Handbook of Behavior Genetics

Yong-Kyu Kim
Editor

Handbook of Behavior Genetics

Springer

Editor
Yong-Kyu Kim
University of Georgia
Athens, GA
USA
yongkyu@arches.uga.edu

ISBN 978-0-387-76726-0   e-ISBN 978-0-387-76727-7
DOI 10.1007/978-0-387-76727-7

Library of Congress Control Number: 2008941695

Printed on acid-free paper

springer.com

Dedicated to my teachers

Preface

Behavior genetics is an interdisciplinary area combining the behavioral sciences and genetics. The study of behavior genetics has become increasingly important as we see growth spurts in finding genes involved in complex behaviors following on advances in molecular genetic techniques. This domain has been growing rapidly since the 1970s and increasingly receives attention from many different disciplines. It has now become a vast common ground for scientists from very diverse fields including psychology, psychiatry, neurology, endocrinology, biochemistry, neuroimaging, and genetics.

When I was invited to organize this book by Springer, I was preparing for a new course, *Behavior Genetics*, at the University of Georgia in fall, 2005. Only a few textbooks were available at that time, but I could not find good references for graduate students and scientists. I thought that we needed to offer research guides to the studies of genetic and environmental influences on a variety of complex behaviors in humans and animals. I had little idea about the proper scope for such a book. I contacted senior colleagues of the Behavior Genetics Association and they gave me excellent advice. I initially invited contributors who were largely members of the Behavior Genetics Association and the handbook was outlined with 14 chapters. As the Handbook developed, it became clear that the first draft was not sufficient to cover all important domains in behavior genetics. In the second meeting with contributors during the BGA meeting in Hollywood, CA, we discussed expanding the handbook to other related domains, such as evolutionary psychology, health behavior, and neurosciences. I invited additional contributors from other disciplines, and added chapters on the history of behavior genetics, quantitative methods and models, as well as more studies of animal models. Now the handbook stands with 34 chapters and integrates many of the basic issues in behavior genetics. *In each chapter, current research and issues on the selected topics are intensively reviewed and directions for future research on these topics are highlighted: new research designs, analytic methods, and their implications are addressed.* It is anticipated that the handbook will contribute to our understanding of behavior genetics and future research endeavors in the 21st century.

Chapter 1 addresses a history of behavior genetics going back to some of Plato's ideas and discusses the nature–nurture controversies on behavior in the modern era which sometimes brought about uproar in our community. In Part I, we address designs and methods in behavior genetic research. Chapters 2 and 4 introduce statistical models and analyses, i.e., biometrical models and multivariate genetic analyses, which explain genetic and environmental causes of covariation between quantitative traits and comorbidity between disorders. In Chapter 3, quantitative trait locus (QTL) analysis is introduced and methods of linkage and association mapping of continuous traits are discussed. Results of the QTL analyses in several quantitative traits are presented throughout this volume. Chapter 5 addresses the importance of animals as models of human behaviors – cognition, personality, and pathology are presented in this volume.

Part II addresses the genetics of cognition in humans and animals with nine selected topics. Chapter 6 discusses genetic and environmental influences on general intelligence using twin studies, followed by new twin research designs, analytic methods, findings, and their implications. In Chapter 7, behavioral genetic research on cognitive aging is reviewed: genetic and environmental contributions to age-related changes in cognitive abilities; contributions of genes and lifestyle variables to dementia, and to the terminal decline in cognitive functioning; and quantitative methods for investigating cognitive aging are presented. Chapter 8 addresses behavioral genetic research on reading, and the genetic and environmental etiologies of reading ability and disability are discussed. Chapter 9 explores behavioral and molecular genetic studies elucidating the role of the genome in the development and manifestation of disorders of speech and language. The human brain continues to show dynamic changes from childhood into adulthood. Genetic and environmental influences in brain volumes are addressed in Chapter 10. Using quantitative magnetic resonance imaging (MRI), brain structures in patients with a clear genetic etiology are reviewed. Genetic approaches to the search for genes associated with brain volume are discussed. Cognitive abilities in animals as models of human behavior are presented in Chapters 11, 12, 13, and 14. Quantitative and molecular genetic approaches to cognition research in rodents are presented in Chapter 11. Cognitive deficits affected by genetic manipulations and mouse models for human cognitive disabilities are discussed. Specifically, Chapter 12 reviews human cognitive impairment associated with chromosomal abnormalities, and mouse models of trisomy 21 are discussed addressing the relationships among genes, brain, and cognitive function. *Drosophila* (fruit fly) models of Alzheimer's disease are introduced in Chapter 13. Pathological roles of Aβ peptides in fly brains, memory defects, and locomotor dysfunctions are discussed. Chapter 14 addresses *Drosophila* courtship songs which are utilized for intersexual selection and species recognition in nature. Quantitative and molecular genetic studies on the phylogenetic patterns of song evolution in different species groups are reviewed.

In Part III, the genetics of personality in humans and animals is addressed with 10 selected topics. Personality is influenced by both genes and environment during development. Chapter 15 explores genotype–environment correlation through a review of the behavioral genetic literature on genetic and environmental influences on family relationships. It is very important that behavioral genetic models that measure behaviors of interest reflect the content of the domains. Chapter 16 reviews behavioral genetic methods and models for personality research and theory, and addresses some methodological issues. Chapter 17 addresses the roles of specific genes, i.e., DRD4 and 5-HTTLPR genes, contributing to the multifaceted dimensions of human personality, including altruism. Temperament, developing early in life and possibly forming the basis for later personality and psychopathology, is explored in Chapter 18 in which quantitative and molecular genetic findings, as well as endophenotypic approaches, are discussed. Sexual orientation is a controversial issue in our communities. A growing body of evidence suggests that familial and genetic factors affect human sexual orientation. Quantitative and molecular genetic studies on sexual orientation are reviewed in Chapter 19. Three chapters introduce animal models of personality and aggression. Chapter 20 explores personality differences in rats widely used in laboratories and discusses anatomical and neurochemical analyses in this endeavor. Behavioral and genetic research on offensive aggression in mice is reviewed and comparative genetic studies of aggression across species are addressed in Chapter 21. Chapter 22 discusses aggressive behavior in fruit flies from the ecological, genetic, neurological, and evolutionary perspectives. Approximately 10% of the population are left-handers. The history, determination, and etiology of handedness are addressed in Chapter 23. Chapter 24 introduces exercise behavior as a new discipline in behavior genetics. A large proportion of adults in the world do not regularly engage in exercise, although benefits of exercise are well documented. Genetic determinants of variability in exercise behavior are discussed.

In Part IV the genetics of psychopathology is represented with nine selected topics. Some psychiatric disorders like ADHD are only diagnosed by questionnaires or psychiatric interviews, rather than by clinical tests, and consequently the genetic studies of the disorders can vary as a function of applied assessment methods and informants. Chapter 25 addresses such behavioral measure issues concerning ADHD. Depression and anxiety have their origins in childhood and arise from genetic and shared environmental effects. Epidemiological and behavior genetic research on childhood depression and anxiety are discussed in Chapter 26. Autism is familial and, thus, relatives of probands with autism are at high risk for depression, anxiety, and personality attributes. Chapter 27 reviews current findings in the genetic epidemiology of autism and its etiological issues concerning the definition of autism phenotypes are discussed. Two chapters address substance abuse behaviors, that is, smoking, drugs, and drinking. Smoking behaviors aggregate in families and in peer networks due to genetic dispositions and common environmental influences. Chapter 28 reviews behavioral genetic research on smoking behavior and nicotine dependence, using Finnish sample studies, and its comorbidities with other substance use, depression, and schizophrenia are discussed. Behavioral and molecular genetic research on the use and abuse of both alcohol and drugs is reviewed in Chapter 29. Substance abuse and substance use disorder co-occur with conduct disorder and antisocial behavior. Chapter 30 gives results of a meta-analysis of twin and adoption studies examining genetic and environmental influences on conduct disorder and antisocial behavior. Association and linkage studies for genes influencing antisocial behavior are discussed. Chapter 31 explores the behavioral and molecular genetic approaches to the origins of two major psychoses: schizophrenia and bipolar mood disorder. The concept of endophenotypes, which are measured intermediate traits or states between genotypes (genetic liability) and phenotypes (disorders), is discussed. Chapter 32 discusses indepth longitudinal "high-risk" studies that intend to identify endophenotypes in the first-degree relatives of schizophrenic probands and to offer putative behavioral predictors of *future* schizophrenia spectrum disorders. Mouse models of cognitive dysfunctions in schizophrenia are explored in Chapter 33 where the role of dopamine in attention and working memory is discussed. Finally, in Chapter 34, future directions for behavior genetics are addressed.

It is not surprising that, at the final publication date of a book like the *Handbook of Behavior Genetics*, research has moved on. In 2008 we saw the publication of genome wide association studies for Bipolar disorder (Ferreira et al., 2008), for five dimensions of personality (Terracciano et al., 2008), ADHD (Neale et al., 2008) and major depressive disorder (Sullivan et al., 2008). Many more GWA studies of complex behavioral and psychiatric phenotypes are expected in the next few years. The landscape of behavior genetics has changed remarkably in a relatively short space of time. The field continues to progress from comparatively small studies to consortia-based efforts that target the inherited components of complex diseases and behaviors and which typically involve thousands of participants (Orr & Chanock, 2008).

References

Ferreira, M. A., O'Donovan, M. C., Meng, Y. A., Jones, I. R., Ruderfer, D. M., Jones, L., et al. (2008). Collaborative genome-wide association analysis supports a role for ANK3 and CACNA1C in bipolar disorder. *Nature Genetics, 40*, 1056–1058.

Neale, B. M., Lasky-Su, J., Anney, R., Franke, B., Zhou, K., Maller, J. B., et al. (2008). Genome-wide association scan of attention deficit hyperactivity disorder. *American Journal of Medical Genetics B Neuropsychiatric Genetics, 147B*, 1337–1344.

Orr, N., & Chanock, S. (2008). Common genetic variation and human disease. *Advances in Genetics, 62*, 1–32.

Sullivan, P. F., de Geus, E. J. C., Willemsen, G., James, M. R., Smit, J. H., Zandbelt, T., et al. (2008). Genome-wide association for major depressive disorder: a possible role for the presynaptic protein piccolo. *Molecular Psychiatry*, 1–17.

Terracciano, A., Sanna, S., Uda, M., Deiana, B., Usala, G., Busonero, F., et al. (2008). Genome-wide association scan for five major depressions of personality. *Molecuar Psychiatry* (in press).

Acknowledgments

I gratefully acknowledge John DeFries, L. Erlenmeyer-Kimling, Irving Gottesman, Jenae Neiderhiser, Nancy Pedersen, and Anita Thapar who provided useful suggestions for shaping the contents of this handbook. I owe a great debt to my colleagues whose contributions create the substance of this exciting book. They have been very patient, and they have promptly responded and revised their chapters in light of reviewers' comments and suggestions. I hope that they are as pleased with the final result as I am. I also acknowledge Janice Stern, my Springer Editor, for her patience on the long delay of the handbook, and I thank her staff for their help in publishing this book.

I am especially grateful to the numerous colleagues who critically reviewed and improved the earlier versions of the chapters with very valuable comments and suggestions. These people include W. Anderson, J. Ando, R. Asarnow, H. Bléaut, D. Blizzard, D. Boomsma, J. Bridle, D. Bruck, T. Button, A. Caspi, S. Cherny, K. Christensen, J. Crabbe, W. Crusio, K. Deater-Deckard, S. de Boer, C. DeCarli, J. DeFries, D. Dick, L. DiLalla, M. Dunne, L. Ehrman, T. Eley, L. Erlenmeyer-Kimling, W. Etges, J.-F. Ferveur, G. Fisch, B. Fish, J. Flint, S. Gammie, X. Ge, M. Geyer, N. Gillespie, J. Gleason, D. Gooding, I. Gottesman, T. Gould, P. A. Gowaty, E. Grigorenko, N. Harlaar, J. Harris, N. Henderson, J. Hewitt, A. Holmes, J. Horn, G. Jackson, J.-M. Jallon, K. Jang, J. Kaprio, W. Kates, J. Kuntsi, C. Kyriacou, R. Landgraf, J.-O. Larsson, L. Leve, J. Loehlin, M. Luciano, H. Maes, T. Markow, L. Matzel, S. Maxson, R. McCrae, S. McGuire, S. Medland, K. Morley, P. Mullineaux, J. Neiderhiser, R. Nelson, Y. Oguma, D. Overstreet, D. Patterson, S. Petrill, Λ. Pike, R. Pillard, R. Plomin, M. Pogue-Geile, G. Riedel, G. Rieger, F. Rijsdijk, J. Ringo, P. Roubertoux, M. Rutter, S. Smith, M. Stallings, T. Suzuki, G. Swan, P. Szatmari, M. Thomis, R. Todd, T. Verson, E. Viding, J. Vink, I. Weiner, H. Welzl, J. Wilson, D. Wolfer, and J. Young.

Contents

Conclusion

Contributors

Judith G. Auerbach Department of Psychology, Ben-Gurion University, Beer-Sheva, Israel; Center for Advanced Studies, Norwegian Academy of Science and Letters, Oslo, Norway,
Email: judy@bgu.ac.il

J. Michael Bailey Department of Psychology, Northwestern University, Evanston, IL 60637, USA,
Email: jm-bailey@northwestern.edu

Patrick F. Bolton Department of Psychologist Medicine, Institute of Psychiatry, London SE5 8AF, UK,
Email: p.bolton@iop.kcl.ac.uk

Dorret I. Boomsma Department of Biological Psychology, Vrije Universiteit, Amsterdam 1081 BT, The Netherlands,
Email: DI.Boomsma@psy.vu.nl

Ulla Broms Department of Public Health, University of Helsinki, Helsinki 00014, Finland; Department of Mental Health and Alcohol Research, National Public Health Institute, Helsinki, Finland,
Email: ulla.broms@helsinki.fi

Michèle Carlier Institut Universitaire de France and Laboratoire Psychologie Cognitive, UMR 6146, CNRS Aix-Marseille 1, Marseille Cedex 20, France,
Email: Michele.Carlier@univ-provence.fr

Stacey S. Cherny Department of Psychiatry, Genome Research Centre, and The State Key, Laboratory of Brain and Cognitive Sciences, University of Hong Kong, Pokfulam, Hong Kong,
Email: cherny@hku.hk

Maria G. Corda Department of Toxicology, University of Cagliari, Cagliari 09124, Italy,
Email: mgcorda@unica.it

Sarah Curran Department of Psychologist Medicine, Institute of Psychiatry, London SE5 8AF, UK,
Email: s.curran@iop.kcl.ac.uk

Khytam Dawood Department of Psychology and Center for Developmental and Health Genetics, Pennsylvania State University, University Park, PA 16802, USA,
Email: Khytam@psu.edu

Eco J.C. de Geus Department of Biological Psychology, Vrije Universiteit, Amsterdam 1081 BT, The Netherlands,
Email: JCN.de.Geus@psy.vu.nl

Eske M. Derks Department of Biological Psychology, Vrije Universiteit, Amsterdam 1081 BT, The Netherlands; University Medical Center Utrecht, Divisie hersenen, Heidelberglaan 100, 3584 CX Utrecht, The Netherlands,
Email: E.M.Derks@umcutrecht.nl

Danielle M. Dick Virginia Institute of Psychiatric and Behavioral Genetics, Virginia Commonwealth University, Richmond, VA 23298, USA,
Email: ddick@vcu.edu

Peter Driscoll Institute for Animal Science, ETHZ, Schwerzenbach 8603, Switzerland,
Email: peter-driscoll@ethz.ch

Richard P. Ebstein Department of Psychology, Scheinfeld Center of Genetic Studies for the Social Sciences, Hebrew University of Jerusalem; Sarah Herzog Memorial Hospital, Givat Shaul, Jerusalem 91905, Israel,
Email: ebstein@mscc.huji.ac.il

Lee Ehrman School of Natural and Social Sciences, Purchase College, State University of New York, Purchase, NY 10577, USA,
Email: lee.ehrman@purchase.edu

L. Erlenmeyer-Kimling Departments of Medical Genetics, New York State Psychiatric Institute, New York, NY 10032, USA; Departments of Psychiatry and of Genetics and Development, Columbia University, New York, NY, USA,
Email: le4@columbia.edu

Alberto Fernández-Teruel Medical Psychology Unit, Department of Psychiatry and Forensic Medicine, University of Barcelona, Bellaterra 08193, Spain,
Email: albert.fernandez.teruel@uab.es

Deborah Finkel Department of Psychology, School of Social Sciences, Indiana University Southeast, New Albany, IN 47150, USA,
Email: dfinkel@ius.edu

Gene S. Fisch Bluestone Clinical Research Center, NYU Colleges of Dentistry & Nursing, Yeshiva University, New York, NY 10010, USA,
Email: gene.fisch@nyu.edu

Barbara Fish Department of Psychiatry and Behavioral Sciences, University of California, Los Angeles, CA 90095, USA,
Email: bfish@ucla.edu

Jeffrey R. Gagne Department of Psychology, University of Wisconsin, Madison, WI 53706, USA,
Email: jgagne@wisc.edu

Michael J. Galsworthy Division of Neuroanatomy and Behavior, Institute of Anatomy, University of Zurich, Zurich CH-8057, Switzerland,
Email: mike_galsworthy@yahoo.co.uk

Osvaldo Giorgi Department of Toxicology, University of Cagliari, Cagliari 09124, Italy,
Email: giorgi@unica.it

H. Hill Goldsmith Department of Psychology, University of Wisconsin, Madison, WI 53706, USA,
Email: hhgoldsm@wisc.edu

Elena L. Grigorenko Child Study Center, Department of Psychology, Department of Epidemiology & Public Health, Yale University, New Haven, CT 06519-1124, USA,
Email: elena.grigorenko@yale.edu

Sydney L. Hans School of Social Service Administration, University of Chicago, Chicago, IL 60637, USA,
Email: s-hans@uchicago.edu

Daniel R. Hanson Departments of Psychiatry & Psychology, University of Minnesota, Minneapolis, MN 55454, USA,
Email: drhanson@umn.edu

Sara A. Hart Department of Human Development and Family Science, The Ohio State University, Columbus, OH 43210, USA,
Email: hart.327@osu.edu

Anneli Hoikkala Department of Biological and Environmental Science, University of Jyväskylä, Jyväskylä, Finland,
Email: anneli.a.hoikkala@jyu.fi

James J. Hudziak Department of Psychiatry and Medicine (Division of Human Genetics), Center for Children, Youth and Families, University of Vermont, Burlington, VT, USA,
Email: James.Hudziak@uvm.edu

Hilleke E. Hulshoff Pol Department of Psychiatry, Rudolf Magnus Institute of Neuroscience, University Medical Center, Utrecht, The Netherlands,
Email: h.e.hulshoff@umcutrecht.nl

Koichi Iijima Laboratory of Neurodegenerative Diseases and Gene Discovery; Farber Institute for Neurosciences; Department of Biochemistry and Molecular Biology, Thomas Jefferson University, Philadelphia PA19107, USA,
Email: Koichi.Iijima@jefferson.edu

Kanae Iijima-Ando Laboratory of Neurogenetics and Pathobiology; Farber Institute for Neurosciences; Department of Biochemistry and Molecular Biology, Thomas Jefferson University, Philadelphia PA 19107, USA,
Email: Kanae.Iijima-Ando@jefferson.edu

Loring J. Ingraham Department of Psychology, The George Washington University, Washington, DC 20037, USA,
Email: ingraham@gwu.edu

Salomon Israel Department of Psychology, Scheinfeld Center of Genetic Studies for the Social Sciences, Hebrew University of Jerusalem, Jerusalem 91905, Israel,
Email: salomon.israel@mail.huji.ac.il

Kerry L. Jang Department of Psychiatry, University of British Columbia, Vancouver, British Columbia, Canada, V6T 2A1.
Email: kjang@interchange.ubc.ca

Christopher Janus Mayo Clinic Jacksonville, Department of Neuroscience, University of Florida, Jacksonville, FL 32224, USA,
Email: Janus.Christopher@mayo.edu

Wendy Johnson Centre for Cognitive Ageing and Cognitive Epidemiology and Department of Psychology, University of Edinburgh, Edinburgh, UK, and Department of Psychology, University of Minnesota, Minneapolis, MN 55455, USA,
Email: john4350@umn.edu

René S. Kahn Department of Psychiatry, Rudolf Magnus Institute of Neuroscience, University Medical Center, Utrecht, The Netherlands,
Email: r.kahn@umcutrecht.nl

Jaakko Kaprio Department of Public Health, University of Helsinki, Helsinki 14, Finland; Department of Mental Health and Alcohol Research, National Public Health Institute, Helsinki, Finland,
Email: jaakko.kaprio@helsinki.fi

Yong-Kyu Kim Department of Genetics, University of Georgia, Athens, GA 30602, USA,
Email: yongkyu@uga.edu

Tellervo Korhonen Department of Public Health, University of Helsinki, Helsinki 14, Finland; Department of Mental Health and Alcohol Research, National Public Health Institute, Helsinki, Finland,
Email: tellervo.korhonen@helsinki.fi

John C. Loehlin Department of Psychology, University of Texas at Austin, Austin, TX 78712, USA,
Email: loehlin@psy.utexas.edu

Joseph Marcus Department of Psychiatry, University of Chicago, Chicago, IL 60637, USA,
Email: jmarcusmd@cox.net

Nicholas G. Martin Department of Genetic Epidemiology, Queensland Institute of Medical Research, Brisbane, Queensland, Australia,
Email: Nick.Martin@qimr.edu.au

Stephen C. Maxson Department of Psychology, University of Connecticut, Storrs, CT 06269-1020, USA,
Email: Stephen.Maxson@Uconn.Edu

Dominique Mazzi Institute of Plant Sciences, Swiss Federal Institute of Technology, Zurich, Switzerland,
Email: dominique.mazzi@ipw.argl.ethz.ch

Matthew McGue Department of Psychology, University of Minnesota, Minneapolis, MN 55455, USA; Department of Epidemiology, Southern Denmark University, Denmark,
Email: mcgue001@tc.umn.edu

Thomas F. McNeil Department of Psychiatric Epidemiology, University Hospital, Lund University, Israel,
Email: Thomas.McNeil@med.lu.se

Michael C. Neale Departments of Psychiatry and Human Genetics, Virginia Institute for Psychiatric and Behavioral Genetics, Virginia Commonwealth University, Richmond, VA 23298, USA,
Email: neale@vcu.edu

Jenae M. Neiderhiser Department of Psychiatry and Behavioral Sciences, Center for Family Research, The Pennsylvania State University, University Park, PA 16802, USA,
Email: jenaemn@psu.edu

Jiska S. Peper Department of Psychiatry, Rudolf Magnus Institute of Neuroscience, University Medical Center, Utrecht, The Netherlands,
Email: j.s.peper@umcutrecht.nl

Ira B. Perelle Department of Psychology, Mercy College, Dobbs Ferry, NY 10522, USA,
Email: iperelle@mercy.edu

Stephen A. Petrill Department of Human Development and Family Science, Ohio State University, Columbus, OH 43210, USA,
Email: spetrill@ehe.osu.edu

Danielle Posthuma Department of Biological Psychology, Section Medical Genomics, and Section Functional Genomics, Vrije Universiteit and Vrije Universiteit Medical Center, Amsterdam 1081BT, The Netherlands,
Email: danielle@psy.vu.nl

Carol Prescott Department of Psychology, University of Southern California, Los Angeles, CA 90089, USA,
Email: cprescot@usc.edu

Chandra A. Reynolds Department of Psychology, University of California, Riverside, CA 92521, USA,
Email: chandra.reynolds@ucr.edu

Soo Hyun Rhee Institute for Behavioral Genetics, University of Colorado, Boulder, CO 80309, USA,
Email: Soo.Rhee@colorado.edu

Frances Rice Child & Adolescent Psychiatry Section, Department of Psychological Medicine, School of Medicine, Cardiff University, Cardiff CF14 4XN, UK,
Email: ricef@cardiff.ac.uk

Richard J. Rose Department of Psychological and Brain Sciences, Indiana University, Bloomington, IN 47405, USA; Department of Public Health, University of Helsinki, Finland, Email: rose@indiana.edu

Pierre L. Roubertoux INSERM UMR910 Génétique Médicale et Génomique Fonctionnelle, Université d'Aix-Marseille 2, Faculté de Médecine, 27 Bvd Jean Moulin, 13385 Marseille Cedex 05, France, Email: pierre.roubertoux@univmed.fr

Claudia Schmauss Department of Psychiatry and Molecular Therapeutics, Columbia University and New York State Psychiatric Institute, New York, NY 10032, USA, Email: cs581@columbia.edu

Erland Schubert Department of Psychiatric Epidemiology, University Hospital, Lund University, Israel, Email: erland.schbert@psychepi.lu.se

Nancy L. Segal Department of Psychology, California State University, Fullerton, CA 92834, USA, Email: nsegal@fullerton.edu

Thierry Steimer Clinical Psychopharmacology Unit, University Hospital of Geneva, 1225 Chêne-Bourg, Switzerland, Email: thierry.steimer@hcuge.ch

Janine H. Stubbe Department of Biological Psychology, Vrije Universiteit, Amsterdam 1081 BT, The Netherlands, Email: janine.stubbe@tno.nl

Anita Thapar Child & Adolescent Psychiatry Section, Department of Psychological Medicine, Cardiff University, Cardiff CF14 4XN, UK, Email: thapar@Cardiff.ac.uk

Jennifer A. Ulbricht Center for Family Research; Department of Psychology, George Washington University, Washington, DC 20037, USA, Email: jau5b@gwu.edu

Matthew K. Vendlinski Department of Psychology, University of Wisconsin, Madison, WI 53706, USA, Email: vendlinski@wise.edu

Irwin D. Waldman Department of Psychology, Emory University, Atlanta, GA 30322, USA, Email: psyiw@emory.edu

Hans Welzl Division of Neuroanatomy and Behavior, Institute of Anatomy, University of Zurich, Zurich CH-8057, Switzerland, Email: welzl@anatom.uzh.ch

David P. Wolfer Division of Neuroanatomy and Behavior, Institute of Anatomy, University of Zurich, Zurich CH-8057, Switzerland, Email: dpwolfer@anatom.uzh.ch

Shinji Yamagata Faculty of Letters, Keio University, Tokyo, 108-8345, Japan,
Email: yamagata@bayes.cu-tokyo.ad.jp

Yi Zhong Cold Spring Harbor Laboratory, Cold Spring Harbor, NY 11724, USA,
Email: zhongyi@cshl.edu

Marcel P. Zwiers F.C. Donders Center for Cognitive Neuroscience, Nijmegen, The
Netherlands; Psychiatry Department, Radboud University, Nijmegen Medical Center,
Nijmegen 6525 HE, The Netherlands,
Email: m.zwiers@fcdonders.ru.nl

Introduction

Chapter 1

History of Behavior Genetics

John C. Loehlin

Hermann Ebbinghaus (1908) said of psychology that it had a long past, but only a short history. The same may be said of behavior genetics. One cannot specify an exact date at which behavior genetics came to be regarded as a distinct scientific discipline, but for convenience let us say 1960, the publication date of Fuller and Thompson's textbook of that title.

This chapter considers both the long past and some aspects of the short history of behavior genetics. We begin with the long past: the recognition since antiquity that behavioral traits are in part inherited, and the controversy concerning the extent to which this is so, a discussion often going under the label of the *nature–nurture controversy*.

The Long Past of Behavior Genetics

From Ancient Times to the Renaissance

Ancient Times

Where does the long past start? Perhaps with the domestication of dogs for behavioral as well as physical traits, a process which probably took place at least 15,000 years ago (Savolainen, Zhang, Luo, Lundeberg, & Leitner, 2002) – although one must suppose that in its early days this was more an evolution of a subgroup of wolves to fit a niche around human habitation than a process deliberately undertaken by man (Morey, 1994). In any case, about 5000 years ago in Egypt and the Near East, it appears that deliberate animal breeding was well established (Brewer, Clark, & Phillips, 2001); several distinctive varieties of cattle and dogs are portrayed in ancient Egyptian art.

Greeks, Romans, Hebrews

By classical times, 3000–1500 years ago, many varieties of dogs with distinctive physical and behavioral characteristics were recognized. More than 50 breeds are named in surviving Greek and Roman documents, falling into such categories as scent- and sight hounds, shepherd dogs, guard dogs, war dogs, and pets (Brewer et al., 2001).

The ancient Greeks held that humans inherited qualities, including behavioral ones, from their ancestors. Thus in Book IV of Homer's *Odyssey*, Menelaus greets two young visiting strangers, "Ye are of the line of men that are sceptred kings ... for no churls could beget sons like you" (Homer, trans. 1909, p. 49). And later (p. 53), to one of them, "Thou has said all that a wise man might say or do, yea, and an elder than thou; – for from such a sire too thou art sprung, wherefore thou dost even speak wisely." A similar notion was expressed in the Hebrew scriptures: "I am the heir of wise men, and spring from ancient kings" (Isaiah 19:11, *New English Bible*).

A few hundred years later, the Greek philosopher Plato in Book V of the *Republic* – his prescription for an ideal state – took both inheritance and instruction into account in the development of the "Guardians," the ruling elite. He begins with the question, "How can marriages be made most beneficial?" He discusses the breeding of hunting dogs and birds, noting that "Although they are all of a good sort, are not some better than others?" "True." "And do you breed from them all indifferently, or do you take care to breed from the best only?" "From the best" (Plato, trans. 1901, p. 149). From there Plato goes on to generalize to the class of elite humans in his ideal state – to the desirability of matching the best with the best, and rearing their offspring with special attention.

Plato recognizes that good ancestry is not infallibly predictive and recommends applying, at least in early youth, a universal education to the citizens of his state; demoting, when inferior, offspring of the elite class of guardians and elevating into the ranks of the guardians offspring of the lower classes who show merit.

J.C. Loehlin (✉)
Department of Psychology, *The University of Texas at Austin*, Austin, TX 78712, USA

Y.-K. Kim (ed.), *Handbook of Behavior Genetics*,
DOI 10.1007/978-0-387-76727-7_1, © Springer Science+Business Media, LLC 2009

We need not debate the pros and cons of Plato's particular social proposals; people have been arguing about them
ever since his day. We only need observe that well over 2000
years ago the interplay of nature and nurture – and its social
implications – was being discussed.

Middle Ages

What of the contrary view, the notion that all men are
born equal? A major impetus to such an idea came from
the medieval Catholic Church (Pearson, 1995). All men are
sons of God, and therefore of equal value in His sight. Or,
from another perspective, as the fourteenth-century English
proverb had it, "When Adam delved and Eve span/Who was
then a gentleman?"

The Renaissance

Ideas concerning the inheritance of behavior were present in
Shakespeare's day. The Countess of Rossilon in *All's Well
That Ends Well* says, about a wise daughter of a wise father,
"Her dispositions she inherits" (Act I:i). The nature–nurture
controversy itself appears to have got its label from Pros-
pero's remark in *The Tempest* about his subhuman creature,
Caliban, "A devil, a born devil, on whose nature nurture will
never stick" (Act IV:i).

The Nature–Nurture Controversy in the Modern Era

Although ideas about the roles of nature and nurture in
human and animal behaviors have been with us for thousands
of years, the modern form of the controversy traces back
fairly directly to the seventeenth-century philosopher John
Locke and the nineteenth-century naturalist Charles Darwin.

John Locke

Locke may be considered to be the chief ideological father
of the nurture side of the controversy. In *An Essay Concerning Human Understanding* (Locke, 1690/1975), he invoked
the metaphor of the mind as a blank sheet of paper upon
which knowledge is written by the hand of experience. In the
opening paragraph of his book *Some Thoughts Concerning
Education*, he said, "I think I may say, that of all the Men
we meet with, nine Parts of ten are what they are, good or
evil, useful or not, by their Education" (Locke, 1693/1913,
Sect. 1). Locke's political view that all men are by nature
equal and independent, and that society is a mutual contract

entered into for the common good, had an immense influence
via Jefferson, Voltaire, Rousseau, and the other theorists of
the American and French revolutions.

Indeed, one may view many of the events of the nature–
nurture controversy since Locke's day as a series of challenges to the prevailing Lockean position, with those steeped
in that tradition rising indignantly to battle what they perceived to be threats to inalienable human rights of liberty and
equality.

Locke himself, however, was not nearly as alien to hered-
itarian concepts as some of his followers have been. He
rejected the concept of inborn ideas, but not of all innate characteristics. In a marginal note on a pamphlet by one Thomas
Burnet, Locke wrote "I think noe body but this Author who
ever read my book [*An Essay Concerning Human Understanding*] could doubt that I spoke only of innate Ideas ...
and not of *innate powers* ... " (see Porter, 1887). Elsewhere
in *Some Thoughts Concerning Education* Locke wrote,

> Some Men by the unalterable Frame of their Constitutions are
> *stout*, others *timorous*, some *confident*, others *modest*, *tractable*,
> or *obstinate*, *curious* or *careless*, *quick* or *slow*. There are not
> more Differences in Men's Faces, or in the outward Lineaments
> of their Bodies, than there are in the Makes and Tempers of their
> Minds. (1693/1913, Sect. 101)

John Stuart Mill

Many of Locke's successors in the English liberal tradition
came out more strongly than Locke did on the side of nurture.
John Stuart Mill wrote in his *Autobiography* (1873, p. 192),

> I have long felt that the prevailing tendency to regard all the
> marked distinctions of human character as innate, and in the
> main indelible, and to ignore the irresistible proofs that by far the
> greater part of these differences, whether between individuals,
> races, or sexes, are such as not only might but naturally would
> be produced by differences in circumstances, is one of the chief
> hindrances to the rational treatment of great social questions, and
> one of the greatest stumbling blocks to human improvement.

Charles Darwin

During roughly the same period as Mill, Charles Darwin
gave the nature side of the controversy its modern form
by placing behavior, including human behavior, solidly in
the framework of biological evolution. In addition to his
major treatise *The Origin of Species* (1859), Darwin in such
works as *The Descent of Man* (1871) and *The Expression
of the Emotions in Man and Animals* (1872) made it clear
that human behavior shared ancestry with that of other animal forms, and was subject to the same evolutionary process of hereditary variation followed by natural selection of
the variants that proved most successful in their particular
environments.

In *The Descent of Man* (1871, pp. 110–111) Darwin wrote,

> So in regard to mental qualities, their transmission is manifest in our dogs, horses, and other domestic animals. Besides special tastes and habits, general intelligence, courage, bad and good temper, etc. are certainly transmitted. With man we see similar facts in almost every family; and we now know through the admirable labours of Mr. Galton that genius, which implies a wonderfully complex combination of high faculties, tends to be inherited; and on the other hand, it is too certain that insanity and deteriorated mental powers likewise run in the same families.

Francis Galton

Darwin's younger cousin Francis Galton agreed with Darwin and disagreed with Mill. In his book *Inquiries into Human Faculty* (1883, p. 241) he concluded,

> There is no escape from the conclusion that nature prevails enormously over nurture when the differences of nurture do not exceed what is commonly to be found among persons of the same rank of society and in the same country.

Galton is not saying that environment never matters. However, he is saying that the ordinary differences we observe among people in the same general social context are mostly due to heredity.

Galton was a central, crystallizing figure in behavior genetics' "long past." His emphasis on the measurement of individual differences and their statistical treatment became a core theme in the development of the field. His studies of "hereditary genius" and "the comparative worth of different races" (Galton, 1869) foreshadowed recent controversies about IQ. He proposed the study of twins as a way of getting at the relative effect of nature and nurture. And his promotion of eugenics – that is, the encouragement of the more useful members of society to have more children and the less useful to have fewer (as in Plato's scheme for an ideal state) – has generated on occasion a good deal of heat. Here is a recent example (Graves, 2001, p. 100): "Galton's scientific accomplishments are sufficient for some still to consider him an intellectual hero. Whereas for others (this author included) he was an intellectual mediocrity, a sham, and a villain."

The Twentieth Century

Vigorous disagreements on the relative impact of nature and nurture on behavior continued into the twentieth century. On the whole, twentieth-century psychology was heavily environmentalistic, emphasizing the crucial role of learning in shaping behavior. The high-water mark of this tradition was the famous claim of John B. Watson (1925, p. 82):

> Give me a dozen healthy infants, well-formed, and my own specified world to bring them up in and I'll guarantee to take any one at random and train him to become any type of specialist I might select – doctor, lawyer, artist, merchant-chief, and yes, even beggar-man and thief, regardless of his talents, penchants, tendencies, abilities, vocations, and race of his ancestors.

The year 1928 saw the publication of the *Twenty-Seventh Yearbook* of the National Society for the Study of Education. It was entitled *Nature and Nurture*, and it contained the reports of two adoption studies of IQ. One, by Barbara Burks, emphasized the effects of nature. The other, by Freeman, Holzinger, and Mitchell, came down on the side of nurture. The nature–nurture controversy continued, but students of the effects of heredity and environment on behavior were gathering data. When enough had been gathered for a textbook to be written, the short history of behavior genetics could begin.

The Short History of Behavior Genetics

Most of the short history of behavior genetics, as it applies to the study of both humans and other animal species, will not be discussed in this chapter. It is a tale of steady scientific progress on a variety of fronts, despite occasional controversies, confusions, and setbacks, and it is a tale told in the other chapters of this handbook. The reader who wants a quick sense of the scope of scientific progress in the field of behavior genetics during the last 40-odd years, and the prospects opening up in it today, can achieve this by scanning through the chapter introductions and summaries, and the editor's final chapter. The reader who aspires to a more solid grasp of this short history will need, of course, to proceed more systematically through the book, as well as following up some of its many references.

The remainder of this chapter addresses two other aspects of behavior genetics' short history. First, we look briefly at some institutional features of the field: its principal scholarly and scientific organization, the *Behavior Genetics Association*; the discipline's key journal, *Behavior Genetics*; and some major centers of behavior genetics research. Following this, we look at the social context of behavior genetics, at instances in which the scientific and scholarly pursuits of the field have become entangled with public political and social concerns. These instances include a series of controversies concerning the genetic or environmental bases of differences in psychological characteristics between groups defined by race, sex, or social class. Controversies about group differences have roots in behavior genetics' long past and have persisted into its short history. They are far from central in the activities of most working behavior geneticists, but they represent an important part of the public face of the field.

The Institutional History of Behavior Genetics

The Behavior Genetics Association

After some informal discussions in the late 1960s, and the circulation of a mailing to a list of persons who had recently published in the area of behavior genetics, an organizational meeting took place at Urbana, Illinois, in March 1970. R. H. Osborne, then editor of the journal *Social Biology*, was chosen to act as president pro tem, and five committees were appointed to lay the groundwork for a Behavior Genetics Association (or Society – there was some argument about a suitable name). In April 1971, the fledgling organization held its first formal meeting, at Storrs, Connecticut. In addition to scientific sessions, a draft constitution was discussed to be submitted to the initial membership via mail ballot for approval. Nominations and an election followed, and at the time of the second annual meeting at Boulder, Colorado, in April 1972, the Behavior Genetics Association (BGA) was officially underway, and its first set of officers took office: Theodosius Dobzhansky was president, John Fuller was president-elect, R. H. Osborne served as past president, the secretary was Elving Anderson, the treasurer was John Loehlin, and the two executive committee members-at-large were Seymour Kessler and L. Erlenmeyer-Kimling.

The association proved viable. Table 1.1 shows the successive presidents of the BGA and the location of its annual meetings. Note that a special extra international meeting was held in Jerusalem in 1981, and that thereafter the regular annual BGA meeting was periodically held in countries outside the USA: in England (twice), the Netherlands (twice), France, Australia, Spain, Canada (twice), and Sweden.

Over time, the association grew in size. Forty-four persons responded to the initial mailing indicating interest in such an association. There were 69 paid-up members at the time of the first annual meeting at Storrs. By the time of the 34th annual meeting in Aix-en-Provence, France, in 2004, the BGA had 270 regular and 109 associate members (the latter chiefly graduate students). Approximately two-thirds were from North America and one-third from other continents.

The Journal *Behavior Genetics*

In 1970, a decade after Fuller and Thompson's textbook, the scientific journal *Behavior Genetics* began with Vol. 1, No. 1. Its founding editors were Steven G. Vandenberg and John C. DeFries. They stated their hopes for the new journal in an editorial (p. 1):

> Research in behavior genetics continues to be undertaken at an accelerating rate. Nevertheless, no single journal has existed heretofore which was dedicated primarily to the publication of papers in this important area. Since manuscripts in behavior

Table 1.1 BGA Presidents and Annual Meetings

Year	President	Site of meeting
1971	R. H. Osborne [pro tem]	Storrs CT
1972	Th. Dobzhansky	Boulder CO
1973	John L. Fuller	Chapel Hill NC
1974	Gerald E. McClearn	Minneapolis MN
1975	J. P. Scott	Austin TX
1976	Irving I. Gottesman	Boulder CO
1977	W. R. Thompson	Louisville KY
1978	Lee Ehrman	Davis CA
1979	V. Elving Anderson	Middletown CT
1980	John C. Loehlin	Chicago IL
1981	Norman D. Henderson	Purchase NY/Jerusalem
1982	John C. DeFries	Ft Collins CO
1983	David W. Fulker	London, England
1984	Steven G. Vandenberg	Bloomington IN
1985	Sandra Scarr	State College PA
1986	Ronald S. Wilson	Honolulu HI
1987	Peter A. Parsons	Minneapolis MN
1988	Leonard L. Heston	Nijmegen, Netherlands
1989	Robert Plomin	Charlottesville VA
1990	Carol B. Lynch	Aussois, France
1991	Lindon J. Eaves	St. Louis MO
1992	David A. Blizard	Boulder CO
1993	Thomas J. Bouchard, Jr.	Sydney, Australia
1994	Glayde Whitney	Barcelona, Spain
1995	James Wilson	Richmond VA
1996	Nicholas G. Martin	Pittsburgh PA
1997	Nicholas G. Martin	Toronto, Canada
1998	Norman D. Henderson	Stockholm, Sweden
1999	Richard Rose	Vancouver, Canada
2000	John Hewitt	Burlington VT
2001	Matt McGue	Cambridge, England
2002	Nancy Pedersen	Keystone CO
2003	Andrew Heath	Chicago IL
2004	Michèle Carlier	Aix-en-Provence, France
2005	H. Hill Goldsmith	Hollywood CA
2006	Laura Baker	Storrs CT
2007	Pierre Roubertoux	Amsterdam, Netherlands

Source: BGA web site (June 27, 2007); http://www.bga.org

genetics have thus been published in widely scattered journals, a clear identification with this discipline has been lacking. It is our hope that *BEHAVIOR GENETICS* will fulfill this need.

The journal has largely lived up to their hopes. It never stood completely alone – for example, at the time there was an existing journal focused on twin research, *Acta Geneticae Medicae et Gemellologiae*, which published many behaviorally oriented papers. The journal *Social Biology* – whose editor, R. H. Osborne, played an important role in founding the Behavior Genetics Association – initially served as the official organ of the BGA. (*Behavior Genetics* assumed that role in 1974.) Other journals have since emerged – for example, the recent journals *Genes, Brains, and Behavior* and *Twin Research*. Many important papers in behavior genetics continue to be published in journals in the neighboring behavioral and biological sciences. Nevertheless, *Behavior Genetics*, as the official organ of the Behavior Genetics Association, remains a major defining force in the field.

It is instructive to compare Vol. 1 (1970) of *Behavior Genetics* with Vol. 35 (2005). The journal became a good deal bigger: from three issues in Vol. 1 (Nos. 3 and 4 were bound together) to six in Vol. 35 from 274 to 854 pages (and nearly twice the number of words per page because of larger pages). In Vol. 1, there were 24 papers, an editorial, and 2 "short communications." In Vol. 35 there were 66 papers, plus 142 abstracts from the Behavior Genetics Association meeting, and various BGA minutes, announcements, etc. *Behavior Genetics* continues to publish both substantively and methodologically oriented papers, featuring various animal species, but the mix changed from Vol. 1 to Vol. 35. In Vol. 1 there were 7 papers (27%) focused on human behavior, 16 papers (62%) involving rodents, mostly inbred mice, 1 paper on another species (*Drosophila*), and 2 papers primarily methodological (statistical) in character. In Vol. 35, there was an increased proportion of substantive papers involving humans, 28 (42%); proportionately fewer involving rodents, 14 (21%); an increase in those involving other animal species, 9 (15%) – mostly *Drosophila*, but one on rainbow trout. For many of the remaining 22% of papers, the species might be described as the computer: These were methodological papers, many involving a heavy dose of computer model-fitting or simulation.

Major Behavior Genetics Centers

Preeminent among academic centers for teaching and research in behavior genetics has been the Institute for Behavioral Genetics (IBG) at the University of Colorado at Boulder. Among the notable behavior geneticists who have served on its faculty are Gregory Carey, John DeFries, David Fulker, John Hewitt, Carol Lynch, Gerald McClearn, Robert Plomin, Steven Vandenberg, and James Wilson. It has also served as home for the journal *Behavior Genetics*, except for 1978–1985 when Jan Bruell edited the journal at the University of Texas and 2000–2002 when Norman Henderson edited it at Oberlin College. The IBG has also hosted several BGA annual meetings and a number of summer training institutes on behavior genetics methods.

Next in line as a center of behavior genetics activity would probably be the University of Minnesota, whose faculty has included important behavior geneticists like Elving Anderson, Thomas Bouchard, Irving Gottesman, Leonard Heston, Gardner Lindzey, David Lykken, Matthew McGue, Sheldon Reed, Sandra Scarr, and Auke Tellegen. A third center, at least in the early days, was the University of Texas at Austin, with Jan Bruell, Joseph Horn, Gardner Lindzey, John Loehlin, Delbert Thiessen, and Lee Willerman. A current major behavior genetics center is at the Virginia Commonwealth University; its faculty includes Lindon Eaves, Kenneth Kendler, Hermine Maes, and Michael Neale. Other important U.S. centers include Washington University in St. Louis (Robert Cloninger, Andrew Heath, & John Rice) and Penn State (David Blizard, Gerald McClearn, & George Vogler). Outside the USA, Kings College, London, has recruited an eminent group of behavior genetics researchers, including Peter McGuffin, Robert Plomin, and Michael Rutter. The Vrije Universiteit in Amsterdam also has a substantial behavior genetics contingent, including Dorret Boomsma and Danielle Postuma. Stable international coalitions are becoming increasingly common, greatly facilitated by the Internet. Notable examples include collaborations between groups at Indiana University and the University of Helsinki, Penn State and the Karolinska Institute in Stockholm, and several U.S. groups with the Queensland Institute for Medical Research in Australia.

Beside the institutions mentioned above, dozens of other universities and research institutes, including many outside the USA, have developed and maintained strong programs in human or animal behavior genetics on the strength of one or two distinguished researchers on their faculties. Almost half the presidents of the BGA, for example, would represent this category. The hosting of an annual BGA meeting (see Table 1.1) also tends to reflect a strong local program.

Public Controversies – Group Differences

The possibility that there might be genetic differences in psychological traits between groups defined by race, sex, or social class has led to a good deal of public uproar and not a little confusion. It has provided an inflammatory intersection between the scientific discipline of behavior genetics and Western attitudes of equality stemming from religious, political, and philosophical roots. Racist, sexist, and classist ideas (as references to such group differences are sometimes called) tend to drive traditional Lockean ideologists up the wall, so that clear thinking has not always prevailed in this area.

A few general points should be noted. First, the main business of behavior geneticists has always been individual differences, not group differences, so that for the day-to-day research of most behavior geneticists, questions about group differences are at best an unwelcome distraction. Second, as Lewontin (1970) made clear, a demonstration that individual differences are due to genes does not imply that group differences are genetic. He used the analogy of genetically varied seeds raised in a greenhouse in two pots under identical regimens, except that one pot lacked a crucial trace nutrient present for the other. The heights of the plants are subsequently measured. The variation of height within each pot, except for random measurement errors, is entirely genetic, since the plants within each pot vary genetically, but are treated exactly the same. The average difference in plant height between the two pots is entirely environmental, because it stems from the presence or absence of the critical nutrient. Clearly, this example implies that group differences may be different in their genetic and environmental origins from individual differences. However, it is sometimes forgotten that *may* does not imply *are*. There remains the empirical question for any particular trait and any particular group difference in any particular population: To what relative extent are genetic and environmental differences between the groups *in fact* involved? There also remains the social question: How much (if at all) does this matter?

The empirical question is not necessarily an easy one to answer. For one thing, it may well have different answers for different traits and different groups (Loehlin, 2000). If one were to demonstrate that profiles of cognitive ability differ for genetic reasons between Asian Americans and European Americans, it would not imply that a difference in average intellectual performance between European Americans and African Americans has a genetic origin. To make matters worse, the social excitement and media hoopla surrounding the issue of group differences has discouraged most behavior geneticists from addressing such matters empirically. It is not as though informative research designs do not exist. One listing of promising areas of research on racial-ethnic ability differences listed ten possible approaches, ranging from studies of race mixtures and cross-racial adoptions to piggy-back studies on educational or nutritional programs which were being undertaken for other reasons (Loehlin, Lindzey, & Spuhler, 1975, pp. 251–254).

Jensen

Less than a decade into behavior genetics' short history, the educational psychologist Arthur Jensen published a long article in the *Harvard Educational Review* entitled "How much can we boost IQ and scholastic achievement?" (Jensen, 1969). Jensen noted the fact that compensatory education programs had not lived up to their advance billing and concluded that this might partly reflect the genetic contribution to IQ, which he estimated at a fairly high 80%. Almost in passing, he noted the possibility that the persistent IQ gap between U.S. blacks and whites might in part be genetic in origin. He did not say that this had been demonstrated to be the case, but suggested that the matter should be looked into empirically. Jensen's article, particularly the suggestion that there might be a genetic contribution to black–white IQ differences, created an immediate furor. There were numerous published critiques, not all judicious and carefully thought out. And this was not just a genteel academic debate – tires were slashed and public

meetings disrupted. A graphic account of the goings-on may be found in Pearson (1991). The controversy about possible racial differences in mental abilities has continued to the present – the interested reader may wish to consult *Race Differences in Intelligence* (Loehlin et al., 1975), *Race, IQ and Jensen* (Flynn, 1980), *The Black–White Test Score Gap* (Jencks & Phillips, 1998), and *The New Know-Nothings* (Hunt, 1999). Rushton and Jensen (2005) provide a recent review emphasizing the genes: "Thirty years of research on race differences in cognitive ability," which, along with a number of critiques from various points of view, fills an issue of *Psychology, Public Policy, and Law* [Vol. 11(2), 2005].

The Bell Curve

Twenty-five years after Jensen's article, a similar uproar arose, this time due to the publication of a book by the psychologist Richard Herrnstein and the sociologist Charles Murray entitled *The Bell Curve* (Herrnstein & Murray, 1994). Although much of the furor focused on race differences in cognitive skills, the authors did not in fact devote a great deal of attention to this topic and took a fairly mild position on it. After emphasizing via a version of Lewontin's metaphor that a genetic basis for individual differences does not imply a genetic basis for group differences, they said of U.S. ethnic differences in average IQ (p. 312):

> They may well include some (as yet unknown) genetic component, but nothing suggests that they are entirely genetic. And, most important, it matters little whether the genes are involved at all.

Their argument in support of the second sentence was that for an appropriate treatment of an individual it is his or her own IQ that is relevant (if IQ is relevant at all), not the average IQs of some group to which the individual may belong. One might add, however, that for long-term social policy, the fact that an average group difference has its source in genes or in the environment can sometimes matter, because it can affect the choice of a remedy to alter that difference – eugenics versus Head Start, for example.

Herrnstein on Social Class and IQ

The Bell Curve did not represent Herrnstein's first engagement with group differences and public controversy. In an article in *The Atlantic* (Herrnstein, 1971) and in a subsequent book, *I.Q. in the Meritocracy* (1973), Herrnstein elaborated on an idea by Cyril Burt (1961) that social class and occupational differences in IQ will be partly genetic in a society that features social mobility. If IQ is partly genetic, and higher IQ individuals tend to move up in social and occupational status, while lower IQ individuals tend to move down, then IQ differences between social classes and occupational

groups will come to be partly genetic. This is not a hereditary aristocracy – far from it – it is a dynamic phenomenon
that depends on continued mobility up and down the social
scale. An important question is, How much? Some evidence
suggests that about 40% of IQ differences in occupation and
income in Western societies are associated with genetic differences (Rowe, Vesterdal, & Rodgers, 1998; Tambs, Sundet,
Magnus, & Berg, 1989). Phenotypically, there are substantial average differences in IQ between different occupational
groups. For example, in the U.S. standardization sample for
the 1981 revision of the Wechsler Adult Intelligence Scale,
there was a 22-point difference between the average IQs of
persons in professional and technical occupations and persons who were unskilled laborers (Reynolds, Chastain, Kaufman, & McLean, 1987). And yet there was nearly as much
variation in IQ within these two occupational groups (standard deviations of 14.4 and 15.2) as in the U.S. population
as a whole (standard deviation of 15.1). It is an interesting
paradox that there may be real and significant differences in
average IQ between different groups, yet individuals vary so
widely within them that an individual's group membership is
of almost no value for predicting his or her IQ.

The Glayde Whitney Affair

In his 1995 presidential address to the Behavior Genetics
Association, Glayde Whitney, whose distinguished research
career had mostly focused on taste sensitivity in mice, turned
to humans and elected to address the topic of black–white
differences in the frequency of criminal behavior. He pointed
out the large discrepancies on the phenotypic level, such as
a ninefold difference in murder rates between blacks and
whites in the USA. Compared to a dozen other industrialized
countries, the USA had the highest overall murder rate. However, based only on its white population, it ranked third from
the bottom, with a lower murder rate than such countries
as Switzerland, Denmark, Finland, and Sweden. Whitney
argued that behavior geneticists should be willing to explore
both genetic and environmental hypotheses about such differences; he also argued that the current intellectual climate
in the USA made such discussion virtually impossible – and
he made some critical remarks about the contribution of the
political Left to this situation (Whitney, 1995).

Whitney's address was perhaps not a model of tact: for
example, in addition to his comments about the Left, he
noted that Richmond, Virginia, the city in which he was
speaking as a guest, was the second-worst large city in the
USA with respect to its murder rate. Nor did he address
the question of how behavior geneticists were to go about
deciding to what extent the group differences in criminality
were genetic or environmental. Subsequent events within the
Behavior Genetics Association proved, however, that he was

clearly right about the difficulty of public discussion of such
questions. An announcement was issued the next day by the
BGA Executive Committee to the effect that Whitney was
not acting as the official spokesman of the association, that
presentations at BGA meetings should be strictly scientific,
and that "members are not encouraged to express their personal political and moral views" (Heath, 1995, p. 590). A
special December meeting of the BGA Executive Committee
was scheduled to consider removing Whitney from the BGA
Board of Directors, of which he was automatically a member
as past president (e-mail announcement to the BGA membership, October 12, 1995). President-elect Pierre Roubertoux
and Wim Crusio, a member-at-large of the Executive Committee, resigned from the association because it was unwill-
ing to adopt sufficiently strong sanctions against Whitney.
The incoming president-elect, Nicholas Martin, took over
for Roubertoux as president, and later served his own term,
accounting for his double appearance in Table 1.1, in 1996
and 1997 (Heath, 1996).

Lawrence Summers and Sex Differences

On January 14, 2005, Harvard President Lawrence H. Summers informally addressed a conference on "Diversifying the
Science and Engineering Workforce" which was considering
the reasons for a shortage of women at the highest levels in
the scientific professions (Summers, 2005). With the avowed
intention of provoking discussion, Summers proposed three
hypotheses for his audience's consideration: (a) Many tal-
ented women prefer devoting some of their time to children
and families rather than undertaking the 80-hour work-weeks
required for reaching the top levels in elite research organi-
zations; (b) there may be biological differences between the
sexes, such as a greater variance for males on many traits,
producing an excess of males at the extremes; and (c) subtle
and not-so-subtle patterns of discrimination may exist that
lead the present elite in these fields, mostly males, to choose
others like them to join them. Summers thought it likely that
all three of these factors contributed, and he guessed that
they might rank in importance in the order given. Summers
is an economist by training, not a behavior geneticist, but he
cited some behavior genetic evidence against an overwhelming role of socialization in producing behavioral differences,
and suggested that the effects in hypotheses (a) and (b) might
have in part a biological basis. Summers' remarks aroused a
firestorm in the press and in feminist circles, which in turn
provoked assorted indignant rejoinders. It is not necessary to
pursue these in detail here – a quick survey on the Internet
will yield an ample sampling of widely varying views about
Summers' remarks – views expressed with widely varying
degrees of heat and light. Pinker (2002, Chap. 18) provides
a readable survey of the considerable evidence that at least

some male–female psychological differences have a biological component – although, presumably, few are exclusively so, and many questions remain open empirically.

The Future?

One take-home lesson from the various controversies concerning group differences is that the nature–nurture controversy is not dead, even though it has been declared moribund on many occasions in recent decades. Although behavior geneticists have had an appreciable impact on public thinking about individual differences, the question of the relative genetic and environmental contributions to group differences has been both more socially explosive and much less successfully addressed empirically.

What does the future hold? This will depend, in part, on future behavior genetics research on these topics – some of it, perhaps, carried out by readers of this book. One may be fairly confident that nature–nurture controversies will not vanish completely anytime soon. However, one may hope that as knowledge expands, the cloud of misunderstandings on which these controversies feed will gradually shrink, and that one day we may have an agreed-upon body of facts on which to base social policy.

Conclusion

Yes, behavior genetics has had a long past, which extends into the nature–nurture controversies of the present day. It has also had a short but solid history of substantive accomplishment and institutional establishment. The date at which the short history will make the long past seem quaint and obsolete in the eyes of the general educated public remains to be determined. Readers of this book will help determine it.

References

Brewer, D. J., Clark, T., & Phillips, A. (2001). *Dogs in antiquity*. Warminster, England: Aris & Phillips.

Burks, B. S. (1928). The relative influence of nature and nurture upon mental development: A comparative study of foster parent-foster child resemblance and true parent-true child resemblance. In *27th Yearbook of the National Society for the Study of Education, Part 1*, pp. 219–316.

Burt, C. (1961). Intelligence and social mobility. *British Journal of Statistical Psychology, 14*, 3–24.

Darwin, C. (1859). *On the origin of species by means of natural selection*. London: John Murray.

Darwin, C. (1871). *The descent of man, and selection in relation to sex*. London: John Murray.

Darwin, C. (1872). *The expression of the emotions in man and animals*. London: John Murray.

Ebbinghaus, H. (1908). *Psychology: An elementary textbook* (M. F. Meyer, Trans.). Boston: D. C. Heath.

Flynn, J. R. (1980). *Race, IQ and Jensen*. London: Routledge & Kegan Paul.

Freeman, F. N., Holzinger, K. J., & Mitchell, B. C. (1928). The influence of environment on the intelligence, school achievement, and conduct of foster children. In *27th Yearbook of the National Society for the Study of Education, Part 1*, pp. 103–217.

Fuller, J. L., & Thompson, W. R. (1960). *Behavior genetics*. New York: Wiley.

Galton, F. (1869). *Hereditary genius: An inquiry into its laws and consequences*. London: Collins.

Galton, F. (1883). *Inquiries into human faculty and its development*. London: Macmillan.

Graves, J. L., Jr. (2001). *The emperor's new clothes: Biological theories of race at the millennium*. New Brunswick, NJ: Rutgers University Press.

Heath, A. C. (1995). Secretary's report: The 25th annual meeting of the Behavior Genetics Association, Richmond, Virginia. *Behavior Genetics, 25*, 589–590.

Heath, A. C. (1996). Secretary's report: The 26th annual meeting of the Behavior Genetics Association, Richmond, Virginia [Pittsburgh, Pennsylvania]. *Behavior Genetics, 26*, 605–606.

Herrnstein, R. J. (1971). I.Q. *Atlantic Monthly 228*(3), 43–64.

Herrnstein, R. J. (1973). *I.Q. in the meritocracy*. Boston: Little, Brown.

Herrnstein, R. J., & Murray, C. (1994). *The bell curve: Intelligence and class structure in American life*. New York: The Free Press.

Homer. (n.d./1909). *The Odyssey* (S. H. Butcher & A. Lang, Trans.). New York: Collier.

Hunt, M. (1999). *The new know-nothings*. New Brunswick, NJ: Transaction Publishers.

Jensen, A. R. (1969). How much can we boost IQ and scholastic achievement? *Harvard Educational Review, 39*, 1–123.

Jencks, C., & Phillips, M. (Eds.) (1998). *The Black-White test score gap*. Washington, DC: Brookings Institution.

Lewontin, R. C. (1970). Race and intelligence. *Bulletin of the Atomic Scientists, 26*(3), 2–8.

Locke, J. (1690/1975). *An essay concerning human understanding* (P. H. Nidditch, Ed.). Oxford: Clarendon Press.

Locke, J. (1693/1913). *Some thoughts concerning education* (R. H. Quick, Ed.). Cambridge: Cambridge University Press.

Loehlin, J. C. (2000). Group differences in intelligence. In R. J. Sternberg (Ed.), *Handbook of intelligence* (pp. 176–193). Cambridge: Cambridge University Press.

Loehlin, J. C., Lindzey, G., & Spuhler, J. N. (1975). *Race differences in intelligence*. San Francisco: Freeman.

Mill, J. S. (1873). *Autobiography*. London: Longmans.

Morey, D. F. (1994). The early evolution of the domestic dog. *American Scientist, 82*, 336–347.

National Society for the Study of Education (1928). *27th yearbook: Nature and nurture*. Bloomington. IL: Public School Publishing.

Pearson, R. (1991). *Race, intelligence and bias in academe*. Washington, DC: Scott-Townsend.

Pearson, R. (1995). The concept of heredity in the history of Western culture: Part One. *The Mankind Quarterly, 35*, 229–266.

Pinker, S. (2002). *The blank slate: The modern denial of human nature*. New York: Viking.

Plato (n.d./1901). *The Republic* (B. Jowett, Trans.). New York: Willey Book Co.

Porter, N. (1887). Marginalia Locke-a-na. *New Englander and Yale Review, 11*, 33–49.

Reynolds, C. R., Chastain, R. L., Kaufman, A. S., & McLean, J. T. (1987). Demographic characteristics and IQ among adults: Analysis of the WAIS-R standardization sample as a function of the stratification variables. *Journal of School Psychology, 25*, 323–342.

Rowe, D. C., Vesterdal, W. J., & Rodgers, J. L. (1998). Herrnstein's syllogism: Genetic and shared environmental influences on IQ, education, and income. *Intelligence, 26*, 405–423.

Rushton, J. P., & Jensen, A. R. (2005). Thirty years of research on race differences in cognitive ability. *Psychology, Public Policy, and Law, 11*, 235–294.

Savolainen, P., Zhang, Y., Luo, J., Lundeberg, J., & Leitner, T. (2002). Genetic evidence for an East Asian origin of domestic dogs. *Science, 298*, 1610–1613.

Summers, L. H. (2005). *Remarks at NBER conference on diversifying science & engineering workforce*. Retrieved September 18, 2005, from http://www.president.harvard.edu/speeches/2005/nber.html

Tambs, K., Sundet, J. M., Magnus, P., & Berg, K. (1989). Genetic and environmental contributions to the covariance between occupational status, educational attainment, and IQ: A study of twins. *Behavior Genetics, 19*, 209–222.

Vandenberg, S. G., & DeFries, J. C. (1970). Our hopes for behavior genetics. *Behavior Genetics, 1*, 1–2.

Watson, J. B. (1925). *Behaviorism*. New York: Norton.

Whitney, G. (1995). Twenty-five years of behavior genetics. *Mankind Quarterly, 35*, 328–342.

Part I
Quantitative Methods and Models

Chapter 2

Biometrical Models in Behavioral Genetics

Michael C. Neale

Introduction

The main goal of this chapter is to describe the research designs and statistical methods that are in popular use in behavioral genetics (BG). We begin with a brief overview of the historical background to BG in general and twin studies in particular. Next, we describe some elementary statistics required for understanding biometrical modeling. Then follows a statistical model for genetic variation, as articulated by Fisher in his classic 1918 paper, in which additive and dominance genetic variance terms are defined. The coefficients of resemblance between relatives derived from this model are then implemented in structural equation models for the analysis of data from twins and other relatives. Overall the intent is to provide a general and extensible infrastructure for the modeling of genetically informative data.

Historical Background

Behavior genetics is the synthesis of two domains: behavior, which is defined as the actions or reactions of an object or organism, and genetics, which is the science of heredity and variation. The primary focus of contemporary BG is variation in behavior, while broader psychological constructs such as internal mental states and cognition are frequently included. Individual differences in this activity are readily observed in virtually all forms of animal life and may also be seen in certain plant species, such as *Dionaea muscipula* (the Venus Fly Trap). The ability to predict behavior in other organisms, be they of the same or a different species – would seem to have substantial survival value. Today, tremendous investment is made by both medical and military agencies in order

to understand the origins of behavioral differences, and with good reason. Many of the most pressing health problems in modern cultures have behavioral components: obesity, cardiovascular disease, cancer, drug abuse and psychopathology are obvious examples. It is also the case that human conflicts, be it a marital dispute, a street fight, or a world war, are primarily behavioral. Thus much of human suffering has behavioral origins. One aim of BG is to identify potential ways to alleviate this distress by correctly identifying both genetic and environmental sources of individual differences in behavior and susceptibility to environmental insults.

Behavioral genetics as a field was perhaps first established by the exceptional 18th century cousins, Charles Darwin and Francis Galton. The former, in Chapter 8 of *On the Origin of Species* (Darwin, 1859) discusses instincts in animals as diverse as dogs, birds, insects, and notes individual differences in behavior within species. In his later work, the Descent of Man (Darwin, 1871), he wrote:

> If no organic being excepting man had possessed any mental power, or if his powers had been of a wholly different nature from those of the lower animals, then we should never have been able to convince ourselves that our high faculties had been gradually developed. But it can be shown that there is no fundamental difference of this kind.

It is well known that different breeds of dog have different average temperaments. Typically, Labradors are affectionate, Border Collies are intelligent and Bull Terriers are aggressive. Even the most ardent critics of behavioral genetics do not seem to quibble with this or any other behavioral differences that are observed in either domesticated species or those in the wild (Mann, 1994). That selection experiments can produce reliable behavioral differences between strains of rats and mice is well established for numerous traits, including mazesolving ability (Tryon, 1941), activity (Defries, Gervais, & Thomas, 1978), brain weight (Fuller & Herman, 1974), and alcohol preference (Li, Lumeng, & Doolittle, 1993). Selective breeding experiments are essentially univariate in design; those with high or low scores on the single trait of interest are used to populate the next generation. However, it is commonly observed that changes in

M.C. Neale (✉)
Departments of Psychiatry and Human Genetics, Virginia Institute for Psychiatric and Behavioral Genetics, Virginia Commonwealth University, Richmond, VA 23298-0126, USA
e-mail: neale@vou.edu

Y.-K. Kim (ed.), *Handbook of Behavior Genetics*,
DOI 10.1007/978-0-387-76727-7_2, © Springer Science+Business Media, LLC 2009

other, secondary, phenotypes occur together with the selected trait. Although many human traits might respond to selection, it is widely considered morally unacceptable to impose reproductive constraints (or to carry out 'ethnic cleansing') on human beings. The author knows of no behavior geneticist who would consider such activities. Rather, the goal of BG is to identify putative genetic and environmental pathways that may prove fruitful targets for the prevention and treatment of mental and physical disorders.

During the past 30 years, the classical twin study has been extensively used to differentiate between genetic and environmental sources of variation in human populations. The idea is to compare the similarity of MZ twin pairs reared together by their parents in the same home to that of DZ twin pairs reared in the same circumstances. Today there are dozens of well-established registries of twins around the world (Busjahn & Hur, 2006). Exactly to whom this very popular research design should be credited is a matter of some debate. While many credit Sir Francis Galton (1875), he did not explicitly propose this comparison. At the time, the distinction between MZ and DZ twins' genetic relationship was not clear. Indeed, Thorndike (1905) doubted the existence of two types of twins, which may have retarded the development of the twin method. As discussed by Rende, Plomin, and Vandenberg (1990) the first published comparison of MZ and DZ twins' similarities appears to be that of Merriman (1924). In the same year, the dermatologist Siemens described the approach in his book *Die Zwillingpathologie* (twin pathology). That period also saw the advent of adoption studies. Gordon (1919) found that sibling resemblance for cognitive ability was approximately the same for pairs reared together in the same home as for those reared apart. It is reasonable to expect that most of the readers of this volume will witness the centennials of Fisher's classic paper, the adoption and the twin study.

The advent of molecular genetics, from the elucidation of the structure of deoxyribonucleic acid (DNA) in the 1950s by Watson and Crick (1953) and others to the development of polymerase chain reaction (PCR) for DNA amplification by Mullis (1990) and subsequent explosive growth of biotechnology has had a dramatic impact on modern behavioral genetics. Early approaches to establish genetic linkage focused on known measurable genetic polymorphisms such as blood groups or the human leukocyte antigen region, as these were the only readily available genetic markers (this is akin to looking under the lamppost for one's keys even though it seems unlikely that that is where they were lost). The identification of microsatellite markers along the genome at approximately 10 million base-pair intervals (the human genome is approximately 3 billion base-pairs in length) permitted linkage studies to provide approximate localization of variants responsible for individual differences. Fine mapping was typically restricted to regions

of linkage or loci that were plausible candidates by knowledge of their function. In the past few years, microarray technology has enabled very large-scale genotyping, of over a million single nucleotide polymorphisms (SNPs) on a single array. This has ushered in an era of whole genome association studies, which holds considerable promise for dissecting the origins of individual differences in behavior. These developments in measuring genomic variation have been paralleled by synergistic advances in statistical methods to analyze data.

For further readings in the history of genetics, the reader is referred to the internet resource Wikipedia (http://www.wikipedia.org). The maxim 'don't believe everything you read' which is a key aspect of scientific method should of course be applied to Wikipedia articles – as well as, e.g., this chapter. At the time of writing, the Behavioral Genetics entries in Wikipedia are in need of revision, something that the Behavioral Genetics Association has noted, and will hopefully address by the time this volume is published.

Measuring Variation and Covariation

Note that whatever the phenotype being studied, it is variation in the phenotype that is the focus of behavior genetics. As in any other scientific domain, measurement is key. One consideration is how we measure the traits of interest. The second is how we measure variation and covariation in those traits, which is a matter for some elementary statistics, which we discuss now.

Summarizing Variation

Most people are familiar with the concept of the mean or average of a set of measurements. This quantity is simply calculated by dividing the sum of a series of measurements by the number, N of measures that have been made. The conventional notation for this operation is:

$$\mu = \frac{\sum_{i=1}^{N} x_i}{N}$$

where the symbol μ is the mean, x_i is the measurement obtained from case i, and $\sum$ denotes the summation. The idea behind measuring variation is to measure the average dispersion or spread of scores from the mean. Several possible measures of dispersion might be taken: the average absolute distances from the mean, the average squared deviation from the mean, or average of any higher even-numbered power (4, 6, 8, etc.) of the deviation from the mean. Statisticians almost always use the squared deviation, because this measure is, on average the most accurate. In other words, if we took a sample of measurements and computed these alternative measures of variance, and repeated this process multiple times, and then compared the variance of these measures of variance, the squared deviations would have the smallest variance. This result was shown by Fisher (1922). The formula for the variance that is most commonly used is:

$$\sigma_x^2 = \frac{\sum_{i=1}^{N}(x_i - \mu)^2}{N - 1}.$$

The $N - 1$ denominator departs slightly from the formula for the average (whose denominator is N) because it is less biased in small samples. On occasion, N may be used in place of $N - 1$; for example, the maximum likelihood estimate of a variance is equivalent to using the divisor N rather than $N - 1$. Asymptotically, as N tends to infinity, these estimates converge. The average-squared deviation from the mean is a general statistic. It does not require that the data follow a particular distribution in the population. Indeed, we employ it below in the context of measuring the variation due to a single diallelic locus in Equation 2.2. At the same time, the mean and the variance are not always *sufficient statistics* in that they do not describe all distributions up to an arbitrary constant. As discussed in Equation 2.2, the mean and the variance are sufficient for the normal distribution, but this is not generally the case.

The next important consideration is how to measure covariation. As discussed in the biometrical genetics below, the model predicts that if genetic factors influence a trait, then MZ twin pairs will (on average) show greater covariation than will DZ twin pairs. Following the same general principle for average variation, we wish to determine the average of the extent to which the deviations from the mean of one measure are similar to those of another measure. Thus the formula for the covariance is:

$$\sigma_{xy} = \frac{\sum_{i=1}^{N}(x_i - \mu_x)(y_i - \mu_y)}{N}.$$

Covariance is maximized when $x_i - \mu_x$ and $y_i - \mu_y$ are equal for all $i = 1 \ldots N$ pairs of data points in the sample. The covariance is then equal to the square root of the product of the variances of x and y, and the correlation (defined below) equals unity. These summary statistics – means, variances and covariances – are commonly used to provide an overview of characteristics of the data. Many genetic models predict that the variances of MZ and DZ twins should be equal. Conversely, certain violations of the assumptions of the twin method predict that these variances would differ; for example, if one's cotwin forms a substantial part of the relevant trait environment, then different total variances of MZ and DZ twins would be expected (Eaves, 1976; Carey, 1986).

Especially valuable in the inspection of data from different classes of relatives is a comparison between their correlation coefficients. The correlation is simply the standardized covariance:

$$r_{xy} = \sigma_{xy}/\sigma_x\sigma_y.$$

Another commonly encountered statistic in biometrical genetic analysis is the regression coefficient. Most readers should be familiar with the simple linear regression formula:

$$Y = a + bX$$

Following some simple algebra (see almost any introductory statistics text, e.g., Edwards, 1979) it can be shown that the least squares solution to this equation yields an estimate of b:

$$\hat{b} = \frac{\sigma_{xy}}{\sigma_x^2}.$$

In essence, the regression of Y on X may be thought of as the covariance between X and Y scaled by the variance of X. When X is standardized to unit variance, the covariance equals the regression coefficient. It is on this simple model that the idea of *explained variance* is based. The residual variance of Y – that not explained by the regression – is $\sigma_y^2 - b^2$ or in the case of standardized X and Y, it is $1 - b^2$ so the variance explained by the regression is b^2. There is a potential pitfall when considering data from relatives in this fashion. Our conceptual model for twin resemblance is that there are certain factors (genetic or environmental) that relatives share. The relatives' phenotypes are regressed on these shared factors, rather than, e.g., the phenotype of twin 1 regressed on that of twin 2. The variance explained by the common factor would equal the correlation, rather than the square of the correlation. This problem, noted by Loehlin (1996) has not stopped the development of a useful regression-based approach to the analysis of twin data, known as DeFries–Fulker regression.

$$P_{T1} = B_1 P_{T2} + B_2 R + K$$

where P_{Tj} is the phenotype of twin j; B_1 is the regression on the cotwin's score, B_2 is the regression on the coefficient of relationship R, which is set to 1 for MZ twins and 0.5 for DZ twins; and K is the regression constant. B_2 therefore reflects differential twin resemblance. The augmented model

$$P_{T1} = B_3 P_{T2} + B_4 R + B_5 P_{T2} R + K$$

directly estimates additive genetic effects via the parameter B_5 and shared environment effects via parameter B_3.

The method has several virtues. First is that it is computationally straightforward, in that it can be implemented in practically any standard statistical package. Second, it can be extended in numerous ways. Fulker and Cardon (1994) showed the incorporation of measured genetic markers to test for the genetic linkage to a measured putative quantitative trait locus. Third, it is particularly convenient for the analysis of data from selected samples (DeFries & Fulker, 1988; LaBuda, DeFries, & Fulker, 1986). It also provides a natural framework for testing for interactions. At the same time there are some limitations. One is that the specification of genetic factor models, which incorporate latent variables which influence several or all of the measures is not straightforward. A second is that the method of double entry of the data (in which both twin 1 and twin 2 are used as dependent variables) does not naturally lead to the analysis of more extended pedigrees. However, some recent developments in this area show considerable promise in making the analysis of more complex pedigree structures practical. McArdle and Prescott (McArdle & Prescott, 2005) show that by judicious coding of dummy variables it is possible to analyze nuclear family structures in a relatively general fashion. An alternative parameterization of this model, which is more efficient for binary variables, was described by Rabe-Hesketh, Skrondal, and Gjessing (2007). It seems reasonable to expect that multivariate extensions of this approach will be developed in the future. However, the explicit modeling of common genetic and environmental factors that influence some or all the observed measures in a multivariate analysis would seem complicated at best in the context of this method.

Binary and Ordinal Measures

Many traits of interest to the behavior geneticist are not measured on a continuous, interval scale. Psychiatric disorders, substance use and abuse, and political affiliation are examples of measures that are inherently categorical in nature. Even traits that approximately follow the normal distribution in the population, such as measures of cognitive ability or personality, are typically derived from the aggregation of a set of binary or ordinal items that are designed to measure the trait in question. Most psychiatric diagnoses are based on the presence of one or more required signs or symptoms, together with a sufficient number of additional criteria (e.g., five from nine possible symptoms of depression).

There is an interesting duality to the specification of genetic models. As we have seen, the models typically begin at the binary level, with SNPs on the genome that affect the expression of genes which, typically through quite convoluted biological pathways, subsequently generate differ-

ences in measured outcome traits. Since Fisher's classic 1918 paper, and perhaps even earlier (Pearson, 1901) the idea that a large number of such elementary factors combine to generate trait variation has had substantial appeal. That many human physical traits follow the normal distribution that is generated (per the central limit theorem) from the aggregation of a large number of factors, increases the appeal of the multifactorial model. The question then becomes of how best to model binary or ordinal data.

The inherent complexity of the systems that generate behavior – most obviously the structure and function the brain – suggests that it is unlikely that any one SNP will have a major outcome on behavior. There are counterexamples, of course, such as the metabolic disorder phenylketonuria, which without appropriate intervention leads to severe cognitive deficits, but these may be relatively rare. Particularly for traits related to reproductive fitness, it seems likely that the systems involved will be inherently redundant, since all other things being equal, the organism with a 'failsafe' redundant system will be more likely to reproduce than one without. It is natural, therefore, to expect that a large number of genetic variations across the human genome will influence behavioral outcomes. The same expectation seems likely to hold for environmental factors and for the interactions within and between these primary sources. Thus even traits that are measured at the binary level – presence or absence of major depressive disorder, for example – may be appropriately modeled by assuming that there is a continuous, underlying normal distribution of liability with a threshold. Those with liability below this threshold would not be affected by the disorder, while those with liability above it would. This situation is shown in Fig. 2.1.

It should be emphasized that the underlying normal distribution of liability model is not be appropriate for all binary

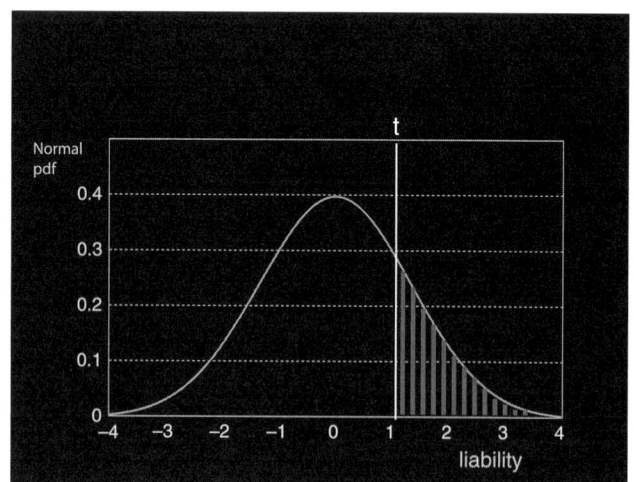

Fig. 2.1 Liability threshold model. Liability is assumed to be normally distributed in the population; those with liability above threshold t are affected (score 1) and those with liability below are unaffected (score 0)

traits. For example, a SNP genotype itself might be regarded as a binary or ordinal trait (zero, one or two copies of a particular SNP allele), but it would not be appropriate to suppose that there is an underlying normal distribution of liability to having this genotype. Likewise, any phenotype purely related to a particular SNP genotype would be better modeled in terms of allele frequencies than as an underlying continuous liability distribution (Risch, 2001). Ultimately, the number of genetic factors that influence a particular trait is an empirical question, and one that is likely to be answered in the relatively near future, due to the advent of SNP chip genotyping technology.

The traditional summary statistics for the liability threshold model closely correspond to the mean and correlation for continuous data. Typically, the mean of the liability distribution is assumed to be zero. The threshold is usually estimated in a z-score metric of standard deviations. The covariance between two liability dimensions is not calculated via the usual Pearson's formula for the covariance because this statistic is biased toward zero to an increasing extent as the ratio of those above vs. below the threshold departs from 50:50 (so variables with a small proportion of 0's or 1's are biased more towards zero). Instead, a tetrachoric correlation coefficient may be used. This statistic can be estimated from the data by computing the proportions of observations that would fall into the 00, 01, 10, and 11 categories for a given pair of thresholds and tetrachoric correlation on the underlying bivariate normal distribution. Thus the tetrachoric correlation may be estimated through a model-fitting method, where the predicted quadrants of the bivariate normal distribution are 'matched' to the observed data through what is generally known as a loss function. Alternative estimates of the thresholds and the correlation are tried iteratively until those that best predict the frequencies of the four outcomes are found. Similar procedures may be used for ordinal-level measurements, such as may be obtained from a Likert scale with ordered response categories (e.g., never, rarely, often or always). The same underlying normal distribution model with thresholds may be used to predict the cell frequencies, and a polychoric correlation may be estimated. An important distinction is that when there are more than two categories of outcome it is possible to test how well the bivariate normal distribution fits the contingency table data. Failure of the model may occur for many reasons, such as un-modeled population heterogeneity, or non-linearity of the relationship between latent trait score and response outcome. While estimation of variance components may be robust to minor violations of the model assumptions, it is always good practice to try to establish the origins of departures from expectations. We now turn to the derivation of the expected resemblance between relatives from genetic theory, before considering the application of model-fitting to behavior genetic data.

Biometrical Genetics

The principles of biometrical and quantitative genetics are central to practically all the statistical models employed in this book. A little knowledge of elementary biometrical genetics can provide valuable insights into the variance components being estimated and the assumptions being made about the mode of gene action in the population. A thorough account of biometrical genetics would easily fill an entire volume. Those interested in deeper study of this area might consult any of the excellent textbooks on the subject, such as Mather and Jinks (1982), Falconer (1990) or Lynch and Walsh (1998). Here we review the derivation of additive and dominance variance components from basic principles. Our treatment follows that of David Fulker, published in the Neale and Cardon (1992) volume.

At the time of writing, genetics is a discipline which is rapidly changing in many respects: genotyping technology, bioinformatics, statistical methods, substantive findings, and even nomenclature are all in a state of flux. For the most part, this situation is excellent, because much scientific progress is being made. However, the nomenclature deserves attention in this section to avoid possible confusion. In what follows, we use the term *gene* to refer to a 'unit factor of inheritance' that influences an observable trait or traits, following the earlier usage by Fuller and Thompson (1978). Measured traits are referred to as *phenotypes*. The position of a gene on a chromosome is known as the *locus*. Any gene may have multiple *alleles*, which are alternative forms at the same locus. The simplest type of allele we consider is a *Single Nucleotide Polymorphism*, or *SNP*, which is where only a single base-pair differs. In earlier works, alleles are frequently written with uppercase vs. lowercase letters, such as A and a, or B and b (letters such as C are not ideal for this purpose because it has the same shape in both cases). An alternative notation is to denote alleles by subscript: A_1 vs. A_2, and it is common in molecular genetic work to simply refer to SNP alleles as 1 or 2. When larger strands of DNA are considered (*haplotypes*, which may span part or all of one or more genes), a larger number of alleles may exist. For statistical purposes, a two-allele system (known as *diallelic* or *biallelic*) may still provide a useful approximation, such as when there is one operational allele and multiple inoperational forms. The *genotype* is the chromosomal complement of aileles for an individual. At a single diallelic locus, the genotype may be symbolized AA, Aa, or aa; when two loci are considered, the genotype may be written $AABB$, $AABb$, $AAbb$, $AaBB$, $AaBb$, $Aabb$, $aaBB$, $aaBb$ or $aabb$. If an individual has the same allele at the same locus on both chromosomes we say they are *homozygous* – genotype AA or aa. If their alleles differ, they are said to be *heterozygous* and would have, e.g., genotype Aa. In a quantitative trait, it is common to consider the average trait value of individuals with a particular genotype, which are referred to as *genotypic values*. The *additive value* of a gene is the sum of the average effects of the individual alleles. A *dominance deviation* refers to the extent to which the heterozygote genotypic value differs from the mean of the genotypic values of the two homozygotes. Genetic variation in a trait is referred to as *polygenic* ('many genes') when many genes influence the trait. It seems likely that almost all traits are polygenic, since the pathway from genotype to phenotype is rarely simple. Disorders (or very large effects on a quantitative trait) that are caused by a single gene may be referred to as *monogenic* or *Mendelian*. However, it is often the case that phenotypic variability remains among individuals who have the same genotype at the *major locus* which has a large effect on the phenotype. *Pleiotropy* occurs when a gene or set of genes influences more than one trait. Again, it is likely that any gene that influences one trait will also influence others. Pleiotropic effects seem especially likely when considering *endophenotypes* (Gottesman & Gould, 2003) which may be defined as being (i) associated with the phenotype in the population, (ii) heritable, (iii) state-independent (e.g., present regardless of the phenotypic value of an individual), and (iv) correlated with the phenotype within families.

Most biometrical models are initially defined in terms of gene action at a single locus, but assume that the system is polygenic. The development of this model was one of the many seminal contributions of Sir Ronald Fisher (Fisher, 1918). Typically it is assumed that many loci influence the trait in question and that each locus has a relatively small effect. This assumption is consistent with the central limit theorem, which states that asymptotically (as the number of factors of small and equal effect tends to infinity) a normal distribution of the trait will emerge. In reality, this assumption has to be incorrect because the genotype of any species consists of a finite number of genes. Nevertheless, it is likely to provide a good approximation to a normal distribution even when the number of genes is a few as 10 (Kendler & Kidd, 1986). The central limit theorem may also apply when the factors involved are of unequal effect, or are non-independent (Lehmann, 1998). Should a phenotype be influenced by a major locus, the remaining genetic variation may be polygenic. As will become clear, the biometrical genetic model is sufficiently general and extensible to cover a wide variety of models of gene action.

In a diallelic system with alleles B and b there are three possible genotypes BB, Bb, and bb. If we were to measure a phenotype from a sample of the population, and then calculated the mean for each genotype, three observed statistics (the means) would be obtained. Three parameters might be used to characterize these means: a grand mean (μ) for the population, the distance between the two homozygotes ($2a$), and the deviation d of the heterozygote from the midpoint

of the two homozygotes. Thus the genotype values would be $\mu - a$ for bb, $\mu + d$ for Bb, and $\mu + a$ for BB. Estimates of the model parameters μ, a, and b could be obtained from an appropriately genotyped and phenotyped sample. It is possible then to test whether d or a may be set to zero, to conduct a simple population-based association study. Statistically, we could conduct this test using one-way analysis of variance or by likelihood ratio test in a model-fitting context. We can also consider how much variation a locus that operates in this fashion would generate. This component of variance could then be compared to the variance in the population as a whole, to obtain a proportion of variance accounted for, which we refer to as *locus heritability*. The variation caused by a diallelic locus depends not only on the a and d parameters but also on the frequencies, p and $q = 1 - p$ of the two alleles.[1]

Note that all three genotypes share the term μ for the population mean. Since adding a constant to a set of observations does not change the variance (which is simply the average squared deviation from the mean), we can calculate the variance generated by the locus with the grand mean parameter μ set to zero. Parameters, a and d are referred to as *genotypic effects*. They are shown graphically in Fig. 2.2.

To make this model concrete, suppose that we are considering genetic effects on a measure of intelligence (IQ) which has a mean of 100 and a standard deviation of 15. Let us assume that there are 50 loci with approximately equal and purely additive effects on the phenotype and that at each locus the alleles are equally frequent in the population. If allele b at each locus contributes -1 IQ point, and allele B contributes $+1$ IQ point, then individuals with genotype bb will score (on average) 98, those with Bb will score 100, and those with BB will score 102. The estimate of the parameter a for the B locus would be $+1$, and parameter d would be

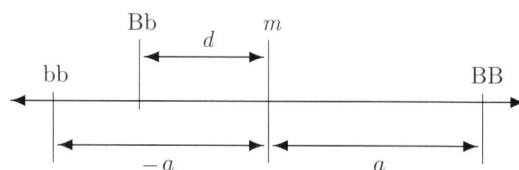

Fig. 2.2 The additive (a) and dominance (d) parameterization of the gene difference $B - b$. Bb may lie on either side of the midpoint, m, of the two homozygotes, and the sign of d will vary accordingly; in the case illustrated d would be negative. Similarly, the mean of the BB homozygote may be greater or less than that of the bb homozygotes and the estimate of a would be positive or negative, respectively (Adapted from Mather & Jinks, 1977, p. 32)

---

[1] The term frequency is technically incorrect here, because frequency refers to count, and p and q are effectively allele *proportions*, but the term allele frequency is in widespread use to mean proportion in this context.

estimated at zero. Individuals with a full set of the decreasing alleles at every locus would be expected to have a mean IQ of zero, while those with a full set of increasing alleles would be expected to average 200. Note that we are assuming here that (i) the 50 genes operate *independently* – their action is influenced neither by genes at other loci nor by environmental factors; (ii) having allele *B* does not influence the probability that one has a particular allele at another locus (i.e., there is no disequilibrium due to e.g., linkage or assortative mating); and (iii) there is no genotype by environment interaction.

Breeding Experiments: Crosses with Inbred Strains

Biometrical genetics originated with the analysis of data on inbred strains, particularly in R. A. Fisher's work on agriculture. Breeding experiments with inbred strains provide a very simple and intuitive framework for understanding genetic contributions to population variation. The analysis of data from outbred populations, such as humans, is compared.

Inbred strains formed by multiple generations of brother–sister or parent–offspring matings are homozygous at almost every locus. Mutation is expected to generate heterozygosity at a small number of loci, but for the most part homozygosity will be observed throughout the genome. When two such strains ('parental strains' P_1 and P_2) are crossed, the resulting offspring strain (known as first-generation filial, or F_1) will be homozygous at all loci at which the parental strains do not differ, and heterozygous (Bb) at all loci at which they do differ. A cross between two F_1 individuals creates what Gregor Mendel termed the 'second filial' generation, or F_2, and it may be shown that this generation would be expected to comprise 1/4 individuals of genotype BB, 1/4 bb, and 1/2 Bb. Mendel's first law, the *Law of Segregation*, states that parents with genotype Bb will produce the gametes B and b in equal proportions. The geneticist Reginald Punnett developed a device known as the *Punnett square*, which gives the proportions of genotypes expected to arise under random mating. A simple example of the Punnett square is shown in Table 2.1 for the mating of two heterozygous parents in a diallelic system. The gamete frequencies in Table 2.1 (shown out-side the box) are known as *gene or allelic frequencies*, and they can be used to calculate the *genotypic frequencies* by a simple product of independent probabilities. This assumption of independence makes the biometrical model tractable yet readily extends to more complex situations, such as random mating in populations where the gene frequencies are unequal. It also forms a simple basis for considering the more complex effects of non-random mating, or *homogamy*, which are known to be important in human populations –

Table 2.1 Punnett square illustrating offspring genotypes (and expected probabilities of their occurrence) that theoretically result from the mating of two heterozygous parents

		Male Gametes	
		$B(\frac{1}{2})$	$b(\frac{1}{2})$
	$B(\frac{1}{2})$	$BB(\frac{1}{4})$	$Bb(\frac{1}{4})$
Female Gametes			
	$b(\frac{1}{2})$	$Bb(\frac{1}{4})$	$bb(\frac{1}{4})$

especially for traits such as substance use, educational attainment, and social attitudes.

In the simple case of equal allele frequencies ($p = q = \frac{1}{2}$), such as in an F_2 population, it is easily shown that random mating over successive generations changes neither the allele nor the genotype frequencies of the population. Male and female gametes of the type B and b from an F_2 population are produced in equal proportions so that random mating may be represented by the same Punnett square as given in Table 2.1, which simply reproduces a population with identical structure to the F_2 from which we started. More importantly this result holds regardless of the allele frequency p, and also when there are more than two alleles. This state is known as *Hardy – Weinberg Equilibrium* and is a fundamental principle of quantitative and population genetics. From this result, the effects of non-random mating and other forces that change populations, such as natural selection, migration, and mutation, may be derived (Gale, 1980; Lynch & Walsh, 1998).

The genotypic frequencies from the Punnett square are important because they allow calculation of the population mean and variance of the phenotype. The genotypes, frequencies, and genotype means for this model in Table 2.1 are as follows:

Genotype (i)	BB	Bb	bb
Frequency (f)	$\frac{1}{4}$	$\frac{1}{2}$	$\frac{1}{4}$
Mean (x)	a	d	$-a$

It is now possible to calculate the mean effect of the B locus, by computing the sum of the products of the frequencies (which are proportions, per the preceding footnote) with their corresponding expected mean, which yields:

$$\begin{aligned}
\mu_B &= \sum f_i x_i \\
&= \frac{1}{4}a + \frac{1}{2}d - \frac{1}{4}a \qquad (2.1) \\
&= \frac{1}{2}d
\end{aligned}$$

The variance of the genetic effects of the B locus is given by the sum of the products of the genotypic frequencies and their squared deviations from the population mean (d/2)

$$\sigma_A^2 = \sum f_i (x_i - \mu_A)^2$$
$$= \frac{1}{4}(a - \frac{1}{2}d)^2 + \frac{1}{2}(d - \frac{1}{2}d)^2 + \frac{1}{4}(-a - \frac{1}{2}d)^2$$
$$= \frac{1}{4}a^2 - \frac{1}{4}ad + \frac{1}{16}d^2 + \frac{1}{8}d^2 + \frac{1}{4}a^2 + \frac{1}{4}ad + \frac{1}{16}d^2$$
$$= \frac{1}{2}a^2 + \frac{1}{4}d^2$$

(2.2)

For this single locus with equal allele frequencies, $\frac{1}{2}a^2$ is known as the additive genetic variance, or V_A, and $\frac{1}{4}d^2$ is known as the dominance variance, V_D. When more than one locus is involved, Mendel's *law of independent assortment* permits the simple summation of the individual effects of separate loci in both the mean and the variance. Thus, for (k) multiple loci,

$$\mu = \frac{1}{2}\sum_{i=1}^{k} d_i,$$

(2.3)

and

$$\sigma^2 = \frac{1}{2}\sum_{i=1}^{k} a_i^2 + \frac{1}{4}\sum_{i=1}^{k} d_i^2$$
$$= V_A + V_D.$$

(2.4)

It is the parameters V_A and V_D that are estimated for inbred animal and outbred animal or human populations elsewhere in this volume.

To appreciate the relationship between these variance components and the estimates derived by fitting a structural equation model to data collected from relatives, we need to consider the joint effect of genes in related individuals. That is, we need to derive, from the genotype frequencies and the parameters a and d, expectations for MZ and DZ covariances in terms of V_A and V_D components. These derived expectations may then be set as fixed parameters in the structural equation model, as described on page 26. This approach is general, in that expectations may be derived for a wide variety of types of relative. Some additional complications arise in the presence of inbreeding, as described by Lynch and Walsh (1998).

Genetic Covariance Between Relatives

In order to use data collected from relatives to estimate variance components or proportions of variance such as heritability, it is necessary to establish the expected covariance between relatives in terms of the variance components of interest. Accordingly, we now consider the derivation of

the additive and dominance genetic contributions to the covariance between relatives. It is our knowledge of the mechanisms by which genetic factors are transmitted across generations which permits precise specification of the expected covariance between relatives. Although genetic models have been precisely formulated for almost a century, the same cannot be said for models of the action of environmental agents. Of this second source of variation we have much less understanding. The physical mechanisms by which environmental events are transmitted from one individual to another, perceived and subsequently encoded in the brain, and thereafter influence the thoughts and actions are not sufficiently characterized to make other than relatively crude approximations to their degree of resemblance between relatives. We return to models of environmental resemblance later in this chapter, but for now note that having at least one source of variation appropriately specified is better than having none.

Twin correlations may be derived in a number of different ways, but the most direct method is to list all possible twin-pair genotypes (taken as deviations from the population mean) and the frequency with which they arise in a random-mating population. Then the expected covariance may be obtained by multiplying the expected mean for twin 1 and twin 2 for each pair type, weighting them by their frequency of occurrence, and summing across all possible pairs. By this method the covariance among pairs is calculated directly. The mean of all pairs is, of course, simply the population mean, $\frac{1}{2}d$, in the case of equal allele frequencies. There are shorter methods for obtaining the same result, but they are less intuitively obvious. In this section, we consider the more general case of unequal allele frequencies. To do this, we need to know the population mean, and the frequencies that pairs of relatives, classified according to their pairwise genotypes, are expected to occur.

The results for equal allele frequencies were known by a number of biometricians shortly after the rediscovery of Mendel's work (Castle, 1903; Pearson, 1904; Yule, 1902). However, it was not until Fisher's remarkable 1918 paper that the full generality of the biometrical model was elucidated. Allele frequencies do not have to be equal, nor do they have to be the same for the various polygenic loci involved in the phenotype for the simple fractions, 1, $\frac{1}{2}$, $\frac{1}{4}$, and 0 to hold, as long as we define V_A and V_D appropriately. The algebra is considerably more complicated with unequal allele frequencies and it is necessary to define carefully what we mean by V_A and V_D. However, the end result is extremely simple, which is perhaps somewhat surprising. The interested reader should refer to the classic texts in this field for further information (Crow & Kimura, 1970; Falconer, 1990; Kempthorne, 1960; Mather & Jinks, 1982). We note that the elaboration of this biometrical model and its power and elegance has been largely responsible for the tremendous strides in economical plant and animal food production throughout the world.

Consider the three genotypes, BB, Bb, and bb, with population genotypic frequencies P, Q, R:

Genotypes	BB	Bb	bb
Frequency	P	Q	R

The allele frequencies (proportions) are

$$\text{allele frequency } (A) = P + \frac{Q}{2} = p$$

$$(a) = R + \frac{Q}{2} = q. \quad (2.5)$$

These expressions derive from the simple fact that the BB genotype contributes only B alleles and the heterozygote, Bb, contributes $\frac{1}{2}B$ and $\frac{1}{2}b$ alleles. A Punnett square showing the allelic form of gametes uniting at random gives the genotypic frequencies in terms of the allele frequencies:

		Male Gametes	
		$p\,A$	$q\,a$
Female Gametes	$p\,A$	p^2BB	$pqBb$
	$q\,a$	$pqBb$	q^2bb

which yields an alternative representation of the genotypic frequencies

Genotypes	BB	Bb	bb
Frequency	p^2	$2pq$	q^2

That these genotypic frequencies are in Hardy–Weinberg equilibrium may be shown by using them to calculate allele frequencies in the new generation showing them to be the same, and then reapplying the Punnett square. Using expression 2.5, substituting p^2, $2pq$, and q^2, for P, Q, and R, respectively, and noting that the sum of allele frequencies is 1 ($p + q = 1.0$), we can see that the new allele frequencies are the same as the old and that genotypic frequencies will not change in subsequent generations

$$p_1 = p^2 + \frac{1}{2}2pq = p^2 + pq = p(p + q) = p$$

$$q_1 = q^2 + \frac{1}{2}2pq = q^2 + pq = q(p + q) = q. \quad (2.6)$$

The biometrical model is developed in terms of these equilibrium frequencies and genotypic effects as

Genotypes	BB	Bb	bb	
Frequency	p^2	$2pq$	q^2	(2.7)
Genotypic effect	a	d	$-a$	

The mean and variance of a population with this composition is obtained in analogous manner to that in 1. The mean is

$$\mu = p^2 a + 2pqd - q^2 a = (p - q)a + 2pqd. \quad (2.8)$$

Because the mean is a reasonably complex expression, it is not convenient to sum weighted deviations to express the variance as in 2.2, instead, we rearrange the variance formula

$$\begin{aligned}
\sigma^2 &= \sum f_i (x_i - \mu)^2 \\
&= \sum f_i (x_i^2 - 2x_i\mu + \mu^2) \\
&= \sum f_i x_i^2 - 2\mu \sum f_i x_i + \mu^2 \\
&= \sum f_i x_i^2 - 2\mu^2 + \mu^2 \\
&= \sum f_i x_i^2 - \mu^2 \quad (2.9)
\end{aligned}$$

Applying this formula to the genotypic effects and their frequencies given in 2.7 above, we obtain

$$\begin{aligned}
\sigma^2 &= p^2 a^2 + 2pqd^2 + q^2 a^2 - [(p - q)a + 2pqd]^2 \\
&= p^2 a^2 + 2pqd^2 + q^2 a^2 - [(p - q)^2 a^2 \\
&\quad + 4pqad(p - q) + 4p^2 q^2 d^2] \\
&= p^2 a^2 + 2pqd^2 + q^2 a^2 - [(p^2 - 2pq - q^2)a^2 \\
&\quad + 4pqad(p - q) + 4p^2 q^2 d^2] \\
&= 2pq[a^2 + 2(q - p)ad + (1 - 2pq)d^2] \\
&= 2pq[a^2 + 2(q - p)ad + (q - p)d^2 + 2pqd^2] \\
&= 2pq[a + (q - p)d]^2 + 4p^2 q^2 d^2. \quad (2.10)
\end{aligned}$$

When the variance is arranged in this form, the first term, $2pv[a + (v - p)d]^2$, defines the additive genetic variance, V_A, and the second term ($4p^2 q^2 d^2$) defines the dominance variance, V_D. Why this particular arrangement is used to define V_A and V_D rather than some other may be seen if we introduce the notion of gene dose and the regression of genotypic effects on this variable, which is how Fisher proceeded to develop the concepts of V_A and V_D.

If B is the increasing allele, then we can consider the three genotypes, BB, Bb, bb, as containing 2, 1, and 0 doses of the B allele, respectively. The regression of genotypic effects on these gene doses is shown in Fig. 2.3. The values that enter into the calculation of the slope of this line are

Genotype	BB	Bb	bb
Genotypic effect (y)	a	d	$-a$
Frequency (f)	p^2	$2pq$	q^2
Dose (x)	2	1	0

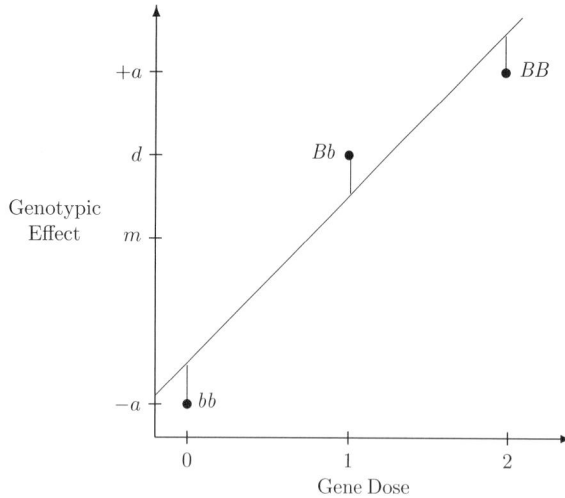

Fig. 2.3 Regression of genotypic effects on gene dosage showing additive and dominance effects under random mating. The figure is drawn to scale for $p = q = \frac{1}{2}$, $d = 1$, and $h = \frac{1}{2}$

From these values the slope of the regression line of y on x in Fig. 2.3 is given by $\beta_{y,x} = \sigma_{x,y}/\sigma_x^2$. In order to calculate σ_x^2 we need μ_x, which is

$$\mu_x = 2p^2 + 2pq$$
$$= 2p(p + q)$$
$$= 2p. \tag{2.11}$$

Then, σ_x^2 is

$$\sigma_x^2 = 2^2 p^2 + 1^2 2pq - 2^2 p^2$$
$$= 4p^2 + 2pq - 4p^2$$
$$= 2pq$$

using the variance formula in 2.9. In order to calculate $\sigma_{x,y}$ we need to employ the covariance formula

$$\sigma_{x,y} = \Sigma f_i x_i y_i - \mu_x \mu_y, \tag{2.12}$$

where μ_y and μ_x are defined as in 2.8 and 2.11, respectively. Then,

$$\sigma_{xy} = 2p^2 a + 2pqd - 2p[(p - q)a + 2pqd]$$
$$= 2p^2 a + 2pqd - 2p^2 a + 2pqa - 4p^2 qd$$
$$= 2pqa + d(2pq - 4p^2 q)$$
$$= 2pqa + 2pqd(1 - 2p)$$
$$= 2pqa + 2pqd(1 - p - p)$$
$$= 2pqa + 2pqd(q - p)$$
$$= 2pq[a + (q - p)d]. \tag{2.13}$$

Therefore, the slope is

$$\beta_{y,x} = \frac{\sigma_{xy}}{\sigma_x^2}$$
$$= 2pq[a + (q - p)d]/2pq$$
$$= a + (q - p)d. \tag{2.14}$$

Following standard procedures in regression analysis, we can partition σ_y^2 into the variance due to the regression and the variance due to residual. The former is equivalent to the variance of the expected y; that is, the variance of the hypothetical points on the line in Fig. 2.3, and the latter is the variance of the difference between the observed y and the expected values.

The variance due to regression is

$$\beta\sigma_{xy} = 2pq[a + (q - p)d][a + (q - p)d]$$
$$= 2pq[a + (q - p)d]^2$$
$$= V_A \tag{2.15}$$

and we may obtain the residual variance simply by subtracting the variance due to regression from the total variance of y. The variance of genotypic effects (σ_y^2) was given in Equation 2.10, and when we subtract the expression obtained for the variance due to regression 2.15, we obtain the residual variances:

$$\sigma_y^2 - \beta\sigma_{x,y} = 4p^2 q^2 d^2$$
$$= V_D. \tag{2.16}$$

In this representation, genotypic effects are defined in terms of the regression line and are known as genotypic values. They are related to a and d, the genotypic effects we defined in Fig. 2.2, but now reflect the population mean and allele frequencies of our random-mating population. Defined in this way, the genotypic value (G) is $G = A + D$, the additive (A) and dominance (D) deviations of the individual.

G	=	A	+	D	frequency
G_{BB}	=	$2q[a + d(q - p)]$	−	$2q^2 d$	p^2
G_{Bb}	=	$(q - p)[a + d(q - p)]$	+	$2pqd$	$2pq$
G_{bb}	=	$-2p[a + d(q - p)]$	−	$2p^2 d$	q^2

In the case of $p = q = \frac{1}{2}$, this table becomes

G	=	A	+	D	frequency
G_{BB}	=	a	−	$\frac{1}{2}d$	$\frac{1}{4}$
G_{Bb}	=			$\frac{1}{2}d$	$\frac{1}{2}$
G_{bb}	=	$-a$	−	$\frac{1}{2}d$	$\frac{1}{4}$

from which it can be seen that the weighted sum of all G's is zero ($\sum f_i G_i = 0$).

The genotypic values A and D that we employ in the structural equation model have precisely the expectations given above in 2.15 and 2.16, but are summed over all polygenic loci contributing to the trait. Thus, the biometrical model gives a precise definition to the latent variables employed in structural equation models for the analysis of twin data.

Coefficients of Resemblance Between Relatives

In order to compute the predicted degree of resemblance between pairs of relatives for a variance component, it is useful to list all possible genotype pairings together with certain other salient pieces of information. The left-most column of Table 2.2 lists all possible pairs of diallelic genotypes that may be shared by a pair of relatives. Additive and dominance effects (the genotype means) of these genotypes are shown in columns two and three for relative one (x_{1i}) and two (x_{2i}), respectively. Next are shown the deviations of these two genotype means from the population mean μ, and then their cross-product, from which we compute the covariance. Finally, the population frequencies of MZ, DZ, and unrelated siblings are listed. For MZ twins, the genotypes must be identical, so there are only three possibilities and these occur with the population frequency of each of the possible genotypes (p^2, $2pq$, and q^2). For unrelated pairs, the population frequencies of the three genotypes are simply multiplied within each pair of siblings since genotypes are paired at random. The frequencies for DZ twins, which are the same as for ordinary siblings, are more difficult to obtain. All possible parental types and the proportion of paired genotypes they can produce must be enumerated, and these categories collected up across all possible parental types. These frequencies and the method by which they are obtained may

be found in standard texts (e.g., Crow & Kimura, 1970, pp. 136–137; Falconer, 1960, pp. 152–157; Mather & Jinks, 1971, pp. 214–215).

The covariances of three types are sibling are given by the frequency-weighted sum of the cross-products:

$$\text{Cov(MZ)} = 2pq[a + (q - p)d]^2 + 4p^2q^2d^2 = V_A + V_D$$
$$\text{Cov(DZ)} = pq[a + (q - p)d]^2 + p^2q^2d^2 \quad = \tfrac{1}{2}V_A + \tfrac{1}{4}V_D$$
$$\text{Cov(U)} \quad = 0 \qquad\qquad\qquad\qquad\qquad = 0$$

By similar calculations, the expectations for half-siblings and for parents and their offspring may be shown to be $\tfrac{1}{4}V_A$ and $\tfrac{1}{2}V_A$, respectively. That is, these relationships do not reflect dominance effects. The MZ and DZ resemblances are the primary focus of this chapter and others in this volume, but the approach is very general and may be used for extended pedigrees and the large number of different relationships found in, e.g., studies of twins, their parents, spouses, and children (Maes et al., 2006).

Given a wider variety of types of relative we can assess the effects of *epistasis*, or non-allelic interaction, since the biometrical model may be extended easily to include such genetic effects. Another important problem we have not considered is that of assortative mating, which one might have thought would introduce insuperable problems for the model. However, once we are working with genotypic values such as A and D, the effects of assortment can be readily accommodated in the model using the Pearson–Aitken formulae for deriving the mean vector and variance–covariance matrix among variables that covary with variables on which selection has occurred (Aitken, 1934). Fulker (1988) describes this approach in the context of Fisher's (1918) model of assortment.

In this section, we have given a brief introduction to the biometrical model that underlies the structural equation modeling of data from twins, and we have shown how additional genetic variance components may be incorporated in the model. However, in addition to genetic influences, we must

Table 2.2 Genotypes, gene effects, deviations, cross-products and frequencies for MZ, DZ, and Unrelated Siblings

Genotype pair		Effects		Deviations		Product	Frequency		
		x_{1i}	x_{2i}	$x_{1i} - \mu$	$x_{2i} - \mu$	$(x_{1i} - \mu)(x_{2i} - \mu)$	MZ	DZ	U
BB	BB	a	a	$a - d/2$	$a - d/2$	$a^2 - ad + d^2/4$	p^2	$p^4 + p^3q + \tfrac{1}{4}p^2q^2$	p^4
BB	Bb	a	d	$a - d/2$	$d/2$	$ad/2 - d^2/4$	0	$p^3q + \tfrac{1}{2}p^2q^2$	$2p^3q$
BB	bb	a	$-a$	$a - d/2$	$-a - d/2$	$-a^2 + d^2/4$	0	$\tfrac{1}{4}p^2q^2$	p^2q^2
Bb	BB	d	a	$d/2$	$a - d/2$	$ad/2 - d^2/4$	0	$p^3q + \tfrac{1}{2}p^2q^2$	$2p^3q$
Bb	Bb	d	d	$d/2$	$d/2$	$d^2/4$	$2pq$	$p^3q + 3p^2q^2 + pq^3$	$4p^2q^2$
Bb	bb	d	$-a$	$d/2$	$-a - d/2$	$-ad/2 - d^2/4$	0	$\tfrac{1}{2}p^2q^2 + pq^3$	$2pq^3$
bb	BB	$-a$	a	$-a - d/2$	$a - d/2$	$-a^2 + d^2/4$	0	$\tfrac{1}{4}p^2q^2$	p^2q^2
bb	Bb	$-a$	d	$-a - d/2$	$d/2$	$-ad/2 - d^2/4$	0	$\tfrac{1}{2}p^2q^2 + pq^3$	$2pq^3$
bb	bb	$-a$	$-a$	$-a - d/2$	$-a - d/2$	$a^2 + ad + d^2/4$	p^4	$\tfrac{1}{4}p^2q^2 + pq^3 + q^4$	q^4

consider the effects of the environment in any phenotype. These may be easily accommodated by defining environmental influences that are common to sib pairs and those that are unique to the individual. If these environmental effects are unrelated to the genotype, then the variances due to these influences simply add to the genetic variances we have just described. If they are not independent of genotype, as in the case of sibling interactions and cultural transmission, both of which are likely to occur in some behavioral phenotypes, then the structural equation model may be suitably modified to account for them.

Modeling of Data from Relatives

In this section we consider the statistical underpinnings of one of the more popular approaches to modeling data from relatives. The treatment here is not by any means intended to be exhaustive. Rather, we wish to give enough information for the reader to appreciate how the coefficients of resemblance between relatives that were calculated in the preceding section can be implemented in a statistical model and used to analyze data. A key feature of the approach we describe is that it is extensible. While there are inevitably some limitations to any selected approach, it will be shown that it is highly versatile. Some alternative methods will be considered at the end of the chapter.

It should be recognized that biometrical modeling is intended to discern between putative causes of variation in the population. That is, it is a part of the study of individual differences. As such, it differs from, e.g., a physiologist's study of the mechanics of a particular organ or cellular signalling pathway, because the focus is on what differs between the individuals in the population rather than on what they have in common. This choice of focus is partly motivated by the desire to understand why certain individuals suffer from physical or psychological disorders, while others do not. In essence, therefore, the approach is partly guided by epidemiological concerns, and the term genetic epidemiology is frequently used to describe this area of study.

Our first concern coincides with almost every area of science: how do we measure the quantities of interest? Usually, a researcher with a substantive interest in a particular trait, disorder, or condition has some idea of how to measure it. In practice, behavior is often measured through self-report, which may be obtained by questionnaire administered by a paper-and-pencil test or via the Internet. These approaches are popular because they are inexpensive. More costly studies may involve personal interviews which may be conducted face-to-face or by telephone (Internet videoconferencing is possible at the time of writing but is not yet sufficiently widespread to be feasible for a population-based survey).

Perhaps the most desirable approach would be to measure behavior by direct observation, as for example are being conducted by Goldsmith and colleagues (Goldsmith, Lemery-Chalfant, Schmidt, Arneson, & Schmidt, 2007). The major drawbacks to such methods are that it is possible that knowledge that one is being observed may change the behavior itself, and that it is very expensive. Such observational methods are typically not well-suited to the assessment of internal states, although it is possible that neuroimaging may provide some non-introspective insights in this area. Behavior genetics may also concern itself with health outcomes that are only partly behavioral in their origin. Physical characteristics such as cardiovascular functioning (which may be assessed with high quality measurements) are likely to reflect behavioral variation such as the amount of physical exercise a person takes. The scope of behavior genetics is therefore very broad, encompassing many different types of measurement and different substantive areas. At times we may be interested in comparing the effects of different sources of variation — genetic vs. environment or more explicit subcomponents thereof, perhaps at different ages. Other studies may focus on how they relate to each other to address questions of comorbidity such as whether the same genetic factors influence variation in two traits. It is therefore important to devise and implement a statistical framework that is sufficiently general to encompass different types of measurements as well as different putative causes of variation.

Structural Equation Modeling: Latent and Observed Variables

Structural equation modeling (Bollen, 1989; Cudeck, Toit, & Sorbom, 2001) is a popular statistical method in many branches of social science. It is essentially an extension of multiple linear regression (Maxwell, 1977) which includes latent as well as observed variables. An especially useful feature of the approach for the non-mathematical audience is that it is possible to draw a graphical representation of the model, known as a path diagram. Structural equation modeling is thus sometimes referred to as path analysis, and the diagrams may be known as path models. This useful device was invented by the population geneticist Sewall Wright around the time that Ronald Fisher was developing biometrical genetics (Wright, 1921, 1934). In their basic form, these diagrams may contain two types of variable: (i) latent (drawn as circles), which have not been directly measured; and (ii) observed (drawn as squares), which have. Two types of relationship (paths) between variables may be depicted: causal, shown as single-headed arrows from cause to effect; and correlational, shown as double-headed arrows. A double-headed arrow from a variable to itself may be used to represent the variance of a variable in a path diagram. Often these variance paths are omitted when the variables are standardized, but it is good practice to include them because it makes the diagrams mathematically complete (i.e., sufficiently precisely specified to enable unambiguous derivation of all predicted variances and covariances between the variables).

There are two general ways of deriving the predicted variances and covariances between the variables in a path diagram. One is to follow a set of tracing rules. For mathematically complete diagrams as described above, a suitable set of rules is:

1. Trace the path chains between the variables, as follows:

   a) Start at one of the variables of interest, trace backwards along any number (zero or more) of single-headed arrows
   b) Change direction at a double-headed arrow
   c) Trace forwards along zero or more single-headed arrows to the second variable of interest

2. Multiply the paths along the path chain to obtain the path effect
3. Repeat the chain procedure for all distinct chains. Chains are distinct if they pass through different nodes, or (in the case of computing a variance only) the order of the paths in the chain is different.
4. Sum all the path effects to obtain the predicted covariance.

It is possible to organize a path diagram into matrices. There are several ways of doing so; the one described here is completely general but is not the most computationally efficient. It has the advantage of being conceptually simple and is due to my colleagues Jack McArdle and Steve Boker (McArdle & Boker, 1990). Two matrices are defined to contain the paths between all the variables (latent and observed) in the model. Both matrices are square with nv rows and nv columns. The first, labelled $\mathbf{S}$, is symmetric and contains all the two-headed paths in the diagram, with zeroes otherwise. The second, labelled A, contains the asymmetric paths or arrows, which are specified between causes in the columns to effects in the rows. The following relatively simple quadratic matrix algebra formula may then be used to derive the predicted covariances between all the variables in the diagram:

$$\Sigma = (\mathbf{I} - \mathbf{A})^{-1}\mathbf{S}(\mathbf{I} - \mathbf{A})^{-1\prime}$$

where $^{-1}$ and $\prime$ denote the inverse and transpose operations, respectively, and $\mathbf{I}$ is the identity matrix. Typically, the predicted variances and covariances of only the observed variables are of use for model-fitting. These can be filtered from Σ using an elementary matrix (one containing 1's and 0's) $\mathbf{F}$ with columns the same as those of matrices $\mathbf{S}$ and $\mathbf{A}$ above and rows containing only the observed variables. All elements of $\mathbf{F}$ are set to zero except those where the column variable and the row variable are identical.

Another useful feature of the path analysis approach is that it is straightforward to use them to derive predicted means as well as covariances. This aspect of path diagrams appears to have been a much later development. Sörböm (1974) presented the general approach to implementing mean structures in structural equation models, in terms of matrix equations. Graphical representation of mean structures in path models appears to have been first described by McArdle and Boker's RAMPATH. Triangles are used to denote mean deviations, and a set of path tracing rules and corresponding matrix algebra permits unambiguous derivation of predicted means.

1. Trace the path chains back from the variable whose expected mean is required to all triangles in the diagram:

   a) Start at the variables of interest, trace backwards along any number (one or more) single-headed arrows to a triangle
   b) Multiply the paths along the path chain to obtain the path effect

2. Repeat the chain procedure for all distinct chains. Chains are distinct if they pass through different nodes.
3. Sum all the path effects to obtain the predicted mean

Using the same organization of matrices $\mathbf{A}$, $\mathbf{S}$ and $\mathbf{I}$ as before, we can derive the expected means under the model using the formula

$$\mu = (\mathbf{I} - \mathbf{A})^{-1}\mathbf{M}\mathbf{U},$$

where matrix $\mathbf{M}$ has rows corresponding to those in $\mathbf{S}$ and columns corresponding to each of the triangles in the diagram. Paths from the triangles to the circles or squares are entries (from column to row) in the matrix $\mathbf{M}$, which contains zeroes where no path is drawn. The matrix $\mathbf{U}$ is a column vector with as many rows as there are triangles in the model, and with every element set to unity.

Alternative matrix specifications. The approach described above is general, but it is not usually the most efficient. The large majority of models in use in behavior genetics do not require use of the matrix inverse to derive the expectations. Often a very succinct and efficient matrix specification can be described. The main advantage of specifying models with matrices is that they are very easy to change when the number of observed variables or the number of factors changes.

Univariate Model for Twin Data

For most variables, the first foray into establishing the relative impact of genetic and environmental factors on trait variation involves a classical twin study. Ideally, random samples of twin pairs are obtained from the population under study,

and data are collected from both members of each pair. Twins
are usually subclassified according to zygosity (MZ or DZ)
and sex (male–male, female–female, or male–female) pairs.
For the present, we consider only the MZ/DZ distinction and
assume that the sample at hand consists of all males or all
females. Note that it is generally inadvisable to collate DZ
pairs when opposite sex pairs are present, since lower correlation between opposite sex pairs are frequently observed
(likely because different genetic or environmental factors are
operating in the two sexes).

While studying pairs of relatives is a substantial advance
over the study of unrelated pairs, it must be remembered
that there are limits to the number of parameters that can be
identified. In the univariate case, the twin study augments the
simple measure of trait variance that would be available from
unrelated persons to include the MZ and DZ covariances.
At its simplest level, this presents us with the opportunity
to estimate two additional parameters. Typically, additive
genetic (V_A) and a common (shared) environment (V_C) component are estimated, and this model will be presented here.
The main alternative model specifies variance due to genetic
dominance (V_D) as derived in Section above. It is important
to recognize that these two components are confounded in
the classical twin study and that while a DZ twin correlation
that is greater than half that of the MZ twins might imply
shared environment variance, and while a model with V_A,
V_C and V_E will fit better than one with V_D in place of V_C,
it does not exclude the possibility that V_D exists for the trait,
but is masked by the effects of V_C.

A path diagram for twin pairs is shown in Fig. 2.4. The
model specifies that the phenotypes of the two twins, $P1$
and $P2$, regress on their respective latent additive genetic
(A), common environment (C), or specific environment variance (E) component. This model is often referred to as an
ACE model, per the acronym for its variance components.
The latent variables are specified to correlate differently for

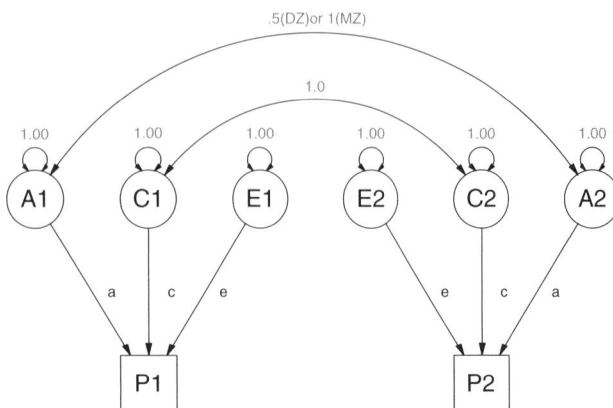

Fig. 2.4 Structural equation model for MZ or DZ twin pairs, showing
additive genetic (A), common environment (C) and specific environment variance components (E)

MZ and DZ twin pairs, as derived on p. 25. Note that the
C components covary 1.0 regardless of zygosity type. This
specification is the embodiment of the equal environments
assumption – that the environmental influences correlate, on
average, the same regardless of zygosity. The free parameters
of this model are the three regression coefficients a, c, and
e, and the mean, m. All other parameters (the variances of,
and covariances between, the latent variables) are fixed to a
priori values according to theory. The free parameters may
in principle be estimated at any value from $-\infty$ to $+\infty$,
though most software algorithms constrain the values somewhat ($-10{,}000$ to $10{,}000$ are the defaults in Mx). Different
values of the parameters predict different means, variances,
and covariances for the twins. Typically, the variances and
means are constrained to be equal not only for the two members of a twin pair but also for both MZ and DZ twins. That
different values of the parameters a and c predict different
patterns of covariances between twins is the essence of the
biometrical twin model.

Specification of this model with the Mx graphical interface is described in detail in Chapter 2 of the Mx manual
(Neale, Boker, Xie, & Maes, 2003), so it will not be repeated
here. Of note, however, is that very simple matrix algebra
expressions for the means and covariance matrix can be written as

$$\mu = (mm); \ \Sigma = \begin{pmatrix} a^2 + c^2 + e^2 & \alpha a^2 + c^2 \\ \alpha a^2 + c^2 & a^2 + c^2 + e^2 \end{pmatrix},$$

where $\alpha = 1$ for MZ paris and 0.5 for DZ pairs.

Model-Fitting and Statistical Considerations

As described in the preceding sections, the biometrical
genetic theory, and the specification of models according
to that theory furnish us with sets of expected covariances
and means in terms of the parameters of the model. It now
remains to describe how these expectations are made to
'match' a set of observed data. This process is known as
model-fitting, and there are a variety of approaches that may
be used. It is instructive to consider the simplest approach,
which can be done 'on the back of an envelope' because it
provides insight into how the information is used. This discussion will be followed by a brief description of maximum
likelihood methods.

Suppose that we have collected data from 600 pairs of
MZ twins, and from 400 pairs of DZ twins, and have computed the two correlations: $r_{MZ} = 0.8$ and $r_{DZ} = 0.5$. The
expected correlations under the ACE model are obtained by
standardizing the variance – covariance matrix Σ in Equation
above, which become

$$\Sigma^*_{MZ} = \begin{pmatrix} 1 & a^{*2} + c^{*2} \\ a^{*2} + c^{*2} & 1 \end{pmatrix}$$

$$\Sigma^*_{DZ} = \begin{pmatrix} 1 & .5a^{*2} + c^{*2} \\ .5a^{*2} + c^{*2} & 1 \end{pmatrix},$$

where a^{*2} and c^{*2} are the standardized variance components. Combining the observed data and these two simultaneous equations we obtain

$$r_{MZ} = .8 = a^{*2} + c^{*2} \tag{2.17}$$

$$r_{DZ} = .5 = .5a^{*2} + c^{*2}. \tag{2.18}$$

Twice the difference between these equations gives $\hat{a}^{*2} = 0.6$. Substituting this value into the equation for r_{MZ} gives an estimate of $\hat{c}^{*2} = 0.2$. A third equation is that the standardized variance components sum to unity: $a^{*2} + c^{*2} + e^{*2} = 1$ and substituting the estimates for a^{*2} and c^{*2} into this equation we obtain

$$0.6 + 0.2 + e^{*2} = 1 \tag{2.19}$$

$$e^{*2} = 1 - 0.6 - 0.2 \tag{2.20}$$

$$e^{*2} = 0.2. \tag{2.21}$$

At this point, one might wonder why behavior geneticists do not routinely exploit this simple algebra to obtain estimates of variance components all the time. As discussed elsewhere (Neale, 2003a), there are at least eight problems with this algorithm for estimating components of variance. First, it is possible to obtain nonsensical estimates of the heritability, either greater than 1.0 or less than zero. Second, it takes no account of the relative precision of the r_{MZ} and r_{DZ} statistics, which may be unequal if the sample sizes or the values of the correlations differ. Third, there is no assessment of whether the correlations are consistent with the ACE model. Fourth, the method does not easily generalize to the multivariate case to permit testing hypotheses concerning why variables correlate with each other. Fifth, it is not easy to correct estimates for the effects of covariates such as age and sex. Sixth, it does not generalize to extended twin studies that involve other relatives. Seventh, it is inefficient when there are missing data, and eighth, it is not suitable for nonrandomly ascertained samples of twins. Accordingly, we now discuss maximum likelihood estimation which provides a solution to all these difficulties.

Maximum likelihood estimation. Early in the 20th century, R. A. Fisher began publishing a seminal series of articles on the statistical properties of maximum likelihood estimates. His first publication using the method was in the context of curve-fitting (Fisher, 1912), and he explored its advantages during the following decade (Fisher, 1922). In many respects, maximum likelihood estimates are ideal. They are invariant to one-to-one transformations; given the maximum likelihood estimate (MLE) of $\hat{a}$, the MLE of $\hat{a}^2$ can be computed directly ($\hat{a} \times \hat{a} = \hat{a}^2$). Second, asymptotically (i.e., as the sample size goes to infinity) MLE's are unbiased. Third, of all estimates that are asymptotically unbiased, MLE's have the smallest variance (that is, they are, on average, the most accurate). Fourth, the distribution of MLE's is asymptotically normal, so the error of an MLE is described simply by a single number (its variance). Of course, nothing is perfect, and there are conditions where alternative estimators are superior; with small sample sizes an estimator with lower bias may be preferred. Perhaps the most significant limitation is that MLE's can be time consuming to compute. While the dramatic advances in computer hardware and software that have occurred in the past 25 years continue to make ever more complex analyses feasible, several areas – including the analysis of ordinal data – still take too long for practical purposes.

The application of maximum likelihood estimation to the analysis of data collected from twins is in widespread use in behavior genetic analysis at this time (2007). Early methods fitted models to previously computed variance – covariance matrices (Fulker, Baker, & Bock, 1983). This approach exploited valuable work by Lawley (1940) and Jöreskog (1967). When data are input as covariance matrices, the fit function is usually

$$\text{ML}_C = df\{\ln|\Sigma| - \ln|S| + \text{tr}S\Sigma^{-1}) - p\}, \tag{2.22}$$

where Σ is the estimated covariance matrix, S is the sample observed covariance matrix, p is the number of observed variables, df is the sample size used to compute observed covariance matrix, and $|\Sigma|$, Σ^{-1} and $\text{tr}\Sigma$ denote the determinant, inverse, and trace of the matrix Σ, respectively. Following the later work by Sörbom, it became popular to fit models jointly to a covariance matrix and a vector of observed sample means, $\mathbf{x}$. In this case the fit function is augmented as

$$\text{ML}_{C+M} = df\{\ln|\Sigma| - \ln|S| + \text{tr}S\Sigma^{-1}) - p \\ + (x - \mu)'\Sigma^{-1}(x - \mu)\}, \tag{2.23}$$

where μ is the vector of predicted means. This approach is highly practical for two main reasons. First, it is computationally efficient, requiring the evaluation of Equation 2.22 or 2.23 only once per set of trial parameter estimates used to iterate toward the maximum. Second, the loss function in fact is not the log-likelihood itself, but is the log of the likelihood ratio of the fitted model to a 'saturated' model in which the observed covariance matrix is substituted for the population matrix. This model is called saturated because it has as many free parameters as there are observed statistics. Under certain regularity conditions (Steiger, 1990), twice the difference between the log-likelihood of a model with u free parameters and a submodel with $v < u$ free parameters is distributed as χ^2 with degrees of freedom equal to $u - v$. Thus when fitting models to covariance matrices, a measure of overall goodness-of-fit of the model to the data is generated automatically. Furthermore, alternative 'nested' models can be compared to one another by this same *likelihood ratio test*.

A disadvantage of this approach is that it is awkward to use when there are missing data, since different covariance matrices must be computed and modeled for each different pattern of missing data. Such problems are commonly encountered in longitudinal studies and with multivariate analysis. The problem can become particularly acute when covariates or moderator variables are used. For example, Visscher et al. (2006) use a direct measure of genotype sharing between DZ twin pairs, and use this statistic to specify a different genetic correlation for each pair in the sample. Analyses of this type are not suited to the above summary statistic approach. They motivate the use of direct 'Full Information Maximum Likelihood' (FIML) to which we now turn.

The principles of ML estimation are somewhat obscured by the use of the summary statistic formulae 2.22 and 2.23. The approach is almost mind-boggling in its simplicity. Any probability density function (pdf) has parameters which govern its general form. The univariate normal pdf, for example, has two parameters, its mean μ and variance σ^2. We can fit a normal distribution to set of independent observations (hypothesized to be sampled from a normal distribution) by varying the estimates $\hat{\mu}$ and $\hat{\sigma}^2$ until the joint likelihood of the data points is maximized. Given that the data points are independent, the joint likelihood is obtained by multiplying the likelihood of the individual data points together (Neale, 2000). Typically, the individual likelihoods consist of a large number of likelihoods, most or all of which take values between zero and one. The product of a large number of such values may be very close to zero, which in turn can cause computational problems. Use of logarithms of the individual likelihoods avoids this difficulty, as these may be summed to yield the log of the joint likelihood.

As discussed above, many models of complex traits assume that individual differences arise as the result of the action of a large number of independent factors, which gives rise to a normal distribution of trait variation. The joint distribution of pairs of relatives is assumed to be bivariate normal, and this assumption may be extended to the multivariate case. The multivariate normal probability density function for a vector of observed variables x_i may be written as

$$|2\pi \boldsymbol{\Sigma}|^{-n/2} \exp \left\{ -\frac{1}{2}(\mathbf{x}_i - \mu_i)' \boldsymbol{\Sigma}^{-1}(\mathbf{x}_i - \mu_i) \right\} .$$

For univariate twin data, the matrix $\boldsymbol{\Sigma}$ is as on p. 28 and hence varies according to the particular set of parameter estimates of $\hat{a}$, $\hat{c}$, and $\hat{e}$ being evaluated during optimization. Typically, the estimates of the elements of the mean vector μ are constrained to be equal across the members of a twin pair and between MZ and DZ twins. This assumption is testable by likelihood ratio test. When there are missing data, the mean vector μ is trimmed to match only those variables that are in the observed data vector x_i. The covariance is similarly filtered to retain only those rows and columns corresponding to the data that were actually observed. The advantage of this approach is that it produces asymptotically unbiased estimates under either or both of two quite general possible causes of missingness. First is when the data are missing completely at random, i.e., the missingness is not related to any of the variables being analyzed. Second is when missing values are predicted by other variables that are observed in the data set. For example, should structural MRI scan data be obtained only from those subjects who are above a certain height, but height is analyzed jointly with the scan data, then the estimates of the mean, the variance, and the covariances between the twins would be asymptotically unbiased. This is despite the fact that the height and structural MRI measures are likely to be correlated.

The application of this model to binary or ordinal data presents no special conceptual difficulties when the threshold model is employed. Essentially, the model states that a particular value on an ordinal scale is observed when an individual's trait score lies between two thresholds t_i and t_j, with $t_i = -\infty$ for the lowest category and $t_j = \infty$ for the highest. However, for high-dimensional problems – when there are large pedigrees or the problem is highly multivariate, analyses become time consuming. High-dimensional numerical quadrature to obtain the multidimensional integrals is computationally intensive, particularly when optimization is required over a large number of parameters of the model. Alternative methods, such as the Gibbs sampler (Geman & Geman, 1984; Casella & George, 1992), offer a promising alternative in such situations.

Conclusion

This chapter provides a brief introduction to the basic genetic theory and statistical methods that are in common use in human behavioral genetics. Using simple algebra and maximum likelihood estimation, it is possible to estimate the contribution of genetic and environmental factors to human variation in quantitative, ordinal, or categorical traits. These contributions have direct counterparts in the variance components estimated from animal and plant breeding experiments, but they do *not* require experimentation with human reproduction. It is especially important to realize that the treatment in this chapter is brief and considers only the simplest of models for genetically informative data. Nevertheless, the platform on which it is built is very general indeed, and many extensions are well established. First, data from many other classes of relative may be encountered in behavior genetic studies. Such additional sources of data prove especially valuable in removing confounds between, e.g., common environment and dominance genetic variance components (Keller & Coventry, 2005), and may resolve alternative models of assortative mating (Heath, 1987). Indeed, the twin study should be considered as only an initial step in the genetic epidemiological study of a trait. Second, extensions to multivariate analysis, which will be discussed in Chapter 4, have great potential for understanding comorbidity between disorders, and covariance between traits. Beyond simple estimation of variance components, it is possible to estimate the relative contributions of these components to the covariance between two or more traits. Factor analysis, which can yield valuable insights to the covariance between multiple phenotypic measures, may be applied to genetic and environmental variance components separately (Martin & Eaves, 1977; McArdle & Goldsmith, 1990). Furthermore, under certain circumstances it is possible to resolve between models that specify different directions of causation between traits (Heath et al., 1993; Neale & Kendler, 1995; Duffy & Martin, 1994; Neale et al., 1994). Third, the models generalize for the analysis of longitudinal data in a variety of ways. Simple Cholesky factorization across time provides an initial view of whether different genetic or environmental factors are operating at different stages of development. More elaborate approaches that exploit Markov chain models (Eaves, Long, & Heath, 1986), growth curves (Neale & McArdle, 2000), dynamical systems modeling (Neale, Boker, Bergeman, & Maes, 2005), and dual change score models (Gillespie et al., 2007) offer great promise for multivariate longitudinal genetic modeling. Third, it is straightforward to carry out univariate or multivariate linkage analyses (Eaves, Neale, & Maes, 1996; Fulker & Cherny, 1996). Linkage analysis may be thought of as partitioning DZ twin pairs into those sharing zero, one, or two alleles identical by descent (IBD) at a specific locus (Nance & Neale, 1989). When marker data unambiguously classify sibling pairs into these IBD groups, the approach is analogous to the MZ vs. DZ vs. unrelated partitioning used to obtain estimates of total heritability. Should the marker data not yield unambiguous classification, then a mixture distribution approach may be used to reflect the uncertainty. The same method is also useful in the context of twin studies in which zygosity diagnosis is subject to error (Neale, 2003b). Similarly, data on specific genetic markers can be included to effect a joint variance components and association analysis. Fourth, models of genotype by environment interaction are undergoing a surge of popularity, partly fueled by the availability of genetic markers. Two main varieties are common: those with measured genotypes (Caspi, Harrington, Moffitt, Milne, & Poulton, 2006) and those with unmeasured variance components (Purcell, 2002). Finally, while the above brief list of extensions may all be approached within the maximum likelihood framework, we note that other applications may be more easily approached via Bayesian or other methods. Cases requiring high-dimensional integration (such as highly multivariate binary or ordinal data) may be tackled with the Gibbs sampler, and models for genotype–environment interaction in the presence of measurement error or genotype–environment covariance have also been successfully analyzed within this framework (Eaves & Erkanli, 2003; Eaves, Foley, & Silberg, 2003). In future we may expect to see a convergence of these methods that will facilitate further analysis of genetically informative data. Hopefully these advances will yield deeper understanding not only of genetic and environmental factors but also of their complex interplay that generates the incredible variety of human behavior.

Acknowledgment Michael Neale is grateful for support from PHS grants DA-18673 and MH-65322.

References

Aitken, A. C. (1934). Note on selection from a multivariate normal population. *Proceedings of the Edinburgh Mathematical Society B, 4*, 106–110.

Bollen, K. A. (1989). *Structural equations with latent variables*. New York: John Wiley.

Busjahn, A., & Hur, Y.-M. (2006). Twin registries: An ongoing success story. *Twin Res Hum Genet, 9*(6), 705.

Carey, G. (1986). A general multivariate approach to linear modeling in human genetics. *American Journal of Human Genetics, 39*, 775–786.

Casella, G., & George, E. I. (1992). Explaining the gibbs sampler. *The American Statistician, 46*, 167–174.

Caspi, A., Harrington, H., Moffitt, T. E., Milne, B. J., & Poulton, R. (2006). Socially isolated children 20 years later: Risk of cardiovascular disease. *Archives of Pediatrics Adolescent Medicine, 160*(8), 805–811.

Castle, W. E. (1903). The law of heredity of Galton and Mendel and some laws governing race improvement by selection. *Proceedings of the American Academy of Sciences, 39*, 233–242.

Crow, J. F., & Kimura, M. (1970). *Introduction to population genetics theory*. New York: Harper and Row.

Cudeck, R., du Toit, S. H. C., & Sorbom, D. (2001). *Structural equation modeling: Present and future*. Chicago: Scientific Software International.

Darwin, C. (1859). *On the origin of species by means of natural selection*. London: John Murray.

Darwin, C. (1871). *The descent of man, and selection in relation to sex*. London: John Murray.

Defries, J., Gervais, M., & Thomas, E. (1978). Response to 30 generations of selection for open-field activity in laboratory mice. *Behavior Genetics, 8*(1), 3–13.

DeFries, J. C., & Fulker, D. W. (1988). Multiple regression analysis of twin data: Etiology of deviant scores versus individual differences. *Acta Genet Med Gemellol (Roma), 37*(3–4), 205–216.

Duffy, D. L., & Martin, N. G. (1994). Inferring the direction of causation in cross-sectional twin data: Theoretical and empirical considerations. *Genetic Epidemiology, 11*(6), 483–502.

Eaves, L., & Erkanli, A. (2003). Markov chain monte carlo approaches to analysis of genetic and environmental components of human developmental change and g x e interaction. *Behavior Genetics, 33*(3), 279–299.

Eaves, L., Foley, D., & Silberg, J. (2003). Has the 'equal environments' assumption been tested in twin studies? *Twin Research, 6*(6), 486–489.

Eaves, L. J. (1976). A model for sibling effects in man. *Heredity, 36*, 205–214.

Eaves, L. J., Long, J., & Heath, A. C. (1986). A theory of developmental change in quantitative phenotypes applied to cognitive development. *Behavior Genetics, 16*, 143–162.

Eaves, L. J., Neale, M. C., & Maes, H. (1996). Multivariate multipoint linkage analysis of quantitative trait loci. *Behavior Genetics, 26*, 519–525.

Edwards, A. L. (1979). *Multiple regression and the analysis of variance and covariance.* San Francisco, CA: W. H. Freeman.

Falconer, D. S. (1960). *Quantitative genetics.* Edinburgh: Oliver and Boyd.

Falconer, D. S. (1990). *Introduction to quantitative genetics* (3rd ed.). New York: Longman Group Ltd.

Fisher, R. A. (1912). On an absolute criterion for fitting frequency curves. *Messenger of Mathematics, 41*, 155–160.

Fisher, R. A. (1918). The correlation between relatives on the supposition of Mendelian inheritance. *Translations of the Royal Society, Edinburgh, 52*, 399–433.

Fisher, R. A. (1922). On the mathematical foundations of theoretical statistics. *Philosophical Transactions of the Royal Society of London, Series A, 222*, 309–368.

Fulker, D. W. (1988). Genetic and cultural transmission in human behavior. In B. S. Weir, E. J. Eisen, M. M. Goodman, & G. Namkoong (Eds.), *Proceedings of the second international conference on quantitative genetics* (pp. 318–340). Sunderland, MA: Sinauer.

Fulker, D. W., Baker, L. A., & Bock, R. D. (1983). Estimating components of covariance using LISREL. *Data Analyst, 1*, 5–8.

Fulker, D. W., & Cardon, L. R. (1994). A sib-pair approach to interval mapping of quantitative trait loci. *American Journal of Human Genetics, 54*(6), 1092–1103.

Fulker, D. W., & Cherny, S. S. (1996). An improved multipoint sib-pair analysis of quantitative traits. *Behavior Genetics, 26*, 527–532.

Fuller, J. L., & Herman, B. H. (1974). Effect of genotype and practice upon behavioral development in mice. *Developmental Psychobiology, 7*, 21–30.

Fuller, J. L., & Thompson, W. R. (1978). *Foundations of behavior genetics.* St. Louis: C. V. Mosby.

Gale, J. S. (1980). *Population genetics.* New York: J. Wiley.

Galton, F. (1875). The history of twins, as a criterion of the relative powers of nature and nurture. *Journal of the Anthropological Institute, 12*, 566–576.

Geman, S., & Geman, D. (1984). Stochastic relaxation, gibbs distributions, and the bayesian restoration of images. *IEEE Transactions on Pattern Analysis and Machine Intelligence, 6*, 721–741.

Gillespie, N. A., Kendler, K. S., Prescott, C. A., Aggen, S. H., Gardner, C. O. J., Jacobson, K., et al. (2007). Longitudinal modeling of genetic and environmental influences on self-reported availability of psychoactive substances: Alcohol, cigarettes, marijuana, cocaine and stimulants. *Psychological Medicine, 37*(7), 947–959.

Goldsmith, H. H., Lemery-Chalfant, K., Schmidt, N. L., Arneson, C. L., & Schmidt, C. K. (2007). Longitudinal analyses of affect, temperament, and childhood psychopathology. *Twin Research and Human Genetics, 10*(1), 118–126.

Gordon, K (1919). Report on psychological tests of orphan children. *Journal of Deliquency, 4*, 45–55.

Gottesman, I. I., & Gould, T. D. (2003). The endophenotype concept in psychiatry: Etymology and strategic intentions. *American Journal of Psychiatry, 160*(4), 636–645.

Heath, A. C. (1987). The analysis of marital interaction in cross-sectional twin data. *Acta Geneticae Medicae et Gemellologiae, 36*, 41–49.

Heath, A. C., Kessler, R. C., Neale, M. C., Hewitt, J. K., Eaves, L. J., & Kendler, K. S. (1993). Testing hypotheses about direction of causation using cross-sectional family data. *Behavior Genetics, 23*(1), 29–50.

Jöreskog, K. G. (1967). Some contributions to maximum likelihood factor analysis. *Psychometrika, 32*, 443–482.

Keller, M. C., & Coventry, W. L. (2005). Quantifying and addressing parameter indeterminacy in the classical twin design. *Twin Research and Human Genetics, 8*(3), 201–213.

Kempthorne, O. (1960). *Biometrical genetics.* New York: Pergammon Press.

Kendler, K. S., & Kidd, K. K. (1986). Recurrence risks in an oligogenic threshold model: The effect of alterations in allele frequency. *Annals of Human Genetics, 50 (Pt 1)*, 83–91.

LaBuda, M. C., DeFries, J. C., & Fulker, D. W. (1986). Multiple regression analysis of twin data obtained from selected samples. *Genetic Epidemiology, 3*, 425–433.

Lawley, D. (1940). The estimation of factor loadings by the method of maximum likelihood. *Proceedings of the Royal Society of Edinburgh, 60*, 64–82.

Lehmann, E. L. (1998). *Elements of large-sample theory.* New York: Springer.

Li, T.-K., Lumeng, L., & Doolittle, D. (1993). Selective breeding for alcohol preference and associated responses. *Behavior Genetics, 23*(2), 163–170.

Loehlin, J. (1996). The cholesky approach: A cautionary note. *Behavior Genetics, 26*, 65–69.

Lynch, M., & Walsh, B. (1998). *Genetics and analysis of quantitative traits.* Sunderland, MA: Sinauer.

Maes, H. H., Neale, M. C., Kendler, K. S., Martin, N. G., Heath, A. C., & Eaves, L. J. (2006). Genetic and cultural transmission of smoking initiation: an extended twin kinship model. *Behavior Genetics, 36*(6), 795–808.

Mann, C. C. (1994). Behavioral genetics in transition. *Science, 264*(5166), 1686–1689.

Martin, N. G., & Eaves, L. J. (1977). The genetical analysis of covariance structure. *Heredity, 38*, 79–95.

Mather, K., & Jinks, J. L. (1971). *Biometrical genetics.* London: Chapman and Hall.

Mather, K., & Jinks, J. L. (1977). *Introduction to biometrical genetics.* Ithaca, New York: Cornell University Press.

Mather, K., & Jinks, J. L. (1982). *Biometrical genetics: The study of continuous variation* (3rd ed.). London: Chapman and Hall.

Maxwell, A. E. (1977). *Multivariate analysis in behavioral research.* New York: John Wiley.

McArdle, J. J., & Boker, S. M. (1990). *Rampath path diagram software.* Denver, CO: Data Transforms Inc.

McArdle, J. J., & Goldsmith, H. H. (1990). Alternative common factor models for multivariate biometric analyses. *Behavior Genetics, 20*(5), 569–608.

McArdle, J. J., & Prescott, C. A. (2005). Mixed-effects variance components models for biometric family analyses. *Behavior Genetics, 35*(5), 631–652.

Merriman, C. (1924). The intellectual resemblance of twins. *Psychological Monographs, 33*, 1–58.

Mullis, K. (1990). The unusual origin of the polymerase chain reaction. *Scientific American, 262*(4), 56–61, 64–5.

Nance, W. E., & Neale, M. C. (1989). Partitioned twin analysis: A power study. *Behavior Genetics, 19*, 143–150.

Neale, M. C. (2000). Individual fit, heterogeneity, and missing data in multigroup sem. In T. D. Little, K. U. Schnabel, & J. Baumert (Eds.), *Modeling longitudinal and multiple-group data: Practical issues, applied approaches, and specific examples*. Hillsdale, NJ: Lawrence Erlbaum Associates.

Neale, M. C. (2003a). Twins studies: Software and algorithms. In D. N. Cooper & N. J. Hoboken (Eds.), *Encyclopedia of the human genome*. Macmillan Publishers Ltd, Nature Publishing Group, London, p.88–96.

Neale, M. C. (2003b). A finite mixture distribution model for data collected from twins. *Twin Research, 6*(3), 235–239.

Neale, M., Boker, S., Bergeman, C., & Maes, H. (2005). The utility of genetically informative data in the study of development. In S. Boker & C. Bergeman (Eds.), *Notre dame quantitative methods in psychology*. New York: Erlbaum.

Neale, M., Boker, S., Xie, G., & Maes, H. (2003). *Mx: Statistical modeling* (6th ed.). Box 980126 Richmond VA: Department of Psychiatry, Virginia Commonwealth University.

Neale, M. C., & Cardon, L. R. (1992). *Methodology for genetic studies of twins and families*. Dordrecht: Kluwer Academic Publishers.

Neale, M. C., & Kendler, K. S. (1995). Models of comorbidity for multifactorial disorders. *American Journal of Human Genetics, 57*(4), 935–953.

Neale, M. C., & McArdle, J. J. (2000). Structured latent growth curves for twin data. *Twin Research, 3*, 165–77.

Neale, M.C., Walters, E., Health, A. C., Kessler, R. C., Perusse, D., Eaves, L. J.; et al. (1994). Depression and parental bonding: cause, consequence, or genetic covariance? *Genetic Epidemiology, 11*(6), 503–522.

Pearson, K. (1901). Mathematical contributions to the theory of evolution. VII. On the correlation of characters not quantitatively measurable. *Proceedings of the Royal Society, 66*, 241–244.

Pearson, K. (1904). On a generalized theory of alternative inheritance, with special references to Mendel's laws. *Philosophical Transactions of the Royal Society A, 203*, 53–86.

Purcell, S. (2002). Variance components models for gene-environment interaction in twin analysis. *Twin Research, 5*(6), 554–571.

Rabe-Hesketh, S., Skrondal, A., & Gjessing, H. (2008). Biometrical modeling of twin and family data using standard mixed model software. *Biometrics, 64*, 280–288.

Rende, R. D., Polmin, R., & Vandenberg, S. G. (1990). Who discovered the twin method? *Behavior Genetics, 20*(2), 277–285.

Risch, N. (2001). The genetic epidemiology of cancer: Interpreting family and twin studies and their implications for molecular genetic approaches. *Cancer Epidemiology Biomakers Prevention, 10*(7), 733–741.

Sörbom, D. (1974). A general method for studying differences in factor means and factor structures between groups. *British Journal of Mathematical and Statistical Psychology, 27*, 229–239.

Steiger, J. H. (1990). Structural model evaluation and modification: An interval estimation approach. *Multivariate Behavioral Research, 25*, 173–180.

Thorndike, E. L. (1905). Measurement of twins. *Archieves of Philosophy, Psychology and Scientific Methods, 1*, 1–64.

Tryon, R. C. (1941). Studies in individual differences in maze ability. x. ratings and other measures of initial emotional responses of rats to novel inanimate objects. *Journal of Comparative Psychology, 32*, 447–473.

Visscher, P. M., Medland, S. E., Ferreira, M. A., Morley, K. I., Zhu, G., Cornes, B. K., et al. (2006). Assumption-free estimation of heritability from genome-wide identity-by-descent sharing between full siblings. *PLoS Genetics, 2*(3), e41.

Watson, J. D., & Crick, F. H. (1953). Genetical implications of the structure of deoxyribonucleic acid. *Nature, 171*(4361), 964–967.

Wright, S. (1921). Correlation and causation. *Journal of Agricultural Research, 20*, 557–585.

Wright, S. (1934). The method of path coefficients. *Annals of Mathematical Statistics, 5*, 161–215.

Wright, S. (1968). *Evolution and the genetics of populations. Volume 1. Genetic and Biometric foundations*. Chicago: University of Chicago Press.

Yule, G. U. (1902). Mendel's laws and their probable relation to intraracial heredity. *New Phytology, 1*, 192–207.

Chapter 3

QTL Methodology in Behavior Genetics

Stacey S. Cherny

Introduction

In the second chapter of this volume, biometrical models in behavioral genetics are presented. Such models provide the foundation for quantitative trait locus (QTL) analysis. The present chapter specifically deals with applying those models to QTL analysis, in both linkage and association contexts. Until relatively recently, linkage was the preferred method for mapping QTLs. The approach has limited power in detecting small effects, unless an extremely large sample size is available. However, linkage extends over large chromosomal regions and so can be used to localize QTLs of relatively large effects to large segments of DNA. In contrast to association mapping, linkage can be detected without actually genotyping a causal variant or a locus that is in linkage disequilibrium with a causal locus. In this case, linkage disequilibrium implies a correlation between a marker and a causal variant within the population as a whole. With the availability of low-cost single nucleotide polymorphism (SNP) chip-based genotyping technologies, the focus has shifted toward association mapping. Currently, for example, Affymetrix has a chip set that includes nearly 1,000,000 SNPs, spaced on average nearly every 3000 base pairs along the genome. Such chips are ideally suited to genomewide association scans. The present chapter will deal with both linkage and association methods, however, since linkage may still be the preferred first-pass analysis in a genomewide scan if one is interested in finding genes of relatively large (perhaps 10% of the total phenotypic variance) effects, including multiple rare variants at a single locus. While coverage with the new SNP chips is relatively complete, if a causal variant is not in linkage disequilibrium with a typed SNP, it will not be detected via association mapping, yet still can be detected using linkage. Therefore, linkage and association can be considered complimentary approaches to gene mapping.

Maximum likelihood variance components methods for estimating and testing genetic and environmental variance contributions to a trait of interest have been the analysis methods of choice in behavior genetics since Jinks and Fulker's seminal 1970 paper (Jinks & Fulker 1970). The approach is extremely flexible and readily lends itself to the analysis of complex pedigrees and phenotypically multivariate datasets. In the years since then, these methods have been refined and extended and are most readily implemented using the Mx statistical modelling package (Neale, Boker, Xie, & Maes, 2003), although other packages can be used as well but tend to involve more cumbersome specifications. A good basic introduction to such model fitting can be found in Neale and Cardon (1992) and the second edition of this text, Neale and Maes (in preparation), in addition to Chapter 2 of this volume. QTL mapping extends these models of estimating overall genomewide genetic influences to estimating genetic effects at any (or all) position(s) along the genome.

Linkage Mapping of QTLs

While studies which estimate the extent to which genetic influences contribute to quantitative traits are still very important as a first step in genetic analysis, with the low cost of typing DNA markers across the genome, we are much more interested in finding the individual genes which contribute to the genetic variance twin and adoption studies can detect. That is, we wish to detect a quantitative trait locus, a gene which has a quantitative effect on a trait. Linkage analysis for complex behavioral genetic traits typically involves the analysis of large numbers of small families, typed on hundreds or thousands of DNA markers. We can extend the basic twin model (see previous chapter) to estimate variance attributable to a single genetic locus, rather than the genomewide contribution of all loci. In estimating genomewide genetic variance, we rely on the difference in

S.S. Cherny (✉)
Department of Psychiatry, Genome Research Centre, and The State Key Laboratory of Brain and Cognitive Sciences, The University of Hong Kong, Pokfulam, Hong Kong
e-mail: cherny@hku.hk

Y.-K. Kim (ed.), *Handbook of Behavior Genetics*,
DOI 10.1007/978-0-387-76727-7_3, © Springer Science+Business Media, LLC 2009

the amount of genetic material shared by MZ vs. DZ twins, on average. When examining a particular locus to determine its effect on a phenotype, we contrast allele sharing between sibling pairs only at that particular locus. A sibling pair can either share 0, 1, or 2 alleles at any position in the genome. Sharing is generally not known with certainty and so must be estimated from marker data, with this estimate being denoted $\hat{\pi}$. While estimating allele sharing for a single marker locus is relatively straightforward (e.g., see Haseman & Elston, 1972), we typically now genotype markers all along each chromosome, necessitating use of more complex algorithms (see e.g., Abecasis, Cherny, Cookson, & Cardon, 2002; Kruglyak, Daly, Reeve-Daly, & Lander, 1996). In order to estimate the effect of a QTL using linkage analysis, we fit the following model to the data:

$$\Sigma_i = \begin{pmatrix} h_q^2 + c^2 + e^2 & \hat{\pi}_i h_q^2 + c^2 \\ \hat{\pi}_i h_q^2 + c^2 & h_q^2 + c^2 + e^2 \end{pmatrix} \quad (3.1)$$

where h_q^2 is an estimate of the variance attributable to the QTL, c^2 is the variance attributable to environmental influences shared by a pair of DZ twins or siblings, confounded with polygenic variance, and e^2 is the variance attributable to environmental influences unique to the individual, which includes measurement error (unless that is explicitly modelled through use of multiple measures of a trait). If these three quantities are standardized to the phenotypic variance, the quantities of QTL heritability, the proportion of variance due to the putative trait locus, shared environmentality (which in this case includes polygenic influences) and non-shared environmentality result, analogous to the basic twin model presented in the previous chapter.

In order to estimate the parameters of this model, we maximize the (log of the) following likelihood of the data across all i sib pairs:

$$L = \prod_i \frac{1}{2\pi |\Sigma_i|} e^{-\frac{1}{2}(y_i - \mu_i)' \Sigma_i^{-1}(y_i - \mu_i)} \quad (3.2)$$

This expression also contains a means model, where y_i is a vector of observed data and μ_i is a vector of expected, or predicted means. We would expect means of the first and second members of a sib pair to be equal, but this is a testable assumption.

In addition to allowing estimation of these parameters, their statistical significance is tested by fitting a model with the parameter of interest (h_q^2 for testing the QTL variance) included and estimating the likelihood of the data (L_1) and comparing it with a model with that parameter fixed to zero (L_0). Twice the difference between these two (natural) log-likelihoods is asymptotically distributed as χ^2:

$$\chi^2 = -2 \ln \frac{L_0}{L_1} \quad (3.3)$$

In practice, we would estimate sharing across the entire genome and repeatedly fit the genetic model at perhaps 1 centi-Morgan (cM) intervals and perform a test of h_q^2 (the heritability due to the particular locus), plotting the obtained test statistic at each position. We would choose an interval of 1 cM because the probability of a recombination event occurring during a meiosis within 1 cM of DNA is approximately 1%, making for a smooth linkage curve. Typically linkage could locate a gene to at best 10 cM, and more often only 20–30 cM. In linkage analysis, the test statistic commonly employed is the LOD score, where

$$LOD = \log_{10} \frac{L_1}{L_0} \quad (3.4)$$

yielding the simple relationship

$$LOD = \frac{\chi^2}{2 \ln(10)} \quad (3.5)$$

Lander and Kruglyak formalized some guidelines as to what size LOD score constitutes a significant linkage finding (E. Lander & Kruglyak 1995). They determined that for sibling-based linkage studies, a LOD of 3.6 is required for statistical significance, where the probability of a positive linkage arising anywhere in the genome is 0.05 under the null hypothesis of no QTL present. They also proposed a LOD of 2.2 for "suggestive linkage", where a single genomewide false-positive would be expected under the null hypothesis.

In the above example of QTL linkage analysis, when we use $\hat{\pi}_i$ in the model, we are looking for a QTL which has an additive effect on the quantitative trait of interest. That is, if the gene contains just two alleles, A_1 and A_2, if an individual had genotype $A_1 A_1$, we would expect an individual's trait score to be a above the mean and if the individual were $A_2 A_2$, we would expect their trait score to be a below the mean. The heterozygote, $A_1 A_2$, would be expected to not deviate from the mean, in the additive case. Nonadditive (dominant or recessive) models can also be fitted, which would then imply a nonlinear relationship between the three possible genotypes and the expected trait scores. Discussion of additive vs. dominance models in linkage analysis can be found in the statistical genetics textbook by Sham (1998).

QTL linkage models can be fitted using structural equation modelling packages (e.g., LISREL, EQS, Calis, Mx). In the field of behavior genetic analysis, Mx (Neale et al. 2003) is a very convenient and powerful structural modelling package, and there is a large archive of Mx scripts for QTL linkage analysis and many other behavior genetic analyses at the Mx Scripts Library (http://www.psy.vu.nl/mxbib/) and at the Mx web site (http://www.vcu.edu/mx/). However, for the most basic variance components linkage analysis for quantitative traits, the Merlin package (Abecasis et al. 2002) can perform these analyses most efficiently across the entire genome.

Multivariate QTL Linkage Mapping

Just as multiple measures can be analyzed to examine genetic and environmental covariance structure using twins and other genetically informative relationships (see Chapter 2), linkage analysis can also be performed in a multivariate manner, both as a means of increasing power and as a way to determine which aspects of a phenotype are linked to particular genetic loci. This can be done through use of the (relatively) unconstrained Cholesky decomposition to model the environmental components of variation and the more restrictive common factor model for the QTL components of the model (see Chapter 2 for a discussion of these models). The common factor model is particularly useful when generalizing the multivariate twin model to the detection of QTLs, since a single genetic locus can influence several traits simultaneously and this influence is necessarily perfectly correlated across the traits. The magnitudes of the effect of the given locus on each trait can differ dramatically, but it still results in a common factor influence, with no specific variances resulting from the QTL.

Extending the Cholesky decomposition to the mapping of QTLs, we still model a full Cholesky on the shared and nonshared environmental covariance structure, but we substitute a single common factor representing a single gene, or QTL, in place of the full Cholesky. This simply involves reducing the Λ_G matrix of the genetic Cholesky parameters (see Chapter 2) down to a single column of parameters, with the QTL loading on each of the observed measures. This allows estimation of the effect of the QTL on each measure in the multivariate model, but ensures these effects are correlated completely across measures, since it is the same QTL influencing multiple measures. This results in an n degree-of-freedom test for the QTL, where n is the number of phenotypic measures. Despite the increased degrees of freedom, we generally see increased power as a result of the multivariate test, both in theory (Schmitz, Cherny, & Fulker, 1998) and in practice (Marlow et al. 2003). We use the Λ_G to model the QTL effect in ordinary siblings or DZ twins, as we would model the polygenic effect in twin pairs. However, if we have MZ twin pairs available for QTL analysis, we would also model the polygenic variance by including a Cholesky decomposition of that component. MZ twins may improve the power of a QTL analysis somewhat over an analysis of DZ twins alone (Schork & Xu 2000). Additionally, a pair of MZ twins along with their sibling can be more powerful

than just using one member of the MZ pair (Evans & Medland 2003).

An example of the added power of multivariate QTL linkage analysis can be found in Marlow et al. (2003). This study presents a multivariate analysis of six measures of reading ability, in an attempt to find a QTL influencing reading disability. While the sample was selected for reading problems, the selection criteria were so weak that all measures were essentially normally distributed and so variance components analysis, without any ascertainment correction, was performed. As can be seen in Fig. 3.1, the multivariate analysis yielded a test statistic similar to that obtained from one of the individual measures (PD) on Chromosome 6, but did a better job of localizing the QTL, as indicated by the sharper peak. However, for Chromosome 18, an individual measure (Read) yielded a higher test statistic. Results such as these are not surprising, however. If all measures in the analysis are equally good at tapping the underlying construct which the QTL influences, a multivariate test will be more powerful. However, when a specific measure better taps what the QTL influences, of course no more power would be expected of a multivariate analysis. In the case of reading disability presented here, perhaps that is the case for the Chromosome 18 region. Nonetheless, multivariate analysis performed better for Chromosome 6, in terms of providing a more precise estimate of the peak location.

Association Mapping

The maximum likelihood variance components framework can easily be adapted to test for genetic association in addition to linkage (Abecasis et al. 2000, Abecasis, Cookson, & Cardon 2000, Fulker, Cherny, Sham, & Hewitt 1999). While linkage to quantitative trait loci involves detection of differences in sibling covariance among phenotypes as a function of differences in the degree of allelic sharing, association mapping involves the detection of mean differences in the phenotype as a function of an individual's genotype. When modelling linkage using sibling pairs, the QTL is modelled in the covariance matrix among the sibling pairs. When modelling association, the QTL is modelled in the mean vector, as one deals with any other covariate. In the simplest case, in order to estimate and detect association of a gene with a trait, each element of μ_i in the likelihood expression would be modelled as

$$\mu_i = \mu + \beta_a g_i \tag{3.6}$$

where the g_i for a given individual is coded as -1 for genotype aa, 0 for the heterozygote Aa, and $+1$ for genotype AA, in the case of modelling an additive effect of a diallelic locus.

A

B

C

D

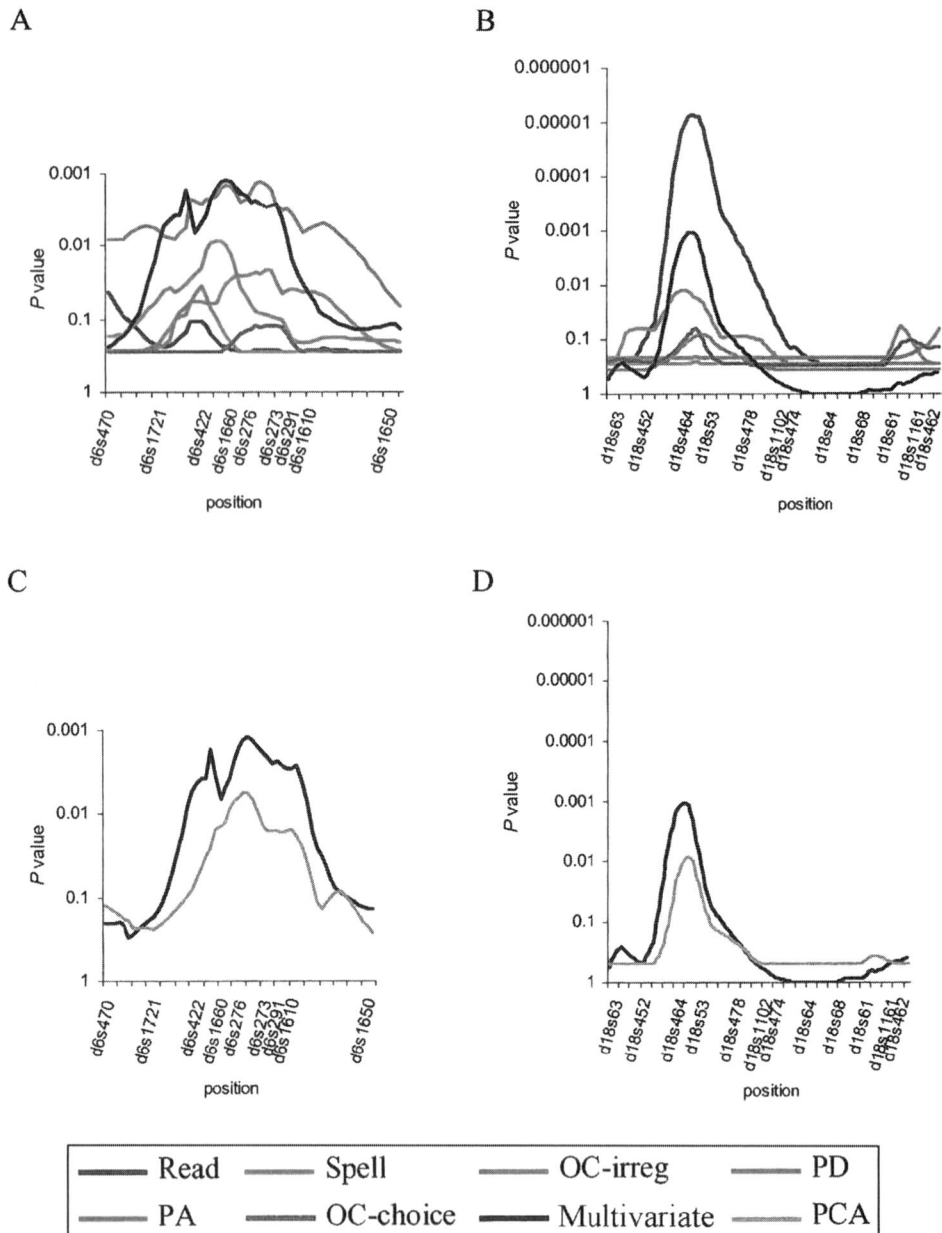

| Read | Spell | OC-irreg | PD |
| PA | OC-choice | Multivariate | PCA |

Fig. 3.1 Multivariate and univariate linkage analysis of the six reading-related measures—on a 54 cM region of chromosome 6p (**A**) and a 137 cM region spanning the whole of chromosome 18 (**B**)—and comparison of multivariate linkage and use of the first factor from a PCA approach as the phenotypic measure for linkage analysis, on chromosomes 6p (**C**) and 18 (**D**). A subset of the markers is shown on the graphs. The significance of the linkage results is reported in all cases as p values. For univariate measures, the p values are empirically derived as described elsewhere (S. Fisher et al. 2002); for multivariate and PCA results, the p values are asymptotic. From Marlow et al. (2003)

β_a would then estimate the additive effect of the locus on the trait and dropping β_a from the model and comparing it with a model that includes β_a in it would provide a test of association.

Fulker et al. (1999) further extended this simple association model to allow one to control for population stratification, a problem of concern in association studies, and proposed further to simultaneously model linkage and asso-

ciation using the maximum likelihood variance components approach. Population stratification refers to the problem of the individuals in the sample coming from different genetic populations. Different populations may have different allelic frequencies for reasons unrelated to the trait of interest (e.g., genetic drift). However, if allele frequencies differ between populations and there are mean differences between populations, association could be detected between those alleles and the trait, when in fact there is no causal relationship. Stratification was controlled for by decomposing the genotype score, g_i, into between-family (b) and within-family (w) components. Since stratification would contribute to between-family differences, the within-family component of association would be free of stratification effects. The means model they proposed is

$$\mu_{ij} = \mu + \beta_b b_i + \beta_w w_{ij} \qquad (3.7)$$

where b_i and w_{ij} are the orthogonal between- and within-family components of g_{ij}, respectively, with j denoting a particular individual in family i. Using the specification of Abecasis et al. (2000) and their extension to deal with parental genotypes,

$$b_i = \frac{\sum_i g_{ij}}{n_i} \qquad (3.8)$$

if parental genotypes are unknown and

$$b_i = \frac{g_{Fi} + g_{Mi}}{2} \qquad (3.9)$$

if parental genotypes are available, with F and M referring to the father's and mother's genotypes in family i. The within-family component is given as

$$w_{ij} = g_{ij} - b_i \qquad (3.10)$$

A robust test for association, controlling for stratification, is obtained by computing the 1-df difference χ^2 between a model with β_w free and a model with β_w set to 0, while β_b is estimated in both models. In the case of a locus which is dominant or multi-allelic, these models are readily extended to allow a between- and within-dominance or multi-allelic parameters. Furthermore, the modelling of multiple alleles is a straightforward extension, although with inclusion of dominance parameters in such a model, the number of df associated with the test of association increases dramatically with more and more alleles.

This test of association, because it is based on maximum likelihood variance components estimation procedures, brings all the advantages of that procedure to it. One major advantage is that linkage can be modelled simultaneously with the association parameters. Linkage is modelled in the covariance structure, as illustrated above, while the association parameters, along with other covariates, if desired, are modelled on the means. One would estimate all parameters as a full model and compare it to various submodels, allowing individual tests of association and linkage. A simple test of the within-association parameter would yield a robust test of association while controlling for stratification. Testing linkage while simultaneously modelling association would

provide a test of whether the putative QTL locus is likely causative or whether it is merely in disequilibrium with a trait locus. If significant linkage is detected while modelling association, one can conclude that the putative locus is not the functional gene, but rather a locus in disequilibrium with a trait locus. Such tests have been explored more formally by Cardon and Abecasis (2000).

The approach not only allows one to control for stratification but readily suggests a test for it. If the β_b parameter cannot be equated to the β_w parameter, one can conclude that at least part of the association observed is a result of population stratification. However, if β_w on its own is still significant, one can still conclude that a true association has been found.

Application of this method is beautifully illustrated by a study by McKenzie et al. (2001). In their study, they measured circulating angiotensin I-converting enzyme (ACE) in two samples and tested for linkage and association with 10 polymorphisms in the ACE gene. The study is both proof of principle that linkage and association could be detected, since it was already known that the ACE gene influences ACE levels, but the primary purpose was to narrow down which particular variant in the ACE gene is responsible for the effect. The Fulker et al. (1999) and Abecasis et al. (2000) model was fitted to these data, whereby linkage was modelled without association parameters in the model, association was modelled without the linkage parameters in the model, and a combined linkage and association model was fitted. As can be seen in the first panel of Fig. 3.2, the evidence for linkage is very strong. However, because the region is quite small, a short segment of DNA in a single gene, linkage cannot be expected to resolve location at all, since the recombination fraction between markers is nearly zero. However, association has much greater resolution capability, since linkage disequilibrium between the markers examined is not complete. And as can be seen in the second panel of Fig. 3.2, the association test statistic is greatest for the four markers on the right-hand side in the figure, already a great improvement on the linkage analysis. The top line represents the combined test of β_w and β_b, while the bottom line is the robust, but less powerful, test of linkage in the presence of stratification, the test of just β_w. However, if both linkage and association are modelled together, we would expect to see the linkage signal disappear to the extent that the marker in association is the causal variant. Indeed, for one of the markers (I/D) in this analysis, the linkage signal disappeared completely when the association parameters were modelled simultaneously.

Just as we can extend the linkage mapping methods to the analysis of multivariate phenotypes, association mapping can also be extended to deal with multivariate traits, both to better understand the genotype–phenotype relationship and to increase power.

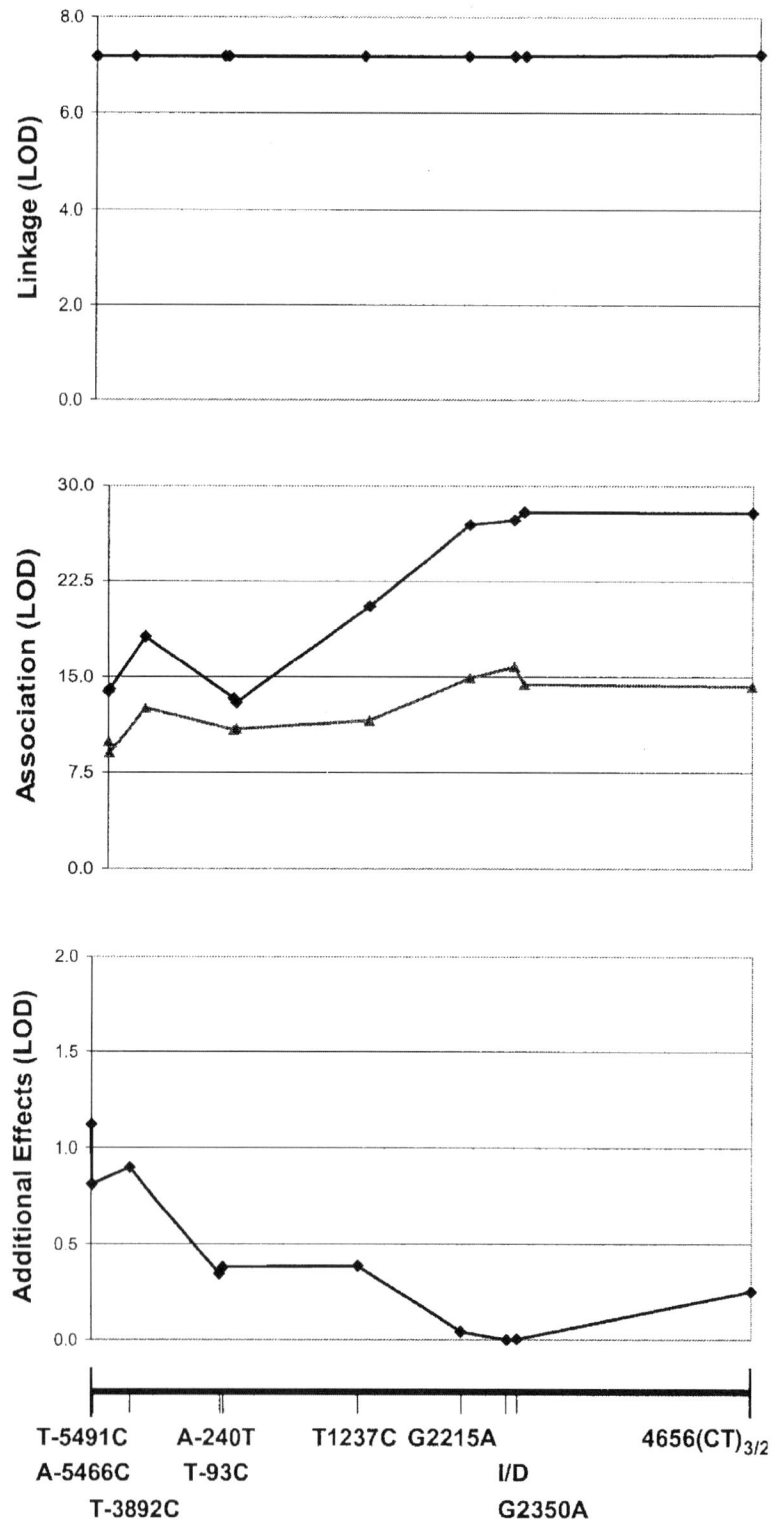

Fig. 3.2 Linkage and association analyses of ACE polymorphism, from McKenzie et al. (2001). See text for details

Regression Models

Thus far this chapter has dealt with maximum likelihood variance components approaches to modelling linkage and association to quantitative traits using sibships. It has been shown that such methods are optimally powerful and given their flexibility in designing complex genetic and environmental models, if it is appropriate to use them, they are preferable. However, these methods are less flexible when dealing with selected samples, where families and individuals within families are not sampled at random. Sham, Zhao, Cherny, and Hewitt (2000) have developed a conditioning on traits approach for modelling linkage using variance components with selected samples, but in dealing with sibships larger than pairs it becomes quite unwieldy. Also, variance components methods can yield inflated test statistics when dealing with non-normal data. Next, a brief history of regression methods is discussed, adapted from Sham, Purcell, Cherny, & Abecasis (2002).

Haseman–Elston

It is commonly known that simple multiple linear regression is more robust than maximum likelihood variance components methods to violations of normality and also deals with selection on independent variables without any modifications of the method. Regression methods for linkage analysis in sib pairs date back to the Haseman–Elston (1972) approach, whereby the phenotypic squared sib pair difference score, $(X - Y)^2$, is regressed on the sib pair's proportion of alleles shared identical by descent (IBD) at a given locus, $\hat{\pi}$:

$$(X - Y)^2 = 2(1 - r) + 2Q(\hat{\pi} - .5) + \epsilon \qquad (3.11)$$

In the above expression, r is the correlation between phenotypic scores of sib 1 (X) and sib 2 (Y), with the intercept of the regression equal to $2(1 - r)$ and the regression slope $\beta = -2Q$, where Q is the variance explained by the additive effect of the putative QTL, if there is no recombination between the marker and the QTL, and ϵ is the residual. A one-tailed t-test, $\hat{\beta}/SE(\hat{\beta})$, provides a test of significance for the QTL. At a given chromosomal location, a pair of siblings can share either none, one, or both alleles IBD from their parents. For example, sibling 1 could have received one allele at a given locus from her maternal grandmother and the other from her paternal grandmother. If her sibling also received those same alleles, they will be IBD 2, or $\hat{\pi} = 1$. However, if sibling 2 received the maternal grandmother's allele, but the paternal grandfathers's allele, the sib pair is IBD 1, or $\hat{\pi} = 0.5$. Finally, the sib pair could be $\hat{\pi} = 0$ if they differ in the origin of both alleles (e.g., sib 2 received

the maternal grandfather allele and the paternal grandfather allele. While the original H–E method dealt with linkage to a single marker, a later multipoint extension by Fulker, Cherny, and Cardon (1995) replaced $\hat{\pi}$ with an estimate of allele sharing at any position in the genome, using all available flanking markers, thereby eliminating the effect of recombination between the marker and QTL on effect size and on the power of the test statistic and providing a more powerful test of linkage. There are several methods available for estimating IBD at any position along a chromosome, given all the genetic information available on the chromosome. When dealing with sib pairs or small families, exact methods such as the Lander–Green algorithm (Lander & Green 1987) and improvements on it (Idury & Elston 1997), as implemented in Merlin (Abecasis et al. 2002), are preferred. The H–E regression test for linkage has the advantages of simplicity and robustness (Allison, Fernandez, Heo, & Beasley 2000), but is less powerful than variance components (VC) models (Fulker & Cherny 1996), which make optimal use of the data.

Extensions to Haseman–Elston

Several authors have proposed modifications to the original H–E method in order to improve its power (Drigalenko 1998, Elston, Buxbaum, Jacobs, & Olson 2000, Forrest 2001, Sham & Purcell 2001, Visscher & Hopper 2001, Wright 1997, Xu, Weiss, Xu, & Wei 2000). The H–E method regresses squared sib pair differences on an estimate of allelic sharing. However, there is more phenotypic information in a sib pair than simply the square of its phenotypic difference score. One proposal was to perform the regression on squared sums instead of squared differences (Drigalenko 1998), which capture different, yet correlated, information in the sib pair:

$$(X + Y)^2 = 2(1 + r) + 2Q(\hat{\pi} - .5) + \epsilon \qquad (3.12)$$

This led to a further extension by Elston et al. (2000) which attempted to incorporate the information contained in both the sums and differences by using the cross-product XY as the dependent variable:

$$XY = r + Q(\hat{\pi} - .5) + \epsilon \qquad (3.13)$$

Xu et al. (2000) proposed a unified H–E method that uses a weighted linear combination of the estimates of Q obtained from the original H–E regression on squared differences and the alternative regression on squared sums. Rijsdijk, Hewitt, and Sham (2001) showed that this method yielded a noncentrality parameter which is identical to that from the VC linkage test. Sham and Purcell (2001) further simplified this unified method such that it could be performed with a single regression:

$$\frac{(X+Y)^2}{(1+r)^2} - \frac{(X-Y)^2}{(1-r)^2} = -\frac{4r}{1-r^2} + \frac{4(1+r^2)}{(1-r^2)^2}Q(\hat{\pi}-.5)+\epsilon$$

(3.14)

with identical results.

Sham and Purcell (2001) examined all of the above regression-based methods, comparing them under various conditions. A variance components (VC) approach (Fulker & Cherny 1996), modelling the full familial covariance structure, can be considered the gold standard, and Sham and Purcell compared these existing regression-based methods with VC, showing that none of the other methods captured all available information in the data. They noted that in the special case of a sib correlation (r) equal to zero, the noncentrality parameters (NCPs) for using squared sums and squared differences are the same, and they sum to the NCP obtained from using the cross-products. As r increases, however, the original squared differences regression test gains in power, while using sums or cross-products loses power; when $r > (2 - \sqrt{3})$, cross-products are less powerful than squared differences, which should be noted is a likely minimum sib correlation (0.27) for typical linkage studies. They further showed that their new regression method retains the advantages of the original H–E regression method in being computationally less demanding than VC and, more importantly, more suited to the analysis of selected samples, which comprise the vast majority of linkage studies.

Sham, Purcell, Cherny, and Abecasis (2002) extended this method to deal with sibships larger than size 2 and more general pedigree configurations, while retaining all the advantages of the other regression-based methods. Within a single pedigree, the analysis is a case of multivariate regression, with as many observations as there are pairs of family members, each contributing an estimated proportion of alleles shared IBD. These estimated IBD-sharing proportions are regressed on an equal number of squared sums and an equal number of squared differences. The regression from the previous methods is reversed, where now the trait values are regressed on $\hat{\pi}$, because sample selection is often through trait values but almost never through marker genotypes. The estimate of a regression coefficient is not biased by sample selection through the independent variable, but can be biased by sample selection through the dependent variable. Therefore, IBD sharing is put on the dependent side of the regression and the squared sums and squared differences are the independent variables.

In this multivariate regression, the estimated IBD sharing of a pair of relatives is modelled not only on the squared sum and squared difference of the same relative pair but also on the squared sums and squared differences of other relative pairs in the pedigree. Since the full distribution of IBD sharing is uncertain under imperfect marker information, a weighted least squares estimation procedure is adopted,

which requires only the covariance matrix of IBD sharing. The weighted least squares estimators of the regression coefficients can be written as a function of three covariance matrices: (1) the covariance matrix of the IBD sharing proportions, (2) the covariance matrix of the squared sums and squared differences, and (3) the covariance matrix between the estimated IBD proportions and the squared sums and squared differences. The last of these matrices is in part determined by the additive variance explained by a linked QTL. The solution of this multivariate regression in a single pedigree provides an estimate of the additive QTL variance, together with its sampling variance. It is then straightforward to combine these estimates across all the pedigrees in a sample, weighting them by the inverse of their variances. This also provides the sampling variance of the combined estimate and a chi-squared test for linkage. The asymptotic distribution of this test statistic in large samples is ensured by the central limit theorem. The method has been shown to provide unbiased LODs and p values under selection Sham et al. (2002).

This method is implemented in Merlin (Abecasis et al. 2002, Sham et al. 2002) and its application is illustrated by an analysis of data from a large population-based sample of sibships assessed for anxiety and depression measures in the GENESiS study (Nash et al. 2004). This study involved collection of questionnaire data on a large community-based sample of 34,371 people, of which 14,807 were members of sibships of sizes 2–7, with the majority sib pairs. Using an algorithm implemented in the SEL (selection for QTL linkage) program (Purcell, Cherny, Hewitt, & Sham 2001), DNA samples were solicited from the most informative 10% of all sibships, based on their composite phenotypic scores across several measures. Informativeness was then recalculated on the 65% of samples who returned DNA, and another round of selection was performed to choose those to be genotyped on 400 microsatellite markers. Sibships which are most discordant or extreme concordant on the phenotype are the most informative, with the ratio of those two groups a function of the sibling correlation on the phenotype. The SEL program ranks the sibships based on their likely contribution to the noncentrality parameter for linkage, using only phenotypic information. Given that this was a selected sample, it was ideally suited for regression-based linkage analysis (Sham et al. 2002). Figure 3.3 shows the results of performing a genome scan using Merlin-regress (Sham et al. 2002) on the neuroticism measure from the Eysenck personality questionnaire (EPQ-N; Eysenck & Eysenck 1975), measured on two occasions in 78% of those subjects genotyped. Using an average of two repeated measures, rather than just a single measurement occasion, will reduce measurement error, thereby increasing heritability (Falconer & Mackay 1996). Merlin-regress was extended to make use of this additional information, allowing the use of averaged

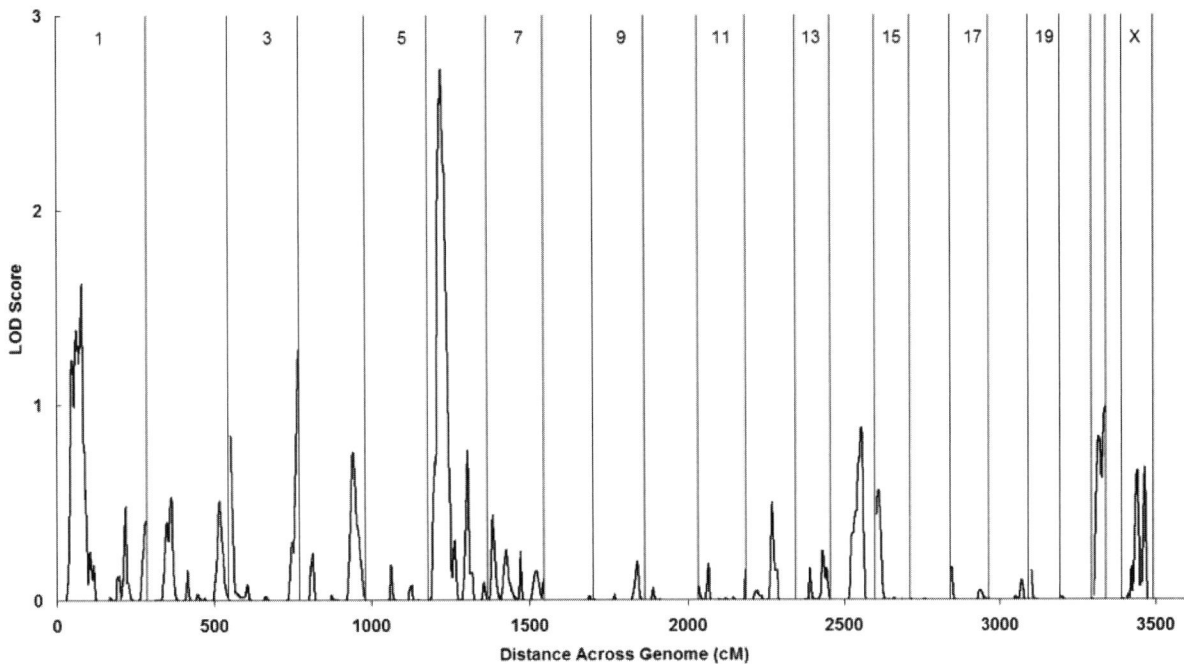

Fig. 3.3 Genomewide linkage scan results for a composite measure of neurotism, from Nash et al. (2004). See text for details

phenotypes from any number of measurement occasions on a per-subject basis, with the method taking account of the differing expected error variance on a per-subject basis (see Appendix in Nash et al. 2004). This analysis resulted in a LOD of nearly 3 on Chromosome 6 for the EPQ-N. This study further refined the analysis by examining males and females separately, yielding a somewhat higher LOD score (Nash et al. 2004).

DeFries–Fulker Regression

An alternative regression model for linkage analysis using sibship data is one based on the DeFries–Fulker (DF) method for analysis of twin data (DeFries & Fulker 1985, 1988, LaBuda, DeFries, & Fulker 1986). The method was developed for estimating genetic and environmental components of variance in both population-based samples of twins and samples where proband members of a twin pair were selected for having extreme scores exceeding (or below) a designated threshold. The method was later extended for analysis of genetic marker data (Fulker et al. 1991, Fulker & Cardon 1994) and the power of the method explored (Cardon & Fulker 1994). It has been used successfully to map a gene for reading disability (Cardon et al. 1994, 1995) which has since been replicated using this methodology (Gayán et al. 1999) and other methods (S. E. Fisher et al. 1999, Grigorenko et al. 1997, Marlow et al. 2003).

The DF basic model (DeFries & Fulker 1985; LaBuda et al. 1986), appropriate for selected sample analysis, when used for linkage mapping in sib pairs, is given by

$$C_i = B_1 P_i + B_2 \hat{\pi}_i \qquad (3.15)$$

where for sib pair i, the co-sib's phenotypic score (C_i) is predicted by the proband's score (P_i) and the sib pair's $\hat{\pi}$, the proportion of alleles shared IBD at a given location along a chromosome. B_2 provides an estimate of the proportion of variance accounted for by the putative QTL (provided the data are suitably transformed prior to analysis (LaBuda et al. 1986)) and the corresponding t statistic provides a test of linkage. In the case of sibships larger than pairs, the method can still be applied by having each sib's score predicted by each proband's score. In the case of multiple probands in a sibship, the probands appear on both sides of the regression equation (once as proband and again as a co-sib). The resulting t statistic is then adjusted by producting it by the square root of the ratio of the number of unique pairings to the number of pairings resulting from this double-entry procedure.

The DF augmented model (DeFries & Fulker 1985, LaBuda et al. 1986) can also be used for linkage mapping, both in selected and population-based samples:

$$C_i = B_3 P_i + B_4 \hat{\pi}_i + B_5 P_i \hat{\pi}_i \qquad (3.16)$$

where B_5 provides an alternative test for linkage (appropriate both in selected and unselected samples; see LaBuda et al. (1986) for an explanation of the distinction between tests of B_2 and B_5 in selected samples). For analysis of unselected samples, all possible pairings of siblings within a family must be formed, with each sib appearing on both sides of the regression equation. The resulting t statistics are then adjusted by producting them by $\sqrt{2}$ to adjust for this double-entry procedure. Double entry of all individuals in an unselected sample acknowledges the intraclass relationship among siblings in a family.

Once the proportions of alleles shared IBD at given chromosomal locations are available, implementation in a statistical analysis package of the DF method of QTL mapping is trivial, since it uses simple linear regression. However, complications arise in merging pairwise IBD information as output from Genehunter (Kruglyak et al. 1996) or Merlin (Abecasis et al. 2002), computing the regression repeatedly across the chromosome, accommodating sibships larger than sib pairs, and accommodating sibships with multiple probands. A macro package for the SAS$^{\circledR}$ statistical analysis software is available which makes the application of the DF approach to sibship data a simple matter for the researcher (Lessem & Cherny 2001).

Conclusions

The regression approach lends itself to various extensions, such as multivariate analysis and tests of association. However, these extensions are not as straightforward to implement as they are in variance components and as yet have not been fully developed. Additionally, it is doubtful a general implementation of the regression approach could be devised so that the researcher can accommodate more complex data structures, using a flexible structural equation modelling package such as Mx (Neale et al. 2003). The variance components approach offers all the flexibility that structural equation modelling brings. Dealing with complex issue such as gene by environment interaction is relatively straightforward, both when estimating genomewide genetic variance, such as in twin studies, and searching for QTLs using genetic marker data. We could stratify our samples into multiple groups, estimating parameters separately for males vs. females, or different ethnicities, or social strata. Or, alternatively, the models could be written, for example, such that the genetic and environmental parameters are linear (or nonlinear) functions of measures of the environment. However, while the regression methods are fast and robust under selected samples and non-normal data, as yet the flexibility offered by structural modelling cannot be matched, as can be seen by the large collection of scripts made available at the Mx Scripts Library (http://www.psy.vu.nl/mxbib/) or the Mx web site (http://www.vcu.edu/mx/).

Acknowledgment Preparation of this chapter was partially supported by National Institutes of Health grant EY-12562.

References

Abecasis, G. R., Cardon, L. R., & Cookson, W. O. (2000, January). A general test of association for quantitative traits in nuclear families. *American Journal of Human Genetics, 66*(1), 279–92.

Abecasis, G. R., Cherny, S. S., Cookson, W. O., & Cardon, L. R. (2002). Merlin—rapid analysis of dense genetic maps using sparse gene flow trees. *Nature Genetics, 30*, 97–101.

Abecasis, G. R., Cookson, W. O., & Cardon, L. R. (2000, July). Pedigree tests of transmission disequilibrium. *European Journal of Human Genetics, 8*(7), 545–551.

Allison, D. B., Fernandez, J. R., Heo, M., & Beasley, T. M. (2000). Testing the robustness of the new haseman-elston quantitative-trait loci-mapping procedure. *American Journal of Human Genetics, 67*, 249–252.

Cardon, L. R., & Abecasis, G. R. (2000). Some properties of a variance components model for fine-mapping quantitative trait loci. *Behavior Genetics, 30*, 235–243.

Cardon, L. R., & Fulker, D. W. (1994). The power of interval mapping of quantitative trait loci, using selected sib pairs. *American Journal of Human Genetics, 55*, 825–833.

Cardon, L. R., Smith, S. D., Fulker, D. W., Kimberling, W. J., Pennington, B. F., & DeFries, J. C. (1994). Quantitative trait locus for reading disability on chromosome 6. *Science, 266*, 276–279.

Cardon, L. R., Smith, S. D., Fulker, D. W., Kimberling, W. J., Pennington, B. F., & DeFries, J. C. (1995). Quantitative trait locus for reading disability: Correction. *Science, 268*, 1553.

DeFries, J. C., & Fulker, D. W. (1985). Multiple regression analysis of twin data. *Behavior Genetics, 15*, 467–473.

DeFries, J. C., & Fulker, D. W. (1988). Multiple regression analysis of twin data: Etiology of deviant scores versus individual differences. *Acta Geneticae Medicae et Gemellologiae, 37*, 205–216.

Drigalenko, E. (1998). How sib pairs reveal linkage. *American Journal of Human Genetics, 63*, 1242–1245.

Elston, R. C., Buxbaum, S., Jacobs, K. B., & Olson, J. M. (2000). Haseman and elston revisited. *Genetic Epidemiology, 19*, 1–17.

Evans, D. M., & Medland, S. E. (2003). A note on including phenotypic information from monozygotic twins in variance components qtl linkage analysis. *Annals of Human Genetics, 67*, 613–617.

Eysenck, H. J., & Eysenck, S. B. G. (1975). *Manual of the eysenck personality questionnaire*. San Diego: Digits.

Falconer, D. S., & Mackay, T. F. C. (1996). *Introduction to quantitative genetics* (pp. 136–142). London: Longman.

Fisher, S., Francks, C., Marlow, A. J., MacPhie, I. L., Newbury, D. F., Cardon, L. R., et al. (2002). Independent genome-wide scans identify a chromosome 18 quantitative-trait locus influencing dyslexia. *Nature Genetics, 30*, 86–91.

Fisher, S. E., Marlow, A. J., Lamb, J., Maestrini, E., Williams, D. F., Richardson, A. J., et al. (1999). A quantitative-trait locus on chromosome 6p influences different aspects of developmental dyslexia. *American Journal of Human Genetics, 64*, 146–156.

Forrest, W. (2001). Weighting improves the "new haseman-elston" method. *Human Heredity, 52*, 47–54.

Fulker, D. W., & Cardon, L. R. (1994). A sib-pair approach to interval mapping of quantitative trait loci. *American Journal of Human Genetics, 54*, 1092–1103.

Fulker, D. W., Cardon, L. R., DeFries, J. C., Kimberling, W. J., Pennington, B. F., & Smith, S. D. (1991). Multiple regression analysis of sib-pair data on reading to detect quantitative trait loci. *Reading and Writing: An Interdisciplinary Journal, 3*, 299–313.

Fulker, D. W., & Cherny, S. S. (1996). An improved multipoint sib-pair analysis of quantitative traits. *Behavior Genetics, 26*, 527–532.

Fulker, D. W., Cherny, S. S., & Cardon, L. R. (1995). Multipoint interval mapping of quantitative trait loci, using sib pairs. *American Journal of Human Genetics, 56*, 1224–1233.

Fulker, D. W., Cherny, S. S., Sham, P. C., & Hewitt, J. K. (1999, January). Combined linkage and association sib-pair analysis for quantitative traits. *American Journal of Human Genetics, 64*(1), 259–267.

Gayán, J., Smith, S. D., Cherny, S. S., Cardon, L. R., Fulker, D. W., Brower, A. M., et al. (1999). Quantitative trait locus for specific language and reading deficits on chromosome 6p. *American Journal of Human Genetics, 64*, 157–164.

Grigorenko, E. L., Wood, F. B., Meyer, M. S., Hart, L. A., Speed, W. C., Schuster, A., et al. (1997). Susceptibility loci for distinct components of developmental dyslexia on chromosomes 6 and 15. *American Journal of Human Genetics, 60*, 27–39.

Haseman, J. K., & Elston, R. C. (1972). The investigation of linkage between a quantitative trait and a marker locus. *Behavior Genetics, 2*, 3–19.

Idury, R. M., & Elston, R. C. (1997). A faster and more general hidden markov model algorithm for multipoint likelihood calculations. *Human Heredity, 47*, 197–202.

Jinks, J. L., & Fulker, D. W. (1970). Comparison of the biometrical genetical, MAVA, and classical approaches to the analysis of human behavior. *Psychological Bulletin, 73*, 311–349.

Kruglyak, L., Daly, M. J., Reeve-Daly, M. P., & Lander, E. S. (1996). Parametric and nonparametric linkage analysis: A unified multipoint approach. *American Journal of Human Genetics, 58*, 1347–1363.

LaBuda, M. C., DeFries, J. C., & Fulker, D. W. (1986). Multiple regression analysis of twin data obtained from selected samples. *Genetic Epidemiology, 3*, 425–433.

Lander, E., & Kruglyak, L. (1995). Genetic dissection of complex traits: Guidelines for interpreting and reporting linkage results. *Nature Genetics, 11*, 241–247.

Lander, E. S., & Green, P. (1987). Construction of multilocus genetic maps in humans. *Proceedings of the National Academy of Sciences, 84*, 2363–2367.

Lessem, J. M., & Cherny, S. S. (2001). Defries-fulker multiple regression of sibship qtl data: A SAS® macro. *Bioinformatics, 17*, 371 372.

Marlow, A. J., Fisher, S. E., Francks, C., MacPhie, I. L., Cherny, S. S., Richardson, A. J., et al. (2003). Use of multivariate linkage analysis for dissection of a complex cognitive trait. *American Journal of Human Genetics, 72*, 561–570.

McKenzie, C. A., Abecasis, G. R., Keavney, B., Forrester, T., Ratcliffe, P. J., Julier, C., et al. (2001). Trans-ethnic fine mapping of a quantitative trait locus for circulating angiotensin i-converting enzyme (ace). *Human Molecular Genetics, 10*, 1077–1084.

Nash, M. W., Huezo-Diaz, P., Williamson, R. J., Sterne, A., Purcell, S., Hoda, F., et al. (2004). Genome-wide linkage analysis of a composite index of neuroticism and mood-related scales in extreme selected sibships. *Human Molecular Genetics, 13*, 2173–2182.

Neale, M. C., Boker, S. M., Xie, G., & Maes, H. H. (2003). *Mx: Statistical modeling* (6th ed). VCU Box 900126, Richmond, VA 23298 USA: Department of Psychiatry, Virginia Commonwealth University.

Neale, M. C., & Cardon, L. R. (1992). *Methodology for genetic studies of twins and families, NATO ASI series*. Dordrecht, The Netherlands: Kluwer Academic Press.

Neale, M. C., & Maes, H. H. (in preparation). *Methodology for genetic studies of twins and families* (2nd ed). Dordrecht, The Netherlands: Kluwer Academic Press.

Purcell, S., Cherny, S., Hewitt, J., & Sham, P. (2001). Optimal sibship selection for genotyping in quantitative trait locus linkage analysis. *Human Heredity, 52*, 1–13.

Rijsdijk, F. V., Hewitt, J. K., & Sham, P. C. (2001). Analytic power calculation for variance-components linkage analysis in small pedigrees. *European Journal of Human Genetics, 9*, 335–340.

Schmitz, S., Cherny, S. S., & Fulker, D. W. (1998). Increase in power through multivariate analyses. *Behavior Genetics, 28*, 357–363.

Schork, N. J., & Xu, X. (2000). The use of twins in quantitative trait locus mapping. In T. D. Spector, H. Snieder, & A. J. MacGregor (Eds), *Advances in twin and sib-pair analysis* (pp. 189–202). London: Greenwich Medical Media.

Sham, P. (1998). *Statistics in human genetics*. London: Arnold.

Sham, P. C., & Purcell, S. (2001). Equivalence between haseman-elston and variance-components linkage analyses for sib pairs. *American Journal of Human Genetics, 68*, 1527–1532.

Sham, P. C., Purcell, S., Cherny, S. S., & Abecasis, G. R. (2002). Powerful regression-based quantitative-trait linkage analysis of general pedigrees. *American Journal of Human Genetics, 71*, 238–253.

Sham, P.C., Zhao, J. H., Cherny, S., & Hewitt, J. (2000). Variance-components qtl linkage analysis of selected and non-normal samples: conditioning on trait value. *Genetic Epidemiology, 19*, S22–S28.

Visscher, P. M., & Hopper, J. L. (2001). Power of regression and maximum likelihood methods to map qtl from sib-pair and dz twin data. *Annals of Human Genetics, 65*, 583–601.

Wright, F. A. (1997). The phenotypic difference discards sib-pair QTL linkage information. *American Journal of Human Genetics, 60*, 740–742.

Xu, X., Weiss, S., Xu, X., & Wei, L. J. (2000). A unified haseman-elston method for testing linkage with quantitative traits. *American Journal of Human Genetics, 67*, 1025–1028.

Chapter 4

Multivariate Genetic Analysis

Danielle Posthuma

Introduction

The main goal of behavior genetics' research is to understand the causes of variation in (human) traits. When single traits are considered, observed trait variation is decomposed into sources of genetic and environmental variation. A genetically informative design, such as the classical twin design, allows estimating the relative contributions of these sources of variation. When multiple traits are considered, genetically informative designs additionally allow investigating the causes of *co-variation* between two or more traits. Such multivariate genetic analyses are usually more powerful than univariate genetic analyses (Schmitz, Cherny, & Fulker, 1998), may aid in understanding underlying biological mechanisms, and may provide a faster route to gene-finding and elucidating environmental factors that influence a trait (Leboyer et al., 1998).

The key source of information in multivariate twin studies is the comparison of MZ and DZ *cross-trait cross-twin* correlations (CTCTs). A CTCT is a correlation between trait A of a twin and trait B of his or her co-twin. A larger MZ CTCT correlation than DZ CTCT correlation implies that the correlation between traits A and B is due to co-variation at a genetic level, while similar MZ and DZ CTCT correlations imply that (shared) environmental factors are responsible for the co-variation between traits A and B. For example, using a sample of MZ and DZ twins for whom data on both brain volume and IQ were available, correlations were calculated between brain volume of a twin and the IQ score of his or her co-twin (Posthuma et al., 2002). It was found that the CTCT correlation was larger in MZ twins than in DZ twins, which indicates that genes must mediate the correlation between brain volume and IQ. In support of this, it was also found that the MZ cross-trait cross-twin correlation was the same as the correlation between brain volume and IQ in the same person, indicating that non-shared environmental influences do not mediate the correlation between brain volume and IQ. The prediction of one's IQ score can thus be made with similar reliability from one's own brain volume as from the brain volume of one's genetically identical co-twin. This finding directs further research aimed at finding genetic determinants for intelligence, as it indicates that some of the genes that influence brain volume may also influence intelligence (and vice versa). Since some genes for brain volume have already been identified (such as *ASPM* and *Microcephalin*), these may pose good candidate genes for intelligence. The identification of such underlying common sources of variation may therefore direct further research aimed at identifying the actual genes (or actual environmental factors).

This chapter is intended as an introductory text explaining the basics of multivariate genetic models. In some cases formal quantification is offered to allow the reader to apply these models using software designed to model genetically informative data, such as Mx (Neale, 1997) or LISREL (Jöreskog & Sörbom, 1986). For more advanced models, a theoretical description and a brief discussion of the limitations of the models are provided. The reader is then directed to other sources in the literature that deal with the formal quantification of these advanced models. Below we will first explain how the univariate twin model can be extended to a multivariate model, using path analysis and matrix algebra. We will then continue with more complex multivariate models.

Multivariate Analysis of Twin Data: Determining Genetic and Environmental Correlations

The univariate model can easily be extended to a multivariate model when more than one trait per subject is measured (Boomsma & Molenaar, 1986, 1987; Eaves & Gale, 1974; Martin & Eaves, 1977). For simplicity we will first discuss a bivariate design, as depicted in Fig. 4.1 (two observed traits

D. Posthuma (✉)
Department of Biological Psychology, Section Medical Genomics, and Section Functional Genomics, Vrije Universiteit and Vrije Universiteit Medical Center, Amsterdam 1081BT, The Netherlands
e-mail: danielle@psy.vu.nl

Y.-K. Kim (ed.), *Handbook of Behavior Genetics*,
DOI 10.1007/978-0-387-76727-7_4, © Springer Science+Business Media, LLC 2009

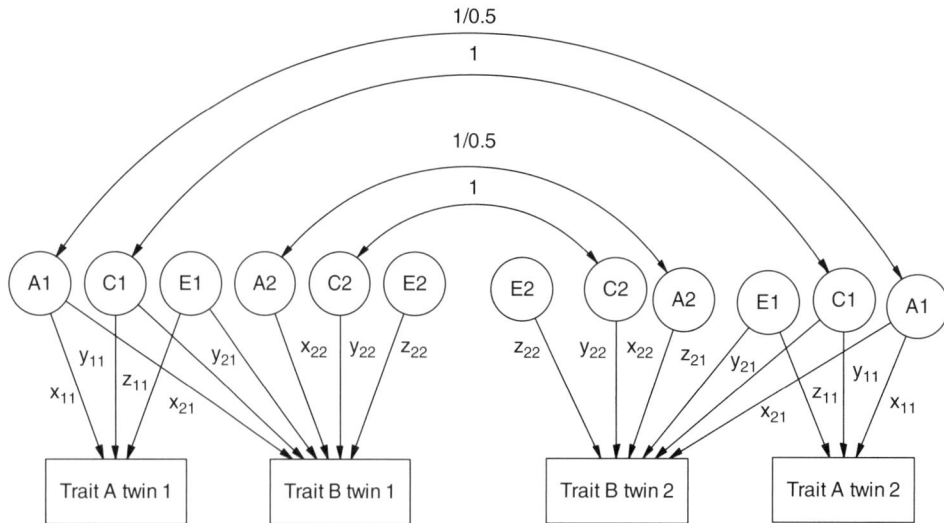

Fig. 4.1 A bivariate twin design: Cholesky factorization. The observed (co-)variance of traits A and B is decomposed into sources of genetic (A), shared (C), and non-shared (E) environmental variation, following a Cholesky factorization. Genetic factors correlate 1 in MZ twins and 0.5 in DZ twins. Shared environmental factors correlate 1
Note that the path coefficients are not included in the figure for estheti-

cal reasons. For the genetic paths, the paths are named x_{11}, x_{21}, and x_{22}, where the subscripts refer to 'goes to the nth measured variable, comes from the nth latent variance component'. The path coefficients for the shared environmental paths are denoted as z_{11}, z_{21}, and z_{22}, and for the non-shared environmental paths z_{11}, z_{21}, and z_{22}

per subject; four observed traits for a pair of twins or siblings). We assume each trait is influenced by additive genetic influences (A), shared environmental influences (C), and non-shared environmental influences (E). The corresponding matrix algebra expressions for the expected MZ or DZ variances and co-variances for traits i and j are similar to those in a univariate model. Let X, Y, and Z be the matrices representing the genetic, non-shared, and shared environmental paths, respectively. In the multivariate case, the dimensions of matrices **X**, **Y**, and **Z** are no longer 1×1, but are a function of the number of traits. An often convenient form for those matrices is lower triangular of dimensions $n \times n$ (where n is the number of traits assessed on a single subject; in Fig. 4.1, $n = 2$). Matrix **X** is thus $\begin{bmatrix} x_{11} & 0 \\ x_{21} & x_{22} \end{bmatrix}$, with each matrix entry corresponding to a path coefficient. Element x_{11} corresponds to the path that goes to the first observed trait and comes from the first genetic factor (A1), while x_{21} corresponds to the path that goes to the second observed trait and comes from the first genetic factor (A1), and x_{22} corresponds to the path that goes to the second trait and comes from the second latent genetic factor. As a rule of thumb, the subscripts are thus in the order '*goes to, comes from*' in path diagrams and '*rows first then columns*' in matrices. Similar reasoning goes for matrices **Y** $\begin{bmatrix} y_{11} & 0 \\ y_{21} & y_{22} \end{bmatrix}$ and **Z** $\begin{bmatrix} z_{11} & 0 \\ z_{21} & z_{22} \end{bmatrix}$.

Multivariate genetic designs allow the decomposition of an observed correlation between two variables into a genetic and an environmental part. This can be quantified by

calculating the genetic and environmental correlations and the genetic and environmental contributions to the observed correlation.

The additive genetic, shared, and non-shared environmental variances and co-variances can be represented as elements of the symmetric matrices $\mathbf{A} = \mathbf{XX}^{\mathrm{T}}$, $\mathbf{C} = \mathbf{YY}^{\mathrm{T}}$, and $\mathbf{E} = \mathbf{ZZ}^{\mathrm{T}}$, where the superscripted T denotes matrix transposition. On the diagonals these matrices contain the additive genetic and (non-)shared environmental variances for variables 1 to n. **X**, **Y**, and **Z** are known as the Cholesky factorization of the matrices **A**, **C**, and **E**, assuring that these matrices are nonnegative definite, which is required for variance–covariance matrices.

The *genetic correlation* between traits i and j (r_{gij}) is derived as the genetic covariance between traits i and j (denoted by element ij of matrix **A**; a_{ij}) divided by the square root of the product of the genetic variances of traits $i(a_{ii})$ and $j(a_{jj})$: $r_{gij} = \frac{a_{ij}}{\sqrt{a_{ii} \times a_{jj}}}$.

The *shared environmental correlation* (r_{cij}) between variables i and j is derived as the environmental covariance between variables i and j divided by the square root of the product of the shared environmental variances of variables i and j: $r_{cij} = \frac{c_{ij}}{\sqrt{c_{ii} \times c_{jj}}}$. Analogously, the *non-shared environmental correlation* (r_{eij}) between variables i and j is derived as the non-shared environmental covariance between variables i and j divided by the square root of the product of the non-shared environmental variances of variables i and j: $r_{eij} = \frac{e_{ij}}{\sqrt{e_{ii} \times e_{jj}}}$.

The phenotypic correlation r is the sum of the product of the genetic correlation and the square roots of the standardized genetic variances (i.e., the heritabilities) of the two traits, and the product of the shared environmental correlation and the square roots of the standardized shared environmental variances of the two traits and the product of the non-shared environmental correlation and the square roots of the standardized non-shared environmental variances of the two traits:

$$r = r_{gij} \times \sqrt{\frac{a_{ii}}{(a_{ii} + c_{ii} + e_{ii})}} \times \sqrt{\frac{a_{jj}}{(a_{jj} + c_{jj} + e_{jj})}}$$
$$+ r_{cij} \times \sqrt{\frac{c_{ii}}{(a_{ii} + c_{ii} + e_{ii})}} \times \sqrt{\frac{c_{jj}}{(a_{jj} + c_{jj} + e_{jj})}}$$
$$+ r_{eij} \times \sqrt{\frac{e_{ii}}{(a_{ii} + c_{ii} + e_{ii})}} \times \sqrt{\frac{e_{jj}}{(a_{jj} + c_{jj} + e_{jj})}}$$

(i.e., observed correlation is the sum of the genetic contribution and the environmental contributions).

The genetic contribution to the observed correlation between two traits is a function of the two sets of genes that influence the traits and the correlation between these two sets. However, a large genetic correlation does not imply a large phenotypic correlation, as the latter is also a function of the heritabilities. If the heritabilities are low, the genetic contribution to the observed correlation will also be low.

If the genetic correlation is 1, the two sets of genetic influences overlap completely. If the genetic correlation is less than 1, at least some genes are a member of only one of the sets of genes. A large genetic correlation, however, does not imply that the overlapping genes have effects of similar magnitude on each trait. The overlapping genes may even act additively for one trait and show dominance for the second trait. In addition, a genetic correlation less than one cannot exclude that all of the genes are overlapping between the two traits (Carey, 1988). Similar reasoning applies to the environmental correlation.

It should be noted that the Cholesky factorization described above is just identified. That is, with data on MZ and DZ twins one can estimate three sources of variation, and for each source there are no more latent factors than observed traits (i.e., two in the bivariate case). In the Cholesky factorization, there is no path specified from the second latent factor to the first observed trait (e.g., element x_{12} is zero) – as an additional path coefficient is not identified (unless this coefficient is restricted to be equal to one of the other path coefficients specified). This *lower* triangular solution, however, is mathematically completely equivalent with an *upper* triangular solution. Or, in other words, the order of the observed variables is completely arbitrary. In practice this means that when two traits (e.g., anxiety and depression; personality and smoking; sports participation and self-reported health) are modeled in a bivariate genetic design, we are restricted to statements on the genetic (or environmental) overlap between the two traits. Statements of the form '*new genes are important for depression, whereas all of the genetic information important for anxiety is also important for depression*' cannot be made. Instead of the lower triangular solution, one can interpret the standardized solution represented by a correlated factors model with the standardized factor loadings (e.g., $\sqrt{h^2_1}$ and $\sqrt{h^2_2}$) and the genetic correlation between them, using $\mathbf{XRX}^T$, where X is diagonal and R is a standardized, correlation matrix containing the genetic correlation between factors A1 and A2. Similar reasoning applies to the shared and non-shared environmental factors (see Fig. 4.2).

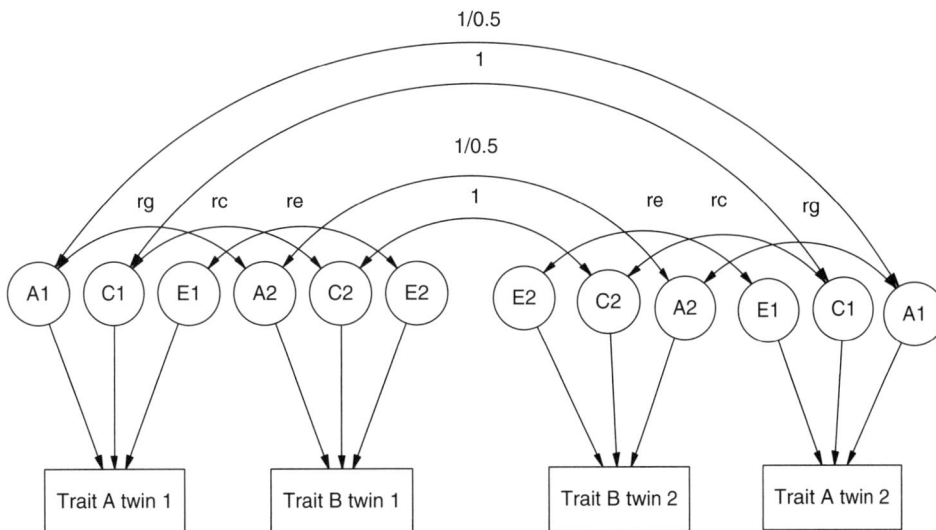

Fig. 4.2 A bivariate twin design: standardized solution. The path coefficients refer to the square roots of the relative proportions of A, C, and E to the observed variation. rg = genetic correlation, rc = shared environmental correlation, re = non-shared environmental correlation

There is one exception to the restrictions described above; when the two traits of the bivariate model are naturally ordered (for example when they concern two measurements over time or two nested conditions), specific hypothesis-driven tests can in fact be carried out. It can then be tested whether new genes come into play (by testing whether the influence of the second genetic factor on the second measurement significantly differs from zero) or whether genetic amplification is in order (when the two path coefficients from the first genetic factor to the two traits are equal). For example, De Geus, Kupper, Boomsma, and Snieder (2007) used bivariate genetic modeling to discriminate between amplification of genetic effects (the same genes influence two traits with different effect sizes) and emergence of new genetic effects by comparing cardiovascular measures in resting and stress conditions. They found that the heritability of these cardiovascular traits increased during stress which was due both to an amplification of genetic effects that are important in a resting condition and to new genetic effects emerging under stress.

Extension of the bivariate design described above to more than two traits is straightforward, allowing the estimation of genetic and environmental correlations between multiple traits at the same time (see Fig. 4.3).

The full Cholesky model provides a complete mathematical description of the observed variables. In other words, it has as many latent factors as there are observed traits. Apart from decomposing the observed (co-)variation into sources of genetic and environmental (co-)variation, it merely provides a description of the observed data. In order to test several different theoretical models to the data, one usually seeks to reduce the number of explanatory, latent factors. One way to do so is via factor analysis.

Genetic Factor Models

The main goal of factor analysis is to explain (co-)variation between and within a set of measured traits by a lesser number of latent factors. The use of factor analysis was pioneered by Spearman (1904) in the context of the measurement of intelligence and different mental abilities. Spearman noted that performance in several different mental abilities correlates highly within subjects. This led him to believe that there is a general mental ability (the 'g'-factor) underlying performance on all other specific mental abilities. The use of factor analytic techniques in the social sciences has since flourished. Factor analysis can either be exploratory or confirmatory. In exploratory factor analysis, there is no a priori hypothesis on the number of latent factors or the nature of the relationship between them. In contrast, confirmatory factor analysis is based on testing hypotheses on the latent factor structure underlying multiple traits. Confirmatory factor analysis typically has fewer factors than observed variables and specifies the presence or absence of correlations between the latent factors. For a full treatment of phenotypic factor analytic techniques we refer the reader to Hotelling (1933), Lawley & Maxwell (1971), Mulaik (1972), and Thurstone (1947). Briefly, in phenotypic factor analysis, an observed trait P in an individual is modeled as a function of that individual's score on a latent factor F and error or unique variance U:

$$P_i = f_i F + U$$

Genetic factor analysis follows similar principles as phenotypic factor analysis, except that the latent factors are

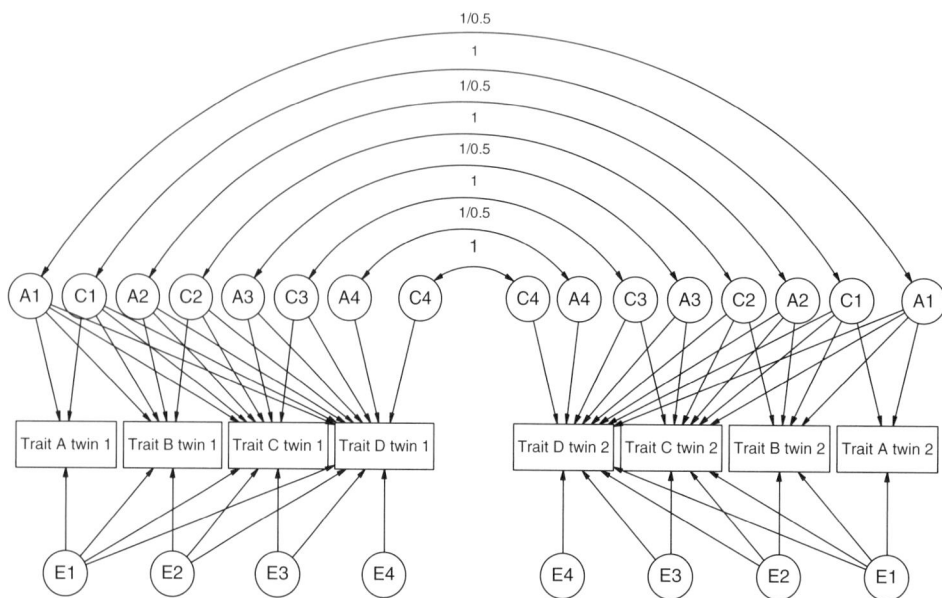

Fig. 4.3 A four-variate twin model: Cholesky factorization

decomposed into genetic and environmental latent factors. A genetic factor model is formally represented as (see Neale & Maes, in press)

$$P_i = a_i A - c_i C - e_i E - U$$

In standard factor analysis the complete factor model is formally represented in one matrix (**B**) containing the common variances (for all sources of variation) and a matrix (**U**) containing the specific variances. In addition, a matrix that specifies the relations between the latent factors (**R**) is included. The formal notation of the factor model is then

$$\sum_{PP} = \mathbf{BRB}^T + \mathbf{U}$$

where the dimensions of matrices **B** and **R** depend on the number of latent sources of variance × the number of factors (columns) and the number of observed traits (for all individuals) (rows). Matrix **R** is a diagonal matrix containing correlations between the latent factors, while matrix **U** is a diagonal and contains the specific variances for each observed trait. When the latent factors are uncorrelated, **R** is an identity matrix and can be omitted from the model. While this notation follows most applications for phenotypic factor analysis, an alternative specification is commonly used in behavior genetics research. Using classical behavior genetic conventions, the different sources of variation and the common and specific variances are specified in separate matrices (see Neale & Cardon, 1992, or Neale & Maes, in press). The formal representation for the factor structure then becomes

$$\sum_{PP} = \mathbf{A} - \mathbf{C} - \mathbf{E}$$

where $\mathbf{A} = \mathbf{XRX}^T$, $\mathbf{C} = \mathbf{YSY}^T$, $\mathbf{E} = \mathbf{ZTZ}^T + \mathbf{UU}^T$. Matrices **X**, **Y**, and **Z** are of dimensions number of traits × number of factors, while matrix **U** is diagonal with dimensions number of traits × number of traits and assumes that all specific variance is due to measurement error. (Note, however, that specific variation may also be decomposed into sources of genetic and environmental variation.) Matrices **R**, **S**, and **T** are diagonal matrices that contain the relationships between the latent genetic, shared, and non-shared environmental factors, respectively.

Two factor models that are often used in behavior genetics are the independent and common pathway models. In the independent pathway model, the genetic and environmental sources of variation independently influence the observed traits, although for each source of variation there is one common factor (see Fig. 4.4).

The formal representation of this model is equivalent to the genetic factor model described above, except that the accompanying matrices **X**, **Y**, and **Z** are of dimensions 4 × 1 (4 traits × 1 latent factor), matrices **S**, **T**, and **U** are identity matrices, and matrix **U** is diagonal and 4 × 4. Again, for simplicity, this model assumes that all specific variance is due to measurement error.

In the common pathway model, all observed traits are indicators of one common latent factor which is influenced by sources of genetic and environmental variation (see Fig. 4.5).

A formal representation of this model is provided by $\sum_{PP} = \mathbf{A} - \mathbf{C} - \mathbf{E}$, where $\mathbf{A} = \mathbf{F}(\mathbf{XRX}^T)\mathbf{F}^T$, $\mathbf{C} = \mathbf{F}(\mathbf{YSY}^T)\mathbf{F}^T$, $\mathbf{E} = \mathbf{F}(\mathbf{ZTZ}^T)\mathbf{F}^T + \mathbf{UU}^T$. Matrices **X**, **Y**, and **Z** are of dimensions 1 (number of latent traits) × 1 (number of latent genetic or environmental factors); matrix **F** is of dimension 4 (number of observed traits) × 1 (number of latent traits); matrices **R**, **S**, and **T** are identity matrices; and

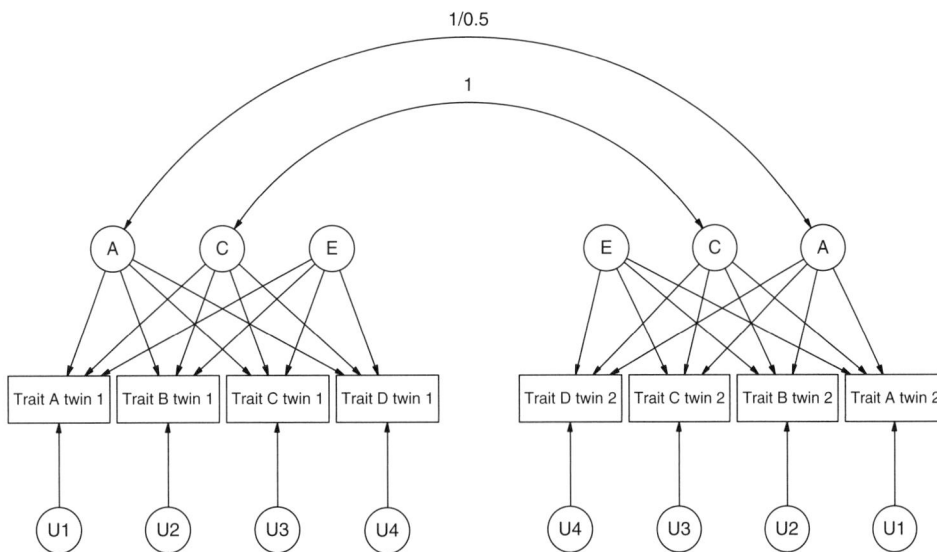

Fig. 4.4 A four-variate twin model: independent pathway model

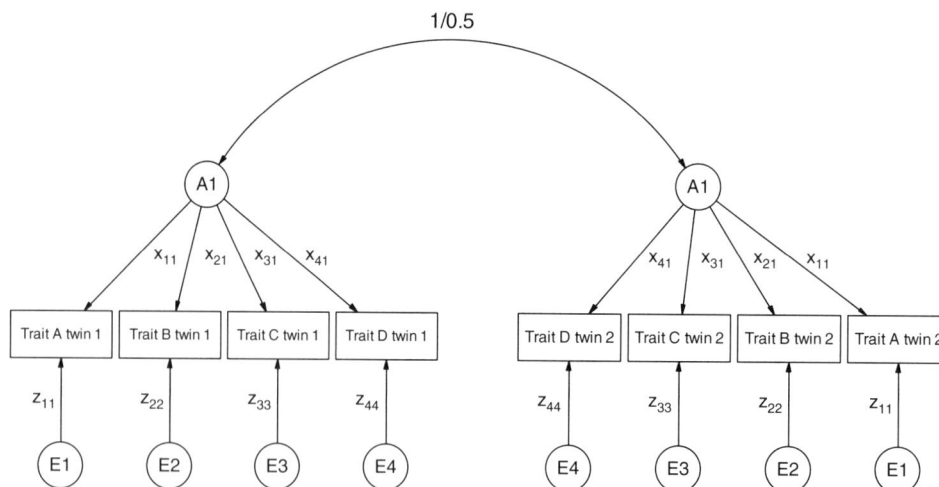

Fig. 4.5 A four-variate twin model: common pathway model

matrix **U** is diagonal and 4×4. Again, for simplicity, this model assumes that all specific variance is due to measurement error. In addition to these specifications, a restriction should be made in the model to assure that the variance of the latent factor F is scaled to 1. Common pathway models are often used when multiple indicators (e.g., multiple raters) are collected of the same underlying trait data.

More complex multivariate genetic models that are extensions of the models described above can be designed. For example, Rijsdijk, Vernon, and Boomsma (2002) tested whether hierarchical genetic factor models would provide a good description of scores on subscales of two intelligence tests: the Wechsler Adult Intelligence Scale (WAIS) and the Raven Standard Progressive Matrices (Raven SPM). They found that at a genetic level the subscales could be described best by a two-level hierarchical model; the first level consisted of three factors (verbal comprehension, perceptual organization, and freedom-from-distractibility), while the second level consisted of one general factor. This general factor was suggested to support the notion of a biological basis for general intelligence, or g (see Fig. 4.6).

Longitudinal Analysis of Twin Data

When the same trait is measured at multiple time points, genetically informative samples can provide information on the sources of the stability and change in a trait over time, using longitudinal analysis of genetic data. The aim of longitudinal analysis of twin data is to consider the genetic and environmental contributions to the dynamics of twin pair responses through time. In this case the trait is measured at several distinct time points for each twin in a pair. To analyze such data one must take the serial correlation between the consequent measurements of the trait into consideration. The

classical genetic analysis methods described above are aimed at the analysis of a trait or traits measured cross-sectionally and provide a way of estimating the time-specific heritability and variability of environmental effects, as well as covariation with other measured traits. However, these methods are not able to handle serially correlated longitudinal data efficiently.

To deal with these issues the classic genetic analysis methods have been extended to investigate the effects of genes and environment on the development of traits over time (Boomsma & Molenaar, 1987; McArdle, 1986). Methods based on the Cholesky factorization of the covariance matrix of the responses treat the multiple trait measurements in a multivariate genetic analysis framework (as discussed above). 'Markov chain' (or 'Simplex') models (Dolan, 1992; Dolan, Molenaar, & Boomsma, 1991) provide an alternative account of change in covariance and mean structure of the trait over time. The Markov model structure implies that future values of the trait depend on the current trait values alone, not on the entire past history. Methods of function-valued quantitative genetics (Pletcher & Geyer, 1999) or the genetics of infinite-dimensional characters (Kirkpatrick & Heckman, 1989) have been developed for situations where it is necessary to consider the time variable on a continuous scale. The aim of these approaches is to investigate to what extent the variation of the traits at different times may be explained by the same genetic and environmental factors acting at different time points and to establish how much of the genetic and environmental variation is time specific.

An alternative approach for the analysis of longitudinal twin data is based on random growth curve models (Neale & McArdle, 2000). The growth curve approach to genetic analysis was introduced by Vandenberg and Falkner (1965) who first fitted polynomial growth curves for each subject and then estimated heritabilities of the components. These methods focus on the rate of change of the phenotype (i.e.,

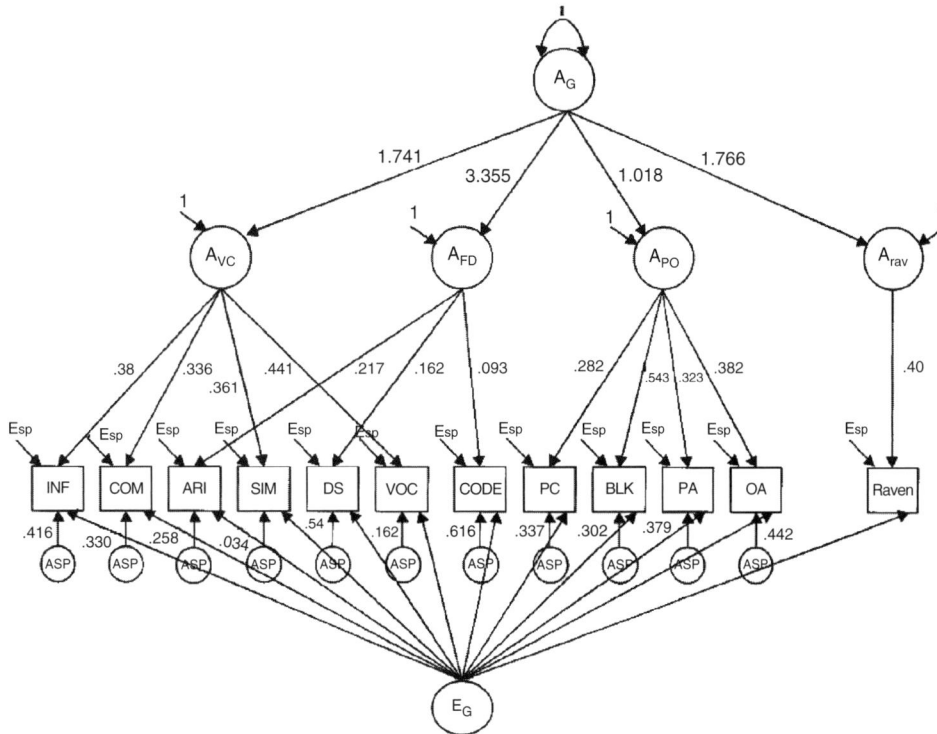

Fig. 4.6 Example of a hierarchical genetic factor model on subtests of the WAIS intelligence test and the Raven SPM. The genetic variance follows a hierarchical, two-level factor model. There is a common non-shared environmental factor that explains part of the correlation between subtests (E_G). Additionally specific genetic and non-shared environmental influences are modeled for each subtest
WAIS subtests: INF = information; COM = comprehension; ARI = arithmetic; SIM = similarities; DS = digit span; VOC = vocabulary; CODE = coding; PC = picture completion; BLK = block design; PA = picture arrangement; OA = object assembly: *Factors:* VC = verbal comprehension; FD = freedom of distractibility; PO = perceptual organization. Asp = specific genetic influences; Esp = specific non-shared environmental influences. Reprinted with permission from Rijsdijk et al. (2002)

its slope or partial derivative) as a way to predict the level at a series of points in time. It is assumed that the individual trait trajectory in time may be described by a parametric growth curve (e.g., linear, exponential, logistic, etc.). The parameters of the growth curve (e.g., intercept and slope, also called latent variables) are assumed to be random and individual specific. However, the random intercepts and slopes may be dependent within a pair of twins because of genetic and shared environmental influences on the random coefficients. The basic idea of the method is that the mean and covariance structure of the latent variables determines the expected mean and covariance structure of the longitudinal phenotype measurements and one may therefore estimate the characteristics of the latent variable distribution based on the longitudinal data.

The random growth curve approach permits to investigate new questions concerning the nature of genetic influence on the dynamic characteristics of the trait, such as the rate of change. If the random parameters of the growth curve would be observed, they might have been analyzed directly using the classical methods of multivariate genetic analysis. However, their latent nature requires a more elaborate statistical

approach. Since the growth curve model may be formulated in terms of the mean and covariance structure of the random parameters one might simply take the specification of the mean and covariance structure of a multivariate trait as predicted by the classical methods of multivariate genetic analysis and transfer that to the growth curve model. The resulting two-level latent variable model would then allow for multivariate genetic analysis of the random coefficients (see Fig. 4.7).

In the following sections we consider the bivariate linear growth curve model applied to longitudinal twin data using age as timescale. The approach may be extended to other parametric growth curves (e.g., exponential, logistic, etc.) using first-order Taylor expansions and the resulting mean and covariance structure approximations (Neale & McArdle, 2000).

The Linear Growth Curve Model

A simple implementation of the random effects approach is carried out using linear growth curve models. In this case each individual is characterized by a random intercept and a random slope, which are considered to be the new traits. In a linear growth curve model the continuous age-dependent trait (Y_{1t}, Y_{2t}) for sib 1 and sib 2 are assumed to follow a linear age trajectory given the random slopes and intercepts with some additive measurement error: $Y_{it} = \alpha_i + \beta_i t + \varepsilon_{it}$, for sibling i, where $i = 1, 2$, t denotes the time point ($t = 1, 2, \ldots, n$), α_i and β_i are the individual (random) intercept and slope of sib i, respectively, and ε_{it} is a zero-mean individual error residual, which is assumed to be independent of α_i and β_i. The aim of the study is then the genetic analysis of the individual intercepts (α_i) and slopes (β_i). The model may easily be extended to include covariates.

Assume the trait (Y_{it}) is measured for the two sibs at $t = 1$, $2, \ldots, n$. The measurements on both twins at all time points may be written in vector form as $\mathbf{Y} = (Y_{11}, \ldots, Y_{1n}, Y_{21}, \ldots, Y_{2n})^{\mathrm{T}}$ (where ^{T} denotes transposition). Furthermore, if $\mathbf{L}$ denotes the vector of the random growth curve parameters, the matrix form of the linear growth curve model is $\mathbf{Y} = \mathbf{DL} + \mathbf{E}$ where

$$L = \begin{pmatrix} \alpha_1 \\ \beta_1 \\ \alpha_2 \\ \beta_2 \end{pmatrix}, D = \begin{pmatrix} F & 0 \\ 0 & F \end{pmatrix}, \quad F = \begin{pmatrix} 1 & 1 \\ 1 & 2 \\ : & : \\ 1 & n \end{pmatrix}, E = \begin{pmatrix} \varepsilon_{11} \\ \vdots \\ \varepsilon_{1n} \\ \varepsilon_{21} \\ \vdots \\ \varepsilon_{2n} \end{pmatrix}.$$

Note that the linear growth curve model actually represents a structural equation model with latent variables α_i and β_i (and ε_{it}s) with loadings of the latent variables on the observed responses Y_{it} given by either 1 or t. This implies that this model may be analyzed using the general structural equation modeling techniques. In particular, parameter estimation may be carried out via the maximum likelihood method under multivariate normality assumptions using the fact that the moment structure ($\boldsymbol{m}_Y$, $\boldsymbol{\Sigma}_Y$) of $\mathbf{Y}$ can be expressed in terms of $\boldsymbol{m}$, $\boldsymbol{\Sigma}$, and Var(ε_{it}), where $\boldsymbol{m} = \mathbf{mean}(\mathbf{L})$, $\boldsymbol{\Sigma} = \mathrm{Cov}(\mathbf{L}, \mathbf{L})$: $\boldsymbol{m}_Y = \mathbf{D}\boldsymbol{m}$, $\boldsymbol{\Sigma}_Y = \mathbf{D}\boldsymbol{\Sigma}\mathbf{D}^{\mathrm{T}} + \boldsymbol{\Sigma}_\varepsilon$, where $\boldsymbol{\Sigma}_\varepsilon = \mathrm{Cov}(\mathbf{E}, \mathbf{E})$.

As described previously in the section on multivariate genetic analysis, the two-dimensional phenotype (α_i, β_i) may be analyzed by modeling the covariance matrix $\boldsymbol{\Sigma}$ for MZ and DZ twins using the Cholesky factorization approach (see Fig. 4.7 above).

The two-level model construction leads to a parameterization of the joint likelihood for the trait in terms of the variance components, the respective mean vectors, and residual variances. This yields estimates of the two heritability values of α_i and β_i (and respective variabilities of the environmental effects) and also estimates of correlations between the genetic and environmental components of α_i and β_i, as described earlier. Posthuma et al. (2003) further describe how the predicted individual random growth curve parameters can be obtained. These predictions may be useful for selection of most informative pairs for subsequent linkage analysis of the random intercepts and slopes.

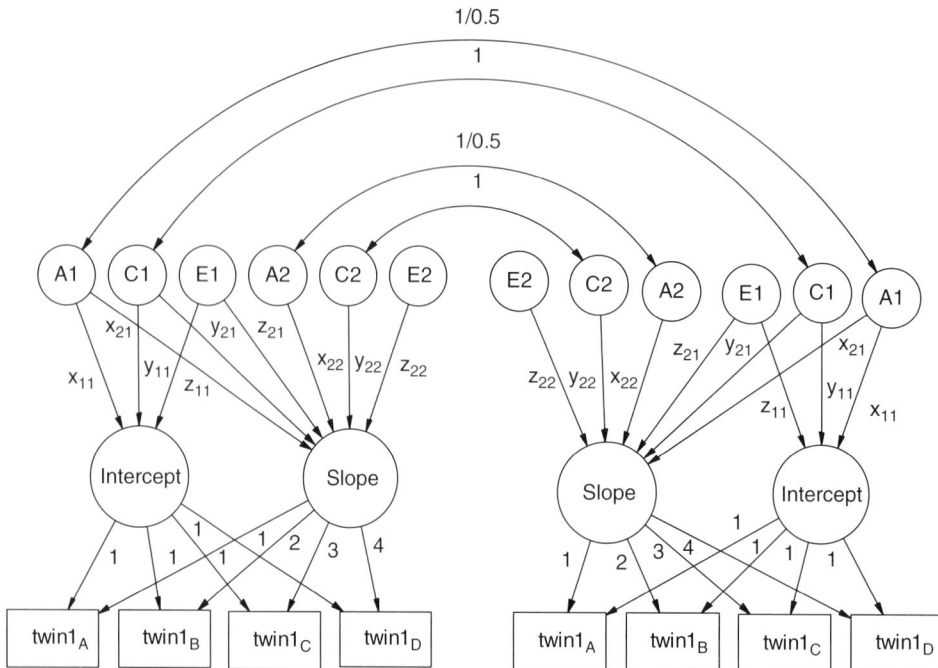

Fig. 4.7 A latent growth curve model, where the latent mean and intercept are decomposed into sources of A, C, and E

Co-morbidity Models

Genetic and environmental correlations provide information on the nature of an observed phenotypic correlation between two traits, whereas longitudinal models provide information on the causes of stability and change over time. However, these models do not provide information on the actual mechanism underlying this correlation. For example, a genetic correlation does not make any inferences on the direction of causation or on the reasons why two traits tend to occur together. In order to understand and investigate the causes of the co-occurrence of two traits, we need to apply models of co-morbidity.

Co-morbidity is usually investigated in the context of (psychiatric) diseases, where a disease is either present or absent. Co-morbidity may also apply to other traits (i.e., not particularly disorders, not particularly categorical), e.g., red hair and blue eyes, although we would then simply call it co-occurrence, or correlation. When patients have two co-occurring disorders, we refer to them as co-morbid for these disorders. Understanding co-morbidity can be of crucial scientific importance to understanding disease etiology, and also has high practical significance (Rutter, 1994). For patients two disorders combine to produce greater impact on normal functioning than each disorder separately. Understanding co-morbidity may further have implications both for diagnosing disorders and treating them.

Two important papers provide an exhaustive overview of co-morbidity models: Klein and Riso (1993) and Neale and Kendler (1995). Klein and Riso (1993) have provided a scholarly overview of many possible models of co-morbidity. Neale and Kendler (1995) have provided a scholarly overview of many possible models of co-morbidity. Neale and Kendler (1995) later extended these models and translated their implications into formal, falsifiable expectations. Table 4.1 summarizes the Klein and Riso (KR) and Neale and Kendler (NK) models. In the following we provide a conceptual overview of the models described in KR and NK models. For a more mathematical explanation of these models we refer the reader to the original papers.

Co-morbidity Due to Sampling and Base Rates

Models 1–3 (in Table 4.1) explain co-morbidity due to sampling bias and base rates. The simplest explanation for co-morbidity between two traits is chance: when trait A has a frequency of 0.20 in the population and trait B has a frequency of 0.30, they are expected to co-occur with a frequency of $0.20 \times 0.30 = 0.06$ (KR1/NK1). When a sample is ascertained through clinical records individuals with two or more diseases may be more likely to be included in the study. Such sampling bias induces a spuriously high rate of

Table 4.1 Models of co-morbidity according to Klein & Riso (1993) and Neale & Kendler (1995)

Explanations based on sampling and base rates
KR1/NK1: Co-morbidity due to chance
KR2/NK2: Co-morbidity due to sampling bias
KR3/NK3: Co-morbidity due to population stratification
Explanations based on artifacts of diagnostic criteria
KR4: Co-morbidity due to overlapping diagnostic criteria
KR5: Co-morbidity due to one disorder encompassing the other
NK7: Submodels of NK3 and 4
Explanations based on drawing boundaries in the wrong places
KR6: Multiformity of the co-morbid condition
NK8: Submodels of NK 3 and 4
KR7: Heterogeneity (of severe form)
NK5: Random multiformity and NK6: Extreme multiformity
KR8: The co-morbid disorder is a third, independent factor
NK9: Three independent disorders
KR9/NK4: The pure and co-morbid conditions are different phases or alternative expressions of the same disorder
Explanations based on etiological relationships
KR10: One disorder is a risk factor for the other; NK11: Causal model
KR11: The two disorders arise from overlapping etiological processes.
NK10: Correlated liabilities
KR-/NK12: Reciprocal causation

KR = Klein & Riso (1993); NK = Neale & Kendler (1995).

co-occurrence between traits (KR2/NK2). Another spurious association between two traits may arise when two independent sets of risk factors for having traits A and B are both elevated in certain subpopulations but not in others. Analyzing all subpopulations as a whole will then show a statistical association between traits A and B (KR3/NK3).

Co-morbidity Due to Artifacts of Diagnostic Criteria

Klein and Riso models 4 and 5 explain co-morbidity in terms of artifacts of diagnostic criteria; when criteria for certain traits or disorders overlap (i.e., similar symptoms are included as criteria for different diagnoses), two traits may be diagnosed at the same time in one person (KR4). Klein and Riso model 5 states that disorder A is really a manifestation of disorder B. This is actually a specific case of KR model 6 (KR5/NK7).

Co-morbidity Due to Drawing Boundaries in the Wrong Places

Klein and Riso models 6–9 concern the concept of liability and drawing boundaries at the wrong places. Although presented as dichotomous traits, many psychiatric diseases are thought to show an underlying liability. This latent liability follows a normal distribution, with a certain threshold above which a disorder becomes manifest. The multiformity explanation of co-morbidity (KR6/NK3 and NK4) states that disorders are manifested in several heterogeneous forms, including symptoms typically associated with other disorders: The co-morbid disorder is regarded as an atypical or more severe form of disorder A and is distinct from disorder B. The heterogeneity explanation (KR7/NK5 and 6) states that the co-morbid disorder is regarded as atypical forms of both disorders A and B. Following the Neale and Kendler formulation this model is named the random multiformity model of liability to disorders A and B. Disorder A arises if individuals are above threshold on the liability to disorder A or with probability r if they are above threshold on the liability to disorder B. Disorder B arises if individuals are above threshold for disorder B or with probability p if they are above threshold for disorder A. In the extreme multiformity model of Neale and Kendler, the underlying liability shows two thresholds, dividing the liability scale into low scorers (below the first threshold), medium scorers (between the first and second threshold) and high scorers (above the second threshold). Disorder A arises if individuals are above either threshold on the liability to disorder A or above the second threshold on the liability to disorder B. Disorder B arises if individuals are above either threshold on the liability to disorder B or if they are above the second threshold on the liability to disorder A.

In Klein and Riso model 8 (Neale and Kendler model 9) the co-morbid disorder is regarded as a completely distinct disorder. In terms of the Neale and Kendler formulation, disorder A arises when individuals are above threshold on either the liability to disorder A or the liability to the third, combined disorder. Disorder B arises when individuals are above threshold on either the liability to disorder B or the liability to the third, combined disorder.

Klein and Riso model 9 further states that the two pure disorders and the co-morbid disorder are all phases or different expressions of the same underlying single disorder. Individuals above threshold on the liability to disorder A express disorder A with probability p and disorder B with probability r. Probabilities p and r are independent, so co-morbid cases arise with frequency pr. This is an extreme form of multiformity (NK model 4).

Co-morbidity Due to Etiological Relationships

The last two models introduced by Klein and Riso explain co-morbidity in terms of etiological processes. Neale and Kendler added an additional model (NK12). KR11 explains co-morbidity in terms of overlapping etiological processes. For example, the same environmental stressor may cause both depression and alcohol abuse. Or, in terms of genetics: one (set of) genes may show pleiotropic effects on multiple disorders. This model relates to the simplest bivariate genetic model as well as to the genetic factor models for multivariate traits. Alternatively, one disorder may be a risk factor for developing another disorder, or vice versa. In this case co-morbidity is explained by disorder A causing disorder B, either unidirectionally (KR10/NK11) or reciprocally (NK12).

The co-morbidity models described above provide theoretical explanations for co-morbidity at a phenotypic level. Applying these models to actual data and comparing how well each model fits the data requires specialized sample designs. Key simulation work in this area has been carried out by Rhee et al. (Rhee, Hewitt, Corley, & Stallings, 2003; Rhee et al., 2004, 2006), as reviewed in Krueger & Markon (2006). Rhee and colleagues showed that different co-morbidity models can be distinguished well in many circumstances, bar certain caveats. As may be expected, similar co-morbidity models, e.g., different subtypes of co-morbidity models, are more difficult to distinguish than co-morbidity models that are structurally very different, such as the alternate forms and directional causation models. In addition, distinguishing between different co-morbidity models becomes difficult when the prevalence of one or both of the disorders is very low or when correlations between liabilities are small. Finally, Rhee et al. have shown that very large samples may be required to obtain adequate power to discriminate between different co-morbidity models. The studies of Rhee et al. (2003, 2004, 2006) stress the importance of study design in distinguishing between different models of co-morbidity. The three possible designs to study co-morbidity are epidemiological designs, longitudinal designs, and family designs. Family designs preclude the need of longitudinal data, and in many instances are more cost-effective than any of the other designs. Many models that are indistinguishable in simple phenotypic cross-sectional design can be distinguished in family designs due to the addition of information about co-morbidity patterns across relatives (Neale & Kendler, 1995; Simonoff, 2000).

Although there is an extremely large amount of literature on bivariate relationships between disorders, relatively few studies have explicitly compared multiple models of co-morbidity within a KR/NK framework, as stated by Krueger and Markon (2006). The co-morbidity between depression and anxiety disorder seems to be the most frequently modeled. Middeldorp, Cath, Van Dyck, and Boomsma (2005) reviewed twin and family studies of depression and anxiety in the framework of KR/NK models and concluded that shared genetic liability can explain much of the co-morbidity between depression and anxiety.

Of all co-morbidity models described above the direction of causation models elegantly show the added value of genetically informative designs in delineating the causes of co-morbidity or correlation between two disorders or traits. The application of direction of causation models in genetic designs is described in further detail below.

Direction of Causation

Heath et al. (1993) reviewed the conditions under which cross-sectional twin data are informative about the direction of causation between two traits. Below we describe the main issues that have been put forward in direction of causation models using genetically informative data sets (see also Duffy & Martin, 1994; Neale, Duffy, & Martin, 1994; Neale, Eaves, Kendler, Heath, & Kessler, 1994; Neale & Walters, et al., 1994).

Four different models of causation can be distinguished: (1) no causation, but pleiotropy between traits A and B; (2) unidirectional causation – trait A causes trait B; (3) unidirectional causation – trait B causes trait A; and (4) reciprocal causation – trait A causes trait B and vice versa. The power to distinguish between the two unidirectional causation models is strongly dependent on the difference in modes of inheritance in the two traits. The most optimal situation

arises when variation in trait A is mainly due to genetic variation and variation in trait B is mainly due to environmental variation. Figure 4.8 depicts the two situations where trait A causes trait B (top) and where trait B causes trait A (bottom). Variation in trait A is due to additive genetic variation and non-shared environmental variation, with the influence of shared environmental variation mediated by trait B. Variation in trait B is influenced by shared and non-shared variation while additive genetic influences are mediated by trait A.

The expected MZ and DZ CTCTs under these two models are

$$A \rightarrow B : \text{MZ CTCT} = b \times a^2; \text{DZ CTCT} = b \times 1/2\,a^2$$
$$B \rightarrow A : \text{MZ CTCT} = b \times c^2; \text{DZ CTCT} = b \times c^2$$

Thus, when trait A causes trait B, the CTCTs are a function of the mode of inheritance of trait A, whereas when trait B causes trait A, the CTCTs are a function of the mode

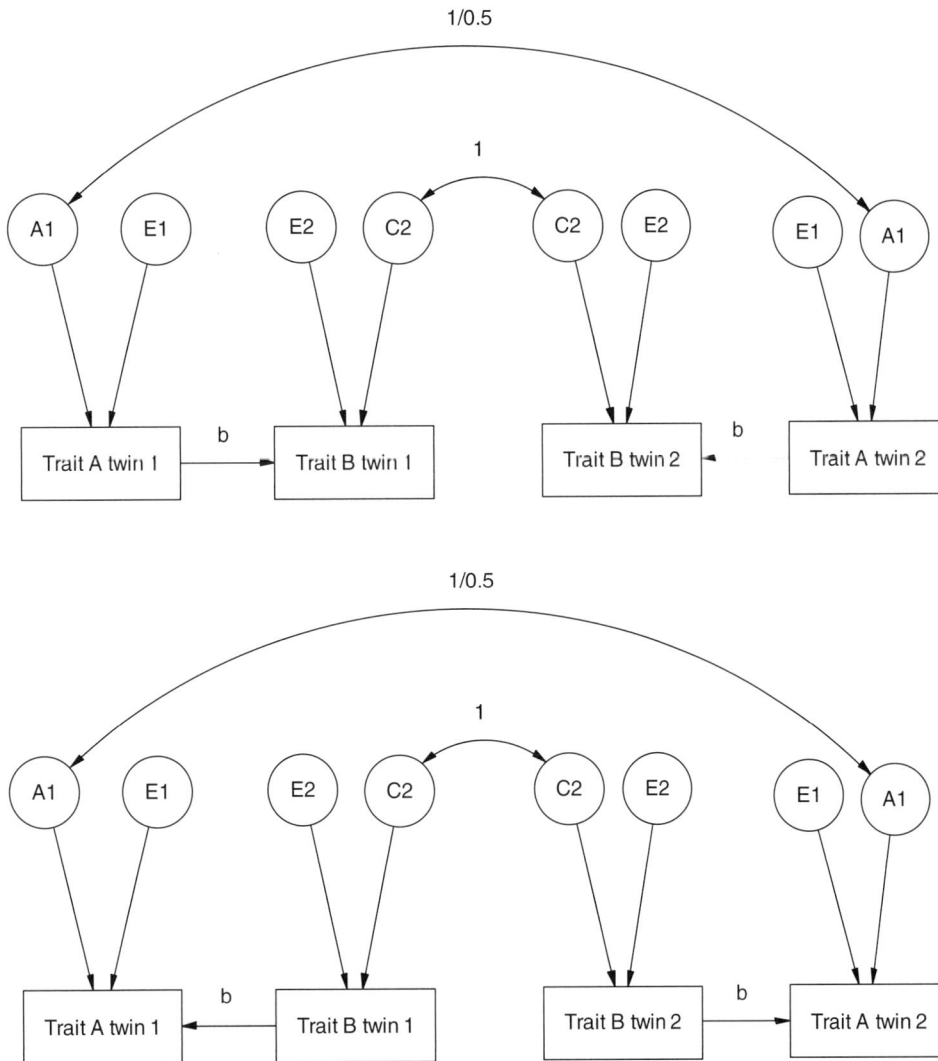

Fig. 4.8 A bivariate twin design: direction of causation model.
Top: trait A causes trait B; *Bottom*: trait B causes trait A

of inheritance of trait B. Formally the direction of causation model for two observed traits in one individual is represented as

$$\Sigma_P = (\mathbf{I} - \mathbf{B})^{-1}(\mathbf{A} + \mathbf{C} + \mathbf{E})((\mathbf{I} - \mathbf{B})^{-1})^T,$$

where $\mathbf{A} = \mathbf{X}\mathbf{X}^T, \mathbf{C} = \mathbf{Y}\mathbf{Y}^T, \mathbf{E} = \mathbf{Z}\mathbf{Z}^T$

Matrices $\mathbf{X}$, $\mathbf{Y}$, and $\mathbf{Z}$ are diagonal matrices of dimensions $2{\times}2$ (number of traits × number of traits). Matrix $\mathbf{I}$ is an identity matrix, whereas matrix $\mathbf{B}$ is a subdiagonal matrix if trait A causes trait B. The total variance is thus pre- and post-multiplied by the inverse of $(\mathbf{I}{-}\mathbf{B})$. Matrix $\mathbf{B}$ is transposed when trait B causes trait A. For reciprocal causation, matrix $\mathbf{B}$ needs to be respecified such that the diagonals are zero whereas the off-diagonals are the reciprocal path coefficients. For a formal derivation of reciprocal causation we refer the reader to Heath et al. (1993) and Neale and Eaves et al. (1994).

Cross-sectional data can thus be informative for direction of causation as long as a genetically informative design is used. However, there are several limitations to be noted. Naturally if the mode of inheritance of trait A is highly similar to the mode of inheritance of trait B, these two models become indistinguishable. In addition to that Heath et al. (1993) note that these unidirectional models can only be tested if there are at least three different sources of variance of importance for the two traits, unless multiple indicators are used. With only two sources of variation, the two unidirectional models become indistinguishable from each other as well as from the general bivariate model, unless we know that both traits are measured without error or we have additional information on the measurement variances (Heath et al., 1993, p. 38).

The reciprocal model cannot be tested at all when only a single indicator of each trait is available. With multiple indicators available, testing the reciprocal causation model is feasible only when at least three sources of variation are present. The use of multiple indicators in testing direction of causation is strongly advised, as it reduces the error variance of each trait. This is crucial to direction of causation models as ignoring measurement error affects the within-person covariance and therefore influences all other parameters in the direction of causation model (Heath et al., 1993). Another limitation is that because the direction of causation models are nested under the general bivariate model, we are restricted to testing whether the (reciprocal) causation is the *only* cause of the observed correlation between two traits. It is not possible to test whether both a causation mechanism and a pleiotropic mechanism influence correlation between two traits at a phenotypic level.

Conclusion

In the above we have aimed to provide a general introduction in the basics of multivariate genetic modeling as well as discuss some of the more advanced multivariate genetic models, such as longitudinal models, genetic factor models, and co-morbidity models. We have discussed that genetically informative designs often provide a cost-effective framework for determining the causes of co-variation between multiple traits. The models described in this chapter do not reflect an exhaustive list of all possible multivariate genetic models, but merely aim to provide a good starting point to gain insight in the theoretic underpinnings as well as putative extensions of multivariate genetic models. Alternative multivariate genetic models that have not been described above include for example multivariate genetic linkage models (see, e.g., J. Liu, Y. Liu, X. Liu, & Deng, 2007; Marlow et al., 2003; Williams, Van Eerdewegh, Almasy, & Blangero, 1999, see also Hottenga & Boomsma, 2007) or models that deal with data obtained from multiple raters (Derks, Hudziak, van Beijsterveldt, Dolan, & Boomsma, 2004; Hartman, Rhee, Willcutt, & Pennington, 2007; Hewitt, Silberg, Neale, Eaves, & Erickson, 1992; Simonoff et al., 1995).

In general, multivariate genetic modeling can be of great value when trying to understand the causes of co-variation between quantitative traits and co-morbidity between disorders. Some of the remaining chapters of this book will provide excellent examples of the application of multivariate genetic modeling.

References

Boomsma, D. I., & Molenaar, P. C. M. (1986). Using LISREL to analyze genetic and environmental covariance structure. *Behavior Genetics, 16*, 237–250.

Boomsma, D. I., & Molenaar, P. C. M. (1987). The genetic analysis of repeated measures I: Simplex models. *Behavior Genetics, 17*, 111–123.

Carey, G. (1988). Inference about genetic correlations. *Behavior Genetics, 18*, 329–338.

De Geus, E. J., Kupper, N., Boomsma, D. I., & Snieder, H. (2007). Bivariate genetic modeling of cardiovascular stress reactivity: does stress uncover genetic variance? *Psychosomatic Medicine, 69*(4), 356–364. Epub 2007 May 17. Erratum in (2007 Jun): *Psychosomatic Medicine, 69*(5), 489.

Derks, E. M., Hudziak, J. J., van Beijsterveldt, C. E., Dolan, C. V., & Boomsma, D. I. (2004). A study of genetic and environmental influences on maternal and paternal CBCL syndrome scores in a large sample of 3-year-old Dutch twins. *Behavior Genetics, 34*(6), 571–583.

Dolan, C. V. (1992). *Biometric decomposition of phenotypic means in human samples*. PhD thesis, University of Amsterdam, The Netherlands.

Dolan, C. V., Molenaar, P. C. M., & Boomsma, D. I. (1991). Simultaneous genetic analysis of longitudinal means and covariance structure in the simplex model using twin data. *Behavior Genetics, 21*, 49–65.

Duffy, D. L., & Martin, N. G. (1994). Inferring the direction of causation in cross-sectional twin data: Theoretical and empirical considerations. *Genetic Epidemiology, 11*(6), 483–502.

Eaves, L. J., & Gale, J. S. (1974). A method for analyzing the genetic basis of covariation. *Behavior Genetics, 4*, 253–267.

Hartman, C. A., Rhee, S. H., Willcutt, E. G., & Pennington, B. F. (2007). Modeling rater disagreement for ADHD: Are parents or teachers biased? *Journal of Abnormal Child Psychology, 35*(4), 536–542.

Heath, A. C., Kessler, R. C., Neale, M. C., Hewitt, J. K., Eaves, L. J., Kendler, K. S. (1993). Testing hypotheses about direction of causation using cross-sectional family data. *Behavior Genetics, 23*(1), 29–50.

Hewitt, J. K., Silberg, J. L., Neale, M. C., Eaves, L. J., & Erickson, M. (1992). The analysis of parental ratings of children's behavior using LISREL. *Behavior Genetics, 22*(3), 293–317.

Hotelling, H. (1933). Analysis of a complex of statistical variables into principal component. *Journal Educational Psychology, 24*, 417–441, 498–520.

Hottenga, J. J., & Boomsma, D. I. (2007). QTL detection in multivariate data from sibling pairs. In M. Ferreira, B. Neale, S. E. Medland, & D. Posthuma (Eds.), *Dissection of complex trait variation through linkage and association* (pp. 239–264.). Taylor & Francis.

Jöreskog, K., & Sörbom, D. (1986). *LISREL: Analysis of linear structural relationships by the method of maximum likelihood*. Chicago: National Education Resources.

Kirkpatrick, M., & Heckman, N. (1989). A quantitative genetic model for growth, shape, reaction norms, and other infinite-dimensional characters. *Journal of Mathematical Biology, 27*, 429–450.

Klein, D. N., & Riso, L. P. (1993). Psychiatric disorders: Problems of boundaries and co-morbidity. In Costello (Ed.), *Basic issues in psychopathology*. New York: Guildford.

Krueger, R. F., & Markon, K. E. (2006). Reinterpreting comorbidity: A model-based approach to understanding and classifying psychopathology. *Annual Review of Clinical Psychology, 2*, 111–133.

Lawley, D. N., & Maxwell, A. E. (1971). *Factor analysis as a statistical method*. London: Butterworths.

Leboyer, M., Bellivier, F., Nosten-Bertrand, M., Jouvent, R., Pauls, D., & Mallet, J. (1998). Psychiatric genetics: Search for phenotypes. *Trends in Neuroscience, 21*(3), 102–105.

Liu, J., Liu, Y., Liu, X., & Deng, H. W. (2007). Bayesian mapping of quantitative trait loci for multiple complex traits with the use of variance components. *American Journal of Human Genetics, 81*(2), 304–320. Epub 2007, July 3.

Marlow, A. J., Fisher, S. E., Francks, C., MacPhie, I. L., Cherny, S. S., Richardson, A. J., et al. (2003). Use of multivariate linkage analysis for dissection of a complex cognitive trait. *American Journal of Human Genetics, 72*(3), 561–570. Epub 2003, February 13.

Martin, N. G., & Eaves, L. J. (1977). The genetical analysis of covariance structure. *Heredity, 38*, 79–95.

McArdle, J. J. (1986). Latent variable growth within behavior genetic models. *Behavior Genetics, 16*, 163–200.

Middeldorp, C. M., Cath, D. C., Van Dyck, R., & Boomsma, D. I. (2005). The co-morbidity of anxiety and depression in the perspective of genetic epidemiology: A review of twin and family studies. *Psychological Medicine, 35*(5), 611–624.

Mulaik, S. A. (1972). *The foundations of factor analysis*. New York: McGraw-Hill Book Company.

Neale, M. C. (1997). *Mx: Statistical modeling* (3rd ed.). Box 980126 MCV, Richmond VA 23298.

Neale, M. C., & Cardon, L. R. (1992). *Methodology for genetic studies of twins and families*. Dordrecht, The Netherlands: Kluwer Academic Publishers.

Neale, M. C., Duffy, D. L., & Martin, N. G. (1994). Direction of causation: Reply to commentaries. *Genetic Epidemiology, 11*(6), 463.

Neale, M. C., Eaves, L. J., Kendler, K. S., Heath, A. C., & Kessler, R. C. (1994). Multiple regression with data collected from relatives: Testing assumptions of the model. *Multivariate Behavioral Research, 29*(1), 33–61.

Neale, M. C., & Kendler, K. S. (1995). Models of co-morbidity for multifactorial disease. *American Journal of Human Genetics, 57*, 935–953.

Neale, M. C., & Maes, H. H. (in press). *Methodology for genetic studies of twins and families*. Dordrecht, The Netherlands: Kluwer Academic Publishers.

Neale, M. C., & McArdle, J. J. (2000). Structured latent growth curves for twin data. *Twin Research, 3*, 165–177.

Neale, M. C., Walters, E., Heath, A. C., Kessler, R. C., Pérusse, D., Eaves, L. J., et al. (1994). Depression and parental bonding: Cause, consequence, or genetic covariance? *Genetic Epidemiology, 11*(6), 503–522.

Pletcher, S. D., & Geyer, C. J. (1999). The genetic analysis of age-dependent traits: Modelling the character process. *Genetics, 153*, 825–835.

Posthuma, D., Beem, A. L., de Geus, E. J., van Baal, G. C., von Hjelmborg, J. B., Iachine, I., et al. (2003). Theory and practice in quantitative genetics. *Twin Research, 6*(5), 361–376.

Posthuma, D., de Geus, E. J., Baare, W. F., Hulshoff Pol, H. E., Kahn, R. S., & Boomsma, D. I. (2002). The association between brain volume and intelligence is of genetic origin. *Nature Neuroscience, 5*(2), 83–84.

Rhee, S. H., Hewitt, J. K., Corley, R. P., & Stallings, M. C. (2003). The validity of analyses testing the etiology of comorbidity between two disorders: Comparisons of disorder prevalences in families. *Behavior Genetics, 33*(3), 257–269.

Rhee, S. H., Hewitt, J. K., Lessem, J. M., Stallings, M. C., Corley, R. P., & Neale, M. C. (2004). The validity of the Neale and Kendler model-fitting approach in examining the etiology of comorbidity. *Behavior Genetics, 34*(3), 251–265.

Rhee, S. H., Hewitt, J. K., Young, S. E., Corley, R. P., Crowley, T. J., Neale, M. C., et al. (2006). Comorbidity between alcohol dependence and illicit drug dependence in adolescents with antisocial behavior and matched controls. *Drug and Alcohol Dependence, 84*(1), 85–92.

Rijsdijk, F. V., Vernon, P. A., & Boomsma, D. I. (2002). Application of hierarchical genetic models to Raven and WAIS subtests: A Dutch twin study. *Behavior Genetics, 32*(3), 199–210.

Rutter, M. (1994). Co-morbidity: Meanings and mechanisms clinical psychology. *Science and Practice, 1*(1), 100–103.

Schmitz, S., Cherny, S. S., & Fulker, D. W. (1998). Increase in power through multivariate analyses. *Behavior Genetics, 28*(5), 357–363.

Simonoff, E. (2000). Extracting meaning from comorbidity: Genetic analyses that make sense. *Journal of Child Psychology and Psychiatry, 41*(5), 667–674.

Simonoff, E., Pickles, A., Hewitt, J., Silberg, J., Rutter, M., Loeber, R., et al. (1995). Multiple raters of disruptive child behavior: Using a genetic strategy to examine shared views and bias. *Behavior Genetics, 25*(4), 311–326.

Spearman, C. (1904). General intelligence, objectively determined and measured. *American Journal of Psychology, 15*, 201–293.

Thurstone, L. L. (1947). *Multiple factor analysis*. Chicago: University of Chicago Press.

Vandenberg, S. G., & Falkner, F. (1965). Hereditary factors in human growth. *Human Biology, 37*, 357–365.

Williams, J. T., Van Eerdewegh, P., Almasy, L., & Blangero, J. (1999). Joint multipoint linkage analysis of multivariate qualitative and quantitative traits. I. Likelihood formulation and simulation results. *American Journal of Human Genetics, 65*(4), 1134–1147.

Chapter 5

Models of Human Behavior: Talking to the Animals

Gene S. Fisch

If we could talk to the animals, learn their languages
Think of all the things we could discuss
If we could walk with the animals, talk with the animals,
Grunt and squeak and squawk with the animals,
And they could squeak and squawk and speak and talk to us.

(from the Musical "Dr. Doolittle")

Prologue and Introduction

Inheritance of behavioral characteristics was known to humankind in prehistoric times and likely came about while domesticating animals. In the Middle East, sheep, goats, and pigs were likely tamed between 6000 and 9000 B.C. There is no written record of the early rise of animal husbandry, but rearing and training of animals were known to the ancient Romans. Well-defined breeding techniques for domesticated livestock were underway in England in the 18th century. At the turn of the 19th century, even rats were bred for their variegated coat colors and behavioral peculiarities (Brush & Driscoll, 2002). Breeders conserved the desired characteristics and controlled for undesired aspects by repeatedly selecting those preferred features in offspring, mating "like with like" and producing increasingly homogeneous strains.

Inheritance of traits in humans stemmed from the belief in "blood theory" – the child is a fusion, or blend, not only in the characteristics of the parent, but of all preceding generations – which was widespread in the 19th century. However, the mechanism of inheritance was not known. Darwin and Galton speculated as to the process, but it remained for William Bateson and the rediscovery of Mendel's published account of his research on peas for the modern concept of genetics – and that of the gene – to emerge as the paradigm for the inheritance of traits.

That animals could be used as models of human behavior, that the heritability of human traits could be replicated in nonhuman creatures, emerged in the mid-to-late 19th century. Until then, Aristotelian taxonomy separated plants from infrahuman animals and humans (cf. Fisch, 2006). That there was continuity in structure and mind between humans and infrahumans, and that mental abilities could be inherited, grew out of the thinking and writings of Charles Darwin and his cousin, Francis Galton.

In *The Origin of Species* (1859), Darwin alluded to the many similarities in structure between animals and humans, and the inheritance of instincts related to behavior. He also made mention of the "acquirement of each mental power and capacity by gradation" (Darwin, 1859, p. 455). Later, in *The Expression of the Emotions in Man and Animals* (1872), Darwin was less circumspect: "not only has the body been inherited by animal ancestors, but there is continuity in respect to mind between animals and humans."

What was the mechanism by which mind was inherited? As noted earlier, the blood theory of inheritance was popular, and Darwin and Galton set off to validate the theory by performing blood transfusions in various species of rabbits. Both obtained negative results. As a consequence, and unaware of Mendel's experiments, Galton redirected his thinking away from a physiological theory of inheritance in favor of a statistical model.

G.S. Fisch (✉)
Bluestone Clinical Research Center, New York University, New York, NY 10010, USA
e-mail: gene.fisch@nyu.edu

Y.-K. Kim (ed.), *Handbook of Behavior Genetics*,
DOI 10.1007/978-0-387-76727-7_5, © Springer Science+Business Media, LLC 2009

Galton and His Statistical Model of Inherited Traits

The statistical model Galton used to develop his ideas about correlation and regression was based on the Gaussian distribution. Galton was aware of the anthropological findings of his time that quantitative characteristics in fossilized plants and animals appeared to approximate the Gaussian distribution. In *Hereditary Genius* (1869), Galton argued that mental abilities were likely normally distributed as well, and provided an outline for a theory of genetics, based on Darwin's theory of pangenesis (cf. Bulmer's (2003) scholarly treatise of Galton's investigations of genetics and statistics). In a presentation to the Royal Society in 1887, Galton stated that heredity must, therefore, follow the statistical laws derived from the Gaussian distribution.

Animal Models of Human Characteristics

While Galton was absorbed with the inheritance of mental ability, Darwin's evolutionary notion of the continuity of structure and mind between animals and humans provided the basis for systematic investigations of animal intelligence and behavior as they might relate to humans, and spawned studies in comparative psychology and animal experimental psychology. One of Darwin's earliest supporters of continuity of mind was his friend, George Romanes, who busily collected anecdotes about animal behavior that he likened to humans, drawing additional inferences about the state of the animal mind.

Not all investigators of animal behavior embraced Romanes' argument. Contrarian views were held by animal experimental psychologists such as E. L. Thorndike and J. B. Watson, who were trained in the 19th century British associationist tradition and saw no need to make additional assumptions about the existence of an animal mind in order to study the behavior. The arguments are now more than a century old, but the gap between behaviorism and mentalism has never been bridged. The rise of cognitive psychology and cognitive neuroscience in the mid-20th century, the advances in mathematics, logic, and computation, coupled with the similarities observed between neural networks and computer parallel processing, plus the discovery of the structure of DNA, all combined to catapult cognitive neuroscience to the forefront of investigations of gene–brain–behavior relationships, leaving unresolved disputes concerning the existence and nature of an animal mind. I will return to the dilemma posed by behaviorism and mentalism later.

Validity of Animal Models

As noted earlier, selective breeding for specific characteristics in animals has been known for centuries, and animal models have taught us much about the pathogenesis of many diseases. Therefore, it would seem logical that inbred mouse strains could provide an effective means by which to optimize the search for genetic factors of complex behavioral phenotypes, as Wehner, Radcliffe, and Bowers (2001) note. Complex phenotypes exhibit continuous variation, from which we may deduce that they are quantitative in nature and, therefore, likely the outcome of polygenic sources. Quantitative traits have, in turn, been mapped to chromosomal regions referred to as quantitative trait loci (QTLs) that contain the gene or genes affecting the phenotype, based on statistical inferences drawn from linkage analysis (Lander & Botstein, 1989). Once a QTL has been identified, congenic mouse strains can be developed using overlapping QTL sections to examine their respective phenotypes for the presence or absence of behaviors. Congenic strains can be created by recursively back-crossing the behavioral phenotype strain onto another strain.

To isolate the effect of a specific gene polymorphism producing a phenotype requires some manipulation of the gene itself. Successful introduction of gene sequences into mouse embryos resulting in the development of transgenic mice was first realized by Gordon, Scangos, Plotkin, Barbosa, and Ruddle (1980). However, the protocol for targeting mutations in any gene that would effectively produce knockout models of genes known to produce disorders was developed by Thomas and Capecchi (1987). Using a specialized construct of the neomycin resistance gene, these researchers effectively produced an *Hprt ko* mouse.

While many types of apparatus used to evaluate rats can be converted into tools to test mice, mice are not small rats and equipment for mice needed to be modified to account for species differences. Crawley & Paylor (1997) were among the first researchers to propose a comprehensive test battery for mice. Animals should first be examined to ascertain that their sensorimotor and neurological functions are within normal limits. To examine cognitive abilities, i.e., learning and memory, several types of apparatus can be used. Chief among them has been the MWM, but as was noted, other techniques, e.g., the radial maze, delayed-matching-to-sample operant conditioning, cued and contextual conditioning, and passive and active avoidance paradigms have all been employed (cf. Crawley & Paylor, 1997). Aggressive behaviors have been assessed using resident–intruder designs; anxiety has been investigated using the elevated plus maze or light–dark exploration; depression has been evaluated using the forced swimming test; schizophrenia has been tested by implementing the acoustic startle reflex to assess prepulse inhibition (PPI); and drug abuse has been examined using self-administration tasks.

The MWM has probably been the most frequently used procedure to ascertain the presence of visual/spatial deficits in learning and memory, deficits associated with lesions in the hippocampus. It has been observed, however, that not all mouse strains perform equally well on different behavior-specific tasks. For example, C57BL/6J mice perform well on the MWM, but DBA/2J mice perform poorly (Upchurch & Wehner, 1988). Moreover, as Wolfer and Lipp (2000) have observed, some mice develop thigmotaxis (wall hugging) in the MWM, while others float passively, and both factors account for significantly greater variance in performance than do visual/spatial variables. In their meta-analysis of swimming behavior, Wolfer and Lipp (2000) also reported that differences in genetic background and mutation status produce a wide variety of outcomes in the MWM. In an attempt to salvage the less-than-adequate performances by some mouse strains and to generalize its validity, Wahlsten, Cooper and Crabbe (2005) modified the MWM apparatus. These researchers then examined performances in variants of the MWM, and the 4-arm version of the MWM, of mice from 21 inbred strains. Wahlsten et al. (2005) observed significant strain differences with respect to speed during pre-trial training, percent trials floating, escape latencies, and swimming distance in the standard MWM. Patterns of escape latency over trials also differed significantly. However, in the 4-arm version of the MWM, they found that several kinds of errors were significantly reduced; and, as measured by latency to find the submerged platform, learning ability improved. Thus, while the MWM has been extensively used to study cognitive deficits, conclusions regarding behavior assessed by it should be tempered by the mouse strain enlisted, the apparatus configuration employed, and any unusual behavioral characteristics exhibited by the animals, before drawing conclusions.

Contextual and cued fear conditioning (Pavlovian conditioning) provide another means by which to examine learning. The association between a visual or auditory stimulus (CS) and a mildly aversive stimulus (UCS) has been related to functioning in the amygdala (LeDoux, 2000) and to some extent, the hippocampus (Logue, Paylor, & Wehner, 1997). Investigations of QTLs in two strains of mice have converged on two chromosomes, one of which – chromosome 1 – is associated specifically with cued fear conditioning (Wehner et al. 2001). Even so, Crawley et al. (1997) caution that there are strain differences in contextual fear conditioning, and that C57BL strains perform better than FVB or DBA strains. Consequently, finding QTLs and genes associated with contextual fear conditioning may prove more problematic than originally thought.

Environment also plays an important role in producing differences in mouse behavior. Crabbe, Wahlsten, and Dudek (1999) found a significant interactive effect between strain and laboratory environment. In addition to examining laboratory differences, Wahlsten, Bachmanov, Finn and Crabbe (2006) also attempted to replicate results from earlier studies by employing the same mouse strain. Wahlsten et al. (2006) found that, for ethanol preference and locomotor activity, results across labs and time could be replicated; but, for studies of anxiety behavior using exploration on the elevated maze, effects of lab differences were found. Previously, enriched cage environments were found to have a positive effect on learning (Fiala, Joyce, & Greenough, 1978). On the other hand, Wolfer et al. (2004) noted that enriched cage environments had no significant effects on a variety of learning tasks presented to C57BL and DBA inbred mouse strains, although strain differences were observed.

Thus it appears that, while there are instances in which laboratory environment affects tests of learning and memory in mice, inbred strain may be a more salient (genetic) component of their behavioral phenotype. Crawley and colleagues (1997) examined the implications of mouse strains on phenotype. They noted that most knockout mice are created from 129 substrains and bred with C57BL/6 females. According to the authors, the phenotype that develops, and the tests used to evaluate the phenotype, will have a significant interactive effect with the mutated genes of interest. Crawley et al. (1997) examined several dozen inbred strains of mice and noted that C57BL/6 and C57BL/10 mice perform well in MWM. However, even within C57BL strains there are differences in responses depending on the cognitive task administered. Deacon, Thomas, Rawlins, and Morley (2007) examined C57BL/6 and C57BL/10 mice performing several learning tasks and found significant differences in percent correct responses of spontaneous alteration in a T-maze, as well as time to reach goal on the Lashley III maze.

Another issue that researchers need to consider when choosing an animal model concerns the process by which the knockout is produced. As noted earlier, knockouts are typically created using the embryonic stem cells from 129 substrains, after which they are mated with females from C57BL strain. By developing congenic strains, as recommended by the Banbury Conference, breeding strategies can be systematized for dealing with genetic background. However, in creating the knockout, genes flanking the one of interest may also be affected. Wolfer, Crusio, and Lipp (2002) propose several different strategies to be employed, including the development of a conditional knockout, to overcome these problems.

Issues involving mouse genetics – inbred strain, knockout creation – and environment – laboratory housing, test apparatus type – obligate researchers to institute standards for testing transgenic and knockout mice. Wahlsten (2001) and Wahlsten, Rustay, Metten, and Crabbe (2003) set about examining the problem of standards, albeit without presenting all-embracing solutions. One recommendation made was that several common strains free of neurological defects and derived from different ancestries be used in any one study of complex behavior. Second, in order to attain consistency in responding, the apparatus used should be standardized and its use maintained over the course of generations of experiments. Unfortunately, the number of variables that need to be considered which affect response outcomes – strain, sex, mouse supplier, lab environment, apparatus, effect size – would require vast numbers of animals to be used in any one study, making a completely justifiable analysis impossible. Nonetheless, any attempt to standardize across these variables will increase the validity of experimental outcomes.

Genetics and Behavior

It has been said that human beings are complex and dangerous creatures. Human behavior manifests itself as a vast, multifaceted hodgepodge of verbal and nonverbal activities, some of which are in response to the immediate surroundings, some of which are in response to other individuals, or groups of individuals, some of which are in response to nothing apparent. Behavior genetics is primarily concerned with the genetics of individual behavior and, as such, it behooves researchers in the field to simplify this complex into comprehensible, analyzable units. To that end, the psychological community has reduced complex behavior into several major components which generally fall into the following categories:

1. learning, cognition, and intelligence;
2. personality and temperament;
3. language.

If there is continuity in mind from animals to humans, in learning and intelligence, personality and temperament, as well as in language, one must first establish heritability of these characteristics of human behavior, then find their analogs in nonhuman animals.

Current Issues

Human Intelligence and Cognitive Function

Galton first measured intelligence using eminence as a surrogate. Later, Spearman developed the super-ordinate concept of monarchic intelligence, "g", composed of specific abilities. Then, with the aid of the statistical machinery developed by L. L. Thurstone, factor analysis was used to load specific variables onto factors to produce a general measure of intelligence, g.

But what is intelligence? E. G. Boring famously said that it was whatever intelligence tests test. Many definitions emphasize the ability to learn, to "educing either relations or correlates" (Spearman, 1927). At the symposium on intelligence convened in 1921 (Thorndike, 1921), then at a second in 1986 (Sternberg & Detterman, 1986), psychologists could not agree on a common definition, but did agree that, at the core of intelligence is the ability to adapt effectively to the environment. Interestingly, IQ tests using different concepts of intelligence produce remarkably similar composite score results.

The link between genetics and intelligence was first systematically articulated by Galton (1883). However, the basis for the modern statistical model, represented mathematically as the sum of genetic and environmental variance components, was devised by Fisher (Kemphorne, 1997). Plomin (2003) commented on the use of g to determine molecular-genetic correlates of behavior, arguing that g is the quantitative feature diverse cognitive abilities have in common, is stable over time and, therefore, a trait of the individual. Twin studies have demonstrated genetic contributions to intelligence in more than 100 published articles, with differing degrees of genetic and family pairings, and across many countries (cf. Bouchard, Lykken, McGue, Segal, & Tellegen, 1990; Bouchard & McGue, 1981; also see Deary, Spinath, & Bates, 2006, for an overview).

Although twin studies of intelligence provide a basis for its heritability, to date no one gene has been implicated directly. Specific gene effects on intellectual function could be determined by association and linkage analysis, as Flint (1999) noted. However, contributions by individual genes may be limited and their effect size small; and, linkage analysis has proven most successful when mapping genetic disorders that produce dichotomous phenotypes, e.g., disease vs. no disease, where the effect size is obvious (Wahlsten, 1999). Wahlsten also observed that linkage analysis has also been less than successful in finding loci for complex disorders.

Kovas and Plomin (2006) acknowledged the problem of finding specific genes for complex behaviors and proposed a "generalist genes" hypothesis: the same genes that produce cognitive disabilities also most affect cognitive abilities. Generalist genes function by utilizing two genetic mechanisms: pleiotropy and polygenicity. Pleiotropy refers to the spread of effect of an individual gene on the phenotype, while polygenicity refers to the effect of many genes to produce a specific phenotype. Accordingly, pleiotropy could justify gradations of intelligence, while polygenicity could account for differences in cognitive profiles.

Animal Intelligence and Cognitive Function

Do Androids Dream of Electric Sheep?
– Philip Dick, 1968

Systematic studies of animal intelligence began in the late 19th century with E. L. Thorndike and his publication, *Animals in a Puzzle Box* (1898). Thorndike employed a variety of experimental devices to demonstrate what he referred to as animal intelligence, without reference to unobserved mental processes. In the century that followed, many other apparatus types – among them, variously configured mazes, Skinner's operant experimental chamber, the Lashley jumping stand, the Morris water maze (MWM) – were developed to study animal learning and memory.

To determine the heritability of intelligence in animals, one could imagine that the techniques of selective breeding might be used. If one could selectively breed for size and speed, one might argue that cognitive ability could be bred selectively as well. Based on their observed skills in maze learning, Tolman (1924) was the first to breed rats selectively for intelligence, with limited results. However, Tryon (1940) who later published on maze learning in 18 generations of rats bred for "maze bright" and "maze dull" characteristics was probably first to demonstrate systematically that there was a genetic basis for animal intelligence. Although not originally bred for their cognitive ability, Sara, Devauges, Biegon, and Blizard (1994) found that Maudsley reactive strain rats performed worse on memory tasks involving a 12-arm radial maze than Maudsley nonreactive (MNR) rats. On the other hand, the reactive strain performed better than the MNR strain on other types of mazes. Their results suggest that there is more than one way to measure cognitive abilities.

As there are differences among psychologists as to what constitutes human intelligence, there are differences among animal experimentalists as to what represents intelligence in nonhuman animals. Bailey, McDaniel, and Thomas (2007) noted that when intelligence and cognitive functioning were based on different kinds of learning skills, a hierarchy of learning tasks could be developed to identify different levels of intelligence. At its pinnacle (level 8) is the ability to use class concepts in bi-conditional relationships. Bailey et al. (2007) reported that many studies demonstrated animals' ability to acquire learning sets – level 5 – which is associated with executive function and decision making in humans. To date, however, no studies have been published in which animals have demonstrated learning at level 8.

Notwithstanding the learning hierarchy described by Bailey et al. (2007), MacPhail (1998) noted that there was no general agreement among psychologists as to the borders of the set of tasks defining animal intelligence. Part of the problem is whether to consider animal intelligence as monarchic – factor loading specific abilities onto a unitary entity – as in human intelligence or whether it should be considered as the interaction of several diverse and distinct areas in brain. Indeed, Gardner (1983) has proposed that there are multiple intelligences in humans, and that the IQ scores measure only a few of these.

If intelligence and cognitive processes are affected by generalist genes in humans and nonhuman animals, and these genes are responsible for most cognitive abilities and disabilities, as Kovas and Plomin (2006) argue, then it should be possible to amass a list of possible candidate genes. To that end, Morley and Montgomery (2001) examined studies of human and nonhuman animals for genes that were associated with various phenotypes associated with cognition, learning, memory and mental retardation, and reported on several found in both humans and mice. In particular, the ADRA2C and dopamine beta-hydroxylase genes have been associated with ADHD and learning disabilities in children; and, that mice which over-express its homologue, *Adra2c*, perform poorly in the MWM (Morley & Montgomery, 2001).

Human Personality and Temperament

Personality measurement developed at about the time factor analysis was devised for intelligence testing. Unlike the notion of *g* in intelligence testing, theories of personality were built around either single-trait (one-dimensional) or multi-trait (multi-dimensional) theories. Like factors in intelligence, the intent of multi-trait theory is to capture all dimensions of personality and to develop measures by which to assess each dimension. Several multi-trait models have been developed, but the one currently preferred is the Five Factor Model (FFM), originally developed by Tupes and Christal (1961), and validated by McCrae and Costa (1987).

To examine the genetic and environmental components of personality, Jang, Livesley, and Vernon (1996) used the revised NEO Personality Inventory to assess MZ and DZ twins. Genetic effects were found to have extensive influence on all FFM dimensions, while nonshared environmental factors accounted for most of the remainder. More recently, Borkenau, Riemann, Spinath, and Angleitner (2006) examined genetic and environmental influences on MZ and DZ twins' personality and found personality profiles among MZ twins were more alike than among DZ twins. Also using the revised NEO Personality Inventory, Yamagata et al. (2006) examined the cross-cultural aspects of personality in twins in Canada, Japan, and Germany, and found that all five factor structures from the FFM were nearly identical with one another across cultures.

Most psychologists consider temperament closely related to personality and a stable trait, as intelligence is to cognition. Temperament is continuously measurable, multi-dimensional, and appears early in the life of the developing infant. Consequently, the origins of temperament are thought to be genetic. Current research suggests that dimensions of temperament are fivefold: emotionality, activity, attentiveness/perseverance, sociability, and reactivity (Saudino, 2005).

Several instruments have been developed to measure temperament. However, they do not all agree as to the number of dimensions of which temperament consists, as Mathiesen and Tambs (1999) noted. Moreover, the psychometric properties of the instruments used to measure temperament are quite modest. Mathiesen and Tambs (1999) sought to cross-validate one of these instruments, comparing factors and factor scores from the EAS Temperament Survey in a large sample of Norwegian children with an earlier study of Dutch children. The factor loadings they obtained for both Dutch and Norwegian samples were remarkably similar, thereby strengthening the external validity of the EAS.

Saudino, McGuire, Reiss, Hetherington, and Plomin (1995) used the EAS to examine adolescents and found moderate intraclass correlations among MZ twins for all dimensions. DZ twins, however, exhibited markedly lower correlations, and only correlations derived from the Emotionality scale were statistically significantly different from zero. Using the updated version measure of temperament, the EASI, Goldsmith, Buss, and Lemery (1997), found intraclass correlation coefficients similar to those obtained by Saudino et al. (1995) for both MZ and DZ twins. Goldsmith et al. (1997) also computed heritability coefficients for each factor and found that they ranged from 0.21 (for the pleasure dimension) to 0.72 (for activity). The heritability coefficients for activity obtained by Saudino et al. (1995) and Goldsmith et al. (1997) were comparable and argue in favor of a heritable component for temperament, albeit one which is smaller than the heritability component of intellectual abilities.

Animal Personality and Temperament

Although Romanes collected only anecdotal evidence for the existence of personality traits in nonhuman animals, scientists such as Pavlov and Yerkes attempted to examine personality systematically in animals. Since animals cannot be expected to fill out survey questionnaires, measuring personality in animals requires that several conditions be met:

1. Can two observers agree on the personality traits measured?
2. Are the inter-observer agreement correlations found in studies of humans comparable to those found in animals?
3. Are the behaviors assessed which are associated with personality consonant with the dimensions of human behavior found in the FFM?
4. Are the measurements of personality valid?
5. Are the traits heritable in humans and animals?

To address many of these issues, Gosling (2001) undertook in a comprehensive review of all published experimental studies of personality in animals. He examined the inter-rater reliability of personality traits across many species and found that, although there were wide ranging values in correlations, the median correlation coefficient for all studies was moderately high ($r = 0.61$). Test–retest reliability was high when the interval was brief, but weak when the interval was long.

Gosling (2001) also found that all the traits examined could be linked to one or another of the dimensions in the FFM for humans. He also noted that the median inter-observer correlations for specific personality traits in animals were greater than those calculated in humans. To explain the discrepancy, Gosling argued that humans learn to restrain themselves from exhibiting socially unacceptable behaviors, e.g., aggression, neuroticism, and so these behaviors tend to be less apparent.

Validity is as much a concern in studies of personality in animals as it is an issue in human personality research. Unfortunately, studies of validity in animal research are few in number. In one study, Feaver, Mendl, and Bateson (1986) found high convergent validity ($r = 0.85$) for aggressive behavior. On the other hand, Capitanio (1999) investigated predictive validity of sociability in monkeys but found only modest to weak correlations. Gosling (2001) noted, however, that these studies of validity were based on post hoc findings, that reliability of the observed behaviors had been assumed rather than tested. In some instances, observers who coded behaviors were the same as those who made personality ratings, thus introducing the possibility of rater bias.

In their review of animal behavior, Réale, Reader, Sol, McDougall, and Dingemanse (2007) did not distinguish between personality traits and temperament. Instead, they assigned temperament traits to FFM categories. Like Gosling, Réale et al. emphasized the importance of developing valid experimental tests to measure temperament traits, comparing measurements of temperament across species, and establishing a connection between trait expression and survival fitness. When tests are valid and precise, Réale et al. expected that results will be repeatable and set an upper bound to the heritability of the trait ascertained. Unlike Gosling (2001), Réale et al. (2007) argue against measuring temperament by direct observation, since such behaviors are often interpreted subjectively. Whenever possible, trait measures should correspond to biological/physiological markers.

One method for identifying personality traits in animals is to correlate phenotypes across species under a specific set of environmental conditions. Nonetheless, it may not be possible to quantify temperamental traits in species with markedly different routines and habits. To resolve this issue, Réale et al. suggested that species with closely related anatomical and physiological features be compared. It should be noted, however, that evidence for heritability of traits varies greatly, depending on the behavior assessed and the organism evaluated. For example, aggressive-like behaviors appear to have higher heritability in land mammals ($r = 0.53 - 0.61$) than sea creatures such as fish ($r = 0.14$) or squid ($r = 0.21$), whereas fearfulness in domesticated animals is weak to moderate ($r = 0.32 - 0.56$) as Réale et al. reported.

Language in Humans

Speech enables man to utter what he does not think.
Thomas Hobbes, *Elements of Law, Natural and Politic*

Language, as opposed to simple communication, is thought to be a multifaceted, particularly human trait. Cognitive psychologists consider language an essential component for thinking. Language is also used in memorizing, reasoning, and socializing. There are cultural differences in language, identifiable by their lexical and grammatical categories, e.g., boundaries for categories of colors, the use (or not) of the definite article. Other parameters of language are the phonological aspects (speech sound and production), syntax (the arrangement of words and how they are related), and pragmatics (the study of the use of language in given situations).

There are two major and opposing theories of language acquisition. One assumes language is innate; that humans are "hard-wired" to develop language (Chomsky, 1965). The opposing view argues that language is acquired solely from one's environment (e.g., Whorf, 1940; Skinner, 1957). Thus, language is conceived as having emerged predominantly from either a genetic or environmental source.

Demonstrating the genetic component in language has been problematic in part due to developmental issues. Stromswold (2001) has shown that the rate of language growth in MZ twins is slower than among singleton births. A substantially higher proportion of variance for phonology and syntax can be accounted for by genetic factors for typically developing MZ twins than for language-impaired MZ twins. Problems associated with perinatal environments enveloping twin fetuses occur more frequently than with singletons. The risk, therefore, of sub-threshold damage to neurological structures associated with language development is greater among twins in utero than among singletons and would account for the slower development of lan-

guage in, and greater discordance between, MZ twins (cf. Stromswold, 2006, for a comprehensive review of language development and genetics.).

Language in Animals

As noted earlier, one feature of language in humans is communication, and communication in animals has been noted and argued over in many species, in bees in particular (Gould, 1975). Another feature of language in humans, vocal learning, is infrequently found in nonhuman animals. One notable instance in common for humans and animals is the gene that encodes forkhead box P2 (FOXP2). Mutations in FOXP2 have been associated with speech and language disorders in humans (Fisher, Vargha-Khadem, Watkins, Monaco, and Pembrey, 1998). The homologous *Foxp2* gene is found in mice, and vocalizations in *Foxp2 wt* and *ko* mice are currently under study by Fisher and his colleagues (White, Fisher, Geschwind, Scharff, & Holy, 2006).

MacPhail (1982) reviewed the literature regarding language abilities in nonhuman animals, particularly among primates, and found that vocalizations approximating human language have been less than successful. Other approaches to language acquisition and production, such as the use of American Sign Language, have also achieved limited results; concerns about how signs were acquired, how to interpret a sign produced by a primate, and whether the sequences in which signs have been presented are grammatically correct, have also been raised.

Cognitive Deficits and Speech/Language Disorders in Humans

According to DSM-IV-TR criteria (American Psychiatric Association, 2000), human cognitive deficits first diagnosed in infancy or early childhood are categorized as mental retardation (MR) or learning disorders (LD). The formal definition for MR requires that both IQ scores and adaptive behavior (DQ) scores be at least 2 standard deviations below the population mean. LD has been variously defined, but according to DSM criteria, is represented by a 2 standard deviation difference between IQ score and achievement score on standardized assessment instruments.

LD and MR affect 5–10% of the general population. MR is further sub-typed as mild, moderate, severe, or profound, according to the IQ score. Moderate-to-severe MR – about 1% of the general population – is mostly caused by genetic factors. Many genetic abnormalities occur spontaneously, e.g., trisomy 21 and the micro-deletion disorders; but some, like the fragile X mutation, are highly heritable. NF1 produces LD, and 50% of individuals with NF1 have inherited the mutation. As noted previously, expressive language disorders have been associated with mutations in the FOXP2 gene. In their review, Lewis et al. (2006) reported that, in studies of speech and language disorders, high concordance rates among MZ twins were found.

Another somewhat related category of DSM-IV-TR childhood disorders is referred to as pervasive developmental disorders (PDD). Chief among PDDs is autism. Autism has been referred to as a triad of dysfunction involving severe deficits in communication, socialization, and restricted or bizarre activities. About 75% of children with autism are also diagnosed as MR and may have no speech, or severely limited speech and expressive language. The ratio of males to females with the disorder is 4:1. Among families in which a first born has been diagnosed as autistic, the probability that a second or subsequent child will also be diagnosed with autism is almost 2 orders of magnitude greater than expected in the general population. Among MZ twins, the correlation of a diagnosis of autism is moderately high ($r = 0.59$), and statistical analysis showed that 57% of the variance could be attributed to the genetic component (Hoekstra, Bartels, Verweij, and Boomsma, 2007).

Animal Models of Mental Retardation and Learning Disability

One of the earliest animal models of MR was created in 1994 by the Dutch–Belgian Fragile X Consortium (DBFXC, 1994). The *fmr1* gene in mice is homologous to the FMR1 gene in humans. A mutation in the FMR1 gene, typically caused by an excess number of CGG repeats in the promoter region combined with hypermethylation at the fragile site, produces the fragile X syndrome in humans. To examine cognitive function in mice, the DBFXC tested their *fmr1 ko* and *wt* mice using the MWM. They observed that, on average, *fmr1 ko* mice performed less well than *wt* littermate controls on the reversal hidden platform task in the MWM, but did not perform significantly differently on other tasks in the MWM. Other researchers report similar, small differences (e.g., D'Hooge et al., 1997). Interestingly, Fisch, Hao, Bakker, and Oostra (1999) found that *fmr1 ko* performed better than *wt* controls on both visual and auditory discriminant operant conditioning tasks, while Paradee et al. (1999) found strain differences had a significant effect. More recently, Fisch and colleagues (Fisch, 2003) performed a follow-up evaluation of *fmr1 ko* and *wt* controls on a delayed-matching-to-sample task and found that *fmr1 ko* mice performed marginally, but not sig-

nificantly, worse compared to *wt* controls. After Fisch (2003) examined the mice DNA and FMR protein (FMRP) in their study, they discovered the presence of FMRP in brain in 5/8 knockouts. As Welzl, D'Adamo, Wolfer, and Lipp (2006) note, the emergence of a subtle phenotype in the *fmr1 ko* mouse will make assessments of future therapies problematic.

In addition to the fragile X mutation, there are other genes located on the X chromosome in which mutations produce MR, i.e., X-linked MR or XLMR. One such is FRAXE, the other fragile X disorder, the mutation site for which is downstream from the FMR1 locus. In general, cognitive deficits associated with FRAXE are mild compared to those produced by mutations in the FMR1 gene. The gene associated with FRAXE, FMR2, has been identified (Gecz, Gedeon, Sutherland, and Mulley, 1996), a *ko* mouse created (Gu et al., 2002), and a battery of behavioral tests administered to examine the phenotype. Except for a test of conditioned fear, in which the *ko* mice exhibited a lower freezing response than *wt* littermate controls, there were no significant differences in behavior between the two groups of animals.

A different X-linked abnormality producing MR is caused by a mutation in GDI1. A knockout mouse for the *Gdi1* gene was created by D'Adamo et al. (2002). These researchers also administered a wide battery of behavioral tests to the mice, most of which found that *ko*s did not differ significantly from littermate controls. However, D'Adamo et al. did find that *Gdi1 ko* mice performed significantly less well than controls on the radial maze learning task, and exhibited less aggressive behavior than controls.

Another form of inherited XLMR, Coffin–Lowry syndrome, is characterized by many craniofacial and skeletal abnormalities, the expression of which is severe in males but mild in females. It is caused by a mutation in the RSK2 gene. Poirier et al. (2007) examined exploratory and emotional reactivity, and learning and memory in *Rsk2* mutant mice, and found that *wt* performed significantly better than *ko*s on visual–spatial activities, and on tasks that involved exploratory behavior.

Three years after the *fmr1 ko* mouse was created, Silva and colleagues (Silva et al. 1997) investigated cognitive function in the *Nf1 ko* mouse, also using the MWM. These researchers observed significantly poorer performance in the *Nf1 ko* compared to its *wt* littermate controls. Costa et al. (2001) later found that a mutation in exon 23a of the *Nf1* gene affected the Ras signaling pathway and increased GABA-mediated inhibition, and was associated with cognitive deficits in the *Nf1 ko* mice. Recently, Donarum, Halperin, Stephan, and Narayanan (2006) performed a gene expression analysis to examine the profiles of expression in the *Nf1* mouse brain. These researchers found a functional relationship between the NF1 protein, neurofibromin, the amyloid beta precursor complex, and the dopamine receptor, Drd3, which suggests that these pathways may also be involved in cognitive dysfunction in humans with NF1.

Animal Models of Autism and Pervasive Developmental Disorders

As noted earlier, autism is a pervasive developmental disorder primarily affecting males and is characterized by a triad of dysfunction involving impairments in communication, socialization, and bizarre and/or restricted interests or activities. There is reason to believe that there is a genetic etiology. For example, a higher proportion of individuals with the fragile X mutation have been diagnosed with autism than in the general population. Researchers have also found genetic mutations associated with autism elsewhere on the X chromosome. Jamain et al. (2003) examined two brothers diagnosed with autism and found mutations in neuroligin genes NLGN3 and NLGN4. Risch et al. (1999) performed a genomic screen on 90 multiplex sibships with autism and concluded that the phenotype may result from more than 15 genes. Consequently, constructing a mouse model for the disorder has been, and will be, problematic.

As noted earlier, the *fmr1 ko* mouse' ability to mimic the fragile X syndrome in humans has produced mixed results. These obstacles notwithstanding, Mineur, Huynh, and Crusio (2006) examined social habituation as an autistic-like behavior in C57BL/6 strains of *fmr1 ko* and *wt* mice, and as compared to the BALB and DBA strains. They found insignificant differences in social behavior among the three strains. They also noted that *fmr1 ko* mice from the C57BL/6 strain exhibit significantly fewer social interactions with females than do *fmr1 wt* littermate controls, suggestive of autistic-like behavior. Moon et al. (2006) examined performance of *fmr1 ko* and *wt* mice on a series of attention tasks and found that *fmr1 ko* mice made a significantly higher proportion of premature responses than *wt* littermate controls, suggestive of impaired control of inhibitory responding. They also found a significantly higher proportion of premature responses to the sustained attention task compared to controls. These researchers commented that, although these dysfunctional behaviors were evocative of attention deficits observed in humans with the fragile X mutation, the problems observed in mice were only temporary, and that permanent deficits were not significant.

As autism is defined by its behavioral phenotype and not associated with specific biological markers, current mouse models of autism are highly dependent upon an accurate depiction of behaviors associated with the disorder. To that end, Bourgeron, Jamain, and Granon (2006) have proposed a

set of behavioral tests for animal models of autism to provide a standardized framework for experimentation. They suggest that two types of social behaviors be investigated: ability to interact with a conspecific and social transmission of food cues. In addition to social behaviors, Bourgeron et al. (2006) also advise that anxiolytical and stereotypical behaviors be assessed. To evaluate anxiety, they recommend investigating exploratory activity in an open-field apparatus. To study stereotypic behavior, an operant lever press task should be employed. Learning and memory should also be evaluated and assessed using the MWM and a radial arm maze, both of which examine visual/spatial skills. Finally, sensory gating should be inspected using the procedures involving the startle reflex to measure PPI.

In their efforts to assess autism in mice, Crawley and her colleagues focused on the sociability aspects – or rather, lack thereof (Moy et al., 2004) – by studying five different mouse strains on tests of social approach to familiar and novel environments. Bolivar, Walters, and Phoenix (2007) also examined social behavior in several inbred strains and found, as did Moy et al. (2004), that BTBR mice exhibit significantly lower social interactions than other strains. They also noted deficits in the corpus callosum in the BTBR strain, similar to *postmortem* findings of individuals with autism (Brambilla et al., 2003). More recently, Moy et al. (2007) developed a series of tasks designed to emulate behavioral deficits observed in autism, and again found the BTBR strain to be a useful model. Taken together, these results suggest that the BTBR strain may be valuable in locating QTLs and genes for autism in humans.

In pursuing the genetic side of autism, Sadakata et al. (2007) identified a candidate gene, CADPS2, residing in the autism susceptibility locus on chromosome 7q31–33 in humans, and associated with autistic-like behavior in the *Cadps2 ko* mouse. Employing a complex test battery to evaluate mouse behavior, these researchers noted abnormal sleep patterns, increased locomotor activity, and decreased responses to a novel object, all of which are suggestive of several aspects of impaired behavior observed in individuals with autism. As for the genetic component, they also found aberrant splicing in CADPS2 mRNA in several patients with autism. Tabuchi et al. (2007) created neuroligin knockin (*ki*) and knockout mice that emulated the genetic mutations noted by Jamain et al. (2003). Tabuchi et al. (2007) found that neuroligin-3 *ki* mice exhibited some social deficits compared to littermate controls, but curiously performed better on the MWM than controls. Also, unlike the results obtained by Sadakata et al. (2007), Tabuchi et al. (2007) found that neuroligin-3 *ki* mice did not differ from controls in their social responses to a novel intimate.

Rett syndrome is another pervasive developmental disability, but predominantly affects females, and results from a mutation in the MECP2 gene (Nan et al., 1998). The phenotype develops early in infancy, resulting in a loss of speech, development of neuromuscular degeneration, and the emergence of hand-wringing in most cases. The *Mecp2 ko* mouse exhibits a phenotype consistent with Rett syndrome: normal development for the first several weeks after birth, followed by progressive neurological dysfunction (Shahbazian et al., 2002). In addition to autistic-like behavior, Rett syndrome is associated with severe cognitive deficits. Moretti et al. (2006) tested *Mecp2 ko* mice and found that, in addition to social interaction deficits, *Mecp2 ko* mice take longer to find the platform in the MWM, thereby exhibiting deficits in learning and memory. Surprisingly, however, these mice do not display the neuronal dysmorphology found in *postmortem* analyses in humans with Rett syndrome.

Personality and Temperamental Disorders in Humans

DSM-IV-TR also defines a variety of other disorders: personality disorders; anxiety disorders; mood disorders; attention deficit/hyperactivity disorder (ADHD); obsessive/compulsive disorder (OCD); excessive aggressiveness, e.g., oppositional defiant (OD) disorder and conduct disorder (CD); as well as the psychoses and schizophrenia, disorders that involve hallucinations and/or delusions. Although certain aspects of these disorders can be quantized and measured, e.g., hyperactivity and attentiveness, other features, such as hallucinations, cannot.

Hudziak et al. (2004) found concordance rates for OCD of about 90% in MZ twins. Ørstavik et al. (2007) examined depression and found heritability estimates of 49% among female MZ twins, and 25% among male MZ twins. Kendler, Gatz, Gardner, and Pedersen (2006) found moderate correlations for major depression between female MZ twins and modest correlations between male MZ twins. Kendler et al. (2006) found heritability of major depression among MZ twins was 0.38. Ehringer, Rhee, Young, Corley, and Hewitt (2006) found modest-to-moderate heritability coefficients among MZ twins for lifetime development of ADHD, CD, OD, generalized anxiety disorder (GAD), and separation anxiety disorder. Thus, personality, anxiety, and mood disorders appear to have a modest-to-moderate heritable component. In their follow-up study, Kendler and colleagues (Kendler, Gardner, Gatz, & Pedersen, 2007) examined comorbidity between GAD and major depression and found high genetic correlations between the two.

Animal Models of Personality and Anxiety Disorders

As Widiger and Trull (2007) argue, a dimensionality approach to personality disorders would be a vast improvement over the limitations of the existing diagnostic categories in DSM-IV-TR. They endorse the use of the FFM, employing a Likert-like scale to assess each dimension of personality. By doing so, personality research would be more in accord with studies of cognitive abilities and intelligence that investigate MR and LD. To that end, Willis-Owen and Flint (2007) suggested that, as a dimension of personality, neuroticism be measured as a gauge of emotionality and emotional stability. To support their argument, Willis-Owen and Flint (2007) report that, in a large sample study of humans, Fullerton and colleagues discovered five QTLs for emotionality on chromosome 1 (Fullerton et al., 2003). At the same time, Willis-Owen and Flint (2007) noted that selective inbreeding techniques – backcrosses, intercrosses – have identified emotional reactivity in mice – as measured by avoidance behaviors, defecation, urination, among others – and mapped QTLs onto chromosomes 1, 15, 18, but primarily on chromosome 1.

As in studies of the genetics of animal intelligence, selective breeding in rats was among the first techniques enlisted to identify genetic factors in anxiety. Using the elevated plus maze to differentiate high- and low-anxiety behaviors, high-anxiety (HAB) and low-anxiety (LAB) rats have been bred for more than a decade (Landgraf & Wigger, 2002). In preliminary studies, these researchers found single nucleotide polymorphisms in a candidate gene, vasopressin, in their HAB rats. Using defecatory responses to an open-field environment as a measure of anxiety, Maudsley rats have also been bred for reactive or nonreactive strains. Maudsley rats have also been bred to study alcohol addiction, contact with novel objects (a measure of fearfulness), and aggressive behavior (Blizard & Adams, 2002).

Mouse models of anxiety, fear, and depression have been extensively studied, and Crawley and Paylor (1997) proposed a battery of behavioral tests to examine anxiety and depression in mice. In humans, the serotonin receptor gene, 5-HT1A, has been associated with anxiety disorder and depression. However, when Ramboz et al. (1998) examined anxiety and depression in *5HT1A*-deficient mice, they found that the *5-HT1A ko* performed as expected on tests of anxiety, but performed better that *wt* littermates on the forced swim test (a measure of depression). These researchers had also noted that, although *5-HT1A ko* develop and breed normally, adult *ko*s are more impulsive (Brunner & Hen, 1997). As a result, they decided to study the maternal affects of and on *ko* and *wt 5-HT1A* mice. Weller et al. (2003) examined ultrasonic vocalizations (USV) in pups tested in isolation and found that the genotype of the mother affected the USV level, a measure of anxiety.

To differentiate anxiety from fear, Holmes and Cryan (2006) suggest that mice be assessed on separate test batteries. To measure anxiety, they note the use of approach/avoidance tasks such as the elevated maze, whereas, to measure fearful behavior, cues associated with aversive stimuli are used. Holmes and Cryan (2006) also report that laboratory environment and genetic strain also have effects on animal responses to experimental conditions, factors that were discussed earlier in this chapter.

In addition to the personality disorders noted, there is evidence to suggest that many of the addiction disorders listed in DSM-IV-TR – alcoholism, drug abuse – also have a strong genetic component. Research into addiction disorders has been quite active of late, and animal models have been developed to examine behaviors associated with addiction (e.g., Blizard, 2007; Haile, Kosten, & Kosten, 2007; Metten et al., 2007).

Animal Models of Mood Disorders

Animal models of mood disorders have been widely investigated. Tests for depression – the forced swim test, tail suspension test, learned helplessness paradigm – and genetic rodent models – rats and mice – have been developed and used to investigate mood disorders (cf. Yacoubi & Vaugeois, 2007). Nonetheless, Urani, Chourbaji, and Gass (2005) argue that not all behavioral despair paradigms are specific to models of depression, since outcomes can be interpreted in more than one way. For example, is floating a measure of despair or a means of conserving energy? Consequently, most researchers agree that standards must be met for an animal model to be considered valid. Nearly four decades ago, McKinney and Bunney (1969) proposed four criteria that must be attained: (1) the animal's behavior be "reasonably analogous" to the human disorder in its features and/or symptomatology, i.e., face validity; (2) behavioral changes be monitored objectively, i.e., eliminate rater bias; (3) that the disordered behavior in the model be reversed by the same treatment modalities which are effective in humans, i.e., predictive validity; and (4) the study be reproducible by other investigators, i.e., external validity. Others note that, in addition to these criteria, etiological factors thought to elicit the disorder in humans should also elicit analogous behaviors in the animal model; and, that the neurobiological and neurophysiological processes involved in humans be implicated in the animal model as well (Newport, Stowe, & Nemeroff, 2002).

Although researchers determined to meet these standards have constructed a variety of prototypes to model depression, they have also encountered many obstacles. Depression, as defined by DSM-IV-TR criteria, is characterized primarily by chronic mood change, anhedonia, large swings in weight and/or sleep habits, psychomotor agitation or retardation, feelings of worthlessness, and diminished cognitive functioning. These features are heterogeneous and vary in appearance and extent within and across individuals, and over time. Psychopharmacological treatment has often been effective, but

then so have been cognitive–behavioral interventions, as well as no interventions at all.

Factors that produce stress leading to depression are also many and varied. They can be neurogenic or psychogenic, controllable or not, organismic (e.g., age, sex), experiential (e.g., early life experiences), and mediated by personal and psychosocial factors. There are also genetic differences in responses to stressors. For example, in mice, the BALB strain is more reactive than the C57BL strain (cf. Anisman & Matheson, 2005). Anisman and Matheson (2005) also acknowledge that many animal prototypes of depression and etiological factors have been investigated, and argue that, while all the symptoms of depression should be studied, they give primacy to anhedonia, which they consider central to the disorder.

Based on the efficacy of certain pharmacological agents to ameliorate the symptoms of depression in humans, i.e., show predictive validity, molecular targets of these antidepressants have been studied. Three important neurophysiological pathways have been investigated: those which involve serotonergic signaling (5HT released into the synaptic cleft affecting transcription factors such as CREB and genes such as BDNF); the neurotropin pathway involving 5HT, CREB and BDNF; and dysregulation of the endocrinological stress system along the hypothalamic–pituitary–adrenal (HPA) axis. When considering each of these pathways, candidate genes have been proposed based on molecular findings in humans, and mutant (transgenic or knockout) mouse models created. For example, the serotonin transporter (SERT) gene plays an important role in the serotonin signaling pathway in humans. Consequently, SERT mutant mice were created to examine the neurobiological pathways and test whether depression-like behaviors develop. Although many neurobiological aspects of the serotonin signaling system were confirmed in SERT mutant mice, Holmes, Yang, Murphy, and Crawley (2002) found that despair-like behavior based on the tail suspension test did not differ significantly in C57BL6 mutant mice, compared to controls. Given the heterogeneous nature of the disorder, and differing pathways that may be involved, Urani et al. (2005) propose that behavioral traits associated with depression be more well defined, and that mutant models reflect more accurately the complexity of the disorder.

The problem of creating a valid mouse model for mood disorders is compounded when attempting to create one for bipolar disorder. Bipolar disorder is characterized by alternating episodes of depression and excessive euphoria. In addition to involvement of the HPA axis associated with intervals of depression, dysregulation of stress responses involving corticotropin releasing factor (CRF) has been implicated in the manic component. Given these alternating phases, Dirks, Groenink, and Olivier (2006) regard the development of an animal model for bipolar disorder to be problematic. However, Dirks et al. (2006) note that transgenic models that over-express neurotransmitters involved in glucocorticoid receptors display a greater lability of positive and negative responses to tests of emotionality, suggesting that these transgenic mice may make good candidates for animal studies of bipolar disorder.

Psychoses and Schizophrenia in Humans

Psychosis and schizophrenia are also defined by DSM-IV-TR. Positive symptoms in the family of psychotic disorders include hallucinations (sensing something that is not present) and/or delusions (beliefs about oneself that do not conform to reality). Twin studies have shown statistically significant genetic components for the psychoses and schizophrenia. Jang, Woodward, Lang, Honer, and Livesley (2005) found that correlation of psychotic features among MZ twins was moderate. Shih, Belmonte, and Zandi (2004) reviewed twin studies of psychopathology in adults and found concordance rates among MZ twins between 20 and 75% for bipolar disorder. For schizophrenia, Shih et al. found concordance rates among MZ twins were between 41 and 79%.

Animal Models of Psychosis and Schizophrenia

Identifying specific genetic factors in schizophrenia has been difficult, due in part to the lack of complete concordance among MZ twins. Also, as in depression, phenotypic heterogeneity has compounded the problem. Therefore, should schizophrenia be considered a unitary disorder or a collection of subtypes? Studies of schizophrenia have established that both dopaminergic and glutamatergic systems are implicated in its pathophysiology. Unfortunately, findings associating schizophrenia with genetic polymorphisms related to dopaminergic or glutamatergic function have been inconsistent (cf. O'Tuathaigh et al., 2006).

Despite these difficulties, several genes have been implicated in the risk for schizophrenia, including NRG1 (Stefansson et al., 2003), RGS4 (Chowdari et al., 2002), Dysbindin (Straub et al., 2002), COMT (Williams, Owen, and O'Donovan, 2007), PRODH (Liu et al., 2002), and DISC1 (Millar et al., 2000), from which mutant mouse models have been created. However, as O'Tuathaigh et al. (2006) and others have cautioned, given the likely polygenic origins of schizophrenia, the loss of function or haploinsufficiency for any one gene will probably not provide researchers with a valid animal model for the disorder. On the other hand, Miyakawa et al. (2003) found that their CN mutant mice displayed behavioral abnormalities consistent with schizophrenic behavior in humans. As a result,

Miyakawa et al. (2003) carried out a series of behavioral tests and observed significant impairment in PPI, locomotor hyperactivity, impaired latent inhibition, and impaired nesting behavior in CN mutants compared to littermate controls. More recently, Rojas, Joodmardi, Hong, Perlmann, and Ogren (2007) found that the transcription factor, Nurr1 (NR4A2), plays an essential role in dopaminergic function in neurons, and developed a heterozygous *Nurr1 ko* mouse. Rojas et al. (2007) reported that mouse behavior was consistent with that of schizophrenia, and with other rodent models of schizophrenia.

Unfortunately, the normal strategy for determining genetic factors associated with complex behavioral disorders, for finding susceptibility genes, specifically for anxiety and depression, has largely been unsuccessful, as Kas, Fernandes, Schalkwyk, and Collier (2007) noted. The problem is due in part to the overlapping features of many psychiatric disorders, e.g., schizophrenia, bipolar disorder, and unipolar depression. They suggest that researchers focus on the genetics of naturally occurring inter-species behaviors rather than the complex-syndrome genetics to optimize the establishment of genotype–phenotype relations. Further, in order to identify shared genotype–phenotype relations, the same gene should evoke analogous phenotypes in both species, a criterion we noted earlier. Consequently, the choice of phenotype will be crucial. One strategy has been to identify an endophenotype, as Gould and Gottesman (2003) have proposed. The endophenotype, which is thought to be dimensionally less complex, may have evolved from simpler genetic bases. Consequently, behaviors associated with the endophenotype that fall along a comparable dimension in mice and humans could be identified more readily. One such endophenotype associated with schizophrenia is impaired sensory gating, which has been observed in humans, and which produces a suppressed auditory-evoked P50 response, i.e., prepulse inhibition (PPI) of the startle response (Freedman, Adler, Waldo, Pachtman, & Franks, 1983).

A propos of the use of endophenotypes, Paylor et al. (2006) developed a knockout mouse which models 22q11 deletion syndrome, a relatively common genetic disorder that produces cognitive and behavioral deficits, and psychiatric dysfunction – primarily schizophrenia. Asperger's syndrome has also been observed in humans with the 22q11 deletion. A decade earlier, Karayiorgou et al. (1995) found interstitial deletions in 22q11 in patients with schizophrenia. The *Df1/+* mice developed by Paylor et al. (2006) display much abnormal behavior, including impaired PPI. Four genes were located in the critical region associated with PPI, including *Tbx1*. Previously, mutations in *TBX1* had been found in patients with 22q11 deletion. Accordingly, Paylor and colleagues examined a family in which 22q11 deletion syndrome was segregating and located mutations in the *TBX1* gene in two brothers diagnosed with Asperger's syndrome.

Finale and Future Directions: A Few Cautionary Notes

What a piece of work is a man! ... the Paragon of Animals.
William Shakespeare, *Hamlet, Act II*

In our discussion of animal models of human behavior, it is important to note that Darwin's thesis regarding the continuity in structure and mind in animals and humans remains an open question. As Welzl et al. (2006) point out, mice are not diminutive humans, nor are they little rats. These researchers also state that, while the hippocampus and amygdala are an integral part of learning and memory in mice and other mammals, they likely serve more diverse neurological functions in mice than the more specialized subcortical regions found only in primates and humans. And, despite its surface appeal, the assumption underlying the generalist genes theory, as expressed by Kovas and Plomin (2006) – that the same genes producing cognitive deficits are those involved in intelligence – may not be a valid hypothesis. Genetic mutations that produce MR and LD, while they may interrupt cognitive processes and produce dysfunction, may not be sufficient to produce the gradations in intelligence observed in the normal distribution of IQ scores for humans, the concept of pleiotropy notwithstanding. An example is the FMR1 gene associated with fragile X mutation. The normal range of CGG polymorphisms in the FMR1 promoter region is unrelated to gradations in intelligence in the general population (Mazzocco & Reiss, 1997).

As for the manner in which intelligence is characterized in animals, should researchers attempt to assess a unitary *g* or examine the specialized abilities within each species? Nearly two decades ago, Thompson and colleagues (Thompson, Crinella, & Yu, 1990) developed a test battery consisting of various problem-solving apparatuses designed to assess a variety of appetitive and aversively motivated activities. Then, using factor analysis, they constructed a general measure of intelligence comparable to *g* in humans that differentiated brain-lesioned rats from controls. Curiously, no other investigators seem to have utilized the Thompson et al. (1990) "psychometric" battery to assess animal intelligence. One might also ask, when researchers decide to examine specialized abilities within each species, what is the human intelligence test equivalent to the MWM? For that matter, what is normal intelligence in mice?

As MacPhail (1982) and others note, any debate about the similarities and differences between human and animal intelligence – as well as other dimensions of behavior –

requires a discussion of language and language acquisition. All activities in which nonhuman animals can demonstrate mastery can be mastered by humans, but not all tasks in which humans demonstrate mastery can be mastered by nonhuman animals. The difference has been accounted for by the fact that humans have language and can solve problems using it, while animals do not and cannot. As noted earlier, MacPhail (1982) surveyed studies of language and communication in nonhuman animals and concluded that there was no compelling evidence to support the conjecture that animals are capable of developing language in the sense that humans understand language.

The problem of language also arises when researchers are intent upon investigating neuropsychiatric disorders. Although the criteria proposed by McKinney and Bunney (1969) have been a useful guide when considering animal models of human psychopathology, they may be less than satisfactory, particularly when invoking the criterion, "reasonably analogous." Hayes and Delgado (2006, 2007) argue that the behavioral repertoires of humans, especially those that are essentially psychopathological, are insufficiently analogous to nonhuman organisms. They point out that the most important dimension of human behavior lacking in nonhuman animals is language, and the effects of this discrepancy between humans and nonhumans are exacerbated along the other dimensions of human behavior, i.e., cognition and intelligence, personality and temperament. The extent to which clinical diagnoses are made are largely contingent upon the verbal reports of patients. Thus, when cognition, intelligence, personality and temperament are assessed in order to make a diagnosis of neuropsychiatric disorder – MR and LD, PDD, psychosis and schizophrenia, depression and bipolar disorder, anxiety and other personality disorders – researchers employing mouse models should be expected to furnish a more precise operational definition of animal behaviors for "reasonably analogous."

Having reflected on the use of animals to study disease in medicine, Lafollette and Shanks (1995) assert that, in biomedical animal experimentation, there are two types of models: (1) causal analog models (CAMs), in which the effects of various causes observed in animals are likely to be the effects observed in humans and (2) hypothetical analog models (HAMs), in which experiments stimulate the creation of hypotheses about similar biological phenomena in humans. If animals are to be considered appropriate CAMs for psychopathology, the causal properties must then be connected to a specific set of outcomes for both humans and animals; and, "there must be no causal *disanalogies* (my italics) between the (animal) model and (human characteristic) being modeled." (p.147). Lafollette and Shanks (1995) also observe that, while investigators may present the Darwinian argument that humans and animals are phylogenetically continuous or phylogenetically close, this does not guarantee causally relevant comparability. According to these authors, evolutionary theory provides researchers with good reasons to think that animals are good HAMs, not that they are necessarily good CAMs.

In my discourse on animal models, I noted that in a very few instances, animals seem to express sufficiently many features of a disorder so that they may engender good CAMs. Perhaps NF1 and the *Nf1 ko* mouse are good examples. However, in other instances cited, there often appear to be what Lafollette and Shanks referred to as disanalogies. The model should then be considered only suggestive, as in the case of mutations in the FMR1 gene and the moderate-to-severe MR detected in humans compared to the mild cognitive deficits and incongruities observed in the *fmr1 ko* mouse. One solution would be to identify comparable biological markers in humans and animals directly correlated with the behaviors or disorders in question that would obviate the need to validate "reasonably analogous" behavior between the species. Otherwise, we as researchers should begin to view our respective animal models in a more conservative light, as HAMs, and perhaps we will better appreciate the use of animals in future studies of the genetic components of behavior. That is to say, animal models are models of, and not isomorphic with, human behavior.

References

Anisman, H., & Matheson, K. (2005). Stress, depression, and anhedonia: caveats concerning animal models. *Neuroscience and Biobehavioral Reviews, 29*, 525–546.

American Psychiatric Association. Diagnostic and Statistical Manual of Mental Disorders: DSM-IV-TR (2000). (4th ed., text revision) Washington, DC: American Psychiatric Association.

Bailey, A. M., McDaniel, W. F., & Thomas, R. K. (2007). Approaches to the study of higher cognitive functions related to creativity in nonhuman animals. *Methods, 42*, 3–11.

Blizard, D. A. (2007). Sweet and bitter taste of ethanol in C57BL/6J and DBA2/J mouse strains. *Behavior Genetics, 37*, 146–159.

Blizard, D. A., & Adams, N. (2002). The Maudsley reactive and nonreactive strains: A new perspective. *Behavior Genetics, 32*, 277–299.

Bolivar, V. J., Walters, S. R., & Phoenix, J. L. (2007). Assessing autism-like behavior in mice: Variations in social interactions among inbred strains. *Behavioural Brain Research, 176*, 21–26.

Borkenau, P., Riemann, R., Spinath, F. M., & Angleitner, A. (2006). Genetic and environmental influences on Person x Situation profiles. *Journal of Personality, 74*, 1451–1480.

Bouchard, T. J., & McGue, M. (1981). Familial studies of intelligence: A review. *Science, 212*, 1055–1059.

Bouchard, T. J., Lykken, D. T., McGue, M., Segal, N. L., & Tellegen, A. (1990). Sources of human psychological differences: The Minnesota study of twins reared apart. *Science, 250*, 223–250.

Bourgeron, T., Jamain, S., & Granon, S. (2006). Animal models of autism: Proposed behavioral paradigms and biological studies. In G. S. Fisch & J. Flint (Eds.), *Transgenic and knockout models of neuropsychiatric disorders* (pp. 151–174). Totowa NJ: Humana Press.

Brambilla, P., Hardan, A., di Nemi, S. U., Perez, J., Soares, J. C., & Barale, F. (2003). Brain anatomy and development in autism: Review of structural MRI studies. *Brain Research Bulletin, 61*, 557–569.

Brunner, D., & Hen, R. (1997). Insights into the neurobiology of impulsive behavior from serotonin receptor knockout mice. *Annals of the NY Academy of Sciences, 836*, 81–105.

Brush, F. R., & Driscoll, P. (2002). Selective breeding program with rats: Introduction. *Behavior Genetics, 32*, 275–276.

Bulmer, M. (2003). *Francis Galton: Pioneer of heredity and biometry*. Baltimore, MD: Johns Hopkins Press.

Capitanio, J. P. (1999). Personality dimensions in adult male rhesus macaques: Prediction of behaviors across time and situation. *American Journal of Primatology, 47*, 299–320.

Chomsky, N. (1965). *Aspects of the theory of syntax*. Cambridge, MA: MIT Press.

Chowdari, K. V., Mirnics, K., Semwal, P., Wood, J., Lawrence, E., Bhatia, T., et al. (2002). Association and linkage analyses of RGS4 polymorphisms in schizophrenia. *Human Molecular Genetics, 11*, 1373–1380.

Costa, R. M., Yang, T., Huynh, D. P., Pulst, S. M., Viskochil, D. H., Silva, A. J., et al. (2001). Learning deficits, but normal development and tumor predisposition, in mice lacking exon 23a of Nf1. *Nature Genetics, 27*, 399–405.

Crabbe, J. C., Wahlsten, D., & Dudek, B. C. (1999). Genetics of mouse behavior: Interactions with laboratory environment. *Science, 284*, 1670–1672.

Crawley, J. N., Belknap, J. K., Collins, A., Crabbe, J. C., Frankel, W., Henderson, N., et al. (1997). Behavioral phenotypes of inbred mouse strains: Implications and recommendations for molecular studies. *Psychopharmacology, 132*, 107–124.

Crawley, J. N., & Paylor, R. (1997). A proposed test battery and constellations of specific behavioral paradigms to investigate the behavioral phenotypes of transgenic and knockout mice. *Hormones and Behavior, 31*, 197–211.

D'Adamo, P., Welzl, H., Papadimitriou, S., Raffaele di Barletta, M., Tiveron, C., Tatangelo, L., et al. (2002). Deletion of the mental retardation gene Gdi1 impairs associative memory and alters social behavior in mice. *Human Molecular Genetics, 11*, 2567–2580.

Darwin, C. R. (1859). *On the origin of species by means of natural selection, or the preservation of the favoured races in the struggle for life*. London: John Murray.

Darwin, C. R. (1872). *The expression of the emotion in man and animals*. London: John Murray.

Deacon, R. M., Thomas, C. L., Rawlins, J. N., & Morley, B. J. (2007). A comparison of the behavior of C57BL/6 and C57BL/10 mice. *Behavioural Brain Research, 179*, 239–247.

Deary, I. J., Spinath, F. M., & Bates, T. C. (2006). Genetics of intelligence. *European Journal of Human Genetics, 14*, 690–700.

D'Hooge, R., Nagels, G., Franck, F., Bakker, C. E., Reyniers, E., Storm, K., et al. (1997). Mildly impaired water maze performance in male Fmr1 knockout mice. *Neuroscience, 76*, 367–376.

Dick, P. (1968). Do Androids Dream of Electric Sheep? New York: Del Ray Books.

Dirks, A., Groenink, L., & Olivier, B. (2006). Mutant mouse models of bipolar disorder: Are there any? In G. S. Fisch & J. Flint (Eds.), *Transgenic and knockout models of neuropsychiatric disorders* (pp. 265–285). Totowa NJ: Humana Press.

Donarum, E. A., Halperin, R. F., Stephan, D. A., & Narayanan, V. (2006). Cognitive dysfunction in NF1 knock-out mice may result from altered vesicular trafficking of APP/DRD3 complex. *BMC Neuroscience, 7*, 22.

Dutch-Belgian Fragile X Consortium. (1994). Fmr1 knockout mice: a model to study fragile X mental retardation. *Cell, 78*, 23–33.

Ehringer, M. A., Rhee, S. H., Young, S., Corley, R., & Hewitt, J. K. (2006). Genetic and environmental contributions to common psychopathologies of childhood and adolescence: A study of twins and their siblings. *Journal of Abnormal Child Psychology, 34,* 1–17.

Feaver, J., Mendl, M., & Bateson, P. (1986). A method for rating the individual distinctiveness of domestic cats. *Animal Behaviour, 34,* 1016–1025.

Fiala, B. A., Joyce, J. N., & Greenough, W. T. (1978). Environmental complexity modulates growth of granule cell dendrites in developing but not adult hippocampus of rats. *Experimental Neurology, 59,* 372–383.

Fisch, G. S. (2003). Transgenic models of complex behavioral phenotypes. Invited Symposium for the annual meeting of the American Society of Human Genetics, Los Angeles, CA, November, 2003.

Fisch, G. S. (2006). Transgenic and knockout models of neuropsychiatric disorders: Introduction, history, assessment. In G. S. Fisch & J. Flint (Eds.), *Transgenic and knockout models of neuropsychiatric disorders* (pp. 3–23). Totowa NJ: Humana Press.

Fisch, G. S., Hao, H. H., Bakker, C., & Oostra, B. A. (1999). Learning and memory in the FMR1 knockout mouse. *American Journal of Medical Genetics, 84,* 277–282.

Fisher, S. E., Vargha-Khadem, F., Watkins, K. E., Monaco, A. P., & Pembrey, M. E. (1998). Localisation of a gene implicated in a severe speech and language disorder. *Nature Genetics, 18,* 168–170.

Flint, J. (1999). The genetic basis of cognition. *Brain, 122,* 2015–2032.

Freedman, R., Adler, L. E., Waldo, M. C., Pachtman, E., & Franks, R. D. (1983). Neurophysiological evidence for a defect in inhibitory pathways in schizophrenia: Comparison of medicated and drug-free patients. *Biological Psychiatry, 18,* 537–551.

Fullerton, J., Cubin, M., Tiwari, H., Wang, C., Bomhra, A., Davidson, S., et al. (2003). Linkage analysis of extremely discordant and concordant sibling pairs identifies quantitative-trait loci that influence variation in the human personality trait neuroticism. *American Journal of Human Genetics, 72,* 879–890.

Galton, F. (1869). *Hereditary genius: An inquiry into its laws and consequences.* London: MacMillan and Co.

Galton, F. (1883). *Inquiries into human faculty and its development.* New York: AMS Press.

Gardner, H. (1983). *Frames of mind: The theory of multiple intelligences.* New York: Basic Books.

Gecz, J., Gedeon, A. K., Sutherland, G. R., & Mulley, J. C. (2006). Identification of the gene FMR2, associated with FRAXE mental retardation. *Nature Genetics, 13,* 105–108.

Goldsmith, H. H., Buss, K. A., & Lemery, K. S. (1997). Toddler and childhood temperament: Expanded content, stronger genetic evidence, new evidence for the importance of environment. *Developmental Psychology, 33,* 891–905.

Gordon, J. W., Scangos, G. A., Plotkin, D. J., Barbosa, J. A., & Ruddle, F. H. (1980). Genetic transformation of mouse embryos by microinjection of purified DNA. *Proceedings of the National Academy of Science, USA, 77,* 7380–7384.

Gosling, S. D. (2001). From mice to men: What can we learn about personality from animal research? *Psychological Bulletin, 127,* 45–86.

Gould, J. L. (1975). Honey bee recruitment: The dance-language controversy. *Science, 189,* 685–693.

Gould, T. D., & Gottesman, I. I. (2003). Psychiatric endophenotypes and the development of valid animal models. *Genes, Brain, and Behavior, 5,* 113–119.

Gu, Y., McIlwain, K. L., Weeber, E. J., Yamagata, T., Xu, B., Antalffy, B. A., et al. (2002). Impaired conditioned fear and enhanced long-term potentiation in Fmr2 knock-out mice. *Journal of Neuroscience, 22,* 2753–2763.

Haile, C. N., Kosten, T. R., & Kosten, T. A. (2007). Genetics of dopamine and its contribution to cocaine addiction. *Behavior Genetics, 37,* 119–145.

Hayes, L. J., & Delgado, D. (2006). If only they could talk: Genetic mouse models for psychiatric disorders. In G. S. Fisch & J. Flint (Eds.), *Transgenic and knockout models of neuropsychiatric disorders* (pp. 69–83). Totowa NJ: Humana Press.

Hayes, L. J., & Delgado, D. (2007). Invited commentary on animal models in psychiatry: Animal models of non-conventional human behavior. *Behavior Genetics, 37,* 11–17.

Hoekstra, R. A., Bartels, M., Verweij, C. J., & Boomsma, D. I. (2007). Heritability of autistic traits in the general population. *Archives of Pediatric and Adolescent Medicine, 161,* 372–377.

Holmes, A., & Cryan, J. F. (2006). Modeling human anxiety and depression in mutant mice. In G. S. Fisch & J. Flint (Eds.), *Transgenic and knockout models of neuropsychiatric disorders* (pp. 237–264). Totowa, NJ: Humana Press.

Holmes, A., Yang, R. J., Murphy, D. L., & Crawley, J. N. (2002). Evaluation of antidepressant-related behavioral responses in mice lacking the serotonin transporter. *Neuropsychopharmacology, 27,* 914–923.

Hudziak, J. J., Van Beijsterveldt, C. E., Althoff, R. R., Stanger, C., Rettew, D. C., Nelson, E. C., et al. (2004). Genetic and environmental contributions to the child behavior checklist obsessive-compulsive scale: A cross-cultural twin study. *Archives of General Psychiatry, 61,* 608–616.

Jamain, S., Quach, H., Betancur, C., Råstam, M., Colineaux, C., Gillberg, I. C., et al. (2003). Mutations of the X-linked genes encoding neuroligins NLGN3 and NLGN4 are associated with autism. *Nature Genetics, 34,* 27–29.

Jang, K. L., Livesley, W. J., & Vernon, P. A. (1996). Heritability of the big five personality dimensions and their facets: A twin study. *Journal of Personality, 64,* 577–591.

Jang, K. L., Woodward, T. S., Lang, D., Honer, W. G., & Livesley, W. J. (2005). The genetic and environmental basis of the relationship between schizotypy and personality: A twin study. *Journal of Nervous and Mental Disease, 193,* 153–159.

Karayiorgou, M., Morris, M. A., Morrow, B., Shprintzen, R. J., Goldberg, R., Borrow, J., et al. (1995). Schizophrenia susceptibility associated with interstitial deletions of chromosome 22q11. *Proceedings of the National Academy of Sciences in the United States of America, 92,* 7612–7616.

Kas, M. J., Fernandes, C., Schalkwyk, L. C., & Collier, D. A. (2007). Genetics of behavioural domains across the neuropsychiatric spectrum; of mice and men. *Molecular Psychiatry, 12,* 324–330.

Kemphorne, O. (1997). Heritability: Uses and abuses. *Genetica, 99,* 109–112.

Kendler, K. S., Gardner, C. O., Gatz, M., & Pedersen, N. L. (2007). The sources of co-morbidity between major depression and generalized anxiety disorder in a Swedish national twin sample. *Psychological Medicine, 37,* 453–462.

Kendler, K. S., Gatz, M., Gardner, C. O., & Pedersen, N. L. (2006). A Swedish national twin study of lifetime major depression. *American Journal of Psychiatry, 163,* 109–114.

Kovas, Y., & Plomin, R. (2006). Generalist genes: Implications for the cognitive sciences. *Trends in Cognitive Sciences, 10,* 198–203.

Lafollette, H., & Shanks, N. (1995). Two models of models in biomedical research. *Philosophical Quarterly, 45,* 141–160.

Lander, E. S., & Botstein, D. (1989). Mapping Mendelian factors underlying quantitative traits using RFLP linkage maps. *Genetics, 121,* 185–199.

Landgraf, R., & Wigger, A. (2002). High vs low anxiety-related behavior rats: an animal model of extremes in trait anxiety. *Behavior Genetics, 32,* 301–314.

LeDoux, J. E. (2000). Emotion circuits in the brain. *Annual Review of Neuroscience, 23,* 155–184.

Lewis, B. A., Shriberg, L. D., Freebairn, L. A., Hansen, A. J., Stein, C. M., Taylor, H. G., et al. (2006). The genetic bases of speech sound disorders: Evidence from spoken and written language. *Journal of Speech, Language, and Hearing Research, 49,* 1294–1312.

Liu, H., Heath, S. C., Sobin, C., Roos, J. L., Galke, B. L., Blundell, M. L., et al. (2002). Genetic variation at the 22q11 PRODH2/DGCR6 locus presents an unusual pattern and increases susceptibility to schizophrenia. *Proceedings of the National Academy of Sciences in the United States of America, 99*, 3717–3722.

Logue, S. F., Paylor, R., & Wehner, J. M. (1997). Hippocampal lesions cause learning deficits in inbred mice in the Morris water maze and conditioned-fear task. *Behavioral Neuroscience, 111*, 104–113.

Macphail, E. M. (1982). *Brain and intelligence in vertebrates.* Oxford: Clarendon Press.

Macphail, E. M. (1998). *The evolution of consciousness.* Oxford: Oxford University Press.

Mathiesen, K. S., & Tambs, K. (1999). The EAS temperament questionnaire – factor structure, age trends, reliability, and stability in a Norwegian sample. *Journal of Child Psychology and Psychiatry, 40*, 431–439.

Mazzocco, M. M. M., & Reiss, A. L. (1997). Normal variation in size of the FMR1 gene is not associated with intellectual performance. *Intelligence, 24*, 355–366.

McCrae, R. R., & Costa, P. T. Jr. (1987). Validation of the five-factor model of personality across instruments and observers. *Journal of Personality and Social Psychology, 52*, 81–90.

McKinney, W. T. Jr., & Bunney, W. E. Jr. (1969). Animal model of depression. I. Review of evidence: implications for research. *Archives of General Psychiatry, 21*, 240–248.

Metten, P., Buck, K. J., Merrill, C. M., Roberts, A. J., Yu, C. H., & Crabbe, J. C. (2007). Use of a novel mouse genotype to model acute benzodiazepine withdrawal. *Behavior Genetics, 37*, 160–170.

Millar, J. K., Wilson-Annan, J. C., Anderson, S., Christie, S., Taylor, M. S., Semple, C. A., et al. (2000). Disruption of two novel genes by a translocation co-segregating with schizophrenia. *Human Molecular Genetics, 9*, 1415–1423.

Mineur, Y. S., Huynh, L. X., & Crusio, W. E. (2006). Social behavior deficits in the Fmr1 mutant mouse. *Behavioural Brain Research, 168*, 172–175.

Miyakawa, T., Leiter, L. M., Gerber, D. J., Gainetdinov, R. R., Sotnikova, T. D., Zeng, H. et al. (2003). Conditional calcineurin knockout mice exhibit multiple abnormal behaviors related to schizophrenia. *Proceedings of the National Academy of Science, USA, 100*, 8987–8992.

Moon, J., Beaudin, A. E., Verosky, S., Driscoll, L. L., Weiskopf, M., Levitsky, D. A., et al. (2006). Attentional dysfunction, impulsivity, and resistance to change in a mouse model of fragile X syndrome. *Behavioral Neurosciences, 120*, 1367–1379.

Moretti, P., Levenson, J. M., Battaglia, F., Atkinson, R., Teague, R., Antalffy, B., et al. (2006). Learning and memory and synaptic plasticity are impaired in a mouse model of Rett syndrome. *Journal of Neuroscience, 26*, 319–327.

Morley, K. I., & Montgomery, G. W. (2001). The genetics of cognitive processes: Candidate genes in humans and animals. *Behavior Genetics, 31*, 511–531.

Moy, S. S., Nadler, J. J., Perez, A., Barbaro, R. P., Johns, J. M., Magnuson, T. R., et al. (2004). Sociability and preference for social novelty in five inbred strains: An approach to assess autistic-like behavior in mice. *Genes, Brain and Behavior, 3*, 287–302.

Moy, S. S., Nadler, J. J., Young, N. B., Perez, A., Holloway, L. P., Barbaro, R. P., et al. (2007). Mouse behavioral tasks relevant to autism: Phenotypes of 10 inbred strains. *Behavioural Brain Research, 176*, 4–20.

Nan, X., Ng, H. H., Johnson, C. A., Laherty, C. D., Turner, B. M., Eisenman, R. N., et al. (1998). Transcriptional repression by the methyl-CpG-binding protein MeCP2 involves a histone deacetylase complex. *Nature, 393*, 386–389.

Newport, D. J., Stowe, Z. N., & Nemeroff, C. B. (2002). Parental depression: Animal models of an adverse life event. *American Journal of Psychiatry, 159*, 1265–1283.

O'Tuathaigh, C. M., Babovic, D., O'Meara, G., Clifford, J. J., Croke, D. T., & Waddington, J. L. (2006). Susceptibility genes for schizophrenia: Characterisation of mutant mouse models at the level of phenotypic behaviour. *Neuroscience and Biobehavioral Reviews, 31*, 60–78.

Ørstavik, R. E., Kendler, K. S., Czajkowski, N., Tambs, K., and Reichborn-Kjennerud, T. (2007). Genetic and environmental contributions to depressive personality disorder in a population-based sample of Norwegian twins. *Journal of Affective Disorders, 99*, 181–189.

Paradee, W., Melikian, H. E., Rasmussen, D. L., Kenneson, A., Conn, P. J., & Warren S. T. (1999). Fragile X mouse: Strain effects of knockout phenotype and evidence suggesting deficient amygdala function. *Neuroscience, 94*, 185–192.

Paylor, R., Glaser, B., Mupo, A., Ataliotis, P., Spencer, C., Sobotka, A., et al. (2006). Tbx1 haploinsufficiency is linked to behavioral disorders in mice and humans: implications for 22q11 deletion syndrome. *Proceedings of the National Academy of Sciences in the United States of America, 103*, 7729–7734.

Plomin, R. (2003). Genetics, genes, genomics and g. *Molecular Psychiatry, 8*, 1–5.

Poirier, R., Jacquot, S., Vaillend, C., Soutthiphong, A. A., Libbey, M., Davis, S., et al. (2007). Deletion of the Coffin-Lowry syndrome gene Rsk2 in mice is associated with impaired spatial learning and reduced control of exploratory behavior. *Behavior Genetics, 37*, 31–50.

Ramboz, S., Oosting, R., Amara, D. A., Kung, H. F., Blier, P., Mendelsohn, M., et al. (1998). Serotonin receptor 1A knockout: An animal model of anxiety-related disorder. *Proceedings of the National Academy of Sciences in the United States of America, 95*, 14476–14481.

Réale, D., Reader, S. M., Sol, D., McDougall, P. T., & Dingemanse, N. J. (2007). Integrating animal temperament within ecology and evolution. *Biological Review of the Cambridge Philosophy Society, 82*, 291–318.

Risch, N., Spiker, D., Lotspeich, L., Nouri, N., Hinds, D., Hallmayer, J., et al. (1999) A genomic screen of autism: Evidence for a multilocus etiology. *American Journal of Human Genetics, 65*, 493–507.

Rojas, P., Joodmardi, E., Hong, Y., Perlmann, T., & Ogren, S.O. (2007). Adult mice with reduced Nurr1 expression: An animal model for schizophrenia. *Molecular Psychiatry, 12*, 756–766.

Sadakata, T., Washida, M., Iwayama, Y., Shoji, S., Sato, Y., Ohkura, T., et al. (2007). Autistic-like phenotypes in Cadps2-knockout mice and aberrant CADPS2 splicing in autistic patients. *The Journal of Clinical Investigation, 117*, 931–943.

Sara, S. J., Devauges, V., Biegon, A., & Blizard, D. A. (1994). The Maudsley rat strains as a probe to investigate noradrenergic-cholinergic interaction in cognitive function. *Journal of Physiology, 88*, 337–345.

Saudino, K. J. (2005). Behavioral genetics and child temperament. *Journal of Developmental and Behavioral Pediatrics, 26*, 214–223.

Saudino, K. J., McGuire, S., Reiss, D., Hetherington, E. M., & Plomin, R. (1995). Parent ratings of EAS temperaments in twins, full siblings, half siblings, and step siblings. *Journal of Personality and Social Psychology, 68*, 723–733.

Shahbazian, M., Young, J., Yuva-Paylor, L., Spencer, C., Antalffy, B., Noebels, J., et al. (2002). Mice with truncated MeCP2 recapitulate many Rett syndrome features and display hyperacetylation of histone H3. *Neuron, 35*, 243–254.

Shih, R. A., Belmonte, P. L., & Zandi, P. P. (2004). A review of the evidence from family, twin and adoption studies for a genetic contribution to adult psychiatric disorders. *International Review of Psychiatry, 16*, 260–283.

Silva, A. J., Frankland, P. W., Marowitz, Z., Friedman, E., Laszlo G. S., Cioffi, D., et al. (1997). A mouse model for the learning and memory deficits associated with neurofibromatosis type I. *Nature Genetics, 15*, 281–284.

Skinner, B. F. (1957). *Verbal behavior*. New York: Appleton-Century-Crofts.

Spearman, C. (1927). *The abilities of man, their nature and measurement*. London: MacMillan and Co.

Stefansson, H., Sarginson, J., Kong, A., Yates, P., Steinthorsdottir, V., Gudfinnsson, E., et al. (2003). Association of neuregulin 1 with schizophrenia confirmed in a Scottish population. *American Journal of Human Genetics, 72*, 83–87.

Sternberg, R. J., & Detterman, D. K. (1986). *What is intelligence? Contemporary viewpoints on its nature and definition*. Norwood, NJ: Ablex.

Straub, R. E., Jiang, Y., MacLean, C. J., Ma, Y., Webb, B. T., Myakishev, M. V., et al. (2002) Genetic variation in the 6p22.3 gene DTNBP1, the human ortholog of the mouse dysbindin gene, is associated with schizophrenia. *American Journal of Human Genetics, 71*, 337–348.

Stromswold, K. (2001). The heritability of language: a review and meta-analysis of twin, adoption and linkage studies. *Language, 77*, 647–723.

Stromswold, K. (2006). Why aren't identical twins linguistically identical? Genetic, prenatal and postnatal factors. *Cognition, 101*, 333–384.

Tabuchi, K., Blundell, J., Etherton, M. R., Hammer, R. E., Liu, X., Powell C. M., et al. (2007). A neuroligin-3 mutation implicated in autism increases inhibitory synaptic transmission in mice. *Science, 318*, 71–76.

Thomas, K. R., & Capecchi, M. R. (1987). Site-directed mutagenesis by gene targeting in mouse embryo-derived stem cells. *Cell, 51*, 503–512.

Thompson, R., Crinella, F. M., & Yu, J. (1990). *Brain mechanisms in problem solving and intelligence*. New York: Plenum Press.

Thorndike, E. L. (1898). Animal intelligence: An experimental study of the associative processes in animals. *The Psychological Review: Monograph Supplements*, 8.

Thorndike, E. L. (1921). Intelligence and its measurement: A symposium. *Journal of Educational Psychology, 12*, 124–127.

Tolman, E. C. (1924). The inheritance of maze-learning ability in rats. *Journal of Comparative Psychology, 4*, 1–18.

Tryon, R. C. (1940). Genetic differences in maze-learning ability in rats. *Yearbook for National Social Studies and Education, 39*, 111–119.

Tupes, E. C., & Christal, R. E. (1961). *Recurrent personality factors based on trait ratings*. USAF ASD Technical Report, No. 61–97.

Upchurch, M., & Wehner, J. M. (1988). Differences between inbred strains of mice in Morris water maze performance. *Behavior Genetics, 18*, 55–68.

Urani, A., Chourbaji, S., & Gass P. (2005). Mutant mouse models of depression: Candidate genes and current mouse lines. *Neuroscience & Biobehavioral Reviews, 29*, 805–828.

Wahlsten, D. (1999). Single-gene influences on brain and behavior. *Annual Review of Psychology, 50*, 599–624.

Wahlsten, D. (2001). Standardizing tests of mouse behavior: Reasons, recommendations, and reality. *Physiology & Behavior, 73*, 695–704.

Wahlsten, D., Bachmanov, A., Finn, D. A., & Crabbe, J. C. (2006). Stability of inbred mouse strain differences in behavior and brain size between laboratories and across decades. *Proceedings of the National Academy of Sciences in the United States of America, 103*, 16364–16369.

Wahlsten, D., Cooper S. F., & Crabbe, J. C. (2005). Different rankings of inbred mouse strains on the Morris maze and a refined 4-arm water escape task. *Behavioural Brain Research, 165*, 36–51.

Wahlsten, D., Rustay, N. R., Metten, P., & Crabbe, J. C. (2003). In search of a better mouse test. *Trends in Neuroscience, 26*, 132–136.

Wehner, J. M., Radcliffe, R. A., & Bowers, B. J. (2001). Quantitative genetics and mouse behavior. *Annual Review in Neuroscience, 24*, 845–867.

Weller, A., Leguisamo, A. C., Towns, L., Ramboz, S., Bagiella, E., Hofer, M., et al. (2003). Maternal effects in infant and adult phenotypes of 5HT1A and 5HT1B receptor knockout mice. *Developmental Psychobiology, 42*, 194–205.

Welzl, H., D'Adamo, P., Wolfer, D. P., & Lipp, H.-P. (2006). Mouse models of hereditary mental retardation. In G. S. Fisch, & J. Flint (Eds.), *Transgenic and knockout models of neuropsychiatric disorders* (pp. 101–126). Totowa NJ: Humana Press.

White, S. A., Fisher, S. E., Geschwind, D. H., Scharff, C., & Holy, T. E. (2006). Singing mice, songbirds, and more: Models for FOXP2 function and dysfunction in human speech and language. *Journal of Neuroscience, 11*, 10376–10379.

Whorf, B. (1940). *Science and linguistics*. reprinted in *Language, thought & reality*. Cambridge, MA: MIT Press.

Widiger, T. A., & Trull, T. J. (2007). Plate tectonics in the classification of personality disorder: Shifting to a dimensional model. *American Psychologist, 62*, 71–83.

Williams, H. J., Owen, M. J., & O'Donovan, M. C. (2007). Is COMT a susceptibility gene for schizophrenia? *Schizophrenia Bulletin, 33*, 635–641.

Willis-Owen, S. A., & Flint, J. (2007). Identifying the genetic determinants of emotionality in humans; insights from rodents. *Neuroscience and Biobehavioural Reviews, 31*, 115–124.

Wolfer, D. P., Crusio, W. E., & Lipp, H. P. (2002). Knockout mice: Simple solutions to the problems of genetic background and flanking genes. *Trends in Neuroscience, 25*, 336–340.

Wolfer, D. P., & Lipp, H. P. (2000). Dissecting the behaviour of transgenic mice: Is it the mutation, the genetic background, or the environment? *Experimental Physiology, 85*, 627–634.

Wolfer, D. P., Litvin, O., Morf, S., Nitsch, R. M., Lipp, H. P., & Würbel, H. (2004). Laboratory animal welfare: Cage enrichment and mouse behaviour. *Nature, 432*, 821–822.

Yacoubi, M., & Vaugeois, J. M. (2007). Genetic rodent models of depression. *Current Opinion in Pharmacology, 7*, 3–7.

Yamagata, S., Suzuki, A., Ando, J., Ono, Y., Kijima, N., Yoshimura, K., et al. (2006). Is the genetic structure of human personality universal? A cross-cultural twin study from North America, Europe, and Asia. *Journal of Personality and Social Psychology, 90*, 987–998.

Chapter 6

Twin Studies of General Mental Ability

Nancy L. Segal and Wendy Johnson

Introduction

Twin studies are a vital source of information about genetic and environmental influences on general mental ability. The classic twin design—comparison of the relative similarity between monozygotic (MZ) and dizygotic (DZ) twins—is a simple and elegant approach to estimating the effects of genes and experience on developmental traits. However, while this method was considered state of the art in behavioral genetics in the 1960s and 1970s, it is now only one of many more sensitive and sophisticated twin designs. Twin research on behavioral and medical traits, in general, and on intelligence, in particular, has advanced at an impressive rate.

The focus of the present chapter is on what twin studies have thus far contributed to our understanding of individual differences in intelligence. The chapter begins by briefly summarizing key events and controversies that have marked the ontogeny of twin research on intellectual development. Subsequent sections examine new twin research designs, analytic methods, findings, and their implications. Topics include recent evidence from studies of twin-singleton differences, twins reared apart and together, virtual twins, longitudinal analyses, prenatal environments, parenting practices, shared environments, epigenetic processes (e.g., DNA methylation), the heritability of relevant endophenotypes, associations between genetic variance and socioeconomic status (SES), and the search for specific genes underlying intelligence. Links between twin studies and other research areas, both within and outside behavioral genetics, are explored.

Key Events and Controversies: A Brief Summary

Twin studies began with Sir Francis Galton's (1875) paper, "The history of twins, as a criterion of the relative powers of nature and nurture." This monograph set forth the simple, but elegant logic that lends twins their vast research potential: "It is, that their history affords means of distinguishing between the effects of tendencies received at birth, and of those that were imposed by the circumstances of their after lives; in other words, between the effects of nature and nurture" (p. 391). At that time the biological bases of twinning had not been elaborated, but Galton correctly surmised that there were two types of twins: those who shared all their heredity (identical twins) and those who shared some of their heredity (fraternal twins). He concluded that greater resemblance between the former, compared with the latter, demonstrated genetic influence on the physical and behavioral traits in question. Galton's paper presented qualitative comparisons between twins, based on material gathered from correspondence sent to him by twins and family members. As such, it departed in significant ways from current quantitative analyses conducted with systematically recruited twin pairs.

The biological bases of twinning were not revealed until the early 1900s. Weinberg (1901) developed the formula for estimating the frequency of one-egg and two-egg twins. Newman and Patterson (1910) discovered how identical quadruplets are produced by armadillos, establishing that identical twinning occurs in mammals. (Mammalian twinning is rare, probably due to the reduced genetic variability among multiple offspring.) The natural identical twinning rate in humans is about 1/250; see Segal, 2000a. Arey (1922) offered the terms monozygotic (MZ) and dizygotic (DZ) as labels for one-egg (identical) and two-egg twins (fraternal), respectively. Further developments in the biology of twinning have revealed numerous subclasses of both MZ and DZ twins, based on placentation and other factors (see Machin & Keith, 1999 for a comprehensive review). Organizing twin samples according to placental structure and other features has been informative with

N.L. Segal (✉)
Department of Psychology, California State University, Fullerton, CA 92834, USA
e-mail: nsegal@fullerton.edu

Y.-K. Kim (ed.), *Handbook of Behavior Genetics*,
DOI 10.1007/978-0-387-76727-7_6, © Springer Science+Business Media, LLC 2009

respect to some traits, including intelligence, as will be demonstrated.

An early twin study by Thorndike (1905) classified twins according to age, rather than twin type, showing that cognitive resemblance did not differ across younger and older pairs. It was not until Merriman's (1924) investigation that the first modern twin study of intelligence appeared. [Note that the first classical MZ–DZ twin comparison was Jablonski's (1992) study of refractive error.] Merriman found greater IQ similarity in MZ than DZ twin children, demonstrating genetic effects. Since then, numerous studies using the classic twin design, as well as variations of that design, have produced evidence consistent with Merriman's report. However, despite the agreement across studies and the credibility generally accorded them, twin research assumptions and findings have been attacked, as well as embraced, over the years.

Some of the charges against twin studies have also been raised against family and adoption studies. For example, evidence of genetic effects on behavior (regardless of the source) has been rejected by those who mistakenly equate genetic effects with biological determinism. The results of twin studies have never indicated biological determinism. All behavioral phenotypes are products of both genes and environments. Genes do not operate in isolation, but are expressed within (or transact with) environments, at both prenatal and/or postnatal levels.

Gene–environment interactions (G × E) refer more specifically to the different expression of different genes in a given environment. For example, a high-IQ child might excel in a classroom rich with learning opportunities, whereas an average-IQ child might be less inspired. G × E also refers to the different expression of a certain genotype depending upon environmental events. For example, a mathematically gifted child will probably display his or her quantitative skills if provided with an appropriate curriculum. However, this same child may not perform to the same degree if he or she is not sufficiently challenged. An excellent design for demonstrating such effects is co-twins control. This involves systematically exposing MZ co-twins to different experiences and assessing the outcome. For example, the effects of extra training in verbal skill could be examined by providing one twin, but not the other, with supplementary classes. Alternatively, different training programs could be administered to each co-twin. Interactions among genes at different loci also affect behavioral phenotypes.

Research shows that genetic effects underlie most measured behaviors and predispositions that are sensitive to environmental influence; however, the extent to which behaviors may be modified by environments and by experience is trait specific. Environmental influences on all behaviors are evident by the fact that no twin study has ever reported a perfect MZ twin correlation for any measured trait. Measurement error and variable gene expression also account, in part, for MZ intraclass correlations of less than 1.00.

Other charges have been specific to twin studies. Critics have questioned the applicability of the equal environments assumption (EEA), the fundamental principle underlying the twin design. The idea here is that environmental influences on specific traits must be the same for MZ and DZ twins if findings are to be valid and generalizable. Some people have argued that MZ twins are treated more alike than DZ twins, thus violating this assumption. This environmental challenge has been evaluated by behavioral genetic investigators and has been found wanting (Bouchard & McGue, 1993; LaBuda, Svikis, & Pickens, 1997). Specifically, it has been found that twins who are treated more alike do not show greater behavioral resemblance than those treated less alike. For example, twins who are dressed alike do not resemble one another in personality more than twins who are dressed differently. It is important to note that if MZ twins are treated more alike than DZ twins, it is most likely associated with their genetically based behavioral similarities. Interestingly, parents who are mistaken about their twins' zygosity tend to treat them or rate them in accordance with their true zygosity (see Segal, 2000a).

Critics have also questioned whether twins' unique prenatal circumstances (e.g., shared intrauterine environment, premature delivery) render twin study findings inapplicable to non-twin populations. Christensen, Vaupel, Holm, and Yashin (1995) reported that after 6 years of age, disease incidence and mortality are comparable for twins and singletons, a finding confirmed for most other behavioral and physical measures.

Twin studies' reputation has additionally suffered from serious misuse of the methodology by some individuals. Dr. Josef Mengele's horrific medical experiments using twins, dwarfs, and individuals with genetic defects, conducted in the Auschwitz concentration camp between 1943 and 1945, are exemplary (see Segal, 1985a, 2005a). Dr. Viola Bernard's intentional separation of adopted infant twins and Dr. Peter Neubauer's (Neubauer & Neubauer, 1990) longitudinal study that took unfair advantage of these twins and their families also hurt the ability of other researchers to make constructive use of twin research (see Perlman, 2005; Segal, 2005b). Another controversy concerned the truthfulness of the reared apart twin IQ data gathered by Cyril Burt, although that situation was ultimately resolved in his favor (Fletcher, 1991; Joynson, 1989). Scientific sources no longer cite Burt's studies, but since his results were consistent with current findings their omission does not affect interpretations or conclusions concerning influences on general intelligence.

Twin research has survived these challenges as evidenced by increased applications of this approach across many behavioral, medical, and social science fields. Its recovery was due, in part, to advances in genetic research and growing disillusionment with environmental explanations of human behavior and development (Vandenberg, 1969). The molecular structure of DNA was identified in 1953, enhancing understanding of the transmission and expression of genetic factors. The genetic underpinning of Down's syndrome (trisomy 21) and the metabolic mechanism associated with the mental retardation caused by phenylketonuria (PKU) drew attention to gene-behavior relationships. Social explanations of abnormal behavior became less satisfying, thus renewing attention to biological components of mental disorder.

This altered research climate was conducive to some landmark studies of general intelligence. Erlenmeyer-Kimling and Jarvik (1963) surveyed the twin and adoption literature and concluded that genes substantially influence mental ability. Since then, more extensive updated analyses have been completed by Bouchard and McGue (1981, 1993), with the same results. In 1988, Snyderman and Rothman showed that the majority of behavioral science researchers endorsed genetic influence on intelligence. Today, some researchers are searching for links between specific genes and mental ability and disability, and some promising leads have been found. For example, Haarla, Butcher, Meaburn, Sham, Craig, and Plomin (2005) found associations (albeit, modest) between DNA markers and general cognitive ability in 7-year-old children.

Behavioral genetics, of which twin research is a critical component, entered the mainstream of psychological research in the 1980s and has stayed there. Much of what is currently known about the bases and progression of general intelligence, special mental abilities, Alzheimer's disease, and associations between earnings and education comes from twin-based analyses. The most important recent advances in twin research include elaboration of twin research designs, greater availability of population-based twin registries, increased sophistication of analytic methods, and new insights on epigenetic processes.

Research Designs and Findings: Using Twins to Find Genetic and Environmental Influences on General Intelligence

Twin-Singleton Differences

The question of possible intellectual differences between twins and singletons is important. That is because it only makes sense to think of twin-based estimates of genetic and environmental effects as applicable to the general population if twins can be considered typical of that population for the trait in question. The older psychological literature

includes a number of studies showing a five- to ten-point IQ disadvantage for twins, relative to non-twins (Bouchard & Segal, 1985; Segal, 2000a). Explanations for this difference refer mostly to twins' lower average birth weight (due to premature delivery) and close social relationship that could restrict their range of learning opportunities. However, twin studies have revealed only modest birth weight-IQ correlations. Interestingly, MZ twins show greater birth weight differences than DZ twins, but this pattern tends to reverse itself by 3 months of age (Wilson, 1986). Wilson (1979) also found that co-twins in ten MZ pairs differing in birth weight by over one and three-quarter pounds did not show pronounced IQ differences at 6 years of age, although they did maintain their size difference. He suggested that "a high degree of buffering for the nervous system against the effects of malnutrition" in viable fetuses may protect against early insult (p. 217). It has also been found that fetuses with modest nutritional deficits may show accelerated development of their lungs and brain (Amiel-Tison & Gluck, 1995), possibly explaining the low birth weight-IQ correlations from twin studies.

Low birth weight in singletons has been linked to later cognitive difficulties (Caravale, Tozzi, Albino, & Vicari, 2005; Davis, Burns, Wilkerson, & Steichen, 2005), but low birth weight in twins may not predict comparable developmental delays in otherwise healthy twins. It is possible that the birth weight difference as a percentage of total weight may be a more meaningful factor in individual pairs. A number of recent studies have explored associations between birth weight and intelligence in twins. Boomsma, van Beijsterveldt, Rietveld, Bartels, and van Baal (2001) found that genetic factors mediate the link between birth weight and IQ in twins until 10 years of age. An association between intrapair differences in birth weight and IQ was detected for DZ twins (who differ genetically), but not for MZ twins. Luciano, Wright, and Martin (2004), in a study of 16-year-old twins, found that genetic variance in birth weight overlapped with genetic variance in verbal IQ, but not with non-verbal or overall IQ. It was suggested that verbal IQ may serve as a proxy for parents' education or intelligence and that higher-IQ mothers may provide more favorable intrauterine environments for their children than lower-IQ mothers.

It is important to note that most studies comparing the intellectual levels of twins and singletons have focused on the IQ scores of young children, comparing them to those of unrelated non-twins. This approach fails to control for biological and experiential family background measures.

The more recent literature presents a more varied picture of twins' intellectual abilities. Posthuma, De Geus, Bleichrodt, and Boomsma (2000) compared the adult IQ scores of MZ and DZ twins with those of their singleton siblings, circumventing some problematic features of earlier studies. No IQ difference between the two groups was found in this study, suggesting that twin studies provide informative estimates of IQ heritability. The twins were part of the Netherlands National Twin Registry and were recruited as adults through City Councils and notices in newsletters. Twins in the study had taken an IQ test as part of a previous study of adult brain function, minimizing effects of self-selection.

Findings contradicting those of Posthuma et al. were recently reported by Scottish investigators. Ronalds, De Stavola, and Leon (2005) also compared IQ scores for twins and their non-twin siblings, born between 1950 and 1956; thus, twins in this sample were younger than those used by the Dutch investigators. In this study twins scored 5.3 and 6.0 points below their singleton siblings at ages 7 and 9, respectively. Adjusting for sex, mother's age, and number of older siblings did not affect the data; however, adjusting for birth weight and gestational age reduced the IQ differences to 2.6 and 4.1 points at 7 and 9 of age, respectively. (See also Deary, Pattie, Wilson, & Whalley, 2005.)

A problematic feature of the Scottish study was failure to differentiate MZ and DZ twins. Given that MZ twins are more likely to be exposed to adverse prenatal factors than DZ twins, it is possible that combining MZ and DZ pairs actually decreased twin-singleton differences. It is also possible that better medical care than was available to twins born in the 1950s would reduce, or eliminate, twin-singleton differences among more recently born twins. However, this cannot be the full explanation because MZ and DZ twins in the Dutch study (who did not differ from their siblings in IQ) were 39.7 and 37.3 years of age, respectively. As such, some twins were born in the 1960s when medical technology was less effective than it is today. Consistent with the Dutch findings are those from a more recent Danish study by Christensen, Petersen, Herskind, and Bingley (2006) that did not detect twin-singleton differences in general intelligence, using school children from a nation-wide population register.

The varied results from the recent twin-singleton studies call for additional twin-singleton comparisons of IQ, using representative samples of twin children and adults. In the area of language development, however, twins' average deficits relative to non-twins have been well documented (Segal, 2000a).

A number of young twins display language delays that are explained mostly by postnatal family influences (e.g., patterns of parent–child communication and interaction), rather than by birth and delivery factors (Rutte, Thorpe, Greenwood, Northstone, & Golding, 2003; Thorpe, Rutter, & Greenwood, 2003; also see Segal, 2000a). Research shows that parents of twins direct less speech to each child and are more controlling in their verbal interactions, relative to mothers of non-twins (Tomasello, Mannle, & Kruger, 1986). Thorpe et al. (2001) have described two language features: *private language* (communication used exclusively within pairs, but which is unintelligible to others) and *shared verbal*

understanding (communication used both within pairs and with others, but which is unintelligible to others). Shared verbal understanding was observed among 50 and 19.7% of twins at 20 and 36 months of age, respectively, and among and 2.5 and 1.3% of non-twins. Private language was observed among 11.8 and 6.6% of twins at 20 and 36 months of age, respectively, and among 2.5 and 1.3% of non-twin pairs. Children showing these speech characteristics scored lower on most cognitive ability measures than those who did not, especially children showing private language at age 36 months. Shared verbal understanding is, however, considered a not uncommon developmental feature in twins and in near-in-age siblings. Children showing such language delays usually recover by age 3 years. However, language delays could partially explain the lower average IQ scores observed among some young twin samples.

The recent dramatic increase in the twinning rate should facilitate additional cognitive comparisons between twins and non-twins. Twins currently occur in approximately 1 in 30 births, as compared with 1 in 50 to 1 in 60 births in 1980 (Center for Disease Control, 2003). Reasons for this increase are important with respect to analyses of general intelligence in twins and non-twins.

The rise in twinning is due mostly to the greater availability of artificial reproductive technologies (ART) that enable infertile couples to conceive, although delayed childbearing (associated with DZ twinning) explains part of the trend. Most twins conceived via ART are DZ, although a minority of MZ twins is thought to result from splitting of the embryo due to its micromanipulation outside the womb (Hecht & Magoon, 1998). Studies of artificially conceived non-twin infants have indicated early delivery, low birth weight, and developmental delays (Stromberg et al., 2002). This raises the possibility that using ART twins in behavioral genetic studies might bias estimates of genetic and environmental influence on measured traits. Studies have yielded mixed findings in this regard.

Some investigators have reported no differences in birth and health outcomes between naturally and artificially conceived twins (Helmerhorst, Perquin, Donker, & Keirse, 2004; Tully, Moffit, & Caspi, 2003). However, other studies have found an increased risk of birth defects (Kuwata et al., 2004) and lower birth weights and reduced co-twin resemblance in birth weight and problem behaviors among artificially conceived twins, relative to naturally conceived twins (Goody et al., 2005). Continued comparison of these twin groups will be important in studies exploring biological and experiential factors affecting mental ability. Clearly, relationships among prenatal factors, health status, and complex behaviors such as general intelligence are not straightforward (Segal, in press, 2009).

Twins and singletons experience different biological and social situations affecting their development. Recall, however, that once children reach the age of 6 years there do not appear to be meaningful differences between twins and singletons that would prevent generalizability of twin studies findings to non-twins (Christensen et al., 1995).

Twin-Family Designs

Families constructed from MZ twins, their spouses, and children yield a range of genetically and environmentally informative relationships. Twin aunts and uncles are genetically equivalent to genetic mothers and fathers, and first cousins are genetically equivalent to half-siblings. This design has been used to study a variety of traits, such as birth weight (Magnus, 1984), non-verbal intelligence (Rose, Harris, Christian, & Nance, 1979), schizophrenia predisposition (Gottesman & Bertelsen, 1989) conduct disorder (Haber, Jacob, & Heath, 2005), and social closeness (Segal, Seghers, Marelich, Mechanic, & Castillo, 2007). Rose (1979) found evidence of genetic effects on non-verbal ability, given the higher parent–child and twin parent–niece/nephew correlations, relative to spouse uncle/aunt–niece/nephew, and spouse–spouse correlations. Maternal effects were not present.

Twins Reared Apart and Together

Studies of the rare sets of MZ twins reared apart (MZA) yield direct estimates of genetic influence on measured traits. If co-twins experience little or no social contact until reunion in adulthood and are raised in uncorrelated environments, their similarity is associated with their shared genes. Bouchard (2005) has asserted that "With monozygotic twins reared apart, correlation is, in fact, an estimate of causation, and the magnitude of the correlation tells you the effect of the genes" (p. 7). DZ twins reared apart (DZA) constitute an important control group, providing opportunities to assess interactions between genotypes and behavior (Segal, 2005c). Unfortunately DZA pairs were not included in studies prior to the 1970s.

Separated twin studies have been conducted, or are underway, in the United States (Bouchard, Lykken, McGue, Segal, & Tellegen, 1990; Newman, Freeman, and Holzinger, 1937), Great Britain (Shields, 1962), Denmark (Juel-Nielsen, 1965), Japan (Hayakawa, Shimizu, Kato, Onoi, & Kobayashi, 2002), Sweden (Pedersen, McClearn, Plomin, & Nesselroade, 1992), and Finland (Kervinen, Kaprio, Koskenvuo, Juntunen, & Kesaniemi, 1998; also see Segal, 2003). Most studies are comprehensive, including a wide array of behavioral and physical measures (Segal, 2000a); however, the present discussion focuses on analyses of general intelligence.

A remarkable level of consistency has been demonstrated in the magnitude of the intraclass correlations for general intelligence reported across studies, with MZA correlations for primary tests ranging from 0.68 to 0.78 (Bouchard et al., 1990). This is especially impressive given that reared apart twin studies span multiple age groups, protocols, countries, and cultures. Organizing the studies according to time of publication yields mean IQ correlations of 0.72 ("old data," $n = 65$; three small reared apart twin studies conducted between 1937 and 1966) and 0.78 ("new data, $n = 93$; two relatively larger reared apart twin studies conducted in 1990 in Minnesota, and in 1992 in Sweden); see Plomin, DeFries, McClearn, and McGuffin (2001). These findings indicate that heritable factors explain 72–78% of the variance in general intelligence. This value exceeds the 50% heritability based upon young twin, sibling, and parent–child pairs. This may be due to the fact that IQ heritability increases with age (discussed below), and MZA twins are studied mostly as adults.

MZA twin similarity in any trait is best viewed against the extant data for other genetically and environmentally informative kinships. These data, displayed in Fig. 6.1, show a trend toward increasing IQ similarity with increasing genetic relatedness. Several features in the graph deserve attention. First, MZ twins reared together (MZT) show slightly greater resemblance (0.86) than MZAs (0.78). One explanation is that growing up together may enhance IQ similarity between MZ co-twins, albeit slightly. However, recall that most studies of reared together twins include young pairs living at home, when the shared environment exerts its greatest effect on development. In contrast, MZA data are typically gathered when twins are reunited as adults. Note that DZ twins reared together (DZT) also show slightly greater resemblance (0.60) than DZA twin pairs (0.52; see Pedersen, McClearn, Plomin, & Friberg, 1985), possibly for the same reasons. Second, both DZA and DZT pairs show greater resemblance than ordinary full siblings (0.47) even though all pairs share half their genes, on average, by descent. This could conceivably reflect shared age (DZAs and DZTs) and/or shared environmental factors (DZTs). A third intriguing effect shown in Fig. 6.1 is the reduced IQ correlation for full siblings reared apart (0.24). It is impossible to rule out differences in rearing as explanatory. Another possibility is that the reared apart siblings include a number of half-siblings, due to multiple paternities.

Explaining MZA twin similarity in intelligence has taken several routes. Taylor (1980) asserted that four classes of environmental similarities (age at separation, age at reunion, rearing by relatives and social environments) explained IQ resemblance among reunited twin pairs in the three earliest studies. Subsequent to Taylor's publication, Bouchard (1983) failed to constructively replicate his findings using the alternate IQ measure in each study. (The investigators had obtained more than one measure of general mental ability.) More recently, examining characteristics of MZA twins' rearing families has failed to yield meaningful associations that would challenge genetic interpretations of mental ability. Specifically, Bouchard et al. (1990) and Johnson et al. (2006) found negligible correlations between twins' IQ similarity and similarity in childhood physical facilities in the home, social status indicators, and parenting practices. Frequency of contact between MZ twins prior to assessment was also unrelated to their IQ similarity.

It would, however, be incorrect to claim that MZA twin studies eliminate a role for experiential influences on intellectual development. Newman, Freeman, and Holzinger (1937) reported correlations of 0.79 and 0.55 between co-twin differences in educational measures and Binet IQ and Otis IQ scores, respectively, and correlations of 0.51 and 0.53 between co-twin differences in social environments and IQ scores. Note that these are within-pair measures so they indicate that educational and social factors can affect individuals' intellectual development. At the same time, the reared apart twins' IQ correlations were 0.67 (Binet IQ) and 0.73 (Otis IQ). These are between-pair measures so they demonstrate genetic effects.

It would seem impossible to explain the intellectual similarities between MZA twins, relative to unrelated siblings reared together, without reference to genetic factors. It is likely that twins' genetically based predispositions partly explain their tendencies to seek similar opportunities and experiences in their separate environments, illustrative of active gene–environment correlation. It is also likely that parents and significant caretakers respond to individual twins' preferences and abilities by providing them with meaningful and relevant opportunities, a process termed reactive gene–environment correlation. Unfortunately, the clearest picture of such processes would be available from prospective longitudinal studies of reared apart twins, research that is not practically and ethically feasible.

IQ findings summarized in Fig. 6.1 also include data on a novel, relatively unstudied kinship called virtual twins, described below.

Virtual Twins

Virtual twins (VT) are same-age unrelated children, reared together since infancy (Segal, 1997, 2000a; Segal & Hershberger, 2005). They fall into two classes: adopted–adopted pairs and adopted–biological pairs. These unique sibships mimic the situations of MZ and DZ twins, but without the genetic link; of course, VTs are somewhat more akin to DZ twins who do not look physically alike. Another way to think about VTs is that they represent the reverse of MZ twins reared apart because the former share environments, not genes, while the latter share genes, not environments. VTs offer researchers a valuable "twin-like" design for studying the extent to which shared environments influence gen-

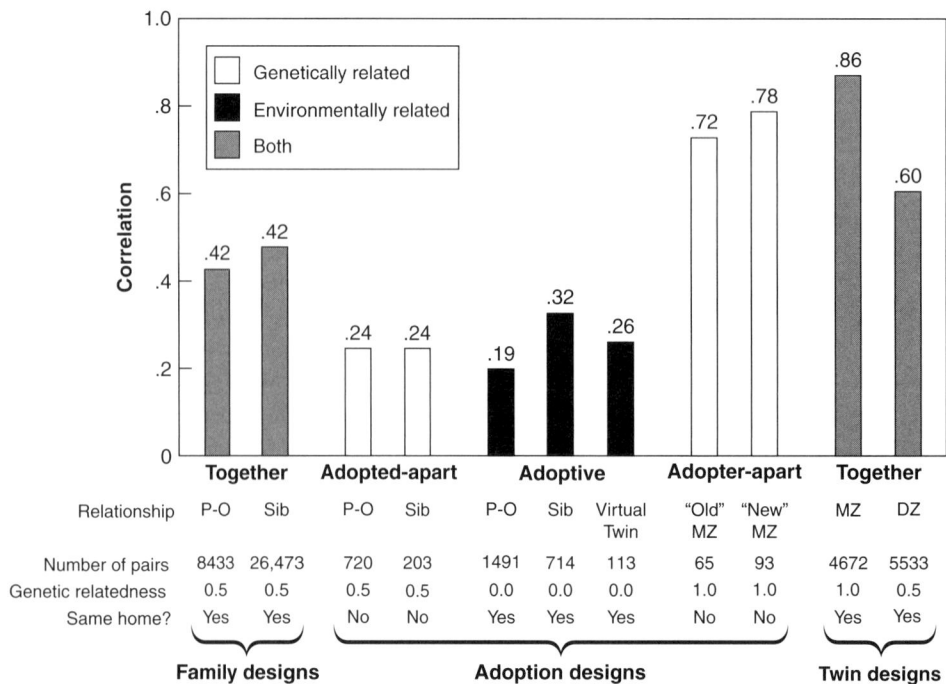

Relationship	P-O	Sib	P-O	Sib	P-O	Sib	Virtual Twin	"Old" MZ	"New" MZ	MZ	DZ
	Together		**Adopted-apart**		**Adoptive**			**Adopter-apart**		**Together**	
Number of pairs	8433	26,473	720	203	1491	714	113	65	93	4672	5533
Genetic relatedness	0.5	0.5	0.5	0.5	0.0	0.0	0.0	1.0	1.0	1.0	0.5
Same home?	Yes	Yes	No	No	Yes	Yes	Yes	No	No	Yes	Yes
	Family designs				**Adoption designs**					**Twin designs**	

Fig. 6.1 IQ correlations for family members varying in genetic and environmental relatedness. Adapted from Plomin et al. (2001). Virtual twin data are from Segal & Hershberger (2005)

eral intelligence and other phenotypes (Segal, 2000a, 2000b; Segal & Allison, 2002). Specifically, they circumvent some problematic features of ordinary adoptive siblings who differ in age, time of entry into the family, and often in placement history. It is possible, for example, that resources in the home might differ for two children adopted several years apart, benefiting one child but not the other.

A recent report of IQ resemblance included 113 VT pairs, with a mean age of 8.10 years (SD = 8.56) and age range of 5–54 years. (About 70% of the pairs were younger than 7 years of age. The remaining 30% included twin children and adolescents, and seven pairs aged 22 years and older.) The mean age difference between the pairs was 3.10 months (SD = 2.80) and ranged between 0 and 9.20 months. Intraclass correlations were 0.26 for full-scale IQ score, 0.23 for verbal IQ score, and 0.21 for performance IQ score. The profile correlation across IQ subtests was 0.07. Thus, shared family environment is associated with modest intellectual similarity among family members during childhood.

These results are best appraised against the backdrop of findings for MZ and DZ twin pairs. IQ correlations for MZ ($n = 4,672$) and DZ twin pairs ($n = 5,533$), averaged across a number of studies, were 0.86 and 0.60, respectively (see Fig. 6.1). A study of 7- to 13-year-old twins (comparable in age to the VTS) reported IQ correlations of 0.85 for MZ twin pairs ($n = 69$) and 0.45 for DZ twin pairs ($n = 35$ pairs), and profile correlations (across subtests) of 0.45 for MZ twin pairs and 0.24 for DZ twin pairs (Segal, 1985b).

It is noteworthy that the VT IQ correlation (0.26) is nearly identical to the weighted average adopted sibling correlation of 0.25 (McGue, Bouchard, Iacono, & Lykken, 1993). It could be argued that efforts should be directed toward studying ordinary adoptive siblings, rather than the rare VT pairs. However, given the controversies still surrounding genetic explanations of intelligence, it is important to control for environmental features that could potentially affect ability to gather the most environmentally informative pairs possible and to identify novel kinships for replication of findings.

Additional IQ analyses are possible using VTs because of the availability of both biological and adoptive children. Biological children of the parents of virtual twin pairs scored significantly higher in full-scale IQ, verbal IQ, and performance IQ, relative to the adoptive children. This finding is consistent with other adoption studies (see, for example, Cardon, 1994; Dumaret & Stewart, 1985), although the bases for this difference are uncertain. The majority of parents in the VT study held professional or managerial positions, so it is possible that their biological children inherited predispositions for advanced intellectual skills. In contrast, the adoptive children may have come from more varied biological family backgrounds. Of course, relationships between adoption and later behavioral development are complex, as shown by a recent meta-analysis by Van Ijzendoorn, Juffer, and Poel-

huis (2005). It was found that adopted children's IQ scores exceeded those of their non-adopted (biological) siblings and their peers raised by the birth family or placed in institutional care. Their school performance was also better. Second, adopted children's IQ scores did not differ from those of their non-adopted siblings (the children with whom they were raised) or those of their current peers; however, they performed less well at school, showed poorer language skills and required more special education referrals. As the investigators concluded, adoption has positive effects on intelligence, but the varied effects of early deprivation, emotional correlates of adoption, and other factors may affect intellectual performance and progress.

IQ assessments continue to be conducted for new VT pairs. In addition, pairs who have already participated are being reassessed. This longitudinal component to the project is part of the TAPS (Twins, Adoptees, Peers, and Siblings) Study, a collaboration between investigators at California State University Fullerton and the University of San Francisco. A new analysis of the IQ similarity of 43 young VT pairs, retested 1.70–8.96 years after their initial assessment, is now available (Segal, McGuire, Havlena, Gill, & Hershberger, 2007). A decrease in IQ resemblance was observed, suggesting increased genetic and/or non-shared environmental effects and decreased shared environmental effects on general intelligence during childhood.

It will be especially interesting to examine VT similarity as siblings approach adolescence, given that previous adoption studies have reported correlations of 0.30 for adopted siblings at age 8, but near zero correlations for adopted siblings in the teenage years (Loehlin, Horn, & Willerman, 1989). Further longitudinal analyses of VTs may shed light on gene–environment interaction effects on IQ if the IQ scores of adoptive children in adoptive–biological pairs begin to approach those of their non-adopted siblings.

Longitudinal and Life Span Twin Studies

Behavioral geneticists interested in developmental issues segued into developmental behavioral geneticists (although some developmental psychologists and others still resist genetic perspectives on behavior; see Pinker, 2002). Questions of the extent to which genes and environments accounted for continuity and change in intelligence, personality, and physical features were addressed via longitudinal twin studies. Combining these studies with data on the twins' singleton siblings added additional informative features.

Probably the most widely cited longitudinal twin analysis of intellectual development is Wilson's (1983) tracking of MZ and DZ twins' intellectual progress from 3 months to 15 years of age. Little difference in the magnitudes of the

MZ and DZ correlations was apparent at 6 months of age, after which MZ twin pairs showed correlations about 0.10 higher than those of DZ twins during the period between 9 and 36 months. MZ correlations increased steadily from 0.67 (9 months) to 0.88 (36 months), while DZ correlations increased from 0.51 (9 months) to 0.73 (24 months), decreased to 0.65 (30 months), and increased again to 0.79 (36 months). Then, the MZ twin correlations remained stable, with a mean correlation of 0.82 for ages 8–15 years. In contrast, the DZ twin correlations declined except for a slight rise at the age 6 years, yielding a mean correlation of 0.50 from 8 to 15 years of age. Thus, heritability increased from early childhood (0.16) to adolescence (0.64).

Another remarkable outcome from Wilson's (1983) study was that the twin correlations exceeded the age-to-age continuity, meaning that twin A's score at a particular age better predicted twin B's score than a previous score of twin B. Furthermore, plots of individual MZ and DZ pairs depicted coordinated and discrepant patterns of "spurts" and "lags," respectively, demonstrating genetic influence on intellectual growth patterns. Finally, the twin-sibling correlations confirmed the findings for the DZ twins, i.e., sibling correlations increased from 0.38 at age 3 years to 0.55 at age 7 years, then declined slightly to 0.50 at age 15 years.

A number of longitudinal twin studies have followed Wilson's classic work, so only selected examples will be presented. Spinath, Ronald, Harlaar, Price, and Plomin (2003) showed that genetic influence increased from early childhood (20–30%) to middle childhood (40%), and again with the approach of adolescence (50%). It was also shown that shared environmental influence on general intelligence declined to near zero across this portion of the life span. These findings have been generally supported by other longitudinal data from twins, as well as from biological siblings and adoptees, gathered from 1 to 12 years of age by Bishop et al. (2003). Two exceptional findings from that study were that (1) non-shared environmental factors were associated with both IQ stability and change in middle childhood and (2) genetic factors were only associated with IQ stability at adolescence. However, Dutch investigators Rietveld, Dolan, van Baal, and Boomsma (2003) found that IQ stability across ages 5, 7, and 10 years was mostly explained by genetic factors and that non-shared environment contributed only to variance that was age specific. Genetically influenced transition times in intellectual development have also been identified via longitudinal twin studies. Fulker, Cherny, and Cardon (1993) reported developmental changes during childhood. Specifically, genetic influence on cognition seemed to stabilize by age 4, with new variation appearing at age 7.

Some studies have restricted IQ analyses to twins at the older end of the life span. These studies are striking in that they also show increasing IQ heritability across the life span. A study of 80-year-old Swedish twins (McClearn

et al., 1997) yielded an IQ heritability of 0.60. Evidence of increasing IQ heritability also comes from cross-sectional analyses, in which the MZ–DZ difference in correlations widens after adolescence, into adulthood (McGue et al., 1993). As in Wilson's (1983) study, the increasing MZ–DZ similarity difference is explained by the growing discordance between DZ co-twins, relative to the stable concordance between MZ co-twins.

Reasons for increasing genetic influence on intelligence across the life span have been considered. It seems likely that two sources of influence are operative. First, small genetic effects present in childhood may become more important over time, leading to larger behavioral effects. Second, shared environmental influence declines and genetic influence increases as individuals become more active in seeking opportunities for learning and for self-expression (Plomin et al., 2001), especially with the end of required schooling. This would be an example of gene–environment correlation—the concept that certain genotypes are selectively found in certain environments. There is, however, some evidence that heritability declines at the very oldest ages. An IQ study of Swedish twins aged 50 years and older indicated a heritability of 0.80 (Pedersen et al., 1992), a figure that fell to 0.60 for twins at age 80. This suggests that environmental factors, including physical health, may gain some importance toward the end of the life span. With this in mind, a recent study examined sources of influence on rate of cognitive change, using two groups of older twins (65 years and younger; older than 65 years) from the Swedish study (Reynolds, Finkel, Gatz, & Pedersen, 2002). Genetic influences were associated with individual differences in ability, whereas environmental factors were more closely tied to rate of change. Trends toward increased longevity will facilitate further efforts along these lines.

Environmental Influences on General Intelligence

Two main classes of environmental influence on general intelligence are of interest with respect to twin studies: prenatal environments and parenting practices. This is because such effects could bias estimates of heritability if they have a meaningful impact on twins' intellectual development. Other potential sources of environmental influence on intelligence (e.g., quality and years of schooling; parental socioeconomic status) have been addressed in the psychological literature and are beyond the scope of the present chapter. However, some comments along these lines will be included at the end of this section.

Prenatal Environments: Truths and Consequences

The behavioral implications of twins' shared intrauterine environment are often misunderstood. A common assumption is that sharing a womb enhances twins' phenotypic similarity because the fetuses are equally affected by the mother's diet, health, medications, and other factors. However, the unique effects of the prenatal environment tend to make twins *less* alike, not more alike, especially in the case of MZ twins. Furthermore, twins' prenatal situation cannot be considered vis-à-vis measured traits without reference to twin type (MZ or DZ) and the presence in MZ twins of separate or shared placentas and fetal membranes. Devlin, Daniels, and Roeder (1997) overlooked these distinctions, incorrectly concluding that twins' shared prenatal environments contribute to their IQ similarity. (It was found that 20% of the covariance between twins and 3% of the covariance between siblings was explained by shared prenatal factors.) Thus, this analysis produced lower estimates of genetic effects than most other studies.

In fact, MZ twins come in several varieties: separate amnions, chorions and placentae or separate amnions, and chorions with a fused placenta (about one-third of MZ twin pairs; zygotic division occurs before the second post-conceptional day); single chorion, separate amnions, and single placenta (about two-thirds of MZ twin pairs; zygotic division occurs between the second and eighth post-conceptional day); single amnion, chorion, and placenta (zygotic division occurs between the eighth and thirteenth post-conceptional day; about one-twentieth of MZ twin pairs). DZ twins come in two varieties: separate amnions, chorions and placentae; separate amnions and chorions and fused placenta (Endres & Wilkins, 2005; Machin & Keith, 1999). Early studies indicated that 43 and 42% of dichorial MZ and DZ twins, respectively, had fused placentas (Bulmer, 1970), but the variable criteria for fusion makes such estimates problematic (Bryan, 1983). The different placental arrangements are depicted in Fig. 6.2.

Several studies have compared cognitive resemblance between MZ co-twins, organized according to placentation. A study using 4- to 6-year-old MZ twins did not detect differences across six mental abilities between one-chorion and two-chorion pairs (Sokol et al., 1995). The investigators did, however, find that one-chorion twins were more alike in some personality measures, such as social competence and self-control. More recently, Jacobs et al. (2003) found greater similarity among one-chorion twins than two-chorion twins for two Wechsler subtests (arithmetic and vocabulary), but not for total IQ score. Other studies have either found the same pattern of difference on selected cognitive measures (Spitz et al., 1996) or have found no difference (Reed, Carmelli, & Rosenman, 1991).

The puzzle that emerges from the foregoing is that greater resemblance between two-chorion twins, not one-chorion twins, would be expected. This is because one-chorion twins are generally at greater physical risk, due to complications from shared fetal circulation. It is possible that one-chorion twins who survive and volunteer for research represent a remarkably healthy subgroup of such sets (Segal, 2000a), but this remains speculative. Perhaps there are, as yet, unidentified features associated with delayed zygotic splitting conducive to co-twins' matched phenotypic development in some domains. One candidate with respect to female twins would be X-inactivation patterns, for which late-splitting twins show greater concordance; this would lead to greater co-twin resemblance in X-linked traits (Trejo et al., 1994). The important point is that placentation should be considered in twin-based analyses of human behavior. Further efforts along these lines may shed light on the mechanisms underlying the differential resemblance between MZ twin types, thus refining estimates of heritability.

Parenting and Twin Studies: Effects on Intelligence

Examining parenting practices can help refine heritability estimates by revealing whether treatment of twins is causal or reactive in nature. As indicated above, the view that similar treatment of MZ twins produces similar behavioral outcomes has caused some individuals to question results from twin research. Of course, the key question is not whether MZ twins receive more similar treatment than DZ twins, but whether any more similar treatment they receive enhances their phenotypic resemblance (Rowe, 1994).

A substantial number of studies have variously assessed the similarity of parenting effects, physical similarity, and childhood experiences on twins' similarity in personality (Borkenau, Riemann, Angleitner, & Spinath, 2002), eating behaviors (Klump, Holly, Iacono, McGue, & Wilson, 2000), behavioral problems (Cronk et al., 2002; Morris-Yates, Andrews, Howie, & Henderson, 1990), and psychiatric illness (Kendler, Neale, Kessler, Heath, & Eaves, 1994). These studies have found that environmental similarity was largely unrelated to twins' similarity in the measured traits, thus affirming the equal environments assumption. Fewer studies have assessed the effects of rearing on intelligence, but those that have done so have drawn the same conclusion.

In their landmark study of 850 twin sets, Loehlin, & Nichols (1976) found negligible correlations between twin differences in NMSQT measures and differential experience measures. The same pattern held for interests. Recall that the twins reared apart data (reviewed above) also indicated little effect of family background variables on twins' IQ scores.

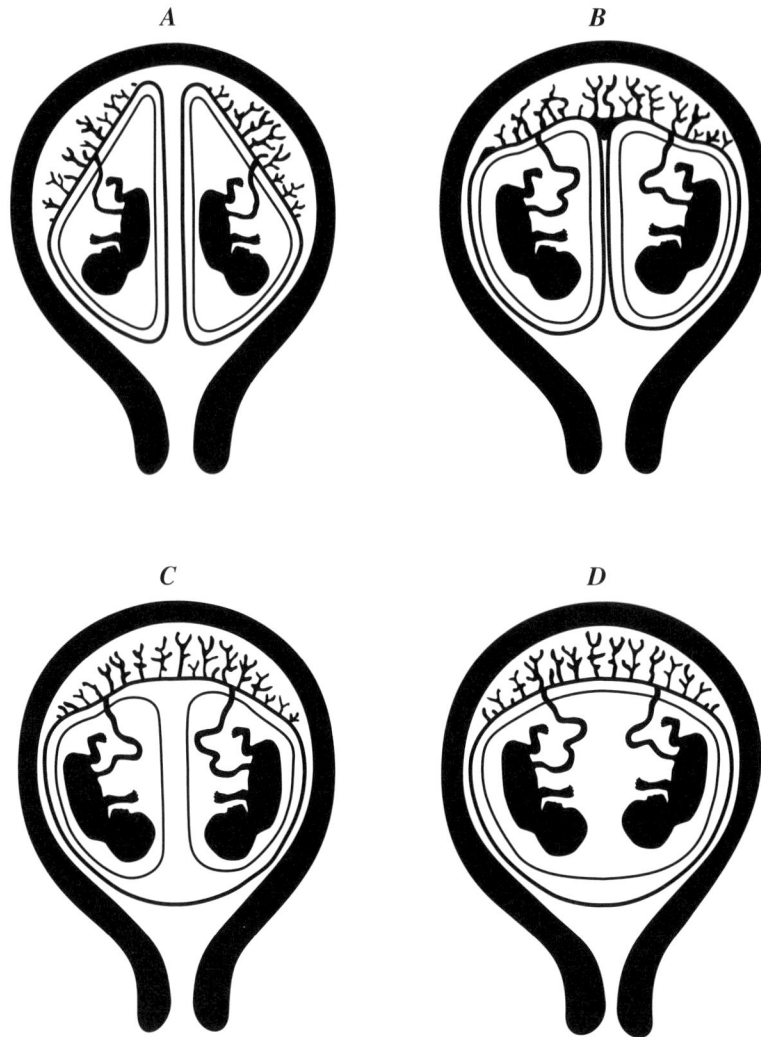

Fig. 6.2 Placental arrangements for MZ and DZ twin pairs. **A**: MZ or DZ twins with separate chorions, amnions and placentae. **B**: MZ or DZ twins with separate chorions and amnions and fused placentae. **C**: MZ twins with a shared chorion and placenta, but separate amnions. **D**: MZ twins with a shared chorion, amnion and placenta. Adapted from Potter (1948) and Stern (1960); see Segal (2000a)

Another informative series of analyses has compared parents' judgments of twins' behaviors when parents were correct and incorrect about twin type. The majority of such studies have found parental ratings to be consistent with true twin type, rather than with assumed twin type (Scarr, 1969; Goodman and Stevenson, 1991). Goodman and Stevenson (1991) did, however, find that the majority of parental warmth and criticism was unrelated to the child's behavior. Such studies have not been conducted with reference to cognitive skills, but there is little reason to suspect that they would deviate from the patterns found for behaviors in other domains.

In concluding this section, a study by Turkheimer, Haley, Waldron, D'Onofrio, and Gottesman (2003) is worth noting. It was found that heritability estimates of general intelligence were close to zero among 7-year-old twin children from impoverished families and that 60% of the IQ vari-

ance was associated with shared environmental factors. In contrast, the reverse was true for twin children from affluent families. As the authors indicated, it would be inappropriate to claim that behavioral differences among children from poor environments are more closely tied to their environments than are outcome differences among children from favorable environments. This is because the genetic influences that varied with socioeconomic status were only those that were independent of socioeconomic status. It is possible, if not likely, that a substantial portion of the genetic influences on general intelligence are common to genetic influences on socioeconomic status (SES). If so, these genetic influences were not measured in this study at all; also see Plomin et al. (2001).

Such common genetic influences as those referenced above would occur, for example, if parents who attain high levels of education and income do so because they have high intelligence and, therefore, pass these genes to their children along with their high SES environment. Samples of reared-apart twins and/or samples allowing for the measurement of intergenerational transmission and the separate measurement of SES, or its effects in co-twins, will be necessary to estimate all genetic and environmental associations involved. This is an important direction for future research.

Endophenotypes: What Can They Tell Us?

General mental ability is typically assessed by evaluating composite performance on a number of mental tasks requiring diverse knowledge, skills, and reasoning (Jensen, 1998). General mental ability is, therefore, a highly abstract concept. It reflects not only the complex behaviors involved in solving cognitive ability problems that we can see, but also the variance common to a variety of these behaviors. Thus, the gap between the gene products and the environments in which they are formed, and the "hidden" genetic and environmental influences on general mental ability we measure through most twin and related studies is large. In order to understand how genes and environments transact to create the mental ability performances we observe as behavior, we need to understand the biological processes lying within this gap. The concept of the endophenotype is potentially useful in this regard. Twin studies can contribute importantly to identifying endophenotypes and investigating their roles in the development of general mental ability.

The term endophenotype was adapted by Gottesman and Shields (1973) from evolutionary theory involving insect biology to describe the "internal phenotypes discoverable by a biochemical test or microscopic examination" involved in schizophrenia (Gottesman and Gould, 2003, p. 637). The term refers to biological mechanisms thought to be closer to the immediate products of genes and, thus, under stronger and perhaps less polygenic genetic influence than are the manifest behaviors they undergird. For example, the inability to synthesize phenylalanine would be considered an endophenotype for PKU (phenylketonuria)-induced mental retardation. The idea that there are neurological and biochemical bases of general mental ability and other psychological features did not originate with Gottesman. However, the term endophenotype allows for clearer characterization of some of the roles played by neurological factors in psychological manifestations.

Gottesman intended the concept of endophenotypes to be specific to genetic influences resulting from DNA sequence variations. He did this because the specific purpose of identifying endophenotypes is to assist in the search for particular genes involved in a behavior. Once accomplished, this can further understanding of how those genes transact with the environment to result in the biological processes involved in the phenotype.

Gottesman and Gould (2003) specified five criteria for the designation of endophenotypes. These criteria were, however, developed to apply to schizophrenia, a behavioral pattern that can be considered an overt disorder and for which irregularities in biological processes have been identified as endophenotypes. In contrast, general cognitive ability is a clearly overtly continuous trait for which the concept of a threshold of disorder is less clearly applicable. Endophenotypes of cognitive ability are also more likely to be continuous. This renders two of Gottesman and Gould's criteria irrelevant. The remaining relevant criteria are (1) the endophenotype is associated with general mental ability in the population; (2) the endophenotype is heritable; and (3) the endophenotype and general mental ability are related within families, as well as throughout the population.[1] This is equivalent to saying that there are common genetic influences on the endophenotype and on general mental ability.

Identification of possible endophenotypes related to general mental ability has proceeded along the lines earmarked by the first two criteria. First, studies have sought to relate brain structure and function to general mental ability. Twin studies are of limited value here, but we review this area because of its obvious importance to the overall process of identifying endophenotypes for general mental ability. Early studies in this area were based on patients with brain damage, and patient–control studies still provide important data. However, more recent studies have focused on assessments of normally functioning groups of participants. These studies have been sufficiently successful such that studies of brain structures and malfunctions involved in disease now routinely correct for estimated mental ability, prior to inception of the disease state (Gray & Thompson, 2004). Magnetic resonance imaging (MRI) studies have revealed structural associations. There are substantial correlations between general mental ability and total brain volume and total volumes of both gray and white brain matters, as well as volumes of gray and white matter in specific brain areas, particularly those in the frontal and parietal lobes involved in language (Haier, Jung, Yeo, Head, & Alkire, 2004; McDaniel, 2005). More provocatively, one study (Pennington et al., 2000) found that the volumes of 13 brain regions were substantially intercorrelated, with a general factor accounting for 48% of the variance. This suggests a general structural factor linked to the general mental ability factor.

In addition, functional studies of neural activity using positron emission tomography (PET) and functional mag-

---

[1] The other two criteria were that the endophenotype is present regardless of whether or not the disorder is currently present, and the endophenotype is present at a higher rate in the unaffected relatives of people with the disorder than in the general population.

netic resonance imaging (fMRI) have revealed areas of the brain involved in intelligent performance. For example, Duncan et al. (2000) found that three tasks (Spatial, Verbal, and Circles) requiring different kinds of mental ability were associated with greater neural activity in several brain regions, but only one area (the lateral prefrontal cortex) was activated during all tasks. This finding of a central location for general mental ability, however, contradicts the findings of several fMRI studies (Esposito, Kirby, Van Horn, Ellmore, & Berman, 1999; Gray, Chabris, & Braver, 2003; Prabhakaran, Smith, Desmond, Glover, & Gabrielli, 1997) that have reported widespread activity throughout the brain during cognitive tasks. The contradiction may reflect the different technologies used and may be more apparent than real with respect to hypotheses regarding unitary general versus multiple intelligences. This is because the functional units of higher cognition may include networks of brain areas as well as single areas (Gray & Thompson, 2004).

Functional studies have also investigated individual differences in the magnitudes of various aspects of brain activity and their associations with general mental ability. Electroencephalograms and event-related potentials have been used to demonstrate an association between the speed and the reliability of neural transmission and general mental ability (Deary, 2000; Neisser et al., 1996). Positron emission tomography (PET) studies indicate that general mental ability is negatively associated with glucose metabolism during mental activity (Haier et al., 1992), suggesting that general mental ability is associated with some level of neural efficiency. In addition, general mental ability appears to be mediated by neural mechanisms that support executive control of attention (Gray, Chabris, & Braver, 2003). That is, controlling for overall performance and for performance on less difficult trials in the same brain region, the association between general mental ability (as measured by Ravens Advanced Progressive Matrices) and task performance on more difficult attention trials can be explained by activity in the lateral prefrontal and parietal regions associated with attentional control. In contrast, the lateral prefrontal region does not show activity correlated with general mental ability while watching videotapes, an activity placing little demand on general mental ability (Haier, White, & Alkire, 2003). This has important implications because any area of the brain involved with mental cognitive ability should show activity when engaged in intelligent performance, but should not when not so engaged.

The second approach related to the identification of endophenotypes has involved assessing the extent to which brain structure is heritable. This is the area in which twin studies are particularly valuable. Assessment of the heritabilty of brain structure is accomplished through the use of a standardized brain atlas template and an algorithm that classifies images of tissue as gray matter, white matter, cerebral spinal fluid, and non-brain material and maps them onto the atlas (Toga & Thompson, 2005). The MRI scans used have sufficient resolution to track individual differences in cortical gray and white matter, thus making estimates of heritability possible. In such studies, total brain volume has shown very high heritability, on the order of 0.90 (Bartley, Jones, & Weinberger, 1997; Tramo et al., 1998). Volumes of gray and white matter in individual brain regions also show much greater similarity in MZ twins than in DZ twins, who in turn show much greater similarity than randomly paired individuals (Baaré et al., 2001; Thompson et al., 2001). The heritability estimates from these studies generally lack precision due to very small sample sizes, but they also are on the order of 0.90. The volumes of some individual brain structures, including the corpus collosum, the ventricles, and the temporal horns adjacent to the hippocampus are heritable as well, though to a lesser degree, about 0.60 (Oppenheim, Skerry, Tramo, & Gazzaniga, 1989; Pfefferbaum, Sullivan, Swan, & Carmelli, 2000). The volume of the hippocampus itself appears to be somewhat less heritable (Sullivan, Pfefferbaum, Swan, & Carmelli, 2001), about 0.40, and gyral patterns are much less heritable (Bartley, Jones, & Weinberger, 1997).

This discussion of work to date shows that we have identified total brain size, regional brain size, and level of brain activity as potential endophenotypes of general mental ability. Of course, the key to identifying these features as actual endophenotypes is that they share common genetic influences with general mental ability. This has been investigated in only one study. Importantly, however, it was a twin study, and this is another area in which twin studies will prove to be important in future research. Posthuma et al. (2002) found that the observed correlations between general mental ability and volumes of gray and white brain matter in a sample of Dutch twins and their siblings were due completely to genetic influences. Thus, volumes of gray and white matter can be considered endophenotypes for general cognitive ability. This means that, as we identify genes controlling the development of gray and white brain matter and come to understand the processes involved in their expression, we should build directly on our understanding of the biological development of general mental ability, as well. Given that the correlations between brain matter volumes and general mental ability are on the order of 0.3, however, we should expect that we will need to identify other endophenotypes and perhaps environmentally based biological processes, as well, before we fully understand the development and manifestation of general mental ability.

Genes for General Mental Ability: Can We Find Them?

The fact that general mental ability is heritable means that its variation in the population arises, at least in part, from variations in DNA. These variations, called polymorphisms, are passed from parents to their offspring. However, two randomly selected individuals will show differences in only 0.1–0.2% of the nucleotides in their genome (Thompson et al., 2001). Many of these differences likely have little relevance to general mental ability. In addition, relatively few actually occur in protein coding regions. This emphasizes an important point in the search for specific genes involved in general mental ability: the brain is a complex organ whose function is absolutely essential to the organism throughout its life. Together, this complexity and vital importance imply the existence of considerable redundancy of genetic control of biological function. This is because they suggest the existence of multiple genetic mechanisms through which any essential brain functions can be maintained.

Evidence for this kind of redundancy of function is routinely provided by the regenerative and compensatory capacities shown by people suffering brain damage due to stroke, injury, or disease (Beatty, 1995). It is also routinely provided by non-human studies involving animals in which particular genes have been removed or rendered inactive. Such animals, nevertheless, effectively function normally despite absence of the inactivated gene. This kind of redundancy clearly complicates the search for specific polymorphisms associated with general mental ability. This is because it suggests that polymorphisms associated with high or low function in one family may not even be present in another family, due to the different possible pathways to high or low function.

At the same time, building a brain is a general process, as the actual and potential endophenotypes that have been identified to date make clear. To the extent that quantitative characteristics (e.g., volumes of total, gray, or white brain matter) or specific kinds of brain activity are associated with general mental ability, there are unlikely to be specific polymorphisms that contribute directly to variation in general mental ability. This is because everyone's brain contains both gray and white matter, and everyone's brain shows activity during task performance. Thus, genes that contribute directly to the formation of brain matter or the elicitation of activity will be unlikely to segregate among individuals. Instead, phenotypic variations from individual to individual may result from differences in expression by these genes. These differences, in turn, could be due to differences in other genes that regulate their expression, to environmental influences on gene expression, or both. Much of the work involved in investigating these kinds of genetic processes is carried out in experimental animals in which genetic background can be closely controlled. In humans, twin studies provide the closest scientific and ethical alternatives.

For example, evidence for differential environmental effects on genetic expression in rats is provided by Weaver et al. (2004), who described long-term differences in offsprings' hypothalamic–pituitary–adrenal response to stress.

These differences appeared to result from differences in DNA methylation of a glucocorticoid receptor gene promoter in the hippocampus, brought on by differences in maternal licking, grooming, and nursing practices. The differences appeared to contribute to parental behavior by the offspring as well, leading to differences in stress response that were transmitted from generation to generation. It is unknown whether the particular genetic and environmental mechanisms involved in this process in rats have directly analogous mechanisms in humans. It is highly likely, however, that this kind of process takes place in humans, contributing to differences in genetic expression that cannot be observed through examination of DNA samples. The fact that genetic expression also appears to be subject to genetic influence (York et al., 2005) complicates this picture further, as it suggests another means by which genetic influences on general mental ability may be rather indirect.

It is also possible that some environmental stresses may elicit expression of genetic variation that has lain dormant in the population. This has been demonstrated in *Drosophila melanogaster* through the generation of phenocopies. Phenocopies are phenotypes that closely match known genetic mutations, usually aberrant. They are, however, generated in genetically wild-type animals by delivering particular stresses during specific developmental periods (Rutherford, 2000). For example, at 21–23 hours of pupal development, 4 hours of heat treatment disrupts the posterior cross-veins in a small percentage of flies (Waddington, 1957). This deleterious phenotype is very similar to the effects of mutations in the cv gene. Heat stress-induced phenocopies do not result from mutations, however, but rather from the interactions of several to many other genes with the stressful environment.

Waddington (1953) demonstrated the heritability of this effect in a series of well-known experiments by crossing the specific animals affected by the heat treatment and subjecting their progeny to it again. In response to this selection, the proportion of affected animals produced each generation increased until nearly all the animals receiving the treatment showed the effect. At this point, some fraction of the control flies from the same selection lines expressed the effect even without receiving the treatment. Thus, previously silent genetic variation revealed by environmental stresses can be selected to the point at which the stress-induced phenotype is reliably expressed even in the absence of the stress. Again, it is highly likely that similar processes take place in humans, generating situations in which general mental ability is possibly associated with certain genes in some individuals or groups, but not in others. Some evidence that this might be the case has been provided by Turkheimer et al. (2003), who found, as noted above, that, in young children, genetic variability in IQ independent of SES increased with socioeconomic status.

In humans, MZ twins do provide information about another mechanism that complicates the search for genes for multigenic traits such as general mental ability, in an interesting twist on the classical twin study method that focuses on twin similarity. Though MZ twins share a common genetic background, which includes genetic influence on gene expression, significant variation in gene expression remains. The extent of this variation increases with age (Fraga et al., 2005), suggesting environmental influences. It tends to be found in genes involving signaling and communication or immune and related functions (Sharma et al., 2005), implicating the involvement of general mental ability due to the general, brain-wide nature of its known endophenotypes.

Comparing the similarity of MZ twins reared together and apart across multiple traits suggests, however, that post-natal environmental experiences are not the only sources of these epigenetic differences. This is because MZ twins tend to be similar to the same degree in many traits regardless of whether they are reared together or apart (Wong, Gottesman, & Petronis, 2005). This "similarity of similarity" may be due to gene–environment correlation, or the tendency for genetically similar individuals to seek out similar environments and experiences. It may also be due to the tendency for additional resemblance due to shared rearing environments to be offset by social differentiation, or the intentional selection of different experiences, by members of twin pairs growing up together (Segal, in press).

In addition, in laboratory animals for which even prenatal environments can be tightly controlled, such epigenetic differences persist in genetically identical organiams, i.e., clones (Gartner, 1990). At the same time, Gartner and Baunack (1981) carried out an interesting experiment using genetically identical cloned mouse pairs, controlling the environment to the same degree. The mouse pairs included MZ and DZ sets, created by transplanting divided and nondivided eight-cell embryos into pseudo-pregnant surrogates. Specifically, MZ pairs were created by artificial separation of embryos, while DZ pairs were created from multiple zygotes produced by inbred mice. Thus, each pair was genetically identical and reared in the same environment. There was more phenotypic variation within the DZ pairs than within the MZ pairs. The investigators termed this variance the "third component" after genes and environment, but its molecular basis remains unknown. It is clear, however, that there is more to epigenetic effects than simply the accumulation of different experiences over a lifetime.

Still another potential complication arises from the operation of standard kinetics in multi-step steady-state systems, illustrated nicely with an example from human and mouse blood pressure, recounted by Smithies (2005). The renin–angiotensin system is generally acknowledged to be one of the most important means through which blood pressure is genetically controlled. In this system, renin (produced in the kidney) acts on angiotensinogen (AGT, produced in the liver) to generate angiotensin I. This is converted by the enzyme angiotensin-converting-enzyme ACE to angiotensin II, which acts through several different receptors to increase blood pressure. Using gene-titration to increase the numbers of copies of the genes for AGT and ACE in mice artificially, Kim et al. (1995) showed that increasing expression of the relevant gene increased the concentration of AGT in the blood and also increased the blood pressure of the mice. This technique also increased the concentration of ACE in the blood, but it had no effect on the blood pressure of the mice (Krege et al., 1997) as ACE effectively acted only as a gatekeeper in the conversion process from angiotensin I to angiotensin II. The point here is that the effects that variations in genetic expression will have on the outcome phenotype depend on the roles in the underlying biochemical processes, played by each specific gene product. Without complete understanding of these roles we may find it difficult to detect the associated genes. This, of course, is where endophenotypes can be helpful, because they suggest candidate genes. In contrast to this, however, the phenylketonuria (PKU) gene that causes a severe form of mental retardation (unless phenylalanine is removed from the childhood diet) was identified in the 1930s and cloned in 1983 (Woo, Lidsky, Guttler, Chandra, & Robson, 1983), yet we still do not understand how the mutated gene damages brain function.

Any consideration of genetic influences on general mental ability has to account for their place in evolution. Two issues are important here. First, general mental ability seems to be part of a very general system of biological processes. This can be seen by studying monogenic cases of mental retardation caused by deleterious mutations. Some 282 monogenic disorders have been reported to involve mental ability (Inlow & Restifo, 2004). Most, however, were originally identified in relation to their associations with other medical conditions. Furthermore, the ranges of mental ability demonstrated by affected individuals are wide, because they reflect the damaging effects of the mutations, but leave the rest of the normal distribution of ability intact. The genes involve general biological processes such as metabolic and signaling pathways, transcription, and aspects of neuronal and glial biology in highly pleiotropic ways (Inlow & Restifo, 2004).

In spite of the fact that such monogenic disorders cause only a small proportion of cases of mental retardation (and only a small proportion of cases of physical illness or disability, as well), they make clear the degree to which general mental ability is integrated into the broader system of biological processes involved in the overall integrity of the organism. This suggests that the effects of more common genes, though less severe, may be similarly pervasive. As such, the complex mix of genes influencing medical conditions such as hypertension, cardiovascular disease, diabetes, asthma, and multiple sclerosis may influence general mental ability, as well. At the same time, genes influencing overall health indices, such as immune response and physical growth, may confer benefits for general cognitive ability, as well. Thus, regardless of the direction of their effects, many genes associated with mental ability may be very general in their effects. Those that are positive may have been subject to natural selection, while those that are negative may survive simply because their individual effects are rather small.

Second, and in contrast (though brain function is clearly strongly general), it is obvious that mental abilities can take specific forms such as musical, artistic, or computational talents. Miller (2000) has hypothesized that both general and specific mental abilities have evolved because they have been subject to sexual selection, meaning that they confer advantages to their holders in mate competition. This suggests that some genes involved in mental ability may have very specific, largely ornamental effects that are difficult to identify, because we have yet to develop tests that accurately measure the presence of those abilities. In addition, some genes that confer cognitive advantages in the heterozygote may have deleterious consequences in the homozygote. One example of this has been suggested by Cochran, Hardy, and Harpending (2006), who note that several genes involved in DNA repair, including BRCA1 (a gene associated with breast and colon cancer and other autoimmune disorders), have deleterious mutations that have reached polymorphic frequency in Ashkenazi Jews, who also average higher IQ's than any other human population. They suggest that these mutations have survived in these populations, despite their damaging consequences, because they benefit general mental ability.

This discussion of complications is not intended to discourage the search for specific genes involved in general mental ability. Most studies involving such searches today, however, make use of genetic linkage, association, and scanning techniques, based on the assumption that there is a one-to-one correspondence between genotype and phenotype. Findings from these studies have been difficult to replicate. The complications discussed here may provide some explanations for these failures. At the same time, the existence of these kinds of phenomena underline the richness and intimacy of transactions between genes and environments we are likely to discover as we continue searching for specific genes involved in general mental ability. This reminds us that even when genes associated with general mental ability have been identified, use of genetic modification techniques may have unintended consequences, both positive and negative.

Conclusions and Future Directions

The preponderance of evidence from classic twin studies, and studies using variant twin designs, is consistent with genetic influence on general intelligence. Given past controversies surrounding this conclusion, it is anticipated that debates over the extent of genetic influence will continue. Capitalizing further on naturally occurring "human experiments," such as twins reared apart and virtual twins, will be a vital part of future research in this area.

It is expected that twins will continue to be used in creative ways in the future, to further address issues and questions concerning the development of general intelligence and its correlates. As an example, Australian investigators used twins to assess the heritability of inspection time (IT) and its covariance with IQ (Luciano et al., 2001). Results yielded a shared genetic factor influencing IT and IQ, separate from the total genetic variance, a finding informative with respect to how individuals process information. Other investigators have looked at co-morbidity between verbal and non-verbal delays in 2-year-old twin children (Purcell et al., 2001). A major finding was that co-morbidity between the two is largely genetic in origin, whereas differences are largely environmental. It was suggested that such efforts can lead to improved diagnostic systems, based on genetic factors rather than observed symptoms. Australian researchers found common genetic influence on a standard test of academic achievement and IQ (particularly verbal IQ) in a study of 256 MZ and 326 DZ twin pairs (Wainwright, Wright, Geffen, Luciano, & Martin, 2005). Clearly, twins bring added perspective to research on general intelligence, given the genetic and family background controls.

Cross-cultural analyses of intelligence using twins would offer insights into the impact of cultural effects on the heritability of general mental ability. A Russian longitudinal twin study reported a decrease in genetic effects as children transitioned from 6 to 7 years of age (Malykh, Zyrianova, & Kuravsky, 2003). At age 7 shared environmental effects had increased substantially. This line of inquiry would profit substantially by studying the rare group of MZ twins reared separately in different cultures. An ongoing prospective study of young Chinese twins, adopted together and apart, will also shed light on genetic and environmental effects on intellectual development (Segal, Chavarria, & Stohs, 2008).

Identifying genes associated with intelligence may also move ahead as a result of twin studies. Recently, Harlaar et al. (2005) found five DNA markers associated with general cognitive ability in a longitudinal study of 7,414 twin pairs, assessed at ages 2, 3, 4, and 7 tears. Of course, effects sizes of the five markers were quite small.

The search for mechanisms to explain how, and why, MZ twins show remarkable similarities in their intellectual development, despite differences in rearing and differences in some brain characteristics, raises intriguing contradictions. Research on the cerebral development of MZ twins reveals both striking similarities and differences. A study of twins, using magnetic resonance imaging (MRI), showed a 94% heritability of the corpus callosum (CC) midsaggital size (Scamvougeras, Kigar, Jones, Weinberger, & Witelson, 2003). It was suggested that correlates of CC size, such as lateralization patterns, cognitive skills, and neuropsychological functions, could be associated with genetic factors affecting CC morphology. In contrast, a recent study showed little MZ twin resemblance in the shape of the planum temporale, a brain structure possibly linked to language (Steinmetz, Herzog, Schlaug, Huang, & Jancke, 1995). Furthermore, development of neural structures proceeds according to dynamic processes, some of which may be stochastic, such that MZ twins may conceivably show "functionally significant variant wiring" (Edelman, 1987, p. 323). Resolution of such disparate findings can enhance understanding of intellectual development in the general population. It is clear that the natural experiment provided by genetically identical twins, who can serve as controls for one another, will continue to provide unique insights into the development and manifestation of mental ability throughout the foreseeable future.

References

Amiel-Tison, C., & Gluck, L. (1995). Fetal brain and pulmonary adaptation in multiple pregnancy. In L. G. Keith, E. Papiernik, D. M. Keith, & B. Luke (Eds.), *Multiple pregnancy: Epidemiology, gestation and perinatal outcome* (pp. 585–597). New York: Parthenon.

Arey, L. B. (1922). Direct proof of the monozygotic origin of human identical twins. *Anatomical Record, 23*, 245–251.

Baaré, W. F. C., van Oel, C. J., Hulshoff Pol, H. E., Schnack, H. G., Durston, S., Sitskoorn, et al. (2001). Volumes of brain structures in twins discordant for schizophrenia. *Archives of General Psychiatry, 58*, 33–40.

Bartley, A. J., Jones, D. W., & Weinberger, D. R. (1997). Genetic variability of human brain size and cortical gyral patterns. *Brain, 120*, 257–269.

Beatty, J. (1995). *Principles of behavioral neuroscience.* Madison: Brown & Benchmark.

Bishop, E. G., Cherny, S. S., Corley, R., Plomin, R., DeFries, J. C., & Hewitt, J. K. (2003). Development genetic analysis of general cognitive ability from 1 to 12 years in a sample of adoptees, biological siblings, and twins. *Intelligence, 31*, 31–49.

Boomsma, D., van Beijsterveldt, C. E. M., Rietveld, M. J. H., Bartels, M., & van Baal, G. C. M. (2001). Genetics mediate relation of birth weight to childhood IQ. *British Medical Journal, 323*, 1426.

Borkenau, P., Riemann, R., Angleitner, A., & Spinath, F. M. (2002). Similarity of childhood experiences and personality resemblance in monozygotic and dizygotic twins: A test of the equal environments assumption. *Personality and Individual Differences, 33*, 261–269.

Bouchard, T. J., Jr. (1983). Do environmental similarities explain the similarity in intelligence of identical twins reared apart? *Intelligence, 7*, 175–184.

Bouchard, T. J., Jr. (2005). 2005 Kistler Prize Recipient: Dr. Thomas J. Bouchard Jr. *Foundation for the Future News*, Winter 2005/2006.

Bouchard, T. J., Jr., Lykken, D. T., McGue, M., Segal, N. L., & Tellegen, A. (1990). Sources of human psychological differences: The Minnesota Study of Twins Reared Apart.*Science, 250*, 223–228.

Bouchard, T. J., Jr., & McGue, M. (1981). Familial studies of intelligence: A review. *Science, 212*, 1055–1059.

Bouchard, T. J. Jr., & McGue, M. (1993). Genetic and environmental influences on human psychological differences. *Journal of Neurobiology, 54*, 4–45.

Bouchard, T. J., Jr., & Segal, N. L. (1985). IQ and environment. In B. B. Wolman (Ed.), *Handbook of intelligence* (pp. 391–464). New York: John Wiley & Sons.

Bryan, E. M. (1983). *The nature and nurture of twins.* London: Baillière Tindall.

Bulmer, M. G. (1970). *The biology of twinning in man.* Oxford: Clarendon.

Caravale, B., Tozzi, C., Albino, G., & Vicari, S. (2005). Cognitive development in low risk preterm infants at 3–4 years of life. *Archives of Diseases in Childhood (Fetal and Neonatal Edition), 90*, F474–479.

Carden, L. (1994). Specific cognitive abilities. In J. C. DeFries, R. Plomin, & D. W. Fulker (Eds.), *Nature and nurture during middle childhood* (pp. 57–76). Oxford:Blackwell.

Center for Disease Control. (2003). Births: Final data for 2002. *National Vital Statistics Reports, 52*, 1–116.

Christensen, K., Petersen, I., Herskind, A.-M., & Bingley, P. (2006). Twin/singleton differences in intelligence? A Danish nation-wide population-based register study of test scores and classroom assessments. *British Medical Journal, 333*, 1095.

Christensen, K., Vaupel, J. W., Holm, N. V., & Yashin, A. I. (1995). Mortality among twins after age 6: Fetal origins hypothesis versus twin method. *British Medical Journal, 310*, 432–436.

Cochran, G., Hardy, J., & Harpending, H. (2006). Natural history of Ashkenazi intelligence. *Journal of Biosocial Science, 38*, 659–693.

Cronk, N. J., Slutske, W. S., Madden,P. A. F., Bucholz, K. K., Reich, W., & Heath, A. C. (2002). Emotional and behavioral problems among female twins: An evaluation of the equal environments assumption. *Journal of the American Academy of Child and Adolescent Psychiatry, 41*, 829–837.

Davis, D. W., Burns, B. M., Wilkerson, S. A., & Steichen, J. J. (2005). Visual perceptual skills in children born with very low birth weights. *Journal of Pediatric Health Care, 19*, 363–368.

Deary, I. J. (2000). *Looking down on human intelligence.* New York: Oxford University Press.

Deary, I. J., Pattie, A., Wilson, V., & Whalley, L. J. (2005). The cognitive cost of being a twin: Two whole-population surveys. *Twin Research and Human Genetics, 8*, 376–383.

Devlin, B., Daniels, M., & Roeder, K. (1997). The heritability of IQ. *Nature, 388*, 468–471.

Dumaret, A., & Stewart, J. (1985). IQ, scholastic performance and behaviour of sibs raised in contrasting environments. *Journal of Child Psychology and Psychiatry, 26*, 553–580.

Duncan, J., Seitz, R. J., Kolodny, J., Bor, D., Herzog, H., Ahmed, A., et al. (2000). A neural basis for general intelligence. *Science, 289*, 457–460.

Edelman, G. (1987). *Neural Darwinism: The theory of neuronal group selection.* New York: Basic Books.

Endres, L., & Wilkins, I. (2005). Epidemiology and biology of multiple gestations. In K. A. Edelman & J. Stone (Eds.), *Clinics in perinatology: Multiple gestations* (pp. 301–314). Philadelphia: Elsevier.

Erlenmeyer-Kimling, L., & Jarvik, L. F. (1963). Genetics and intelligence: A review. *Science, 142*, 1477–1479.

Esposito, G., Kirby, B. S., Van Horn, J. D., Ellmore, T. M., & Berman, K. F. (1999). Context-dependent, neural-system-specific neurophysiological concomitants of ageing: Mapping PET correlates during cognitive activation. *Brain, 122*, 963–979.

Fletcher, R. (1991). *Science, ideology and the media: The Cyril Burt scandal.* London: Transaction Publishers.

Fraga, M. F., Ballestar, E., Paz, M. F., Ropero, S., Setien, F., Ballestar, M. L., et al. (2005). Epigenetic differences arise during the lifetime of monozygotic twins. *Proceedings of the National Academy of Sciences, 102*, 10604–10609.

Fulker, D. W., Cherny, S. S., & Cardon, L. R. (1993). Continuity and change in cognitive developement. In R. Plomin & G. E. McClearn (Eds.), *Nature, nurture, and psychology* (pp. 77–97). Washington, D. C.:American Psychological Association.

Galton, F. (1875). The history of twins as a criterion of the relative powers of nature and nurture. *Journal of the Anthropological Institute, 5*, 391–406.

Gartner, K. (1990). A third component causing random variability beside environment and genotype: A reason for the limited success of a 30 year long effort to standardize laboratory animals? *Laboratory Animals, 24*, 71–77.

Gartner, K., & Baunack, E. (1981). Is the similarity of monozygotic twins due to genetic factors alone? *Nature, 292*, 646–647.

Goodman, R., & Stevenson, J. (1991). Parental criticism and warmth toward unrecognized monozygotic twins. *Behavioral and Brain Sciences, 14*, 394–395.

Goody, A., Rice, F., Bolvin, J., Harold, G. T., Hay, D. F., & Thapur, A. (2005). Twins born following fertility treatment: Implications for quantitative genetic studies. *Twin Research and Human Genetics, 8*, 337–345.

Gottesman, I. I., & Bertelsen, A. (1989). Confirming unexpressed genotypes for schizophrenia. *Archives of General Psychiatry, 46*, 867–872.

Gottesman, I. I., & Gould, T. D. (2003). The endophenotype concept in psychiatry: Etymology and strategic intentions. *American Journal of Psychiatry, 160*, 636–645.

Gottesman, I. I., & Shields, J. (1973). Genetic theorizing and schizophrenia. *British Journal of Psychiatry, 122*, 15–30.

Gray, J. R., Chabris, C. F., & Braver, T. S. (2003). Neural mechanisms of general fluid intelligence. *Nature Neuroscience, 6*, 316–322.

Gray, J. R., & Thompson, P. M. (2004). The neurobiology of intelligence: Science and ethics. *Nature Neuroscience, 5*, 471–482.

Haber, J. R., Jacob, T., & Heath, A. C. (2005). Paternal alcoholism and offspring conduct disorder: Evidence for the 'common genes' hypothesis. *Twin Research and Human Genetics, 8*, 120–131.

Haier, R. J., Jung, R. E., Yeo, R. A., Head, K., & Alkire, M. T. (2004). Structural brain variation and general intelligence. *NeuroImage, 23*, 425–433.

Haier, R. J., Siegel, B. V., MacLachlan, A., Soderling, E., Lottenberg, S., & Buchsbaum, M. S. (1992). Regional glucose metabolic changes after learning a complex visual-spatial motor task: A positron emission tomography study. *Brain Research, 570*, 134–143.

Haier, R. J., White, N. S., & Alkire, M. T. (2003). Individual differences in general intelligence correlate with brain function during nonreasoning tasks. *Intelligence, 31*, 429–441.

Harlaar, N., Butcher, L. M., Meaburn, E., Sham, P., Craig, I. W., & Plomin, R. (2005). A behavioural genomic analysis of DNA markers associated with general cognitive ability in 7-year-olds. *Journal of Child Psychology and Psychiatry, 46*, 1097–1107.

Hayakawa, K., Shimizu, T., Kato, K., Onoi, M., & Kobayashi, Y. (2002). A gerontological cohort study of aged twins: The Osaka University Aged Twin Registry. *Twin Research, 5*, 387–388.

Hecht, B. R., & Magoon, M. W. (1998). Can the epidemic of iatrogenic multiples be conquered? *Clinical Obstetrics and Gynecology, 41*, 126–137.

Helmerhorst, F. M., Perquin, D. A. M., Donker, D., & Keirse, M. J. N. C. (2004). Perinatal outcome of singletons and twins after assisted conception: A systematic review of controlled studies. *British Medical Journal, 328*, 261–265.

Inlow, J. K., & Restifo, L. L. (2004). Molecular and comparative genetics of mental retardation. *Genetics, 166*, 835–881.

Jablonski, W. (1992). A contribution to the heredity of refraction in human eyes. *Archiv Augenheilk, 91*, 308–328.

Jacobs, N., Van Gestel, S., Derom, C., Thiery, E., Vernon, P., Derom, R., et al. (2003). Heritability estimates of intelligence in twins: Effect of chorion type. *Behavior Genetics, 31*, 209–217.

Jensen, A. (1998). *The g factor.* Westport, CT: Praeger.

Johnson, W., Bouchard, T. J., Jr., McGue, M., Segal, N. L., Tellegen, A., Keyes, M., et al. (2006). Genetic and environmental influences on the verbal-perceptual-image rotation (VPR) model of the structure of mental abilities in the Minnesota Study of Twins Reared Apart. *Intelligence, 35*, 452–462.

Joynson, R. B. (1989). *The Burt affair.* New York: Routledge.

Juel-Nielsen, N. (1965). *Individual and environment: Monozygotic twins reared apart.* New York: International Universities Press.

Kendler, K. S., Neale, M. C., Kessler, R. C., Heath, A. C., & Eaves, L. J. (1994). Parental treatment and the equal environment assumption in twin studies of psychiatric illness. *Psychological Medicine, 23*, 579–590.

Kervinen, K., Kaprio, J., Koskenvuo, M., Juntunen, J., & Kesaniemi, Y. A. (1998). Serum lipids and apolipoprotein E phenotypes in identical twins reared apart. *Clinical Genetics, 53*, 191–199.

Kim, H., Krege, J. H., Kluckman, K. D., Hagaman, J. R., Hodgin, J. B., Best, C. F., et al. (1995). Genetic control of blood pressure and the angiotensinogen locus. *Proceedings of the National Academy of Sciences, 92*, 2735–2739.

Klump, K. L., Holly, A., Iacono, W. G., McGue, M., & Wilson, L. E. (2000). Physical similarity and twin resemblance for eating attitudes and behaviors: A test of the equal environments assumption. *Behavior Genetics, 30*, 51–58.

Krege, J. H., Kim, H. S., Moyer, J. S., Jennette, J. C., Peng, L., Hiller, S.K., et al. (1997). Angiotensin-converting-enzyme gene mutations, blood pressures, and cardiovascular homeostasis. *Hypertension, 29*, 150–157.

Kuwata, T., Matsubara, S., Ohkuchi, A., Watanabe, T., Izumi, A., Honma, Y., et al. (2004). The risk of birth defects in dichorionic twins conceived by assisted reproductive technology. *Twin Research and Human Genetics, 7*, 223–227.

LaBuda, M., Svikis, D. S., & Pickens, R. W. (1997). Twin closeness and co-twin risk for substance use disorders: assessing the impact of the equal environment assumption. *Psychiatric Research, 70*, 155–164.

Loehlin, J. C., Horn, J. M., & Willerman, L. (1989). Modeling IQ change: Evidence from the Texas Adoption Project. *Child Development, 60*, 993–1004.

Loehlin, J. C., & Nichols, R. C. (1976). *Heredity, environment, and personality: A study of 850 sets of twins.* Austin: University of Texas Press.

Luciano, M., Smith, G. A., Wright, M. J., Geffen, G. M., Geffen, L. B., & Martin, N. G. (2001). On the heritability of inspection time and its covariance with IQ: a twin study. *Intelligence, 29*, 443–457.

Luciano, M., Wright, M. J., & Martin, N. G. (2004). Exploring the etiology of the association between birthweight and IQ in an adolescent twin sample. *Twin Research, 7*, 62–71.

Machin, G. A., & Keith, L. G. (1999). *An atlas of multiple pregnancy: Biology and pathology.* New York: Parthenon.

Magnus, P. (1984). Causes of variation in birth weight: A study of offspring of twins. *Clinial Genetics, 25*, 15–24.

Malykh, S. B., Zyrianova, N. M., & Kuravsky, L. S. (2003). Longitudinal genetic analysis of childhood IQ in 6- and 7-year-old Russian twins. *Twin Research, 6*, 285–291.

McClearn, G. E., Johansson, B., Berg, S., Pedersen, N. L., Ahern, F., Petrill, S. A., et al. (1997). Substantial genetic influence on cognitive abilities in twins 80+ years old. *Science, 276*, 1560–1563.

McDaniel, M. A. (2005). Big-brained people are smarter: A meta-analysis of the relationship between in vivo brain volume and intelligence. *Intelligence, 33*, 337–346.

McGue, M., Bouchard, T. J., Jr., Iacono, W. G., & Lykken, D. T. (1993). Behavioral genetics of cognitive ability: A life-span perspective. In R. Plomin & G. E. McClearn (Eds.), *Nature, nurture and psychology* (pp. 59–76). Washington, DC: APA Press.

Merriman, C. (1924). The intellectual resemblance of twins. *Psychological Monographs, 33*, 1–58.

Miller, G. F. (2000). *The mating mind: How sexual choice shaped the evolution of human nature.* New York: Doubleday.

Morris-Yates, A., Andrews, G., Howie, P., & Henderson, S. (1990). Twins; A test of the equal environments assumption. *Acta Psychiatrica Scandinavica, 81*, 322–326.

Neisser, U., Boodoo, G., Bouchard, T. J., Jr., Boykin, A. W., Brody, N., Ceci, S. J., et al. (1996). Intelligence: Knowns and unknowns. *American Psychologist, 51*, 77–101.

Neubaur, P. B., & Neubaur, A. (1990). *Nature's thumbprint: The new genetics of personality.* New York: Addison-Wesley.

Newman, H. N., Freeman, F. N., & Holzinger, K. J. (1937). *Twins: A study of heredity and environment.* Chicago: University of Chicago Press.

Newman, H. H., & Patterson, J. T. (1910). The development of the nine-banded armadillo from the primitive streak stage to birth, with especial reference to the question of specific polyembryony. *Journal of Morphology, 21*, 359.

Oppenheim, J. S., Skerry, J. E., Tramo, M. J., & Gazzaniga, M. S. (1989). Magnetic resonance imaging of the corpus callosum in monozygotic twins. *Annual Neurology Review, 26*, 100–104.

Pedersen, N. L., McClearn, G. E., Plomin, R., & Friberg, L. (1985). Separated fraternal twins: Resemblance for cognitive abilities. *Behavior Genetics, 15*, 407–419.

Pedersen, N. L., McClearn, G. E., Plomin, R., & Nesselroade, J. R. (1992). Effects of early rearing environment on twin similarity in the last half of the life span. *British Journal of Developmental Psychology, 10*, 255–267.

Pennington, B. F., Filipek, P. A., Lefly, D., Chhabildas, N., Kennedy, D. N., Simon, J. H., et al. (2000). A twin MRI study of size variations in human brain. *Journal of Cognitive Neuroscience, 12*, 223–232.

Perlman, L. M. (2005). Memories of the Child Development Center study of adopted monozygotic twins reared apart: An unfulfilled promise. *Twin Research and Human Genetics, 8*, 271–275.

Pfefferbaum, A., Sullivan, E. V., Swan, G. E., & Carmelli, D. (2000). Brain structure in men remains highly heritable in the seventh and eighth decades of life. *Neurobiology of Aging, 21*, 63–74.

Pinker, S. (2002). *The blank slate: The modern denial of human nature.* New York: Viking.

Plomin, R., DeFries, J. C., McClearn, G. E., & McGuffin, P. (2001). *Behavioral genetics* (4th ed.). New York: Worth Publishers.

Posthuma, D., De Geus, E. J., Baare, W. F. C., Pol, H. E. H., Kahn, R. S., & Boomsma, D. I. (2002). The association between brain volume and intelligence is of genetic origin. *Nature Neuroscience, 5*, 83–84.

Posthuma, D., De Geus, E. J., Bleichrodt, N., & Boomsma, D. L. (2000). Twin-singleton differences in intelligence? *Twin Research, 3*, 83–87.

Potter, C. E. L. (1948). *Fundamentals of human reproduction.* New York: McGraw-Hill.

Prabhakaran, V., Smith, J. A., Desmond, J. E., Glover, G. H., & Gabrielli, J. D. (1997). Neuronal substrates of fluid reasoning: An fMRI study of neocortical activation during performance of the Raven's Progressive Matrices Test. *Cognitive Psychology, 33*, 43–63.

Purcell, S., Eley, T. C., Dale, P. S., Oliver, B., Petrill, S. A., Price, T. S., et al. (2001). Comorbidity between verbal and non-verbal cognitive delays in 2-year-olds: A bivariate twin analysis. *Developmental Science, 4*, 194–207.

Reed, T., Carmelli, D., & Rosenman, R. H. (1991). Effects of placentation on selected Type A behaviors in adult males in the National Heart, Lung, and Blood Institute (NHLBI) twin study. *Behavior Genetics, 21*, 9–19.

Reynolds, C. A., Finkel, D., Gatz, M., & Pedersen, N. L. (2002). Sources of influence on rate of cognitive change over time in Swedish twins: An application of latent growth models. *Experimental Aging Research, 28*, 407–433.

Rietveld, M. J. H., Dolan, C. V., van Baal, C. M., & Boomsma, D. I. (2003). A twin study of differentiation of cognitive abilities in childhood. *Behavior Genetics, 33*, 367–381.

Ronalds, G. A., De Stavola, B. L., & Leon, D. A. (2005). The cognitive cost of being a twin: Evidence from comparisons within families in the Aberdeen children of the 1950s cohort. *British Medical Journal, 331*, 1306–1310.

Rose, R. J., Harris, E. L., Christian, J. C., & Nance, W. E. (1979). Genetic variance in nonverbal intelligence: Data from the kinships of identical twins. *Science, 205*, 1153–1155.

Rowe, D. C. (1994). *The limits of family influence.* New York: Guiford Press.

Rutherford, S. L. (2000). From genotype to phenotype: Buffering mechanisms and the storage of genetic information. *BioEssays, 22*, 1095–1105.

Rutter, M., Thorpe, K., Greenwood, R., Northstone, K., & Golding, J., (2003). Twins as a natural experiment to study the causes of mild language delay: I: Design; twin-singleton differences in language, and obstetric risks. *Journal of Child Psychology and Psychiatry, 44*, 326–341.

Scamvougeras, A., Kigar, D. L., Jones, D., Weinberger, D. R., & Witelson, S. F. (2003). Size of the human corpus callosum is genetically determined: An MRI study in mono and dizygotic twins. *Neuroscience Letters,338*, 91–94.

Scarr, S. (1969). Environmental bias in twin studies. In M. Manosevits, G. Lindzey, & D. D. Thiessen (Eds.), *Behavioral genetics: Method and theory* (pp. 597–605). New York: Appleton-Century-Crofts.

Segal, N. L. (1985a). Holocaust twins: Their special bond. *Psychology Today, 19*, 52–58.

Segal, N. L. (1985b). Monozygotic and dizygotic twins: A comparative analysis of mental ability profiles. *Child Development, 56*, 1051–1058.

Segal, N. L. (1997). Twin research perspective on human development. In N. L. Segal, G. E. Weisfeld, & C. C. Weisfeld (Eds.), *Uniting psychology and biology: Integrative perspectives on human development* (pp. 145–173). Washington, DC: APA Press.

Segal, N. L. (2000a). *Entwined lives: Twins and what they tell us about human behavior.* New York: Plume.

Segal, N. L. (2000b). Virtual twins: New findings on within-family environmental influences on intelligence. *Journal of Education Psychology, 92*, 442–448.

Segal, N. L. (2003). Spotlights (Reared apart twin researchers); research sampling; literature, politics, photography and athletics. *Twin Research, 6*, 72–81.

Segal, N. L. (2005a). *Indivisible by two: Lives of extraordinary twins.* Cambridge, MA: Harvard University Press.

Segal, N. L. (2005b). More thoughts on the Child Development Center Twin Study. *Twin Research and Human Genetics, 8*, 276–281.

Segal, N. L. (2005c). Twins reared apart design. In B. Everitt & D. C. Howell (Eds.), *Encyclopedia of statistics in behavioral science. 4* (pp. 2072–2076). Chichester, UK: John Wiley & Sons.

Segal, N. L. (In press, 2009). *Multiple births: Developmental perspectives.* Chicago Companion to the Child, Chicago: University of Chicago Press.

Segal, N. L., & Allison, D. B. (2002). Twins and virtual twins: Bases of relative body weight revisited. *International Journal of Obesity, 26*, 437–441.

Segal, N. L., Chavarria, K. A., & Stohs, J. H. (2008). Twin research: Evolutionary perspective on social relations. In T. Shackelford & C. D. Salmon (Eds.), *Family relationships: An evolutionary perspective* (pp. 312–333). Oxford, England: Oxford University Press.

Segal, N. L., & Hershberger, S. L. (2005). Virtual twins and intelligence: Updated and new analyses of within-family environmental influences. *Personality and Individual Differences, 39*, 1061–1073.

Segal, N. L., McGuire, S. A., Havlena, J., Gill, P., & Hershberger, S. L. (2007). Intellectual similarity of virtual twin pairs: Developmental trends. *Personality and Individual Differences, 42*, 1209–1219.

Segal, N. L., Seghers, J. P., Marelich, W. D., Mechanic, M., & Castillo, R. (2007). Social closeness of monozygotic and dizygotic twin parents toward their nieces and nephews. *European Journal of Personality, 21*, 487–506.

Sharma, A., Sharma, V. K., Horn-Saban, S., Lancet, D., Ramachandran, S., & Brahmachari, S. K. (2005). Assessing natural variations in gene expression in humans by comparing with monozygotic twins using microarrays. *Physiological Genomics, 21*, 117–123.

Shields, J. (1962). *Monozygotic twins: Brought up apart and together.* London: Oxford University Press.

Smithies, O. (2005). Many little things: One geneticist's view of complex diseases. *Nature Reviews – Genetics, 6*, 419–425.

Snyderman, M., & Rothman, S. (1988). *The IQ controversy, the media and publication.* New Brunswick, NJ: Transaction.

Sokol, D. K., Moore, C. A., Rose, R. J., Williams, C. J., Reed, T., & Christian, J. C. (1995). Intrapair differences in personality and cognitive ability among young monozygotic twins distinguished by chorion type. *Behavior Genetics, 25*, 457–466.

Spinath, F. M., Ronald, A., Harlaar, N., Price, T. S., & Plomin, R. (2003). Phenotypic g early in life: On the etiology of general cognitive ability in a large population sample of twin children aged 2–4 years. *Intelligence, 31*, 194–210.

Spitz, E., Carlier, M., Vacher-Lavenu, M.-C., Reed, T., Moutier, R., Busnel, M.-C., et al. (1996). Long term effect of prenatal heterogeneity among monozygotes. *Current Psychological Cognition, 15*, 283–308.

Steinmetz, H., Herzog, A., Schlaug, G., Huang, Y., & Jancke, L. (1995). Brain (A)symmetry in monozygotic twins. *Cerebral Cortex, 5*, 296–300.

Stern, K. (1960). *Principles of human genetics.* (2nd ed.). San Francisco: W.H. Freeman.

Stromberg, B., Dahlquist, G., Ericson, A., Finnstrom, O., Koster, M., & Stjernqvist, K. (2002). Neurological sequelae in children born after in-vitro fertilisation: A population-based study. *Lancet, 359*, 461–465.

Sullivan, E. V., Pfefferbaum, A., Swan, G. E., & Carmelli, D. (2001). Heritability of hippocampal size in elderly twin men: Equivalent influences from genes and environment. *Hippocampus, 11*, 754–762.

Taylor, H. F. (1980). *The IQ game: A methodological inquiry into the heredity-environment controversy.* New Brunswick, NJ: Rutgers University Press.

Thompson, P. M., Cannon, T. D., Narr, K. L., Erp, T. V., Poutanen, V. P., Huttunen, M., et al. (2001). Genetic influences on brain structure. *Nature Neuroscience, 4*, 1–6.

Thorndike, E. L. (1905). Measurement of twins. *Journal of Philosophy, Psychology and Scientific Methods, 1*, 1–64.

Thorpe, K., Greenwood, R., Eivers, A., & Rutter, M. (2001). Prevalence and developmental course of 'secret language.' *International Journal of Language and Communication Disorders, 36*, 43–62.

Thorpe, K., Rutter, M., & Greenwood, R. (2003). Twins as a natural experiment to study the causes of mild language delay: II: Family interaction risk factors. *Journal of Child Psychology and Psychiatry, 44*, 342–355.

Toga, A. W., & Thompson, P. M. (2005). Genetics of brain structure and intelligence. *Annual Review of Neuroscience, 28*, 1–5.

Tomasello, M., Mannle, S., & Kruger, A. C. (1986). Linguistic environment of 1- to 2-year-old twins. *Developmental Psychology, 22*, 169–176.

Tramo, M. J., Loftus, W. C., Stukel, T. A., Green, R. L., Weaver, J. B., & Gazzaniga, M. S. (1998). Brain size, head size, and intelligence quotient in monozygotic twins. *Neurology, 50*, 1246–1252.

Trejo, V., Derom, C., Vlietinck, R., Ollier, W., Silman, A., Ebers, G., et al. (1994). X chromosome inactivation patterns correlate with fetal-placental anatomy in monozygotic twin pairs: Implications for immune relatedness and concordance for autoimmunity. *Molecular Medicine, 1*, 62–70.

Tully, L. A., Moffit, T. B., & Caspi, A. (2003). Maternal adjustment, parenting and child behavior in families of school-aged twins conceived after IVF and ovulation induction. *Journal of Child Psychology and Psychiatry, 44*, 316–325.

Turkheimer, E., Haley, A., Waldron, M., D'Onofrio, B., & Gottesman, I. I. (2003). Socioeconomic status modified heritability of IQ in young children. *Psychological Science, 14*, 623–628.

Vandenberg, S. G. (1969). Contributions of twin research to psychology. In M. Manosevits, G. Lindzcy, & D. D. Thiessen (Eds.), *Behavioral genetics: Method and theory* (pp. 145–164). New York: Appleton-Century-Crofts.

Van Ijzendoorn, M. H., Juffer, F., & Poelhuis, C. W. K. (2005). Adoption and cognitive development: A meta-analytic comparison of adopted and nonadopted children's IQ and school performance. *Psychological Bulletin, 131*, 301–316.

Waddington, C. H. (1953). Genetic assimilation of an acquired character. *Evolution, 7*, 118–126.

Waddington, C. H. (1957). *The strategy of the genes.* New York: Macmillan.

Wainwright, M. A., Wright, M. I., Geffen, G. M., Luciano, M., & Martin, N. G. (2005). The genetic basis of academic achievement on the Queensland Core Skills Test and its shared genetic variance with IQ. *Behavior Genetics, 35*, 133–145.

Weaver, I. C. G., Cervoni, N., Champagne, F. A., D'Alessio, A. C., Sharma, S., Seckl, J. R., et al. (2004). Epigenetic programming by maternal behavior. *Nature Neuroscience, 7*, 847–854.

Weinberg, W. (1901). Beitrage zur physiologie und pathologie der mehrlingsgeburten beim menschen. *Pflugers Archiv fur die Gesamte Physiologie de Menschen und der Tiere, 88*, 346–430.

Wilson, R. S. (1986). Twins: Genetic influence on growth. In R. M. Malina & C. Bouchard (Eds.), *Sports and human genetics* (pp. 1–21). Champaign, IL: Human Kinetics.

Wilson, R. S. (1979). Twin growth: Initial deficit, recovery, and trends in concordance from birth to nine years. *Human Biology, 6*, 205–220.

Wilson, R. S. (1983). The Louisille Twin Study: Developmental synchronies in behavior. *Child Development, 54*, 298–316.

Wong, A. H. C., Gottesman, I. I., & Petronis, A. (2005). Phenotypic differences in genetically identical organisms: The epigenetic perspective. *Human Molecular Genetics, 14*, R11–R18.

Woo, S. L., Lidsky, A. S., Guttler, F., Chandra, T., & Robson, K. J. (1983). Cloned human phenylalanine hydroxylase gene allows prenatal diagnosis and carrier detection of classical phenylketonuria. *Nature, 306*, 151–155.

York, T. P., Miles, M. F., Kendler, K. S., Jackson-Cook, C., Bowman, M. L., & Eaves, L. (2005). Epistatic and environmental control of genome-wide gene expression. *Twin Research, 8*, 5–15.

Chapter 7

Behavioral Genetic Investigations of Cognitive Aging

Deborah Finkel and Chandra A. Reynolds

Introduction

One of the universal concerns of human beings is the issue of aging: How will I meet the changes and challenges that occur with age? What impact can I have on my own aging process? The first modern twin study designed to investigate genetic and environmental influences on the aging process was the New York State Psychiatric Study of Aging Twins begun in 1946. In the last two decades, there has been an upsurge in the number of behavioral genetic studies of the various facets of the aging process. A recent summary reports over two dozen twin studies investigating physical, psychological, and social aspects of aging (Bergeman, 2007). In this review, we focus on the three components of cognitive aging. Primary aging is the normal and pervasive changes in cognitive abilities that occur with age. In contrast, secondary aging is typified by changes in cognitive functioning that result from disease or pathological processes. The distinction between primary and secondary aging is of paramount importance; some changes that were once thought to be an inevitable part of aging (e.g., senile dementia) have been shown to be the outcome of disease processes that can be diagnosed and potentially treated. Finally, the acceleration in decline of cognitive functioning that occurs in the years immediately preceding death is termed tertiary aging or terminal decline. Distinguishing primary, secondary, and tertiary aging is paramount to understanding the nature of cognitive aging.

Behavioral genetic research on primary aging has focused both on general intelligence and specific cognitive abilities, as well as covariates of intellectual functioning that may be the sources of genetic and environmental contributions to cognitive aging. In investigations of secondary aging, the contributions of both measured genotypes and mea-sured lifestyle variables to forms of dementia are discussed. Methods for investigating tertiary aging are presented, including estimating trajectories of change from age at death. The issues that will face future investigators and the quantitative methodologies that will be required to address these issues are also discussed.

Primary Aging

Primary aging encompasses general age-related changes due to ontogenetic or inherent processes (Berger, 2005; Busse, 2002). Consistent findings of a loss (or gain) in cognitive ability systematic with chronological age in healthy adults would be a suggestive of primary aging effects where confounding factors such as education or socioeconomic background are ruled out. Systematic loss with age has been observed most prominently for perceptual speed and fluid abilities (Horn, 1988; Schaie, 1996) while relatively more conserved abilities tend to show losses later in the second half of the lifespan, e.g., memory loss (Horn, 1988; Schaie, 1996).

General Cognitive Ability

Original results from cross-sectional studies of cognitive aging suggested higher levels of heritability for general cognitive ability than is typically observed in young adulthood (cf. Chapter 6, this volume). As evidenced by the data summarized in Table 7.1, heritability estimates for general cognitive ability are about 0.80 in adulthood. The remaining variance is primarily nonshared environmental variance. The large age ranges (45–68 years) included in these studies, however, may be masking age changes in heritability as a result of successive phases of the aging process. Evidence from cohort sequential and longitudinal analyses, for example, provide a more complex image of the

D. Finkel (✉)
Department of Psychology, School of Social Sciences, Indiana University Southeast, New Albany, IN 47150, USA
e-mail: dfinkel@ius.edu

Y.-K. Kim (ed.), *Handbook of Behavior Genetics*,
DOI 10.1007/978-0-387-76727-7_7, © Springer Science+Business Media, LLC 2009

Table 7.1 Results from cross-sectional twin studies of primary cognitive aging

Variable	Study	Age range	Heritability
General cognitive ability	GOSAT	18–70	0.81
	MTSADA	27–95	0.75
	NTR	42–87	0.81
Verbal ability			
Information	MTSADA	27–95	0.77
Vocabulary	GOSAT	18–70	0.64–0.68
Verbal measures	NTR	42–87	0.59–0.86
Word fluency	GOSAT	18–70	0.53
Spatial ability			
Block design	MTSADA	27–95	0.73
Block design	OKUT	50–78	0.60
Spatial measures	GOSAT	18–70	0.39–0.57
Spatial measures	NTR	42–87	0.33–0.58
Memory			
Digit span	MTSADA	27–95	0.55
Figure memory	MTSADA	27–95	0.60
Memory factor	NHLBI	59–80	0.56
Memory	NTR	42–87	0.51
Text recall	MTSADA	27–95	0.53
Processing speed			
Digit symbol	MTSADA	27–95	0.62
Digit symbol	NHLBI	59–80	0.67
Digit symbol	OKUT	50–78	0.22
Perceptual speed	NTR	42–87	0.49–0.75

Note: GOSAT, German Observational Study of Adult Twins (Neubauer et al., 2000); MTSADA, Minnesota Twin Study of Adult Development and Aging (Finkel, Pedersen, & McGue, 1995a; Finkel, Pedersen, McGue, & McClearn, 1995b); NHLBI, National Heart Lung Blood Institute Twin Study (Swan et al., 1990, 1999); NTR, Norwegian Twin Register (Sundet, Tambs, Harris, Magnus, & Torjussen, 2005); OKUT, Osaka/Kinki University Twin Study (Hayakawa, Shimizu, Ohba, & Tomioka, 1992).

nature of genetic influences on cognitive abilities across the adult lifespan. Using latent growth curve analyses (McArdle, Prescott, Hamagami, & Horn, 1998; Chapter 2, this volume), genetic and environmental influences on static (intercept) and dynamic (rates of change) measures of cognitive aging can be investigated. In addition, changes with age in genetic and environmental components of variance can be calculated from the latent growth curve parameters. In other words, both the heritability of change and the change in heritability can be estimated.[1]

---

[1] It is important to note the challenges inherent in applying latent growth curve models to studies of aging twins. As with all studies of aging, attrition due to both nonresponse and death has a significant impact on sample size and assumptions about missing data. Furthermore, in studies of cognitive aging, the reasons for nonresponse (e.g., dementia or terminal decline) may be integrally related to the phenotype in question. In twin studies of aging we have the additional issue of twinness: in order for a twin pair to contribute fully to the investigation of genetic and environmental influences, both twins must participate. Finally, to ensure stability of parameter estimates in the latent growth curve model, at least three waves of measurement are necessary, and estimate stability increases with additional measurement occasions, see McArdle et al. (1998) for a more complete discussion of these issues.

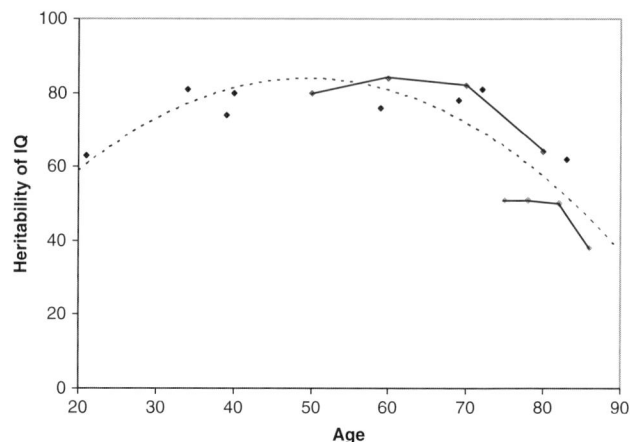

Fig. 7.1 Summary of heritability estimates for general cognitive ability across the adult lifespan. Single-point estimates are from the following cross-sectional studies: Bouchard, Lykken, McGue, Segal, and Tellegen (1990), Finkel et al. (1995b), McClearn et al. (1997), Neubauer et al. (2000), Sundet et al. (2005), Tambs, Sundet, and Magnus (1986) and Tambs et al. (1989). Longitudinal data are from McGue and Christensen (2002) and Reynolds et al. (2005). The *dotted line* is the polynomial regression line fitted to all the points ($R^2 = 0.71$)

A growing body of evidence from longitudinal investigations of cognitive aging indicates a decrease in heritability in late adulthood. A summary of the data on general cognitive ability is presented in Fig. 7.1: evidence from both cross-sectional and longitudinal studies converges on the conclusion that heritability for general cognitive ability increases from young adulthood, plateaus in adulthood, and decreases late in life. The dotted line represents the polynomial regression line fitted to all points, and even through the data come from eight different studies, the regression line explains 71% of the variability among the data points. When the data are considered in terms of raw variance, instead of proportion of variance, it becomes clear that the decrease in heritability results from fairly constant genetic variance and increasing nonshared environmental variance (e.g., Reynolds et al., 2005). Thus we can interpret the results as an indication of an accumulation of unique environmental influences that begin to have a greater impact on individual differences in cognitive ability in late life.

The implications of late life changes in heritability are supported by investigations of the heritability of change in general intelligence. It is possible that the genetic and environmental factors influencing *level* of cognitive functioning are not the same as the genetic and environmental factors that affect *change* with age. In fact, results from longitudinal twin studies indicate greater genetic influences on the intercept, or level of cognitive performance, than for either linear or quadratic components of cognitive decline (McGue & Christensen, 2002; Reynolds, Finkel, Gatz, & Pedersen, 2002; Reynolds et al., 2005). Results from several longitudinal twin studies of aging are summarized in

Table 7.2 Results from longitudinal twin studies of primary cognitive aging

Variable	Study	Age range	Change in heritability		Heritability of change		
			Young–old	Old–old	Intercept	Linear slope	Quadratic
General cognitive ability	LSADT	70–97	0.51	0.38	0.76	0.06	
	SATSA	50–92	0.80	0.64	0.91	0.01	0.43
	OCTO	80–95		0.62[a]			
Verbal ability							
Analogies	SATSA	50–92	0.87	0.22	0.78	0.19	0.09
Information	OCTO	80–95		0.55[a]	0.68[b]	–[c]	
Information	SATSA	50–92	0.65	0.54	0.70	0.09	0.42
Synonyms	OCTO	80–95		0.55[a]	0.46[b]	–[c]	
Synonyms	SATSA	50–92	0.78	0.46	0.75	0.17	0.13
Vocabulary	NYT	58–90	0.73	0.71	0.88	0.33	
Spatial ability							
Block design	NYT	58–90	0.63	0.35	0.96	0.70	
Block design	OCTO	80–95		0.32[a]	0.62[b]	–[c]	
Block design	SATSA	50–92	0.89	0.31	0.80	0.35	0.56
Card rotations	SATSA	50–92	0.81	0.33	0.74	0.35	0.38
Figure logic	OCTO	80–95		0.32[a]	0.70[b]	–[c]	
Figure logic	SATSA	50–92	0.57	0.03	0.67	0.14	0.33
Memory							
Digit span forward	OCTO	80–95		0.27[a]	0.36[b]	–[c]	
Digit span backward	OCTO	80–95		0.49[a]	0.84[b]	–[c]	
Digit span	SATSA	50–92	0.36	0.63	0.52	0.35	
Picture memory	OCTO	80–95	0.47[a]	0.14[b]	–[c]		
Picture memory	SATSA	50–92	0.50	0.33	0.84	0.06	0.70
Prose recall	OCTO	80–95		0.04[a]	0.12[b]	–[c]	
Processing speed							
Digit symbol	NHLBI	59–80				–[d]	
Digit symbol	OCTO	80–95		0.62[a]	0.72[b]	–[c]	
Symbol digit	SATSA	50–92	0.67	0.64	0.85	0.03	0.75
Figure identification	SATSA	50–92	0.58	0.40	0.78	0.15	0.39

[a] The full OCTO-Twin sample is in the old–old age range; therefore, no young–old estimate of heritability is reported.
[b] Calculated from twin correlations provided in Johansson et al. (2004).
[c] Twin correlations for slope estimates were unstable, but indicated largely environmental influences (Johannson et al., 2004).
[d] MZ concordance for decline = 45% and DZ concordance for decline = 8%.
Note: NYT, New York Twin Study (McArdle et al., 1998); SATSA, Swedish Adoption/Twin Study of Aging (Reynolds et al., 2005); NHLBI, National Heart Lung Blood Institute Twin Study (Swan, LaRue, Carmelli, Reed, & Fabsitz, 1992); LSADT, Longitudinal Study of Aging Danish Twins (McGue & Christensen, 2002); OCTO-Twin, Origins of Variance in the Oldest Old (Johansson et al., 1999, 2004; McClearn, Johansson, Berg, & Pedersen, 1997).

Table 7.2, including data on heritability of change in general cognitive ability. The evidence suggests that individual differences in the level of cognitive performance reflect primarily genetic influences. Although linear change results almost entirely from nonshared environmental influences, both genetic and nonshared environmental factors impact accelerating decline. Therefore, over half of the variance in general cognitive decline reflects person-specific environmental influences lending support to theories of stochastic or chance processes in aging (Finch & Kirkwood, 2000).

Specific Cognitive Abilities

Even though measures of general cognitive ability provide an overall view of changes in functioning that occur with age, intelligence is not a unitary construct and neither is cognitive aging. Decades of gerontological research indicate longitudinal decline for spatial, fluid, and memory abilities, with relatively smaller age changes noted for verbal ability (e.g., Carmelli, Swan, LaRue, & Eslinger, 1997; Korten et al., 1997; Schaie, 1994; Singer, Verhaeghen, Ghisletta, Lindenberger, & Baltes, 2003; Small, Stern, Tang, & Mayeux, 1999). Two-component theories of intelligence (e.g., Lindenberger, 2001) predict greater cultural impact on aging trajectories for aging-resilient (i.e., crystallized) abilities and a more biological foundation for age changes in age-sensitive (i.e., fluid) abilities. On the whole, results from twin studies of specific cognitive abilities in older adults support these predictions.

Cross-sectional twin data report higher heritability estimates for measures of verbal and spatial abilities and lower heritability estimates for memory and processing speed (see Table 7.1). The differences in heritability estimates reported in Table 7.1, both between- and within-cognitive variables, can be attributed to differences in the age ranges covered by the studies and to the multi-faceted nature of the constructs, especially as they relate to the aging brain. These issues can be addressed by including multiple measures of the constructs and by investigating longitudinal changes in heritability across the age ranges in question. Results of longitudinal twin studies of specific cognitive abilities are presented in Table 7.2. Measures of verbal ability from three different studies indicate high genetic influences on the level of performance, but limited genetic variance for either linear or quadratic rates of decline. In contrast, as predicted by two-component theories of intelligence, results for spatial abilities show high heritability for the intercept and at least moderate heritability for both linear and quadratic declines. The third pattern is evident for processing speed: high heritability for level of performance, low heritability for linear decline, and high heritability for accelerating decline. We will consider the genetic influences on processing speed further when we discuss the possible mediational role of processing speed in cognitive aging. Results for measures of memory performance are mixed. With the exception of digit span backward, OCTO-Twin reports minimal heritability for the intercept and consistently low heritability for linear decline; however, twin data can become particularly unstable in the latest part of the lifespan, limiting our ability to make strong inferences (Johansson et al., 2004). In addition, SATSA reports only linear trajectories for digit span and quadratic trajectories for picture memory; although in combination the results support at least moderate genetic influences on decline.

Relationships Among Specific Cognitive Abilities

Specific cognitive abilities are not independent of each other and, in fact, it is possible that age changes in one cognitive ability may mediate or drive age changes in other components of cognition. Mediational theories of age-related cognitive have identified processing speed as a factor that may underlie demonstrated declines in a variety of cognitive tasks (e.g., Birren, 1964; Salthouse, 1996). Using data from twin studies of older adults, we can investigate not only the extent of the relationship between age changes in speed and age changes in cognition but also the underlying nature of that relationship. Genetic and environmental influences on the relationship between processing speed and cognitive aging have been investigated on three levels. First, cross-sectional studies demonstrate that correlations between processing speed and various measures of cognitive ability in middle to late adulthood are almost entirely genetically mediated (Finkel & Pedersen, 2000; Neubauer, Spinath, Riemann, Angleitner, & Borkenau, 2000; Posthuma, de

Geus, & Boomsma, 2001; Posthuma, Mulder, Boomsma, & de Geus, 2002). Thus, a significant proportion of the genetic influences on cognitive ability in the second half of the lifespan arises from genetic factors affecting processing speed. Second, analysis of longitudinal twin data indicates that with age, an increasing proportion of genetic variance for cognitive ability can be attributed to genetic influences on processing speed (Finkel & Pedersen, 2004). Finally, by analyzing longitudinal twin data with sufficient time points to estimate quadratic patterns of decline, we find that it is not the linear age changes but the accelerating age changes in cognitive performance that share genetic variance with processing speed, at least for fluid abilities (Finkel, Reynolds, McArdle, & Pedersen, 2005).

Applying standard behavioral genetic and growth curve methods, we can investigate the genetic and environmental contributions to the covariance among specific cognitive abilities. To extend our understanding of these longitudinal relationships, however, it would be informative to be able to determine whether one cognitive variable is the leading indicator of subsequent changes in cognitive performance, as well as the extent of genetic and environmental influences driving the system. By using latent difference scores, instead of latent growth curves, McArdle and Hamagami (2003) have proposed a model that taps the dynamic interaction between age changes in specific cognitive abilities. In other words, the bivariate dual change score model allows for the identification of leading indicators of cognitive change: the extent to which changes in one variable drive changes in a related variable. They applied the model to vocabulary and block design measures from the New York Twin Study. They found small but significant coupling from block design to vocabulary scores, indicating that age changes in block design lead to age changes in vocabulary. Estimates of genetic variance for vocabulary changed only slightly over the age range when the dynamic coupling with block design was included in the model: heritability for vocabulary was stable instead of declining when the impact of block design was removed. The analyses need to be repeated with variables are more strongly correlated (McArdle & Hamagami, 2003).

Covariates of Primary Cognitive Aging

In addition to cognitive measures that mediate cognitive aging, we can also consider environmental and biological variables that may account for some of the genetic and environmental contributions to age changes in cognitive performance. A review of the evidence for environmental measures as sources of environmental variance in cognitive abilities in adulthood reported mixed success (Finkel & Pedersen, 2001). Moreover, it is important to distinguish between environmental phenotypes and environmental etiology. Behavioral genetic methods can be used to investigate whether the relationships between cognitive aging and its covariates are explained via genetic or environmental pathways. For example, researchers have questioned the extent to which apparently environmental variables truly reflect environmental influences, as opposed to genetic influences (e.g., Plomin, 1994; Rowe, 1994). Occupation, education, and attitudes toward education explained most of the shared environmental variance in cognitive performance. However, only a very limited proportion of the nonshared environmental variance could be explained by measured environmental variables, primarily social class, occupation, and smoking history. More success has been achieved in identifying environmental variables that account for a portion of the genetic variation in cognitive abilities. For example, including education as a covariate in latent growth curve models of cognitive ability indicated that education shared genetic variance with the model intercept, but environmental variance with rate of change (Reynolds et al., 2002).

Biological variables have also been proposed as candidates for sources of genetic variance in cognitive abilities. Analyses of data from the Minnesota Twin Study of Adult Development and Aging reported only small and nonsignificant genetic correlations between physical activity or health factors and measures of cognitive performance (Finkel & McGue, 1993, 1998). In contrast, investigations using data from the Swedish Adoption/Twin Study of Aging reported that pulmonary function shared significant genetic variance with four measures of cognitive performance: information, digit symbol, block design, and digit span backward (Emery, Pedersen, Svartengren, & McClearn, 1998). Similarly, using growth curve models, researchers found that the covariance between pulmonary function and level of performance was primarily genetically mediated, although the covariance with decline in cognitive performance was mediated by nonshared environmental factors (Reynolds et al., 2002).

Identifying Genes Related to Primary Aging

Moving beyond anonymous components of genetic and environmental variance to measurable genes and environment has become one of the goals of behavioral genetic research. The apolipoprotein E *(APOE)* gene, coding for the brain cholesterol transporter apolipoprotein E, has been most often studied with respect to normal cognitive aging. The *APOE* e4 allele has been identified as a potential risk variant for normative cognitive change based on studies reporting consistent association with Alzheimer's disease risk (Strittmatter et al., 1993). In addition to well-known links of *APOE* e4 and AD-associated neuropathology [e.g., senile plaques containing Aβ deposits, neurofibrillary tangles (Ohm

et al., 1995; Takahashi, Nam, Edgar, & Gouras, 2002)], studies indicate that the apoE protein may be involved in neuronal development and plasticity (Nathan et al., 2002; Ohm et al., 2003; Teter et al., 2002). While several population-based or community-based studies of individuals have indicated a positive association with *APOE* e4 and cognitive change (Deary et al., 2002; Hofer et al., 2002; Mayeux, Small, Tang, Tycko, & Stern, 2001; Mortensen & Hogh, 2001), findings are not entirely consistent (Anstey & Christensen, 2000; Pendleton et al., 2002; Small et al., 2000). The memory domain evidences the most consistent findings (Anstey & Christensen, 2000; Bretsky, Guralnik, Launer, Albert, & Seeman, 2003; Nilsson, Nyberg, & Backman, 2002; Reynolds, Jansson, Gatz, & Pedersen, 2006a; Wilson et al., 2002), even in studies removing participants with slight indications of dementia (Mayeux et al., 2001). A recent twin-based association study added to the consensus by finding significant association of APOE with change in working memory (i.e., digit span) in nondemented Swedish twins (Reynolds, Prince, Feuk, Gatz, & Pedersen, 2006b).

Findings for other gene candidates have begun to appear relatively recently though with few if any replications published to date (e.g., de Frias et al., 2005; Harris et al., 2005; Reynolds, Jansson, et al., 2006a; Reynolds, Prince, et al., 2006b). The genes encoding serotonin 2A receptors *(HTR2A)* and catechol-*O*-methyltransferase *(COMT)* are of particular interest given their involvement in brain regions associated with learning and memory. In particular, working memory, episodic, and semantic performance have been associated with serotonin function and/or proteins involved in the catabolism or breakdown of dopamine, i.e., catechol-*O*-methyltransferase. Evidence of their potential biological importance has been marshaled across a variety of designs including neuroimaging studies (Egan et al., 2001; Sheline, Mintun, Moerlein, & Snyder, 2002), mRNA (Akil et al., 2003; Amargos-Bosch et al., 2004; but see Bray, Buckland, Hall, Owen, & O'Donovan, 2004), and, for serotonin 2A, in vivo receptor activity manipulations in primates (Williams, Rao, & Goldman-Rakic, 2002).

Serotonin 2A *(HTR2A)* may play a role in memory-related formation of synaptic connections (Kandel, 2001). Furthermore, age-related decreases in serotonin 2A receptors have been noted in the hippocampus and prefrontal cortex (Sheline et al., 2002) with evidence of downregulated gene expression in the frontal cortex beginning at midlife (Lu et al., 2004). Episodic memory performance in young adults (de Quervain et al., 2003) and recognition memory change in nondemented Swedish twins (Reynolds et al., 2006a) have been associated with the gene encoding serotonin 2A *(HTR2A)*.

The Val108/158 Met *COMT* functional variant has been associated with prefrontal cortex ERP P300 latencies (Tsai et al., 2003), cognitive stability and flexibility in schizophrenics (Nolan, Bilder, Lachman, & Volavka, 2004), working memory performance in schizophrenics and healthy adults (Bruder et al., 2005; Egan et al., 2001; Malhotra et al., 2002), and declarative memory performance (de Frias et al., 2004; Harris et al., 2005). Most recently the COMT variant was associated with cognitive change over 5 years in nondemented Swedish adults ranging from 35 to 85 years at baseline with respect to executive function (de Frias et al., 2005).

Secondary Aging

Secondary aging effects are a result of environmental factors, such as those due to lifestyle or many disease processes (Berger, 2005). Examples of disease that result in loss of cognitive abilities are the dementias, such as vascular dementia and Alzheimer's disease (AD). All-cause dementia prevalence is approximately 6% at age 60 years and older (Bowler, 2005; Ferri et al., 2005). Alzheimer's disease is by far the most common form of dementia representing approximately two-thirds of dementia cases (Ferri et al., 2005). Prevalence rates for AD increase with age reaching 45% by age 95 years (Nussbaum & Ellis, 2003). Whereas vascular dementia appears to be influenced by shared and nonshared environmental factors without evidence of genetic influence (Bergem, Engedal, & Kringlen, 1997), late-onset AD is due to multifactorial causes and is highly heritable (Gatz et al., 1997, 2006a). The largest population-based study to date estimates heritability as high as 79% for prevalent AD with remaining variation due to nonshared environmental effects, and with no significant evidence for sex differences (Gatz et al., 2006a). Thus, individual differences in risk for Alzheimer's disease is largely due to genetic differences. One might argue that Alzheimer's disease (AD) represents a primary aging process given the high heritability and the reflection that if one lived long enough one would eventually evidence characteristics of the disease (Ebly, Parhad, Hogan, & Fung, 1994). However, evidence suggests that AD is not inevitable given that elderly into their 90s do not necessarily show clinical evidence of AD even in the case where subsequent postmortem autopsies indicate AD-like neuropathology (Morris, 1999). Additionally, while the prevalence of AD increases with age, a recent retrospective family-history study indicates that familiality of AD likely decreases with age based on negative relationship between AD risk and age of onset of family members with AD (Silverman, Ciresi, Smith, Marin, & Schnaider-Beeri, 2005).

Candidate Genes and AD

Consistent association has been observed for *APOE* e4 and Alzheimer's disease risk both in family-based studies

(Strittmatter et al., 1993) and numerous case–control studies (Rubinsztein, & Easton, 1999). No other AD gene candidate has achieved such coherent findings. Indeed a compendia of 62 candidate gene makers was recently published indicating after correcting for multiple tests, only *APOE* remained significantly associated with AD risk as well as AD-relevant biomarkers (i.e., CSF Aβ; CSF tau) (Blomqvist et al., 2006). That said, it is apparent that APOE does not fully account for the genetic bases of AD with indications that there may be four to five major genes yet to locate (Warwick Daw et al., 2000). Additional studies indicate potential genomic regions where additional candidates may lie, including regions on chromosomes 1, 9, 10, and 19 that achieved genome-wide significance (Blacker et al., 2003; Kehoe et al., 1999).

Head Injury

Head injury with loss of consciousness is an oft-reported environmental event that increases the risk for Alzheimer's disease (AD) based primarily on case–control studies (Fleminger, Oliver, Lovestone, Rabe-Hesketh, & Giora, 2003) but the effect is also found in community-based samples (Schofield et al., 1997). A compelling population-based cohort study of World War II military veterans indicates that moderate head injury with loss of consciousness during military service significantly increased the risk of dementia as well as AD (Plassman et al., 2000). Head injury may also synergistically interact with *APOE* e4 status in the risk of AD as described below.

Antioxidants

Antioxidants have been touted as potentially decreasing the risk of cognitive decline and AD (Kontush & Schekatolina, 2004). Specifically, taking vitamin supplements or eating food rich in vitamins C and E may reduce free radical damage or oxidative stress and thus stave off or delay cognitive decline (Kontush & Schekatolina, 2004; Maxwell, Hicks, Hogan, Basran, & Ebly, 2005; M. C. Morris, Evans, Bienias, Tangney, & Wilson, 2002). Community-based and population-based prospective studies suggest that both dietary and supplemental vitamin E and/or vitamin C intake may be related to lessened cognitive change (M. C. Morris et al., 2002; Maxwell et al., 2005), though not all have found a reduced risk of incident dementia or AD (Maxwell et al., 2005). Antioxidant use in the context of behavior genetic designs has yet to be explored though such analyses may prove useful in determining the nature of the antioxidant use/cognitive decline relationship, particularly in discordant twins.

"Use It or Lose It"

Education may be protective against cognitive decline and risk of AD (Anstey & Christensen, 2000). The nature of the protective association is not clearly understood as educational attainment is influenced by both genetic and environmental factors (Baker, Treloar, Reynolds, Heath, & Martin, 1996; Heath & Berg, 1985; Lichtenstein & Pedersen, 1997; Tambs, Sundet, Magnus, & Berg, 1989). Higher education levels may lead to greater cognitive reserve resulting in greater hardiness that may delay a clinical presentation of AD, or education may serve as a proxy for health and nutritional habits such that as those with higher education may have better access to resources leading to a healthier lifestyle. For example, a recent study suggests that exercise frequency is associated with reduced risk of or delayed onset of dementia (Larson et al., 2006). Consistent with the reserve capacity hypothesis, education, cognitive ability, and mental status performance in the SATSA twin study were associated phenotypically due to a common genetic factor (Pedersen, Reynolds, & Gatz, 1996). Epidemiological studies are also supportive: educational attainment remains associated with cognitive change even when relevant health variables are controlled (Lee, Kawachi, Berkman, & Grodstein, 2003). That said, there may be evidence of an environmental explanation as well: cotwin control or discordant twin pair analyses of Swedish twins provide evidence for "use it or lose it" as a protective factor against dementia or AD including the cognitive complexity of one's lifetime occupations (Andel et al., 2005) and participation in leisure activities (Crowe, Andel, Pedersen, Johansson, & Gatz, 2003).

Early Life SES-Related Factors

Lower SES in early life may be associated with a cascade of influences relevant to later cognitive aging, including poorer health and nutrition and lower rates of educational and occupational achievements. Epidemiological studies have reported that having a higher number of siblings, lower paternal occupational status and residing in urban locales before 18 may increase the risk of AD (Moceri et al., 2001; Moceri, Kukull, Emanuel, van Belle, & Larson, 2000). However, the Religious Orders Study reported that early childhood SES and community-level SES were predictive of cognitive ability but not cognitive decline or an increased risk of AD (Wilson et al., 2005). In AD-discordant twin pairs, early adult tooth loss before the age of 35 years was a significant risk factor for AD in the HARMONY study that includes twins from the Swedish Twin Registry aged 65 years and older. Adult tooth loss before the age of 35 years may be a marker of poorer early life health and/or an indicator of

inflammation processes detrimental to neuronal health (see Gatz et al., 2006b).

Tertiary Aging

Longitudinal studies have indicated that cognitive performance in those 3–6 years from death is lower than those who survive (Johansson & Berg, 1989; Small, Fratiglioni, von Strauss, & Backman, 2003; Wilson, Beckett, Bienias, Evans, & Bennett, 2003). There may be an attenuation of effect when controlling for cardiovascular disease and stroke (Hassing et al., 2002), while others have not found an association with cause of death (Small et al., 2003) suggesting a common process may be at play. Results from an investigation of genetic influences on low cognitive functioning in late adulthood indicated little or no heritability (Petrill et al., 2001). The authors suggested that nonsignificant heritability at the low end of cognitive ability may be attributable to the processes of terminal decline. Their interpretation highlights the question: To what extent is terminal decline in cognitive functioning influenced by genetic and environmental factors? Regardless of whether there is a genetic component to terminal decline of cognitive abilities, it is possible that the timing of entry into terminal decline results from environmental factors. Consequently, twin similarity for cognitive performance may decrease in late adulthood as the abilities of each member of a twin pair begin to decline at a slightly different time, even though the decline itself may be genetically influenced. In fact, analyses of cognitive data from the same sample of twins aged 80 and older as presented by Petrill et al. (2001) support this idea. As one or both members of a twin pair approached death, twin similarity for cognitive performance decreased, indicating both heterogeneity in timing of decline and a decreasing role for genetic influences on cognitive performance with approaching mortality (Johansson et al., 2004). An investigation including imputation of age at death reached a similar conclusion: genetic variance for the cognitive task was significantly lower when age at death was included in the model (Pedersen et al., 2003). It will require closely spaced assessments of cognitive performance in twins both before and during the period of terminal decline to provide sufficient data to differentiate genetic and environmental influences on timing of terminal decline versus rates of decline.

Future Directions

Dynamic Models

To best capture the nature of cognitive aging it would be ideal to capture the cognitive aging process in real time considering the push and pull relationships of the multiple processes leading up to the points of change. As that is not possible, dynamic change models may provide a fruitful approach to better understand the nature of cognitive change (see McArdle & Nesselroade, 2003; and Chapter 2, this volume). Simply correlating rate of change for two traits does not capture dynamic change but rather indicates how rates of static change are associated (McArdle & Nesselroade, 2003), e.g., if one is changing more rapidly on perceptual speed do they tend to change more rapidly on working memory? One cannot tell if change in a particular trait precedes change in the other because there is no information about the timing of change in each trait. Only one application of dynamic models to longitudinal twin data of cognitive abilities has been published to date (McArdle & Hamagami, 2003). Using dynamic approaches and identifying lead–lag relationships between two or more traits across age may move the field closer to identifying the mechanistic nature of the associations, in addition to understanding whether the relationship is due to common genetic and/or environmental influences (McArdle & Hamagami, 2003; McArdle & Nesselroade, 2003).

Distinguishing Primary, Secondary, and Tertiary Cognitive Aging

Whether one can clearly distinguish primary, secondary, and tertiary cognitive aging is important to understanding the nature(s) of cognitive aging. The search for relevant candidate genes and environments indeed bear upon this. On the one hand, if primary cognitive aging is universal then one would expect heritability to be zero as there would be no individual differences. Rather perhaps we should search for what affects the entry or timing into decline as it is clear that chronological age is only a proxy: some age cognitively at a quicker pace than others despite the same chronological age. As to secondary aging, it would be expected that environmental factors would be largely important though it is clear that even age-associated diseases such as Alzheimer's are highly heritable. Indeed, some consider AD to represent the extreme end of the continuum of cognitive change, even suggesting that the search for genes for IQ is likely to turn up genes relevant to AD and vice versa (Plomin & Spinath, 2004). The candidate gene *APOE* fits neatly within the continuum view, given its relationship both with cognitive decline and AD. Finally, distinguishing tertiary cognitive aging from normative aging and that due to disease/environmental factors is not likely to be any easier. There is no clear agreement as to when tertiary aging begins. More studies that follow participants to the end and applying growth models where the time metric is years to death may provide invaluable evidence for distinguishing tertiary cognitive aging from the others.

Identifying Genes Associated with Cognitive Aging

Given the increase in heritability indicated during the majority of adulthood (as indicated in Fig. 7.1), it is surprising that researchers have not experienced more success in identifying genes that are associated with cognitive aging. To date, research has demonstrated the impact of only handful of genes on aspects of primary, secondary, and tertiary aging. It may be that genes involved in cognitive aging account for such a small proportion of individual differences in aging trajectories that current studies lack sufficient power to detect them. In contrast, the presence of gene-by-environment interactions (see our discussion below) or complex gene–environment pathways is limiting the ability of researchers to detect cognitive aging genes. As the search progresses, it will be vital to investigate the differential role, if any, the identified genes have with regard to primary, secondary, and tertiary aging processes.

Gene–Environment Interactions

More and more studies are reporting the presence of gene–environment interactions for human behavioral traits (e.g., Caspi et al., 2002, 2003). Little is known with respect to cognitive aging, save for the findings of a possible synergistic effect of head injury coupled with positive *APOE* e4 status on the risk of AD (Jellinger, 2004; Tang et al., 1996). An indicator of possible gene–environment interaction in the normative cognitive aging literature is the pattern of increasing nonshared environmental variance with age (e.g., Reynolds et al., 2005); any unspecified variance due to gene–environment interaction will be included as nonshared environmental variance (Falconer, 1989). Future studies should examine the possibility of G × E interaction given the tools and knowledge available, including examination of within-pair differences for cognitive change in MZ twins stratified by measured genotype (Martin, 2000).

References

Akil, M., Kolachana, B. S., Rothmond, D. A., Hyde, T. M., Weinberger, D. R., & Kleinman, J. E. (2003). Catechol-O-Methyltransferase genotype and dopamine regulation in the human brain. *Journal of Neuroscience, 23*, 2008–2013.

Amargos-Bosch, M., Bortolozzi, A., Puig, M. V., Serrats, J., Adell, A., Celada, P., et al. (2004). Co-expression and in vivo interaction of serotonin1A and serotonin2A receptors in pyramidal neurons of prefrontal cortex. *Cereb Cortex, 14*, 281–299.

Andel, R., Crowe, M., Pedersen, N. L., Mortimer, J., Crimmins, E., Johansson, B., et al. (2005). Complexity of work and risk of

Alzheimer's disease: A population-based study of Swedish twins. *Journals of Gerontology, Series B: Psychological and Social Sciences, 60*, P251–P258.

Anstey, K., & Christensen, H. (2000). Education, activity, health, blood pressure and apolipoprotein E as predictors of cognitive change in old age: A review. *Gerontology, 46*, 163–177.

Baker, L. A., Treloar, S. A., Reynolds, C. A., Heath, A. C., & Martin, N. G. (1996). Genetics of educational attainment in Australian twins: Sex differences and secular changes. *Behavior Genetics, 26*, 89–102.

Bergem, A. L., Engedal, K., & Kringlen, E. (1997). The role of heredity in late-onset Alzheimer disease and vascular dementia: A twin study. *Archives of General Psychiatry, 54*, 264–270.

Bergeman, C. S. (2007). Behavioral genetics. In Birren, J. E. (Ed.), *Encyclopedia of gerontology* (2nd ed.). New York: Academic Press.

Berger, K. S. (2005). *The developing person through the lifespan* (6th ed.). New York: Worth Publishers.

Birren, J. E. (1964). *The psychology of aging.* Englewood Cliffs, NJ: Prentice-Hall.

Blacker, D., Bertram, L., Saunders, A. J., Moscarillo, T. J., Albert, M. S., Wiener, H., et al. (2003). Results of a high-resolution genome screen of 437 Alzheimer's disease families. *Human Molecular Genetics, 12*, 23–32.

Blomqvist, M. E., Reynolds, C., Katzov, H., Feuk, L., Andreasen, N., Bogdanovic, N., et al. (2006). Towards compendia of negative genetic association studies: An example for Alzheimer disease. *Human Genetics, 119*, 29–37.

Bouchard, T. J., Jr., Lykken, D. T., McGue, M., Segal, N. L., & Tellegen, A. (1990). Sources of human psychological differences: The Minnesota Twin Study of Twins Reared Apart. *Science, 250*, 223–228.

Bowler, J. V. (2005). Vascular cognitive impairment. *Journal of Neurology, Neurosurgery, and Psychiatry, 76*(Suppl 5), v35–v44.

Bray, N. J., Buckland, P. R., Hall, H., Owen, M. J., & O'Donovan, M. C. (2004). The serotonin-2A receptor gene locus does not contain common polymorphism affecting mRNA levels in adult brain. *Molecular Psychiatry, 9*, 109–114.

Bretsky, P., Guralnik, J. M., Launer, L., Albert, M., & Seeman, T. E. (2003). The role of APOE-epsilon4 in longitudinal cognitive decline: MacArthur Studies of Successful Aging. *Neurology, 60*, 1077–1081.

Bruder, G. E., Keilp, J. G., Xu, H., Shikhman, M., Schori, E., Gorman, J. M., et al. (2005). Catechol-O-methyltransferase (COMT) genotypes and working memory: Associations with differing cognitive operations. *Biological Psychiatry, 58*, 901–907.

Busse, E. W. (2002). Scope and development in the twentieth century. In M. S. John Pathy, A. J. Sinclair, & J. E. Morley (Eds.), *Principles and practice of geriatric psychiatry* (2nd ed., pp. 7–8). New York: John Wiley & Sons Ltd.

Carmelli, D., Swan, G. E., LaRue, A., & Eslinger, P. J. (1997). Correlates of change in cognitive function in survivors from the Western Collaborative Group Study. *Neuroepidemiology, 16*, 285–295.

Caspi, A., McClay, J., Moffitt, T. E., Mill, J., Martin, J., Craig, I. W., et al. (2002). Role of genotype in the cycle of violence in maltreated children. *Science, 297*, 851–854.

Caspi, A., Sugden, K., Moffitt, T. E., Taylor, A., Craig, I. W., Harrington, H., et al. (2003). Influence of life stress on depression: Moderation by a polymorphism in the 5-HTT gene. *Science, 301*, 386–389.

Crowe, M., Andel, R., Pedersen, N. L., Johansson, B., & Gatz, M. (2003). Does participation in leisure activities lead to reduced risk of Alzheimer's disease? A prospective study of Swedish twins. *Journals of Gerontology, Series B: Psychological and Social Sciences, 58*, P249–P255.

de Frias, C. M., Annerbrink, K., Westberg, L., Eriksson, E., Adolfsson, R., & Nilsson, L. G. (2004). COMT gene polymorphism is associated with declarative memory in adulthood and old age. *Behavior Genetics, 34*, 533–539.

de Frias, C. M., Annerbrink, K., Westberg, L., Eriksson, E., Adolfsson, R., & Nilsson, L. G. (2005). Catechol O-methyltransferase Val158Met polymorphism is associated with cognitive performance in nondemented adults. *Journal of Cognitive Neuroscience, 17*, 1018–1025.

de Quervain, D. J. F., Henke, K., Aerni, A., Coluccia, D., Wollmer, M. A., Hock, C., et al. (2003). A functional genetic variation of the 5-HT2a receptor affects human memory. *Nature Neuroscience, 6*, 1141–1142.

Deary, I. J., Whiteman, M. C., Pattie, A., Starr, J. M., Hayward, C., Wright, A. F., et al. (2002). Cognitive change and the APOE epsilon 4 allele. *Nature, 418*, 932.

Ebly, E. M., Parhad, I. M., Hogan, D. B., & Fung, T. S. (1994). Prevalence and types of dementia in the very old: Results from the Canadian Study of Health and Aging. *Neurology, 44*, 1593–1600.

Egan, M. F., Goldberg, T. E., Kolachana, B. S., Callicott, J. H., Mazzanti, C. M., Straub, R. E., et al. (2001). Effect of COMT Val108/158 Met genotype on frontal lobe function and risk for schizophrenia. *PNAS, 98*, 6917–6922.

Emery, C. F., Pedersen, N. L., Svartengren, M. S., & McClearn, G. E. (1998). Longitudinal and genetic effects in the relationship between pulmonary function and cognitive performance. *Journal of Gerontology: Psychological Sciences, 53B*, P311–P317.

Falconer, D. S. (1989). *Introduction to quantitative genetics* (3rd ed.). New York: John Wiley & Sons.

Ferri, C. P., Prince, M., Brayne, C., Brodaty, H., Fratiglioni, L., Ganguli, M., et al. (2005). Global prevalence of dementia: A Delphi consensus study. *The Lancet, 366*, 2112–2117.

Finch, C. E., & Kirkwood, T. B. L. (2000). *Chance, development, and aging.* New York: Oxford University Press.

Finkel, D., & McGue, M. (1993). The origins of individual differences in memory among the elderly: A behavior genetic analysis. *Psychology and Aging, 8*, 527–537.

Finkel, D., & McGue, M. (1998). Age differences in the nature and origin of individual differences in memory. *International Journal of Aging and Human Development, 47*, 217–239.

Finkel, D., & Pedersen, N. L. (2000). Contribution of age, genes, and environment to the relationship between perceptual speed and cognitive ability. *Psychology and Aging, 15*, 56–64.

Finkel, D., & Pedersen, N. L. (2001). Sources of environmental influence on cognitive abilities in adulthood. In E. L. Grigorenko & R. J. Sternberg (Eds.), *Family environment and intellectual functioning: A life-span perspective* (pp. 173–194). Mahwah, NJ: Lawrence Erlbaum Associates.

Finkel, D., & Pedersen, N. L. (2004). Processing speed and longitudinal trajectories of change for cognitive abilities: The Swedish Adoption/Twin Study of Aging. *Aging, Neuropsychology, and Cognition, 11*, 325–345.

Finkel, D., Pedersen, N. L., & McGue, M. (1995a). Genetic influences on memory performance in adulthood: Comparison of Minnesota and Swedish twin studies. *Psychology and Aging, 10*, 437–446.

Finkel, D., Pedersen, N. L., McGue, M., & McClearn, G. E. (1995b). Heritability of cognitive abilities in adult twins: Comparison of Minnesota and Swedish data. *Behavior Genetics, 25*, 421–431.

Finkel, D., Reynolds, C. A., McArdle, J. J., & Pedersen, N. L. (2005). The longitudinal relationship between processing speed and cognitive ability: Genetic and environmental influences. *Behavior Genetics, 35*, 535–549.

Fleminger, S., Oliver, D. L., Lovestone, S., Rabe-Hesketh, S., & Giora, A. (2003). Head injury as a risk factor for Alzheimer's disease: The evidence 10 years on; a partial replication. *Journal of Neurology, Neurosurgery, and Psychiatry, 74*, 857–862.

Gatz, M., Pedersen, N. L., Berg, S., Johansson, B., Johansson, K., Mortimer, J. A., et al. (1997). Heritability for Alzheimer's disease:

The study of dementia in Swedish twins. *Journal of Gerontology: Medical Sciences, 52*, M117–M125.

Gatz, M., Reynolds, C. A., Fratiglioni, L., Johansson, B., Mortimer, J. A., Berg, S., et al. (2006a). Role of genes and environments for explaining Alzheimer's disease. *Archives of General Psychiatry, 63*, 168–174.

Gatz, M., Mortimer, J. A., Fratiglioni, L., Johansson, B., Berg, S., Reynolds, C. A., & Pedersn, N. L. (2006b). Potentially modifiable risk factors for dementia in identical twins. *Alzheimer's & Dementia, 2*, 110–117.

Harris, S. E., Wright, A. F., Hayward, C., Starr, J. M., Whalley, L. J., & Deary, I. J. (2005). The functional COMT polymorphism, Val 158 Met, is associated with logical memory and the personality trait intellect/imagination in a cohort of healthy 79 year olds. *Neuroscience Letters, 385*, 1–6.

Hassing, L. B., Johansson, B., Berg, S., Nilsson, S. E., Pedersen, N. L., Hofer, S. M., et al. (2002). Terminal decline and markers of cerebro- and cardiovascular disease: Findings from a longitudinal study of the oldest old. *Journals of Gerontology, Series B: Psychological and Social Sciences, 57*, P268–P276.

Hayakawa, K., Shimizu, T., Ohba, Y., & Tomioka, S. (1992). Risk factors for cognitive aging in adult twins. *Acta Geneticae Medicae Gemellologiae, 41*, 187–195.

Heath, A. C., & Berg, K. (1985). Effects of social policy on the heritability of educational achievement. *Progress in Clinical and Biological Research, 177*, 489–507.

Hofer, S. M., Christensen, H., Mackinnon, A. J., Korten, A. E., Jorm, A. F., Henderson, A. S., et al. (2002). Change in cognitive functioning associated with apoE genotype in a community sample of older adults. *Psychology and Aging, 17*, 194–208.

Horn, J. (1988). Thinking about human abilities. In J. R. Nesselrode & R. B. Cattell (Eds.), *Handbook of multivariate experimental psychology*. New York: Plenum Press.

Jellinger, K. A. (2004). Head injury and dementia. *Current Opinion in Neurology, 17*, 719–723.

Johansson, B., & Berg, S. (1989). The robustness of the terminal decline phenomenon: Longitudinal data from the Digit-Span Memory Test. *Journal of Gerontology, 44*, P184–P186.

Johansson, B., Hofer, S. M., Allaire, J. C., Maldonado-Molina, M. M., Piccinin, A. M., Berg, S., et al. (2004). Change in cognitive capabilities in the oldest old: The effects of proximity to death in genetically related individuals over a 6-year period. *Psychology and Aging, 19*, 145–156.

Johansson, B., Whitfield, K., Pedersen, N. L., Hofer, S. M., Ahern, F., & McClearn, G. E. (1999). Origins of individual differences in episodic memory in the oldest-old: A population-based study of identical and same-sex fraternal twins aged 80 and older. *Journal of Gerontology: Psychological Sciences, 54B*, P173–P179.

Kandel, E. R. (2001). The molecular biology of memory storage: A dialogue between genes and synapses. *Science, 294*, 1030–1038.

Kehoe, P., Wavrant-De Vrieze, F., Crook, R., Wu, W. S., Holmans, P., Fenton, I., et al. (1999). A full genome scan for late onset Alzheimer's disease. *Human Molecular Genetics, 8*, 237–245.

Kontush, A., & Schekatolina, S. (2004). Vitamin E in neurodegenerative disorders: Alzheimer's disease. *Annals of the New York Academy of Sciences, 1031*, 249–262.

Korten, A. E., Henderson, A. S., Christensen, H., Jorm, A. F., Rodgers, B., Jacomb, P., et al. (1997). A prospective study of cognitive function in the elderly. *Psychological Medicine, 27*, 919–930.

Larson, E. B., Wang, L., Bowen, J. D., McCormick, W. C., Teri, L., Crane, P., et al. (2006). Exercise is associated with reduced risk for incident dementia among persons 65 years of age and older. *Annals of Internal Medicine, 144*, 73–81.

Lichtenstein, P., & Pedersen, N. L. (1997). Does genetic variance for cognitive abilities account for genetic variance in educational achievement and occupational status? A study of twins reared apart and twins reared together. *Social Biology, 44*, 77–90.

Lee, S., Kawachi, I., Berkman, L. F., & Grodstein, F. (2003). Education, other socioeconomic indicators, and cognitive function. *American Journal of Epidemiology, 157*, 712–720.

Lichtenstein, P., & Pedersen, N. L. (1997). Does genetic variance for cognitive abilities account for genetic variance in educational achievement and occupational status? A study of twins reared apart and twins reared together. *Social Biology, 44*, 77–90.

Lindenberger, U. (2001). Lifespan theories of cognitive development. In N. J. Smelser & P. B. Baltes (Eds.), *International encyclopedia of the social and behavior sciences* (pp. 8848–8854). Oxford, England: Elsevier.

Lu, T., Pan, Y., Kao, S. Y., Li, C., Kohane, I., Chan, J., et al. (2004). Gene regulation and DNA damage in the ageing human brain. *Nature, 429*, 883–891.

Malhotra, A. K., Kestler, L. J., Mazzanti, C., Bates, J. A., Goldberg, T., & Goldman, D. (2002). A functional polymorphism in the COMT gene and performance on a test of prefrontal cognition. *American Journal of Psychiatry, 159*, 652–654.

Martin, N. G. (2000). Gene–environment interaction and twin studies. In T. D. Spector, H. Snieder, & A. J. MacGregor (Eds.), *Advances in twin and sib-pair analysis* (pp. 144–150). London: Greenwich Medical Media.

Maxwell, C. J., Hicks, M. S., Hogan, D. B., Basran, J., & Ebly, E. M. (2005). Supplemental use of antioxidant vitamins and subsequent risk of cognitive decline and dementia. *Dementia and Geriatric Cognitive Disorders, 20*, 45–51.

Mayeux, R., Small, S. A., Tang, M., Tycko, B., & Stern, Y. (2001). Memory performance in healthy elderly without Alzheimer's disease: Effects of time and apolipoprotein-E. *Neurobiology and Aging, 22*, 683–689.

McArdle, J. J., & Hamagami, F. (2003). Structural equation models for evaluating dynamic concepts within longitudinal twin analyses. *Behavior Genetics, 33*, 137–159.

McArdle, J. J., & Nesselroade, J. R. (2003). Growth curve analysis in contemporary psychological research. In J. A. Schinka & W. F. Velicer (Eds.), *Handbook of psychology: Research methods in psychology* (Vol. 2, pp. xxiii, 711pp.). New York, NY: John Wiley & Sons, Inc.

McArdle, J. J., Prescott, C. A., Hamagami, F., & Horn, J. L. (1998). A contemporary method for developmental-genetic analyses of age changes in intellectual abilities. *Developmental Neuropsychology, 14*, 69–114.

McClearn, G. E., Johansson, B., Berg, S., & Pedersen, N. L. (1997). Substantial genetic influences on cognitive abilities in twins 80 or more years old. *Science, 276*, 1560–1563.

McGue, M., & Christensen, K. (2002). The heritability of level and rate-of-change in cognitive functioning in Danish twins aged 70 years and older. *Experimental Aging Research, 28*(4), 435–451.

Moceri, V. M., Kukull, W. A., Emanuel, I., van Belle, G., & Larson, E. B. (2000). Early-life risk factors and the development of Alzheimer's disease. *Neurology, 54*, 415–420.

Moceri, V. M., Kukull, W. A., Emanual, I., van Belle, G., Starr, J. R., Schellenberg, G. D., et al. (2001). Using census data and birth certificates to reconstruct the early-life socioeconomic environment and the relation to the development of Alzheimer's disease. *Epidemiology, 12*, 383–389.

Morris, J. C. (1999). Is Alzheimer's disease inevitable with age?: Lessons from clinicopathologic studies of healthy aging and very mild Alzheimer's disease. *Journal of Clinical Investigation, 104*, 1171–1173.

Morris, M. C., Evans, D. A., Bienias, J. L., Tangney, C. C., & Wilson, R. S. (2002). Vitamin E and cognitive decline in older persons. *Archives of Neurology, 59,* 1125–1132.

Mortensen, E. L., & Hogh, P. (2001). A gender difference in the association between APOE genotype and age-related cognitive decline. *Neurology, 57,* 89–95.

Nathan, B. P., Jiang, Y., Wong, G. K., Shen, F., Brewer, G. J., & Struble, R. G. (2002). Apolipoprotein E4 inhibits, and apolipoprotein E3 promotes neurite outgrowth in cultured adult mouse cortical neurons through the low-density lipoprotein receptor-related protein. *Brain Research, 928,* 96–105.

Neubauer, A. C., Spinath, F. M., Riemann, R., Angleitner, A., & Borkenau, P. (2000). Genetic and environmental influences on two measures of speed of information processing and their relation to psychometric intelligence: Evidence form the German Observational Study of Adult Twins. *Intelligence, 28,* 267–289.

Nilsson, L. G., Nyberg, L., & Backman, L. (2002). Genetic variation in memory functioning. *Neuroscience and Biobehavioral Reviews, 26,* 841–848.

Nolan, K. A., Bilder, R. M., Lachman, H. M., & Volavka, J. (2004). Catechol O-methyltransferase Val158Met polymorphism in schizophrenia: Differential effects of Val and Met alleles on cognitive stability and flexibility. *American Journal of Psychiatry, 161,* 359–361.

Nussbaum, R. L., & Ellis, C. E. (2003). Alzheimer's disease and Parkinson's disease. *New England Journal of Medicine, 348,* 1356–1364.

Ohm, T. G., Glockner, F., Distl, R., Treiber-Held, S., Meske, V., & Schonheit, B. (2003). Plasticity and the spread of Alzheimer's disease-like changes. *Neurochemical Research, 28,* 1715–1723.

Ohm, T. G., Kirca, M., Bohl, J., Scharnagl, H., Gross, W., & Marz, W. (1995). Apolipoprotein E polymorphism influences not only cerebral senile plaque load but also Alzheimer-type neurofibrillary tangle formation. *Neuroscience, 66,* 583–587.

Pedersen, N. L., Reynolds, C. A., & Gatz, M. (1996). Sources of covariation among Mini-Mental State Examination scores, education, and cognitive abilities. *Journals of Gerontology: Series B: Psychological Sciences and Social Sciences, 51B,* P55–P63.

Pedersen, N. L., Ripatti, S., Berg, S., Reynolds, C. A., Hofer, S. M., Finkel, D., et al. (2003). The influence of mortality on twin models of change. *Behavior Genetics, 33,* 161–169.

Pendleton, N., Payton, A., van den Boogerd, E. H., Holland, F., Diggle, P., Rabbitt, P. M., et al. (2002). Apolipoprotein E genotype does not predict decline in intelligence in healthy older adults. *Neuroscience Letters, 324,* 74–76.

Petrill, S. A., Johansson, B., Pedersen, N. L., Berg, S., Plomin, R., Ahern, F., et al. (2001). Low cognitive functioning in nondemented 80+-year-old twins is not heritable. *Intelligence, 29,* 75–83.

Plassman, B. L., Havlik, R. J., Steffens, D. C., Helms, M. J., Newman, T. N., Drosdick, D., et al. (2000). Documented head injury in early adulthood and risk of Alzheimer's disease and other dementias. *Neurology, 55,* 1158–1166.

Plomin, R. (1994). *Genetics and experience.* Newbury Park, CA: Sage Publications.

Plomin, R., & Spinath, F. M. (2004). Intelligence: Genetics, genes, and genomics. *Journal of Personality & Social Psychology, 86,* 112–129.

Posthuma, D., de Geus, E. J. C., & Boomsma, D. I. (2001). Perceptual speed and IQ are associated through common genetic factors. *Behavior Genetics, 31,* 593–602.

Posthuma, D., Mulder, E. J. C. M., Boomsma, D. I., & de Geus, E. J. C. (2002). Genetic analysis of IQ, processing speed and stimulus-response incongruency effects. *Biological Psychology, 61,* 157–182.

Reynolds, C. A., Finkel, D., Gatz, M., & Pedersen, N. L. (2002). Sources of influence on rate of cognitive change over time in Swedish twins: An application of latent growth models. *Experimental Aging Research, 28,* 407–433.

Reynolds, C. A., Finkel, D., McArdle, J. J., Gatz, M., Berg, S., & Pedersen, N. L. (2005). Quantitative genetic analysis of latent growth curve models of cognitive abilities in adulthood. *Developmental Psychology, 41,* 3–16.

Reynolds, C. A., Jansson, M., Gatz, M., & Pedersen, N. L. (2006a). Longitudinal change in memory performance associated with HTR2A polymorphism. *Neurobiology of Aging, 27,* 150–154.

Reynolds, C. A., Prince, J. A., Feuk, L., Gatz, M., & Pedersen, N. L. (2006b). Longitudinal memory performance during normal aging: Twin association models of APOE and other Alzheimer candidate genes. *Behavior Genetics, 36,* 185–194.

Rowe, D. C. (1994). *The limits of family influence: Genes, experience, and behavior.* New York: Guilford.

Rubinsztein, D. C., & Easton, D. F. (1999). Apolipoprotein E genetic variation and Alzheimer's disease: A meta-analysis. *Dementia and Geriatric Cognitive Disorders, 10,* 199–209.

Salthouse, T. A. (1996). The processing-speed theory of adult age differences in cognition. *Psychological Review, 103,* 403–428.

Schaie, K. W. (1994). The course of intellectual development. *American Psychologist, 49,* 304–313.

Schaie, K. W. (1996). Intellectual development in adulthood. In J. E. Birren & K. W. Schaie (Eds.), *Handbook of the psychology of aging* (pp. 266–286). San Diego, CA, US: Academic Press, Inc.

Schofield, P. W., Tang, M., Marder, K., Bell, K., Dooneief, G., Chun, M., et al. (1997). Alzheimer's disease after remote head injury: an incidence study. *Journal of Neurology, Neurosurgery and Psychiatry, 62,* 119–124.

Sheline, Y. I., Mintun, M. A., Moerlein, S. M., & Snyder, A. Z. (2002). Greater loss of 5-HT(2A) receptors in midlife than in late life. *American Journal of Psychiatry, 159,* 430–435.

Silverman, J. M., Ciresi, G., Smith, C. J., Marin, D. B., & Schnaider-Beeri, M. (2005). Variability of familial risk of Alzheimer disease across the late life span. *Archives of General Psychiatry, 62,* 565–573.

Singer, T., Verhaeghen, P., Ghisletta, P., Lindenberger, U., & Baltes, P. B. (2003). The fate of cognition in very old age: Six-year longitudinal findings in the Berlin Aging Study (BASE). *Psychology and Aging, 18,* 318–331.

Small, B. J., Fratiglioni, L., von Strauss, E., & Backman, L. (2003). Terminal decline and cognitive performance in very old age: does cause of death matter? *Psychology and Aging, 18,* 193–202.

Small, B. J., Graves, A. B., McEvoy, C. L., Crawford, F. C., Mullan, M., & Mortimer, J. A. (2000). Is APOE–epsilon4 a risk factor for cognitive impairment in normal aging? *Neurology, 54,* 2082–2088.

Small, S. A., Stern, Y., Tang, M., & Mayeux, R. (1999). Selective decline in memory function among healthy elderly. *Neurology, 52,* 1392–1396.

Strittmatter, W. J., Saunders, A. M., Schmechel, D., Pericak-Vance, M., Enghild, J., Salvesen, G. S., et al. (1993). Apolipoprotein E: High-avidity binding to beta-amyloid and increased frequency of type 4 allele in late-onset familial Alzheimer disease. *Proceedings of the National Academy of Sciences of the United States of America, 90,* 1977–1981.

Sundet, J. M., Tambs, K., Harris, J. R., Magnus, P., & Torjussen, T. M. (2005). Resolving the genetic and environmental sources of the correlation between height and intelligence: A study of nearly 2600 Norwegian male twin pairs. *Twin Research and Human Genetics, 8*(4), 307–311.

Swan, G. E., Carmelli, D., Reed, T., Harshfield, G. A., Fabsitz, R. R., & Eslinger, P. J. (1990). Heritability of cognitive performance in aging twins. The National Heart, Lung, and Blood Institute Twin Study. *Archives of Neurology, 47,* 259–262.

Swan, G. E., LaRue, A., Carmelli, D., Reed, T. E., & Fabsitz, R. R. (1992). Decline in cognitive performance in aging twins: Heritability and biobehavioral predictors from the National Heart, Lung, and Blood Institute Twin Study. *Archives of Neurology, 49*, 476–481.

Swan, G. E., Reed, T., Jack, L. M., Miller, B. L., Markee, T., Wolf, P. A., et al. (1999). Differential genetic influence for components of memory in aging adult twins. *Archives of Neurology, 56*, 1127–1132.

Takahashi, R. H., Nam, E. E., Edgar, M., & Gouras, G. K. (2002). Alzheimer beta-amyloid peptides: Normal and abnormal localization. *Histology and Histopathology, 17*, 239–246.

Tambs, K., Sundet, J. M., & Magnus, P. (1986). Genetic and environmental contributions to the covariation between the Wechsler Adult Intelligence Scale (WAIS) subtests: A study of twins. *Behavior Genetics, 16*, 475–491.

Tambs, K., Sundet, J.M., Magnus, P., & Berg, K. (1989). Genetic and environmental contributions to the covariance between occupational status, educational attainment, and IQ: A twin study. *Behavior Genetics, 19*, 209–222.

Tang, M. X., Maestre, G., Tsai, W. Y., Liu, X. H., Feng, L., Chung, W. Y., et al. (1996). Effect of age, ethnicity, and head injury on the association between APOE genotypes and Alzheimer's disease. *Annals of the New York Academy of Sciences, 802*, 6–15.

Teter, B., Xu P.-T., Gilbert, J. R., Roses, A. D., Galasko, D., & Cole, G. M. (2002). Defective neuronal sprouting by human apolipoprotein

E4 is a gain-of-negative function. *Journal of Neuroscience Research, 68*, 331–336.

Tsai, S.-J., Yu, Y. W.-Y., Chen, T.-J., Chen, J.-Y., Liou, Y.-J., Chen, M.-C., et al. (2003). Association study of a functional catechol-*O*-methyltransferase-gene polymorphism and cognitive function in healthy females. *Neuroscience Letters, 338*, 123–126.

Warwick Daw, E., Payami, H., Nemens, E. J., Nochlin, D., Bird, T. D., Schellenberg, G. D., et al. (2000). The number of trait loci in late-onset Alzheimer disease. *American Journal of Human Genetics, 66*, 196–204.

Williams, G. V., Rao, S. G., & Goldman-Rakic, P. S. (2002). The physiological role of 5-HT2A receptors in working memory. *Journal of Neuroscience, 22*, 2843–2854.

Wilson, R. S., Beckett, L. A., Bienias, J. L., Evans, D. A., & Bennett, D. A. (2003). Terminal decline in cognitive function. *Neurology, 60*, 1782–1787.

Wilson, R. S., Scherr, P. A., Hoganson, G., Bienias, J. L., Evans, D. A., & Bennett, D. A. (2005). Early life socioeconomic status and late life risk of Alzheimer's disease. *Neuroepidemiology, 25*, 8–14.

Wilson, R. S., Schneider, J. A., Barnes, L. L., Beckett, L. A., Aggarwal, N. T., Cochran, E. J., et al. (2002). The apolipoprotein E epsilon 4 allele and decline in different cognitive systems during a 6-year period. *Archieves of Neurology, 59*, 1154–1160.

Chapter 8

The Genetics and Environments of Reading: A Behavioral Genetic Perspective

Sara A. Hart and Stephen A. Petrill

Introduction

Reading is a complex skill involving the interaction of letter and word recognition, grapheme–phoneme correspondence, deciphering the meaning of a given word and, finally, the understanding of the text in its entirety (Adams, 1990). Learning to read is also a process, one that spans a wide age range, beginning in the home and continuing through formal education. Given the complexity of this construct, it is not surprising that debate continues concerning not only the processes governing the development of reading, but more fundamentally, the etiological factors that influence the development of reading. This chapter reviews the behavioral genetic research on reading in the context of the theoretical debates related to the constructs of reading. Consequently, this chapter is divided into two distinct parts. First, we will describe the current state of the literature on behavioral genetics and reading. Within this section will be a discussion on the etiology of reading ability in general, as well as the sub skills that influence reading. Following this, we will describe the genetic and environmental etiology of the relationships between different reading aspects. We will then examine molecular genetic findings, as well as research describing the impact of measures of the environment within genetically sensitive designs. The second and final section will be a brief examination of future research directions.

Learning to read begins at a very young age, even prior to beginning formal reading instruction, and continues well into the late school years (see Dale & Crain-Thoreson, 1999, for review). The "Emergent Literacy" period is one describing the interaction of a young child with literacy, prior to formal reading instruction (i.e., birth to kindergarten; Kamhi & Catts, 2005). This time is defined by the skills, knowledge, and attitudes gained by a young child in the process of learning the conventions of reading and writing (Teale & Sulzby, 1986; Whitehurst et al., 1988; Whitehurst & Lonigan, 1998). During early reading *instruction*, the main requirement for successful reading is the ability to recognize words that are already known to the child orally, also called decoding, described as "learning to read" (Chall, 1983). Decoding is measured by content-based tasks, such as (but not limited to) phonological awareness and orthographic skills. As children get older and master decoding, they begin to "read to learn", in that they use skills already learned to identify and comprehend new words not already part of their oral vocabulary (Chall, 1983). Therefore, skills such as oral vocabulary, letter knowledge, and concepts of print may be most important to younger children. However, as children age, these skills are replaced by school-based comprehension skills, a shift that occurs around the third or forth grade (Curtis, 1980). This is when a child begins to use text to learn about new content subjects that were previously unknown to the child (Snow, Burns & Griffin, 1998). The cause of this developmental shift is not completely understood. School instruction is tailored to teach reading to children in this matter, but whether this is the cause of the shift, or a social response to a biologically routed cognitive change, is not clear.

Like many other complex cognitive abilities, decades of research have suggested a strong familial component for reading and reading disability (Bakwin, 1973; DeFries, Singer, Foch, & Lewitter, 1978; Foch, DeFries, McClearn, & Singer, 1977; Matheny, Dolan, & Wilson, 1976; Stevenson, Graham, Fredman, & McLoughlin, 1987; Zerbin-Rüdin, 1967). Bakwin (1973) found an 84% concordance rate between monozygotic (MZ) twins for reading disability, and only 29% concordance in dizygotic (DZ) twins, suggesting genetic effects. Twin and adoption studies have further elucidated the genetic, shared environmental, and non-shared environmental (including error) influences on reading (Plomin, DeFries, McClearn, & McGuffin, 2001).

S.A. Petrill (✉)
Department of Human Development and Family Science, Ohio State University, Columbus, OH 43210, USA
e-mail: spetrill@ehe.osu.edu

Y.-K. Kim (ed.), *Handbook of Behavior Genetics*,
DOI 10.1007/978-0-387-76727-7_8, © Springer Science+Business Media, LLC 2009

Studies of Reading

The following section reviews the contributions of the major twin and adoption projects that have examined reading ability and disability. We will first describe the studies, and then describe their results. The first projects were cross-sectional in nature, examining reading and reading disability across a wide age range.

Colorado Reading Study: The Colorado Reading Study through the Colorado Learning Disabilities Research Center was one of the first large-scale twin projects to examine reading (Plomin & DeFries, 1983; Gayán & Olson, 2003; Olson, Forsberg, & Wise, 1994). It was a cross-sectional study that was interested in reading difficulties in older children (mean age around 11 years old, with a range of 8–20 years; Davis, Knopik, Olson, Wadsworth, & DeFries, 2001; Gayán & Olson, 2003; Olson, et al., 1994). Twins were recruited throughout Colorado, with most of the twin pairs having at least one twin with reading difficulties (459 pairs with disability, 297 without).

Register for Child Twins: Another important cross-sectional study in reading was conducted by Stevenson and colleagues in the UK from the initial Register of Child Twins (Stevenson et al., 1987; Eley, Lichtenstein, & Stevenson, 1999). This project was interested in looking at the etiology of reading in a normal population. Reading skills were assessed in 218 monozygotic twin pairs and 173 pairs of same-sex dizygotic twin pairs, aged 8–16 (mean age of around 12 years).

Western Reserve Twin Project: A further cross-sectional project was the Western Reserve Twin Project, which was interested in measuring cognitive ability and academic achievement in 132 same-sex dizygotic (DZ) twins, and 146 monozygotic (MZ) twins, aged 6–13 (Thompson, Detterman, & Plomin, 1991).

Queensland Institute of Medical Research Twin Project: A cross-sectional twin project in Australia focusing on medical histories, but also included some measures of reading and spelling (Bates et al., 2004). In total, 470 twin pairs (125 MZ; 345 DZ), mean age of 18.5 years, have been reported on.

These cross-sectional studies have led to important findings concerning reading skills across a broad age range. In particular, these studies have led to a general acceptance of the existence of *both* genetic and environmental influences on reading in childhood. However, more recently, behavioral genetic studies have begun to examine the development of reading skills from a behavioral genetic perspective. Specifically, several longitudinal projects have or are currently being conducted to examine the etiology of the development of readings skills in prospective samples.

Colorado Adoption Project: One of the first longitudinal studies interested in cognitive abilities in general was the Colorado Adoption project (CAP; DeFries, Plomin, Vandenberg, & Kuse, 1981). The investigators for this project were interested in looking at the genetic and environmental influences on cognitive development, using a sample of 245 adopted children and their biological and adoptive parents, as well as a matched sample of 245 biological control families. The children were tested at 1, 2, 3, 4, 7, 12, and 16 years of age. Reading ability was not the main focus of the project, but CAP did include one reading recognition measure, the Peabody Individual Achievement Test (PIAT; Markwardt, 1998) assessed at 7, 12, and 16 years of age (Wadsworth, Corley, Hewitt, & DeFries, 2001).

Twins Early Development Study: The Twins Early Development Study (TEDS) is a population-based sample of 7500 pairs of twins in the UK (Trouton, Spinath, & Plomin, 2002). The TEDS project was originally designed to measure the early development of communication disorders, mental impairment, and behavior problems. Twins were measured at 2, 3, 4, 7, and 10 years of age via parent testing, phone testing, teacher report, and Internet testing, with reading ability being one of the major areas of assessment at ages 7 and 10.

The International Longitudinal Twin Study: Byrne et al. (2005) examined the etiology of early reading skills in children from preschool to kindergarten across three countries. This is an ongoing project, with 325 sets monozygotic (MZ) and same-sex dizygotic (DZ) twins examined presently, 99 MZ and 107 DZ pairs of twins from the United States, and 73 MZ and 46 DZ pairs of twins from Australia. This comparison across countries allowed for increased generalizability to different cultures and school environments. The larger project described by Byrne et al. (2002) and Samuelsson et al. (2005) also includes twins from Scandinavia, but Byrne et al. (2005) is the first publication including the kindergarten data (the others are only preschool) to date.

The Western Reserve Reading Project: Petrill. Deater-Deckard, Thompson, DeThorne, and Schatschneider (2006a) examined a slightly older age range (kindergarteners and first graders, mean age = 6.1 years) than Byrne et al. (2005), and involved in-depth tester administered assessments. The ongoing project, The Western Reserve Reading Project (WRRP), has data presently from 128 MZ and 175 same-sex DZ twins, mostly from Ohio. The investigators of WRRP intend to measure the sample annually for 6 years, beginning in kindergarten or the first grade of a total sample of 450 twin pairs. A large variety of environmental measures are assessed as well.

The following section will review some of the initial behavioral genetic results of reading ability and disability as a whole, as well as the more recent results examining the characteristics of the different constructs of reading.

Behavioral Genetic Findings

Primary results from genetically sensitive designs within reading focused on univariate estimates of reading ability. "Univariate genetic estimates" define the mathematically derived point estimates of genetic and environmental influences based on the variance of just *one* trait (e.g., any one reading construct). A cautionary note should be made here. There is no method for testing the differences between various genetic and environment point estimates, so care should be taken if attempting to do so. Although overall magnitudes can be contrasted for descriptive purposes, valid comparisons can only be made by differences in the significance of the estimates.

Univariate Analysis of Reading Skills

Overall Reading Ability: Research has strongly suggested that genetic influences are important to reading ability and disability (see Pennington & Smith, 1983, 1988, for review; Stevenson et al., 1987). Some of the first point estimates of "reading ability" (based on a single measurement of reading accuracy and reading comprehension) suggested genetic influences of $h^2 = 0.44$ on average (Stevenson et al.). In contrast, shared environmental influences were lower, at around $c^2 = 0.27$ (Stevenson et al.). Pennington and Smith (1988) concluded after a lengthy and inclusive review of the literature, that "dyslexia is familial, heritable, and genetically heterogeneous" (p. 820), a remark that is still valid today.

More recently, the literature has moved away from the initial interest in simply measuring reading ability with psychometric tests, and has recognized the importance of studying the gene and environmental influences on the constructs of reading, in addition to overall reading skills.

Beginning with the Colorado Reading Project, subsequent research has suggested that the individual constructs of reading, such as phonology, rapid naming, orthography, and reading comprehension, show modest to moderate genetic influences, with varying estimates of shared environmental influences (see Olson, Gillis, Rack, DeFries, & Fulker, 1991; Gayán & Olson, 2001, 2003; Harlaar, Spinath, Dale, & Plomin, 2005; Knopik, Alarcón, & DeFries, 1998; Olson, et al., 1994; Petrill et al., 2006a). These are described in more detail below:

Phonological Awareness is one of the most important predictors of early reading (see National Reading Panel (NRP), 2000, for a review). Phonological awareness (PA) is the ability to isolate and manipulate phonemes in speech. A "phoneme" is the smallest component of language, and is the sound that a letter, or a group of letters, makes. There are approximately 41 phonemes in English, and these combine together to make syllables and whole words (NRP, 2000). There are many ways to measure phonological awareness, and one such measurement assesses "phoneme isolation", which is the ability to separate and identify phonemes in whole words (e.g., responding that /k/ is the first sound in the word "cat"). Despite the numerous measurement options, confirmatory factor analysis suggests that phonological awareness is a unitary construct (see Schatschneider, Francis, Foorman, Fletcher, & Mehta, 1999).

Cross-sectional studies across a wide age range have suggested moderate and statistically significant heritability estimates for phonological awareness, $h^2 = 0.59 - 0.77$ (Olson, et al., 1994; Gayán and Olson, 2003). Shared environment estimates were lower, and only significant when individual measures of phonological awareness were used (Olson, et al., 1994), rather than a latent factor (Gayán & Olson, 2003), $c^2 = 0.14$–0.27. Similarly, developmental studies of young readers suggest that genetics play an important role in phonological awareness. Heritability estimates are moderate and significant, ranging from $h^2 = 0.38$ to 0.60 across studies (Byrne et al., 2005; Kovas et al., 2005; Petrill et al., 2006a). Interestingly, estimates of shared environment ranged greatly in these younger samples, from $c^2 = 0.06$ to 0.43. The Petrill et al. (2006a) study was the only one to report a significant estimate (representing the highest estimate) within the developmental studies. Therefore, these results suggest that phonological awareness has moderate genetic influences, and possible shared environmental influences.

Rapid Automatized Naming (RAN) is the ability to name a series of familiar visual objects (usually letters, numbers, colors, and simple pictures of objects) as quickly as possible with the fewest amounts of errors (Wolf, Bowers, & Biddle, 2000). Initial research on rapid naming showed that the ability of the task to predict reading ability is in the *speed* at which the task is completed, not the accuracy of the task (Denckla & Rudel, 1976). The theoretical importance of RAN as an individual reading construct, separate from phonological awareness, is a continuing source of debate (Torgesen, Wagner, Rashotte, Burgess, & Hecht, 1997; Wolf & Bowers, 1999). Although RAN is moderately correlated with phonological awareness ($r = 0.38$), it also contributes significant unique variance to reading ability beyond phonological awareness, suggesting that it is an important, and separate from phonological awareness, concept of reading (Swanson, Trainin, Necoechea, & Hammill, 2003; Wolf et al., 2000).

Behavioral genetic estimates for RAN are highly consistent. Cross-sectional results suggest a moderate genetic influence, $h^2 = 0.62$, with a very low and non-significant estimate for shared environmental influence (Davis et al., 2001). Similar to cross-sectional studies, genetic influences were significant ranging from $h^2 = 0.60$

(Byrne et al., 2005) to 0.77 (Petrill et al., 2006a). Shared environment influences were not significant at any age range, with estimates ranging from $c^2 = 0.01$ to 0.17.

Orthographic Coding is another reading construct that is similar to phonological decoding. Orthographic coding is the ability to distinguish between homophones, which are words that sound the same, but differ in meaning, and sometimes spelling. Simple phonology would not help with the irregular spelling seen frequently in English, such as choosing whether "anser" or "answer" is the correct word (Gayán & Olson, 2003). Further, in some cases, homophones must be memorized as specific spelling patterns for each individual word, such as the difference between the words "their", "they're", and "there" (Gayán & Olson, 2003). This skill allows for seamless reading, so that words are immediately recognized, rather than having to be continually sounded out using phonological processes (Kamhi & Catts, 2005). The two processes of orthographic coding and phonology are moderately correlated to one another after orthogonal rotation in factor analysis ($r = 0.43$), but they also each load onto separate factors (Olson, et al., 1994).

Few behavioral genetic studies have examined orthographic coding, and only one sample has published data that could be found. Using the Colorado Reading Study sample, Gayán & Olson (2003) found a significant point heritability estimate ($h^2 = 0.87$) and a non-significant shared environmental estimate ($c^2 = 0.01$).

Reading Comprehension is another important reading skill, most relevant to older readers (NRP, 2000). Comprehension is a complex process, without an agreed upon definition or a method of measurement, hindered by the fact that it is not a unitary construct (i.e., "comprehension" does not mean any one specific ability, but rather requires many abilities; Duke, 2005; Sweet, 2005). However, what is accepted in the literature is that in addition to efficient decoding, in order to comprehend text, the reader must also actively engage the text, purposely looking for keys to understanding (NRP, 2000). The behavioral genetic literature on reading comprehension is sparse, which is reflective of the overall reading literature, which has tended to concentrate on early reading skills (e.g., decoding), and not later comprehension skills. Therefore, the only available data to present here is again from the Colorado group project of older readers, and the WRRP project in early readers. Davis et al. (2001) reported estimates of a composite measure of "reading" including measures of reading recognition, spelling, and reading comprehension. In the normal range twins, it was suggested that genetic influences were $h^2 = 0.76$ (no significance values given for all estimates) and shared environment $c^2 = 0.09$ (Davis et al., 2001). Similar effect sizes were obtained in the developmentally assessed Western Reserve Reading Project (Petrill et al., 2007). Similar to other reported reading constructs, there is a strong genetic aspect to reading comprehension, and little to no influence of shared environment.

Taken together, one common theme across the differing aspects of reading is that genetics play a significant role in reading irrespective of age. In contrast, the significance of shared environmental influences is less consistent (Byrne et al., 2002; Gayán & Olson, 2001, 2003; Petrill, Deater-Deckard, Schatschneider, & Davis, 2005; Petrill et al., 2006a). In particular, the magnitude, and therefore, in some cases, the significance, of the shared environment on reading ability and disability may vary depending on age, and amount of formal reading instruction required for mastery. Shared environmental effects are moderate and significant in preschool- and early school-aged twins for the "content-based" measures of reading (such as knowing letters, $c^2 = 0.55$, and grammar rules, $c^2 = 0.49$; Byrne et al., 2002; Petrill et al., 2006a). In contrast, "process-based" measures, such as the ability to rapidly identify and name colors ($c^2 = 0.00$), etc., do not show significant effects of shared environment (Byrne et al., 2005; Petrill et al., 2006a). This is in contrast to older children (i.e., middle-to-late grades school and high school) where the magnitude of the shared environment is negligible (Gayán & Olson, 2003). Although much more work is needed before conclusions can be drawn, these results may be reflective of the importance of variability in direct formal instruction in early education, with genetic effects increasing as the majority of students meet basic reading criteria. In the first few years of formal instruction, individual differences among children is reflective of a wider variability in the skills first learnt at home (such as the beginnings of phonology instruction, being read to, etc.), and therefore represents shared environmental influences. As more children learn the fundamental aspects of reading through formal instruction, the variability related to these shared environmental effects may diminish, leading to larger genetic estimates. This is not to imply that environments are not important for the development of reading – fifth graders have better reading skills on average than third graders. However, those individual differences in reading outcomes appear to be more heavily influenced by genetics in older children than younger children.

Classification of the Extremes of Reading Ability/Disability

In addition to examining reading skills across the range of ability, others have examined the genetic and environmental etiology of reading disability. There have been several methods employed to measure and classify disability, which is usually treated as a categorical variable (e.g., either affected or not affected). The first, and most fundamental, is to compare proband concordance rates (i.e., where both members of a pair are concordant for reading disability), giving an initial estimate of twin similarity in reading disability. If the monozygotic (MZ) twins' concordance rates for disability are higher than the dizygotic (DZ) twins, there is suggestion of genetic influences. The more the two groups are similar in concordance, the greater the suggestion of shared environmental influences.

Another method for measuring and analyzing disability is based on the idea that there is an underlying normally distributed liability continuum that contains a certain threshold. Above the threshold, an individual is in one category (e.g., not reading disabled), and below the threshold, an individual is in another category (e.g., reading disabled; Falconer, 1960). The proportion of individuals in each category is defined by the area under the normal curve compared to the area defined by the threshold curve. In twins, this technique allows for tetrachoric correlations to be calculated based on zygosity, and from these, genetic and environmental estimates can be made.

The third method for analyzing disability is the DeFries–Fulker (DF; DeFries & Fulker, 1985) analysis, which compares co-twins means, rather than variances, to determine the etiology of the extremes of reading ability. When a twin is selected for low ability, the co-twin's score will regress toward the mean. Greater dizygotic twins' regression to the mean over the monozygotic twins' is suggestive of genetic influences. Equal regression to the mean by both groups is suggestive of shared environmental effects.

DeFries–Fulker analyses for reading disability suggest that group correlations (analogous to an interclass correlation, but in this context an indicator of how far a co-twin regresses toward the population mean; Plomin et al., 2001) of reading disability between MZ twins are 0.90, and 0.65 for DZ twins. This suggests that there is a group heritability for reading disability, estimated at $h_g^2 = (0.90–0.65) \times 2 = 0.50$ (DeFries & Gillis, 1993; Plomin et al., 2001). Therefore, from this it can be suggested that 50% of the mean difference between reading-disabled and non-disabled groups is due to genetic differences between the groups. Moreover, because the magnitude of genetic effects are similar to those in the normal range, these data are consistent with the hypothesis that the genes responsible for reading disability are the same genes responsible for reading ability (Plomin & Kovas, 2005).

Similarly, liability threshold analyses have suggested that genetic effects are significant for word recognition (Harlaar, Dale, & Plomin, 2005). Tetrachoric correlations from this study are quite similar to the correlations given above, $r_{MZ} = 0.92$ and $r_{DZ} = 0.67$ (Harlaar, Dale et al., 2005). Continuous ability analyses (i.e., comparing reading-disabled groups to non-disabled groups, with both groups being treated as a normal population) have also resulted in similar results,

with heritability estimates from one study suggestive of high genetic influences on reading disability measured by a test of word recognition ($h_g^2 = 0.42$; Light, DeFries, & Olson, 1998).

As behavioral genetic studies have become more accepted within the field of reading, research has moved away from the simple univariate estimates of various reading aspects to examining the relationships among them. These constructs can be measured together with "multivariate genetic estimates", which is the operation of looking for genetic and environmental overlap, as well as uniqueness, from the covariance between multiple traits (e.g., between two reading concepts).

Multivariate Analysis of Reading – Relationships Between Components

Before our discussion on the covariance between various reading components, it is important to first mention the literature examining the gene–environment contribution to the moderate to strong correlations between general cognitive ability, "g", and individual reading components (e.g., Conners & Olson, 1990). In their cross-sectional sample, Gayán and Olson (2003) reported a moderate correlation between the genetic influences in phonological awareness and the genetics of g ($r_g = 0.56$), in a group of non-reading-disabled older twins. Estimates of shared environmental influences were all non-significant. Moreover, the influence of g on reading (reading defined here as word recognition fluency and phonemic decoding) was examined in 7-year-old twins by Harlaar, Hayiou-Thomas, and Plomin (2005) in their longitudinal sample. Approximately 17% of the total variance in reading was accounted for by genetic influences related to g. Put another way, a large portion of the genetic variance in reading was associated with genetic influences of g, as was exhibited by a genetic correlation of $r_g = 0.50$ (representing about 25% of the variance). Similarly, shared environmental influences of g accounted for about 8% of the variance in reading, and shared environmental effects also correlated highly ($r_c = 0.68$). In other words, although there was only a low influence of shared environment on reading in this sample, that amount was highly shared with g (Harlaar, Thomas et al., 2005). This would suggest that the shared environmental influences for both reading and g are almost the same in this sample. In the Western Reserve Twin Project sample, Luo, Thompson, and Detterman (2003) found similar results in very different measures of general cognitive ability and reading ability. The authors found that about 30% of the variation within a scholastic achievement measure of reading was accounted for by a mental speed component of an IQ test, and this covariation was largely due to genetic effects.

Beyond "g", there have been studies examining the unique and shared etiology between the different constructs of reading. In their cross-sectional study, Gayán & Olson (2001) examined the genetic covariance between phonological awareness and orthographic coding. Results suggested that genetic influences on orthographic coding deficits correlated 0.28–0.39 with deficits in phonological awareness measures. In a similar study in twin normal readers, Gayán and Olson (2003) found that genetics in phonological awareness correlated $r_g = 0.55$ with the genetics in orthographic coding. No shared environmental estimates were reported in the 2001 study, and none of the estimates were significant in the 2003 study, and were therefore also not reported. Bates et al. (2007) found analogous results in their sample of adolescent twins when tested for "irregular word" and "nonword" reading (representing orthographic coding and phonological coding, respectively). The authors found that 59% of the variance between the two factors was due to common genetic influences. However, there were also independent genetic effects, representing 14% of the variance in irregular word reading, and 11% of the variance in nonword reading. Shared environmental variance had no effect in this model. It appears that although there is similarity within orthographic coding and phonological awareness accounted for by overlapping genetics (perhaps because phonology skills are necessary for orthography), the two processes also show independence when measuring deficits within them.

Moreover, there has been literature examining the relationship between reading and fluency. Davis et al. (2001) were interested in the correlation between rapid automatized naming (RAN; split here into two components, numbers and letters, and colors and objects) and various reading aspects (phonological decoding and orthographic coding) in the Colorado Twin Study. Bivariate results suggested significant genetic effects, with $h^2{}_g = 0.09$–0.25 (colors and objects being the lower estimate) between phonological decoding and RAN, and $h^2{}_g = 0.21$–0.45 (colors and objects being the lower estimate) between orthographic coding and RAN.

Another ongoing debate concerns the association between rapid automatized naming (RAN) and phonological awareness (PA). As mentioned above, the two constructs are correlated ($r = 0.39$), but are factorially distinct (Swanson et al., 2003). Byrne et al. (2002) found that there was substantial shared genetic influence between RAN and PA, as well as independent genetic effects on RAN affecting reading outcomes in a longitudinal sample of young readers. Petrill, Deater-Deckard, Thompson, DeThorne, and Schatschneider (2006b) also found that RAN and PA independently predicted reading outcomes in a young reader longitudinal sample, and that there was a general genetic factor influencing the covariance between RAN, PA, and

reading outcomes (measures of reading ability that are not necessarily defined by only one construct, such as the ability to recognize and define a whole word). However, Petrill et al. (2006b) did not find significant unique genetic effects for RAN independent from phonological awareness. Moreover, Petrill et al. (2006b) did not find evidence for shared environmental effects on RAN related to PA (just independent effects for PA), but Byrne et al. did find overlap. Despite the lack of complete overlap in results, there is evidence for a genetic overlap between rapid automatized naming and phonological awareness across both studies.

The results from the univariate and multivariate analyses are suggestive of two important points. First, reading ability is a complex skill that involves multiple genetic and environmental influences – there is no one gene or environment "for" reading. Second, different aspects of reading (e.g., phonology, rapid naming) overlap and this overlap is heavily influenced by genes. However, it is important to note that statistically significant genetic estimates do not mean that environments do not matter. Reading is a learned skill, and gene–environment effects are included in heritability estimates.

Identifying Genetic and Environmental Effects

To this point, we have described methodologies that indirectly estimate genetic and environmental influences by comparing the covariance among family members with different levels of genetic relatedness (MZ vs. DZ twins, biological siblings vs. adoptive siblings). While these methods remain useful, in particular, multivariate genetic methods (as this technique allows for a more general and complete understanding of the relationship between all aspects of reading), researchers have also begun to attempt to identify specific genetic and environmental factors that affect reading ability.

Molecular Genetics

Molecular genetic research within reading has informed the literature on the location of possible candidate loci that may explain a portion of the heritability of reading and related outcomes (see Grigorenko, 2005, for review; Smith, Kimberling, Pennington, & Lubs, 1983). Linkage analysis has been used for the majority of the reports on quantitative trait loci (QTLs) in reading disability. Briefly, a "QTL" is a region of the genome that is attributed to a phenotypic trait, and is a stretch of DNA that is related to genes that explain variance in that trait. Linkage is established when a genetic marker(s) co-occurs with a particular phenotype (in this case, reading disability) within a family of known ancestry. When examining just the children of a family for linkage analysis, siblings with the given phenotype are expected to show the same version of the marker. Siblings without the phenotype are expected to show a different version of the marker related to that of the phenotype in question. Siblings in which one sib has the phenotype and the other does not are expected to show different versions of the marker.

Replicated results of QTLs via linkage studies have been reported for chromosome 15 (15q), chromosome 6 (6p21.2-23 and 6q), chromosome 2 (2p15-p16), chromosome 3 (3cen), chromosome 18 (18p), chromosome 11 (11p), chromosome 1 (1p34-p36), and the sex chromosome X (Xq27; see Fisher & DeFries, 2002; Grigorenko, 2005 for review).

The 6p21.2-23 region has attracted considerable interest. This region of chromosome 6 was first investigated because it contains the human leukocyte antigen (HLA) region, and phenotypically, there is common co-occurrence of reading disorders and autoimmune disorders within families (Cardon et al., 1994). Various studies interested in looking for reading QTLs at this site for various reading constructs have been differentially successful. Gayán et al. (1999) found evidence of a QTL for orthographic coding and phonological decoding on chromosome 6 (6p22.3B-6p22.1). Alternatively, meta-analysis has suggested that RAN is not associated with any region on chromosome 6 (Grigorenko, 2005).

Despite some successes, QTLs are generally difficult to replicate, so while markers have been identified, the field is not close to finding all of the markers associated with reading ability/disability. It has been suggested that this may be due to differential linkage by the different reading phenotypes, which are expressed at varying regions of the genome (Grigorenko et al., 1997), which the results above would seem to suggest is true. In other words, component QTLs, such as for phonological awareness, may be located in different areas in the genome than comprehension components, and may also be influenced by how that construct was measured.

More recently, association studies have been brought to bear to identify specific genes within the general chromosomal regions identified by linkage studies. Association studies use a between-subjects design, where a large group of unrelated people are studied, and any co-occurrence of a phenotype and a known marker in a large proportion of the sample suggests that a marker is responsible for the phenotype. These types of studies require large sample sizes, and are just beginning to be utilized within the reading literature (for example, Francks et al., 2004). Using association methodology based on results from previous linkage literature, Meng et al. (2005) examined the DYX2 gene on the 6p22 QTL as a possible candidate gene for reading disability. They discovered that the gene, which codes for modulation of neuronal development areas of the brain already associated with reading fluency, is a good candidate gene

associated with reading disability. Researchers have also identified another gene on the 6p chromosome (KIAA0319) that appears to be a candidate gene for reading disability, although the function of this gene has yet to be identified (Cope et al., 2005). In general, as more large-scale genetically sensitive designs (e.g., twin projects) become available for association studies, then the ability for replication of previous results within a developmental framework, as well as the opportunity to identify more candidate genes for reading disability, will come to fruition.

The Environment

Although evidence for genetic influences is substantial, this does not mean that the environment is not important for the development of reading. Reading is clearly a learned skill and intervention studies suggest that the environment has important effects. In particular, the National Reading Panel (NRP, 2000) conducted a review of instruction of phonology, looking at the effect of instruction on subsequent measures of phonological awareness, as well as on reading outcomes. The group found that instruction in phonological awareness resulted in a large effect on subsequent measures of phonological awareness (0.86, where 0.00 is no effect, and 0.80 and above is a large effect), and had a moderate effect on reading outcomes (0.53), with results lasting over a period of months. Moreover, numerous researchers have explored reading intervention techniques using scientific methodology to determine the effectiveness of such early reading skills (see Foorman & Ciancio, 2005, for review). Foorman and colleagues suggested in one study where they looked at the effect of increasing teaching effectiveness in low-income schools that "reading development was not immutable; students were improving in reading" (Foorman et al., 2006, p. 23). There is substantial evidence that reading interventions have an effect on subsequent reading outcomes, even above a students' initial reading level (Foorman et al., 2006).

Additionally, the reading socialization literature has described various characteristics of the home that are associated with reading ability. For example, parental involvement and expectations for their children are influential to early reading success, even beyond the child's own interest in reading (Briggs & Elkind, 1977; Thompson, Alexander, & Entwisle, 1988). It has been shown that parents having a high opinion of the value of education and having high expectations for their children are also related to good school outcomes (Stevenson & Lee, 1990). Further, some evidence for the importance of parental reading behaviors has been suggested. The "Family Literacy Environment", exemplified by shared reading with parents and children, trips to the library, and supervised homework time, has been shown to be positively associated with reading outcomes (Christian, Morrison, & Bryant, 1998; Griffin & Morrison, 1997). Moreover, there is the suggestion that this influence of the Family Literacy Environment is only important in future reading outcomes in pre-kindergarten, and is not influential in children's scores after kindergarten (Rebello Britto, 2001).

However, because these environmental variables are measured within biological families living together, it is not possible to separate the genetic and environmental etiology of these experiences. "Passive gene–environment correlation" occurs when children are in environments that are correlated with their parent's genes, such as if parents are avid readers, their children are more likely to grow up in a house with many books. Therefore, the possibility of passive gene–environmental effects cannot be ruled out (as the effects of genes and environments within a related family cannot be separated). Therefore, throughout the above discussion of environmental variables, the results could be, in fact, reflective of a passive gene–environment correlation. This issue must also be recognized in further studies by twin projects attempting to identify specific environmental mediators in a behavioral genetic perspective.

Adoption studies, however, are able to determine if an identified environmental variable truly is a shared environmental influence, and not a passive gene–environment correlation, as the methodology involves non-genetically related families living together. One such project has identified a specific shared environmental variable within an adoption study. In young readers, Petrill et al. (2005) found that number of books read by mothers accounted for 5% of the variance in children's vocabulary scores, suggesting that parental modeling of reading is important in reading outcomes for children (Taylor, 1983). Additionally, Petrill et al. (2005) found that the number of books read by the child accounted for 5% of the variance in phonological awareness scores.

However, there are aspects of learning to read that we yet do not understand. First, most longitudinal studies to this point have focused on the development of early reading. This means that the results so far from these twin projects have been intimately tied with the development of decoding skills only. What we do not know is if the reported effects will persist once these children master decoding and move into comprehension-based skills. As the importance of a different skill set becomes clearer, a genetic discontinuity may be seen. Secondly, we do not know how the etiology of the environment will change during this developmental shift. As children mature and complete early reading instruction, they may become more in control of their reading and their reading environments. As this occurs, shared environmental effects may lessen increasing the importance of genetic effects.

Apart from the generalized influences of genetics and environments on the development of reading, we also do not know how different aspects of reading (e.g., language, decoding, and phonology) vary as reading develops. One next step in the multivariate analyses of reading is to examine how genes and environments impact the development of reading using broader assessment strategies across a broader age range. Presently, there is considerable genetic overlap between many aspects of reading, but this overlap may be due to the fact that measures of reading used in behavioral genetic studies overemphasize decoding. Moreover, many of the research projects done to date have focused on a narrow developmental timeframe – so it is unclear how genes related to decoding in early reading are related to later reading, which may be more language based as comprehension skills become more important. In other words, are there common genes shared between comprehension and language that are not overlapping with decoding, or, are the genetic influences the same between these constructs throughout reading development.

Future Directions

Developmental Genetic Studies of Reading

Behavioral genetic studies have only just begun to assess reading as a developmental outcome. What has been found from these studies, as well as the initial results from the longitudinal projects, is threefold. First, there is a strong suggestion that genetic influences are important, even in young readers. Second, shared environments are important, but they may be limited to measures of direct instruction, as well as to young readers only. Last, although the influences may be small, it is possible to identify specific aspects of the environment that impact early reading acquisition.

Better Understanding Identified Environments and Genes

Although initial research has attempted to highlight measures of the environments associated with reading within genetically sensitive designs, there are many more unanswered questions, in particular the effect of gene–environment processes in reading. There are three types of gene–environment (GE) correlation (Plomin, DeFries, & Loehlin, 1977). Again, passive GE correlation describes how children are in environments that are correlated with their parent's genes (e.g., number of books in the home). Reactive or evocative GE correlation occurs when an environment is provided for a child that is in reaction to that child's genes (e.g., children who show early reading precocity may be given more opportunity to read harder books by their parents). Finally, active GE correlation is when a child actively searches for environmental situations as influenced by their genes. For example, children with high reading ability may ask their parents to read to them more often. Gene × environment interaction occurs when there is a nonlinear association between genetic and environmental influences. For example, the heritability of IQ is lower in low-SES families compared to children from high-SES families (Rowe, Jacobson, & Van den Oord, 1999; Scarr, 1971; Turkheimer, Haley, Waldron, D'Onofrio, & Gottesman, 2003).

Genetically sensitive designs examining reading are in a unique position to comment on gene–environment processes, and much research needs to be done. The importance of such a research is great. In particular, there is a significant influence of genetics on reading ability and disability. How these influences impact the probability of coming into contact with environments associated with reading remains an important question. Understanding these effects will allow us not only to better understand reading and its components, but also the environments associated with reading. Both univariate and multivariate analyses will be useful in this regard, allowing us to understand how specific environmental influences interact with genetic effects.

With respect to integrating molecular genetic, there is potential to examine the relationship between specific DNA polymorphisms and measured environments. Put another way, there is potential for studies of possible genotype × environment interactions, which describe genetic susceptibility to specific environments. Although this type of research within reading has not yet been conducted, there has been notable success in the area of socioemotional development (see Moffitt, Caspi, & Rutter, 2005, for a review) and thus, may be extended to reading outcomes. These efforts may allow for better and earlier identification of reading problems and lead to targeted environmental interventions.

References

Adams, M. J. (1990). *Learning to read: Thinking and learning about print.* Cambridge, MA: MIT Press.

Bakwin, H. (1973). Reading disability in twins. *Developmental Medicine and Child Neurology, 15*, 184–187.

Bates, T. C., Castles, A., Coltheart, M., Gillespie, N., Wright, M., & Martin, N. G. (2004). Behaviour genetics analyses of reading and spelling: A component processes approach. *Australian Journal of Psychology, 56*(2), 115–126.

Bates, T. C., Castles, A., Luciano, M., Wright, M. J., Coltheart, M., & Martin, N. G. (2007). Genetic and environmental bases of reading and spelling: A unified genetic dual route model. *Reading and Writing, 20*(1–2), 147–171.

Briggs, C., & Elkind, D. (1977). Characteristics of early readers. *Perceptual and Motor Skills, 44*, 1231–1237.

Byrne, B., Delaland, C., Fielding-Barnsley, R., Quain, P., Samuelsson, S., Høien, T., et al. (2002). Longitudinal twin study of early reading development in three countries: Preliminary results. *Annals of Dyslexia, 52*, 49–73.

Byrne, B., Wadsworth, S., Corley, R., Samuelsson, S., Quain, P., & DeFries, J. C. (2005). Longitudinal twin study of early literacy development: Preschool and kindergarten phases. *Scientific Studies of Reading, 9*(3), 219–235.

Cardon, L. R., Smith, S. D., Fulker, D. W., Kimberling, W. J., Pennington, B. F., & DeFries, J. C. (1994). Quantitative trait locus for reading disability on chromosome 6. *Science, 266*(5183), 276–279.

Chall, J. S. (1983). *Stages of reading development.* New York: McGraw-Hill.

Christian, K., Morrison, F. J., & Bryant, F. B. (1998). Predicting kindergarten academic skills: Interactions among child care, maternal education, and family literacy environments. *Early Child Research Quarterly, 13*(3), 501–521.

Conners, F., & Olson, R. K. (1990). Reading comprehension in dyslexic and normal readers: A component-skills analysis. In D. A. Balota, G. B. Flores-d'Arcais, & K. Rayner (Eds.), *Comprehension processes in reading* (pp. 557–579). Hillsdale, NJ: Erlbaum.

Cope, N., Harold, D., Hill, G., Moskvina, V., Stevenson, J., Holmans, P., et al. (2005). Strong evidence that KIAA0319 on chromosome 6p is a susceptibility gene for developmental dyslexia. *American Journal of Human Genetics, 76*(4), 581–591.

Curtis, M. E. (1980). Development of components of reading skill. *Journal of Educational Psychology, 72*, 656–669.

Dale, P. S., & Crain-Thoreson, C. (1999). Language and literacy developmental perspective. *Journal of Behavioral Education, 9*(1), 23–33.

Davis, C. J., Knopik, V. S., Olson, R. K., Wadsworth, S. J., & DeFries, J. C. (2001). Genetic and environmental influences on rapid naming and reading ability: A twin study. *Annals of Dyslexia, 51*, 231–247.

DeFries, J. C., & Fulker, D. W. (1985). Multiple regression analysis of twin data. *Behavior Genetics, 15*, 467–473.

DeFries, J. C., & Gillis, J. J. (1993). Genetics and reading disability. In R. Plomin & G. E. McClearn (Eds.), *Nature, nurture, and psychology* (pp. 121–145). Washington, DC: American Psychological Society.

DeFries, J. C., Plomin, R., Vandenberg, S. G., & Kuse, A. R. (1981). Parent-offspring resemblance for cognitive abilities in the Colorado Adoption Project Biological adoptive and control parents and one year-old children. *Intelligence, 5*, 245–277.

DeFries, J. C., Singer, S. M., Foch, T. T., & Lewitter, F. I. (1978). Familial nature of reading disability. *British Journal of Psychiatry, 132*, 361–367.

Denckla, M. B., & Rudel, R. G. (1976). Rapid automatized naming (R.A.N.): Dyslexia differentiated from other learning disabilities. *Neuropsychologia, 14*, 471–479.

Duke, N. K. (2005). Comprehension of what for what: Comprehension as a nonunitary construct. In S. G. Paris & S. A. Stahl (Eds.), *Children's reading: Comprehension and assessment* (pp. 93–104). Mahwah, NJ: Lawrence Erlbaum.

Eley, T. C., Lichtenstein, P., & Stevenson, J. (1999). Sex differences in the etiology of aggressive and nonaggressive antisocial behavior: Results from two twin studies. *Child Development, 70*, 155–168.

Falconer, D. S. (1960). *Introduction to quantitative genetics.* London: Oliver and Boyd London.

Fisher, S. E., & DeFries, J. C. (2002). Developmental dyslexia: Genetic dissection of a complex cognitive trait. *Nature Reviews: Neuroscience, 3*, 767–780.

Foch, T. T., DeFries, J. C., McClearn, G. E., & Singer, S. M. (1977). Familial patterns of impairment in reading disability. *Journal of Educational Psychology, 69*(4), 316–329.

Foorman, B. R., & Ciancio, D. J. (2005). Screening for secondary intervention: Concept and context. *Journal of Learning Disabilities, 38*(6), 494–499.

Foorman, B. R., Schatschneider, C., Eakin, M. N., Fletcher, J. M., Moats, L. C., & Francis, D. J. (2006). The impact of instructional practices in grades 1 and 2 on reading and spelling achievement in high poverty schools. *Contemporary Educational Psychology, 31*, 1–29.

Francks, C., Paracchini, S., Smith, S. D., Richardson, A. J., Scerri, T. S., Cardon, L. R., et al. (2004). A 77-kilobase region of chromosome 6p22.2 is associated with dyslexia in families from the United Kingdom and from the United States. *American Journal of Human Genetics, 75*(6), 1046–1058.

Gayán, J., & Olson, R. K. (2001). Genetic and environmental influences on orthographic and phonological skills in children with reading disabilities. *Developmental Neuropsychology, 20*(2), 487–511.

Gayán, J., & Olson, R. K. (2003). Genetic and environmental influences on individual differences in printed word recognition. *Journal of Experimental Child Psychology, 84*, 97–123.

Gayán, J., Smith, S. D., Cherny, S. S., Cardon, L. R., Fulker, D. W., Brower, A. M., et al. (1999). Quantitative-trait locus for specific language and reading deficits on chromosome 6p. *American Journal of Human Genetics, 118*, 27–39.

Griffin, E. A., & Morrison, F. J. (1997). The unique contribution of home literacy environment to differences in early literacy skills. *Early Child Development and Care, 127–128*, 233–243.

Grigorenko, E. L. (2005). A conservative meta-analysis of linkage and linkage-association studies of developmental dyslexia. *Scientific Studies of Reading, 9*(3), 285–316.

Grigorenko, E. L., Wood, F. B., Meyer, M. S., Pauls, J. E. D., Hart, L. A., Speed, W. C., et al. (1997). Susceptibility loci for distinct components of developmental dyslexia on chromosome 6 and 15. *American Journal of Human Genetics, 66*, 715–723.

Harlaar, N., Dale, P. S., & Plomin, R. (2005). Telephone testing and teacher assessment of reading skills in 70 year-olds: II. Strong genetic overlap. *Reading and Writing, 18*, 401–423.

Harlaar, N., Hayiou-Thomas, M. E., & Plomin, R. (2005). Reading and general cognitive ability: A multivariate analysis of 7-year-old twins. *Scientific Studies of Reading, 9*(3), 197–218.

Harlaar, N., Spinath, F. M., Dale, P. S., & Plomin, R. (2005). Genetic influences on word recognition abilities and disabilities: A study of 7 year old twins. *Journal of Child Psychology and Psychiatry and Allied Disciplines, 46*(4), 373–384.

Kamhi, A. G., & Catts, H. W. (2005). Reading development. In H. W. Catts & A. G. Kamhi (Eds.), *Language and reading disabilities* (pp. 26–49). Boston: Pearson.

Knopik, V. S., Alarcón, M., & DeFries, J. C. (1998). Common and specific gender influences on individual differences in reading performance: A twin study. *Personality and Individual Differences, 25*(2), 269–277.

Kovas, Y., Hayiou-Thomas, M. E., Oliver, B., Dale, P. S., Bishop, D. V. W., & Plomin, R. (2005). Genetic influences in different aspects of language development: The etiology of language skills in 4.5-year-old twins. *Child Development, 76*(3), 632–651.

Light, J. G., DeFries, J. C., & Olson, R. K. (1998). Multivariate behavioral genetic analysis of achievement and cognitive measures in reading-disabled and control twin pairs. *Human Biology, 70*, 215–237.

Luo, D., Thompson, L. A., & Detterman, D. K. (2003). Phenotypic and behavioral genetic covariation between elemental cognitive components and scholastic measures. *Behavior Genetics, 33*(3), 221–246.

Markwardt, F. C., Jr. (1998). *Peabody individual achievement test-Revised/Normative update (PIAT-R/NU) manual*. Circle Pines, MN: American Guidance Service.

Matheny, A. P., Jr., Dolan, A. B., & Wilson, R. S. (1976). Twins with academic learning problems: Antecedent characteristics. *American Journal of Orthopsychiatry, 46*(3), 464–469.

Meng, H., Smith, S. D., Hager, K., Held, M., Liu, J., Olson, R. K., et al. (2005). DCDC2 is associated with reading disability and modulates neuronal development in the brain. *Proceedings of the National Academy of Sciences of the United States of American, 102*(51), 18763.

Moffitt, T. E., Caspi, A., & Rutter, M. (2005). Strategy for investigating interactions between measured genes and measured environments. *Archives of General Psychiatry, 62*(5), 473–481.

National Reading Panel. (2000). *Teaching children to read: An evidence-based assessment of the scientific research literature on reading and its implications for reading instruction*. (NIH Publication No. 00-4769). Washington, DC: U.S. Government Printing Office.

Olson, R. K., Forsberg, H., & Wise, B. (1994). Genes, environment, and the development of orthographic skills. In V. W. Berninger (Ed.), *The varieties of orthographic knowledge 1: Theoretical and developmental issues* (pp. 27–71). Dordrecht, The Netherlands: Kluwer Academic Publishers.

Olson, R. K., Gillis, J. J., Rack, J. P., DeFries, J. C., & Fulker, D. W. (1991). Confirmatory factor analysis of word recognition and process measures in the Colorado Reading Project. *Reading and Writing, 3*, 235–248.

Pennington, B. F., & Smith, S. D. (1983). Genetic influences on learning disabilities and speech and language disorders. *Child Development, 54*(2), 369–387.

Pennington, B. F., & Smith, S. D. (1988). Genetic influences on learning disabilities: An update. *Journal of Consulting and Clinical Psychology, 56*(6), 817–823.

Petrill, S. A., Deater-Deckard, K., Schatschneider, C., & Davis, C. (2005). Measured environmental influences on early reading: Evidence from an adoption study. *Scientific Studies of Reading, 9*(3), 237–259.

Petrill. S. A., Deater-Deckard, K., Thompson, L. A., DeThorne, L. S., & Schatschneider, C. (2006a). Reading skills in early readers: Genetic AND shared environmental influences. *Journal of Learning Disabilities, 39*(1), 48–55.

Petrill. S. A., Deater-Deckard, K., Thompson, L. A., DeThorne, L. S., & Schatschneider, C. (2006b). Genetic and environmental effects of serial naming and phonological awareness on early reading outcomes. *Journal of Educational Psychology, 98*(1), 112–121.

Petrill, S. A., Deater-Deckard, K., Thompson, L., Schatschneider, C., DeThorne, L., & Vandenberg, D. (2007). Longitudinal genetic analysis of early reading: The Western Reserve Reading Project. *Reading and Writing, 20*(1–2), 127–146.

Plomin, R., & DeFries, J. C. (1983). The Colorado Adoption Project. *Child Development, 54*(2), 276–289.

Plomin, R., DeFries, J. C., & Loehlin, J. (1977). Genotype-environment interaction and correlation in the analysis of human behavior. *Psychological Bulletin, 84*, 309–322.

Plomin, R., DeFries, J. C., McClearn, G. E., & McGuffin, P. (2001). *Behavioral Genetics* (4th ed.). New York: W. H. Freeman.

Plomin, R., & Kovas, Y. (2005). Generalist genes and learning disabilities. *Psychological Bulletin, 131*(4), 592–617.

Rebello Britto, P. (2001). Family literacy environments and young children's emerging literacy skills. *Reading Research Quarterly, 36*(4), 346–347.

Rowe, D. C., Jacobson, K. C., & Van den Oord, E. J. C. G. (1999). Genetic and environmental influences on vocabulary IQ: Parent education as moderator. *Child Development, 70*(5), 1151–1162.

Samuelsson, S., Byrne, B., Quain, P., Wadsworth, S., Corley, R., DeFries, J. C., et al. (2005). Environmental and genetic influences on prereading skills in Australia, Scandinavia, and the United States. *Journal of Educational Psychology, 97*(4), 705–722.

Scarr, S. (1971). Race, social class, and IQ. *Science, 174*(4016), 1285–1295.

Schatschneider, C., Francis, D. J., Foorman, B. R., Fletcher, J. M., & Mehta, P. (1999). The dimensionality of phonological awareness: An application of item response theory. *Journal of Educational Psychology, 91*(3), 439–449.

Smith, S. D., Kimberling, W. J., Pennington, B. F., & Lubs, H. A. (1983, March 18). Specific reading disability: Identification of an inherited form through linkage analysis. *Science, 219,* 1345–1347.

Snow, C. E., Burns, M. S., & Griffin, P. (1998). *Preventing reading difficulties in young children.* Washington, DC: National Academy Press.

Stevenson, J., Graham, P., Fredman, G., & McLoughlin, V. (1987). A twin study of genetic influences on reading and spelling ability and disability. *Journal of Child Psychology and Psychiatry, 28*(2), 229–247.

Stevenson, H. W., & Lee, S. Y. (1990). Contexts of achievement. *Monographs of the Society for Research in Child Development, 55(3–4),* Serial #221.

Swanson, H. L., Trainin, G., Necoechea, D. M., & Hammill, D. D. (2003). Rapid naming, phonological awareness, and reading: A meta-analysis of the correlation evidence. *Review of Educational Research, 73*(4), 407–440.

Sweet, A. P. (2005). Assessment of reading comprehension: The RAND reading study group vision. In S. G. Paris & S. A. Stahl (Eds.), *Children's reading: Comprehension and assessment* (pp. 3–12). Mahwah, NJ: Lawrence Erlbaum.

Taylor, D. (1983). *Family literacy: Young children learning to read and write.* Exeter, NH: Heineman.

Teale, W. H., & Sulzby, E. (Eds.). (1986). *Emergent literacy: Writing and reading.* Norwood, NJ: Ablex.

Thompson, M. S., Alexander, K. L., & Entwisle, D. R. (1988). Household composition, parental expectations, and school environment. *Social Forces, 67*(2), 424–451.

Thompson, L. A., Detterman, D. K., & Plomin, R. (1991). Associations between cognitive abilities and scholastic achievement: Genetic overlap but environmental differences. *Psychological Science, 2*(3), 158–165.

Torgesen, J. K., Wagner, R. K., Rashotte, C. A., Burgess, S., & Hecht, S. (1997). Contributions of phonological awareness and rapid automic naming to the growth of word-reading skills in second grade to fifth-grade children. *Scientific Studies of Reading, 1,* 161–155.

Trouton, A., Spinath, F. M., & Plomin, R. (2002). Twins' Early Development Study (**TEDS**): A multivariate, longitudinal genetic investigation of language, cognition and behaviour problems in childhood. *Twin Research,* 5, 444–448.

Turkheimer, E., Haley, A., Waldron, M., D'Onofrio, B., & Gottesman, I. I. (2003). Socioeconomic status modifies heritability of IQ in young children *Psychological Science, 14*(6), 623–628.

Wadsworth, S. J., Corley, R. P., Hewitt, J. K., & DeFries, J. C. (2001). Stability of genetic and environmental influences on reading performance at 7, 12, and 16 years of age in the Colorado Adoption Project. *Behavior Genetics, 31*(4), 353–359.

Whitehurst, G. J., Falco, F., Lonigan, C. J., Fischel, J. E., DeBaryshe, B. D., Valdez-Menchaca, M. C., et al. (1988). Accelerating language development through picture-book reading. *Developmental Psychology, 24,* 552–558.

Whitehurst, G. J., & Lonigan, C. J. (1998). Child development and emergent literacy. *Child Development, 69*(3), 848–872.

Wolf, M., & Bowers, P. G. (1999). The double-deficit hypothesis for the development dyslexics. *Journal of Educational Psychology, 91,* 415–438.

Wolf, M., Bowers, P. G., & Biddle, K. (2000). Naming-speed processes, timing, and reading: A conceptual review. *Journal of Learning Disabilities, 33*(4), 387–407.

Zerbin-Rüdin, E. (1967). Congenital word-blindness. *Bulletin of the Orton Society, 17,* 47–56.

Chapter 9

Behavior-Genetic and Molecular Studies of Disorders of Speech and Language: An Overview

Elena L. Grigorenko

Introduction

As of September 2, 2007 when this chapter was written, there were an estimated 994,638 words in the English language (http://www.languagemonitor.com/). Everyone agrees that we are not born with the knowledge of these words and the rules by which they are strung together. A newborn child does not know or use any words. However, a child entering school has a lexicon of ~13,000 words which, by the end of schooling, expands to ~60,000 words and during adulthood reaches ~120,000 words (Pinker, 1999).

How this ascent from 0 to 120,000 (and for a tiny minority, all the way to 994,638!) occurs and how we acquire the rules for assembling endless combinations of words into meaningful sentences have been central questions in the fields of developmental and cognitive psychology and linguistics since their emergence. This field of study within and across disciplines is typically referred to as language acquisition. Although everyone agrees that language acquisition has distinct developmental stages, there is also consensus on the high number of individual differences that characterize this process. Mounting evidence indicates that a very important source of these individual differences is the genome. The goal of this chapter is to review the behavior- and molecular-genetic studies elucidating the role of the genome in the atypical acquisition of language that manifests itself in disorders of speech and language.

Defining the Phenotypes for Studies of DSL

The first and, arguably, most essential step in behavior- and molecular-genetic studies of complex behaviors is to define

the phenotypes to be investigated. It is never a trivial task, and this certainly holds true for studies of typical and atypical language acquisition. Below I briefly review a number of issues that are important to consider in this context.

Concepts, Terminology, and Theories

Before reviewing the behavior- and molecular-genetic studies of disorders of speech and language (DSL), it is critical to introduce a number of concepts and terms that will be used throughout the chapter. These terms pertain to both the psychological and linguistic "texture" of language and the deficiencies observed in children and adults with DSL.

The *phonetic* (or *phonological*) representation of a word pertains to its pronunciation and understanding, whether spoken or written. *Phonemes* are the smallest units of sound that form the texture of meaningful speech, whether oral or printed, in a given language. Every word in a language is characterized by a particular internal structure. *Morphology* pertains to studies of word structure or word formation and relationships between words; correspondingly, *morphological* and *lexical* representations capture this type of word-related information. When words are connected together in a manner specific to particular languages, they form sentences. Internal structures of sentences and rules of word connectedness are captured by *syntactic* representations. Every real word in a language has a meaning or several meanings associated with it. This aspect of words and languages is captured by *semantic* representations. Finally, the knowledge of the appropriate use of words, phrases, and sentences in the context of discourse or conversation is captured by *pragmatic* representation.

A fundamental question in studies of language acquisition is how all these psychological representations are acquired and what goes wrong in the formation of these representations that leads to the development of a DSL. There are many competing theories regarding these issues and here I mention these theoretical debates only briefly and selectively.

E.L. Grigorenko (✉)
Child Study Center, Department of Psychology, Department of Epidemiology & Public Health, Yale University, New Haven, CT 06519-1124, USA
e-mail: elena.grigorenko@yale.edu

Y.-K. Kim (ed.), *Handbook of Behavior Genetics*,
DOI 10.1007/978-0-387-76727-7_9, © Springer Science+Business Media, LLC 2009

For example, there are competing accounts of the acquisition and production of regular and irregular forms of past tense and plural nouns in English (McClelland & Patterson, 2002; Pinker & Ullman, 2002). Specifically, a dual-mechanism theory (Pinker, 1991) postulates that the generation of regular forms (e.g., *worked* and *flowers*) is mechanistically different from the generation of irregular forms (e.g., *sang* and *cacti*). In detail, the generation of the first type of words is driven by rules and procedural knowledge, whereas the generation of the second type of words is explained by rule learning and declarative knowledge (Pinker, 1999); the two mechanisms have different brain correlates (Ullman et al., 1997). This framework is one of the number of specific realizations of the general approach that emphasizes the primacy of learning mechanisms that involve the abstraction of categorical, symbolic rules and the memorization of arbitrary facts that constitute exceptions to those rules (Pinker, 1999).

An opposing position, known as the distributed processing (PDP) or connectionist model, argues for the similarities between generating rule-driven and exceptional forms and views them as products of a single connectionist lexical processing system whose construction is driven by statistical learning (Rumelhart & McClelland, 1987). In this parallel processing model, different forms of linguistic representations are "graded functions of multiple probabilistic constraints" (Haskell, MacDonald, & Seidenberg, 2003, p. 130). In other words, in typical development, language (both spoken and written)-related information is acquired by experiencing properties of different linguistic elements (e.g., phonemes, graphemes, morphemes, words, sentences) in different contexts, and, thus, establishing the characteristics and constraints of permissible and impermissible linguistic uses. When trained, the model generates performance patterns with some variability in outcome that, arguably, can be viewed as analogs of individual differences (Zevin & Seidenberg, 2006). Similarly, an analog of an impairment has been modeled by introducing random damage to the learning connectionist network (Plaut, 1996).

Although this debate originated in the domain of spoken languages, it has been extended into the domain of reading studies as well [see (Coltheart, Curtis, Atkins, & Haller, 1993; Coltheart, Rastle, Perry, Langdon, & Ziegler, 2001) vs. (Seidenberg, 2005; Seidenberg & McClelland, 1989), for dual-route and parallel processing/triangle models, respectively].

This debate also has an etiological twist. The first model is identified as the nativist or innate model of language acquisition, whereas the second model is typically viewed as the "learned from experience" model. Yet, in spite of such differentiation, it is important to note that, in fact, both models provide ample room for the role of genes in model acquisition by assuming that genes might influence the architecture, functional properties, and/or learning capacities of the models.

Typologies of Disorders of Speech and Language

As mentioned earlier, there is substantial evidence that a sizable group of children have difficulties mastering language acquisition in the typical way. These difficulties have been observed and studied in children for a long time, but, presently, there is no clear classification of the related difficulties (Verhoeven & van Balkom, 2004). In fact, there are many alternative classifications, some of which are institutionalized in diagnostic manuals or conventionally used in research and clinical practice. For example, below are categories used in the world's two leading classification manuals, published by the World Health Organization (ICD-10, http://www.who.int/classifications/icd/en/) and the American Psychiatric Association (DSM-IV, http://www.psychiatryonline.com/resourceTOC.aspx?resourceID=1), respectively. Specifically, the following categories are listed in the corresponding manuals.

- ICD-10: ICD-10's Chapter V (*Mental and Behavioural Disorders*)

  - *Disorders of Psychological Development*

    - *Specific Developmental Disorder of Speech and Language* (Specific Speech Articulation Disorder, Expressive Language Disorder, Receptive Language Disorder, Acquired Aphasia with Epilepsy, Other Developmental Disorders of Speech and Language, and Developmental Disorder of Speech and Language, Unspecified).

- DSM-IV: *Disorders Usually First Diagnosed in Infancy, Childhood, or Adolescence*

  - *Communication Disorders* (Expressive Language Disorder, Mixed Receptive–Expressive Language Disorder, Phonological Disorder, Stuttering, and Communication Disorder NOS).

Even a cursory examination of these lists suggests substantial uncertainty and disagreement in the field. Moreover, as mentioned earlier, the research literature does not directly map itself on either diagnostic manual. In this chapter, I capitalize on the distinctions made in the research literature between disorders of speech and sound (Speech and Sound Disorders, SSD) and disorders of language (Specific Language Impairment, SLI). In presenting behavior- and molecular-genetic findings, I use these two categories broadly, including in SSD such disorders as apraxia and stuttering, and in the SLI category both receptive and expressive language disorders.

Two additional remarks are in order before I review the behavior- and molecular-genetic findings on DSL.

First, as is the case for many other complex common disorders, there is an ongoing debate about how to define DSL: through their categorical distinctness as defined by a bimodal distribution of some underlying liability trait, or through their representation of the extreme lower end of normal developmental variation, where the distribution does not separate typical and atypical language development (Bishop, 1994b, 1995; Cole, Schwartz, Notari, Dale, & Mills, 1995). This debate is central to assumptions about the specificity vs. generality of the genetic mechanisms affecting typical and atypical (disordered) language acquisition. Specifically, if the distributions of typical and atypical skills are assumed to be discrete, then different genetic mechanisms[1] might be assumed to contribute to the acquisition of different skills. If disordered skills are viewed as the lower end of the normal distribution of language acquisition skills, then the same genetic mechanisms can be assumed to influence the manifestation of both typical and atypical skills.

Second, although researchers may disagree on the characterization of the underlying linguistic–cognitive problems, their findings on the manifestations of the disorders are quite consistent. Thus, children with SSD demonstrate inadequate mastery of sound production and perception (Shriberg, 2002). Children with SLI demonstrate great difficulty acquiring and using grammatical markers that express purely structural relations (the so-called functional categories), such as various tense and agreement markers, including the past tense -ed, the third person singular -s, the auxiliary verb do, and so on (Cleave & Rice, 1997; Marshall & van der Lely, 2006; Redmond & Rice, 2001; Rice, Cleave, & Oetting, 2000; Rice, Tomblin, Hoffman, Richman, & Marquis, 2004; Rice, Wexler, & Cleave, 1995; Rice, Wexler, Marquis, & Hershberger, 2000; Rice, Wexler, & Redmond, 1999; van der Lely, 2005; van der Lely, Rosen, & Adlard, 2004; van der Lely & Stollwerck, 1996, 1997). Yet, there are multiple theories attempting to explain specific

disorders. For example, intending to account for SLI, researchers have suggested that this impairment is (1) a specific distortion of syntax acquisition (Rice et al., 1995); (2) a result of poor auditory perception (Bates, 2004); (3) a consequence of phonological deficits (Joanisse, 2004); or (4) a deficit in processing capacity (Bishop, 1994a), or some combination of all these factors (Leonard, 1998). Of interest to this discussion is that behavior-genetic research has been used to clarify the theoretical and phenomenological work surrounding DSL. For example, in one study, researchers (Bishop, Adams, & Norbury, 2006) attempted a direct comparison of selected theories by measuring indices of phonological processing (short-term phonological memory) and grammaticality (a combined index of verb inflections) in a set of twins, ~66% of whom were selected because of their language difficulties and ~33% of whom were chosen at random from a large sample with no evidence of a language disorder. Both measures provided high estimates of heritability for poor performance (0.61 and 0.74, for phonological and grammatical indicators, respectively). Bivariate analyses revealed no indication of genetic overlap between the two measures, suggesting that they were probably driven by different genetic etiologies. This finding is challenging to explain within the context of theories that view SLI as an outcome of a phonological deficiency of some kind.

In summary, given the debates in the field regarding general mechanisms of language acquisition and specifics of definition, manifestation, typologies, and theories of different DSL, the task of defining phenotypes in behavior- and molecular-genetic studies of DSL is critical. Therefore, in the discussion below I distinguish between specific phenotypes when these distinctions are important for the discussion.

Current Research and Issues

An Overview of Behavior- and Molecular-Genetic Studies of DSL

This main section of the chapter is subdivided into subsections covering studies of familiality and heritability as well as molecular-genetic studies of DSL.

Familiality and Heritability of DSL

There is substantial evidence that DSL cluster in families. Investigations of this clustering have been carried out using a number of methodologies, including biological (Choudhury & Benasich, 2003; Kidd, Heimbuch, Records, Oehlert, & Webster, 1980; Lahey & Edwards, 1995; Lewis, Ekelman, & Aram, 1989; Neils & Aram, 1986; Rice,

---

[1] In fact, this "line of reasoning" can be also applied, hypothetically, to different mutations within a single gene, with these mutations resulting in different outcomes. For example, a null mutation (i.e., a mutation that leads to the nonsynthesis of the protein coded by a mutated gene or to the synthesis of a defective protein lacking its typical functional properties) in a gene contributing to genetic variation in language acquisition might result in the formation of a dichotomous distribution with individuals who possess such a mutation not able to acquire language at all. A less deleterious mutation (i.e., one that challenges some but preserves other functions of a protein) might be associated with a less severe phenotype. Correspondingly, if many mutations of different degrees of deleteriousness exist within one gene, these various mutations (when, of course, considered, in a very large sample of individuals) might result in a more continuous distribution.

Haney, & Wexler, 1998; Tallal et al., 2001; Tallal, Ross, & Curtiss, 1989; Tomblin, 1989) and adoptive family (Felsenfeld & Plomin, 1997) and twin designs (Bishop, North, & Donlan, 1995; Felsenfeld et al., 2000; Howie, 1981; Lewis & Thompson, 1992; Tomblin & Buckwalter, 1998). Evidence from these studies has clearly demonstrated the importance of genetic factors in the development of various forms of DSL (e.g., stuttering, SSD, SLI).

Specifically, Stromswold (1998) reviewed a number of studies of familial aggregation of DSL. These studies indicated a higher than expected presence of DSL among the relatives of DSL probands: the median incidence rate for language difficulties in the families of children with language impairment was 35% compared with a median incidence rate of 11% in control families (Stromswold, 1998). The reported prevalence estimates range from 24 to 78%, ~5–10 times higher than the 2–7% observed in the general population (Ambrose, Yairi, & Cox, 1993; Barry, Yasin, & Bishop, 2006; Bishop & Edmundson, 1986; Kidd, 1984; Neils & Aram, 1986; Tallal et al., 1989).

Twin studies have consistently reported a substantive difference in concordance between monozygotic and dizygotic twins, with the former showing higher concordance rates than the latter (ranging from 25 to 100% in the published literature; for a review, see Stromswold, 2001), thus, indicating the presence of genetic influences (Felsenfeld et al., 2000; Howie, 1981; Lewis & Thompson, 1992; Tomblin & Buckwalter, 1998). In addition, heritability estimates are high for a variety of psychometric indicators capturing psychological processes thought to be deficient in DSL (DeThorne et al., 2006; Kovas et al., 2005). Of note is that heritability estimates appear to be high whether ascertained directly from the child through psychometric assessment or clinical interview or through parent or teacher report (Bishop, Laws, Adams, & Norbury, 2006).

Thus, considered holistically, the accumulated evidence points to the substantial role of genes and relatively small role of environment in the development and manifestation of DSL. Yet, collectively, these studies have not revealed particular replicable modes of transmission for any of the DSL, although multiple attempts have been made (e.g., Ambrose et al., 1993; Cox, Kramer, & Kidd, 1984; Lewis, Cox, & Byard, 1993; Viswanath, Lee, & Chakraborty, 2004). Presently, it is hypothesized that, as is the case for many complex common disorders, the etiology of DSL is probably related to linear and interactive influences of combinations of genetic and environmental factors (Smith, 2007).

The KE Family

A very special and interesting case is a large, three-generation pedigree from the United Kingdom (referred to as KE), in which about 50% of the members suffer from a severe complex disorder that distorts speech, language, and cognitive functioning and is passed through the generations in a simple Mendelian pattern of dominant transmission (i.e., the disorder is present in each generation at a particular rate). Since the family was first presented in the literature (Hurst, Baraitser, Auger, Graham, & Norell, 1990), many subsequent steps have been undertaken to track the manifestation (Alcock, Passingham, Watkins, & Vargha-Khadem, 2000a, 2000b; Gopnik & Crago, 1991; Gopnik & Goad, 1997; Vargha-Khadem, Watkins, Alcock, Fletcher, & Passingham, 1995; Watkins, Dronkers, & Vargha-Khadem, 2002), brain correlates (Belton, Salmond, Watkins, Vargha-Khadem, & Gadian, 2003; Liégeois et al., 2003; Vargha-Khadem et al., 1998; Watkins, Vargha-Khadem et al., 2002), and genetic etiology of the disorder (Fisher, Vargha-Khadem, Watkins, Monaco, & Pembrey, 1998; Lai, Fisher, Hurst, Vargha-Khadem, & Monaco, 2001; MacDermot et al., 2005).

As mentioned earlier, this family's phenotype is complex, with deficiencies manifesting as verbal dyspraxia (i.e., speech articulation problems) and difficulties in receptive and expressive vocabulary and language morphology and syntax. Of note also is that, in the KE family, difficulties penetrate all types of linguistic processing (i.e., oral and written language) in aspects of both reception and production. Finally, as a group, when compared with the unaffected individuals in the pedigree, affected individuals are characterized by lower nonverbal IQ, although the cognitive difficulties do not co-segregate "true" with speech and language difficulties; this might suggest that two independent disorders segregate in this family. Thus, the phenotypic texture of difficulties transmitted in this pedigree is complex. Moreover, a number of deficits (e.g., cognitive deficits) are observed among unaffected individuals. Yet, there appears to be a single phenotype that best differentiates affected and unaffected individuals – individual performance on a nonword repetition task (Watkins, Dronkers et al., 2002).

Given the dominant pattern of transmission of this complex disorder in the KE family, researchers expected that the disorder was controlled by a single major gene (for a more complete review of the genetic etiology of the disorder in the KE family, see Fisher, Lai, & Monaco, 2003). To identify this gene, a genome-wide screen of the pedigree was carried out; the maximum-strength signal (LOD = 6.62) was localized to a small interval on the long arm of chromosome 7 (7q31). This interval was reported to have, at that point, 20 known genes and 50 unknown transcripts. Researchers narrowed it down by genotyping a number of additional newly generated polymorphism markers; this resulted in the exclusion of four genes. They then prepared to launch a systematic sift through the unexcluded genes and transcripts in the region (Lai et al., 2000). This exercise, which could have been extremely lengthy given the size of the interval of interest, was shortened substantially by the identification of a patient (named CS) unrelated to the KE family who was diagnosed with verbal dyspraxia and language impairment and had a balanced exchange of genetic material between chromosomes 5 and 7 (a balanced translocation). On chromosome 7, the translocation involved region 7q31; further investigation of the breakpoint led to the identification of a partially characterized gene, then called *CAGH44*. Using bioinformatics, researchers assembled the entire coding region of this gene and confirmed the assembly by analyzing the messenger RNA derived from the sequence. This analysis resulted in the discovery that the gene encoded a DNA-binding motif, a three-dimensional structural element or fold within the DNA sequence, which also appeared in a variety of other molecules, the so-called forkhead/winged-helix domain (Kaufmann & Knöchel, 1996). Correspondingly, the gene was renamed *FOXP2* (forkhead box P2). This family of genes codes for proteins that specifically bind to promoter or enhancer regions on DNA and thereby control the transcription of other genes. It was shown that the breakpoint in patient CS disrupted one copy of the *FOXP2* gene, right between exons 3 and 4. Following up on the information from this patient, the researchers decided to sequence the coding region of the *FOXP2* gene in the KE family. This exercise divulged a G-to-A nucleotide change in exon 14. This mutation, which substituted the amino acid histidine for the amino acid arginine, was not found in a single individual from a group of 364 screened control subjects without DSL, but co-segregated perfectly with the affected status in the KE family and, thus, was recognized as a cause of the disorder transmitted in this family (Lai et al., 2001). Subsequent investigations of other patients with similar phenotypes revealed additional mutations in *FOXP2* that caused deficiencies in these patients (MacDermot et al., 2005). It was hypothesized that, like other transcription factors, the FOXP2 protein might be important in embryogenesis, especially at the stage of brain development (Fisher et al., 2003). However, the specifics of this causal chain have yet to be determined (French et al., 2007; White, Fisher, Geschwind, Scharff, & Holy, 2006), although new promising evidence indicates the *FOXP2*-based regulation of the expression of other genes (e.g., *CNTNAP2*) whose function may be challenged in disorders of speech and language (Vernes et al., 2008).

Molecular-Genetic Studies of DSL

The identification of *FOXP2* as a causal gene for DSL in the KE family and four additional cases of complex DSL generated significant attention and enthusiasm for this gene. A number of researchers began to investigate genetic associations between *FOXP2* and different types of DSL. These investigation unfolded for more common cases of speech and language impairments (Meaburn, Dale, Craig, & Plomin, 2002; Newbury et al., 2002; O'Brien, Zhang, Nishimura, Tomblin, & Murray, 2003) and autism (Bartlett et al., 2002; Newbury et al., 2002). At present, there is no consistent evidence supporting the involvement of *FOXP2* in more common forms of DSL than those in the KE family and a few interesting cases.

Conscientious of the apparent noninvolvement of *FOXP2* in DSL in the general population, researchers carried out a number of molecular-genetic studies of DSL of two different types: (1) whole-genome screens and (2) regional studies of various forms of DSL. Using whole-genome scans, researchers attempted to identify regions of the genome that might contain susceptibility genes for DSL anew, whereas in the regional studies the task was to test for linkage with DSL those regions that have been implicated in comorbid developmental disorders such as autism and developmental dyslexia. Here both types of studies are illustrated briefly.

Genome-Wide Screens of DSL Phenotypes

For example, five large extended Canadian families of Celtic ancestry meeting criteria of SLI were genotyped with a set of highly polymorphic markers spaced at ~9 cM apart (Bartlett et al., 2002). Three categorical (discrete, diagnostic) phenotypes were used in these analyses: (1) language impairment (spoken language quotient standard score of 0.85); (2) reading impairment (1 SD discrepancy between performance IQ and nonword reading ability); and (3) clinical impairment (poor performance on language or reading subtests and/or at least 2 years of therapeutic intervention). The outcomes of this analysis identified a number of regions of interest, in order of the significance of the findings: 13q21 (for the phenotype of reading impairment), 2p22 (language impairment), and 17q23 (reading impairment).

Another whole-genome scan (density at ~8 cM) was carried out with 219 sibpairs from 98 small UK nuclear families ascertained through probands with SLI (The SLI Consortium, 2002). In this study, the phenotypes were captured by quantitative traits, specifically the receptive and expressive scales of a standardized clinical assessment battery of language functioning and a test of nonword repetition. The results revealed two susceptibility regions, 16q24 (nonword repetition) and 19q13 (expressive language), in order of significance. Each of the identified regions spans ~30 cM and potentially contains ~100 genes (Haines & Camarata, 2004). To follow up on previously reported findings, the researchers ascertained an additional sample of 86 new families; both regions were linked in this sample as well [however, both regions 16q and 19q appeared to be linked to nonword repetition (The SLI Consortium, 2004)].

Furthermore, when both samples were amalgamated (393 sibpairs from 184 families), the 16q peak reached the threshold referred to as highly significant (MLS = 7.46 and 4.41 for multipoint and single point, respectively). However, the 19q results were not replicated in the amalgamated sample, which might not be that surprising given both samples' link to two different phenotypes. It is also curious that, although smaller in magnitude, linkage in the same region on 16q was established with three reading-related measures (basic reading, spelling, and reading comprehension).

Five whole-genome scans have been carried out to detect regions harboring genes controlling genetic variation in stuttering, a developmental disorder of speech characterized by disfluency of verbal expression manifested in involuntary repetitions of sounds or syllables. Two whole-genome scans were completed at the density of ~10 cM with highly polymorphic markers (STRPs). One screen involved 68 families of European ancestry; modest evidence for linkage (nonparametric LOD = 1.51) was established for a region on 18q (Shugart et al., 2004). The second scan was done using samples from a large Cameroonian family (Levis, Ricci, Lukong, & Drayna, 2004); this study provided evidence of linkage to 1q21–1q22 (LOD = 2.27). The third scan included 46 highly inbred Pakistani families (Riaz et al., 2005). Two susceptibility regions were identified, one at 12q (LOD = 4.61) and the second at 1q (LOD = 2.93). The fourth whole-genome scan used a panel of 10,000 SNPs genotyped in 100 families from The Illinois International Genetics of Stuttering Project and revealed general linkage peaks on chromosomes 9q and 15q for the whole sample and sex-specific signals on chromosomes 7q for male-only and 21q for female-only data subsets, as well as evidence for gene × gene interactions involving chromosomes 15q and 13q, 15q and 20q, 12q and 7q, 2q and 7q, and 2q and 9q (Suresh et al., 2006). Finally, a genome screen was conducted on a sample of individuals with stutters connected in a single 232-person Hutterite family: ~1,200 markers, both STRPs and SNPs, were used (Wittke-Thompson et al., 2007). Three linkage signals were detected (3q, 13q, and 15q), although none of them were particularly strong. All five investigations used only categorical diagnoses-based phenotypes, although more than one (e.g., narrow and broad, susceptibility and persistence) diagnostic category was used.

In summary, although each particular study produced an array of results, there were no obvious overlaps of chromosomal regions in the primary linkage analyses. Such a pattern of results might indicate that (1) many different genes are involved in the development and manifestation of DSL; (2) DSL are very heterogeneous disorders and each disorder might have its own genetic etiology; (3) no study, when considered separately, has enough power to produce consistent results and some of these signals might still be false positives; (4) a higher density of genetic markers (e.g.,

as in the study by Suresh et al., 2006 and higher) is needed to detect "weak" signals consistently; and (5) the phenotypic definitions of deficits segregating in families of the probands of DSL might not be precise enough. All these possible explanations of the inconsistency of the current pattern of results might be valid; thus, further molecular-genetic studies of DSL are needed to explain and understand this pattern.

Regional Studies of DSL Phenotypes

In addition to genome-wide screens, a number of region-specific studies with a variety of DSL phenotypes have been carried out. For example, in a large sample of families of probands with SLI (O'Brien et al., 2003), the 7q31 region was covered with six markers in and around the *FOXP2* gene. Although no genetic association with *FOXP2* was established, strong association was found with a marker within the *CFTR* (cystic fibrosis transmembrane conductance regulator) gene and the marker D7S3052, both adjacent to *FOXP2*.

A different research group investigated a set of susceptibility regions identified in studies of autism. Capitalizing on genetic findings in the field of autism and on a set of results from their own whole-genome screen (Bartlett et al., 2002), the researchers studied two SLI samples (one from Canada and one from the USA) for linkage to the previously identified regions on chromosomes 2, 7, and 13 (Bartlett et al., 2004). The formerly registered linkage to 13q21 was replicated in both samples. In addition, although the overall evidence for linkage signals on chromosomes 2 and 7 was weak, markers in the *CFTR* gene (see above) demonstrated genetic association with the diagnosis of SLI.

Following up on a lead obtained through studies of developmental dyslexia, a different group of researchers explored linkage to the pericentrometric region of chromosome 3 in 77 families ascertained through a child with SSD (Stein et al., 2004). In this study, phenotypes were defined as quantitative traits capturing a number of linguistic processes (i.e., phonological memory, phonological representation, articulation, receptive and expressive vocabulary, and reading decoding and comprehension skills). Overall, the results indicated the linkage of a number of the investigated phenotypes to the region on chromosome 3, with the highest linkage score found for measures of phonological memory, followed by scores for single-word decoding for both real and pseudowords. Using the same but enlarged sample (151 families), the same group of investigators also studied the 15q14–q21 region that had been previously implicated in a number of studies of developmental disorders including autism and developmental dyslexia. A categorical phenotype of SSD and a number of SSD-related continuous phenotypes were used in these analyses. Positive linkage signals were obtained for the 15q14 region for the phenotypes of SSD, and quantitative phenotypes of oral motor function, articulation, and phonological memory (Stein et al., 2006). Yet another region, 1p34–36, previously implicated as a candidate region for developmental dyslexia, was investigated in the same sample with two categorical diagnoses (SSD and SLI) and a number of DSL-related phenotypes. Globally speaking, the linkage to 1p34–36 was replicated, although the specific peaks varied for different phenotypes. The strongest signals were obtained for articulation and listening comprehension, although the precise locations of these signals did not overlap (Miscimarra et al., 2007).

Still another set of region-specific studies was carried out on another sample of 86 probands with SSD and their siblings from 65 families (Smith, Pennington, Boada, & Shriberg, 2005). Seven phenotypes were used, among which were five indicators of speech development and two spoken language measures of phonological processing (nonword repetition and phonological awareness). These phenotypes were investigated for linkage with markers on chromosomes 1p36, 6p22, and 15q21; all regions had been previously reported as susceptibility regions for developmental dyslexia. The results of this linkage investigation indicated the presence of linkage for the speech and nonword reading phenotypes with the regions on chromosomes 6 and 15, with suggestive results for the chromosome 1 locus.

Chromosomal Abnormalities and DSL

Similar to the description of patient CS, whose chromosomal abnormality assisted in the identification of the *FOXP2* gene, other studies of isolated patients with various chromosomal abnormalities that result in DSL have been published (e.g., Somerville et al., 2005 for 7q11.23; Tyson, McGillivray, Chijiwa, & Rajcan-Separovic, 2004 for 7q31.3). Yet, the results of these single cases have not been generalized to any family or population data.

Concluding Comments and Future Directions

The review of the evidence in this chapter permits an unequivocal conclusion that the development and manifestation of DSL is driven, at least partially and substantially, by genetic variation. Yet, in addition to this general statement, only a few remarks can be made with any degree of certainty.

First, it appears that the genetic mechanisms involved in DSL are complex. At the beginning of this chapter, the issue of "general" vs. "specific" mechanisms unifying or differentiating typical and atypical language acquisition was discussed. In fact, the review of the literature carried out in this chapter suggests that this differentiation might be inadequate, because both types of mechanisms might be at work. The

example of the KE family indicates that there might be major genes that differentiate families like KE and individuals like CS from the general population by the impact of a gene such as *FOXP2*. Given the multidimensional nature of their phenotypes, KE family members and CS truly form a subgroup of individuals different from the general population – rare distinct examples of DSL. Yet, the mutations in *FOXP2* do not appear to generalize to "common" forms of DSL and account for only a very small number of cases. Thus, it is plausible that other genes, which may or may not be functionally related to *FOXP2*, account for common DSL.

Second, the landscape of findings discussed in this chapter underscores even more the importance of carefully defining the phenotype to be studied. For example, many if not all theories of DSL include deficiencies in phonological representation as an element of the clinical and psychological profiles of DSL. Latent deficits in phonological representation can manifest in multiple constructs, which in turn can be captured by observed variables. One such construct is phonological memory, a type of memory that ensures storage and processing of phonemes and their sequences. Phonological memory can be captured and quantified through a number of processes. For example, as discussed earlier, one measure of phonological memory is the processing of nonwords – a skill of repeating orally presented nonsense words of varying levels of difficulty.

Correspondingly, familiality and heritability estimates can be obtained for such a skill. For example, in one twin study the estimate of heritability for the indicator of nonword repetition was close to 1.0, suggesting that genes are an important source of individual differences for this trait (Bishop, North, & Donlan, 1996). Based on this and other findings, the phenotype of nonword repetition has been used as a phenotype in molecular-genetic research, and linkage was established to this phenotype (The SLI Consortium, 2004). In addition, the nonword repetition task appears to be the single phenotype that best differentiates affected and unaffected individuals (Watkins, Dronkers et al., 2002). It is also of interest that when samples of parents of probands with SLI and typically developing children were compared on a number of psychological indicators relevant to linguistic functioning (Barry et al., 2006), the two groups differed significantly in their performance on nonword repetition, oral motor, and digit span tasks. Although all three tasks highlighted important information that differentiated the two groups, the nonword repetition task alone discriminated 75.8% of the cases. The authors concluded that nonword repetition can be viewed as a reliable marker of family risk for language impairment. Finally, it has been shown in longitudinal studies of SLI that twins with SLI had deficits in nonword repetition even when the holistic, syndromatic presentation of the disorder was not present (Bishop et al., 1996). Thus, it is possible that there are particular "dimensions" of all DSL that capture deficits common to all of them. Yet, given the diversity in the presentation and course of different DSL, it is likely that each disorder will have its unique dimensions.

Third, this "partial overlap/partial specificity" of multidimensional phenotypes of different DSL might lead to what is now referred to as the retrospective and prospective comorbidity of these disorders. For example, the typically observed onset of stuttering is between the ages of 3 and 6 years, with reported rates of natural, unassisted recovery of ~75% (Yairi & Ambrose, 1999, 2005). Thus, the prevalence of stuttering as a lifetime disorder is much lower than its incidence (0.5–1% vs. 4–5%, respectively, Bloodstein, 1995; Felsenfeld, 2002). Yet, childhood stuttering is a significant risk factor for other DSL that develop later in life, even if stuttering stops manifesting (Ardila et al., 1994). It is possible that this "continuity" in the context of developmental transformation is due to the presence of particular dimensions of the complex phenotypes common to all DSL.

Finally, the issue of pleiotropy, or the impact of the same genes on multiple phenotypes, is also a theme that has been discussed multiple times in the literature on DSL, given the substantive overlap in regions of linkage for a variety of developmental disorders, such as SSD and developmental dyslexia (Miscimarra et al., 2007; Stein et al., 2006, 2004), and SLI and autism (Tager-Flusberg & Joseph, 2003). Once again, whether these are true examples of pleiotropy or outcomes of the imprecision of phenotype definitions is yet to be determined. In this context, molecular-genetic studies of DSL can be instrumental in re-categorizing probands into those having particular genetic defects (e.g., particular mutations) and those not having such defects and exploring their presentational features. Such classifications of probands based on their genetic etiology might result in the establishment of more precise phenotypes and, possibly, the re-design of clinical classifications.

Nevertheless, although the field of behavior- and molecular-genetic studies of DSL today presents a variegated quilt of findings with only a limited number of common threads, it has been enormously productive in generating exciting findings and providing leads and examples for studies of the genetic etiology of complex human behaviors.

Acknowledgment Preparation of this article was supported by Grants R21 TW006764-02 from the Fogarty Program as administered by the National Institutes of Health, Department of Health and Human Services and R01 DC007665 as administered by the National Institute of Deafness and Communication Disorders (PI: Grigorenko). Grantees undertaking such projects are encouraged to express their professional judgment freely. Therefore, this chapter does not necessarily reflect the position or policies of the National Institutes of Health, and no official endorsement should be inferred. The content of this chapter partially overlaps with Grigoranko (in press). I am thankful to Ms. Robyn Rissman and Ms. Mei Tan for their editorial assistance and to Ms. Jodi Reich and anonymous reviewers for their helpful comments.

References

Alcock, K. J., Passingham, R. E., Watkins, K. E., & Vargha-Khadem, F. (2000a). Oral dyspraxia in inherited speech and language impairment and acquired dysphasia. *Brain & Language, 75*, 17–33.

Alcock, K. J., Passingham, R. E., Watkins, K. E., & Vargha-Khadem, F. (2000b). Pitch and timing abilities in inherited speech and language impairment. *Brain & Language, 75*, 34–46.

Ambrose, N., Yairi, E., & Cox, N. J. (1993). Genetic aspects of early childhood stuttering. *Journal of Speech & Hearing Research, 36*, 701–706.

Ardila, A., Bateman, J. R., Nino, C. R., Pulido, E., Rivera, D. B., & Vanegas, C. J. (1994). An epidemiologic study of stuttering. *Journal of Communication Disorders, 27*, 37–48.

Barry, J. G., Yasin, I., & Bishop, D. V. (2006). Heritable risk factors associated with language impairments. *Genes, Brain & Behavior, 6*, 66–76.

Bartlett, C. W., Flax, J. F., Logue, M. W., Smith, B. J., Vieland, V. J., Tallal, P., et al. (2004). Examination of potential overlap in autism and language loci on chromosomes 2, 7, and 13 in two independent samples ascertained for specific language impairment. *Human Heredity, 57*, 10–20.

Bartlett, C. W., Flax, J. F., Logue, M. W., Vieland, V. J., Bassett, A. S., Tallal, P., et al. (2002). A major susceptibility locus for specific language impairment is located on 13q21. *American Journal of Human Genetics, 71*, 45–55.

Bates, E. A. (2004). Explaining and interpreting deficits in language development across clinical groups: where do we go from here? *Brain & Language, 88*, 248–253.

Belton, E., Salmond, C. H., Watkins, K. E., Vargha-Khadem, F., & Gadian, D. G. (2003). Bilateral brain abnormalities associated with dominantly inherited verbal and orofacial dyspraxia. *Human Brain Mapping, 18*, 194–200.

Bishop, D. V. (1994a). Grammatical errors in specific language impairment: Competence or performance limitations? *Applied Psycholinguistics, 15*, 507–550.

Bishop, D. V. (1994b). Is specific language impairment a valid diagnostic category? Genetic and psycholinguistic evidence. *Philosophical Transactions of the Royal Society of London – Series B: Biological Sciences, 346*, 105–111.

Bishop, D. V. (2005). Developmental cognitive genetics: how psychology can inform genetics and vice versa. *Quarterly Journal of Experimental Psychology, 59*, 1153–1168.

Bishop, D. V., Adams, C. V., & Norbury, C. F. (2006). Distinct genetic influences on grammar and phonological short-term memory deficits: Evidence from 6-year-old twins. *Genes, Brain & Behavior, 5*, 158–169.

Bishop, D. V., & Edmundson, A. (1986). Is otitis media a major cause of specific developmental language disorders? *British Journal of Disorders of Communication, 21*, 321–338.

Bishop, D. V., Laws, G., Adams, C., & Norbury, C. F. (2006). High heritability of speech and language impairments in 6-year-old twins demonstrated using parent and teacher report. *Behavior Genetics, 36*, 173–184.

Bishop, D. V., North, T., & Donlan, C. (1995). Genetic basis of specific language impairment: evidence from a twin study. *Developmental Medicine & Child Neurology, 37*, 56–71.

Bishop, D. V., North, T., & Donlan, C. (1996). Nonword repetition as a behavioural marker for inherited language impairment: evidence from a twin study. *Journal of Child Psychology & Psychiatry & Allied Disciplines, 37*, 391–403.

Bloodstein, O. (1995). *A handbook on stuttering*. Chicago, IL: National Easter Seal Society.

Choudhury, N., & Benasich, A. A. (2003). A family aggregation study: the influence of family history and other risk factors on language development. *Journal of Speech Language & Hearing Research, 46,* 261–272.

Cleave, P. L., & Rice, M. L. (1997). An examination of the morpheme BE in children with specific language impairment: the role of contractibility and grammatical form class. *Journal of Speech Language & Hearing Research, 40,* 480–492.

Cole, K. N., Schwartz, I. S., Notari, A. R., Dale, P. S., & Mills, P. E. (1995). Examination of the stability of two methods of defining specific language impairment. *Applied Psycholinguistics, 16,* 103–123.

Coltheart, M., Curtis, B., Atkins, P., & Haller, M. (1993). Models of reading aloud: Dual-route and parallel-distributed-processing approaches. *Psychological Review, 100,* 589–608.

Coltheart, M., Rastle, K., Perry, C., Langdon, R., & Ziegler, J. (2001). DRC: A dual route cascaded model of visual word recognition and reading aloud. *Psychological Review, 108,* 204–256.

Cox, N. J., Kramer, P., & Kidd, K. (1984). Segregation analyses of stuttering. *Genetic Epidemiology, 1,* 245–253.

DeThorne, L. S., Hart, S. A., Petrill, S. A., Deater-Deckard, K., Thompson, L. A., Schatschneider, C., et al. (2006). Children's history of speech-language difficulties: genetic influences and associations with reading-related measures. *Journal of Speech Language & Hearing Research, 49,* 1280–1293.

Felsenfeld, S. (2002). Finding susceptibility genes for developmental disorders of speech: the long and winding road. *Journal of Communication Disorders, 35,* 329–345.

Felsenfeld, S., Kirk, K., Zhu, G., Statham, M., Neale, M., & Martin, N. (2000). A study of the genetic and environmental etiology of stuttering in a selected twin sample. *Behavior Genetics, 30,* 359–366.

Felsenfeld, S., & Plomin, R. (1997). Epidemiological and offspring analyses of developmental speech disorders using data from the Colorado Adoption Project. *Journal of Speech, Language and Hearing Research, 41,* 778–791.

Fisher, S. E., Lai, C. S., & Monaco, A. P. (2003). Deciphering the genetic basis of speech and language disorders. *Annual Review of Neuroscience, 26,* 57–80.

Fisher, S. E., Vargha-Khadem, F., Watkins, K. E., Monaco, A. P., & Pembrey, M. E. (1998). Localisation of a gene implicated in a severe speech and language disorder. *Nature Genetics, 18,* 168–170.

French, C. A., Groszer, M., Preece, C., Coupe, A. M., Rajewsky, K., & Fisher, S. E. (2007). Generation of mice with a conditional Foxp2 null allele. *Genesis: the Journal of Genetics & Development, 45,* 440–446.

Gopnik, M., & Crago, M. B. (1991). Familial aggregation of a developmental language disorder. *Cognition, 39,* 1–50.

Gopnik, M., & Goad, H. (1997). What underlies inflectional error patterns in genetic dysphasia? *Journal of Neurolinguistics, 10,* 109–137.

Grigorenko, E. L. (in press). Speaking genes or genes for speaking? Deciphering the genetics of speech and language. *Journal of Child Psychology and Psychiatry.*

Haines, J., & Camarata, S. (2004). Examination of candidate genes in language disorder: a model of genetic association for treatment studies. *Mental Retardation & Developmental Disabilities Research Reviews, 10,* 208–217.

Haskell, T. R., MacDonald, M. C., & Seidenberg, M. S. (2003). Language learning and innateness: Some implications of Compounds Research. *Cognitive Psychology, 47,* 119–163.

Howie, P. M. (1981). Concordance for stuttering in monozygotic and dizygotic twin pairs. *Journal of Speech & Hearing Research, 24,* 317–321.

Hurst, J. A., Baraitser, M., Auger, E., Graham, F., & Norell, S. (1990). An extended family with a dominantly inherited speech disorder. *Developmental Medicine and Child Neurology, 32,* 352–355.

Joanisse, M. F. (2004). Specific Language Impairments in children phonology, semantics, and the English past tense. *Current Directions in Psychological Science, 13,* 156–160.

Kaufmann, E., & Knöchel, W. (1996). Five years on the wings of fork head. *Mechanisms of Development, 57,* 3–20.

Kidd, K. K. (1984). Stuttering as a genetic disorder. In R. Curlee & W. Perkins (Eds.), *Nature and treatment of stuttering* (pp. 149–169). San Diego, CA: College Hill.

Kidd, K. K., Heimbuch, R. C., Records, M. A., Oehlert, G., & Webster, R. L. (1980). Familial stuttering patterns are not related to one measure of severity. *Journal of Speech & Hearing Research, 23,* 539–545.

Kovas, Y., Hayiou-Thomas, M. E., Oliver, B., Dale, P. S., Bishop, D. V., & Plomin, R. (2005). Genetic influences in different aspects of language development: The etiology of language skills in 4.5-year-old twins. *Child Development, 76,* 632–651.

Lahey, M., & Edwards, J. (1995). Specific language impairment: Preliminary investigation of factors associated with family history and with patterns of language performance. *Journal of Speech and Hearing Research, 38,* 643–657.

Lai, C. S., Fisher, S. E., Hurst, J. A., Levy, E. R., Hodgson, S., Fox, M., et al. (2000). The SPCH1 region on human 7q31: genomic characterization of the critical interval and localization of translocations associated with speech and language disorder. *American Journal of Human Genetics, 67,* 357–368.

Lai, C. S., Fisher, S. E., Hurst, J. A., Vargha-Khadem, F., & Monaco, A. P. (2001). A forkhead-domain gene is mutated in a severe speech and language disorder. *Nature, 413,* 519–523.

Leonard, L. B. (1998). *Children with specific language impairment.* Cambridge, MA: MIT Press.

Levis, B., Ricci, D., Lukong, J., & Drayna, D. (2004). Genetic linkage studies in a large West African kindred. *American Journal of Human Genetics, 75,* S2026.

Lewis, B. A., Cox, N. J., & Byard, P. J. (1993). Segregation analysis of speech and language disorders. *Behavior Genetics, 23,* 291–297.

Lewis, B. A., Ekelman, B. L., & Aram, D. M. (1989). A familial study of severe phonological disorders. *Journal of Speech and Hearing Research, 32,* 713–724.

Lewis, B. A., & Thompson, L. A. (1992). A study of developmental speech and language disorders in twins. *Journal of Speech and Hearing Research, 35,* 1086–1094.

Liégeois, F., Baldeweg, T., Connelly, A., Gadian, D. G., Mishkin, M., & Vargha-Khadem, F. (2003). Language fMRI abnormalities associated with FOXP2 gene mutation. *Nature Neuroscience, 6,* 1230–1237.

MacDermot, K. D., Bonora, E., Sykes, N., Coupe, A. M., Lai, C. S., Vernes, S. C., et al. (2005). Identification of FOXP2 truncation as a novel cause of developmental speech and language deficits. *American Journal of Human Genetics, 76,* 1074–1080.

Marshall, C. R., & van der Lely, H. K. (2006). A challenge to current models of past tense inflection: the impact of phonotactics. *Cognition, 100,* 302–320.

McClelland, J. L., & Patterson, K. (2002). Rules or connections in past-tense inflections: What does the evidence rule out? *Trends in Cognitive Sciences, 6,* 465–472.

Meaburn, E., Dale, P. S., Craig, I. W., & Plomin, R. (2002). Language-impaired children: No sign of the FOXP2 mutation. *Cognitive Neuroscience and Neuropsychology, 13,* 1075–1077.

Miscimarra, L. E., Stein, C. M., Millard, C., Kluge, A., Cartier, K. C., Freebairn, L. A., et al. (2007). Further evidence of pleiotropy influencing speech and language: Analysis of the DYX8 region. *Human Heredity, 63,* 47–58.

Neils, J., & Aram, D. M. (1986). Family history of children with developmental language disorders. *Perceptual Motor Skills, 63,* 655–658.

Newbury, D. F., Bonora, E., Lamb, J. A., Fisher, S. E., Lai, C. S. L., Baird, G., et al. (2002). FOXP2 is not a major susceptibility gene for autism or specific language impairment. *American Journal of Human Genetics, 70,* 1318–1327.

O'Brien, E. K., Zhang, X., Nishimura, C., Tomblin, J. B., & Murray, J. C. (2003). Association of Specific Language Impairment (SLI) to the region of 7q31. *American Journal of Human Genetics, 72,* 1536–1543.

Pinker, S. (1991). Rules of language. In P. Bloom (Ed.), *Language acquisition: Core readings.* New York, NY: Harvester Wheatsheaf.

Pinker, S. (1999). *Words and rules: The ingredients of language.* New York, NY, US: Basic Books.

Pinker, S., & Ullman, M. T. (2002). The past and future of the past tense. *Trends in Cognitive Sciences, 6,* 456–463.

Plaut, D. C. (1996). Relearning after damage in connectionist networks: Toward a theory of rehabilitation. *Brain and Language, 52,* 25–82.

Redmond, S. M., & Rice, M. L. (2001). Detection of irregular verb violations by children with and without SLI. *Journal of Speech Language & Hearing Research, 44,* 655–669.

Riaz, N., Steinberg, S., Ahmad, J., Pluzhnikov, A., Riazuddin, S., Cox, N. J., et al. (2005). Genomewide significant linkage to stuttering on chromosome 12. *American Journal of Human Genetics, 76,* 647–651.

Rice, M. L., Cleave, P. L., & Oetting, J. B. (2000). The use of syntactic cues in lexical acquisition by children with SLI. *Journal of Speech Language & Hearing Research, 43,* 582–594.

Rice, M. L., Haney, K. R., & Wexler, K. (1998). Family histories of children with SLI who show extended optional infinitives. *Journal of Speech Language & Hearing Research, 41,* 419–432.

Rice, M. L., Tomblin, J. B., Hoffman, L., Richman, W. A., & Marquis, J. (2004). Grammatical tense deficits in children with SLI and nonspecific language impairment: relationships with nonverbal IQ over time. *Journal of Speech Language & Hearing Research, 47,* 816–834.

Rice, M. L., Wexler, K., & Cleave, P. L. (1995). Specific language impairment as a period of extended optional infinitive. *Journal of Speech & Hearing Research, 38,* 850–863.

Rice, M. L., Wexler, K., Marquis, J., & Hershberger, S. (2000). Acquisition of irregular past tense by children with specific language impairment. *Journal of Speech Language & Hearing Research, 43,* 1126–1145.

Rice, M. L., Wexler, K., & Redmond, S. M. (1999). Grammaticality judgments of an extended optional infinitive grammar: evidence from English-speaking children with specific language impairment. *Journal of Speech Language & Hearing Research, 42,* 943–961.

Rumelhart, D. E., & McClelland, J. L. (1987). On learning the past tenses of English verbs. In J. L. McClelland, D. E. Rumelhart, & t. P. R. Group (Eds.), *Parallel distributed processing: Explorations in the microstructure of cognition* (Vol. 2). Cambridge, MA: MIT Press.

Seidenberg, M. S. (2005). Connectionist models of word reading. *Current Directions in Psychological Science, 14,* 238–242.

Seidenberg, M. S., & McClelland, J. L. (1989). A distributed, developmental model of word recognition and naming. *Psychological Review, 96,* 523–568.

Shriberg, L. D. (2002, November 22–24). *Classification and misclassification of child speech sound disorders.* Paper presented at The Annual Conference of the American Speech-Language-Hearing Association, Atlanta.

Shugart, Y. Y., Mundorff, J., Kilshaw, J., Doheny, K., Doan, B., Wanyee, J., et al. (2004). Results of a genome-wide linkage scan for stuttering. *American Journal of Medical Genetics, 124,* 133–135.

Smith, S. D. (2007). Genes, language development, and language disorders. *Mental Retardation & Developmental Disabilities Research Reviews, 13,* 96–105.

Smith, S. D., Pennington, B. F., Boada, R., & Shriberg, L. D. (2005). Linkage of speech sound disorder to reading disability loci. *Journal of Child Psychology and Psychiatry, 46,* 1057–1066.

Somerville, M. J., Mervis, C. B., Young, E. J., Seo, E. J., del Campo, M., Bamforth, S., et al. (2005). Severe expressive-language delay related to duplication of the Williams-Beuren locus. *New England Journal of Medicine, 353,* 1694–1701.

Stein, C. M., Millard, C., Kluge, A., Miscimarra, L. E., Cartier, K. C., Freebairn, L. A., et al. (2006). Speech sound disorder influenced by a locus in 15q14 region. *Behavior Genetics, 36,* 858–868.

Stein, C. M., Schick, J. H., Taylor, G. H., Shriberg, L. D., Millard, C., Kundtz-Kluge, A., et al. (2004). Pleiotropic effects of a chromosome 3 locus on speech-sound disorder and reading. *American Journal of Human Genetics, 74,* 283–297.

Stromswold, K. (1998). Genetics of spoken language disorders. *Human Biology, 70,* 297, 291–212.

Stromswold, K. (2001). The heritability of language: A review and meta-analysis of twin, adoption, and linkage studies. *Language, 77,* 647–723.

Suresh, R., Ambrose, N., Roe, C., Pluzhnikov, A., Wittke-Thompson, J. K., Ng, M. C.-Y., et al. (2006). New complexities in the genetics of stuttering: Significant sex-specific linkage signals. *American Journal of Human Genetics, 78,* 554–563.

Tager-Flusberg, H., & Joseph, R. M. (2003). Identifying neurocognitive phenotypes in autism. *Philosophical Transactions of the Royal Society of London – Series B: Biological Sciences, 358,* 303–314.

Tallal, P., Hirsch, L. S., Realpe-Bonilla, T., Miller, S., Brzustowicz, L. M., Bartlett, C. W., et al. (2001). Familial aggregation in Specific Language Impairment. *Journal of Speech, Language and Hearing Research, 44,* 1172–1182.

Tallal, P., Ross, R., & Curtiss, S. (1989). Familial aggregation in specific language impairment. *Journal of Speech and Hearing Disorders, 54,* 167–173.

The SLI Consortium. (2002). A genomewide scan identifies two novel loci involved in SLI. *American Journal of Human Genetics, 70,* 384–398.

The SLI Consortium. (2004). Highly significant linkage to the SLI1 locus in an expanded sample of individuals affected by SLI. *American Journal of Human Genetics, 74,* 1225–1238.

Tomblin, J. B. (1989). Familial concentration of developmental language impairment. *Journal of Speech and Hearing Disorders, 54,* 287–295.

Tomblin, J. B., & Buckwalter, P. R. (1998). Heritability of poor language achievement among twins. *Journal of Speech and Hearing Research, 41,* 188–199.

Tyson, C., McGillivray, B., Chijiwa, C., & Rajcan-Separovic, E. (2004). Elucidation of a cryptic interstitial 7q31.3 deletion in a patient with a language disorder and mild mental retardation by array-CGH. *American Journal of Medical Genetics. Part A., 129,* 254–260.

Ullman, M. T., Corkin, S., Coppola, M., Hickok, G., Growdon, J. H., Koroshetz, W. J., et al. (1997). A neural dissociation within language: Evidence that the mental dictionary is part of declarative memory, and that grammatical rules are processed by the procedural system. *Journal of Cognitive Neuroscience, 9,* 266–276.

van der Lely, H. K. (2005). Domain-specific cognitive systems: insight from Grammatical-SLI. *Trends in Cognitive Sciences, 9,* 53–59.

van der Lely, H. K., Rosen, S., & Adlard, A. (2004). Grammatical language impairment and the specificity of cognitive domains: Relations between auditory and language abilities. *Cognition, 94,* 167–183.

van der Lely, H. K., & Stollwerck, L. (1996). A grammatical specific language impairment in children: an autosomal dominant inheritance? *Brain and Language, 52,* 484–504.

van der Lely, H. K., & Stollwerck, L. (1997). Binding theory and grammatical specific language impairment in children. *Cognition, 62,* 245–290.

Vargha-Khadem, F., Watkins, K., Alcock, K. J., Fletcher, P., & Passingham, R. E. (1995). Praxic and nonverbal cognitive deficits in a large family with a genetically transmitted speech and language disorder. *Proceedings of the National Academy of Sciences of the United States of America, 92*, 930–933.

Vargha-Khadem, F., Watkins, K. E., Price, C. J., Ashburner, J., Alcock, K. J., Connelly, A., et al. (1998). Neural basis of an inherited speech and language disorder. *Proceedings of the National Academy of Sciences of the United States of America, 95*, 12695–12700.

Verhoeven, L., & van Balkom, H. (Eds.). (2004). *Classification of developmental language disorders*. Mahwah, NJ: Lawrence Erlbaum.

Vernes, S. C., Newbury, D. F. Abrahsms., B. S., Winchester, L., Nicod, J., Groszer, M., et al. (2008). A functional genetic link between distinct developmental language disorders. *The New England Journal of Medicine, 359*, 2337–2345.

Viswanath, N., Lee, H. S., & Chakraborty, R. (2004). Evidence for a major gene influence on persistent developmental stuttering. *Human Biology, 76*, 401–412.

Watkins, K. E., Dronkers, N. F., & Vargha-Khadem, F. (2002). Behavioural analysis of an inherited speech and language disorder: comparison with acquired aphasia. *Brain, 125*, 452-464.

Watkins, K. E., Vargha-Khadem, F., Ashburner, J., Passingham, R. E., Connelly, A.,Friston, K. J., et al. (2002). MRI analysis of an inherited speech and language disorder: structural brain abnormalities. *Brain, 125*, 465–478.

White, S. A., Fisher, S. E., Geschwind, D. H., Scharff, C., & Holy, T. E. (2006). Singing mice, songbirds, and more: models for FOXP2 function and dysfunction in human speech and language. *Journal of Neuroscience, 26*, 10376–10379.

Wittke-Thompson, J. K., Ambrose, N., Yairi, E., Roec, C., Cook, E. H., Ober, C., et al. (2007). Genetic studies of stuttering in a founder population. *Journal of Fluency Disorders, 32*, 33–50.

Yairi, E., & Ambrose, N. G. (1999). Early childhood stuttering I: Persistency and recovery rates. *Journal of Speech, Language, and Hearing Research, 42*, 1097–1112.

Yairi, E., & Ambrose, N. G. (2005). *Early childhood stuttering*. Austin, TX: Pro-Ed.

Zevin, J. D., & Seidenberg, M. S. (2006). Simulating consistency effects and individual differences in nonword naming: A comparison of current models. *Journal of Memory and Language, 54*, 145–160.

Chapter 10

Human Brain Volume: What's in the Genes?

Jiska S. Peper, Marcel P. Zwiers, Dorret I. Boomsma, René S. Kahn, and Hilleke E. Hulshoff Pol

Introduction

The human brain continues to grow considerably after birth. Compared to measurements taken at birth (mean, SD was 34.9, 1.1 cm), head circumference was found to increase by more than 30% in the first year (46.6, 1.3 cm); between 1 and 4 years of age it increased by another 9% (50.9, 1.4 cm) and between 4 and 8 years by an additional 4% (53.4, 1.4 cm) in a normal cohort (Gale, O'Callaghan, Bredow, & Martyn, 2006). Magnetic resonance imaging (MRI) research has shown that at 6 years of age total cerebral volume has reached 95% of its adult volume (Giedd et al., 1999). However, the brain continues to show dynamic changes from childhood into adulthood in overall gray and white matter and in subcortical structures. In early adolescence gray matter starts to decrease (Giedd et al., 1999), whereas overall white matter volume still increases (Bartzokis et al., 2001; Giedd et al., 1999; Paus et al., 1999). Also, subcortical structures show developmental changes after childhood. For instance, the thalamus and caudate nucleus decrease with age (Sowell, Trauner, Gamst, & Jernigan, 2002) and the posterior hippocampus increases with age, whereas the anterior hippocampus decreases with age (Gogtay et al., 2006) (for a review on brain maturation, see Toga, Thompson, & Sowell, 2006).

The contribution of specific genes and environmental factors to these developmental brain changes is currently not understood. However, it is known that in adulthood, the extent of variation in human brain volume is highly heritable, with estimates between 80 and 90% (Baaré et al., 2001; Pennington et al., 2000; Pfefferbaum, Sullivan, Swan, & Carmelli, 2000). Most heritability estimates of brain volumes are based on data from monozygotic twin pairs (MZ, who are nearly always genetically identical) and dizygotic twin

pairs (DZ, who share on average 50% of their segregating genes). If brain volumes of monozygotic twin pairs resemble each other more closely than those of dizygotic twin pairs, it can be inferred that variation of brain volumes is under genetic control. These findings from twins can be generalized to the general (singleton) population, particularly after correcting for head size or intracranial volume (Hulshoff Pol et al., 2002).

Importantly, the high heritability of brain volume is functionally relevant. For instance, the association between brain volumes and intelligence was found to be of genetic origin (Posthuma et al., 2002) and the association between frontal gray matter volume and intelligence is suggested to be due to genetic factors (Thompson et al., 2001; Toga & Thompson, 2004). Recently, the association of intelligence with frontal, occipital, and parahippocampal gray matter and connecting white matter was found to be influenced by genes common to brain structure and intelligence (Hulshoff Pol et al., 2006). These findings demonstrate that a common set of genes may underly the association between brain structure and cognitive functions. However, in elderly twins, the associations between fronto-temporal brain volumes and executive function were found to be due to common environmental influences shared by twins from the same family (Carmelli, Reed, & DeCarli, 2002). These results point to the possibility that overlapping sets of genes or common environmental influences cause variation in two distinct phenotypes. However, other, causal, models are also consistent with the findings. It might be, for example, that a higher level of cognitive function leads a person to select an environment that also increases brain size. The genetic influence on brain size then simply reflects the genetic influences on cognition. Thus, the specific mechanism, pathways, and genes that are involved in human brain morphology and its association with cognitive functions remain elusive.

A few studies have been published in which particular genetic polymorphisms (a gene with at least two relatively common variants, also called alleles) are associated with variation in brain structure. However, without some prior assumptions about which genes are good candidates,

J.S. Peper (✉)
Department of Psychiatry, University Medical Center, Utrecht, The Netherlands
e-mail: j.s.peper@umcutrecht.nl

Y.-K. Kim (ed.), *Handbook of Behavior Genetics*,
DOI 10.1007/978-0-387-76727-7_10, © Springer Science+Business Media, LLC 2009

this may be comparable with searching for a needle in a haystack. Another approach to search for genes involved in human brain volume might be to study subjects with a known genetic makeup or a known genetic abnormality, i.e., groups in which the genetic variant is known, and to search for abnormalities in their brain volumes. This approach has also been applied to study genes involved in cognitive impairments in subjects with mental retardation (Nokelainen & Flint, 2002).

Here we review the studies on the influence of genes onto human brain volumes using quantitative magnetic resonance imaging (MRI). To this end, twin studies are reviewed to assess the heritability of human brain volume variation in the general population. In addition, brain structures in patients with diseases caused by mutations in genes located on autosomal chromosomes are discussed. For this purpose, MRI brain studies on diseases with a clear genetic etiology were included, i.e., Huntington's disease (expansion of triplet repeat on chromosome 4), Down syndrome (21-trisomy), Williams syndrome (hemideletion on chromosome 7q11.23), and Velocardiofacial syndrome (deletion on chromosome 22q11). Finally, other genetic approaches to the search of genes in brain structure are discussed. These other approaches include studies on brain volume of families with a particular genetic makeup, studies searching for genes in subjects with brain morphological abnormalities, and studies on the association of brain volumes with genetic polymorphisms in candidate genes in healthy subjects.

Current Research

Methods

A PubMed indexed search was carried out for each of the three different approaches, with a limitation for human subjects and the following keywords: (1) (brain volume) or (white/gray matter) and ((twin) or (heritability)); (2) (brain volume) or (white/gray matter) and ((Huntington's Disease) or (Down syndrome) or (Williams Syndrome) or (Velocardiofacial Syndrome)); and (3) (Brain volume/abnormality) or (white/gray matter) and ((polymorphism) or (genes)). All the abstracts were inspected ($n=260$), and papers written in English, using structural magnetic resonance imaging (MRI) were selected. These included volumetric MRI (both global and focal measures), voxel-based morphometry (VBM) and diffusion tensor imaging (DTI) (for information on white matter integrity). Case studies or qualitative studies were not included. These selection criteria resulted in 90 papers coming from the following topics: twin studies ($n=18$), Huntington's disease ($N=20$), Down syndrome ($N=13$), Williams

syndrome ($n=14$), Velocardiofacial syndrome ($N=14$), and other genetic approaches ($N=11$).

If available, information on the number of subjects, p-values/effect sizes, age of the sample, and heritability estimates was extracted from the papers.

Results

Twin Studies and Human Brain Morphology

To determine the relative contribution of genetic, common, and unique environmental influences on variation in brain structures, the (extended) twin model is a powerful approach (Posthuma & Boomsma, 2000). For genetic influences (heritability), the extent to which brain structures of monozygotic (MZ) twin pairs resemble each other more than in the case for dizygotic (DZ) twin pairs is the determining factor. However, in addition to genetic influences, common (or shared) environmental influences may play a role in explaining resemblances. The presence of shared environmental factors is suggested when correlations in DZ twins are larger than half the MZ correlation (Boomsma, Busjahn, & Peltonen, 2002). A first impression of the importance of unique environmental factors is obtained from the extent to which MZ twins do not resemble each other.

Brain structure in healthy MZ and DZ twin pairs was first quantitatively studied using computed tomography (CT) (Reveley, Reveley, Chitkara, & Clifford, 1984) (Table 10.1). In this study it was found that lateral ventricle variation was mostly explained by genetic factors. Later studies using MRI found high heritability estimates of global brain measures including intracranial volume ($>81\%$) (Baaré et al., 2001; Carmelli et al., 1998; Pfefferbaum et al., 2000) and total brain volume (66–97%) (Baaré et al., 2001; Bartley, Jones, & Weinberger, 1997; Pennington et al., 2000; Wright, Sham, Murray, Weinberger, & Bullmore, 2002). The first twin-sibling study to measure the genetic contributions to variation in global gray and white matter found heritability estimates of 82% for gray matter and 88% for white matter (Baaré et al., 2001). The volumes of each cerebral hemisphere showed 65% heritability (Geschwind, Miller, DeCarli, & Carmelli, 2002). For variation in cerebellar volume a heritability of 88% was reported (Posthuma et al., 2000).

A number of global brain areas seem to be mainly under environmental control. For example, this was found for the overall gyral patterning of the cortex (Bartley et al., 1997; Eckert et al., 2002). Common and unique environmental factors explained the individual variation in lateral ventricle volumes (Baaré et al., 2001; Wright et al., 2002). However, individual differences in lateral ventricle size were mainly of genetic origin in a study consisting of elderly

Table 10.1 Twin studies on brain volume

Brain area	Authors	Group size	Age (years)	Heritability
Lateral ventricles Total brain, gyral patterns	Reveley et al. (1984)* Bartley et al. (1997)	18 MZ, 18 DZ 10 MZ (6 males), 9 DZ (3 males)	NA MZ: 31 (19–54) DZ: 23 (18–29)	82–85% TB: 94% Gyral patterns: 7–17%
Intracranial volume (IC) and white matter hyperintensities (WMH)	Carmelli et al. (1998)	74 MZ, 71 DZ (males)	68–79 years	IC: 91% WMH: 71%
Total cerebral volume	Pennington et al. (2000)	RD: 25 MZ (12 males), 23 DZ (16 males) Non-RD: 9 MZ (4 males), 9 DZ (4 males)	RD: MZ 17.1, DZ 16.8 Non-RD: MZ 19.4, DZ 18.7	97% (two factors: neocortex 56%, subcortex 70%)
Intracranial volume (IC), midsagittal corpus callosum (CC) and mids lat ventricles (LV) size	Pfefferbaum et al. (2000)	45 MZ, 40 DZ (males)	MZ 72.2 DZ 71.4 68–78	IC: 81 % CC: 79%\$ LV: 79%
Intracranial volume, total brain, gray and white matter, lateral ventricles	Baaré et al. (2001)	54 MZ (33 males) 58 DZ (17 males, 21 OS) 34 sibs	MZM: 31.2, MZF: 34.1, DZM: 30.3, DZF: 30.6, OS: 30.3, sibs: 29	IC: 88% TB: 90% GM: 82% WM: 87% LV: C: 59%, E 41%
Microstructure corpus callosum (with DTI)	Pfefferbaum et al. (2001)	15 MZ, 18 DZ	NA	Relative proportion G:E in CC size: 5:1, microstructure: 3:1
Hippocampus	Sullivan et al. (2001)	45 MZ, 40 DZ (males)	MZ 72.2 DZ 71.4 68–78	40%
Frontal lobes, Broca and Wernicke's areas Planum temporale asymmetry	Thompson et al. (2001) Eckert et al. (2002)	10 MZ (5 males), 10 DZ (5 males) 27 MZ, 12 DZ (all males)	48.2 (±3.4) MZ: 6.9–16.4 DZ: 6.1–15.0	Frontal lobe 90–95% NA
Cerebral hemispheres Cerebral asymmetry/handedness	Geschwind et al. (2002)	72 MZ, 67 DZ (males)	MZ: 72.3 DZ: 71.8	Both hemispheres: 65% In left handers: less genetic control of hemispheres. More c^2 influence on left hemisphere
Cerebellum	Posthuma et al. (2000)	54 MZ (33 males) 58 DZ (17 males, 21 OS) 34 sibs	MZM: 31.2, MZF: 34.1,DZM: 30.3, DZF: 30.6, OS: 30.3, sibs: 29	88%
Total brain, lateral ventricles (LV) and regional GM (with factor analysis)	Wright et al. (2002)	10 MZ (6 males), 10 DZ (4 males). (nb: same sample as Bartley et al., 1997)	MZ: 31 (19–54) DZ: 23 (18–29)	TB: 66% LV: C 48% and E 50% GM subregions (>30) >50% heritability

Table 10.1 (continued)

Brain area	Authors	Group size	Age (years)	Heritability
TB, GM, WM, cerebellum (CB), caudate (cau), putamen (put), thalamus (thal), cortical depth	White et al. (2002)	12 MZ (6 males), 12 control pairs (6 males) no DZ	MZ: 24.5 ± 7.2 Cont: 24.4 ± 7.2	MZ correlations: CB, TB, GM, WM: $r>0.90$ Cau, put, thal: $r>.75$ cort depth low correlations
Corpus callosum area White matter hyperintensities (WMH)	Scamvougeras et al. (2003) Atwood et al. (2004)	14 MZ, 12 DZ 1330 Individuals in genetically useful relationships	Mean 27 years 60.99±9.61	94% WMH: 55%
Longitudinal: genetic stability (4-year follow up). Corpus callosum (CC) and LV	Pfefferbaum et al. (2004)	34 MZ, 37 DZ	T1:68–80 years T2: 72–84 years	CC: a increase from 89 to 92% LV: a decrease from 92 to 88% New environmental influences
Density of focal brain gray and white matter. Parahippocampal gyrus	Hulshoff Pol et al. (2006)[#]	54 MZ (33 males) 58 DZ (17 males, 21 OS) 34 sibs	MZM: 31.2, MZF: 34.1, DZM: 30.3, DZF: 30.6, OS: 30.3, sibs: 29	occ-front-temp and white matter pathways: 76–85% parahippocampal gyrus: 69%

* Manual segmentation; #VBM; $midsagittal surface measurement; NA, not available.
MZ, monozygotic; DZ, dizygotic; MZM, monozygotic male; MZF, monozygotic female; DZM, dizygotic male; DZF, dizygotic female; OS, opposite sex; A, additive genetic; C, common environment; E, unique environment; FA, fractional anisotropy; GM, gray matter; WM, white matter; TB, total brain; IC, intracranial volume; occ-front-temp, occipito-fronto temporal.

subjects (Pfefferbaum et al., 2000; Pfefferbaum, Sullivan, & Carmelli, 2004).

A few studies have examined possible genetic effects on more specific brain areas. Volumes of frontal and temporal gray matter (GM) are particularly influenced by genetic factors (Thompson et al., 2001; Wright et al., 2002). Furthermore, brain density of the medial and superior frontal, superior temporal and occipital gray matter and connecting white matter of the superior occipito-frontal fasciculus and corpus callosum are particularly influenced by genetic factors (Hulshoff Pol et al., 2006).

Area measurements of the corpus callosum revealed heritability estimates between 79 and 94% (Pfefferbaum et al., 2000; Scamvougeras, Kigar, Jones, Weinberger, & Witelson, 2003). Variation in hippocampus volume was found to have a lower heritability with estimates of 40% (Sullivan, Pfefferbaum, Swan, & Carmelli, 2001) and 69% (Hulshoff Pol et al., 2006). Unique environmental factors influenced vast gray matter and white matter areas surrounding the lateral ventricles (up to 50%) (Hulshoff Pol et al., 2006).

In a study that did not include DZ twin pairs, MZ twin pair correlations were high (>0.90 for cerebellum, total brain, gray, and white matter and >0.75 for caudate nucleus, putamen, thalamus, and cortical depth) as compared to a healthy comparison group (White, Andreasen, & Nopoulos, 2002).

In the only study to date that measured heritability estimates of changes in brain volumes over time, genetic contributions to variability in intracranial volume, corpus callosum, and lateral ventricles were found to be high in healthy elderly (Pfefferbaum et al., 2000) remained high at longitudinal follow-up of 4 years (Pfefferbaum et al., 2004).

Next to twin studies, other designs can also be applied to yield estimates on genetic and environmental influences. For example, a family-based study reported heritability estimates of white matter hyperintensities of 55%. These estimates increased with age (Atwood et al., 2004).

In summary, human brain volume is considerably heritable. Moreover, it remains to be largely explained by genetic factors, even in old age. Individual variation in lateral ventricles is mainly explained by environmental factors, suggesting that surrounding brain tissue is at least partly influenced by environmental factors. Genetic effects were shown to vary regionally, with high heritabilities of frontal and temporal lobe volumes and densities, but moderate estimates in the hippocampus, and environmental influences on several medial brain areas. Areas that show high heritability for volume emphasize the relevance of these brain areas when searching for genes influencing brain structure.

It should be noted that only one longitudinal twin study is carried out in elderly subjects. Moreover, twin studies in children and/or adolescents are currently lacking. Therefore, no conclusions can be drawn about the stability of genetic

influences on brain volume. Studies are under way to determine the influence of genetic and environmental factors on brain changes with age.

Autosomal Genetic Abnormalities and Human Brain Morphology

Huntington's Disease

Huntington's disease (HD) is an autosomal-dominant neurodegenerative disease, which is associated with increases in the length of a CAG triplet repeat present in a gene called "huntingtin" located on chromosome 4p16.3 (Rosas et al., 2001). Cognitively, HD patients suffer from attention impairments and problems with executive functioning as well as psychomotoric functions, whereas semantic memory and delayed recall memory seem to be intact (Ho et al., 2003).

Several MRI studies have demonstrated that, compared to healthy controls, HD is associated with global loss in volumes of total brain, total cerebrum, cerebral cortex (Table 10.2a) (Paulsen et al., 2006; Rosas et al., 2003). Also white matter reductions (Jernigan, Salmon, Butters, & Hesselink, 1991; Rosas et al., 2003) and cortical thinning (Rosas et al., 2005) have been reported. Focal atrophy in the basal ganglia is an often found abnormality in HD patients (Aylward et al., 1994, 1997, 1998, 2000, 2004; Beglinger et al., 2005; Fennema-Notestine et al., 2004; Harris et al., 1992; Jernigan et al., 1991; Kassubek, Juengling, et al., 2004; Kassubek, Juengling, Ecker, & Landwehrmeyer, 2005; Kipps et al., 2005; Mascalchi et al., 2004; Paulsen et al., 2006; Peinemann et al., 2005; Rosas et al., 2001, 2003; Thieben et al., 2002). Other structures in HD that were smaller as compared to healthy subjects are the following: the hypothalamus (Kassubek, Juengling, et al., 2004), thalamus (Kassubek et al., 2005; Paulsen et al., 2006) amygdala, hippocampus, brainstem, cerebellum (Rosas et al., 2003), insula, dorsal midbrain (Thieben et al., 2002), and the frontal lobe (Aylward et al., 1998).

Interestingly, the major brain abnormality in HD, i.e., basal ganglia atrophy, was positively correlated with CAG repeat length, symptom severity (Aylward et al., 1997; Kassubek, Bernhard, et al., 2004; Rosas et al., 2001) as well as age of onset of the disease symptoms (Aylward et al., 1997).

In pre-clinical Huntington patients (who do not have symptoms yet, but who test positively for the "Huntingtin gene"), decreased volumes of the striatum, insula, and dorsal forebrain were detected when compared to healthy control subjects (Thieben et al., 2002). Furthermore, more progressive atrophy in the basal ganglia was found in clinical patients in a follow-up measurement as compared to pre-clinical patients (Kipps et al., 2005). Finally, striatal

Table 10.2 Autosomal genetic diseases

Syndrome	Brain area	Authors	Group size	Age (years)	p-Value/ES
(a) Huntington's disease					
	Total brain, total cerebrum, (GM and WM) ↓	Rosas et al. (2003)* 18 HD, 18 HC	HD: HC: matched	27–63 (34.8±10.41)	GM: p<0.02 WM: p<0.0001
		Paulsen et al. (2006)*	HD: 24 HC: 24	HD: 36.8 (±9.7) HC: 36.6 (±9.1)	P<0.001
	White matter ↓	Jernigan et al. (1991)*	11 HD 55 HC	HD: 49±12.5 HC: 54±14.1	P<0.01
		Rosas et al. (2003)*	18 HD, 18 HC	HD: 27–63 (34.8±10.41) HC: matched	P<0.0001
	Cerebellum ↓	Fennema-Notestine et al. (2004)*	15 HD, 22 HC	HD: 46.7 (±11.0) HC: 47.1 (±9.8)	WM: p<0.001 GM: p<0.01
	Basal ganglia ↓	Jernigan et al. (1991)	11 HD 55 HC	HD: 49±12.5 HC: 54±14.1	Caudate: P=0.000
		Harris et al. (1992)*[1]	15 HD, 19 HC	HD: 29–72 (43.2±12.9) HC: 23–73 (45.2±15.9)	Putamen: P<0.000001 Caudate: P<0.004
		Aylward et al. (1994)* (tb not diff)	10 HD+ 18 HD–	HD+: 34.4±5.44 HD–: 39.72±8.99	P<0.001 (total basal ganglia)
		Aylward et al. (1997)*[2] Longitudinal	23 HD	HD: 42.57±13.02 Interval: 1.73 years	P<0.001 (change)
		Aylward et al. (1998)*[1]	20 HD (10 mild, 10 mod), 20 HC	HD-mild: 44.8±15.4 HD-mod: 42.7±11.3 HC-mild: 43.9±16 HC-mod: 42.6±11.5	P<0.004
		Aylward et al. (2000)*[1] Aylward et al. (2004)*[1]	30 HD 19 pre-HD, 17 HD	HD (total): 43.6 (±10.2) HD: 32.1(±6.6) HC: 32.5(±6.7)	P<0.001 P=0.05
		Rosas et al. (2001)*	27 HD, 24 HC	HD: 25–63 43.8 (±10.3) HC: 29–62 41.2 (±9.8)	P<0.0001
		Rosas et al. (2003)*[1]	18 HD, 18 HC	HD: 27–63 (34.8±10.41) HC: matched	P<0.0001
		Thieben et al. (2002)#	18 HD (from 34 tested)	HD–: 38.04±11.43 HD+: 43.58±12.02	Left-putamen: p=0.009 Caud: p=0.003
		Kassubek et al. (2004b)#	44 HD , 22 HC	HD: 23–66 44.7 (±10.7) HC: 25–68 44.1 (±16.9)	P<0.001
		Mascalchi et al. (2004)	21 HD, 21 HC	HD: 58±11 HC: 54±13	P<0.001

Table 10.2 (continued)

Syndrome	Brain area	Authors	Group size	Age (years)	p-Value/ES
		Fennema-Notestine et al. (2004)*,1	15 HD, 22 HC	HD: 46.7 (±11.0) HC: 47.1 (±9.8)	
		Peinemann et al. (2005)#	25 HD 25 HC	HD: 43.8 (±7.7) HC: 42.9 (±9.8)	Caudate/putamen P<0.05
		Paulsen et al. (2006)*,1	HD: 24 HC: 24	HD: 36.8 (±9.7) HC: 36.6 (±9.1)	Cau: p<0.002 Put: p<0.0001
		Kipps et al. (2005) TBM Progression of atrophy	17 HD+ 13 HD−	HD+: 43.8 (±10.0) HD−: 42.0 (±11.4)	p<0.005
		Beglinger et al. (2005)*,1	10 HD 10 HC	HD: 54.3 (±8.2) HC: 53.9 (±9.0)	Cau: p<0.0002 Put: p<0.0001
		Kassubek et al. (2005)#	44 HD, 22	HD: 23–66 44.7 (±10.7) HC: 25–68 44.1 (±16.9)	Cau and put: p<0.001
	Thalamus ↓	Kassubek et al. (2005)# 44 HD, 22	HD: 23–66 44.7 (±10.7)	HC: 25–68 44.1 (±16.9)	P<0.001
		Paulsen et al. (2006)*,1	HD: 24 HC: 24	HD: 36.8 (±9.7) HC: 36.6 (±9.1)	p<0.019
	Hypothalamus ↓	Kassubek et al. (2004b)#	44 HD, 22	HD: 23–66 44.7 (±10.7) HC: 25–68 44.1 (±16.9)	p<0.001
	Amygdala (amyg), hippocampus (hip), brainstem ↓	Rosas et al. (2003)*,1	18 HD, 18 HC	HD: 27–63 (34.8±10.41) HC: matched	Amyg: p<0.001 Hip: p<0.01 Brainstem: p<0.0001
	Insula, dorsal midbrain ↓	Thieben et al. (2002)#	18 HD (from 34 tested)	HD−: 38.04±11.43 HD+: 43.58±12.02	Insula: R: p<0.032 L: p<0.010
	Frontal lobe ↓	Aylward et al. (1998)*,1	20 HD (10 mild, 10 mod), 20 HC	HD-mild: 44.8±15.4 HD-mod: 42.7±11.3 HC-mild: 43.9±16 HC-mod: 42.6±11.5	NA
(b) Down syndrome					
	Cerebrum ↓	Raz et al. (1995)*	13 DS, 12 HC	DS: 22–50 (35.2±8.32) HC: 23–49 (35.3±8.12)	P<0.002
		Weis et al. (1991)* (cerebral cortex)	7 DS, 7 HC	DS: 30–45 (38) HC: 36–44 (38)	P=0.003
	Cerebellum ↓	White et al. (2003)#	19 DS, 11 HC	DS: 34–52 41.9(±6.0) HC: 37–56 45.6(±6.1)	P<0.05
		Pinter et al. (2001a)*,1	16 DS, 15 HC	DS: 5–23.8 11.3 (±5.2) HC: 5.4–23.2 11.9 (±4.7)	P<0.0001

Table 10.2 (continued)

Syndrome	Brain area	Authors	Group size	Age (years)	p-Value/ES
		Raz et al. (1995)*,[3],	13 DS, 12 HC	DS: 22–50 (35.2±8.32) HC: 23–49 (35.3±8.12)	P <0.002
		Jernigan et al. (1993)*	6 DS, 21 HC	DS: 10–20 (15.5±3.4) HC: 10–24 (14.5±3.8)	P <0.001
	White matter ↓	Weis et al. (1991)*	7 DS, 7 HC	DS: 30–45 (38) HC: 36–44 (38)	P = 0.05 (controlled for TB: p=0.06)
		Weis et al. (1991)*	7 DS, 7 HC	DS: 30–45 (38) HC: 36–44 (38)	P=0.004
	Cingulate gyrus ↓	White et al. (2003)#	19 DS, 11 HC	DS: 34–52 41.9 (±6.0) HC: 37–56 45.6 (±6.1)	P <0.05
	Amygdala ↔	Pinter et al. (2001b)*,[1]	16 DS, 15 HC	DS: 11.3 (±5.2) HC: 11.9 (±4.7)	R: p<0.32 L: p<0.27
	Hippocampus ↓	Pinter et al. (2001b)*,[1]	16 DS, 15 HC	DS: 11.3 (±5.2) HC: 11.9 (±4.7)	R: p<0.03 L: p<0.002
		Raz et al. (1995)*,[3]	13 DS, 12 HC	DS: 22–50 (35.2±8.32) HC: 23–49 (35.3±8.12)	P <0.002
		Kesslak et al. (1994)*,[2]	13 DS, 10 HC	DS: 23–51 HC: matched	P=0.0285
		Teipel et al. (2003)*,[1]	34 DS, 31 HC	DS: m: 41.6 years HC: 41.8 years	P <0.001
	Corpus callosum ↓ (area)	Teipel et al. (2003)*,[1]	34 DS, 31 HC	DS: m: 41.6 years HC: 41.8 years	P=0.05
	Planum temporale ↓	Frangou et al. (1997)*,[1]	17 DS, 17 HC	DS: 39.2 (±8.7) HC: 39.5 (±9.1)	P <0.0007
	Mammillary bodies ↓	Raz et al. (1995)*,[3]	13 DS, 12 HC	DS: 22–50 (35.2±8.32) HC: 23–49 (35.3±8.12)	P <0.005
	Parahippocampal gyrus ↑	White et al. (2003)#	19 DS, 11 HC	DS: 34–52 41.9 (±6.0) HC: 37–56 45.6 (±6.1)	P <0.05
		Raz et al. (1995)*,[3]	13 DS, 12 HC	DS: 22–50 (35.2±8.32) HC: 23–49 (35.3±8.12)	P <0.008
		Kesslak et al. (1994)*,[2]	13 DS, 10 HC	DS: 23–51 HC: matched	P=0.0022
	Parahippocampal gyrus →	Teipel et al. (2004)# (with age)	27 DS	DS: 41.1 (±9.1)	P <0.05
		Krasuski et al. (2002)*,[1] (with age)	31 DS, 33 HC	DS: 41.6 (±9.1) HC: 41.6 (±10.7)	R: p<0.04 L: p<0.009

Table 10.2 (continued)

Syndrome	Brain area	Authors	Group size	Age (years)	p-Value/ES
(c) Williams syndrome					
	Total brain ↓	Reiss et al. (2000)*	14 WS, 14 HC	WS: 19–44 (28.7±8.9) HC: 20–48 (29.0±9.0)	P<0.001
		Schmitt et al. (2001a)*,[1]	20 WS, 20 HC	WS: 19–44 28.5 (±8.3) HC: 19–48 28.5 (±8.2)	P<0.001
		Cherniske et al. (2004)*	20 WS, Normative data comparison	WS: 30–51 38.8	NA
	Cortical complexity ↑	Thompson et al. (2005)	42 WS 40 HC	WS: 29.2 (±9.0) HC27.5 (±7.4)	p<0.0015
	Parieto-occipital sulcus ↓	Meyer-Lindenberg et al. (2004)[#]	13 WS, 11 HC	HC: 30.8 (±7.6) WS: 28.3 (±9.6)	p<0.001
		Boddaert et al. (2006)[#]	9 WS 11 HC	WS: 11.6 (±3.1) HC: 11.8 (±2.2)	p<0.05
	Intraparietal sulcus	Meyer-Lindenberg et al. (2004)[#]	13 WS, 11 HC	WS: 28.3 (±9.6) HC: 30.8 (±7.6)	p<0.001
		Kippenhan et al. (2005) (sulcal depth)	14 WS 13 HC	WS: 27.6 (±9.6) HC: 31.2 (±7.1)	p<0.01
	Hypothalamus ↓	Meyer-Lindenberg et al. (2004)[#]	13 WS, 11 HC	WS: 28.3 (±9.6) HC: 30.8 (±7.6)	p<0.001
	Superior and inferior parietal lobe ↓	Eckert et al. (2005)*,[1]	17 WS, 17 HC	WS: 28.9 HC: 27.1	p<0.05
	Occipital lobe and thalamus ↓	Reiss et al. (2004)*,[1]	43 WS, 40 HC	WS: 12–50 28.9 (±9.2) HC: 18–49 27.5 (±7.4)	Occ: p<0.003 Thal: p<0.0001
	Corpus callosum ↓	Tomaiuolo et al. (2002)*,[1] (corrected Talairach stereotaxic space)	12 WS, 12 HC	WS-M: 13–20 20 (±7.2) WS-F: 13–20 16.6 (±3.01) HC-M: 13–29 20 (±6.9) HC-F: 13–19 16 (±3.46)	p<0.015
		Schmitt et al. (2001b)	20 WS, 20 HC	WS: 19–44 28.5 (±8.3) HC: 19–48 28.5 (±8.2)	p = 0.04
	Cerebellum ↑	Jones et al. (2002)	9 WS, 9 HC	WS: 0.6–3.5 (mean: 1.75) HC: 1.7–3.5 (mean: 2.42)	NA

Table 10.2 (continued)

Syndrome	Brain area	Authors	Group size	Age (years)	p-Value/ES
		Schmitt et al. (2001a)*,[1]	20 WS, 20 HC	WS: 19–44 28.5 (±8.3) HC: 19–48 28.5 (±8.2)	p<0.001
		Reiss et al. (2000)*,[1]	14 WS, 14 HC	WS:19–44 (28.7±8.9) HC: 20–48 (29.0±9.0)	p<0.02
		Wang et al. (1992)*	11 WS, 18 HC	WS: 10–20 (14.7) HC: 10–24 (15.4)	p<0.05
	Amygdala (Am), orbito and orbitofrontal cortex (OFC), anterior cingulated (AC) and insular cortex ↑	Reiss et al. (2004)*,[1]	43 WS, 40 HC	WS: 12–50 28.9 (±9.2) HC: 18–49 27.5 (±7.4)	Am: p<0.001 AC: p<0.005 OFC: p<0.0001
	Superior temporal gyrus ↑	Reiss et al. (2000)*,[1]	14 WS, 14 HC	WS:19–44 (28.7±8.9) HC: 20–48 (29.0±9.0)	p=0.05
		Reiss et al. (2004)	43 WS, 43 HC	WS: 12–50 28.9 (±9.2) HC: 18–49 27.5 (±7.4)	p<.001
(d) Velocardiofacial syndrome					
	Total brain ↓	Eliez et al. (2001a&b)*	23 VCFS, 23 HC	5.8–21.0 VC: 12.7(±3.9) HC: 12.9(±4.1)	P<0.0001
		Simon et al. (2005)#	18 VCFS 18 HC	VC: 7.3–14 9.88(±1.4) HC: 7.5–14.1 10.42(±1.98)	P<0.01
		Kates et al. (2006)*,[1]	47 VCFS, 15 sibs, 18 HC	VC: 11.7 (±2.1), sibs: 11.5 (± 1.8), HC: 11.5 (±2.0)	P<0.03
	WM>GM ↓	Eliez et al. (2000)*	15 VCFS, 15 HC	VC: 10.5(±3.1) HC: 10.8(±2.7)	NA
		Kates et al. (2001)* (non-frontal wm)	10 VCFS, 10 HC	7.9–14.5 VC: 10.1(±1.8) HC: 10.1(±1.9)	P<0.02
		Simon et al. (2005) 18 VCFS	18 HC VC: 7.3–14	9.88(±1.4) HC: 7.5–14.1	10.42(±1.98) NA
		Campbell et al. (2006)#	39 VCFS 26 HC	VC: 11 ± 3 HC: 11 ± 3	P<0.008
	Cerebellum ↓	V Amelsvoort et al. (2004)#,*,[1]	12 VCFS (no schizos) 14 HC	VC: 31(±10) HC: 36(±10)	P<0.001
		Eliez et al. (2001a)*,[1] (lobules VI–VII)	24 VCFS, 24 HC	VC: 12.5(±4) HC: 12.7(±4.2)	P<0.01 Pons P <0.0001

Table 10.2 (continued)

Syndrome	Brain area	Authors	Group size	Age (years)	p-Value/ES
		Campbell et al. (2006)[#] 39 VCFS	26 HC VC: 11 ± 3	HC: 11 ± 3	P <0.008
		Bish et al. (2006)[$]	31 VCFS 23 HC	VC: 9 (±.9) HC: 10 (±.6)	P <0.013
	Hippocampus (hip), temporal lobe (tem), superior temporal gyrus (Sup) ↓	Eliez et al. (2001b)[*,1] (when controlled for TB: no difference!)	23 VCFS, 23 HC	5.8–21.0 VC: 12.7(±3.9) HC: 12.9(±4.1)	Hip: P <0.006 Tem: p <0.0001 Sup: p <0.02
	Left caudate nucleus ↓ (ratio r-cau:TB larger in controls!)	Sugama et al. (2000)[*,1]	17 VC 15 HC	VC: 0.7–21 Mean: 5.41 HC: 1.1–25 Mean: 6.0	P <0.05
	Left parietal lobe ↓	Eliez et al. (2000)[*,1]	15 VCFS, 15 HC	VC: 10.5 (±3.1) HC: 10.8 (±2.7)	P <0.05
	FA frontal, parietal, and occipital lobe ↓	Barnea-Goraly et al. (2003)[#](&dti)	19 VCFS, 19 HC	7.2–21.8 VC: 12.2 (±3.9) HC: 14.4 (±4.2)	Right-midfrontal gyrus: Z=5.07 Left-superior parietal gyrus: Z=3.7
	Posterior thalamus ↓	Bish et al. (2004)	18 VCFS 18 HC	VC: 9.8 (±1.4) HC: 10.4 (±1.9)	P < 0.031
	FA in corpus callosum ↓	Simon et al. (2005)[#]	18 VCFS 18 HC	VC: 7.3–14 9.88 (±1.4) HC: 7.5–14.1 10.42 (±1.98)	Z=4.74
	FA cingulate gyrus ↑	Simon et al. (2005)[#]	18 VCFS 18 HC	VC: 7.3–14 9.88 (±1.4) HC: 7.5–14.1 10.42 (±1.98)	Z=4.78
	Right and left amygdale ↑	Kates et al. (2006)[*,1]	47 VCFS, 15 sibs, 18 HC	VC: 11.7 (±2.1), sibs: 11.5 (±1.8), HC: 11.5 (±2.0)	Left: p <0.002 Right: p <0.01
	Right caudate nucleus ↑	Kates et al. (2004) Campbell et al. (2006)[*,1]	10 VCFS, 10 HC 39 VCFS 26 HC	NA VC: 11 ± 3 HC: 11 ± 3	NA P <0.04
	Frontal lobe (not wm) ↑	Eliez et al. (2000)[*,1]	15 VCFS, 15 HC	VC: 10.5 (±3.1) HC: 10.8 (±2.7)	P <0.001
		Simon et al. (2005)[#]	18 VCFS 18 HC	VC: 7.3–14 9.88 (±1.4) HC: 7.5–14.1 10.42 (±1.98)	Z = 5.1

* Manual segmentation; [#] VBM; [$] midsagittal surface measurement; [1] controlled for total brain/intracranium; [2] no control TB/IC; [3] controlled for height

Abbreviations 2a–d: HD, Huntington's disease; DS, Down syndrome; WS, William's syndrome; VCFS, Velocardiofacial syndrome; HC, healthy controls; FA, fractional anisotropy; GM, gray matter; WM, white matter; TB, total brain; IC, intracranial volume

decline in pre-clinical HD patients was predictive of the time of occurrence of the first clinical symptoms (Aylward et al., 2004).

Down Syndrome

Down syndrome (DS) is caused by a third copy of chromosome 21 (trisomy). DS is associated with mental retardation, and after the age of 40, individuals with DS suffer from cognitive decline or dementia (Lott & Head, 2001). A rapidly growing number of MRI studies have investigated brain atrophy in DS (Table 10.2b). When adult DS patients are compared to healthy subjects, they have smaller volumes of total cerebrum (Raz et al., 1995; Weis, Weber, Neuhold, & Rett, 1991), cerebellum (Jernigan & Bellugi, 1990; Raz et al., 1995; Weis et al., 1991; White, Alkire, & Haier, 2003), and total white matter (Weis et al., 1991). Regional decreases in volume in DS patients have been observed in the cingulate gyrus (White et al., 2003), hippocampus (Kesslak, Nagata, Lott, & Nalcioglu, 1994; Pinter, Brown, et al., 2001; Raz et al., 1995), planum temporale (Frangou et al., 1997), and mammillary bodies (Raz et al., 1995).

Cross-sectional studies carried out in DS patients show significantly more atrophy in patients than healthy controls with advancing age, mainly in the hippocampus (Krasuski, Alexander, Horwitz, Rapoport, & Schapiro, 2002; Teipel et al., 2003), amygdala (Krasuski et al., 2002), parahippocampal gyrus (Krasuski et al., 2002; Teipel et al., 2004), corpus callosum (Teipel et al., 2003), and frontal, parietal, and occipital gyrus (Teipel et al., 2004). However, in an earlier follow-up study no evidence was found for progressive changes in the hippocampus and amygdala of DS patients (Aylward et al., 1999).

Furthermore, children with DS also show brain abnormalities in the cerebellum (Jernigan, Bellugi, Sowell, Doherty, & Hesselink, 1993; Pinter, Eliez, Schmitt, Capone, & Reiss, 2001) and amygdala (Pinter, Brown, et al., 2001) compared to age-matched controls. When a direct distinction is made between DS children and adults, it appears that DS children already have a decreased volume of the cerebellum and hippocampus, although the amygdala and parietal gray matter seem to be preserved (Pinter, Eliez, et al., 2001).

When demented and non-demented DS patients are compared, demented DS patients show more pronounced atrophy with age (Pearlson et al., 1998). The amygdala showed no volumetric differences between demented and non-demented DS patients (Aylward et al., 1999).

A structure that has been reported to be enlarged in DS is the parahippocampal gyrus (Kesslak et al., 1994; Raz et al., 1995; White et al., 2003). Other studies, however, could not replicate this finding (Krasuski et al., 2002; Teipel et al., 2004).

Williams Syndrome

Williams syndrome (WS) patients have a well-defined hemideletion on chromosome 7q11.23. WS patients are characterized by selective preservation of certain complex faculties (language, music, face processing, and sociability) in contrast to marked and severe deficits in nearly every other cognitive domain (Levitin et al., 2003).

Morphometric MRI studies have both investigated adult subjects as well as children and adolescents. In adults, studies demonstrated a decreased total brain volume in WS patients as compared to healthy control subjects (Table 10.2c) (Cherniske et al., 2004; Reiss et al., 2000; Schmitt, Eliez, Warsofsky, Bellugi, & Reiss, 2001a). Furthermore, taken the smaller total brain volume into account, reductions in parieto-occipital (Kippenhan et al., 2005; Meyer-Lindenberg et al., 2004) and intraparietal sulcus (Kippenhan et al., 2005; Meyer-Lindenberg et al., 2004), hypothalamus (Meyer-Lindenberg et al., 2004), superior parietal lobe (Eckert et al., 2005), gray matter of the occipital lobe, thalamus (Reiss et al., 2004), and corpus callosal area (Schmitt, Eliez, Warsofsky, Bellugi, & Reiss, 2001b) were found.

Studies in WS children and adolescents showed reductions in parieto-occipital sulcus (Boddaert et al., 2006) and corpus callosal area (Tomaiuolo et al., 2002).

Some brain regions are found to be increased in WS adult patients when compared to healthy subjects. These include the cerebellum (Reiss et al., 2000; Schmitt et al., 2001a), amygdala, orbitofrontal and medial prefrontal cortex, anterior cingulate, insular cortex (Reiss et al., 2004), and superior temporal gyrus (Reiss et al., 2000, 2004). Furthermore, increased overall cortical complexity was found in WS (Thompson et al., 2005) as well as increased cortical gyrification in the right parietal and occipital lobes and in the left frontal lobe (Schmitt et al., 2002).

Enlargements in the cerebellum were found in WS infants (Jones et al., 2002) and adolescents (Wang, Hesselink, Jernigan, Doherty, & Bellugi, 1992).

In WS, as opposed to HD where one specific gene (i.e., the "huntingtin" gene) seems to be involved, only the locus of the deletion on the chromosome is known and knowledge of specific genes and their working mechanism(s) in the deleted region is scarce. Animal studies suggest involvement of the LIMK1-gene in abnormal brain development, which is located in the deleted region at chromosome 7q11.23 (Table 10.3) (Hoogenraad et al., 2002; Meng et al., 2002). Other genes mapped to region 7q11.23 and linked to abnormal brain development are CYLN2 (Hoogenraad et al., 1998), GTF21 (Danoff, Taylor, Blackshaw, & Desiderio, 2004; Morris et al., 2003), and WBSCR14 (Cairo, Merla, Urbinati, Ballabio, & Reymond, 2001).

Table 10.3 Genes related to brain volumetric changes, discussed in this review[1]

Brain area	Associated genes	Candidate genes	Disease	Gene map locus	Number of studies
Cerebral cortex (TB)		ASPM*,**	Microcephaly	1q31	3
Prefrontal cortex, hippocampus	BDNF*,**			11p13	4
Prefrontal cortex		COMT*,**	VCFS	22q11.2	2
Hippocampus	ApoE*,**			19q13.2	10
Limbic system, orbitofrontal cortex	MAOA*,**			Xp11.23	4
Caudate nucleus	FOXP2*,**			7q31	3
Basal ganglia		Huntingtin*	HD	4p16.3	17
Synaptic connections		ProDH**	VCFS	22q11.2	2
Synaptic connections		TBX1**	VCFS	22q11.2	2
Brain development		LIMK1**	WS	7q11.23	2
		CYLN2**	WS	7q11.23	2
		WBSCR14**	WS	7q11.23	1

[1] For references: see text
* Has been associated with brain volume changes in humans
** Has been associated with brain volume changes in animals
HD, Huntington's disease; DS, Down syndrome; WS, William's syndrome; VCFS, Velocardiofacial syndrome; TB, total brain volume

Velocardiofacial Syndrome

Velocardiofacial syndrome (VCFS) is a neurogenetic disorder caused by deletions on chromosome 22q11.2. Patients with VCFS are characterized by learning disabilities (Swillen et al., 1999) and are often diagnosed with schizophrenia (Bassett et al., 2003; Murphy, Jones, & Owen, 1999).

Most of the structural MRI studies on VCFS are carried out in children and adolescents. In these studies, abnormalities have been found in several brain areas (Table 10.2d). A smaller total brain volume was reported (Eliez, Schmitt, White, & Reiss, 2000; Eliez, Schmitt, White, Wellis, & Reiss, 2001; Eliez, Blasey, et al., 2001; Simon et al., 2005) with (non-frontal) white matter relatively more affected than gray matter (Eliez et al., 2000; Kates et al., 2001; Simon et al., 2005; Campbell et al., 2006). More focal areas that appeared smaller in VCFS as compared to control subjects included the cerebellum (Bish et al., 2006), vermal lobules VI–VII, pons (Eliez, Schmitt, et al., 2001), temporal lobe, superior temporal gyrus, hippocampus (Eliez, Blasey, et al., 2001), left and right amygdala (Kates, Miller, et al., 2006), left caudate nucleus (Sugama et al., 2000), posterior thalamus (Bish, Nguyen, Ding, Ferrante, & Simon, 2004), and left parietal lobe (Eliez et al., 2000). Moreover, DTI studies investigating fractional anisotropy (FA) in white matter, an index of white matter coherence and integrity, found lower FA values in frontal, parietal, and temporal cortex, in connections between the frontal and temporal lobes (Barnea-Goraly et al., 2003) and corpus callosum (Simon et al., 2005). However, increased FA values were reported for the cingulate gyrus (Simon et al., 2005).

In the one study carried out in adult VCFS patients, a reduction in cerebellum density was found (van Amelsvoort et al., 2004). In the same study, adult VCFS patients with and without schizophrenia were compared. It was shown that VCFS patients with schizophrenia had larger ventricles and less overall white matter as compared to VCFS patients without schizophrenia (van Amelsfoort et al., 2004).

A brain area that is enlarged in VCFS is the right caudate nucleus (Kates et al., 2004; Campbell et al., 2006). Frontal lobe volumes seem to be relatively preserved (Eliez et al., 2000; Simon et al., 2005), although this does not seem to hold for the frontal white matter (Kates et al., 2004).

Similar to WS, in VCFS the locus of the deletion on the chromosome is known (22q11.2), but knowledge of specific genes and their working mechanism(s) in the deleted region is limited. Recently, the COMT (catechol-*O*-methyltransferase) low-activity genotype was identified as a risk factor for decline in prefrontal cortical volume (Gothelf et al., 2005). Furthermore, this finding showed an interaction with sex (Kates, Antshel, et al., 2006). In animal studies, the ProDH and TBX1 genes are also mapped to region 22q11 and are thought to be involved in refinement and stabilization of synaptic connections in the adolescent mouse brain (Rakic, Bourgeois, & Goldman-Rakic, 1994).

Other Genetic Approaches

Next to studying brain volume in specific genetic abnormalities, there are other genetic approaches that may elucidate genes involved in brain variation. These include studying the brains of families with a particular genetic makeup, searching for genes in subjects with brain morphological abnormalities, and associating brain volumes with genetic polymorphisms in candidate genes in healthy subjects.

In the so-called "KE family", half the family in three generations is affected by a severe speech and language disorder, which is transmitted as an autosomal-dominant monogenic trait (Vargha-Khadem, Watkins, Alcock, Fletcher, & Passingham, 1995). Genetic linkage studies identified a locus, designated SPCH1, on chromosome 7q31 (Fisher, Vargha-Khadem, Watkins, Monaco, & Pembrey, 1998). A point mutation was identified in the affected family members which alters an invariant amino acid residue in the DNA-binding domain of a forkhead/winged helix transcription factor, encoded by the gene FOXP2 (Table 10.3) (Lai, Fisher, Hurst, Vargha-Khadem, & Monaco, 2001). The affected family members have a reduction in volume of the caudate nucleus bilaterally, as well as changes in gray matter in other mostly motor- and speech-related brain areas, as compared to the unaffected members and healthy control subjects (Watkins et al., 2002). The discovery of the responsible gene in the "KE family" led to further research into the FOXP2 gene and its role in brain development. For example, the expression pattern of the FOXP2 mRNA has been found in the developing brain of mouse (Ferland, Cherry, Preware, Morrisey, & Walsh, 2003; Lai, Gerrelli, Monaco, Fisher, & Copp, 2003) and human, including the basal ganglia, thalamus, and cerebellum (Lai et al., 2003).

The search for genes in subjects with particular morphological changes in the brain was successful in autosomal recessive primary microcephaly (MCPH). MCPH is characterized by shrinkage of nearly 70% of the cortex. Involvement of the ASPM gene (Bond et al., 2002; Mekel-Bobrov et al., 2005) and the microcephalin gene (MCPH1) (Evans et al., 2004) was suggested in the determination of cerebral cortex size. The ASPM gene is the human ortholog (i.e., evolved from) of the *Drosophila melanogaster* abnormal spindle gene (asp), which is essential for normal mitotic spindle function in embryonic brain development. Mutations in the ASPM gene associated with MCPH suggest that regulation of mitotic spindle orientation may be an important evolutionary mechanism controlling brain size. However, in healthy subjects, recently no associations of allelic variants of the ASPM gene and MCPH1 gene and total brain volume were found (Woods et al., 2006). It was argued that outside the context of the microcephalic state, it is misleading to refer to the ASPM gene and/or MCPH1 as regulating or controlling brain size (Woods et al., 2006).

Another genetic approach that may elucidate genes involved in brain variation is studying polymorphisms of specific genes in healthy subjects. A polymorphism is defined as the existence of multiple alleles of a gene within a population. It is a naturally occurring variation in the sequence of genetic information on a segment of DNA among individuals. Those variations are considered normal (not to be confused with true mutations, which are alterations of the original genetic material, often being harmful).

The few studies on polymorphisms in healthy subjects have revealed associations with brain volumes or densities. For example, Val/met (i.e., valine/methionine amino acids) variant carriers of the brain-derived neurotrophic factor (BDNF) gene (a gene involved in reducing the amount of naturally occurring neuronal cell death) were found to have a reduced size of the prefrontal cortex (Pezawas et al., 2004) and hippocampus compared to val/val carriers (Bueller et al., 2006; Pezawas et al., 2004; Szeszko et al., 2005). In addition, in met-BDNF carriers a negative relation was found between dorsolateral prefrontal cortex volume and age, which was not present in the val-BDNF carriers (Nemoto et al., 2006).

A study of allelic variants of the apolipoprotein (ApoE) gene – thought to be involved in cell growth and regeneration of nerves – showed that healthy elderly subjects who were homozygous for the Epsilon4 allele, i.e., e4-e4 genotype had smaller hippocampal volumes than subjects heterozygous for that allele and than e4 non-carriers (Lemaitre et al., 2005; Lind et al., 2006). Also, the presence of a single ApoE-epsilon4 allele is associated with an increased rate of hippocampal volume loss in healthy women (Cohen, Small, Lalonde, Friz, & Sunderland, 2001).

Two variants of the X-linked monoamine oxidase A gene (MAOA) were recently associated with brain volumes in healthy subjects. The low expression variant predicted volume reductions in cingulate gyrus, amygdala, insula, and hypothalamus, whereas the high expression variant was associated with changes in orbitofrontal volume (Meyer-Lindenberg et al., 2006).

Studies of polymorphisms and brain volumetric variation in psychiatric populations also found genes associated with alterations in brain volume. For example, in schizophrenia, a reduction in BDNF production and availability in the dorsolateral prefrontal cortex (DLPFC) was found (Weickert et al., 2003). Furthermore, the disrupted-in-schizophrenia 1 (DISC1) gene was associated with prefrontal gray matter loss (Cannon et al., 2005) and hippocampus decrease (Callicott et al., 2005).

In a study on attention-deficit hyperactivity disorder (ADHD) it was shown that homozygosity for the 10R-allele of the dopamine transporter 1 (DAT1) gene was associated with smaller caudate nucleus volumes and homozygosity of 4R-allele of the dopamine receptor D4 (DRD4) gene with smaller prefrontal gray matter (Durston et al., 2005).

Overall, studying polymorphisms in healthy subjects yields valuable information on specific genes that may be involved in brain volume. However, as it is a newly developing area of research, the robustness of the findings needs to be pointed out and therefore replication is warranted.

Discussion and Conclusion

In this chapter, the influences of genes on human brain volume were reviewed. For this purpose, twin studies were included to assess the heritability of human brain volumetric variation in the general population. In addition, brain structures in patients with a clear genetic etiology were reviewed. Finally, other genetic approaches to the search of genes involved in brain volume were discussed. These other approaches included studies on brain volumes of families with a particular genetic makeup, studies that search for genes in subjects with brain morphological abnormalities, and studies examining genetic polymorphisms in healthy subjects.

Twin studies showed high heritability estimates for specific brain structures and for overall brain size in adulthood (between 66 and 97%). Both global gray and global white matter are largely determined by genes. However, individual variation in lateral ventricles is mainly explained by environmental factors, suggesting that surrounding brain tissue is at least partly influenced by environmental factors. Genetic effects were shown to vary regionally, with high heritability estimates of frontal lobe volumes (90–95%), but moderate estimates of the hippocampus (40–69%), and environmental influences on several medial brain areas. Areas that show a high heritability for volume emphasize the relevance of these brain areas when searching for genes influencing brain structure. For focal structures heritability estimates differ, suggesting that different genes influence focal brain structures differentially.

The study of diseases with a clear genetic etiology yielded specific information on changes in brain volumes, densities, and fractional anisotropy. In patients with Huntington's disease, decreased volumes of the basal ganglia were found. Moreover, age at onset of the first symptoms was significantly related to the amount of atrophy in the basal ganglia. Also, the larger the CAG repeat length in Huntington's disease, the more atrophy in the basal ganglia was found. In Down syndrome, a decreased cerebellum and increased parahippocampal gyrus volume and density were found. In Williams syndrome, an increased amygdala, superior temporal gyrus, and cerebellum were reported. Finally in Velocardiofacial syndrome a decreased parietal lobe volume was found. Interestingly, across all disorders, pronounced decreases in white matter volume and hippocampus volume were revealed, irrespective of the genes and/or chromosomes involved. Furthermore, in all brain imaging studies of autosomal abnormalities, a decreased total brain volume was consistently found. It must be noted that although most studies found decreases in brain volume associated with autosomal abnormalities, there are also genetic disorders in which an enlarged brain is present. These include Sotos syndrome (haplo-insufficiency of the NSD1 gene on 5q35) (Kurotaki et al., 2002), also known as cerebral gigantism. However, no quantitative MRI studies in Sotos syndome have been carried out, and therefore these data were not included in this review.

Results of other genetic approaches, such as investigating allelic variation in the healthy population, have revealed information on specific genes that may be involved in human brain volume. Polymorphisms of the brain-derived neurotrophic factor (BDNF) and apolipoprotein (ApoE) genes have been associated with prefrontal cortex and hippocampus volumes. More specifically, met-BDNF carriers showed reduced prefrontal cortex and hippocampus volumes compared to val-BDNF carriers. Homozygous carriers of the Epsilon 4-allele showed smaller hippocampus volumes than heterozygous carriers. In addition, high- and low-expression variants of the monoamine oxidase A gene (MAOA) resulted in structural differences in limbic and frontal areas. The study of polymorphisms in healthy subjects is a rapidly developing area of research which allows direct investigation of genetic influence (not confounded by disease).

Establishing the extent to which brain morphology is influenced by genes (and environment) contributes both to our understanding of healthy functioning as well as to elucidating the causes of brain disease. More specifically, it enhances our knowledge of individual variation in brain functioning and facilitates the interpretation of the morphological changes found in psychiatric disorders such as schizophrenia. Also, it allows future efforts to find particular genes responsible for brain structures to be concentrated in areas that are under considerable genetic influence (Hulshoff Pol et al., 2006).

Taken together, studies have shown that adult human brain volume is highly genetically determined. Since brain volume changes dynamically throughout life, longitudinal twin studies in childhood as well as in adulthood are needed to investigate the stability of genetic (and environmental) influences. During different age ranges, genes may exert different effects. Studies carried out in autosomal pathologies were reviewed to search for genes or chromosomal regions which are involved in volumetric changes. The genes that have been discovered in these areas might serve as a model for the genes being implicated in healthy individuals; however, direct evidence of the influence of specific genes on the (maintenance of) human brain volume (throughout life) is currently lacking. Polymorphism research on these candidate genes might be helpful in enhancing our knowledge on their influence in healthy human brain volume.

There are a number of limitations to the reviewed approaches of studying genes involved in human brain volume. These limitations need to be taken into account when interpreting findings of studies into the genes involved in human brain structure. First, it must be noted that twin studies in children and adolescents have not been carried out so far. Therefore, no conclusions can be drawn as to the genetic influences on brain volume during childhood. Moreover, only one longitudinal MRI study in twins pairs (in older adults) was completed up to now, and therefore conclusions as to the stability of genetic influences onto brain volume throughout life await further study. Furthermore, it has been argued that the twin method may yield an inflated estimation of heritabilities compared to family and/or adoption studies. On the other hand, family studies might give lower heritability estimations as different ages within families are compared (for a discussion on this topic, see Martin, Boomsma, & Machin, 1997).

Limitations in studying genetic disorders include the presence of co-morbidity in some disorders. In Down syndrome patients who also suffer from dementia, global volumetric reductions are more pronounced with age and particularly present in the amygdala. Also, in Velocardiofacial patients diagnosed with schizophrenia larger ventricles and less white matter are found as compared to Velocardiofacial patients without schizophrenia. However, this finding does not necessarily mean that reductions in white matter results from genetic expression associated with brain morphology. In Velocardiofacial, the reduction of (the integrity of) white matter may well be a secondary to the vascular risk of these patients, i.e., heart defects. Vascular risk factors have been related to white matter lesions (de Leeuw et al., 2004; DeCarli et al., 2005). A second limitation includes the possible confounding effects of the pathology on brain volume. For example, it can be argued that being in a disadvantageous environment (a disease–environment interaction) might lead to decreases in brain morphology. However, brain volumetric changes can be directly associated with the genetic abnormality, which is suggested in Huntington's disease: decreased caudate nucleus volumes were reported prior to disease onset in subjects having the mutation in the huntingtin gene (Thieben et al., 2002). Here, it is important to mention that while Huntington's disease is the only neurodegenerative disorder discussed in this chapter, it offers valuable information on the effects of a single gene in subjects with and without having symptoms. Third, the relative small number of subjects usually involved in the studies may have limited its statistical power. Fourth, different types of medication of the subjects might have confounded the results. For example, Huntington's disease patients often use antipsychotics and/or antidepressants (Bonelli, Wenning, & Kapfhammer, 2004), which have been found to affect brain morphology (Bremner & Vermetten, 2004; Lieberman et al., 2005). Fifth, it is difficult to form a well-matched control group to diseases as Down syndrome, where mental retardation should be taken into account. A limitation in the section of polymorphism studies is that psychiatric and neurological disorders such as Alzheimer's disease and schizophrenia were not discussed in this paper. These conditions can also give more insight into the genetic mechanisms influencing brain volume.

Finally, a limitation which applies to all the reviewed studies is the MRI methodology. Intracranial or total brain volume was not always corrected for, which limits conclusions regarding the influence of a particular gene on small brain structures. Also, methodology of quantification of small structures in the brain can differ across the reviewed studies. For instance, manual segmentation of a structure versus region-of-interest (ROI) analysis with voxel-based morphometry might not lead to completely overlapping findings.

Future Directions

The studies that were discussed in this review have revealed several genes to be associated with the regulation of human brain structure. However, at this point it seems too early to draw general conclusions about which genes are implicated in human brain morphology. Future studies, with other genetic approaches and new MRI methodology may enhance our understanding of the genes involved in human brain structure.

Without specific knowledge of candidate genes, linkage studies are now employed with the goal to localize a gene that influences a phenotype. This approach can be used when genetic marker data (based on DNA polymorphisms of known location in the genome) are available in extended families or in sibling pairs. Linkage studies are often called a theoretical ("blind" search for genes) in contrast to association studies which require knowledge of candidate genes (Vink & Boomsma, 2002). Linkage studies require data collection in related individuals (e.g., siblings or large pedigrees). Also, if the location of a certain polymorphism is not known, a linkage study of the whole genome can be carried out. To our knowledge, only one genome-wide linkage study in healthy subjects has been performed, in relation to brain volume. For white matter hyper-intensity volumes one linkage peak was identified on chromosome 4p (DeStefano et al., 2006). This is the region where the gene responsible for Huntington's disease, i.e., huntingtin, is located. The area of genome-wide research deserves further study as it allows identifying candidate genes involved in human brain volume.

A newly emerging field of genetic research is the study of epigenetics. Epigenetics comprises mechanisms of inheritance, which are not the consequences of changes in DNA structure. They affect gene transcription with environmental factors acting as modulators or inducers of epigenetic factors. One such (important) factor is DNA methylation (see Santos, Mazzola, & Carvalho, 2005 for a review on the working mechanisms). The genome-wide pattern of DNA methylation was found to be more alike within monozygotic young than in monozygotic adult and elderly twin pairs (Fraga et al., 2005). It is therefore important to investigate which environmental factors have an influence on the expression of genes (as found in DNA methylation). Consequently, the study of interaction between genes and environmental factors is warranted. Furthermore, the simultaneous effects of multiple genes and possibly the interaction among genes, also need investigation as one could argue that a single gene polymorphism cannot explain morphological changes in the brain.

New brain imaging methods, such as diffusion tensor imaging (DTI)-fiber tracking, allow study of the connections and/or coherence of white matter fibers. Since white matter was found to be affected in most genetic diseases, future attention could therefore be focused on genes involved in neural networks. Considering the changes in brain structure throughout development in both childhood and adulthood, the study of genes involved in the plasticity of brain structure throughout life is warranted. Indeed, longitudinal studies in (pre)adolescent twin pairs and their siblings are underway to study these effects (Peper et al., 2004).

In summary, it can be concluded that adult human brain volume is highly determined by genetic factors. Specific genes have already been associated with volumes of several brain structures. Particularly white matter and hippocampus volumes are associated with a number of these candidate genes. Many more genes and their interaction with environmental factors that are involved in brain volume in childhood, adolescence, and adulthood are expected to be found in the coming years.

Acknowledgment This work was supported in part by a grant from the Dutch Science Organization for Medical Research NWO-ZonMw (HEH, DIB, RSK, 051.02.060: HEH 051.02.061).

References

Atwood, L. D., Wolf, P. A., Heard-Costa, N. L., Massaro, J. M., Beiser, A., D'Agostino, R. B., et al. (2004). Genetic variation in white matter hyperintensity volume in the Framingham Study. *Stroke, 35*, 1609–1613.

Aylward, E. H., Brandt, J., Codori, A. M., Mangus, R. S., Barta, P. E., & Harris, G. J. (1994). Reduced basal ganglia volume associated with the gene for Huntington's disease in asymptomatic at-risk persons. *Neurology, 44*, 823–828.

Aylward, E. H., Li, Q., Stine, O. C., Ranen, N., Sherr, M., Barta, P. E., et al. (1997). Longitudinal change in basal ganglia volume in patients with Huntington's disease. *Neurology, 48*, 394–399.

Aylward, E. H., Anderson, N. B., Bylsma, F. W., Wagster, M. V., Barta, P. E., Sherr, M., et al. (1998). Frontal lobe volume in patients with Huntington's disease. *Neurology, 50*, 252–258.

Aylward, E. H., Li, Q., Honeycutt, N. A., Warren, A. C., Pulsifer, M. B., Barta, P. E., et al. (1999). MRI volumes of the hippocampus and amygdala in adults with Down's syndrome with and without dementia. *American Journal of Psychiatry, 156*, 564–568.

Aylward, E. H., Codori, A. M., Rosenblatt, A., Sherr, M., Brandt, J., Stine, O. C., et al. (2000). Rate of caudate atrophy in presymptomatic and symptomatic stages of Huntington's disease. *Movement Disorders, 15*, 552–560.

Aylward, E. H., Sparks, B. F., Field, K. M., Yallapragada, V., Shpritz, B. D., Rosenblatt, A., et al. (2004). Onset and rate of striatal atrophy in preclinical Huntington disease. *Neurology, 63*, 66–72.

Baaré, W. F., Hulshoff Pol, H. E., Boomsma, D. I., Posthuma, D., de Geus, E. J., Schnack, H. G., et al. (2001). Quantitative genetic modeling of variation in human brain morphology. *Cerebral Cortex, 11*, 816–824.

Barnea-Goraly, N., Menon, V., Krasnow, B., Ko, A., Reiss, A., & Eliez, S. (2003). Investigation of white matter structure in velocardiofacial syndrome: a diffusion tensor imaging study. *American Journal of Psychiatry, 160*, 1863–1869.

Bartley, A. J., Jones, D. W., & Weinberger, D. R. (1997). Genetic variability of human brain size and cortical gyral patterns. *Brain, 120*(Pt 2), 257–269.

Bartzokis, G., Beckson, M., Lu, P. H., Nuechterlein, K. H., Edwards, N., & Mintz, J. (2001). Age-related changes in frontal and temporal lobe volumes in men: a magnetic resonance imaging study. *Archives of General Psychiatry, 58*, 461–465.

Bassett, A. S., Chow, E. W.C., AbdelMalik, P., Gheorghiu, M., Husted, J., & Weksberg, R. (2003). The schizophrenia phenotype in 22q11 deletion syndrome. *American Journal of Psychiatry, 160*, 1580–1586.

Beglinger, L. J., Nopoulos, P. C., Jorge, R. E., Langbehn, D. R., Mikos, A. E., Moser, D. J., et al. (2005). White matter volume and cognitive dysfunction in early Huntington's disease. *Cognitive and Behavioral Neurology, 18*, 102–107.

Bish, J. P., Nguyen, V., Ding, L., Ferrante, S., & Simon, T. J. (2004). Thalamic reductions in children with chromosome 22q11.2 deletion syndrome. *Neuroreport, 15*, 1413–1415.

Bish, J. P., Pendyal, A., Ding, L., Ferrante, H., Nguyen, V., Donald-McGinn, D., et al. (2006). Specific cerebellar reductions in children with chromosome 22q11.2 deletion syndrome. *Neuroscience Letters, 399*, 245–248.

Boddaert, N., Mochel, F., Meresse, I., Seidenwurm, D., Cachia, A., Brunelle, F., et al. (2006). Parieto-occipital grey matter abnormalities in children with Williams syndrome. *Neuroimage, 30*, 721–725.

Bond, J., Roberts, E., Mochida, G. H., Hampshire, D. J., Scott, S., Askham, J. M., et al. (2002). ASPM is a major determinant of cerebral cortical size. *Nature Genetics, 32*, 316–320.

Bonelli, R. M., Wenning, G. K., & Kapfhammer, H. P. (2004). Huntington's disease: present treatments and future therapeutic modalities. *International Clinical Psychopharmacology, 19*, 51–62.

Boomsma, D., Busjahn, A., & Peltonen, L. (2002). Classical twin studies and beyond. *Nature Reviews Genetics, 3*, 872–882.

Bremner, J. D., & Vermetten E. (2004). Neuroanatomical changes associated with pharmacotherapy in posttraumatic stress disorder. *Biobehavioral Stress Response: Protective and Damaging Effects, 1032*, 154–157.

Bueller, J. A., Aftab, M., Sen, S., Gomez-Hassan, D., Burmeister, M., & Zubieta, J. K. (2006). BDNF Val66Met allele is associated with reduced hippocampal volume in healthy subjects. *Biological Psychiatry, 59*, 812–815.

Cairo, S., Merla, G., Urbinati, F., Ballabio, A., & Reymond A. (2001). WBSCR14, a gene mapping to the Williams–Beuren syndrome deleted region, is a new member of the Mlx transcription factor network. *Human Molecular Genetics, 10*, 617–627.

Callicott, J. H., Straub, R. E., Pezawas, L., Egan, M. F., Mattay, V. S., Hariri, A. R., et al. (2005). Variation in DISC1 affects hippocampal structure and function and increases risk for schizophrenia.*Proceedings of the National Academy of Sciences of the United States of America, 102*, 8627–8632.

Campbell, L. E., Daly, E., Toal, F., Stevens, A., Azuma, R., Catani, M., et al. (2006). Brain and behaviour in children with 22q11.2 deletion syndrome: A volumetric and voxel-based morphometry MRI study. *Brain, 129*, 1218–1228.

Cannon, T. D., Hennah, W., van Erp, T. G., Thompson, P. M., Lonnqvist, J., Huttunen, M., et al. (2005). Association of DISC1/TRAX haplotypes with schizophrenia, reduced prefrontal gray matter, and impaired short- and long-term memory. *Archives of General Psychiatry, 62,* 1205–1213.

Carmelli, D., DeCarli, C., Swan, G. E., Jack, L. M., Reed, T., Wolf, P. A., et al. (1998). Evidence for genetic variance in white matter hyperintensity volume in normal elderly male twins. *Stroke, 29,* 1177–1181.

Carmelli, D., Reed, T., & DeCarli C. (2002). A bivariate genetic analysis of cerebral white matter hyperintensities and cognitive performance in elderly male twins. *Neurobiology of Aging, 23,* 413–420.

Cherniske, E. M., Carpenter, T. O., Klaiman, C., Young, E., Bregman, J., Insogna, K., et al. (2004). Multisystem study of 20 older adults with Williams syndrome. *American Journal of Medical Genetics, 131A,* 255–264.

Cohen, R. M., Small, C., Lalonde, F., Friz, J., & Sunderland T. (2001). Effect of apolipoprotein E genotype on hippocampal volume loss in aging healthy women. *Neurology, 57,* 2223–2228.

Danoff, S. K., Taylor, H. E., Blackshaw, S., & Desiderio S. (2004). TFII-I, a candidate gene for Williams syndrome cognitive profile: Parallels between regional expression in mouse brain and human phenotype. *Neuroscience, 123,* 931–938.

de Leeuw, F. E., Richard, F., de Groot, J. C., van Duijn, C. M., Hofman, A., van Gijn, J., et al. (2004). Interaction between hypertension, apoE, and cerebral white matter lesions. *Stroke, 35,* 1057–1060.

DeCarli, C., Massaro, J., Harvey, D., Hald, J., Tullberg, M., Au, R., et al. (2005). Measures of brain morphology and infarction in the Framingham heart study: establishing what is normal. *Neurobiology of Aging, 26,* 491–510.

DeStefano, A. L., Atwood, L. D., Massaro, J. M., Heard-Costa, N., Beiser, A., Au, R., et al. (2006). Genome-wide scan for white matter hyperintensity: The Framingham Heart Study. *Stroke, 37,* 77–81.

Durston, S., Fossella, J. A., Casey, B. J., Hulshoff Pol, H. E., Galvan, A., Schnack, H. G., et al. (2005). Differential effects of DRD4 and DAT1 genotype on fronto-striatal gray matter volumes in a sample of subjects with attention deficit hyperactivity disorder, their unaffected siblings, and controls. *Molecular Psychiatry, 10,* 678–685.

Eckert, M. A., Leonard, C. M., Molloy, E. A., Blumenthal, J., Zijdenbos, A., & Giedd, J. N. (2002). The epigenesis of planum temporale asymmetry in twins. *Cerebral Cortex, 12,* 749–755.

Eckert, M. A., Hu, D., Eliez, S., Bellugi, U., Galaburda, A., Korenberg, J., et al. (2005). Evidence for superior parietal impairment in Williams syndrome. *Neurology, 64,* 152–153.

Eliez, S., Schmitt, J. E., White, C. D., & Reiss, A. L. (2000). Children and adolescents with velocardiofacial syndrome: a volumetric MRI study. *American Journal of Psychiatry, 157,* 409–415.

Eliez, S., Blasey, C. M., Schmitt, E. J., White, C. D., Hu, D., & Reiss, A. L. (2001). Velocardiofacial syndrome: are structural changes in the temporal and mesial temporal regions related to schizophrenia? *American Journal of Psychiatry, 158,* 447–453.

Eliez, S., Schmitt, J. E., White, C. D., Wellis, V. G., & Reiss, A. L. (2001). A quantitative MRI study of posterior fossa development in velocardiofacial syndrome. *Biological Psychiatry, 49,* 540–546.

Evans, P. D., Anderson, J. R., Vallender, E. J., Gilbert, S. L., Malcom, C. M., Dorus, S., et al. (2004). Adaptive evolution of ASPM, a major determinant of cerebral cortical size in humans. *Human Molecular Genetics, 13,* 489–494.

Fennema-Notestine, C., Archibald, S. L., Jacobson, M. W., Corey-Bloom, J., Paulsen, J. S., Peavy, G. M., et al. (2004). In vivo evidence of cerebellar atrophy and cerebral white matter loss in Huntington disease. *Neurology, 63,* 989–995.

Ferland, R. J., Cherry, T. J., Preware, P. O., Morrisey, E. E., & Walsh, C. A. (2003). Characterization of Foxp2 and Foxp1 mRNA and protein in the developing and mature brain. *Journal of Comparative Neurology, 460,* 266–279.

Fisher, S. E., Vargha-Khadem, F., Watkins, K. E., Monaco, A. P., & Pembrey, M. E. (1998). Localisation of a gene implicated in a severe speech and language disorder. *Nature Genetics, 18,* 168–170.

Fraga, M. F., Ballestar, E., Paz, M. F., Ropero, S., Setien, F., Ballestart, M. L., et al. (2005). Epigenetic differences arise during the lifetime of monozygotic twins. *Proceedings of the National Academy of Sciences of the United States of America, 102,* 10604–10609.

Frangou, S., Aylward, E., Warren, A., Sharma, T., Barta, P., & Pearlson G. (1997). Small planum temporale volume in Down's syndrome: a volumetric MRI study. *American Journal of Psychiatry, 154,* 1424–1429.

Gale, C.R., O'Callaghan, F. J., Bredow, M., & Martyn, C. N. (2006). The influence of head growth in fetal life, infancy, and childhood on intelligence at the ages of 4 and 8 years. *Pediatrics, 118,* 1486–1492.

Geschwind, D. H., Miller, B. L., DeCarli, C., & Carmelli D. (2002). Heritability of lobar brain volumes in twins supports genetic models of cerebral laterality and handedness. *Proceedings of the National Academy of Sciences of the United States of America, 99,* 3176–3181.

Giedd, J. N., Blumenthal, J., Jeffries, N. O., Castellanos, F. X., Liu, H., Zijdenbos, A., Paus, T., et al. (1999). Brain development during childhood and adolescence: A longitudinal MRI study. *Nature Neuroscience, 2,* 861–863.

Gogtay, N., Nugent, T. F., III, Herman, D. H., Ordonez, A., Greenstein, D., Hayashi, K. M., et al. (2006). Dynamic mapping of normal human hippocampal development. *Hippocampus, 16,* 664–672.

Gothelf, D., Eliez, S., Thompson, T., Hinard, C., Penniman, L., Feinstein, C., et al. (2005). COMT genotype predicts longitudinal cognitive decline and psychosis in 22q11.2 deletion syndrome. *Nature Neuroscience, 8,* 1500–1502.

Harris, G. J., Pearlson, G. D., Peyser, C. E., Aylward, E. H., Roberts, J., Barta, P. E., et al. (1992). Putamen volume reduction on magnetic resonance imaging exceeds caudate changes in mild Huntington's disease. *Annals of Neurology, 31,* 69–75.

Ho, A. K., Sahakian, B. J., Brown, R. G., Barker, R. A., Hodges, J. R., Ane, M. N., et al. (2003). Profile of cognitive progression in early Huntington's disease. *Neurology, 61,* 1702–1706.

Hoogenraad, C. C., Eussen, B. H., Langeveld, A., van Haperen, R., Winterberg, S., Wouters, C. H., et al. (1998). The murine CYLN2 gene: genomic organization, chromosome localization, and comparison to the human gene that is located within the 7q11.23 Williams syndrome critical region. *Genomics, 53,* 348–358.

Hoogenraad, C. C., Koekkoek, B., Akhmanova, A., Krugers, H., Dortland, B., Miedema, M., et al. (2002). Targeted mutation of Cyln2 in the Williams syndrome critical region links CLIP-115 haploinsufficiency to neurodevelopmental abnormalities in mice. *Nature Genetics, 32,* 116–127.

Hulshoff Pol, H. E., Posthuma, D., Baare, W. F., de Geus, E. J., Schnack, H. G., van Haren, N. E., et al. (2002). Twin-singleton differences in brain structure using structural equation modelling. *Brain, 125,* 384–390.

Hulshoff Pol, H. E., Schnack, H. G., Posthuma, D., Mandl, R. C.W., Baare, W. F., van Oel, C., et al. (2006). Genetic contributions to human brain morphology and intelligence. *Journal of Neuroscience, 40,* 10235–10242.

Jernigan, T. L., & Bellugi U. (1990). Anomalous brain morphology on magnetic resonance images in Williams syndrome and Down syndrome. *Archives of Neurology, 47,* 529–533.

Jernigan, T. L., Salmon, D. P., Butters, N., & Hesselink, J. R. (1991). Cerebral structure on MRI, Part II: Specific changes in Alzheimer's and Huntington's diseases. *Biological Psychiatry, 29,* 68–81.

Jernigan, T. L., Bellugi, U., Sowell, E., Doherty, S., & Hesselink, J. R. (1993). Cerebral morphologic distinctions between Williams and Down syndromes. *Archives of Neurology, 50,* 186–191.

Jones, W., Hesselink, J., Courchesne, E., Duncan, T., Matsuda, K., & Bellugi U. (2002). Cerebellar abnormalities in infants and toddlers with Williams syndrome. *Developmental Medicine and Child Neurology, 44,* 688–694.

Kassubek, J., Bernhard, L. G., Ecker, D., Juengling, F. D., Muche, R., Schuller, S., et al. (2004). Global cerebral atrophy in early stages of Huntington's disease: Quantitative MRI study. *Neuroreport, 15,* 363–365.

Kassubek, J., Juengling, F. D., Kioschies, T., Henkel, K., Karitzky, J., Kramer, B., et al. (2004). Topography of cerebral atrophy in early Huntington's disease: A voxel based morphometric MRI study. *Journal of Neurology Neurosurgery and Psychiatry, 75,* 213–220.

Kassubek, J., Juengling, F. D., Ecker, D., & Landwehrmeyer, G. B. (2005). Thalamic atrophy in Huntington's disease co-varies with cognitive performance: A morphometric MRI analysis. *Cerebral Cortex, 15,* 846–853.

Kates, W. R., Burnette, C. P., Jabs, E. W., Rutberg, J., Murphy, A. M., Grados, M., et al. (2001). Regional cortical white matter reductions in velocardiofacial syndrome: A volumetric MRI analysis. *Biological Psychiatry, 49,* 677–684.

Kates, W. R., Burnette, C. P., Bessette, B. A., Folley, B. S., Strunge, L., Jabs, E. W., et al. (2004). Frontal and caudate alterations in velocardiofacial syndrome (deletion at chromosome 22q11.2). *Journal of Child Neurology, 19,* 337–342.

Kates, W. R., Antshel, K. M., Abdulsabur, N., Colgan, D., Funke, B., Fremont, W., et al. (2006). A gender-moderated effect of a functional COMT polymorphism on prefrontal brain morphology and function in velo-cardio-facial syndrome (22q11.2 deletion syndrome). *American Journal of Medical Genetics part B: Neuropsychiatric Genetics, 141,* 274–280.

Kates, W. R., Miller, A. M., Abdulsabur, N., Antshel, K. M., Conchelos, J., Fremont, W., et al. (2006). Temporal lobe anatomy and psychiatric symptoms in velocardiofacial syndrome (22q11.2 deletion syndrome). *Journal of the American Academy of Child and Adolescent Psychiatry, 45,* 587–595.

Kesslak, J. P., Nagata, S. F., Lott, I., & Nalcioglu O. (1994). Magnetic resonance imaging analysis of age-related changes in the brains of individuals with Down's syndrome. *Neurology, 44,* 1039–1045.

Kippenhan, J. S., Olsen, R. K., Mervis, C. B., Morris, C. A., Kohn, P., Lindenberg, A. M., et al. (2005). Genetic contributions to human gyrification: Sulcal morphometry in Williams syndrome. *Journal of Neuroscience, 25,* 7840–7846.

Kipps, C. M., Duggins, A. J., Mahant, N., Gomes, L., Ashburner, J., & McCusker, E. A. (2005). Progression of structural neuropathology in preclinical Huntington's disease: A tensor based morphometry study. *Journal of Neurology Neurosurgery and Psychiatry, 76,* 650–655.

Krasuski, J. S., Alexander, G. E., Horwitz, B., Rapoport, S. I., & Schapiro, M. B. (2002). Relation of medial temporal lobe volumes to age and memory function in nondemented adults with Down's syndrome: Implications for the prodromal phase of Alzheimer's disease. *American Journal of Psychiatry, 159,* 74–81.

Kurotaki, N., Imaizumi, K., Harada, N., Masuno, M., Kondoh, T., Nagai, T., et al. (2002). Haploinsufficiency of NSD1 causes Sotos syndrome. *Nature Genetics, 30,* 365–366.

Lai, C. S., Fisher, S. E., Hurst, J. A., Vargha-Khadem, F., & Monaco, A. P. (2001). A forkhead-domain gene is mutated in a severe speech and language disorder. *Nature, 413,* 519–523.

Lai, C. S., Gerrelli, D., Monaco, A. P., Fisher, S. E., & Copp, A. J. (2003). FOXP2 expression during brain development coincides with adult sites of pathology in a severe speech and language disorder. *Brain, 126,* 2455–2462.

Lemaitre, H., Crivello, F., Dufouil, C., Grassiot, B., Tzourio, C., Alperovitch, A., et al. (2005). No epsilon4 gene dose effect on hippocampal atrophy in a large MRI database of healthy elderly subjects. *Neuroimage, 24,* 1205–1213.

Levitin, D. J., Menon, V., Schmitt, J. E., Eliez, S., White, C. D., Glover, G. H., et al. (2003). Neural correlates of auditory perception in Williams syndrome: An fMRI study. *Neuroimage, 18,* 74–82.

Lieberman, J. A., Tollefson, G. D., Charles, C., Zipursky, R., Sharma, T., Kahn, R. S., et al. (2005). Antipsychotic drug effects on brain morphology in first-episode psychosis. *Archives of General Psychiatry, 62,* 361–370.

Lind, J., Larsson, A., Persson, J., Ingvar, M., Nilsson, L. G., Backman, L., et al. (2006). Reduced hippocampal volume in non-demented carriers of the apolipoprotein E epsilon4: relation to chronological age and recognition memory. *Neuroscience Letters, 396,* 23–27.

Lott, I. T., & Head E. (2001). Down syndrome and Alzheimer's disease: A link between development and aging. *Mental Retardation and Developmental Disabilities Research Reviews, 7,* 172–178.

Martin, N. G., Boomsma, D. I., & Machin, G. (1997). A twin-pronged attack on complex traits. *Nature Genetics, 17,* 387–392.

Mascalchi, M., Lolli, F., Della, N. R., Tessa, C., Petralli, R., Gavazzi, C., et al. (2004). Huntington disease: volumetric, diffusion-weighted, and magnetization transfer MR imaging of brain. *Radiology, 232,* 867–873.

Mekel-Bobrov, N., Gilbert, S. L., Evans, P. D., Vallender, E. J., Anderson, J. R., Hudson, R. R., et al. (2005). Ongoing adaptive evolution of ASPM, a brain size determinant in Homo sapiens. *Science, 309,* 1720–1722.

Meng, Y., Zhang, Y., Tregoubov, V., Janus, C., Cruz, L., Jackson, M., et al. (2002). Abnormal spine morphology and enhanced LTP in LIMK-1 knockout mice. *Neuron, 35,* 121–133.

Meyer-Lindenberg, A., Kohn, P., Mervis, C. B., Kippenhan, J. S., Olsen, R. K., Morris, C. A., et al. (2004). Neural basis of genetically determined visuospatial construction deficit in Williams syndrome. *Neuron, 43,* 623–631.

Meyer-Lindenberg, A., Buckholtz, J. W., Kolachana, B., Hariri, R., Pezawas, L., Blasi, G., et al. (2006). Neural mechanisms of genetic risk for impulsivity and violence in humans. *Proceedings of the National Academy of Sciences of the United States of America, 103,* 6269–6274.

Morris, C. A., Mervis, C. B., Hobart, H. H., Gregg, R. G., Bertrand, J., Ensing, G. J., et al. (2003). GTF2I hemizygosity implicated in mental retardation in Williams syndrome: genotype-phenotype analysis of five families with deletions in the Williams syndrome region. *American Journal of Medical Genetics A, 123,* 45–59.

Murphy, K. C., Jones, L. A., & Owen, M. J. (1999). High rates of schizophrenia in adults with velo-cardio-facial syndrome. *Archives of General Psychiatry, 56,* 940–945.

Nemoto, K., Ohnishi, T., Mori, T., Moriguchi, Y., Hashimoto, R., Asada, T., et al. (2006). The Val66Met polymorphism of the brain-derived neurotrophic factor gene affects age-related brain morphology. *Neuroscience Letters, 397,* 25–29.

Nokelainen, P. & Flint, J. (2002). Genetic effects on human cognition: lessons from the study of mental retardation syndromes. *Journal of Neurology Neurosurgery and Psychiatry, 72,* 287–296.

Paulsen, J. S., Magnotta, V. A., Mikos, A. E., Paulson, H. L., Penziner, E., Andreasen, N. C., et al. (2006). Brain structure in preclinical Huntington's disease. *Biological Psychiatry, 59,* 57–63.

Paus, T., Zijdenbos, A., Worsley, K., Collins, D. L., Blumenthal, J., Giedd, J. N., et al. (1999). Structural maturation of neural pathways in children and adolescents: in vivo study. *Science, 283,* 1908–1911.

Pearlson, G. D., Breiter, S. N., Aylward, E. H., Warren, A. C., Grygorcewicz, M., Frangou, S., et al. (1998). MRI brain changes in subjects with Down syndrome with and without dementia. *Developmental Medicine and Child Neurology, 40,* 326–334.

Peinemann, A., Schuller, S., Pohl, C., Jahn, T., Weindl, A., & Kassubek J. (2005). Executive dysfunction in early stages of Huntington's disease is associated with striatal and insular atrophy: a neuropsychological and voxel-based morphometric study. *Journal of the Neurological Sciences, 239,* 11–19.

Pennington, B. F., Filipek, P. A., Lefly, D., Chhabildas, N., Kennedy, D. N., Simon, J. H., et al. (2000). A twin MRI study of size variations in human brain. *Journal of Cognitive Neuroscience, 12*, 223–232.

Peper, J. S., Zwiers, M. P., van Leeuwen, M., Van den Berg, S. M., Boomsma, D. I., Kahn, R. S., et al. (2004). Genetic modeling of cognitive brain maturation in pre-adolescence: A longitudinal study in twins. *Twin Research and Human Genetics, 7*, S172.

Pezawas, L., Verchinski, B. A., Mattay, V. S., Callicott, J. H., Kolachana, B. S., Straub, R. E., et al. (2004). The brain-derived neurotrophic factor val66met polymorphism and variation in human cortical morphology. *Journal of Neuroscience, 24*, 10099–10102.

Pfefferbaum, A., Sullivan, E. V., Swan, G. E., & Carmelli D. (2000). Brain structure in men remains highly heritable in the seventh and eighth decades of life. *Neurobiology of Aging, 21*, 63–74.

Pfefferbaum, A., Sullivan, E. V., & Carmelli D. (2004). Morphological changes in aging brain structures are differentially affected by time-linked environmental influences despite strong genetic stability. *Neurobiology of Aging, 25*, 175–183.

Pinter, J. D., Brown, W. E., Eliez, S., Schmitt, J. E., Capone, G. T., & Reiss, A. L. (2001). Amygdala and hippocampal volumes in children with Down syndrome: a high-resolution MRI study. *Neurology, 56*, 972–974.

Pinter, J. D., Eliez, S., Schmitt, J. E., Capone, G. T., & Reiss, A. L. (2001). Neuroanatomy of Down's syndrome: a high-resolution MRI study. *American Journal of Psychiatry, 158*, 1659–1665.

Posthuma, D. & Boomsma, D. I. (2000). A note on the statistical power in extended twin designs. *Behavioral Genetics, 30*, 147–158.

Posthuma, D., de Geus, E. J.C., Neale, M. C., Pol, H. E.H., Baare, W. E.C., Kahn, R. S., et al. (2000). Multivariate genetic analysis of brain structure in an extended twin design. *Behavior Genetics, 30*, 311–319.

Posthuma, D., de Geus, E. J., Baare, W. F., Hulshoff Pol, H. E., Kahn, R. S., & Boomsma, D. I. (2002). The association between brain volume and intelligence is of genetic origin. *Nature Neuroscience, 5*, 83–84.

Rakic, P., Bourgeois, J. P., & Goldman-Rakic, P. S. (1994). Synaptic development of the cerebral cortex: implications for learning, memory, and mental illness. *Progression in Brain Research, 102*, 227–243.

Raz, N., Torres, I. J., Briggs, S. D., Spencer, W. D., Thornton, A. E., Loken, W. J., et al. (1995). Selective neuroanatomic abnormalities in Down's syndrome and their cognitive correlates: evidence from MRI morphometry. *Neurology, 45*, 356–366.

Reiss, A. L., Eliez, S., Schmitt, J. E., Straus, E., Lai, Z., Jones, W., et al. (2000). IV. Neuroanatomy of Williams syndrome: a high-resolution MRI study. *Journal of Cognitive Neuroscience, 12*(Suppl 1), 65–73.

Reiss, A. L., Eckert, M. A., Rose, F. E., Karchemskiy, A., Kesler, S., Chang, M., et al. (2004). An experiment of nature: brain anatomy parallels cognition and behavior in Williams syndrome. *Journal of Neuroscience, 24*, 5009–5015.

Reveley, A. M., Reveley, M. A., Chitkara, B. & Clifford, C. (1984). The genetic basis of cerebral ventricular volume. *Psychiatry Research, 13*, 261–266.

Rosas, H. D., Goodman, J., Chen, Y. I., Jenkins, B. G., Kennedy, D. N., Makris, N., et al. (2001). Striatal volume loss in HD as measured by MRI and the influence of CAG repeat. *Neurology, 57*, 1025–1028.

Rosas, H. D., Koroshetz, W. J., Chen, Y. I., Skeuse, C., Vangel, M., Cudkowicz, M. E., et al. (2003). Evidence for more widespread cerebral pathology in early HD: An MRI-based morphometric analysis. *Neurology, 60*, 1615–1620.

Rosas, H. D., Hevelone, N. D., Zaleta, A. K., Greve, D. N., Salat, D. H., & Fischl B. (2005). Regional cortical thinning in preclinical Huntington disease and its relationship to cognition. *Neurology, 65*, 745–747.

Santos, K. F., Mazzola, T. N., & Carvalho, H. F. (2005). The prima donna of epigenetics: The regulation of gene expression by DNA methylation. *Brazilian Journal of Medical and Biological Research, 38*, 1531–1541.

Scamvougeras, A., Kigar, D. L., Jones, D., Weinberger, D. R., & Witelson, S. F. (2003). Size of the human corpus callosum is genetically determined: an MRI study in mono and dizygotic twins. *Neuroscience Letters, 338*, 91–94.

Schmitt, J. E., Eliez, S., Warsofsky, I. S., Bellugi, U., & Reiss, A. L. (2001a). Enlarged cerebellar vermis in Williams syndrome. *Journal of Psychiatric Research, 35*, 225–229.

Schmitt, J. E., Eliez, S., Warsofsky, I. S., Bellugi, U., & Reiss, A. L. (2001b). Corpus callosum morphology of Williams syndrome: Relation to genetics and behavior. *Developmental Medicine and Child Neurology, 43*, 155–159.

Schmitt, J. E., Watts, K., Eliez, S., Bellugi, U., Galaburda, A. M., & Reiss, A. L. (2002). Increased gyrification in Williams syndrome: Evidence using 3D MRI methods. *Developmental Medicine and Child Neurology, 44*, 292–295.

Simon, T. J., Ding, L., Bish, J. P., McDonald-McGinn, D. M., Zackai, E. H., & Gee J. (2005). Volumetric, connective, and morphologic changes in the brains of children with chromosome 22q11.2 deletion syndrome: an integrative study. *Neuroimage, 25*, 169–180.

Sowell, E. R., Trauner, D. A., Gamst, A., & Jernigan, T. L. (2002). Development of cortical and subcortical brain structures in childhood and adolescence: A structural MRI study. *Developmental Medicine and Child Neurology, 44*, 4–16.

Sugama, S., Bingham, P. M., Wang, P. P., Moss, E. M., Kobayashi, H., & Eto Y. (2000). Morphometry of the head of the caudate nucleus in patients with velocardiofacial syndrome (del 22q11.2). *Acta Paediatrica, 89*, 546–549.

Sullivan, E. V., Pfefferbaum, A., Swan, G. E., & Carmelli D. (2001). Heritability of hippocampal size in elderly twin men: equivalent influence from genes and environment. *Hippocampus, 11*, 754–762.

Swillen, A., Devriendt, K., Legius, E., Prinzie, P., Vogels, A., Ghesquiere, P., et al. (1999). The behavioural phenotype in velocardio-facial syndrome (VCFS): From infancy to adolescence. *Genetic Counseling, 10*, 79–88.

Szeszko, P. R., Lipsky, R., Mentschel, C., Robinson, D., Gunduz-Bruce, H., Sevy, S., et al. (2005). Brain-derived neurotrophic factor val66met polymorphism and volume of the hippocampal formation. *Molecular Psychiatry, 10*, 631–636.

Teipel, S. J., Schapiro, M. B., Alexander, G. E., Krasuski, J. S., Horwitz, B., Hoehne, C., et al. (2003). Relation of corpus callosum and hippocampal size to age in nondemented adults with Down's syndrome. *American Journal of Psychiatry, 160*, 1870–1878.

Teipel, S. J., Alexander, G. E., Schapiro, M. B., Moller, H. J., Rapoport, S. I., & Hampel H. (2004). Age-related cortical grey matter reductions in non-demented Down's syndrome adults determined by MRI with voxel-based morphometry. *Brain, 127*, 811–824.

Thieben, M. J., Duggins, A. J., Good, C. D., Gomes, L., Mahant, N., Richards, F., et al. (2002). The distribution of structural neuropathology in pre-clinical Huntington's disease. *Brain, 125*, 1815–1828.

Thompson, P. M., Cannon, T. D., Narr, K. L., van Erp, T., Poutanen, V. P., Huttunen, M., et al. (2001). Genetic influences on brain structure. *Nature Neuroscience, 4*, 1253–1258.

Thompson, P. M., Lee, A. D., Dutton, R. A., Geaga, J. A., Hayashi, K. M., Eckert, M. A., et al. (2005). Abnormal cortical complexity and thickness profiles mapped in Williams syndrome. *Journal of Neuroscience, 25*, 4146–4158.

Toga, A. W., & Thompson, P. M. (2004). Genetics of Brain Structure and Intelligence. *Annual Review of Neuroscience, 28*, 1–23.

Toga, A. W., Thompson, P. M., & Sowell, E. R. (2006). Mapping brain maturation. *Trends in Neurosciences, 29*, 148–159.

Tomaiuolo, F., Di Paola, M., Caravale, B., Vicari, S., Petrides, M., & Caltagirone C. (2002). Morphology and morphometry of the corpus callosum in Williams syndrome: A T1-weighted MRI study. *Neuroreport, 13*, 2281–2284.

van Amelsvoort, T., Daly, E., Henry, J., Robertson, D., Nguyen, V., Owen, M., et al. (2004). Brain anatomy in adults with velocardiofacial syndrome with and without schizophrenia: preliminary results of a structural magnetic resonance imaging study. *Archives of General Psychiatry, 61*, 1085–1096.

Vargha-Khadem, F., Watkins, K., Alcock, K., Fletcher, P., & Passingham, R. (1995). Praxic and nonverbal cognitive deficits in a large family with a genetically transmitted speech and language disorder.*Proceedings of the National Academy of Sciences of the United States of America, 92*, 930–933.

Vink, J. M., & Boomsma, D. I. (2002). Gene finding strategies. *Biological Psychology, 61*, 53–71.

Wang, P. P., Hesselink, J. R., Jernigan, T. L., Doherty, S., & Bellugi, U. (1992). Specific neurobehavioral profile of Williams' syndrome is associated with neocerebellar hemispheric preservation. *Neurology, 42*, 1999–2002.

Watkins, K. E., Vargha-Khadem, F., Ashburner, J., Passingham, R. E., Connelly, A., Friston, K. J., et al. (2002). MRI analysis of an inherited speech and language disorder: structural brain abnormalities. *Brain, 125*, 465–478.

Weickert, C. S., Hyde, T. M., Lipska, B. K., Herman, M. M., Weinberger, D. R., & Kleinman, J. E. (2003). Reduced brain-derived neurotrophic factor in prefrontal cortex of patients with schizophrenia. *Molecular Psychiatry, 8*, 592–610.

Weis, S., Weber, G., Neuhold, A., & Rett, A. (1991). Down syndrome: MR quantification of brain structures and comparison with normal control subjects. *AJNR American Journal of Neuroradiology, 12*, 1207–1211.

White, N. S., Alkire, M. T., & Haier, R. J. (2003). A voxel-based morphometric study of nondemented adults with Down Syndrome. *Neuroimage, 20*, 393–403.

White, T., Andreasen, N. C., & Nopoulos P. (2002). Brain volumes and surface morphology in monozygotic twins. *Cerebral Cortex, 12*, 486–493.

Woods, R. P., Freimer, N. B., De Young, J. A., Fears, S. C., Sicotte, N. L., Service, S. K., et al. (2006). Normal Variants of Microcephalin and ASPM Do Not Account for Brain Size Variability. *Human Molecular Genetics, 15*, 2025–2029.

Wright, I. C., Sham, P., Murray, R. M., Weinberger, D. R., & Bullmore, E. T. (2002). Genetic contributions to regional variability in human brain structure: methods and preliminary results. *Neuroimage, 17*, 256–271.

Chapter 11

Cognition in Rodents

Christopher Janus, Michael J. Galsworthy, David P. Wolfer, and Hans Welzl

Introduction

Cognition is a *loosely defined term* with divergent meanings in different disciplines and species. In human psychology, 'cognition' is often used in reference to concepts such as 'mind' or 'higher mental functions'. However, in more general terms, 'cognition' is regularly used to refer to all manner of information organization by the brain: from collection, to processing, to storage and recognition or recall. Whereas 'cognition' would seem to permeate all mental functions, including subjective perception and innate responses, 'cognitive ability' has a slightly more specific connotation – something more akin to intelligence or information-processing ability. Thus, 'cognition' deals with mental process structure and 'cognitive abilities' with natural variations impinging upon functioning at the higher end of that structure. Although the term 'cognition' sometimes subsumes or substitutes 'cognitive ability' in the literature, understanding this methodological distinction allows us to read across the two fields without the misunderstandings that classical cognitive psychologists have sometimes shown for cognitive ability research.

All aspects of cognition in rodents can only be studied indirectly by collecting behavioral data within suitable experimental environments. In this domain, there is a lesser distinction between cognitive processes and cognitive abilities as the predominant model centers on the genetic, pharmacological, or lesion manipulation of rodents. Resultant changes in learning, memory, or problem-solving paradigms then indicate the effect of the gene, drug, or locus on information processing. There are no introspective reports to monitor the associated thought contents. Nevertheless, an increasing number of psychologists, neuroscientists, and geneticists study cognitive processes in animals as a way to increase knowledge of neural and genetic mechanisms influencing cognitive functioning in normal and diseased states.

Cognitive abilities vary in populations of humans as well as animals. Behavior genetic studies have provided ample evidence that variability in behaviors reflecting cognition is – like almost all types of behavior – to a lesser or greater extent genetically influenced (Plomin, DeFries, McClearn, & McGuffin, 2001). Knowing the *genetic contribution* to behavior is also essential to determine the extent of environmental influences on behavioral variability. However, investigating a genotype–behavior relationship is difficult since many genes have an impact on a single type of behavior, and each gene affects several different types of behavior. The genotype–environment complexity is yet another reason why animals such as rodents provide experimental opportunities otherwise unavailable if humans were the only subjects studied.

Research investigating the genetic contribution to cognition in rodents has followed *three major lines*. In the first line of research, a handful of laboratories have used a quantitative (or 'psychometric') approach to look for genetic or neuronal correlates of general cognitive ability in rats and mice. General cognitive ability in humans, or 'g' as it was labeled by Spearman (1904), represents the core performance in a battery of cognitive tests. Similarly, the goal of the equivalent animal research is to develop a battery of cognitive task for rats or mice from which a general cognitive performance, or 'g' factor, can be extracted. Such a battery will then help in the search for alleles or brain properties that predict or are associated with cognitive performance level across a range of circumstances. The second line of research focuses on the contribution of specific genes to cognitive abilities, normally by manipulating genes and recording the resultant changes in learning and memory tasks. Based on the results, models are constructed explaining the molecular and cellular mechanisms crucial for learning and memory. A third line of research attempts to unravel the basic functioning of human mental retardation and senile dementia genes via mouse models. Mouse models carrying similar gene defects, or transgenically overexpressing mutated

H. Welzl (✉)
Division of Neuroanatomy and Behavior, Institute of Anatomy, University of Zurich, Zurich CH-8057, Switzerland
e-mail: welzl@anatom.uzh.ch

Y.-K. Kim (ed.), *Handbook of Behavior Genetics*,
DOI 10.1007/978-0-387-76727-7_11, © Springer Science+Business Media, LLC 2009

human genes implicated in a particular disease, are tools to investigate the etiology and neuropathological changes responsible for reduced cognitive abilities. The following sections deal with these three lines of research in detail.

Quantitative Genetic Approaches to 'General Cognitive Ability'

In humans, the almost universally accepted model of cognitive abilities is a statistical one. The essence of the model was first articulated just over 100 years ago when Charles Spearman noticed that people who did well on one mental task tended to do better on other mental tasks, even if those tasks seemed quite different in cognitive demand (Spearman, 1904). Expressed differently, Spearman found that all cognitive tasks tended to correlate positively with each other. Spearman deduced that this must mean that they all give, in part, the same basic information. Beyond that, they have their own specific information. Spearman called this common core in performance 'g', short for 'general cognitive ability'.

A century of intensive work including intelligence testing in fields of academia, military, schools, health, and industry has provided overwhelming support for the practical and explanatory utility of this approach. The current model is a slightly more complex hierarchy, describing performance on a cognitive task as being produced by general cognitive ability variance + specific module variance (e.g., verbal fluency) + variance specific to the task (Fig. 11.1). For a while, there were various academics who gained popular mileage by refusing the existence of 'g' and instead attempt-

ing to split cognitive abilities into many categories, thus implicitly denouncing the idea that some people could be 'generally' more intelligent than others (e.g., Gardner, 1983). However, the data repeatedly pointed to commonality alongside different factors, and this commonality is not a philosophical position, but rather a reality of the data and a key aspect in understanding important general cognitive problems such as those encountered in mild and severe mental retardation. It is also the 'general' background against which specific impairments, such as dyslexia or memory impairments, can be more clearly delineated.

Quantitative genetic studies in humans strongly support a genetic influence on 'g' as well as on specific cognitive abilities and specific tasks (Plomin, 2001), thus indicating that genes influencing cognitive performance measures could have their effects at any level, from the specific task parameters right up (or down) to processes fundamental to all cognition. This last point is particularly pertinent to the interpretation of 'lower scores' on cognitive tasks in animal research, as will be discussed later. In short, this is why animal cognitive research, as with human, greatly benefits from employing batteries of tasks.

Quantitative genetic research of cognitive abilities, or 'intelligence', in rodents began in the 1920s, when human intelligence research was also in its infancy. Edward Tolman attempted to explore the genetics of cognitive ability differences by selectively breeding 'bright' and 'dull' rats. The work was continued by his student Robert Tryon and they found that after eight generations of selective breeding for performance on a T-maze, there was no population overlap (original figure presented in Plomin & Galsworthy, 2003). Such data evidenced genetic contributions to cognitive task

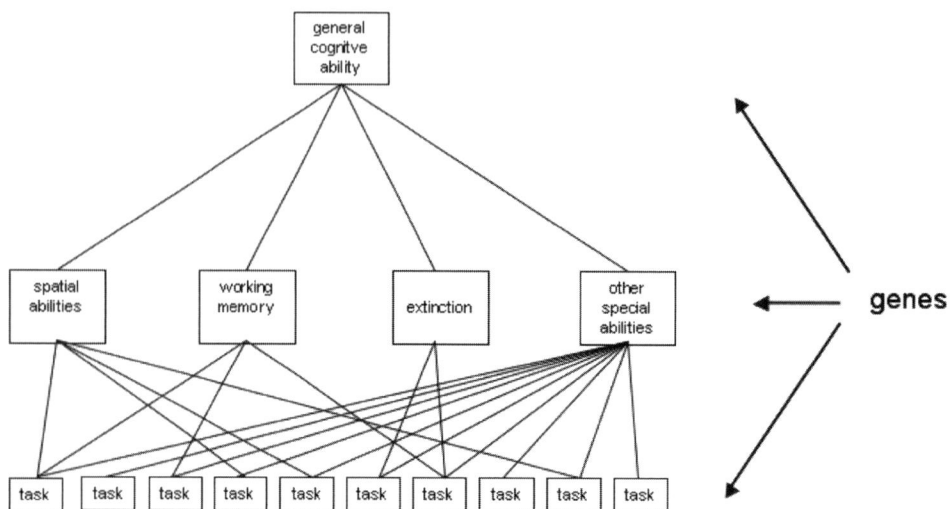

Fig. 11.1 A hypothetical description of a possible clustering hierarchy is depicted. Performance in different learning and memory tasks (*bottom level*) is affected by specific cognitive abilities (*middle level*), and all specific cognitive abilities are influenced by a general cognitive ability (*top level*). Genes have an effect on performance by modulating abilities on all three levels (adapted from Plomin et al., 2001)

performance, whether it be via 'general' ability or something altogether different. Although there was some evidence for 'g' in rats from other authors exploring correlations between tasks (e.g., Thorndike, 1935), this literature lay dormant until very recently.

The exploration of 'g' in mice and rats uses a population of animals with genetic variation (e.g., outbred strains or F2s of inbred strains) and subjects them to a *battery of tasks*. If general cognitive ability considerably influences performance in all cognitive tasks, then an individual should maintain its rank score within a group over all cognitive tasks. Note that 'g' in mice does not equate with 'g' in humans; merely the data-reduction method is the same. Simply, the approach asks how the tasks overlap in information, and if they all have a common overlap, then this is called 'g' to denote that there are some general (genetic or environmental) elements that have a general influence on cognitive performance.

During the last decade a few laboratories have *attempted to establish a battery of tasks measuring 'g' in mice* (Galsworthy et al., 2005; Galsworthy, Paya-Cano, Monleón, & Plomin, 2002; Locurto, Fortin, & Sullivan, 2003; Matzel et al., 2003). These studies used different batteries of cognitive tasks with varying types of motivations and stress levels, and all but one of them (Locurto et al., 2003) found evidence for 'g' in mice, with 'g' accounting for approximately 30–60% of the variance in performance. Nevertheless, these studies have also helped demonstrate that 'cognitive' tasks are strongly influenced by non-cognitive factors that overshadow the influence of 'g' in most individual tasks. Galsworthy et al. (2002) showed a lack of correlation between the extracted 'g' factor and activity and anxiety measures, thus evidencing that the commonality across cognitive tasks was probably not due to these motivational elements. However, there is clearly substantial random or non-cognitive variance both within and between cognitive tasks as the correlations for both of these are surprisingly low (Galsworthy et al., 2005), offering ample opportunity for non-cognitive factors to influence 'cognitive tasks' more than cognitive processes do. Furthermore, confounding factors do not necessarily have uniform influence on all tasks. Higher activity can generate shorter latencies in problem-solving tasks, but also higher 'error-zone' entries if uninhibited. Similarly, stress can motivate – but beyond a certain level can also impair performance, with freezing and panic behaviors seen. Therefore, although confounding traits may persist *across* tasks, they may show different expressions *within* tasks of different environment or motivation.

In a further step, the variability in 'g' can be correlated with allele variability. Defined outbred lines of mice or the F2 generation of inbred strains can be used in studies looking for a *linkage* between high or low 'g' scores and specific genetic markers. This theoretically straightforward approach is somewhat hampered by the large number of genes that very likely influence 'g', but still feasible. Many mice subjected to several tasks would be necessary to detect individual gene influences. Such a study would be time consuming and costly, and genes contributing only modestly to 'g' might not be detectable at all. Correlation with gross physiological or neural parameters might provide a more productive entry point to the natural variability.

This leads directly to the very fundamental question of what might be the *neural basis* of commonality across cognitive task performance. Research in humans suggests that variability in brain structure, or changes in brain structure during development (Shaw, Greenstein, Lerch, Clasen, Lenroot, Gogtay, Evans, Rapoport, & Giedd, 2006), might be related to 'g' (for review, see Toga & Thompson, 2005). Tentative data for a similar relationship between brain size and performance in cognitive tasks also exist for rats (Anderson, 1993). Variability in processes involved in neural development and plasticity are plausible candidates influencing 'g'. These processes are complex and difficult to access; so detecting a relationship between such nebulous entities and 'g' will not be an easy task.

Although the quantitative approach to cognition in rodents has not yet isolated specific genes that quantitatively contribute to the variability in 'g', the concept of what 'g' is and how it could be measured with a test battery is highly *relevant for all other studies looking for genetic influences* on cognition. The search for batteries of cognitive tasks that measure 'g' has served to sharply highlight limitations of different learning and memory tests. A low score of genetically manipulated mice in learning and memory tests could be due to disturbed cognitive or due to changes in non-cognitive processes. Continued efforts to improve batteries of cognitive tasks will reveal the extent to which cognitive as well as non-cognitive processes influence individual tasks; and hopefully they will result in an increasingly improved batteries of tasks that reliably and efficiently measure core elements of cognition.

Molecular Genetic Approaches to Cognition Research in Rodents

Whereas quantitative genetic approaches look at the big picture of how much genes are influencing different tasks and the natural structures of genetic influence, most molecular genetic approaches have a very narrow focus, namely the contribution of a single gene. Thus, molecular genetic studies investigate whether a specific gene of interest is essential for, or modifies, learning and memory in mice, and this is usually traced through associated brain mechanisms. This bottom-up approach starts by *manipulating specific genes* and then compares the behavior of genetically engineered mice with that of their wild-type littermates in *learning and memory tasks*. Through gene targeting and transgenesis, genes can be disabled, reduced, or increased in expression (for review, see Müller, 1999). Whereas the so-called constitutive mutant mice carry the mutation in all cells and throughout their life, 'conditionally' mutant mice carry the mutation only in a part of the brain and/or only during a restricted time in development. Furthermore, mutated forms of a gene can replace the native form so that the gene product loses or changes some of its properties.

This line of research is less concerned with whether variability in alleles correlates with variability in cognitive task performance. As Plomin & Kosslyn (2001) remarked, '... although knocking out a gene can have major effects, such experiments do not imply that the gene has anything to do with the variation responsible for hereditary transmission of individual differences within a species'. However, research with mutant mice has significantly helped to dissect the intricate interrelationship between molecules involved in synaptic plasticity, the cellular basis of learning and memory.

Genetically engineered mice are *only one way to investigate molecular mechanisms* of synaptic plasticity. For several decades, synapse structure and function have been the subject of countless in vitro and in vivo studies using different neuroanatomical, neurochemical, neuropharmacological, and neurophysiological methods. Thanks to the high conservation of synaptic plasticity mechanisms, invertebrate research has also provided important clues to which molecules might be critically involved (for review, see, e.g., Waddell & Quinn, 2001; Kandel, 2001; Crow, 2004). Thus, in most cases, one should be able to predict whether a manipulated mouse gene affects learning and memory. Disabling practically any of the hundreds of molecules involved in the modification of synaptic transmission could potentially impair learning and memory. It is, therefore, surprising how often inactivation of a specific gene does not abolish or even attenuate learning and memory. This suggests that the processes of synaptic plasticity have some potential to compensate for a loss of its parts.

When a mutation-induced deficit in synaptic plasticity is suspected, mice are usually subjected to learning and memory tests to prove that there is a real cognitive effect of the gene. Initially the task of choice was the water maze; later studies included fear conditioning, object recognition, and other learning and memory tasks. However, the interpretation of learning and memory deficits in genetically engineered mice is limited in studies which employ only one or a few tasks. Performance in learning and memory tasks cannot be uncritically equated with cognition, and alternative explanations suggesting non-cognitive deficits as the cause of learning and memory deficits cannot be ruled out. Including more cognitive as well as non-cognitive tasks to the test battery improves the validity of results. To date, the selection of tasks, task design, and procedures are still highly variable from laboratory to laboratory.

The *multitude of genes* whose genetic manipulation affects learning and memory mirrors the complexity of synaptic processes. Genes having an impact on learning and memory processes span a wide range. They may code for proteins involved in exocytosis, hormones, receptors, protein members of signaling cascades, proteins involved in transcription and translation, and membrane-bound proteins such as cell adhesion molecules or postsynaptic density proteins (for review, see Bolivar, Cook, & Flaherty, 2000; Chen & Tonegawa, 1997; Morley & Montgomery, 2001). Below we present a small selection of studies (with a more detailed review of glutamate receptor mutants) that have found learning and memory deficits in genetically engineered mice.

The *first two articles* describing a learning and memory deficit in mice after inactivation of a gene appeared in 1992. Mice lacking the expression of the fyn-type tyrosine kinase were impaired in the water maze (Grant et al., 1992); and mice defective in αCaMKII had similar problems in spatial navigation (Silva, Paylor, Wehner, & Tonegawa, 1992). These studies subjected mice to variations of only one learning and memory task which limited the validity of behavioral results. However, subsequent studies with mutant mice confirmed the crucial role of kinases such as the αCaMKII for learning and memory (for review, see Deutsch, 1993; Elgersma, Sweatt, & Giese, 2004; Shors & Matzel, 1997).

The last two decades have greatly improved our knowledge of *presynaptic proteins* involved in neurotransmitter release. To study release mechanisms, mice carrying homo- or heterozygous mutations in genes coding for release-related proteins were created, and a few mutant lines underwent learning and memory tasks. Although almost exclusively spatial learning in the water maze and fear conditioning were used as learning and memory paradigms, results indicate that αCaMKII, ataxin I, complexin II, GAP-43, PAC1, synapsin I and II, Rab3A, RIM1α, and synaptotagmin mutations impaired performance (for review, see Powell, 2006).

Hormone receptors whose manipulation affected spatial learning and memory include receptors for mineralcorticoids (Berger et al., 2006) and for glucocorticoids (e.g., Oitzl, de Kloet, Joels, Schmid, & Cole, 1997; Rousse et al., 1997; Steckler, Weis, Sauvage, Mederer, & Holsboer, 1999). With glucocorticoid receptors, a point mutation that prevents DNA binding has been associated with impaired spatial memory (Oitzl, Reichardt, Joels, & de Kloet, 2000). Genetic manipulations of growth hormones and their receptors have some of the strongest impacts on cognitive abilities out of the studies to date. Deletion of TrkB receptors restricted to the forebrain and occurring only during postnatal development was shown to impair spatial learning in the water maze (Minichiello et al., 1999). However, deficits were especially prominent in stressful tasks, indicating that the key deficit might not be purely cognitive in nature. Impaired spatial learning in the water maze was also seen with mutant mice carrying a heterozygous deletion of BDNF (Linnarsson, Bjorklund, & Ernfors, 1997) and GDNF (Gerlai et al., 2001). Heterozygous BDNF mice had increased striatal dopamine concentrations and their behavioral responses involving the nigrostriatal dopaminergic system were altered (Dluzen et al., 2001). Thus, deletion of a gene might disturb the complex make-up of biological processes controlling behavior in a way that is not always predictable. The role of the dopaminergic system and BDNF in cognition has recently been reviewed (Savitz, Solms, & Ramesar, 2006).

Two decades ago, Morris, Anderson, Lynch, & Baudry (1986) demonstrated the importance of *NMDA receptors* in spatial learning using a pharmacological tool to selectively inactivate the receptors. Using a variety of neuropharmacological tools and cognitive tasks, the importance of NMDA receptors for many forms of learning and memory appears well established (for review, see Nakazawa, McHugh, Wilson, & Tonegawa, 2004; Robbins & Murphy, 2006). Thus, it is not surprising that knocking out genes coding for one of the five subunits of the ionotropic NMDA receptor in the forebrain caused deficits in different forms of learning and memory. Deletion of the NR1 subunit in the CA1 region of the hippocampus impaired spatial learning (Tsien, Huerta, & Tonegawa, 1996) novel object recognition, context but not cued fear conditioning, and social transmission of food preference (Rampon et al., 2000). Deletions can be brought under the control of a food additive. In such mice, deletions of the NR1 subunit in the CA1 region during training as well as following training impaired spatial memory and context fear conditioning (Shimizu, Tang, Rampon, & Tsien, 2000); and when the NR1 subunit was deleted in the forebrain, also cued fear conditioning and taste aversion were impaired (Cui, Lindl, Mei, Zhang, & Tsien, 2005; Cui et al., 2004). When the NR1 subunit was switched off for a longer time between training and retrieval, spatial memory, fear memory, and taste memory were disrupted, suggesting a role for the NMDA receptor in maintenance of the memory trace (Cui et al., 2005, 2004; Shimizu et al., 2000; however, see also the critical comments on these puzzling results by Day & Morris, 2001).

Another study using mice in which the *NR1 subunit was selectively deleted in part of the hippocampus* suggests a role of the hippocampus in complex but not simple forms of learning and memory. These mice had no problem with discriminating two different odors. However, they failed to use the relationship between odor stimuli as cues (Rondi-Reig et al., 2001). NR1 heterozygous mice and wild types acquired context fear conditioning equally well (Frankland, Cestari, Filipkowski, McDonald, & Silva, 1998; Huerta, Sun,

Wilson, & Tonegawa, 2000). However, they performed worse than wild types in trace fear conditioning (conditioned and unconditioned stimulus separated by a time interval; Huerta et al., 2000) or when a context discrimination was required (Frankland et al., 1998). The impairments described in the latter study were partly reversed by environmental enrichment. Finally, already a single point mutations in the glycine binding site of the NR1 subunit moderately impaired spatial learning during the first few sessions (Kew et al., 2000).

Manipulating *NR2 subunits* also affected learning and memory. Targeted disruption of the NR2A subunit impaired spatial learning (Sakimura et al., 1995), latent learning in a water finding task (Miyamoto et al., 2001), auditory fear conditioning (Moriya et al., 2000), and eye blink conditioning (Kishimoto et al., 1997; Kishimoto, Kawahara, Mori, Mishina, & Kirino, 2001a; Kishimoto, Kawahara, Suzuki, Mori, Mishina, & Kirino, 2001b). NR2C-deficient mice showed no deficit in auditory fear conditioning (Moriya et al., 2000). Mice with targeted deletion of the intracellular domain of all NR2 subunits failed to acquire a step-down avoidance response; but they had also motor disturbances and probably other unmeasured behavioral defects (Sprengel et al., 1998). Furthermore, the constitutively expressed mutation makes it difficult to separate acute from developmental disturbances.

Findings of *improved performance* in NMDA receptor mutant mice (Tang et al., 1999) were widely publicized and discussed. Mice whose NR2B subunits were overexpressed performed slightly better in a spatial task, contextual and cued fear conditioning acquisition and extinction, and novel object recognition. Keeping mice of both genotypes in an enriched environment improved learning and memory in wild types but had no further effect on learning and memory in mutants (Tang, Wang, Feng, Kyin, & Tsien, 2001). These data suggest a more complex and indirect effect of the mutation on performance enhancement.

Besides the NMDA receptor, *other ionotropic and metabotropic glutamate receptors* play a role in learning and memory (Malinow & Malenka, 2002; Riedel, Platt, & Micheau, 2003). Mice lacking the AMPA receptor GluR1 (GluRA) subunit were unimpaired in spatial learning (Zamanillo et al., 1999). Only a more detailed analysis of spatial learning using various paradigms revealed selective working memory deficits (Reisel et al., 2002; Schmitt, Deacon, Seeburg, Rawlins, & Bannerman, 2003). Mice lacking the metabotropic glutamate receptor mGluR1 were slightly impaired in spatial learning, eye blink conditioning, and context fear conditioning (Aiba et al., 1994a,b; Conquet et al., 1994). Part if not all of the deficits might have been due to the simultaneously observed severe motor disturbances. Mice with deleted mGluR5 receptors were impaired in a spatial task and context fear conditioning (Lu et al., 1997), and mice with deletions of the gene for mGluR7 receptors committed more working memory errors in the radial maze and did not learn a conditioned taste aversion (Hölscher et al., 2004; Masugi et al., 1999). In the latter mice, avoidance learning was intact (Cryan et al., 2003).

Non-glutamate receptor genes whose manipulation affected learning and memory tasks include the GABAergic receptors (Crestani et al., 2002, 1999), nicotinic and muscarinic cholinergic receptors (for review, see Drago, McColl, Horne, Finkelstein, & Ross, 2003; Matsui, Yamada, Oki, Manabe, Taketo, & Ehlert, 2004), adrenergic receptors (e.g., Kobayashi & Kobayashi, 2001; Spreng, Cotecchia, & Schenk, 2001), and 5HT1B serotonergic receptors (for review, see Buhot, Wolff, Benhassine, Costet, Hen, & Segu, 2003). Whether or not deletion of a receptor gene impaired learning and memory depended on the subunit or receptor type deleted, the type of task, and the task protocol with deficits showing up more readily when cognitive demands were high.

Learning and memory is based on plasticity in synaptic transmission which involves pre- and/or postsynaptic enzymes and *signaling cascades* (for review, see Sweatt, 2004; Thomas & Huganir, 2004; Waltereit & Weller, 2003). For long-term memory formation, signaling cascades activate transcription factors which then initiate protein synthesis (for review, see Davis & Squire, 1984; Stork & Welzl, 1999). Thus, it is not surprising that deletion of genes that code for links in the Ras-MAP kinase signaling cascade differentially affected learning and memory. However, results from different laboratories were not always consistent. Mice lacking Ras-GRF1 in one laboratory were impaired in spatial learning (Giese et al., 2001), whereas those from another laboratory were not (Brambilla et al., 1997). The two different lines of Ras-GRF1 mutant mice differed also in other tasks. Such inconsistencies might depend on the presence of partially active truncated proteins, differences in genetic background, differences in task procedures, and/or differences in neo gene insertions (for discussion, see Giese et al., 2001). Similarly, two different lines of mice lacking the ERK1 isoform of MAP kinase either performed well (Selcher, Nekrasova, Paylor, Landreth, & Sweatt, 2001) or were impaired (Mazzucchelli et al., 2002) in a passive avoidance task. Similar arguments to the above could explain the discrepancy.

Interfering with the *transcription factor CREB* in *Drosophila* and *Aplysia* impaired their long-term memory formation in different learning tasks (for review, see Kandel, 2001; Waddell & Quinn, 2001). In rodents, evidence for a role for CREB in long-term memory came from a number of sources, including from experiments with mutant mice (for review, see Carlezon, Duman, & Nestler, 2005; Lonze & Ginty, 2002). As in experiments investigating the role of the Ras-MAP kinase signaling cascade in learning and memory, CREB mutant mice displayed selective long-term memory

deficits. However, results from different laboratories did not always agree to what extent and in what ways CREB facilitates long-term memory in intact animals or when an inactivation of CREB impairs learning and memory.

CREB comes in several different isoforms, and the type(s) of isoforms deleted might influence results. Mice with a targeted mutation in the two main isoforms (α and δ) of CREB showed intact short-term but defunct long-term memory for contextual fear conditioning and spatial learning (Bourtchuladze et al., 1994). Forebrain-specific and induced dominant negative repression of all CREB isoforms also impaired spatial learning and object recognition but not context fear conditioning (Pittenger et al., 2002). Other studies did not find deficits in contextual fear conditioning or spatial learning in conditional CREB-deficient mice (Balschun et al., 2003; Rammes et al., 2000). Gene dosage, background genetics, procedural differences in behavioral tasks, and/or differences in neo gene insertions might influence the extent of memory deficits in mutant mice with disabled CREB function (Balschun et al., 2003; Gass et al., 1998; Kogan et al., 1996). Furthermore, non-cognitive performance abnormalities such as thigmotaxis in the water maze could contribute to the learning and memory deficits observed in CREB mutants (Balschun et al., 2003).

Sensory stimulation or electrical stimulation of specific brain sites alters synaptic strength accompanied by changes in *spine density and spine motility*. It has been suggested that similar changes on a smaller scale might take place during long-term memory formation (for review, see Bailey & Kandel, 1993; Nimchinsky, Sabatini, & Svoboda, 2002). Evidence for this suggestion is sparse. However, indirect evidence for an involvement of spine motility in long-term memory formation comes from studies interfering with synthesis and function of proteins involved in cytoskeletal dynamics such as actin-regulating proteins or cell adhesion molecules. Mice lacking *beta-adducin*, a protein that promotes the binding of the two cytoskeletal proteins actin and spectrin, are impaired in fear conditioning as well as in a spatial learning task (Rabenstein et al., 2005). Mice with inactivated *N-CAM* gene, a neural cell adhesion molecule involved in maintaining synapse structure, were impaired in spatial learning (Cremer et al., 1994; Stork et al., 2000), contextual fear conditioning, and to a lesser extent cued fear conditioning (Stork et al., 2000) when compared to their wild-type littermates.

In summary, the use of genetically engineered mice confirmed previous results and has expanded our knowledge of various mechanisms of synaptic plasticity which underlie all forms of learning and memory. Several aspects of this line of research should be emphasized. (1) Mutant mice subjected to a battery of tasks have repeatedly revealed unexpected *pleiotropic effects* of genes. Specifically, manipulating a 'cognitive' gene also may change anxiety, locomotor activity, aggression, or other behavioral or physiological properties alongside (or instead of) the learning and memory change. Such effects appeared even when gene manipulation was restricted to certain brain areas and in time. (2) Related to the first aspect, manipulating genes may *up- or downregulate* expression of other genes, and it may lead to a *readjustment* in various systems not directly targeted by the genetic manipulation. To give just two examples, deleting the CREB gene in mice upregulated a novel CREB mRNA isoform that has not been described before (Blendy, Kaestner, Schmid, Gass, & Schütz, 1996); and in mice lacking the α4 nicotinic receptor subunit dopamine transporter function was impaired (Parish et al., 2005). (3) *Genetic background* may exacerbate or compensate for a deleted gene. Even when careful breeding schedules are employed and mutant and wild-type littermates are used for testing, a mutation-induced defect may show up with one genetic background but not with another. Mouse strains greatly differ in cognitive (see, e.g., Brooks, Pask, Jones, & Dunnett, 2005) as well as in non-cognitive behaviors (see, e.g., Kim, Chae, Lee, Yang, & Shin, 2005). (4) Lines of mice with the same gene targeted but *created in different laboratories* sometimes differ in their behavioral phenotype (see, e.g., Brambilla et al., 1997 versus Giese et al., 2001). Partially active truncated proteins and/or differences in neo gene insertions might be responsible for such discrepancies. (5) Performance in learning and memory tasks *cannot be uncritically equated with cognition*. As quantitative genetic studies convincingly demonstrated each individual task depends on non-cognitive factors, on specific cognitive abilities, and on general cognitive ability. Testing mutant mice in just one or two cognitive tasks tells us little about the manipulated gene's impact on cognition. For a valid conclusion mice have to undergo a battery that includes several cognitive as well as non-cognitive tasks. Only with a full behavioral profile can the gene effect be sensibly characterized.

Modeling Human Cognitive Disabilities in Mice

Human cognitive impairments with known or suspected genetic causes or genetic risk factors occur mainly during early development as mental retardation or later in life as senile dementias. When the genes causing or promoting dementia are known, an attempt can be made to model the disease in mice by introducing similar mutations, or selecting mutations that replicate part of the pathology. The goal of this line of research is to understand disease etiology at the most fundamental biochemical levels and, with the help of the model, to develop therapeutic strategies. The approach could be called a top-down–bottom-up approach because first the

genetic locus has to be detected in affected humans before the genetic defect can be modeled in mice employing similar techniques as described in the previous section.

Mental Retardation

Clinical research has identified several dozens of single genes or chromosomal regions whose mutation causes mental retardation (for review, see Inlow & Restifo, 2004; Weeber, Levenson, & Sweatt, 2002; see also Online Mendelian Inheritance in Man (OMIM) http://www.ncbi.nlm.nih.gov/entrez/query.fcgi). Most forms of mental retardation are syndromic, i.e., besides cognitive deficits the clinical picture includes other anatomical, physiological, and behavioral symptoms. Genetic background and environmental influences modulate cognitive deficits as well as other symptoms to a larger or smaller degree. These characteristics of mental retardation caused by single gene mutation, i.e., pleiotropic effects of genes and dependence of the phenotype on the genetic background, resemble the observations made in other mutant mice as described in the previous section.

To model mental retardation, the homologues of human genes known to cause mental retardation were deleted or replaced by mutated forms in mice (for review, see, e.g., Branchi, Bichler, Berger-Sweeney, & Ricceri, 2003; Welzl, D'Adamo, Wolfer, & Lipp, 2006). The best investigated mouse model is that for *fragile X syndrome*, the most common form of hereditary mental retardation (for review, see Bakker & Oostra, 2003; Kooy, 2003). Fragile X syndrome in humans is due to a massive triple repeat expansion in the *Fmr1* gene on the X chromosome causing hypermethylation and silencing that gene (Jin & Warren, 2003). Besides mental retardation, the syndrome includes anatomical features such as elongated faces and large testicles alongside behavioral changes such as hyperkinesia. When a homologous triple repeat expansion was introduced into mice, however, only moderate instability was observed (Bontekoe et al., 2001). Mice with deleted *Fmr1* gene displayed some but not all anatomical features (for review, see Bakker & Oostra, 2003; Kooy, 2003). In more than 90% of adult mutant mice, testes were enlarged but facial features were normal. Dendritic spines of mutant mice showed anomalies similar to those in human patients, and mutants were only slightly impaired in learning and memory tasks. The mild behavioral phenotype seemed to be dependent on the genetic background.

Mouse models for *other forms of hereditary mental retardation* yielded similar results, i.e., symptoms were usually weaker than in human patients or even completely absent in the mouse model. Relatively mild impairments in learning and memory tasks accompanied models for Coffin–Lowry syndrome (Poirier et al., 2006), GDI 1 mental retardation (D'Adamo et al., 2002), Rett syndrome (Shah-bazian et al., 2002), Agtr2 mutation with X-linked mental retardation (Sakagawa et al., 2000), L1 mutation and CRASH syndrome (Fransen et al., 1998; Law et al., 2003; Wolfer, Mohajeri, Lipp, & Schachner, 1998), and neurofibromatosis type 1 (for review, see Costa & Silva, 2003). Most of these 'mental retardation' genes belonged to the class of genes listed in the previous section, i.e., genes coding for proteins involved in synaptic plasticity.

Critical interpretation problems similar to those listed at the end of the previous section apply also for the present section (pleiotropic effects, mutation-induced changes in non-targeted systems, genetic background effects, differences between lines targeting the same gene, need for improved cognitive test batteries). In addition, syndromes of mouse models only partly replicate the clinical picture. One possible explanation for that observation could be a better compensatory mechanism for deleted genes in mice. Another explanation might be the lack of good tasks to measure 'g' in mice compared to the available tasks measuring 'g' in humans. Most of the tasks tapping into cognition in mice are probably too simple and can be mastered even after a reduction in cognitive abilities. Furthermore, cognition in humans is highly dependable on language. It is possible that language in humans is more sensitive to disruption than learning and memory tasks such as spatial learning or conditioning in mice.

decline in cognitive performance with compromised learning, memory, and speed of problem solving (Albert, 1996), sometimes accompanied by delusions, depression, agitation, and aggressive behavior (Victoroff, Zarow, Mack, Hsu, & Chui, 1996). Although cognitive evaluation is indicative of AD, a definitive diagnosis is only reached postmortem based on neuropathological changes consisting of senile plaques, predominantly containing amyloid-β (Aβ) protein and neurofibrillary tangles, composed of tau protein (Fig. 11.2; Price et al., 1991). These changes are accompanied by neuronal damage and death mainly in brain regions critical for learning and memory such as the neocortex, hippocampus, amygdala, anterior thalamus, and basal forebrain (Arnold, Hyman, Flory, Damasio, & Van Hoesen, 1991; Horn et al., 1996; Hyman, Van Hoesen, Damasio, & Barnes, 1984; Morrison & Hof, 1997; Samuel, Terry, Deteresa, Butters, & Masliah, 1994; Whitehouse et al., 1982). In addition, the functionality of the monoaminergic and cholinergic systems is reduced (Braak & Braak, 1994; Jope, Song, & Powers, 1997; Mattson & Pedersen, 1998; Tong & Hamel, 1999). Early onset AD cases are mostly familial and linked to the presence of autosomal dominant mutations. Mutations causing familial AD affect at least one of three different genes that encode the amyloid precursor protein (APP) or the presenilins (PS1 or PS2). Furthermore, the ApoE4 allele of apoliprotein (apoE) gene is known to be a potent susceptibility factor for late-onset idiopathic ('sporadic') form of AD (reviewed in Selkoe, 1997).

Senile Dementia of the Alzheimer Type

Aging, in conjunction with genetic, epigenetic, and environmental factors, can compromise brain function, eventually leading to brain degeneration. Although the decline is typically a gentle one, the general consensus is that no form of memory is completely spared during aging (Fratiglioni, Small, Winblad, & Bäckman, 2001). Dementias are characterized by progressive and accelerated decline in cognitive function that results from loss of the underlying neuronal architecture. Patients suffering from advanced Alzheimer's disease (AD), fronto-temporal dementia and Parkinsonism linked to chromosome 17, vascular dementia, corticobasal degeneration, or Pick's disease show cognitive impairment that is indicative of widespread neuronal damage. Yet each disease possesses a unique cognitive phenotype that emerges from a patterned destruction of specific neuronal architectonics (for review, see Lee, Goedert, & Trojanowski, 2001).

AD is the most prominent of the dementias and based on cognitive tests accounts for more than 75% of patients suffering from dementia (Price, Davies, Morris, & White, 1991). The major clinical symptom of AD is a progressive

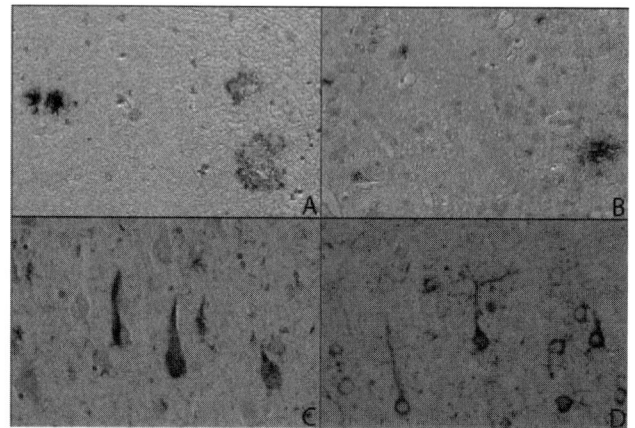

Fig. 11.2 Comparison of plaques and tangles in human AD patients (*above*) and transgenic mouse models (*below*). (**A**) Senile plaques in cortex of a human AD case, stained for pan Ab species. (**B**) Senile plaques in cortex of Tg2576 mice, stained for pan Ab species. (**C**) Tangles in the cortex of human patient, stained with MC1 antibody (detects paired helical filaments). (**D**) Tangles in the cortex of Tgr4510 mice, stained with MC1 antibody [picture supplied by one of the authors (Christopher Janus)]

The identification of gene mutations associated with AD and other tauopathies allowed the creation of *mice carrying either mutated or wild-type forms of human AD-causing genes*. These transgenic mice should replicate the relevant features of these diseases, including neuropathology and associated cognitive deficits. In light of what is known about AD in humans, a mouse model should (1) replicate at least one and preferably more pathologic hallmarks of AD; (2) should exhibit cognitive deficits in different behavioral paradigms targeting the same memory system; (3) in models employing genetic mutation(s), phenotypic changes described in 1 and 2 should be associated with the presence of a mutation(s) (these phenotyping effects should be absent or less pronounced in age-matched mice expressing wild-type (wt) gene alleles expressed at equal (or greater) steady-state levels as the mutated allele(s)). Although transgenic mice might never recapitulate all facets of the human disease, transgenic mouse models for AD present unique and important systems for in vivo study of the pathophysiology of a gene of interest.

Most of the *mouse models* which replicate amyloid pathology *express high levels of APP* and its metabolite Aβ peptide, with temporal and spatial expression patterns of the transgene depending on the promoter used. Commonly used promoters include the APP promoter (Lamb et al., 1993), the brain-enriched Prion protein promoter (Chishti et al., 2001; Hsiao et al., 1995, 1996), the platelet-derived growth factor b-chain (PDGFb) promoter (Games et al., 1995) (both PrP and PDGF promoters resulting in a transgene expression also outside of the CNS), or the neuronal-specific Thy-1 promoter (Sturchler-Pierrat et al., 1997).

Learning and memory in mouse models of AD. Interestingly, mice overexpressing human *PS1* wild-type and mutated genes developed no overt amyloid plaque pathology (Citron et al., 1997; Duff et al., 1996). Intensive behavioral characterization of these mouse lines revealed no sensorimotor or spatial reference memory deficits (Holcomb et al., 1999, 1998; Janus et al., 2000a). The transgenic models which most convincingly replicated AD-related neuropathology overexpressed human mutated *APP* (Indiana mutation: Games et al., 1995; Swedish mutation: Andra et al., 1996; Hsiao et al., 1996; Swedish and Indiana mutations: Chishti et al., 2001; Li et al., 2004; Janus et al., 2000b; J20 mice: Mucke et al., 2000; London mutation: Moechars et al., 1999; *APP* and *PS1* genes: Borchelt et al., 1996; Dineley, Xia, Bui, Sweatt, & Zheng, 2002). A detailed description of all transgenic lines is presented in reviews which chronicle the complete history of APP transgenesis and the characteristic of specific lines (Ashe, 2001; Dodart, Mathis, Bales, & Paul, 2002b; Greenberg et al., 1996; Higgins & Jacobsen, 2003; Janus, Phinney, Chishti, & Westaway, 2001; Janus & Westaway, 2001; Seabrook &

Rosahl, 1999; van Leuven, 2000). Most of the transgenic APP mice rapidly increased Aβ levels and amyloid plaque deposition with age. These mice also recapitulated other neurological features of AD (astrogliosis, microgliosis, cytokinine production, oxidative stress, dystrophic neurites) but no overt neuronal loss. Behavioral studies from different laboratories found that these mice were significantly impaired in spatial reference memory in the water maze, in working memory evaluated in a radial arm water maze, in the T-maze test, Y-maze, object recognition test, and contextual and trace fear conditioning.

Another main pathological feature of AD, apart from Aβ plaques, is the formation of *neurofibrillary tangles*. Interestingly, no detectable neurofibrillary tangles developed in transgenic mice overexpressing APP and/or PS1 on different genetic backgrounds. A recently introduced triple transgenic mouse model addressed the relationship between amyloid and tau pathology. The triple transgenic AD mice consist of APP (Swedish), PS1, and tau mutations (Oddo, Caccamo, Kitazawa, Tseng, & LaFerla, 2003a). These mice develop Aβ plaques in an age-dependent fashion, first in the neocortex and then later in the hippocampus. The development of neurofibrillary tangles followed Aβ pathology. Tangles first appear in the hippocampus and later spread to the cortex (Oddo et al., 2003b). The triple transgenic AD mice also develop an age-dependent synaptic dysfunction which preceded plaque and tangle formation (Oddo et al., 2003b), and an age-progressing memory impairment that correlated with the accumulation of intraneuronal Aβ (Billings, Oddo, Green, McGaugh, & LaFerla, 2005). The regional and temporal patterns of pathology development observed in triple transgenic AD mice is reminiscent of the development of human AD pathology. Also, the appearance of Aβ pathology before tau pathology supports the amyloid cascade hypothesis of AD pathogenesis (Hardy & Selkoe, 2002), making this model particularly suitable for the validation of this hypothesis.

In conclusion, the development of mouse models of tauopathies has brought research closer to an accurate replication of neurodegenerative processes in AD, and this ultimately should help elucidate the causes of neuronal death and cognitive decline in this type of dementia. Understanding the interplay between genetic, epigenetic, and environmental risk factors underlying dementia is a *conditio sine qua non* for the development of preventive and curative therapies. In this respect, the present mouse models have proved to be robust and have already contributed to our understanding of basic biology and pathogenesis of AD. One major problem when investigating mouse models of AD is how well the complex cognitive characteristics of human patients can be replicated in mice and what would be the most sensitive tests to assess cognitive deficits in mice that resemble the cognitive deficits in human patients.

Perspectives and Future Directions for the Study of Cognition in Rodents

A genetic influence on cognitive abilities in health and disease in mice and man is well supported by the literature as outlined above. However, increased knowledge about genetic mechanisms and expression throughout development has brought with it appreciations of the complexity of those gene effects on behavior. The old presumptions of a few key genes influencing one behavior regardless of background factors are long obsolete. We are now also aware how difficult it is to select behavioral tasks and interpret behavioral data when studying a gene–behavior relationship. The wealth of available data and the multitude of techniques on hand to investigate gene effects on cognition should provide ample material for more refined future research, if only the appropriate care and attention are taken. Progress is likely to appear along all three lines of research discussed here.

Continuing efforts in *psychometric research* will hopefully provide a better understanding as to what extent behavioral tasks and their multiple parameters inform on general cognitive, specific cognitive, or confounding traits. The results should provide us with improved behavioral test batteries which reliably profile both cognitive and noncognitive processes so that genetic effects or other manipulations can be correctly categorized and quantified. These test batteries will not necessarily be more extensive; but they will include only those tasks and their parameters with known high loadings for specific behavioral aspects. Thus the psychometric research is also providing the critical service of selecting and refining individual tasks with a view to maximizing key information. Improved behavioral test batteries will certainly be the basis of the quantitative genetic approach as well as proving necessary to other approaches looking for accurate explanations of the relationship between their gene of interest and cognition.

A *quantitative genetic* approach to cognition in mice will benefit from better cognitive test batteries as well as new resources and techniques. Whole genome single nucleotide polymorphism (SNP) panels for mice are available which potentially allow a more comprehensive search for quantitative trait loci associated with cognitive abilities and other behavioral traits (Moran et al., 2006; Petkov et al., 2004; Tsang et al., 2005; Yan, Wang, Lemon, & You, 2004). These SNP panels very likely will be expanded in the future. In addition to more mouse SNPs, pooled DNA analysis and microarrays could potentially reduce costs in the future; these techniques have been introduced in human research to detect phenotype-specific quantitative trait loci (for review, see Butcher, Kennedy, & Plomin, 2006). Finally, RNA microarrays can help to indicate genes whose expression level correlates with performance level in learning and memory paradigms (see, e.g., Cavallaro, D'Agata, Manickam, & Alkon, 2002; D'Agata & Cavallaro, 2003; Leil, Ossadtchi, Cortes, Leahy, & Smith, 2002; Leil, Ossadtchi, Nichols, Leahy, & Smith, 2003; Letwin et al., 2006; Paratore et al., 2006; Robles et al., 2003), thus really examining the gene contributions in a quantitative way.

A *molecular genetic approach* to finding genes for cognitive abilities, or more specifically for learning and memory, will profit from a number of recent technical developments. Genetically engineered mice have been developed in which gene expression can be switched on and off at any time during development (Michalon, Koshibu, Baumgartel, Spirig, & Mansuy, 2005; Uchida et al., 2006). This technique has already been used to investigate, for example, the role of calcineurin in learning and memory (Genoux et al., 2002). Gene delivery to specific brain areas can be achieved through injection of a compound consisting of a viral vector and specific gene in mice as well as other species. Such viral transfection has been successfully applied to deliver human apolipoprotein E isoforms into the brains of mice modeling AD (Dodart et al., 2005). The application of gene transfer to limbic system research has been discussed and reviewed by Robert Sapolsky (2003). Another more recent approach that might help to elucidate the role of genes in learning and memory is gene silencing by small interfering RNAs injected into the brain (for review, see McManus & Sharp, 2002; Novina & Sharp, 2004).

In the field of genetically determined or predisposed *cognitive disabilities*, clinical research and investigation of animal models will continue to influence each other. Clinical studies can detect further genes causing or promoting different forms of cognitive impairments. Modeling gene defects in mice can help to understand the role gene products play in synaptic or cellular processes. Mouse models may also help to test potential therapeutic approaches. This field is rapidly moving and it is impossible to provide an exhaustive account of this research area. For readers who are interested in learning more, we would like to mention several key papers (Austin et al., 2003; Dodart et al., 2002a; Janus et al., 2000b; Kotilinek et al., 2002; Sigurdsson, Scholtzova, Mehta, Frangione, & Wisniewski, 2001) and reviews (Bush, 2001; Dodart et al., 2002b; Duff, 1999; Janus, 2003) that chronicle the history of immunization against Aβ in mouse models, and the subsequent results of clinical trials (Gilman et al., 2005; Hock et al., 2002, 2003; Orgogozo et al., 2003).

In summary, the study of cognitive processes and cognitive abilities in rodents does not suffer from lack of usefulness to the understanding of human brain function and disease; nor is there a lack of genetic or neuroscience technology to back it up; nor is there a lack of scientists trying to find genes associated with cognition. What is sorely lacking, however, is the behavioral technology to give clear interpretations of the genetic manipulation effects. Huge numbers of genes have been 'characterized' in terms of 'cognitive' tasks, but the interpretation in most of these results is dubious by virtue of very limited behavior information interpreted with simplistic or wishful thinking. The field of rodent psychometrics, which initially faded somewhere in the 1960s before it was really born, has recently surged back into the spotlight. Simply put; with all the bucketful of candidate genes pouring into different tests, we need to know the structure of rodent cognition and we need to know what our tasks really measure. It is already observed that as no one task measures one thing purely, combinations ('batteries') of tasks are the only way to fully profile the difference between a wild type and a manipulated group. Psychometrics in animals, as with humans, offers the best data-driven way to clean up and collate our cognitive tasks into the informative batteries that we should have had a long time ago.

Acknowledgment This work was supported by the NCCR 'Plasticity and Repair'.

References

Aiba, A., Chen, C., Herrup, K., Rosenmund, C., Stevens, C. F., & Tonegawa, S. (1994a). Reduced hippocampal long-term potentiation and context-specific deficit in associative learning in mGluR1 mutant mice. *Cell, 79*, 365–375.

Aiba, A., Kano, M., Chen, C., Stanton, M. E., Fox, G. D., Herrup, K., et al. (1994b). Deficient cerebellar long-term depression and impaired motor learning in mGluR1 mutant mice. *Cell, 79*, 377–388.

Albert, M. S. (1996). Cognitive and neurobiological markers of early Alzheimer disease. *Proceedings of the National Academy of Sciences of the United States of America, 93*, 13547–13551.

Anderson, B. (1993). Evidence from the rat for a general factor that underlies cognitive performance and that relates to brain size: Intelligence? *Neuroscience Letters, 153*, 98–102.

Andra, K., Abramowski, D., Duke, M., Probst, A., Wiederhold, K. H., Burki, K., et al. (1996). Expression of APP in transgenic mice: A comparison of neuron-specific promoters. *Neurobiology of Aging, 17*, 183–190.

Arnold, S. E., Hyman, B. T., Flory, J., Damasio, A. R., & Van Hoesen, G. W. (1991). The topographical and neuroanatomical distribution of neurofibrillary tangles and neuritic plaques in the cerebral cortex of patients with Alzheimer's disease. *Cerebral Cortex, 1*, 103–116.

Ashe, K. (2001). Learning and memory in transgenic mice modelling Alzheimer's disease. *Learning & Memory, 8*, 301–308.

Austin, L., Arendash, G. W., Gordon, M. N., Diamond, D. M., DiCarlo, G., Dickey, C., et al. (2003). Short-term ß-amyloid vaccinations do not improve cognitive performance in cognitively impaired APP+PS1 mice. *Behavioral Neuroscience, 117*, 478–84.

Bailey, C. H., & Kandel, E. R. (1993). Structural changes accompanying memory storage. *Annual Review of Physiology, 55*, 397–426.

Bakker, C. R., & Oostra, B. A. (2003). Understanding fragile X syndrome: Insights from animal models. Cytogenet. *Genome Research, 100*, 111–123.

Balschun, D., Wolfer, D. P., Gass, P., Mantamadiotis, T., Welzl, H., Schutz, G., et al. (2003). Does cAMP response element-binding protein have a pivotal role in hippocampal synaptic plasticity and hippocampus-dependent memory? *The Journal of Neuroscience, 23*, 6304–6314.

Berger, S., Wolfer, D. P., Selbach, O., Alter, H., Erdmann, G., Reichardt, H. M., et al. (2006). Loss of the limbic mineralocorticoid receptor impairs behavioral plasticity. *Proceedings of the National Academy of Sciences of the United States of America, 103*, 195–200.

Billings, L. M., Oddo, S., Green, K. N., McGaugh, J. L., & LaFerla, F. M. (2005). Intraneuronal Abeta causes the onset of early Alzheimer's disease-related cognitive deficits in transgenic mice. *Neuron, 45*, 675–688.

Blendy, J. A., Kaestner, K. H., Schmid, W., Gass, P., & Schütz, G. (1996). Targeting of the CREB gene leads to up-regulation of a novel CREB mRNA isoform. *EMBO Journal, 15*, 1098–1106.

Bolivar, V., Cook, M., & Flaherty, L. (2000). List of transgenic and knockout mice: Behavioral profiles. *Mammalian Genome, 11*, 260–274.

Bontekoe, C. J., Bakker, C. E., Nieuwenhuizen, I. M., van der Linde, H., Lans, H., de Lange, D., et al. (2001). Instability of a $(CGG)_{98}$ repeat in the *Fmr1* promotor. *Human Molecular Genetics, 10*, 1693–1699.

Borchelt, D. R., Thinakaran, G., Eckman, C. B., Lee, M. K., Davenport, F., Ratovitsky, T., et al. (1996). Familial Alzheimer's disease-linked presenilin 1 variants elevate Abeta1-42/1-40 ratio in vitro and in vivo. *Neuron, 17*, 1005–1013.

Bourtchuladze, R., Frenguelli, B., Blendy, J., Cioffi, D., Schutz, G., & Silva, A. J. (1994). Deficient long-term memory in mice with a targeted mutation of the cAMP-responsive element-binding protein. *Cell, 79*, 59–68.

Braak, H., & Braak, E. (1994). Pathology of Alzheimer's disease. In D. B. Calne (Ed.), *Neurodegenerative Diseases* (pp. 585–613). Philadelphia: Saunders.

Brambilla, R., Gnesutta, N., Minichiello, L., White, G., Roylance, A. J., Herron, C. E., et al. (1997). A role for the Ras signalling pathway in synaptic transmission and long-term memory. *Nature, 390*, 281–286.

Branchi, I., Bichler, Z., Berger-Sweeney, J., & Ricceri, L. (2003). Animal models of mental retardation: From gene to cognitive function. *Neuroscience & Biobehavioral Reviews, 27*, 141–153.

Brooks, S. P., Pask, T., Jones, L., & Dunnett, S. B. (2005). Behavioral profiles of inbred mouse strains used as transgenic backgrounds. II: Cognitive tests. *Genes, Brain and Behavior, 4*, 307–17.

Buhot, M.-C., Wolff, M., Benhassine, N., Costet, P., Hen, R., & Segu, L. (2003). Spatial learning in the 5-HT1B receptor knockout mouse: Selective facilitation/impairment depending on the cognitive demand. *Learning & Memory, 10*, 466–477.

Bush, A. I. (2001). Therapeutic targets in the biology of Alzheimer's disease. *Current Opinion in Psychiatry, 14*, 341–348.

Butcher, L. M., Kennedy, J. K. J., & Plomin, R. (2006). Generalist genes and cognitive neuroscience. *Current Opinion in Neurobiology, 16*, 145–151.

Carlezon, W. A., Jr., Duman, R. S., & Nestler, E. J. (2005). The many faces of CREB. *Trends in Neurosciences, 28*, 436–445.

Cavallaro, S., D'Agata, V., Manickam, P., & Alkon, D. L. (2002). Memory-specific temporal profiles of gene expression in the hippocampus. *Proceedings of the National Academy of Sciences of the United States of America, 99*, 16279–16284.

Chen, C., & Tonegawa, S. (1997). Molecular genetic analysis of synaptic plasticity, activity-dependent neural development, learning, and memory in the mammalian brain. *Annual Review of Neuroscience,* *20,* 157–184.

Chishti, M. A., Yang, D. S., Janus, C., Phinney, A. L., Horne, P., Pearson, J., et al. (2001). Early-onset amyloid deposition and cognitive deficits in transgenic mice expressing a double mutant form of amyloid precursor protein 695. *The Journal of Biological Chemistry,* *276,* 21562–21570.

Citron, M., Westaway, D., Xia, W., Carlson, G., Diehl, T., Levesque, G., et al. (1997). Mutant presenilins of Alzheimer's disease increase production of 42-residue amyloid ß-protein in both transfected cells and transgenic mice. *Nature Medicine, 3,* 67–72.

Conquet, F., Bashir, Z. I., Davies, C. H., Daniel, H., Ferraguti, F., Bordi, F., et al. (1994). Motor deficit and impairment of synaptic plasticity in mice lacking mGluR1. *Nature, 372,* 237–243.

Costa, R. M., & Silva, A. J. (2003). Mouse models of neurofibromatosis type I: Bridging the GAP. *Trends in Molecular Medicine, 9,* 19–23.

Cremer, H., Lange, R., Christoph, A., Plomann, M., Vopper, G., Roes, J., et al. (1994). Inactivation of the NCAM gene in mice results in size-reduction of the olfactory bulb and deficits in spatial learning. *Nature, 367,* 455–459.

Crestani, F., Keist, R., Fritschy, J. M., Benke, D., Vogt, K., Prut, L., et al. (2002). Trace fear conditioning involves hippocampal alpha5 GABA(A) receptors. *Proceedings of the National Academy of Sciences of the United States of America, 99,* 8980–8985.

Crestani, F., Lorez, M., Baer, K., Essrich, C., Benke, D., Laurent, J. P., et al. (1999). Decreased GABAA-receptor clustering results in enhanced anxiety and a bias for threat cues. *Nature Neuroscience, 2,* 833–839.

Crow, T. (2004). Pavlovian conditioning of Hermissenda: Current cellular, molecular, and circuit perspectives. *Learning & Memory, 11,* 229–38.

Cryan, J. F., Kelly, P. H., Neijt, H. C., Sansig, G., Flor, P. J., & van der Putten, H. (2003). Antidepressant and anxiolytic-like effects in mice lacking the group III metabotropic glutamate receptor mGluR7. *European Journal of Neuroscience, 17,* 2409–2417.

Cui, Z., Lindl, K. A., Mei, B., Zhang, S., & Tsien, J. Z. (2005). Requirement of NMDA receptor reactivation for consolidation and storage of nondeclarative taste memory revealed by inducible NR1 knockout. *European Journal of Neuroscience, 22,* 755–763.

Cui, Z., Wang, H., Tan, Y., Zaia, K. A., Zhang, S., & Tsien, J. Z. (2004). Inducible and reversible NR1 knockout reveals crucial role of the NMDA receptor in preserving remote memories in the brain. *Neuron, 41,* 781–793.

D'Adamo, P., Welzl, H., Papadimitriou, S., Raffaele di Barletta, M., Tiveron, C., Tatangelo, L., et al. (2002). Deletion of the mental retardation gene Gdi1 impairs associative memory and alters social behavior in mice. *Human Molecular Genetics, 11,* 2567–2580.

D'Agata, V., & Cavallaro, S. (2003) Hippocampal gene expression profiles in passive avoidance conditioning. *European Journal of Neuroscience, 18,* 2835–2841.

Davis, H. P., & Squire, L. R. (1984). Protein synthesis and memory: A review. *Psychological Bulletin, 96,* 518–559.

Day, M., & Morris, R. G. M. (2001). Memory consolidation and NMDA receptors: Discrepancy between genetic and pharmacological approaches. *Science, 293,* 755a.

Deutsch, J. A. (1993). Spatial learning in mutant mice. *Science, 262,* 760–761.

Dineley, K. T., Xia, X., Bui, D., Sweatt, J. D., & Zheng, H. (2002). Accelerated plaque accumulation, associative learning deficits, and up-regulation of alpha 7 nicotinic receptor protein in transgenic mice co-expressing mutant human presenilin 1 and amyloid precursor proteins. *The Journal of Biological Chemistry, 277,* 22768–22780.

Dluzen, D. E., Gao, X., Story, G. M., Anderson, L. I., Kucera, J., & Walro, J. M. (2001). Evaluation of nigrostriatal dopaminergic func-

tion in adult +/+ and +/− BDNF mutant mice. *Experimental Neurology, 170,* 121–8.

Dodart, J. C., Bales, K. R., Gannon, K. S., Greene, S. J., DeMattos, R. B., Mathis, C., et al. (2002a). Immunization reverses memory deficits without reducing brain Abeta burden in Alzheimer's disease model. *Nature Neuroscience, 5,* 452–457.

Dodart, J.-C., Marr, R. A., Koistinaho, M., Gregersen, B. M., Malkani, S., Verma, I. M., et al. (2005). Gene delivery of human apolipoprotein E alters brain Aβ burden in a mouse model of Alzheimer's disease. *Proceedings of the National Academy of Sciences of the United States of America, 102,* 1211–1216.

Dodart, J. C., Mathis, C., Bales, K. R., & Paul, S. M. (2002b). Does my mouse have Alzheimer's disease? *Genes, Brain and Behavior, 1,* 142–155.

Drago, J., McColl, C. D., Horne, M. K., Finkelstein, D. I., & Ross, S. A. (2003). Neuronal nicotinic receptors: Insights gained from gene knockout and knockin mutant mice. *Cellular and Molecular Life Sciences, 60,* 1267–1280.

Duff, K. (1999). Curing amyloidosis: Will it work in humans? *Trends in Neurosciences, 22,* 485–486.

Duff, K., Eckman, C., Zehr, C., Yu, X., Prada, C.-M., Perez-tur, J., et al. (1996). Increased amyloid-ß42(43) in brains of mice expressing mutant presenilin 1. *Nature, 383,* 710–713.

Elgersma, Y., Sweatt, J. D., & Giese, K. P. (2004). Mouse Genetic Approaches to Investigating Calcium/Calmodulin-Dependent Protein Kinase II Function in Plasticity and Cognition *The Journal of Neurosciences, 24,* 8410–8415.

Frankland, P. W., Cestari, V., Filipkowski, R. K., McDonald, R. J., & Silva, A. J. (1998). The dorsal hippocampus is essential for context discrimination but not for contextual conditioning. *Behavioral Neuroscience, 112,* 863–874.

Fransen, E., D'Hooge, R., Van Camp, G., Verhoye, M., Sijbers, J., Reyniers, E., et al. (1998). L1 knockout mice show dilated ventricles, vermis hypoplasia and impaired exploration patterns. *Human Molecular Genetics, 7,* 999–1009.

Fratiglioni, L., Small, B. J., Winblad, B., & Bäckman, L. (2001). The Transition from Normal Functioning to Dementia in the Aging Population. In K. Iqbal, S. Sisodia, & B. Winblad (Eds.), *Alzheimer's disease: Advances in etiology, pathogenesis and therapeutics* (pp. 3–10). Chichester: John Wiley & Sons. Ltd.

Galsworthy, M. J., Paya-Cano, J. L., Liu, L., Monleón, S., Gregoryan, G., Fernandes, C., et al. (2005). Assessing reliability, heritability and general cognitive ability in a battery of cognitive tasks for laboratory mice. *Behavior Genetics, 35,* 675–692.

Galsworthy, M. J., Paya-Cano, J. L., Monleón, S., & Plomin, R. (2002). Evidence for general cognitive ability (g) in heterogeneous stock (HS) mice and an analysis of potential confounds. *Genes, Brain and Behavior, 1,* 88–95.

Games, D., Adams, D., Alessandrini, R., Barbour, R., Berthelette, P., Blackwell, C., et al. (1995). Alzheimer-type neuropathology in transgenic mice overexpressing V717F beta-amyloid precursor protein. *Nature, 373,* 523–527.

Gardner, H. (1983) *Frames of Mind: The theory of multiple intelligences.* New York: Basic Books.

Gass, P., Wolfer, D. P., Balschun, D., Rudolph, D., Frey, U., Lipp, H. P., et al. (1998). Deficits in memory tasks of mice with CREB mutations depend on gene dosage. *Learning & Memory, 5,* 274–288.

Genoux, D., Haditsch, U., Knobloch, M., Michalon, A., Storm, D., & Mansuy, I. M. (2002). Protein phosphatase 1 is a molecular constraint on learning and memory. *Nature, 418,* 929–930.

Gerlai, R., McNamara, A., Choi-Lundberg, D. L., Armanini, M., Ross, J., Powell-Braxton, L., et al. (2001). Impaired water maze learning performance without altered dopaminergic function in mice heterozygous for the GDNF mutation. *European Journal of Neuroscience, 14,* 1153–63.

Giese, K. P., Friedman, E., Telliez, J. B., Fedorov, N. B., Wines, M., Feig, L. A., et al. (2001). Hippocampus-dependent learning and memory is impaired in mice lacking the Ras-guanine-nucleotide releasing factor 1 (Ras-GRF1). *Neuropharmacology, 41*, 791–800.

Gilman, S., Koller, M., Black, R. S., Jenkins, L., Griffith, S. G., Fox, N. C., et al. (2005). Clinical effects of Abeta immunization (AN1792) in patients with AD in an interrupted trial. *Neurology, 64*, 1553–1562.

Grant, S. G., O'Dell, T. J., Karl, K. A., Stein, P. L., Soriano, P., & Kandel, E. R. (1992). Impaired long-term potentiation, spatial learning, and hippocampal development in fyn mutant mice. *Science, 258*, 1903–10.

Greenberg, B. D., Savage, M. J., Howland, D. S., Ali, S. M., Siedlak, S. L., Perry, G., et al. (1996). APP transgenesis: Approaches toward the development of animal models for Alzheimer disease neuropathology. *Neurobiology of Aging, 17*, 153–171.

Hardy, J., & Selkoe, D. J. (2002). The amyloid hypothesis of Alzheimer's disease: Progress and problems on the road to therapeutics. *Science, 297*, 353–356.

Higgins, G. A., & Jacobsen, H. (2003). Transgenic mouse models of Alzheimer's disease: Phenotype and application. *Behavioural Pharmacology, 14*, 419–438.

Hock, C., Konietzko, U., Papassotiropoulos, A., Wollmer, A., Streffer, J., von Rotz, R., et al. (2002). Generation of antibodies specific for ß-amyloid by vaccination of patients with Alzheimer disease. *Nature Medicine, 8*, 1270–1275.

Hock, C., Konietzko, U., Streffer, J. R., Tracy, J., Signorell, A., Muller-Tillmanns, B., et al. (2003). Antibodies against beta-amyloid slow cognitive decline in Alzheimer's disease. *Neuron, 38*, 547–554.

Holcomb, L. A., Gordon, M. N., Jantzen, P., Hsiao, K., Duff, K., & Morgan, D. (1999). Behavioral changes in transgenic mice expressing both amyloid precursor protein and presenilin-1 mutations: Lack of association with amyloid deposits. *Behavior Genetics, 29*, 177–185.

Holcomb, L., Gordon, M. N., McGowan, E., Yu, X., Benkovic, S., Jantzen, P., et al. (1998). Accelerated Alzheimer-type phenotype in transgenic mice carrying both mutant amyloid precursor protein and presenilin 1 transgenes. *Nature Medicine, 4*, 97–100.

Hölscher, C., Schmid, S., Pilz, P. K. D., Sansig, G., van der Putten, H., & Plappert, C. F. (2004). Lack of the metabotropic glutamate receptor subtype 7 selectively impairs short-term working memory but not long-term memory. *Behavioural Brain Research, 154*, 473–481.

Horn, R., Ostertun, B., Fric, M., Solymosi, L., Steudel, A., & Möller, H.-J. (1996). Atrophy of hippocampus in patients with Alzheimer's Disease and other diseases with memory impairment. *Dementia, 7*, 182–186.

Hsiao, K. K., Borchelt, D. R., Olson, K., Johannsdottir, R., Kitt, C., Yunis, W., et al. (1995). Age-related CNS disorder and early death in transgenic FVB/N mice overexpressing Alzheimer amyloid precursor proteins. *Neuron, 15*, 1203–1218.

Hsiao, K., Chapman, P., Nilsen, S., Eckman, C., Harigaya, Y., Younkin, S., et al. (1996). Correlative memory deficits, a-beta elevation, and amyloid plaques in transgenic mice. *Science, 274*, 99–102.

Huerta, P. T., Sun, L. D., Wilson, M. A., & Tonegawa, S. (2000). Formation of temporal memory requires NMDA receptors within CA1 pyramidal neurons. *Neuron, 25*, 473–80.

Hyman, B. T., Van Hoesen, G. W., Damasio, A. R., & Barnes, C. L. (1984). Alzheimer's Disease: Cell-specific pathology isolates the hippocampus formation. *Science, 225*, 1168–1170.

Inlow, J. K., & Restifo, L. L. (2004). Molecular and comparative genetics of mental retardation. *Genetics, 166*, 835–881.

Janus, C. (2003). Vaccines for Alzheimer's disease: how close are we? *CNS Drugs, 17*, 457–538.

Janus, C., D'Amelio, S., Amitay, O., Chishti, M. A., Strome, R., Fraser, P., et al. (2000a). Spatial learning in transgenic mice expressing human presenilin 1 (PS1) transgenes. *Neurobiology of Aging, 21*, 541–549.

Janus, C., Pearson, J., McLaurin, J., Mathews, P. M., Jiang, Y., Schmidt, S. D., et al. (2000b). A beta peptide immunization reduces behavioural impairment and plaques in a model of Alzheimer's disease. *Nature, 408*, 979–982.

Janus, C., Phinney, A. L., Chishti, M. A., & Westaway, D. (2001). New developments in animal models of Alzheimer's disease. *Current Neurology and Neuroscience Reports, 1*, 451–457.

Janus, C., & Westaway, D. (2001). Transgenic mouse models of Alzheimer's disease. *Physiology & Behavior, 73*, 873–886.

Jope, R. S., Song, L., & Powers, R. E. (1997). Cholinergic activation of phosphoinositide signaling is impaired in Alzheimer's disease brain. *Neurobiology of Aging, 18*, 111–120.

Jin, P., & Warren, S. T. (2003). New insights into fragile X syndrome: From molecules to neurobehaviors. *Trends in Biochemical Sciences, 28*, 152–158.

Kandel, E. R. (2001). The molecular biology of memory storage: A dialogue between genes and synapses. *Science, 294*, 1030–1038.

Kew, J. N.-C., Koester, A., Moreau, J.-L., Jenck, F., Ouagazzal, A.-M., Mutel, V., et al. (2000). Functional consequences of reduction in NMDA receptor glycine affinity in mice carrying targeted point mutations in the glycine binding site. *The Journal of Neuroscience, 20*, 4037–4049.

Kim, D., Chae, S., Lee, J., Yang, H., & Shin, H. S. (2005). Variations in the behaviors to novel objects among five inbred strains of mice. *Genes, Brain and Behavior, 4*, 302–6.

Kishimoto, Y., Kawahara, S., Kirino, Y., Kadotani, H., Nakamura, Y., Ikeda, M., et al. (1997). Conditioned eyeblink response is impaired in mutant mice lacking NMDA receptor subunit NR2A. *Neuroreport, 8*, 3717–3721.

Kishimoto, Y., Kawahara, S., Mori, H., Mishina, M., & Kirino, Y. (2001a). Long-trace interval eyeblink conditioning is impaired in mutant mice lacking the NMDA receptor subunit $\varepsilon 1$. *European Journal of Neuroscience, 13*, 1221–1227.

Kishimoto, Y., Kawahara, S., Suzuki, M., Mori, H., Mishina, M., & Kirino, Y. (2001b). Classical eyeblink conditioning in glutamate receptor subunit delta 2 mutant mice is impaired in the delay paradigm but not in the trace paradigm. *European Journal of Neuroscience, 13*, 1249–53.

Kobayashi, K., & Kobayashi, T. (2001). Genetic evidence for noradrenergic control of long-term memory consolidation. *Brain and Development, 23*, S16–S23.

Kogan, J. H., Frankland, P. W., Blendy, J. A., Coblentz, J., Marowitz, Z., Schutz, G., et al. (1996). Spaced training induces normal long-term memory in CREB mutant mice. *Current Biology, 7*, 1–11.

Kooy, R. F. (2003). Of mice and the fragile X syndrome. *Trends in Genetics, 19*, 148–154.

Kotilinek, L. A., Bacskai, B., Westerman, M., Kawarabayashi, T., Younkin, L., Hyman, B. T., et al. (2002). Reversible memory loss in a mouse transgenic model of Alzheimer's disease. *The Journal of Neuroscience, 22*, 6331–6335.

Lamb, B. T., Sisodia, S. S., Lawler, A. M., Slunt, H. H., Kitt, C. A., Kearns, W. G., et al. (1993). Introduction and expression of the 400 kilobase amyloid precursor protein gene in transgenic mice. *Nature Genetics, 5*, 22–30.

Law, J. W. S., Lee, A. Y. W., Sun, M., Nikonenko, A. G., Chung, S. K., Dityatev, A., et al. (2003). Decreased anxiety, altered place learning, and increased CA1 basal excitatory synaptic transmission in mice with conditional ablation of the neural cell adhesion molecule L1. *The Journal of Neuroscience, 23*, 10419–10432.

Lee, V. M., Goedert, M., & Trojanowski, J. Q. (2001). Neurodegenerative tauopathies. *Annual Review of Neuroscience, 24*, 1121–1159.

Leil, T. A., Ossadtchi, A., Cortes, J. S., Leahy, R. M., & Smith, D. J. (2002). Finding new candidate for learning and memory. *Journal of Neuroscience Research, 68*, 127–137.

Leil, T. A., Ossadtchi, A., Nichols, T. E., Leahy, R. M., & Smith, D. J. (2003). Genes regulated by learning in the hippocampus. *Journal of Neuroscience Research, 71*, 763–768.

Letwin, N. E., Kafkafi, N., Benjamini, Y., Mayo, C., Frank, B. C., Luu, T., et al. (2006). Combined application of behavior genetics and microarray analysis to identify regional expression themes and gene–behavior associations. *The Journal of Neuroscience, 26*, 5277–87.

Li, F., Calingasan, N. Y., Yu, F., Mauck, W. M., Toidze, M., Almeida, C. G., et al. (2004). Increased plaque burden in brains of APP mutant MnSOD heterozygous knockout mice. *Journal of Neurochemistry, 89*, 1308–1312.

Linnarsson, S., Bjorklund, A., & Ernfors, P. (1997). Learning deficit in BDNF mutant mice. *European Journal of Neuroscience, 9*, 2581–7.

Locurto, C., Fortin, E., & Sullivan, R. (2003). The structure of individual differences in Heterogeneous Stock mice across problem types and motivational systems. *Genes, Brain and Behavior, 2*, 40–55.

Lonze, B. E., & Ginty, D. D. (2002). Function and regulation of CREB family transcription factors in the nervous system. *Neuron, 35*, 605–623.

Lu, Y.-M., Jia, Z., Janus, C., Henderson, J. T., Gerlai, R., Wojtowicz, J. M., et al. (1997). Mice lacking metabotropic glutamate receptor 5 show impaired learning and reduced CA1 long-term potentiation (LTP) but normal CA3 LTP. *The Journal of Neuroscience, 17*, 5196–5205.

Malinow, R., & Malenka, R. C. (2002). AMPA receptor trafficking and synaptic plasticity. *Annual Review of Neuroscience, 25*, 103–126.

Masugi, M., Yokoi, M., Shigemoto, R., Muguruma, K., Watanabe, Y., Sansig, G., et al. (1999). Metabotropic glutamate receptor subtype 7 ablation causes deficit in fear response and conditioned taste aversion. *The Journal of Neuroscience, 19*, 955–963.

Matsui, M., Yamada, S., Oki, T., Manabe, T., Taketo, M. M., & Ehlert, F. J. (2004). Functional analysis of muscarinic acetylcholine receptors using knockout mice. *Life Sciences, 75*, 2971–2981.

Mattson, M. P., & Pedersen, W. A. (1998). Effects of amyloid precursor protein derivatives and oxidative stress on basal forebrain cholinergic systems in Alzheimer's disease. *International Journal of Developmental Neuroscience, 16*, 737–753.

Matzel, L. D., Han, Y. R., Grossman, H., Karnik, M. S., Patel, D., Scott, N., et al. (2003). Individual differences in the expression of a 'general' learning ability in mice. *The Journal of Neuroscience, 23*, 6423–6433.

Mazzucchelli, C., Vantaggiato, C., Ciamei, A., Fasano, S., Pakhotin, P., Krezel, W., et al. (2002). Knockout of ERK1 MAP kinase enhances synaptic plasticity in the striatum and facilitates striatal-mediated learning and memory. *Neuron, 34*, 807–820.

McManus, M. T., & Sharp, P. A. (2002). Gene silencing in mammals by small interfering RNAs. *Nature Reviews Genetics, 3*, 737–747.

Michalon, A., Koshibu, K., Baumgartel, K., Spirig, D. H., & Mansuy, I. M. (2005). Inducible and neuron-specific gene expression in the adult mouse brain with the rtTA2S-M2 system. *Genesis, 43*, 205–12.

Minichiello, L., Korte, M., Wolfer, D., Kuhn, R., Unsicker, K., Cestari, V., et al. (1999). Essential role for TrkB receptors in hippocampus-mediated learning. *Neuron, 24*, 401–14.

Miyamoto, Y., Yamada, K., Noda, Y., Mori, H., Mishina, M., & Nabeshima, T. (2001). Hyperfunction of dopaminergic and serotonergic neuronal systems in mice lacking the NMDA receptor 1 subunit. *The Journal of Neuroscience, 21*, 750–757.

Moechars, D., Dewachter, I., Lorent, K., Reverse, D., Baekelandt, V., Naidu, A., et al. (1999). Early phenotypic changes in transgenic mice that overexpress different mutants of amyloid precursor protein in brain. *Journal of Biological Chemistry, 274*, 6483–6492.

Moran, J. L., Bolton, A. D., Tran, P. V., Brown, A., Dwyer, N. D., Manning, D. K., et al. (2006). Utilization of a whole genome SNP panel for efficient genetic mapping in the mouse. *Genome Research, 16*, 436–440.

Moriya, T., Kouzu, Y., Shibata, S., Kadotani, H., Fukunaga, K., Miyamoto, E., et al. (2000). Close linkage between cal-cium/calmodulin kinase II α/β and NMDA-2A receptors in the lateral amygdala and significance for retrieval of auditory fear conditioning. *European Journal of Neuroscience, 12*, 3307–14.

Morley, K. I., & Montgomery, G. W. (2001). The genetics of cognitive processes: Candidate genes in humans and animals. *Behavior Genetics, 31*, 511–531.

Morris, R. G., Anderson, E., Lynch, G. S., & Baudry, M. (1986). Selective impairment of learning and blockade of long-term potentiation by an N-methyl-D-aspartate receptor antagonist, AP5. *Nature, 319*, 774–776.

Morrison, J. H., & Hof, P. R. (1997). Life and death of neurons in the aging brain. *Science, 278*, 412–419.

Mucke, L., Masliah, E., Yu, G. Q., Mallory, M., Rockenstein, E. M., Tatsuno, G., et al. (2000). High-level neuronal expression of abeta 1-42 in wild-type human amyloid protein precursor transgenic mice: Synaptotoxicity without plaque formation. *The Journal of Neuroscience, 20*, 4050–4058.

Müller, U. (1999). Ten years of gene targeting: Targeted mouse mutants, from vector design to phenotype analysis. *Mechanisms of Development, 82*, 3–21.

Nakazawa, K., McHugh, T. J., Wilson, M. A., & Tonegawa, S. (2004). NMDA receptors, place cells and hippocampal spatial memory. *Nature Review Neuroscience, 5*, 361–372.

Nimchinsky, E. A., Sabatini, B. L., & Svoboda, K. (2002). Structure and function of dendritic spines. *Annual Review of Physiology, 64*, 313–53.

Novina, C. D., & Sharp, P. A. (2004). The RNAi revolution. *Nature, 430*, 161–164.

Oddo, S., Caccamo, A., Kitazawa, M., Tseng, B. P., & LaFerla, F. M. (2003a). Amyloid deposition precedes tangle formation in a triple transgenic model of Alzheimer's disease. *Neurobiology of Aging, 24*, 1063–1070.

Oddo, S., Caccamo, A., Shepherd, J. D., Murphy, M. P., Golde, T. E., Kayed, R., et al. (2003b). Triple-transgenic model of Alzheimer's disease with plaques and tangles: Intracellular Abeta and synaptic dysfunction. *Neuron, 39*, 409–421.

Oitzl, M. S., de Kloet, E. R., Joels, M., Schmid, W., & Cole, T. J. (1997). Spatial learning deficits in mice with a targeted glucocorticoid receptor gene disruption. *European Journal of Neuroscience, 9*, 2284–96.

Oitzl, M. S., Reichardt, H. M., Joels, M., & de Kloet, E. R. (2000). Point mutation in the mouse glucocorticoid receptor preventing DNA binding impairs spatial memory. *Proceedings of the National Academy of Sciences of the United States of America, 98*, 12790–12795.

Orgogozo, J. M., Gilman, S., Dartigues, J. F., Laurent, B., Puel, M., Kirby, L. C., et al. (2003). Subacute meningoencephalitis in a subset of patients with AD after Abeta42 immunization. *Neurology, 61*, 46–54.

Paratore, S., Alessi, E., Coffa, S., Torrisi, A., Mastrobuono, F., & Cavallaro, S. (2006). Early genomics of learning and memory: A review. *Genes, Brain and Behavior, 5*, 209–221.

Parish, C. L., Nunan, J., Finkelstein, D. I., McNamara, F. N., Wong, J. Y., Waddington, J. L., et al. (2005). Mice Lacking the 4 Nicotinic Receptor Subunit Fail to Modulate Dopaminergic Neuronal Arbors and Possess Impaired Dopamine Transporter Function. *Molecular Pharmacology, 68*, 1376–1386.

Petkov, P. M., Cassell, M. A., Sargent, E. E., Donnelly, C. J., Robinson, P., Crew, V., et al. (2004). Development of a SNP genotyping panel for genetic monitoring of the laboratory mouse. *Genomics, 83*, 902–911.

Pittenger, C., Huang, Y. Y., Paletzki, R. F., Bourtchouladze, R., Scanlin, H., Vronskaya, S., et al. (2002). Reversible inhibition of CREB/ATF transcription factors in region CA1 of the dorsal hippocampus disrupts hippocampus-dependent spatial memory. *Neuron, 34*, 447–462.

Plomin, R. (2001). The genetics of g in human and mouse. *Nature Review Neuroscience, 2*, 136–141.

Plomin, R., DeFries, J. C., McClearn, G. E., & McGuffin, P. (2001). *Behavioral genetics* (4th ed.). New York: Worth Publishers.

Plomin, R., & Kosslyn, S. M. (2001). Genes, brain and cognition. *Nature Neuroscience, 4*, 1153–55.

Plomin, R., & Galsworthy, M.J. (2003). Intelligence and Cognition. In D. N. Cooper (Ed.),*Nature Encyclopedia of the Human Genome* (Vol. 3, pp. 508–514). London: Nature Publishing Group.

Poirier, R., Jacquot, S., Vaillend, C., Soutthiphong, A. A., Libbey, M., Davis, S., et al. (2006). Deletion of the Coffin–Lowry syndrome gene *Rsk2* in mice is associated with impaired spatial learning and reduced control of exploratory behavior. *Behavior Genetics, 37*, 31–50.

Powell, C. M. (2006). Gene targeting of presynaptic proteins in synaptic plasticity and memory: Across the great divide. *Neurobiology of Learning and Memory, 85*, 2–15.

Price, J. L., Davies, P. B., Morris, J. C., & White, D. L. (1991). The distribution of tangles, plaques and related immunohistochemical markers in healthy aging and Alzheimer's disease. *Neurobiology of Aging, 12*, 295–312.

Rabenstein, R. L., Addy, N. A., Caldarone, B. J., Asaka, Y., Gruenbaum, L. M., Peters, L. L., et al. (2005). Impaired synaptic plasticity and learning in mice lacking β-adducin, an actin-regulating protein. *The Journal of Neuroscience, 25*, 2138–2145.

Rammes, G., Steckler, T., Kresse, A., Schutz, G., Zieglgansberger, W., & Lutz, B. (2000). Synaptic plasticity in the basolateral amygdala in transgenic mice expressing dominant-negative cAMP response element-binding protein (CREB) in forebrain. *European Journal of Neuroscience, 12*, 2534–2546.

Rampon, C., Tang, Y. P., Goodhouse, J., Shimizu, E., Kyin, M., & Tsien, J. Z. (2000). Enrichment induces structural changes and recovery from nonspatial memory deficits in CA1 NMDAR1-knockout mice. *Nature Neuroscience, 3*, 238–244.

Reisel, D., Bannerman, D. M., Schmitt, W. B., Deacon, R. M. J., Flint, J., Borchardt, T., et al. (2002). Spatial memory dissociations in mice lacking GluR1 *Nature Neuroscience, 5*, 868–873.

Riedel, G., Platt, B., & Micheau, J. (2003). Glutamate receptor function in learning and memory. *Behavioural Brain Research, 140*, 1–47.

Robbins, T. W., & Murphy, E. R. (2006). Behavioral pharmacology: 40+ years of progress, with a focus on glutamate receptors and cognition. *Trends in Pharmacological Sciences, 27*, 141–148.

Robles, Y., Vivas-Mejía, P. E., Ortiz-Zuazaga, H. G., Félix, J., Ramos, X., & Peña de Ortiz, S. (2003). Hippocampal gene expression profiling in spatial discrimination learning. *Neurobiology of Learning and Memory, 80*, 80–95.

Rondi-Reig, L., Petit, G. H., Tobin, C., Tonegawa, S., Mariani, J., & Berthoz, A. (2001). Impaired Sequential Egocentric and Allocentric Memories in Forebrain-Specific-NMDA Receptor Knock-Out Mice during a New Task Dissociating Strategies of Navigation. *The Journal of Neuroscience, 26*, 4071–4081.

Rousse, I., Beaulieu, S., Rowe, W., Meaney, M. J., Barden, N., & Rochford, J. (1997). Spatial memory in transgenic mice with impaired glucocorticoid receptor function. *Neuroreport, 8*, 841–845.

Sakagawa, T., Okuyama, S., Kawashima, N., Hozumi, S., Nakagawasai, O., Tadano, T., et al. (2000).Pain threshold, learning and formation of brain edema in mice lacking the angiotensin II type 2 receptor. *Life Sciences, 67*, 2577–2585.

Sakimura, K., Kutsuwada, T., Ito, I., Manabe, T., Takayama, C., Kushiya, E., et al. (1995). Reduced hippocampal LTP and spatial learning in mice lacking NMDA receptor 1 subunit. *Nature, 373*, 151–155.

Samuel, W., Terry, R. D., Deteresa, R., Butters, N., & Masliah, E. (1994). Clinical Correlates of Cortical and Nucleus Basalis Pathology in Alzheimer Dementia. *Archives of Neurology, 51*, 772–778.

Sapolsky, R. M. (2003). Altering behavior with gene transfer in the limbic system. *Physiology & Behavior, 79*, 479–486.

Savitz, J., Solms, M., & Ramesar, R. (2006). The molecular genetics of cognition: Dopamine, COMT and BDNF. *Genes, Brain and Behavior, 5*, 311–328.

Schmitt, W. B., Deacon, R. M. J., Seeburg, P. H., Rawlins, J. N. P., & Bannerman, D. M. (2003). A within-subjects, within-task demonstration of intact spatial reference memory and impaired spatial working memory in glutamate receptor-A-deficient mice. *The Journal of Neuroscience, 23*, 3953–3959.

Seabrook G. R., & Rosahl, T. W. (1999). Transgenic animals relevant to Alzheimer's disease. *Neuropharmacology, 38*, 1–17.

Selcher, J. C., Nekrasova, T., Paylor, R., Landreth, G. E., & Sweatt, J. D. (2001). Mice lacking the ERK1 isoform of MAP kinase are unimpaired in emotional learning. *Learning & Memory, 8*, 11–19.

Selkoe, D. J. (1997). Alzheimer's Disease: Genotypes, phenotypes, and treatments. *Science, 275*, 630–631.

Shahbazian, M. D., Young, J. I., Yuva-Paylor, L. A., Spencer, C. M., Antalffy, B. A., Noebels, J. L., et al. (2002). Mice with truncated MeCP2 recapitulate many Rett syndrome features and display hyperacetylation of histone H3. *Neuron, 35*, 243–254.

Shaw, P., Greenstein, D., Lerch, J., Clasen, L., Lenroot, R., Gogtay, N., et al. (2006). Intellectual ability and cortical development in children and adolescents. *Nature, 440*, 676–679.

Shimizu, E., Tang, Y. P., Rampon, C., & Tsien, J. Z. (2000). NMDA receptor-dependent synaptic reinforcement as a crucial process for memory consolidation. *Science, 290*, 1170–1174.

Shors, T. J., & Matzel, L. D. (1997). Long-term potentiation: What's learning got to do with it? *Behavioral and Brain Sciences, 20*, 597–655.

Sigurdsson, E. M., Scholtzova, H., Mehta, P. D., Frangione, B., & Wisniewski, T. (2001). Immunization with a nontoxic/nonfibrillar amyloid-beta homologous peptide reduces Alzheimer's disease-associated pathology in transgenic mice. *American Journal of Pathology, 159*, 439–447.

Silva, A. J., Paylor, R., Wehner, J. M., & Tonegawa, S. (1992). Impaired spatial learning in alpha-calcium-calmodulin kinase II mutant mice. *Science, 257*, 206–211.

Spearman, C. (1904). 'General intelligence' objectively determined and measured. *American Journal of Psychology, 15*, 201–293

Spreng, M., Cotecchia, S., & Schenk, F. (2001). A behavioral study of alpha-1b adrenergic receptor knockout mice: Increased reaction to novelty and selectively reduced learning capacities. *Neurobiology of Learning and Memory, 75*, 214–229.

Sprengel, R., Suchanek, B., Amico, C., Brusa, R., Burnashev, N., Rozov, A., et al. (1998). Importance of the intracellular domain of NR2 subunits for NMDA receptor function in vivo. *Cell, 92*, 279–289.

Steckler, T., Weis, C., Sauvage, M., Mederer, A., & Holsboer, F. (1999). Disrupted allocentric but preserved egocentric spatial learning in transgenic mice with impaired glucocorticoid receptor function. *Behavioural Brain Research, 100*, 77–89.

Stork, O., & Welzl, H. (1999). Memory formation and the regulation of gene expression. *Cellular and Molecular Life Sciences, 55*, 575–592.

Stork, O., Welzl, H., Wolfer, D., Schuster, T., Mantei, N., Stork, S., et al. (2000). Recovery of emotional behavior in neural cell adhesion molecule. (NCAM) null mutant mice through transgenic expression of NCAM180. *European Journal of Neuroscience, 12*, 3291–3306.

Sturchler-Pierrat, C., Abramowski, D., Duke, M., Wiederhold, K.-H., Mistl, C., Rothacher, S., et al. (1997). Two amyloid precursor protein transgenic mouse models with Alzheimer disease-like pathology. *Proceedings of the National Academy of Sciences of the United States of America, 94*, 13287–13292.

Sweatt, J. D. (2004). Mitogen-activated protein kinases in synaptic plasticity and memory. *Current Opinion in Neurobiology, 14*, 311–317.

Tang, Y. P., Shimizu, E., Dube, G. R., Rampon, C., Kerchner, G. A., Zhuo, M., et al. (1999). Genetic enhancement of learning and memory in mice. *Nature, 401*, 63–69.

Tang, Y. P., Wang, H., Feng, R., Kyin, M., & Tsien, J. Z. (2001). Differential effects of enrichment on learning and memory function in NR2B transgenic mice. *Neuropharmacology, 41*, 779–790.

Thomas, G. M., & Huganir, R. L. (2004). MAPK cascade signalling and synaptic plasticity. *Nature Reviews Neuroscience, 5*, 173–83.

Thorndike, R. L. (1935). Organization of behavior in the albino rat. *Genetic Psychology Monographs, 17*, 1–70.

Toga, A. W., & Thompson, P. M. (2005). Genetics of brain structure and intelligence. *Annual Review of Neuroscience, 28*, 1–23.

Tong, X. K., & Hamel, E. (1999). Regional cholinergic denervation of cortical microvessels and nitric oxide synthase-containing neurons in Alzheimer's disease. *Neuroscience, 92*, 163–175.

Tsang, S., Sun, Z., Luke, B., Stewart, C., Lum, N., Gregory, M., et al. (2005). A comprehensive SNP-based genetic analysis of inbred mouse strains. *Mammalian Genome, 16*, 476–80.

Tsien, J. Z., Huerta, P. T., & Tonegawa, S. (1996). The essential role of hippocampal CA1 NMDA receptor-dependent synaptic plasticity in spatial memory. *Cell, 87*, 1327–1338.

Uchida, S., Sakai, S., Furuichi, T., Hosoda, H., Toyota, K., Ishii, T., et al. (2006). Tight regulation of transgene expression by tetracycline-dependent activator and repressor in brain. *Genes, Brain and Behavior, 5*, 96–106.

van Leuven, F. (2000). Single and multiple transgenic mice as models for Alzheimer's disease. *Progress in Neurobiology, 61*, 305–312.

Victoroff, J., Zarow, C., Mack, W. J., Hsu, E., & Chui, H. C. (1996). Physical aggression is associated with preservation of substantia nigra pars compacta in Alzheimer disease. *Archives of Neurology, 53*, 428–434.

Waddell, S., & Quinn, W. G. (2001). Flies, genes, and learning. *Annual Review of Neuroscience, 24*, 1283–1309.

Waltereit, R., & Weller, M. (2003). Signaling from cAMP/PKA to MAPK and synaptic plasticity. *Molecular Neurobiology, 27*, 99–106.

Weeber, E. J., Levenson, J. M., & Sweatt, J. D. (2002). Molecular genetics of human cognition. *Molecular Interventions, 2*, 376–391.

Welzl, H., D'Adamo, P., Wolfer, D. P., & Lipp, H. P. (2006). Mouse models of hereditary mental retardation. In G. S. Fisch & J. Flint (Eds.), *Transgenic and knockout models of neuropsychiatric disorders*. R. Lydic & H. A. Baghdoyan (Series Eds.), *Contemporary clinical neuroscience* (pp. 101–125). Totowa, New Jersey USA: Humana Press.

Whitehouse, P. J., Price, D. L., Struble, R. G., Clark, A. W., Coyle, J. T., & Delon, M. R. (1982). Alzheimer's disease and senile dementia: Loss of neurons in the basal forebrain. *Science, 215*, 1237–1239.

Wolfer, D. P., Mohajeri, H. M., Lipp, H. P., & Schachner, M. (1998). Increased flexibility and selectivity in spatial learning of transgenic mice ectopically expressing the neural cell adhesion molecule L1 in astrocytes. *European Journal of Neuroscience, 10*, 708–717.

Yan, Y., Wang, M., Lemon, W. J., & You, M. (2004). Single nucleotide polymorphism (SNP) analysis of mouse quantitative trait loci for identification of candidate genes. *Journal of Medical Genetetics, 41*, e111.

Zamanillo, D., Sprengel, R., Hvalby, Ø., Jensen, V., Burnashev, N., Rozov, A., et al. (1999). Importance of AMPA Receptors for Hippocampal Synaptic Plasticity But Not for Spatial Learning. *Science, 284*, 1805–1811.

Chapter 12

Neurogenetic Analysis and Cognitive Functions in Trisomy 21

Pierre L. Roubertoux and Michèle Carlier

Introduction

Trisomy 21 (TRS21) is also known as "Down's syndrome" and for a long time was called "mongolism". At the beginning of the third millennium, TRS21 remains the most frequent genetic cause of mental deficiency in Western society. According to estimates in recent studies by Roizen and Patterson (2003), TRS21 occurs once in every 800 or 1,000 births. TRS21 is a syndrome, defined by a complex set of cardiac, immune, bone, brain, and cognitive disorders, most being highly variable in expression. Not all persons with TRS21 have leukemia, although it is more prevalent in the TRS21 population compared to the general population. Cardiac disorders are responsible for approximately 60% of perinatal mortality in neonates with TRS21. Immune disorders are more common in TRS21 (30% of TRS21 persons have abnormal levels of T-lymphocytes – Ugazio, Maccario, Notarangelo, & Burgio, 1990); bone anomalies are also more common. The characteristic morphology is short and stocky with virtually no neck because of skeletal abnormalities. Facial features of persons with TRS21 typically include oblique eye fissures, epicanthic eye-folds, a flat nasal bridge, the mouth permanently open and the tongue protruding. The limbs are malformed, and hands are short and broad with a single transverse palmar crease and shortened, incurved fifth finger. Mental deficiency, while of variable severity, is the most constant feature of persons with TRS21 (Antonarakis, Lyle, Dermitzakis, Reymond, & Deutsch, 2004; Patterson & Costa, 2005).

Mental deficiency is heterogeneous, varying between individuals, and appears as a complex cognitive pattern rather than an overall deficit. Some skills remain preserved, while others are impaired to different degrees (Nadel, 1995, 1999). The complexity of the impairment is not only seen in the psychological patterns, but also seen in brain structure volumes and histological characteristics. TRS21 may thus reflect the complexity of brain structure volumes that are affected unequally and in complex patterns. TRS21 offers potential scope for correlating anatomic defects in the brain with cognitive characteristics (Teipel & Hampel, 2006). Even more interesting prospects are available for behavior-genetic analysis as the human 21 chromosome (HSA21) has been sequenced (Hattori et al., 2000). The triplicated region implicated in TRS21 encompasses 225 genes but a recent estimate of the HSA21 genes indicates that HAS 21 carries 283 protein-encoding genes (Watanabe et al., 2004). This is not a very large number and hopes of establishing correlations between genes, the brain and psychological variants do not appear to be utopian. Comparative genomics have identified many synthesis between the human species and other mammalian species; these have been thoroughly documented in the mouse, thus providing an opportunity to develop mouse models of TRS21 and investigate the function of genes or groups of genes.

Chromosomal abnormalities have been widely reported in relation to cognitive disorders. The hemizygous deletions of large chromosomal fragments, 7q11.23 in Williams–Beuren (Morris & Mervis, 2000) and 22q.11 syndromes (Liu et al., 2002) are associated with a typical profile of cognitive dysfunction or psychotic disorders. Several phenotypes are known to be the result of trisomies affecting only one gene. Recent deciphering of gene–brain–behavior relationships has seen considerable progress in investigations of TRS21 and could serve as a model for other chromosomal anomalies.

Knowledge of gene–brain–behavior relationships can offer prospects for uncovering treatments to improve brain functions in TRS21, but needs to be extended to gain an understanding of the underlying processes involved in the cognitive disorders of this syndrome.

P.L. Roubertoux (✉)
INSERM UMR910 Génétique Médicale et Génomique Fonctionnelle, Université d'Aix-Marseille 2, Faculté de Médecine, 27 Bvd Jean Moulin, 13385 Marseille Cedex 05, France
e-mail: pierre.roubertoux@univmed.fr

Y.-K. Kim (ed.), *Handbook of Behavior Genetics*,
DOI 10.1007/978-0-387-76727-7_12, © Springer Science+Business Media, LLC 2009

The First Identified Chromosomal Bases of Cognitive Defects

J.E.D. Esquirol (1838) devoted a large part of one of his books to "idiocy"; idiocy was what we refer to as "mental deficiency", "mental retardation" or "feeble mindedness". Several observations were of what we now call TRS21, and the description clearly tallies with the modern symptomatology of the disorder. Esquirol described a particular category of patient characterized by oblique eye fissures, epicanthic eye-folds, a flat nasal bridge and protruding tongue. He noted their short, stocky stature with virtually no neck, malformed limbs and mental deficiency. One illustration by Ambroise Tardieu in the book by Esquirol shows some of the physical characteristics of persons with trisomy 21 (see Roubertoux & Kerdelhué, 2006, for more details). Eight years later, Edouard Séguin (1846) adopted this description of the symptomatological group, adding a detailed description of the small nose and open mouth; he described the morphology of the tongue, that was thick and cracked, and the susceptibility of the lungs and integuments to infection. In a later paper, first published in English (Séguin, 1856, 1866), Séguin wrote that in spite of "profound idiocy", these "good kids" had language and were able to gain some basic knowledge. He described the mental pathology reported by Esquirol in greater detail and named it "furfuraceous cretinism" because of the bran-like appearance of the skin of persons with TRS21. Six years after the pioneering description by Séguin, the British alienist, John Langdon Haydon Down, classified patients in the mental hospital where he was working, assigning them to so-called "ethnic groups". "*A very large number of congenital idiots are typical Mongols. So marked is this that, when placed side by side, it is difficult to believe the specimens compared are not children of the same parents.*" (Down, 1862). Several astute observations taken from texts by Esquirol and in particular by Séguin (1846, 1956) can be recognized in Down's three-page note (1862) and in the even shorter note published later (1867). The term "Down's syndrome" was substituted for "mongolism" in the mid-20th century, but this new terminology is unsatisfactory as it infers that the pioneering work was done by J. H. Down, which is obviously inaccurate, as we know it was done by Esquirol and Séguin. The term "Trisomy 21" (TRS21) should therefore be preferred for these reasons.

What was the origin of the syndrome? Several hypotheses were formulated during the first part of the 20th century, several implicating the chromosomal arrangement, with an extra copy of one chromosome. Two elements were needed to prove the chromosomal basis of the syndrome still called "mongolism" at the time: the exact number of human chromosomes had to be established, and Tijo and Levan did not discover that the human genome has 46 chromosomes until the middle of the century (Tijo & Levan, 1956), and individual cytological characterization of each chromosome still had to be performed. The technique was only available in a small number of laboratories at the forefront of research into cytogenetics. Two years after Tijo and Levan published their paper, Lejeune observed a "Mongoloid" patient with 47 chromosomes instead of 46. He came up with an "either/or" hypothesis: either a translocation on chromosome 4 or an extra chromosome. Morphological analysis of the chromosomes producing this anomalous karyotype confirmed the finding of an extra chromosome 21, first in two patients, then in nine "mongoloid" patients (Lejeune, Gauthier, & Turpin, 1959; Lejeune, Turpin, & Gautier, 1958). The observation of a triple chromosome 21 in patients with the same clinical diagnosis was confirmed in independent groups.

Cognitive and Brain Phenotypes of Patients with TRS21

Questions raised with TRS21 are

- Is the neuropsychological profile of TRS21 persons the result of greater developmental instability?
- Is the neuropsychological profile the consequence of a dosage effect?

In other words, are the brain and psychological features observed in TRS21 specific to the genetic disorder or are they non-specific consequences of mental deficiency? The issue at stake is important. If the neuropsychological profile is the consequence of mental deficiency, similar profiles should be found in other genetic diseases, and the deciphering of relationships between HSA21 genes and cognitive impairment will be irrelevant to any understanding of the interface between the genome and the highest integrative functions. But if the profiles are specific and vary from one genetic disorder to another, this specificity should lead to genes involved in the different facets of cognition being identified. We will present a brief synopsis of cognitive and brain characteristics of TRS21 and will then show that the characteristics are different when compared to Williams–Beuren syndrome due to another chromosomal aberration with mental deficiency.

Patterns of Cognition in TRS21

TRS21 cognitive processes have been well documented (Brown et al., 2003; Clark & Wilson, 2003; Krinsky-McHale, Devenny, & Silverman, 2002; Laws, 2002; Pennington, Moon, Edgin, Stedron, & Nadel, 2003; Vicari, 2006).

Cognitive impairment is the salient feature of the syndrome, with IQ ranging from 30 to 70 and averaging around 45 (Chapman & Hesketh, 2000; Carr, 2005). Carr (2005) reported that cognitive decline in adult age affected verbal skills more than performance ability, whereas the reverse is observed in typically developing adults. Persons with TRS21 have a normal performance level for simple tasks (operant conditioning at 3 months, imitation learning at 55 months and classical conditioning at adult age), but difficulties with spatial memory, poor long-term memory performances (Brown et al., 2003; Clark & Wilson, 2003; Hodapp, Ewans, & Gray, et al., 1999; Raz et al., 1995) and very poor performance in language abilities (Nadel, 1999; Rondal, 1999) are observed.

Persons with TRS21 have difficulty in acquiring new skills (Raz et al., 1995; Hodapp, 1999 et al.), mostly because of the persistent use of old strategies to solve new problems (Wishart, 1993; Raz et al., 1995; Hodapp, 1999 et al.), and given the small size of the hippocampus (see below), they learn less efficiently than typically developing children (compared when matched for mental age) as tested with different tasks: word list, a computer-generated virtual Morris water maze and pattern recognition (Pennington et al., 2003).

Brain Characteristics of TRS21

Figure 12.1 summarizes concordant studies of volumes of different brain structures. Observing the results in published papers, we can see that the size of the brain structures is generally smaller in TRS21 persons; in particular there is a dramatic reduction in the size of the hippocampus. Few biochemical studies are available on TRS21 persons. More stringent age controls and replications are needed before reaching the state of the art.

TRS21 and William–Beuren Syndrome: Comparative Neuropsychology

Neuropsychological studies have shown that mental deficiency from different genetic etiologies can be characterized by very different quantitative and qualitative profiles of cognitive impairment. The focus here will be on studies comparing TRS21 and Williams–Beuren syndrome (WBS) as there is extensive literature on the topic. WBS is a rare genetic syndrome (between 1/7,500 and 1/25,000 live births), resulting from a hemizygous deletion of contiguous genes on the long arm of chromosome 7 at 7q11.23, including the gene for *elastin* (*ELN*). In 95% of affected individuals, the size of the deletion is 1.6 Mb encompassing at least 25 genes (Bayés et al., 2003). The specific contribution of most of the deleted genes to psychological traits is still unknown (see, for example, Gray, Karmiloff-Smith, Funnell, & Tassebehji 2006).

The TRS21 and WBS phenotypes are characterized by mental deficiency. The mean level of cognitive performance is lower in TRS21 than in WBS; the difference is at least 10 IQ points, but this figure has to be considered with caution as intellectual level depends on the tool used to measure IQ, and particularly so in the low range. The psychological profiles in TRS21 and WBS are complex, and they differ.

The reader must not forget that when differences are shown between psychological profiles of persons with TRS21 and WBS, it cannot be concluded that one group performed in the normal range and that the other achieved a low performance level, but simply that one is relatively higher than the other. Moreover, when a TRS21 group and a WBS group are matched for mental age, the groups are not representative of the whole group of persons with the genetic disorder concerned; and as the mean IQ is lower for TRS21 than for WBS, individuals with very low scores in the TRS21 group have to be excluded for the purposes of statistical

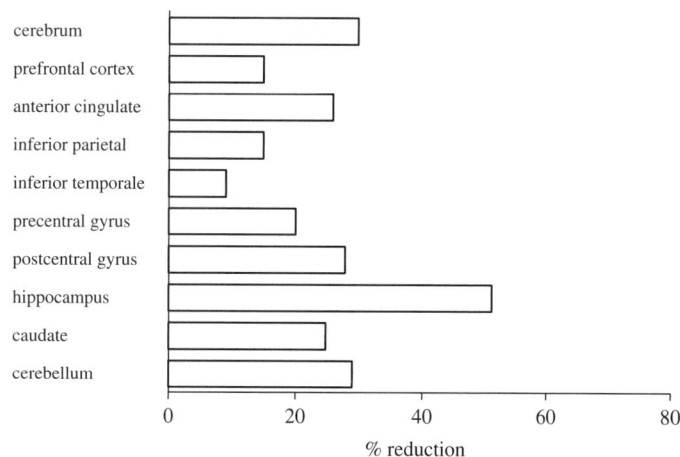

Fig. 12.1 Reduction of the volume of brain structures in trisomic 21 persons compared to non-trisomics

analysis, and individuals scoring at the top of the WBS IQ range, also have to be excluded (Carlier & Ayoun, 2007).

Language development and auditory rote memory are relatively good in WBS syndrome, while visuospatial processing skills are severely impaired. The reverse is observed in persons with TRS21 who display a severe deficit in language development and relative strength in visuospatial skills (Vicari, 2006). Studies of adolescents and adults with TRS21 and WBS, when matched for age and full-scale IQ, show clear differences in language skills at a number of levels: phonological, lexical, morphological, syntactic, plus prosody, discourse and narrative (Bellugi, Lichtenberger, Mills, Galaburda & Korenberg, 1999; Mervis, 2003). Different patterns of impairment in long-term visual and spatial memory have also been observed. When compared to a group of mental-age-matched typically developing children, individuals with WBS displayed specific difficulties in visual–spatial tasks, but not with visual-object and verbal memory tasks, while persons with TRS21 performed poorly on both tasks (Vicari, Belluci, & Carlesimo, 2005).

Aspects of the psychological profile other than cognitive traits have been investigated. Children with WBS were described as hyper-sociable compared to typically developing individuals and individuals with TRS21 (Doyle, Bellugi, Korenberg, & Graham, 2004). A recent study of lateral preference for hand and foot compared samples of children and adolescents: WBS, TRS21 and typically developing persons (Carlier et al., 2006; Gérard-Desplanches et al., 2006). Carlier and her group found more left-handers and more left-foot preferences in the group with TRS21 than in the other two groups. Manual and foot inconsistencies (i.e., preference varying within tasks) were observed in both the TRS21 and the WBS groups, but was very rare in the age-matched typically developing group. With a card-reaching task, age differences were observed in the typically developing group, but not in either the TRS21 or the WBS group. In short, there were different patterns of laterality, between the TRS21 and the WBS groups and the typically developing group, and also between the TRS21 group and the WBS group. Animal models should help describe the physiological pathways between the laterality and the TRS21 and WBS genes (Roubertoux et al., 2005).

Differences in brain development and brain morphologies between TRS21 and WBS have been reported. Teipel & Hampel (2006) presented a review of neuroanatomical investigations of TRS21 and some conclusions are summarized here. Postmortem analyses of fetuses and infants have described characteristics such as lower brain weight, brachycephaly, a small cerebellum, small frontal and temporal lobes, shallow depth of the cerebral sulci and a narrow superior temporal gyrus. Even after correcting for body size, brain volume is smaller in persons with TRS21 compared to age-matched typically developing individuals (see Fig. 12.1).

Volumetric studies using nuclear magnetic resonance observing young, non-demented adults with TRS21 and comparing them to age-matched non-TRS21 individuals have confirmed these data: the overall brain volume is smaller, and this includes the cerebellum and cerebral gray and white matter. To understand the results obtained with animal models of TRS21, it is important to note that the volume of the hippocampus is dramatically small, whereas the volume of amygdala is smaller, but in proportion to the overall smaller size of the brain.

Children with WBS display marked atrophy of the posterior regions of the brain and of the basal ganglia. Recent studies using nuclear magnetic resonance techniques evidenced abnormal surface complexity and thickness of the cortex (Thompson et al., 2005), hypoactivation in the parietal portion of the dorsal stream (Meyer-Lindenberg et al., 2004) and a decrease in gray matter concentration in the left parieto-occipital region (Boddaert et al., 2005). In the hippocampus of patients with WBS, the normal volume is intact, but the metabolism is impaired (reduction in blood flow, in N-acetyl aspartate metabolism and in the responsiveness of the anterior formation – Meyer-Lindenberg et al., 2005). Abnormal development of the corpus callosum has been reported, with it being more convex, particularly in the splenium and the caudal part of the callosal body (Tomaiuolo et al., 2002).

Human Chromosome 21

The publication of the full sequence of the long arm of chromosome 21 (HSA21) (Hattori et al., 2000) inaugurated a new era in understanding the relationships between chromosome 21 (HSA21) and cognition and, more generally, between genes and brain processes involved in mental deficiency. What was even more important for understanding these processes was the description of synthesis between HSA21 and mouse chromosomes 16 (MMU16), 17 (MMU17) and 10 (MMU10).

Properties of HSA21 Genes

HSA21 is one of the smallest human chromosomes. With its 33,827,477 bp it accounts for 1% of the human genome. The long arm was initially considered as carrying 225 genes. Recent estimates suggest that the number should be higher and could reach 283 (Watanabe et al., 2004). No empirical estimates of the number of the genes carried by the short arm are available yet. Triplication of HSA21 can be caused by non-disjunction during meiosis, by translocation to another chromosome or mosaicism. The first case is the most frequent, and examination of the extra chromosome DNA shows that it is always maternal in origin. The extra chromosome is from maternal origin in 90% of the cases, while 10% are from paternal (Hassold & Sherman, 2000)

The phenotypes of TRS21 patients depend on the allelic forms carried by not only the pair of chromosomes from the father and mother but also the extra chromosome. The allelic forms that differ from one patient to another and intra- or inter-locus interactions contribute substantially to individual differences between TRS21 persons. The triple copy of an allelic form of a gene does not necessarily mean triple expression of the gene. This has been shown with the Ts65Dn mouse, one model for TRS21 (see below). The genes carried by the triplicated fragment of MMU16 (syntenic to HSA21) do not produce the same gene dosage effects. Lyle, Gehrig, Neergaard-Henrichsen, Deutsch, & Antonarakis (2004) noted that not all the genes present in the three copies are overexpressed. The expression of some of the HSA21 genes in three copies is no greater than would be expected with two genes copied (Antonarakis et al., 2004). The normalized expression value for a gene with two alleles is 1.00, so the normalized value expected for a gene with three allelic copies would be 1.50. Of the 78 genes investigated, 18% showed expression greater than 1.5, for 37%, expression was 1.5, for 45% it was between 1 (normal) and 1.5, while 9% had the same level of expression as in diploid cells. Gitton et al. (2002), Kahlem et al. (2004) and Kahlem (2006) concluded that differences in overexpression of triplicated genes of HSA21 were tissue dependent. Reymond et al. (2002) and Lyle et al. (2004), studying mice, observed that the level of expression was age dependent, varying from embryonic to aerial life. The Usp16 gene provides a good illustration of variations in expression that can be observed between cells. The expression of the three copies gives a normalized value of 1.44 in the brain and 1.11 in the kidney at day 30, but increases from 1.11 to 1.52 in the kidney between day 30 and month 11.

The effect of triplicated copies of HSA21 genes can therefore be direct or indirect. The genes reacting to the triple dosage effect may induce modifications to expression in the genome and, specifically in the chromosome 21 genes. These variations in expression may contribute to the TRS21 phenotype. Depending on the allelic forms carried by the non-HSA21 genes, the TRS21 phenotype can vary.

Although HSA21 is one of the smallest chromosomes, it carries a large number of genes. The discovery of partial trisomies seemed to pave the way for deciphering the genes involved in the mental deficiency observed in TRS21. The first partial trisomy was reported by Aula, Leisti, & von Koskull (1973) and was related to 21q22.1 and 21q22.2 bands in a person with the TRS21 phenotype. Partial trisomies have proven invaluable in investigating the function of genes carried by chromosome 21. Delabar et al. (1993) and Korenberg et al. (1994) suggested that certain regions of chromosome 21 were linked with most of the signs observed

by Jackson, North III, and Thomas (1976) in their description of TRS21. These regions are bounded by *D21S17* and *ETS2*. Persons with a triple *D21S17-ETS2* region were recognized as more severely mentally retarded than persons carrying a triple copy of another HSA21 region. Unfortunately, in both studies, assessments of cognitive capacities were often approximate, with an overall score on an intelligence test, or sometimes mental performance was simply inferred from "examinations or discussions with the family and patient" (Korenberg et al., 1994). The idea of a "critical region" or of a "minimal region" has been challenged recently (Olson, Richtsmeier, Leszl, & Reeves, 2004); these findings will be discussed in the section on animal models.

Cases of partial TRS21 are very rare, accounting for no more than 1% of living persons with TRS21. The scientific community therefore turned to animal models to establish correlations between HSA21 and phenotypes. The mouse was chosen as the syntenies between MMU16 and HAS21 have been well documented over an extended period.

Mouse Models of TRS21

The selection of a model gives rise to a dilemma. Mice can be produced with extra copies of large synthetic fragments of HSA21, and even with the whole HSA21. These mice can be used to see how the mouse phenotype fits the human syndrome, but they cannot be used to identify individual functions of genes and their role in TRS21 mental retardation. Other limitations are encountered with single mice overexpressing a single gene. The hypothesis implicit in the use of transgenic mice with single genes from HSA21 is that one gene may have a substantial effect. The effect should be detectable by comparison with diploid mice and the effect would be more difficult to detect if a large number of genes were involved or if the effect of a single gene was produced by epistasis. Mouse models of TRS21 include segmental trisomies and trisomies for a single gene. The extra fragment can be of mouse origin, in which case it is caused by translocation, or it can be of human origin, through transfection of HSA21 genes or chromosomal fragments (Fig. 12.2). The discrepancies between an extra copy produced by translocation and one produced by transgenesis could be determined by the different degrees of homology between the species, and also by the efficiency of the promoter selected for the transfection.

Extra Copy of Mouse Origin

Gropp, Kolbus, and Giers (1975) used spontaneous Robertsonian translocations to generate aneuploidy for MMU16 in mice. The model was labeled Ts16. However, the embryos

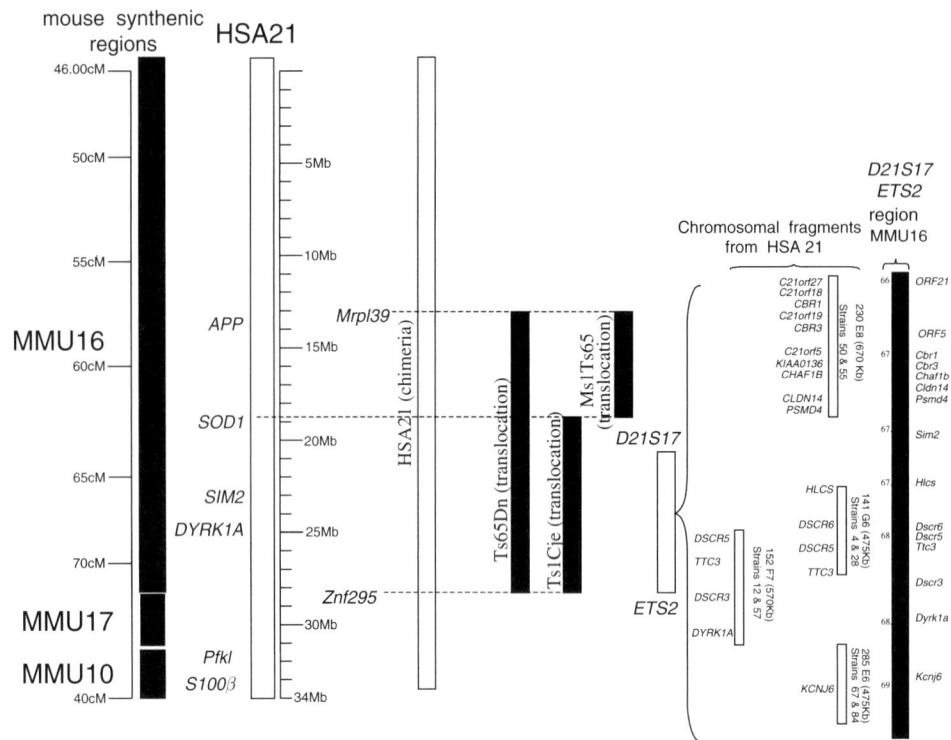

Fig. 12.2 Available mouse models of trisomy 21. Only models used for cognitive studies are shown. *Black* = mouse origin, *white* = human origin. The main genes transfected are shown on the human chromo-some (HSA21). The genes carried by human chromosomal fragments covering *D21S17-ETS2* are shown on the *right*

died in utero and, with the presence of other syntenies between MMU16 and HSA3, HSA8, HSA16 and HSA21, the model proved to be of limited value for experimental purposes. Davisson, Schmidt, and Akeson (1990), Davisson et al. (1993) developed a second MMU16 aneuploidy for a single region of chromosome 16 syntenic with human chromosome 21. The model was labeled Ts65Dn. The extra region encompasses 132 genes from *Mrpl39* to *Znf295* (Baxter, Moran, Richtsmeier, Troncoso, & Reeves, 2000). In the course of a targeted gene experiment, Sago et al. (1998) accidentally generated a third aneuploidy on the Ts65Dn background for the region between Mrpl39 and Znf295. A partial trisomy of MMU16 occurred when the *Sod-1* gene was targeted. This included 85 genes from *Sod-1* to *Znf295*, *Sod-1* not being overexpressed. The new model was labeled Ts1Cje. A fourth aneuploidy (labeled Ms1Cje) was also generated on the same background, but covered the centromeric part of Ts65Dn; Ms1Cje includes the complementary region covered by Ts65Dn; the extra-chromosomal fragment encompasses 46 genes with *App* and *Sod-1* as boundaries.

Extra Copy of Human Origin

A second category of mouse model for TRS21 was produced by inserting all or part of HSA21. These genes

are from the human genome and are cited in capitals and italics. Recent studies (O'Doherty et al., 2005; Shinohara et al., 2001) incorporated almost the entire HSA21 chromosome into the mouse genome, using two different groups of techniques. Smith, Zhu, Zhang, Cheng, and Rubin (1995) and Smith et al. (1997) selected short contiguous fragments of HSA21 covering the *D21S17-ETS2* region. The fragments were inserted into Yeast Artificial Chromosomes (YACs) or phages and transfected into the mouse genome to produce transgenic mice. The chromosomal fragments cover a region encompassing the 21q22.2 band and located between the human *D21S55* (*D21S55* Site Targeted Sequence – STS) and the *ZNF295* genes, with one gap corresponding to the *SIM2* gene and its neighboring region.

A recent paper challenged the concept of a "critical" or "minimal" region of TRS21 (Olson et al., 2004). Mice with a duplicated region between *D21S17* and *ETS2* did not present the same cranial abnormalities as mice with the extra Ts65Dn fragment. These mice were not tested for cognitive performance, but other studies on mice carrying extra fragments encompassed in the *D21S17-ETS2* region have reported specific neural and cognitive impairments (Smith et al., 1997; Chabert et al., 2004; Roubertoux et al., 2005; Roubertoux et al., 2006; Sérégaza, Roubertoux, Jamon, & Soumireu-Mourat, 2006). The idea of a "minimal" or "critical" region for TRS21 for all the Jackson signs does not stand up to experimental evidence. The crucial contribution of the *D21S17-ETS2* region cannot, however, be eliminated for cognitive processes and brain morphology.

MMU10 and MMU17 Syntenies

As stated above, HSA21 is syntenic in mice to MMU16, MMU10 and MMU17. Triple copies of MMU16 have been extensively analyzed but no trisomic mice with MMU10 or MMU17 syntenies are available. Chromosomal engineering has been developed for these regions by combining gene-targeting techniques using mouse embryo stem cells and the Cre/*loxP* site-specific recombination system, and may lead to the emergence of chromosomal-engineering technology in mice (Yu & Bradley, 2001). The authors suggest that "chromosomal engineering could also be used to generate small overlapping duplications in mouse chromosomal regions conserved with the trisomic human chromosome 21 regions to identify the crucial genomic domain(s) and causative gene(s) that are responsible for the clinical characteristics of the disorder."

Over-Expression of Single Human Genes

Extra copies of single human genes carried by HSA21 have been inserted into the mouse genome and have generated single gene trisomy. Several mice with single gene trisomy have been used to investigate gene involvement in cognitive impairment: *SIM2, S100β, SOD1, APP and DYRK1A*.

Chromosome-Phenotype Correlation and Partial Trisomy 21

The different TRS21 models have been investigated to varying degrees through behavioral tests and brain examinations. Behavioral examinations have focused on cognitive functioning. Learning performances were tested with the Morris water maze, radial maze and T-maze, as well as active and passive avoidance tests, fear conditioning and operant conditioning procedures. Exploration and reaction to novelty were also studied. Unfortunately, "home-made" adaptations of the tasks limit benefits that could have been gained from this apparent abundance. Many cognitive functions remain unexplored in mouse models of TRS21. Non-declarative memory was not investigated in either of the two studies on the development of transgenic mice carrying the entire HSA21. Neither procedural nor non-associative categories of non-declarative memories were investigated in Ms1Ts65 or Ts1Cje (Sérégaza et al., 2006, Table 2).

A recent review of the literature on the correlation between the extra-chromosomal fragment of MMU16 and cognitive phenotypes (Sérégaza et al., 2006) was conducted from two perspectives. The first was to establish whether mouse models matched features observed in persons with TRS21. The second was to analyze the pathways between behavioral and neuronal differences by reanalyzing these correlations using the terminology of Milner, Squire, & Kandel (1998). The well-known classification of these authors using two types of memory, i.e., declarative vs non-declarative, is based on distinct brain systems with non-declarative memory subdivided into three categories (procedural, priming and associative).

How do Mouse Models of TRS21 Match the Cognitive Features Reported in TRS21 Persons?

The prevalence of cardiac and immune disorders in TRS21 is variable, but the low IQ is a dominant feature. As stated above, not all skills are affected in all persons or to the same extent (Brown et al., 2003; Crnic & Pennington, 2000; Krinsky-McHale et al., 2002; Laws, 2002; Clark & Wilson, 2003; Pennington et al., 2003; Vicari, 2006). Persons with TRS21 can perform simple learning tasks in spite of their low IQ. We found 8 of 11 studies showing that segmental TRS21 mice succeeded in performing simple tasks (fear conditioning: percentage of freezing in identical context; Morris water maze: the time to reach visible platform; passive avoidance: latency to step through).

Persons with TRS21 have difficulties when spatial cues need to be used (Vicari, 2006). Recourse to spatial cues increased the number of failures for most TRS21 models tested in the Morris water maze or other mazes when the non-proximal cue version was used. Ts65Dn mice generally failed to perform the tasks with the hidden or virtual platform and radial maze. Partial aneuploid TslCJe mice, which carry only one extra fragment of the translocated Ts65Dn region, needed more time to reach the hidden platform than diploid mice (Sago et al., 1998), whereas the complementary fragment in Ms1Cje, the equivalent of the centromeric region of Ts65Dn, produced minor learning impairment; the size of the effect was considered as negligible by the authors themselves (Sago et al., 2000). The 152F7 strain, which carries a small fragment of the D21S17-ETS2 region, needed more time to reach the hidden platform than diploid mice (Chabert et al., 2004; Smith et al., 1997). The DYRK1A gene (Altafaj et al., 2001) which is located on the region covered by the 152F7 fragment, confirms the crucial role of the D21S17-ETS2 region in spatial learning. A recent report by Harris-Cerruti et al. (2004) observed an interaction between

APP and SOD1 (missing in Ts1Cje), although the size of the effect was small compared to the effect in 152F7.

Cognitive flexibility is lower in persons with TRS21 than in typically developing individuals. The reversing of the position of the platform in the Morris water maze or changing of the rules in the radial maze was more difficult for both Ts65Dn and Ts1Cje, but this was not the case for Ms1Ts65. Lower flexibility was also observed with 152F7, covering part of the *D21S17-ETS2* region (Chabert et al., 2004), and with *DYRK1A* transgenic mice (Altafaj et al., 2001). The *S100β* homolog is on MMU10 and the difficulty that appears in mice overexpressing this gene indicates (1) that the *D21S17-ETS2* region does not play an exclusive role in cognitive impairment and (2) urges us to investigate the MMU10 and 17 syntenic regions.

Long-term memory deficits in persons with TRS21 have been reported (see above). Diploid mice performed better than Ts65Dn aneuploid when re-tested several days later. Unfortunately there is no data for TsCje and Ms1Ts65. Winocur, Roder, & Lobaugh (2001) highlighted the role of the MMU10 syntenic region when obtaining similar results with *S100β* transgenic mice.

Attention impairment is a crucial component of mental deficiency and was frequently detected in persons with TRS21 (Brown et al., 2003; Krinsky-McHale et al., 2002; Laws, 2002). Intensive exploration of a new environment may reveal alterations to hippocampal function involved in recognizing new or familiar objects. Excessive exploration

of environments could be indicative of an attention deficit. Exploration in both an open-field and hole-board activity must be considered as multifactorial, even if the novelty seeking associated with attention deficits could contribute to the behavior patterns observed with these tasks.

In a wide variety of situations and for 7 of 9 measurements, Ts65DN explored more than diploids. *DYRK1A* transgenic mice explored less, even though the gene is on the MMU16 fragment carried in triplicate by Ts65Dn mice. The same result was found in the two papers studying *DYRK1A* (Altafaj et al., 2001; Fotaki et al., 2002).

Cognitive Processes in Mouse Models of TRS21

Performance scores for separate memory categories can also be measured in mice. Declarative memory is involved in (1) fear conditioning, with fewer freezing episodes caused when subjecting the mouse to changes in the environment; (2) in the radial maze, with non-repeated visits to a reinforced arm and (3) in the Morris water maze, reversal difficulties. Declarative memory cannot be divided into categories, but non-declarative memory is comprised of three categories: (1) procedural memory or formation of habits and acquisition of skills – reaching the platform under proximal cue conditions, (2) priming and (3) associative, with classical conditioning

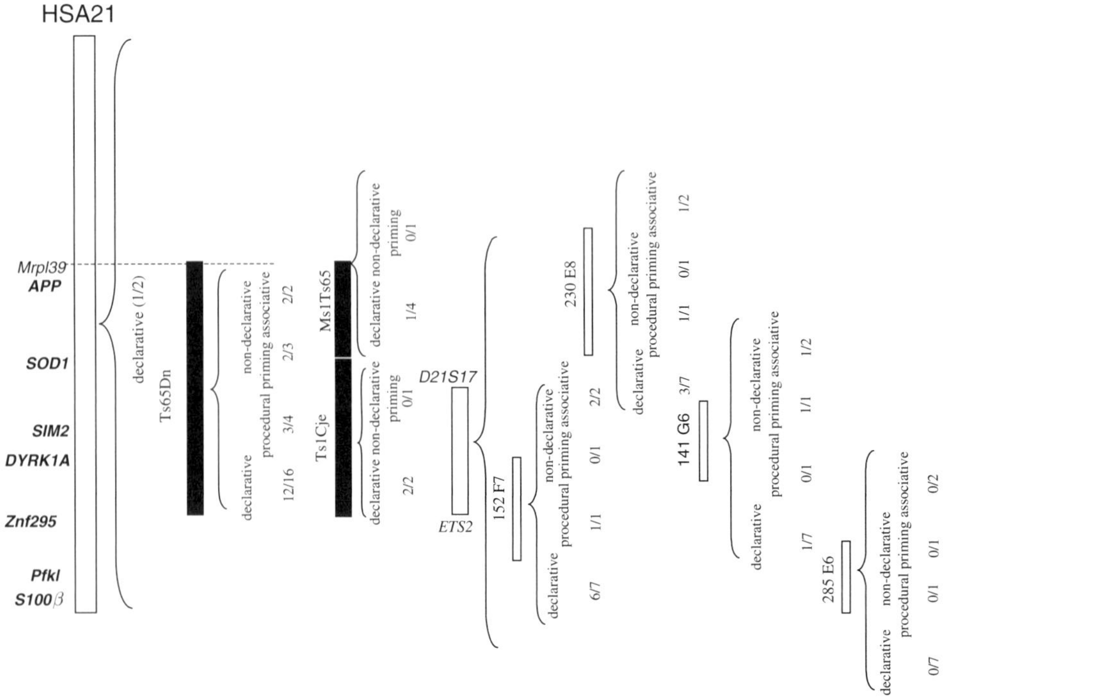

Fig. 12.3 Impairment in categories of memory in mouse models of TRS21. Number of studies indicating memory impairment/number of studies performed (adapted from Sérégaza et al., 2006)

measured as output. Correspondences to learning situations in mice have been illustrated by Sérégaza et al. (2006).

Declarative memory impairment occurred in Ts65Dn. Fragment 152F7 (encompassing *DYRK1A*) in the *D21S17-ETS2* sub-region and, to a lesser extent, 230E8, another fragment from this sub-region were associated with declarative memory impairment. 152F7 is also associated with declarative memory dysfunction. Poor performance on tasks requiring procedural non-declarative memory observed in Ts65Dn was also found with 152F7. Priming only affected Ts65Dn and 141G6. Associative learning impairment was seen in Ts65Dn and also in chromosomal fragments covering the *D21S17-ETS2* region; the impairment was greater in 152F7 than in 230E8 and 141G6. General impairment therefore characterizes the largest triplication (Ts65Dn), including the fragment 152F7 which is included in the *D21S17-ETS2* region and DYRK1A, one of the genes present in 152F7 (see Fig. 12.3).

Long-term potentiation is a correlate of synaptic plasticity and reduced long-term potentiation is a sign of hippocampal dysfunction. Ts65Dn and 152F7 have impaired declarative memory and exhibited low long-term potentiation. Abnormalities of the synaptic structure were observed in the fascia dentata and other regions of the diencephalon in Ts65Dn (Belichenko et al., 2004).

Conclusion

Correlations between the HAS21 genes and the neurocognitive measurements are relevant for different arguments.

It should be noted that there is high prevalence of TRS21 in western society and increasing prevalence related to the age of the mother, as pregnancy today occurs at a later age than it did only 20 years ago. With the identification of genes involved in mental deficiency, it should be possible to conduct proteomic investigations of cell functions of the gene product. Knowledge of these functions should pave the way for therapeutic approaches, not to cure the syndrome, but to improve some of the skills affected. Knowledge is therefore needed on the physiological pathways between genes and behavior. This will mean investigating gene expression in nerve tissue and will require a bank of trisomic brains.

In addition to the public health aspect, studies of cognitive disorders in TRS21 can be used as a guide for further basic research investigating genes involved in cognition. Initial findings show the need to move from the field of differential psychology, or from the psychology of skills, toward the analysis of the processes deciphered by the burgeoning cognitive sciences. The proteomic approach has also proven to be important. We have seen that some HSA21 genes are not "dosage sensitive". An extra copy of a gene can be present without inducing a triple dose of the primary product of the gene. Extra copies of HSA21 genes might have direct effect plus indirect effect by modifying the expression of non-HSA21 genes (Roubertoux & Carlier, 2007).

Chromosomal abnormalities relevant to behavior–genetic analysis extend beyond TRS21. Other syndromes, such as Williams–Beuren, include cognitive disorders. Some deletions that will be discovered by the fine sequencing of the human chromosomes could uncover the genetic bases for developmental or psychiatric disorders. A short hemizygous deletion of HSA22 located on 22q11 has been associated with schizophrenic disorders affecting young adults (Liu et al., 2002), and the triplication of individual genes may be responsible for certain pathologies, as was shown recently with Alzheimer's disease.

Acknowledgment The present study was funded by INSERM U 910, Génétique Médicale, Génomique Fonctionnelle, UMR 6146 Aix-Marseille Université CNRS, Laboratoire Psychologie cognitive, France, and Institut Universitaire de France to Michéle Carlier. We also wish to express our gratitude to the Board of the Fondation Jérôme Lejeune for both the financial support to Michéle Carlier and to Pierre Roubertoux, and fruitful scientific discussions.

References

Altafaj, X., Dierssen, M., Baamonde, C., Marti, E., Visa, J., Guimera, J., et al. (2001). Neurodevelopmental delay, motor abnormalities and cognitive deficits in transgenic mice overexpressing Dyrk1A (minibrain), a murine model of Down's syndrome. *Human Molecular Genetics, 10*, 1915–1923.

Antonarakis, S. E., Lyle, R., Dermitzakis, E. T., Reymond, A., & Deutsch, S. (2004). Chromosome 21 and Down syndrome: from genomics to pathophysiology. *Nature Review Genetics, 5*, 725–738.

Aula, P., Leisti, J., & von Koskull, H. (1973). Partial trisomy 21. *Clinical Genetics, 4*, 241–251.

Baxter, L. L., Moran, T. H., Richtsmeier, J. T., Troncoso, J., & Reeves, R. H. (2000). Discovery and genetic localization of down syndrome cerebellar phenotypes using the Ts65Dn mouse. *Human Molecular Genetics, 9*, 195–202.

Bayés, M., Magano, L. F., Rivera N., Flores R., & Pérez Jurado L. A. (2003). Mutational mechanisms of Williams-Beuren syndrome deletions. *American Journal of Human Genetics, 73*, 131–151.

Belichenko, P. V., Masliah, E., Kleschevnikov, A. M., Villar, A. J., Epstein, C. J., Salehi, A., et al. (2004). Synaptic structural abnormalities in the Ts65Dn mouse model of Down syndrome. *Journal of Comparative Neurology, 480*, 281–298.

Bellugi, U., Lichtenberger, L., Mills, D., Galaburda A., & Korenberg, J. R. (1999). Bridging cognition, the brain and molecular genetics: Evidence from Williams syndrome. *Trends in Neuroscience, 22*, 197–207.

Boddaert, N., Mochel, F., Meresse, I., Seidenwurm, D., Cachia, A., Brunelle, F., et al. (2005). Parieto-occipital grey matter abnormalities in children with Williams syndrome. *Neuroimage, 30*, 721–725.

Brown, J. H., Johnson, M. H., Paterson, S. J., Gilmore, R., Longhi, E., & Karmiloff-Smith, A. (2003). Spatial representation and attention in toddlers with Williams syndrome and Down syndrome. *Neuropsychology, 41*, 1037–1046.

Carlier, M., & Ayoun, C. (2007). *Déficiences intellectuelles et intégration sociale*. Wavre (Belgique): Mardaga.

Carlier, M., Stefanini, S., Deruelle, C., Volterra, V., Doyen, A.-L., Lamard, C., et al. (2006). Laterality in Persons with Intellectual Disability. I. – Do Patients with Trisomy 21 and Williams-Beuren syndrome differ from typically developing persons? *Behavior Genetics, 36*, 365–376.

Carr, J. (2005). Stability and change in cognitive ability over the life span: A comparison of population with and without Down syndrome.*Journal of Intellectual Disability Research, 49*, 915–928.

Chabert, C., Jamon, M., Cherfouh, A., Duquenne, V., Smith, D. J., Rubin, E., et al. (2004). Functional analysis of genes implicated in Down syndrome: 1. Cognitive abilities in mice transpolygenic for Down syndrome chromosomal region-1 (DCR-1). *Behavior Genetics, 34*, 559–569.

Chapman, R. S., & Hesketh, L. J. (2000). Behavioral phenotype of individuals with Down syndrome. *Mental Retardation and Developmental Disabilities Research Reviews, 6*, 84–95.

Clark, D., & Wilson, G.N. (2003). Behavioral assessment of children with Down syndrome using the Reiss psychopathology scale. *American Journal Medical Genetics, 118*, 210–216.

Crnic, L. S., & Pennington, B. F. (2000). Down syndrome: Neuropsychology and animal models. *Progress in Infancy Research, 1*, 69–111.

Davisson, M. T., Schmidt, C., & Akeson, E. C. (1990). Segmental trisomy of murine chromosome 16: A new model system for studying Down syndrome. *Progress in Clinical and Biological Research, 360*, 263–280.

Davisson, M. T., Schmidt, C., Reeves, R. H., Irving, N. G., Akeson, E. C., Harris, B. S., & Bronson, R. T. (1993). Segmental trisomy as a mouse model for Down syndrome. *Progress in Clinical and Biological Research, 384*, 117–133.

Delabar, J. M., Theophile, D., Rahmani, Z., Chettouh, Z., Blouin, J. L., Prieur, M., et al. (1993). Molecular mapping of twenty-four features of Down syndrome on chromosome 21. *European Journal of Human Genetics, 1*, 114–124.

Down, J. L. H. (1862). Observations on an Ethnic Classification of Idiots. *London Hospital Report, 3*, 259–262.

Down, J. L. H. (1867). Observations on an ethnic classification of idiots. *Journal of Mental Science, 13*, 121–123.

Doyle, T. F., Bellugi, U., Korenberg, J. R., & Graham, J. (2004). "Everybody in the world is my friend" hypersociability in young children with Williams syndrome. *American Journal of Medical Genetics, 124*, 263–273.

Esquirol, J. E. D. (1838). *Des maladies mentales considérées sous les rapports médical, hygiénique et médico-légal*. Paris J.-B. Baillères.

Fotaki, V., Dierssen, M., Alcantara, S., Martinez, S., Marti, E., Casas, C., et al. (2002). Dyrk1A haploinsufficiency affects viability and causes developmental delay and abnormal brain morphology in mice. *Molecular and Cellular Biology, 22*, 6636–6647.

Gérard-Desplanches, A., Deruelle, C., Stefanini, S., Ayoun, C., Volterra, V., Vicari, S., et al. (2006). Laterality in Persons with Intellectual Disability. II. Hand, foot, ear and eye laterality in persons with Trisomy 21 and Williams-Beuren syndrome. *Developmental Psychobiology, 4*, 482–497.

Gitton, Y., Dahmane, N., Baik, S., Ruiz i Altaba, A., Neidhardt, L., Scholze, M., et al. (2002). A gene expression map of human chromosome 21 orthologues in the mouse. *Nature, 420*, 586–590.

Gray, V., Karmiloff-Smith, A., Funnell, E., & Tassebehji, M. (2006). In-depth analysis of spatial cognition in Williams syndrome: A critical assessment of the role of the LIMK1 gene. *Neuropsychologia, 44*, 679–685.

Gropp, A., Kolbus, U., & Giers, D. (1975). Systematic approach to the study of trisomy in the mouse II. *Cytogenetics and Cell Genetics, 14*, 42–62.

Harris-Cerruti, C., Kamsler, A., Kaplan, B., Lamb, B., Segal, M., & Groner, Y. (2004). Functional and morphological alterations in compound transgenic mice overexpressing Cu/Zn superoxide dismutase and amyloid precursor protein. *European Journal of Neuroscience, 19*, 1174–1190.

Hassold, T., & Sherman, S. (2000). Down syndrome: Genetic recombination and the origin of the extra chromosome 21. *Clinical Genetics, 57*, 95–100.

Hattori, M., Fujiyama, A., Taylor, T. D., Watanabe, H., Yada, T., Park, H. S., et al. (2000). The DNA sequence of human chromosome 21. *Nature, 405*, 311–319.

Hodapp, R. M., Ewans, D. E., & Gray, F. L. (1999). Intellectuel development in children with Down syndrome. In J.-A. Rondal, J. Perera, & L. Nadel (Eds.), *Down Syndrome: A Review of Current Knowledge* (pp. 124–132). Whurr Publisher: London, U.K.

Jackson, J. F., North III, E. R., & Thomas, J. G. (1976). Clinical diagnosis of Down's syndrome. *Clinical Genetics, 9*, 483–487.

Kahlem, P. (2006). Gene dosage effect on chromosome 21 transcriptome in trisomy 21: Implication in Down's syndrome cognitive disorders. *Behavior Genetics, 36*, 416–428.

Kahlem, P., Sultan, M., Herwig, R., Steinfath, M., Balzereit, D., Eppens, B., et al. (2004). Transcript level alterations reflect gene dosage effects across multiple tissues in a mouse model of Down syndrome. *Genome Research, 14*, 1258–1267.

Korenberg, J. R., Chen, X. N., Schipper, R., Sun, Z., Gonsky, R., Gerwehr, S., et al. (1994). Down syndrome phenotypes: The consequences of chromosomal imbalance. *Proceedings of the National Academy of Sciences USA, 91*, 4997–5001.

Krinsky-McHale, S. J., Devenny, D. A., & Silverman, W. P. (2002). Changes in explicit memory associated with early dementia in adults with Down's syndrome. *Journal of Intellectual Disability Research, 46*, 198–208.

Laws, G. (2002). Working memory in children and adolescents with Down syndrome: evidence from a colour memory experiment. *Journal of Child Psychology and Psychiatry, 43*, 353–364.

Lejeune, J., Turpin, R., & Gautier, M. (1958/1959). Le mongolisme. Premier exemple d'aberration autosomique humaine. *Annales de génétique, Paris, 1*, 41–49.

Lejeune, J., Gauthier, M., & Turpin, R. (1959). Etude des chromosomes somatiques de neuf enfants mongoliens. *Comptes Rendus de l'Académie des Sciences Paris, 248*, 1721–1722.

Liu, H., Abecasis, G. R., Heath, S. C., Knowles, A., Demars, S., Chen,Y. J., et al. (2002). Genetic variation in the 22q11 locus and susceptibility to schizophrenia. *Proceedings of the National Academy of Sciences USA, 99*, 16859–16864.

Lyle, R., Gehrig, C., Neergaard-Henrichsen, C., Deutsch, S., & Antonarakis, S. E. (2004). Gene expression from the aneuploid chromosome in a trisomy mouse model of Down syndrome. *Genome Research, 14*, 1268–1274.

Mervis, C. B. (2003). Williams syndrome: 15 years of psychological research. *Develomental Neuropsychology, 23*, 1–12.

Meyer-Lindenberg, A., Kohn, P., Mervis, C. B., Kippenhan, J. S., Olsen, R. K., Morris, C. A., et al. (2004). Neural basis of genetically determined visuospatial construction deficit in Williams syndrome. *Neuron, 43*, 623–631.

Meyer-Lindenberg, A., Mervis, C. B., Sarpal, D., Koch, P., Steele, S., Kohn, P., et al. (2005). Functional, structural, and metabolic abnormalities of the hippocampal formation in Williams syndrome. *Journal of Clinical Investigation, 115*, 1888–1895.

Milner, B, Squire, L. R., & Kandel, E. R. (1998). Cognitive neuroscience and the study of memory. *Neuron, 20*, 445–468.

Morris C. A., & Mervis, C. B. (2000). Williams syndrome and related disorders. *Annual Review of Genomics and Human Genetics, 1*, 461–484.

Nadel, L. (1995). Neural and cognitive development in Down syndrome. In L. Nadel & D. Rosenthal (Eds.), *Down syndrome Living and learning in the community* (pp. 107–114). New-York: J. Wiley.

Nadel, L. (1999). Learning and memory in Down syndrome. In J. Rondal, J. Perera, & L. Nadel (Eds.), *Down syndrome A review of current knowledge* (pp.133–142). London: Whurr Publishers.

O'Doherty, A., Ruf, S., Mulligan, C., Hildreth, V., Errington, M. L., Cooke, S., et al. (2005). An aneuploid mouse strain carrying human chromosome 21 with Down syndrome phenotypes. *Science, 309*, 2033–2037.

Olson, L. E., Richtsmeier J. T., Leszl, J., & Reeves R. H. (2004). A chromosome 21 critical region does not cause specific Down syndrome phenotypes. *Science, 306*, 687–690.

Patterson, D., & Costa, A. C. (2005). Down syndrome and genetics – a case of linked histories. *Nature Review Genetics, 6*, 137–147.

Pennington, B. F., Moon, J., Edgin, J., Stedron, J., & Nadel, L. (2003). The neuropsychology of Down syndrome: Evidence for hippocampal dysfunction. *Child Development, 74*, 75–93.

Raz, N., Torres, I. J., Briggs, S. D., Spencer, W. D., Thornton, A. E., Loken, W. J., et al. (1995). Selective neuroanatomic abnormalities in Down's syndrome and their cognitive correlates: evidence from MRI morphometry. *Neurology, 45*, 356–366.

Reymond, A., Marigo, V., Yaylaoglu, M. B., Leoni, A., Ucla, C., Scamuffa, N., et al. (2002). Human chromosome 21 gene expression atlas in the mouse. *Nature, 420*, 582–586.

Roizen, N. J., & Patterson, D. (2003). Down's syndrome. *Lancet, 361*, 1281–1289.

Rondal, J. A. (1999) Language in Down syndrome: Current perspectives. In J. Rondal, J. Perera, & L. Nadel (Eds.), *Down syndrome A review of current knowledge* (pp. 143–149). London: Whurr Publishers.

Roubertoux, P. L., Bichler, Z., Pinoteau, W., Sérégaza, Z., Fortes, S., Jamon, M., et al. (2005). Functional analysis of genes implicated in Down syndrome: 2. Laterality and corpus callosum size in mice transpolygenic for Down syndrome chromosomal region-1 (DCR-1). *Behavior Genetics, 35*, 333–341.

Roubertoux, P. L., Bichler, Z., Pinoteau, W., Jamon, M., Sérégaza, Z., Smith, D. J., et al. (2006). Pre-weaning Sensorial and Motor Development in Mice Transpolygenic for the Critical Region of Trisomy 21. *Behavior Genetics, 36*, 377–386.

Roubertoux, M., & Carlier M. (2007). From DNA to the mind. *EMBO Reports*, 8, Science & Society, Special Issue, S7–S11.

Roubertoux, P. L., & Kerdelhué, B. (2006). Trisomy 21: From chromosomes to mental retardation. *Behavior Genetics, 36*, 434–345.

Sago, H., Carlson, E. J., Smith, D. J., Kilbridge, J., Rubin, E. M., Mobley, W. C., et al. (1998). Ts1Cje, a partial trisomy 16 mouse model for Down syndrome, exhibits learning and behavioral abnormalities. *Proceedings of the National Academy of Sciences USA, 95*, 6256–6261.

Sago, H., Carlson, E. J., Smith, D. J., Rubin, E. M., Crnic, L. S., Huang, T. T., et al. (2000). Genetic dissection of region associated with behavioral abnormalities in mouse models for Down syndrome. *Pediatric Research, 48*, 606–613.

Séguin, E. (1846). *Traitement moral, hygiène et éducation des idiots et autres enfants arriérés ou retardés dans leurs mouvements, agités de mouvements volontaires*. Paris: J.-B. Ballières.

Séguin, E. (1856). Origin of the treatment and training of idiots. *American Journal of Educ*ation, 2, 145–152.

Séguin, E. (1866) *Idiocy and its treatment by the physiological method*. New York: William Wood & Co.

Sérégaza, Z., Roubertoux, P. L., Jamon, M., & Soumireu-Mourat, B. (2006). Mouse Models of Cognitive Disorders in Trisomy 21: A Review. *Behavior Genetics, 36*, 387–404.

Shinohara, T., Tomizuka, K., Miyabara, S., Takehara, S., Kazuki, Y., Inoue, J., et al. (2001). Mice containing a human chromosome 21 model behavioral impairment and cardiac anomalies of Down's syndrome. *Human Molecular Genetics, 10*, 1163–1175.

Smith, D. J., Zhu, Y., Zhang, J., Cheng, J. F., & Rubin, E. M. (1995). Construction of a panel of transgenic mice containing a contiguous 2-Mb set of YAC/P1 clones from human chromosome 21q22.2. *Genomics*, 27, 425–434.

Smith, D. J., Stevens, M. E., Sudanagunta, S. P., Bronson, R. T., Makhinson, M., Watabe, A. M., et al. (1997). Functional screening of 2 Mb of human chromosome 21q22.2 in transgenic mice implicates minibrain in learning defects associated with Down syndrome. *Nature Gene*tics, *16*, 28–36.

Teipel, S. J., & Hampel, H. (2006). Neuroanatomy of Down syndrome in vivo: A model of preclinical Alzheimer's disease. *Behavior Gene*tics, 36, 405–415.

Thompson, P. M., Lee, A. D., Dutton, R. A., Geaga, J. A., Hayashi, K. M., Eckert, M. A., et al. (2005). Abnormal cortical complexity and thickness profiles mapped in Williams syndrome. *Journal of Neuroscience, 25*, 4146–4158.

Tijo, H., & Levan, A. (1956). The chromosomes of man. *Hereditas, 42*, 1–6.

Tomaiuolo, F., Di Paola, M., Caravale, B., Vicari, S., Petrides, M., & Caltagirone, C. (2002). Morphology and morphometry of the corpus callosum in Williams Syndrome: A magnetic resonance imaging analysis. *NeuroReport, 13*, 1–5

Ugazio, A. G, Maccario, R., Notarangelo, L. D., & Burgio, G. R. (1990). Immunology of Down syndrome: A review. *American Journal of Medical Genetics*, (Suppl.) 7, 204–212.

Vicari, S. (2006). Motor development and neuropsychological patterns in persons with Down syndrome.*Behavior Genetics, 36*, 355–364.

Vicari, S., Belluci, S., & Carlesimo G. A. (2005). Visual and spatial long-term memory: Differential pattern of impairments in Williams and Down syndromes. *Developmental Medicine & Child Neurol*ogy, *47*, 305–311.

Watanabe, H., Fujiyama, A., Hattori, M., Taylor, T. D., Toyoda, A., Kuroki, Y., et al. (2004). DNA sequence and comparative analysis of chimpanzee chromosome 22, *Nature, 429*, 382–388.

Winocur, G., Roder, J., & Lobaugh, N. (2001). Learning and memory in *S100-beta* transgenic mice: an analysis of impaired and preserved function. *Neurobiology of Learning and Memory, 75*, 230–243.

Wishart, J. G. (1993). The development of learning difficulties in children with Down's syndrome. *Journal of Intellectual Disability Research, 37*, 389–403.

Yu, Y., & Bradley, A. (2001). Engineering chromosomal rearrangements in mice. *Nature Review Genetics, 10*, 780–790.

Chapter 13

Evolution of Complex Acoustic Signals in *Drosophila* Species

Anneli Hoikkala and Dominique Mazzi

Introduction

Males of most *Drosophila* species produce complex acoustic cues, so-called courtship songs, while pursuing a female. In most of the over 100 species studied so far (see the list of these species in Hoikkala, 2005), such cues are produced by wing vibration. Other mechanisms of song production include abdomen purring (Hoy, Hoikkala, & Kaneshiro, 1988) and rapid vibrations of the whole body (Ritchie & Gleason, 1995). The carrier frequency of songs produced through any of these actions ranges from 150 to 500 Hz. A hitherto unknown mechanism enables males of some Hawaiian species to generate songs of up to 15,000 Hz (Hoikkala, Hoy, & Kaneshiro, 1989).

The structure of the courtship songs of closely related species often reflects phylogenetic relationships of the species, implying that the songs may be diverging as a by-product of genetic divergence and/or drift. The course of song evolution may also be driven or constrained by various forms of selection pressure, causing or reinforcing song divergence at population and species level or, conversely, inhibiting differentiation. Darwin (1871) proposed that sexual selection promotes the origin and exaggeration of precopulatory male displays which enable females to choose the best among an array of potential mates. However, sexual selection need not be the exclusive, nor the predominant, selection pressure affecting song evolution. Courtship song is to some extent species-specific and can as such contribute to reproductive isolation and hence, ultimately, to speciation. The courtship songs of *Drosophila* species may thus be an example of signals evolved under the conflicting pressures of directional sexual selection toward the exaggeration of the targets of female preference on the one hand and stabilizing natural selection acting to maintain the consistency within species required for species recognition on the other hand.

The genetic architecture of song traits as well as the physical constraints in male song production and female song perception may set boundaries to song evolution. The evolution of species-specific songs may have involved novel genetic processes, such as changes in gene regulation and/or fixation of alleles with a large effect, or may have come about through the steady accumulation of minor genetic changes. Furthermore, the location of a gene, whether it is on the X-chromosome or an autosome, may affect the probability of fixation of new alleles in populations. Correlations between the song traits may restrict song evolution, especially if the traits are evolving under different selection pressures. Also, in species where the males produce songs by wing vibration, the contraction power of wing muscles may prevent song evolution beyond certain limits (Ewing, 1979).

The evolution of complex behavioral traits such as male signals and female preferences for these signals requires coordinated changes in both sexes, i.e., coevolution of trait and preference, a matter that has proven hard to validate empirically (Pomiankowski & Sheridan, 1994). Female song perception may constrain male song evolution, if the females are not capable of hearing the male song or of discerning specific patterns. In *Drosophila melanogaster*, hearing is supported by mechanosensory neurons transducing sound-induced vibrations of the antennae (Göpfert & Robert, 2003). Antennal hearing organs mediate the detection of conspecific songs in females, the aristal tips of the antennae being moderately tuned to frequencies around 425 Hz (Göpfert & Robert, 2002). The high-frequency songs of some Hawaiian *Drosophila* species would be inaudible to such 'ears', thus requiring alternative means of sensory transduction. When producing high-frequency songs, males usually stroke the abdomen of their partner with their forelegs, thereby possibly favoring the perception of the sounds as vibrational signals (Hoikkala et al., 1989).

Courtship songs are not the only cues that *Drosophila* flies produce during the mating rituals. In some species they are an essential part of the courtship, while in other species the

A. Hoikkala (✉)
Department of Biological and Environmental Science, University of Jyväskylä, Finland
e-mail: anneli.a.hoikkala@jyu.fi

Y.-K. Kim (ed.), *Handbook of Behavior Genetics*,
DOI 10.1007/978-0-387-76727-7_13, © Springer Science+Business Media, LLC 2009

courtship relies more heavily on visual or olfactory cues (see Markow & O'Grady, 2005). During the course of evolution, less-important signals may have been lost from the courtship repertoire. For example, in *D. subobscura*, males do not produce any courtship song (Ewing & Bennet-Clark, 1968), but rather rely exclusively on exceptionally complex visual displays (Brown, 1965).

Current Topics

Phylogenetic Patterns in Song Evolution

Studies of the phylogenetic patterns of song evolution in different species groups, or preferably entire clades, are essential for the understanding of the past history and the evolution of different song types. In *Drosophila* species, the songs consist of trains of uni- or polycyclic sound pulses, arranged to form a variety of song patterns (Fig. 13.1). In addition to the pulse song, the males of several species produce a song consisting of modified sine waves (a sine song; Cowling & Burnet, 1981). The males of some species produce several types of songs during courtship. The homology of different song types and/or song characters in different species is often difficult to trace back. In closely related interfertile species, the song homologies can be studied by analyzing the songs of interspecific hybrids (e.g., Hoikkala & Lumme, 1987). Also, analyses of courtship interactions in the presence and absence of auditory cues (e.g., Liimatainen, Hoikkala, Aspi, & Welbergen, 1992), studies of the effects of heterospecific songs in mating (e.g., Tomaru, Doi, Higuchi, & Oguma, 2000) and playback experiments (e.g., Ritchie, Halsey, & Gleason, 1999) may help to trace song homologies, though it should be reminded that the song function can change during the speciation processes.

Comparative studies of the auditory cues of closely related species are essential for unraveling sequential evolution of different song types. Mapping the songs onto a phylogeny can reveal which states of the songs are ancestral and which are derived. Largely because of difficulties in establishing the homology of song characters in different species, this approach has been only rarely applied to *Drosophila* groups. Ewing & Miyan (1986) reconstructed the evolution of song types in the *D. repleta* group by assessing song variation among 22 species of the group on a phylogeny based on chromosomal inversions. They suggested the archetypical song of this group to be composed of two distinct components produced either at the early ('A song', with short pulse trains) or at the late ('B song', with longer pulse trains) stages of courtship. During speciation processes, some species lost the A song, others the B song and in many species the B song

D. melanogaster

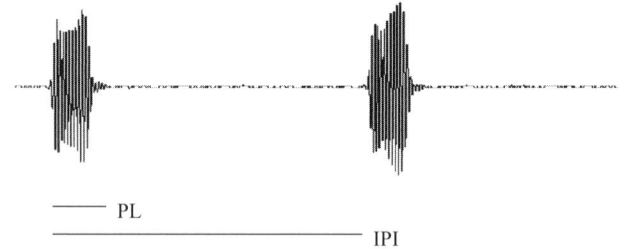

D. littoralis

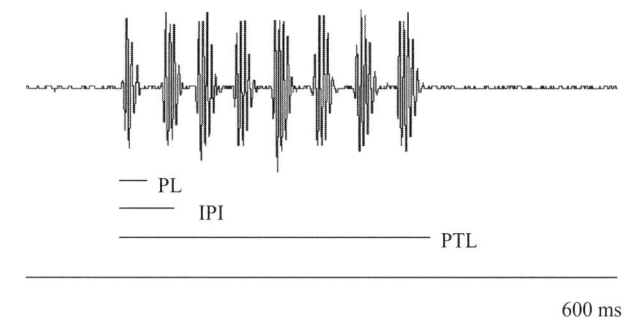

D. montana

600 ms

Fig. 13.1 Oscillograms of the male courtship songs of *D. melanogaster*, *D. littoralis* and *D. montana*. PL = length of the sound pulse, IPI = time from the beginning of one pulse to the beginning of the next one and PTL = length of a pulse train

became less regular and more complex. Later on, Alonso-Pimentel, Spangler, & Heed (1995) and Etges (2002) confirmed that song evolution in the *D. repleta* group shows a complex pattern of diversification, character loss and reverse evolution. The courtships of the *D. willistoni* group species differ from each other by the presence or absence of different song types such as rasps, pulses and trembles and by the quantitative traits of the songs (Ritchie & Gleason, 1995). Gleason & Ritchie (1998) found no phylogenetically informative pattern when imposing different song types on the molecular phylogeny of the group or a correlation between the quantitative song characters and the genetic distances between the species. The *funebris* group, where the males of most species produce single cycle sound pulses near the minimum interpulse interval of 10 ms, shows a different evolutionary trend. Ewing (1979) suggests that in this group, song evolution has undergone a bottleneck with regard to the song pattern, leading to arrangement of the minimum interval pulses into bursts with species-specific interburst intervals.

In the *D. melanogaster, D. obscura* and *D. virilis* groups, where the inheritance of song characters has been studied most extensively, male courtship songs have clearly retained some phylogenetic information. In the *melanogaster* group the males of most species produce a simple mono- or polycyclic pulse song and/or a sine song during courtship (e.g., Cowling & Burnet, 1981). In the species of the *bipectinata* complex of the *ananassae* subgroup of this group the males may produce different kinds of songs at the beginning and at the end of the courtship (Crossley, 1986), while in some species of the *montium* subgroup, the males produce songs mainly or solely during copulation (Hoikkala & Crossley, 2000; Tomaru & Oguma, 1994). In the *obscura* group, the males of some species, such as *D. persimilis* and *D. pseudoobscura*, produce two kinds of songs consisting of polycyclic sound pulses, while males of the other species produce only one song type consisting of single or polycyclic sound pulses, or no song at all (Ewing & Bennet-Clark, 1968). In the *virilis* group, the songs of the species belonging to the *virilis* subgroup consist of dense pulse trains without pauses between successive pulses, whereas the songs of the species belonging to more rapidly evolving lineages of the *montana* subgroup have distinct species-specific songs with shorter or longer intervals between consecutive pulses (Hoikkala & Lumme, 1987).

The best example of a species group with drastic changes in song production mechanisms is the Hawaiian picture-winged species group, including several subgroups and nearly 100 species. In the *planitibia* subgroup, the males of most species produce simple pulse songs, rhythmic phrase songs and/or sound bursts by vibrating their wings (Hoikkala, Kaneshiro, & Hoy, 1994). In one intermediate species, *D. neoperkinsi*, males produce all three song types, while the males of the remaining species produce one song, two songs or no song at all. Males of the most derived species of the subgroup (e.g., *D. silvestris* and *D. heteroneura*) produce only one wing song, namely sound bursts, and a novel song produced by abdomen purring (Hoy et al., 1988). In the species of the more derived *pilimana, vesciseta* and *grimshawi* subgroups, the males may vibrate their wings slowly at the same courtship stage as the *planitibia* subgroup males, but they fail to produce any audible sound. Thereafter, the males may, however, fold their wings back and make small 'scissoring' movements generating loud high-frequency clicking sounds of up to 15,000 Hz through a still unidentified mechanism (Hoikkala et al., 1989).

Selective Forces Affecting Song Evolution

Male courtship songs, as well as other cues emitted during the mating rituals, may play a role in sexual selection exercised by females on conspecific males and/or in species recognition. As different song characters may be involved in both processes, their variation may carry signs of different selection pressures.

Sexual Selection Within the Species

While the predicted pattern of song changes can be compared to a phylogeny, the selective pressures affecting song traits can best be studied at the population level. Notwithstanding the adequacy of *D. melanogaster* as a model organism for genetic studies, it is not the best suited species for investigating sexual selection based on song traits, as its song is quite conserved and female preferences are not straightforwardly quantified. Species of the *virilis* group prevail in this respect, thanks to large intra- and interspecific genetic and phenotypic variability in male song traits and the ease with which female preference functions for synthetic song stimuli can be assessed experimentally.

Four *virilis* group species (*D. montana, D. ezoana, D. littoralis* and *D. lummei*) occurring sympatrically in northern Finland share a short mating period in spring. Soon after the snow has melted, the flies gather on food patches of rotting plant material, where courtship and mating take place. This form of 'lekking' as well as a comparatively large body size of the flies enable the studies of reproductive ecology and mating behavior in the wild (Aspi, Lumme, Hoikkala, & Heikkinen, 1993). Courting males of these sympatric species produce species-specific songs, which effectively prevent interspecific matings (Liimatainen & Hoikkala, 1998). Occasional crowding on food patches potentially also allows females to choose their mating partner among numerous competing males.

Mechanisms of intraspecific sexual selection are most thoroughly studied in *D. montana*, a species belonging to the *montana* subgroup of the *virilis* group. In this species, the courtship song is an obligate prerequisite for mating (Liimatainen et al., 1992), to which receptive females respond with a wing-spreading acceptance gesture (Vuoristo, Isoherranen, & Hoikkala, 1996). Aspi & Hoikkala (1995) found that in the field, the songs of the males caught in copula had shorter and denser sound pulses (i.e., higher carrier frequency) than the songs of the average males of the same population. Playback experiments in the laboratory verified that song frequency is a target of sexual selection through female choice (Ritchie, Townhill, & Hoikkala, 1998). The preference function for male song frequency, though stabilizing in shape, is effectively directional, as it peaks at values in the upper range of the distribution of the male trait (Ritchie, Saarikettu, Livingstone, & Hoikkala, 2001).

Pulse traits of the song (including the carrier frequency) are very sensitive to environmental effects and may convey information about the condition of a courting male (Hoikkala & Isoherranen, 1997). Condition dependency is typical for traits evolving under sexual selection (e.g., Rowe & Houle, 1996), potentially providing females with a reliable criterion in their choice of partners. Indeed, *D. montana* females derive an indirect benefit from choosing males that sing at higher frequency, as egg-to-adult survival is positively associated with the carrier frequency of the sire's song (Hoikkala, Aspi, & Suvanto, 1998).

Coevolution of Male Songs and Female Song Preferences

In spite of the wealth of available evidence on male ornaments and displays, and on female preferences for them, empirical investigations of the mechanisms and the genetic basis of coevolution are scarce (Butlin & Ritchie, 1989). Inspired by Fisher's runaway process, which requires the linked inheritance of genes for both the preferred trait and the preference in the offspring of attractive males and choosy females, genetic covariances between trait and preference have sometimes been demonstrated (Bakker, 1993; Wilkinson & Reillo, 1994, but see also Gilburn & Day, 1994). Genetic covariance may arise through an intrinsic genetic association between genes caused by pleiotropy or tight genetic linkage, or, provided that there is sufficient genetic variation for trait and preference, through assortative mating and gametic phase linkage disequilibrium.

Studies of coevolution are of importance, as the variance and covariance between male displays and female preferences for them determine the form and strength of sexual selection and can thus ultimately affect population divergence and speciation. Traits such as the male courtship song and the female preference for certain characteristics of this song in various *Drosophila* species offer a valuable opportunity to study the evolution of traits requiring coordinated changes in both sexes. Acoustic signals are relatively easy to quantify, and the relationship between parameters of the male song and the probability of female acceptance provides the shape of the female preference (Ritchie, 1996). Accordingly, the extent to which sexual selection acts on specific song traits in different populations can be determined by examining the shape of the female preference function and the degree of its overlap with the song trait distribution. Comparisons of male and female traits in different species and strains with established phylogenies can also supply information as to which of the two traits is driving the coevolution. Therefore, acoustic communication systems may prove well suited to study the factors affecting display evolution as well as to evaluate the dynamics of coevolution between male displays and female preferences.

The male trait and the female preference may not always evolve in concert. The concept of 'sensory bias' assumes that an arbitrary female preference predates the evolution of a corresponding male trait, and males evolve in response so as to match the signals females are predisposed to perceive as attractive. A delay in the males' exploitation of the preexisting female preference may explain why traits and preferences are sometimes found to be out of step (Basolo, 1990). Even more radically with respect to the evolutionary dynamics, the relatively recent 'chase-away hypothesis' (reviewed in Pizzari & Snook, 2003) states that sexual conflict promotes antagonistic – rather than mutualistic – coevolution, whereby manipulative reproductive strategies in one sex are counteracted by the evolution of resistance to such strategies in the other sex.

In a recent study on the variation of male songs and female preference functions and their covariance in a natural population of *D. montana*, Ritchie, Saarikettu, & Hoikkala (2005) failed to find the expected genetic covariances between male traits and female preferences. In spite of significant within- and among-families variation in most components of mate choice, all females preferred the songs with the highest carrier frequency. The females varied only in their responsiveness (readiness to mate), a fact that may have counteracted the build-up of a positive genetic covariance between trait and preference. Furthermore, Klappert, Mazzi, Hoikkala, & Ritchie (2007) found considerable variation in male song frequency and female preference for this trait within and between geographically and phylogenetically distinct *D. montana* populations, but no positive covariation of the traits among populations.

Song as a Species Recognition Signal

Henderson & Lambert (1982) have argued that species recognition signals should be relatively uniform over the species' distribution range, if they are under stabilizing selection. Nonetheless, recent studies on song variation among populations of different *Drosophila* species have shown that the songs may vary over a species' distribution range. Species-specific mating signals can significantly contribute to sexual isolation, provided that differences between species are matched by narrowly tuned differences in female preferences (Ritchie et al., 1999). For example, the songs of the cosmopolitan (non-African) strains of *D. melanogaster* exhibit smaller variation in interpulse interval (IPI) than the songs of *D. simulans* (Kawanishi & Watanabe, 1980), and, accordingly, *D. melanogaster* females discriminate against heterospecific songs more accurately than *D. simulans* females (Ritchie et al., 1999). African strains of *D. melanogaster*, which are genetically different and partially reproductively isolated from the cosmopolitan strains, have unusually short interpulse intervals in their song, which, however, seem to have little direct influence on sexual isolation (Colegrave, Hollocher, Hinton, & Ritchie, 2000).

There is abundant indirect evidence of the role of male songs in causing and maintaining sexual isolation between *Drosophila* species, while compelling direct evidence is difficult to obtain. Kyriacou & Hall (1982) have shown that *D. melanogaster* and *D. simulans* females readily mate when hearing a simulated song with the correct combination of IPI and song rhythm, and Williams, Blouin, & Noor (2001) have shown male song IPI to be strongly associated with mating success in hybrids between *D. pseudoobscura* and *D. persimilis*. Yamada, Matsuda, & Oguma (2002) have suggested the male song to play a crucial role in sexual isolation also between *D. ananassae* and *D. pallidosa*, the genes with a major effect on the control of the song characters associated with sexual isolation being located on the second chromosome. Etges, Over, Cardoso de Oliveira, & Ritchie (2006) have found significant variation in song characters among geographically isolated *D. mojavensis* populations, in agreement with the documented differences in sexual isolation between populations. In the *virilis* group, sympatric species have unique songs, suggesting that the songs may function in species recognition (Hoikkala, Lakovaara, & Romppainen, 1982). The observation that interspecific courtship episodes occurring in the wild break down when the male produces an inadequate song (Liimatainen & Hoikkala, 1998) corroborates the role of male song as a species recognition signal. Saarikettu, Liimatainen, & Hoikkala (2005) demonstrated that sexual isolation between two sympatric species, *D. montana* and *D. lummei*, can be overcome by playing back to *D. montana* females synthetic courtship song with their own species-specific interpulse interval (IPI) while they are being courted by a mute (wingless) *D. lummei* male.

Genetic Analysis of Song Variation

Studies of the genetic basis of song trait variation at the intra- and interspecific level reveal the evolutionary mechanisms behind song divergence. Quantitative genetic analyses involve the production of a F1 or F2 generation or of backcross hybrids between the parental species/strains. Such crosses (and their modifications) allow to determine the magnitude of additive and dominance variation, the direction of dominance, the existence and extent of interactions (epistasis) between nonallelic genes and the approximate number and location of genes affecting the examined traits. The genetic architecture of song traits may also prompt conjectures about the selection pressures driving their evolution (Wagner & Bürger, 1985), provided that the current lev-

els of interaction, dominance, etc., correspond to those prevailing while the songs were diverging. The candidate gene approach can give further information on song evolution. Here, information on genes found to affect song traits in a model species, usually *D. melanogaster*, is used to determine whether the same genes play a role in determining the species specificity of songs. Candidate genes can be used as markers in gene localization (QTL) studies. The analysis of the DNA sequences of the coding regions of these genes provides the chance to look for mutations with a potential effect on song variation and to trace signs of selection at the level of DNA (e.g., Tajima, 1989). The function of the candidate genes in different species can be traced, for example, in gene transfer experiments or in microarray-based gene expression studies.

Quantitative Genetic Analyses

Variation in quantitative traits provides the raw material on which selection acts to produce the observed phenotypic diversity. Knowledge of the genetic architecture of quantitative traits is required to address issues concerning for example the maintenance of genetic variation for adaptive traits or the consistency of the loci at which variation occurs across populations and species. Recent advances in the development of neutral polymorphic markers and of sophisticated statistical methods have boosted the increasingly detailed characterization of the genetics underlying quantitative traits.

Quantitative genetic analysis, for instance, successfully uncovered the mechanisms of inheritance of song traits in the species of the *D. virilis* group. In such studies, the magnitude and direction of dominance are of special interest, as they tell about the direction of past evolution (Wagner & Bürger, 1985). In selection theory, traits under directional selection are presumed to retain relatively little additive genetic variation and much (mainly directional) dominance variation. Hence, the direction of dominance is expected to match the direction of trait evolution. In the *virilis* subgroup species, only the number of pulses in a pulse train (PN) and the length of the pulse trains (PTL) vary to a great extent among conspecific strains or species (Hoikkala & Lumme, 1987). A diallelic cross between all species and subspecies of the subgroup (*D. virilis, D. lummei, D. americana americana, D. a. texana* and *D. novamexicana*) showed that both song traits are autosomally inherited and polygenic, the direction of dominance suggesting that song evolution has proceeded toward denser sound pulses and longer pulse trains (Hoikkala & Lumme, 1987). A biometrical analysis of song differences between two *D. virilis* strains representing extreme phenotypes in PN and PTL revealed significant additive and dominance components, the direction of dominance being toward shorter and denser sound pulses (Huttunen & Aspi, 2003). The genetic basis of intraspecific song variation has also been studied in *D. montana*, another representative of the *virilis* group. Diallelic crosses between four inbred *D. montana* strains revealed additive genetic variation in four out of five song traits, the carrier frequency of the song showing unidirectional dominance toward higher values (Suvanto, Liimatainen, Tregenza, & Hoikkala, 2000). Aspi (2000) found evidence of additivity for the pulse train characters (pulse number, pulse train length and interpulse interval) and additivity, dominance and epistasis for the pulse characters (pulse length, the number of sound cycles in a pulse and carrier frequency) using partly different sets of inbred strains.

Interspecific crosses are harder to create than intraspecific ones. They are essential for tracking the evolution of species-specific signals, but they may not always give reliable information on the initial steps of song evolution. In the *virilis* group species, one or more X-chromosomal song genes largely determine the interpulse interval (IPI) and pulse length (PL) of species-specific songs. Hoikkala & Lumme (1987) have suggested that an X-chromosomal change increasing variation in these two traits has occurred during the separation of the *virilis* and *montana* subgroups. The gene (or a group of genes) with a major effect on song differences between *D. virilis* and *D. littoralis* has been localized on the proximal end of the X-chromosome. Unfortunately, a large inversion in *D. littoralis* prevents more precise localization by conventional crossing methods (Päällysaho, Huttunen, & Hoikkala, 2001). The genetic basis of interspecific song differences has also been studied by crossing *D. virilis* (a *virilis* subgroup species) and *D. flavomontana* (a *montana* subgroup species) females with males of four different species with species-specific song. This study confirmed a central role of the X-chromosomal factors in determining species-specific song characters and in controlling the action of some autosomal genes (Päällysaho, Aspi, Liimatainen, & Hoikkala, 2003).

Ewing (1969) has argued that X-linked genes are more likely to be involved in behavioral differences between closely related species than autosomes, thanks to immediate exposure of favorable partial recessives to selection, and thus faster evolution. Sex linkage may, however, be easily overestimated because of differential expression of the genes in males (i.e., under hemizygous conditions) and because of the lack of recombination in QTL studies. As Ritchie & Phillips (1998) have stated, evidence of a disproportionate role of the sex chromosomes requires the difference between reciprocal hybrid males to be greater than expected if the genes were distributed randomly throughout the karyotype, taking into account doubled expression of the X-linked genes. The genetic background of male song traits does not always involve single genes of large effect and/or a disproportionate contribution of the sex chromosome or epistasis (e.g., Pugh & Ritchie, 1996).

Quantitative Trait Loci (QTL) Mapping

A comprehensive review of QTL mapping theory and methodology as well as an exhaustive list of references is given in part I of the present volume and in Mackay (2001). QTL mapping is essentially a search for a phenotype–genotype association in individuals of a mapping population. It requires two parental strains fixed for different alleles at loci affecting the variation in the trait of interest (the power of the tests is greatly reduced if the parental strains are not fixed) and a polymorphic molecular marker linkage map. A mapping population of backcross, F2, recombinant inbred lines or other segregating generation derived from the two divergent parental strains is generated, and the phenotype and multi-locus genotype of each individual are compared.

It is worth emphasizing that the number of QTL found to affect variation in a quantitative trait is always a minimum estimate of the actual number, as the power of QTL studies crucially depends on sample size and on whether a map has sufficient resolution to single out linked QTL. Such techniques are therefore biased to detect genes with large and unconditional effects on the phenotype. It should also be reminded that locations and effects of QTL are peculiar to the conditions of any one experiment and may not replicate across sexes and environments, as genotype-by-sex, genotype-by-environment and epistatic interactions are common and may challenge the generality of reported findings.

Drosophila, especially *D. melanogaster*, has featured prominently since the first pioneering mapping attempts, thanks to a long history of genetic manipulability in the laboratory. However, while several studies mapped QTL affecting morphological (e.g., bristle numbers) or life history traits (e.g., longevity), comparatively few studies have dealt with the genetic architecture of courtship song parameters. Gleason & Ritchie's (2004) attempt to uncover the genetic basis of differences in song interpulse interval (IPI) between *D. simulans* and *D. sechellia* revealed six autosomal QTL explaining 40.7% of the observed variation. Yet, none of these QTL coincided with those formerly found to affect variation for the same trait in recombinant inbred lines of *D. melanogaster* (Gleason, Nuzhdin, & Ritchie, 2002), suggesting that different genes contribute to intra- and interspecific variation. Williams et al. (2001) identified significant QTL involved in differences in the carrier frequency (FRE) and interpulse interval (IPI) between *D. pseudobscura* and *D. persimilis* on the second as well as on the X-chromosome, which may well cover large fractions of the entire genome. Huttunen, Aspi, Hoikkala, & Schlötterer (2004) mapped strains of *D. virilis* representing extreme phenotypes for the length of the pulse train (PTL) and the number of pulses in a train (PN) and found two significant QTL on the third chromosome, each of them explaining slightly more than 10% of the variation in PN and even less than 10% in PTL.

QTL regions identified in mapping studies can encompass from one up to numerous loci affecting the trait in question. Depending on the distribution of the markers on the linkage map and on recombination frequencies, localized QTL can span large physical distances of the genome. In *D. melanogaster*, the availability of deficiency stocks has proven useful for refining the map location of recessive mutations of large impact (quantitative deficiency mapping, quantitative complementation mapping), and thus to nominate candidate genes for further studies. Mutagenesis (e.g., single P-element transposon insertions) and linkage disequilibrium mapping can help in the quest for the crucial genes within QTL regions. The Berkeley Drosophila Gene Disruption Project is currently creating mutations with transposable P-elements in *D. melanogaster*, thus preparing the grounds for a more thorough search for mutations, also in genes affecting male courtship song. The candidate gene approach can be applied also to little-known *Drosophila* species, thereby taking advantage of the fact that gene function is commonly conserved across distant lineages, despite millions of years of evolutionary divergence.

Candidate Gene Approach

A number of mutations are known to distort song production in *D. melanogaster* (see Gleason, 2005 for a complete listing), and the target genes of these mutations are obvious candidate genes affecting natural song variation in *Drosophila* species. Such mutations include the *dissonance* (*diss*) mutation of the *no-on-transient-A* (*nonA*) locus, the *cacophony* (*cac*) mutation of the *DmcalA* locus and several mutant alleles of the *period* and *fruitless* loci. None of those mutations exclusively affect song production (for instance, *per* mutants have altered circadian rhythm in addition to abnormal length of the cycle of mean IPI; Kyriacou & Hall, 1980), implying that exclusive song genes may be rare or not exist at all. Peixoto & Hall (1998) suggest that the genes involved in ion-channel function (as many of the genes mentioned above are) might be a source of genetic variation in courtship songs. Such pleiotropy may generate trade-offs and/or cause mating signal divergence as a side effect of local adaptation.

Candidate song genes have been used as markers in QTL analyses to check for their correspondence with the song QTL regions. In general, the overlap is poor. This either points to a real phenomenon indicative of the contribution of different genes to song variation within and among species or is an artifact due to insufficient resolution. In a QTL study on *D. melanogaster* recombinant inbred strains, only one candidate song gene (*tipE*) coincided with a significant song QTL (Gleason et al., 2002), while in a study on hybrids between *D. simulans* and *D. sechellia*, three candidate genes (*mle, cro* and *fru*) co-localized with significant QTL (Gleason &

Ritchie, 2004). Neither of these studies detected QTLs on the X-chromosome. Among the *D. melanogaster* song genes, *nonA* and *DmcalA* are, however, located on the same end of the X-chromosome as the major X-linked genes causing species differences in male song in the *virilis* group species (Päällysaho et al., 2001).

Transformation of *D. melanogaster* with orthologous candidate song genes (genes that have evolved from the same ancestral gene) offers an alternative approach to investigate the role of these genes in controlling the species specificity of songs. The courtship song of *D. melanogaster* males transformed with the *per* gene of *D. simulans* (Wheeler et al., 1991), or with the *nonA* gene of *D. virilis* (Campesan, Dubrova, Hall, & Kyriacou, 2001), assumed characteristics typical of the species from which the gene originated, suggesting that these genes carry species-specific information on song structure. By examining the song pattern of males expressing a chimeric gene (the regulatory region and the coding region of the gene derived from different species), or by causing gene knock-out by means of homologous recombination (Rong & Golic, 2001), it could be possible to determine which part (down to which amino acid) of the gene product is crucial for species-specific patterns of the songs.

An important goal of studies of DNA sequence variation is to identify genes that have been targets of natural or sexual selection, and thus contribute to individual differences in fitness within populations. Regions of the genome that have been affected by selection may show an excess of functionally important molecular changes, which may be detected through both intra- and interspecific DNA sequence comparisons (McDonald & Kreitman, 1991; Tajima, 1989; Yang, 2000). Examining the DNA sequence of candidate song genes can reveal whether the genes have been subject to directional selection, such as female choice, in their past evolution, and whether polymorphisms detected in gene sequences are associated with song variation. Huttunen, Campesan, & Hoikkala (2002) screened the DNA sequence of the coding region of the *nonA* gene for fixed polymorphisms and/or signs of selection explaining species specificity of the songs of the *virilis* group species. They sequenced the coding region of this gene in *D. littoralis* and compared it to that of *D. virilis* and *D. melanogaster*, but failed to find any nonsynonymous mutations which could have given rise to song variation or any signs of selection on gene sequences. It cannot be excluded, however, that species-specific information lies in the regulatory region of the gene.

Directions to the Future Research in the Field

Tracing the past history and evolution of male courtship songs in the genus *Drosophila* requires information on the selection pressures affecting song evolution and on the genetic changes underlying song divergence in various *Drosophila* species. Mapping the songs and their characters on phylogenies comprising different species groups would help to trace the ancestral and derived status of different song types (Markow & O'Grady, 2005), and mapping the female preferences for the songs in the same phylogenies would give further information on coevolution of male songs and female song preferences. This kind of studies requires, however, song homologies to be clarified.

One of the central issues in the study of song evolution in *Drosophila* species is concerned with the genetic changes underlying the diversification of song characters and the extent to which these changes have involved modifications in regulatory and structural genes. QTL studies are especially suited to answer these questions, as they give insight into the relative relevance of major (possibly regulatory) and minor genes affecting the traits as well as on gene interactions. This technique is, however, quite limited in its power to identify the actual genes that are responsible for quantitative variation. Studies on the function of candidate genes falling within a QTL interval increase the possibilities to find the relevant song genes. Identification of the genes responsible for intra- and/or interspecific variation in male song would open up several avenues of research, such as the comparison of gene structure and function across the species of the genus *Drosophila* (see Matthews, Kaufman, & Gelbart, 2005), the evaluation of extant variation in wild populations and the estimation of the relationship between variation in song characters and DNA polymorphism.

Progressing from the study of the location and structure of behavioral genes to the study of their function may not be easy, but it may prove rewarding, as shown by Riedl, Neal, Robichon, Westwood, & Sokolowski (2005) for *Drosophila* foraging behavior. Species and population divergence at the transcriptional level can be studied using various molecular genetic techniques such as microarrays, Northern blot, SAGE, real-time PCR and tissue in situ hybridization (e.g., Thomas & Klaper, 2004). The primary application of microarrays has so far been in the identification of the sets of genes that respond in an extreme manner to some treatment or that differentiate two or more tissues. Quantitative genomics using microarray analyses also enables an estimate of fundamental parameters of gene expression variation, including the additivity, dominance and heritability of transcription (Gibson, 2002). While microarray analyses offers new prospects in evolutionary research, its use in behavioral studies is still rudimentary and usually needs complementation with studies on the function of numerous genes with altered transcription (e.g., Mackay, Heinsohn, Lyman, Moehring, Morgan, & Rollman, 2005).

The synthesis of ecological, evolutionary and developmental biology has yielded new insights in evolutionary research, emphasizing the interplay of ecological factors, molecular variation and gene expression. While the new genomics technologies offer precious tools for use in evolutionary research, studies on male song evolution should still be made in the quantitative genetics framework, including examinations of wild populations whenever possible.

References

Alonso-Pimentel, H., Spangler, H. G., & Heed, W. B. (1995). Courtship sounds and behaviour of the two saguaro-breeding *Drosophila* and their relatives. *Animal Behaviour, 50,* 1031–1039.

Aspi, J. (2000). Inbreeding and outbreeding depression in male courtship song characters in *Drosophila montana. Heredity, 84,* 273–282.

Aspi, J., & Hoikkala, A. (1995). Male mating success and survival in the field with respect to size and courtship song characters in *Drosophila littoralis* and *D. montana* (Diptera, Drosophilidae). *Journal of Insect Behavior, 8,* 67–87.

Aspi, J., Lumme, J., Hoikkala, A., & Heikkinen, E. (1993). Reproductive ecology of the boreal riparian guild of *Drosophila. Ecography, 16,* 65–72.

Bakker, T. C. M. (1993). Positive genetic correlation between female preference and preferred male ornament in sticklebacks. *Nature, 363,* 255–257.

Basolo, A. L. (1990). Female preference predates the evolution of the sword in swordtail fish. *Science, 250,* 808–810.

Butlin, R. K., & Ritchie, M. G. (1989). Genetic coupling in mate recognition systems: What is the evidence? *Biological Journal of the Linnean Society, 37,* 237–246.

Brown, R. G. B. (1965). Courtship behaviour in the *Drosophila obscura* group. Part II. Comparative studies. *Behaviour, 25,* 281–323.

Campesan, S., Dubrova, Y., Hall, J. C., & Kyriacou, C. P. (2001). The *nonA* gene in *Drosophila* conveys species-specific behavioral characteristics. *Genetics, 158,* 1535–1543.

Colegrave, N., Hollocher, H., Hinton, K., & Ritchie, M. G. (2000). The courtship song of African *Drosophila melanogaster. Journal of Evolutionary Biology, 13,* 143–150.

Cowling, D. E., & Burnet, B. (1981). Courtship songs and genetic control of their acoustic characteristics in sibling species of the *Drosophila melanogaster* subgroup. *Animal Behaviour, 29,* 924–935.

Crossley, S. A. (1986). Courtship sounds and behaviour in the four species of the *Drosophila bipectinata* complex. *Animal Behaviour, 34,* 1146–1159.

Darwin, C. (1871). *The descent of man, and selection in relation to sex.* London: J. Murray.

Etges, W. J. (2002). Divergence in mate choice systems: does evolution play by rules? *Genetica, 116,* 151–166.

Etges, W. J., Over, K. F., Cardoso de Oliveira, C., & Ritchie, M. G. (2006). Inheritance of courtship song variation among geographically isolated populations of *Drosophila mojavensis. Animal Behaviour, 71,* 1205–1214.

Ewing, A. W. (1969). The genetic basis of sound production in *Drosophila pseudoobscura* and *D. persimilis. Animal Behaviour, 17,* 555–560.

Ewing, A. W. (1979). Complex courtship songs in the *Drosophila funebris* species group: escape from an evolutionary bottleneck. *Animal Behaviour, 27,* 343–349.

Ewing, A. W, & Bennet-Clark, H. C. (1968). The courtship songs of *Drosophila. Behaviour, 31,* 288–301.

Ewing, A. W., & Miyan, J. A. (1986). Sexual selection, sexual isolation and the evolution of song in the *Drosophila repleta* group of species. *Animal Behaviour, 34*, 421–429.

Gibson, G. (2002). Microarrays in ecology and evolution: a preview. *Molecular Ecology, 11*, 17–24.

Gilburn, A. S., & Day, T. H. (1994). Evolution of female choice in seaweed flies: fisherian and good genes mechanisms operate in different populations. *Proceedings of the Royal Society of London. Series B, 255*, 159–165.

Gleason, J. M. (2005). Mutations and natural genetic variation in the courtship song of *Drosophila*. *Behavior Genetics, 35*, 265–277.

Gleason, J. M., & Ritchie, M. G. (1998). Evolution of courtship song and reproductive isolation in the *Drosophila willistoni* species complex: Do sexual signals diverge the most quickly? *Evolution, 52*, 1493–1500.

Gleason, J. M., & Ritchie, M. G. (2004). Do quantitative trait loci (QTL) for a courtship song difference between *Drosophila simulans* and *D. sechellia* coincide with candidate genes and intraspecific QTLs? *Genetics, 166*, 1303–1311.

Gleason, J. M., Nuzhdin, S. V., & Ritchie, M. G. (2002). Quantitative trait loci affecting a courtship signal in *Drosophila melanogaster*. *Heredity, 89*, 1–6.

Göpfert, M. C., & Robert, D. (2002). The mechanical basis of *Drosophila* audition. *Journal of Experimental Biology, 205*, 1199–1208.

Göpfert, M. C., & Robert, D. (2003). Motion generation by *Drosophila* mechanosensory neurons. *Proceedings of the National Academy of Sciences USA, 100*, 5514–5519.

Henderson, N. R., & Lambert, D. M. (1982). No significant deviation from random mating of worldwide populations of *Drosophila melanogaster*. *Nature, 300*, 437–440.

Hoikkala, A. (2005). Inheritance of male sound characteristics in *Drosophila* species. In S. Drosopoulos & M. F. Claridge (Eds.), *Insect sounds and communication: Physiology, behaviour, ecology and evolution* (pp. 167–177). Boca Raton: CRC Taylor & Francis.

Hoikkala, A., & Crossley, S. (2000). Copulatory courtship in *Drosophila*: Behaviour and songs of *D. birchii* and *D. serrata*. *Journal of Insect Behavior, 13*, 71–86.

Hoikkala, A., & Isoherranen, E. (1997). Variation and repeatability of courtship song characters among wild-caught and laboratory-reared *Drosophila montana* and *D. littoralis* males (Diptera: Drosophilidae). *Journal of Insect Behavior, 10*, 193–202.

Hoikkala, A., & Lumme, J. (1987). The genetic basis of evolution of male courtship sounds in the *Drosophila virilis* group. *Evolution, 4*, 827–845.

Hoikkala, A., Lakovaara, S., & Romppainen, E. (1982). Mating behavior and male courtship sounds in the *Drosophila virilis* group. In S. Lakovaara (Ed.), *Advances in genetics, development, and evolution of Drosophila* (pp. 407–421). New York: Plenum Publishing Corporation.

Hoikkala, A., Hoy, R., & Kaneshiro, K. (1989). High-frequency clicks of Hawaiian picture-winged *Drosophila* species. *Animal Behaviour, 37*, 927–934.

Hoikkala, A., Kaneshiro, K., & Hoy, R. (1994). Courtship songs of the picture-winged *Drosophila planitibia* subgroup species. *Animal Behaviour, 47*, 1363–1374.

Hoikkala, A., Aspi, J., & Suvanto, L. (1998). Male courtship song frequency as an indicator of male genetic quality in an insect species, *Drosophila montana*. *Proceedings of the Royal Society of London. Series B, 265*, 503–508.

Hoy, R., Hoikkala, A., & Kaneshiro, K. (1988). Hawaiian courtship songs: Evolutionary innovation in communication signals of *Drosophila*. *Science, 240*, 217–219.

Huttunen, S., & Aspi, J. (2003). Complex inheritance of male courtship song characters in *Drosophila virilis*. *Behavior Genetics, 33*, 17–24.

Huttunen, S., Campesan, S., & Hoikkala, A. (2002). Nucleotide variation at the *no-on-transient A* gene in *Drosophila littoralis*. *Heredity, 88*, 39–45.

Huttunen, S., Aspi, J., Hoikkala, A., & Schlötterer, C. (2004). QTL analysis of variation in male courtship song characters in *Drosophila virilis*. *Heredity, 92*, 263–269.

Kawanishi, M., & Watanabe, T. K. (1980). Genetic variation of courtship song of *Drosophila melanogaster* and *D. simulans*. *Japanese Journal of Genetics, 55*, 235–240.

Klappert, K., Mazzi, D., Hoikkala, A. & Ritchie, M. (2007). Male courtship song and female preference variation between phylogeographically distinct populations of *Drosophila montana*. *Evolution, 61*, 1481–1488.

Kyriacou, C. P., & Hall, J. C. (1980). Circadian rhythm mutations in *Drosophila melanogaster* affect short-term fluctuations in the male's courtship song. *Proceedings of the National Academy of Sciences USA, 77*, 6729–6733.

Kyriacou, C. P., & Hall, J. C. (1982). The function of courtship song rhythms in *Drosophila*. *Animal Behaviour, 30*, 794–801.

Liimatainen, J. O., & Hoikkala, A. (1998). Interactions of the males and females of three sympatric *Drosophila virilis* group species, *D. montana*, *D. littoralis*, and *D. lummei* (Diptera: Drosophilidae) in intra- and interspecific courtships in the wild and in the laboratory. *Journal of Insect Behavior, 11*, 399–417.

Liimatainen, J. O., Hoikkala, A., Aspi, J., & Welbergen, P. (1992). Courtship in *Drosophila montana*: The effects of male auditory signals on the behaviour of the flies. *Animal Behaviour, 43*, 35–48.

Mackay, T. F. C. (2001). Quantitative trait loci in *Drosophila*. *Nature Reviews Genetics, 2*, 11–20.

Mackay, T. F. C., Heinsohn, S. L., Lyman, R. F., Moehring, A. J., Morgan, T. J., & Rollman, S. M. (2005). Genetics and genomics of *Drosophila* mating behavior. *Proceedings of the National Academy of Sciences USA, 102*, 6622–6629.

Markow, T. A., & O'Grady, P. M. (2005). Evolutionary genetics of reproductive behavior in *Drosophila*: connecting the dots. *Annual Review of Genetics, 39*, 263–291.

Matthews, K. A., Kaufman, T. C., & Gelbart, W. M. (2005). Research resources for *Drosophila*: the expanding universe. *Nature Reviews Genetics, 6*, 179–193.

McDonald, J. H., & Kreitman, M. (1991). Adaptive protein evolution at the *Adh* locus in *Drosophila*. *Nature, 351*, 652–654.

Peixoto, A. A., & Hall, J. C. (1998). Analysis of temperature-sensitive mutants reveals new genes involved in the courtship song of *Drosophila*. *Genetics, 148*, 827–838.

Pizzari, T., & Snook, R. (2003). Perspective: Sexual conflict and sexual selection: Chasing away paradigm shifts. *Evolution, 57*, 1223–1236.

Pomiankowski, A., & Sheridan, L. (1994). Linked sexiness and choosiness. *Trends in Ecology and Evolution, 9*, 242–244.

Pugh, A. R. G. & Ritchie, M. G. (1996). Polygenic control of a mating signal in *Drosophila*. *Heredity, 77*, 378–382.

Päällysaho, S., Huttunen, S., & Hoikkala, A. (2001). Identification of X chromosomal restriction fragment length polymorphism markers and their use in a gene localization study in *Drosophila virilis* and *D. littoralis*. *Genome, 44*, 242–248.

Päällysaho, S., Aspi, J., Liimatainen, J. O., & Hoikkala, A. (2003). Role of X chromosomal song genes in the evolution of species-specific courtship songs in *Drosophila virilis* group species. *Behavior Genetics, 33*, 25–32.

Riedl, C. A. L., Neal, S. J., Robichon, A., Westwood, J. T., & Sokolowski, M. B. (2005). *Drosophila* soluble guanylyl cyclase mutants exhibit increased foraging locomotion: behavioral and genomic investigations. *Behavior Genetics, 35*, 231–244.

Ritchie, M. G. (1996). The shape of female mating preferences. *Proceedings of the National Academy of Sciences USA, 93,* 14628–14631.

Ritchie, M. G., & Gleason, J. M. (1995). Rapid evolution of courtship song pattern in *Drosophila willistoni* sibling species. *Journal of Evolutionary Biology, 8,* 463–479.

Ritchie, M. G., & Phillips, S. D. F. (1998). The genetics of sexual isolation. In D. J. Howard & S. H. Berlocher (Eds.), *Endless forms and speciation* (pp. 291–308). Oxford: Oxford University Press.

Ritchie, M. G., Townhill, R. M., & Hoikkala, A. (1998). Female preference for fly song: playback experiments confirm the targets of sexual selection. *Animal Behaviour, 56,* 713–717.

Ritchie, M. G., Halsey, E. J., & Gleason, J. M. (1999). *Drosophila* song as a species-specific mating signal and the behavioural importance of Kyriacou & Hall cycles in *D. melanogaster* song. *Animal Behaviour, 58,* 649–657.

Ritchie, M. G., Saarikettu, M., Livingstone, S., & Hoikkala, A. (2001). Characterization of female preference functions for *Drosophila montana* courtship song and a test of the temperature coupling hypothesis. *Evolution, 55,* 721–727.

Ritchie, M. G., Saarikettu, M., & Hoikkala, A. (2005). Variation, but no covariance, in female preference functions and male song in a natural population of *Drosophila montana*. *Animal Behaviour, 70,* 849–854.

Rong, Y. S., & Golic, K. G. (2001). A targeted gene knockout in *Drosophila*. *Genetics, 157,* 1307–1312.

Rowe, L., & Houle, D. (1996). The lek paradox and the capture of genetic variance by condition-dependent traits. *Proceedings of the Royal Society of London. Series B, 263,* 1415–1421.

Saarikettu, M., Liimatainen, J. O., & Hoikkala, A. (2005). The role of male courtship song in species recognition in *Drosophila montana*. *Behavior Genetics, 35,* 257–263.

Suvanto, L., Liimatainen, J. O., Tregenza, T., & Hoikkala, A. (2000). Courtship signals and mate choice of the flies of inbred*Drosophila montana* strains. *Journal of Evolutionary Biology, 13,* 583–592.

Tajima, F. (1989). Statistical method for testing the neutral mutation hypothesis by DNA polymorphism. *Genetics, 123,* 585–595.

Thomas, M. A., & Klaper, R. (2004). Genomics for the ecological toolbox. *Trends in Ecology and Evolution, 19,* 439–445.

Tomaru, M., & Oguma, Y. (1994). Differences in courtship song in the species of the *Drosophila auraria* complex. *Animal Behaviour, 47,* 133–140.

Tomaru, M., Doi, M., Higuchi, H., & Oguma, Y. (2000). Courtship song recognition in the *Drosophila melanogaster* complex: heterospecific song make females receptive in *D. melanogaster*, but not in *D. sechellia*. *Evolution, 54,* 1286–1294.

Vuoristo, M., Isoherranen, E., & Hoikkala, A. (1996). Female wing spreading as acceptance signal in the *Drosophila virilis* group of species. *Journal of Insect Behavior, 9,* 505–516.

Wagner, G. P., & Bürger, R. (1985). On the evolution of dominance modifiers II: a nonequilibrium approach to the evolution of genetic systems. *Journal of theoretical Biology, 113,* 475–500.

Wheeler, D. A., Kyriacou, C. P., Greenacre, M. L., Yu, Q., Rutila, J. E., Rosbash, M., et al. (1991). Molecular transfer of a species-specific behavior from *Drosophila simulans* to *Drosophila melanogaster*. *Science, 251,* 1082–1085.

Wilkinson, G. S., & Reillo, P. R. (1994). Female choice response to artificial selection on an exaggerated male trait in a stalk-eyed fly. *Proceedings of the Royal Society of London. Series B, 255,* 1–6.

Williams, M. A., Blouin, A. G., & Noor, M. A. F. (2001). Courtship songs of *Drosophila pseudodobscura* and *D. persimilis*. II. Genetics of species differences. *Heredity, 86,* 68–77.

Yamada, H., Matsuda, M., & Oguma, Y. (2002). Genetics of sexual isolation based on courtship song between two sympatric species: *Drosophila ananassae* and *D. pallidosa*. *Genetica, 116,* 225–237.

Yang, Z. (2000). Phylogenetic analysis by maximum likelihood (PAML). Version 3.0. London: University College London.

Chapter 14

Drosophila Model of Alzheimer's Amyloidosis

Koichi Iijima, Kanae Iijima-Ando, and Yi Zhong

Introduction

The establishment of animal models of human diseases is crucial for understanding disease pathogenesis as well as for the discovery and evaluation of potential therapies. In the last decades, numerous models of human neurodegenerative diseases have been established in various laboratory organisms. The mouse has been the most popular choice for this purpose and has been used to test many hypotheses derived from in vitro and in vivo observations. In addition to these hypothesis-driven approaches, many groundbreaking discoveries in various biological contexts have been made by non-biased and systematic genome-wide genetic screenings using simple organisms.

Yeast (*Saccharomyces cerevisiae*), nematodes (*Caenorhabditis elegans*) and fruit flies (*Drosophila melanogaster*) are today the most sophisticated genetic models available for the study of almost any field of biology. During the past several years, scientists have attempted to model human neurodegenerative diseases in these simpler organisms (Bilen & Bonini, 2005; Link, 2005; Outeiro & Giorgini, 2006). These models recapitulate many important aspects of the human diseases in question and have greatly enhanced our understanding of the pathogenic mechanisms behind them.

In this chapter, we will focus our discussion solely on Alzheimer's disease (AD) and our fly model of this disease, since a number of informative reviews about modeling neurodegenerative diseases in the fly have already been published elsewhere (Bilen & Bonini, 2005; Marsh & Thompson, 2004; Sang & Jackson, 2005; Shulman, Shulman, Weiner, & Feany, 2003).

Current Research or Issues

Alzheimer's Disease (AD)

AD is a fatal disorder. In its later stage, global cognitive functions are disturbed and the associated motor disability leads patients to become bedridden (Cummings, 2003). Short-term memory impairment is detected in the early stage of the disease along with other psychiatric problems such as sleep disorders and increased agitation, which distinguishes AD from other neurodegenerative conditions such as Parkinson's disease, tauopathies or Huntington's disease (Cummings, 2003; Selkoe, 2002).

At the pathological level, extensive neuronal loss and two characteristic hallmarks, senile plaques (SPs) and neurofibrillary tangles (NFTs), are observed in AD brains (Cummings, 2003) (Fig. 14.1). SPs are extracellularly deposited protein aggregates that are called amyloid deposits. Biochemical studies have revealed that their major components are 40- or 42-amino acid peptides termed amyloid-β 40 or 42 (Aβ40 or Aβ42) (Glenner & Wong, 1984). Although a low number of SPs can be detected in normal-aged brains, this lesion is relatively specific to AD. NFTs, which are intracellular protein inclusions composed of hyperphosphorylated microtubule-associated tau protein (Lee, Balin, Otvos, & Trojanowski, 1991), are also observed in many other neurological diseases, suggesting its more general role in neurodegenerative process.

What is the cause of AD? The majority of AD cases are sporadic with disease onset after 65 years. Less than 10% of all AD cases are inherited in an autosomal dominant manner (Bertram & Tanzi, 2005). Genetic analysis of these families identified causative mutations in the *amyloid precursor protein, APP* (chromosome 21), *Presenilin 1* (chromosome 14) and *Presenilin 2* (chromosome 1) genes (Tanzi & Bertram, 2005). Recently, APP duplication was found in a French pedigree, which is similar to the situation of Down syndrome patients who carry three copies of APP and

K. Iijima (✉)
Laboratory of Neurodegenerative Diseases and Gene Discovery, Farber Institute for Neurosciences, Department of Biochemistry and Molecular Biology, Thomas Jefferson University, Philadelphia, PA19107, USA
e-mail: Koichi.Iijima@jefferson.edu

Y.-K. Kim (ed.), *Handbook of Behavior Genetics*,
DOI 10.1007/978-0-387-76727-7_14, © Springer Science+Business Media, LLC 2009

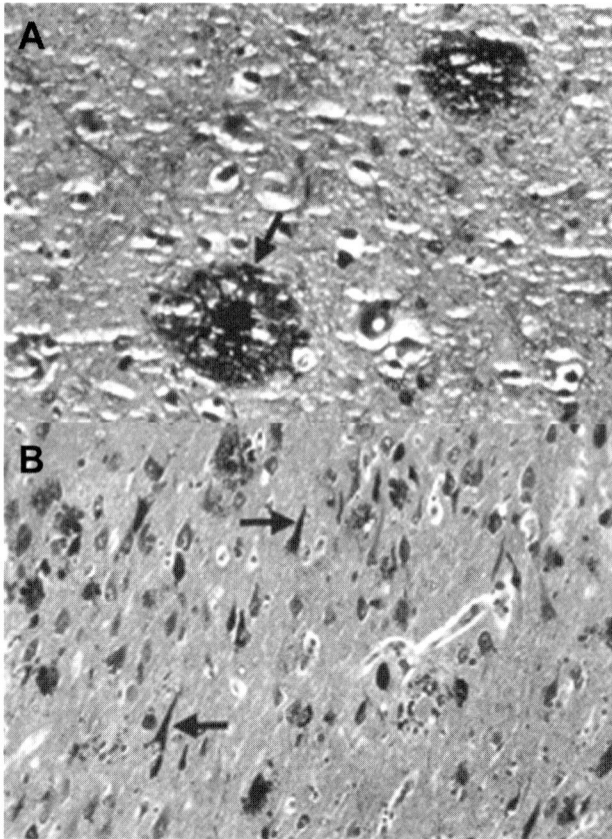

Fig. 14.1 Neuritic (senile) plaque and neurofibrillary tangle. (**A** and **B**) Neuritic (senile) plaque (**A**, magnification 435 × 220; Bielschowsky stain) and neurofibrillary tangle (**B**, magnification 435 × 220; Bielschowsky stain). (Modified and adapted from The Neuropsychiatry of Alzheimer's Disease and Related Dementias, Martin Dunitz Ltd, Taylor and Francis Group plc., with permission)

develop AD-like pathology (Rovelet-Lecrux et al., 2006). These familial AD (FAD) patients show much earlier onset than sporadic cases with excessive accumulation of Aβ peptide in their brains (Price, Tanzi, Borchelt, & Sisodia, 1998).

Aβ peptide is derived from a type 1 transmembrane protein termed amyloid precursor protein (APP) by sequential cleavage of β- and γ-secretases. Heterogeneity of γ-secretase cleavage gives rise to a series of Aβ peptides that include Aβ40 and Aβ42. The APP mutations found in FAD patients are clustered around or within the Aβ sequence of APP and increase Aβ42 production or Aβ aggregation. The aforementioned presenilin 1 or 2 is a critical component of the γ-secretase complex; mutations in these genes also cause overproduction of Aβ42 (see recent review, Gandy, 2005). Therefore, it is generally accepted that accumulation of Aβ42 in the brains causes the development of AD (Hardy & Selkoe, 2002). However, it remains elusive how Aβ42 peptide can initiate complex AD pathogenesis.

Advantages of Drosophila *as a Model for the Study of Neurodegenerative Diseases*

Analysis of the *Drosophila* genome revealed that approximately 70% of human disease-related genes have homolog genes in *Drosophila* (Fortini, Skupski, Boguski, & Hariharan, 2000; Reiter, Potocki, Chien, Gribskov, & Bier, 2001), suggesting that molecular mechanisms underlying many human diseases may be conserved in fly.

One of the most important tools that *Drosophila* provides is the ability to carry out large-scale genetic screens for mutations that affect a given process. Since the fly does not possess as many redundant gene families as mammals in general, studying the consequence of a single gene disruption is much easier in the fly. A number of forward genetic screens using chemical or transposon mutagenesis, in which genes were randomly disrupted, have been carried out to unravel the molecular mechanisms underlying various biological processes ranging from development to behavior. These efforts have resulted in the availability of an amazing collection of loss of function mutations and transgenic lines that are freely available in the fly community. Moreover, the completion of whole genome analysis in *Drosophila* (Adams et al., 2000) enables us to perform systematic RNAi-based knockout of all fly genes (Ueda, 2001; Dietzl et al., 2007).

Overexpression of exogenous proteins in the fly can be achieved by transgenic strategies. The Gal4/UAS system is widely used to express the protein of interest through spatial control (Brand & Perrimon, 1993) (Fig. 14.2). Briefly, one group of fly lines carry the transgenes in which the genes of interest are placed under the control of a yeast upstream activating sequence (UAS). Another group of flies have insertions of the yeast Gal4 transgene under the control of endogenous or exogenous promoters that express Gal4 protein in specific cell types or structures. By crossing these UAS- and Gal4-transgenic lines, it is possible to readily express exogenous protein in a specific location. Gal4 lines with a variety of expression patterns are available. Furthermore, collections of random insertion lines of P-element-containing UAS sequences have been established; these collections can be used to target the endogenous genes that are adjacent to P-elements for Gal4-induced gain of function screening (Rorth et al., 1998; Toba et al., 1999). Recently, use of the temperature-sensitive yeast Gal80 protein, which is a repressor of Gal4, has allowed us to control a transgene both temporally and spatially (McGuire, Le, Osborn, Matsumoto, & Davis, 2003).

Drosophila possesses a well-organized brain and provides a unique opportunity for studying complex behaviors such as learning and memory (Heisenberg, 2003; Waddell & Quinn, 2001). Although the fly brain has many fewer cells (around 10^5) than the human brain (approximately 10^{11}),

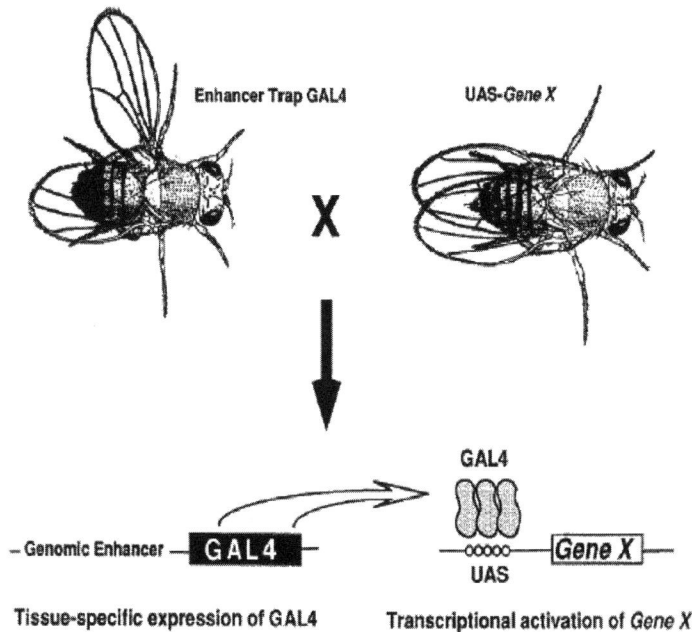

Fig. 14.2 Directed gene expression in *Drosophila*. To generate transgenic lines expressing GAL4 in various cell- and tissue-specific patterns, the GAL4 gene is inserted randomly into the genome and GAL4 expression is driven by a variety of genomic enhancers. A GAL4-dependent target gene can then be constructed by subcloning the sequence behind the GAL4-binding sites. The target gene is silent in the absence of GAL4. To activate the target gene in a cell- or tissue-specific pattern, flies carrying the target (*UAS-Gene X*) are crossed with flies that express GAL4 (enhancer trap GAL4). In the progeny of this cross, it is possible to activate *UAS-Gene X* in cells where GAL4 is expressed and to observe the effect of this directed misexpression on development. (Adapted from Brand, A. H., & Perrimon, N. (1993). *Development, 118*, 401–415, with permission)

and its structural organization is very different, the basic cell biology and neurotransmitter system are remarkably conserved. Several methods using different cues such as olfactory, visual, taste and courtship have been established to study memory in the fly (Heisenberg, 2003; Waddell & Quinn, 2001). Of these, classical Pavlovian olfactory conditioning is a particularly sensitive and reliable assay to measure associative learning and memory in a quantitative manner (Quinn, Harris, & Benzer, 1974; Tully & Quinn, 1985). In this conditioning assay, flies learn to avoid one of two odors because it is associated with an electric shock (Fig. 14.3). Thus, during the training phase, flies are exposed to one odor together with an electric shock after which the second odor is delivered without electric shock. Temporally distinct memories such as learning, short-term memory and middle-term memory can be tested at various time points after the training by allowing flies to choose between the two odors presented without electric shock (Tully & Quinn, 1985). The first genetic screening with chemical mutagenesis using this assay led to the discovery of four mutants, namely, dunce (dnc) (Dudai, Jan, Byers, Quinn, & Benzer, 1976), rutabaga (rut) (Livingstone, Sziber, & Quinn, 1984), amnesiac (amn) (Quinn, Sziber, & Booker, 1979) and radish (rsh) (Folkers, Drain, & Quinn, 1993). The gene products of dnc, rut and amn

all participate in the same biochemical pathway (cyclic adenosine 3',5'-monophosphate; cAMP). Repetitive training with interspersed rest intervals induces long-term memory (LTM) that can persist for days and the transcription factor cAMP-response-element-binding protein (CREB) is critical for LTM formation (Yin et al., 1994). The involvement of the cAMP pathway and CREB in learning and memory has been shown to be conserved across animal phyla (Yin & Tully, 1996; Guo, Tong, Hannan, Luo, & Zhong, 2000). Since then, many memory genes have been identified by transposon-induced mutant screening (Grotewiel, Beck, Wu, Zhu, & Davis, 1998) or DNA microarray analyses (Dubnau et al., 2003).

The short life span of flies (2–3 months) is suitable for studying late-onset disorders, including neurodegenerative diseases (Grotewiel, Martin, Bhandari, & Cook-Wiens, 2005). While the first attempt to model prion disease in *Drosophila* in 1995 was not successful (Raeber, Muramoto, Kornberg, & Prusiner, 1995; see also Deleault, Dolph, Feany, & Cook, 2003), a number of human neurodegenerative conditions, including polyglutamine diseases (Jackson et al., 1998; Warrick et al., 1998), Parkinson's disease (Feany & Bender, 2000), tauopathies (Jackson et al., 2002; Wittmann et al., 2001) and Alzheimer's disease (AD) (Crowther et al., 2005; Finelli, Kelkar, Song, Yang, &

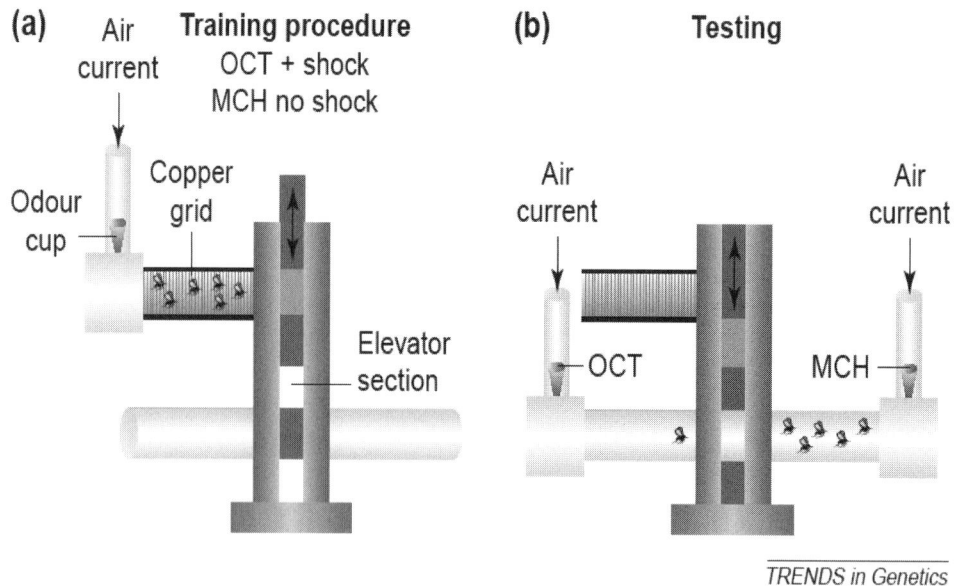

Fig. 14.3 Odor-aversion learning in *Drosophila* using electric-shock reinforcement. During training, flies experience an odor in conjunction with an electric shock. During subsequent testing, the flies avoid the shock-associated odor. A group of ~100 flies are trained in a chamber (**a**) with an inner surface covered with an electrifiable printed-circuit grid. Odors are delivered into the chamber by an air current. The flies are exposed to one odor (e.g., 3-octanol; OCT) while the walls of the chamber are electrified (CS+). They then experience another odor (e.g., 4-methylcyclohexanol; MCH) without shock (CS−). The flies are then tested for learning or memory performance by transporting them to a choice point between converging air currents suffused with the two odors (**b**). After 2 minutes the flies that have run into each odor arm are trapped, anesthetized and counted. A score is calculated as (number of flies avoiding CS+ minus number of flies avoiding CS−) divided by the total number of flies. A single learning index is the average from two groups of flies trained to avoid each of the two odors. Here 'learning' denotes a measure of performance within 2 minutes after training whereas 'memory' is a measure of the same performance at longer times after training. (Adapted from Waddell, S., & Quinn, W. G. (2001). *TRENDS in Genetics, 17*, 719–726, with permission)

Konsolaki, 2004; Greeve et al., 2004; Iijima et al., 2004), have been generated in flies that overexpress the causative proteins of these diseases. Extensive genetic and pharmacological screenings are underway in all of these models (Marsh & Thompson, 2006), and the treatment strategy discovered using polyglutamine disease models has now entered clinical trials (Steffan et al., 2001).

Drosophila Model of Alzheimer's Amyloidosis

Establishment of Transgenic Flies Expressing Human Aβ40 and Aβ42 Peptides

We were interested in studying the pathogenic effects of Aβ40 and Aβ42 in *Drosophila*. Our primary question was whether the fly could be a model to study the toxicity of Aβ peptides. Simultaneously, we wished to examine the toxicities of Aβ40 and Aβ42 in the brains separately, which had not yet been achieved in any animal model.

Drosophila does not produce Aβ peptide endogenously. A fly counterpart of APP termed dAPPL exists (Luo, Tully, & White, 1992) but the Aβ sequence is not conserved between fly and human. Moreover, the Aβ peptide cannot be produced from human APP in the fly, because *Drosphila* lacks β-secretase activity (Fossgreen et al., 1998). Therefore, to overproduce Aβ40 or Aβ42 in the secretory pathway, each Aβ peptide was directly fused with a secretion signal peptide at the N-terminus. Using Gal4/UAS transgene expression system, we confirmed that this artificial construct could produce intact Aβ40 or Aβ42 peptide in the fly brains (Iijima et al., 2004).

Formation of Amyloid Deposits in Aβ42 but Not in Aβ40 Fly Brains

Both Aβ40 and Aβ42 peptides accumulated during aging in the fly brains, but only Aβ42 formed clear amyloid deposits (Iijima et al., 2004: Fig. 14.4). The number and size of these deposits increased during aging. This result is consistent with the previous observations of amyloid deposits in AD or Down syndrome patient brains that revealed Aβ42 first accumulates into amyloid plaques in the brain parenchyma (Iwatsubo et al., 1994).

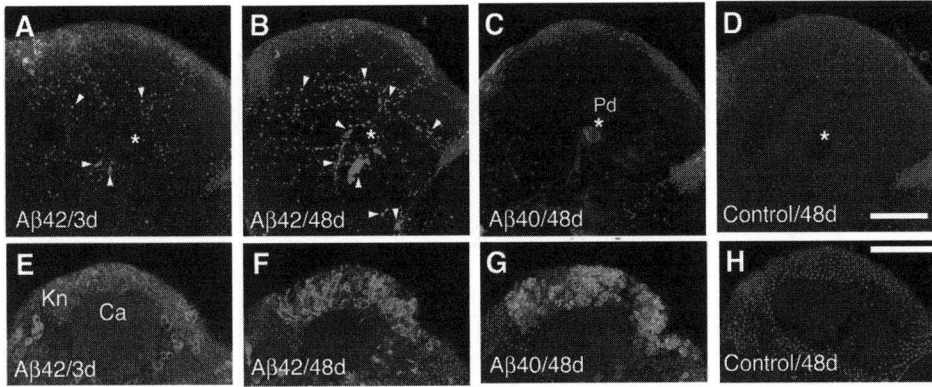

Fig. 14.4 Detection of Aβ deposits in the fly brain. (**A–H**) Whole-mount Aβ immunostaining in the neuropil region (**A–D**) and the Kenyon cell layer (**E–H**). *Arrowheads*, deposited Aβ42 (**A** and **B**); asterisks, *the peduncle structure (an axon bundle of Kenyon cells)*. Scale bar: 50 μm. (Adapted from Iijima, K. et al. (2004). *PNAS 17*, 6623–6628, with permission)

Age-Dependent Behavioral Defects Induced by Aβ40 and Aβ42

To examine if these flies show short-term memory defects, we used the Pavlovian olfactory classical conditioning assay (Quinn et al., 1974; Tully & Quinn, 1985). A short-term memory defect that was expressed in an age-dependent manner was observed in Aβ42 flies but not in the control flies (Fig. 14.5). When Aβ40 was expressed at much higher levels (five times more), it also caused short-term memory defects (Fig. 14.5), which suggests that excessive accumulation of Aβ40 may be toxic to synaptic plasticity. The sensory motor activity was not significantly affected in Aβ flies (Iijima et al., 2004), which indicates that the observed defects were attributable to short-term memory defects (Fig. 14.5).

We used a climbing assay to measure locomotor defects in Aβ42 flies. Briefly, about 20 flies were placed in empty plastic vials and were lightly tapped to the bottom. Naïve flies immediately climbed the wall, an activity known as negative geotaxis. Aβ42 flies started to show locomotor disability around 20 days and completely lost climbing ability around 30 days (Fig. 14.6). In contrast, Aβ40 flies did not show any defects in this assay (Fig. 14.6). In addition to this locomotor defect, the life span of Aβ42 flies was much shorter than that of the control flies, while the life span of Aβ40 flies was not affected (Fig. 14.6).

Late-Onset Progressive Neuronal Cell Loss Induced by Aβ42 but Not Aβ40

Extensive neuronal cell loss was observed in aged Aβ42 flies but not in Aβ40 or control flies (Fig. 14.7). This cell loss was of late onset and progressive since neurodegeneration could be only detected after 30 days of age but not at younger

stages. Electron microscopic analysis revealed that most of the cell death could be classified as necrotic type (Fig. 14.7).

In summary, both Aβ40 and Aβ42 induce progressive learning defects but only Aβ42 is capable of causing the formation of amyloid deposits, locomotor dysfunction, severe neuronal loss and premature death. Our study demonstrated that excessive accumulation of Aβ42 is sufficient to cause memory defects and neurodegeneration and that the molecular basis underlying Aβ toxicity may be conserved in *Drosophila*.

Comparison of Our Fly Model to Mouse Models of Alzheimer's Disease

There are many animal models of AD ranging from nematodes to rodents (Link, 2005; McGowan, Eriksen, & Hutton, 2006). Each model recapitulates somewhat different features of AD and thus it would be useful to review these models. Since the mouse models have been the most frequently used and best-characterized models, we will compare our fly model with mouse models that overproduce Aβ from human APP. Although there are many AD mouse models expressing human tau protein (Lee, Kenyon, & Trojanowski, 2005), we will not discuss them in this chapter.

It was more than a decade ago that the first two Aβ mouse models of AD were established. In 1995, Games et al. reported a transgenic mouse model (PDAPP) expressing human APP carrying the familial V717I mutation found in a British family (Games et al., 1995). This mouse overproduced human Aβ40 and Aβ42 and recapitulated some of the key pathologies such as the production of extracellular amyloid plaques, dystrophic neurites, gliosis, loss of synapse density and age-dependent memory impairment. However, these lesions were not accompanied by neurofibrillary

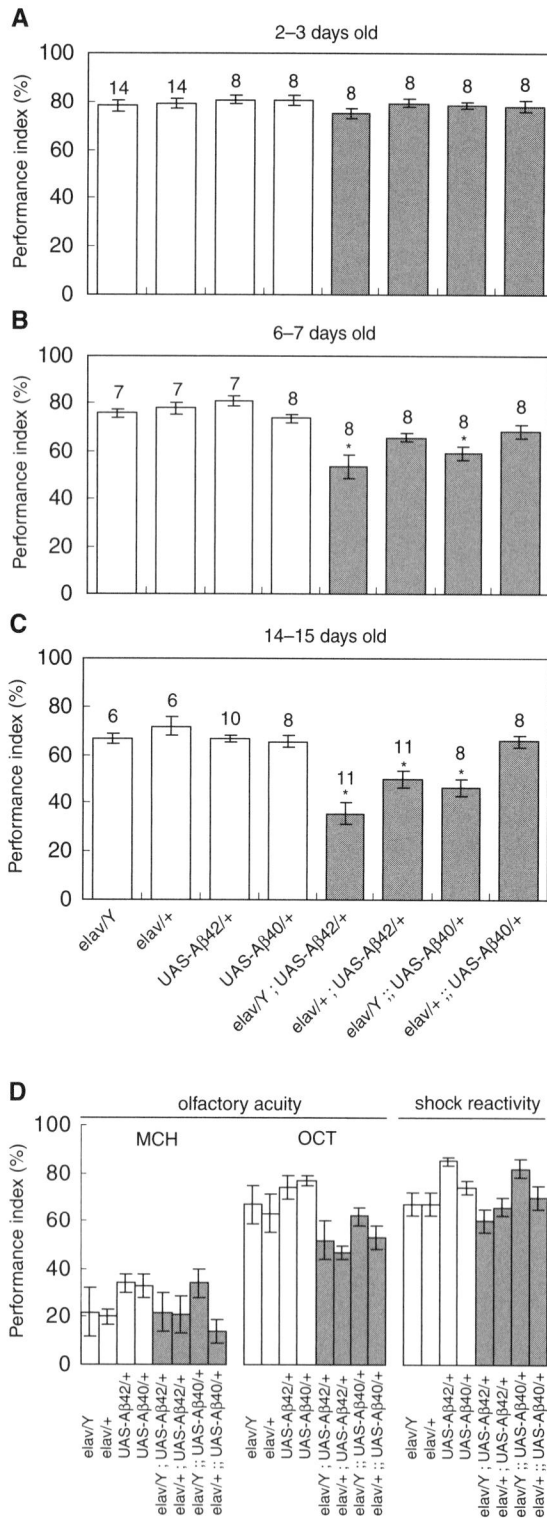

Fig. 14.5 Progressive loss of learning ability in Aβ flies. (**A–C**) The learning abilities at the ages of 2–3 (**A**), 6–7 (**B**) and 14–15 days (**C**) are presented as means ± SEM. The numbers of experiments are indicated on top of the bars. Asterisks show statistical differences relative to controls ($\alpha < 0.05$, Tukey-Kramer HSD). (**D**) There is no statistical difference in the olfactory acuity and shock reactivity of experimental genotype and appropriate control genotypes at the ages of 14–15 days ($\alpha < 0.05$, Tukey–Kramer HSD). (Adapted from Iijima, K. et al. (2004). *PNAS 17*, 6623–6628, with permission)

tangle (NFT) formation, and, more importantly, no global neuronal loss was observed. A year later, Hsiao et al. reported another transgenic mouse model (Tg2576) carrying the FAD mutation found in a Swedish family (Hsiao et al., 1996). This mouse model showed age-dependent memory deficits in addition to extracellular amyloid deposits, dystrophic neurites, and gliosis, but also failed to show NFT formation or loss of synapses or neurons. Since then, a number of mouse models with different combinations of promoters, transgenes and mutations have been established (for recent reviews, see McGowan et al., 2006; Spires & Hyman, 2005). However, although most transgenic mice that overproduce Aβ successfully recapitulate several pathological lesions and memory defects similar to the PDAPP and Tg2576 mice, they also do not show NFT formation or global neuronal loss. It is not clear why these mouse models do not develop those pathologies of AD. One possible explanation may be that the mouse life span is too short to see cell loss, and/or only human tau, but not mouse tau, and can cause the NFT formation and cell death induced by Aβ (McGowan et al., 2006).

Senile plaques have been the characteristic hallmark of AD for almost a century, and recapitulation of extracellular amyloid deposits have been one of the key criteria for the successful generation of mouse models of AD (McGowan et al., 2006). However, the pathological roles of these Aβ42 deposits have been questioned because of a low correlation between affected brain areas and senile plaque loads in AD brains (Terry et al., 1991). Recently, extracellular soluble oligomers, but not deposited forms of Aβ, have been proposed to be responsible for memory defects in the rodent model of AD (Cleary et al., 2005; Lesne et al., 2006; Walsh et al., 2002). The involvement of those oligomers in neurodegeneration remains to be elucidated.

In contrast to mouse models, extensive cell loss was induced in transgenic Aβ42 flies. What causes this difference between the mouse models and our flies? One remarkable difference between these two systems is that extensive accumulation of Aβ42 occurs within neurons in our fly model. The human AD brains show significant accumulation of intraneuronal Aβ42 (Gouras et al., 2000; Takahashi et al., 2002), and very recently, a few mouse models of AD showed intraneuronal accumulation of Aβ42 in association with memory defects (Oddo et al., 2003), significant neuron loss (Casas et al., 2004; Oakley et al., 2006) and axonopathy (Wirths, Weis, Szczygielski, Multhaup, & Bayer, 2006). Thus, it is possible that intraneuronal Aβ42 may contribute to neuron loss in AD brains.

Another unique feature of our fly model is that the toxicity of Aβ40 and Aβ42 can be dissected in vivo. Since most of the mouse models of AD overexpress APP, which produces a series of Aβ species including Aβ40 and Aβ42, it is impossible to study the toxicity of either species individually.

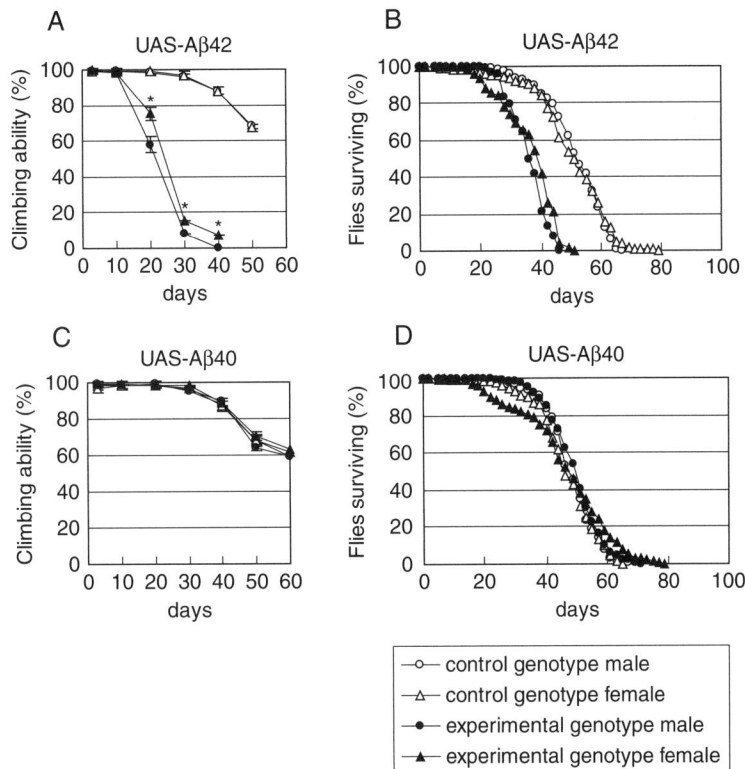

Fig. 14.6 Progressive climbing disability and shortened life span of Aβ42 flies. (**A** and **C**) Climbing ability of Aβ42 flies (**A**, *asterisks*, p<0.001, student's *t* test) and Aβ40 flies (**C**). The SDs of 10 trials are within the symbols. (**B** and **D**) Survival rate of Aβ42 flies (**B**) and Aβ40 flies (**D**). (Adapted from Iijima, K. et al. (2004). *PNAS 17*, 6623–6628, with permission)

We found that while both Aβ40 and Aβ42 could cause memory defects, only Aβ42 caused neurodegeneration with amyloid deposits. Recently, transgenic mouse models separately expressing Aβ40 or Aβ42 have been reported (McGowan et al., 2005). Consistent with our results, this study showed that only Aβ42 forms amyloid deposits in the mouse brain. However, no neuronal loss is observed in the Aβ42 mice. This may be because Aβ42 accumulates only extracellularly in these mice. Our fly model thus provides a unique tool for the study of 'intraneuronal' Aβ42 toxicity and may help in exploring features of AD.

Future Directions

Unlike cancer where biopsy samples are sometimes available to study disease progression in humans, most information about pathological changes of AD at the molecular level depends entirely on autopsy samples, which probably only represent the end stage of AD. Therefore, establishment of animal models is crucial to understand the early stage of AD pathogenesis, which is essential to the development of therapeutic strategies.

Accumulating evidence demonstrates that many human diseases can be modeled in simple organisms such as *Drosophila* (Marsh & Thompson, 2006). Since AD occurs in human brain, which is unique to humans, it is impossible to observe the entire spectra of the disease conditions in any animal model. Therefore, detailed characterization and precise understanding of each animal model are crucial; this information allows us to select the model system that is appropriate for addressing specific questions. As described above, our Aβ42 flies recapitulated important aspects of AD that are rarely seen in existing mouse models. This indicates that this fly model will contribute to AD research not only as a powerful genetic screening system but also as a novel model with which to study AD from new angles.

It is vital to corroborate discoveries made by using simple organisms, such as the ones we made with our fly AD studies, in higher organisms up to humans. The genes discovered in our fly model can be tested in the mouse AD models or may be reanalyzed in genetic association studies of AD phenotypes to identify new AD-related genes. Certainly, the effects of candidate genes identified by the genetic association studies with humans on the toxicity of Aβ42 can be systematically evaluated in our fly models.

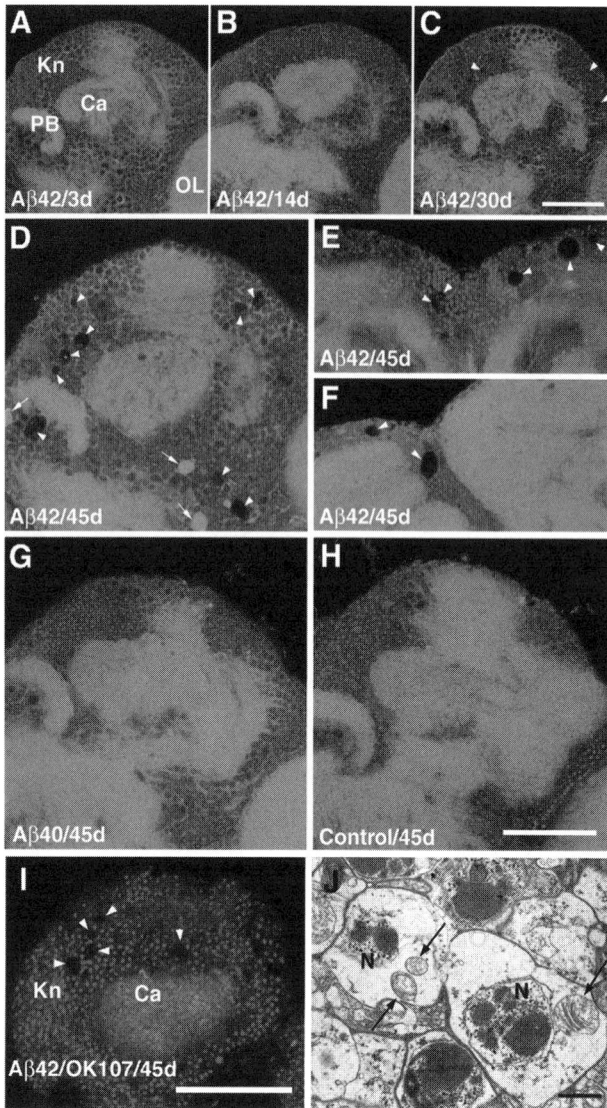

Fig. 14.7 Late-onset progressive neurodegeneration in Aβ42 brains. (**A–H**) Progressive neuronal loss was observed in the Aβ42 brains (**A–F**, *arrow heads*) but not in the Aβ40 or control brains (**G** and **H**). (**A–D**, **G** and **H**) The Kenyon cell region. (**E**) The medial brain. (**F**) The lateral brain. White, neuropil structure; dark gray, cell bodies. *Arrows* in (**D**) indicate the aggregates, which are presumably amyloid deposits. Kn, Kenyon cell layer; Ca, calyx; PB, protocerebral bridge; OL, optic lobe. Scale bars in C and H: 50 μm. (**I**) Neuronal loss induced by a different Gal4 line (OK107). Scale bar: 50 μm. (**J**) Ultrastructural analysis of the degenerating neurons that reveals digested cytoplasm (electron-lucent) and swollen mitochondria (*arrows*). N, nucleus. Scale bar: 1 μm. (Adapted from Iijima, K. et al. (2004). *PNAS 17*, 6623–6628, with permission)

In addition to memory defects, other cognitive dysfunctions can be observed in AD patients. To what extent can the fruit fly recapitulate human behavior and cognitive functions of a human? A recent study revealed that flies can also be used to study sleep (Hendricks et al., 2001), aggression (Chen, Lee, Bowens, Huber, & Kravitz, 2002), attention (van

Swinderen & Greenspan, 2003) and fear-related behaviors (Besson & Martin, 2005). These new findings may allow us to use flies to model the psychiatric aspects of AD in the future.

A precise understanding of the anatomical organization of the fly brain is also important for modeling neurodegenerative diseases. While the characterization of the fly brain is still behind that of the human or rodent brain, the development of unique genetic tools such as the Gal4-UAS system has significantly accelerated this area of research. Many cell types have been identified and the neuronal circuits involved in the visual (Morante & Desplan, 2004) or olfactory (Jefferis & Hummel, 2006) systems have been extensively analyzed. For example, it has been reported that the dominant excitatory system in the fly brain is composed of cholinergic neurons (Yasuyama, Kitamoto, & Salvaterra, 1995), which are the most vulnerable neurons in AD (Davies, 1983). Another example is that the dopaminergic neuron in the fly brain can be used to study the toxicity of α-synuclein, which causes Parkinson's disease (Auluck, Chan, Trojanowski, Lee, & Bonini, 2002; Feany & Bender, 2000). Moreover, the spatial and temporal manipulation of neuronal activity in specific subsets of neurons in the fly brain by using temperature-sensitive genetic tools can enable us to map the functional anatomy of the fly brain (Kitamoto, 2001). In addition, in vivo imaging of neuronal activity using a genetically encoded GFP-based Ca2+ indicator in combination with two-photon microscopy will allow us to visualize the functional abnormalities that are induced in the fly brain by Aβ42 (Wang et al., 2004). Thus, the ongoing characterization of the fly brain at the molecular, cellular, neural circuit, neural activity and behavioral levels will facilitate the use of the fly model in studies of human disease conditions.

In summary, *Drosophila* can be an excellent model system to study complex human diseases such as AD. Along with the state-of-the-art genetic tools that are available and are constantly being developed, we believe that *Drosophila* studies will make as significant a contribution to AD research as they did in other biological contexts.

References

Adams, M. D., Celniker, S. E., Holt, R. A., Evans, C. A., Gocayne, J. D., Amanatides, P. G., et al. (2000). The genome sequence of Drosophila melanogaster. *Science, 287*(5461), 2185–2195.

Auluck, P. K., Chan, H. Y., Trojanowski, J. Q., Lee, V. M., & Bonini, N. M. (2002). Chaperone suppression of alpha-synuclein toxicity in a Drosophila model for Parkinson's disease. *Science, 295*(5556), 865–868.

Bertram, L., & Tanzi, R. E. (2005). The genetic epidemiology of neurodegenerative disease. *The Journal of Clinical Investigation, 115*(6), 1449–1457.

Besson, M., & Martin, J. R. (2005). Centrophobism/thigmotaxis, a new role for the mushroom bodies in Drosophila. *Journal of Neurobiology, 62*(3), 386–396.

Bilen, J., & Bonini, N. M. (2005). Drosophila as a model for human neurodegenerative disease. *Annual Review of Genetics, 39*, 153–171.

Brand, A. H., & Perrimon, N. (1993). Targeted gene expression as a means of altering cell fates and generating dominant phenotypes. *Development, 118*(2), 401–415.

Casas, C., Sergeant, N., Itier, J. M., Blanchard, V., Wirths, O., van der Kolk, N., et al. (2004). Massive CA1/2 neuronal loss with intraneuronal and N-terminal truncated Abeta42 accumulation in a novel Alzheimer transgenic model. *The American Journal of Pathology, 165*(4), 1289–1300.

Chen, S., Lee, A. Y., Bowens, N. M., Huber, R., & Kravitz, E. A. (2002). Fighting fruit flies: a model system for the study of aggression. *Proceedings of the National Academy of Sciences of the USA, 99*(8), 5664–5668.

Cleary, J. P., Walsh, D. M., Hofmeister, J. J., Shankar, G. M., Kuskowski, M. A., Selkoe, D. J., et al. (2005). Natural oligomers of the amyloid-beta protein specifically disrupt cognitive function. *Nature Neuroscience, 8*(1), 79–84.

Crowther, D. C., Kinghorn, K. J., Miranda, E., Page, R., Curry, J. A., Duthie, F. A., et al. (2005). Intraneuronal Abeta, non-amyloid aggregates and neurodegeneration in a Drosophila model of Alzheimer's disease. *Neuroscience, 132*(1), 123–135.

Cummings, J. L. (2003). *The neuropsychiatry of Alzheimer's disease and other dementias*. London: Martin Dunitz.

Davies, P. (1983). An update on the neurochemistry of Alzheimer disease. *Advances in Neurology, 38*, 75–86.

Deleault, N. R., Dolph, P. J., Feany, M. B., Cook, M. E., Nishina, K., Harris, D. A., et al. (2003). Post-transcriptional suppression of pathogenic prion protein expression in Drosophila neurons. *Journal of Neurochemistry, 85*(6), 1614–1623.

Dietzl, G., Chen, D., Schnorrer, F., Su, K. C., Barinova, Y., & Fellner, M., et al. (2007) *Nature, 448*(7150), 151–156.

Dubnau, J., Chiang, A. S., Grady, L., Barditch, J., Gossweiler, S., McNeil, J., et al. (2003). The staufen/pumilio pathway is involved in Drosophila long-term memory. *Current Biology, 13*(4), 286–296.

Dudai, Y., Jan, Y. N., Byers, D., Quinn, W. G., & Benzer, S. (1976). dunce, a mutant of Drosophila deficient in learning. *Proceedings of the National Academy of Sciences of the USA, 73*(5), 1684–1688.

Feany, M. B., & Bender, W. W. (2000). A Drosophila model of Parkinson's disease. *Nature, 404*(6776), 394–398.

Finelli, A., Kelkar, A., Song, H. J., Yang, H., & Konsolaki, M. (2004). A model for studying Alzheimer's Abeta42-induced toxicity in Drosophila melanogaster. *Molecular and Cellular Neurosciences, 26*(3), 365–375.

Folkers, E., Drain, P., & Quinn, W. G. (1993). Radish, a Drosophila mutant deficient in consolidated memory. *Proceedings of the National Academy of Sciences of the USA, 90*(17), 8123–8127.

Fortini, M. E., Skupski, M. P., Boguski, M. S., & Hariharan, I. K. (2000). A survey of human disease gene counterparts in the Drosophila genome. *The Journal of Cell Biology, 150*(2), F23–F30.

Fossgreen, A., Bruckner, B., Czech, C., Masters, C. L., Beyreuther, K., & Paro, R. (1998). Transgenic Drosophila expressing human amyloid precursor protein show gamma-secretase activity and a blistered-wing phenotype. *Proceedings of the National Academy of Sciences of the USA, 95*(23), 13703–13708.

Games, D., Adams, D., Alessandrini, R., Barbour, R., Berthelette, P., Blackwell, C., et al. (1995). Alzheimer-type neuropathology in transgenic mice overexpressing V717F beta-amyloid precursor protein. *Nature, 373*(6514), 523–527.

Gandy, S. (2005). The role of cerebral amyloid beta accumulation in common forms of Alzheimer disease. *The Journal of Clinical Investigation, 115*(5), 1121–1129.

Glenner, G. G., & Wong, C. W. (1984). Alzheimer's disease and Down's syndrome: sharing of a unique cerebrovascular amyloid fibril protein. *Biochemical and Biophysical Research Communications, 122*(3), 1131–1135.

Gouras, G. K., Tsai, J., Naslund, J., Vincent, B., Edgar, M., Checler, F., et al. (2000). Intraneuronal Abeta42 accumulation in human brain. *The American Journal of Pathology, 156*(1), 15–20.

Greeve, I., Kretzschmar, D., Tschape, J. A., Beyn, A., Brellinger, C., Schweizer, M., et al. (2004). Age-dependent neurodegeneration and Alzheimer-amyloid plaque formation in transgenic Drosophila. *The Journal of Neuroscience, 24*(16), 3899–3906.

Grotewiel, M. S., Beck, C. D., Wu, K. H., Zhu, X. R., & Davis, R. L. (1998). Integrin-mediated short-term memory in Drosophila. *Nature, 391*(6666), 455–460.

Grotewiel, M. S., Martin, I., Bhandari, P., & Cook-Wiens, E. (2005). Functional senescence in Drosophila melanogaster. *Ageing Research Reviews, 4*(3), 372–397.

Guo, H. F., Tong, J., Hannan, F., Luo, L., & Zhong, Y. (2000) *Nature, 403*(6772), 895–898.

Hardy, J., & Selkoe, D. J. (2002). The amyloid hypothesis of Alzheimer's disease: progress and problems on the road to therapeutics. *Science, 297*(5580), 353–356.

Heisenberg, M. (2003). Mushroom body memoir: from maps to models. *Nature Reviews Neuroscience, 4*(4), 266–275.

Hendricks, J. C., Williams, J. A., Panckeri, K., Kirk, D., Tello, M., Yin, J. C., et al. (2001). A non-circadian role for cAMP signaling and CREB activity in Drosophila rest homeostasis. *Nature Neuroscience, 4*(11), 1108–1115.

Hsiao, K., Chapman, P., Nilsen, S., Eckman, C., Harigaya, Y., Younkin, S., et al. (1996). Correlative memory deficits, Abeta elevation, and amyloid plaques in transgenic mice. *Science, 274*(5284), 99–102.

Iijima, K., Liu, H. P., Chiang, A. S., Hearn, S. A., Konsolaki, M., & Zhong, Y. (2004). Dissecting the pathological effects of human Abeta40 and Abeta42 in Drosophila: a potential model for Alzheimer's disease. *Proceedings of the National Academy of Sciences of the USA, 101*(17), 6623–6628.

Iwatsubo, T., Odaka, A., Suzuki, N., Mizusawa, H., Nukina, N., & Ihara, Y. (1994). Visualization of A beta 42(43) and A beta 40 in senile plaques with end-specific A beta monoclonals: evidence that an initially deposited species is A beta 42(43). *Neuron, 13*(1), 45–53.

Jackson, G. R., Salecker, I., Dong, X., Yao, X., Arnheim, N., Faber, P. W., et al. (1998). Polyglutamine-expanded human huntingtin transgenes induce degeneration of Drosophila photoreceptor neurons. *Neuron, 21*(3), 633–642.

Jackson, G. R., Wiedau-Pazos, M., Sang, T. K., Wagle, N., Brown, C. A., Massachi, S., et al. (2002). Human wild-type tau interacts with wingless pathway components and produces neurofibrillary pathology in Drosophila. *Neuron, 34*(4), 509–519.

Jefferis, G. S., & Hummel, T. (2006). Wiring specificity in the olfactory system. *Seminars in Cell & Developmental Biology, 17*(1), 50–65.

Kitamoto, T. (2001). Conditional modification of behavior in Drosophila by targeted expression of a temperature-sensitive shibire allele in defined neurons. *Journal of Neurobiology, 47*(2), 81–92.

Lee, V. M., Balin, B. J., Otvos, L., Jr., & Trojanowski, J. Q. (1991). A68: a major subunit of paired helical filaments and derivatized forms of normal Tau. *Science, 251*(4994), 675–678.

Lee, V. M., Kenyon, T. K., & Trojanowski, J. Q. (2005). Transgenic animal models of tauopathies. *Biochimica et Biophysica Acta, 1739*(2–3), 251–259.

Lesne, S., Koh, M. T., Kotilinek, L., Kayed, R., Glabe, C. G., Yang, A., et al. (2006). A specific amyloid-beta protein assembly in the brain impairs memory. *Nature, 440*(7082), 352–357.

Link, C. D. (2005). Invertebrate models of Alzheimer's disease. *Genes, Brain, and Behavior, 4*(3), 147–156.

Livingstone, M. S., Sziber, P. P., & Quinn, W. G. (1984). Loss of calcium/calmodulin responsiveness in adenylate cyclase of rutabaga, a Drosophila learning mutant. *Cell, 37*(1), 205–215.

Luo, L., Tully, T., & White, K. (1992). Human amyloid precursor protein ameliorates behavioral deficit of flies deleted for Appl gene. *Neuron, 9*(4), 595–605.

Marsh, J. L., & Thompson, L. M. (2004). Can flies help humans treat neurodegenerative diseases? *Bioessays, 26*(5), 485–496.

Marsh, J. L., & Thompson, L. M. (2006). Drosophila in the study of neurodegenerative disease. *Neuron, 52*(1), 169–178.

McGowan, E., Eriksen, J., & Hutton, M. (2006). A decade of modeling Alzheimer's disease in transgenic mice. *Trends in Genetics, 22*, 281–289.

McGowan, E., Pickford, F., Kim, J., Onstead, L., Eriksen, J., Yu, C., et al. (2005). Abeta42 is essential for parenchymal and vascular amyloid deposition in mice. *Neuron, 47*(2), 191–199.

McGuire, S. E., Le, P. T., Osborn, A. J., Matsumoto, K., & Davis, R. L. (2003). Spatiotemporal rescue of memory dysfunction in Drosophila. *Science, 302*(5651), 1765–1768.

Morante, J., & Desplan, C. (2004). Building a projection map for photoreceptor neurons in the Drosophila optic lobes. *Seminars in Cell & Developmental Biology, 15*(1), 137–143.

Oakley, H., Cole, S. L., Logan, S., Maus, E., Shao, P., Craft, J., et al. (2006). Intraneuronal beta-amyloid aggregates, neurodegeneration, and neuron loss in transgenic mice with five familial Alzheimer's disease mutations: potential factors in amyloid plaque formation. *The Journal of Neuroscience, 26*(40), 10129–10140.

Oddo, S., Caccamo, A., Shepherd, J. D., Murphy, M. P., Golde, T. E., Kayed, R., et al. (2003). Triple-transgenic model of Alzheimer's disease with plaques and tangles: intracellular Abeta and synaptic dysfunction. *Neuron, 39*(3), 409–421.

Outeiro, T. F., & Giorgini, F. (2006). Yeast as a drug discovery platform in Huntington's and Parkinson's diseases. *Biotechnology Journal, 1*(3), 258–269.

Price, D. L., Tanzi, R. E., Borchelt, D. R., & Sisodia, S. S. (1998). Alzheimer's disease: genetic studies and transgenic models. *Annual Review of Genetics, 32*, 461–493.

Quinn, W. G., Harris, W. A., & Benzer, S. (1974). Conditioned behavior in Drosophila melanogaster. *Proceedings of the National Academy of Sciences of the USA, 71*(3), 708–712.

Quinn, W. G., Sziber, P. P., & Booker, R. (1979). The Drosophila memory mutant amnesiac. *Nature, 277*(5693), 212–214.

Raeber, A. J., Muramoto, T., Kornberg, T. B., & Prusiner, S. B. (1995). Expression and targeting of Syrian hamster prion protein induced by heat shock in transgenic Drosophila melanogaster. *Mechanisms of Development, 51*(2–3), 317–327.

Reiter, L. T., Potocki, L., Chien, S., Gribskov, M., & Bier, E. (2001). A systematic analysis of human disease-associated gene sequences in Drosophila melanogaster. *Genome Research, 11*(6), 1114–1125.

Rorth, P., Szabo, K., Bailey, A., Laverty, T., Rehm, J., Rubin, G. M., et al. (1998). Systematic gain-of-function genetics in Drosophila. *Development, 125*(6), 1049–1057.

Rovelet-Lecrux, A., Hannequin, D., Raux, G., Le Meur, N., Laquerriere, A., Vital, A., et al. (2006). APP locus duplication causes autosomal dominant early-onset Alzheimer disease with cerebral amyloid angiopathy. *Nature Genetics, 38*(1), 24–26.

Sang, T. K., & Jackson, G. R. (2005). Drosophila models of neurodegenerative disease. *NeuroRx, 2*(3), 438–446.

Selkoe, D. J. (2002). Alzheimer's disease is a synaptic failure. *Science, 298*(5594), 789–791.

Shulman, J. M., Shulman, L. M., Weiner, W. J., & Feany, M. B. (2003). From fruit fly to bedside: translating lessons from Drosophila models of neurodegenerative disease. *Current Opinion in Neurology, 16*(4), 443–449.

Spires, T. L., & Hyman, B. T. (2005). Transgenic models of Alzheimer's disease: learning from animals. *NeuroRx, 2*(3), 423–437.

Steffan, J. S., Bodai, L., Pallos, J., Poelman, M., McCampbell, A., Apostol, B. L., et al. (2001). Histone deacetylase inhibitors arrest polyglutamine-dependent neurodegeneration in Drosophila. *Nature, 413*(6857), 739–743.

Takahashi, R. H., Milner, T. A., Li, F., Nam, E. E., Edgar, M. A., Yamaguchi, H., et al. (2002). Intraneuronal Alzheimer abeta42 accumulates in multivesicular bodies and is associated with synaptic pathology. *The American Journal of Pathology, 161*(5), 1869–1879.

Tanzi, R. E., & Bertram, L. (2005). Twenty years of the Alzheimer's disease amyloid hypothesis: a genetic perspective. *Cell, 120*(4), 545–555.

Terry, R. D., Masliah, E., Salmon, D. P., Butters, N., DeTeresa, R., Hill, R., et al. (1991). Physical basis of cognitive alterations in Alzheimer's disease: synapse loss is the major correlate of cognitive impairment. *Annals of Neurology, 30*(4), 572–580.

Toba, G., Ohsako, T., Miyata, N., Ohtsuka, T., Seong, K. H., & Aigaki, T. (1999). The gene search system. A method for efficient detection and rapid molecular identification of genes in Drosophila melanogaster. *Genetics, 151*(2), 725–737.

Tully, T., & Quinn, W. G. (1985). Classical conditioning and retention in normal and mutant Drosophila melanogaster. *Journal of Comparative Physiology A, 157*(2), 263–277.

Ueda, R. (2001). Rnai: a new technology in the post-genomic sequencing era. *Journal of Neurogenetics, 15*(3–4), 193–204.

van Swinderen, B., & Greenspan, R. J. (2003). Salience modulates 20–30 Hz brain activity in Drosophila. *Nature Neuroscience, 6*(6), 579–586.

Waddell, S., & Quinn, W. G. (2001). What can we teach Drosophila? What can they teach us? *Trends in Genetics, 17*(12), 719–726.

Walsh, D. M., Klyubin, I., Fadeeva, J. V., Cullen, W. K., Anwyl, R., Wolfe, M. S., et al. (2002). Naturally secreted oligomers of amyloid beta protein potently inhibit hippocampal long-term potentiation in vivo. *Nature, 416*(6880), 535–539.

Wang, Y., Guo, H. F., Pologruto, T. A., Hannan, F., Hakker, I., Svoboda, K., et al. (2004). Stereotyped odor-evoked activity in the mushroom body of Drosophila revealed by green fluorescent protein-based Ca2+ imaging. *The Journal of Neuroscience, 24*(29), 6507–6514.

Warrick, J. M., Paulson, H. L., Gray-Board, G. L., Bui, Q. T., Fischbeck, K. H., Pittman, R. N., et al. (1998). Expanded polyglutamine protein forms nuclear inclusions and causes neural degeneration in Drosophila. *Cell, 93*(6), 939–949.

Wirths, O., Weis, J., Szczygielski, J., Multhaup, G., & Bayer, T. A. (2006). Axonopathy in an APP/PS1 transgenic mouse model of Alzheimer's disease. *Acta Neuropathologica (Berl), 111*(4), 312–319.

Wittmann, C. W., Wszolek, M. F., Shulman, J. M., Salvaterra, P. M., Lewis, J., Hutton, M., et al. (2001). Tauopathy in Drosophila: neurodegeneration without neurofibrillary tangles. *Science, 293*(5530), 711–714.

Yasuyama, K., Kitamoto, T., & Salvaterra, P. M. (1995). Immunocytochemical study of choline acetyltransferase in Drosophila melanogaster: an analysis of cis-regulatory regions controlling expression in the brain of cDNA-transformed flies. *The Journal of Comparative Neurology, 361*(1), 25–37.

Yin, J. C., & Tully, T. (1996). CREB and the formation of long-term memory. *Current Opinion in Neurobiology, 6*(2), 264–268.

Yin, J. C., Wallach, J. S., Del Vecchio, M., Wilder, E. L., Zhou, H., Quinn, W. G., et al. (1994). Induction of a dominant negative CREB transgene specifically blocks long-term memory in Drosophila. *Cell, 79*(1), 49–58.

Part III
Genetics of Personality

Chapter 15

Genotype–Environment Correlation and Family Relationships

Jennifer A. Ulbricht and Jenae M. Neiderhiser

Introduction

Traditionally, the unit for analysis in considering adult mental health has been the individual. Research, assessment, and treatment of psychological and behavioral issues have most often focused on identifying or changing certain characteristics of the recognized target individual. When considering development and children, the focus often widens to include parents, though research, intervention, and treatments again tend to address behaviors, emotions, or cognitions of the identified "problem child." Exceptions to this general trend can be found in empirically supported preventive and intervention programs that involve children, parents, schools, and communities (e.g., Hawkins, Catalano, & Arthur, 2002; Olds, Hill, O'Brien, Rache, & Mortiz, 2003; Reid & Webster-Stratton, 2001; Robbins, Alexander, & Turner, 2000). These programs address the systems (such as school, workplace, peer groups, and families) in which individuals act and interact and the influences that these systems can have on psychological and behavioral functioning. Though evidence indicates that these systems-based interventions are effective, it is not yet clear what the mechanisms are through which individual factors influence and are influenced by others.

One of the most universal systems in human development is the family. There have been numerous correlational studies linking various parent attributes to child behavior and developmental outcomes (e.g., Belsky, Crnic, & Woodworth, 1995; Bornstein, 2002; Maccoby, 2002). For instance, parental depression – particularly maternal depression – and marital conflict are risk factors for various forms of psychopathology in children, including both behavioral and emotional problems (Whiffen, 2005). In addition, aggressive parents are often found to have aggressive children who grow into aggressive adults (Huesmann, Eron, Lefkowitz, & Walder, 1984; Patterson, Capaldi, & Bank, 1991). Further studies of the mechanisms of these associations have focused on parenting behavior as the probable link between parent psychopathology and child outcome (e.g., Fleming, Kim, Harachi, & Catalano, 2002; Kaczynski, Lindahl, Malik, & Laurenceau, 2006). Additionally, research has moved toward conceptualizing parenting not as something that happens to a child but as a relational process that is dependent on the influence of individual traits of both parent and child (e.g., Ge, Conger, Cadoret, Neiderhiser, Yates, & Troughton, 1996; Simons, Whitbeck, Conger, & Melby, 1990). In this way, the parent–child relationship environment is partially a product of the traits (many that are genetically influenced) of each person involved and can be impacted by the various other systems in which the individual members of the family participate. In this chapter, the interplay between genes and environment (specifically genotype–environment correlations) will be explored through a review of the behavioral genetic literature on genetic and environmental influences on parenting, sibling relationships, marital relationships, and general family environment. There is now accumulating evidence that genotype–environment correlation plays an important role in the associations among these family subsystems as well as in the links between family relationships and adjustment. These topics will be considered, as will the current and future directions for the study of genotype–environment correlation.

Genotype–Environment Correlation

A central question regarding genetic influences on family relationships is concerned with the mechanisms by which these influences occur. One explanation is genotype–environment (GE) correlation. GE correlation refers to the correlation between an individual's genetically influenced behavior and their exposure to certain environmental situations (e.g., Neiderhiser et al., 2004; Rutter, 2005; Silberg & Eaves, 2004). Typically three types of GE

J.M. Neiderhiser (✉)
The Pennsylvania State University, Department of Psychology,
222 Moore Building, University Park, PA 16802, USA
e-mail: jenaemn@psu.edu

correlation have been described: passive, active, and evocative (Plomin, DeFries, & Loehlin, 1977; Scarr, 1992; Scarr & McCartney, 1988). Passive GE correlation arises in biological families because parents and children share both genes and environment. For example, passive GE correlation in parenting could be conceptualized as a parent's genetically influenced traits impacting their parenting behavior, which is part of the environment for their children. Therefore, due to the fact that parents and their biological children also share genes, the environments that parents provide may correlate with the child's genotype due, at least in part, to those shared genes. In the parent–child relationship, it has been hypothesized that personality may be a good candidate for a genetically influenced trait shared by parent and child that can explain passive GE correlation. For example, a parent's genetically influenced warm and sociable personality makes their parenting more warm and responsive while, simultaneously, the child is warm and sociable due to inherited traits and a positive environment. There is some evidence that parents' personalities do influence their parenting, at least in part (e.g., Belsky et al., 1995; Clark, Kochanska, & Ready, 2000), although few studies have examined these associations within a genetically informed design (see Spinath & O'Connor, 2003, for an exception).

Active GE correlation is the result of an individual actively selecting environments correlated with their genetically influenced characteristics. A good example of active genotype–environment correlation for an environmental measure concerns peers. When genetic influences on peer relationships are found (e.g., Manke, McGuire, Reiss, Hetherington, & Plomin, 1995) one explanation is that adolescents are selecting peers who are like them and the characteristics that they are selecting for are genetically influenced. The classic example of active GE correlation in childhood is with musical ability: a child has a genetically influenced proclivity for music and thus chooses to take band class (an environment providing musical enrichment).

Finally, evocative GE correlation may be most interesting for understanding family process. An evocative GE correlation occurs when genetically influenced traits of the individual evoke a certain type of response from others (the environment). GE correlations are especially relevant for understanding genetic influences on family process because they provide an explanation of possible mechanisms. For example, a child that demonstrates high levels of negative emotionality due to genetically influenced traits may elicit harsher or more rejecting behavior from parents. In peer relationships, the evocative process is also easily conceptualized. For example, an adolescent who exhibits genetically influenced antisocial traits such as rule breaking or drug use may elicit rejection from more prosocial peers.

It is clear that genetic influences on family relations, like parenting or marital conflict, are not due to an individual's genes directly influencing the environment. Instead, indirect processes, involving mechanisms like GE correlations, genotype × environment interaction[1], or most likely a combination of the two, must be involved. Therefore, in order to better understand family process, it is important to also understand GE correlation (and G × E interaction) and to identify which type of GE correlation may be operating. Finally, GE correlation also helps to explain why heritability estimates for some traits tend to increase with age. It may be that individuals – and their genetically influenced tendencies – have increasing impact on their environments throughout the course of development which serves to reinforce and strengthen those tendencies (Bouchard & McGue, 2003).

Parenting Behavior

As noted above, parenting is a family process that has received extensive attention in the developmental psychology literature, although the majority of developmental studies have examined one child per family. There is now a sizable literature reporting evidence of genetic and environmental influences on parenting across the life span. These studies have used both child-based (children vary in genetic relatedness) and parent-based (parents vary in genetic relatedness) designs and have considered a wide variety of perspectives, reporters, and parenting constructs as well as global measures of family environment. At least two detailed reviews of the literature examining genetic and environmental influences on parenting have been published (e.g., McGuire, 2003; Towers, Spotts, & Neiderhiser, 2002). These reviews conclude that genetic and environmental influences tend to differ based upon the parenting construct, the reporter, and the developmental stage or age of the children in the study. In addition, findings indicate that parent- and child-based designs are valuable for detecting different mechanisms for genetic influence on the parent–child relationship. Because child-based twin and family designs ostensibly detect the influence of the child's genes, it might be tempting to interpret these findings as support for the child's heritable characteristics as the main vehicle for genetic influences on the parent–child relationship. However, because children receive 50% of their genes from each parent, a child-based design is unable to decisively disentangle passive from evocative GE correlation for parenting. Interestingly, the different designs show a somewhat different pattern of

---

[1] Rutter and Silberg (2002) define G × E interaction as the impact of genetically influenced individual differences in the sensitivity to certain environmental features. For example, an individual with genotype A may develop depressive symptoms in response to a job loss while an individual with genotype B may respond with more resilience.

genetic and environmental influences; considering these differences, we can begin to clarify mechanisms of GE correlation. A detailed discussion of how to use parent-based and child-based designs to disentangle passive, evocative, and active GE correlation can be found elsewhere (Neiderhiser et al., 2004; Rutter, Moffitt, & Caspi, 2006) and a brief description specific to parenting is provided here. Passive GE correlation is indicated by genetic influences on parenting in a parent-based design, while it may emerge as shared environmental influences in child-based studies. Evocative GE correlations are indicated by genetic influences on parenting in a child-based design and will emerge as shared and/or nonshared environmental influences in a parent-based design. Genetic and environmental influences on parenting in child-based and parent-based designs will be described below with an emphasis on the type of GE correlation that is indicated by the findings.

In general, studies have converged on the finding that most parenting constructs show significant genetic and shared environmental influences in child-based designs (see Fig. 15.1 for a conceptual summary of the findings for different family relationships). Parental warmth, support, and negativity typically are influenced by genetic and shared environmental influences (e.g., Elkins, McGue, & Iacono, 1997; Lau, Rijsdijk, & Eley, 2006). The findings for warmth, support, and negativity are generally similar in parent-based designs, indicating a possible passive GE mechanism (Neiderhiser et al., 2004). There is also some evidence of evocative GE correlation because the child-based designs show significant and substantial genetic influences. The

construct of parental control shows shared environmental and little to no genetic influence in most child-based studies. In contrast, parent-based studies of parental controlling behavior have found genetic influences on parental control (Kendler, 1996; Losoya, Callor, Rowe, & Goldsmith, 1997; Neiderhiser et al., 2004; Perusse, Neale, Heath, & Eaves, 1992). These findings, taken together, suggest that the parents' genes may be the largest contributing genetic factor for parent control, while children's genetically influenced traits have little impact on this construct. Therefore, parental control is best explained by passive GE correlation. Taken together, these studies support the conclusion that genes and environment influence different parenting constructs through different mechanisms.

There is some evidence that patterns of genetic and environmental influences change throughout the life span, most using child-based designs. Shared environmental influences are larger and genetic influences smaller for parenting of infants and young children (Boivin et al., 2005). Genetic influences on parenting tend to increase during middle childhood and adolescence although shared environmental effects are typically still important (Elkins et al., 1997; McGue, Elkins, Walden, & Iacono, 2005; Reiss, Neiderhiser, Hetherington, & Plomin, 2000; Rende, Slomkowski, Stocker, Fulker, & Plomin, 1992; Rowe, 1981, 1983). Although there have been only preliminary analyses exploring the parent–child relationships when the children are young adults, the findings suggest that genetic influences are about the same as they were during adolescence, with little to no evidence of shared environmental influences and substantial nonshared

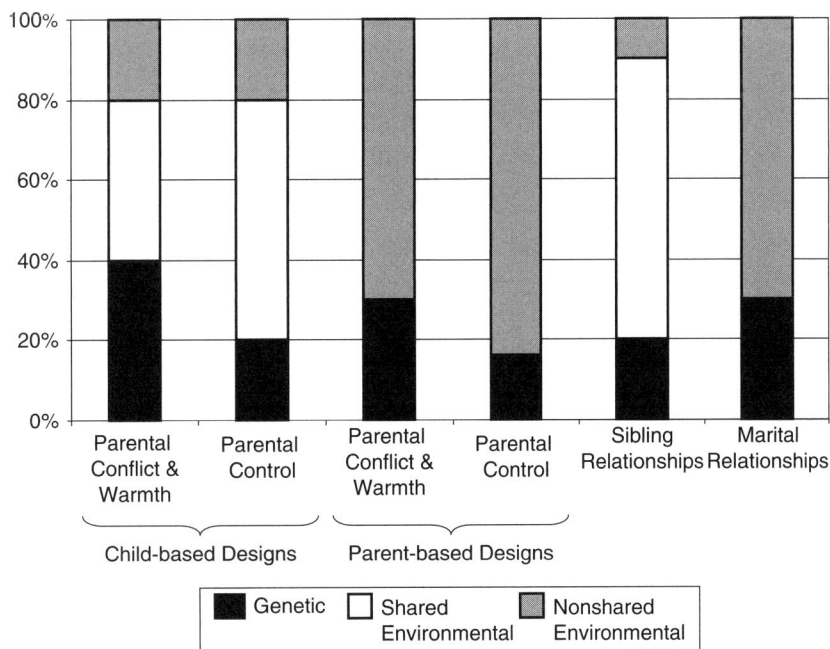

Fig. 15.1 General patterns of genetic and environmental influences on different family relationships

environmental influences (Neiderhiser, 2003). Interestingly, retrospective reports of parental warmth, collected from adults during middle age, show patterns more similar to those for adolescents than for young adults. Specifically, genetic, shared environmental, and nonshared environmental influences make nearly equal contributions for twin women's retrospective reports of warmth from their mother and father (Lichtenstein et al., 2003). It may be that when asked to remember how warm their parents were, these twin women focused more on adolescence than at their more proximal relationship experiences with their parents. As illustrated in Fig. 15.2 genetic influences on parental warmth and conflict tend to increase from infancy to adolescence, when they flatten out and may decrease slightly in young adulthood. Shared environmental influences, on the other hand, show a slow but steady decrease from infancy to adolescence and then abruptly drop off for young adults. Finally, nonshared environmental influences on parenting are modest until young adulthood when they show a sudden increase. The sharp decrease in shared environmental influences and increase in nonshared environmental influences are most likely due to the fact that the children are no longer living in the same household.

As noted earlier, genetic and environmental influences on parenting also tend to differ depending on the reporter (e.g., Neiderhiser & Reiss, 2004; O'Connor, Hetherington, Reiss, & Plomin, 1995; Plomin, Reiss, Hetherington, & Howe, 1994). Observer ratings of parenting have consistently shown relatively modest genetic influences and sizable shared and nonshared environmental influences (e.g., Leve, Winebarger, Fagot, Reid, & Goldsmith, 1998; O'Connor et al., 1995), while estimates of genetic influences are larger when parent and adolescent reports of parenting are examined (e.g., McGue et al., 2005; Neiderhiser et al., 2004; Reiss et al., 2000). It is worth noting that shared environmental influences tend to be at least moderate for most constructs of parenting in child-based designs, regardless of the reporter.

Generally, reports compiled across multiple measures of parenting behavior as well as parent, child, and observer ratings offer a view of parenting behavior that may be more consistent across time and settings. Interestingly, these composites also tend to be influenced more by genetic factors than single-rater constructs (e.g., Feinberg, Neiderhiser, Howe, & Hetherington, 2001; Reiss et al., 2000).

There have been a series of reports attempting to specify the type of GE correlation operating for parenting. These reports have used two very different designs and have converged on similar findings. First, two adoption studies examined associations between biological parents' disorders (an estimation of genotype) and adoptive parents' parenting (Ge et al., 1996; O'Connor, Deater-Deckard, Fulker, Rutter, & Plomin, 1998). Both studies suggested that evocative GE correlation was important, as indicated by the finding that child behavior acted as a mediator of the association between the biological parents' characteristics and parenting. A second set of reports have relied on the comparison of a sample of twin parents (parent based) with a sample of adolescent twins and siblings (child based). The Twin Moms study (Reiss et al., 2001) and its extension to a second cohort and twin fathers, the Twin/Offspring Study in Sweden (TOSS), was designed in part as a parent-based complement to the child-based NEAD study. Analyses comparing these two samples on mothering, and more recently on fathering, have found evidence for passive and evocative GE correlation for all of the parenting constructs examined, although the patterns of findings tended to vary by reporter, parent, and construct (Neiderhiser, Reiss, Lichtenstein, Spotts, & Ganiban, 2007; Neiderhiser et al., 2004). Generally, the pattern of findings suggested passive GE correlation for mothers' positivity and monitoring and evocative GE correlation for mothers' negativity and control. In other words, a mother may exert more control or demonstrate more negativity *in response* to her child's genetically influenced traits or behaviors. Fathering showed

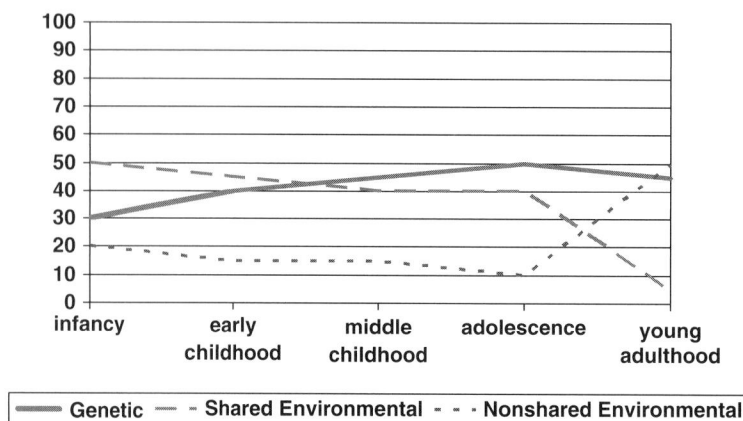

Fig. 15.2 General trends in parenting behavior (warmth and negativity) throughout the life span

a similar pattern of findings with one exception, evocative GE correlation was suggested for father's positivity. This suggests that while mothers' and children's positivity is shaped by shared genetics, fathers parent more positively *in response* to their children's genetically influenced traits or behaviors. The next step in understanding the mechanisms of GE correlation is to examine these constructs within the same sample in order to reduce the likelihood of differences in findings emerging due to sample differences rather than because of real differences in the mechanisms of effects.

Understanding how genetic factors influence parent–child processes beyond parenting has been examined in a number of recent reports. Genetic and environmental influences on dyadic mutuality – defined as emotional reciprocity, co-responsiveness, and cooperation in parent–child dyads – were examined in two reports (Deater-Deckard & O'Connor, 2000; Deater-Deckard & Petrill, 2004). The first sampled same-sex twins and their families at age 3 and found that mutuality varied both between and within families and that similarity between siblings' relationships with their parent was due largely to genetic and nonshared environmental influences (Deater-Deckard & O'Connor, 2000). The second report extended this analysis to include child outcome using a sample of adoptive siblings (mean age = 8.96) and found that variation in mutuality in the parent–child relationship was due to evocative GE correlation (the child's genetically influenced traits evoke certain levels of mutuality) and unspecified environmental mechanisms that may include external influences on the parent such as stressors (Deater-Deckard & Petrill, 2004). These two sets of findings provide additional support to the notion that children directly contribute to their parents' behavior toward them and that this contribution is, at least in part, parental responses to children's genetically influenced characteristics.

Sibling Relationships

Despite the fact that most behavioral genetic studies involve siblings, relatively few have examined genetic and environmental influences on the sibling relationship. One of the first, a child-based adoption design, found that positive and negative behaviors between adoptive and non-adoptive siblings during middle childhood were mostly influenced by shared environmental, and moderate genetic, influences (Rende et al., 1992). A study examining the same sample during early adolescence using the social relations model found evidence for genetic, shared environmental, and nonshared environmental influences on aspects of sibling relationships such as conflict, warmth, and self-disclosure (Manke & Plomin, 1997). In the adolescent NEAD sample

shared environmental influences were most important in explaining individual differences in sibling relationships, with only modest genetic influences for most of the sibling relationship constructs (Bussell et al., 1999; Reiss et al., 2000). Interestingly, the findings from NEAD indicate a high level of reciprocity in sibling relationships, indicating that adolescents both contributed to and "got something" out of the sibling relationship (Reiss et al., 2000). A different study of adolescent twins also found that sibling relationships were due to mostly genetic and shared environmental influences, with modest nonshared environmental influences (Pike & Atzaba-Poria, 2003). All of these findings underscore the high degree of reciprocity in sibling relationships as indicated by the significant and substantial shared environmental influences. Because these large shared environmental influences are found regardless of whether child report, parent report, or behavioral observations are used, the likelihood of a sibling rater bias as an explanation for shared environmental influences is low. Instead, it appears that the shared environmental influences found for sibling relationships are due to the fact that siblings are closely attuned to one another resulting in highly reciprocal relationships. It is noteworthy that the reciprocal nature of the sibling relationship seems to increase during adolescence, suggesting that adolescents are especially sensitive to the behavior of their sibling and tend to respond in a similar way.

Marital Relationships

The first studies to examine marital relations focused on divorce, pair bonding (maintaining stable heterosexual marriage), and mate diversification (remarriage following divorce). In the first of these studies, genetic and environmental influences on divorce were examined in a large sample of adult twins (McGue & Lykken, 1992). This study found that divorce was influenced by primarily genetic and non-shared environmental factors. A later report using the same sample found that most of this genetic influence on divorce could be explained by genetic influences on personality traits correlated with divorce (Jocklin, McGue, & Lykken, 1996). For example, the personality factors of positive emotionality and negative emotionality were positively linked to divorce, while the factor of constraint was linked negatively. The most recent report from this group found that one's propensity to be married was also substantially genetically influenced and the genetic influences on propensity to marry were correlated with genetic influences on personality (Johnson, McGue, Krueger, & Bouchard, 2004). Genetic and environmental influences on pair bonding and mate diversification were examined on a sample of male veteran twin pairs (Trumbetta & Gottesman, 2000). Both constructs were heritable, but with nonshared environmental influences accounting for the bulk of the variance.

Although these studies provide interesting insight into the factors that influence one's tendency to marry or divorce, they do not provide any information on how genetic and environmental factors may influence the quality of a marriage or each partner's satisfaction. In an effort to do just this, marital quality was examined in a sample of adult female twin pairs and their partners (Spotts et al., 2004). Genetic influences were found for marital quality as reported by the twin women. Interestingly, genetic influences were also found for the husbands' reports of their own marital satisfaction. Because only the wives were twins, this suggests that genetically influenced characteristics of the women set a tone for the marriage that then influences the husbands' reports of their own satisfaction. In other words, evocative GE correlation was suggested. Using the same sample of twin women and their partners, personality characteristics of both spouses were examined as potential sources of genetic influences on marital quality (Spotts et al., 2005). The personality characteristics of the women accounted for genetic and nonshared environmental influences on their own marital satisfaction, husbands' marital satisfaction, and the covariation between wives and husbands on marital satisfaction. Husbands' personality characteristics did not account for additional variance in marital satisfaction. This finding suggests that there may be different genetic and environmental mechanisms operating for marriage in men and women. A study examining this question using a large unselected sample of adult male and female twin pairs found evidence for small differences between genetic and environmental influences on marital quality for men and women (Spotts, Prescott, & Kendler, 2006). Specifically, marital warmth and conflict were influenced by the same genetic factors in men and by different genetic factors in women.

In sum, genetic influences have been found for all marriage-related constructs that have been examined. Nonshared environmental influences explain the bulk of the variance for all constructs; genetic influences are substantially smaller for marital relationships than has been reported for parent–child relationships. This suggests that genetically influenced traits may operate differently within different relationship contexts and highlights the importance of the interplay between genes and environment in influencing behavior (Spotts et al., 2004).

General Family Environment

In addition to examining genetic and environmental influences on specific relationships within the family, studies have also examined measures of the global family environment.

One measure that has been examined in a number of studies is the Family Environment Scale (FES; Moos & Moos, 1981) which assesses the overall family environment in terms of cohesion, conflict, and control. The first of these used a sample of adult twins reared together and apart who retrospectively rated their perceptions of the family environment *in which they were reared* (Plomin, McClearn, Pedersen, Nesselroade, & Bergeman, 1988). The findings indicated significant genetic influences on all of the FES scales. A second study using the same sample assessed the twins' reports of their *current* family environment and also found moderate genetic influences on all of the FES scales (Plomin, McClearn, Pedersen, Nesselroade, & Bergeman, 1989). Similar patterns of genetic and environmental influences have been found for child and parent reports on the FES during the transition from late childhood to early adolescence, although there was more variation between reporters, and the pattern of genetic and environmental influence was different for parent report and child report (Deater-Deckard, Fulker, & Plomin, 1999). In most studies examining genetic and environmental influences on FES, control/structure tends to show less genetic influence than acceptance/support. One exception comes from a recent analysis of current FES using an adolescent sample of male twins which found genetic influences on all FES constructs, including control/structure (Herndon, McGue, Krueger, & Iacono, 2005).

Another widely used measure of family environment is the Home Observation for Measurement of the Environment (HOME; Caldwell & Bradley, 1978). The HOME is a semi-structured interview that yields a score reflecting the observed family environment. Researchers using the Colorado Adoption Project sample (Plomin, DeFries, & Fulker, 1988), a sibling adoption design, found evidence for substantial genetic influences on the HOME at ages 1 and 2 (Braungart, Fulker, & Plomin, 1992). In addition, a number of studies have also detected genetic influences on measures of socioeconomic status (e.g., Lichtenstein, Harris, Pederson, & McClearn, 1993). Finally, family connectedness, a construct designed to assess feelings of general closeness or connectedness within the family, was examined using the Add Health sample, which includes twins and siblings of varying degrees of genetic relatedness, and it also shows genetic influences (Jacobson & Rowe, 1999). Gender differences emerged, however, for shared environmental influences which were significant for male adolescents, but not for females. From the studies described above, it appears that constructs traditionally conceptualized as "environment" or "nurture" variables are, in fact, influenced in part by the genetic makeup of the individuals involved. It is likely that GE correlations and interactions play a substantial role in the heritability of family environment measures.

Associations Among Family Subsystems

Within the family systems theory framework, we might expect that these different family relationships may influence each other and that genetic factors are likely to be involved. "Spillover" is a term often used in discussing family systems and processes referring to situations where marital discord impacts the relationship of children and parents. While there have been many phenotypic studies of how these processes operate together (e.g., Belsky, Youngblade, & Pensky, 1989; Conger, Patterson, & Ge, 1995; Shamir, Du Rocher Schudlich, & Cummings, 2001) few studies have examined the genetic and environmental influences on associations between marital quality and parent–child relationships.

At least one study, however, focused on associations among other family relationships. Bussell and colleagues (1999) examined associations between mother–child relationships and sibling relationships in the NEAD sample using composites of observer ratings, mother reports, and child reports. Findings indicated that although both mother–child and sibling relationships are significantly influenced by genetic factors, the two relationships covaried for primarily shared environmental reasons. Additional analyses examining genetic and environmental influences on the covariation of parenting and sibling relationships consistently found that shared environmental influences accounted for the bulk of the covariation (Reiss et al., 2000). This finding held for mother–child and father–child relationships and across positivity and negativity in both relationships. These findings of significant and substantial shared environmental influences across the family subsystems of parent–child and sibling relationships suggest that there may be a common style of interaction across family relations within a family. Parents may play a role in shaping sibling interactions through modeling a preferred relational style or intervening and controlling sibling interactions, thereby both directly and indirectly influencing the sibling relationship via shared environmental mechanisms.

The covariation between marital conflict about each adolescent child with parenting and with sibling relationships has also been examined using the NEAD sample (Reiss et al., 2000). The findings were somewhat different than those for parenting and sibling relationships, however. The covariation between marital conflict and parenting was due in near equal parts to genetic and shared environmental influences. In other words, genetic influences of the child had an important impact on both marital conflict about the adolescent children and on parent–child relationships. On the other hand, the pattern of findings for associations between sibling negativity and marital conflict about the children was consistent with the findings for parent–child and sibling relationships. Specifically, shared environmental influences accounted for the bulk of this covariation.

These findings suggest that different processes are operating for different sets of family relationships. Specifically, sibling relationships are highly reciprocal and may be more responsive to the family climate. Although the associations between marital conflict about the children and parent–child relationships are also influenced by shared environmental influences, they are nearly equally due to genetic influences. In other words, they may be influenced by the general family climate, but they are also influenced by the child's genetically influenced characteristics. Unfortunately, these findings are based on only a single sample assessed during adolescence. Additional studies using different aged samples may help to clarify the mechanisms involved in influencing different types of family relationships.

Family Subsystems and Adjustment

Parenting

There is an extensive literature indicating that interpersonal relationships can have an important impact on mental health. In particular, there are a number of studies linking parent–child relationships to various child outcomes, and results indicate that the associations are complex. For instance, reports from the NEAD study have indicated that genetic influences on adolescent perceptions of parenting behavior mediate the relationship between parenting behavior and adolescent adjustment (Neiderhiser, Pike, Hetherington, & Reiss, 1998). This indicates that it is not necessarily the actions of the parents that impact child outcome, but how the child perceives their parents' actions. Interestingly, this perception is influenced by the child's genetic makeup and therefore illustrates how genes can influence the environment's effects on behavior (Neiderhiser, 1998). Genetic and environmental influences on associations between parenting and child and adolescent adjustment have been examined in a number of studies. Associations among parental negativity, positivity, monitoring and control, and seven composites of adolescent adjustment were systematically examined in the NEAD sample (Pike, McGuire, Hetherington, & Reiss, 1996; Reiss et al., 2000). Typically, the bulk of the association between parenting and adolescent adjustment was due to genetic influences on the covariation.

These associations were also examined longitudinally from middle to later adolescence in the NEAD project (Neiderhiser, Reiss, Hetherington, & Plomin, 1999; Reiss et al., 2000). Using a full cross-lagged design to estimate genetic and environmental contributions to the associations between parenting in middle adolescence and child adjustment in later adolescence, a number of interesting conclusions emerged. First, associations between parenting at time 1 and adolescent adjustment 3 years later were accounted for primarily by genetic factors. This indicates that genetic influences play an important role in how parenting in middle adolescence impacts adjustment in later adolescence. In this child-based design, the genetic influence measured is that of the *child's* genes, which suggest that GE correlation (either passive, evocative, or both) accounts for the observed associations, though these models were not tested explicitly (Neiderhiser et al., 1999). Second, there were differences in the associations between parenting and adolescent adjustment during middle and later adolescence. Most often genetic influences were larger for these associations during middle than later adolescence. This finding and the findings for the cross-lagged associations emphasize that genetic influences *change over time*. Finally, shared environment was slightly more important for fathering than for mothering. It is possible that this finding is due to a father's tendency to treat the adolescents more similarly and, perhaps, to be less sensitive to individual differences than are mothers.

Given the implications of these results, it is not surprising that the focus of such studies has shifted from models involving the *direct* influence of genes or parenting on problem behaviors to models that consider GE correlation as a mediator of the associations among genes, parenting, and antisocial behavior. One report using a sample of adolescent adoptees found evidence of evocative GE correlation for adoptive parent hostility (Ge et al., 1996). Specifically, the psychiatric disorders of the biological parents were related to the adolescent's antisocial behavior *and* with adoptive parents' behavior. A subsequent report using the CAP found that children at genetic risk for behavior problems received more negative efforts of control from their adoptive parents, even after controlling for selective placement (O'Connor, Deater-Deckard, Fulker, Rutter, & Plomin, 1998). In both of these studies, the presence of evocative GE correlation was supported and the results were consistent with established theories of coercive family processes (Patterson, 1982), highlighting the importance of considering genetic influences within the context of the family environment.

More recently, Burt, McGue, Krueger, and Iacono (2005) used the Minnesota Twin Family Study to examine the relationship between parent–child conflict and externalizing symptoms in adolescent boys over a 3 year period using a cross-lagged approach. This study was designed, in part, to clarify the source and direction of the relationship between child externalizing and conflictual parent–child relationships. Three primary findings emerged from this report. Stability in both parent–child conflict and externalizing could be explained by primarily genetic influences. The relationship between parent–child conflict and adolescent externalizing is bidirectional, both cross-sectionally and longitudinally. Finally, the role of evocative GE correlation was supported. Specifically, parent–child conflict was, at least in part, a

response to externalizing behavior and this association could be explained by genetic influences. At the same time, parent–child conflict contributed to child externalizing behavior via shared and nonshared environmental influences. These findings extend and support those reported earlier using a different sample and a slightly different model to explore parent conflict–negativity and adolescent antisocial behavior (Neiderhiser et al., 1999). Taken together, these findings emphasize the importance of considering longitudinal effects of parenting on child and adolescent adjustment and for considering the possibility of reciprocal influences and differing genetic and environmental mechanisms.

Interestingly, findings from this line of research have also indicated that there are some gender differences in the connections between genes, parenting, and child outcome, for both child and parent gender (e.g., Jacobson & Rowe, 1999; Neiderhiser et al., 1999). Typically, separate models for mothers and fathers or sons and daughters consistently provide the "best fit" for the data, although clear differences in the *patterns* of influence have not been indicated for most constructs (e.g., Neiderhiser et al., submitted; Reiss et al., 2000). In addition, there is clear evidence from a wide range of different studies that parents treat their children differently (e.g., Conger & Conger, 1994; Feinberg & Hetherington, 2001; McGuire & Dunn, 1994; McHale, Crouter, McGuire, & Updegraff, 1995; Reiss, Howe, Simmens, & Bussell, 1996) and that this difference in treatment is linked to the child's genetically influenced traits (Deater-Deckard et al., 2001). Though family process and relational theories of reciprocity predict that the quality of the parent–child relationship should influence the parent's mental health, there is little genetically informed research to support this idea to date.

Marriage

Marital quality has been associated with both positive and negative effects on mental health for both members of the dyad, though women show stronger associations between marital quality and mental health (e.g., Gove, Hughes, & Style, 1983; Prigerson, Maciejewski, & Rosenheck, 1999). Traditionally, these associations have been explained through psychosocial mechanisms with little attention given to the possible contribution of genetic factors to the correlations, despite evidence for genetic influence on measures of marital quality and mental health. Spotts et al. (2003) found that interpersonal relationships accounted for a moderate amount of the variance (18–31%) in women's depressive symptoms, using a sample of Swedish twin women and their husbands. The associations among measurements of marital quality, social support, and depressive symptoms were accounted for by genetic influences when husband's reports were used. Genetic and environmental influences on the associations between marital relationships, social support, and women's *positive* mental health have also been examined using the same sample of twin women (Spotts et al., 2005). Results indicated that while genetic influences are partially responsible for associations among marital quality, adequacy of social support, and mental health, nonshared environmental influences are of equal, if not greater, importance.

A few studies have examined genetic influences on the relationship between marital conflict and child outcome as well. O'Connor, Caspi, DeFries, and Plomin (2000) examined associations between divorce and child achievement, social adjustment, behavior problems, and substance use in the CAP sample. Genetic and environmental contributions tended to vary based on child outcome. Environmental factors mediated the links between divorce and child adjustment, but results for social adjustment and achievement indicated passive GE correlation (O'Connor, Caspi, DeFries, & Plomin, 2000). This work was extended by examining a moderation model for the association among genetic risk, adopted parent separation, and child adjustment (O'Connor, Caspi, DeFries, & Plomin, 2003). Findings indicated that in the absence of marital discord in the adoptive home, genetic risk (estimated from the birth parents) was not correlated with child adjustment. However, in the presence of marital discord genetic risk was significantly correlated with adjustment. These results suggest that genetic vulnerability may play a role in the causal pathway between divorce and child psychopathology.

Connections between marital instability and child drug and alcohol use, internalizing problems, and behavior problems have also been examined using a Children of Twins Design as part of a large study of Australian twins (D'Onofrio et al., 2005). With this method, the children of adult twins who were discordant for marital difficulty were compared. In other words, a child whose parent was experiencing marital instability was compared to their parent's twin (their aunt or uncle) who was *not* having trouble in their marriage. This design is useful in identifying the type of GE correlation operating, particularly passive GE correlation theorized to exist between marital conflict and child psychopathology. Results indicated that environmental influences linked to divorce explained the greater rates of psychological difficulty among the children rather than common genetically influenced traits that impact both marriage and mental health (D'Onofrio et al., 2005). While specific environmental factors were not examined in this study, they were hypothesized to include decline in effective parenting, conflict between parents, loss of contact or inadequate parenting from non-residential parent, and increased economic pressures.

A number of non-genetic research studies have found support for the hypothesized environmental processes proposed by D'Onofrio and colleagues (2005). For instance, Mann and MacKenzie (1996) found that marital dissatisfaction and overt marital conflict predicted oppositional behavior problems for school-aged boys. Most studies have concluded, however, that the connections between marital conflict and child behavior problems are mediated by parenting behavior such as rejection and inept discipline. These pathways help to explain why divorce is related to child difficulties in some, but not all children: essentially, parents who can maintain positive interaction styles and consistent discipline insulate their children from the negative effects of conflict in the marital relationship (Bullock & Forgatch, 2005).

In addition to general marital conflict and dissatisfaction, associations between adolescent adjustment and marital conflict *specifically* about each child were examined in the Nonshared Environment in Adolescent Development (NEAD; Reiss et al., 2000) study, which included families with adolescent twins, nontwins, half siblings, and stepsiblings. Findings indicate that the bulk of the association between parental conflict about each child was due to genetic influences on both constructs. This suggests that the marital conflict about a child may be, at least in part, a response to genetic influences on the child's behavior.

Siblings

While the studies discussed previously have established a framework for genetic and environmental influence on sibling relationships, research on processes involved in sibling influence on child outcome is rarer and these associations remain unclear (e.g., Dunn, 2002). Sibling conflict has been established as a risk factor for adjustment difficulties and problem behaviors, but the genetic and environmental mechanisms of this association remain largely unexplored. Sibling conflict and negativity have been related positively to measures of antisocial behavior; conversely, sibling warmth and support were positively associated with positive measures of adjustment like social responsibility and self-worth (Reiss et al., 2000). One study, using the Add Health sample, sought to explore sibling effects in smoking behavior (Slomkowski, Rende, Novak, Lloyd-Richardson, & Niaura, 2005). Both shared environmental and genetic factors were found to have main effects on smoking behavior, but social connectedness between siblings moderated the shared environmental influences, even after controlling for parent and peer smoking. These findings indicate that the quality of the sibling relationship may change the manner in which genetic and environmental factors influence adolescent smoking behavior. Future research would benefit from examining the genetic and environmental influences on processes like collusion and sibling deviancy training because identification of contributing factors would aid in development of targeted interventions as suggested by Bullock and Dishion (2002). Additionally, research has suggested that sibling conflict and negativity may contribute to increased negativity in the overall family environment and may also change the way genes and environment influence other relationships (parent–child or interparental) within the family (Feinberg, Reiss, Neiderhiser, & Hetherington, 2005).

Family Environment

Measures of global family environment have, not surprisingly, been linked to positive and negative child outcome as well. For instance, in the previously discussed Braungart, Fulker, and Plomin (1992) study of the CAP sample, the association between observed family environment at age 2 and child IQ at age 7 was genetically mediated. Another study of full- and half siblings found that genetic factors accounted for 25% of the correlation between HOME scores and child achievement, while shared environmental factors accounted for the remaining majority of the association (Cleveland, Jacobson, Lipinski, & Rowe, 2000). Furthermore, associations between family environment and child problem behaviors such as aggression and drug use have been attributed to both genes and shared environment (e.g., Braungart-Rieker, Rende, Plomin, DeFries, & Fulker, 1995; Meyer et al., 2000; Jang, Vernon, Liversley, Stein, & Wolf, 2001). Gender differences were found in a number of these studies. Interestingly, while genetic influences account for a greater portion of the association between family environment and *depressive* symptoms for *females*, the association between family environment and *externalizing behaviors* appears to be more strongly genetically influenced for *males* (Braungart-Rieker, Rende, Plomin, DeFries, & Fulker, 1995; Jacobson & Rowe, 1999). The patterns of these findings suggest a complex interplay between genes and environment that can vary both by construct and by gender.

Future Directions

Although the number of studies examining genetic and environmental influences on "environmental" measures have exploded since the seminal report by Plomin and Bergeman (1991) we are only just beginning to disentangle the mechanisms involved. Examining measured environments within the context of a traditional behavioral genetic model is just the first step. A number of studies have extended this

work to examine associations among the environment and the adjustment of individuals, to examine the associations longitudinally, and to use novel designs in an attempt to better understand processes. There are at least two promising approaches currently underway. First, the use of the Children of Twins design has a number of advantages over the traditional child-based or parent-based approaches (e.g., Silberg & Eaves, 2004). Most notably, this design combines both approaches within a single sample allowing comparisons between twin parents *and* between their children. Unfortunately, there are also disadvantages to this approach and to date it has yet to be fully utilized in teasing apart passive from evocative GE correlation (Eaves, Silberg, & Maes, 2005).

A second approach, that is not new but has been somewhat limited by measurement constraints like limited assessment of birth parents, is the adoption design. When birth parents, adoptive parents, and adopted children are all well assessed and followed longitudinally inferences about the relative roles of genetic and environmental influences, as well as about evocative GE correlation, can be made. Currently, there are a number of studies focused on understanding shared environmental influences using adoption samples (e.g., McGue & Sharma, 1995; Petrill & Deater-Deckard, 2004; Rutter, O'Connor, & English and Romanian Adoptees Study Team, 2004) and one study that also includes longitudinal assessment of birth parents (Leve et al., 2007). In all cases, environmental factors are measured in some detail and the findings of these studies will help to advance our understanding of genetic influences on environmental measures and on our understanding of GE correlation.

Finally, studies examining measured genes and measured environment have just begun to find evidence of GE correlation (Dick et al., 2006). The number of studies including both measured genes and measured environment are growing and this is likely to result in more findings of this kind. A number of advances in the fields of developmental and family psychology and quantitative and molecular genetics have enabled a clearer understanding of how genes and the environment operate together in influencing development. As these fields continue to expand and as approaches and strategies are combined we will be able to better understand the processes and mechanisms involved in the interplay between genes and the environment. As we better understand the mechanisms, targets for prevention and intervention efforts can be identified, thereby advancing the translation of this basic science to application.

References

Belsky, J., Crnic, K., & Woodworth, S. (1995). Personality and parenting: Exploring the mediating role of transient mood and daily hassles. *Journal of Personality, 63*(4), 905–929.

Belsky, J., Youngblade, L., & Pensky, E. (1989). Childrearing history, marital quality, and maternal affect: Intergenerational transmission in a low-risk sample. *Development & Psychopathology, 1*(4), 291–304.

Boivin, M., Perusse, D., Dionne, G., Saysset, V., Zoccolillo, M., Tarabulsy, G. M., et al. (2005). The genetic-environmental etiology of parents' perceptions and self-assessed behaviours toward their 5-month-old infants in a large twin and singleton sample. *Journal of Child Psychology and Psychiatry, 46*(6), 612–630.

Bornstein, M. H. (2002). *Handbook of parenting*: Vol. 1: Children and parenting (2nd ed). Mahwah, NJ, US: Lawrencce Erlbaum Associates Publishers.

Braungart-Rieker, J., Rende, R. D., Plomin, R., DeFries, J. C., & Fulker, D. W. (1995). Genetic mediation of longitudinal associations between family environment and childhood behavior problems. *Development and Psychopathology, 7*, 233–245.

Bussell, D. A., Neiderhiser, J. M., Pike, A., Plomin, R., Simmens, S., Howe, G. W., et al. (1999). Adolescents' relationships to siblings and mothers: A multivariate genetic analysis. *Developmental Psychology, 35*(5), 1248–1259.

Caldwell, B. M., & Bradley, R. H. (1978). *Home Observation for Measurement of the Environment*. Little Rock: University of Arkansas.

Clark, L. A., Kochanska, G., & Ready, R. (2000). Mothers' personality and its interaction with child temperament as predictors of parenting behavior. *Journal of Personality & Social Psychology, 79*(2), 274–285.

Conger, K. J., & Conger, R. D. (1994). Differential parenting and change in sibling differences in delinquency. *Journal of Family Psychology, 8*, 287–302.

Conger, R. D., Patterson, G. R., & Ge, X. (1995). It takes two to replicate: A mediational model for the impact of parents' stress on adolescent adjustment. *Child Development, 66*, 80–97.

Deater-Deckard, K., Fulker, D. W., & Plomin, R. (1999). A genetic study of the family environment in the transition to early adolescence. *Journal of Child Psychology and Psychiatry, 40*(5), 769–775.

Deater-Deckard, K., & O'Connor, T. G. (2000). Parent-child mutuality in early childhood: Two behavioral genetic studies. *Developmental Psychology, 36*(5), 561–570.

Deater-Deckard, K., & Petrill, S. A. (2004). Parent-child dyadic mutuality and child behavior problems: An investigation of gene-environment processes. *Journal of Child Psychology and Psychiatry, 45*(6), 1171–1179.

Dick, D. M., Agrawal, A., Schuckit, M. A., Bierut, L., Hinrichs, A., Fox, L., et al. (2006). Marital status, alcohol dependence, and GABRA2: Evidence for gene-environment correlation and interaction. *Journal of Studies on Alcohol, 67*(2), 185–194.

Dunn, J. (2002). Sibling relationships. In P. K. Smith & C. H. Hart (Eds.), *Blackwell handbook of childhood social development* (pp. 223–237). Malden, MA: Blackwell Publishing.

Eaves, L. J., Silberg, J. L., & Maes, H. H. (2005). Revisiting the children of twins: Can they be used to resolve the environmental effects on dyadic parental treatment on child behavior? *Twin Research and Human Genetics, 8*(4), 283–290.

Elkins, I. J., McGue, M., & Iacono, W. G. (1997). Genetic and environmental influences on parent-son relationships: Evidence for increasing genetic influence during adolescence. *Developmental Psychology, 33*(2), 351–363.

Feinberg, M., & Hetherington, E. (2001). Differential parenting as a within-family variable. *Journal of Family Psychology, 15*(1), 22–37.

Feinberg, M., Neiderhiser, J., Howe, G., & Hetherington, E. (2001). Adolescent, parent, and observer perceptions of parenting: Genetic and environmental influences on shared and distinct perceptions. *Child Development, 72*(4), 1266–1284.

Feinberg, M. E., Reiss, D., Neiderhiser, J. M., & Hetherington, E. M. (2005). Differential association of family subsystem negativity on siblings' maladjustment: using behaviour genetic methods to test process theory. *Journal of Family Psychology, 19*(4), 601–610.

Ge, X., Conger, R. D., Cadoret, R. J., Neiderhiser, J. M., Yates, W., & Troughton, E. (1996). The developmental interface between nature and nurture: A mutual influence model of child antisocial behavior and parent behaviors. *Developmental Psychology, 32*(4), 574–589.

Herndon, R. W., McGue, M., Krueger, R. F., & Iacono, W. G. (2005). Genetic and environmental influences on adolescents' perceptions of current family environment. *Behavior Genetics, 35*(4), 373–380.

Huesmann, L. R., Eron, L. D., Lefkowitz, M. M., & Walder, L. O. (1984). Stability of aggression over time and generations. *Developmental Psychology, 20*, 1120–1134.

Jacobson, K. C., & Rowe, D. C. (1999). Genetic and environmental influences on the relationships between family connectedness, school connectedness, and adolescent depressed mood: Sex differences. *Developmental Psychology, 35*(4), 926–939.

Jocklin, V., McGue, M., & Lykken, D. T. (1996). Personality and divorce: A genetic analysis. *Journal of Personality and Social Psychology, 71*(2), 288–299.

Johnson, W., McGue, M., Krueger, R. F., & Bouchard, T. J., Jr. (2004). Marriage and personality: A genetic analysis. *Journal of Personality and Social Psychology, 86*(2), 285–294.

Kendler, K. S. (1996). Parenting: A genetic-epidemiologic perspective. *American Journal of Psychiatry, 153*, 11–20.

Lau, J. Y. F., Rijsdijk, F., & Eley, T. C. (2006). I think, therefore I am: A twin study of attributional style in adolescents. *Journal of Child Psychology and Psychiatry, 47*(7). 696–703.

Leve, L. D., Neiderhiser, J. M., Ge, X., Scaramella, L. V., Conger, R. D., & Reid, J. B., et al. (2007). The Early Growth and Development Study: A prospective adoption design. *Research and Human Genetics, 10*(1), 84–95.

Leve, L. D., Winebarger, A. A., Fagot, B. I., Reid, J. B., & Goldsmith, H. (1998). Environmental and genetic variance in children's observed and reported maladaptive behavior. *Child Development, 69*(5), 1286–1298.

Lichtenstein, P., Ganiban, J., Neiderhiser, J. M., Pedersen, N. L., Hansson, K., Cederblad, M., et al. (2003). Remembered parental bonding in adult twins: Genetic and environmental influences. *Behavior Genetics, 33*(4), 397–408.

Losoya, S. H., Callor, S., Rowe, D. C., & Goldsmith, H. H. (1997). Origins of familial similarity in parenting: A study of twins and adoptive siblings. *Developmental Psychology, 33*(6), 1012–1023.

Maccoby, E. E. (2002). Parenting effects: Issues and controversies. In J. G. Borkowski, S. L. Ramey, & M. Bristol-Power (Eds.), *Parenting and the child's world: Influences on academic, intellectual, and social-emotional development*. Mahwah, NJ: Lawrence Erlbaum Associates.

Manke, B., & Plomin, R. (1997). Adolescent familial interactions: A genetic extension of the Social Relations Model. *Journal of Social and Personal Relationships, 14*(4), 505–522.

McGue, M., Elkins, I., Walden, B., & Iacono, W. G. (2005). Perceptions of the parent-adolescent relationship: A longitudinal investigation. *Developmental Psychology, 41*(6), 971–984.

McGue, M., & Lykken, D. T. (1992). Genetic influence on risk of divorce. *Psychological Science, 3*(6), 368–373.

McGue, M., & Sharma, A. (1995). Parent and sibling influences on adolescent alcohol use and misuse: Evidence from a U.S. adoption cohort. *Journal of Studies on Alcohol, 57*(1), 8–18.

McGuire, S. (2003). The heritability of parenting. *Parenting: Science and Practice, 3*(1), 73–94.

McGuire, S., & Dunn, J. (1994). Nonshared environment in middle childhood. In J. C. Defries, R. Plomin, & D. W. Fulker (Eds.), *Nature and Nurture During Middle Childhood*. Oxford: Blackwell.

McHale, S. M., Crouter, A. C., McGuire, S. A., & Updegraff, K. A. (1995). Congruence between mothers' and fathers' differential treatment of siblings: Links with family relations and children's well-being. *Child Development, 66*(1), 116–128.

Moos, R. H., & Moos, B. S. (1981). *Family Environment Scale*. Palo Alto, CA: Consulting Psychologists Press.

Neiderhiser, J. M. (2003). Genetic and environmental influences on change and continuity in family relationships from adolescence to young adulthood. *Behavior Genetics, 33*(6), 713.

Neiderhiser, J. M., Pike, A., Hetherington, E. M., & Resiss, D. (1998). Adolescent perceptions as mediators of parenting: Genetic and environmental contributions. *Developmental Psychology, 34*(6), 1459–1469.

Neiderhiser, J. M., & Reiss, D. (2004). Family investment and child and adolescent adjustment: The role of genetic research. In A. Kalil & T. DeLeire (Eds.), *Family investments in children's potential: Resources and parenting behaviors that promote success Monographs in parenting* (pp. 33–47). Mahwah, NJ: Lawrence Erlbaum Associates, Publishers.

Neiderhiser, J. M., Reiss, D., Hetherington, E., & Plomin, R. (1999). Relationships between parenting and adolescent adjustment over time: Genetic and environmental contributions. *Developmental Psychology, 35*(3), 680–692.

Neiderhiser, J. M., Reiss, D., Lichtenstein, P., Spotts, E. L. & Ganiban, J. (2007). Father-adolescent relationship and the role of genotype-environment correlation. *Journal of Family Psychology, 21(4)*, 560–571.

Neiderhiser, J. M., Reiss, D., Pedersen, N. L., Lichtenstein, P., Spotts, E. L., Hansson, K., et al. (2004). Genetic and environmental influences on mothering of adolescents: A comparison of two samples. *Developmental Psychology, 40*(3), 335–351.

O'Connor, T. G., Deater-Deckard, K., Fulker, D., Rutter, M., & Plomin, R. (1998). Genotype-environment correlations in late childhood and early adolescence: Antisocial behavioral problems and coercive parenting. *Developmental Psychology, 34*(5), 970–981.

O'Connor, T. G., Hetherington, E., Reiss, D., & Plomin, R. (1995). A twin-sibling study of observed parent-adolescent interactions. *Child Development, 66*(3), 812–829.

Patterson, G., Capaldi, D., & Bank, L. (1991). An early starter model for predicting delinquency. In D. J. Pepler & K. H. Rubin (Eds.), *The development and treatment of childhood aggression* (pp. 139–168). Hillsdale, NJ: Lawrence Erlbaum Associates, Inc.

Perusse, D., Neale, M. C., Heath, A. C., & Eaves, L. J. (1992). Human parental behavior: Evidence for genetic influence and implication for gene-culture transmission (abstract). *Behavior Genetics, 22*(6), 744.

Petrill, S. A., & Deater-Deckard, K. (2004). The heritability of general cognitive ability: A within-family adoption design. *Intelligence, 32*(4), 403–409.

Pike, A., & Atzaba-Poria, N. (2003). Do sibling and friend relationships share the same temperamental origins? A twin study. *Journal of Child Psychology and Psychiatry, 44*(4), 598–611.

Pike, A., McGuire, S., Hetherington, E., & Reiss, D. (1996). Family environment and adolescent depressive symptoms and antisocial behavior: A multivariate genetic analysis. *Developmental Psychology, 32*(4), 590–604.

Plomin, R., & Bergeman, C. S. (1991). The nature of nurture: Genetic influence on "environmental" measures. *Behavioral and Brain Sciences, 14*, 373–427.

Plomin, R., DeFries, J., & Loehlin, J. C. (1977). Genotype-environment interaction and correlation in the analysis of human behavior. *Psychological Bulletin, 84*(2), 309–322.

Plomin, R., DeFries, J. C., & Fulker, D. W. (1988). *Nature and Nurture During Infancy and Early Childhood*. Cambridge, England: Cambridge University Press.

Plomin, R., McClearn, G., Pedersen, N. L., Nesselroade, J. R., & Bergeman, C. S. (1989). Genetic influences on adults' ratings of their current family environment. *Journal of Marriage and the Family, 51*, 791–803.

Plomin, R., McClearn, G. E., Pedersen, N. L., Nesselroade, J. R., & Bergeman, C. S. (1988). Genetic influence on childhood family environment perceived retrospectively from the last half of the life span. *Developmental psychology, 24*(5), 738–745.

Plomin, R., Reiss, D., Hetherington, E., & Howe, G. W. (1994). Nature and nurture: Genetic contributions to measures of the family environment. *Developmental Psychology, 30*(1), 32–43.

Reiss, D., Howe, G. W., Simmens, S. J., & Bussell, D. A. (1996). Genetic questions for environmental studies: Differential parenting and psychopathology in adolescence. *Annual Progress in Child Psychiatry & Child Development, 206–235.*

Reiss, D., Neiderhiser, J. M., Hetherington, E., & Plomin, R. (2000). *The relationship code: Deciphering genetic and social influences on adolescent development*. Cambridge, MA: Harvard University Press.

Reiss, D., Pedersen, N. L., Cederblad, M., Lichtenstein, P., Hansson, K., Neiderhiser, J. M., et al. (2001). Genetic probes of three theories of maternal adjustment: I. Recent evidence and a model. *Family Process, 40*(3), 247–259.

Rende, R., Slomkowski, C. L., Stocker, C., Fulker, D. W., & Plomin, R. (1992). Genetic and environmental influences on maternal and sibling interaction in middle childhood: A sibling adoption study. *Developmental Psychology, 17*, 203–208.

Rowe, D. C. (1981). Environmental and genetic influences on dimensions of perceived parenting: A twin study. *Developmental Psychology, 17*(2), 203–208.

Rowe, D. C. (1983). A biometrical analysis of perceptions of family environment: A study of twin and singleton sibling kinships. *Child Development, 54*(2), 416–423.

Rutter, M. (2005). Environmentally mediated risks for psychopathology: Research strategies and findings. *Journal of the American Academy of Child and Adolescent Psychiatry, 44*(1), 3–18.

Rutter, M. & Silberg, J. (2002). Gene-environment interplay in relation to emotional and behavioral disturbance. *Annual Review of Psychology, 53*, 463–490.

Rutter, M., Moffitt, T. E., & Caspi, A. (2006). Gene-environment interplay and psychopathology: Multiple varieties but real effects. *Journal of Child Psychology and Psychiatry, 47*(3–4), 226–261.

Rutter, M., O'Connor, T. G., & English and Romanian Adoptees Study Team, T. (2004). Are there biological programming effects for psychological development? Findings from a study of Romanian adoptees. *Developmental Psychology, 40*(1), 81–94.

Scarr, S. (1992). Developmental theories for the 1990's: Development and individual differences. *Child Development, 63*, 1–19.

Scarr, S., & McCartney, K. (1988). How people make their own environments: A theory of genotype -> environment effects. *Child Development, 54*, 424–435.

Shamir, H., Du Rocher Schudlich, T., & Cummings, E. (2001). Marital conflict, parenting styles, and children's representations of family relationships. *Parenting: Science & Practice, 1*(1–2), 123–151.

Silberg, J. L., & Eaves, L. J. (2004). Analysing the contributions of genes and parent-child interaction to childhood behavioural and emotional problems: A model for the children of twins. *Psychological Medicine, 34*(2), 347–356.

Simons, R. L., Whitbeck, L. B., Conger, R. D., & Melby, J. N. (1990). Husband and wife differences in determinants of parenting: A social learning and exchange model of parental behavior. *Journal of Marriage & the Family, 52*(2), 375–392.

Slomkowski, C., Rende, R., Novak, S., Lloyd-Richardson, E., & Niaura, R. (2005). Sibling effects on smoking in adolescence: Evidence for social influence from a genetically informative design. *Addiction, 100(4)*, 430–438.

Spinath, F. M., & O'Connor, T. G. (2003). A Behavioral Genetic Study of the Overlap Between Personality and Parenting. *Journal of Personality, 71*(5), 785–808.

Spotts, E. L., Lichtenstein, P., Pedersen, N., Neiderhiser, J. M., Hansson, K., Cederblad, M., et al. (2005). Personality and marital satisfaction: A behavioural genetic analysis. *European Journal of Personality, 19*(3), 205–227.

Spotts, E. L., Neiderhiser, J. M., Towers, H., Hansson, K., Lichtenstein, P., Cederblad, M., et al. (2004). Genetic and environmental influences on marital relationships. *Journal of Family Psychology, 18*(1), 107–119.

Spotts, E. L., Prescott, C., & Kendler, K. (2006). Examining the origins of gender differences in marital quality: A behaviour genetic analysis. *Journal of Family Psychology, 20(4)*, 605–613.

Towers, H., Spotts, E. L., & Neiderhiser, J. M. (2002). Genetic and environmental influences on parenting and marital relationships: Current findings and future directions. *Marriage & Family Review, 33*(1), 11–29.

Trumbetta, S., & Gottesman, I. (2000). Endophenotypes for marital status in the NAS-NRC twin registry. In J. L. Rogers & D. C. Rowe (Eds.), *Genetic influences on human fertility and sexuality*, (pp. 253–269). Boston: Kluwer Academic.

Chapter 16

Personality

Kerry L. Jang and Shinji Yamagata

Introduction

Gordon Allport (1937) has penned some of the most influential lines in the history of personality research. He defined personality as the "... dynamic organization within the individual of those psychophysical systems that determine his unique [sic] adjustments to this environment" (p. 48) and that "... personality is something and personality does something" (p. 48). Together, these lines neatly summarize the primary mission of personality research: (1) the characterization of enduring qualities that give rise to regularities and consistencies in behaviour and the organization of these qualities and (2) how they achieve coherent functioning to actively adapt to the social environment (Livesley & Jang, 2005). As a result, much of mainstream personality research has been directed towards determining the number of basic traits, their organization, how they can be measured reliably, and the relationship between normal personality function and personality disorder.

The sheer volume of research on these questions has converged to show neuroticism (N), extraversion (E), openness to experience (O), agreeableness (A), and conscientiousness (C), popularly referred to as the "Big Five" (e.g. John & Srivastava, 1999), delineate the basic traits of normal personality. Although it would be incorrect to consider the Big Five as the definitive model of personality, its value cannot be underestimated because it serves as a useful framework in which to study personality. This is because the Big Five is related to competing personality models, such as Eysenck and Eysenck's (1975) psychoticism, extraversion, neuroticism model (PEN model) and Zuckerman, Kuhlman, Thornquist, and Kiers's (1991) alternative five-factor model in generally predictable ways. Moreover, it is of sufficient depth and breadth to subsume the ideas of alternate personality models (e.g. Aluja, Garcia, & Garcia, 2004; Larstone,

Jang, Livesley, Vernon, & Wolf, 2002; Markon, Krueger, & Watson, 2005).

Given reliability of the major personality measures and their well-understood relationship between one another, personality has been a target of behavioural geneticists for decades. Classic twin studies have yielded what is perhaps one of the most stable findings reported in the social sciences – additive genetic influences (h^2_A) account for between 40 and 50% of the total variability in personality, with nonshared environmental influences (e^2) accounting for the remainder and shared family effects (c^2) accounting for a negligible portion (e.g. Ando et al., 2002; Bouchard & Loehlin, 2001; Jang, Livesley, & Vernon, 1996; Livesley, Jang, Jackson, & Vernon, 1993). These findings have been consistently replicated across most of the popular inventories of normal and abnormal personality, awarding the psychological entity known as "personality" the status of a biologically based anatomical structure. As a result, personality inventories became the basis of several studies designed to identify putative loci for personality.

The Problem

However, something unexpected happened. Despite the stability of the phenotype and consistency of the heritability estimates, years of intense molecular genetic research has been unable to consistently identify the loci underlying any of the major personality traits (e.g. Munafo et al., 2003). One of the most famous examples is Ebstein et al.'s (1996) report that novelty seeking scores from the Temperament and Character Inventory (TCI) which was developed to operationalize Cloninger's "psychobiological model of temperament and character" (Cloninger, Svrakic, & Przybeck, 1993) was associated with the long form of the dopamine receptor D4 (DRD4) allele as predicted by the model. This association was followed by several independent replications but also non-replications by the same research group (e.g. Ebstein, Segman, & Benjamin, 1997; Ebstein,

K.L. Jang (✉)
Department of Psychiatry, University of British Columbia, Vancouver, British Columbia, V6T 2A1, Canada

Y.-K. Kim (ed.), *Handbook of Behavior Genetics*,
DOI 10.1007/978-0-387-76727-7_16, © Springer Science+Business Media, LLC 2009

Gritsenko, & Nemanov, 1997). Several other research groups also attempted to replicate this association, and two meta-analyses of these reports concluded that the association was negligible (Kluger, Siegfried, & Ebstein, 2002; Schinka, Letsch, & Crawford, 2002). This inconsistency has been reported for other dopaminergic genes and personality measures. For example, no association between DRD4 or the D2 dopamine receptor (DRD2) gene and novelty seeking measured by the Multidimensional Personality Questionnaire (MPQ; Tellegen, 1982, unpublished manuscript) was found whereas a positive association was reported between DRD2 and a measure of novelty seeking (Berman, Ozkaragoz, McDonald, Young, & Noble, 2002).

Similar inconsistencies have been reported between genes regulating the serotonergic system and neuroticism. Several studies have reported an association between the short form of the 5HTT allele (that produces less serotonin transporter mRNA) and significantly increased NEO-PI-R neuroticism scores (e.g. Lesch et al., 1996) and related traits, such as harm avoidance (Osher, Hamer, & Benjamin, 2000) accompanied by a number of non-replications (e.g. Flory et al., 1998). Even when meta-analytic techniques are used, the results are mixed. Two meta-analyses concluded that the association was significant when neuroticism was assessed with the NEO-PI-R but not with TCI or related scales (Schinka, Busch, & Robichaux-Keene, 2004; Sen, Burmeister, & Ghosh, 2004), whereas another reached the opposite conclusion (Munafo, Clark, & Flint, 2005a; but see also Munafo, Clark, & Flint, 2005b). Presently, these inconsistencies have been attributed to methodological issues, including those of sample size and power. Current wisdom in the field is to overcome these limitations with brute force methods, such as designing studies that take advantage of different family relationships, testing increasingly more loci that are located closer together, and to recruit huge samples of subjects in the hope of aggregating effect size and averaging out error (see Plomin, DeFries, Craig, & McGuffin, 2003 for a discussion).

Studies with the purpose of identifying the unique experiences, milieu, or conditions that account for individual differences in personality have also largely come up empty-handed. This has been particularly surprising given that non-shared environmental factors consistently account for over half of the overall variability observed in most personality measures and their centrality in psychological and psychiatric theory. This failure is highlighted by Turkheimer and Waldron's (2000) extensive meta-analysis of the environmental research on personality. They found that differences in family constellation variables on average accounted for 1.1% of the variance; maternal and paternal behaviour accounted for 2.3 and 1.6%, respectively; sibling interaction accounted for 2.4% on average; and peer–teacher interactions accounted for 5.3% on average. In short, measured nonshared

environmental variables leave about 90% of the variance in personality and temperament unaccounted for. Furthermore, it should be noted that when they limited their analyses to studies that used a genetically informative design (e.g. twin and family studies), the effect size of measured nonshared environmental variables, on average, was halved!

The search for explanations for this state of affairs and their resolution will likely drive personality behavioural genetics for the next decade. A critical examination of some of the current issues in mainstream personality research and behavioural genetics suggests that perhaps the failure to identify putative loci and environment is because we may have "put the cart before the horse". That is, the molecular genetic research was undertaken on perhaps the naïve assumption that the personality phenotype was well understood. How might we have misunderstood personality? Gottesman and Gould (2003) pointed out that the more complex behaviour is the more genes are involved. Flipping the problem over, it also suggests that even if it is accepted that given the convergence of research that the Big Five represents the basic traits of personality, current measures of personality do not adequately capture its complexity.

For example, there remain disagreements as to which behaviours actually define each trait. Depue and Collins (1999) reviewed the definition of extraversion as assessed by the major scales. They found that all of the scales recognized sociability and affiliation but not all recognized agency (e.g. surgency, exhibitionism), activation (e.g. activity level), impulsivity – sensation seeking (e.g. novelty seeking, monotony avoidance), positive emotions (e.g. enthusiasm, cheerfulness), or optimism. Similarly, the NEO-PI-R neuroticism scale (Costa & McCrae, 1992) contains items assessing impulsive behaviours whereas these are not measured by the EPQ-R neuroticism (Eysenck & Eysenck, 1992), indicating that the definition of trait neuroticism is fundamentally different in each model. Thus, each measure can be reflecting the action of quite different genes.

Behavioural geneticists are now faced with two broad challenges to coming to understand this complexity. First, at the level of the phenotype, do current personality measures actually have sufficient bandwidth and fidelity to reflect personality? Second, do behavioural genetic methods or how they have been used sufficiently capture and reflect this complexity? In order to meet these challenges, we must first examine some of the recent findings in the behavioural genetics of personality and the issues they raise. Given the long association of behavioural genetics and personality there are many issues to consider that no short chapter such as this can adequately cover. As such, we have chosen to broadly discuss a few of the issues that we feel will be the source of some head-scratching over the next decade.

Current Topics

Using Genetics to Understand the Personality Phenotype

There is no doubt that personality is a complex phenotype that has required a huge effort to develop a systematic and rational approach to scale development to make personality accessible for research and clinical use. A good example of this systematic and rational approach is "the lexical models of personality". This approach takes advantage of the natural language concepts used to describe behaviour. One source of such adjectives used in the past is the dictionary. Individuals are asked to rate themselves on each adjective (typically using a Likert scale of 1 "not like me at all" to 5 "very much like me") and these ratings are subjected to factor analysis, with each extracted factor representing a personality "trait". In fact, the "Big Five" basic personality traits mentioned earlier is the result of convergence of results from several lexical studies, more properly known as the five-factor model (FFM).

One of the most popular measures of the Big Five is the revised NEO personality inventory (NEO-PI-R; Costa & McCrae, 1992). Research using this inventory has shown that the five-factor structure and its psychometric properties are remarkably consistent across gender, age, race, and when translated into different languages, across cultures as well (e.g. Costa, McCrae, & Dye, 1991; McCrae & Allik, 2002; McCrae, Terracciano, & 78 Members of the Personality Profiles of Cultures Project, 2005). Given this stability, McCrae and Costa (1999) and McCrae (2004) proposed the five-factor theory (FFT) where they argued that the FFM structure is universal because the five personality traits are not affected by culture but solely shaped by biology that is common to human species.

Behavioural genetic methods, especially multivariate approaches, are well suited to test this hypothesis. Multivariate genetic analysis presumes that the observed correlation (or covariation) between two variables is mediated by both genetic and environmental factors shared by the two variables. The extent to which the two variables share genetic (i.e. pleiotropy) and environmental influences is indexed by the genetic (r_G) and environmental (r_E) correlation coefficients, respectively. Both r_G and r_E yield a coefficient that varies from -1.0 to $+1.0$ and is interpreted as any other correlation coefficient (e.g. Pearson's r) and can be subjected to factor analysis to determine the degree to which variables share a common genetic and environmental basis.

To test the basic premise of FFT, Yamagata et al. (2006) factored genetic and environmental correlations computed between all 30 of the NEO-PI-R facets using twin data from Canada, Germany, and Japan. For each sample, the congruence coefficients between the genetic, environmental, and phenotypic factors ranged from 0.95 to 0.99, indicating that both genetic and environmental factors are responsible for the patterns of trait covariation observed in phenotypic analyses of trait structure. Comparison of genetic and environmental factors across three samples also revealed that the influence of genetic and environmental factors on trait covariance is similar across cultures.

These findings can be taken as supporting the validity of FFT, although it is unclear from this study what is genetically universal – the specific structure of the FFM or covariation of personality traits in general. We raise this issue because similar studies using different personality models also showed that by and large, no matter the phenotypic structure of traits (e.g. the model under study posits that two, three, four, or five factors) the number of genetic and environmental factors and patterns of loadings remain highly congruent (e.g. Carey & DiLalla, 1994; Livesley, Jang, & Vernon, 1998; Loehlin, 1982, 1987; Krueger, 2000; McCrae, Jang, Livesley, Riemann, & Angleitner, 2001). The fact that aetiological structure almost always resembles phenotypic structure no matter what that structure may be raises questions about how basic the Big Five are – revisiting the classic "number of factors question".

An added complexity is that when one examines the personality literature most of the current research uses measures that have been developed using the concepts and language of one culture that was then exported to another. For example, the Yamagata et al. (2006) study used the NEO-PI-R which was developed using American usage of the English language and translated into German and Japanese. However, an imported scale may not capture all of the important personality characteristics found in another culture. There is a growing body of personality research using the lexical approach to analyse personality in other languages, such as Korean, Chinese, Japanese, Tagalog, and Croatian. A series of papers by Ashton and Lee (Ashton et al., 2004; Boies, Yoo, Ebacher, Lee, & Ashton, 2004; Hahn, Lee, & Ashton, 1999) has suggested a robust sixth factor describing honesty–humility or truthfulness. Also, Cheung et al. (2001) have shown that indigenous Chinese personality scales typically extract a sixth factor called "ren qing" (relationship orientation), reflecting harmony and face, personality features that do not generally appear in the English language, which were not well captured by the NEO-PI-R.

To add to the mix are the studies that fail to find the general congruence between phenotypic and aetiological structure. For example, Heath, Eaves, and Martin (1989) examined genetic and environmental structure of Eysenck's PEN model and found that the items comprising extraversion, neuroticism (and a fourth validity scale called Lie) scales were genetically and environmentally influenced via single common latent factor per scale, whereas psychoticism factor showed unusual finding of zero heritability. In a subsequent study, Heath and Martin (1990) also reported that the phenotypic structure of psychoticism corresponded well to shared and nonshared environmental structure, but not to the genetic one, suggesting that psychoticism is an aetiologically heterogeneous construct. One obvious explanation is that the difference in findings simply reflects the natural characteristics of the population under study. A second more serious possibility is questions regarding the precision of the personality phenotype in general.

It is the latter question that is of most current interest to personality researchers. For example, let us return to look at Cloninger et al.'s (1993) "Psychobiological Model of Temperament and Character" and the measures developed to operationalize its tenets (the TCI and the earlier Tridimensional Personality Questionnaire; Cloninger, Przybeck, & Svrakic, 1991) in a little detail. This model has become very popular with behavioural geneticists because it has provided substantial guidance in the selection of candidate genes since the scales were developed to reflect the influence of specific biological processes. This model hypothesizes that personality is composed of four temperament traits and three character traits. The model posits that temperament traits manifest early in life, functioning as preconceptual biases in perceptual memory and habit formation. Each trait is hypothesized to be controlled by a unique genetically based neurotransmitter system: the dopaminergic system for novelty seeking (NS); the serotonergic system for harm avoidance (HA); and the noradrenergic system for reward dependence (RD). A fourth dimension labelled "persistence" has been suggested (Cloninger et al., 1993) but no putative neurotransmitter system has been hypothesized.

The three character dimensions are self-directedness (SD), cooperativeness (CO), and self-transcendence (ST), which are hypothesized to be traits that reflect learned, maturational variations in goals, values, and self-concepts that develop in adulthood through conceptual or insight-based learning. As such, character traits should show little heritable influence in contrast to temperament traits. Despite the model's theoretical strength, psychometric examination of the TCI scales themselves has raised several questions regarding their reliability (Gana & Trouillet, 2003; Stewart, Ebmeier, & Deary, 2004). Straightforward behavioural genetic analyses have also questioned the theory and scales. First, contrary to predictions, the character traits have been found to be substantially heritable[1] (Ando et al., 2002; Gillespie, Cloninger, Heath, & Martin, 2003), and molecular

genetic studies have found that 5-HTTLPR gene was associated with cooperativeness, but not with temperament traits (e.g. Kumakiri et al., 1999).

Ando et al. (2004) examined genetic and environmental factor structure of TCI and used the findings of multivariate genetic analyses to reorganize the content of the TCI scales. Each of the TCI dimensions, like most personality scales, is composed of several narrow sub- or "facet" traits. For example, NS is defined by exploratory excitability, impulsiveness, extravagance, and disorderliness. The genetic correlation between all of the facet traits defining NS, HA, and RD dimensions on a sample of 414 pairs of MZ and 203 DZ twin pairs from Japan was factored (see Table 16.1). Factor analysis of the genetic intercorrelations yielded factors that did not quite resemble the phenotypic structure of NS, HA, and RD. As shown in Table 16.1, only the subtraits defining reward dependence (factor II) lined up as originally designed. Using this information, harm avoidance (r-HA) was revised to consist of (low) exploratory excitability, anticipatory worry, fear of uncertainty, shyness, and fatigability. Novelty seeking (r-NS) was revised to consist of impulsiveness, extravagance, and disorderliness, and RD was unchanged. The genetic and environmental correlations between r-NS, r-HA, and RD were very small (ranging from −0.02 to 0.11), indicating that the revised temperament scales were rendered genetically homogeneous and independent. What is important about these results is that they suggest that the genotyping research based on TCI dimensions as originally designed is reflecting several, possibly competing, influences. Thus, if the primary phenotype was the total dimension score, it would be unclear as to what specific personality trait is actually associated with the gene.

This issue highlights one research agenda for behavioural genetics – the revision of scales to be more genetically homogeneous. Genetic correlations are not only useful in revising

Table 16.1 Varimax rotated principal factor analysis loading matrix of the genetic correlations estimated between the TCI temperament subscales

	I	II	III	IV
Novelty seeking				
Exploratory excitability	0.62	0.28	0.25	0.29
Impulsiveness	−0.10	0.03	0.03	0.76
Extravagance	0.02	0.16	0.00	0.72
Disorderliness	0.03	−0.16	0.15	0.74
Harm avoidance				
Anticipatory worry	−0.87	0.01	−0.03	−0.01
Fear of uncertainty	−0.51	0.28	−0.32	−0.43
Shyness	−0.76	−0.16	−0.22	−0.05
Fatigability	−0.72	−0.26	−0.06	−0.10
Reward dependence				
Sentimentality	−0.06	0.56	0.60	−0.04
Attachment	0.04	0.70	0.00	0.22
Dependence	−0.09	0.86	−0.19	0.13

Source: Ando et al. (2004).

[1] Cloninger conceded that character traits are as heritable as temperamental traits, yet have different biological base than temperament traits (personal communication, July, 2005; see also Gillespie et al., 2003).

content, but they can also be used to address basic psychometric issues such as reliability. For example, this approach can be easily applied to the correlations between items to index the internal consistency of personality scales as well as test–retest reliability. Genetic (and environmental) correlations between scores of a personality scale measured at different time intervals (e.g. longitudinal twin study) provide information regarding the extent to which a trait is consistently influenced by the same genes across time (e.g. Gillespie, Evans, Wright, & Martin, 2004). Genetically reliable personality measures are suitable for linkage/association study because it provides more statistical power to detect genes. It should be also noted that similar logic could also be applied to the use of environmental correlations, which can help study searching for the specific environmental factors influencing personality.

Another recommendation for future research is to use genetic/environmental factor scores for linkage/association studies. Sham et al. (2001) recently described a basic process that enables a score on any measure, like neuroticism, to be split into two scores: to reflect pattern of genetic influences and the other environmental influences. The basic approach is to derive a weight for each of the genetic and environmental effects that can be applied to the phenotypic score of an existing inventory. These weights function much like weights used to compute factor scores but instead of being derived from the phenotypic correlations between variables, they are derived from matrices of genetic and environmental correlations.

Theoretically, it is possible even to break the genetic score further into separate scores that reflect the variability in neuroticism due to different sets of genes. These genetically and environmentally indexed scales would reduce genetic/environmental noise in a measure of a behaviour when searching for specific environment/genes influencing the trait and increase the power to find them (Lander & Botstein, 1989), as was reported in several simulation studies and studies using actual data (e.g. Eaves & Meyer, 1994; Cardon, Smith, Fulker, Kimberling, Pennington, & DeFries, 1994; Boomsma, 1996; Boomsma & Dolan, 1998). For clinicians, such scales would be also extremely useful as they could reflect the level at which psychotherapeutic or pharmacological treatments are acting.

What Is Inherited?

Genetic and environmental factor analyses discussed above are simple forms of what is called the "independent pathway (IP) model" illustrated in Fig. 16.1. In this model, higher-order constructs (e.g. neuroticism) simply reflect sum total of the pleiotropic action of genes and environmental influences shared by all lower-order traits rather than the effects of a phenotypic entity. Thus, under an IP model, higher-order constructs are, in effect, relegated to the status of a convenient heuristic device to label the covariance of traits.

A more stringent multivariate genetic model is the "common pathway (CP) model" illustrated in Fig. 16.2. The centre section of this figure is identical to the form of the contemporary factor analysis model. The addition of genetic and environmental influences to the latent variable P (note that the existence of P is unmeasured and is inferred by the degree to which the measured variables appear together), which represents a higher-order trait, transforms the latent variable into a veridical entity that has a basis in biology. The most important aspect of this figure is that P mediates 100% of the covariance of the lower-order traits that define it and that each of the lower-order traits is best understood as exemplars of the higher-order construct. Although both IP and CP models hypothesize that personality trait facets share a common genetic basis, the primary difference is that in the IP model, *no* independently inherited P is required to explain the covariation between personality measures. Knowing whether personality is structured like an IP or CP model is fundamentally important because each suggests quite different approaches to the search for putative genes. For example, if the CP model is correct, this implies that there are specific genes for the higher-order construct or domain. As such, as in current approaches, the phenotypic factor score could be used because each facet is an exemplar of the domain that has a common genetic basis with all other aspects of the domain. In contrast, if the IP model is correct, it would mean that either the total summative score across facet traits or phenotypic factor score would not be ideal because it could reflect different, possibly competing, aetiological influences. Rather, a phenotypic factor score based on only those subsets of facets that share a common genetic basis can be associated with various candidate genes (or genetic factor score as described above can be used).

It is important to note that the IP and CP models illustrated in Figs. 16.1 and 16.2 appear equivalent because every facet trait is shown to share a common genetic and environmental basis with the others defining each domain. However, in the CP model it must be remembered that because of the latent phenotype P, *all* facets must share a common genetic and environmental basis. When this is not the case, for example, when one of the facets may not share a common genetic basis with the other facets, the CP model is effectively reduced to the IP model in which that path between the common genetic factor G and the affected facet trait is set to zero. The observed covariance of the affected facet with the others is maintained by the environmental influences in common. Thus, it should be clear that the CP and IP models provide quite different explanations for why facet traits covary.

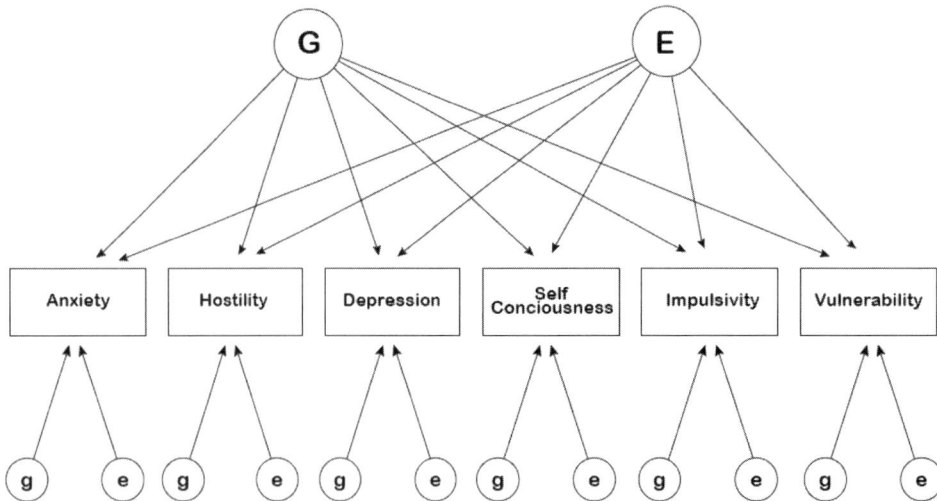

Fig. 16.1 Independent pathways model

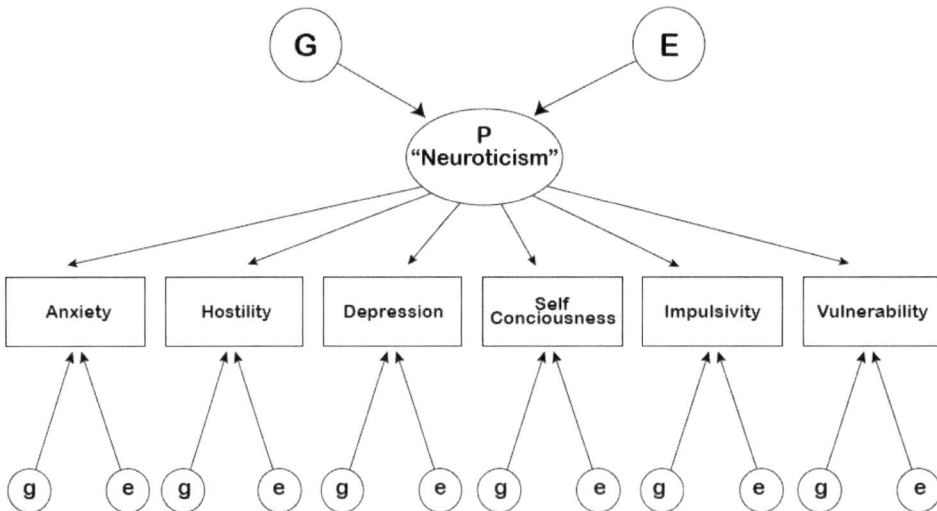

Fig. 16.2 Common pathways model

The relative correctness of the IP or CP models has important implications for personality theory. It speaks of the validity of higher-order traits such as N, E, O, A, and C as veridical entities.

Despite the importance of these structural models for personality research and theory, current research using IP and CP models has yet to provide a clear indication as to how personality is organized. For example, Jang et al. (2002) fit both CP and IP models to the six facets defining each of the NEO-PI-R domains and the CP model was rejected for all domains. Instead, an IP model specifying two additive genetic factors and two nonshared environmental factors provided the best fit. This finding suggests that there are *no* genes for N, E, O, A, and C per se. Rather, as noted earlier, higher-order traits are simply a convenient heuristic to describe the action of genes. This conclusion is consistent with the lex-

ical view of Saucier and Goldberg (1996), who argued that the five domains are merely a convenient way of organizing lower-order traits. In contrast, Johnson and Krueger (2005a) reported that for the adjectives describing neuroticism and extraversion, a CP model provided a good fit, and similar to Jang et al. (2001), for the adjectives describing agreeableness, conscientiousness, and openness, the IP model provided the best fit. This finding supports the traditional view that N and E are inherited entities and can be treated as such.

However, it should be noted that in these two papers, the decision to retain the IP or CP model was based on what could be called "flexible" statistical grounds. For example, by accepting a slightly different cut-off point for a fit index, such as 0.08 as opposed to 0.05, either the CP or IP model provides a satisfactory fit. Similarly, both papers used different fit indices to make their decisions. It is quite possible that electing to place emphasis on one fit index over the other (e.g. Bayesian information criterion versus the root mean square residual) can yield quite different conclusions.

In fact, in current research it is quite common to find that *both* the CP and IP models provide a satisfactory explanation for the data. To illustrate the phenomenon, consider Young, Stallings, Corley, Krauter, and Hewitt's (2000) genetic analysis of adolescent behavioural disinhibition. In this study of DSM-IV symptom counts for conduct disorder, attention deficit disorder, substance experimentation, and novelty seeking, a one-factor CP model and a one genetic factor IP provided a good fit to the data. The reported $\chi^2_{difference}$ between the models was a non-significant ($p > 0.05$) 1.51. As a result, Young et al. selected the common pathways model because "...this is a more parsimonious model than the independent pathway model and shows no significant decrement in fit by χ^2 difference test..." (pp. 690–691). Because statistical guidance was lacking, the authors picked the simplest model. However, is the CP model actually the simplest? It adds complexity in that it trades off simplicity for stringency because it specifies that the trait covariation is entirely due to the mediating action of a higher-order variable. It can be argued that being able to drop the requirement of having to invoke an unmeasured higher-order construct renders a simpler explanation for the data.

At the present time, it appears that when statistical guidance is lacking, the choice of model can become rather arbitrary. In the meantime, the easiest way to help resolve the problem is to conduct replication studies, use of larger sample sizes, and power calculation to help gauge the veracity of the results.

Role of Personality Theory

Theories of personality and personality development in mainstream personality research have tended to take a backseat to efforts focused on developing reliable measures that had tended to use an empirical approach (e.g. lexical approach). However, this has caused a disconnect between trait models of personality and concepts used to describe personality and its development (Digman, 1997). For example, do any of the major personality domains assessed by any of the major inventories capture the concepts from Freud's theories on psychosocial development?

Why is this important? Some may argue that these theories are only of historical interest to personality psychologists who would not expect behavioural genetic or molecular genetic studies to address the constructs of classical personality theory. We disagree. Personality theory is the source of testable hypotheses regarding personality development,

the relationship between different traits, its structure, and the role personality plays in everyday living and mental illness. Human beings strive for "personal growth" or "self-actualization" – not just to be more or less extraverted. Such striving implies the interplay of personality traits from several domains, such as extraversion and openness to experience. The real question for personality research is how do trait concepts come together (genetic? environmental?) to uncover the mechanisms (genetic? environmental? interplay?) through which "self-actualizing" tendencies develop and are expressed? Behavioural genetics can help address these basic questions just as they did for trait models. Moreover, in light of our previous discussion on model selection, theory is a means to guide the selection of model when statistical criteria are ambiguous.

There is a growing body of research along these lines. For example, phenotypically, Digman (1997) has shown that two higher-order factors have been consistently extracted from the Big Five personality factors (Digman, 1997). The first he called alpha (α) which was typically defined by factor loadings from agreeableness, conscientiousness, and neuroticism which encompass aggression, hostility, impulse restraint, and neurotic defence akin to Freud's theories on psychosocial development or Adler's "social interest". Similarly, the emergence of extraversion and openness to experience as a single factor (β) is a far more comprehensive reflection of constructs such as Rodger's "personal growth", Adler's "superiority striving", or Maslow's "self-actualization" than is possible by either trait alone. He argued that such higher-order traits are important because they provide a tangible link between psychometric models used to develop reliable taxonomies and measures to theories of personality development.

Despite the apparent increase in conceptual clarity afforded by α and β, their actual relationship to developmental concepts has not been evaluated empirically. Such work cannot be conducted until the stability of these higher-order traits is better established. A recent paper by Jang et al. (2006) has begun to address this issue by conducting multivariate genetic analyses of the five NEO-PI-R domain inducting to determine if the α and β constructs can be reliably reproduced across a diverse range of independent samples. They report that two CP models could explain the covariance of N, E, O, A, and C from twin samples drawn from Canada, Germany, and Japan. The first model broadly represented α and the second resembled β.

Of particular importance is the finding that although the domains share some genetic and environmental influences to differing degrees, a great deal of the variability of domains is due to genetic and environmental factors *unique* to each facet. For example, Jang, McCrae, Angleitner, Riemann, and Livesley (1998) estimated the heritability of the 30 NEO-PI-R facet traits after all genetic influence due to the higher-order traits was removed using regression techniques. Substantial residual heritability was found for each trait that accounted for between 25% (competence) and 65% (dutifulness) of the variance. Livesley et al. (1998) also reported similar findings for personality disorder traits. These findings illustrate the oft forgotten fact that not all domains play equally important roles in personality when in practice, the opposite is often assumed. Personality research often tends to focus on what is common between traits when in fact, genetic and environmental factors specific to each facet or domain account for a greater proportion of the total variability. As such, when working with personality scale scores, it is not clear at which level the genes and environmental influences are exerting their influences; certain genes may contribute to facet-specific genetic variance, perhaps even item-specific variance, whereas other genes contribute to the variability of the domains or higher-order factors such as α. A simple recommendation for association study is to examine effects of specific genes/environment at all levels.

Consistency of Heritability Estimates

The chapter opened with a pronouncement that additive genetic and nonshared environmental influences accounted for all of the variability in personality measures. However, there exist several studies which suggest that not all of the genetic variability is additive (e.g. Loehlin, 1986, 1992; Loehlin, Horn, & Willermann, 1997; Loehlin & Nicholls, 1976) but non-additive (dominance or additive-to-additive epistasis effects; symbolized as d^2). There are a number of potential explanations for these findings, some of which are methodological and some of which causes one to pause and think about the inclusiveness of the major models of personality. For example, one possibility is that the presence of these effects is a reflection of content unique to specific personality scales such as the MPQ (Waller & Shaver, 1994), the TCI (Cloninger et al., 1993), or the Revised Eysenck Personality Questionnaire (EPQ-R; Eysenck & Eysenck, 1975), as reported in Keller, Coventry, Heath, and Martin (2005), in contrast to studies using the NEO-PI-R (e.g. Jang et al., 1996, 1998; Riemann, Angleitner, & Strelau, 1997) in which d^2 effects are rarely, if ever, found. This suggests that any one personality inventory alone does not necessarily provide a comprehensive assessment of personality.

The presence of non-additive effects may not be a personality measurement issue per se, but rather attributable to the possibility that behavioural genetic methods are not sensitive enough to reliably detect these effects. This is suggested by the reports of non-additive effects inconsistently found on the same measures (e.g. Ando et al., 2002; Eaves, Eysenck, & Martin, 1989; Gillespie et al., 2003;

Heiman, Stallings, Hofer, & Hewitt, 2003). This is because additive and non-additive genetic effects (both epistasis and dominance) are confounded in a twin-reared-together design, and only when large sample size is used and other genetically informative relationships (e.g. parents or non-twin siblings) are incorporated into the twin-reared-together design are non-additive effects reported more consistently (Eaves et al., 1999; Eaves & Carbonneau, 1998; Finkel & McGue, 1997; Keller et al., 2005; Lake, Eaves, & Maes, 2000). Several association studies showing that only combination of several genes can explain variance in personality suggest that non-additive effects are an important influence in personality (e.g. Benjamin, Osher, Kotler, et al., 2000; Benjamin Benjamin, Osher, Lichtenberg, et al., 2000; Strobel, Lesch, Jatzke, Paetzold, & Brocke, 2003).

Methodological issues also impact the magnitude of genetic effects. Although heritability in the 30–50% is typically observed for self-report measures, when multiple peer ratings or both self- and peer ratings are subjected to behavioural genetic analysis, h^2_A has been found to account for between 60 and 80% of the total variability (e.g. Heath, Neale, Kessler, Eaves, & Kendler, 1992; Riemann et al., 1997; Wolf, Angleitner, Spinath, Riemann, & Strelau, 2004). Multiple peer ratings can control rater-specific bias, which are included in estimates of e^2, and thus are considered more objective than self-report. Thus, inflated rates of measurement error may be an important contributing factor for the disappointing results of studies designed to find sources of e^2 because it is possible that a significant proportion (50–70%) of nonshared environmental variance typically observed for classical twin studies may largely reflect measurement error and not actual environmental variation.

Another issue regarding the consistency of the heritability estimates is gene–environment interaction. If one thinks about what the heritability coefficient represents, one could say it represents (1) a snapshot of the magnitude of genetic effects on any sample and (2) the average genetic effect over all conditions. Thus, it could be that all research is correct, and the differences in results – be they variations in values of h^2 or type of genetic effect, h^2_A, h^2_D, c^2 – simply reflect the fact that different genes are in action at different times or in the face of certain environments. Personality theory has always stressed that events in childhood are important in personality development and future psychopathology. Despite the ubiquity of primacy of early experience in many of the theories of psychopathology, the primacy of early experience has a weak evidence base (e.g. Paris, 2001). For example, in the personality disorders a large body of research (e.g. Garmezy & Masten, 1994) shows that negative childhood experiences need not necessarily lead to psychopathological outcomes in adult life. As a result, theorists suggest that adversities in combination with genetic liabilities during development increase the risk for mental disorders. Within the medicine and the social sciences this process of gene–environment interplay is called the diathesis-stress model of illness. This model in some form is usually invoked to explain why despite the fact that many people may carry the genes for mental illness, or may have experienced some kind of horrible trauma, not all of them will develop mental illness. In its simplest form, the model postulates that adverse experiences interact with an underlying genetic liability that leads to disease.

However, this explanatory model is very broad and does not specify the mechanisms of gene–environment interplay. Behavioural genetics has provided the means to test and delineate the mechanism of action. One such mechanism is genetic control of exposure to the environment in which genetic factors influence the probability of exposure to adverse events (Kendler & Eaves, 1986). In some fields, this phenomenon is referred to as an "amplification effect" (e.g. Paris, 1994, 1996) but within behavioural genetics it is called genotype by environment correlation – the extent to which individuals are exposed to environments as a function of their genetic propensities. Three general types of genotype–environment correlation have been hypothesized (see Plomin, DeFries, & Loehlin, 1977; Scarr & McCartney, 1983). The first is passive genotype–environment correlation that occurs because children share heredity and environments with members of their family and can thus passively inherit environments correlated with their genetic propensities. The second is called reactive in which the experiences of the child are derived from reactions of other people to the child's genetic propensities. The third is the active type that occurs when children actively select or create environments commensurate with their underlying genetic propensities.

The other interplay effect is environmental moderation of genetic and environmental variability referred to as gene–environment interaction (Plomin, DeFries, and McClearn, 1990). The importance of gene–environment interaction has been shown by several molecular genetic studies. For example, Caspi, McClay, and Moffitt's (2002) study of the development of antisocial behaviour is particularly salient to this discussion. Clinical research has identified childhood abuse, such as erratic, coercive, and punitive parenting, as one of the major risk factors for the development of antisocial behaviour in boys, and the risk for conduct disorder increases the earlier the abuse begins. However, there is often little 1:1 correspondence between environmental conditions and phenotype, so the presence of a genetic liability for the disorder must be involved. In the case of antisocial behaviour, the monoamine oxidase A gene (MAOA gene Xp11.23-11.4) was selected because it has been associated with aggressive behaviour in mice as well as some human studies.

This sample consisted of 1,037 children who had been assessed at 9 different ages for levels of maltreatment (no, probable, and severe maltreatment) and MAOA activity (low or high). They found that the effect of maltreatment was significantly weaker among males with high MAOA activity than those with low activity. Moreover, the probable and high maltreatment group did not differ in MAOA activity, indicating that the genotype did not influence exposure to maltreatment. These results demonstrate that the MAOA gene modifies the influence of maltreatment. Similar dramatic gene–environment interaction effects have been shown for genes believed implicated in depression (e.g. Caspi et al., 2003; but see also Eaves, 2006, for possibility of statistical artefacts in these studies).

For classical twin studies, there is a growing body of literature demonstrating that heritability varies over environmental condition. For example, Jang et al. (2005) showed that perceived levels of family conflict and maternal indulgence moderated the genetic influences underlying emotional instability, a trait delineating personality disorder, and is related to neuroticism. Specifically, this study found that the estimates of h^2_A varied between 70 and 44% over levels of maternal indulgence and 92 and 15% over levels of family conflict. Moderation was not limited to genetic effects, shared family environment effects (c^2), and ranged from 23 to 40% and 30 to 44% over levels of maternal overprotection and paternal care, respectively. This final set of results suggests that the apparent lack of c^2 is the result of its effects being averaged out over different conditions and that they might be cumulative. Moreover, it also highlights the fact that environment–environment interaction or experience by environment interaction is just as important. Such models allow us to test hypotheses about how some people can live in the most adverse conditions (e.g. extreme poverty) but display no ill effects. Is it because of the presence of another environmental factor, such as a caring mother who attends to the emotional needs of a child, cancelling out the influence of poverty?

There are an increasing number of studies reporting gene–environment interactions. Recently, Button, Scourfield, Martin, Purcell, and McGuffin (2005) observed that increasing genetic influences on antisocial behaviours with a concomitant decrease in c^2 as level of family dysfunction increased in samples of 5- to 18-year-old children. Boomsma, de Geus, van Baal, and Koopmans (1999) observed that genetic influences on disinhibition were smaller and shared environmental influences were larger for male adolescents raised in a religious family than for those raised in a non-religious family. Turkheimer, Haley, Waldron, D'Onofrio, and Gottesman (2003) observed that among 7-year-old children, genetic influences on intelligence were larger and shared environmental influences were smaller for those in more impoverished families (see also

Dick, Rose, Viken, Kaprio, & Koskenvuo, 2001; Johnson & Krueger, 2005a, 2005b). It is clear that the study of interplay effects is vitally important for personality research. Studies of this kind may be the key to understanding how personality develops and explain the inconsistencies in the current behavioural genetic literature.

What Does Personality Do?

Allport (1937) also noted that "personality does something". Basically, personality is a general disposition that governs the way individuals normally behave across situations, and current research in this area has tended to focus on the role of personality on mental illness. Clinical studies consistently find that personality features are correlated (i.e. comorbid) with virtually all forms of common psychopathology. For example, individuals with depression frequently report being anxious (Kendler, 1996; Zimmerman, Chelminski, & McDermut, 2002). Further, personality features have been delineated as diagnostic criteria in many forms of psychopathology. For example, the diagnostic criteria for DSM-IV major depressive episode (APA; 1994) includes "feelings of guilt and worthlessness", item content typically found in measures of trait neuroticism.

A review of the personality literature suggests that personality impacts psychopathology in three ways (Jang et al., 2006). The first hypothesizes that personality factors increase the risk for developing psychiatric disorder. In this model, both personality and psychopathology are qualitatively distinct entities; however, certain personality dimensions alone or in combination with others increase the likelihood of developing psychiatric disorder (e.g. Metalsky, Halberstadt, & Abramson, 1987). The second hypothesizes that personality and psychopathology occupy a single domain or spectrum and psychopathology is simply a display of the extremes of normal personality function (Eysenck, 1994). For example, Trull, Waudby, and Sher (2004) reported that DSM-IV cluster B personality disorder symptoms (particularly antisocial and borderline symptoms) were consistently associated with alcohol use disorders in a non-clinical sample of 395 young adults but also significantly associated with alcohol use disorders above and beyond what was accounted for by normal personality traits. This suggests that personality disorder symptoms predict unique variance in substance use disorders that clearly reflects maladaptive aspects of personality. The third hypothesis is that personality plays a minor role in the development of disorder and changes in observed personality are simply the result of disorder. For example, although minor personality changes are frequently observed before the onset of major depressive disorder, major personality

changes occur after onset (e.g. Chien & Dunner, 1996; see also Goldsmith, Lemery, & Essex, 2004; Widiger, Verheul, & van den Brink, 1999, for a similar classification).

Little, if any, current behavioural genetics research has directly investigated these possibilities. Rather, research has largely been descriptive, tending focus on showing that the comorbidity of personality and psychopathology is partially due to a common genetic basis. However, current findings provide some circumstantial evidence for the spectrum model of disorder. For example, Krueger (1999) reported that the phenotypic comorbidity of 10 DSM-III-R common mental disorders can be arranged into 2 higher-order constructs that describe disorders directed inward towards oneself ("internalizing" such as depression) as opposed to disorders that are directed outward ("externalizing" such as antisocial personality). In a subsequent study, Krueger et al. (2002) fitted a variety of genetic models to the data and found that a one-factor common pathway genetic model (similar in form to Figs. 16.1 and 16.2) provided the best explanation for the covariation of these measures, suggesting that the externalizing disorders are inherited as a single genetically based syndrome. However, despite Krueger et al.'s (2002) report that the fit of a one-factor common pathways model provided a superior fit to any model (e.g. models that did not include P), as discussed earlier, some lingering doubts remain as to whether this model truly provides the best explanation. For example, Kendler, Prescott, Myers, and Neale (2003) found that the internalizing and externalizing factors are not inherited as a single genetically based syndrome at all. They showed that an independent pathways model (of the form illustrated in Fig. 16.1) in which multiple genetic and environmental factors directly influenced each disorder provided a far superior fit to their data.

So which is correct? Are syndromes inherited as a unitary entity or are they convenient names for frequently comorbid disorders that share a common genetic basis? Once again, these kinds of results highlight important strengths and weaknesses of the behavioural genetic models and approaches used to study comorbidity. On the one hand, the models explain why comorbidity exists and provides, in principle, the means to study the organization of variables. As discussed earlier, what the current research has highlighted is our reliance on statistics and fit indices to make our decisions. This is not a new idea, because as shown quite often in the literature, such indices alone are insufficient to make a decision.

Directions for the Future

The current behavioural genetics of personality research has been extremely important in bringing to light a number of issues concerning what personality is and what personality does. We have also highlighted broad directions where more research needs to be spent. The remainder of this chapter will present some potential ideas to resolve these issues. A theme of this chapter is the complexity of the personality phenotype. Another approach that might provide some insight into disentangling this complexity is to develop personality endophenotypes (neurobiological correlates of personality) and use these to validate existing scales (see Gottesman & Gould, 2003). This approach has been useful in alcohol research. For example, Hesselbrock, Begleiter, Porjesz, O'Connor, and Bauer (2001) noted that a difficulty with genotyping studies is that the clinical heterogeneity of the disorder results in a poorly defined phenotype for genetic analysis and that better results may be obtained by switching to diagnostic endophenotypes.

One endophenotype that has received considerable attention is the P300 event-related brain potential (ERP) waveform. ERPs are recordings of neuro-electrical activity in response to stimuli recorded by electrodes on the scalp. Common stimuli include flashing lights, reactions to a short emotionally laden video clip, or noises (beeps, etc.). This endophenotype has been shown to be a valid neurobiological correlate of alcoholism in both males and females (Prabhu, Porjesz, Chorlian, Wang, Stimus, & Begleitner, 2001). Hesselbrock et al. (2001) found significant reductions in P300 amplitude between alcoholics and non-alcoholics, between unaffected relatives of alcoholics and relatives of controls, and between unaffected offspring of alcoholic fathers and offspring of controls. Almasy et al. (2001) conducted a genome-wide scan of P300 responses to a semantic priming task on 604 individuals in 100 pedigrees. They showed that the P300 waveform was significantly heritable (40–50% range) and reported significant evidence of linkage to chromosome 5 and suggestive evidence of linkage to chromosome 4.

Within personality research, a promising endophenotype suitable for linkage/association study is frontal electroencephalographic (EEG) asymmetry (Allen & Kline, 2004; Davidson, Jackson, & Larson, 2000). Frontal EEG asymmetry represents balance between cortical activity in left and right frontal regions; more specifically, a frontal EEG asymmetry score is obtained by subtracting log-transformed power density value in the alpha band (8–13 Hz) in the left frontal cite (e.g. F3) from the right frontal cite (e.g. F4). Because alpha power is inversely associated with cortical activity, the positive score indicates greater left-sided activity (see Tomarken, Davidson, Wheeler, & Doss, 1992). Previous research has shown that greater left-sided activity was consistently associated with stronger behavioural activation system (i.e. sensitivity to reward; Gray, 1987) and weaker behavioural inhibition system (i.e. sensitivity to punishment; Gray, 1987; Coan & Allen, 2003; Harmon-Jones & Allen, 1998; Sutton & Davidson, 1997), more positive and

less negative mood (Tomarken et al., 1992), higher reactivity to positive stimuli and lower reactivity to negative stimuli (Tomarken, Davidson, & Henriques, 1992; Wheeler, Davidson, & Tomarken, 1993), and reactivity to maternal separation among infants (Davidson & Fox, 1989). Lastly, a pilot study observed that frontal EEG asymmetry was heritable and genetically associated with negative emotionality among 66 female twin pairs (Coan, Allen, Malone, & Iacono, 2003).

For those interested in what personality may do, behavioural genetic methods provide the means to determine the role(s) any specific personality variable plays in a disorder. As discussed earlier, does variation in trait anxiety increase the vulnerability to depression, or is depression the extremes of normal anxiousness, or are changes in anxiety simply a consequence of depression? Current behavioural genetic methods provide some avenues to explore these alternatives. A straightforward way to investigate if personality acts as a risk factor is to apply models of gene–environment correlation and gene–environment interaction. The validity of a personality-as-risk-factor model rests on demonstrating comorbidity that is not contemporaneous but veridical by being caused by a shared aetiology estimated by r_G and r_E. Second, it depends on some form of gene–environment correlation – that is, genetically based personality traits help create or modify environments suspected to increase risk for a particular psychopathology. The third step is to demonstrate that the environment that personality helped shape "triggers" for the onset of another genetically based disorder. For example, people with high levels of genetically based sensation seeking prefer an urban as opposed to rural lifestyle where alcohol and other substances are readily available. The accessibility of alcohol could trigger the onset of genetically based alcoholism via the mechanism of gene–environment interaction.

A test for a personality-psychopathology spectrum is that personality and disorder share a common aetiological basis. However, one possible way to differentiate the risk and spectrum models is the degree to which personality and the other disorder share a common genetic basis and how this shared aetiology is structured. What those conditions are would be a fascinating research process. Finally, it is important that time is included in these tests. It is clearly the most crucial element to determine if personality is the cause or effect of psychopathology. To test if personality is simply a consequence of other mental illnesses, it must be shown that significant personality changes do not predate the onset of disorder. However, this may not be enough. Recall that quantitative genetic theory states that the phenotype represents the sum of genetic and environmental action, and it should be clear that genetic factors underlying personality can substantially increase the risk for the development of another disorder, although not reflected in the phenotype. As such, a significant genetic correlation can exist between two variables, but because environmental factors can work in the opposite direction (that is, r_E is negative) the sum of these effects as reflected in the phenotypic correlation is zero (Carey, 2003), suggesting that the two variables are unrelated when in fact they are related.

Thus, it becomes extremely important to study genetic and environmental influences in the context of time. Current longitudinal behavioural genetic research has shown that different aetiological factors can operate at different times during development as shown by the well-known Nonshared Environment in Adolescent Development project (Neiderhiser, Reiss, & Hetherington, 1996) which has shown that genetic influences were important for both change and stability in antisocial behaviour.

In summary, the ability of behavioural genetic methods to move beyond the phenotype to focus on aetiology has been extremely important in supporting long-held assumptions or has shaken the foundations of issues long thought resolved in traditional personality research. It has also certainly minimized the centrality of some long-debated issues, such as the number of domains debate, but has renewed issues over content and definition of the domains. It has also begun to force an integration of personality measurement and personality theory. The goal of behavioural geneticists working in personality is to continue to apply these methods to resolve these issues. That is, these methods should be applied to issues important in personality research such as to aid in scale development and revision as opposed to uncritically accepting current conceptions of personality. It is hoped that the ideas in this chapter will stimulate thought and provoke a more thorough integration of personality and behavioural genetics.

Acknowledgment Special thanks to the three anonymous reviewers for their comments and suggestions. Many of these are incorporated in this chapter. Given the volume of their comments, it is clear that behavioural genetics will continue to play a huge role in personality research for some time to come.

References

Allen, J. J. B., & Kline, J. P. (2004). Frontal EEG asymmetry, emotion, and psychopathology: The first, and the next 25 years. *Biological Psychology, 67*(1–2), 1–5.

Allport, G. W. (1937). *Personality: A psychological interpretation* (p. 48). New York: Holt, Rinehart & Winston.

Almasy, L., Porjesz, B., Blangero, J., Goate, A., Edenberg, H.J., Chorlian, D.B., et al. (2001). Genetics of event-related brain potentials in response to a semantic priming paradigm in families with a history of alcoholism. *American Journal of Human Genetics, 68*, 128–135.

Aluja, A., Garcia, O., & Garcia, L. F. (2004). Replicability of the three, four and five Zuckerman's personality super-factors: exploratory and confirmatory factor analysis of the EPQ-RS, ZKPQ and NEO-PI-R. *Personality and Individual Differences, 36*(5), 1093–1108.

Ando, J., Ono, Y., Yoshimura, K., Onoda, N., Shinohara, M., Kanba, S., et al. (2002). The genetic structure of Cloninger's seven-factor model of temperament and character in a Japanese sample. *Journal of Personality, 70*(5), 583–609.

Ando, J., Suzuki, A., Yamagata, S., Kijima, N., Maekawa, H., Ono, Y., et al. (2004). Genetic and environmental structure of Cloninger's temperament and character dimensions. *Journal of Personality Disorders,18*(4), 379–393.

Ashton, M. C., Lee, K., Perugini, M., Szarota, P., de Vries, R. E., DiBlas, L., et al. (2004). A six-factor structure of personality-descriptive adjectives: Solutions from psycholexical studies in seven languages. *Journal of Personality and Social Psychology, 86*(2), 356–366.

Benjamin, J., Osher, Y., Kotler, M., Gritsenko, I., Nemanov, L., Belmaker, R. H., et al. (2000). Association between tridimensional personality questionnaire (TPQ) traits and three functional polymorphisms: dopamine receptor D4 (DRD4), serotonin transporter promoter region (5- HTTLPR) and catechol O-methyltransferase (COMT). *Molecular Psychiatry, 5*(1), 96–100.

Benjamin, J., Osher, Y., Lichtenberg, P., Bachner-Melman, R., Gritsenko, I., Kotler, M., et al. (2000). An interaction between the catechol O-methyltransferase and serotonin transporter promoter region polymorphisms contributes to Tridimensional Personality Questionnaire persistence scores in normal subjects. *Neuropsychobiology, 41*(1), 48–53.

Berman, S.A., Ozkaragoz, T., Young, R.M., & Noble, E.P. (2002). D2 dopamine receptor gene polymorphism discriminates two kinds of novelty seeking. *Personality and Individual Differences, 33*, 867–882.

Boies, K., Yoo, T. Y., Ebacher, A., Lee, K., & Ashton, M. C. (2004). Psychometric properties of scores on the French and Korean versions of the HEXACO Personality Inventory.*Educational and Psychological Measurement, 64*(6), 992–1006.

Boomsma, D. I. (1996). Using multivariate genetic modeling to detect pleiotropic quantitative trait loci. *Behavior Genetics, 26*, 161–166.

Boomsma, D. I., & Dolan, C. V. (1998). A comparison of power to detect a QTL in sib-pair data using multivariate phenotypes, mean phenotypes, and factor scores. *Behavior Genetics, 28*, 329–340.

Boomsma. D. I., de Geus, E. J. C., van Baal, G. C. M., & Koopmans, J. R. (1999). A religious upbringing reduces the influence of genetic factors on disinhibition: Evidence for interaction between genotype and environment on personality. *Twin Research, 2*, 115–125.

Bouchard, T. J., Jr., & Loehlin, J. C. (2001). Genes, evolution, and personality. *Behavior Genetics, 31*(3), 243–273.

Button, T. M. M., Scourfield, J., Martin, N., Purcell, S., & McGuffin, P. (2005). Family dysfunction interacts with genes in the causation of antisocial symptoms. *Behavior Genetics, 35*(2), 115–120.

Cardon, L. R., Smith, S. D., Fulker, D. W., Kimberling, W. J., Pennington, B. F., & DeFries, J. C. (1994). Quantitative trait locus for reading disability on chromosome 6. *Science, 266*, 276–279.

Carey, G. (2003). *Human genetics for the social sciences.* Thousand Oaks: Sage Publications.

Carey, G., & DiLalla, (1994). D.L. Personality and psychopathology: Genetic perspectives. *Journal of Abnormal Psychology, 103*, 32–43.

Caspi, A., McClay, J., & Moffitt, T. (2002). Role of genotype in the cycle of violence in maltreated children. Science, 297(5582), 851–854.

Caspi, A., Sugden, K., Moffitt, T. E., Taylor, A., Craig, I. W., Harrington, H., et al. (2003). Influence of life stress on depression: Moderation by a polymorphism in the 5-HTT gene. *Science, 301*(5631), 386–389.

Cheung, F. M., Leung, K., Zhang, J. X., Sun, H. F., Gan, Y. Q., Song, W. Z., et al. (2001). Indigenous Chinese personality constructs – Is the five-factor model complete? *Journal of Cross-Cultural Psychology, 32*(4), 407–433.

Chien, A. J., & Dunner, D. L. (1996). The tridimensional personality questionnaire in depression: State versus trait issues. *Journal of Psychiatric Research, 30*(1), 21–27.

Cloninger, C. R., Przybeck, T., & Svrakic, D. (1991). The Tridimensional Personality Questionnaire: US normative data. *Psychological Reports, 69*, 1047–1057.

Cloninger, C. R., Svrakic, D. M., & Przybeck, T. R. (1993). A Psychobiological Model of Temperament and Character. *Archives of General Psychiatry, 50*(12), 975–990.

Coan, J. A., & Allen, J. J. B. (2003). Frontal EEG asymmetry and the behavioral activation and inhibition systems. *Psychophysiology, 40*(1), 106–114.

Coan, J. A., Allen, J. J. B., Malone, S., & Iacono, W. G. (2003). The heritability of trait midfrontal EEG asymmetry and negative emotionality: Sex differences and genetic nonadditivity. *Psychophysiology, 40*, S34–S34.

Costa, P.T., & McCrae, R.R. (1992). *The Revised NEO Personality Inventory (NEO-PI-R and the NEO Five-Factor Inventory (NEO-FFI) professional manual.* Odessa, FL: Psychological Assessment Resources.

Costa, P. T., McCrae, R. R., & Dye, D. A. (1991). Facet scales for agreeableness and conscientiousness: A revision of the NEO Personality Inventory. *Personality and Individual Differences, 12*(9), 887–898.

Davidson, R. J., & Fox, N. A. (1989). Frontal brain asymmetry predicts infants response to maternal separation. *Journal of Abnormal Psychology, 98*(2), 127–131.

Davidson, R. J., Jackson, D. C., & Larson, C. L. (2000). Human electroencephalography. In J.T. Cacioppo, L. G. Tassinary, & G.G. Berntson, (Eds.), *Handbook of psychophysiology* (2nd ed., pp. 27–52). New York, NY, US: Cambridge University Press,.

Depue, R. A. & Collins, P. F. (1999). Neurobiology of the structure of personality: Dopamine, facilitation of incentive motivation, and extraversion. *Behavioral and Brain Sciences, 22*, 491–569.

Dick, D. M., Rose, R. J., Viken, R. J., Kaprio, J., & Koskenvuo, M. (2001). Exploring gene-environment interactions: Socioregional moderation of alcohol use. *Journal of Abnormal Psychology, 110*(4), 625–632.

Digman, J. M. (1997). Higher-order factors of the Big Five. *Journal of Personality & Social Psychology, 73*(6), 1246–1256.

Eaves, L., & Meyer, J. (1994). Locating human quantitative trait loci: guidelines for the selection of sibling pairs for genotyping. *Behavior Genetics, 24*, 443–455.

Eaves, L. J. (2006). Genotype × environment interactions: Fact or artifact? *Twin Research & Human Genetics, 9*, 1–8.

Eaves, L. J., & Carbonneau, R. (1998).Recovering components of variance from differential ratings of behavior and environment in pairs of relatives. *Developmental Psychology, 34*(1), 125–129.

Eaves, L. J., Eysenck, H. J., & Martin, N. G. (1989). *Genes, culture and personality: An empirical approach.* London: Oxford University Press.

Eaves, L. J., Heath, A. C., Martin, N. G., Neale, M. C., Meyer, J. M., Silberg, J. L., et al. (1999). Biological and cultural inheritance of stature and attitudes. In C. R. Cloninger (Ed.), *Personality and psychopathology* (pp. 269–308). Washington, DC, US: American Psychiatric Association.

Ebstein, R. P., Gritsenko, I. & Nemanov, L., (1997). No association between the serotonin transporter gene regulatory region polymorphism and the tridimensional personality questionnaire (TPQ) temperament of harm avoidance. *Molecular Psychiatry, 2*, 224–226.

Ebstein, R. P., Novick, O., Umansky, R., Priel, B., Osher, Y., Blaine, D., et al. (1996). Dopamine D4 receptor (D4DR) exon III polymorphism associated with the human personality trait of novelty seeking. *Nature Genetics, 12*(1), 78–80.

Ebstein, R. P., Segman, R., & Benjamin, J. (1997). 5-HT2C (HTR2C) serotonin receptor gene polymorphism associated with the human personality trait of reward dependence: Interaction with dopamine D4 receptor (D4DR) and dopamine D3 (D3DR) polymorphisms. *American Journal of Medical Genetics, 74*, 65–72.

Eysenck, H. J. (1994).Normality-abnormality and the three-factor model of personality. In S. Strack & M. Lorr (Eds.), *Differentiating normal and abnormal personality* (pp. 3–25). New York, NY, US: Springer Publishing Co.

Eysenck, H. J., & Eysenck, S. B. G. (1975). *Manual of the Eysenck personality questionnaire.* London: Hodder and Stoughton.

Eysenck, H. J., & Eysenck, S. B. G. (1992). *Manual for the Eysenck personality questionnaire-revised.* San Diego, CA: Educational and Industrial Testing Service.

Flory, J. D., Mann, J. J., Manuck, S. B., & Muldon, M. F. (1998). Recovery from major depression is not associated with normalization of serotonergic function. *Biological Psychiatry, 43*, 320–326.

Gana, K., & Trouillet, R. (2003). Structure invariance of the temperament and character inventory (TCI).*Personality and Individual Differences, 35*(7), 1483–1495.

Garmezy, N., & Masten, A. S. (1994). Chronic adversities. In M. Rutter & L. Hersov (Eds.), *Child and adolescent psychiatry: Modern approaches* (3rd ed., pp. 191–208). London: Blackwell.

Gray, J. A. (1987). The neuropsychology of emotion and personality. In S. M. Stahl & S.D. Iversen (Eds.), *Cognitive neurochemistry* (pp. 171–190). Oxford, England: Oxford University Press.

Gillespie, N. A., Cloninger, C. R., Heath, A. C., & Martin, N. G. (2003). The genetic and environmental relationship between Cloninger's dimensions of temperament and character. *Personality and Individual Differences, 35*(8), 1931–1946.

Gillespie, N. A., Evans, D. E., Wright, M. M., & Martin, N. G. (2004). Genetic simplex modeling of Eysenck's dimensions of personality in a sample of young Australian twins. *Twin Research, 7*(6), 637–648.

Goldsmith, H. H., Lemery, K. S., & Essex, M. J. (2004). Temperament as a liability factor for childhood behavioral disorders: The concept of liability. In L.F. DiLalla (Ed.), *Behavior genetics principles: Perspectives in development, personality, and psychopathology* (pp. 19–39). Washington, DC, US: American Psychological Association.

Gottesman, I. I., & Gould, T. D. (2003). The endophenotype concept in psychiatry: Etymology and strategic intentions. *American Journal of Psychiatry, 160*(4), 636–645.

Hahn, D. W., Lee, K., & Ashton, M. C. (1999). A factor analysis of the most frequently used Korean personality trait adjectives. *European Journal of Personality, 13*(4), 261–282.

Harmon-Jones, E., & Allen, J. J. B. (1998). Anger and frontal brain activity: EEG asymmetry consistent with approach motivation despite negative affective valence. *Journal of Personality and Social Psychology, 74*(5), 1310–1316.

Heath, A. C., Eaves, L. J., & Martin, N. G. (1989). The genetic structure of personality: III. Multivariate genetic item analysis of the EPQ scales. *Personality and Individual Differences, 10*(8), 877–888.

Heath, A. C., & Martin, N. G. (1990). Psychoticism as a dimension of personality: A multivariate genetic test of Eysenck and Eysenck's psychoticism construct. *Journal of Personality and Social Psychology, 58*, 111–121.

Heath, A. C., Neale, M. C., Kessler, R. C., Eaves, L. J., & Kendler, K. S. (1992). Evidence for genetic influences on personality from self-reports and informant ratings. *Journal of Personality and Social Psychology, 63*, 85–96.

Heiman, N., Stallings, M. C., Hofer, S. M., & Hewitt, J. K. (2003). Investigating age differences in the genetic and environmental structure of the tridimensional personality questionnaire in later adulthood. *Behavior Genetics, 33*(2), 171–180.

Hesselbrock,V., Begleiter, H., Porjesz, B., O'Connor, S., & Bauer, L. (2001). Event-related potential amplitude as an endophenotype of alcoholism: Evidence from the collaborative study on the genetics of alcoholism. *Journal of Biomedical Science, 8*, 77–82.

Jang, K. L. (2005). *The Behavioral Genetics of Psychopathology: A Clinical Guide.* Mahwah, New Jersey: Lawrence Erlbaum and Associates.

Jang, K. L., Dick, D. M., Taylor, S., Stein, M. B., Wolf, H., Vernon, P. A., et al. (2005). Psychosocial Adversity and Emotional Instability: An Application of Gene-Environment Interaction Models. *European Journal of Personality, 19*(4), 359–372. Special issue: Personality and Personality Disorders.

Jang, K. L., Hu, S., Livesley, W. J., Angleitner, A., Riemann, R., Ando, J., et al. (2001). The covariance structure of neuroticism and agreeableness: A twin and molecular genetic analysis of the role of the serotonin transporter gene. *Journal of Personality and Social Psychology, 81*(2), 295–304.

Jang, K. L., Livesley, W. J., Ando, Y., Yamagata, S., Suzuki, A., Angleitner, A., et al. (2006). Behavioral genetics of the higher-order factors of the Big Five. *Personality and Individual Differences, 41,* 261–272.

Jang, K. L., Livesley, W. J., Angleitner, A., Riemann, R., & Vernon P.A. (2002). Genetic and environmental influences on the covariance of facets defining the domains of the five-factor model of personality. *Personality and Individual Differences, 33*(1), 83–101.

Jang, K. L., Livesley, W. J., & Vernon, P. A. (1996). Heritability of the big five personality dimensions and their facets: A twin study. *Journal of Personality, 64*(3), 577–591.

Jang, K. L., McCrae, R. R., Angleitner, A., Riemann, R., & Livesley, W. J. (1998). Heritability of facet-level traits in a cross-cultural twin sample: Support for a hierarchical model of personality. *Journal of Personality and Social Psychology, 74*, 1556–1565.

John, O. P., & Srivastava, S. (1999). The big five trait taxonomy: History, measurement, and theoretical perspectives. In L. A. Pervin & O. P. John (Eds.), *Handbook of personality* (pp. 102–138). New York, NY: Guilford.

Johnson, W., & Krueger, R. F. (2005a). Genetic effects on physical health: Lower at higher income levels. *Behavior Genetics, 35*(5), 579–590.

Johnson, W., & Krueger, R. F. (2005b). Higher perceived life control decreases genetic variance in physical health: Evidence from a national twin study. *Journal of Personality and Social Psychology, 88*(1), 165–173.

Keller, M. C., Coventry, W. L., & Heath, A. C. (2005). Widespread evidence for non-additive genetic variation in Cloninger's and Eysenck's personality dimensions using a twin plus sibling design. *Behavior Genetics, 35*(6), 707–721.

Kendler, K. S. (1996). Major depression and generalised anxiety disorder same genes, (Partly) different environments Revisited.*British Journal of Psychiatry, 168*(Suppl 30), 68–75.

Kendler, K. S., & Eaves, L. J. (1986). Models for the joint effect of genotype and environment on liability to psychiatric illness. *American Journal of Psychiatry, 143*(3),279–289.

Kendler, K. S., Prescott, C. A., Myers, J., & Neale, M. C. (2003). The structure of genetic and environmental risk factors for common psychiatric and substance use disorders in men and women. *Archives of General Psychiatry, 60*(9), 929–937.

Kluger, A. N., Siegfried, Z., & Ebstein, R. P. (2002). A meta-analysis of the association between DRD4 polymorphism and novelty seeking. *Molecular Psychiatry, 7*(7), 712–717.

Krueger, R. F. (1999). The structure of common mental disorders. *Archives of General Psychiatry, 56*, 921–926.

Krueger, R. F. (2000). Phenotypic, genetic, and nonshared environmental parallels in the structure of personality: A view from the Multidimensional Personality Questionnaire. *Journal of Personality and Social Psychology, 79*(6), 1057–1067.

Krueger, R. F., Hicks, B. M., Patrick, C. J., Carlson, S. R., Iacono, W. G., & McGue, M. (2002). Etiologic connections among substance dependence, antisocial behavior, and personality: modeling the externalizing spectrum. *Journal of Abnormal Psychology, 111*, 411–424.

Kumakiri, C., Kodama, K., Shimizu, E., Yamanouchi, N., Okada, S., Noda, S., et al. (1999). Study of the association between the serotonin transporter gene regulatory region polymorphism and personality traits in a Japanese population. *Neuroscience Letters, 263*(2–3), 205–207.

Lake, R. I. E., Eaves, L. J., & Maes, H. M. (2000). Further evidence against the environmental transmission of individual differences in neuroticism from a collaborative study of 45,850 twins and relatives on two continents. *Behavior Genetics, 30*(3), 223–233.

Lander, E. S., & Botstein, D. (1989). Mapping Mendelian factors underlying quantitative traits using RFLP linkage maps. *Genetics, 121,* 185–199.

Larstone, R. M., Jang, K. L., Livesley, W. J., Vernon, P. A., & Wolf, H. (2002). The relationship between Eysenck's P-E-N model of personality, the five-factor model of personality, and traits delineating personality disorder. *Personality and Individual Differences, 33*(1), 25–37.

Lesch, K. P., Bengel, D., Heils, A., Sabol, S. Z., Greenberg, B. D., Petri, S., et al. (1996). Selective alteration of personality and social behavior by serotonergic intervention. *Science, 274*, 1527–1531.

Livesley, W. J., & Jang, K. L. (2005). Genetic Contributions to Personality Structure. In S. Strack (Ed.), *Handbook of Personology and Psychopathology* (pp. 103–119). New York: Wiley.

Livesley, W. J., Jang, K. L., Jackson, D. N., & Vernon, P. A. (1993). Genetic and Environmental Contributions to Dimensions of Personality Disorder. *American Journal of Psychiatry, 150*(12), 1826–1831.

Livesley, W. J., Jang, K. L., & Vernon, P. A. (1998). The phenotypic and genetic architecture of traits delineating personality disorder. *Archives of General Psychiatry, 55*, 941–948.

Loehlin, J. C. (1982). Are personality traits differentially heritable? *Behavior Genetics, 12*, 417–428.

Loehlin, J. C. (1986). Are California Psychological Inventory items differentially heritable? *Behavior Genetics, 16*, 599–603.

Loehlin, J.C. (1987). Heredity, environment, and the structure of the California Psychological Inventory. *Multivariate Behavioral Research, 22*, 137–148.

Loehlin, J.C. (1992). *Genes and environment in personality development.* Newbury Park, CA: Sage.

Loehlin, J. C., Horn, J. M., & Willermann, L. (1997). Heredity, environment, and IQ in the Texas adoption study. In R. J. Sternberg & E. L. Grigorenko (Eds.), *Heredity, environment, and intelligence* (pp. 105–125). New York: Cambridge University Press.

Loehlin, J. C., & Nicholls, R. C. (1976). *Heredity, environment, and personality.* Austin: University of Texas Press.

Markon, K. E., Krueger, R. F., & Watson, D. (2005). Delineating the structure of normal and abnormal personality: An integrative hierarchical approach. *Journal of Personality and Social Psychology, 88*(1), 139–157.

McCrae, R. R. (2004). Human nature and culture: A trait perspective. *Journal of Research in Personality, 38*(1), 3–14.

McCrae, R. R., & Allik, J. (2002). *The Five-Factor model of personality across cultures.* New York, NY, US: Kluwer Academic/Plenum Publishers.

McCrae, R. R., Costa, P. T., Jr. (1999). A five-factor theory of personality. In L.A. Pervin & O.P. John (Eds.), *Handbook of personality: Theory and research* (2nd ed., pp. 139–153). New York, NY, US: Guilford Press.

McCrae, R. R., Jang, K. L., Angleitner, A., Riemann, R., & Livesley, W.J. (2001). Sources of structure: genetic, environmental, and artifactual influences on the covariation of personality traits. *Journal of Personality, 69*(4), 511–535.

McCrae, R. R., Terracciano, A., & 78 Members of the Personality Profiles of Cultures Project. (2005). Universal features of personality traits from the observer's perspective: Data from 50 cultures. *Journal of Personality and Social Psychology, 88*(3), 547–561.

Metalsky, G. I., Halberstadt, L. J., & Abramson, L. Y. (1987). Vulnerability to depressive mood reactions – toward a more powerful test of the diathesis-stress and causal mediation components of the reformulated theory of depression. *Journal of Personality and Social Psychology, 52*(2), 386–393.

Munafo, M. R., Clark, T., & Flint, J. (2005a). Does measurement instrument moderate the association between the serotonin transporter gene and anxiety-related personality traits? A meta-analysis. *Molecular Psychiatry, 10*(4), 415–419.

Munafo, M. R., Clark, T., & Flint, J. (2005b). Promise and pitfalls in the meta-analysis of genetic association studies: A response to Sen and Schinka. *Molecular Psychiatry, 10*(10), 895–897.

Munafo, M. R., Clark, T. G., Moore, L. R., Payne, E., Walton, R., & Flint, J. (2003). Genetic polymorphisms and personality in healthy adults: A systematic review and meta-analysis. *Molecular Psychiatry, 8*(5), 471–484.

Neiderhiser, J. M., Reiss, D., & Hetherington, E. M. (1996). Genetically informative designs for distinguishing developmental pathways during adolescence: Responsible and antisocial behavior. *Development and Psychopathology, 8*(4), 779–791.

Osher, Y., Hamer, D., & Benjamin, J. (2000). Association and linkage of anxiety-related traits with a functional polymorphism of the serotonin transporter gene regulatory region in Israeli sibling pairs. *Molecular Psychiatry, 5*(2), 216–219.

Paris, J. (1994). *Borderline personality disorder: A multidimensional approach.* Washington, DC: American Psychiatric Press.

Paris, J. (1996). *Social factors in the personality disorders.* New York: Cambridge University Press

Paris, J. (2001). Psychosocial adversity. In W. J. Livesley (Ed.), *Handbook of personality disorders* (pp. 231–241). New York: Guilford Press.

Plomin, R., DeFries, J. C., Craig, I. W., &, McGuffin, P. (Eds). (2003). *Behavioural genetics in the post-genomic era.* Washington, DC: American Psychological Association.

Plomin, R., DeFries, J. C., & Loehlin, J. C. (1977).Genotype-environment interaction and correlation in the analysis of human behavior. *Psychological Bulletin, 84*(2), 309–322.

Plomin, R., DeFries, J. C., & McClearn, G. E. (1990). *Behavioral genetics: A primer* (2nd ed.). New York, NY, USA: W. H. Freeman and Co, Publishers.

Prabhu, V., Chorlian, D. B., Wang, K., Stimus, A., & Begleitner, H. (2001). Visual P3 in female alcoholics. *Alcoholism: Clinical and Experimental Research, 25*(4), 531–539.

Riemann, R., Angleitner, A., & Strelau, J. (1997). Genetic and environmental influences on personality: A study of twins reared together using the self- and peer-report NEO-FFI scales. *Journal of Personality, 65*(3), 449–475.

Saucier, G., & Goldberg, L.R. (1996). The language of personality: Lexical perspectives on the Five-Factor Model. In J.S. Wiggins (Ed.), *The five factor model of personality* (pp. 21–50). New York: Guilford Press.

Scarr, S., & McCartney, K. (1983). How people make their own environments: A theory of genotype —> environment effects. *Child Development, 54*, 424–435.

Schinka, J. A., Busch, R. M., & Robichaux-Keene, N. (2004). A meta-analysis of the association between the serotonin transporter gene polymorphism (5-HTTLPR) and trait anxiety. *Molecular Psychiatry, 9*(2), 197–202.

Schinka, J. A., Letsch, E. A., & Crawford, F. C. (2002). DRD4 and novelty seeking: Results of meta-analyses. *American Journal of Medical Genetics, 114*(6), 643–648.

Sen, S., Burmeister, M., & Ghosh, D. (2004). Meta-analysis of the association between a serotonin transporter promoter polymorphism (5-HTTLPR) and anxiety-related personality traits. *American Journal of Medical Genetics Part B-Neuropsychiatric Genetics, 127B*(1), 85–89.

Sham, P.C., Sterne, A., Purcell, S., Webster, M., Rijsdijk, F., Asherson P., et al. (2001). GENESiS: Creating a composite index of the vulnerability to anxiety and depression in a community-based sample of siblings. *Twin Research, 3*, 316–322.

Stewart, M.E., Ebmeier, K. P., & Deary, I.J. (2004). The structure of Cloninger's Tridimensional Personality Questionnaire in a British sample. *Personality and Individual Differences, 36(6)*, 1403–1418.

Strobel, A., Lesch, K. P., Jatzke, S., Paetzold, F., & Brocke, B. (2003). Further evidence for a modulation of Novelty Seeking by DRD4 exon III, 5-HTTLPR, and COMT val/met variants. *Molecular Psychiatry, 8*(4), 371–372.

Sutton, S. K., & Davidson, R. J. (1997). Prefrontal brain asymmetry: A biological substrate of the behavioral approach and inhibition systems. *Psychological Science, 8*(3), 204–210.

Tellegen, A. (1982). *Brief manual for the Multidimensional Personality Questionnaire.* Unpublished manuscript, University of Minnesota, Minneapolis.

Tomarken, A. J., Davidson, R. J., & Henriques, J. B. (1990). Resting frontal brain asymmetry predicts affective responses to films. *Journal of Personality and Social Psychology, 59*(4), 791–801.

Tomarken, A. J., Davidson, R. J., Wheeler, R. E., & Doss, R. C. (1992). Individual-differences in anterior brain asymmetry and fundamental dimensions of emotion. *Journal of Personality and Social Psychology, 62*(4), 676–687.

Tomarken, A. J., Davidson, R. J., Wheeler, R. E., & Kinney, L. (1992). Psychometric properties of resting anterior EEG asymmetry - temporal stability and internal consistency. *Psychophysiology, 29*(5), 576–592.

Trull, T. J., Waudby, C. J., & Sher, K. J. (2004). Alcohol, tobacco, and drug use disorders and personality disorder symptoms. *Experimental and Clinical Psychopharmacology, 12*(1), 65–75.

Turkheimer, E., Haley, A., Waldron, M., D'Onofrio, B., & Gottesman, II. (2003). Socioeconomic status modifies heritability of IQ in young children. *Psychological Science, 14*(6), 623–628.

Turkheimer, E., & Waldron, M. (2000). Nonshared environment: A theoretical, methodological, and quantitative review. *Psychological Bulletin, 126*(1), 78–108.

Waller, N. G, & Shaver, P. R. (1994). The importance of nongenetic influences on romantic love styles: A twin-family study. *Psychological Science, 5*(5), 268–274.

Wheeler, R. E., Davidson, R. J., & Tomarken, A. J. (1993). Frontal Brain Asymmetry and Emotional Reactivity - a Biological Substrate of Affective Style. *Psychophysiology, 30*(1), 82–89.

Widiger, T. A., Verheul, R., & van den Brink, W. (1999). Personality and psychopathology. In L.A. Pervin, & O.P. John (Eds.), *Handbook of personality: Theory and research* (2nd ed., pp. 347–366). New York, NY, US: Guilford Publications.

Wolf, H., Angleitner, A., Spinath, F. M., Riemann, R., & Strelau, J. (2004). Genetic and environmental influences on the EPQ-RS scales: A twin study using self- and peer reports. *Personality and Individual Differences, 37*(3), 579–590.

Yamagata, S., Suzuki, A., Ando, J., Ono, Y., Kijima, N., Yoshimura, K., et al. (2006). Is the genetic structure of human personality universal? A cross-cultural twin study from North America, Europe, and Asia. *Journal of Personality and Social Psychology, 90*(6), 987–998.

Young, S. E., Stallings, M. C., Corley, R. P., Krauter, K. S., & Hewitt, J. K. (2000). Genetic and environmental influences on behavioral disinhibition. *American Journal of Medical Genetics, 96*, 684–695.

Zimmerman, M., Chelminski, I., & McDermut, W. (2002). Major depressive disorder and axis I diagnostic comorbidity. *Journal of Clinical Psychiatry, 63*(3), 187–193.

Zuckerman, M., Kuhlman, D. M., Thornquist, M., & Kiers, H. (1991). 5 (or 3) Robust questionnaire scale factors of personality without culture. *Personality and Individual Differences, 12*(9), 929–941.

Chapter 17

Molecular Genetics of Personality: How Our Genes can Bring Us to a Better Understanding of Why We Act the Way We Do

Richard P. Ebstein and Salomon Israel

Introduction

In the 4th century, St. Augustine of Hippo used the Biblical parable of Jacob and Esau, twin brothers who displayed remarkably different characters, to disprove prevailing notions that astrology dictated personality. The search for underlying factors that contribute to individual differences in personality and character has continued to capture the interest and intrigue of scientists as well as the common man looking for a little personal insight. Advancements in behavioral genetics, by expounding on the molecular biological inputs of personality, have added a particularly fresh face to our understanding of what makes us the way we are and what makes us act different than our neighbors.

Defining and Measuring Personality

The term "personality", and the one we will use throughout the remainder of this chapter, is defined as *the individual psychological aspects of people that make them "recognizable," which is to say different from each other*.[1] In this sense we describe Hamlet's personality as brooding and indecisive, Bart Simpson as impulsive and James Bond as suave. We can say that a friend is gregarious and kindhearted and that a colleague is arrogant – we can even say that a potential love interest while maybe not the most attractive candidate, nevertheless "has a great personality". In this sense, personalities

describe the characteristic *manner* of our behavior, separate from other aspects such as intelligence, appearance or ability that may influence but not compose our personalities.

These descriptions tend to provide an intuitive grasp of personality based on a few conspicuous traits; however, they are not systematically interconnected through a singular framework. A general empirical theory of personality should include a manageable set of dimensions that account for most of the variance in personality. These dimensions should relate to universal and fundamental aspects of personality and demonstrate stability over time. Virtually all personality assessments in genetic research are evaluated either through pencil-and-paper questionnaires that inventory dimensional personality traits or via direct laboratory manipulation and observation. The questionnaire approach has the advantage of replicability over extended periods, but loses out on some of the objectivity offered by the laboratory approach. The lab method, on the other hand, is both labor intensive and perhaps more sensitive to context and time effects.

Personality dimensions in pencil-and-paper inventories assume that an accurate rating imparts a reasonable and economic description of an individual's personality. The dimensions are typically developed in one of the three ways. *Approaches based on psychopathology* derive personality traits from psychiatric illnesses. These take the view that psychiatric illnesses are extreme variants and therefore the clearest expressions of normal personality. For example, the Minnesota Multiphasic Personality Inventory (Butcher, 2001) grades people on degrees of hysteria, obsessive-compulsive disorder and schizophrenia. While this approach is reflected in popular discourse, it is not widely used in genetic research.

Theoretical approaches are based on biological models drawn from associations between personality traits and nervous system function. For example, Cloninger's Tridimensional Personality Questionnaire (TPQ) (Cloninger, 1986) employs animal and human findings from lesioning, imaging and pharmacological studies involving neurotransmitter pathways to show connections between monoamines dopamine, noradrenaline and serotonin – and temperament.

Richard P. Ebstein (✉)
Scheinfeld Center of Genetic Studies for the Social Sciences, Psychology Department, Hebrew University, Mt. Scopus; Sarah Herzog Memorial Hospital, Givat Shaul, Jerusalem 91905, Israel
e-mail: ebstein@mscc.huji.ac.il

[1] Other approaches to defining personality, of which Freud is the classic example, consider more universal aspects of personality such as the unconscious, interpersonal relations and self-actualization. While undoubtedly having a major influence on 20th century culture, much of the field is psychoanalytic, metaphoric and impractical for empirical analysis.

Y.-K. Kim (ed.), *Handbook of Behavior Genetics*,
DOI 10.1007/978-0-387-76727-7_17, © Springer Science+Business Media, LLC 2009

Dopaminergic (DA) pathways are implicated in drug use, sensation seeking and explorative behaviors, and emotions like curiosity and recklessness, which the TPQ calls "novelty seeking" (NS). Individuals high in NS are risk takers and impulsive while those with low NS are deliberate and frugal. Noradrenergic (NE) pathways are implicated in approach behaviors, which the TPQ calls "reward dependence" (RD). Human beings high on RD are sentimental and affectionate; those with low RD are tough and pragmatic. Serotonin (5-HT) pathways are implicated in behaviors devoted to avoiding harm or escaping punishment, and to the associated emotion of anxiety (called "harm avoidance" (HA) in the TPQ). Human subjects with high HA are classified as "neurotic" on many other instruments; those with low HA are "stable" or "healthy" or "well adjusted" (see Table 17.1 for sample questions from the TPQ). The eight possible combinations of the two extremes of the distributions of these three dimensions yield personality constellations held to reflect clinically recognized personality disorders like antisocial psychopathy and obsessive–compulsive personality disorder. A later version of the TPQ is the Temperament and Character Inventory (TCI).

The third approach to personality assessment is derived from *everyday speech*. Natural language is considered to have proved itself during evolution to be a reasonably accurate and adaptive way of describing people. Personality inventories have been constructed by listing thousands of personality descriptors culled from dictionaries of everyday language and then reduced to sizable numbers via factor analysis. The most popular such inventory in clinical and academic use is the NEO (Costa & McCrae, 1997), which assesses five main personality factors (neuroticism, extraversion, openness to experience, agreeableness and conscientiousness), also known as "the Big Five". Neuroticism resembles TPQ HA. Extraversion is the degree of interest in other people and in social dominance. Openness to new ideas and experiences is the opposite of traditionalism. Agreeableness is a measure of how likeable an individual is. Conscientiousness assesses conscience, thoughtfulness, planning and order, and their opposites: spontaneity and lack of prudence. Extraversion and conscientiousness (with opposite signs) correlate

with TPQ NS. While details of dimensions differ between instruments, a number of recent factor analyses concur that five main factors adequately describe human personality differences.

Given the inherent uncertainty in measuring personality traits, simultaneously studying more than one instrument and/or laboratory assessment improves the robustness of a measure. Good correlations between apparently similar traits in two or more assessments, and converging findings of associations between given polymorphisms and those traits in multiple assessments, enhances confidence in those findings. This is all the more necessary, given the challenges of replication in molecular genetic findings for personality genetics, similar to the problems encountered in other complex phenotypes.

The Molecular Genetics of Personality

The debut of molecular genetic studies of personality was ushered in with the simultaneous publication of two articles in *Nature Genetics* in 1996 showing an association between Tridimensional Personality Questionnaire (TPQ) (Cloninger, 1987; Cloninger, Svrakic, & Przybeck, 1993) Novelty Seeking (Ebstein, Novick, et al., 1996) and NEO-PI-R Extraversion (Benjamin, Patterson, Greenberg, Murphy, & Hamer, 1996) and the dopamine D4 exon III (D4.7) seven repeat (Van Tol, Wu, et al., 1991; Van Tol, Bunzow, et al., 1992). These two seminal reports were promptly followed by a study (Lesch et al., 1996) showing an association between the short *SLC6A4* (serotonin transporter) promoter 44 bp repeat deletion/insertion (5-HTTLPR) and NEO-PI-R Neuroticism (Costa & McCrae, 1997). These reports spurred a series of investigations that have resulted in both successful and unsuccessful efforts at replicating these first studies. Molecular personality genetics has evidently not escaped the conundrum of non-replication that continues to plague the genetics of complex human phenotypes (see Mayeux, 2005 for review) in the issue of *Journal of Clinical Investigation* dedicated to complex disorders. Nevertheless, despite apparent difficulties in confirming earlier findings,

Table 17.1 Sample questions from Cloninger's TPQ

Cloninger's TPQ	
The TPQ is a 100-question instrument that requires a yes/no answer. Most subjects complete the form in about a half an hour	
Sample Novelty Seeking Questions	I often try new things just for fun or thrills, even if most people think it is a waste of time (yes)
	I often do things based on how I feel at the moment without thinking about how they were done in the past (yes)
	I am much more controlled than most people (no)
Sample Harm Avoidance Questions	I usually am confident that everything will go well, even in situations that worry most people (no)
	I often feel tense and worried in unfamiliar situations, even when others feel there is no danger at all (yes)
	I often have to stop what I am doing because I start worrying about what might go wrong (yes)

the enthusiasm for personality genetic studies continues to swell testifying to the allure of this subject for many behavioral scientists.

Reviews and Meta-Analyses

The status of personality genetics has been the focus of many reviews (Benjamin, Ebstein, & Belmaker, 2000a, 2000b; Ebstein & Kotler, 2002; Ebstein, Zohar, Benjamin, & Belmaker, 2002; Jang, Vernon, & Livesley, 2001; Noblett & Coccaro, 2005; Savitz & Ramesar, 2004; Torgerson, 2005) and several meta-analyses have been carried out for *DRD4* (Novelty Seeking/Extraversion) and the serotonin transporter (Harm Avoidance/Neuroticism) (Kluger, Siegfried, & Ebstein, 2002; Munafo, Clark, & Flint, 2005; Munafo, Clark, Moore, et al., 2003; Schinka, Busch, & Robichaux-Keene, 2004; Schinka, Letsch, & Crawford, 2002; Sen, Burmeister, & Ghosh, 2004; Willis-Owen et al., 2005). Overall, meta-analyses compiled from self-report questionnaires failed to provide definitive evidence either that the *DRD4* D4.7 repeat is significantly contributing to Novelty Seeking or that the 5-HTTLPR short allele is significantly contributing to Neuroticism/Harm Avoidance (HA). The evidence appears strongest in the transporter-Neuroticism association and the *DRD4* C-521T promoter region SNP and novelty seeking. A cardinal feature of complex phenotypes, small effect size ($d = 0.23$ for Neuroticism and 5-HTTLPR (Schinka, Busch, & Robichaux-Keene, 2004); $d = 0.34$ for Novelty Seeking and DRD4 SNP C-521T (Schinka, et al., 2002); where effect size "d" was calculated as the difference in personality measure means between genotype groups divided by the pooled standard deviation of the two groups), is likely an important reason for challenges in replicating the first findings. Subtle differences in the inventories can easily obscure effect sizes of polymorphisms that only explain a few percent of the variance between subjects. Additionally, as noted in all the meta-analyses, considerable heterogeneity was observed between studies indicating that there is greater variation among outcomes than expected by chance (Jang et al., 2001; Sen, Burmeister, et al., 2004).

The Worry Wart – Serotonin Transporter SLC6A4

In comparison to the highly polymorphic *DRD4* gene the serotonin transporter has a more conserved genotype. 5-HTTLPR is a variable number tandem repeat (VNTR) polymorphism consisting of 14–16 copies of 22 bp imperfect repeat sequences, also known as 44 bp Ins/Del because the most common polymorphisms are 16 repeats (long) and 14 repeats (short) (Heils et al., 1996; Lesch et al., 1996). The short allele showed lower expression in a dominant manner. The presence of one or two copies of the short allele significantly reduced the rate of transporter transcription, which was about 65% lower in brain and about 35% lower in lymphoblasts. Most studies of personality genetics have only genotyped 5-HTTLPR; however, an intron 2 VNTR (17 bp repeat) is also of interest (Lesch, Balling, et al., 1994). Linkage disequilibrium (LD) between the two loci (5-HTTLPR and intron 2 VNTR) was found in most of the studied populations: it ranged from moderate (e.g., in Europeans) to very strong (e.g., in Native Americans), while completely absent in some others (e.g., Chinese; Gelernter, Cubells, Kidd, Pakstis, & Kidd, 1999). Allele-dependent differential enhancer activity of the polymorphic region in the second intron was demonstrated as different levels of reporter gene expression in embryonic stem cells (Fiskerstrand, Lovejoy, & Quinn, 1999) and in mouse embryo (MacKenzie & Quinn, 1999). A thorough study by Wildenauer and his colleagues (Hranilovic et al., 2004) showed separate and combined effects of the intron 2 VNTR and 5-HTTLPR polymorphisms on the rate of transporter mRNA transcription in lymphoblasts.

Interestingly, a single nucleotide variant (A to G) was detected in the sixth motif, present only in the long variant of 5-HTTLPR, creating an AP2-binding site (De Luca, Zai, et al., 2006; Goldman, Hu, Zhu, Lipsky, & Murphy, 1994). In European American and African American populations, the frequency of the polymorphism is low but perhaps polymorphic enough if the size of the sample is large enough to provide reasonable statistical power. The long "G" and long "A" alleles are functionally distinct, in fact only the long A is associated with high levels of transporter mRNA expression. The function of the long variant when the G nucleotide is present is more similar to the 5-HTTLPR short allele, with a low level of mRNA. Carriers of the combination of long A and intron 2 VNTR 12 repeat alleles were at risk for suicide attempt (De Luca, Zai, et al., 2006), consistent with a role for the purported role of the gene in depression, anxiety-related personality traits and similar phenotypes.

Animal Studies

Murphy et al. (2001) reviewed the efficacy of animal models in exploring the complexities of human behavioral phenotypes, with a specific focus on their group's development of an *SLC6A4* knockout mouse. This bottom-up approach (animal gene → human phenotype) suggested increased anxiety and fearfulness as a major behavioral phenotype of the serotonin transporter mutant mouse, findings consistent with many of the human studies of this gene.

Ansorge and his colleagues skillfully illustrated the value of animal studies in elucidating how human behavioral genes operate. Although much of the evidence on the SLC6A4 appears to support the contention that reduced expression generated by the presence of the 5-HTTLPR short allele is associated with abnormal affective and anxiety-like symptoms in humans, the neurochemical mechanism of this effect remains obscure. Especially confounding is that while SSRIs, by inhibiting the serotonin transporter, are effective anxiolytic agents the short 5-HTTLPR promoter allele (equivalent to an *endogenous* SSRI) is purported to predispose to anxiety. Ansorge, Zhou, Lira, Hen, & Gingrich (2004) resolved this paradox in a well-designed developmental study. They inhibited transporter expression during a limited period in early mouse development with fluoxetine which led to abnormal emotional behaviors in adult mice. This effect mimicked the behavioral phenotype of mice genetically deficient in transporter expression which points to a critical role of serotonin in the maturation of brain systems that modulate emotional function in the adult. These results additionally suggest a developmental mechanism to explain how the short 5-HTTLPR promoter allele increases vulnerability to anxiety-related disorders.

Imaging

Weinberger and his colleagues pioneered the use of functional neuroimaging in the analysis of genotype–phenotype relationships in healthy individuals and have coined the nascent field "imaging genomics" (Hariri & Weinberger, 2003). Their studies strengthened the connection between serotonin transporter promoter region polymorphisms and anxiety-related personality traits by showing increased amygdala activation in subjects presented with fearful faces possessing the short 5-HTTLPR allele (Hariri, Drabant, et al., 2005; Hariri, Mattay, et al., 2002). Interestingly, these genotype effects are consistent with a dominant short allele effect and are equally prominent in men and women. However, neither 5-HTTLPR genotype, amygdala reactivity, nor genotype-driven variability in this reactivity was reflected in Harm Avoidance scores underscoring the apparent need for large numbers of subjects when the pencil-and-paper self-report strategy is employed. Although the imaging studies from the Weinberger group appear to be robust, a note of caution is appropriate since to our knowledge these studies have yet to be replicated by other investigators.

The Thrill Seeker – Dopamine D4 Receptor

Numerous DRD4 polymorphisms have been identified in the genomic (5′ upstream) coding and intronic regions adding

a source of complexity in analyzing the role of the gene in behavior. LD between these polymorphisms is surprisingly weak (Bachner-Melman, Gritsenko, et al., 2005; Lowe et al., 2004) sometimes even between adjacent SNPs, stressing the importance of testing all genomic loci for association toward extracting the maximum amount of genetic information. The importance of fully assessing allelic heterogeneity was illustrated by the study of Grady et al. (2003) that explored the unusual degree of variation within the exon III repeat and its role in contributing risk to ADHD.

A host of studies have detailed the functional role of the DRD4 exon III repeat region. Studies of G protein coupling (Ashgari et al., 1994), cyclic AMP synthesis (Ashgari et al., 1995), in vitro expression (Schoots & Van Tol, 2003) and chaperone-induced folding (Van Craenenbroeck et al., 2005), provide increasingly solid evidence that the shorter exon III repeats code for a more efficient gene at the level both of transcription, translation and second messenger generation compared to the D4.7 repeat.

The first evidence that *DRD4* promoter region polymorphisms contribute to personality was the report by a Japanese group (Okuyama et al., 2000) who showed that a C-521T promoter region SNP is both functional and associated with Novelty Seeking. C/C homozygotes had higher TPQ Novelty Seeking scores. Additionally, a transient expression method revealed that the T variant of the C-521T polymorphism reduces transcriptional efficiency, findings that have been confirmed by other studies (Bookman, Taylor, Adams-Campbell, & Kittles, 2002; Eichhammer et al., 2005; Golimbet, Gritsenko, Alfimova, & Ebstein, 2005; Ronai et al., 2001) with one exception (Mitsuyasu et al., 2001).

A second functional polymorphism in the *DRD4* promoter region is a 120 base-pair tandem duplication first identified by Seaman et al. (1999) located 1.2 kb upstream of the initiation codon. The duplicated region contains consensus sequences of binding sites for several known transcription factors, suggesting differences in transcriptional activity in long and short repeats. The polymorphism has been associated with attention deficit (Arcos-Burgos et al., 2004; McCracken et al., 2000; Xing et al., 2003) and TPQ Novelty Seeking (Rogers et al., 2004). Subsequently, the longer allele (240 bp) was shown to display lower transcriptional activity than the short allele (120 bp) (D'Souza et al., 2004).

Animal Studies

Spurred by the first human studies suggesting an association between Novelty Seeking and *DRD4*, a mouse knockout model of the DRD4 receptor was employed to explore a potential role for the gene in behavioral responses to novelty (Dulawa, Grandy, Low, Paulus, & Geyer, 1999). *DRD4* knockout mice exhibit reduced behavioral responses to new environments as well as a supersensitivity to alcohol and cocaine (Rubinstein et al., 1997) consistent with some human studies of this gene. Other studies have found associations between the exon III and aggressive canine behavior (Inoue-Murayama et al., 2002; Ito et al., 2004; Niimi, Inoue-Murayama, Kato, et al., 2001; Niimi, Inoue-Murayama, Murayama, Ito, & Iwasaki, 1999), as well as curiosity and vigilance in horses (Momozawa, Takeuchi, Kusunose, Kikusui, & Mori, 2005).

A Related Phenotype Fibromyalgia

Seen predominantly in women, fibromyalgia syndrome (FM) is an underdiagnosed musculoskeletal condition of unknown etiology affecting more than 10% of patients attending general medical clinics and 15–20% attending rheumatology clinics (Buskila, 2001). Patients complain that they ache all over and report symptoms such as fatigue, morning stiffness, sleep disturbance, paresthesias and headaches. In 1999, a German group reported an association between this syndrome and the short 5-HTTLPR allele (Offenbaecher et al., 1999) consistent with the observed psychosocial profile of these women who score high on anxiety-related personality traits. Later studies demonstrated a marked increase in the frequency of the short/short 5-HTTLPR genotype and a decrease in the frequency of the DRD4 exon III 7 repeat allele among fibromyalgia patients compared to the general population (Cohen, Buskila, Neumann, & Ebstein, 2002). They also demonstrated significantly higher Harm Avoidance and Persistence scores and significantly lower Novelty Seeking scores. These results can be interpreted as a "proof of principle" of the small effect sizes of common polymorphisms in a more extreme phenotype as well as demonstrating the benefits of expanding the evaluation of a phenotype via different measures, a perhaps necessary step in the identification and appraisal of complex traits such as personality.

The Social Personality – AVPR1a

Although progress has been made in unraveling the molecular genetic architecture of individual personality traits (Bouchard & Loehlin, 2001; Ebstein, Benjamin, & Belmaker, 2000; Kluger et al., 2002; Schinka, Letsch, et al., 2002), little is known regarding the genetic basis of social behaviors (interaction between at least two individuals) in humans. Much more has been elucidated on the genetic influence of social behavior in animals, including insects and lower vertebrates (Giraud, Pedersen, & Keller, 2002; Keller & Parker, 2002; Krieger & Ross, 2002). In particular, research over the last two decades has revealed the molecular mechanisms by which two peptide hormones,

vasopressin and oxytocin, shape social behavior from fish to rodents (Young, Lim, Gingrich, & Insel, 2001). However, despite speculation (Taylor et al., 2000), little evidence has been forthcoming linking these hormones to corresponding human behaviors.

Arginine-vasopressin (AVP) in rodents has been associated with the modulation of a broad range of behavioral phenotypes including social recognition and learning, affiliative behaviors, aggression, dominant–subordinate relationships, parental behavior, grooming and categories of pair bonding such as monogamy (Ferguson, Young, & Insel, 2002). Of particular interest are affiliative behaviors that interact with, but are distinct from reproductive pair bonding. For example, in squirrel monkeys vasopressin has been shown to modulate male–male interactions (Winslow & Insel, 1991). Extrapolating from these studies, vasopressin might modulate a range of human behaviors distinct from romantic pair bonding. Strengthening this notion are recent studies showing linkage between the Arginine-vasopressin promoter region 1a receptor (AVPR1a) and complex social behavior such as self-presentation, sibling relationships (Bachner-Melman, Zohar, et al., 2005) and autism, a behavior disorder of which dysfunctional social interaction is a core symptom (Kim et al., 2002).

Gene–Environment Interactions

An important event of considerable impact in human behavioral genetics were publications by Caspi and his colleagues demonstrating that environmental effects were contingent upon certain genotypes (Caspi, Moffitt, et al., 2005; Caspi, Sugden, et al., 2003; Moffitt, Caspi, & Rutter, 2005). These studies showed what was widely hypothesized but rarely demonstrated, that nature and nurture (gene × environment interactions) jointly contribute to the determination of behavioral phenotypes. Remarkably, similar studies in non-human primates also show an interaction between the serotonin transporter polymorphism, depression/anxiety and alcoholism (Barr et al., 2003; Barr, Newman, Lindell, et al., 2004; Barr, Schwandt, Newman, & Higley, 2004).

G × E studies, an emerging trend in behavioral genetics, are helping to elucidate the confluence of genetic effects and environmental experiences. For example, Finnish investigators have exploited the availability of early environmental information in their cohort study to test G × E interactions for DRD4 and TPQ Novelty Seeking (Keltikangas-Jarvinen, Raikkonen, Ekelund, & Peltonen, 2004) in an enriched genetic model. When the childhood-rearing environment was more hostile (emotionally distant, low tolerance of the child's normal activity and strict discipline), the participants carrying D4.2 or D4.5 alleles had a significantly greater risk of exhibiting high Novelty Seeking scores as adults. Similarly when the father, but not the mother, reported more frequent alcohol consumption or drunkenness, there was an association between D4.2 and D4.5 and extreme Novelty Seeking scores (Lahti et al., 2005). Other interesting experiments explored the interaction of genetic effects on cigarette and cannabis smoking (Caspi, Moffitt, et al., 2005; Lerer, Kanyas, Kami, Ebstein, & Lerer, 2006).

Linkage Studies

Four studies (Cloninger, Van Eerdewegh, Goate, Edenberg, & Blangero, 1998; Dina et al., 2005; Fullerton et al., 2003; Zohar et al., 2003) have now identified a broad region on chromosome 8p that harbors a locus that contributes to individual differences in a personality trait that is a measure of emotional liability. Two of the studies found linkage to TPQ harm avoidance whereas the Fullerton et al. report used Eysenck's Psychoticism scale (Eysenck, 1952) and found linkage to neuroticism.

The Dina et al. results suggest the possibility that the same locus near the neuroregulin 1 gene on chromosome 8p confers risk for both an anxiety-related personality trait as well as schizophrenia. This common genetic factor may contribute to emotional liability during early development which constitutes a predisposition for major psychosis.

Evolutionary Considerations

An intriguing feature of the *DRD4* exon III polymorphism is its recent evolutionary history revealed by the investigations of Kidd and his colleagues (Ding et al., 2002; Wang et al., 2004). It was estimated that the D4.7 allele arose prior to the upper Paleolithic era (~40,000–50,000 years ago), a crucial period in the history of our species when in one key wave of migration, there was a spread of modern humans out of Africa. Kidd and his colleagues hypothesized the emergence of the D4.7 as a rare mutational event (or events) that nevertheless increased to high frequency in human populations by positive selection (Ding et al., 2002; Wang et al., 2004). Ding et al. hypothesize that the currently observed population frequencies of the D4.4 and D4.7 alleles are an example of balanced selection. The balanced maintenance of both the D4.4 and D4.7 repeats is related to the need for diverse behavioral phenotypes in human populations partially determined by this gene, altruistic and prosocial (D4.4) versus a more aggressive, novelty seeking or perhaps even antisocial type (D4.7). Indeed, models suggest that populations composed of entirely altruistic individuals would be unstable (Sigmund & Hauert, 2002) since they would be prone to invasion and exploitation by defectors.

A second and perhaps complimentary phenotype that we suggest is also determined for in part by the *DRD4* gene is altruism (Bachner-Melman, Gritsenko, et al., 2005). The paradox of human altruism, helping others and thereby reducing one's own fitness, has confounded evolutionary biologists since the days of Darwin. Not only is altruism a puzzle for evolutionary biologists but the trait has also perplexed psychologists who have questioned the merits of an altruistic personality (Batson, 1991; Gergen, Gergen, & Meter, 1972). Nevertheless, altruistic behavior is a commonplace and a unique feature of human altruism is that it extends beyond Hamilton's concept of "inclusive fitness" (Hamilton, 1964), which explains altruistic behavior by virtue of genetic relatedness, and even beyond other concepts such as reciprocal altruism and reputation-based altruism (Sigmund & Hauert, 2002). Out of all the animals, only humans practice wholesale mutual aid among genetically unrelated individuals.

Several twin studies (Emde et al., 1992; Loat, Asbury, Galsworthy, Plomin, & Craig, 2004; Loehlin & Nichols, 1976; Matthews, Batson, Horn, & Rosenman, 1981; Rushton, Fulker, Neale, Nias, & Eysenck, 1986; Zahn-Waxier, Robinson, & Emde, 1992) reported significant heritability of prosocial attitudes and as expected of a trait that is partially influenced by genes, prosocial or altruistic dispositions show individual differences in early childhood and stability over developmental time (Eisenberg, Guthrie, Cumberland, et al., 2002; Eisenberg, Guthrie, Murphy, et al., 1999). Although the theoretical basis for the evolutionary (Fehr & Fischbacher, 2003) and psychological mechanisms (Eisenberg, 2003) that underlie altruism has greatly increased in the past several decades, an exploration into specific genes contributing to this behavior is almost entirely lacking.

Locus Heterogeneity, Gene × Gene Interactions and Epistasis

How many genes are estimated to contribute to complex behavioral phenotypes including personality? A discussion of this question using rodent genetics as a background is found in the studies by Flint and his colleagues (Flint, 2003; Flint & Mott, 2001). For the pessimists among us it is notable that although over the past 15 years, of the more than 2,000 quantitative trait loci (QTLs) that have been identified in crosses between inbred strains of mice and rats, less than 1% have been characterized at the molecular level (Flint, Valdar, Shifman, & Mott, 2005). Even more sobering is that even if no more QTLs are mapped in rodent studies, at the present rate of progress (20 genes identified in 15 years) it will take 1,500 years to find all the genes that underlie

Table 17.2 Gene × gene interactions

What is the evidence for specific gene × gene interactions including epistasis in personality studies?	
Genes	Linked to...
HTR2C × DRD4	...Persistence and reward dependence (Ebstein, Segman, et al., 1997; Kuhn et al., 1999)
DRD4 × COMT × 5-HTTLPR	...Novelty seeking (Benjamin, Osher, et al., 2000; Strobel, Lesch, Jatzke, Paetzold, & Brocke, 2003)
DRD4 × SLC6A4	...Temperament in infants (Lakatos et al., 2003; De Luca, Rizzardi, Buccino, et al., 2003; De Luca, Rizzardi, Torrente, et al., 2001; Gervai et al., 2004; Auerbach, Faroy, Ebstein, Kahana, & Levine, 2001; Auerbach, Geller, et al., 1999; Ebstein, Levine, et al., 1998)
SLC6A4 × GABA(A)	...Neuroticism (Sen, Villafuerte, et al., 2004)
HTR2A × COMT	...Altered states of consciousness (Ott et al., 2005) which correlates to hypnotizability (Lichtenberg, Bachner-Melman, Ebstein, & Crawford, 2004; Lichtenberg, Bachner-Melman, Gritsenko, & Ebstein, 2000)
5-HTTLPR × AVPR1a	...Creative dance and reward dependence (Bachner-Melman, Zohar, et al., 2005)

known QTLs. Flint is nevertheless optimistic and notes that new analytical tools, including probabilistic ancestral haplotype reconstruction in outbred mice, Yin-Yang crosses and in silico analysis of sequence variants in many inbred strains, could make QTL cloning tractable. Flint notes that further high-resolution information at other QTLs might drive the estimate of average effect size below 5%. Altogether, finding genes for behavioral QTLs even in mice is not an easy task and suggests that the road ahead in human personality genetics is likely to be long and undoubtedly frustrating on occasion.

Incorporating environmental information in human behavioral genetic models might result in far fewer genes contributing to these traits than expected but such genes, including crucial environmental information, would be predicted to display larger effect sizes (Caspi, McClay, et al., 2002). This notion remains to be demonstrated but deserves serious consideration. A good bet is that the use of a robust phenotype definition such as an imaging paradigm coupled with early and reliable environmental information, more informative genotyping (accomplished by haplotype analysis across the genomic region) and epistatic interactions at different genetic loci might account for a large fraction of the variance of individual differences in personality traits.

Until recently it was assumed that alleles at separate QTLs contribute to most behavioral phenotypes additively. Recently, interest has shifted to the role of non-additive or epistatic interactions in contributing to complex phenotypes (Carlborg & Haley, 2004) and both Cloninger's and Eysenck's Personality Dimensions show widespread evidence for non-additive genetic variation (Keller, Coventry, Heath, & Martin, 2005). In its broadest sense, epistasis implies that the effect of a particular genotype on the phenotype depends on the genetic background. In its simplest form, this refers to an interaction between a pair of loci, in which the phenotypic effect of one locus depends on the genotype at the second locus. In the case of QTLs, epistasis describes a situation in which the phenotype of a given genotype cannot be predicted by the sum of its component single-locus effects

(see Table 17.2 for a summary of gene × gene personality studies).

Beyond Self-Report Temperament Measures

In addition to its stand-alone appeal, personality genetics, by parceling up complex disorders into more bite-sized traits, serves as an appealing method to better understand complex psychiatric disorders such as schizophrenia or ADHD. Indeed, we and others have suggested that personality traits are endophenotypes for mental illness (Cloninger, Adolfsson, & Svrakic, 1996; Gottesman & Gould, 2003; Benjamin, Ebstein, et al., 2002) and the linkage findings of Harm Avoidance and a schizophrenia locus at 8p strengthens this notion.

However, the multiple challenges posed by small effect sizes, polygenic inheritance, environmental influences and the limitations of self-report questionnaires make it unlikely that future studies based solely on a single instrument or experimental paradigms will substantially improve our understanding of how genes such as *DRD4* and *SLC6A4* impact personality. The current state of affairs regarding questionnaire-based temperament studies may be as good as it gets, and it is not very good. Future research should adopt a broader, multidisciplinary study of genes likely to contribute to a range of related phenotypes toward arriving at a comprehensive picture of how genes contribute to personality.

Traits such as altruism (Bachner-Melman, Gritsenko, et al., 2005), spirituality (Ott, Reuter, Hennig, & Vaitl, 2005), social communication (Bachner-Melman, et al., 2005; Bachner-Melman, Zohar, et al., 2005), empathy and shyness (Arbelle et al., 2003) to name just a few novel "personality" phenotypes are likely to be studied in greater depth in the future. Moreover, many of these phenotypes might best be studied by laboratory-based paradigms. Restricting conclusions to evidence derived from meta-analyses that are based on self-report questionnaires is suggested to be too conservative a strategy and it might be time to change the game plan and look at the forest and not only the trees. GeneCard (http://www.genecards.org/index.shtml) lists 23 OMIM *phenotypes* associated with the *DRD4* gene and 44 phenotypes associated with the *SLC6A4* gene. Evaluating the role of these two genes toward a broad understanding of human personality surely needs to consider all the available evidence based on all these phenotypes.

We suggest that some novel experimental paradigms such as brain imaging (Passamonti et al., 2006), evoked potentials (Leonard et al., 2002; Yu, Tsai, Hong, Chen, & Yang, 2004), prepulse inhibition (Swerdlow, Wasserman, et al., 2003) and some computer game models (Hollander et al., 2005) may also prove useful in unraveling the role of specific genes in contributing to the multifaceted dimensions of human personality. Knowledge of a gene's action at a lower level of brain organization such as those mediated by prepulse inhibition (Braff, Geyer, & Swerdlow, 2001; Swerdlow, Filion, Geyer, & Braff, 1995; Swerdlow, Martinez, et al., 2000) and evoked potentials (Gurrera, Salisbury, O'Donnell, Nestor, & McCarley, 2005; Strobel et al., 2004) may well clarify the role of particular polymorphisms in higher order behaviors such as personality constructs.

Acknowledgment This research was partially supported by the Israel Science Foundation founded by the Israel Academy of Sciences and Humanities (RPE).

References

Ansorge, M. S., Zhou, M., Lira, A., Hen, R., &Gingrich, J. A. (2004). Early-life blockade of the 5-HT transporter alters emotional behavior in adult mice. *Science, 306*(5697), 879–881.

Arbelle, S., Benjamin, J., Golin, M., Kremer, I., Belmaker, R. H., & Ebstein, R. P. (2003). Relation of shyness in grade school children to the genotype for the long form of the serotonin transporter promoter region polymorphism. *The American Journal of Psychiatry, 160*(4), 671–676.

Arcos-Burgos, M., Castellanos, F. X., Konecki, D., Lopera, F., Pineda, D., Palacio, J. D., et al. (2004). Pedigree disequilibrium test (PDT) replicates association and linkage between DRD4 and ADHD in multigenerational and extended pedigrees from a genetic isolate. *Molecular Psychiatry, 9*(3), 252–259.

Asghari, V., Sanyal, S., Buchwaldt, S., Paterson, A., Jovanovic, V., & Van Tol, H. H. (1995). Modulation of intracellular cyclic AMP levels by different human dopamine D4 receptor variants. *Journal of Neurochemistry, 65*(3), 1157–1165.

Asghari, V., Schoots, O., van Kats, S., Ohara, K., Jovanovic, V., Guan, H. C., et al. (1994). Dopamine D4 receptor repeat: Analysis of different native and mutant forms of the human and rat genes. *Molecular Pharmacology, 46*(2), 364–373.

Auerbach, J. G., Faroy, M., Ebstein, R., Kahana, M., & Levine, J. (2001). The association of the dopamine D4 receptor gene (DRD4) and the serotonin transporter promoter gene (5-HTTLPR) with temperament in 12-month-old infants. *Journal of Child Psychology and Psychiatry, 42*(6), 777–783.

Auerbach, J., Geller, V., Lezer, S., Shinwell, E., Belmaker, R. H., Levine, J., et al. (1999). Dopamine D4 receptor (D4DR) and

serotonin transporter promoter (5-HTTLPR) polymorphisms in the determination of temperament in 2-month-old infants. *Molecular Psychiatry, 4*(4), 369–373.

Bachner-Melman, R., Dina, C., Zohar, A. H., Constantini, N., Lerer, E., Hoch, S., et al. (2005). AVPR1a and SLC6A4 Gene Polymorphisms Are Associated with Creative Dance Performance. *PLoS Genetics, 1*(3), e42.

Bachner-Melman, R., Gritsenko, I., Nemanov, L., Zohar, A. H., Dina, C., & Ebstein, R. P. (2005). Dopaminergic polymorphisms associated with self-report measures of human altruism: A fresh phenotype for the dopamine D4 receptor. *Molecular Psychiatry, 10*(4), 333–5.

Bachner-Melman, R., Zohar, A. H., Bacon-Shnoor, N., Elizur, Y., Nemanov, L., Gritsenko, I., et al. (2005). Linkage between vasopressin receptor AVPR1A promoter region microsatellites and measures of social behavior in humans. *Journal of Individual Differences, 26*(1), 2–10.

Barr, C. S., Newman, T. K., Lindell, S., Shannon, C., Champoux, M., Lesch, K. P., et al. (2004). Interaction between serotonin transporter gene variation and rearing condition in alcohol preference and consumption in female primates. *Archives of General Psychiatry, 61*(11), 1146–1152.

Barr, C. S., Schwandt, M. L., Newman, T. K., & Higley, J. D. (2004). The use of adolescent nonhuman primates to model human alcohol intake: Neurobiological, genetic, and psychological variables. *Annals of the New York Academy of Sciences, 1021*, 221–233.

Barr, C. S., Newman, T. K., Becker, M. L., Parker, C. C., Champoux, M., Lesch, K. P., et al. (2003). The utility of the non-human primate; model for studying gene by environment interactions in behavioral research. *Genes, Brain, and Behavior, 2*(6), 336–340.

Batson, C. D. (1991). *The altruism question: Toward a social psychological answer.* Hillsdale, NJ: L. Erlbaum, Associates, 257 p.

Benjamin, J., Ebstein, R. P., & Belmaker, R. H. (2002). Genes for Human Personality Traits: Endophenotypes of Psychiatric Disorders. In: J. Benjamin, R. P. Ebstein, R. H. Belmaker (Eds.), *Molecular Genetics and the Human Personality* (1st ed., pp. 333–344). Washington, DC: American Psychiatric Publications, Inc.

Benjamin, J., Ebstein, R. P., & Belmaker, R. H. (2002). *Molecular genetics and the human personality* (1st ed.). Washington, DC: American Psychiatric Publications.

Benjamin, J., Osher, Y., Kotler, M., Gritsenko, I., Nemanov, L., Belmaker, R. H., et al. (2000). Association between tridimensional personality questionnaire (TPQ) traits and three functional polymorphisms: Dopamine receptor D4 (DRD4), serotonin transporter promoter region (5-HTTLPR) and catechol *O*- methyltransferase (COMT). *Molecular Psychiatry, 5*(1), 96–100.

Benjamin, J., Li, L., Patterson, C., Greenberg, B. D., Murphy, D. L., & Hamer, D. H. (1996). Population and familial association between the D4 dopamine receptor gene and measures of Novelty Seeking. *Nature Genetics, 12*(1), 81–84.

Bookman, E. B., Taylor, R. E., Adams-Campbell, L., & Kittles, R. A. (2002). DRD4 promoter SNPs and gender effects on Extraversion in African Americans. *Molecular Psychiatry, 7*(7), 786–789.

Bouchard, T. J., Jr., & Loehlin, J. C. (2001). Genes, evolution, and personality. *Behavior Genetics, 31*, 243–273.

Braff, D. L., Geyer, M. A., & Swerdlow, N. R. (2001). Human studies of prepulse inhibition of startle: Normal subjects, patient groups, and pharmacological studies. *Psychopharmacology (Berl), 156*(2–3), 234–258.

Buskila, D. (2001). Fibromyalgia, chronic fatigue syndrome, and myofascial pain syndrome. *Current Opinion in Rheumatology, 13*, 117–27.

Butcher, J. N. (2001). *MMPI-2: Minnesota Multiphasic Personality Inventory-2: manual for administration, scoring, and interpretation.* Minneapolis: University of Minnesota Press, 212 p.

Carlborg, O., & Haley, C. S. (2004). Epistasis: Too often neglected in complex trait studies? *Nature Reviews Genetics, 5*(8), 618–625.

Caspi, A., Moffitt, T. E., Cannon, M., McClay, J., Murray, R., Harrington, H., et al. (2005). Moderation of the effect of adolescent-onset cannabis use on adult psychosis by a functional polymorphism in the Catechol-O-Methyltransferase gene: Longitudinal evidence of a Gene X Environment interaction. *Biological Psychiatry, 57,* 1117–1127.

Caspi, A., Sugden, K., Moffitt, T. E., Taylor, A., Craig, I. W., Harrington, H., et al. (2003). Influence of life stress on depression: Moderation by a polymorphism in the 5-HTT gene. *Science, 301*(5631), 386–389.

Caspi, A., McClay, J., Moffitt, T. E., Mill, J., Martin, J., Craig, I. W., et al. (2002). Role of genotype in the cycle of violence in maltreated children. *Science, 297*(5582), 851–854.

Cloninger, C. R., Van Eerdewegh, P., Goate, A., Edenberg, H. J., Blangero, J., Hesselbrock, V., et al. (1998). Anxiety proneness linked to epistatic loci in genome scan of human personality traits. *American Journal of Medical Genetics, 81*(4), 313–317.

Cloninger, C. R., Adolfsson, R., & Svrakic, N. M. (1996). Mapping genes for human personality [news]. *Nature Genetics, 12*(1), 3–4.

Cloninger, C. R., Svrakic, D. M., & Przybeck, T. R. (1993). A psychobiological model of temperament and character. *Archives of General Psychiatry, 50*(12), 975–990.

Cloninger, C. R. (1987). A systematic method for clinical description and classification of personality variants. A proposal. *Archives of General Psychiatry, 44*(6), 573–588.

Cloninger, C. R. (1986). A unified biosocial theory of personality and its role in the development of anxiety states. *Psychiatric Developments, 4,* 167–226.

Cohen, H., Buskila, D., Neumann, L., & Ebstein, R. P. (2002). Confirmation of an association between fibromyalgia and serotonin transporter promoter region (5- HTTLPR) polymorphism, and relationship to anxiety-related personality traits. *Arthritis Rheum, 46,* 845–847.

Costa, P. T., Jr., & McCrae, R. R. (1997). Stability and change in personality assessment. The revised NEO Personality Inventory in the year (2000). *Journal of Personality Assessment, 68,* 86–94.

De Luca, V., Zai, G., Tharmalingam, S., de Bartolomeis, A., Wong, G., & Kennedy, J. L. (2006). Association study between the novel functional polymorphism of the serotonin transporter gene and suicidal behaviour in schizophrenia. *European Neuropsychopharmacology, 16,* 268–271.

De Luca, A., Rizzardi, M., Buccino, A., Alessandroni, R., Salvioli, G. P., Filograsso, N., et al. (2003). Association of dopamine D4 receptor (DRD4) exon III repeat polymorphism with temperament in 3-year-old infants. *Neurogenetics, 4*(4), 207–212.

De Luca, A., Rizzardi, M., Torrente, I., Alessandroni, R., Salvioli, G. P., Filograsso, N., et al. (2001). Dopamine D4 receptor (DRD4) polymorphism and adaptability trait during infancy: A longitudinal study in 1- to 5-month-old neonates. *Neurogenetics, 3*(2), 79–82.

Dina, C., Nemanov, L., Gritsenko, I., Rosolio, N., Osher, Y., & Heresco-Levy, U., et al. (2005). Fine mapping of a region on chromosome 8p gives evidence for a QTL contributing to individual differences in an anxiety-related personality trait, TPQ harm avoidance. *Neuropsychiatric Genetics, 132B,* 104–108.

Ding, Y. C., Chi, H. C., Grady, D. L., Morishima, A., Kidd, J. R., Kidd, K. K., et al. (2002). Evidence of positive selection acting at the human dopamine receptor D4 gene locus. *Proceedings of the National Academy of Sciences of the USA, 99*(1), 309–314.

D'Souza, U. M., Russ, C., Tahir, E., Mill, J., McGuffin, P., Asherson, P. J., et al. (2004). Functional effects of a tandem duplication poly-

morphism in the 5'flanking region of the DRD4 gene. *Biological Psychiatry, 56*(9), 691–697.

Dulawa, S. C., Grandy, D. K., Low, M. J., Paulus, M. P., & Geyer, M. A. (1999). Dopamine D4 receptor-knock-out mice exhibit reduced exploration of novel stimuli. *The Journal of Neuroscience, 19*(21), 9550–9556.

Ebstein, R. P., Zohar, A. H., Benjamin, J., & Belmaker, R. H. (2002). An update of molecular genetic studies of human personality traits. *Applied Bioinformatics 1*(2), 57–68.

Ebstein, R. P., & Kotler, M. (2002). Personality, Substance Abuse, and Genes. In J. Benjamin, R. P. Ebstein, R. H. Belmaker (Eds.), Molecular Genetics and the Human Personality (1st ed., pp. 151–163), Washington, DC: American Psychiatric Publishing, Inc.

Ebstein, R. P., Benjamin, J., & Belmaker, R. H. (2000). Personality and polymorphisms of genes involved in aminergic neurotransmission. *European Journal of Pharmacology, 410,* 205–214.

Ebstein, R. P., Levine, J., Geller, V., Auerbach, J., Gritsenko, I., & Belmaker, R. H. (1998). Dopamine D4 receptor and serotonin transporter promoter in the determination of neonatal temperament. *Molecular Psychiatry, 3*(3), 238–246.

Ebstein, R. P., Segman, R., Benjamin, J., Osher, Y., Nemanov, L., & Belmaker, R. H. (1997). 5-HT2C (HTR2C) serotonin receptor gene polymorphism associated with the human personality trait of reward dependence: Interaction with dopamine D4 receptor (D4DR) and dopamine D3 receptor (D3DR) polymorphisms. *American Journal of Medical Genetics Neuropsychiatric Genetics, 74*(1), 65–72.

Ebstein, R. P., Novick, O., Umansky, R., Priel, B., Osher, Y., Blaine, D., et al. (1996). Dopamine D4 receptor (D4DR) exon III polymorphism associated with the human personality trait of Novelty Seeking. *Nature Genetics, 12*(1), 78–80.

Eichhammer, P. Sand, P. G., Stoertebecker, P., Langguth, B., Zowe, M., & Hajak, G. (2005). Variation at the DRD4 promoter modulates extraversion in Caucasians. *Molecular Psychiatry, 10*(6), 520–2.

Eisenberg, N. (2003). Prosocial behavior, empathy, and sympathy. In M. H. Bornstein & L. Davidson (Eds.), Well being: Positive development across the life course. Crosscurrents in contemporary psychology (pp. 253–265). Mahwah, NJ: Lawrence Erlbaum Associates, Publishers.

Eisenberg, N., Guthrie, I. K., Cumberland, A., Murphy, B. C., Shepard, S. A., Zhou, Q., & Carlo, G. (2002). Prosocial development in early adulthood: A longitudinal study. *Journal of Personality and Social Psychology, 82,* 993–1006.

Eisenberg, N., Guthrie, I. K., Murphy, B. C., Shepard, S. A., Cumberland, A., & Carlo, G. (1999). Consistency and development of prosocial dispositions: A longitudinal study. *Child Development, 70,* 1360–1372.

Emde, R. N., Plomin, R., Robinson, J. A., Corley, R., DeFries, J., & Fulker, D. W., et al. (1992). Temperament, emotion, and cognition at fourteen months: The MacArthur Longitudinal Twin Study. *Child Development, 63,* 1437–1455.

Eysenck, H. J. (1952). *The scientific study of personality.* London: Routledge & Kegan Paul.

Fehr, E., & Fischbacher, U. (2003). The nature of human altruism. *Nature, 425,* 785–791.

Ferguson, J. N., Young, L. J., & Insel, T. R. (2002). The neuroendocrine basis of social recognition. *Front Neuroendocrinol, 23,* 200–224.

Fiskerstrand C. E., Lovejoy E. A., & Quinn J. P. (1999). An intronic polymorphic domain often associated with susceptibility to affective disorders has allele dependent differential enhancer activity in embryonic stem cells. *FEBS Letters, 458*(2), 171–4.

Flint, J., Valdar, W., Shifman, S., & Mott, R. (2005). Strategies for mapping and cloning quantitative trait genes in rodents. *Nature Reviews Genetics, 6*(4), 271–286.

Flint, J. (2003). Analysis of quantitative trait loci that influence animal behavior. *Journal of Neurobiology, 54*(1), 46–77.

Flint, J., & Mott, R. (2001). Finding the molecular basis of quantitative traits: Successes and pitfalls. *Nature Reviews Genetics, 2*(6), 437–45.

Fullerton, J., Cubin, M., Tiwari, H., Wang, C., Bomhra, A., Davidson, S., Miller, S., Fairburn, C., Goodwin, G., Neale, M. C., Fiddy, S., Moll, R., Allison, D. B., & Flinl, J. (2003). Linkage analysis of extremely discordant and concordant sibling pairs identifies quantitative-trait Loci that influence variation in the human personality trait neuroticism. *American Journal of Human Genetics, 72*, 879–890.

Gelernter, J., Cubells, J. F., Kidd, J. R., Pakstis, A. J., & Kidd, K. K. (1999). Population studies of polymorphisms of the serotonin transporter protein gene. *American Journal of Medical Genetics, 88*(1), 61–6.

Gergen, K.-J., Gergen, M.-M., & Meter, K. (1972). Individual orientations to prosocial behavior. *Journal of Social Issues, 28*, 105–130.

Gervai, J., Nemoda, Z., Lakatos, K., Ronai, Z., Toth, I., Ney, K., et al. (2004). Transmission disequilibrium tests confirm the link between DRD4 gene polymorphism and infant attachment. *American Journal of Medical Genetics B Neuropsychiatric Genetics, 132B*(1), 126–130.

Giraud, T., Pedersen, J. S., & Keller, L. (2002). Evolution of supercolonies: The Argentine ants of southern Europe. *Proceedings of the National Academy of Sciences of the USA, 99*, 6075–6079.

Goldman, D., Hu, X., Zhu, G., Lipsky, R., & Murphy, D. (1994). The Serotonin Transporter: New Alleles, Function and Phenotype. In: 3rd Annual Pharmacogenetics in Psychiatry Meeting, p. 16–17.

Golimbet, V. E., Gritsenko, I. K., Alfimova, M. V., & Ebstein, R. P. (2005). [Polymorphic markers of the dopamine D4 receptor gene promoter region and personality traits in mentally healthy individuals from the Russian population]. *Genetika, 41*(7), 966–972.

Gottesman, I. I., & Gould, T. D. (2003). The endophenotype concept in psychiatry: Etymology and strategic intentions. *The American Journal of Psychiatry, 160*(4), 636–645.

Grady, D. L., Chi, H. C., Ding, Y. C., Smith, M., Wang, E., Schuck, S., et al. (2003). High prevalence of rare dopamine receptor D4 alleles in children diagnosed with attention-deficit hyperactivity disorder. *Molecular Psychiatry, 8*(5), 536–545.

Gurrera, R. J., Salisbury, D. F., O'Donnell, B. F., Nestor, P. G., & McCarley, R. W. (2005). Auditory P3 indexes personality traits and cognitive function in healthy men and women. *Psychiatry Research, 133*(2–3), 215–228.

Hamilton, W. D. (1964). The genetical evolution of social behaviour. *Journal of Theoretical Biology, 7*, 1–16.

Hariri, A. R., Drabant, E. M., Munoz, K. E., Kolachana, B. S., Mattay, V. S., Egan, M. F., et al. (2005). A susceptibility gene for affective disorders and the response of the human amygdala. *Archives of General Psychiatry, 62*(2), 146–152.

Hariri, A. R., & Weinberger, D. R. (2003). Imaging genomics. *British Medical Bulletin, 65*, 259–270.

Hariri, A. R., Mattay, V. S., Tessitore, A., Kolachana, B., Fera, F., Goldman, D., et al. (2002). Serotonin transporter genetic variation and the response of the human amygdala. *Science, 297*(5580), 400–403.

Heils, A., Teufel, A., Petri, S., Stober, G., Riederer, P., Bengel, D., et al. (1996). Allelic variation of human serotonin transporter gene expression. *Journal of Neurochemistry, 66*(6), 2621–2624.

Hollander, E., Pallanti, S., Baldini R. N., Sood, E., Baker, B. R., & Buchsbaum, M. S. (2005). Imaging monetary reward in pathological gamblers. *The World Journal of Biological Psychiatry, 6*(2), 113–120.

Hranilovic, D., Stefulj, J., Schwab, S., Borrmann-Hassenbach, M., Albus, M., Jernej, B., et al. (2004). Serotonin transporter promoter and intron 2 polymorphisms: Relationship between allelic variants and gene expression. *Biological Psychiatry, 55*(11), 1090–1094.

Inoue-Murayama, M., Matsuura, N., Murayama, Y., Tsubota, T., Iwasaki, T., Kitagawa, H., et al. (2002). Sequence comparison of the dopamine receptor D4 exon III repetitive region in several species of the order Carnivora. *The Journal of Veterinary Medical Science, 64*(8), 747–749.

Ito, H., Nara, H., Inoue-Murayama, M., Shimada, M. K., Koshimura, A., Ueda, Y., et al. (2004). Allele frequency distribution of the canine dopamine receptor D4 gene exon III and I in 23 breeds. *The Journal of Veterinary Medical Science, 66*(7), 815–820.

Jang, K. L., Vernon, P. A., & Livesley, W. J. (2001). Behavioural-genetic perspectives on personality function. *Canadian Journal of Psychiatry, 46*(3), 234–244.

Keller, M. C., Coventry, W. L., Heath, A. C., & Martin, N. G. (2005). Widespread Evidence for Non-Additive Genetic Variation in Cloninger's and Eysenck's Personality Dimensions Using a Twin Plus Sibling Design. *Behavior Genetics, 35*, 707–721.

Keller, L. & Parker, J. D. (2002). Behavioral genetics: A gene for supersociality. *Current Biology, 12*, R180–1.

Keltikangas-Jarvinen, L., Raikkonen, K., Ekelund, J., & Peltonen, L. (2004). Nature and nurture in novelty seeking. *Molecular Psychiatry, 9*(3), 308–11.

Kim, S. J., Young, L. J., Gonen, D., Veenstra-VanderWeele, J., Courchesne, R., Courchesne, E., et al. (2002). Transmission disequilibrium testing of arginine vasopressin receptor 1A (AVPR1A) polymorphisms in autism. *Molecular Psychiatry, 7*, 503–507.

Kluger, A. N., Siegfried, Z., & Ebstein, R. P. (2002). A meta-analysis of the association between DRD4 polymorphism and novelty seeking. *Molecular Psychiatry, 7*(7), 712–717.

Krieger, M. J., & Ross, K. G. (2002). Identification of a major gene regulating complex social behavior. *Science, 295*, 328–332.

Kuhn, K. U., Meyer, K., Nothen, M. M., Gansicke, M., Papassotiropoulos, A., & Maier, W. (1999). Allelic variants of dopamine receptor D4 (DRD4) and serotonin receptor 5HT2c (HTR2c) and temperament factors: Replication tests. *American Journal of Medical Genetics, 88*(2), 168–172.

Lahti, J., Raikkonen, K., Ekelund, J., Peltonen, L., Raitakari, O. T., & Keltikangas-Jarvinen, L. (2005). Novelty seeking: Interaction between parental alcohol use and dopamine D4 receptor gene exon III polymorphism over 17 years. *Psychiatric Genetics, 15*(2), 133–139.

Lakatos, K., Nemoda, Z., Birkas, E., Ronai, Z., Kovacs, E., Ney, K., et al. (2003). Association of D4 dopamine receptor gene and serotonin transporter promoter polymorphisms with infants' response to novelty. *Molecular Psychiatry, 8*(1), 90–97.

Leonard, S., Gault, J., Hopkins, J., Logel, J., Vianzon, R., Short, M., et al. (2002). Association of promoter variants in the alpha7 nicotinic acetylcholine receptor subunit gene with an inhibitory deficit found in schizophrenia. *Archives of General Psychiatry, 59*(12), 1085–1096.

Lerer, E., Kanyas, K., Karni, O., Ebstein, R. P., & Lerer, B. (2006). Why do young women smoke? II Role of traumatic life experience, psychological characteristics and serotonergic genes. *Molecular Psychiatry, 11*, 771–781.

Lesch, K. P., Bengel, D., Heils, A., Sabol, S. Z., Greenberg, B. D., Petri, S., et al. (1996). Association of anxiety-related traits with a polymorphism in the serotonin transporter gene regulatory region. *Science, 274*(5292), 1527–1531.

Lesch, K. P., Bengel, D., Heils, A., Sabol, S. Z., Greenberg, B. D., Petri, S., et al. (1996). Association of anxiety-related traits with a polymorphism in the serotonin transporter gene regulatory region [see comments]. *Science, 274*(5292), 1527–1531.

Lesch, K. P., Balling, U., Gross, J., Strauss, K., Wolozin, B. L., Murphy, D. L., et al. (1994). Organization of the human serotonin transporter gene. *Journal of Neural Transmission. General Section, 95*(2), 157–162.

Lichtenberg, P., Bachner-Melman, R., Ebstein, R. P., & Crawford, H. J. (2004). Hypnotic susceptibility: Multidimensional relationships with Cloninger's Tridimensional Personality Questionnaire, COMT polymorphisms, absorption, and attentional characteristics. *The International Journal of Clinical and Experimental Hypnosis, 52*(1), 47–72.

Lichtenberg, P., Bachner-Melman, R., Gritsenko, I., & Ebstein, R. P. (2000). Exploratory association study between catechol-O-methyltransferase (COMT) high/low enzyme activity polymorphism and hypnotizability [In Process Citation]. *American Journal of Medical Genetics, 96*(6), 771–774.

Loat, C. S., Asbury, K., Galsworthy, M. J., Plomin, R., & Craig, I. W. (2004). X inactivation as a source of behavioural differences in monozygotic female twins. *Twin Research, 7*, 54–61.

Loehlin, J. C. & Nichols, R. C. 1976. Heredity, environment, & personality: A study of 850 sets of twins. Austin: University of Texas Press,. 202 p.

Lowe, N., Kirley, A., Mullins, C., Fitzgerald, M., Gill, M., & Hawi, Z. (2004). Multiple marker analysis at the promoter region of the DRD4 gene and ADHD: Evidence of linkage and association with the SNP -616. *American Journal of Medical Genetics B Neuropsychiatric Genetics, 131*(1), 33–37.

MacKenzie, A., & Quinn, J. (1999). A serotonin transporter gene intron 2 polymorphic region, correlated with affective disorders, has allele-dependent differential enhancer- like properties in the mouse embryo. *Proceedings of the National Academy of Sciences of the USA, 96*(26), 15251–15255.

Matthews, K.-A., Batson, C. D., Horn, J., & Rosenman, R.-H. (1981). Principles in his nature which interest him in the fortune of others: The heritability of empathic concern for others. *Journal of Personality, 49*, 237–247.

Mayeux, R. (2005). Mapping the new frontier: Complex genetic disorders. *Journal of Clinical Investigation, 115*(6), 1404–1407.

McCracken, J. T., Smalley, S. L., McGough, J. J., Crawford, L., Del'Homme, M., Cantor, R. M., et al. (2000). Evidence for linkage of a tandem duplication polymorphism upstream of the dopamine D4 receptor gene (DRD4) with attention deficit hyperactivity disorder (ADHD). *Molecular Psychiatry, 5*(5), 531–536.

Mitsuyasu, H., Hirata, N., Sakai, Y., Shibata, H., Takeda, Y., Ninomiya, H., et al. (2001). Association analysis of polymorphisms in the upstream region of the human dopamine D4 receptor gene (DRD4) with schizophrenia and personality traits. *Journal of Human Genetics, 46*(1), 26–31.

Moffitt, T. E., Caspi, A., & Rutter, M. (2005). Strategy for investigating interactions between measured genes and measured environments. *Archives of General Psychiatry, 62*(5), 473–481.

Momozawa, Y., Takeuchi, Y., Kusunose, R., Kikusui, T., & Mori, Y. 2005. Association between equine temperament and polymorphisms in dopamine D4 receptor gene. *Mammalian Genome, 16*(7), 538–544.

Munafo, M. R., Clark, T., & Flint, J. (2005). Does measurement instrument moderate the association between the serotonin transporter gene and anxiety-related personality traits? A meta-analysis. *Molecular Psychiatry, 10*(4), 415–419.

Munafo, M. R., Clark, T. G., Moore, L. R., Payne, E., Walton, R., & Flint, J. (2003). Genetic polymorphisms and personality in healthy adults: A systematic review and meta-analysis. *Molecular Psychiatry, 8*(5), 471–484.

Murphy, D. L., Li, Q., Engel, S., Wichems, C., Andrews, A., Lesch, K. P., et al. (2001). Genetic perspectives on the serotonin transporter. *Brain Research Bulletin, 56*(5), 487–494.

Niimi, Y., Inoue-Murayama, M., Kato, K., Matsuura, N., Murayama, Y., Ito, S., et al. (2001). Breed differences in allele frequency of the dopamine receptor D4 gene in dogs. *The Journal of Heredity, 92*(5), 433–436.

Niimi, Y., Inoue-Murayama, M., Murayama, Y., Ito, S., & Iwasaki, T. (1999). Allelic variation of the D4 dopamine receptor polymorphic region in two dog breeds, Golden retriever and Shiba. *The Journal of Veterinary Medical Science, 61*(12), 1281–1286.

Noblett, K. L., & Coccaro, E. F. (2005). Molecular genetics of personality. *Current Psychiatry Reports, 7*(1), 73–80.

Offenbaecher, M., Bondy, B., de Jonge, S., Glatzeder, K., Kruger, M., Schoeps, P., & Ackenheil, M. (1999). Possible association of fibromyalgia with a polymorphism in the serotonin transporter gene regulatory region. *Arthritis Rheum, 42*, 2482–2488.

Okuyama, Y., Ishiguro, H., Nankai, M., Shibuya, H., Watanabe, A., & Arinami, T. (2000). Identification of a polymorphism in the promoter region of DRD4 associated with the human novelty seeking personality trait. *Molecular Psychiatry, 5*(1), 64–69.

Ott, U., Reuter, M., Hennig, J., & Vaitl, D. (2005). Evidence for a common biological basis of the absorption trait, hallucinogen effects, and positive symptoms: Epistasis between 5-HT2a and COMT polymorphisms. *American Journal of Medical Genetics B Neuropsychiatric Genetics, 137*(1), 29–32.

Passamonti, L., Fera, F., Magariello, A., Cerasa, A., Gioia, M. C., Muglia, M., et al. (2006). Monoamine Oxidase-A Genetic Variations Influence Brain Activity Associated with Inhibitory Control: New Insight into the Neural Correlates of Impulsivity. *Biological Psychiatry, 59*, 334–340.

Rogers, G., Joyce, P., Mulder, R., Sellman, D., Miller, A., Allington, M., et al. (2004). Association of a duplicated repeat polymorphism in the 5'-untranslated region of the DRD4 gene with novelty seeking. *American Journal of Medical Genetics, 126B*(1), 95–98.

Ronai, Z., Szekely, A., Nemoda, Z., Lakatos, K., Gervai, J., Staub, M., et al. (2001). Association between Novelty Seeking and the -521 C/T polymorphism in the promoter region of the DRD4 gene. *Molecular Psychiatry, 6*(1), 35–38.

Rubinstein, M., Phillips, T. J., Bunzow, J. R., Falzone, T. L., Dziewczapolski, G., Zhang, G., et al. (1997). Mice lacking dopamine D4 receptors are supersensitive to ethanol, cocaine, and methamphetamine. *Cell, 90*(6), 991–1001.

Rushton, J. P., Fulker, D. W., Neale, M. C., Nias, D. K., & Eysenck, H. J. (1986). Altruism and aggression: The heritability of individual differences. *Journal of Personality and Social Psychology, 50*, 1192–1198.

Savitz, J. B., & Ramesar, R. S. (2004). Genetic variants implicated in personality: A review of the more promising candidates. *American Journal of Medical Genetics B Neuropsychiatric Genetics, 131*(1), 20–32.

Schinka, J. A., Busch, R. M., & Robichaux-Keene, N. (2004). A meta-analysis of the association between the serotonin transporter gene polymorphism (5-HTTLPR) and trait anxiety. *Molecular Psychiatry, 9*(2), 197–202.

Schinka, J. A., Letsch, E. A., & Crawford, F. C. (2002). DRD4 and novelty seeking: Results of meta-analyses. *American Journal of Medical Genetics, 114*(6), 643–648.

Schoots, O., & Van Tol, H. H. (2003). The human dopamine D4 receptor repeat sequences modulate expression. *The Pharmacogenomics Journal, 3*(6), 343–348.

Seaman, M. I., Fisher, J. B., Chang, F., & Kidd, K. K. (1999). Tandem duplication polymorphism upstream of the dopamine D4 receptor gene (DRD4). *American Journal of Medical Genetics, 88*(6), 705–709.

Sen, S., Burmeister, M., & Ghosh, D. (2004). Meta-analysis of the association between a serotonin transporter promoter polymorphism (5-HTTLPR) and anxiety-related personality traits. *American Journal of Medical Genetics, 127B*(1), 85–89.

Sen, S., Villafuerte, S., Nesse, R., Stoltenberg, S. F., Hopcian, J., Gleiberman, L., et al. (2004). Serotonin transporter and GABAA alpha 6 receptor variants are associated with neuroticism. *Biological Psychiatry, 55*(3), 244–249.

Sigmund, K., & Hauert, C. (2002). Primer: Altruism. *Current Biology, 12*, R270-2.

Strobel, A., Debener, S., Anacker, K., Muller, J., Lesch, K. P., & Brocke, B. (2004). Dopamine D4 receptor exon III genotype influence on the auditory evoked novelty P3. *Neuroreport, 15*(15), 2411–2415.

Strobel, A., Lesch, K. P., Jatzke, S., Paetzold, F., & Brocke, B. (2003). Further evidence for a modulation of Novelty Seeking by DRD4 exon III, 5-HTTLPR, and COMT val/met variants. *Molecular Psychiatry, 8*(4), 371–372.

Swerdlow, N. R., Wasserman, L. C., Talledo, J. A., Casas, R., Bruins, P., & Stephany, N. L. (2003). Prestimulus modification of the startle reflex: Relationship to personality and physiological markers of dopamine function. *Biological Psychology, 62*(1), 17–26.

Swerdlow, N. R., Martinez, Z. A., Hanlon, F. M., Platten, A., Farid, M., Auerbach, P., et al. (2000). Toward understanding the biology of a complex phenotype: Rat strain and substrain differences in the sensorimotor gating-disruptive effects of dopamine agonists. *Journal of Neuroscience, 20*(11), 4325–4336.

Swerdlow, N. R., Filion, D., Geyer, M. A., & Braff, D. L. (1995). "Normal" personality correlates of sensorimotor, cognitive, and visuospatial gating. *Biological Psychiatry, 37*(5), 286–99.

Taylor, S. E., Klein, L. C., Lewis, B. P., Gruenewald, T. L., Gurung, R. A., & Updegraff, J. A. (2000). Biobehavioral responses to stress in females: Tend-and-befriend, not fight-or-flight. *Psychological Review, 107*, 411–429.

Torgersen, S. (2005).Behavioral genetics of personality. *Current psychiatry reports, 7*(1), 51–56.

Van Craenenbroeck, K., Clark, S. D., Cox, M. J., Oak, J. N., Liu, F., & Van Tol, H. H. (2005). Folding efficiency is rate-limiting in dopamine D4 receptor biogenesis. *The Journal of Biological Chemistry, 280*(19), 19350–19357.

Van Tol, H. H., Wu, C. M., Guan, H. C., Ohara, K., Bunzow, J. R., Civelli, O., et al. (1992). Multiple dopamine D4 receptor variants in the human population [see comments].*Nature, 358*(6382), 149–152.

Van Tol, H. H., Bunzow, J. R., Guan, H. C., Sunahara, R. K., Seeman, P., Niznik, H. B., et al. (1991). Cloning of the gene for a human dopamine D4 receptor with high affinity for the antipsychotic clozapine. *Nature, 350*(6319), 610–614.

Wang, E., Ding, Y. C., Flodman, P., Kidd, J. R., Kidd, K. K., Grady, D. L., et al. (2004). The Genetic Architecture of Selection at the Human Dopamine Receptor D4 (DRD4) Gene Locus. *American Journal of Human Genetics, 74*(5), 931–944.

Willis-Owen, S. A., Turri, M. G., Munafo, M. R., Surtees, P. G., Wainwright, N. W., Brixey, R. D., et al. (2005). The Serotonin Transporter Length Polymorphism, Neuroticism, and Depression: A Comprehensive Assessment of Association. *Biological Psychiatry, 58*, 451–456.

Winslow, J., & Insel, T. R. (1991). Vasopressin modulates male squirrel monkeys' behavior during social separation. *European Journal of Pharmacology, 200*, 95–101.

Xing, Q. H., Wu, S. N., Lin, Z. G., Li, H. F., Yang, J. D., Feng, G. Y., et al. (2003). Association analysis of polymorphisms in the upstream region of the human dopamine D4 receptor gene in schizophrenia. *Schizophrenia research, 65*(1), 9–14.

Young, L. J., Lim, M. M., Gingrich, B., & Insel, T. R. (2001). Cellular mechanisms of social attachment. *Hormones and Behavior, 40*, 133–138.

Yu, Y. W., Tsai, S. J., Hong, C. J., Chen, T. J., & Yang, C. W. (2004). Association analysis for MAOA gene polymorphism with long-latency auditory evoked potentials in healthy females. *Neuropsychobiology, 50*(4), 288–291.

Zahn-Waxier, C., Robinson, J.-L., & Emde, R.-N. (1992). The development of empathy in twins. *Developmental Psychology, 28*, 1038–1047.

Zohar, A. H., Dina, C., Rosolio, N., Osher, Y., Gritsenko, I., Bachner-Melman, R., Benjamin, J., Belmaker, R. H., & Ebstein, R. P. (2003). Tridimensional personality questionnaire trait of harm avoidance (anxiety proneness) is linked to a locus on chromosome 8p21. *American Journal of Medical Genetics, 117B*, 66–69.

Chapter 18

The Genetics of Childhood Temperament

Jeffrey R. Gagne, Matthew K. Vendlinski, and H. Hill Goldsmith

Introduction

The field of temperament research stands at the intersection of various disciplines of biobehavioral research. Temperament links personality to psychological development. Temperament links normative variation to psychopathology. Temperament links human and animal research traditions in behavior genetics. Temperament, as we now understand it, was one of the earliest areas of behavior genetic research. For example, temperament in dogs was one focus of research by Scott and Fuller (1965).

In this chapter, we first describe the concept of temperament, with an emphasis on five of its components. Then we review quantitative genetic findings for these five dimensions. The quantitative genetic findings are followed by selected molecular genetic findings relevant to human temperament. Next, the concepts of gene–environment interaction and correlations are defined and illustrated for temperament. After discussing endophenotypic approaches to temperament, we conclude the chapter with potential directions for new research on the genetics of childhood temperament.

Definition of Temperament

Temperamental traits can be conceptualized as behavioral dimensions that develop early in life and form the basis for later personality (Goldsmith et al., 1987). Early emerging temperamental differences tend to change over development and combine with experience to give rise to personality traits in adulthood (Caspi, 2000; Rothbart & Bates, 1998). Generally, temperament refers to emotional or affective aspects of the personality (Goldsmith & Campos, 1982). The link between early temperamental traits and later person-

ality has become an area of intense study during the last decade (Caspi, Roberts, & Shiner, 2005). Like temperamental traits, most personality traits show modest genetic influence (Bouchard & Loehlin, 2001), and thus genetics has not proven to differentiate temperament and personality.

Temperament has not proven to be a precise term. It has ancient origins, tracing at least to Greek physician Galen, and being influential in the writing of various European philosophers. In the late 19th and early 20th century, the Russian physiologist Pavlov studied "transmarginal inhibition" in dogs. He and various neo-Pavlovians probed the "strength" of the nervous system, including reactive and regulatory aspects, phenomena that correspond reasonably well with certain current temperament concepts. In the early 20th century, Germans such as Kretschmer conceived of constitutional groupings of individuals, which were also called temperaments. Thus, the temperament concept has rich history but has accumulated much excess baggage. It also suffers from being a lay term as well as a scientific term; in the common vernacular, an irritable person might be called "temperamental." Within contemporary science, temperament can be viewed as a concept composed of types (categories of individuals) or of dimensional traits. Despite these and other ambiguities in its definition and connotations, temperament research has blossomed over the past 25 years. Behavioral genetic studies have played a significant role in this substantial body of recent research.

Identifying the Dimensions of Temperament

Our review focuses on five of the most commonly examined dimensions: activity level, anger/frustration, behavioral inhibition/fear, effortful control, and positive affect. We do not intend for these dimensions to be representative of any single theoretical framework regarding the structure of temperament. Instead, we review these particular components largely for pragmatic reasons. First, a relative consensus has emerged among temperament scholars about

H.H. Goldsmith (✉)
Department of Psychology, University of Wisconsin, Madison, WI 53706, USA
e-mail: hhgoldsm@wisc.edu

Y.-K. Kim (ed.), *Handbook of Behavior Genetics*,
DOI 10.1007/978-0-387-76727-7_18, © Springer Science+Business Media, LLC 2009

the importance of these dimensions within the structure of temperament. Second, these dimensions have been extensively studied, often within genetically informative designs. Our review is obviously not comprehensive because several well-recognized and well-studied dimensions of temperament have been excluded (e.g., sadness, rhythmicity, soothability, sociability). Our goal is not to provide a summary of all genetic studies of temperament. We aim instead to review genetic studies of temperament focusing on dimensions that illustrate both reactive and regulatory features.

Temperament dimensions often overlap and thus can be organized into supraordinate families of related traits. With the possible exception of activity level, the dimensions discussed in this chapter can be conceptualized as belonging to such families. Below, we define the dimensions of temperament discussed in this chapter and recognize the different aspects of temperament that belong to the same families.

Activity level refers to movement of the arms, legs, head, or trunk with movement being characterized by both intensity and frequency. As a temperament trait, activity level refers to individual differences in proneness to higher versus lower characteristic degrees, or vigor, of movement. More difficult to discern is whether any unitary motivational process accompanies a given level of activity.

Anger/frustration describes the individual's propensity toward experiencing an approach-related type of negative affect within a challenging context, where goals may be blocked or attack may be perceived. Anger is characterized by both intensity and frequency of behaviors such as crying, protesting, hitting, and pouting (Goldsmith & Rothbart, 1991). Concepts in the anger family include hostility, irritability, and aggression. In addition, anger can be divided into interpersonal and task-related subtypes.

Behavioral inhibition/fear can be defined as the individual's tendency to react with distress or wariness toward novel or threatening objects, persons, or situations. This dimension can also be expressed through avoidance of or prolonged latency to approach a novel stimulus (Goldsmith & Rothbart, 1991). Behavioral inhibition/fear shares characteristics with shyness and social inhibition.

Effortful control is a multidimensional concept referring to multiple capabilities in motor, vocal, and cognitive domains. In general, effortful control refers to one's ability to inhibit a prepotent response and instead activate and execute a subdominant response (Murray & Kochanska, 2002; Posner & Rothbart, 2003; Rothbart & Bates, 1998). Effortful control overlaps with the constructs of executive function, cognitive control, and attention regulation and fits within the broader concept of emotion regulation.

Positive affect refers to a propensity to experience and express enjoyment (Bates, 1989) and to be responsive to reward. It can be expressed through smiling, laughter, or playful activity. Within the positive affect family are concepts such as contentment and exuberance that are the subjects of research during childhood. Other family members are less-studied concepts such as awe and relief. Within a motivational framework, positive affect can be classified as occurring pre-goal (anticipatory pleasure) versus postgoal attainment (consumatory pleasure). Within an interpersonal context, positive affect is related to sociability. Within personality research, sociability is a component of the broad factor of extraversion, and within the rubric of the "Big 5" personality factors, it is closely linked to agreeableness.

Significance of Temperament Dimensions

An extensive literature documents the importance of infant and child temperament for developmental outcomes. We provide a few examples of this relevance that best illustrate the practical importance of temperament dimensions highlighted in this chapter.

Several aspects of temperament have been reliably implicated in cognitive development and education. Effortful control is positively associated with cognitive skills such as working memory, planning, and attentional flexibility (Chiappe, Hasher, & Siegel, 2000; Conway, Harries, Noyes, Racsma'ny, & Frankish, 2000; Pallodino, Mammarella, & Vecchi, 2003; Passolunghi & Siegel, 2001; Wolfe & Bell, 2003) as well as learning and education more generally (Posner & Rothbart, 2007). The related dimensions of attention span, persistence, and task orientation have been consistently associated with adjustment and achievement in preschool and grade school, and longitudinally from preschool through adolescence (Galejs, King, & Hegland, 1987; Martin, Olejnik, & Gaddis, 1994; Nakamura & Finck, 1973; Palisin, 1986; Schoen & Nagle, 1994). Activity level is also considered one of the best predictors of daycare and school adjustment (Gallerani, O'Regan, & Reinherz, 1982; Klein, 1980) and is positively associated with mental development in infancy (Matheny, 1989a; Saudino & Eaton, 1991), but negatively associated with IQ after the age of 2 (Halverson & Waldrop, 1976; Matheny, 1989a).

Temperament is also closely tied to social development and relationships. Emotion regulation, effortful control, and positive affect are associated with secure relationships and irritability/anger with insecure relationships (Kopp, 1982; Rothbart, 1989; Rothbart & Bates, 1998). Self-regulation is central to social cognition generally, from regulation of behavior leading to better peer relations (Hughes, White, Sharpen, & Dunn, 2000; Lengua, 2003; Nigg, Quamma,

Greenberg, & Kusche, 1999) to the role of inhibitory control in the development of a theory of others' minds (Frye, Zelaso, & Palfai, 1995; Perner, Leekam, & Wimmer, 1987; Russell, Mauthner, Sharpe, & Tidswell, 1991). Children with higher activity levels are typically more socially interactive, but have more negative interactions with peers and are judged as less popular and socially competent by peers, parents, and teachers (Billman & McDevitt, 1980; Bramlett, Scott, & Lowell, 2000; Eaton, 1994; Gurlanick & Groom, 1990). Conversely, behavioral inhibition and shyness are related to social wariness and withdrawal as well as lower social status across age (Burgess, Rubin, Cheah, & Nelson, 2005; Eisenberg, Shepard, Fabes, Murphy, & Guthrie, 1998; Fox, 2004; Gest, 1997).

Childhood temperament is also related to behavioral adjustment and health outcomes. Temperament can be a liability factor for childhood psychopathology, and this liability can be general or specific (Goldsmith, Lemery, & Essex, 2004). General patterns of fear, shyness, and inhibition most often predict internalizing behavior problems, while negative affect, activity, and low effortful control behaviors predict externalizing (Eisenberg et al., 2001; Rothbart & Bates, 1998). Negative affect and behavioral inhibition are recognized as vulnerability factors for anxiety, and low positive affect is associated with depression (Biederman et al., 2001; Brown, 2007; Clark, Watson, & Mineka, 1994; Kagan, 2001). The dimensions of activity level and effortful and inhibitory control are implicated in the development of ADHD (Campbell & Ewing, 1990; Hall, Halperin, Schwartz, & Newcorn, 1997; Pliszka, Borcherding, Spratley, Leon, & Irick, 1997; Schachar, Tannock, Marriott, & Logan, 1995), and anger proneness is related to conduct problems (Cole, Teti, & Zahn-Waxler, 2003; Deater-Deckard, Petrill, & Thompson, 2007). In addition, effortful control and impulsivity are linked to the development of moral regulation and conscience (Kochanska, DeVet, Goldman, Murray, & Putnam, 1994), and this relation may mediate the likelihood of criminal conviction (Henry, Caspi, Moffitt, Harrington, & Silva, 1999). High activity level and negative emotionality predict higher levels of drug and alcohol use (Rothbart & Bates, 1998), and high activity level is also associated with increased risk for injury due to accidents (Langley, McGee, Silva, & Williams, 1983; Matheny, 1987).

Current Research

The literature on the genetics of temperament can be roughly divided into four categories. First and most extensive are the mainstream twin, family, and adoption studies of temperament, many of which are longitudinal. Another category is allelic association and linkage studies that test candidate genes for their relevance to specific dimensions of temperament. The third category, which has recently increased in frequency, is studies of gene by environment interaction. Finally, with the rise of cognitive and affective neuroscience, genetically informative studies of endophenotypes associated with temperament have become more common. Fortunately, larger projects are beginning to integrate all four of these approaches into research on the same samples.

Selected Twin and Adoption Studies of Reactive and Regulatory Temperaments

Activity Level

One of the earliest twin studies of childhood temperament to examine activity level was the Louisville Twin Study (LTS; Matheny, 1980). The LTS used observational measures of child temperament based on factor analyses of items from Bayley's infant behavior record (IBR), and activity was designated as one of the major factors. This longitudinal study assessed children across 3, 6, 9, 12, 18, and 24 months of age and sample sizes ranged from 72 to 94 for MZ twin pairs and 35 to 54 for DZ twin pairs. Excepting the data at 3 months of age, all MZ twin correlations exceeded DZ twin correlations for the IBR activity factor, indicating that genetic influences were operating on activity in infancy (Matheny, 1980). Later analyses of the 3-month-old twins from the LTS that incorporated more sophisticated model-fitting techniques replicated the finding of no genetic influences on activity level in the neonatal period (Riese, 1990).

Longitudinal investigations of the LTS sample yielded phenotypic age-to-age correlations that showed a simplex pattern across age (Matheny, 1983). To examine genetic influences on the stability of activity level, correlational twin analyses were divided into a 6, 12, and 18 month group and a 12, 18, and 24 month group. The MZ twin correlations exceeded the DZ twin correlations for both age groups and were larger and significantly different from the DZ correlation in the 12–24 month group. This pattern of findings indicated that individual differences of change across development occurred partly as a function of genetic similarity and that genetic influences on activity increased with age. A multi-method approach to activity in the LTS that utilized structural equation models incorporated playroom observations, parent report, and IBR ratings at 24 months of age (Phillips & Matheny, 1997). Interestingly, results suggested that individual differences in responding to specific situations (e.g., a testing situation, play, home) were genetically influenced. This finding is notable in that behavioral scientists typically attribute situational effects to the environment.

Colorado Adoption Project (CAP; Plomin, DeFries, & Fulker, 1988) analyses of parent-rated activity level assessed using the emotionality–activity–sociability–impulsivity rating scales (EASI; Buss & Plomin, 1975, 1984) annually from 1 to 7 years of age found no evidence of genetic influences (Plomin, Coon, Carey, DeFries, & Fulker, 1991). In contrast, previous twin study results indicated substantial genetic influences for all EASI dimensions, with an average of 0.59 and −0.01 for MZ and DZ twin correlations, respectively (Buss & Plomin, 1975, 1984). It was posited that contrast effects could be operating on these parent-rated twin data resulting in inflated MZ and deflated DZ twin correlations that contribute to overestimates of heritability (Plomin et al., 1991). In a unique investigation that utilized both CAP and LTS data, 95 pairs of nonadoptive siblings, 80 pairs of adoptive siblings, 85 MZ twins, and 50 DZ twins were combined in model-fitting analyses for IBR activity (Braungart, Plomin, DeFries, & Fulker, 1992). The combined data yielded heritability estimates of 38% at age 1 and 57% at age 2, with shared environmental influences of 1% at both ages. Ensuing analyses of CAP data using teacher and observer ratings of activity level assessed with the CCTI at 7 years of age found significant genetic influences and nonsignificant shared environmental influences for both (Schmitz, Saudino, Plomin, Fulker, & DeFries, 1996). Therefore, findings of genetic influences on activity in the CAP sample were somewhat mixed; genetic factors were present when temperament was rated by teachers or observers, but not parents.

Genetic and environmental sources of continuity and change in infant activity were also examined in the MacArthur Longitudinal Twin Study (MALTS; Plomin et al., 1993). A MALTS analysis that utilized the IBR included 163 MZ and 138 DZ same-sex twin pairs assessed at 14, 20, and 24 months of age (Saudino, Plomin, & DeFries, 1996). Cross-age correlations for activity were moderately low, suggesting developmental change, and heritability estimates were significant and similar across development; heritable factors contributed to stability (Saudino et al., 1996). Model-fitting results indicated that no new genetic influences emerged at 20 or 24 months, shared environment was nonsignificant across age, and nonshared environment was significant. Therefore, continuity in activity level was due to genetic factors, and change was due to the nonshared environment. MALTS also employed parent CCTI ratings, and results showed substantial stability across 14, 20, 24, and 36 months of age and high heritabilities with near-zero or negative DZ correlations (Saudino & Cherny, 2001). These findings dovetail with the CAP results, possibly reflecting the presence of parental expectations of infant behavior and contrast effects with parent ratings of temperament, and lend further support to multi-method approaches that emphasize unbiased forms of assessment.

An important contribution to this multi-method perspective on activity level measurement was a twin study that employed motion recorders (actometers) attached to all four limbs of 39 MZ and 21 DZ twin pairs at approximately 7 months of age. Intraclass MZ twin correlations exceeded DZ twin correlations, supporting evidence of genetic influences (Saudino & Eaton, 1991). In a follow-up on this sample at 35 months of age, genetic factors continued to be present in activity level assessed with motion recorders, regardless of low phenotypic stability from age to age (Saudino & Eaton, 1995). Moreover, MZ–DZ concordance patterns from 7 to 35 months reflected genetic influences on developmental change (Saudino & Eaton, 1995). A more recent investigation of actometer-assessed activity across situations recruited 7- to 9-year-old twins (463 pairs) from the Twins' Early Development Study (TEDS; Trouton, Spinath, & Plomin, 2002). Model-fitting analyses of testing and break-from-testing situations yielded heritability estimates of 24% and 30%, respectively, with shared environment contributing 27% and 42% of the variance (Wood, Saudino, Rogers, Asherson, & Kuntsi, 2007). The genetic correlation between the test and break situation scores was 1.0, indicating that the same genes were operating on activity level across situations (Wood et al., 2007). The results of twin studies that used objective motion recorders have thus bolstered the findings of those that employed parent, teacher, and observational measures of activity.

Anger/Frustration

Goldsmith and Gottesman (1981) conducted one of the earliest childhood twin studies of an anger/frustration-related temperament dimension, i.e., irritability/negative mood. The authors used data on 110 MZ and 206 DZ twin pairs who had participated in the Collaborative Perinatal Project. Psychologists made ratings of irritability/negative mood after observing child behavior during mental and motor testing and during a free play period that occurred when children were 4 years old. The irritability/negative mood dimension was derived from a factor analysis of behavioral ratings and included observations of emotional reactivity, degree of irritability, degree of cooperation, and degree of dependency. The MZ correlation for this factor (0.45) exceeded the DZ correlation (0.17), indicating that genetic influences played a significant role in the development of irritability/negative mood.

Genetic factors associated with anger proneness have also been examined in a sample of 89 MZ and 95 DZ (62 same-sex) 17- to 36-month-old toddler pairs (Goldsmith, Buss, & Lemery, 1997). Anger proneness was measured using parental report on the Toddler Behavioral Assessment Questionnaire (TBAQ; Goldsmith, 1996). The MZ twin correlation on the anger proneness subscale (0.72) was higher than the DZ correlation (0.55) and estimates of additive (A), common environmental (C), and unique environmental (E) variance components were obtained using DeFries–Fulker regression analyses (DeFries & Fulker, 1985, 1998). The full ACE model fits the data significantly better than either the CE or AE models. Analyses indicated that both genetic and environmental influences played a significant role in the development of anger proneness ($h^2 = 0.34$, $c^2 = 0.38$).

In the same study, Goldsmith et al. (1997) also examined genetic influences on anger among older children (34–99-month-olds). This older sample included 55 MZ and 64 DZ (36 same-sex) twin pairs, and anger was assessed via parent report on the Children's Behavior Questionnaire (CBQ; Rothbart, Ahadi, Hershey, & Fisher, 2001). Because the DZ correlations for same-sex (−0.29) and opposite-sex (0.33) pairs were significantly different from each other and the same-sex DZ correlation was negative, heritability estimates were not calculated for the anger subscale. Instead, anger was combined with discomfort, soothability, fear, and sadness subscales using principal components analysis to form a negative affectivity composite variable, and a heritability estimate was calculated based on this composite score. Analysis using DeFries–Fulker regression demonstrated that an AE model provided the best fit to the data and that genetic influences played a significant role in the development of negative affectivity ($h^2 = 0.42$, $c^2 = 0.00$).

The MALTS project assessed anger using both mother report and laboratory and home observations in a sample of 190 MZ and 161 DZ twin pairs who were assessed across time when they were 14, 20, and 24 months old (Emde, Robinson, Corley, Nikkari, & Zahn-Waxler, 2001). Mothers reported how frequently their children had angry outbursts, described the initiation of physical fights between co-twins, and reported their children's expressions of anger on the Differential Emotions Scale (DES; Izard et al., 1972). Observations of child anger were based on ratings of child protesting behavior while being restrained in four situations (two in the lab and two in the home). Ratings of anger during the four restraint procedures were combined into a single principal component and estimates of h^2 and c^2 were derived. Heritability estimates of mother-reported anger were significant across all three measures and during each of the three time points (range 0.31–0.71). Shared environmental influences were significant for mother report on the DES at age 24 months ($c^2 = 0.37$), for mother report of initiating physical fights at all three time points (range 0.30–0.45), and for mother-reported outbursts at 14 and 24 months ($c^2 = 0.42$ at both ages). Heritability of observed anger was significant only when children were 24 months old ($h^2 = 0.32$). Shared environmental influences on observed anger were significant at 20 and 24 months ($c^2 = 0.34$ and 0.45, respectively).

Although findings from this study were somewhat mixed, mother-reported anger was consistently found to be heritable. This work also provided some evidence for the importance of the shared environment in both mother-reported and observer-rated anger.

Another twin study of anger/frustration used a sample of 105 MZ and 154 same-sex DZ twin pairs ranging in age from 4 to 8 years (Deater-Deckard, Petrill, & Thompson, 2007). Anger/frustration was rated by research assistants based on child behavior observed during a 3-hour home visit. The visit included two tasks completed by the mother together with each twin separately; these tasks were designed to elicit anger/frustration. The MZ correlation for anger/frustration (0.50) was larger than the DZ correlation (0.33) implicating genetic influences. Estimates of genetic, shared environmental, and nonshared environmental variance were computed using path estimates produced by a Cholesky decomposition ($A = 0.25$, $C = 0.23$, $E = 0.51$). Estimates indicated a significant role of genetics and the nonshared environment in the development of anger/frustration.

In addition to calculating variance estimates of anger/frustration, Deater-Deckard et al. (2007) also estimated the influence of genes and environments on child conduct problems and tested for genetic, shared environmental, and nonshared environmental correlations between anger/frustration and child conduct problems. Conduct problems were measured using the Teacher Report Form (TRF; Achenbach, 1991a) and significant genetic, shared, and nonshared environmental variance estimates were detected ($A = 0.29$, $C = 0.38$, $E = 0.32$). Further, a large genetic correlation (0.80) was detected between anger/frustration and conduct problems. This correlation indicated that many of the same genetic influences that contribute to the development of anger/frustration may also impact the emergence of childhood conduct problems.

Overall, the literature on anger/frustration provides quite consistent evidence for heritability. However, findings related to the importance of the shared environment were more equivocal but intriguing given that such findings are not typical. Finally, evidence suggesting that many of the same genetic influences that impact the development of anger/frustration also contribute to the manifestation of child conduct problems is emerging.

Behavioral Inhibition/Shyness/Fear

Because the temperament dimensions of behavioral inhibition, shyness, and fear overlap theoretically, twin study results for all three constructs are included in this section. Behavioral inhibition was first investigated in a study of 33 MZ and 32 DZ twin pairs from the LTS at 12, 18, 24, and

30 months of age (Matheny, 1989b). Parents rated behavioral inhibition with the approach/withdrawal items on the Toddler Temperament Scale (Fullard, McDevitt, & Carey, 1984), and observational measures were based on laboratory play-room vignettes that tapped emotional tone and a fearfulness scale derived from IBR ratings. All MZ twin correlations exceeded DZ twin correlations across measure and age, suggesting genetic influences and stability on behavioral inhibition from 12 to 30 months. The presence of several "too low" DZ correlations may reflect low sample sizes or contrast effects. Age-to-age changes in behavioral inhibition were attributed to a genetically influenced effect, due to the higher concordance rates for MZ twins when these changes occurred.

A second behavioral genetic study of behavioral inhibition investigated a larger sample of 157 24-month-old MALTS twins (DiLalla, Kagan, & Reznick, 1994). They assessed temperament by observing two pairs of unfamiliar twins in a laboratory play situation, and the intraclass correlation comparison between MZ and DZ twins indicated high heritability for behavioral inhibition. In addition, "extreme inhibition" was examined by using a DeFries–Fulker regression model that predicted co-twin's behavioral inhibition scores based on the proband's score and the degree of genetic relatedness. Extreme levels of behavioral inhibition are often used as an analogue to social anxiety in studies of temperament and psychopathology. Results of these analyses indicated that the degree to which extreme behavioral inhibition scores of probands were due to genetic influences was statistically significant. The heritability of extreme inhibition was slightly larger than the estimate from the total sample, although differences were not statistically significant. Therefore, individual differences in both behavioral inhibition and extreme levels of the temperament were due to genetic factors.

Another MALTS analysis of shyness focused on 163 MZ and 138 DZ same-sex twin pairs assessed with observational measures at home and in the laboratory at 14 and 20 months of age (Cherny, Fulker, Corley, & Plomin, 1994). Shyness was indexed by infants' initial responding to strangers in brief videotaped observations. Hierarchical longitudinal model-fitting provided estimates of the genetic and environmental influences that contributed to continuity and change in shyness across age. Findings show that developmental change from 14 to 20 months as well as situation-specific effects for the home and the lab were due to environmental factors, while stability across age and phenotypic associations between the home and lab settings were due to overlapping genetic variance. In a follow-up to this study, analyses that included cross-age within-situation and cross-situation within-age comparisons replicated the findings that stability of shyness across age and situation was a result of common genetic influences (Cherny et al., 2001). The same genetic

effects are involved in shyness at both ages and in both situations.

Studies of common and specific fears tend to focus on older children and adults. An early investigation of common fears used a fear survey in 354 pairs of same-sex twins aged 14–34 years (Rose & Ditto, 1983). Hierarchical multiple regression analyses were employed to predict twin fear from that of the co-twin, as well as age and zygosity effects. For all factors, co-twin fear and zygosity significantly contributed to twin fear, and for two fear factors, age effects were also a significant predictor. This pattern of results indicates that genes do moderate the development of common fears. A second study of 144 MZ and 106 DZ twins sampled in late childhood and adolescence and their parents also utilized a fear survey questionnaire (Neale & Fulker, 1984). Model-fitting analyses yielded significant genetic and environmental variance in two fear factors, as well as a small amount of genetic–environmental covariance explained by the parents' phenotypes influencing the twins' (Neale & Fulker, 1984). Both of these seminal studies used parent- and/or self-ratings that might reflect reporting biases.

Many of the remaining behavioral genetic findings in this area focus on behaviors that are either described as extreme levels of temperament or measure associations at the intersection of temperament and psychopathology. An analysis of several common fears as well as extreme fearfulness was conducted on 144 MZ and 175 DZ same-sex twin pairs (8–18 years of age) who completed a fear survey questionnaire (Stevenson, Batten, & Cherner, 1992). The heritabilities of many fears were significant; however, there was no evidence of increased heritability for extreme fearfulness. In the MALTS, mothers completed the CCTI at 14, 20, 24, and 36 months and assessed problem behaviors with the Child Behavior Checklist (CBCL; Achenbach, 1991b) at 4 years of age (Schmitz et al., 1999). Shyness at all four earlier ages was significantly correlated with the internalizing scale on the CBCL at age 4, and longitudinal genetic analyses revealed that these predictions were largely due to genetic effects. In an investigation of 4-year-olds from the TEDS project, mothers assessed anxiety-related behaviors in 4564 pairs of twins. Genetic influences were found for the fears and shyness/inhibition factors (Eley et al., 2003). In general, these findings implicate genetic variance and suggest that extreme levels of these related areas of temperament might reflect genetically overlapping psychopathology.

All of the findings in this area point to genetic influences on behavioral inhibition, shyness and fear, with very little evidence of shared environment. The MALTS results on developmental change indicated that stability in temperament was due to genetic factors. In contrast, the LTS results implicated genes in age-to-age changes. It is likely that the MALTS analyses, from a larger sample, are more representative; however, both outcomes warrant attention. Neither of the two studies that focused on extreme temperament indicated that heritability for extremes was significantly higher than for the full range of behavior. Lastly, the literature suggests a genetic basis to the overlap between fearful temperament and relevant behavior problems.

Effortful Control

Although it is a relatively new dimension in temperament theories, several behavioral genetic findings in the literature focus on effortful control and related concepts. The first twin study to include effortful control used a small sample of 55 MZ and 64 DZ preschool-age twin pairs (34–99 months of age) pooled from studies in Oregon, Colorado, Texas, and Wisconsin (Goldsmith, Buss, & Lemery, 1997). Parents rated temperament with the Children's Behavior Questionnaire (CBQ; Rothbart, Ahadi, Hershey, & Fisher, 2001), and an effortful control factor was formed with the CBQ attentional focusing, inhibitory control, low pleasure, and perceptual sensitivity scales. The MZ twin correlation exceeded the DZ twin correlation for this factor. Subsequent DeFries–Fulker regression analyses indicated that the ACE ($h^2 = 0.43$) and AE ($h^2 = 0.58$) models fit equally well, and the CE model had a significantly poorer fit than the full model. Therefore, genetic factors accounted for a significant proportion of the variance in effortful control in preschool. In addition, support for genetic variance was strengthened by the fact that the DZ correlation was not "too low," reflecting no evidence of contrast effects often found in parent-rated measures of temperament.

Effortful control was also examined in a Japanese twin study that utilized the self-administered Adult Temperament Questionnaire (ATQ; Rothbart, Ahadi, & Evans, 2000) in a sample of 152 MZ and 73 DZ twin pairs 17–32 years of age (Yamagata et al., 2005). Both intraclass correlations and model-fitting results for the overall effortful control factor and subscales indicated significant genetic influences (AE models fit best). Heritability for effortful control was 49% and ranged from 32% to 45% for the subscales. Multivariate analyses showed that the subscales were highly genetically correlated and moderately environmentally correlated. The self-ratings did not reflect contrast effects, and the heritabilities were consistent with the Goldsmith et al. (1997) preschool-age study, indicating the genetic stability from preschool to young adulthood.

Another study assessed both inhibitory control and attentional focusing as components of effortful control. Lemery-Chalfant, Doelger, and Goldsmith (2008) assessed both of these constructs in 214 MZ and 349 DZ twin pairs with a mean age of 7.6 years and a subsample that was composed of 112 MZ and 171 DZ twin pairs at 5.5 years of age. The CBQ was used at both ages to assess parent ratings of inhibitory control and attentional focusing, and an effortful control composite was formed based on mother and father ratings for both. An observational measure of attentional focusing was derived from items on the Bayley Rating Scale (BRS; Bayley, 1993) for use with the younger sample. MZ twin correlations exceeded DZ correlations for all assessments, indicating genetic influences. The ADE model was the best fitting for the CBQ ratings at both ages with 68–79% of the variance due to broad sense heritability (both additive genetic and dominance effects combined) and the remainder due to nonshared environment. For the observer-rated attentional control measure, the AE model fit best, and genetic factors contributed to 83% of the variance. This study yielded higher heritabilities than those found in other studies, mainly due to the inclusion of dominance effects. It is important to note that CBQ ratings at both ages produced DZ correlations that were very low, which might suggest parental contrast effects. However, the observational results also showed this pattern, and the observations were not susceptible to contrast effects.

Two recent twin studies have focused on inhibitory control in early childhood. The first examined 24-month-olds and included 146 pairs of twins (66 MZ and 80 DZ) rated by parents using the Toddler Behavior Assessment Questionnaire – Revised (TBAQ-R; Goldsmith, 1996) and 96 pairs assessed with the Laboratory Temperament Assessment Battery (Lab-TAB; Goldsmith, Reilly, Lemery, Longley, & Prescott, 1995) as administered by trained observers (Gagne & Saudino, 2005). MZ twin correlations exceeded DZ twin correlations for both measures suggesting the importance of genetic factors. Model-fitting results produced heritability estimates of 54% for TBAQ IC (ACE model is best fitting) and 36% for Lab-TAB IC (AE model fits best) at 24 months of age. The genetic correlation between the two measures was approximately 0.50, and the phenotypic correlation was entirely accounted for by genetic factors.

The second study that examined inhibitory control in early childhood used parent temperament ratings in a longitudinal sample of twins (Gagne & Goldsmith, 2007). Specifically, 129 MZ and 255 DZ twins were assessed with the TBAQ at 22 months, and 144 MZ and 263 DZ twins were assessed with the CBQ at 36 months. Twin correlations suggested genetic influences on inhibitory control, and heritabilities were 50% and 59% for the TBAQ (ACE model) and the CBQ (AE model), respectively. The genetic correlation between the TBAQ and the CBQ inhibitory control subscales was 0.60, indicating genetic stability across age, and the phenotypic correlation was due to overlapping genetic influences. Contrast effects were not apparent in either of these twin studies of inhibitory control during toddlerhood.

In general, research findings on effortful control and related variables support significant genetic influences on these constructs, including some evidence of non-additivity.

Shared environment was negligible and typically only present in results based on parental ratings. Heritability estimates were moderate to high and showed some stability across age. In addition, genetic correlations across age and measure were typically high, and phenotypic correlations were entirely due to genetic influences that overlapped. This indicated that the same genetic factors were operating from age to age and that there is a genetic coherence to effortful control as evaluated by different assessments. Lastly, although contrast effects appeared in one of the studies, many of the parent-rated findings did not produce "too low" DZ twin correlations. In summary, a genetic basis for effortful control seems clear.

Positive Affect

The twin literature on positive affect in children is not extensive, and it does not explore all of the nuanced meanings of positive affect that we summarized above. Among the widely used childhood temperament questionnaires, only the ones authored by Rothbart or Goldsmith contain scales that explicitly assess positive affect. For instance, in the infant twin study by Goldsmith, Lemery, Buss, & Campos (1999), described above, the MZ twin correlation for a smiling and laughter scale was 0.75, with a DZ correlation of 0.59. In the same study, twin correlations for soothability (which was correlated with positive affect) were 0.52 (MZ) and 0.61 (DZ). The correlational patterns imply an effect of the shared environment, which is unusual in temperament research. In Goldsmith et al.'s (1997) study of toddler-aged twins, also described above, a parental report pleasure scale also showed a moderate shared environmental effect, with an MZ correlation of 0.69 and a DZ correlation of 0.64. A variety of earlier evidence supports a role for shared environmental effects on positive affect-related variables, such as person interest (Goldsmith and Gottesman, 1981), laboratory measures of smiling, questionnaire measures of smiling/laughter and pleasure (Goldsmith & Campos, 1986), observations of positive activity (Lytton, 1980), and questionnaire measures of zestfulness (Cohen, Dibble, & Graw, 1977).

Some studies focus on broader constructs that probably encompass children's positive affect. For instance, Deater-Deckard & O'Connor (2000) studied parent–child mutuality, which was influenced by heritable child effects. It seems reasonable that positive affect would be a component of mutuality during interpersonal interaction. Similarly, Knafo & Plomin (2006) studied prosocial behavior in a very large longitudinal sample of young English twins. Again, it seems reasonable that positive affect would be a component of prosocial behavior. Shared environmental effects were relatively strong (47% of the age 2 variance, on average) on prosocial ratings by parents but decreased over age. This longi-

tudinal finding is consistent with the cross-sectional change toward decreased shared environmental effects for positive affect reported by Goldsmith et al. (1997).

What might be the substantive interpretation of this finding? The leading candidate is perhaps some feature of maternal behavior or personality common to the twins. A reasonable speculation is that the primary caregiver might be the source of the common environment, as seen in some other cases (Leve, Winebarger, Fagot, Reid & Goldsmith, 1998). Even more speculatively, twin similarity for mother–child attachment security might play a role in shared environmental effects on positive affectivity. On the other hand, other features of maternal personality such as extraversion could be influential as a source of common experience for the twins. We must also acknowledge the possibility of a social desirability explanation because individual differences in parents' tendency to report their children as happy might mimic a shared environmental effect (Goldsmith, 1996). In any case, the evidence for shared environmental effects on positive affectivity appears to diminish as children grow older.

Linkage and Association Studies

Concepts, methods, and limitations of linkage and association studies are described in other chapters of this handbook and in sources such as Sham (1998). In general, these methods were developed to study the genetic basis of diseases, for which an individual can be considered as affected or not. However, most of the methods can be adapted to study trait variation that is not "all or none," as exemplified by temperamental dimensions, which are analyzed using quantitative trait loci (QTL).

The most common method used for studying temperament is the allelic association study. These studies compare the frequency of a genetic marker in two samples (e.g., an inhibited versus a non-inhibited group). It might seem straightforward to compare extreme temperamental groups (e.g., very shy versus non-shy children) for different frequencies of a candidate gene or QTL. However, high shyness in one subset of individuals might have different roots than similar high shyness in another subset. This trait heterogeneity problem haunts current attempts to associate behavior with specific genes (Cardon, 2006). There are other practical difficulties in this genre of research. For instance, individuals at the extreme of the distribution might possess several alleles promoting higher trait values. The effect of any one allele may be more difficult to discern against this "high-value" genetic background than in more typical genetic backgrounds.

On the whole, however, genetic association studies have yielded inconsistent findings. A recent meta-analysis of studies investigating associations between personality traits and candidate genes concluded that there were few replicable associations (Munafo et al., 2003).

It is important to realize that genes that have only small effects on the phenotype might be functionally related to genes of large effect. That is, if a major gene leading to qualitative dysfunction is identified, other alleles at the same locus could affect more subtle quantitative variation. The allele that leads to qualitative dysfunction could code for an inactive protein whereas a different allelic variant of the same gene might lead to an active but physiologically non-optimal protein.

A prototypic example of an allelic association with a temperamental trait in humans concerned a polymorphism in a promoter of the serotonin transporter gene on chromosome 17q, which accounted for 3–4% of the variance in self-reported anxiety proneness. This finding held in two samples totaling 505 individuals, and the allele in question also differentiated affected from unaffected siblings (Lesch et al., 1996). Serotonin is, of course, involved in neurotransmission in regions of the limbic system and cortex associated with anxious behavior, and some antidepressant and anti-anxiety drugs inhibit uptake of serotonin. The serotonin transporter is a protein that helps "fine-tune" this neurotransmission. The allele of the serotonin transporter promoter that leads to decreased transporter activity—at least in lymphoblast cell lines—occurred at a high frequency, 43%, in the samples studied (Lesch et al., 1996). Persons with this allele would be expected to show increased serotonergic transmission, and they report higher neuroticism scores (but are not different on the other factors of the Big Five personality traits). Lesch et al. (1996) cautioned that their results, from a normal sample, might not generalize as a cause of clinically significant levels of anxiety in patients. Although replications have been reported, some comprehensive failures to replicate these results have also been reported (Willis-Owen et al., 2005).

Another issue in this line of research is the non-specificity of findings from candidate gene studies where the gene affects functioning in key pathways regulating neurotransmitters. Such a lack of specificity is perhaps unsurprising for genes mediating neurotransmitter function in circuitry central to affective processing. Population stratification is also a concern in these studies because Gelernter, Cubells, Kidd, Pakstis, & Kidd (1999) showed that the frequency of the serotonin transporter promoter varies widely (from 0.29 to 0.89) in eight geographically distinct ethnic groups.

In addition to alleles related to serotonin function, genes for dopamine receptors and transporters have been examined for association with emotion-related temperament traits.

For instance, two studies in 1996 showed associations of the personality trait of novelty seeking with the seven repeat alleles of the dopamine receptor D4 (DRD4) gene (Benjamin et al., 1996; Ebstein, Novick, Umansky, Priel, & Osher, 1996). However, a meta-analysis that includes eight subsequent studies, with varying characteristics, casts doubt on the validity or at least the generalizability of the association (Wahlsten, 1999). The literature on this association has grown, with both replications and nonreplications (Strobel, Spinath, Angleitner, Riemann, & Lesch, 2003); more DRD4 polymorphisms have been examined, and interactions both with other genes and with other behaviors have been considered.

Of course, DRD4 is not the only gene that has been examined for association with temperamental approach behaviors. For example, genetic variants in noradrenergic receptors have been examined for association with irritability, hostility, and impulsivity (Comings et al., 2000). In this study, as in others, variation in a single gene accounted for only a few percent of the phenotypic variance.

An obvious conclusion from this line of research is that all findings require replication and tests of generalization. Small samples, multiple tests, genetic heterogeneity, and "cheap phenotyping" via questionnaire are likely culprits in accounting for inconsistencies. In allelic association studies, the behavioral side of the investigation is often not as sophisticated as the genetic side.

Gene–Environment Interaction

Gene–environment interaction refers to a situation in which the association between an environment and a phenotype is moderated by an individual's genotype. Like all statistical interactions, gene–environment interactions can also be conceptualized conversely, i.e., the environment moderates the relationship between the individual's genotype and phenotype.

Specific Allele by Environment Interactions

Fox et al. (2005) used molecular genetic analyses to test for gene–environment interaction in the development of behavioral inhibition among 7-year-olds. The authors examined whether mother-reported social support moderated the association between child genotype (5-HTT allele status) and behavioral inhibition. The authors chose a functional polymorphism in the promoter region of the serotonin transporter gene (5-HTT), which has both long and short alleles, as their polymorphism of interest for multiple reasons. First, several functions of this polymorphism have been identified and these functions including the regulation of transporter levels, serotonin uptake, and 5-HTT transcription (Hariri et al., 2002) could logically be related to behavioral inhibition. Second, possessing the short allele of 5-HTT has been linked to elevated anxiety, negative emotionality, and amygdala activation in response to fearful facial expressions among adults (Hariri et al., 2002; Munafo et al., 2003). Finally, this polymorphism has been shown to interact with stressful life events in the prediction of depression in adults (Caspi et al., 2003). The authors used mother-reported social support as an indicator of environmental stress as it has been shown to reflect child stress levels. Behavioral inhibition was rated by observers when children were 7 years old and was based on the child's onlooking and unoccupied behavior during a group play session. A gene–environment interaction was detected in which the association between genotype and behavioral inhibition was stronger at low levels of social support. Children who carried the short 5-HTT allele and whose mothers reported low social support exhibited higher levels of behavioral inhibition than their counterparts who were homozygous for the long allele.

Genetic Risk by Environment Interaction

Although studies of gene–environment interaction using measured genes provide quite specific information regarding the mechanisms involved in development, they are limited in that single genes typically account for a very small portion of the variance in complex human phenotypes. Quantitative genetic methods address this limitation by estimating an individual's overall genetic risk based on the phenotypes of genetic relatives. One such method first employed by Kendler et al. (1995) uses the phenotypes of co-twins as well as the zygosity of the twin pair to create a 4-point ordinal genetic risk variable. Because MZ twins share 100% of their genes while DZ twins share only 50% of their segregating genes on average, the MZ co-twin of a child exhibiting a phenotype (affected) is at higher genetic risk for developing the same phenotype than the DZ co-twin of an affected child. Extending the logic, a DZ co-twin of a child not exhibiting a particular phenotype (unaffected) is at even less genetic risk whereas the MZ co-twin of an unaffected child is at the lowest genetic risk of all. One recent study implementing this method tested for genetic risk–environment interaction in the development of conduct problems in a representative cohort sample of British 5-year-olds (Jaffee et al., 2005). An interaction was detected in which maltreatment and genetic risk interacted to predict continuously distributed conduct problem scores. The association between genetic risk and child conduct problems was stronger for those children who had been maltreated in comparison to their nonmaltreated counterparts.

Heritability by Measured Environment Interaction

This approach to studying gene–environment interplay involves estimating the heritability of a phenotype across different environmental conditions. Recent studies have found that indeed, for some phenotypes, the portion of variance that is attributable to genetic influences fluctuates across different levels of measured environmental variables. For example, Button, Scourfield, Martin, Purcell, and McGuffin (2005) found that the heritability of child conduct problems varied depending on the level of family dysfunction. Additive genetic effects on conduct problems were found to decrease as the level of family dysfunction increased. Further, shared and nonshared environmental effects increased as family dysfunction increased.

Terminology and Conceptualization Regarding "Environment"

Thus far, we have described extant work in this area as testing for interaction between genetic influences and environments. Within this context, the use of the term "environment" might be too general and may lead to confusion among the consumers of behavior genetic literature. Turkheimer, D'Onofrio, Maes, & Eaves (2005) discussed how confusion regarding the nature of "measured environment" and "shared environment" could lead to the misstatement and misinterpretation of results obtained from twin studies. When study designs include a measured environmental variable that also represents an aspect of the objectively shared environment (see Goldsmith, 1988, for a discussion of objectively versus effectively shared environments) such as SES, researchers sometimes report the effects of that variable to be independent of genetic influences. This is often not the case, however, as the majority of "environmental" variables contain some proportion of genetic variability (Plomin & Bergeman, 1991). Of course, it is not possible to determine if environmental variables measured at the family level are heritable within a traditional twin design. A major implication of the non-independence between genetic influences and measured environments is that studies reporting gene–environment interaction may in fact be mislabeling what is actually a gene–gene interaction. That is, risk genes for a phenotype may interact with the genes influencing a measured environment to predict an outcome. Given the potential for misclassifying interaction effects, it has been suggested that researchers cease reporting their findings as gene–environment interaction unless they can rigorously demonstrate that their measured environmental variable is not heritable. Studies might reduce confusion if they simply replace the term "environment" with a description of the variable that was actually measured (e.g., gene–social support interaction).

Gene–Environment Correlation

Gene–environment correlation describes a situation in which genetic variation leads individuals to be differentially exposed to environments. These environments, in turn, influence the development of heritable phenotypes. Gene–environment correlations are typically divided into three categories: passive, reactive, and active (Plomin, DeFries, & Loehlin, 1977). Passive gene–environment correlation results when a heritable phenotype found in parents contributes to the environment that those parents create for their children. Maternal depression provides a useful example as depression is both a heritable phenotype and is associated with risky family contexts (e.g., compromised parenting). Depressed mothers may pass risk genes to their children and these genes may be correlated with the quality of parenting provided by those mothers. Reactive (also known as evocative) gene–environment correlation occurs when a heritable child phenotype acts to elicit particular behaviors from parents, teachers, siblings, or other relevant people. Reactive gene–environment correlation may occur in the development of oppositional defiant behavior. Oppositional defiant behavior is heritable and may evoke the use of corporal punishment from parents. The use of corporal punishment may then in turn contribute to the maintenance or exacerbation of oppositional defiance. Active gene–environment correlation describes processes in which the child selects an environment that will serve to foster the development of some heritable phenotype. For example, an aggressive child may befriend other aggressive children and such friendships may serve to further develop the child's aggressive tendencies. Scholars have suggested that the nature of gene–environment correlations may change across development since the individual's ability to select environments may increase with age (Scarr & McCartney, 1983). Thus, gene–environment correlations might be largely passive when children are very young but shift to reactive and then to active correlations as children age.

Work on parent–child mutuality (Deater-Deckard & O'Connor, 2000) represents a useful example of gene–environment correlation. Parent–child mutuality refers to aspects of parent–child interaction, including shared positive affect, responsiveness, and cooperation. Central to this construct is the idea that healthy parent–child interaction is characterized by emotional reciprocity and bi-directionality. In this study, trained observers coded 20 min of videotaped parent–child interaction to generate scores for mutuality exhibited between mothers and their 3- to 4-year-old children. Pairs of MZ twins, DZ twins, full siblings, and genetically unrelated adopted siblings were included in the sample. Correlations between siblings on parent–child mutuality (MZ r = 0.61, DZ r = 0.26, full siblings r = 0.25, adopted siblings r = −0.04) suggest that variation in this construct is influenced by genetics and that these genetic influences might be linked to child phenotypes, i.e., reactive gene–environment correlation. More recent work has shown that dyadic mutuality is specific to each child in a family and that child behavior problems are linked to mutuality. As child behavior problems increased, dyadic mutuality was compromised (Deater-Deckard & Petrill, 2004).

Temperament and Endophenotypes

The Endophenotype Concept

Another approach to the genetics of child temperament involves the identification of endophenotypes. An endophenotype is an intermediate link between a genotype and an individual's observable behavior (e.g., a temperament dimension). Such links are implied by the long-standing recognition that a genetically influenced behavioral trait is not encoded directly by the genes that contribute to its variance (Anastasi, 1958; Fuller & Thompson, 1960). Genes relevant to a behavioral trait most likely affect the central nervous system (CNS; e.g., structure, physiology, and/or function of the brain). Thus, an endophenotype may be a measure of the CNS activity involved in the physiology of a behavioral trait or disorder. An endophenotype may be identified at different levels of analysis. Some levels may be more proximal to the gene (e.g., level of a protein) whereas others may be more distal (e.g., level of cognitive functioning). Thus, an endophenotype may be a chemical measure, a measure of brain structure, a measure of neuroendocrine functioning, a measure of cognitive functioning, or perhaps even a different measure of behavior than the phenotype of interest.

Gottesman & Gould (2003) updated the endophenotype concept and suggested criteria for identifying candidate endophenotypes. The first criterion is that the endophenotype must be associated with the trait under study. Those individuals who exhibit the trait must also possess the candidate endophenotype with a greater than chance frequency. Second, individual differences in the endophenotype must be at least partly due to genetic factors (i.e., the endophenotype must be heritable). The endophenotype should also be state independent. In other words, if an individual exhibits different levels of a behavioral trait or it goes through different phases (e.g., elevated levels of normal activity versus hyperactivity), the endophenotype continues to be present. Additionally, the endophenotype must co-segregate with the trait. That is, the endophenotype and the disorder should show a tendency to be transmitted together in family pedigrees. Therefore, family members who do not exhibit the trait will show the endophenotype at a lower rate than family members who do. The final criterion indicates that the endophenotype will present in non-affected family members at a higher rate than in the general population. Taken together, the last two criteria dictate that the endophenotypes will appear in genetic relatives at a rate between the rate in affected persons and the rate in the general population.

Cortisol Function as a Candidate Endophenotype for Behavioral Inhibition

We examine an example of one endophenotypic indicator of a temperamental dimension. Schreiber, Goldsmith, and Gottesman (2007) reviewed the evidence for activity of the hypothalamus–pituitary–adrenal axis, as reflected in cortisol secretion, as an endophenotype for the temperamental dimension of behavioral inhibition. Both basal and reactive cortisol levels have been associated with inhibited behavior, and therefore, both types of measures can be considered candidate endophenotypes. Both basal cortisol during the awakening period (Bartels, de Geus, Kirschbaum, Sluyter, & Boomsma, 2003; Kupper et al., 2005; Wüst, Federenko, Hellhammer, & Kirschbaum, 2000), and measures of reactive cortisol response to a social stress test are heritable (Federenko, Nagamine, Hellhammer, Wadhwa, & Wust, 2004; Kirschbaum, Wust, Faig, & Hellhammer, 1992).

Some evidence supports the stability criterion for an endophenotype. Reactive cortisol shows specific habituation patterns after repeated stress (Schommer, Hellhammer, & Kirschbaum, 2003; Wust, Federenko, van Rossum, Koper, & Hellhammer, 2005). When exposed to a social stress test three times with 4-week intervals between exposures, "high responders" demonstrated elevated cortisol responses after all three social stress test sessions. Thus, there is some evidence for the stability of patterns of response in cortisol reactivity although the issue remains open to investigation.

Currently, no data directly demonstrate associations between cortisol levels and behavioral inhibition among family members or demonstrate shared genetic variance between measures of cortisol activity and behavioral inhibition. One study examined the association of cortisol levels between mother and child (2–4 years of age) in the context of the child's exposure to a novel and challenging task and found that mothers high on ratings of maternal sensitivity had significantly correlated cortisol responses with their children (Sethre-Hofstad, Stansbury, & Rice, 2002). Future twin and family studies are needed to investigate these criteria more extensively. Genes related to the corticotrophin releasing hormone (CRH), its receptors, and peptides have been implicated in animal models of behavior inhibition (Bakshi & Kalin, 2000), and links between a CRH candidate gene and behavioral inhibition in children of a parent with panic disorder have been established (Smoller et al., 2003, 2005). Thus, extant research indicates that cortisol activity has met some of the criteria for being a candidate endophenotype for behavioral inhibition; however, gaps in our knowledge remain.

Additional Temperament Endophenotypes

In addition to cortisol, there are other potential endophenotypes related to child temperament, although research is in the relatively early stages. Behavioral inhibition has been considered as an endophenotype for anxiety disorders. Preliminary data suggest that the central nucleus of the amygdala modulates behavioral inhibition, and positron emission tomography (PET) studies have indicated functional associations between the amygdala and prefrontal cortical regions that are related to anxiety (Kalin & Shelton, 2003). Frontal EEG asymmetry is associated with individual differences in affective style, inhibition, and liability to mood disorders and has been nominated as an endophenotype in genetic studies of temperament and mood and anxiety disorders (Anokhin, Heath, & Myers, 2006; Goldsmith & Lemery, 2000). However, results of a recent twin study indicate that heritability for frontal EEG asymmetry is relatively modest, somewhat dampening enthusiasm for endophenotype status (Anokhin et al., 2006). A potential endophenotype for effortful control is the executive attention system that involves the anterior cingulate and lateral prefrontal brain areas (Rueda, Rothbart, McCandliss, Saccomanno, & Posner, 2005). Several aspects of executive attention have shown modest to high heritability (Doyle et al., 2005; Fan, Wu, Fossella, & Posner, 2001; Goldsmith et al., 1997; Goldsmith, Lemery, Buss, & Campos, 1999; Lemery-Chalfant, Doelger, & Goldsmith, 2008). The evidence suggests that these traits meet some of the standards for qualification as endophenotypes, although more extensive research is required.

Future Directions

Genetic research on temperament is clearly trending toward larger samples with more extensive family configurations, multi-source assessment, and multidisciplinary integration. In the area of sampling, the field has embraced epidemiologically defined samples, with convenience samples of twins becoming dated. We still perceive the need to integrate adoption and stepfamily evidence more fully with the widely available twin results. Sampling has also become longitudinal and more strategically focused on key developmental transitions such as puberty.

A continuing need in quantitative—and especially molecular—genetic studies is for assessment other than self- or parental report of temperament. Some laboratory-based and observational assessment of temperament/emotionality has occurred in twin and adoption studies, but much more is needed.

Genetic research on temperament clearly needs to be synthesized with other disciplinary approaches. As one example, behavioral geneticists have produced massive evidence of genetic input to common clinical phenotypes such as schizophrenia, bipolar affective disorder, unipolar depression, antisocial personality disorder, and attention deficit hyperactivity disorder. Some of these disorders appear to have at least partial dimensional underpinnings, such as poor impulse control, activity level, anger and emotionally motivated aggressiveness, inhibition, and sadness. The latter all map onto temperament. Moreover, developmental epidemiologists have produced substantial evidence for the importance of myriad risk factors (social class, family functioning, crucial life events), but temperamental underpinnings are only beginning to be integrated with the broader field of risk factor research to understand psychopathology (Essex et al., 2006; Schreiber, Krause, Schmidt, Lemery-Chalfant, & Goldsmith, 2008). This research needs to be placed within a genetically informative framework.

Another widely recognized need in this area is integration of assessment of specific genetic and experiential processes into the twin and adoption studies to "actualize" the genetic and environmental components of the observed variation. Synthetic data analytic approaches exist for measuring specific genes within such classic quantitative designs. Measures of individuality for specific social environmental processes that might relate to temperament need to be refined and incorporated into genetically informative studies.

There is a question of whether genetic investigations of temperament will more profitably be gene centered (wherein the behavioral correlates of identified genes are sought) or behavior centered (wherein the investigator begins with an assessment of temperament and searches for associated genetic markers). In either case, temperament research would benefit from a new synergism with animal research, where gene function in the neurophysiological sense is more easily investigated. In either case, an appreciation of developmental plasticity of temperament will be required. As the field advances, it will be crucially important that it not become paradigm-bound if we wish to discern the entire panorama of genetic influences on temperament.

References

Achenbach, T. M. (1991a). *Manual for the Child Behavior Checklist/4–18 and 1991 profile*. Burlington, VT: University of Vermont, Department of Psychiatry.

Achenbach, T. M. (1991b). *Manual for the Teacher's Report Form and 1991 profile*. Burlington, VT: University of Vermont, Department of Psychiatry.

Anastasi, A. (1958). Heredity, environment, and the question "how?" *Psychological Review, 65*, 197–208.

Anokhin, A. P., Heath, A. C., & Myers, E. (2006). Genetic and environmental influences on frontal EEG asymmetry: A twin study. *Biological Psychology, 71*, 289–295.

Bakshi, V. P., & Kalin, N. H. (2000). Corticotropin-releasing hormone and animal models of anxiety: Gene-environment interactions. *Biological Psychiatry, 48*, 1175–1198.

Bartels, M., de Geus, E. J., Kirschbaum, C., Sluyter, F., & Boomsma, D. I. (2003). Heritability of daytime cortisol levels in children. *Behavior Genetics, 33*, 421–433.

Bates, J. E. (1989). Concepts and measures of temperament. In G. A. Kohnstramm, J. E. Bates, & M. K. Rothbart (Eds.), *Temperament in childhood* (pp. 3–26). Chichester, UK: Wiley.

Bayley, N. (1993). *Bayley Scales of Mental Development* (2nd ed.). San Antonio, TX: The Psychological Corporation.

Benjamin, L., Li, L., Patterson, C., Greenberg, B. D., Murphy, D. L., & Hamer D. H. (1996). Population and familial association between the D4 dopamine receptor gene and measures of novelty seeking. *Nature Genetics, 12*, 81–84.

Biederman, J., Hirshfeld-Becker, D. R., Rosenbaum, J. F., Herot, C., Friedman, D., Snidman, et al. (2001). Further evidence of association between behavioral inhibition and social anxiety in children. *American Journal of Psychiatry, 158*, 1673–1679.

Billman, J., & McDevitt, S. C. (1980). Convergence of parent and observer ratings of temperament with observations of peer interaction in nursery school. *Child Development, 51*, 395–400.

Bouchard, T. J. Jr., & Loehlin, J. C. (2001). Genes, evolution, and personality. *Behavior Genetics, 31*, 243–273.

Bramlett, R. K., Scott, P., & Lowell, R. K. (2000). A comparison of temperament and social skills in predicting academic performance in first grade. *Special Services in the Schools, 16*, 147–158.

Braungart, J. M., Plomin, R., DeFries, J. C., & Fulker, D. W. (1992). Genetic influence on tester-rated temperament as assessed by Bayley's Infant Behavior Record; Nonadoptive and adoptive siblings and twins. *Developmental Psychology, 28*, 40–47.

Brown, T. (2007). Temporal course and structural relationships among dimensions of temperament and DSM-IV anxiety and mood disorder constructs. *Journal of Abnormal Psychology, 116*, 313–328.

Burgess, K. B., Rubin, K. H., Cheah, C. S. L., & Nelson, L. J. (2005). Behavioral inhibition, social withdrawal, and parenting. In W. R. Crozier & L. E. Alden (Eds.), *The essential handbook of social anxiety for clinicians* (pp. 99–120), New York, Wiley.

Buss, A. H., & Plomin, R. (1975). *A temperament theory of personality development*. Oxford, UK: Wiley.

Buss, A. H., & Plomin, R. (1984). *Temperament: Early developing personality traits*. Hillsdale, NJ: Erlbaum.

Button, T. M., Scourfield, J., Martin, N., Purcell, S., & McGuffin, P. (2005). Family dysfunction interacts with genes in the causation of antisocial symptoms. *Behavior Genetics, 35*, 115–120.

Campbell, S. B., & Ewing, L. J. (1990). Follow-up of hard-to-manage preschoolers: Adjustment at age 9 and predictors of continuing symptoms. *Journal of Child Psychology and Psychiatry, 31*, 871–889.

Cardon, L. R. (2006). Genetics: Delivering new disease genes. *Science, 314*, 1403–1405.

Caspi, A. (2000). The child is father of the man: Personality continuities from childhood to adulthood. *Journal of Personality and Social Psychology, 78*, 158–172.

Caspi, A., Roberts, B. W., & Shiner, R. L. (2005). Personality development: Stability and change. *Annual Review of Psychology, 56*, 453–484.

Caspi, A., Snugden, K., Moffitt, T. E., Taylor, A., Craig, I. W., Harrington, H., et al. (2003). Influence of life stress on depression: Moderation by a polymorphism in the 5-HTT gene. *Science, 301*, 386–389.

Cherny, S. S., Fulker, D. W., Corley, R. P., & Plomin, R. (1994). Continuity and change in infant shyness from 14 to 20 months. *Behavior Genetics, 24,* 365–379.

Cherny, S. S., Saudino, K. J., Fulker, D. W., Corley, R. P. Plomin, R., & DeFries, J. C. (2001). The development of observed shyness from 14 to 20 months: Shyness in context. In R. N. Emde & J. K. Hewitt (Eds.), *Infancy to early childhood: Genetic and environmental influences on developmental change* (pp. 73–88). New York: Oxford University Press.

Chiappe, P., Hasher, L., & Siegel, L. S. (2000). Working memory, IC, and reading disability. *Memory and Cognition, 28,* 8–17.

Clark, L. A., Watson, D., & Mineka, S. (1994). Temperament, personality, and the mood and anxiety disorders. *Journal of Abnormal Psychology, 103,* 103–116.

Cohen, D J., Dibble, E., & Graw, J. M. (1977). Fathers' and mothers' perceptions of children's personality. *Archives of General Psychiatry, 34,* 480–487.

Cole, P. M., Teti, L. O., & Zahn-Waxler, C. (2003). Mutual emotion regulation and the stability of conduct problems between preschool and early school age. *Development and Psychopathology, 15,* 1–18.

Comings, D. E., Johnson, J. P., Gonzalez, N. S., Huss, M., Saucier, G., McGue, M., et al. (2000). Association between the adrenergic alpha 2A receptor gene (ADRA2A) and measures of irritability, hostility, impulsivity and memory in normal subjects. *Psychiatric Genetics, 10,* 39–42.

Conway, M. A., Harries, K., Noyes, J., Racsma'ny, M., & Frankish, C. R. (2000). The disruption and dissolution of directed forgetting: IC of memory. *Journal of Memory and Language, 43,* 409–430.

Deater-Deckard, K., & O'Connor, T. G. (2000). Parent-child mutuality in early childhood: Two behavioral genetic studies. *Developmental Psychology, 36,* 561–570.

Deater-Deckard, K., & Petrill, S. A. (2004). Parent-child dyadic mutuality and child behavior problems: An investigation of gene-environment processes. *Journal of Child Psychology and Psychiatry, 45,* 1171–1179.

Deater-Deckard, K., Petrill, S. A., & Thompson, L. A. (2007). Anger/frustration, task persistence, and conduct problems in childhood: A behavior genetic analysis. *Journal of Child Psychology and Psychiatry, 48,* 80–87.

DeFries, J. C., & Fulker, D. W. (1985). Multiple regression analysis of twin data. *Behavior Genetics, 15,* 467–473.

DeFries, J. C., & Fulker, D. W. (1988). Multiple regression analysis of twin data: Etiology of deviant scores versus individual differences. *Acta Geneticae Medicae et Gemellolgiae, 37,* 205–216.

DiLalla, L. F., Kagan, J., & Reznick, J. S. (1994). Genetic etiology of behavioral inhibition among 2-year-old children. *Infant Behavior and Development, 17,* 405–412.

Doyle, A. E., Faraone, S. V., Seidman, L. J., Willcutt, E. G., Nigg, J. T., Waldman, I. D., et al. (2005). Are endophenotypes based on measures of executive functions useful for molecular genetic studies of ADHD? *Journal of Child Psychology and Psychiatry, 46,* 774–803.

Eaton, W. O. (1994). Activity level, development, and personality. In C. F. Halverston, G. A. Kohnstamm, & R. P. Martin (Eds.), *The developing structure of temperament and personality from infancy to adulthood* (pp. 173–187). New Jersey: Erlbaum.

Ebstein, R. P., Novick, O., Umansky, R., Priel, B., & Osher, Y. (1996). Dopamine D4 receptor (D4DR) exon III polymorphism associated with the human personality trait of Novelty Seeking. *Nature Genetics, 12,* 78–80.

Eisenberg, N., Cumberland, A., Spinrad, T. L., Fabes, R. A., Shepard, S. A., Reiser, M., et al. (2001). The relations of regulation and emotionality to children's externalizing and internalizing problem behavior. *Child Development, 72,* 1112–1134.

Eisenberg, N., Shepard, S. A., Fabes, R. A., Murphy, B. C., & Guthrie, I. K. (1998). Shyness and children's emotionality, regulation, and coping: Contemporaneous, longitudinal, and across-context relations. *Child Development, 69,* 767–790.

Eley, T. C., Bolton, D., O'Connor, T. G., Perrin, S., Smith, P., & Plomin, R. (2003). A twin study of anxiety-related behaviours in preschool children. *Journal of Child Psychology and Psychiatry, 44,* 945–960.

Emde, R. N., Robinson, J. L., Corley, R. P., Nikkari, D., & Zahn-Waxler, C. (2001). Reactions to restraint and anger-related expressions during the second year. In R. N. Emde & J. K. Hewitt (Eds.), *Infancy to early childhood: Genetic and environmental influences on developmental change* (pp. 127–140). New York: Oxford University Press.

Essex, M. J., Kraemer, H. C., Armstrong, J. M., Boyce, W. T., Goldsmith, H. H., Klein, M. H., et al. (2006). Exploring risk factors for the emergence of children's mental health problems. *Archives of General Psychiatry, 63,* 1246–1256.

Fan, J., Wu, Y., Fossella, J. A., & Posner, M. I. (2001). Assessing the heritability of attentional networks. *BMC Neuroscience, 2,* 14.

Federenko, I. S., Nagamine, M., Hellhammer, D. H., Wadhwa, P. D., & Wust, S. (2004). The heritability of hypothalamus pituitary adrenal axis responses to psychosocial stress is context dependent. *Journal of Clinical Endocrinology and Metabolism, 89,* 6244–6250.

Fox, N. A. (2004). Temperament and early experience form social behavior. In S. G. Kaler & O. M. Rennert (Eds.), *Understanding and optimizing human development: From cells to patients to populations* (pp. 171–178), New York: New York Academy of Sciences.

Fox, N. A., Nichols, K. E., Henderson, H. A., Rubin, K., Schmidt, L., Hamer, D., et al. (2005). Evidence for a gene-environment interaction in predicting behavioral inhibition in middle childhood. *Psychological Science, 16,* 921–926.

Frye, D., Zelaso, P. D., & Palfai, T. (1995). Theory of mind and rule-based reasoning. *Cognitive Development, 10,* 483–527.

Fullard, W., McDevitt, S. C., & Carey, W. B. (1984). Assessing temperament in one to three year old children. *Journal of Pediatric Psychology, 9,* 205–217.

Fuller, J. L., & Thompson, W. R. (1960). *Behavior genetics.* New York: Wiley.

Gagne, J. R. & Goldsmith, H. H. (2007, June). A twin analysis of inhibitory behavior from 12–36 months of age. Paper presented at the Behavior Genetics Association, Amsterdam, The Netherlands.

Gagne, J. R., & Saudino, K. J. (2005). A behavioral genetic analysis of inhibitory control at 24 months of age. *Behavior Genetics, 35,* 801.

Galejs, I., King, A., & Hegland, S. M. (1987). Antecedents of achievement motivation in preschool children. *Journal of Genetic Psychology, 148,* 333–348.

Gallerani, D., O'Regan, M., & Reinherz, H. (1982). Prekindergarten screening: How well does it predict readiness for first grade? *Psychology in the Schools, 19,* 175–182.

Gelernter, J., Cubells, J. F., Kidd, J. R., Pakstis, A. J., & Kidd, K. K. (1999). Population studies of polymorphisms of the serotonin transporter protein gene. *American Journal of Medical Genetics, 88,* 61–66.

Gest, S. D. (1997). Behavioral inhibition: Stability and associations with adaptation from childhood to early adulthood. *Journal of Personality and Social Psychology, 72,* 467–475.

Goldsmith, H. H. (1988). Human developmental behavioral genetics: Mapping the effects of genes and environments. *Annals of Child Development, 5,* 187–227.

Goldsmith, H. H. (1996). Studying temperament via construction of the Toddler Behavior Assessment Questionnaire. *Child Development, 67,* 218–235.

Goldsmith, H. H., Buss, A. H., Plomin, R., Rothbart, M. K., Thomas, A., Chess, S., et al. (1987). Roundtable: What is temperament? Four approaches.*Child Development, 58,* 505–529.

Goldsmith, H. H., Buss, K. A., & Lemery, K. S. (1997). Toddler and childhood temperament: Expanded content, stronger genetic evidence, new evidence for the importance of environment. *Developmental Psychology, 33*, 891–905.

Goldsmith, H. H., & Campos, J. J. (1982). Toward a theory of infant temperament. In R. N. Emde & R. J. Harmon (Eds.), *The development of attachment and affiliative systems* (pp. 161–193). New York: Plenum Press.

Goldsmith, H. H., & Campos, J. J. (1986). Fundamental issues in the study of early temperament: The Denver Twin Temperament Study. In M. E. Lamb, A. L. Brown, & B. Rogoff (Eds.), *Advances in developmental psychology* (pp. 231–283). Hillsdale, NJ: Erlbaum.

Goldsmith, H. H., & Gottesman, I. I. (1981). Origins of variation in behavioral style: A longitudinal study of temperament in young twins. *Child Development, 52*, 91–103.

Goldsmith, H. H., & Lemery, K. S. (2000). Linking temperamental fearfulness and anxiety symptoms: A behavior-genetic perspective. *Biological Psychiatry, 48*, 1199–1209.

Goldsmith, H. H., Lemery, K. S., Buss, K. A., & Campos, J. (1999). Genetic analyses of focal aspects of infant temperament. *Developmental Psychology, 35*, 972–985.

Goldsmith, H. H., Lemery, K. S., & Essex, M. J. (2004). Temperament as a liability factor for childhood behavioral disorders: The concept of liability. In L. F. DiLalla (Ed.), *Behavior genetics principles: Perspectives in development, personality, and psychopathology* (pp. 19–39). Washington, DC: American Psychological Association.

Goldsmith, H. H., Reilly, J., Lemery, K. S., Longley, S., & Prescott, A. (1995). *The laboratory temperament assessment battery-preschool version: Description of procedures.* Technical report, Department of Psychology, University of Wisconsin–Madison.

Goldsmith, H. H., & Rothbart, M. K. (1991). Contemporary instruments for assessing early temperament by questionnaire and in the laboratory. In J. Strelau & A. Angleitner (Eds.), *Explorations in temperament: International perspectives on theory and measurement* (pp. 249–272). New York: Plenum.

Gottesman, I. I., & Gould, T. D. (2003). The endophenotype concept in psychiatry: Etymology and strategic intentions. *American Journal of Psychiatry, 160*, 636–645.

Gurlanick, M. J., & Groom, J. M. (1990). The correspondence between temperament and peer interactions for normally developing and mildly delayed preschool children. *Child: Care, Health and Development, 16*, 165–175.

Hall, S. J., Halperin, J. M., Schwartz, S. T., & Newcorn, J. H. (1997). Behavioral and executive functions in children with attention-deficit hyperactivity disorder and reading disability. *Journal of Attention Disorders, 1*, 235–247.

Halverson, C .F., & Waldrop, M. F. (1976). Relations between preschool activity and aspects of intellectual and social behavior at age $7^{1/2}$. *Developmental Psychology, 12*, 107–112.

Hariri, A. R., Mattay, V. S., Tessitore, A., Kolachana, B., Fera, F., Goldman, D., et al. (2002). Serotonin transporter genetic variation and the response of the human amygdale. *Science, 297*, 400–403.

Henry, B., Caspi, A., Moffitt, T. E., Harrington, H., & Silva, P. A. (1999). Staying in school protects boys with poor self-regulation in childhood from later crime: A longitudinal study. *International Journal of Behavioral Development, 23*, 1049–1073.

Hughes, C., White, A., Sharpen, J., & Dunn, J. (2000). Antisocial, angry, and unsympathetic: "Hard-to-manage" preschoolers' peer problems and possible cognitive influences. *Journal of Child Psychology and Psychiatry, 41*, 169–179.

Izard, C. E. (1972). *Patterns of emotions.* San Diego, CA: Academic Press.

Jaffee, S. R., Caspi, A., Moffitt, T. E., Dodge, K. A., Rutter, M., Taylor, A. et al. (2005). Nature X nurture: Genetic vulnerabilities interact with physical maltreatment to promote conduct problems. *Development and Psychopathology, 17*, 67–84.

Kagan, J. (2001). Temperamental contributions to affective and behavioral profiles in childhood. In S. G. Hofmann & P. M. DiBartolo (Eds.), *From social anxiety to social phobia: Multiple perspectives* (pp. 216–234), Needham Heights, MA: Allyn & Bacon.

Kalin, N. H., & Shelton, S. E. (2003). Nonhuman primate models to study anxiety, emotion regulation, and psychopathology. In J. A. King, C. F. Ferris & I. I. Lederhendler (Eds.), *Roots of mental illness in children* (pp. 189–200). New York: New York Academy of Sciences.

Kendler, K., Kessler, R. C., Walters, E. E., MacLean, C., Neale, M. C., & Eaves, L. J. (1995). Stressful life events, genetic liability, and onset of an episode of major depression in women. *The American Journal of Psychiatry, 152*, 833–842.

Kirschbaum, C., Wust, S., Faig, H. G., & Hellhammer, D. H. (1992). Heritability of cortisol responses to human corticotropin-releasing hormone, ergometry, and psychological stress in humans. *Journal of Clinical Endocrinology and Metabolism, 75*, 1526–1530.

Klein, H. A. (1980). Early childhood group care: Predicting adjustment from individual temperament. *Journal of Genetic Psychology, 137*, 125–131.

Knafo, A., & Plomin, R. (2006). Prosocial behavior from early to middle childhood: Genetic and environmental influences on stability and change. *Developmental Psychology, 42*, 771–786.

Kochanska, G., DeVet, K., Goldman, M., Murray, K., & Putnam, S. P. (1994). Maternal reports of conscience development and temperament in young children. *Child Development, 65*, 852–868.

Kohnstamm, J. A. Bates, & M. K. Rothbart (Eds.), *Temperament in childhood* (pp. 187–247). New York: Wiley.

Kopp, C. B. (1982). Antecedents of self-regulation: A developmental perspective. *Developmental Psychology, 18*, 199–214.

Kupper, N., de Geus, E. J., van den Berg, M., Kirschbaum, C., Boomsma, D. I., & Willemsen, G. (2005). Familial influences on basal salivary cortisol in an adult population. *Psychoneuroendocrinology, 30*(9), 857–868.

Langley, J., McGee, R., Silva, P. A., & Williams, S. (1983). Child behavior and accidents. *Journal of Pediatric Psychology, 8*, 181–189.

Lemery-Chalfant, K., Doelger, L., & Goldsmith, H. H. (2008). Genetic relations between effortful and attentional control and symptoms of psychopathology in middle childhood. *Infant and Child Development, 17*, 365–427.

Lengua, L. J. (2003). Associations among emotionality, self-regulation, adjustment problems, and positive adjustment in middle childhood. *Applied Developmental Psychology, 24*, 595–618.

Lesch, K-P, Bengel, D., Heils, A., Sabol, S. Z., Greenberg, B. D., Petri, S., et al. (1996). Association of anxiety-related traits with a polymorphism in the serotonin transporter gene regulatory region. *Science, 274*, 1527–1531.

Leve, L. D., Winebarger, A. A., Fagot, B. I., Reid, J. B., & Goldsmith, H. H. (1998). Environmental and genetic variance in children's observed and reported maladaptive behavior. *Child Development, 69*, 1286–1298.

Lytton, H. (1980). *Parent-child interaction: The socialization process observed in twin and singleton families.* New York: Plenum.

Martin, R. P., Olejnik, S., & Gaddis, L. (1994). Is temperament an important contributor to schooling outcomes in elementary school? Modeling effects of temperament and scholastic ability on academic achievement. In W. B. Carey & S. C. McDevitt (Eds.), *Prevention and early intervention: Individual differences as risk factors for the mental health of children: A festschrift for Stella Chess and Alexander Thomas*, (pp. 59–68). Philadelphia: Brunner/Mazel.

Matheny, A. P., Jr. (1980). Bayley's Infant Behavior Record: Behavioral components and twin analyses. *Child Development, 51*, 1157–1167.

Matheny, A. P., Jr. (1983). A longitudinal twin study of stability of components from Bayley's Infant Behavior Record. *Child Development, 54*, 356–360.

Matheny, A. P., Jr. (1987). Psychological characteristics of childhood accidents. *Journal of Social Issues, 43*, 45–60.

Matheny, A. P., Jr. (1989a). Temperament and cognition: Relations between temperament and mental test scores. In G. A. Kohnstamm, J. E. Bates, & M. K. Rothbart (Eds.), *Temperament in Childhood* (pp. 263–282). New York: Wiley.

Matheny, A. P., Jr. (1989b). Children's behavioral inhibition over age and across situations: Genetic similarity for a trait during change. *Journal of Personality, 57*, 215–235.

Munafo, M. R., Clark, T. G., Moore, L. R., Payne, E., Walton, R., & Flint, J. (2003). Genetic polymorphisms and personality in healthy adults: A systematic review and meta-analysis. *Molecular Psychiatry, 8*, 471–484.

Murray, K. T., & Kochanska, G. (2002). Effortful control: Factor structure and relation to externalizing and internalizing behaviors. *Journal of Abnormal Child Psychology, 30*, 503–514.

Nakamura, C. Y., & Finck, D. (1973). Effect of social or task orientation and evaluative or nonevaluative situations on performance. *Child Development, 44*, 83–93.

Neale, M. C., & Fulker, D. W. (1984). A bivariate path analysis of fear data on twins and their parents. *Acta Geneticae Medicae et Gemellologiae: Twin Research, 33*, 273–286.

Nigg, J. T., Quamma, J. P., Greenberg, M. T., & Kusche, C. A. (1999). A two-year longitudinal study of neuropsychological and cognitive performance in relation to behavioral problems and competencies in elementary school children. *Journal of Abnormal Child Psychology, 27*, 51–63.

Palisin, H. (1986). Preschool temperament and performance on achievement tests. *Developmental Psychology, 22*, 766–770.

Pallodino, P., Mammarella, N., & Vecchi, T. (2003). Modality-specific effects in inhibitory mechanisms: The interaction of peripheral and central components in working memory. *Brain and Cognition, 53*, 263–267.

Passolunghi, M. C., & Siegel, L. S. (2001). Short-term memory, working memory, and IC in children with difficulties in arithmetic problem solving. *Journal of Experimental Child Psychology, 80*, 44–57.

Perner, J., Leekam, S. R., & Wimmer, H. (1987). Three-year-olds' difficulty with false belief: The case of conceptual deficit. *British Journal of Developmental Psychology, 5*, 125–137.

Phillips, K, & Matheny, A. P., Jr. (1997). Evidence for genetic influence on both cross-situation and situation-specific components of behavior. *Journal of Personality and Social Psychology, 73*, 129–138.

Pliszka, S. R., Borcherding, S. H., Spratley, K., Leon, S., & Irick, S. (1997). Measuring IC in children. *Developmental and Behavioral Pediatrics, 18*, 254–259.

Plomin, R. & Bergeman, C. S. (1991). The nature of nurture: Genetic influence on 'environmental' measures. *Behavioral and Brain Sciences, 14*, 373–427.

Plomin, R., Coon, H., Carey, G., DeFries, J. C., & Fulker, D. W. (1991). Parent-offspring and sibling adoption analyses of parental ratings of temperament in infancy and childhood. *Journal of Personality, 59*, 705–732.

Plomin, R., DeFries, J. C., & Fulker, D. W. (1988). *Nature and nurture during infancy and early childhood*. New York: Cambridge University Press.

Plomin, R., DeFries, J. C., & Loehlin, J. C. (1977). Genotype-environment interaction and correlation in the analysis of human behavior. *Psychological Bulletin, 84*, 309–322.

Plomin, R. Emde, R. N., Braungart, J. M., Campos, J., Corley, R., Fulker, D. W. et al. (1993). Genetic change and continuity from fourteen to twenty months: The MacArthur Longitudinal Twin Study. *Child Development, 64*, 1354–1376.

Posner, M. I., & Rothbart, M. K. (Eds.). (2003). *Developing mechanisms of self-regulation*. New York: Brunner-Routledge.

Posner, M. I., & Rothbart, M. K. (2007). Temperament and learning. In M. I. Posner, M. K. Rothbart (Eds.), *Educating the human brain* (pp. 121–146). Washington, DC: American Psychological Association.

Riese, M. L. (1990). Neonatal temperament in monozygotic and dizygotic twin pairs. *Child Development, 61*, 1230–1237.

Rose, R. J., & Ditto, W. B. (1983). A developmental-genetic analysis of common fears from early adolescence to early adulthood. *Child Development, 54*, 361–368.

Rothbart, M. K. (1989). Temperament in childhood: A framework. In G. A. Kohnstamm, J. E. Bates, & M. K. Rothbart (Eds.), *Temperament in childhood* (pp. 59–73). Chichester, UK: Wiley.

Rothbart, M. K., Ahadi, S. A., & Evans, D. E. (2000). Temperament and personality: Origins and outcomes. *Journal of Personality and Social Psychology, 78*, 122–135.

Rothbart, M. K., Ahadi, S. A., Hershey, K. L., & Fisher, P. (2001). Investigations of temperament at 3–7 years: The Children's Behavior Questionnaire. *Child Development, 72*, 1394–1408.

Rothbart, M. K., & Bates, J. E. (1998). Temperament. In N. Eisenberg (Vol. Ed.), *Handbook of child psychology: Social, emotional, & personality development* (pp. 105–176). New York: Wiley.

Rueda, M. R., Rothbart, M. K., McCandliss, B. D., Saccomanno, L., & Posner, M. I. (2005). Training, maturation, and genetic influences on the development of executive attention. *Proceedings of the National Academy of Sciences USA, 102*, 14931–14936.

Russell, J., Mauthner, N., Sharpe, S., & Tidswell, T. (1991). The 'windows tasks' as a measure of strategic deception in preschoolers and autistic subjects. *British Journal of Developmental Psychology, 9*, 331–349.

Saudino, K. J., & Cherny, S. S. (2001). Parental ratings of temperament in twins. In R. N. Emde & J. K. Hewitt (Eds.), *Infancy to early childhood: Genetic and environmental influences on developmental change* (pp. 73–88). New York: Oxford University Press.

Saudino, K. J., & Eaton, W. O. (1991). Infant temperament and genetics: An objective twin study of motor activity level. *Child Development, 62*, 1167–1174.

Saudino, K. J., & Eaton, W. O. (1995). Continuity and change in objectively assessed temperament: A longitudinal twin study of activity level. *British Journal of Developmental Psychology, 13*, 81–95.

Saudino, K. J., Plomin, R., & DeFries, J. C. (1996). Tester-rated temperament at 14, 20, and 24 months: Environmental change and genetic continuity. *British Journal of Developmental Psychology, 14*, 129–144.

Scarr, S., & McCartney, K. (1983). How people make their own environments: A theory of genotype greater than environment effects. *Child Development, 54*, 424–435.

Schachar, R., Tannock, R., Marriott, M., & Logan, G. (1995). Deficient IC in attention deficit hyperactivity disorder. *Journal of Abnormal Child Psychology, 23*, 411–437.

Schmitz, S., Fulker, D. W., Plomin, R., Zahn-Waxler, C., Emde, R. N., & DeFries, J. C. (1999). Temperament and problem behavior during early childhood. *International Journal of Behavioral Development, 23*, 333–355.

Schmitz, S., Saudino, K. J., Plomin, R., Fulker, D. W., & DeFries, J. C. (1996). Genetic and environmental influences on temperament in middle childhood: Analyses of teacher and tester ratings. *Child Development, 67*, 409–422.

Schoen, M. J., & Nagle, R. J. (1994). Prediction of school readiness from kindergarten temperament scores. *Journal of School Psychology, 32*, 135–147.

Schommer, N. C., Hellhammer, D. H., & Kirschbaum, C. (2003). Dissociation between reactivity of the hypothalamus-pituitary-adrenal axis and the sympathetic-adrenal-medullary system to repeated psychosocial stress. *Psychosomatic Medicine, 65*, 450–460.

Schreiber, J. E., Goldsmith, H. H., & Gottesman, I. I. (2008). Translating the endophenotype concept for dimensional traits. Manuscript in preparation, Department of Psychology, University of Wisconsin–Madison.

Schreiber, J. E., Krause, K. L., Schmidt, N. L., Lemery-Chalfant, K. S., & Goldsmith, H. H. (2008). The associations of temperament and emotion regulation with anxiety and depression in middle childhood. Under editorial review.

Scott, J. P., & Fuller, J. L. (1965). *Genetics and the social behavior of the dog*. Chicago: University of Chicago Press.

Sethre-Hofstad, L., Stansbury, K., & Rice, M. A. (2002). Attunement of maternal and child adrenocortical response to child challenge. *Psychoneuroendocrinology, 27*, 731–747.

Sham, P. (1998). *Statistics in human genetics*. London: Arnold.

Smoller, J. W., Rosenbaum, J. F., Biederman, J., Kennedy, J., Dai, D., Racette, S. R., et al. (2003). Association of a genetic marker at the corticotropin-releasing hormone locus with behavioral inhibition. *Biological Psychiatry, 54*, 1376–1381.

Smoller, J. W., Yamaki, L. H., Fagerness, J. A., Biederman, J., Racette, S., Laird, N. M., et al. (2005). The corticotropin-releasing hormone gene and behavioral inhibition in children at risk for panic disorder. *Biological Psychiatry, 57*, 1485–1492.

Stevenson, J., Batten, N., & Cherner, M. (1992). Fears and fearfulness in children and adolescents: A genetic analysis of twin data. *Journal of Child Psychology and Psychiatry, 33*, 977–985.

Strobel, A., Spinath, F. M., Angleitner, A., Riemann, R., & Lesch, K. P. (2003). Lack of association between polymorphisms of the dopamine D4 receptor gene and personality. *Neuropsychobiology, 47*, 52–56.

Trouton, A., Spinath, F. M., & Plomin, R. (2002). Twins Early Development Study (TEDS): A multivariate, longitudinal genetic investigation of language, cognition and behavior problems in childhood. *Twin Research, 5*, 444–448.

Turkheimer, E., D'Onofrio, B. M., Maes, H. M., & Eaves, L. J. (2005). Analysis and interpretation of twin studies including measures of the shared environment. *Child Development, 76*, 1217–1233.

Wahlsten, D. (1999). Single-gene influences on brain and behavior. *Annual Review of Psychology, 50*, 599–624.

Willis-Owen, S. A., Turri, M. G., Munafo, M. R., Surtees, P. G., Wainwright, N. W., Brixey, R. D., et al. (2005). The serotonin transporter length polymorphism, neuroticism, and depression: A comprehensive assessment of association. *Biological Psychiatry, 58*, 451–466.

Wolfe, C. D., & Bell, M. A. (2003). Working memory and IC in early childhood: Contributions from physiology, temperament, and language. *Developmental Psychobiology, 44*, 68–83.

Wood, A. C., Saudino, K. J., Rogers, H., Asherson, P., & Kuntsi, J. (2007). Genetic influences on mechanically-assessed activity level in children. *Journal of Child Psychology and Psychiatry, 48*, 695–702.

Wust, S., Federenko, I., Hellhammer, D. H., & Kirschbaum, C. (2000). Genetic factors, perceived chronic stress, and the free cortisol response to awakening. *Psychoneuroendocrinology, 25*, 707–720.

Wust, S., Federenko, I. S., van Rossum, E. F., Koper, J. W., & Hellhammer, D. H. (2005). Habituation of cortisol responses to repeated psychosocial stress—further characterization and impact of genetic factors. *Psychoneuroendocrinology, 30*, 199–211.

Yamagata, S., Takahashi, Y., Kijima, N., Maekawa, H., Ono, Y., & Ando, J. (2005). Genetic and environmental etiology of effortful control. *Twin Research and Human Genetics, 8*, 300–306.

Chapter 19

Genetic and Environmental Influences on Sexual Orientation

Khytam Dawood, J. Michael Bailey, and Nicholas G. Martin

Introduction

The primary focus of this chapter is to provide an overview of the evidence to date on the quantitative genetics of sexual orientation, including family and twin studies. The bulk of the available evidence suggests moderate heritability for male sexual orientation. Female sexual orientation has been studied much less extensively, but current studies are consistent with a genetic contribution for women as well (Kirk, Bailey, Dunne, & Martin, 2000; Pattatucci & Hamer, 1995). Familial aggregation has been reported in several family studies of both male and female homosexuality (Dawood & Bailey, 2000), although the genetic and environmental influences on this familial clustering have not been clearly defined by the largest twin studies published thus far, which have produced contradictory results. Recent molecular genetic studies will also be reviewed, including the two main strategies that have been used to date – linkage and association analysis. We will also discuss the implications of recent advances in molecular genetic studies.

Prevalence and Distribution of Sexual Orientation

Sexual orientation describes what is erotically attractive to an individual and is usually consistent with sexual identity which refers to an individual's labeling of self as heterosexual, homosexual, or bisexual; both are typically not consolidated until adolescence or later.

Three large surveys of sexual behavior from UK (Johnson, Wadsworth, Wellings, Bradshaw, & Field, 1992), France (ACFS Investigators, 1992), and the USA (Billy, Tanfer, Grady, & Klepinger, 1993; Laumann, Gagnon, Michael, & Michaels, 1994) have provided current estimates of the frequency of male and female adult homosexual behavior. These estimates vary with the stringency of the respective definitions. For men, the least stringent definition examined in the studies, any homosexual experience ever, yielded an estimate of 4.1, 6.1, and 7.1% for the three samples, respectively. A much more stringent criterion, same-sex activity during the year preceding the survey, yielded rates of 1.1, 1.1, and 2.7% for the three samples, respectively. The criterion closest to that of self-identification as "gay" or "bisexual" was employed only in the American study and applied to 2.4% of men. In general, the rates for female homosexuality appear to be about half that for males; for example, female same-sex activity during the year preceding the survey yielded a rate of 1.3% in the American survey. In addition, male and female sexual orientation appear to be distributed differently in the general population. For men, sexual orientation appears to be bimodally distributed, with most men clustered at the heterosexual end of a continuum, many fewer at the homosexual end, and hardly any in the bisexual middle. For women, however, it tapers gradually from exclusively heterosexual to exclusively homosexual (Bailey, Dunne, & Martin, 2000). Whereas older studies tended to define sexual orientation behaviorally, most current researchers, ourselves included, define sexual orientation psychologically. Because sexual attraction and fantasy are less likely than behavior to be constrained by societal pressures, psychological sexual orientation is thought to be a more stable and fundamental trait. A psychological definition of sexual orientation tends to produce lower prevalence figures compared to a behavioral one and is used in most of the studies demonstrating familiality and heritability.

Behavior Genetic Studies

As currently practiced, behavior genetic research proceeds in three main stages.

K. Dawood (✉)
Department of Psychology and Center for Developmental and Health Genetics, Pennsylvania State University, University Park, PA, USA
e-mail: Khytam@psu.edu

Y.-K. Kim (ed.), *Handbook of Behavior Genetics*,
DOI 10.1007/978-0-387-76727-7_19, © Springer Science+Business Media, LLC 2009

Family studies are initially conducted to determine whether a trait or characteristic (often called a phenotype) runs in families by comparing rates in families of probands (i.e., individuals possessing the trait) vs. families of controls (that typically represent the base rate in the general population). If there is a heritable genetic contribution to a trait, one expects to find familial aggregation. However, the mere demonstration of clustering of a trait in families does not prove a genetic influence, because some traits run in families for environmental reasons, biological or otherwise.

Twin studies are conducted in order to separate genetic from familial environmental effects by comparing the similarity of monozygotic (MZ, also called identical) and same-sex dizygotic (DZ, also called fraternal) twins who have been reared together. MZ twins share all their genes in common whereas DZ twins share 50% of their segregating genes on average, just like any other pair of full siblings. Since both twins share the same pregnancy, it is assumed that they experience essentially the same prenatal environment. Because both twins have also been reared together, postnatal environmental similarity is assumed to be approximately equal. Thus, if MZ twins are more similar (i.e., concordant) for a trait than DZ twins, this is assumed to reflect their greater genetic similarity and is evidence that genetic factors influence the phenotype. Twin studies allow one to estimate heritability, defined as the proportion of the variance in expression of the trait due to all genetic influence(s) combined. Results from twin studies support a significant environmental contribution to most behavioral traits since MZ twins are not 100% concordant.

Adoption studies are a further method used to separate the effects of genes from the environment. Adoption produces family members who share family environment but are not genetically related, and vice versa. Thus, this method allows one to estimate the contribution of family environment to family resemblance. Studies exist from all three research designs regarding both male and female sexual orientation:

Family Studies

The first contemporary family-genetic study of sexual orientation was conducted by Pillard and Weinrich (1986). They recruited homosexual and heterosexual male probands (index subjects) using newspaper advertisements that did not mention the nature of the study. Probands were interviewed about their own sexuality as well as their siblings' sexual orientations. The researchers obtained permission to contact, and successfully contacted, the large majority of probands' siblings in order to verify proband reports. Results suggested that homosexual probands were quite accurate at assessing their siblings' sexual orientations (provided that they expressed a high degree of confidence, which they were typically able to do). Most importantly, gay male probands had an excess of gay brothers (22%) compared to heterosexual male probands' brothers (4%).

Subsequent studies have used similar methodologies, with one exception. A study with a very different methodology (Bailey et al., 1999) recruited gay and bisexual men from consecutive admissions at an HIV outpatient center. (In most studies, both gay and bisexual men have been included as "homosexual" probands.) The most important aspect of such an ascertainment strategy is that it is more systematic than advertising for volunteers, and it may be less subject to self-selection biases. All available studies have focused on the rate of homosexuality in siblings rather than other first-degree relatives (e.g., parents or offspring), due to the decreased reproduction of homosexual individuals.

Both male and female homosexuality appear to run in families. Table 19.1 contains the results of recent family studies. The rate of homosexuality among brothers of homosexual males has been around 9%. These results have exceeded those for heterosexual controls as well as the prevalence estimates from recent large-scale epidemiological surveys, suggesting that male homosexuality is familial. Homosexual women also appear to have more homosexual sisters than do

Table 19.1 Rates of homosexuality for nontwin siblings in recent studies

Study	Criterion	Brothers		Sisters	
		Homosexual	Heterosexual	Homosexual	Heterosexual
Male probands					
Pillard & Weinrich, 1986	Kinsey 2–6	0.22	0.04	0.08	0.09
Bailey et al., 1991	Subject's rating	0.10	0.00	0.02	0.00
Bailey & Pillard, 1991	Subject's rating	0.09	NA	0.06	NA
Bailey & Bell, 1993	Subject's rating	0.09	0.04	0.03	0.01
Bailey et al., 1999	Subject's rating	0.09	NA	0.04	NA
Female probands					
Pillard, 1990	Kinsey 2–6	0.13	0.00	0.25	0.11
Bailey & Benishay, 1993	Subject's rating	0.07	0.01	0.12	0.02
Bailey et al., 1993	Subject's rating	0.05	NA	0.14	NA
Bailey & Bell, 1993	Subject's rating	0.12	0.00	0.06	0.01

Table entries are proportions. NA entries indicate that studies did not assess the respective rate.

heterosexual controls, though the familiality estimates have varied more widely for women. Table 19.1 also contains information concerning the cofamiliality of male and female homosexuality. There is a trend for gay male probands to have more gay brothers than lesbian sisters and for the opposite pattern to be seen for lesbian probands, suggesting that at least some of the familial factors influencing male homosexuality differ from those influencing female homosexuality. However, the largest study to date (Bailey & Bell, 1993) did not find this pattern. Thus, the degree of cofamiliality (reflecting common familial influences) of male and female homosexuality remains inconclusive.

Twin Studies

The most common methodology used by contemporary human behavior geneticists to disentangle genetic and environmental determinants compares the similarity of monozygotic and dizygotic (DZ) twins who have been reared together. Because both kinds of twins have been reared together, environmental similarity is assumed to be equal (more about this assumption later). Thus, if MZ twins are more similar than DZ twins, this reflects their greater genetic similarity and is evidence that genetic factors influence the phenotype.

The first twin study of (male) homosexuality, by Kallmann (1952), ascertained homosexual twins in the "homosexual underworld" and correctional/mental institutions of New York city. Remarkably, 100% of 37 MZ twin pairs were concordant compared to 15% of 26 DZ pairs. Kallmann's (1952) study had a number of methodological defects, including its over reliance on (evidently) mentally ill gay men, lack of information on zygosity diagnosis, and especially its anomalously high rate of MZ concordance compared to other studies (Rosenthal, 1970). Still, it is remarkable that despite its promising results, nearly 40 years passed before another large twin study of male homosexuality was attempted.

Several additional twin studies have been conducted in the past two decades, and their results are given in Table 19.2. These studies have been generally consistent in detecting moderate to large heritabilities for both male and female sexual orientation. However, there have been methodological limitations, in particular, most of the large twin studies of sexual orientation recruited probands via advertisements in gay or lesbian publications. Such sampling is likely to result in volunteer bias that affects twin concordances and heritability analyses, though in most scenarios this would be more likely to lead to a false negative study. The largest twin study of sexual orientation to date (Bailey et al., 2000) recruited twins systematically from a twin registry and reported lower

Table 19.2 Concordance rates for twin studies of homosexuality

Study	MZ concordance	DZ concordance
Male studies		
Kallmann, 1952	1.00	0.15
Heston & Shields, 1968	0.60	0.14
Bailey & Pillard, 1991	0.52	0.22
Buhrich, Bailey, & Martin, 1991	0.47	0.00
Bailey et al., 2000	0.20	0.00
Female studies		
Bailey et al., 1993	0.48	0.16
Bailey et al., 2000	0.24	0.15
Combined male and female		
King & McDonald, 1992	0.25	0.12
Whitam, Diamond, & Martin, 1993	0.66	0.30
Kendler et al., 2000	0.32	0.13

twin concordances for homosexuality than in prior studies, although their findings were also consistent with moderate to large heritabilities for male and female sexual orientation. A further analysis of these data using multivariate structural equation modeling estimated heritability of the latent variable of male homosexuality around 30% and for female homosexuality around 50% (Kirk et al., 2000).

In the discussion below, we focus on some of the largest twin studies of male and female sexual orientation (Bailey et al., 2000; Bailey & Benishay, 1993; Bailey & Pillard, 1991; Kendler, Thornton, Gilman, & Kessler, 2000) conducted to date.

In Bailey and Pillard (1991) as well as in Bailey et al. (1993), homosexual probands were recruited via advertisements in gay or lesbian publications (e.g., "Do you have a twin or an adoptive brother?"). Two kinds of probands were recruited: probands with twins or probands with adoptive brothers or sisters (for the male and female studies, respectively). Adoptive siblings were raised with the probands but are genetically unrelated to them. Probands were interviewed, especially concerning the sexual orientations of their twins. Probands' twins were also contacted when possible and confirmed that probands were quite accurate in assessing their twins' sexual orientations.

In the male study (Bailey & Pillard, 1991), 52% of the MZ cotwins were also gay or bisexual, compared to 22% of the DZ cotwins and 11% of the adoptive brothers. In the female study (Bailey et al., 1993), 48% of the MZ cotwins were also lesbian or bisexual, compared to 16% of the DZ cotwins and 6% of adoptive sisters. Thus, for both men and women, the rates conformed to a partially genetic model, with highest concordance in the most genetically similar (MZ) group and lowest concordance in the least similar (adoptive) group.

In order to calculate heritability from this study, several assumptions were made. First, although sexual orientation was measured on a dichotomous scale (i.e., heterosexual vs.

homosexual), the underlying causal structure was dimensional. That is, genetic influences are polygenic (i.e., numerous genes each with small effect), and environmental influences are similarly multifactorial. This corresponds to a multifactorial threshold model (Reich, Cloninger, & Guze, 1975). Second, a population base rate was assumed for homosexual orientation. The possibilities considered ranged from 4 to 10% for males and from 2 to 10% for females. Third, since ascertainment methods are frequently viewed as leading to concordance-dependent bias, different degrees of such bias were assumed, from none at all to the case in which probands from concordant pairs were three times more likely to be ascertained than probands from discordant pairs. For both men and women, heritability estimates ranged from approximately 0.30 to approximately 0.70.

Kendler et al. (2000) recruited 756 twin and nontwin sibling pairs from a US national probability sample and found that 32% of MZ twins were concordant for homosexual orientation vs. 13% of DZ same-sex twins. The concordance rates for MZ twins were lower than those reported in previous studies which recruited via advertisements in gay publications, suggesting that twin pairs concordant for sexual orientation may be more likely to respond to such advertisements than are twin pairs discordant for sexual orientation. It is also worth noting that unlike previous studies in this area, this study assessed sexual orientation by a single item on a self-report questionnaire.

Bailey et al. (2000) have conducted the largest twin study to date. They recruited male and female twin pairs systematically from the Australian Twin Registry and assessed their sexual orientation as well as two related traits: childhood gender nonconformity and continuous gender identity. Twin concordances for homosexual orientation were lower than in prior studies with 20% of male MZ twins concordant vs. 0% of DZ twins, and 24% of female MZ twins vs. 10.5% of DZ twins. Univariate analyses showed that familial factors were important for all traits, but were less successful in distinguishing genetic from shared environmental influences. Multivariate analyses suggested that the causal architecture differed between men and women and, for women, provided significant evidence for the importance of genetic factors to the traits' covariation.

Adoption Studies

A few of the family and twin studies listed in Tables 19.1 and 19.2 have included adoptive siblings who were reared in the same household as the homosexual probands in their samples. Hence, rates of homosexuality have been estimated for genetically unrelated adoptive siblings. Consistent with a genetic hypothesis, for both sexes the proportion of homosexuals (and bisexuals) was significantly greater for MZ cotwins than for either DZ same-sex cotwins or adoptive siblings.

Molecular Genetic Studies

Once a solid foundation of support for significant genetic influence on a trait has been built by means of behavior genetics (family, twin, and adoption studies), as has been the case for sexual orientation especially with males, molecular genetic studies are the next logical step. The two primary varieties of these studies are linkage and association designs. Linkage analysis exploits the key biological phenomenon during generation of sperm and eggs of meiotic recombination, or crossing over, during which both the maternally and paternally derived chromosomes lie in close proximity and undergo exchange of genetic material between the homologous chromosomes, e.g., between a paternally derived chromosome and its maternally derived counterpart. The chance of crossing over between two loci (locations on a chromosome) is referred to as the recombination fraction. Genes and other genetic markers (DNA sequence variations known as polymorphisms) that are close together are less likely to be separated by this process than are those that are farther apart. Therefore, they are usually inherited together by the progeny cells and are genetically linked. Due largely to the complexity of the genetic contributions to male sexual orientation and uncertainty regarding key parameters (mode of inheritance, number of relevant genes, etc.), the type of linkage analyses preferred are nonparametric allele-sharing methods and, more specifically, the affected sibling pair (ASP) method. ASP designs measure the frequency with which a genetic marker allele (or variant) is inherited from a particular parent (referred to as IBD, meaning identical by descent) in a pair of siblings both manifesting the trait. The presence of a trait-influencing gene is revealed when the IBD allele sharing between affected siblings exceeds the expected 50%.

Association studies are based on linkage disequilibrium (LD). This means that a gene variant influencing a trait was initially associated with specific alleles of nearby polymorphic loci. As generations (and the meioses that produce sperms and eggs) pass, the trait-influencing gene and marker allele may remain statistically associated because their proximity reduces the number of recombinations or crossing over that occurs between them. An advantage of association tests is that the chromosomal region examined is usually much smaller than the region examined by testing for linkage in families. Association is often more powerful than linkage in that a valid association may be detected in a sample when linkage is not detectable, even when the gene is playing only a modest role. Most association studies in the past were the population-based type where the allele frequencies of a group of unrelated cases were compared against those of a group of unrelated controls, and this is the only type of association study done with male sexual orientation. A potential pitfall of population-based case–control studies is that some populations, although they appear homogeneous to superficial examination, are in reality composed of different ancestral human groups, each one potentially with a different allele distribution at the studied loci. If one or more such groups is represented in a largely different proportion in one of the samples of an association dataset (i.e., either in the controls or in the cases), false negative or false positive association findings may easily arise due to methodological artifact.

Linkage Studies

The findings to date from behavior and molecular genetic studies predict that the genetics of male homosexuality will not be simple, and this prediction is consistent with the results of linkage research thus far. Several linkage analyses of male homosexuality to the X chromosome have previously been reported (Hamer, Hu, Magnuson, Hu, & Pattatucci, 1993; Hu et al., 1995; Rice, Anderson, Risch, & Ebers, 1999; Sanders et al., 1998). These studies have been largely based on the assumption that oligogenic (a "few" genes contribute) transmission was most likely, and therefore relied on the ASP method of linkage analysis. See Table 19.3 for a comparison and contrast of the samples examined. While the Xq28 chromosomal region in Hamer et al.'s (1993) study showed a significant linkage signal, with supporting data in a second dataset from the same group in

a follow-up study (Hu et al., 1995), it is still indeterminate (as in many complex traits) whether this finding represents a true positive. An independent group (Sanders et al., 1998) found inconclusive evidence of linkage to Xq28 in 1998, and a third group (Rice et al., 1999) found no support for linkage to Xq28 in 1999. Combining all four linkage studies, with respective affected sibling pair (ASP) sample sizes of 40 (1993), 33 (1995), 54 (1998), and 52 (1999), yields a "multiple scan probability" (MSP) of 0.00003 which is a suggestive p-value when considering all of the chromosomes, i.e., the entire genome (Sanders & Dawood, 2003). The replication MSP (excluding the original positive report from 1993) of 0.07 is at the level of a "trend" and thus not quite statistically significant. This pattern of results is one that has been predicted for complex traits with oligogenic inheritance on the basis of simulation studies: stochastic variation in the degree of co-segregation of any one locus with a trait, which produces variation in the magnitude of linkage findings across samples. Of course, the sample size of the individual linkage studies should be considered a major factor.

In the discussion below, we discuss the first major linkage study of male sexual orientation (Hamer et al., 1993) in greater detail, including a review of the main criticisms of this finding.

Hamer et al.'s (1993) report consisted of two major analyses: a pedigree study and a linkage study. First, they examined family pedigrees in a "randomly ascertained" sample of homosexual probands. As reported in Table 19.1, the probands had a high rate of gay brothers, 13.5%. Furthermore, their pedigrees showed an excess of gay uncles and male first cousins on the maternal side compared to the paternal side, though the difference was not significant. This excess was more pronounced in a subsequent analysis of an

Table 19.3 Sample characteristics of linkage male homosexuality samples

Study	Subject sources	ASPs	DNA	Tools	Inclusion	Exclusion
Hamer et al., 1993	Local clinics, local homophile organizations, homophile publications	40	Proband, homosexual brothers, parents, other siblings	Interview, Kinsey scale, family history	2 (exactly) homosexual brothers	Maximum of one lesbian per family, no male to male transmission
Hu et al., 1995	Local clinics, local homophile organizations, homophile publications	33	Proband, homosexual brothers, parents, other siblings	Interview, Kinsey scale, family history	2 (exactly) homosexual brothers	Maximum of two lesbians per family, no male to male transmission, no bisexuals
Sanders et al., 1998	Homophile organizations, homophile media	54	Proband, homosexual brothers, parents	Kinsey scale, family history	2 (or more) homosexual brothers	No known evidence of male to male transmission
Rice et al., 1999	Homophile media	52	Probands, homosexual brothers	Interview, family history	2 (or more) homosexual brothers	None stated

Number of ASPs are calculated by the n − 1 method for independent ASPs where n is the number of homosexual brothers per sibship. DNA refers to the family members from whom blood was sought for genetic analyses. Tools refer to the clinical methods used to assess the trait of sexual orientation. Major inclusion and exclusion criteria for families are listed.

additional sample of 38 families with two gay brothers. If there are genes for male homosexuality, then these families should be especially rich with them. The probands' maternal uncles and cousins (through maternal aunts; no gay cousins were sons of maternal uncles) had rates of male homosexuality of 10.3 and 12.9%, respectively, compared to rates of 1.5 and 3.1% for paternal uncles and cousins. This pattern of results is precisely what one would expect if an X-linked gene influenced male sexual orientation. In X-linked inheritance, males with the trait inherited the gene from their mothers, and hence have more maternal than paternal relatives with the trait.

Because of the suggestion of X-linkage, Hamer et al. (1993) then searched the X chromosome using linkage analysis. Specifically, they looked at the pairs of gay brothers without evidence of paternal transmission (e.g., they excluded a few cases in which the father may have been gay; the pairs analyzed in the linkage study included all eligible pairs analyzed in both pedigree studies). They examined 22 genetic markers distributed across the X chromosome in order to see if brothers concordant for homosexuality were also concordant for the markers. For chromosomal region Xq28 at the tip of the long arm of the X chromosome, 33 of 40 pairs of gay brothers shared all the markers. This was statistically different from the expected rate (20 of 40), suggesting that a gene influencing male sexual orientation lies within that chromosomal region.

Some skepticism has derived from concerns about Hamer et al.'s (1993) study. Risch, Squires-Wheeler, and Keats (1993) raised three main issues. First, they suggested that the pedigree finding, that gay men had an excess of gay maternal relatives, could be due to bias. They speculated that people may know more about their mothers' side of the family (presumably because mothers are more socially oriented, on average). Second, they argued that even if the finding of an increased rate of gay maternal relatives were true, it could be due to fertility patterns. Even if a gene for male homosexuality were autosomal (i.e., not X-linked) gay men are unlikely to have inherited it from their fathers, because men with the gene tend to be gay and gay men tend not to have children. Third, Risch et al. argued that Hamer et al. overestimated an important parameter, l, that reflects the increased prevalence of a trait in first-degree relatives compared to the background, or general population, rate. This parameter, which affects probability estimates, has not yet been precisely estimated in a large and careful study. Risch et al. chose values from available studies to yield the lowest plausible value of l, which would have rendered the linkage analysis statistically nonsignificant. In our view, the concerns raised by Risch et al. are worth the attention of future research, but are not fatal flaws in Hamer et al.'s study.

Other reasons for skepticism have less to do with Hamer et al.'s study than more general concerns about linkage analysis of genetically complex traits. By genetically complex traits, we mean those whose transmission patterns do not fit classic Mendelian patterns such as autosomal dominant or recessive, or X-linked dominant or recessive. All evidence suggests that male sexual orientation is inherited, if at all, in a complex manner. Linkage analysis has provided important breakthroughs for Mendelian traits, but it has also provided some false leads, especially for genetically complex behavioral traits (Risch & Merikangas, 1993). Indeed, to date not a single molecular finding concerning behavior has been widely accepted as valid by the scientific community, and several highly publicized findings have failed to replicate. This is in part because the number of studies examining any one trait has been relatively small. But it could also reflect the likely possibility that genes underlying behavior variation are typically of small effect and thus difficult to detect. Thus, it is especially important that linkage findings be replicated.

To date, at least three major replication attempts have been reported, including Hu et al. (1995) from Hamer's own research group. Rice et al. (1999) obtained pedigree information from 182 families with at least two gay brothers. They failed to find a significant excess of gay maternal uncles or cousins. In a subset of 41 sibling pairs, they also failed to replicate the finding of linkage to Xq28. Unlike Hamer et al. (1993), however, Rice et al. did not exclude brother pairs with strong evidence of paternal transmission, and it is unclear how many of their subjects would have met Hamer et al.'s (1993) inclusion criteria. Nevertheless, the failure to replicate either of Hamer et al.'s (1993) key findings surely diminishes the probability that they are correct.

In contrast, Hu et al. (1995) from Hamer's lab have reported a successful replication. In this second study, Hu et al. included data from heterosexual brothers as well as gay brothers and found that brothers' similarity for sexual orientation was statistically related to the sharing of Xq28 markers. That is, not only did gay brothers tend to share the markers, but gay-straight pairs tended not to share the markers. The magnitude of the genetic effect was smaller in Hamer's second study, however, and the result was barely statistically significant.

Because of the conflicting replication results, the status of the Xq28 linkage finding is unresolved. When studies are small, replications count more than failures to replicate. Nevertheless, larger studies will be needed to determine whether male sexual orientation is influenced by a gene in Xq28. Currently, the largest linkage study to date of male sexual orientation is underway with DNA samples being collected from a target sample of 1000 families with two or more siblings concordant for homosexuality (Sanders et al., 2005), and researchers in the field eagerly await the results of this large-scale study.

Association Studies

Association studies explore the relation between genetic variation at a specific locus and phenotypic variation. Association studies require that one has a very specific hypothesis, in contrast to linkage studies, which may search the entire genome and examine genetic markers rather than genes. Two association studies have been performed to date for male homosexuality. Macke et al. (1993) used a population-based case–control method to examine DNA sequence variation in the androgen receptor gene, reasoning that some variants may affect sexual differentiation of the brain. This study employed a sample of 197 homosexual males and about 213 unselected (with respect to sexual orientation) male controls with variants of the androgen receptor located on the X chromosome (but not at Xq28) and found no significant differences in the distributions of mutations in homosexual and heterosexual men (i.e., no evidence for association). This gene was selected for examination partly due to its location on the X chromosome since there is some evidence for excess maternal transmission of male homosexuality, which would be consistent with X-linked transmission. However, using linkage analysis, the authors showed that sibling pairs concordant for homosexuality were no more likely than chance to share the same androgen receptor allele.

Another reason the androgen receptor was chosen for examination was not due to its position but rather due to its function, which is to transduce messages from androgens ("male" hormones) to the nucleus of the cell, thus affecting other genes responsive to androgens. In general, many different "candidate" genes may be nominated for examination by means of association testing, but the strength of their candidacy is often in question (relative to any other gene expressed at some point in the brain, which most are). For example, studies in animal models where gene variation may be introduced and the effects systematically examined may propose genes to examine in humans, but there is an unresolved question regarding the validity of extrapolating complex behaviors from species sometimes as different from humans as fruit flies.

Dupree, Mustanski, Bocklandt, Nievergelt, and Hamer (2004) conducted a more recent association study of male sexual orientation to investigate whether differences in the gene encoding the aromatase enzyme influence sexual orientation in men. Aromatase cytochrome P450 (CYP19) is necessary for the conversion of androgens to estrogens and plays an important role in the sexual differentiation of the brain in rodents. This study found no differences between heterosexual and homosexual men in the expression of aromatase mRNA by microarray analysis, suggesting that variation in this gene is not likely to be a major factor in the development of individual differences in male sexual orientation (Dupree et al., 2004).

Additional Molecular Genetic Studies

Two recent studies (Bocklandt, Horvath, Vilain, & Hamer, 2006; Mustanski et al., 2005) have provided additional evidence supporting a heritable component to male and female sexual orientation, and we discuss both studies in further detail below.

Mustanski et al. (2005) conducted the first full-genome scan of sexual orientation in men by genotyping 456 individuals from 146 families with two or more gay brothers with 403 microsatellite markers at 10 cM intervals. They failed to replicate linkage to Xq28 in the full sample; however, they reported three new regions which approached the criteria for near significance (7q36) and for suggestive linkage (8p12 and 10q26). These chromosomal regions may be used in future replication studies with new samples and denser linkage maps.

Bocklandt et al. (2006) measured X chromosome inactivation in a sample of 97 mothers of homosexual men and 103 age-matched control women without gay sons. They reported that extreme skewing of X-inactivation was significantly higher in mothers of gay men (13%) compared to controls (4%) and increased in mothers with two or more gay sons (23%). These findings support a role for the X chromosome in regulating sexual orientation in some homosexual men (Bocklandt et al., 2006), although it is unclear whether these unusual X-inactivation patterns influence the development of sexual orientation in sons via a direct mechanism such as the fraternal birth order effect (Blanchard, 1997) or whether it is the indirect result of a different mechanism. These results have yet to be replicated.

Methodological Issues

The most common criticism aimed at the studies listed in Table 19.2 concerns the "equal environments assumption" that the trait-relevant environment is no more similar for MZ twins than for DZ twins or adoptive siblings. A frequent objection to human twin studies (e.g., Lewontin, Rose, & Kamin, 1984) is that parents treat MZ twins especially similarly and that this similar treatment, rather than the twins' similar genotype, could explain their similar behavior. Indeed, MZ twins are more likely to have been dressed alike and to have shared the same room as children, among other things. The question is whether such treatment makes them more similar, and the evidence suggests that this is not the case, at least for traits studied so far (Plomin, Defries, & McClearn, 1990). For example, MZ twins whose parents make an effort to treat them alike do not behave more similarly than do MZ twins whose parents make an effort to treat them differently. MZ twins whose parents mistakenly believe that they are DZ twins are as similar as they should be based on their true zygosity. It is true (and unfortunate) that the equal environments assumption has not been directly studied in the context of sexual orientation, but it is also true that existing evidence does not contradict the equal environments assumption.

A more serious potential problem concerns ascertainment bias. Ideally, one could recruit probands by interviewing every member of a well-defined population of, say, gay men and asking them if they were twins. Psychiatric genetics has been able to ascertain twins systematically by interviewing consecutive psychiatric admissions, but this strategy was obviously unavailable to those studying homosexuality. The problem with ascertaining twins via advertisements is that self-selection factors are likely to distort results. The most likely way in which this would occur is that gay men whose twins are also gay would be more willing to volunteer than gay men with heterosexual twins (e.g., because the latter might fear conflict from their twins). Kendler and Eaves (1989) have called this kind of bias "concordance-dependent ascertainment bias." This type of bias inflates concordances compared to the population rates, though it does not lead to spurious findings of heritability. Spurious findings could be obtained, however, if concordance-dependent bias was stronger for MZ than for DZ twins. Although there is no evidence that this is so, it cannot presently be excluded. A systematic ascertainment strategy is the most crucial methodological goal for future population genetic studies of sexual orientation.

One final limitation of twin and family studies is worth emphasizing. Even accepting its validity, the evidence reviewed so far is uninformative regarding proximate etiological mechanisms. Genetic evidence does not necessarily support a neuroendocrine explanation, for example. One could envision a host of other genetic pathways to homosexuality. But available studies cannot distinguish among them. Molecular strategies that can identify specific genes for sexual orientation will be much more useful in elaborating the developmental pathways from genes (if they exist) to behavior.

Environmental Influences

The environmental contribution to phenotypic variance is directly comparable to heritability and may be broken down into two subcomponents or parameters: the proportion of variance that is "shared" by family members and the other "nonshared" or "unique". Shared environmental influences are those that make members of a family similar to each other (such as having the same parents and growing up in the same house), while the rest of the variance is described as everything else that siblings do not share.

The most convincing result of the twin studies to date is that environment is sometimes an important determinant of sexual orientation. If it were not so, MZ twins would always have the same sexual orientations, but about half the time, homosexual probands have heterosexual twins (and this is probably an underestimate). Experiences or developmental antecedents that differed between homosexual and heterosexual cotwins would be promising candidates to illuminate relevant environmental factors. Thus, perhaps the most intriguing possible application of the twin method concerns the study of discordant MZ twins (Reiss, Plomin, & Hetherington, 1991). For example, we found in both our male and female studies (Bailey et al., 2000; Dawood, Pillard, Horvath, Revelle, & Bailey, 2000) that discordant MZ twins and nontwin siblings also reported quite different childhood experiences. On questionnaire measures, the homosexual twins reported more gender-atypical behavior, and often in interviews twins mentioned this as an early indication of an important difference between them. This suggests that in many cases, relevant environmental factors operate during childhood. This would contrast, for example, with factors such as adult sexual experiences that have sometimes been alleged to be important in determining sexual orientation (Dawood et al., 2000).

It is important to emphasize that "environment" as construed by behavior geneticists differs from "environment" as it is understood by most laypeople. Environment comprises all causes of variation that are not genetic, where genetic is understood in the specific sense encoded in germline DNA. (Even somatic mutations, which are not typically shared by close relatives, are environmental in this sense.) There is a biological environment – random developmental and intrauterine factors, illness, diet, injury, etc. – as well as a psychosocial environment. Distinguishing between these kinds of environmental factors for sexual orientation will require that specific testable theories be offered.

Results of studies in Table 19.2 suggest that the most important environmental factors are those that typically differ even between MZ twins reared in the same family. Thus, for example, cold, distant fathers are unlikely to be important, because it is unlikely that a father would behave in a cold and distant manner toward only one MZ twin. In contrast, and contrary to intuition, biological differences between MZ twins are not uncommon (Torrey, 1994; Turner, 1994). For example, the twin transfusion syndrome, in which twins receive unequal blood supply, can cause substantial differences in MZ twins' health (and indeed mortality of one twin is common). MZ twins with congenital brain anomalies typically have normal cotwins. Molenaar, Boomsma, and Dolan (1993) have argued that much of apparent within-family environmental variation may be attributable to poorly understood and effectively random developmental processes. There is now considerable speculation, and some evidence, that epigenetic phenomena including differential DNA methylation might be one such class of random process (Oates et al., 2006).

The Fraternal Birth Order Effect

Several excellent reviews of biological research on human sexual orientation have recently been published (Mustanski, Chivers, & Bailey, 2002; Rahman, 2005) which can provide more comprehensive reviews of neuroendocrine and neurodevelopmental theories of sexual orientation. We will note here, however, that perhaps the most replicated finding in sexual orientation research is the fraternal birth order effect in homosexual men (Blanchard, 1997) whereby homosexual men have a greater number of older brothers than heterosexual men do, in diverse community and population-level samples. The estimated odds of being homosexual increase by approximately 33% with each older brother, and statistical modeling using epidemiological procedures suggest that one in seven homosexual men may owe their sexual orientation to the fraternal birth order effect (Cantor, Blanchard, Paterson, & Bogaert, 2002). While purely genetic factors could not explain this effect, recent evidence suggests that the effect is related to prenatal events (Blanchard, 2004; Bogaert, 2006). Currently, a maternal immune response to male-specific, Y-linked antigens is the most plausible explanation for this effect (Blanchard & Bogaert, 1996) which becomes stronger with each male pregnancy. However, empirical studies supporting this hypothesis have yet to be conducted.

Directions for Future Research

During the past two decades, a growing body of evidence has accumulated suggesting that familial and genetic factors affect both male and female sexual orientation. The genetic evidence is substantially stronger for male than for female sexual orientation, and multiple genes could well contribute significant influences on the development of sexual orientation.

Although ongoing studies investigating genetic sources of variation in sexual orientation will contribute a critical aspect for understanding the origins and development of sexual orientation, perhaps the most interesting topic for future research in this area lies in studying nonshared environmental sources of variation and the epigenetic relationship between environmental and genetic factors.

Plomin (1994) has suggested that longitudinal studies and behavior genetic methodologies will be useful in studying the effects of nonshared environments. The study of discordant monozygotic twins, who are genetically identical, can be particularly important since nonshared environmental factors are responsible for monozygotic twins discordant for sexual orientation. For example, future molecular genetic studies could examine epigenetic modifications of DNA between twins using a genomewide screen of differentially methylated regions to identify potential discrete differences between homosexual and heterosexual MZ twins.

Discordant MZ twins could also be used to examine specific environmental influences on the development of sexual orientation. For example, there is strong evidence from both retrospective and prospective studies (Bailey & Zucker, 1995) supporting an association between childhood gender nonconformity and adult sexual orientation. Though significantly larger for males, the effect sizes reported for both sexes are among the largest ever reported in the realm of sex-dimorphic behaviors. Male and female identical twins discordant for sexual orientation might well differ in other gender-related traits, such as childhood gender nonconformity which is also significantly heritable for both men and women (Bailey et al., 2000). If so, it suggests that the nonshared or unique environmental influences that lead to different sexual orientations may also contribute to the development of other gender-related traits, including childhood gender nonconformity.

At this stage, few conclusions can be drawn with certainty regarding genetic and environmental determinants of sexual orientation. Important methodological research innovations hold the most potential for furthering our knowledge on the origins and development of human sexual orientation. Future research should also attempt to integrate different biological approaches in order to provide valuable information about the specific pathways by which genes exert their influence on sexual orientation and its correlates.

References

ACSF Investigators. (1992). AIDS and sexual behaviour in France. *Nature, 360,* 407–409.
Bailey, J. M., & Bell, A. P. (1993). Familiality of male and female homosexuality. *Behavior Genetics, 23,* 313–322.
Bailey, J. M., & Benishay, D. S. (1993). Familial aggregation of female sexual orientation. *American Journal of Psychiatry, 150,* 272–277.
Bailey, J. M., Dunne, M. P., & Martin, N. G. (2000). Genetic and environmental influences on sexual orientation and its correlates in an Australian twin sample. *Journal of Personality and Social Psychology, 78,* 524–536.
Bailey, J. M., & Pillard, R. C. (1991). A genetic study of male sexual orientation. *Archives of General Psychiatry, 48,* 1089–1096.
Bailey, J. M., Pillard, R. C., Dawood, K., Miller, M. B., Farrer, L. A., Trivedi, S., et al. (1999). A family history study of male sexual orientation using three independent samples. *Behavior Genetics, 29,* 79–86.
Bailey, J. M., Pillard, R. C., Neale, M. C., & Agyei Y. (1993). Heritable factors influence sexual orientation in women. *Archives of General Psychiatry, 50,* 217–223.

Bailey, J. M., Willerman, L., & Parks, C. (1991). A test of the maternal stress theory of human male homosexuality. *Archives of Sexual Behavior, 20*, 277–293.

Bailey, J. M, & Zucker, K. J. (1995). Childhood sex-typed behavior and sexual orientation: A conceptual analysis and quantitative review. *Developmental Psychology, 31*, 43–55.

Billy, J. O. G., Tanfer, K., Grady, W. R., & Klepinger, D. H. (1993). The sexual behavior of men in the United States. *Family Planning Perspectives, 25*, 52–60.

Blanchard, R. (1997). Birth order and sibling sex ratio in homosexual versus heterosexual males and females. *Annual Review of Sex Research, 8*, 27–67.

Blanchard, R. (2004). Quantitative and theoretical analyses of the relation between older brothers and homosexuality in men. *Journal of Theoretical Biology, 230*, 173–187.

Blanchard, R., & Bogaert, A. F. (1996). Homosexuality in men and number of older brothers. *American Journal of Psychiatry, 153*, 27–31.

Bocklandt, S., Horvath, S., Vilain, E., & Hamer, D. H. (2006). Extreme skewing of X chromosome inactivation in mothers of homosexual men. *Human Genetics, 118*, 691–694.

Bogaert, A. F. (2006). Biological versus nonbiological older brothers and men's sexual orientation. *Proceedings of the National Academy of Sciences USA, 103*, 10531–10532.

Buhrich, N., Bailey, J. M., & Martin, N. G. (1991). Sexual orientation, sexual identity, and sex-dimorphic behaviors in male twins. *Behavior Genetics, 21*, 75–96.

Cantor, J. M., Blanchard, R., Paterson, A. D., & Bogaert, A. F. (2002). How many gay men owe their sexual orientation to fraternal birth order? *Archives of Sexual Behavior, 31*, 63–71.

Dawood, K., & Bailey, J. M. (2000). The genetics of human sexual orientation. In: J. L. Rodgers, D. C. Rowe, & W. B. Miller (Eds.), *Genetic Influences on Human Fertility and Sexuality: Theoretical and Empirical Contributions from the Biological and Behavioral Sciences*. Boston, MA: Kluwer Academic Publishers.

Dawood, K., Pillard, R. C., Horvath, C., Revelle, W., & Bailey, J. M. (2000). Familial aspects of male homosexuality. *Archives of Sexual Behavior, 29*, 155–163.

Dupree, M. G., Mustanski, B. S., Bocklandt, S., Nievergelt, C., & Hamer, D. H. (2004). A candidate gene study of CYP19 (aromatase) and male sexual orientation. *Behavior Genetics, 34*, 243–250.

Hamer, D. H., Hu, S., Magnuson, V. L., Hu, N., & Pattatucci, A. M. L. (1993). A linkage between DNA markers on the X chromosome and male sexual orientation. *Science, 261*, 321–327.

Hu, S., Pattatucci, A. M. L., Patterson, C., Li, L., Fulker, D., Cherny, S., et al. (1995). Linkage between sexual orientation and chromosome Xq28 in males but not females. *Nature Genetics, 11*, 248–256.

Heston, L. L., & Shields, J. (1968). Homosexuality in twins: A family study and a registry study. *Archives of General Psychiatry, 18*, 149–160.

Johnson, A. M., Wadsworth, J., Wellings, K., Bradshaw, S., & Field, J. (1992). Sexual lifestyles and HIV risk. *Nature, 360*, 410–412.

Kallmann, F. J. (1952). Twin and sibship study of overt male homosexuality. *American Journal of Human Genetics, 4*, 136–146.

Kendler, K. S., & Eaves L. J. (1989). The estimation of probandwise concordance in twins: the effect of unequal ascertainment. *Acta Geneticae et Medicae Gemellologiae, 38*, 253–270.

Kendler, K. S., Thornton, L. M., Gilman, S. E., Kessler, R. C. (2000). Sexual orientation in a U.S. national sample of twin and nontwin sibling pairs. *American Journal of Psychiatry, 157*, 1843–1846.

King, M., & McDonald, E. (1992). Homosexuals who are twins: A study of 46 probands. *British Journal of Psychiatry, 160*, 407–409.

Kirk, K. M., Bailey, J. M., Dunne, M. P., & Martin, N. G. (2000) Measurement models for sexual orientation in a community twin sample. *Behavior Genetics, 30*, 345–356.

Laumann, E. O., Gagnon, J. H., Michael, R. T., & Michaels, S. (1994). *The social organization of sexuality: Sexual practices in the United States.* Chicago: University of Chicago Press.

Lewontin, R. C., Rose, S., & Kamin, L. J. (1984). *Not in our genes.* New York: Pantheon Books.

Macke, J. P., Bailey, J. M., King, V., Brown, T., Hamer, D., & Nathans, J. (1993). Sequence variation in the androgen receptor gene is not a common determinant of male sexual orientation. *American Journal of Human Genetics, 53*, 844–852.

Molenaar, P. C., Boomsma, D. I., & Dolan, C. V. (1993). A third source of developmental differences. *Behavior Genetics, 23*, 519–524.

Mustanski, B. S., Chivers, M. L., & Bailey, J. M. (2002). A critical review of recent biological research on human sexual orientation. *Annual Review of Sex Research, 13*, 89–140.

Mustanski, B. S., Dupree, M. G., Nievergelt, C. M., Bocklandt, S., Schork, N. J., & Hamer, D. H. (2005). A genomewide scan of male sexual orientation. *Human Genetics, 116*, 272–278.

Oates, N. A., van Vliet, J., Duffy, D. L., Kroes, H. Y., Martin, N. G., Boomsma, D. I., et al. (2006). Increased DNA methylation at the *AXIN1* gene in a monozygotic twin from a pair discordant for a caudal duplication anomaly. *American Journal of Human Genetics, 79*, 155–162.

Pattatucci, A. M. L., & Hamer, D. H. (1995). Development and familiarity of sexual orientation in females. *Behavior Genetics, 25*, 407–420.

Pillard, R. C. (1990). The Kinsey Scale: Is it familial? In Homosexuality/Heterosexuality: Concepts of sexual orientation. The Kinsey Institute series, Vol. 2, New York, USA: Oxford University Press.

Pillard, R. C., & Weinrich, J. D. (1986). Evidence of familial nature of male homosexuality. *Archives of General Psychiatry, 43*, 808–812.

Plomin, R. (1994). Genetic research and identification of environmental influences. *Journal of Child Psychology and Psychiatry, 35*, 817–834.

Plomin, R., Defries, J. C., & McClearn, G. E. (1990). Behavioral Genetics. A primer (2nd ed.). New York, W. H. Freeman & Company.

Rahman, Q. (2005). The neurodevelopment of human sexual orientation. *Neuroscience and Biobehavioral Reviews, 29*, 1057–1066.

Reich, T., Cloninger, C. R., & Guze, S. B. (1975). The multifactorial model of disease transmission: I. Description of the model and its use in psychiatry. *British Journal of Psychiatry, 127*, 1–10.

Reiss, D., Plomin, R., & Hetherington, E. M. (1991). Genetics and psychiatry: An unheralded window on the environment. *American Journal of Psychiatry, 148*, 283–291.

Rice, G., Anderson, C., Risch, N., & Ebers, G. (1999). Male homosexuality: absence of linkage to micro satellite markers at Xq28. *Science, 284*, 665–667.

Risch, N., & Merikangas, K. R. (1993). Linkage studies of psychiatric disorders. *European Archives of Psychiatry and Clinical Neuroscience, 243*, 143–149.

Risch, N., Squires-Wheeler, E., & Keats, B. J. B. (1993). Male sexual orientation and genetic evidence. *Science, 262*, 2063–2065.

Rosenthal, D. (1970). Genetic theory and abnormal behavior. New York: McGraw-Hill.

Sanders, A. R, Cao, Q., Zhang, J., Badner, J. A., Goldin, L. R., Guroff, J. J., et al. (1998). Genetic linkage study of male homosexual orientation. Poster presentation at the 151st Annual Meeting of the American Psychiatric Association . Toronto, Ontario, Canada.

Sanders, A. R., & Dawood, K. (2003). Sexual orientation. In *Nature Encyclopedia of Life Sciences*. London: Nature Publishing Group.

Sanders, A. R., Dawood, K., Krishnappa, R., Kolundzija, A., Murphy, T. F., & Bailey, J. M. (2005). Molecular Genetic Study of Sexual Orientation. Poster presentation at the International Behavioral Development Symposium: Biological Basis of Sexual Orientation, Gender Identity, and Sex-Typical Behavior. Minot, ND.

Torrey, E. F. (1994). Are identical twins really identical? *Parabola, 19*, 18–21.

Turner, W. J. (1994). Comments on discordant monozygotic twinning in homosexuality. *Archives of Sexual Behavior, 23*, 115–119.

Whitam, F. L., Diamond, M., & Martin, J. (1993). Homosexual orientation in twins: A report on 61 pairs and three triplet sets. *Archives of Sexual Behavior, 22*, 187–206.

Chapter 20

Some Guidelines for Defining Personality Differences in Rats

Peter Driscoll, Alberto Fernández-Teruel, Maria G. Corda, Osvaldo Giorgi, and Thierry Steimer

Introduction

Whereas many behavioral studies with rats have been traditionally concerned with tests and/or models attempting to deal with subjects such as anxiety, depression, hyperactivity, alcoholism and drug abuse, as they pertain to the human conditions, critical and integrated analyses of the vast amounts of information which have been accumulated, and an application of the same to the realm of personality traits, are long overdue. One such application, for example, might deal with the etiology of substance use and abuse toward which different animal models, considered together, can undoubtedly play a decisive role, especially as genetic factors are known to be extensively involved in the temperament differences underlying these phenomena in both rats and humans. The major contributions of such models toward this goal will, of course, pertain to unraveling the fundamental neurochemical processes involved and to deciphering their genetic origins.

Selected rat lines are particularly suitable to this endeavor, as the integration of all components studied is kept intact in such models (as compared to knockouts and knock-ins, surgical/chemical lesions, etc.). This non-invasive approach reveals naturally selected regulators of behavior by focusing on (e.g., brain) traits which are, basically, evolutionary. Many studies of selective breeding for behavioral divergence show a rapid response, indicating that there must be hereditary factors through which a differential set of alleles can bias behavioral traits, without affecting biological fitness (Lipp et al., 1989). At the same time, research done with non-selected stocks of rodents has, at least until recently, been prevalent in providing much basic data. The information furnished by these studies can be especially valuable when commercially available stocks are compared within the same study or when one takes into account the different suppliers of the animals used by the various laboratories. A major problem which limits the usefulness of most review articles, even those dealing with genetics, is that they have not paid attention to this last point.

The history of psychogenetically selected rat lines/strains goes back many decades (Broadhurst, 1960; Hall, 1938; Tryon, 1929), finally establishing a firm foothold in the 1960s with the Roman (Bignami, 1965) and Syracuse (Brush, 1979) high- and low-avoidance rat lines and the Maudsley reactive and nonreactive rat strains (Broadhurst, 1975). Due to the many similarities between the Roman and Syracuse stocks (for these as well as for the few exceptions, see Brush, 1991), the latter will not be considered further here, due to the limited space available. For a detailed description of those rats, the reader should consult a recent review article (Brush, 2003). The Maudsley rats, which will only be considered briefly later in this chapter, have also been recently reviewed by Blizard and Adams (2002). We will concentrate at the outset on the contributions of the Roman high- (RHA) and low-avoidance (RLA) rat lines/strains to subjects mentioned in the opening sentence, plus that of novelty (sensation) seeking. Reference to these rats will, where appropriate, periodically follow throughout the chapter, during the ensuing description of many of the other rat models being actively utilized at present, or at least recently, in projects which are meaningful for the personality concept in rats. Behavioral tests, neuroanatomical/neurochemical comparisons, physiological parameters and differences/similarities among laboratories and suppliers will be discussed as they relate to the subjects at hand, within this concept.

RHA and RLA Rats: Background

The Swiss sublines of RHA and RLA rats were established in 1972, being derived from the original Roman stock. They have been selected and bred for, respectively, the rapid vs. non-acquisition of two-way, active avoidance

P. Driscoll (✉)
Institute for Animal Science, ETHZ, Schorenstrasse 16, 8603 Schwerzenbach, Switzerland
e-mail: peter-driscoll@ethz.ch

Y.-K. Kim (ed.), *Handbook of Behavior Genetics*,
DOI 10.1007/978-0-387-76727-7_20, © Springer Science+Business Media, LLC 2009

behavior in the shuttle box. It soon became evident that avoidance acquisition was not due to "learning ability" but due, primarily, to emotional factors (Driscoll & Bättig, 1982; reviewed in detail by Escorihuela, Tobeña, Driscoll & Fernández-Teruel, 1995). Indeed, as has been similarly demonstrated with other colonies of high- and low-avoidance rats (Brush, 1991; Willig, M'Harzi, Bardelay, Viet, & Delacour, 1991a; Willig, M'Harzi, & Delacour, 1991b), RLA rats are actually better passive avoiders and perform better in several tasks not involving electric shock, such as spatial learning, object discrimination and lever pressing for food reward than do RHA rats (Aguilar, Escorihuela, Gil, Tobeña, & Fernández-Teruel, 2002a; Driscoll & Bättig, 1982; Driscoll et al., 1995; Fernández-Teruel, Escorihuela, Castellano, González, & Tobeña, 1997a; Nil & Bättig, 1981; Zeier, Baettig, & Driscoll, 1978). An increased emotionality of RLA as compared to RHA rats has been shown on many occasions in relation to higher levels of stressor-induced freezing and grooming behavior, increased shock-induced suppression of drinking, hyponeophagia and higher stressor-induced, acute increases in plasma levels of ACTH, prolactin, renin, aldosterone and, at times, corticosterone (Aubry et al., 1995; Castanon, Dulluc, LeMoal & Mormède, 1992; Castanon & Mormède, 1994; Corda, Piras, Valentini, Scano, & Giorgi, 1998; D'Angio, Serrano, Driscoll, & Scatton, 1988; Driscoll, 1986; Ferré et al., 1995; Gentsch, Lichtsteiner, & Feer, 1981; Gentsch, Lichtsteiner, Driscoll, & Feer, 1982; Imada, 1972; Roozendaal, Wiersma, Driscoll, Koolhaas, & Bohus, 1992; Shephard & Broadhurst, 1983; Steimer, Fleur, & Schulz, 1997a; Steimer & Driscoll, 2003, 2005; Walker, Rivest, Meaney, & Aubert, 1989; Walker, Aubert, Meaney, & Driscoll, 1992). Whereas RLA, but not RHA, rats have shown bradycardia in response to a conditioned emotional stressor (Roozendaal et al., 1992), they have shown tachycardia of greater magnitude and duration than RHA rats when submitted to several unconditioned stressors (D'Angio et al., 1988).

All of the evidence in favor of RLA rats being considered as an "anxiety" model does not preclude the possibility, however, of RHA rats actually being the more relevant model, i.e., as novelty or "sensation" seekers. This eventuality may indeed apply to all laboratories questing after the "gene for anxiety" (e.g., Fernández-Teruel et al., 2002b; Landgraf et al., 2007; Ramos, Moisan, Chaouloff, Mormède, & Mormède, 1999), as "every coin has two sides". A composite model for RLA/RHA rats proposed by Steimer et al. (1997a) and recently elaborated upon (Steimer & Driscoll, 2003) postulates that anxiety and impulsiveness (novelty seeking) lie at opposite poles, the former being associated with a combination of increased emotional reactivity and a passive (reactive) coping style and the latter with a more active (proactive) coping style and decreased emotional reactivity. Both styles of coping, of course, are aimed at successful control of

the environment by opposing personality types (Koolhaas et al., 1999).

Sensation seeking (SS) in humans and cats is characterized by high levels of exploratory and risk-taking behaviors, impulsiveness, aggression and (in humans) a tendency to experiment with a variety of drugs such as ethanol (ETH), opiates and stimulants (Siegel, 1997). For example, SS cats were less successful in controlling bar pressing behavior during an inhibitory DRL task, displaying more exploratory activity and impulsive (quicker) bar pressing (Saxton, Siegel, & Lukas, 1987). That RHA rats are more active and impulsive (including in DRL behavior), more inquisitive, more aggressive and show an increased preference for ETH than RLA rats has been well documented over the years (e.g., Castanon, Dulluc, LeMoal, & Mormède, 1994; Driscoll, Woodson, Fuemm, & Bättig, 1980; Driscoll & Martin, 1987; Driscoll, Cohen, Fackelman, & Bättig, 1990a; Escorihuela et al., 1999; Fernández-Teruel et al., 1992, 1997a; Fernández-Teruel et al., 2002a; Gentsch, Lichtsteiner, & Feer, 1991; Giorgi et al., 1996; Guitart-Masip et al., 2006a; Haney, Castanon, Cador, LeMoal, & Mormède, 1994; Meerlo, Overkamp, & Koolhaas, 1997; Pisula, 2003; Razafimanalina, Mormède, & Velley, 1996; Zeier et al., 1978). A rather spectacular discovery was that RHA rats were also shown to be visual evoked potential (VEP) augmenters, whereas RLA rats (as well as the Wistar rats used for comparison in those studies) were VEP reducers (Siegel, Sisson, & Driscoll, 1993). A subsequent study demonstrated that this difference occurs at the cortical level (Siegel, Gayle, Sharma, & Driscoll, 1996). Furthermore, VEPs recorded from humans, cats and rats have essentially the same waveform, and the same early component (P1) augments or reduces in all three species (reviewed by Siegel, 1997).

RHA and RLA Rats: Foreground

As recently summarized by Giorgi, Piras, and Corda (2007), there is considerable evidence that individuals with high scores for sensation/novelty seeking are at increased risk for using drugs of abuse (Bardo, Donohew, & Harrington, 1996; Verheul & van den Brink, 2000) and that behavioral sensitization (the progressive augmentation of the motor activation induced by psychostimulants and opiates resulting from repeated drug administration) is critically involved in some of the persistent features of addiction, such as drug craving, compulsive drug-seeking behavior and the propensity to relapse (Robinson & Berridge, 2001; Everitt & Wolf, 2002). Genetic factors account for 40–60% of risk in alcoholism, and similar rates of heritability occur in addiction to opiates and psychostimulants (Nestler, 2000; Reich, Hinrichs, Culverhouse, & Bierut, 1999). Most recently, in this connection, it has been shown that RHA rats display a higher propensity to self-administer cocaine than do RLA rats, this being consistent with the views mentioned above regarding behavioral sensitization (Fattore, Piras, Corda, & Giorgi, 2008).

Brain microdialysis studies have shown that the acute administration of morphine (MOR – 0.5 mg/kg s.c.), cocaine (COC – 5.0 mg/kg i.p.) or amphetamine (AMP – 0.15 mg/kg i.p.) induces a larger increase in dopamine (DA) output in the nucleus accumbens (NAc) shell than in the NAc core of RHA rats, whereas no significant differences in DAergic responses are observed between NAc compartments of RLA rats. This line-related difference in the responsiveness of mesolimbic DAergic projections is associated with a more robust, drug-induced increase in ambulatory and stationary (rearing, sniffing and licking/gnawing) activities in RHA than in RLA rats (Giorgi et al., 1997; Giorgi, Piras, Lecca, & Corda, 2005a; Lecca, Piras, Driscoll, Giorgi, & Corda, 2004). The role of the NAc compartments and their projections in motivated behaviors, as well as in the rewarding properties of psychostimulants and opiates and drug-stimulated locomotion, is described in detail by Giorgi et al. (2007; see also Bardo et al., 1996).

Behavioral sensitization to opiates and psychostimulants in rodents, also covered by Giorgi et al. (2007), is characterized by the progressive increase in ambulatory activity and in the frequency of more focused, non-ambulatory behaviors following repeated drug treatments, all of which are believed to reflect long-lasting adaptations in neural circuits involved in motivation and reward (Everitt & Wolf, 2002; Li, Acerbo, & Robinson, 2004; Robinson & Berridge, 2001). In RHA rats, a MOR challenge 21 days after withdrawal from repeated MOR treatments produced a significantly larger motor activation than both RHA control groups and all RLA groups (Piras, Lecca, Corda, & Giorgi, 2003). Similar results have been obtained with COC (Giorgi, Piras, Lecca, & Corda, 2005b; Haney et al., 1994) and AMP (Corda et al., 2005). In keeping with these findings, it was also seen that repeated treatment with MOR or COC produced differential modifications in neurochemical responses to a subsequent drug challenge, with RHA rats being thus handled showing a larger increment in DA output in the NAc core (but not shell) than both RHA control groups and all RLA groups, for both drugs (Giorgi et al., 2007). Once again, similar results were obtained with AMP (Giorgi et al., 2005a).

It was concluded that experimental subjects that are more responsive to the acute effects of addictive drugs, such as RHA rats, are also more susceptible to develop behavioral sensitization upon repeated drug exposure (Giorgi et al., 2007), a situation that resembles clinical observations in humans (O'Brien & McLellan, 1996). Given the proposed role of sensitization in compulsive drug intake (Robinson & Berridge, 2001), the results with RHA and RLA rats,

including the most recent ones (Fattore et al., 2008), support the view that individual susceptibility to succumb to drug addiction is influenced by genetically determined, functional patterns of the mesocortical and mesolimbic DAergic system and associated neural circuits encoding brain reward and goal-directed behaviors. Giorgi et al. (2007) illustrate this by presenting an elaborate graphic model depicting putative, adaptive changes in the DA- and GABAergic neural circuitry of the NAc core and shell, including associated projections to and from other brain regions, some of which will be dealt with later in the chapter.

Spontaneously Hypertensive, and Related, Rats

The spontaneously hypertensive rat (SHR) has long been considered by some to be an adequate model for attention-deficit hyperactivity disorder (AD/HD), showing similarities in deficient sustained attention, overactivity, impulsiveness and other symptoms (Sagvolden, 2000), although some reservations have been expressed regarding therapeutic dosage effects and the production of stereotypic behaviors, for example, when animals are compared to humans (Solanto, 2000). Davids, Zhang, Tarazi, and Baldessarini (2003; see also Hendley, 2000) were concerned with both the hypertension of SHRs and possible beneficial effects of drugs on blood pressure as potential confounding factors as well as a lack of sex differences in that model (in humans, AD/HD is more prevalent in males). Another problem is that the clinical disorder is far from being fully understood in humans. Although, as we have seen, hyperactivity, impulsivity and certain "learning deficits" also exist in RHA rats and, even in accordance with the requirements of Solanto (2000), a low dose of AMP (2.0 mg/kg) has actually been noted to reduce shuttle box avoidance and activity in that line (Driscoll, 1986), we have not (yet) considered pursuing the AD/HD gambit further with the RHA/RLAs. Additional similarities have been found between RHA and SHR rats (see also the next paragraph). RHAs have, actually, often shown a more pronounced hyperactivity vs. RLAs as well as other groups of rats in the open field (OF) test (see e.g., Haney et al., 1994) than have SHRs vs. WKYs, in regard to both "reactive" and "spontaneous" locomotor activity (reviewed by Gentsch, Lichtsteiner, & Feer, 1988).

Despite the hypertension factor (which often even leads to early death), SHRs have also had further application in the field of behavior genetics. For example, Gentsch et al. (1988) found that SHR resembled RHA rats, vs. RLA and Wistar–Kyoto (WKY) rats from Roche-Füllinsdorf, in several parameters relating to reduced "emotionality", such as rapidly acquiring two-way, active avoidance behavior and being more active in an OF, in which both RHAs and SHRs also defecated less and displayed an attenuated elevation of corticosterone. RHA and SHR rats also had heavier thymus glands than their RLA and WKY counterparts, it being known that increased levels of circulating adrenal hormones diminish thymus size (e.g., Steimer & Driscoll, 2003). Castanon, Hendley, Fan, & Mormède (1993) added to these findings, also using WKHA rats, which Hendley (2000, for review) had developed by crossing SHR and WKY rats, thus producing rats behaviorally similar to SHRs but without elevated blood pressure. As they had reported previously with the Roman lines, i.e., RLA > RHA (Castanon et al., 1992), they found that the prolactin response to novelty most clearly distinguished those strains, i.e., WKY > SHR > WKHA (Castanon et al., 1993). In addition, using both sexes of each of these three strains, Cailhol and Mormède (2000) demonstrated that female SHR rats consumed the highest amounts of ETH, as is also the case with RHA vs. RLA rats.

SHR and Other Commercially Available Rats as Models

The same laboratory subsequently examined SHR, WKY, Brown Norway (BN), Wistar–Furth (W–F), Fischer(F)344 and Lewis rat strains, all from IFFA-CREDO, and 12 rats per sex/strain in four different tests. A factor analysis revealed three factors explaining 85% of total variance: approach/avoidance toward aversive stimuli, general activity in novel environments and defecation/time of social interaction (Ramos, Berton, Mormède, & Chaouloff, 1997). Interestingly, in one of the tests which was included in the "aversive stimuli" factor, i.e., the OF, dim lighting of 7 lux was used in order to decrease aversiveness. The most contrasting strains for this factor were the SHR and Lewis rats, although on the elevated plus maze (EPM) this was true only for males in regard to open arm entries, with no female-strain effects being noted for either open arm entries or time spent on the open arms. The best example of contrasting results for males vs. females, however, was for locomotion in the inner OF, where W–F males were first among the males whereas W–F females were last among the female groups (Ramos et al., 1997). This gender effect, so well illustrated by that study, is often, and unfortunately, ignored in this type of research when only one sex is used (usually males). Although there is no room to go into detailed comparisons here, it should be briefly mentioned that a factor analysis later performed, using about 800 males and females from an F2 generation derived from inbred RHA and RLA rats and seven different tests, yielded three factors: learned fear, emotional reactivity and fear of heights (Aguilar et al., 2002b).

Having apparently found the results they had sought with SHR and Lewis rats, Ramos, Mellerin, Mormède, and Chaouloff (1998) crossed them to produce F1 and F2 generations, which were tested in the OF and on the EPM. Inner locomotion in the OF was found to be the most heritable of all traits considered but, unlike previously (Ramos et al., 1997), it was not associated with any EPM variables, suggesting that the OF and EPM traits were independently inherited. Further crossings of later generations produced two lines showing that inhibition of locomotion in the OF was directly related to the aversiveness of the situation (70 lux lighting vs. 7 lux). Because of discrepancies with the EPM, the OF results were now given the term "behavioral inhibition trait" (as opposed to "classical anxiety"), i.e., an "inhibition of motor behavior in aversive places" which was recommended not to be compatible with "emotional behavior" (Mormède, Moneva, Bruneval, Chaouloff, & Moisan, 2002). The EPM will be discussed in more detail later in this chapter. (In the meantime, SHRs apparently went back to being a model for AD/HD.) In conjunction with the multiple-test study previously described (Ramos et al., 1997), hormonal comparisons of BN and F344 rats were subsequently conducted, purposely avoiding tests which involved differences in illumination (BNs were the only pigmented strain used in the earlier study, which may have influenced some of the results and which precluded their replacing SHRs in the subsequent studies together with Lewis rats). They showed a hypersecretion of corticosterone following stress in F344 rats which was, interestingly, negatively correlated with the size of the adrenal glands (Sarrieau & Mormède, 1998). These results have also often been noted in RHA/RLA line/strain comparisons (e.g., Aubry et al., 1995; Castanon et al., 1994; Castanon, Perez-Diaz, & Mormède, 1995; Driscoll & Käsermann, 1977; Fernández-Teruel, Escorihuela, Tobeña, & Driscoll, 1997b; Walker et al., 1989). Similar results (F344 > BN) were found for prolactin and renin secretion (Sarrieau, Chaouloff, Lemaire, & Mormède, 1998), which had also been seen, in the same direction (RLA>RHA), in the Roman lines (Castanon et al., 1992, 1994; Steimer, Driscoll, & Schulz, 1997b; Steimer, Escorihuela, Fernández-Teruel, & Driscoll, 1998).

The main problem confronted by using non-selected, commercially available rats is that even rats bearing the same designation probably vary, depending on the supplier. Compounded by methodological variations among laboratories, as we shall see later, this can lead to (usually undetected) problems in attempting to replicate, or follow-up on, previously performed experiments. Although studies conducted explicitly to check out discrepancies among suppliers have been all too rarely performed, those that have been have regularly revealed potential problems in this direction. For example, substantial differences in social interaction, exploration, motor activity, defecation and adrenal function were found among Lister-hooded rats from three different suppliers in UK (File & Velucci, 1979), differences in stress responses (OF, water restraint) were found among WKY rats from three different suppliers in the eastern USA (Paré & Kluczynski, 1997) and striking behavioral differences (swimming immobility, voluntary ETH intake, EPM activity) have been noted when fawn-hooded (FH) rats from two sources have been compared (Overstreet & Rezvani, 1996). Investigating results which apparently conflicted with 15 years of research on sensorimotor gating, i.e., prepulse inhibition (PPI) of the startle reflex, Swerdlow et al. (2000) showed that opposite effects were obtained when PPI-disruptive effects of the D1/D2 agonist, apomorphine (APO), were compared in Harlan Sprague-Dawley (SD) vs. Harlan Wistar rats than when they were compared in Bantin–Kingman SD vs. Bantin–Kingman Wistar rats.

Several more examples should be mentioned in reference to the widely used "normal control" rat, the SD. It has been shown that SDs from four different vendors in the USA displayed significantly different severity of gastric ulcers following starvation (Paré, Glavin, & Vincent, 1977), that SDs from seven different US suppliers showed a 6–7 fold difference in spontaneous nocturnal rotations, as well as (three suppliers) different effects of AMP (Glick, Shapiro, Drew, Hinds, & Carlson, 1986), that SDs from two different suppliers showed significant differences in heart weight and myocyte number (Campbell & Gerdes, 1987) and even that Hilltop (PA) SDs showed dramatically different incidences of cerebellar abnormalities than did Charles River (QUE) SDs, with the differences also being seen over extended breeding cycles (Ezerman & Kromer, 1985). Even more spectacular was a series of publications dealing with the noradrenergic innervation of the spinal cord in two stocks of SD rats (Sasco, WI and Harlan, MD) prompted by conflicting results which had been published by two different laboratories. It was found, using female rats, that A7 cell group neurons in Sasco SDs innervated the dorsal horn whereas in Harlan SDs, it was innervated primarily from the locus coeruleus (A6). In addition, the A6 region provided most of the ventral horn innervation in Sasco SDs, whereas the source in Harlan SDs was mostly A5/A7 (Clark, Yeomans, & Proudfit, 1991). Using male rats, a different method and Harlan, TX vs. Sasco, MO SDs, Sluka and Westlund (1992) confirmed those findings and determined that data from cats, opossums and monkeys appear to conform more closely to that from Sasco SDs, perhaps making the latter more useful for motor control studies and the Harlan SDs more useful for studies concerned with locus coeruleus/spinal cord interactions in sensory systems. Clark and Proudfit (1992) additionally showed that A6 neurons in Sasco SD innervated cervical spinal cord segments more densely than they did lumbar segments, while the reverse was true for Harlan SD rats.

Profiling Wistar and SD Rats

It should be mentioned at this time that the outbred, Wistar-derived Swiss sublines of RHA/RLA rats, after spending many years as "RHA/Verh" and "RLA/Verh" in Zurich, Lausanne, Geneva, Basel and Zurich again (P.D.), are presently maintained in Cagliari (O.G.) and Geneva (T.S.). Inbred strains were developed from that stock in the mid-1990s and have been maintained in Barcelona (A.F.-T.) since the late 1990s, originally for purposes of a genomic project (see Fernández-Teruel et al., 2002b).

A publication by Cools, Brachten, Heeren, Willemen, and Ellenbroek (1990) was instrumental in pioneering the study of personality differences in rats. With their Wistar-derived APO-susceptible (APO-SUS) and -unsusceptible (APO-UNSUS) rats, based on inbreeding for a gnawing response to 1.5 mg/kg s.c. APO, they demonstrated, using a defeat test, a bimodal shape of variation in "fleeing" (APO-SUS) and "freezing" (APO-UNSUS) rats. These results, at least, paralleled the results of Gentsch, Lichtsteiner, and Driscoll (1989) and Cools (unpublished) that outbred RHA and RLA rats were comparable to the APO-SUS and APO-UNSUS rats, respectively, in both tests. Additional studies with both outbred (Driscoll, Dedek, Fuchs, & Gentsch, 1985) and inbred (Giménez-Llort, Cañete, Guitart-Masip, Fernández-Teruel, & Tobeña, 2005) RHA/RLA males have shown that RHA rats display more pronounced, but less longer-lasting, stereotypy following APO injections.

Some other similarities between RHA and APO-SUS rats were detected by Cools et al. (1990), such as heavier adrenals and more novelty-induced ambulation in a large OF as well as an attenuated prolactin response to a stressor (Rots et al., 1996) and increased sensitization to AMP (Gingras & Cools, 1997). However, several similarities between RLA and APO-SUS rats have also been found, such as poorer, two-way active avoidance behavior and increased ACTH release in response to a CRH challenge (Cools et al., 1993; see also Castanon et al., 1994; Walker et al., 1989, 1992). In general, it is very difficult to compare other selection programs with these rats, as they are interchangeably named and selected by three different methods (APO-SUS/APO-UNSUS, FLEE/NONFLEE and HR/LR), most of the studies have been conducted with relatively young animals and the selection methods appear to be easily reversed by such environmental events as maternal deprivation on only the third, or ninth, postnatal day or being raised by a foster mother of the opposing phenotype (Cools & Gingras, 1998; Ellenbroek, Sluyter, & Cools, 2000). ETH consumption studies (which will be discussed in more detail later) with HR (APO-SUS)/LR (APO-UNSUS) rats were problematic in that both phenotypes appear to have an aversion to ETH (Gingras & Cools, 1995; Sluyter, Ellenbroek, Degen, & Cools, 2000), and OF analyses may have been confounded by the uncommon practice of regularly isolating the rats for a few days before testing (e.g., Gingras & Cools, 1995, 1997; Saigusa, Tuinstra, Koshikawa, & Cools, 1999). In any case, these rats have recently been considered for use as a model for schizophrenia (Ellenbroek, Geyer, & Cools, 1995; Ellenbroek & Cools, 2002; Van Loo & Martens, 2007), which may be more advantageous than attempts made in that direction with non-selected, heterogeneous Wistar rats.

High responder (HR) and low responder (LR) rats originated by dividing SD rats, from IFFA-CREDO, on the basis of locomotion on an open, circular corridor. It was found that rats with a higher "novelty-induced" locomotor activity readily acquired AMP self-administration, as compared to rats with a lower response (Piazza, Deminière, LeMoal, & Simon, 1989). This procedure has become widely used and even improved upon (e.g., Kabbaj, Devine, Savage, & Akil, 2000; Stead et al., 2006), proving to be also useful as pertaining to the anxiety vs. novelty-seeking model mentioned previously (Steimer et al., 1997a; Steimer & Driscoll, 2003). As reactive coping (i.e., freezing behavior) is considered to be indicative of increased emotional reactivity, and to be opposed to proactive coping (Koolhaas et al., 1999), it was inexpedient that the terms "responsiveness" and "reactivity" were used synonymously early in the development of the HR/LR procedure, especially as no evidence was offered that the test actually deals with novelty (Piazza et al., 1990). "Novelty-induced activity" can, and should, be differentiated from other forms of locomotor activity (e.g., habituation) in this type of work. While it is not the intention of this chapter to go into greater detail on this subject, it might be mentioned that other reviews (e.g., Gentsch et al., 1991) have previously discussed the subjects of reactive (i.e., responsive) and spontaneous (i.e., habituated) activity rather thoroughly.

Measuring plasma corticosterone levels after 120 min of exposure to that environment was also unprecedented, especially since no measurements were apparently conducted between 0 and 30, or between 30 and 120, min. Differences were only found at 120 min, i.e., long after the "novelty" of the testing procedure had worn off, making the stated conclusion that HR rats thereby showed "a greater release of corticosterone in a novel environment" (Piazza et al., 1990; Piazza, et al., 1991), which has (unfortunately) been widely cited since, rather unpropitious, to say the least. Did the hormone level remain high in HR rats during the entire 90 min or did it diminish and then recoup? On the basis of what has been learned more recently about HR rats, it appears more likely that after 120 min of exposure to the same, increasingly boring environment, the latter may have been the case, i.e., the proactive HRs may have become more worried than the passively coping LR rats about the outcome of the procedure. Kabbaj et al. (2000) have actually verified that HR rats will seek out novel and varied environments when they can choose between them and environments to which they have become habituated. Finally, although some researchers are still following the original procedure of dividing all rats tested on the circular corridor into HRs and LRs on the basis of being above or below the median, which would obviously produce some similar animals in both supposedly opposed groups, others have decided to use only the top and bottom thirds of the tested population (e.g., Kabbaj et al., 2000) or have actually crossed outbred SD rats from three different colonies (Charles River: NY, MI and QUE) to create HR and LR selection lines (Stead et al., 2006).

In breeding for HR and LR rats, Stead et al. (2006) use a medium-sized acrylic cage in which locomotor activity is measured by photocells every 5 min for 60 min, rather than a circular corridor, for the selection process. After eight generations of breeding, they have shown a fundamental difference in emotionality between their HRs and LRs, utilizing OF (coupled with the administration of chlordiazepoxide), EPM and light–dark (L/D) box tests. Cross-fostering revealed that responses to novelty were largely unaffected by maternal influences, although there were effects on "anxiety-like behavior" (Stead et al., 2006). Kabbaj et al. (2000), using SD rats from Charles River (MA), have labeled their HR rats as "novelty seekers" (as indicated in the previous paragraph – see also the discussion in reference to RHA rats earlier in this chapter) and had results similar to those of Stead et al. (2006) with the EPM and L/D box tests. It was interesting to note that single housing (the social isolation referred to earlier in this section), eliminated the HR/LR differences seen with the L/D box test (Kabbaj et al., 2000). RHA and HR rats are not only more active, but they also show a greater COC-induced elevation of extracellular DA in the NAc than do RLA and LR rats, respectively. In this connection, another study should be mentioned here, one which also pertains to the previous discussion concerning commercially available rats. Ambrosio, Goldberg, and Elmer (1995) investigated the predictive value of spontaneous locomotor activity in the acquisition of drug-reinforced behavior, using male LEW/SsN (Lewis), F344N, ACI/Seg and BN/SsN rats, all from Harlan SD (Inc.). Lewis rats showed both high rates of MOR self-administration and the highest levels of activity. The latter finding contrasts with those for the IFFA-CREDO Lewis rats discussed earlier and may, once again, illustrate potential differences among suppliers. F344 rats, being the least active, also self-administered MOR at the lowest rate, although this increased gradually over the 7 days of testing. A genetic correlation was found between drug intake during the first 5 days and the spontaneous locomotion response to a novel environment for 60 min, although this effect was only seen in rats with cannulas, cannulation apparently having altered the rank order of locomotor activity among the groups, at least in some individuals (a genetic predisposition to stress effect?). Incidentally, Lewis rats from Harlan

SD (IN) are considered to be addiction prone, i.e., self-administering MOR, COC and ETH at high levels, and also exhibit a defect in the biosynthesis of CRH (Beitner-Johnson, Guitart, & Nestler, 1991). If these two phenomena are genetically bound in these rats, that would provide a further indication that stress does not induce increased ETH consumption (Bell et al., 1998), which will be discussed later on.

Fear/Anxiety I: Acoustic Startle

A major problem in the literature, and one which may never be solved satisfactorily, is the way the terms fear, anxiety, emotionality and fearfulness are, more or less, used interchangeably (Steimer & Driscoll, 2003). Compounding this are the questions of whether certain behavioral tests actually represent one or more of these terms and whether or not they should be extended to other domains, e.g., depression. The proposition that anxiety symptoms might be a vulnerability factor in the development of major depressive disorder, with elevated startle magnitude in threatening contexts being a marker for anxiety disorder (Grillon et al., 2005), for example, is hardly reconcilable with the proposition that the distinguishing feature of clinical anxiety is that its occurrence is not attributable to real danger and that analyses based on reaction to danger thereby omit a central feature of that condition (see Stephens, 1997). Attempts are presently being made to distinguish between fear and anxiety based on the assumption that anxiety, unlike fear, can be considered to be a sustained state of apprehension unrelated to immediate environmental threats (Walker, Toufexis, & Davis, 2003). This definition obviously assumes the development and utilization of models that have been sufficiently studied and which remain consistent over time, preferably as a result of selective breeding (Landgraf et al., 2007).

Two types of behavioral tests have proven to be very valuable in this endeavor, those directly involving fear conditioning, such as the acoustic startle (AS) response or shock avoidance acquisition (which crucially involves an initial phase of fear conditioning; see Conti, MacIver, Ferkany, & Abreu, 1990; Escorihuela et al., 1995; Fernández-Teruel et al., 1991; Wilcock & Fulker, 1973), and those which do not, such as the OF or EPM. AS represents a reflexive reaction to a stressful, environmental stimulation. Glowa and Hansen (1994) tested 46 inbred and outbred rat strains/lines from NIH colonies for AS, revealing profound differences among them in both mean amplitude of the startle response and rate of habituation to startle stimuli over repeated trials (the results for both criteria being, in addition, significantly related to one another). Inbred RLA and RHA rats have been compared for AS on a number of occasions, with RLAs showing an increased startle response to acoustic stimuli, i.e.,

a stronger emotional reaction, which was even stronger after footshock presentation (Aguilar, Gil, Tobeña, Escorihuela, & Fernández-Teruel, 2000; López-Aumatell, et al., 2005; Schwegler, et al., 1997; Yilmazer-Hanke, Faber-Zuschratter, Linke, & Schwegler, 2002) compared to RHA rats. Other fear conditioning tests, such as classical fear conditioning (also involving emotional response to footshock presentation) or a "sudden-noise" test, have shown differences in the same direction in both inbred (Escorihuela et al., 1997) and outbred (Roozendaal et al., 1992; Steimer & Driscoll, 2003) RLA/RHA comparisons.

This type of fear response is detected and organized by the amygdala, relayed via basolateral nuclei through it's central nucleus (CEA) to behavioral and autonomic systems, as well as to the hypothalamic–pituitary–adrenal (HPA) axis (LeDoux, 1998). The basolateral amygdala is also a primary source of innervation for the bed nuclei of the stria terminalis (BNST), which is a direct recipient of the startle enhancement effect of CRH (no amygdalar involvement) and which appears to be more vital for fear-eliciting effects on the EPM (Walker & Davis, 1997). Walker et al. (2003) subsequently suggested that the BNST mediates long-duration responses, and the CEA short-duration responses, to conditioned or unconditioned, threatening or aversive, stimuli, finding this to be more suitable than the hypothesis that the BNST mediates unconditioned fear reactions and the CEA mediates conditioned fear reactions. They refer to the sustained response as anxiety and to the short-lasting one as fear. The CEA has often been a subject for study in RHA/RLA rats. Similar electrophysiological activity recorded from the CEA during 4 h of restraint stress, as well as increased stomach pathology, united RLA rats with Wistar rats judged to be "more emotional", as compared to RHA rats and "less-emotional" Wistar rats (Henke, 1988). The knowledge that CEA lesions reduced the fear reactions of rats in an OF and the findings of Roozendaal, Koolhaas, and Bohus (1991) that CEA lesions, also in Wistar rats, not only abolished the immobility response normally seen after footshock but also attenuated all measured hormone responses (including corticosterone and, especially, prolactin) set the stage for their subsequent revelations regarding the CEA in RHA/RLA rats. After showing that RLA rats displayed profound bradycardia and immobility when placed in the same compartment in which they had experienced a 3 sec footshock 24 h previously, and that RHA rats showed no such reactions, Roozendaal et al. (1992) demonstrated that a low dose of arginine-8-vasopressin (AVP) infused into the CEA 30 min before the test enhanced these responses in RLAs, whereas a high dose of AVP or oxytocin (OXT) attenuated them. Once again, there were no effects in RHA rats following the infusions. It was concluded that CEA differences in AVP and OXT innervation and/or receptor densities may contribute to the differences seen in coping strategies between RHA and RLA rats. It was further suggested that RLA rats may have a higher density of AVP innervation in the CEA than RHA rats do (much of which may originate from the BNST). In a recent publication which, unfortunately, failed to discuss both those results and those of Wiersma, Konsman, Knollema, Bohus, and Koolhaas (1998), which had dealt with the effects of CRH infusions directly into the CEA on heart rate and exploration/immobility in RHA and RLA rats, Yilmazer-Hanke et al. (2002) found that inbred RLA rats, which showed a more pronounced fear potentiation of the AS response, had more projection neurons immunoreactive for (anxiogenic) CRH in the CEA than inbred RHA rats did. No differences between the strains were found in basolateral amygdalar neurons. Three different subpopulations of GABAergic neurons were also examined in various brain nuclei.

Returning to possible AVP and OXT receptor-density differences in the CEA between RHA and RLA rats, it has recently been shown that these two receptor types may exert opposing actions on anxiety and levels of stress, for example, and that they are found in different areas of the CEA (lateral and capsular division vs. medial part), based on examining "rat" brain sections (Huber, Veinante, & Stoop, 2005). These observations may support the suggestion by Roozendaal et al. (1992) that the similarity between the effects of the high dose of AVP and those of OXT in RLA rats may have been caused by a mutual OXT receptor stimulation. Windle et al. (2006) have additionally reported that intracerebroventricularly (i.c.v.) infused OXT reduced anxious behavior on the EPM, as well as stressor-induced corticosterone responses, in female SD (Bantin & Kingman, GB) rats, although others have reported that infusions of OXT directly into the CEA increased exploration of the central area of an OF without having any effect on the EPM. Steimer et al. (1997b) measured brain metabolites of progesterone (tetra- and dihydroprogesterone – THP and DHP), which act on GABA-A receptors and have anxiolytic properties, in outbred RHA and RLA rats. The formation of both metabolites was significantly higher in the frontal cortex, and that of DHP in the BNST, of RHA rats, with no differences being found between the rat lines in the CEA or hippocampus. Finally, Giorgi et al. (1994) reported that the stimulating effect of GABA on $^{36}Cl^-$ uptake was less pronounced in the cerebral cortex of RLA than of RHA rats. Progesterone metabolites, especially THP, are precisely capable of increasing GABA-induced Cl^- uptake.

Fear/Anxiety II: The Notorious EPM

The other type of test to be considered in detail (and not involving fear conditioning) is well represented by the EPM.

Although the test is widely used, and perhaps because of this, there is probably no other procedure which shows such extreme inter-laboratory variation, even when genetically identical animals from the same supplier are utilized (e.g., Wahlsten et al., 2003). Intra-laboratory differences (in this case, when the same laboratory is moved to another site within a university) have also recently been reported for the EPM, even when the same stocks of mice, apparatus and procedures have been used (Wahlsten, Bachmanov, Finn, & Crabbe, 2006). Several reviews have listed potential sources of inter-laboratory variation for rats, some of which are strain/stock, gender, housing conditions, previous handling and injection experience, previous EPM experience, previous exposure (or not) to an OF or hole board (HB), light cycle period and time of testing, light intensity, presence or not of the experimenter, presence or not of ledges (raised edges) on the open arms and measures scored, including their definitions (Fernandes & File, 1996; Griebel, Moreau, Jenck, Martin, & Misslin, 1993; Hogg, 1996; Rodgers, 1997; Steimer & Driscoll, 2003). More attention should certainly be paid to the level of lighting of the maze in regard to the effects of potential anxiolytic drugs (Griebel et al., 1993; Hogg, 1996) and to compare the results found among different laboratories (Steimer & Driscoll, 2003). The type and intensity of lighting is not only omitted in many EPM publications, but many investigators even use very dim lighting in order to increase activity on the apparatus to measurable levels (within the apparently mandatory "5 min" limit) which, of course, emphatically reduces it's value as an "anxiety test".

Regarding the effects of anxiolytic drugs, both false positives (e.g., AMP) and false negatives (e.g., buspirone) have been reported (e.g., Dawson, Crawford, Collinson, Iversen, & Tricklebank, 1995; Rodgers, 1997), and Dawson et al. (1995) went so far as to attribute the effects of chlordiazepoxide more to increases in locomotor activity than to fear reduction. They recommended that speed of locomotion and distance traveled on the arms should also be measured rather than just arm entries (which is also subject to definition). We have followed this recommendation and, as a few others have done (e.g., Setem, Pinheiro, Motta, Morato, & Cruz, 1999), have additionally measured rearing and grooming behavior on the EPM (Driscoll et al., 1998; Escorihuela et al., 1999). The more frequent rearing within the enclosed arms often displayed by RHA rats indicates to us that they may prefer to spend more time where they can explore (including the walls, in this event), which would obviously reduce the time they spend on the open arms. As if to emphasize this point, tests with an elevated zero maze, which avoids the closed ends of the closed arms as well as the "uninteresting" central square (at lest to researchers only interested in arm entries), have recently shown that male, inbred RHA rats spent twice as much time on the open sectors of the maze as RLA rats (A.F.-T., unpublished). In any case, in most comparisons made with both the inbred strains (Driscoll et al., 1998; Escorihuela et al., 1999) and outbred lines (Fernández-Teruel et al., 1997a; Steimer & Driscoll, 2003) on the EPM, but not all (see the end of this section), RHA rats showed a significantly higher percentage of open arm entries than RLAs did, as well as more distance traveled on the open arms, when this was measured. In addition, RHA rats have also consistently spent less time self-grooming on the EPM and have shown longer latencies before starting the initial grooming bout.

Self-grooming is a valuable adjunct to the estimation of rodent personality which, unfortunately, is too often ignored in planning and evaluating behavioral studies, particularly those involving the EPM. As indicated above, the information obtained is limited when only total arm entries, open arm entries and/or time spent on the open arms are registered, as is commonly the case, presumably in order to gather the data most easily collected by automated means. Self-grooming behavior, on the other hand, is a typical displacement activity manifesting the conflict between fear and approach behavior in rodents, and it's duration, frequency and commencement latency can be measured in almost all testing situations (Steimer & Driscoll, 2003). It has been shown to be reduced by diazepam injections (Crawley & Moody, 1983) as well as being elicited by infusions of CRH, either i.c.v. (Dunn & Berridge, 1990) or directly into the paraventricular nucleus (PVN) of the hypothalamus (Mönnikes, Heymann-Mönnikes, & Taché, 1992), which may make it a direct behavioral expression of increased CRH secretion and/or that of higher sensitivity to CRH effects, e.g., in RLA rats (Walker et al., 1989; 1992). At any rate, RLA rats invariably show more and longer grooming bouts, as well as shorter latencies to start grooming, than RHA rats do, whenever this behavior is measured in conjunction with tests measuring anxiety and/or fear, such as the EPM, hyponeophagia (Ferré et al., 1995; Steimer et al., 1998), black/white (B/W) box and dark/light OF (Steimer & Driscoll, 2003), tail-pinch (Corda, Lecca, Piras, DiChiara, & Giorgi, 1997; D'Angio et al., 1988), shock-induced suppression of drinking (Corda, Piras, Valentini, Scano, & Giorgi, 1998), exposure to a novel environment, e.g., OF (Aubry et al., 1995; Castanon et al., 1994; D'Angio et al., 1988; Escorihuela et al., 1999; Steimer et al., 1998; etc.) and, especially, the HB test (Guitart-Masip et al., 2006a).

As if in answer to the call for an effective genetic solution to ensure the future of the EPM (Rodgers, 1997), a successful selective breeding program was established at about that time with the high (HAB) and low (LAB) anxiety-related behavior rats (Landgraf & Wigger, 2002; 2003), selected and bred for opposing behavior on the EPM. Entries onto open and closed arms, the time spent on both and latency to the first entry onto an open arm, in all of which HABs differ from LABs, are measured. Whereas females are more

active than males, no difference in total distance traveled on the EPM has been found between the rat lines (Liebsch, Montkowski, Holsboer, & Landgraf, 1998b). The similarities between RHAs and LABs, as compared to RLAs and HABs, respectively, are remarkable. Both HAB and RLA rats (references for the latter have already been given earlier in this chapter, for almost all of the following) are less active in an OF and explore the central area less (Liebsch et al., 1998b; Salomé et al., 2004), display similar results in the B/W box (Henninger et al., 2000) when initially placed in the black (dark) side (see Steimer et al., 1997b, for a discussion of this important point), visit fewer holes on the HB test (Landgraf & Wigger, 2002), generally show less rearing behavior and prefer passive coping strategies (Landgraf & Wigger, 2003), register the same elevated levels of plasma prolactin, ACTH and corticosterone, at least with EPM and OF exposure for HABs (Landgraf, Wigger, Holsboer, & Neumann, 1999; Liebsch et al., 1998a; Salomé et al., 2004), show a blunted activation of the medial prefrontal cortex (mPFC) upon anxiogenic challenge (Kalisch et al., 2004), exhibit increased expression/release of AVP in the PVN (Landgraf & Wigger, 2002; see also Aubry et al., 1995) and even assume the arched-back (active) nursing posture more frequently, the side (passive) nursing posture less frequently and spend more time with their young in maternal behavior studies (Neumann et al., 2005; see also Driscoll, Fümm, & Bättig, 1979; Fuemm & Driscoll, 1981), when compared to their LAB/RHA counterparts. Furthermore, not only does extended neonatal stress (daily separations from the mother) affect certain behaviors and hormonal responses later in life in the same direction in LAB/RHA vs. HAB/RLA rats (Driscoll et al., 1998; Fernández-Teruel et al., 1997a; Neumann et al., 2005), but results in dexamethasone/CRH testing are also similar in both line comparisons, making HAB and RLA rats interesting candidates for some facets of human depression (Keck et al., 2002; Steimer, Python, Schulz, & Aubrey, 2007). The breeding origins of HAB/LAB rats have been described in detail by Landgraf & Wigger (2002) as well as have new directions of research dealing with CRH in the BNST and, in particular, AVP in the PVN (Landgraf et al., 2007; Wigger et al., 2004).

Differences found in comparing HAB/LAB with RLA/RHA rats, other than that HAB/LABs both do not show any particular affinity for ETH (Landgraf & Wigger, 2002), have mainly arisen in connection with the two tests with which these last two sections have been primarily concerned, AS and the EPM. Whereas neuroanatomical analyses demonstrated more projection neurons immunoreactive for (anxiogenic) CRH in the CEA, as well as differences in neurons in other amygdalar nuclei in RLA, as compared to RHA rats (Yilmazer-Hanke et al., 2002), none of those differences were found between HAD and LAD rats (Yilmazer-Hanke, Wigger, Linke, Landgraf, &

Schwegler, 2004). In contrast to the four EPM studies mentioned previously in this section, using both inbred or outbred RHA and RLA rats, Yilmazer-Hanke et al. (2002) found, as Chaouloff, Castanon, & Mormède (1994) had with outbred RHA/RLAs, no differences, or even differences in the opposite direction, between inbred RHA and RLA rats on the EPM. Both of those studies (Chaouloff et al., 1994; Yilmazer-Hanke et al., 2002) measured only number of entries and time spent on the open and closed arms. Lighting differences may have also been a factor here (see Steimer & Driscoll, 2003) as, for example, Chaouloff et al. (1994) used "dim illumination", without stating the actual lux value. The usual differences were found for HAB/LAB rats on the EPM in the Yilmazer-Hanke et al. (2004) study, but it should be noted that, in contrast to Yilmazer-Hanke et al. (2002), the EPM part of that (2004) study had been conducted at the home laboratory, before shipment for the AS testing and neuroanatomical analysis. The AS test, on the other hand, showed the traditional differences for RHA/RLAs, with RLA rats displaying a more pronounced fear potentiation of AS (Yilmazer-Hanke et al., 2002), whereas LABs unexpectedly showed a higher baseline and an increased, fear-sensitized AS, compared to HABs (Yilmazer-Hanke et al. 2004). Anyway, at least in our case, Yilmazer-Hanke et al. (2002) failed to establish further contact with us after obtaining our rats, thereby resulting in such (incorrect) statements in their 2002 publication as, e.g., (a) regarding the "degeneration" of inbred RHA/RLA rats or (b) that Chaouloff et al. (1994) had also studied inbred RHA/RLA rats.

Rats May Drink for Different Reasons . . .

The association between sensation seeking and ETH consumption (e.g., Siegel, 1997) was introduced earlier. Although this association has been recognized for many years in humans (e.g., Cloninger, Sigvardsson, & Bohman, 1988; Schwarz, Burkhart, & Green, 1982), it was only about a decade ago that it was determined that whereas ETH may cause anxiety in some individuals, there was little evidence for ETH consumption being caused by anxiety (e.g., Schuckit & Hesselbrock, 1994). This was confirmed in rats soon thereafter, e.g., when Bell et al. (1998) noted that i.c.v.-infused CRH exerted stressor-like effects while, at the same time, reducing ETH intake in outbred, Long-Evans rats. Differences in ETH consumption were noted even before the initiation of genetically selected rat lines for that behavior, with RHA rats, for example, consistently consuming more ETH than RLA rats from the early years of their development (Drewek & Broadhurst, 1979; Satinder, 1975) up to the present (Fernández-Teruel et al., 2002a). The most commonly used, selectively bred rat lines studied for this behav-

ior in recent years have been the ETH-high preferring (P) vs. low preferring (NP) lines, the high and low ETH-drinking (HAD vs. LAD) replicate lines and the Alko alcohol (AA) vs. Alko non-alcohol (ANA) lines. The P/NP rats were derived from Wistar stock and the HAD/LAD lines from N/Nih heterogeneous stock which were, in turn, produced by an intercross among eight different strains (Spuhler & Deitrich, 1984). N/Nih rats have found frequent application in ETH-related research, e.g., being used to demonstrate that impulsivity predicts individual susceptibility to high levels of ETH consumption (Poulos, Le, & Parker, 1995).

In a recent review dealing with P/NP and HAD/LAD rats, P rats were considered to be the only true model for alcoholism, even simulating "binge drinking" or "loss of control" (Murphy et al., 2002). Although RHA rats are rather "social drinkers" by comparison (Driscoll et al., 1998), both reviews help identify some behavioral similarities between P and RHA rats, as compared to NP and RLA rats, such as in DRL operant responding, OF activity and novelty-seeking behavior. Differences also exist, however, e.g., in AS (Jones et al., 2000; McKinzie et al., 2000). In general, tests of "anxiety" have shown conflicting results in both directions when P and NP rats have been compared, due to differences among testing laboratories, gender differences, etc. (see Stewart, Gatto, Lumeng, Li, & Murphy, 1993 for review). More recent work has indicated that the relationship between heightened response to novelty and ETH consumption was modestly associated, and observed under specific conditions, at least in P/HAD vs. NP/LAD rats, whereas a positive relationship has often been found in other experiments with rats and mice between novelty-seeking behavior and the self-administration of drugs of abuse, such as COC, AMP and ETH (Nowak et al., 2000). Although in inbred mice, high levels of activity in an OF are usually associated with high ETH intake, it has been suggested that some ETH-drinking lines/strains of rats may be drinking ETH for different reasons. Badishtov et al., (1995), comparing several stocks which voluntarily drink ETH, i.e., P, AA, FH and Maudsley non-reactive (MNRA), with their non-drinking counterparts, have shown that P rats were the most active in an OF and Maudsley reactive (MR) rats the least active, but that NP rats were also very active, and that AA and ANA rats were indistinguishable. There was, however, a consistently negative relationship between increased defecation, an often used but controversial indicator of heightened emotionality (see, e.g., Enck, Merlin, Erckenbrecht, & Weinbeck, 1989) and ETH intake. In particular, the ETH-preferring AA rats showed none and the ANAs a high level of defecation, as did the MRs. In a separate set of experiments using the same strains/lines, no relationship was found between ETH intake and activity on the EPM (with only time spent on the open arms being recorded) or in the forced swim test (FST), thereby indicating to the authors that ETH-preferring rats did not drink ETH to reduce high anxiety states (Viglinskaya et al., 1995). In particular, a previous report that P rats were more emotional on the EPM than were NP rats was not confirmed, and it was noted that the original authors were also having difficulty in replicating their original EPM findings. An interesting proposal put forth by Kurtz, Stewart, Zweifel, Li, and Froehlich (1996) suggests that as P rats are less sensitive to the initial behavior-impairing effects of ETH than NP rats are and develop tolerance rather than becoming sensitized with repeated injections of ETH, this may contribute to the differences in ETH consumption between them. We have also noted this phenomenon in preliminary studies with RHA and RLA rats.

Another aspect of ETH preference relates to saccharin (SAC) and quinine (QIN) consumption. Sinclair, Kampov-Polevoy, Stewart, and Li (1992) found that both ANA and NP rats drank much less SAC in solution than AA and P rats did, both as single concentration or in an ascending series. ANAs also drank less than AAs from QIN (bitter), saline or citric acid (sour) solutions, with both P and NP rats showing very low levels of consumption of all three. Basically, however, none of the four groups liked QIN, and it was considered possible that the same preferences for ETH and SAC in P and AA rats may have reflected similar mechanisms mediating reinforcement from preference for sweets and from systemic ETH. In a more detailed study, Overstreet et al. (1993) compared P, FH and MNRA rats with their non-ETH preferring counterparts, also finding that ETH and SAC intakes were highly correlated over all comparisons. FH rats drank the most of both solutions, but they also drank the most fluids, with P rats actually drinking more ETH than FH rats when this was corrected for. Once again, QIN intake was very low for all groups, and there were no significant differences among them. Conversely, in rats selected and bred from Holtzman stock, for high (HiS) vs. low (LoS) SAC consumption, Dess, Badia-Elder, Thiele, Kiefer, and Blizard (1998) showed that HiS drank more ETH than LoS rats did, using 1–10% ETH solutions. They had also previously seen RHA/RLA-like differences in AS testing and, in a detailed study dealing with emotionality indices, had found that LoS appeared to be more emotional than HiS rats, showing longer emergence latencies and more defecation in an OF which included a startbox, and being more affected by electric shock, as assessed by stressor-induced anorexia (Dess & Minor, 1996). Although the preferences for ETH and QIN by the less emotional HiS rats were not overwhelming, they went in the same direction as those seen with RHA rats in a study by Razafimanalina et al., (1996), who showed a strong preference by RHA rats for ETH, SAC and QIN over water, as compared to RLA rats, which actually demonstrated a total aversion to all three substances. Such an affinity for QIN by RHAs is rather unique, even among selected rat populations

(see above), further recommending their being considered as novelty seekers (Driscoll et al., 1998).

. . . And Depression is Not One of Them

Attempts to develop a genetic model for depression in humans originated decades ago with the Flinders sensitive (FSL) and resistant (FRL) lines of rats (see Overstreet, Friedman, Mathé, & Yadid, 2005), which were selected and bred for sensitivity/resistance to the organophosphate anticholinesterase agent, diisopropyl fluorophosphate. Human studies showing more sensitivity to cholinergic agonists for affective disorder patients than normal controls (Janowsky, Risch, Parker, Huey, & Judd, 1980) raised the possibility of FSL rats being a model for depression. The main problems since have been the uncertainty as to which neurochemical changes are responsible for the behavioral alterations seen in depressed individuals and a comparative lack of suitable behavioral tests in the laboratory (see Landgraf et al., 2007). As previously mentioned, results in dexamethasone/CRH testing have been similar for HAB and RLA rats, making them potential models for some facets of human depression. In comparison to RLA vs. RHA rats, FSL rats were also more sensitive than FRLs to the hypothermic effects of oxotremorine, a direct-acting muscarinic agonist (Martin, Driscoll, & Gentsch, 1984; Rezvani, Overstreet, Ejantkar, & Gordon, 1994), poorer in two-way, active avoidance and better in passive avoidance (Overstreet, Rezvani, & Janowsky, 1990) and more sensitive to the hypothermic effect of APO but less sensitive to its stereotypy-inducing effects (Crocker & Overstreet, 1991; Driscoll et al., 1985; Giménez-Llort et al., 2005) and also showed an attenuated (mesocortical) DAergic response following exposure to a stressor (Yadid, Overstreet, & Zangen, 2001). The RHA/RLA references for the latter response and for two-way avoidance, as well as OF, AS, etc. (below) have been cited previously. In addition, FSL rats are also less active in an OF and show more freezing behavior in response to electric shocks, as well as a greater hypothermic response to ETH than FRL rats do (Overstreet et al., 2005), as has been seen for RLA vs. RHA rats. On the other hand, FSL and FRL rats showed no differences in SAC preference (Ayensu et al., 1995) and FSLs showed rather opposing results in regard to AS thresholds (Markou, Matthews, Overstreet, Koob, & Geyer, 1994) and natural killer cell activity (Overstreet et al., 2005; Sandi, Castanon, Vitiello, Neveu, & Mormède, 1991) than RLA rats did.

Because of the evidently higher relevance of the serotonergic than the cholinergic system in the FST, FSL rats metamorphosed to HDS and FRL to LDS by selective breeding for differential hypothermic responses to the serotonin

(5-HT)-1A receptor agonist, 8-OH-DPAT (Overstreet, 2002), with HDS remaining more susceptible to stressor-induced behavioral disturbances than LDS rats (e.g., Commissaris, Ardayfio, McQueen, Gilchrist, & Overstreet, 2000). In addition, another selection for HDS/LDS animals was started from the heterogeneous N/Nih rats mentioned earlier. HDS rats showed a higher intake of SAC but, unusually, this increase was not associated with increased ETH intake (Overstreet, 2002). As also mentioned earlier, Overstreet & Rezvani (1996) had found that differences in FST immobility, EPM activity and voluntary ETH intake surfaced when FH rats from two sources were compared. The FH/Wjd rats have been determined to be not only highly immobile in the FST but also show a high level of ETH consumption, almost like P rats. Rezvani, Parsian, & Overstreet (2002) have favorably compared FH/Wjd and P rats with human alcoholics on the basis of face, construct and predictive validity and have crossbred FH/Wjd with ACI/N rats. Subsequent analyses of the F1 and F2 generations revealed that, although depressive-like behavior and ETH drinking co-occur in FH/Wjd rats, they appear to be under independent genetic control (Overstreet, Rezvani, Djouma, Parsian, & Lawrence, 2007; Rezvani et al., 2002).

Neurochemical ODDS and END(ING)S

As we have seen, experimental subjects which are more responsive to the acute effects of addictive drugs (MOR, COC, AMP), e.g., RHA (vs. RLA) rats, are also more susceptible to develop behavioral sensitization upon repeated drug exposure (Giorgi et al., 2007). These results have also been observed for COC and MOR in another line which drinks ETH, the AA (vs. ANA) rats (Honkanen et al., 1999). The findings that those drugs also induced an increase in DA output in the NAc, both acutely (NAc shell) and in sensitized (NAc core of) RHA, but not RLA, rats (Giorgi et al., 2007), were also obtained for MOR and COC in AA, vs. ANA, rats (Honkanen et al., 1999; Ojanen et al., 2003), although those studies did not differentiate between NAc compartments. Murphy et al. (2002) reported that the mesolimbic DAergic system in ETH-preferring P rats may also be more sensitive to the reinforcing (rewarding) actions of ETH. At the same time, a high initial tolerance to ETH, as also displayed by RHA rats, has been suggested to have good predictive value, being a known risk factor for alcoholism in humans (Guitart-Masip et al., 2006a). Those authors also suggested that the lower sensitivity to ETH of RHA rats joins novelty seeking as a trait seemingly not related to DA function, although further investigations with the two strains (on DA receptor binding, etc.) have indicated their usefulness as a tool to identify DA-

related mechanisms predisposing to drug and alcohol dependence (Guitart-Masip et al., 2006b).

As was also mentioned earlier, Giorgi et al. (1994) reported that the stimulating effect of GABA on ^{36}Cl$^-$ uptake was less pronounced in the cerebral cortex of RLA, than of RHA, rats. This result, confirmed by Bentareha et al. (1998), together with the previously discussed findings of Steimer et al. (1997b), which dealt with metabolites of progesterone, could have interesting implications for RHA/RLA differences in emotional reactivity and/or stress susceptibility, but probably not for differences in ETH consumption. The region of interest (recently illustrated in detail by Giorgi et al., 2007) for GABA and drugs is probably the VTA-NAc. Hwang, Lumeng, Wu, and Li (1990) showed that the GABAergic terminal density in the NAc was greater in P/HAD than in NP/LAD rats and suggested that GABA/DA axonal interactions may be altered in the NAc of ETH-preferring rats. Nowak, McBride, Lumeng, Li, and Murphy (1998) later showed that blocking GABA-A receptors in the VTA actually attenuated ETH intake in P rats, but not SAC intake, suggesting that the VTA mechanisms regulating ETH drinking behavior are under tonic GABA inhibition mediated by GABA-A receptors. On the other hand, Thielen, McBride, Lumeng, and Li (1998), who had also investigated GABA-A receptor function in the cerebral cortex of P and NP rats, detected no differences between them there, thus confirming that the properties of ETH were probably not due to differences in cortical GABA-A receptor function. Actually, they had assumed beforehand that ETH would potentiate GABA-stimulated Cl$^-$ influx in the NP line, which was more sensitive to ETH and as had been seen with other ETH-sensitive mice and rats, but that was not the case.

An area of the cortex which is probably involved in drug responses, however, is the mPFC. It has been shown in Wistar rats (Schwerzenbach) that COC (20 mg/kg i.p.) decreased levels of DA in the mPFC while increasing them in the NAc shell much more than in the NAc core. The locomotor response to COC increased as DA increased in the NAc shell. Response strategies organized in the mPFC were suggested to be translated into actions via the NAc (Hédou, Feldon, & Heidbreder, 1999). Also examining both regions simultaneously, Giorgi, Valentini, Piras, DiChiara, and Corda (1999) found that a highly palatable food produced larger increases in extracellular DA in the mPFC and in the NAc shell of both inbred and outbred RHA, than RLA, rats. RLAs showed longer approach and eating latencies as well as more episodes of freezing behavior. D'Angio et al. (1988) found a higher turnover of DA in the mPFC of RHA vs. RLA rats upon exposing them to various stressors, which they believed to represent a heightened attention or activation of cognitive processes in attempting to cope with the stressors. That interpretation was quite radical at the time, when many were still defining such a DA response in the mPFC to stressors as a sign of heightened emotionality, but there had already been indications of support for the hypothesis from other sectors (reviewed by Driscoll, Dedek, D'Angio, Claustre, & Scatton, 1990b), which subsequently included further verification in experiments performed with RHA/RLA rats (Driscoll et al., 1990b).

Anyway, it has been suggested that the mPFC modulates responses to psychological, rather than physical, stressors, probably doing so by indirect projections to the CRH cells of the PVN via the CEA and/or BNST (Crane, Ebner, & Day, 2003). The CEA, implicated in conditioned fear responses (see the earlier AS section), may be additionally involved in aspects of reward-related behaviors, especially those subjected to modulation by drug-induced increases in NAc DA function (Howes, Dalley, Morrison, Robbins, & Everitt, 2000). P rats display lower levels of CRH in both the CEA and mPFC than NP rats do (Murphy et al., 2002) and, as we have seen, RLAs have displayed more CRH projection neurons in the CEA than RHAs have (Yilmazer-Hanke et al., 2002). Bell et al. (1998) have indicated that CRH may inhibit the mesolimbic DA projections implicated in ETH-mediated reward, as CRH cell bodies and binding sites are closely associated with mesolimbic DA cell bodies and their forebrain terminals. Getting back to the PVN and CRH, however, with the knowledge that AVP and OXT are also released from the hypothalamus, together with CRH (e.g., see Walker et al., 1992), and that CRH and AVP exert a synergistic action on ACTH release from corticotropic cells in the anterior pituitary, Aubry et al. (1995) examined the parvicellular neurons of the PVN in RHA and RLA rats. Whereas no basal differences were found for CRH mRNA, RLAs showed higher basal levels of AVP labeling. OF exposure increased CRH labeling in both rat lines, but more in RLAs. Incidentally, it has also been suggested that paraventricular 5-HT also plays a role in controlling the release of CRH (Feldman, Conforti, & Melamed, 1987) and, in line with this proposal, an increased activity of hypothalamic 5-HT after a repeated footshock stressor in RLA rats only, vs. RHA rats, has been observed (Driscoll, Dedek, Martin, & Zivkovic, 1983).

As hinted at previously, Landgraf et al. (2007) have noted a correlation between the behavioral phenotype of HAB/LAB rats and mice, and the expression of AVP in the PVN, with the resulting "HPA axis overdrive" likely contributing to anxiety- and depression-like behavior. No significant differences were found for OXT. Landgraf et al. (2007) implied that the increased expression of AVP and the more pronounced AVP reaction to stressors in HAB rats may be initially beneficial in adjusting short-term behavior, but that the high levels and/or duration of central AVP release may lead to psychopathology in the long run. They believe that the key lies with an AVP gene. Whether or not this is true, it was appropriate for them to state that the behavioral, anatom-

ical and physiological continuities between rats, in particular, and humans, are considerable (as can be seen from most of the subjects covered in this chapter). Research with rats along these lines is particularly advantageous, considering their intelligence, behavioral skills and complex personalities, all of which are attributes they share with humans (Whishaw, 1999).

In conclusion, this chapter has intended to provide some insight into the realm of exploring personality differences in rats, by offering the reader a candid appraisal of some widely (and less widely) used behavioral tests, as well as some of the types of neurochemical/anatomical analyses considered to be important in this endeavor. Concurrently, and necessarily, most of the pertinent selection programs in rats have been reviewed and similarities, as well as deviations, among them have been considered. This is no easy task, considering the methodological differences to be found among laboratories such as source of the rats, the varied usage of tests and interpretation of results, etc., many of which have been indicated here. We have endeavored to introduce these subjects as a concept within the scope of which one may start to understand personality differences in these animals. Much reference material has additionally been provided for background information to assist the reader along this journey.

References

Aguilar, R., Gil, L., Tobeña, A., Escorihuela, R. M., & Fernández-Teruel, A. (2000). Differential effects of cohort removal stress on the acoustic startle response of the Roman/Verh rat strains. *Behavior Genetics, 30,* 71–75.

Aguilar, R., Escorihuela, R. M., Gil, L., Tobena, A., & Fernández-Teruel, A. (2002a). Differences between two psychogenetically selected lines of rats in a swimming pool matching-to-place task: Long term effects of infantile stimulation. *Behavior Genetics, 32,* 127–134.

Aguilar, R., Gil, L., Flint, J., Gray, J. A., Dawson, G. R., Driscoll, P.,et al. (2002b). Learned fear, emotional reactivity and fear of heights: A fear analytic map from a large F2 intercross of Roman rat strains. *Brain Research Bulletin, 57,* 17–26.

Ambrosio, E., Goldberg, S. R., & Elmer, G. I. (1995). Behavior genetic investigation of the relationship between spontaneous locomotor activity and the acquisition of morphine self-administration behavior. *Behavioral Pharmacology, 6,* 229–237.

Aubry, J.-M., Bartanusz, V., Driscoll, P., Schulz, P., Steimer, T., & Kiss, J. Z. (1995). Corticotropin-releasing factor and vasopressin mRNA levels in Roman high- and low-avoidance rats: Response to open-field exposure. *Neuroendocrinology, 61,* 89–97.

Ayensu, W. K., Pucilowski, O., Mason, G. A., Overstreet, D. H., Rezvani, A. H., & Janowsky, D. S. (1995). Effects of chronic mild stress on serum complement activity, saccharin preference, and corticosterone levels in Flinders lines of rats. *Physiology & Behavior, 57,* 165–169.

Badishtov, B. A., Overstreet, D. H., Kashevskaya, O. P., Viglinskaya, I. V., Kampov-Polevoy, A. B., Seredenin, S. B., et al. (1995). To drink or not to drink: Open field behavior in alcohol-preferring and nonpreferring rat strains. *Physiology & Behavior, 57,* 585–589.

Bardo, M. T., Donohew, R. L., & Harrington, N. G. (1996). Psychobiology of novelty seeking and drug seeking behavior. *Behavioural Brain Research, 77*, 23–43.

Beitner-Johnson, D., Guitart, X., & Nestler, E. J. (1991). Dopaminergic brain reward regions of Lewis and Fischer rats display different levels of tyrosine hydroxylase and other morphine- and cocaine-regulated phosphoproteins. *Brain Research, 561*, 147–150.

Bell, S. M., Reynolds, J. G., Thiele, T. E., Gan, J., Figlewicz, D. P., & Woods, S. C. (1998). Effects of third intracerebroventricular injections of corticotrophin-releasing factor (CRF) on ethanol drinking and food intake. *Psychopharmacology, 139*, 128–135.

Bentareha, R., Araujo, F., Ruano, D., Driscoll, P., Escorihuela, R. M., Tobeña, A., et al. (1998). Pharmacological properties of the GABA-A receptor complex from brain regions of (hypoemotional) Roman high- and (hyperemotional) low-avoidance rats. *European Journal of Pharmacology, 354*, 91–97.

Bignami, G. (1965). Selection for high rates and low rates of avoidance conditioning in the rat. *Animal Behaviour, 13*, 221–227.

Blizard, D. A., & Adams, N. (2002). The Maudsley reactive and nonreactive strains: A new perspective. *Behavior Genetics, 32*, 277–299.

Broadhurst, P. L. (1960). Experiments in psychogenetics: Application of biometrical genetics to the inheritance of behaviour. In H. J. Eysenck (Ed.), *Experiments in Personality: Psychogenetics and Psychopharmacology*, (Vol. 1, pp. 1–102). London: Routledge and Kegan Paul.

Broadhurst, P. (1975). The Maudsley reactive and nonreactive strains of rats: A survey. *Behavior Genetics, 5*, 299–319.

Brush, F. R. (1991). Genetic determinants of individual differences in avoidance learning: Behavioral and endocrine characteristics. *Experientia, 47*, 1039–1050.

Brush, F. R. (2003). The Syracuse strains, selectively bred for differences in active avoidance learning, may be models of genetic differences in trait and state anxiety. *Stress, 6*, 77–85.

Brush, F. R., Froehlich, J. C., & Sakellaris, P. C. (1979). Genetic selection for avoidance behavior in the rat. *Behavior Genetics, 9*, 309–316.

Cailhol, S., & Mormède, P. (2000). Effects of cocaine-induced sensitization on ethanol drinking: Sex and strain differences. *Behavioural Pharmacology, 11*, 387–394.

Campbell, S. C. & Gerdes, M. (1987). Regional differences in cardiac myocyte dimensions and number in Sprague-Dawley rats from different suppliers. *Proceedings of the Society for Experimental Biology and Medicine, 186*, 211–217.

Castanon, N., & Mormède, P. (1994). Psychobiogenetics: Adapted tools for the study of the coupling between behavioral and neuroendocrine traits of emotional reactivity. *Psychoneuroendocrinology, 19*, 257–282.

Castanon, N., Dulluc, J., LeMoal, M., & Mormède, P. (1992). Prolactin as a link between behavioral and immune differences between the Roman rat lines. *Physiology & Behavior, 51*, 1235–1241.

Castanon, N., Hendley, E. D., Fan, X.-M., & Mormède, P. (1993). Psychoneuroendocrine profile associated with hypertension or hyperactivity in spontaneously hypertensive rats. *American Journal of Physiology, 265*, R1304–R1310.

Castanon, N., Dulluc, J., LeMoal, M., & Mormède, P., (1994). Maturation of the behavioral and neuroendocrine differences between the Roman rat lines. *Physiology & Behavior, 55*, 775–782.

Castanon, N., Perez-Diaz, F., & Mormède, P. (1995). Genetic analysis of the relationships between behavioral and neuroendocrine traits in Roman high and low avoidance rat lines. *Behavior Genetics, 25*, 371–384.

Chaouloff, F., Castanon, N., & Mormède, P. (1994). Paradoxical differences in animal models of anxiety among the Roman rat lines. *Neuroscience Letters, 182*, 217–221.

Clark, F. M., & Proudfit, H. K. (1992). Anatomical evidence for genetic differences in the innervation of the rat spinal cord by noradrenergic locus coeruleus neurons. *Brain Research, 591*, 44–53.

Clark, F. M., Yeomans, D. C., & Proudfit, H. K. (1991). The noradrenergic innervation of the spinal cord: Differences between two substrains of Sprague-Dawley rats determined by using retrograde tracers combined with immunocytochemistry. *Neuroscience Letters, 125*, 155–158.

Cloninger, C. R., Sigvardsson, S., & Bohman, M. (1988). Childhood personality predicts alcohol abuse in young adults. *Alcoholism: Clinical & Experimental Research, 12*, 494–505.

Commissaris, R. L., Ardayfio, P. A., McQueen, D. A., Gilchrist, G. A., & Overstreet, D. H. (2000). Conflict behavior and the effects of 8-OH DPAT treatment in rats selectively bred for differential 5-HT1A-induced hypothermia. *Pharmacology, Biochemistry & Behavior, 67*, 199–205.

Conti, L. H., MacIver, C. R., Ferkany, J. W., & Abreu, M. E. (1990). Footshock-induced freezing behavior in rats as a model for assessing anxiolytics. *Psychopharmacology, 102*, 492–497.

Cools, A. R., & Gingras, M. A. (1998). Nijmegen high and low responders to novelty: A new tool in the search after the neurobiology of drug abuse liability. *Pharmacology, Biochemistry & Behavior, 60*, 151–159.

Cools, A. R., Brachten, R., Heeren, D., Willemen, A., & Ellenbroek, B. (1990). Search after neurobiological profile of individual-specific features of Wistar rats. *Brain Research Bulletin, 24*, 49–69.

Cools, A., Dierx, J., Coenders, C., Heeren, D., Ried, S., Jenks, B. G., et al. (1993). Apomorphine-susceptible and apomorphine-unsusceptible Wistar rats differ in novelty-induced changes in hippocampal dinorphin B expression and two-way active avoidance: A new key in the search for the role of the hippocampal-accumbens axis. *Behavioural Brain research, 55*, 213–221.

Corda, M. G., Lecca, D., Piras, G., DiChiara, G., & Giorgi, O. (1997). Biochemical parameters of dopaminergic and GABAergic neurotransmission in the CNS of Roman high-avoidance and Roman low-avoidance rats. *Behavior Genetics, 27*, 527–536.

Corda, M. G., Piras, G., Valentini, V., Scano, P., & Giorgi, O. (1998). Differential sensitivity to shock-induced suppression of drinking in the Roman/Verh lines and strains of rats. *Society of Neuroscience Abstracts, 24*, 1182.

Corda, M. G., Piras, G., Lecca, D., Fernández-Teruel, A., Driscoll, P., & Giorgi, O. (2005). The psychogenetically selected Roman rat lines differ in the susceptibility to develop amphetamine sensitization. *Behavioural Brain Research, 157*, 147–156.

Crane, J. W., Ebner, K., & Day, T. A. (2003). Medial prefrontal cortex suppression of the hypothalamic-pituitary-adrenal axis response to a physical stressor, systematic delivery if interleukin-1B. *European Journal of Neuroscience, 17*, 1473–1481.

Crawley, J. N., & Moody, T. W. (1983). Anxiolytics block excessive grooming behavior induced by ACTH and Bombesin. *Brain Research Bulletin, 10*, 399–401.

Crocker, A. D., & Overstreet, D. H. (1991). Dopamine sensitivity in rats selectively bred for increases in cholinergic function. *Pharmacology, Biochemistry & Behavior, 38*, 105–108.

D'Angio, M., Serrano, A., Driscoll, P., & Scatton, B. (1988). Stressful environmental stimuli increase extracellular DOPAC levels in the prefrontal cortex of hypoemotional (Roman high-avoidance) but not hyperemotional (Roman low-avoidance) rats. An in vivo voltametric study. *Brain Research, 451*, 237–247.

Davids, E., Zhang, K., Tarazi, F. I., & Baldessarini, R. J. (2003). Animal models of attention-deficit hyperactivity disorder. *Brain Research Reviews, 42*, 1–21.

Dawson, G. R., Crawford, S. P., Collinson, N., Iversen, S. D., & Tricklebank, M. D. (1995). Evidence that the anxiolytic-like effects of chlordiazepoxide on the elevated plus maze are confounded by increases in locomotor activity. *Psychopharmacology, 118*, 316–323.

Dess, N. K., & Minor, T. R. (1996). Taste and emotionality in rats selectively bred for high versus low saccharin intake. *Animal Learning & Behavior, 24*, 105–115.

Dess, N. K., Badia-Elder, N. E., Thiele, T. E., Kiefer, S. W., & Blizard, D. A. (1998). Ethanol consumption in rats selectively bred for differential saccharin intake. *Alcohol, 16*, 275–278.

Drewek, K. J., & Broadhurst, P. L. (1979). Alcohol selection by strains of rats selectively bred for behavior. *Journal of Studies on Alcohol, 40*, 723–728.

Driscoll, P., (1986). Roman high- and low-avoidance rats: Present status of the Swiss sublines, RHA/Verh and RLA/Verh, and effects of amphetamine on shuttle-box performance. *Behavior Genetics, 16*, 355–364.

Driscoll, P., & Bättig, K. (1982). Behavioral, emotional and neurochemical profiles of rats selected for extreme differences in active, two-way avoidance performance. In I. Lieblich (Ed.), *Genetics of the Brain* (pp. 95–123). Amsterdam: Elsevier.

Driscoll, P., & Käsermann, H. P., (1977). Differences in the response to pentobarbital sodium of Roman high- and low-avoidance rats. *Arzneimittel-Forschung, 27*, 1582–1584.

Driscoll, P., & Martin, J. R. (1986). The relationship between genetic selection for extreme differences in two-way avoidance and two types of "aggressive" behavior in the rat. *International Journal of Neuroscience, 32*, 318–319.

Driscoll, P., Fümm, H., & Bättig, K. (1979). Maternal behavior in two rat lines selected for differences in the acquisition of two-way avoidance. *Experientia, 35*, 786–788.

Driscoll, P., Woodson, P., Fuemm, H., & Bättig, K. (1980). Selection for two-way avoidance deficit inhibits shock-induced fighting in the rat. *Physiology & Behavior, 24*, 793–795.

Driscoll, P., Dedek, J., Martin, J. R., & Zivkovic, B., (1983). Two-way avoidance and acute shock stress induced alterations of regional noradrenergic, dopaminergic and serotonergic activity in Roman high- and low-avoidance rats. *Life Sciences, 33*, 1719–1725.

Driscoll, P., Dedek, J., Fuchs, A., & Gentsch, C. (1985). Stereotypic, hypothermic, and central dopaminergic effects of apomorphine in Roman high-avoidance (RHA/Verh) and Roman low-avoidance (RLA/Verh) rats. *Behavior Genetics, 15*, 591–592.

Driscoll, P., Cohen, C., Fackelman, P., & Bättig, K. (1990a). Differential ethanol consumption in Roman high- and low-avoidance (RHA and RLA) rats, body weight, food intake, and the influence of pre- and post-natal exposure to nicotine and/or injection stress. *Experientia, 46, supplement*, A60.

Driscoll, P., Dedek, J., D'Angio, M., Claustre, Y., & Scatton, B. (1990b). A genetically-based model for divergent stress responses: Behavioral, neurochemical and hormonal aspects. *Advances in Animal Breeding & Genetics, 5*, 97–107.

Driscoll, P., Ferré, P., Fernández-Teruel, A., Levi de Stein, M., Wolfman, C., Medina, J., et al. (1995). Effects of prenatal diazepam on two-way avoidance behavior, swimming navigation and brain levels of benzodiazepine-like molecules in male Roman high- and low-avoidance rats. *Psychopharmacology, 122*, 51–57.

Driscoll, P., Escorihuela, R. M., Fernández-Teruel, A., Giorgi, O., Schwegler, H., Steimer, T., et al. (1998). Genetic selection and differential stress responses. The Roman lines/strains of rats. *Annals of the New York Academy of Sciences, 851*, 501–510.

Dunn, A. J., & Berridge, C. W. (1990). Physiological and behavioral responses to corticotrophin-releasing factor administration: Is CRF a mediator of anxiety or stress responses? *Brain Research Reviews, 15*, 71–100.

Ellenbroek, B. A., & Cools, A. R. (2002). Apomorphine susceptibility and animal models for psychopathology: Genes and environment. *Behavior Genetics, 32*, 349–361.

Ellenbroek, B. A., Geyer, M. A., & Cools, A. R. (1995). The behavior of APO-SUS rats in animal models with construct validity for schizophrenia. *Journal of Neuroscience, 15*, 7604–7611.

Ellenbroek, B. A., Sluyter, F., & Cools, A. R. (2000). The role of genetic and early environmental factors in determining apomorphine susceptibility. *Psychopharmacology, 148*, 124–131.

Enck, P., Merlin, V., Erckenbrecht, J. F., & Weinbeck, M. (1989). Stress effects on gastrointestinal transit in the rat. *Gut, 30*, 455–459.

Escorihuela, R. M., Tobeña, A., Driscoll, P., & Fernández-Teruel, A. (1995). Effects of training, early handling, and perinatal flumazenil on shuttle box acquisition in Roman low-avoidance rats: Toward overcoming a genetic deficit. *Neuroscience & Biobehavioral Reviews, 19*, 353–367.

Escorihuela, R. M., Fernández-Teruel, A., Tobeña, A., Langhans, W., Bättig, K., & Driscoll, P. (1997). Labyrinth exploration, emotional reactivity, and conditioned fear in young Roman/Verh inbred rats. *Behavior Genetics, 27*, 573–578.

Escorihuela, R. M., Fernández-Teruel, A., Gil, L., Aguilar, R., Tobeña, A., & Driscoll, P. (1999). Inbred Roman high- and low-avoidance rats: Differences in anxiety, novelty-seeking, and shuttlebox behaviors. *Physiology & Behavior, 67*, 19–26.

Everitt, B. J., & Wolf, M. E. (2002). Psychomotor stimulant addiction: A neural systems perspective. *Journal of Neuroscience, 22*, 3312–3320.

Ezerman, E. B., & Kromer, L. F. (1985). Outbred Sprague-Dawley rats from two breeders exhibit different incidences of neuroanatomical abnormalities affecting the primary cerebellar fissure. *Experimental Brain Research, 59*, 625–628.

Fattore, L., Piras, G., Corda, M. G., & Giorgi, O. (2008). The Roman high- and low-avoidance rat lines differ in the acquisition, maintenance, extinction, and reinstatement of intravenous cocaine self-administration. *Neuropsychopharmacology, in the press*.

Feldman, S., Conforti, N., & Melamed, E. (1987). Paraventricular nucleus serotonin mediates neurally stimulated adrenocortical secretion. *Brain Research Bulletin, 18*, 165–168.

Fernandes, C., & File, S. E. (1996). The influence of open arm ledges and maze experience in the elevated plus-maze. *Pharmacology, Biochemistry & Behavior, 54*, 31–40.

Fernández-Teruel, A., Escorihuela, R. M., Nuñez, J. F., Zapata, A., Boix, F., Salazar, W., et al. (1991). The early acquisition of two-way (shuttle-box) avoidance is an anxiety-mediated behavior: Psychopharmacological validation. *Brain Research Bulletin, 26*, 173–176.

Fernández-Teruel, A., Escorihuela, R. M., Nuñez, J. F., Goma, M., Driscoll, P., & Tobeña, A. (1992). Early stimulation effects on novelty-induced behavior in two psychogenetically-selected rat lines with divergent emotionality profiles. *Neuroscience Letters, 137*, 185–188.

Fernández-Teruel, A., Escorihuela, R. M., Castellano, B., González, B., & Tobeña, A. (1997a). Neonatal handling and environmental enrichment effects on emotionality, novelty/reward seeking, and age-related cognitive and hippocampal impairments: Focus on the Roman rat lines. *Behavior Genetics, 27*, 513–526.

Fernández-Teruel, A., Escorihuela, R. M., Tobeña, A., & Driscoll, P. (1997b). The inbred Roman rat strains: Similarities in morphological and pharmacological findings to the outbred Roman lines. *Behavior Genetics, 27*, 589.

Fernández-Teruel, A., Driscoll, P., Gil, L., Aguilar, R., Tobeña, A., & Escorihuela, R. M. (2002a). Enduring effects of environmental enrichment on novelty seeking, saccharin and ethanol intake in two rat lines (RHA/Verh and RLA/Verh) differing in incentive-seeking behavior. *Pharmacology, Biochemistry & Behavior, 73*, 225–231.

Fernández-Teruel, A., Escorihuela, R. M., Gray, J. A., Aguilar, R., Gil, L., Giménez-Llort, L., et al. (2002b). A quantitative trait locus influencing anxiety in the laboratory rat. *Genome Research, 12*, 618–626.

Ferré, P., Fernández-Teruel, A., Escorihuela, R. M., Driscoll, P., Corda, M. G., Giorgi, O., et al. (1995). Behavior of the Roman/Verh high- and low-avoidance rat lines in anxiety tests: Relationship with defecation and self-grooming. *Physiology & Behavior, 58,* 1209–1213.

File, S. E., & Velucci, S. V. (1979). Behavioural and biochemical measures of stress in hooded rats from different sources. *Physiology & Behavior, 22,* 31–35.

Fuemm, H., & Driscoll, P. (1981). Litter size manipulations do not alter maternal behaviour traits in selected lines of rats. *Animal Behaviour, 29,* 1267–1269.

Gentsch, C., Lichtsteiner, M., & Feer, H. (1981). Locomotor activity, defecation score and corticosterone levels during an open-field exposure: A comparison among individually and group-housed rats, and genetically selected rat lines. *Physiology & Behavior, 27,* 183–186.

Gentsch, C., Lichtsteiner, M., Driscoll, P., & Feer, H. (1982). Differential hormonal and physiological responses to stress in Roman high- and low-avoidance rats. *Physiology & Behavior, 28,* 259–263.

Gentsch, C., Lichtsteiner, M., & Feer, H. (1988). Genetic and environmental influences on behavioral and neurochemical aspects of emotionality in rats. *Experientia, 44,* 482–490.

Gentsch, C., Lichtsteiner, M., & Driscoll, P. (1989). Apomorphine-induced gnawing and licking: A comparison between RHA/Verh and RLA/Verh rats. *European Journal of Neuroscience, S2,* 315.

Gentsch, C., Lichtsteiner, M., & Feer, H. (1991). Genetic and environmental influences on reactive and spontaneous locomotor activities in rats. *Experientia, 47,* 998–1008.

Giménez-Llort, L., Cañete, T., Guitart-Masip, M., Fernández-Teruel, A., & Tobeña, A. (2005). Two distinctive apomorphine-induced phenotypes in the Roman high- and low-avoidance rats. *Physiology & Behavior, 86,* 458–466.

Gingras, M. A., & Cools, A. R. (1995). Differential ethanol intake in high and low responders to novelty. *Behavioural Pharmacology, 6,* 718–723.

Gingras, M. A., & Cools, A. R. (1997). Different behavioral effects of daily or intermittent dexamphetamine administration in Nijmegen high and low responders. *Psychopharmacology, 132,* 188–194.

Giorgi, O., Orlandi, M., Escorihuela, R. M., Driscoll, P., Lecca, D., & Corda, M. G. (1994). GABAergic and dopaminergic transmission in the brain of Roman high-avoidance and Roman low-avoidance rats. *Brain Research, 638,* 133–138.

Giorgi, O., Corda, M. G., Orlandi, M., Valentini, V., Carboni, G., Frau, V., et al. (1996). Differential ethanol (ETH) consumption in Roman high-avoidance (RHA) and low-avoidance (RLA) rats. *Society of Neuroscience Abstracts, 22,* 700.

Giorgi, O., Corda, M. G., Carboni, G., Frau, V., Valentini, V., & DiChiara, G. (1997). Effects of cocaine and morphine in rats from two psychogenetically selected lines: A behavioral and brain dialysis study. *Behavior Genetics, 27,* 537–546.

Giorgi, O., Valentini, V., Piras, G., DiChiara, G., & Corda, M. G. (1999). Palatable food differentially activated dopaminergic function in the CNS of Roman/Verh lines and strains of rats. *Society of Neuroscience Abstracts, 25,* 2152.

Giorgi, O., Piras, G., Lecca, D., & Corda, M. G. (2005a). Differential activation of dopamine release in the nucleus accumbens core and shell after acute or repeated amphetamine injections: A comparative study in the Roman high- and low-avoidance rat lines. *Neuroscience, 135,* 987–998.

Giorgi, O., Piras, G., Lecca, D., & Corda, M. G. (2005b). Behavioural effects of acute and repeated cocaine treatments: A comparative study in sensitization-prone RHA rats and their sensitization-resistant RLA counterparts. *Psychopharmacology, 180,* 530–538.

Giorgi, O., Piras, G., & Corda, M. G. (2007). The psychogenetically selected Roman high- and low-avoidance rat lines: A model to study the individual vulnerability to drug addiction. *Neuroscience & Biobehavioral Reviews, 31,* 148–163.

Glick, S. D., Shapiro, R. M., Drew, K. L., Hinds, P. A., & Carlson, J. N. (1986). Differences in spontaneous and amphetamine-induced rotational behavior, and in sensitization to amphetamine, among Sprague-Dawley derived rats from different sources. *Physiology & Behavior, 38,* 67–70.

Glowa, J. R., & Hansen, C. T. (1994). Differences in response to an acoustic startle stimulus among forty-six rat strains. *Behavior Genetics, 24,* 79–84.

Griebel, G., Moreau, J.-L., Jenck, F., Martin, J. R., & Misslin, R. (1993). Some critical determinants of the behaviour of rats in the elevated plus-maze. *Behavioural Processes, 29,* 37–48.

Grillon, C., Warner, V., Hille, J., Merikangas, K. R., Bruder, G. E., Tenke, C. E., et al. (2005). Families at high and low risk for depression: A three-generation startle study. *Biological Psychiatry, 57,* 953–960.

Guitart-Masip, M., Giménez-Llort, L., Fernández-Teruel, A., Cañete, T., Tobeña, A., Ögren, S. O., et al. (2006a). Reduced ethanol response in the alcohol-preferring RHA rats and neuropeptide mRNAs in relevant structures. *European Journal of Neuroscience, 23,* 531–540.

Guitart-Masip, M., Johansson, B., Fernández-Teruel, A., Cañete, T., Tobeña, A., Terenius, L., et al. (2006b). Divergent anatomical pattern of D1 and D3 binding and dopamine- and cyclic AMP-regulated phosphoprotein of 32 kDa mRNA expression in the Roman rat strains: Implications for drug addiction. *Neuroscience, 142,* 1231–1243.

Hall, C. S. (1938). The inheritance of emotionality. *American Scientist, 26,* 17–27.

Haney, M., Castanon, N., Cador, M., LeMoal, M., & Mormède, P. (1994). Cocaine sensitivity in Roman high and low avoidance rats is modulated by sex and gonadal hormone status. *Brain Research, 645,* 179–185.

Hédou, G., Feldon, J., & Heidbreder, C. A. (1999). Effects of cocaine on dopamine in subregions of the rat prefrontal cortex and their efferents to subterritories of the nucleus accumbens. *European Journal of Pharmacology, 372,* 143–155.

Hendley, E. D. (2000). WKHA rats with genetic hyperactivity and hyperreactivity to stress: A review. *Neuroscience and Biobehavioral Reviews, 24,* 41–44.

Henke, P. G. (1988). Electrophysiological activity in the central nucleus of the amygdale: Emotionality and stress ulcers in rats. *Behavioral Neuroscience, 102,* 77–83.

Henninger, M. S. H., Ohl, F., Hölter, S. M., Weissenbacher, P., Toschi, N., Löracher, P., et al. (2000). Unconditioned anxiety and social behaviour in two rat lines selectively bred for high and low anxiety-related behaviour. *Behavioural Brain Research, 111,* 153–163.

Hogg, S. (1996). A review of the validity and variability of the elevated plus-maze as an animal model of anxiety. *Pharmacology, Biochemistry & Behavior, 54,* 21–30.

Honkanen, A., Mikkola, J., Korpi, E. R., Hyytiä, P., Seppälä, T., & Ahtee, L. (1999). Enhanced morphine- and cocaine-induced behavioral sensitization in alcohol-preferring AA rats. *Psychopharmacology, 142,* 244–252.

Howes, S. R., Dalley, J. W., Morrison, C. H., Robbins, T. W., & Everitt, B. J. (2000). Leftward shift in the acquisition of cocaine self-administration in isolation-reared rats: Relationship to extracellular levels of dopamine, serotonin and glutamate in the nucleus accumbens and amygdala-striatal FOS expression. *Psychopharmacology, 151,* 55–63.

Huber, D., Veinante, P., & Stoop, R. (2005). Vasopressin and oxytocin excite distinct neuronal populations in the central amygdala. *Science, 308,* 245–248.

Hwang, B. H., Lumeng, L., Wu, J.-Y., & Li, T.-K. (1990). Increased number of GABAergic terminals in the nucleus accumbens is associated with alcohol preference in rats. *Alcoholism: Clinical & Experimental Research, 14*, 503–507.

Imada, H. (1972). Emotional reactivity and conditionability in four strains of rats. *Journal of Comparative & Physiological Psychology, 79*, 474–480.

Janowsky, D. S., Risch, S. C., Parker, D., Huey, L. Y., & Judd, L. L. (1980). Increased vulnerability to cholinergic stimulation in affective disorder patients. *Psychopharmacology Bulletin, 16*, 29–31.

Jones, A. E., McBride, W. J., Murphy, J. M., Lumeng, L., Li, T.-K., Shekhar, A., et al. (2000). Effects of ethanol on startle responding in alcohol-preferring and –non-preferring rats. *Pharmacology, Biochemistry & Behavior, 67*, 313–318.

Kabbaj, M., Devine, D. P., Savage, V. R., & Akil, H. (2000). Neurobiological correlates of individual differences in novelty-seeking behavior in the rat: Differential expression of stress-related molecules. *Journal of Neuroscience, 20*, 6983–6988.

Kalisch, R., Salomé, N., Platzer, S., Wigger, A., Czisch, M., Sommer, W., et al. (2004). High trait anxiety and hyperactivity to stress of the dorsomedial prefrontal cortex: A combined ph MRI AND FOS study in rats. *NeuroImage, 23*, 382–391.

Keck, M. E., Wigger, A., Welt, T., Müller, M. B., Gesing, A., Reul, J. M. H. M., et al. (2002). Vasopressin mediates the response of the combined dexamethasone/CRH test in hyper-anxious rats: Implications for pathogenesis of affective disorders. *Neuropsychopharmacology, 26*, 94–105.

Koolhaas, J. M., Korte, S. M., DeBoer, S. F., van der Vegt, B. J., van Reenen, C. G., Hopster, H., et al. (1999). Coping styles in animals: Current status in behavior and stress-physiology. *Neuroscience & Biobehavioral Reviews, 23*, 925–935.

Kurtz, D. L., Stewart, R. B., Zweifel, M., Li, T.-K., & Froehlich, J. C. (1996). Genetic differences in tolerance and sensitization to the sedative/hypnotic effects of alcohol. *Pharmacology, Biochemistry & Behavior, 53*, 585–591.

Landgraf, R., & Wigger, A. (2002). High vs low anxiety-related behavior in rats: An animal model of extremes in trait anxiety. *Behavior Genetics, 32*, 301–314.

Landgraf, R., & Wigger, A. (2003). Born to be anxious: Neuroendocrine and genetic correlates of trait anxiety in HAB rats. *Stress, 6*, 111–119.

Landgraf, R., Wigger, A., Holsboer, F., & Neumann, I. D. (1999). Hyper-reactive hypothalamo-pituitary-adrenocortical axis in rats bred for high anxiety-related behaviour. *Journal of Neuroendocrinology, 11*, 405–407.

Landgraf, R., Kessler, M. S., Bunck, M., Murgatroyd, C., Spengler, D., Zimbelmann, M., et al. (2007). Candidate genes of anxiety-related behavior in HAB/LAB rats and mice: Focus on vasopressin and glyoxalase-1. *Neuroscience & Biobehavioral Reviews, 31*, 89–102.

Lecca, D., Piras, G., Driscoll, P., Giorgi, O., & Corda, M. G. (2004). A differential activation of dopamine output in the shell and core of the nucleus accumbens is associated with the motor responses to addictive drugs: A brain dialysis study in Roman high- and low-avoidance rats. *Neuropharmacology, 46*, 688–699.

LeDoux, J. (1998). Fear and the brain: Where have we been, and where are we going? *Biological Psychiatry, 44*, 1229–1238.

Li, Y., Acerbo, M. J., & Robinson, T. E. (2004). The induction of behavioural sensitization is associated with cocaine-induced structural plasticity in the core (but not shell) of the nucleus accumbens. *European Journal of Neuroscience, 20*, 1647–1654.

Liebsch, G., Linthorst, A. C. E., Neumann, I. D., Reul, J. M. H. M., Holsboer, F., & Landgraf, R. (1998a). Behavioral, physiological and neuroendocrine stress responses and differential sensitivity to diazepam in two Wistar rat lines selectively bred for high- and low-anxiety-related behavior. *Neuropsychopharmacology, 19*, 381–396.

Liebsch, G., Montkowski, A., Holsboer, F., & Landgraf, R. (1998b). Behavioural profiles of two Wistar rat lines selectively bred for high and low anxiety-related behaviour. *Behavioural Brain Research, 94*, 301–310.

Lipp, H.-P., Schwegler, H., Crusio, W. E., Wolfer, D. P., Leisinger-Trigona, M.-C., Heimrich, B., et al. (1989). Using genetically-defined rodent strains for the identification of hippocampal traits relevant for two-way avoidance behavior: A non-invasive approach. *Experientia, 45*, 845–859.

López-Aumatell, R., Blazquez, G., Giménez-Llort, L., Gil, L., Aguilar, R., Tobeña, A., et al. (2005). Differences in classical fear conditioning and fear-potentiated startle between the Roman rat strains. Presented at the 11th EBPS meeting in Barcelona, September, 2005.

Markou, A., Matthews, K., Overstreet, D. H., Koob, G. F., & Geyer, M. A. (1994). Flinders resistant hypocholinergic rats exhibit startle sensitization and reduced startle thresholds. *Biological Psychiatry, 36*, 680–688.

Martin, J. R., Driscoll, P., & Gentsch, C. (1984). Differential response to cholinergic stimulation in psychogenetically selected rat lines. *Psychopharmacology, 83*, 262–267.

McKinzie, D. L., Sajdyk, T. J., McBride, W. J., Murphy, J. M., Lumeng, L., Li, T.-K., et al. (2000). Acoustic startle and fear-potentiated startle in alcohol-preferring (P) and –nonpreferring (NP) lines of rats. *Pharmacology, Biochemistry & Behavior, 65*, 691–696.

Meerlo, P., Overkamp, G. J. F., & Koolhaas, J. (1997). Behavioural and physiological consequences of a single social defeat in Roman high- and low-avoidance rats. *Psychoneuroendocrinology, 22*, 155–168.

Mönnikes, H., Heymann-Mönnikes, I., & Taché, Y. (1992). CRF in the paraventricular nucleus of the hypothalamus induces dose-related behavioral profile in rats. *Brain Research, 574*, 70–76.

Mormède, P., Moneva, E., Bruneval, C., Chaouloff, F., & Moisan, M.-P. (2002). Marker-assisted selection of a neuro-behavioural trait related to behavioural inhibition in the SHR strain, an animal model of ADHD. *Genes, Brain & Behavior, 1*, 111–116.

Murphy, J. M., Stewart, R. B., Bell, R. L., Badia-Elder, N. E., Carr, L. G., McBride, W. J., et al. (2002). Phenotypic and genotypic characterization of the Indiana University rat lines selectively bred for high and low alcohol preference. *Behavior Genetics, 32*, 363–388.

Nestler, E. J. (2000). Genes and addiction. *Nature Genetics, 26*, 277–281.

Neumann, I. D., Wigger, A., Krömer, S., Frank, E., Landgraf, R., & Bosch, O. J. (2005). Differential effects of periodic maternal separation on adult stress coping in a rat model of extremes in trait anxiety. *Neuroscience, 132*, 867–877.

Nil, R., & Bättig, K. (1981). Spontaneous maze ambulation and Hebb-Williams learning in Roman high-avoidance and Roman low-avoidance rats. *Behavioral & Neural Biology, 33*, 465–475.

Nowak, K. L., McBride, W. J., Lumeng, L., Li, T.-K., & Murphy, J. M. (1998). Blocking GABA-A receptors in the anterior ventral tegmental area attenuates ethanol intake of the alcohol-preferring P rat. *Psychopharmacology, 139*, 108–116.

Nowak, K. L., Ingraham, C. M., McKinzie, D. L., McBride, W. J., Lumeng, L., Li, T.-K., et al. (2000). An assessment of novelty-seeking behavior in alcohol-preferring and nonpreferring rats. *Pharmacology, Biochemistry & Behavior, 66*, 113–121.

O'Brien, C. P., & McLellan, A. T. (1996). Myths about the treatment of addiction. *Lancet, 347*, 237–240.

Ojanen, S., Koistinen, M., Bäckström, P., Kankaanpää, A., Tuomainen, P., Hyytiä, P., et al. (2003). Differential behavioural sensitization to intermittent morphine treatment in alcohol-preferring AA and alcohol-avoiding ANA rats: Role of mesolimbic dopamine. *European Journal of Neuroscience, 17*, 1655–1663.

Overstreet, D. H. (2002). Behavioral characteristics of rat lines selected for differential hypothermic responses to cholinergic or serotonergic agonists. *Behavior Genetics, 32*, 335–348.

Overstreet, D. H., & Rezvani, A. H. (1996). Behavioral differences between two inbred strains of Fawn-Hooded rat: A model of serotonin dysfunction. *Psychopharmacology, 128*, 328–330.

Overstreet, D. H., Rezvani, A. H., & Janowsky, D. S. (1990). Impaired active avoidance responding in rats selectively bred for increased cholinergic function. *Physiology & Behavior, 47*, 787–788.

Overstreet, D. H., Kampov-Polevoy, A. B., Rezvani, A. H., Murrelle, L., Halikas, J. A., & Janowsky, D. S. (1993). Saccharin intake predicts ethanol intake in genetically heterogeneous rats as well as different rat strains. *Alcoholism: Clinical & Experimental Research, 17*, 366–369.

Overstreet, D. H., Friedman, E., Mathé, A. A., & Yadid, G. (2005). The Flinders sensitive line rat: A selectively bred putative animal model of depression. *Neuroscience & Biobehavioral Reviews, 29*, 739–759.

Overstreet, D. H., Rezvani, A. H., Djouma, E., Parsian, A., & Lawrence, A. J. (2007). Depressive-like behavior and high alcohol drinking co-occur in the FH/Wjd rat but appear to be under independent genetic control. *Neuroscience & Biobehavioral Reviews, 31*, 103–114.

Paré, W. P., & Kluczynski, J. (1997). Differences in the stress response of Wistar-Kyoto (WKY) rats from different vendors. *Physiology & Behavior, 62*, 643–648.

Paré, W. P., Glavin, G. B., & Vincent, G. P. (1977). Vendor differences in starvation-induced gastric ulceration. *Physiology & Behavior, 19*, 315–317.

Piazza, P. V., Deminière, J.-M., LeMoal, M., & Simon, H. (1989). Factors that predict individual vulnerability to amphetamine self-administration. *Science, 245*, 1511–1513.

Piazza, P. V., Deminière, J.-M., Maccari, S., Mormède, P., LeMoal, M., & Simon, H. (1990). Individual reactivity to novelty predicts probability to amphetamine self-administration. *Behavioral Pharmacology, 1*, 339–345.

Piazza, P. V., Maccari, S., Deminière, J.-M., LeMoal, M., Mormède, P., & Simon, H. (1991). Corticosterone levels determine individual vulnerability to amphetamine self-administration. *Proceedings of the National Academy of Sciences USA, 88*, 2088–2092.

Piras, G., Lecca, D., Corda, M. G., & Giorgi, O. (2003). Repeated morphine injections induce behavioral sensitization in Roman high- but not in Roman low-avoidance rats. *Neuroreport, 14*, 2433–2438.

Pisula, W. (2003). The Roman high- and low-avoidance rats respond differently to novelty in a familiarized environment. *Behavioral Processes, 63*, 63–72.

Poulos, C. X., Le, A. D., & Parker, J. L. (1995). Impulsivity predicts individual susceptibility to high levels of alcohol self-administration. *Behavioral Pharmacology, 6*, 810–814.

Ramos, A., Berton, O., Mormède, P., & Chaouloff, F. (1997). A multiple-test study of anxiety-related behaviours in six inbred rat strains. *Behavioural Brain Research, 85*, 57–69.

Ramos, A., Mellerin, Y., Mormède, P., & Chaouloff, F. (1998). A genetic and multifactorial analysis of anxiety-related behaviours in Lewis and SHR intercrosses. *Behavioural Brain Research, 96*, 195–205.

Ramos, A., Moisan, M.-P., Chaouloff, F., Mormède, C., & Mormède, P. (1999). Identification of female-specific QTLs affecting an emotionality-related behavior in rats. *Molecular Psychiatry, 4*, 453–462.

Razafimanalina, R., Mormède, P., & Velley, L. (1996). Gustatory preference-aversion profiles for saccharin, quinine and alcohol in Roman high- and low-avoidance lines. *Behavioral Pharmacology, 7*, 78–84.

Reich, T., Hinrichs, A., Culverhouse, R., & Bierut, L. (1999). Genetic studies of alcoholism and substance dependence. *American Journal of Human Genetics, 65*, 599–605.

Rezvani, A,H., Overstreet, D. H., Ejantkar, A., & Gordon, C. J. (1994). Autonomic and behavioral responses of selectively bred hypercholinergic rats to oxotremorine and diisopropyl fluorophosphate. *Pharmacology, Biochemistry & Behavior, 48*, 703–707.

Rezvani, A. H., Parsian, A., & Overstreet, D. H. (2002). The Fawn-Hooded (FH/Wjd) rat: A genetic animal model of comorbid depression and alcoholism. *Psychiatric Genetics, 12*, 1–16.

Robinson, T. E., & Berridge, K. C. (2001). Incentive-sensitization and addiction. *Addiction, 96*, 103–114.

Rodgers, R. J. (1997). Animal models of "anxiety": Where next? *Behavioral Pharmacology, 8*, 477–496.

Roozendaal, B., Koolhaas, J. M., & Bohus, B. (1991). Attenuated cardiovascular, neuroendocrine, and behavioral responses after a single footshock in central amygdaloid lesioned male rats. *Physiology & Behavior, 50*, 771–775.

Roozendaal, B., Wiersma, A., Driscoll, P., Koolhaas, J. M., & Bohus, B. (1992). Vasopressinergic modulation of stress responses in the central amygdala of the Roman high-avoidance and low-avoidance rat. *Brain Research, 596*, 35–40.

Rots, N. Y., Cools, A. R., Oitzl, M. S., de Jong, J., Sutanto, W., & de Kloet, E. R. (1996). Divergent prolactin and pituitary-adrenal activity in rats selectively bred for different dopamine responsiveness. *Endocrinology, 137*, 1678–1686.

Sagvolden, T. (2000). Behavioral validation of the spontaneously hypertensive rat (SHR) as an animal model of attention-deficit/hyperactivity disorder (AD/HD). *Neuroscience & Biobehavioral Reviews, 24*, 31–39.

Saigusa, T., Tuinstra, T., Koshikawa, N., & Cools, A. R. (1999). High and low responders to novelty: Effects of a catecholamine synthesis inhibitor on novelty-induced changes in behaviour and release of accumbal dopamine. *Neuroscience, 88*, 1153–1163.

Salomé, N., Salchner, P., Viltart, O., Sequeira, H., Wigger, A., Landgraf, R., et al. (2004). Neurobiological correlates of high (HAB) versus low anxiety-related behavior (LAB): Differential Fos expression in HAB and LAB rats. *Biological Psychiatry, 55*, 715–723.

Sandi, C., Castanon, N., Vitiello, S., Neveu, P. J., & Mormède, P. (1991). Different responsiveness of spleen lymphocytes from two lines of psychogenetically selected rats (Roman high and low avoidance). *Journal of Neuroimmunology, 31*, 27–33.

Sarrieau, A., & Mormède, P. (1998). Hypothalamic-pituitary-adrenal axis activity in the inbred Brown Norway and Fischer 344 rat strains. *Life Sciences, 62*, 1417–1425.

Sarrieau, A., Chaouloff, F., Lemaire, V., & Mormède, P. (1998). Comparison of the neuroendocrine responses to stress in outbred, inbred and F1 hybrid rats. *Life Sciences, 63*, 87–96.

Satinder, K. P. (1975). Interactions of age, sex and long-term alcohol intake in selectively bred strains of rats. *Journal of Studies on Alcohol, 36*, 1493–1507.

Saxton, P. M., Siegel, J., & Lukas, J. H. (1987). Visual evoked potential augmenting/reducing slopes in cats – 2. correlations with behavior. *Personality & Individual Differences, 8*, 511–519.

Schuckit, M. A., & Hesselbrock, V. (1994). Alcohol dependence and anxiety disorders: What is the relationship? *American Journal of Psychiatry, 151*, 1723–1734.

Schwarz, R. M., Burkhart, B. R., & Green, S. B. (1982). Sensation-seeking and anxiety as factors in social drinking by men. *Journal of Studies on Alcohol, 43*, 1108–1114.

Schwegler, H., Pilz, P. K. D., Koch, M., Fendt, M., Linke, R., & Driscoll, P. (1997). The acoustic startle response in inbred Roman high- and low-avoidance rats. *Behavior Genetics, 27*, 579–582.

Setem, J., Pinheiro, A. P., Motta, V. A., Morato, S., & Cruz, A. P. M. (1999). Ethopharmacological analysis of 5-HT ligands on the rat elevated plus maze. *Pharmacology, Biochemistry & Behavior, 62*, 515–521.

Shephard, R. A., & Broadhurst, P. L. (1983). Hyponeophagia in the Roman rat strains: Effects of 5-methoxy-N,N-dimethyltryptamine, diazepam, methysergide and the stereoisomers of propranol. *European Journal of Pharmacology, 95*, 177–184.

Siegel, J. (1997). Augmenting and reducing of visual evoked potentials in high- and low-sensation seeking humans, cats and rats. *Behavior Genetics, 27*, 557–563.

Siegel, J., Sisson, D. F., & Driscoll, P. (1993). Augmenting and reducing of visual evoked potentials in Roman high- and low-avoidance rats. *Physiology & Behavior, 54*, 707–711.

Siegel, J., Gayle, D., Sharma, A., & Driscoll, P. (1996). The locus of origin of augmenting and reducing of visual evoked potentials in rat brain. *Physiology & Behavior, 60*, 287–291.

Sinclair, J. D., Kampov-Polevoy, A., Stewart, R., & Li, T.-K. (1992). Taste preferences in rat lines selected for low and high alcohol consumption. *Alcohol, 9*, 155–160.

Sluka, K. A., & Westlund, K. N. (1992). Spinal projections of the locus coeruleus and the nucleus subcoeruleus in the Harlan and the Sasco Sprague-Dawley rat. *Brain Research, 579*, 67–73.

Sluyter, F., Hof, M., Ellenbroek, B. A., Degen, S. B. & Cools, A. R. (2000). Genetic, sex and early environmental effects on the voluntary alcohol intake in Wistar rats. *Pharmacology, Biochemistry & Behavior, 67*, 801–808.

Solanto, M. V. (2000). Clinical pharmacology of AD/HD: Implications for animal models. *Neuroscience & Biobehavioral Reviews, 24*, 27–30.

Spuhler, K., & Deitrich, R. A. (1984). Correlative analysis of ethanol-related phenotypes in rat inbred strains. *Alcoholism: Clinical & Experimental Research, 8*, 480–484.

Stead, J. D. H., Clinton, S., Neal, C., Schneider, J., Jama, A., Miller, S., et al. (2006). Selective breeding for divergence in novelty-seeking traits: Heritability and enrichment in spontaneous anxiety-related behaviors. *Behavior Genetics, 36*, 697–712.

Steimer, T., & Driscoll, P. (2003). Divergent stress responses and coping styles in psychogenetically selected Roman high- (RHA) and low- (RLA) avoidance rats: Behavioural, neuroendocrine and developmental aspects. *Stress, 6*, 87–100.

Steimer, T., & Driscoll, P. (2005). Inter-individual vs line/strain differences in psychogenetically selected Roman high- (RHA) and low- (RLA) avoidance rats: Neuroendocrine and behavioural aspects. *Neuroscience & Biobehavioral Reviews, 29*, 99–112.

Steimer, T., laFleur, S., & Schulz, P. E. (1997a). Neuroendocrine correlates of emotional reactivity and coping in male rats from the Roman high (RHA/Verh)- and low (RLA/Verh)-avoidance lines. *Behavior Genetics, 27*, 503–512.

Steimer, T., Driscoll, P., & Schulz, P. E. (1997b). Brain metabolism of progesterone, coping behaviour and emotional reactivity in male rats from two psychogenetically selected lines. *Journal of Neuroendocrinology, 9*, 169–175.

Steimer, T., Escorihuela, R. M., Fernández-Teruel, A., & Driscoll, P. (1998). Long-term behavioural and neuroendocrine changes in Roman high- (RHA/Verh) and low- (RLA/Verh) avoidance rats following neonatal handling. *International Journal of Developmental Neuroscience, 16*, 165–174.

Steimer, T., Python, A., Schulz, P. E., & Aubrey, J.-M. (2007). Plasma corticosterone, dexamethasone (DEX) suppression and DEX/CRH tests in a rat model of genetic vulnerability to depression. *Psychoneuroendocrinology, 32*, 575–579.

Stephens, D. N. (1997). Animal models of anxiety; grounds for depression? Commentary on Rodgers, "Animal models of anxiety: Where next?". *Behavioural Pharmacology, 8*, 497–501.

Stewart, R. B., Gatto, G. J., Lumeng, L., Li, T.-K., & Murphy, J. M. (1993). Comparison of alcohol-preferring (P) and nonpreferring (NP) rats on tests of anxiety and for the anxiolytic effects of ethanol. *Alcohol, 10*, 1–10.

Swerdlow, N. R., Martinez, Z. A., Hanlon, F. M., Platten, A., Farid, M., Auerbach, P., et al. (2000). Toward understanding the biology of a complex phenotype: Rat strain and substrain differences in the sen-sorimotor gating-disruptive effects of dopamine agonists. *Journal of Neuroscience, 20*, 4325–4336.

Thielen, R. J., McBride, W. J., Lumeng, L., & Li, T.-K. (1998). GABA-A receptor function in the cerebral cortex of alcohol-naive P and NP rats. *Pharmacology, Biochemistry & Behavior, 59*, 209–214.

Tryon, R. C. (1929). The genetics of learning ability in rats. Preliminary report. *University of California Psychology Publications, 4*, 71–89.

Van Loo, K. M. J., & Martens, G. J. M. (2007). Identification of genetic and epigenetic variations in a rat model for neurodevelopmental disorders. *Behavior Genetics, 37*, 697–705.

Verheul, R., & van den Brink, W. (2000). The role of personality pathology in the etiology and treatment of substance use disorders. *Current Opinions in Psychiatry, 13*,163–169.

Viglinskaya, I. V., Overstreet, D. H., Kashevskaya, O. P., Badishtov, B. A., Kampov-Polevoy, A. B., Seredenin, S. B., et al. (1995). To drink or not to drink: Tests of anxiety and immobility in alcohol-preferring and alcohol-nonpreferring rat strains. *Physiology & Behavior, 57*, 937–941.

Wahlsten, D., Metten, P., Phillips, T. J., Boehm, S. L., Burkhart-Kasch, S., Dorow, J., et al. (2003). Different data from different labs: Lessons from studies of gene-environment interaction. *Journal of Neurobiology, 54*, 283-311.

Wahlsten, D., Bachmanov, A., Finn, D. A., & Crabbe, J. C. (2006). Stability of inbred mouse strain differences in behavior and brain size between laboratories and across decades. *Proceedings of the National Academy of Sciences USA, 103*, 16364–16369.

Walker, C.-D., Rivest, R. W., Meaney, M. J., & Aubert, M. L. Differential activation of the pituitary-adrenocortical axis after stress in the rat: Use of two genetically selected lines (Roman low- and high-avoidance rats) as a model. *Journal of Endocrinology, 123*, 477–485.

Walker, C.-D., Aubert, M. L., Meaney, M. J., & Driscoll, P. (1992). Individual differences in the activity of the hypothalamus-pituitary-adrenocortical system after stressors: Use of psychogenetically selected rat lines as a model. In P. Driscoll (Ed.), *Genetically Defined Animal Models of Neurobehavioral Dysfunctions* (pp. 276–296). Boston: Birkhäuser.

Walker, D. L., & Davis, M. (1997). Double dissociation between the involvement of the bed nucleus of the stria terminalis and the central nucleus of the amygdala in startle increases produced by conditioned versus unconditioned fear . *Journal of Neuroscience, 17*, 9375–9383.

Walker, D. L., Toufexis, D. J., & Davis, M. (2003). Role of the bed nucleus of the stria terminalis versus the amygdala in fear, stress, and anxiety. *European Journal of Pharmacology, 463*, 199–216.

Whishaw, I. Q. (1999). The laboratory rat, the Pied Piper of twentieth century neuroscience. *Brain Research Bulletin, 50*, 411.

Wiersma, A., Konsman, J. P., Knollema, S., Bohus, B., & Koolhaas, J. M. (1998). Differential effects of CRH infusion into the central nucleus of the amygdala in the Roman high-avoidance and low-avoidance rats. *Psychoneuroendocrinology, 23*, 261–274.

Wigger, A., Sánchez, M. M., Mathys, K. C., Ebner, K., Frank, E., Liu, D., et al. (2004). Alterations in central neuropeptide expression, release, and receptor binding in rats bred for high anxiety: Critical role of vasopressin. *Neuropsychopharmacology, 29*, 1–14.

Wilcock, J., & Fulker, D. W. (1973). Avoidance learning in rats: Genetic evidence for two distinct behavioral processes in the shuttlebox. *Journal of Comparative & Physiological Psychology, 82*, 247–253.

Willig, F., M'Harzi, M., Bardelay, C., Viet, D., & Delacour, J. (1991a). Roman strains as a psychogenetic model for the study of working memory: Behavioral and biochemical data. *Pharmacology, Biochemistry & Behavior, 40*, 7–16.

Willig, F., M'Harzi, M., & Delacour, J. (1991b). Contribution of the Roman strains of rats to the elaboration of animal models of memory. *Physiology & Behavior, 50,* 913–919.

Windle, R. J., Gamble, L. E., Kershaw, Y. M., Wood, S. A., Lightman, S. L., & Ingram, C. D. (2006). Gonadal steroid modulation of stress-induced hypothalamo-pituitary-adrenal activity and anxiety behavior: Role of central oxytocin. *Endocrinology, 147,* 2423–2431.

Yadid, G., Overstreet, D. H., & Zangen, A. (2001). Limbic dopaminergic adaptation to a stressful stimulus in a rat model of depression. *Brain Research, 896,* 43–47.

Yilmazer-Hanke, D. M., Faber-Zuschratter, H., Linke, R., & Schwegler, H. (2002). Contribution of amygdala neurons containing peptides and calcium-binding proteins to fear-potentiated startle and exploration-related anxiety in inbred Roman high- and low-avoidance rats. *European Journal of Neuroscience, 15,* 1206–1218.

Yilmazer-Hanke, D. M., Wigger, A., Linke, R., Landgraf, R., & Schwegler, H. (2004). Two Wistar rat lines selectively bred for anxiety-related behavior show opposite reactions in elevated plus maze and fear-sensitized acoustic startle tests. *Behavior Genetics, 34,* 309–318.

Zeier, H., Baettig, K., & Driscoll, P. (1978). Acquisition of DRL-20 behavior in male and female, Roman high- and low-avoidance rats. *Physiology & Behavior, 20,* 791–793.

Chapter 21

The Genetics of Offensive Aggression in Mice

Stephen C. Maxson

Introduction

Male aggression was the first behavior studied in inbred strains of mice. Differences were found by Scott (1942) and by Ginsburg and Allee (1942) across the same three inbred strains. Here was the first evidence that genetic variants may have an effect on individual differences in male mouse aggression. These two studies also showed the first strain by environment interaction for a mouse behavior. When C57BL/10 mice were transferred from cage to cage by picking them up by forceps on the tail, they were more aggressive than when they were transferred from cage to cage in a small box or allowed to do so on their own. This treatment had no effect on the aggressive behaviors of the other strains (C3H and Bagg albino [BALB/c]). This finding was replicated (Ginsburg & Jummonville, 1967). Also, the study of Ginsburg and Allee showed for the first time that individual difference in aggression suspected to be due to genes could be modified by experience. Mice of the most pacific strain could be rendered aggressive by helping them to win fights, and mice of the most pugnacious strain could be rendered pacific by subjecting them to a series of defeats.

This chapter considers (1) the description and measurements of two kinds of aggression: offense and defense, (2) experiential parameters for tests of male offense, (3) the effect of the interactions between genes and the experimental parameters on male offense and on the underlying biological systems, (4) whether the same genes are involved in male offense and female aggression, and (5) a comparative genetics of aggression in which the genetics of offense in mice and other animals can be related to human aggression.

S.C. Maxson (✉)
Department of Psychology, The University of Connecticut, Storrs, CT 06269-1020, USA
e-mail: Stephen.Maxson@Uconn.Edu

Description and Measurement of Offense and Defense

Two types of aggression in adult male and female mice are offense and defense. They differ in motor patterns, attack or bite targets, adaptive function, and physiology (Adams, 1980). The motor patterns of offense are chase, sideways offensive posture, upright offensive posture, and attack whereas the motor pattern of defense are flight, sideways defensive posture, upright defensive posture, and attack. Tail rattle also occurs in aggressive encounters but it is difficult to classify as either an offense or defense motor pattern. The attacks in offense are directed primarily at the flanks, rump, and base of tail, whereas the attacks in defense are directed primarily at the face and shoulders. Injuries from offensive attack are rarely lethal whereas injuries from defensive attacks often are. Offense has the adaptive function of obtaining and retaining resources such as space, food, and mates, whereas defense has the adaptive function of defending oneself from injury by others. Defense may also be involved in protecting progeny and mates from injury.

It has been usual in genetic studies of offense to use composite or single scores. Composite scores often index the frequency, severity, or duration of fighting. Single scores may index the latency, frequency, or duration of one of the motor components. Neither provides an adequate description of offense that is suitable for its complete genetic analysis. For example, the frequency of chase and attack is partially but not fully correlated across 11 inbred strains of mice (Roubertoux, Le Roy, Mortaud, Perez-Diaz, & Tordjman, 1999). This implies that some genetic variants affect each motor component and others affect only some of the motor components. Composite and single scores can miss gene effects on one or more motor components. Similarly, there may be gene effects on one or more of the measurements (latency, frequency, or duration) of a motor pattern. For example, across inbred strains, the frequency of chase or attack is partially but not fully correlated with latency to attack (Roubertoux et al., 1999). Reliance on single measures of offense can miss gene effects on other measures. For these reasons, I strongly recommend that after a genetic variant has been shown to affect some aspect of offense, the latency, frequency, and duration for each of the motor components of an agonistic behavior be described and scored for that genetic variant. For some variants, it may also be useful to analyze the temporal sequencing of each motor component.

Life History and Test Parameters

Whether a genetic variant has effects on offense can also depend on life history and test parameters (Roubertoux et al., 1999). The former includes being singly housed versus pair housed for a period before the aggression test. Single housing is often referred to as isolation. The test parameters include the test area (Maxson, 1992a). Two types of test areas are commonly used. These are the resident cage and a neutral cage. The resident-intruder test occurs in the home cage of one of the contestants (the resident). In the resident-intruder test, the offensive behavior of the resident is scored. The neutral cage test occurs in an arena that is not the home cage of either contestant. In the neutral cage test, offense of the subject mouse is scored.

The test parameters also include the type of opponent (Maxson, 1992a; Mizcek, Maxson, Fish, & Faccidomo, 2002). If the genotype of the two opponents is the same, this is referred to as a homogeneous set test. This can only be used with isogenic populations. If the genotype of the opponent is the same for all experimental subjects, this is known as a standard opponent test. It can be used with isogenic and heterogenic populations. Sometimes the opponent is experimentally modified to reduce or eliminate its attacks against the experimental subject. This includes for the opponent either a series of prior defeats, olfactory bulbectomy, or gonadectomy.

There are urinary and other chemosignals from one opponent male that affect the occurrence and intensity of offense in the other opponent male. Some of these signals enhance and others inhibit offense, and some of each kind of signal are testosterone and/or melanocortin dependent. Testosterone-dependent and melanocortin-dependent signals come from the preputial gland (Morgan & Cone, 2006). These signals from the preputial gland consist of alpha and beta farnesene and are found in male urine. Other testosterone-dependent signals come from the bladder. Recent genetic analyses indicate that these urinary pheromones from one male stimulate offense in another male by acting on that male's vomeronasal organ (VNO) and main olfactory epithelium (MOE). The Trp2 ion channel is expressed in the VNO, and mutant male mice lacking this ion channel do not attack intruder males (Leypold

et al., 2002). The genes for the CNGA2 ion channel and 3 adenylate cyclase are expressed in the MOE, and mutant male mice lacking either of these fail to fight other males (Mandiyan, Coats, & Shah, 2005, Wang, Balet Sindreu, Li, Nudelman, Chan, & Storm, 2006). These and other findings are the basis, in part, for a model of the interaction between inputs from the VNO and MOE with effects on aggression (Dulac & Wagner, 2006).

Also, there are strain differences in the urinary chemosignals affecting male offense (Kessler, Harmatz, & Gerling, 1975). Urine from CBA/J, C57BR/cdJ, and DBA/2J strains was daubed on the anogenital region of gonadectomized DBA/2 strain males. Males with DBA/2 strain urine elicited more offense from a DBA/2 strain male than did males with CBA strain urine, and males with CBA strain urine elicited more offense than those with C57BR strain urine. Offense was indexed by latency to attack, number of attacks, and accumulated attack time. Thus, the urinary odors of mice from some strains may elicit more offense than others. There are also testosterone-independent signals. Some of these are individual recognition odor types that are affected by variants of the Y chromosome (Schellinck, Monahan, Brown, & Maxson, 1993) or other genes (Maxson, 1992). It has been suggested that mice with the same Y chromosome and therefore identical odor types are more aggressive than those with different Y chromosomes (Monahan & Maxson, 1998). Thus, mice in a homogenous set test may be more aggressive than in a standard opponent test.

Genetic Effects on Offensive Aggression in Males

Ginsburg (1958) proposed that gene variants could be used as tools to study the biology of behaviors. Benzer (1971) also proposed that this would be best done with single gene mutants having large effects on the behavior. To date, variants of 56 genes, mostly identified from transgenic or knockout studies, have been shown to affect offense in male mice (Table 21.1). Some of these act on olfactory, endocrine, neurotransmitter, or second messenger systems. Still others have effects on other biological systems. Some effects of genetic variants on male offense depend on life history and test parameters, and others are independent of these. Elsewhere, I have suggested that the biology of offense may depend on these parameters and that there may be at least two different biologies of aggression with each dependent on different life history and test parameters (Maxson & Canastar, in press). Roubertoux et al. (1999) have made a similar suggestion. This is the topic of this section.

In a study of offense in 11 inbred strains, there were 4 combinations of life history and test parameters (Roubertoux

Table 21.1 Genes with effects on male offense

Systems	Name (Symbol)	Chromosome
Olfaction		
	Adenylate cyclase-activating polypeptide 1 receptor (*Adcyap1r1*)	6
	Beta 2-microglobin (*B2m*)	12
	Cyclic-nucleotide-gated channel α2 (*Cnga2*)	9
	Guanine nucleotide-binding protein, alpha inhibiting 2 (*Gnai2*)	9
	Melanocortin-5 receptor (Mc5r)	19
	TRP ion channel (*Trpc2*)	7
	Type 3 adenylate cyclase (*Adcy3*)	12
Hormones		
	Androgen receptor (*Ar*)	X
	Aromatase (*Cyp19*)	9
	Corticotropin-releasing hormone receptor 2 (*Crhr2*)	6
	α-Estrogen receptor (*Esr1*)	10
	β-Estrogen receptor (*Esr2*)	12
	Steroid sulfatase (*Sts*)	X/Y pseudoautosomal region
Neurotransmitters		
	Acetylcholine esterase (*Ache*)	5
	Adenosine 1 receptor (*Adora1*)	1
	Adenosine 2a receptor (*Adora2a*)	10
	Adrenergic alpha 2c receptor (*Adra2c*)	5
	Arginine vasopressin 1 b receptor (*Avpr1b*)	1
	Cannabinoid receptor 1 (*Cnr1*)	4
	Catechol-*O*-methyl transferase (*Comt*)	16
	Dopamine DRD2 receptor 1 (Drd2)	9
	Dopamine β hydroxylase (*Dbh*)	2
	Dopamine transporter (*Slc6a3*)	13
	Enkephalin (*Penk1*)	4
	Glutamic acid decarboxylase 65 (*Gad2*)	2
	Glutamate receptor, ionotropic, AMPA1 (alpha 1) (*Gria1*)	11
	Glutamate receptor, ionotropic, AMPA3 (alpha 3) (*Gria3*)	X
	Glutamate receptor, ionotropic, NMDA1 (zeta 1) (*Grin1*)	2
	Histamine 1 receptor (*Hrh1*)	6
	Monoamine oxidase A (*Maoa*)	X
	Membrane metal endopeptidase (*Mm2*)	3
	Nitric oxide synthase 1, neuronal (*Nos1*)	5
	Nitric oxide synthase 3, endothelial (*Nos3*)	5
	Neuropeptide Y1 receptor (*Npy1r*)	8
	Oxytocin (*Oxt*)	2
	Oxytocin receptor (Oxtr)	6
	Pet-1 ETS factor (*Pet-1*)	1
	Phenylethanolamine-*N*-methyltransferase (*Pnmt*)	11
	Serotonin transporter (*Slc6a4*)	11
	Serotonin 1b receptor (*Htrlb*)	9
	Tachykinin-1 receptor (*Tacr1*)	6
	Tryptophan hydroxylase 2 (*Tph2*)	10
Signaling		
	α-Calcium/calmodulin kinase II (*CamK2a*)	18

Table 21.1 (Continued)

Systems	Name (Symbol)	Chromosome
	Breakpoint cluster region (*Bcr*) (regulation of Rho family small GTPase proteins)	10
	Guanine diphosphate dissociation inhibitor (*Gdi-1*)	X
	Regulator of G-protein signaling 2 (*Rgs2*)	1
Other		
	Amyloid beta (A4) precursor *(App)* protein	16
	Brain creatine kinase (Ckb)	12
	Brain-derived neurotrophic factor (*Bdnf*)	2
	Engrailed 2 transcription factor (*En2*)	5
	Fyn tyrosine kinase (*Fyn*)	10
	Gene trap ROSA-b-Geo22 (*Gtrgeo22*)	10
	Huntington's disease gene homolog (*Hdh*)	5
	Interleukin-6 (*Il-6*)	5
	Neural cell adhesion molecule (*Ncam*)	9
	Nuclear receptor family 2 group E member 1 (*Nr3e1*)	10
	Transforming growth factor alpha (*Tgfa*)	6

Except for the following references for each gene may be found in Miczek, Maxson, Fish, and Faccidomo (2001), Maxson, Roubertoux, Guillot, and Goldman (2001), or Maxson and Canastar (2003, 2006): Wang et al., 2006 (*Ac3*), Nicot, Otto, Brabet, and Dicicco-Bloom, 2004 (*Adcyap1r*), Moechars, Gilis, Kuipéri, Laenen, and Van Leuven, 1999 (*App*), Mandiyan et al., 2005 (*Cnga2*), Martin, Ledent, Parmentier, Maldonado, and Valverde, 2002 (*Cnr1*), Coste, Heard, Phillips, and Stenzel-Poore, 2006 (*Crhr2*), Marino, Bourdelat-Parks, Cameron Liles, and Weinshenker (2005 (*Dbh*), Vukhac, Sankoorikal, and Wang (2001 (*Drd2*), Cheh et al., 2006 (*En*), Norlin, Gussing, and Berghard, 2003 (*Gnai2*), Vekovischeva et al., 2004 (*Gria3*), Shimshek et al., 2006 (*Gria3*), Duncan et al., 2004 (*Grin1*), Shelbourne et al., 2006 (*Hdh*), Morgan & Cone, 2006 (*Mc5r*), Karl et al., 2004 (*Npy1r*), Takayanagi et al., 2005 (*Oxtr*), Sorensen et al., 2005 (*Pnmt*), Kulikov, Osipova, Naumenko, and Popova, 2005 (*Tph2*)

et al., 1999). These were (1) non-isolated males, neutral cage test, and A/J opponent; (2) males isolated for 1 day, neutral cage test, and A/J opponent; (3) males isolated for 13 days, resident-intruder test, and A/J opponent; and (4) males isolated for 13 days, resident-intruder test, and homogeneous set test. The behavioral index was the percent of males fighting in a strain. The rank order correlations of the strains between any two conditions were always positive but always less than one. This finding is consistent with some genes acting on offense in all four conditions and some having effects in one or more but not all conditions. Some of these strains have also been used to assess correlations between brain phenotypes and offense in males.

One of these is brain levels of the enzyme, steroid sulfatase or STS (Roubertoux & Carlier, 2003). The steroid sulfatase enzyme is coded for by a gene on the pseudoautosomal region (PAR) of the X and Y chromosomes. This enzyme is expressed in glial cells, and it regulates the levels of sulfated and non-sulfated neurosteroids. Free or sulfated neuros-

teroids appear to have opposite effects on neurotransmitter systems. For example, DHEAS and pregnalone sulfate act as antagonists of the GABA receptor whereas allopregnanalone acts as an agonist of the GABA receptor

The association between brain concentration and initiation of attacks was studied in 11 inbred strains of mice (LeRoy et al., 1999). There was a high (r = 0.89) correlation across the strains for STS concentration (pmols STS/mg protein) in brain and proportion of males attacking. This association was only found for males housed with a female from weaning, tested in the neutral cage, and tested with an A/J opponent. It was not found with other kinds or rearing, test situation, and opponent such as isolated males, a resident-intruder test, and same strain opponents.

In both an F2 and advanced intercross lines derived from crosses of NZB and CBA/H mice, DNA markers on the PAR were associated with number of attacks in tests with non-isolated males, a neutral arena, and A/J opponents (Roubertoux et al., 2005). Also, in an F2 derived from NZB and C57BL/6 mice, DNA markers on the PAR were associated with latency to attack and number of attacks in tests with non-isolated males, a neutral arena, and A/J opponents. However, this association was not found for isolated males in a resident-intruder test with A/J opponents.

Another brain phenotype correlated with offensive aggression is the size of the hippocampal mossy fibers. This tract connects the granule cell of the dentate gyrus to the pyramidal cells of the CA3 area of the hippocampus. This association between the size of the IIPMF (infra- and infrapyramidal mossy fiber fields) and offensive aggression was initially shown across seven inbred strains (Guillot, Roubertoux, & Crusio, 1994). The IIPMF and the proportion of mice attacking were measured in each strain. The strain correlation for the two traits was r = −0.82. Aggression was assessed in a resident-intruder test with an A/J intruder. The resident had been housed with a female until 13 days before the resident-intruder test. This correlation is not observed when aggression is assessed in a neutral cage, when the resident had not been housed with a female and then isolated for 13 days, or when the opponent was of the same strain.

This association has also been shown for a pair of strains selected for latency to attack. These are the short attack latency (SAL) and long attack latency (LAL) mice (van Oortmerssen & Bakker, 1981). The SAL mice have a smaller IIPMF in comparison to the LAL mice (Sluyter, Jamot, van Oortmerssen, & Crusio, 1994). The mice are pair housed before the experiment begins. A test subject is exposed for 2 days to three compartments of a four-compartment arena. One compartment is the resident's home area; it lives there alone for 5 days. For the first 2 days, it is allowed to explore the two adjacent compartments for an hour each day. Then, it encounters an albino opponent male of the MAS-GRO strain. Thus, the SAL and LAL mice differ in attack latency for offensive aggression in a resident-intruder test when the males have been isolated for more than 1 day, and the opponent is a standard strain.

There is also a single gene mutation with pleiotropic effects on both offensive aggression and size of the IIPMF. There is a pair of C57BL/6 strains that differ in a single gene (Jamot, Bertholet, & Crusio, 1994). The N strain has smaller IIPMF than the K strain. In both resident-intruder and neutral cage tests, males of the N strain had fewer tail rattles, made fewer attack, and had a longer attack latency than males of the K strain (Sluyter, Marican, & Crusio, 1999). Also, the K strain had more males attacking than the N strain. From weaning, the males had been housed with a female prior to the neutral cage test but isolated for 14–16 days before the resident-intruder test. The opponent in both tests was a DBA/2 male.

Thus, depending on interaction with other genes, the association between size of the IIPMF and offensive aggression

can occur only in the resident-intruder test after isolation or in the neutral cage test with no or limited isolation or both. In contrast, the association between STS activity and offensive aggression occurs only in the neutral cage test with no isolation and with a standard opponent. Consequently, it may be that brain mechanisms for offensive aggression in mice are not the same for different life histories and test parameters and that some genetic variants will act in one and not the other condition. However, some may act in both. Both dependence and independence of life history and test conditions have been shown in the following research on factor analysis of male offense and biological systems across inbred strains and on association of DNA markers and offensive aggression in F2s of two inbred strains.

In the factor analysis study, there were 12 inbred strains (Roubertoux et al., 2005). The four combinations of life history and test conditions were (1) non-isolated, neutral cage, and A/J opponent with one test; (2) isolated for 13 days, neutral cage, and A/J opponent with two test; (3) isolated for 13 days, resident's cage, and A/J opponent with one test; and (4) isolated for 13 days, resident's cage, and homogenous set test with one test. The aggression index was percent of attacking males in a strain. The biological measures were paired testis weight, plasma testosterone concentration, steroid sulfatase concentration in whole brain, serotonin concentration in plasma, serotonin concentration in whole brain, met-enkephalin concentration in whole brain, dynorphin A concentration in whole brain, β-endorphin concentration in whole brain, and ACTH concentration in whole brain. Three factors were obtained. Factor 1 weighted positively on aggression in test conditions 1 and 2 as well as positively with paired testes weight, plasma testosterone concentration, steroid sulfatase concentration and negatively with brain serotonin, β-endorphin, and ACTH concentration. Factor 2 weighted positively on aggression in test conditions 3 and 4, weakly on aggression in test condition 2, and positively on paired testis weight, and negatively on brain serotonin concentration. Factor 3 weighted positively on aggression in test condition 1, plasma testosterone, β-endorphin, and ACTH concentrations and negatively on met-enkephalin concentration. It is again obvious that the association between gene-based biology and offense in males depends on life history and test parameters. In this study, the main experimental variant is whether the males had been isolated or not.

In a study to identify quantitative trait loci (QTLs), offensive aggression was tested in F2 males from reciprocal crosses of NZB/BlNJ and C57BL/6 inbred strains (Roubertoux et al., 2005). There were two contrasting life history and testing conditions. For one group (Group 1), the males were isolated from weaning, and the encounters occurred in the resident's cage. For the other group (Group 2), the males were housed with a female from weaning, and the encounters occurred in a neutral cage. For both groups, the opponent was an A/J male. The phenotypes were latency to first tail rattle, frequency of tail rattles, latency to first attack, and frequency of attacks. The cosegregation of these phenotypes with SSLP (simple sequence length polymorphisms) DNA markers was assessed. The DNA markers were spaced about 22.5 cM apart across the mouse genome.

For Group 1, QTLs were reported for tail rattling latency at Chromosome 8 (47 $\pm$ 2.8 cM) and Chromosome 9 (41 $\pm$ 27.9 cM), for tail rattling frequency on Chromosome 11 (24 $\pm$ 13.9 cM) and on Chromosome 12 (26 $\pm$ 18.6 cM), for attack latency on Chromosome 12 (19 $\pm$ 11.9 cM) and on the X chromosome (28 $\pm$ 4.5 cM), and for attack frequency on Chromosome 11 (45 $\pm$ 4.5 cM) and on Chromosome 12 (24 $\pm$ 15.1 cM). For Group 2, QTLS were reported for tail rattling latency on Chromosome 8 (44 $\pm$ 26.3 cM), on Chromosome 9 (39 $\pm$ 12.9 cM), and on the X chromosome (75 $\pm$ 12.4 cM), for latency to attack on Chromosome 9 (47 $\pm$ 37.2 cM) and on the X chromosome (75 $\pm$ 5.1 cM), and for attack frequency at Chromosome 8 (45 $\pm$ 15.9 cM), on Chromosome 11 (39 $\pm$ 12.4 cM), on Chromosome 12 (17 $\pm$ 27.9 cM), and on the X chromosome (75 $\pm$ 9 cM). There are three things of note in these findings. First, some but not all QTLs are the same between groups. Second, within a group, some but not all QTLs are the same for each measure of aggression. Third, these QTLs accounted for the same amount of phenotypic variance for latency to tail rattle and for latency to attack for both groups but very different amounts for attack frequency in each group. It was 28% for Group 1 and 87% for Group 2.

Candidate genes have been proposed for the following QTLs: for Group 1, *Got1* (glutamic oxaloacetate transaminase) on latency to tail rattling, *Gabra1* (gamma amino acid receptor, subunit α1) and *Gria1* (glutamate receptor, ionotrophic AMPA1) on frequency of tail rattling, *Ar* (androgen receptor) and *Gabra3* (gamma amino acid receptor, subunit α3) on latency to attack, and *Gria1* and *Esr2* (β estrogen receptor) on frequency of attacks; for Group 2, *Got1*, *Htr1b* (5-hydroxytryptamine receptor, 1b), and *Sts* (steroid sulfatase) on latency to tail rattling, *Sts* on latency to attack, and *Gabra3*, *Esr2*, and *Sts* on frequency of attack. There were two rationale for selecting these as candidate genes. These are the following: (1) The gene is within the QTL region. (2) Other evidence is consistent with a role of this gene or its biology in Group 1 or Group 2 offense. All of these, except *Sts*, need further confirmation.

The studies described so far have with certainty identified only one gene with effects on male offensive aggression. This is *Sts*. The effects of *Sts* variants on offense are only detected when the males have not been isolated and when tested in a neutral cage. Variants of the other 55 genes known to affect offense in mice have been detected when the males have been isolated and tested in the resident's cage. The 56

genes are listed in Table 21.1. Each of the 56 genes has allelic variants with differential effects on one or measures of male offense. Only three of these are listed in Group 1 from the study of Roubertoux et al. (2005). These are *Ar*, *Gria1*, and *Esr1*. So far, effects on male offense of knockouts for *Got1*, *Gabr1*, and *Gabr3* have not been reported. Also, knockouts of two genes listed in Group 2 of Roubertoux et al. (2005) have been shown to have effects with isolation before a resident-intruder test. These are *Esr2* and *Htr1b*. It may be that these are not the genes that underlie the mapped QTLs for test condition 2. Regardless, I recommend, as have Roubertoux et al. (1999 and 2005), that there be a standard set of test conditions (life history, test area, and opponent) as well as aggression measures used in all new and replicate studies of male offense in mice. I recommend that these be isolation before testing, resident-intruder test, and A/J opponent as well as frequency of each motor component. Also, since it appears that gene and neural effects on male offense depend on life history and test arena, follow-up studies of gene effects on male offense should be done with no isolation, neutral cage test, and A/J opponent with the same measures of offense.

For now, it may be possible to use the known biological effects of the 56 genes listed in Table 21.1 to identify some aspects of the neural mechanism of offense in males that have been isolated prior to testing and that are tested in the resident's cage regardless of opponent type or behavioral measure. Under these life history and test conditions, these are as follows: (1) olfactory input from both the main olfactory epithelium and the vomeronasal organ and from these to amygdala and other brain structures is required for offense in males; (2) input from the granule cells of the dentate gyrus to pyramidal cells of CA3 in the hippocampus has a role in male offense; (3) the neurotransmitter systems of acetylcholine, adenosine, argeninevasopressin, cannabinoids, dopamine, enkephalin, GABA, glutamic acid, histamine, nitric oxide, norepinepherin, neuropeptide Y, oxytocin, serotonin, and substance p have a role in male offense; (4) the conversion of testosterone to estradiol and the action of estradiol on the α and β estrogen receptor have a role in male offense. There may also be direct effects of testosterone on the *Ar* receptor. Regardless, item 3 is consistent with pharmacological studies of offense in male mice (Miczek & Fish, 2006), and item 4 is consistent with endocrinological studies of offense in male mice (Simon & Lu, 2006). Further progress will be made in understanding the role of these genes in male offense, after it is known when and where each gene has its primary effect and after we know fully the neural circuits of male offense. To some degree the latter is known from studies in rats (Halasz, Liposits, Meelis, Kruk, & Haller, 2002; Hrabovsky et al., 2005), hamsters (Delville, DeVries, & Ferris, 2000; Ferris, 2006), and mice (Haller, Toth, Halasz, & DeBoer, 2006).

Genetic Effects on Aggression in Female Mice

Until the late 1960s or early 1970s, it was thought with one exception that female mice were not aggressive in encounters similar to those used with males. However, it was shown later that wild female mice would attack female mice in a resident-intruder test (Ebert, 1976). Later, Ogawa and Makino (1981) showed that females of the AKR/J strain but not five other strains would attack males in a resident-intruder test. Female mice will also attack lactating female mice (Haug, Johnson, & Brain, 1992). Also, female mice that are homozygous for a knockout of the *Esr1* gene will attack male mice more than those homozygous for the wild-type allele (Ogawa et al., 1998). Much earlier, it had also been shown that in competitive food encounters female mice are as aggressive as male mice (Fredericson, 1950). In all of these studies, the females were neither pregnant nor lactating. Also, it appears that in all these cases, the females are displaying offensive aggression with the same motor patterns as seen for male offense.

About the same time as Ebert's studies on aggression in non-pregnant and non-lactating wild mice, it was shown that lactating mice attack both male and female mice (St. John & Corning, 1973). This is often referred to as maternal aggression (Gammie & Lonstein, 2006). For lactating females, it has been suggested that offensive aggression occurs with female opponents and defensive aggression occurs with male opponents (Parmigiani, Brain, Mainardi, & Brunoni, 1988; Parmiginai, Palanza, Rodgers, & Ferrari, 1999). Although pups are usually removed just prior to a test of maternal aggression, the expression of maternal aggression is dependent on recent exposure of the dam to pups and on sucking of the dam by the pups. It also depends on olfactory stimuli from the intruder acting on both main olfactory and vomeronasal olfactory systems. For example, mice with the *Trp2* ion channel knockout (Leypold et al., 2002) do not show maternal aggression. It can occur just after birth until about 20 days postpartum. For most strains, there is a peak of maternal aggression from 4 to 12 days after birth. In mice, exogenous testosterone and estradiol inhibit the occurrence of maternal aggression. The development of aggression in lactating females is associated with changes in circulating prolactin. It is maximal when prolactin is high.

In mice, female aggression also occurs during pregnancy. Some studies indicate that this is characterized by lunges with few if any bites against male opponents (Svare, 1989). Other studies show that pregnant female mice show the full range of offensive aggression including bites against male opponents (Ogawa & Makino, 1984). Also, both pregnant and non-pregnant females are attacked by pregnant females. For most strains, female aggression begins about 2–3 days postconception, reaches a maximum about 6–11 days postconception, and declines slightly from then to birth. The development of aggression during pregnancy is associated with changes in circulating progesterone and estradiol during pregnancy. It is maximal when progesterone is high and estradiol low.

There is ample evidence suggesting that genes may affect individual differences in each type of female aggression (Ebert, 1983; Fredericson & Birnbaum, 1954; Haug, Johnson, & Brain, 1992; Ogawa & Makino, 1984; Svare, 1989). Here, I will consider whether the same genes may be involved in each kind of female aggression and then whether the same genes may be involved in both male and female offensive aggression.

Wild mice were trapped from two locations around Bowling Green, Ohio. From this mixed wild population, mice were selectively bred for interfemale aggression (Ebert, 1983). All females were non-pregnant and non-lactating. The test subject females were isolated from 28 to 56 days of age. A C57BL/6 female was introduced into the home cage of the test female. The behavior of the resident mouse was rated on a seven-point scale, and from this an aggression score was derived. Within-group selection was used to reduce inbreeding. There were two high aggression lines, two control lines, and two low aggression lines. Females within a line were bred to untested males which as appropriate were the brother of a high or low aggression-scoring female. After 11 generations of selection, the two high lines had mean aggression scores of 6 and the two low lines had mean aggression scores of 3; there was a non-zero heritability for both the high and low lines.

These lines have been tested for maternal aggression (Hyde & Sawyer, 1979). For these tests a non-lactating C57BL/6 female was introduced into the home cage of a lactating female with her pups. There was a test every day from 2 to 20 days postpartum. Females of each high line had higher aggression scores than females of each control line, and females of each control line had higher aggression scores than females of the low line. Thus, the aggression scores of non-lactating females and of lactating females are correlated and some of the same genetic variants have effects on both types of female aggression.

Aggression in non-pregnant and non-lactating, pregnant, and lactating females was later studied in five inbred stains (Ogawa and Makino, 1981). Non-pregnant, non-lactating females were tested three times. Pregnant females were tested at 9, 15, and 17 days postconception. Lactating mice were tested at 1, 3, 5, 8, 10, 15, 17, and 19 days postpartum. The opponent for all tests was a male mouse. Since only one strain was aggressive when the female was non-pregnant and non-lactating, and most strains were aggressive when lactating, it appears that for these strains few, if any, genes effect both kinds of female aggression. However, more strains were aggressive when either pregnant or lactating. Thus, it may be that some genes have effects on both of these types of female aggression.

I now consider whether or not some of the same genes may affect both male offense and one or more kinds of female aggression. Initially, no association was found between male and female offense in two selection studies. In one study, females were selected for aggression with female opponents in a resident-intruder test (Ebert & Hyde, 1983). The high and low lines did not differ in male offense in a resident-intruder test with male opponents. In another study, males were selected for aggression with male opponents in a resident-intruder test (van Oortmerssen & Bakker, 1981). The high and low lines did not differ in female offense in a resident-intruder test with female opponents. Also, lactating females of each line had equally short attack latencies against a male intruder (Benus, 2001). Thus, in these studies, there appears to be no common effect of genes on male offense and female aggression.

Similarly, when non-pregnant, non-lactating females are housed in groups of three, the dominant female will attack lactating, intruder females whereas males in this situation do not attack lactating females (Haug, Johnson, & Brain, 1992). There are differences in this aggressive behavior for female but not male mice. However, gonadectomized males are aggressive in this situation, and females treated with testosterone are not. Thus, in this situation, there does not appear to be common genetic effects on male and female aggression. This may reflect the action of testosterone on gene transcription in males and its absence in females. However, there are other studies with inbred and selected strains suggesting that some genes affect both male offense and at least one type of female aggression.

There are also studies of strain differences of so-called competitive fighting in male/male and female/female pairs. The mice are food deprived. In one study (Fredericson & Birnbaum, 1954), both male and female C57BL/10 mice fought vigorously to maintain possession of a piece of chow. In contrast, both male and female BALB/c mice shared the food without fights. In general, although there are strain differences for this type of aggression, there are no within-strain sex differences. This implies that in these strains and tests, the same genes may be involved in male and female competitive aggression.

In an early study, St. John and Corning (1973) compared the aggression of males attacking males and of lactating females attacking females in four inbred strains. In their tests DBA/2 and BALB/c males and lactating females were more aggressive than C3H and C57BL/6 males and lactating females. Similarly, Jones and Brain (1987) found a strain correlation for aggression of isolated or female-paired males and aggression of lactating females. The strains were Tuck Ordinary (TO), Swiss-Webster, NZW/Ola, BALB/c, C57BL/10,

DBA/2, CBA/Ca, and C3H/He. In these strains and tests, it does seem that some of the same genes affect both male offense and female aggression.

The TA (Turku Aggressive) and TNA selected (Turku non-aggressive) lines differ in male offense in a neutral cage test with male opponents (Lagerspetz, 1964). Males of the TA line are more aggressive than males of the TNA line. Females of these lines also differ in aggression when lactating but not when pregnant or when neither pregnant nor lactating. Female aggression was tested on the 9th and 18th day of gestation and on the 1st, 3rd, 6th, 9th, and 12th day postpartum. The largest difference in aggression occurred on the 3rd day of lactation. The opponents were males. With these strains and tests, it appears that the some genes may affect both male offense and aggression in lactating but not pregnant females and that the same genes do not affect aggression in pregnant and lactating females (Sandnabba, 1992).

The NC900 and NC100 selected lines also differ in offense of isolated males tested in neutral cage with male opponents (Cairns, MacCombie, & Hood, 1983). Males of the NC900 line are more aggressive than males of the NC100 line. Females of these lines also differ in aggression when lactating (Hood & Cairns, 1988). Female aggression was tested with male opponents on the 3rd, 6th, 9th, and 18th day postpartum. Lactating females of the NC900 line were more aggressive than those of the NC100 line on all test days. Also, after group but not individual housing, non-pregnant and non-lactating females of the NC900 line were more aggressive with male opponents than females of the NC100 line (Hood & Cairns, 1988). Here, it appears that some of the same genes affect not only male offense but also two types of female aggression.

It seems reasonable to conclude from these studies of inbred and selected strains that depending on the test conditions, some genes affect male offense, some genes affect only female offense, and some genes affect both. This conclusion is supported by studies on individual genes and aggression in males and females. Table 21.2 lists transgenics and knockouts that have been studied for effects on male offense and one or more types of female aggression. The following genes affect only male offense: *Nos3* (endothelial nitric oxide synthase), *Gria3* (AMPA-type glutamate receptor), *Grin1* (NMDA-type glutamate receptor), and *Pnmt* (phenylethanolamine *N*-methyl transferase). The following genes affect only female aggression: *Cck2* (cholecystokinin neuropeptide receptor) and *V1r* (vomeronasal receptor 1). The following genes affect male and female aggression in opposite directions: *Crhr2* (corticotrophin-releasing factor receptor), *Esr1* (α-estrogen receptor), *Esr2* (β-estrogen receptor), *Nos1* (neural nitric oxide synthase), and *Tgfa* (transforming growth factor alpha). The following genes have the same effect on male and female aggression: *Avpr1b* (arginine vasopressin 1 b receptor), *Gnai2* (guanine

Table 21.2 Comparison of gene effects on male offense and female aggression

Effect	Gene name (Symbol)	Chromosome
Males only		
	Nitric oxide synthase, endothelial (*Nos3*)	5
	Glutamate receptor, ionotropic, AMPA3 (alpha 3) (*Gria3*)	X
	Glutamate receptor, ionotropic, NMDA1 (zeta 1) (*Grin1*)	2
	Phenylethanolamine-*N*-methyltransferase (*Pnmt*)	11
Females only		
	Cholecystokinin neuropeptide receptor (*Cck2*)	7
	Vomeronasal receptors 1 (V1r)	6
Same for males and females		
	Arginine vasopressin 1 b receptor (*Avpr1b*)	1
	Guanine nucleotide-binding protein, alpha-inhibiting 2 (*Gnai2*)	9
	5-HT1B receptor (*Htr1b*)	9
	Nuclear receptor family 2 group E member 1 (*Nr3e1*)	10
	TRP ion channel (*Trp2*)	7
Opposite for males and females		
	Corticotropin-releasing hormone receptor 2 (*Crhr2*)	6
	α-Estrogen receptor (*Esr1*)	10
	β-Estrogen receptor (*Esr2*)	12
	Nitric oxide synthase 1, neuronal (*Nos1*)	5
	Transforming growth factor alpha (*Tgfa*)	6

Except for the following references for each gene may be found in Miczek et al. (2001), Maxson (1998), Maxson et al. (2001), or Maxson and Canastar (2003, 2006): Wersinger, Caldwell, Christiansen, & Young, 2007 (*Avpr1b*), Abramov et al., 2004 (*Cck2*), Coste et al., 2006 (*Crhr2*), Gammie, Hasen, Stevenson, Bale, & D'Anna, 2005 (*Crhr2*), Shimshek et al., 2006 (*Gria3*), Duncan et al., 2004 (*Grin1*), Sorensen et al., 2005 (*Pnmt*), Leypold et al., 2002 (*Trp2*), Del Punta et al., 2002 (*V1r*)

nucleotide-binding protein alpha inhibiting 2), *Htr1b* (serotonin 1B receptor), *Nre3e1* (nuclear receptor family 2 group E member 1), *Oxt* (oxytocin), and *Trp2* (TRP ion channel 2).

Further progress will be made in understanding the role of these genes in male offense and female aggression, after it is known when and where each gene has its primary effect and after we know fully the neural circuits of male offense in comparison to the neural circuits of the types of female aggression. Some aspects of the neural circuitry of maternal aggression in rodents are discussed by Gammie (2005) and by Gammie and Lonstein (2006).

Comparative Genetics of Offensive Aggression

The first studies on the genetics of aggression in mice were published more than 60 years ago (Ginsburg & Allee, 1942; Scott, 1942). Both studies were conduced not only to understand the biology of mouse aggression in males but also to explore hypotheses about the causes of aggression in human males. This section considers how the findings of genetic

effects on mouse aggression might be related to genetic effects on human aggression.

One approach is to consider the findings in mice as hypothesis generators for human aggression. They would be hypothesis generators in two ways. First, genetic variants in the mouse with effects on offense or other types of aggression have homologues in the human genome. Currently, 56 genes have been shown to affect male mouse offense (Table 21.1). Variants of the human homologues could be assessed for effects on one or more types of human aggression. Second, as has been shown in this article, genetic variants can be used to dissect the neural and other mechanisms of offense in male mice. Some aspects of these gene-based neural mechanisms for offense in mice can potentially generate testable hypotheses about the neural and other mechanisms of one or more types of human aggression. This approach does not require that the type of aggression in mice is isomorphic with that in humans.

Another approach does require that there is some similarity in the aggression of mice and humans. Here one speaks of animal models of aggression. Several types of adaptive aggression are recognized in mice and other animals. These include offense, defense, paternal, and infanticide. Blanchard and Blanchard (2006) have cogently argued that offense and defense occur in humans and that there are functional and mechanistic similarities between these types of aggression in animals and humans. In this approach, one would generate genetic hypotheses about human offense but not defense from genetic data on offense in mice. As I have already discussed, most studies of aggression in male mice as well as female mice have been on offense. Others have sought similarities for non-adaptive or pathological aggression in humans and animals (Haller & Kruk, 2006; Miczek & Fish, 2006). In general, these look for test of aggression in animals where there is very short latency to attack, a very high frequency of attack, inappropriate targets for attacks, or a breakdown in the behavioral sequence of the motor patterns of a kind of aggression. These may be models of extreme or impulsive violence in humans. As far as I know, none of the animal models of non-adaptive or pathological aggression have been the subject of genetic studies in mice. When they are, they can be used to generate genetic hypotheses about one or more types of non-adaptive or pathological aggression in humans.

The first two approaches are primarily concerned with using mice to generate hypotheses about the biology and context of human aggression. A comparative genetic approach is concerned with finding general principles about genes and aggression across animal species including humans. In this approach, the effects of DNA variants of the same gene are studied in similar if not identical types of aggression across species or the correlation in brain expression of the same gene or genes is studied in similar if not identical types of aggression across species. Here some background to comparative genetics will be discussed, then an example of it will be described for foraging behavior, and then it will be applied to the Maoa (monoamine oxidase A) gene, the 5HTT (serotonin transporter) gene and aggression across mice, monkeys, and humans.

The biology of behavior is firmly rooted in the writings of Charles Darwin. Darwin considers behavior and mind in Chapter 8 of the *Origin of Species*, in Chapters 3 and 4 in the *Descent of Man* as well as in numerous chapters on sexual selection, and in *The Expression of Man and Animals*. Throughout these, Darwin sought general principles that apply to both animals and humans. This was accomplished by Darwin through comparative studies but not genetic studies.

Others have also argued, as did Darwin, for studies across species as a way to relate animal and human aggression. Paul Scott (1984, 1989) suggested that no animal species can serve as an exact model of human aggression, but rather that some aspects of animal behavior in this or that species will be relevant to human aggression. Consequently, he proposed that information should be accumulated on various kinds of aggression in a wide range of species. Similarly Bob and

Caroline Blanchard recently wrote, "The importance of such organismic, social, and environmental factors in instigating and modulating offensive and defensive aggression across species highlights the need to analyze aggression in as wide range of species as possible" (2006).

On the basis of his research and studies to date as well as the current state of genetics, Benson Ginsburg wrote and published *Genetics as a Tool in the Study of Behavior* (1958). This was 5 years after the publication of Watson and Crick's model of the structure of DNA and its implications for gene replication, mutation, and function. In this paper, Ginsburg cogently argued that in the context of evolution, genetics is a way of (1) defining natural units of behavior, (2) getting at their underlying mechanisms of behavior, and (3) studying the effects of non-genetic variables on behavior. Applied across species, this approach could find general principles relating genes to behavior.

A comparative genetics approach to behavior has been successfully applied to circadian rhythms, foraging, and learning (Robinson, Grozinger, & Whitfield, 2005). Here, the effect of the same gene, *for*, will be described for foraging behaviors in fruit flies, nematodes, honeybees, and harvester ants. The *for* gene codes for a cGMP-dependent protein kinase (PKG) and is involved in the foraging behaviors of each species. There are two alleles of this gene in fruit flies. These are the high- and low-brain enzyme alleles. Larval and adult flies with higher brain enzyme activity are more likely to shift food patch and thereby travel further for food than those with lower brain enzyme activity. There are also two alleles of the gene in nematodes. But genetic variants in PKG enzyme activity have an opposite effect on foraging behavior in nematodes. Nematodes with the lower enzyme activity travel further than those with the higher activity. Brain expression of this enzyme has been studied in honeybees and in harvester ants. Worker honeybees have essentially two life phases. From birth to about 2 weeks of age, they are nurses and do not leave the hive. From about 2 weeks of age to death, they are foragers and do leave the hive. Nurse bees have very much lower levels of PKG transcripts in the brain than do forager bees. Conversely, in the harvester ant, foragers have a lower level of the PKG transcript in brains than do those working in the nest. Here, then the same gene has a role in the foraging behavior of four species but the relationship between being a forager and brain level of transcript is not the same in each species.

Effects of variants of the MAOA gene on male aggression have been studied in mice, macaque monkeys, and humans. The MAOA gene is on the X chromosome of each. When a transgene was inserted into the MAOA gene of mice, the gene and its protein became inactive (Cases et al., 1995). This was due to replacement of exons two and three of the MAOA gene by the transgene. The MAOA protein of wild-type mice degrades synaptic serotonin and catecholamines. Since the

MAOA protein from the null mutant had essentially no enzymatic activity, there was a higher level in brain of serotonin and norepinephrine but not of dopamine and a lower level of 5HIAA (5-hydroxy indole acetic acid) in the null mutant than in wild-type mice. Aggressive behaviors of adult mice were assessed in two tests. The first was between cage mates in the home cage. Here, skin wounds were assessed in 2-, 3-, 4-, and 7-month-old males. Wounds were found in the null mutant cage mates but not in the wild-type cage mates. The second was in a resident-intruder test. Prior to the aggression test, the resident had either a long period of breeding or had a long period of isolation. For both conditions, the null mutants had lower latency to first attack than did the wild types. Recently, a forebrain-specific (neocortex, amygdala, and hippocampus) *Maoa* transgene has been shown to rescue the effects of the *Maoa* mutant on neurotransmitter levels and offensive aggression (Chen et al., 2007). Also, the effect of the *Maoa* mutant on neurotransmitter levels and offensive aggression is reversed by injection of para-chlorophenylalanine on postnatal day 9 (Chen & Shih, 2006). Para-chlorophenylalanine is an inhibitor of tryptophan hydroxylase and blocks the synthesis of serotonin. Taken together, these findings suggest that the *Maoa* mutant has a developmental rather than physiological effect on offensive aggression and that the elevated levels of serotonin act on the development of forebrain structures involved in offensive aggression.

In macaque monkeys there is a polymorphism for a repeat sequence in the promoter of the MAOA gene. The MAOA gene of monkeys with five or six repeats in the promoter has more MAOA transcript in neuroblastoma cells than the one with seven repeats. Male monkeys were either reared by their mother or a foster mother or in peer groups with the continuous or limited access to the peer group of three to four monkeys. In each rearing category, there were males with the five, six, or seven promoter repeat alleles of the *Maoa* gene. Aggressive behavior was tested when the monkeys were 3–5 years old. There were two tests of aggression. These were aggressive behavior during food competition with the wins and loses recorded and home cage social aggression with the total duration for threats, displacement, contact aggression, and other aggressive acts recorded. There was no main effect of genotype or rearing on either of these, but there was a genotype by rearing interaction for both of these. On both measures of aggression, mother-reared but not peer-reared monkeys with the low-activity allele were more aggressive than those with the high-activity alleles (Newman et al., 2005).

There are macaque species with both the short and long promoter variants of MAOA and those that have only the long promoter variant (Wendland et al., 2006). In a small sample of each of seven species, all but one species monomorphic for MAOA have relaxed dominance, high levels of conciliation, and tolerant societies. The exception was Tonkean macaques. Conversely, all species polymorphic for MAOA have intolerant and hierarchical societies with considerable aggressive social behaviors. These findings need to be replicated with a larger sample.

The first finding of an effect of MAOA variants on aggression was reported in humans (Brunner, 1996). In a Dutch kindred, males with a nonsense mutation of the MAOA gene had higher urinary 5HT and lower urinary levels of 5HIAA and other MAOA metabolites. Also, they were impulsive, made unprovoked attacks resulting in injury, attempted arson, and attempted rape. This is known as Brunner's Syndrome. The nonsense mutant of MAOA has only been found in this family. However, there is in humans as in monkeys a repeat polymorphism in the promoter. The repeat polymorphism in humans and monkey arose independently in each lineage. In humans, individuals with the 3.5 or 4 repeat alleles have higher MAOA expression and activity than those with 2 or 3 repeat alleles. Males from the Dunedin Longitudinal Study in New Zealand were classified by genotype into low- and high-activity MAOA groups (Caspi et al., 2002). Most (95.7%) of the sample had either the three or four repeat alleles. Males in this study were also classified into no, possible, and severe maltreatment groups based on assessments of parental or caregiver behaviors at 3, 5, 7, 9, 11, 13, 15, 18, and 21 years of age. There were assessments for conduct disorder at 11, 13, 15, and 18 years of age, antisocial personality disorder at age 26, dispositions toward violence on the Multidimensional Personality Questionnaire Aggression scale at age 26, and court records of convictions for violent offenses. On a composite measure of antisocial behavior, there were no main effects of genotype, but there were main effects of maltreatment and a genotype by maltreatment interaction. Severely maltreated individuals with the low MAOA activity alleles had a higher composite index of antisocial behavior than those with the high MAOA activity alleles. There was no genotypic difference for those with no or probable maltreatment. A similar pattern is seen for each of the individual measures of antisocial behavior. A metanalysis of five studies on MAOA, maltreatment, and antisocial behavior found significant effect size for the interaction of MAOA genotype and maltreatment on antisocial behavior (Kim-Cohen et al., 2006). A sixth study has also replicated the effect of an interaction between MAOA alleles and maltreatment on conduct disorder (Prom et al., 2006). These genotype by environmental interactions may be mediated by effects of MAOA variant on forebrain structure and function (Meyer-Lindenberg et al., 2006). Males with the low-activity alleles have smaller amygdalas and cingulated cortex and larger orbital frontal cortex in comparison to males with the high-activity alleles. During aversive recall, the amygdala and hippocampus were more reactive in those with the low-activity alleles than in those with the high-activity alleles. Also, during cognitive inhibition, there was greater activation of the cingulate in those with the high than with the low MAOA activity allele.

Across species, individuals with no or low MAOA activity are more prone to aggressive behavior. The studies with mice and humans indicate that low MAOA activity may also effect forebrain development. Also, in monkeys and humans, the effect on aggression of the MAOA variants depends on the social environment. Although this has not been studied in mice, there are anecdotal reports that *Maoa* mutant males receive poor maternal care which may be due to the abnormal motor behaviors of *Maoa* mutant pups.

In mice, there is a knockout mutant of *Slc6a4* which codes for the serotonin transporter (Holmes, Murphy, & Crawley, 2002). The serotonin transporter regulates the reuptake of serotonin after it is released into the synaptic space. As a consequence, in the knockout mouse, there is an increase in extracellular serotonin levels and compensatory desensitization of the serotonin 1A and 1B receptors. In a resident-intruder test, offensive aggression but not social investigation is eliminated in the *Slc6a4* homozygotes and greatly reduced in the heterozygotes. On the first and second encounter, the homozygous knockouts had longer latency to first attack and fewer attacks than the homozygous wild types. Also, there was an increase in aggressive behavior from day 1 to 2 for the wild types but not the homozygous knockouts.

Although the *Maoa* and *Slc6a4* have opposite effects on offensive aggression, they have the same effects on brain levels of serotonin and the development of neocortical barrel fields. Neither have distinct barrel fields which receive projections from the vibrissae. This finding implies that the effects of these two genes on offensive aggression are not mediated by their effects on brain levels of serotonin or on the development of the neocortical barrel fields. Regardless, there is also an opposite effect of the *Maoa* and the *Slc6a4* knockout on expression of the adenosine 2A receptor that may account for the opposite effects of the two knockouts on offense (Mossner et al., 2000). It is downregulated in the *Maoa* knockout and upregulated in the *Slc6a4* knockout. A knockout of the adenosine 2A gene increases aggression in a resident-intruder test (Ledent et al., 1997).

In macaque monkeys there is a polymorphism for a repeat sequence length variation in the promoter of the serotonin transporter gene. The l allele has a 44 bp sequence in the promoter, and the s allele has a 22 bp sequence in the promoter. In human placental carcinoma cells, the l allele produces more serotonin transporter transcript than the short allele. Monkeys were either reared by their mother or a foster mother or in peer groups with the continuous or limited access to the peer group of three to four monkeys. In each rearing category, there were males with the l/l or l/s genotype for the serotonin promoter. In one study CSF 5HIAA was measured (Bennett et al., 2002). In parent-reared mon-

keys, there was no difference in concentration of CSF 5HIAA for those with l/s and l/l genotypes, whereas in peer-reared monkey, there was a lower concentration of CSF 5HIAA in those with the l/s genotype than those with the l/l genotype. In another study aggressive behavior was measured in 2-year-old monkeys (Barr et al., 2003). An aggression score was based on the occurrence of bites, hairpulls, chases, hitting, or slapping. In parent-reared monkeys, aggression scores were the same for both those with the l/l and l/s genotypes, whereas in peer-reared monkeys, aggression scores were higher in those with the l/s than with the l/l genotype.

There are macaque species with both the short and long promoter variants of the serotonin transporter (5HTT) gene and those that have only the long promoter variant (Wendland et al., 2006). In a small sample of each of seven species, all species monomorphic for 5HTT have relaxed dominance, high levels of conciliation, and tolerant societies. Conversely, all species polymorphic for 5HTT have intolerant and hierarchical societies with considerable aggressive social behaviors. These findings need to be replicated with a larger sample.

There is also a repeat sequence length polymorphism in the human 5HTT promoter. The long variant is 16 copies and the short variant is 14 copies of a 20–23 bp sequence. In brain, the short allele is transcribed less than the long allele, and in human lymphoblasts, there is less 5HTT and less 5HT uptake. Some studies suggest an association between the 5HTT promoter variants and aggression. For example, in a laboratory test of aggression, men but not women with the short allele genotype (s/s) had increased aggression when stressed (Verona, Joiner, Johnson, & Bender, 2006). There was no genotypic effect when they were not stressed. There was no effect of stress on aggression in males or females with the long allele genotype (l/l). In children, there is an association between the s allele and aggression as measured in the Child Behavior Check List and Teacher Report Form (Haberstick, Smolen, & Hewitt, 2006). Similarly, the frequency of the s/s or s/l genotype was higher in children with a 2-year history of aggression in comparison to a healthy group (Beitchman et al., 2006). However, some studies have not found an association of the 5HTT promoter variants and aggression in children or in adults.

The involvement, if any, of the promoter variant of 5HTT in human aggression may be a consequence of developmental effects on brain structures. As shown in imaging studies, individuals with the l/s genotype had smaller supragenual anterior cingulated and amygdala than those with the l/l genotype (Pezawas et al., 2005).

In mice, individuals with no 5HTT activity are not aggressive, whereas in monkeys and humans, individuals with low 5HTT activity are prone to aggression. Also, in monkeys and perhaps humans, the effect on aggression of the 5HTT variants depends on the environment. The different effects of low 5HTT activity on aggression between mice and primates and the similar effects of low 5HTT activity on aggression in monkeys and humans may be due to the differential roles of serotonin in brain development of mice versus primates.

Comparative studies can and should also be done for the other 54 genes that affect offensive aggression in mice. It may be that this is best done in species with complete genome sequences known. These include round worms (two species), sea squirts (two species), fruit fly (two species), honeybee, mosquito, mouse, dog, rhesus monkeys, chimpanzees, and humans. Also, the following are being sequenced: jelly fish, worms, sea urchins, wasps, aphids, zebra fish, stickleback fish, frogs, chickens, wallaby, rats, voles, guinea pigs, rabbits, shrew, hedgehog, tenrec, armadillos, pigs, horses, cattle, elephant, cats (small and big), marmosets, and orangutans.

In summary, research on the genetics of offensive aggression in male mice began with two studies of three inbred strains. These were followed by many more inbred strain studies as well as three selective breeding studies. It is clear from these that genes effect individual differences in offensive aggression of male mice and that there are genotype by environment interactions for the development of offensive aggression in male mice. Later breeding, transgenic, and knockout studies have identified 56 genes that are involved in offensive aggression of male mice. Also, the effects of these genes on offensive aggression in male mice often depend on life history, test parameters, and test measure. Some of the genes with effects on offensive aggression in male mice also have either no effect or the same effect or opposite effect on aggression in female mice. Lastly, the genes identified for aggression in either male or female mice can be used to develop a comparative genetics of aggression across species. This may be a basis for extrapolating from findings on genetics of aggression from animals to humans.

References

Abramov, U., Raud, S., Koks, S., Innos J., Kurrikoff, K., Matsui, T., et al. (2004). *Targeted mutation of CCK(2) receptor gene antagonises behavioural changes induced by social isolation in female, but not in male mice. Behavioural Brain Research, 155*, 1–11.
Adams, D. B. (1980). Motivational systems of agonistic behavior in muroid rodents: A comparative review and neural model. *Aggressive Behavior, 4*, 295–346.
Barr, C. S., Newman, T. K., Becker, M. L., Parker, C. C., Champoux, M., Lesch, K. P., et al. (2003). The utility of the non-human primate model for studying gene by environment interactions in behavioral research. *Genes, Brain, and Behavior, 2*, 336–340.
Bennett, A. J., Lesch, K. P., Heils, A., Long, J. C., Lorenz, J. G., Shoaf, S. E., et al. (2002). Early experience and serotonin transporter gene variation interact to influence primate CNS function. *Molecular Psychiatry, 7*, 118–122.
Benus, R. F. (2001). Coping in female mice from lines bidirectionally selected for male aggression. *Behaviour, 138*, 997–1008.

Benzer, S. (1971). From gene to behavior. *Journal of the American Medical Association, 218*, 1015–1022.
Beitchman, J. H., Baldassarra, L., Mik., H., De Luca, V., King, N., Bender, D., et al. (2006). Serotonin transporter polymorphisms and persistent, pervasive childhood aggression. *American Journal of Psychiatry, 163*, 1103–1105.
Blanchard, D. C., & Blanchard, R. J. (2006). Stress and aggressive behaviors. In R. J. Nelson (Ed.), *Biology of Aggression* (pp. 275–291). New York: Oxford University Press.
Brunner, H. G. (1996). MAOA deficiency and abnormal behaviour: perspectives on an association. In: G. R. Bock & J. A. Goode (Eds.), *Genetics of Criminal and Antisocial Behaviour* (pp. 155–164) New York: Wiley.
Cairns, R. B., MacCombie, D. J., & Hood, K. E. (1983). A developmental-genetic analysis of aggressive behavior in mice I: Behavioral outcomes. *Journal of Comparative Psychology, 97*, 69–89.
Cases, O., Seif, I., Grimsby, J., Gaspar, P., Chen, K., Pournin, S., et al. (1995). Aggressive behavior and altered amounts of brain serotonin and norepinephrine in mice lacking MAOA. *Science, 268*, 1763–1766.
Caspi, A., McClay, J., Moffitt, T. E., Mill, J., Martin, J., Craig, I. W., et al. (2002). Role of genotype in the cycle of violence in maltreated children. *Science, 297*, 851–854.
Cheh, M. A., Millonig, J. H., Roselli, L. M., Ming, X., Jacobsen, E., Kamdar, S., et al. (2006). *En2 knockout mice display neurobehavioral and neurochemical alterations relevant to autism spectrum disorder. Brain Research, 1116*, 166–176.
Chen, K., & Shih, J. C. (2006) MAO A knock-out (KO), MAOA AB KO and forebrain specific MAOA knock-in transgenic mice as models for studying aggressive behavior. XVII World Meeting of the International Society for Research on Aggression. Minneapolis, MN. July 26, 2006.
Chen, K., Cases, O., Rebrin, I., Wu, W., Gallaher, T. K., Seif, I., et al. (2007). Forebrain-specific expression of monoamine oxidase A reduces neurotransmitter levels, restores the brain structure, and rescues aggressive behavior in monoamine oxidase A-deficient mice. *Journal of Biological Chemistry, 282*, 115–123.
Coste, S. C., Heard, A. D., Phillips. T. J., & Stenzel-Poore, M. P. (2006). Corticotropin-releasing factor receptor type 2-deficient mice display impaired coping behaviors during stress. *Genes, Brain, and Behavior, 5*, 131–138.
Del Punta, K., Leinders-Zufall, T., Rodriguez, I., Jukam, D, Wysocki, C. J., Ogawa, S., et al. (2002). Deficient pheromone responses in mice lacking a cluster of vomeronasal receptor genes. *Nature, 419*, 70–74.
Delville, Y., De Vries, G. J., & Ferris, C. F. (2000). Neural connections of the anterior hypothalamus and agonistic behavior in golden hamsters. *Brain, Behavior, and Evolution, 55*, 53–76.
Dulac, C., & Wagner, S. (2006). Genetic analysis of brain circuits underlying pheromone signaling. *Annual Review of Genetics, 40*, 449–467.
Duncan, G. E., Moy, S. S., Perez, A., Eddy, D. M., Zinzow, W. M., Lieberman, J. A., et al. (2004) Deficits in sensorimotor gating and tests of social behavior in a genetic model of reduced NMDA receptor function. *Behavioural Brain Research, 53*, 507–519.
Ebert, P. D. (1976). Agonistic behavior in wild and inbred *Mus musculus. Behavioral Biology, 18*, 291–294.
Ebert, P. D. (1983). Selection for aggression in a natural population. In E. C. Simmel, M. E. Hahn & J. K. Walters (Eds.), *Aggressive behavior: Genetic and neural Approaches* (pp. 103–127). Hillsdale, NJ: Lawrence Erlbaum Associates.
Fredericson, E. (1950). The effect of food deprivation upon competitive and spontaneous combat in C57 black mice. *Journal of Psychology, 29*, 89–100.

Fredericson, E., & Birnbaum, E. A. (1954). Competitive fighting between mice with different hereditary backgrounds. *Journal of Genetic Psychology, 85*, 271–280.

Gammie, S. C. (2005). Current models and future directions for understanding the neural circuitries of maternal behaviors in rodents. *Behavioral and Cognitive Neuroscience Reviews, 4*, 119–135.

Gammie, S. C., & Lonstein, J. S. (2006). Maternal aggression. In R. J. Nelson (Ed.), *Biology of Aggression* (pp. 250–274). New York: Oxford University Press.

Gammie, S. C., Hasen, N. S., Stevenson, S. A., Bale, T. L., & D'Anna, K. L. (2005). Elevated stress sensitivity in corticotropin-releasing factor receptor 2 deficient mice decreases maternal, but not intermale aggression. *Behavioural Brain Research, 160*, 169–177.

Ginsburg, B. E. (1958). Genetics as a tool in the study of behavior. *Perspectives in Biology and Medicine, 1*, 397–424.

Ginsburg, B. E., & Allee, W. C. (1942). Some effects of conditioning on social dominance and subordination in inbred strains of mice. *Physiology and Zoology, 15*, 485–506.

Ginsburg, B. E., & Jummonville, J. E. (1967). Genetic variability in response to early stimulation viewed as an adaptive mechanism in population ecology. *American Zoologist, 7*, 795.

Guillot, P. V., Roubertoux, P. L., & Crusio, W. E. (1994). Hippocampal mossy fiber distributions and intermale aggression in seven inbred mouse strains. *Brain Research, 660*, 167–169.

Haberstick, B. C., Smolen, A., & Hewitt, J. K.. (2006). Family-based association test of the 5HTTLPR and aggressive behavior in a general population sample of children. *Biological Psychiatry, 9*, 836–843.

Haller, J., & Kruk, M. R. (2006). Normal and abnormal aggression: Human disorders and novel laboratory models. *Neuroscience and Biobehavioral Reviews, 30*, 292–303.

Haller, J., Toth, M., Halasz, J., & De Boer, S. F. (2006). Patterns of violent aggression-induced brain c-fos expression in male mice selected for aggressiveness. *Physiology and Behavior, 88*, 173–182.

Halasz, J., Liposits, Z., Meelis, W., Kruk, M. R., & Haller, J. (2002). Hypothalamic attack area-mediated activation of the forebrain in aggression. *Neuroreport, 13*, 1267–1270.

Haug, M., Johnson, F. J., & Brain, P. F. (1992). Biological correlates of attack on lactating intruders by female mice: A topical review. In K. Bjorkqvist & P. Nielmela (Eds.), *Of Mice and Women: Aspects of Female Aggression* (pp. 381–393). New York: Academic Press.

Hensbroek, R. A., Sluyter, F., Guillot, P. V., Van Oortmerssen, G. A., & Crusio, W. E. (1995). Y chromosomal effects on hippocampal mossy fiber distributions in mice selected for aggression. *Brain Research, 682*, 203–206.

Holmes, A., Murphy, D. L., & Crawley, J. N. (2002). Reduced aggression in mice lacking the serotonin transporter. *Psychopharmacology, 161*, 160–167.

Hood, K. E., & Cairns, R. B. (1988). A developmental-genetic analysis of aggressive behavior in mice. II. Cross-sex inheritance. *Behavior Genetics, 18*, 605–619.

Hrabovsky, E., Halasz, J., Meelis, W., Kruk, M. R., Liposits, Zs., & Haller, J. (2005). Neurochemical characterization of hypothalamic neurons involved in attack behavior: Glutamatergic dominance and co-expression of thyrotropin-releasing hormone in a subset of glutamatergic neurons. *Neuroscience, 133*, 657–666.

Hyde, J. S., & Sawyer, T. F. (1979). Correlated response to selection for aggressiveness in female mice. II Maternal aggression. *Behavior Genetics, 9*, 571–577.

Jamot, L., Bertholet, J.-Y., & Crusio, W. E. (1994). Neuroanatomical divergence between two substrains of C57BL/6J inbred mice entails differential radial-maze learning. *Brain Research, 644*, 352–356.

Jones, S. E., & Brain P. F. (1987). Performances of inbred and outbred laboratory mice in putative tests of aggression. *Behavior Genetics, 17*, 87–96.

Karl, T., Lin, S., Schwarzer, C., Sainsbury, A., Couzens, M., Wittmann, W., et al. (2004). Y1 receptors regulate aggressive behavior by modulating serotonin pathways. *Proceedings of the National Academy of Science USA, 101*, 12742–12747.

Kessler, S., Harmatz, P., & Gerling, S. A. (1975). The genetics of pheromonally mediated aggression in mice. I. Strain difference in the capacity of male urinary odors to elicit aggression. *Behavior Genetics, 5*, 233–238.

Kim-Cohen, J., Caspi, A., Taylor, A., Williams, B., Newcombe, R., Craig, I. W., et al. (2006). MAOA, maltreatment, and gene-environment interaction predicting children's mental health: new evidence and a meta-analysis. *Molecular Psychiatry, 11*, 903–913.

Kulikov, A. V., Osipova, D. V., Naumenko, V. S., & Popova, N. K. (2005). Association between Tph2 gene polymorphism, brain tryptophan hydroxylase activity and aggressiveness in mouse strains. *Genes, Brain, and Behavior, 4*, 482–485.

Lagerspetz, K. M. J. (1964). Studies on the aggressive behavior in mice. *Annales Academiae Scientiarum Fenniae, Series B, 131*, 1–131.

Ledent, C., Vaugeois. J. M., Schiffmann, S. N., Pedrazzini, T., El Yacoubi, M., Vanderhaeghen, J, J., et al. (1997). Aggressiveness, hypoalgesia and high blood pressure in mice lacking the adenosine A2a receptor. *Nature, 388*, 674–678.

LeRoy, I., Mortaud, S., Tordjman, S., Donsez-Darcel, E., Carlier, M., Degrelle, H., et al. (1999). Genetic correlation between steroid sulfatase concentration and initiation of attack behavior in mice. *Behavior Genetics, 29*, 131–136.

Leypold, B. G., Yu, C. R., Leinders-Zufall, T., Kim, M. M., Zufal, F., & Axel, R. (2002). Altered sexual and social behaviors in trp2 mutant mice. *Proceedings of the National Academy of Sciences U S A, 99*, 6376–6381.

Mandiyan, V. S., Coats, J. K., & Shah, N. M. (2005). Deficits in sexual and aggressive behaviors in Cnga2 mutant mice. *Nature Neuroscience, 8*, 1660–1662.

Marino, M. D., Bourdelat-Parks, B. N., Cameron Liles, L., & Weinshenker, D. (2005). Genetic reduction of noradrenergic function alters social memory and reduces aggression in mice. *Behavioural Brain Research, 161*, 197–203.

Martin, M., Ledent, C., Parmentier, M., Maldonado, R., & Valverde, O. (2002). Involvement of CB1 cannabinoid receptors in emotional behaviour. *Psychopharmacology (Berl), 159*, 379–387.

Maxson, S. C. (1992a). Methodological issues in genetic analyses of an agonistic behavior (offense) in male mice. In D. Goldowitz, D. Wahlsten, & R. E. Wimer (Eds.), *Techniques for the genetic analysis of brain and behavior: Focus on the mouse.* (pp. 349–373). Amsterdam: Elsevier.

Maxson, S. C. (1992b). MHC genes, chemosignals, and genetic analyses of murine social behaviors. In R. L. Doty & D. Muller-Schwarze (Eds.), *Chemical Signals in Vertebrates 6* (pp. 197–203). New York: Plenum Press.

Maxson, S. C. (1998) Homologous genes, aggression and animal models. *Developmental Neuropsychology, 14*, 143–156.

Maxson, S. C., & Canastar, A. (2003). Conceptual and methodological issues in the genetics of mouse agonistic behavior. *Hormones & Behavior, 44*, 258–262.

Maxson, S. C., & Canastar, A. (2006). Aggression: Concepts and methods relevant to genetic analyses in mice and humans. In B. Jones & P. Mormede (Eds.), *Neurobehavioral Genetics: Methods and Applications, Second Edition.* (pp. 281–289) Boca Ratan: CRC Press.

Maxson, S. C., & Canastar, A. (2007). The genetics of aggression in mice. In D. J. Flannery, A. Vazsony, & I. Waldman (Eds.), *The Cambridge Handbook of Violent Behavior* (pp. 91–110). New York: Cambridge University Press

Maxson, S. C., Roubertoux, P. L., Guillot, P., & Goldman, D. (2001). The genetics of aggression: From mice to humans. In M. Martinez (Ed.), *Prevention and Control of Aggression and the Impact on its Victims*, (pp. 71–81), New York: Kluwer Academic.

Meyer-Lindenberg, A., Buckholtz,J. W., Kolachana, B. R., Hariri, A., Pezawas, L., Blasi, G., et al. (2006). Neural mechanisms of genetic risk for impulsivity and violence in humans. *Proceedings of the Naional Academy of Science USA, 103*, 6269–6274

Miczek, K. A., Maxson, S. C., Fish, E. W., & Faccidomo, S. (2001). Aggressive behavioral phenotypes in mice. *Behavioural Brain Research, 125*, 167–181.

Mickck, K. A., & Fish, E. W. (2006). Monoamines, GABA, glutamate and aggression. In R. J. Nelson (Ed.), *Biology of Aggression* (pp 114–149). New York: Oxford University Press.

Moechars, D., Gilis, M., Kuipéri, C., Laenen, I., & Van Leuven, F. (1999). Aggressive behaviour in transgenic mice expressing APP is alleviated by serotonergic drugs. *Neuroreport, 9*, 3561–3564.

Monahan, E. J., & Maxson, S. C. (1998). Y chromosome, urinary chemosignals, and an agonistic behavior (offense) of mice. *Physiology and Behavior, 64*, 123–132.

Morgan, C., & Cone, R. D. (2006). Melanocortin-5 receptor deficiency in mice blocks a novel pathway influencing pheromone-induced aggression. *Behavior Genetics, 36*, 291–300.

Mossner, R., Albert, D., Persico, A. M., Hennig, T., Bengel, D., Holtman, B., et al. (2000). Differential regulation of adenosine A(1) and A(2A) receptors in serotonin transporter and monoamine oxidase A-deficient mice. *European Journal of Neuropsychopharmacology, 10*, 489–493.

Newman, T. K., Syagailo, Y. V., Barr, C. S., Wendland, J. R., Champoux, M., Graessle, M, et al. (2005). Monoamine oxidase A gene promoter variation and rearing experience influences aggressive behavior in rhesus monkeys. *Biological Psychiatry, 57*, 167–172.

Nicot, A., Otto, T., Brabet, P., & Dicicco-Bloom, E. M.. (2004). Altered social behavior in pituitary adenylate cyclase-activating polypeptide type I receptor-deficient mice. *Journal of Neuroscience, 24*, 8786–8795.

Norlin, E. M., Gussing, F., & Berghard, A. (2003). Vomeronasal phenotype and behavioral alterations in Gαi2 mutant mice *Current Biology, 13*, 1214–1219

Ogawa, S., & Makino, J. (1981). Maternal aggression in inbred strains of mice: Effects of reproductive state. *The Japanese Journal of Psychology, 52*, 78–84.

Ogawa, S., & Makino, J. (1984). Aggressive behavior in inbred strains of mice during pregnancy. *Behavioral and Neural Biology, 40*, 195–204.

Ogawa, S., Eng, V., Taylor, J., Lubahn, D. B., Korach, K. S., & Pfaff, D. W. (1998). Roles of estrogen receptor-alpha gene expression in reproduction-related behaviors in female mice. *Endocrinology, 139*, 5070–5081.

Parmigiani, S., Brain, P. F., Mainardi, D., & Brunoni, V. (1988) Different patterns of biting attack employed by lactating female mice (*Mus domesticus)* in encounters with male and female conspecific intruders. *Journal of Comparative Psychology, 102*, 287–293.

Parmiginai, S., Palanza, P. S., Rodgers, J., & Ferrari, P. F. (1999). Selection, evolution of behavior and animal models in behavioral neuroscience. *Neuroscience and Biobehavioral Reviews, 23*, 957–969.

Pezawas, L., Meyer-Lindenberg, A., Drabant, E. M., Verchinski, B. A., Munoz, K. E., Kolachana, B. S., et al. (2005). 5-HTTLPR polymorphism impacts human cingulate-amygdala interactions: a genetic susceptibility mechanism for depression. *Nature Neuroscience, 8*, 828–834.

Prom, E. C., Eaves, L. J., Foley, D. L., Gardner, C. O., Wormley, B. K., Riley, B. P., et al. (2006, July 26). Gender differences in the interaction of monoamine oxidase-A and childhood adversity as risk factor in conduct disorders. XVII World Meeting of the International Society for Research on Aggression. Minneapolis, MN.

Ragnauth, A. K., Devidze, N., Moy, V., Finley, K., Goodwillie, A., Kow, L. M., et al. (2005). Female oxytocin gene-knockout mice, in a semi-natural environment, display exaggerated aggressive behavior. *Genes, Brain, and Behavior, 4*, 229–239.

Robinson, G. E., Grozinger, C. M., & Whitfield, C. W. (2005), Sociogenomics: social life in molecular terms. *Nature Reviews Genetics, 6*, 257–270.

Roubertoux, P. L., & Carlier, M. (2003). Y chromosome and antisocial behavior. In M. P. Mattson (Ed.), *Neurobiology of Aggression: Understanding and Preventing Violence* (pp. 119–134). Totowa, NJ: Humana Press.

Roubertoux, P. L., Le Roy, I., Mortaud, S., Perez-Diaz, F., & Tordjman, S. (1999) Measuring aggression in the mouse. In W. E. Crusio & R. T. Gerlai (Eds.), *Handbook of Molecular-Genetic Techniques for Brain and Behavior Research* (pp. 696–709). Amsterdam: Elsevier.

Roubertoux, P. L., Guillot, P. V., Mortaud, S., Pratte, M., Jamon, M., Cohen-Salmon, C., et al. (2005). Attack behaviors in mice: From factorial structure to quantitative trait loci mapping. *European Journal of Pharmacology, 526*, 172–185.

Sandnabba, N. K. (1992). Aggressive behavior in female mice as a correlated characteristic in selection for aggressiveness in male mice. In K. Bjorkqvist & P. Niemela (Eds.), *Of Mice and Women: Aspects of Female Aggression*. (pp. 367–379). New York: Academic Press.

St John, R. D., & Corning, P. A. (1973). Maternal aggression in mice. *Behavioral Biology, 9*, 635–639.

Schellinck, H. M., Monahan, E., Brown, R. E., & Maxson, S. C. (1993). A comparison of the contribution of the major histocompatibility complex (MHC) and Y chromosomes to the discriminability of individual urine odors of mice by Long-Evans rats. *Behavior Genetics, 23*, 257–263.

Scott, J. P. (1942). Genetic differences in the social behavior of inbred strains of mice. *Journal of Heredity, 33*, 11–15.

Scott, J. P. (1984). The dog as a model for human aggression. In K. J. Flannelly, R. J. Blanchard, & D. C. Blanchard (Eds.), *Biological Perspectives on Aggression* (pp. 97–107). New York: Alan R. Liss.

Scott, J. P. (1989). *The Evolution of Social Systems*. New York: Gordon and Breach.

Shelbourne, P. F., Killeen, N., Hevner, R. F., Johnston, H. M., Tecott, L., Lewandoski, M., et al. (2006). A Huntington's disease CAG expansion at the murine *Hdh* locus is unstable and associated with behavioural abnormalities in mice. *Human Molecular Genetics, 8*, 763–774

Shimshek, D. R., Bus, T., Grinevich, V., Single, F. N., Mack, V., Sprengel, R., et al. (2006). Impaired reproductive behavior by lack of GluR-B containing AMPA receptors but not of NMDA receptors in hypothalamic and septal neurons. *Molecular Endocrinology, 20*, 219–231.

Simon, N. G., & Lu, S.-F. (2006). Androgens and aggression. In R. J. Nelson (Ed.), *Biology of Aggression* (pp. 211–230). New York: Oxford University Press

Sluyter, F., Jamot, L., van Oortmerssen, G. A., & Crusio, W. E. (1994). Hippocampal mossy fiber distributions in mice selected for aggression. *Brain Research, 16*, 145–148.

Sluyter, F., Marican, C. C., & Crusio, W. E. (1999). Further phenotypical characterization of two substrains of C57BL/6J inbred mice differing by a spontaneous single-gene mutation. *Behavioural Brain Research, 98*, 39–43.

Sorensen, D. B., Johnsen, P. F., Bibby, B. M., Bottner A., Bornstein, S. R., Eisenhofer, G., et al. (2005). PNMT transgenic mice have an aggressive phenotype *Hormones and Metabolic Research, 37*, 159–163.

Svare, B. (1989). Recent advances in the study of female aggressive behavior in mice. In P. F. Brain, D. Mainardi, & S. Parmigiani (Eds.) *House Mouse Aggression* (pp. 135–159). New York: Harwood Academic Press.

Takayanagi, Y., Yoshida, M., Bielsky, I. F., Ross, H. E., Kawamata, M., Onaka, T., et al. (2005). Pervasive social deficits, but normal parturition, in oxytocin receptor-deficient mice. *Proceedings of the National Academy Science USA, 102*, 16096–16110

van Oortmerssen, G. A., & Bakker, Th. C. M. (1981). Artificial selection for short and long attack latencies in wild *Mus musculus domesticus. Behavior Genetics, 11*, 115–126.

Vekovischeva, O. Y., Aitta-Aho, T., Echenko, O., Kankaanpaa, A., Seppala, T., Honkanen, A., et al. (2004). Reduced aggression in AMPA-type glutamate receptor GluR-A subunit-deficient mice. *Genes, Brain, and Behavior, 3*, 253–265.

Verona, E., Joiner, T. E., Johnson, F., & Bender, T. W. (2006). Gender specific gene-environment interactions on laboratory-assessed aggression. *Biological Psychiatry, 71*, 33–41.

Vukhac, K.-L., Sankoorikal, E.-B., & Wang, Y. (2001). Dopamine D2L receptor- and age-related reduction in offensive aggression. *Neuroreport, 12*, 1035–1038.

Wang, Z., Balet Sindreu, C., Li, V., Nudelman, A., Chan, G. C.-K., & Storm, D. R. (2006). Pheromone detection in male mice depends on signaling through the type 3 adenylyl cyclase in the main olfactory epithelium. *Journal of Neuroscience, 26*, 7375–7379.

Wendland, J. R., Lesch, K. P., Newman, T. K., Timme, A., Gachot-Neveu, H., Thierry, B., et al. (2006). Differential functional variability of serotonin transporter and monoamine oxidase a genes in macaque species displaying contrasting levels of aggression-related behavior. *Behavior Genetics, 36*, 163–172.

Wersinger, S. R., Caldwell, H. K., Christiansen, M., & Young, W. S., III. (2007). Disruption of the vasopressin 1b receptor gene impairs the attack component of aggressive behavior in mice. *Genes Brain & Behavior, 6*, 653–660.

Chapter 22

Sexual Selection and Aggressive Behavior in *Drosophila*

Yong-Kyu Kim

There are many other structures and instincts which must have been developed through sexual selection – such as the weapons of offence and the means of defense – of the males for fighting with and driving away their rivals – their courage and pugnacity – their various ornaments – their contrivances for producing vocal or instrumental music – and their glands for emitting odors, most of these latter structures serving only to allure or excite the female. It is clear that these characters are the result of sexual and not of ordinary selection, since unarmed, unornamented, or unattractive males would succeed equally well in the battle for life and in leaving a numerous progeny, but for the presence of better endowed males.

Darwin (1871, p. 278)

Introduction

In animals, females make large but few nutritious eggs and males produce small but many mobile sperm. There are tremendous amounts of competition between males over access to females, and females discriminate among their mating partners. Sexual selection arises from differences in reproductive success caused by competition for mates. Sexual selection is a mechanism by which conspicuous traits such as large body size, bright colors, songs, weapons as well as behaviors are highly favored to attract more mates and the traits enhance fitness of individuals (Andersson, 1994; Fisher, 1930; Kirkpatrick, 1987; Lande, 1981). There are two types of sexual selection: (1) *intrasexual selection*, a competition within the same sex, usually males, for mates and (2) *intersexual selection*, mate selection by females. Intrasexual selection can occur in the form of competition for females without fighting with other males or in the form of contest between males. Morphological traits such as large body size, weaponry, and armor as well as aggressiveness are favored in the form of male–male competition. When males, however, are unable to monopolize either females or any resource vital to females, males advertise themselves for mates by displaying courtship, territory, songs, and ornaments. Intersexual selection occurs by choosing individuals with the best displays. Sexually dimorphic traits result from

sexual selection by female choice. Males vigorously compete with each other for mates, and females show preference for males with the conspicuous traits. Female preferences are therefore evolved as a correlated response to selection on male traits, and genes both for attractive male traits and for preferences are inherited together from generation to generation. The two sets of genes are genetically correlated and evolution of the male traits drives a further change in the preference. This process leads to rapid coevolution of trait and preference until the exaggerated traits are opposed by natural selection due to selective pressure on high predation (Fisher, 1930; Lande, 1981) or until the traits are common in frequency in the population (O'Donald, 1983).

Aggressive behavior has been reported in many *Drosophila* species and is often observed at food resources (Dow & von Schilcher, 1975; Hoffmann, 1987a; Jacob, 1960, 1978; Ringo, Kananen, & Wood, 1983; Spieth, 1966, 1968). It serves for the acquisition or defense of food resources as well as in access to mates in nature. Despite its importance, little is known about genetic and neural mechanisms that underlie aggressive behavior, other than that hormones play roles (e.g., Baier, Wittek, & Brembs, 2002). Aggressive behavior in *Drosophila* has recently received attention as aspects of evolutionary biology and neuroscience and several studies have demonstrated genetic and environmental influences (Chen, Lee, Bowens, Huber, & Kravitz, 2002; Dierick & Greenspan, 2006, 2007; Hoffmann, 1990; Kamyshev et al., 2002; Kim, Alvarez, Barber, Brock, & Jeon, 2007; Kim & Ehrman, 1998; Kravitz & Huber, 2003; Lee & Hall, 2000; Nilsen, Chan, Huber, & Kravitz, 2004; Robin, Daborn, & Hoffmann, 2006; Svetec, & Ferveur, 2005;

Y.-K. Kim (✉)
Department of Genetics, University of Georgia, Athens, GA 30602, USA
e-mail: yongkyu@uga.edu

Y.-K. Kim (ed.), *Handbook of Behavior Genetics*,
DOI 10.1007/978-0-387-76727-7_22, © Springer Science+Business Media, LLC 2009

Ueda & Kidokoro, 2002; Vrontou, Nilsen, Kravitz, & Dickson, 2006; Yurkovic, Wang, Basu, & Kravitz, 2006).

In this chapter, studies of aggressive behavior in *Drosophila* will be reviewed from ecological, evolutionary, neurological, and genetic points of view. I will discuss (1) how ecological or behavioral interactions between individuals that influence aggressive behavior in *Drosophila*; (2) relationships between social learning and the *Drosophila* central nervous system; and (3) the genetic architecture of aggressive behavior using both single gene analysis of induced mutations and quantitative genetic analysis. Recommended future research directions on aggressive behavior will be addressed.

Current Research

Ecological and Evolutionary Perspectives

Insects, including fruit flies, search for food and mates for survival and reproduction daily. Food is unevenly distributed in nature. Males mate with multiple females and females discriminate among males. Such ecological and environment conditions lead to a development of aggressive behavior in *Drosophila*. Aggressive behavior functions to defend territories against intruders per se but is also given during courtship. Male mating success is influenced by aggression as well as by courtship. Aggressive behavior may also contribute to reproductive isolation between species.

Territoriality

Most *Drosophila* matings occur at feeding and oviposition sites (see reviews by Wilkinson & Johns, 2005), and there is a wide range of resources used by different *Drosophila* species (see reviews by Markow & O'Grady, 2005). The distribution of food resources attracts females to feed and lay eggs which, in turn, influences the distribution of males at these food resources; they visit to feed and obtain mates. Flies are usually observed near food patches or baits in nature as well as in the laboratory, and they interact with each other during territorial defenses of food resources. Territorial behavior has been reported for several *Drosophila* species (for *Drosophila melanogaster*, Jacob, 1960, 1978; Dow & von Schilcher, 1975; Hoffmann, 1987a; for *D. pinicola*, Spieth & Heed, 1975; for Hawaiian flies, Spieth, 1966, 1974, 1981). For example, *D. melanogaster* males establish and defend patches of food as territories against intruding males (Hoffmann, 1987a; Hoffmann & Cacoyianni, 1990). A single male can defend a maximum diameter of 55–75 mm of food (Hoffmann & Cacoyianni, 1990;

Skrzipek, Kroner, & Hager, 1979). Old males are more successful at holding territories than younger males, although territorial behavior is initiated by as young as 2-day-old flies (Hoffmann, 1989). Body size is positively correlated with territoriality (Hoffmann, 1987b, 1991). In the *D. pinicola* species group, utilizing mushrooms for oviposition sites, *D. melanderi* males vigorously defend their territories and single males individually position on a mushroom cap (Spieth & Heed, 1975). Males of the picture-winged Hawaiian *Drosophila* species occupy their mating territories within leks which consist of 5–10 males each and advertise their sexual readiness by displaying nonresource-based behavior, and females periodically visit these territories (see review by Spieth, 1982). Territorial males in *D. melanogaster* consequently have more encounters and are more successful in mating than non-territorial males (Dow & von Schilcher, 1975; Hoffmann, 1987a, 1991; Jacob, 1960, 1978; Skrzipek et al., 1979). However, Ringo, Kananen, and Wood (1983) found no evidence of territoriality in three species they studied: *D. virilis, D. americana*, and *D. novamexicana*. Territoriality has been shown to be heritable in artificial selection experiments, and progeny of territorial parents have mating advantages (Hoffmann, 1988, 1991; Hoffmann & Cacoyianni, 1989). The incidence of territoriality is influenced by defended food types, male density in territories, female receptivity, as well as age. Higher territoriality is observed in natural breeding sites than in laboratory medium and is also detected when females are present in the territory regardless of their reproductive status (Hoffmann & Cacoyianni, 1990). Territorial defense, however, ceases when benefits from mating advantages are overweighed by costs. Males are less likely to defend territories when food resources are abundant or territorial males are more common in density (Hoffmann & Cacoyianni, 1990).

Resource defense is associated with fighting between individuals, usually males (see *Aggressive Behavior* for details). *Drosophila melanogaster* males vigorously defend their feeding or oviposition sites by displaying a series of aggressive behaviors: chasing, fencing, lunging, tussling, wing threat, boxing, holding, or head-to-head butting (Fig. 22.1). Males of most Hawaiian *Drosophila* species – *heteroneura, silvestris*, and their close relatives – patrol their leks and vigorously defend mating territories exhibiting ritualized fighting such as curling and slashing which involve wing, leg, and body movements (Spieth, 1966, 1968; see Ringo & Hodosh, 1978 for *D. grimshawi* subgroup). *Drosophila melanderi* males approach and attempt to court any individuals invading their territories (Spieth & Heed, 1975), and *D. subobscura* males are reported to display wing threat toward intruders (Milani, 1956). It is hypothesized that forced copulation and patrolling at emergence sites are favored in *Drosophila* species possessing more male-biased operational sex ratio and intense

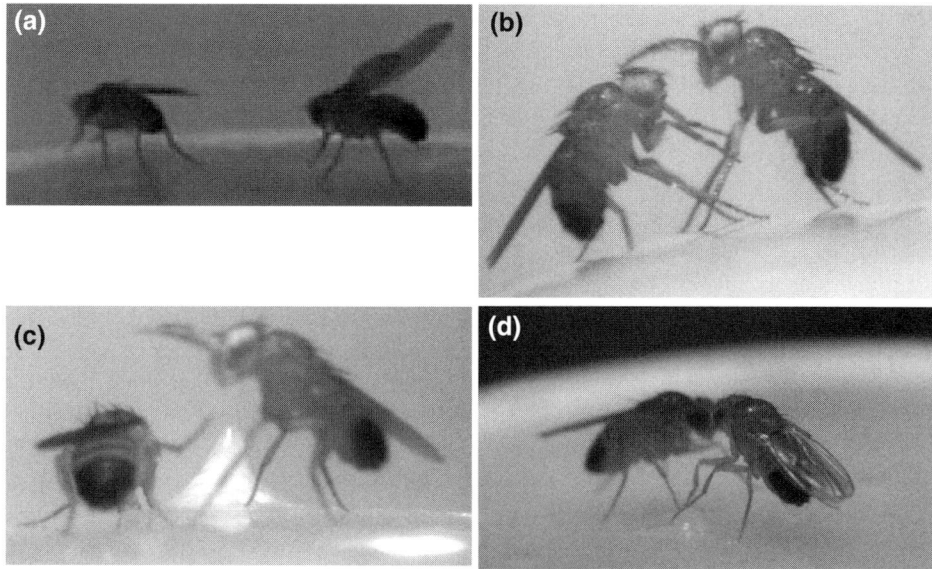

Fig. 22.1 *Drosophila* aggressive behaviors: (**A**) Wing threat; (**B**) Boxing; (**C**) Lunging; (**D**) Head-butting. Photos Courtsey of Dr. Edward Kravitz

competition for mates (Markow, 1996). For example, Markow (2000) reported that *D. melanogaster* and *D. simulans* males patrolled pupation sites, waited for emerging females, and copulated with teneral females that were incapable of displaying rejection behavior; in addition, these mated teneral females successfully produced fertile offspring.

Mate Competition

Individuals of the same sex vigorously compete with each other to acquire mates. Males may pugnaciously fight with other males or have contests over females by exhibiting extravagant ornaments or conspicuous courtship signals toward females or elaborate courtship or mating behavior toward females. Traits that increase reproductive fitness will be favored by sexual selection (Andersson, 1994; Fisher, 1930; Kirkpatrick, 1987; Lande, 1981). Considerable variation in body size, head size, songs, pheromones, and courtship are observed among flies collected in nature and contribute to reproductive success. Adult body size is influenced by a variety of environmental factors such as nutrition, temperature, and larval density on food (David, Allemand, van Herrewege, & Cohert, 1983), and it is heritable in laboratory populations (Robertson & Reeve, 1952). Female *Drosophila* are usually larger than males, and large females have fecundity advantages (Partridge & Farquhar, 1983). Males consequently struggle over females who also compete for larger males. Larger males offer more courtship activities as well as agonistic behavior, and they win aggressive encounters

(Markow, 1988; Partridge, Ewing, & Chandler, 1987; Partridge, Hoffmann, & Jones, 1987; Santos, Ruiz, Barbadilla, Hasson, & Fontdevila, 1988; Spieth, 1982; but see Zamudio, Huey, & Crill, 1995). Small males are less likely than are large males to hold territory (Hoffmann, 1987b, 1991). Hawaiian *Drosophila* species also show considerable variation in morphology, sexual behavior, and pheromones (Alves et al., 2008; Carson, 1997; Carson, Hardy, Spieth, & Stone, 1970; Hoy, Hoikkala, & Kaneshiro, 1988; Kaneshiro & Boake, 1987; Ringo, 1997; Ringo & Hodosh, 1978; Spieth, 1966). These sexually dimorphic traits are evolved under sexual selection via female choice. For example, highly pugnacious *D. heteroneura* males have hammer-shaped heads with stalk eyes. Males with broader heads are more successful in male–male competition and there is a highly significant correlation between male mating success and head width (Boake, DeAngelis, & Andreadis, 1997; Spieth, 1981). In *D. silvestris* males, however, body size is positively correlated with aggressive success but not with mating success (Boake, 1989; Boake & Konigsberg, 1998).

Drosophila males display elaborate courtship behavior to attract mates (e.g., see Spieth, 1966, 1982 for the picture-winged Hawaiian flies) and females are subjected during courtship to a variety of stimuli – visual, acoustic, olfactory, and mechanical elements (Greenspan & Ferveur, 2000; Hall, 1994; Spiess, 1987). Courtship is species specific and plays a major role in mate recognition (Paterson, 1978, 1985) but males show individual variation in quantities of courtship, an indicator of their physical status. Females prefer to mate with males that vigorously display courtship. For example, Kim and Ehrman (1998) have demonstrated that *D. paulistorum* females mated more frequently with males reared in isolation than those reared in groups, because the former were sexually more active in courting. During courtship males also exhibit high levels of agonistic behavior to their sexual partners. Males intersperse between courtship behavior and aggressive behavior when they encounter females. When females are not receptive or display rejection behaviors, males often threaten females by raising both wings up in front of females or lunge toward females (Jacob, 1960), and females repel males by decamping and kicking. Unsuccessful males in mating are often observed to interrupt or terminate courtship of other males by attacking them.

Courtship songs influence male mating success and are sexually selected traits (Aspi & Hoikkala, 1995; Hoikkala, Aspi, & Suvanto, 1998; Ritchie, Saarikettu, Livingstone, & Hoikkala, 2001; Ritchie, Townhill, & Hoikkala, 1998; Saarikettu, Liimatainen, & Hoikkala, 2005; Snook, Robertson, Crudgington, & Ritchie, 2005; Williams, Blouin, & Noor, 2001). During courtship, *Drosophila* males produce courtship song via wing vibration; it has two types of songs: sine song and pulse song (see Tauber & Eberl, 2003 for research approach). It functions as a communication from males to females and it serves to stimulate female receptivity and to identify the species of the courting male. The wing vibration of many *Drosophila* species has been studied (see Gleason, 2005; Hoikkala, 2005 for review). Each species is unique in its acoustic characteristics such as interpulse intervals (IPI); IPI is regarded as the critical parameter for discrimination between species. Such variable courtship songs may contribute to reproductive isolation between sympatric, closely related sibling species (Kyriacou & Hall, 1982; Ritchie & Gleason, 1995; Williams et al., 2001). There is considerable variation in courtship songs among *D. melanogaster* males (Wheeler, Fields, & Hall, 1988). Females prefer to mate with males presenting high frequencies of IPI and courtship song traits are heritable (Ritchie et al., 1998; Ritchie & Kyriacou, 1996). In competitive situations, however, males interrupt other males (Wallace, 1974) by producing a rejection song linked to flicking behavior (Paillette, Ikeda, & Jallon, 1991) or exhibiting aggressive behavior such as wing treat or lunging (Jacob, 1960) and later even shorten the duration of courtship song as well as courtship duration or latency to copulation (Crossley & Wallace, 1987; Ewing & Ewing, 1984; Tauber & Eberl, 2002).

During courtship, males and females exchange olfactory signals as well. Cuticular hydrocarbons present on insect cuticles primarily protect them from desiccation, but they also contribute to intraspecific mate recognition as well as to sexual isolation between species (Blows & Allan, 1998; Cobb & Jallon, 1990; Coyne & Charlesworth, 1997; Coyne, Crittenden, & Mah, 1994; Jallon, 1984; Marcillac, Bousquet,

Alabouvette, Savarit, & Ferveur, 2005; Savarit, Sureau, Cobb, & Ferveur, 1999). Cuticular hydrocarbons are not too volatile and are perceived at a relatively short distance (1–2 mm) by olfactory organs or by contact with taste organs. Quantitative and qualitative variations in cuticular hydrocarbons exist according to sexes, species, and geographic populations (see Ferveur, 2005 for review). Several studies have proven the roles of cuticular hydrocarbons in sexual selection. In *D. melanogaster*, females mate faster and more often with males carrying greater levels of 7-tricosene, a principal male hydrocarbon (Grillet, Dartevelle, & Ferveur, 2006). *Drosophila serrata* females show significant sexual selection according to quantities of the male hydrocarbons (Blows, 2002; Chenoweth & Blows, 2005; Howard, Jackson, Banse, & Blows, 2003). Similarly, males that mated multiply in an array of *D. pseudoobscura* possessed higher relative abundances of 5,9-pentacosadiene, a male-predominant hydrocarbon, than did males that were never accepted by females. They further mated more quickly with females (Kim, Basset, Laverentz, & Anderson, 2005). There is, however, no direct evidence of the role of male hydrocarbons in *Drosophila* aggressive behavior, although aggressive males tend to produce greater amounts of male-predominant hydrocarbons than do non-aggressive males (Kim, Phillips, Chao, & Ehrman, 2004).

Mate choice is influenced both intrinsically and extrinsically. For example, social constraints, such as intrasexual interference and intersexual coercion, significantly influence mate choice. Mate preference tests, where social or ecological constraints were removed, were designed to identify freely chosen preferred and non-preferred partners, and the viability of offspring from matings to preferred and non-preferred partners was measured (Anderson, Kim, & Gowaty, 2007; Gowaty et al., 2007). Matings with preferred partners, when either males or females were choosing, produced more offspring than matings with non-preferred partners. During these trials, individual choosers were permitted to compare both visual and olfactory information concerning two partners of the opposite sex. Preferred partners were bigger in body sizes and produced greater quantities of male- or female-predominant hydrocarbons than non-preferred partners (Kim, Gowaty, & Anderson, 2005). Quantities of hydrocarbons are also influenced by rearing conditions. In the *D. paulistorum* species complex, males reared in isolation produced greater quantities of male hydrocarbons than did males reared in groups, and, as a result, they were more often accepted as mates (Kim and Ehrman, 1998).

Males often mate with several females per day in nature, and mating frequency among males varies. Mating frequency is positively correlated with mating speed, an important component of *Drosophila* fitness and one that is genetically controlled (Fulker, 1966; Kaul & Parsons, 1965; Kessler, 1969; Manning, 1961; Partridge, Mackay, &

Aitken, 1985; Spiess & Langer, 1964). Carson (2002) demonstrated that approximately 30% of Hawaiian *Drosophila* males did not mate in female choice tests. Kim, Basset, et al. (2005) reported similar results with *D. melanogaster, D. pseudoobscura,* and *D. hydei:* Males that mated multiply during observation periods showed significantly faster mating speeds than did males that were not accepted by females during the same intervals. The former also were more aggressive than the latter. Recently, Moehring and Mackay (2004) identified QTLs affecting *D. melanogaster* male mating speed and showed that seven candidate genes are associated with variations in mating behavior. Microarray analysis showed that 21% of the genome is involved in regulating mating speed and that such genes also are likely to be genes involved in neurogenesis, metabolism, development, and cellular processes (Mackay et al., 2005).

Aggressive Behavior

Aggression is an adaptive behavior expressed for access to more mates, among other goals. Individuals of the same sex, usually males, compete and fight for mates. They perform complex and stereotyped aggressive behaviors. Aggressive behavior has been reported in many *Drosophila* species (Dow & von Schilcher, 1975; Hoffmann, 1987a; Jacob, 1960, 1978; Ringo, Kananen, & Wood, 1983; Spieth, 1966, 1968). In *D. melanogaster,* males at feeding sites chase each other, push off with one of their middle legs, elevate both wings for a sustained period of time, rise on their hind legs and tussle with forelegs, hold other flies with forelegs in the back, raise the front part of body, and lunge toward the other (Chen et al., 2002; Hoffmann, 1987b; Jacob, 1960; see Table 22.1). Females also display aggressive though limited fighting behaviors (Kamyshev et al., 2002; Nilsen et al., 2004; Ueda & Kidokoro, 2002; Vrontou et al., 2006). Hierarchical relationships are formed among *D. melanogaster* males, and winners lunge more while losers retreat more (Chen et al., 2002; Nilsen et al., 2004; Yurkovic et al., 2006). Thus, more aggressive males are shown to have a greater mating success than less aggressive males. In the majority of lekking Hawaiian species, males frequently intrude upon one another's leks, leading to aggressive behaviors (Boake, 1989; Spieth, 1966, 1968, 1982): Both males engage in head-on, wing waving, and bobbing. They elevate their heads and forebodies, then fully extend and slash their forelegs downward against the intruders. If the encounter continues, additional bobbing and slashing movements occur and terminate the encounter. *Drosophila heteroneura* in the *planitibia* subgroup of the Hawaiian species, however, display a unique fighting pattern (Spieth, 1981). When they approach head-on, they horizontally keep their bodies depressed and face and push against each other, attempting to force the other backwards. Males with broader heads are more successful at mating success as well as at aggressive success (Boake et al., 1997). Such fighting behavior provides a selection pressure that led to the evolution of secondary sexual characters such as enhanced head size that attract more mates (Boake et al., 1997; Ringo, 1977; Templeton, 1977). However, aggression is not always positively correlated with mating success. Unlike *D. melanogaster* group species, Ringo, Kananen, and Wood (1983) found that courtship and aggressive behavior were largely independent in the *D. virilis* group species; courtship was positively related to mating success, and aggressive males were not more successful at

Table 22.1 Aggressive behavior in *Drosophila melanogaster*

Behavior	Description
Chasing	A male vigorously follows another male in the short distance; often the male vibrates one wing during chasing (Jacob, 1960)
Fencing	A male extends one middle leg on the body and wards off another male (Jacob, 1960)
Wing threat	Both wings are extended to 30° and raised up to 30–40° in front of another male. This behavior is often observed in the front of females (Dow & von Schilcher, 1975)
Boxing	Two males approach each other, stand up and exchange their forelegs; it is infrequently displayed during fighting (Dow & von Schilcher, 1975)
Holding	A male stays behind another male, grabs and holds both wings of another male; the male often display this behavior in front of the opponent (Jacob, 1960)
Lunging	A male raises the front part of his body and lunges down onto his opponent (Hoffmann, 1987a)
Head-butting	Two males face each other in a straight-line position and push against each other using their heads (Lee and Hall, 2000)
Tussling	Two males raise the front part of the body extending their hindlegs and tussle with forelegs (Hoffmann, 1987a)

mating than unaggressive males. Boake (1989) also demonstrated that aggressive success in the picture-winged Hawaiian species, *D. silvestris*, was positively and highly correlated with body size but not correlated with mating success (see also Boake & Konigsberg, 1998). Laboratory strains that have been reared under constant environments for many generations are less aggressive than field-caught flies (Dierick and Greenspan, 2006). These authors also found that flies were less aggressive when food is not available in the observation chamber (see also Hoffmann & Cacoyianni, 1989; Ueda & Kidokoro, 2002).

Age affects aggressiveness in *Drosophila* too. The age at which males become sexually mature differs among species (see Markow, 1996 for details). Old flies of both sexes are more territorial and aggressive than young ones (Hoffmann, 1990; Papaj & Messing, 1998). Shortly after eclosion young males are more often courted by mature males, but aggressiveness develops as they age along with the production of inhibitory pheromones (Jallon, Antony, & Benemar, 1981; Kim, Phillips, et al., 2004; Mane, Tompkins, & Richmond, 1983).

Roles of Aggressiveness in Reproductive Isolation

Both sexes of *D. melanogaster* display aggressive behaviors, although females tend to be less aggressive (Chen et al., 2002; Nilsen et al., 2004). When two or more species are placed together in mating chambers or in population cages, males are often observed to chase each other vigorously. Sometimes males fight with one another to get access to mates or interrupt couples in copulation. A role for aggressiveness in sexual isolation has not been reported because the frequency of aggressive behavior during observations is very low. There is, however, an aggression index which measures durations of observed aggressive behaviors, AI <10% in this instance. Price and Boake (1995) reported that males of the two Hawaiian picture-wing species, *D. silvestris* and *D. heteroneura*, did not fight each other during observation periods and barely displayed aggressive behavior (but see Boake, Price, & Andreadis, 1998). With sympatric and allopatric pairs of *D. pseudoobscura* and *D. persimilis*, aggressive behavior between heterospecific males occurred (Kim, Abramowicz, & Anderson, 2008). *Drosophila pseudoobscura* males, in sympatry with *D. persimilis*, were more aggressive, although not significantly so, than *D. pseudoobscura* males in allopatry. These results suggest potential roles of aggressive behavior on reproductive isolation and are consistent with an increase in female *D. pseudoobscura* discrimination in sympatry, where fewer heterospecific matings occurred (Noor, 1995; Noor & Ortíz-Barrientos, 2006). Both male aggressiveness and female discrimination may contribute to sexual isolation between two species in sympatry.

Neurological and Genetic Perspectives

Aggressive behavior depends on social experience, and therefore entails learning. The *Drosophila* mushroom bodies (MBs), which continue to grow in adults and whose growth is stimulated by social experience, are the site of olfactory learning (Fig. 22.2). The products of several specific genes necessary for the acquisition and storage of olfactory memories have been identified. Therefore, we expect to find that the MBs are the site of learning during aggressive encounters and that the genes involved in learning and memory are necessary for acquisition of some aspects of aggression, too.

Social Experience and Learning

Social interactions or experiences with conspecifics affect *Drosophila* courtship and aggressive behavior (Hoffmann, 1990; Kim, Ehrman, & Koepfer, 1992, 1996; Kim and Ehrman, 1998; McRobert et al., 2003; Papaj & Messing, 1998; Svetec, Cobb, & Ferveur, 2005; Svetec, & Ferveur, 2005; Yurkovic et al., 2006). Individuals receive visual, olfactory, acoustic, and tactile signals during pre- and post-adult stages. These signals will be used as templates and compared at future encounters when adult (Lacy & Sherman, 1983). But when flies are raised in isolation from conspecifics, they do not obtain any useful template information. In a series of investigations into the development of discriminatory abilities in *D. paulistorum*, Kim et al. (1992, 1996) demonstrated that discriminatory abilities increased when flies had social contacts with conspecifics during development from egg to adult stages. However, when individuals were reared in total isolation during development, their discriminatory abilities were minimized and more heterogamic matings occurred. Socially isolated flies did not discriminate efficiently between sexes and homosexuality was frequently observed (Kim and Ehrman, 1998).

Males reared in isolation have been shown to be more aggressive than ones reared in groups (Hoffmann, 1990; Kamyshev et al., 2002; Kim and Ehrman, 1998; Svetec, Cobb, & Ferveur, 2005; see Ueda & Kidokoro, 2002 for females): These isolates displayed lunging and wing threat more frequently than males held in groups. They also established territories more quickly (Hoffmann, 1990). Consequently, the isolated males were more successful at mating than males in groups (Ellis & Kessler, 1975; Kim and Ehrman, 1998). Previous experience in fighting modified their social status when new hierarchical relationships are reestablished (Hoffmann, 1990; Yurkovic et al., 2006): Former males who were more territorial and aggressive tended to establish territories readily and to attack more than do loser

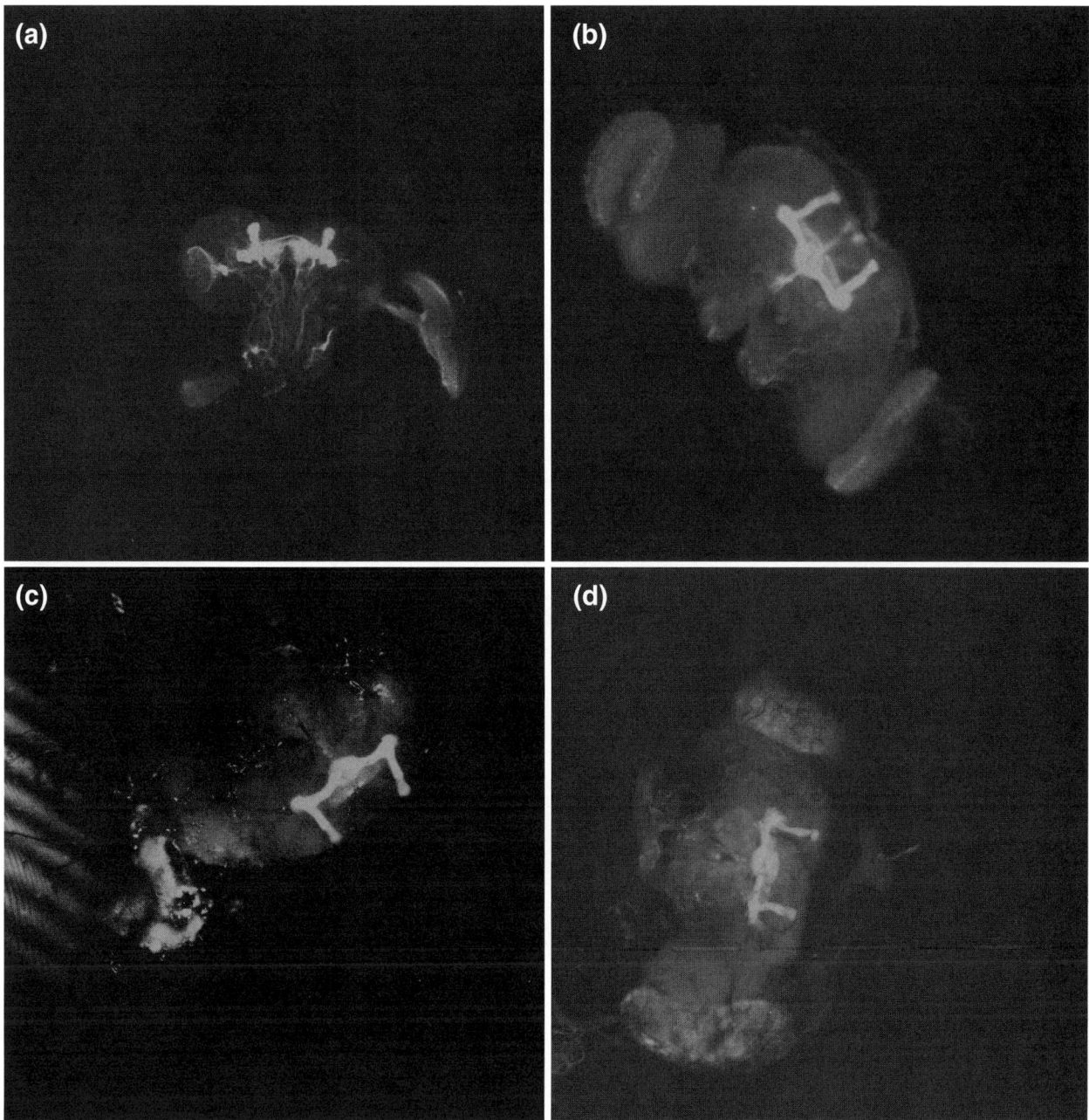

Fig. 22.2 P-Gal4 expression patterns in *Drosophila* mushroom bodies at (**A**) third instar larva; (**B**) pupa; (**C**) 5-day-old adult reared in group; (**D**) 5-day-old adult reared in isolation since egg stage. Kim (2008)

males. Former losers lost in subsequent fights when paired with familiar or unfamiliar winners.

Such results indicate that *Drosophila* learn to recognize conspecifics through social experience, and learning about conspecifics is reinforced by the presence of heterospecifics (Irwin & Price, 1999; Kim, Koepfer, & Ehrman, 1996). For example, when *D. pseudoobscura* and *D. persimilis* coexist in sympatry, the discriminatory abilities of *D. pseudoobscura* females increase (Noor, 1995; Noor & Ortíz-Barrientos, 2006) and more aggressive behaviors by males against heterospecifics were observed (Kim et al., 2008).

Mushroom Bodies as a Site for Associative Learning

Flies receive and process multisensory information, including olfactory stimuli in the brain. They detect odors through sensory receptors on antennae and maxillary palps, and olfactory inputs reach antennal lobes of the brain via the antennal nerve. Larvae also have olfactory organs, the antenno-maxillary complex, at the anterior tip of the animal (Singh & Singh, 1984). Mushroom bodies (MBs) of the brain receive olfactory inputs from antennal lobes and forward outputs to their central complex and the lateral protocerebrum. The MBs are bilateral clusters of about 2,500 Kenyon cells per hemisphere. They are crucial for olfactory associative learning and memory, but are not required for visual, tactile, or motor learning (Davis, 1993; de Belle & Heisenberg, 1994; Dubnau & Tully, 2001; Heisenberg, 2003; Heisenberg, Borst, Wagner, & Byers, 1985; Pascual & Préat, 2001; Roman & Davis, 2001; Strausfeld, Hansen, Li, Gomez, & Ito, 1998; Waddell & Quinn, 2001; Zars, Fischer, Schulz, & Heisenberg, 2000). Flies whose MBs were disrupted chemically or genetically were impaired in olfactory associative learning (Connolly et al., 1996; de Belle & Heisenberg, 1994; Heisenberg et al., 1985). As attempts to dissect the roles of MBs at different phases of memory processing – *acquisition, consolidation, and retrieval* (Dubnau, Chiang, & Tully, 2003) – McGuire, Le, and Davis (2001) inactivated neurotransmission in this region and demonstrated that MB signaling is needed for olfactory memory retrieval. The MBs are also involved in sustaining courtship suppression after males were conditioned with mated females (McBride et al., 1999; but see Kido & Ito, 2002).

Genetic screening for learning-defective mutant flies identified *dunce, rutabaga, amnesiac,* and *radish* mutants (Dubai, Jan, Byers, Quinn, & Benzer, 1976; Folkers, Drain, & Quinn, 1993; Quinn, Sziber, & Booker, 1979). The *dunce* mutant has a defect in the gene coding for cAMP-specific phosphodiesterase, and the *rutabaga* mutant has abnormal adenylyl cyclase, an enzyme activated by Ca^{2+}/calmodulin and $G_s\alpha$ proteins. These genes are expressed in the MBs. The *amnesiac* gene, encoding a pre-proneuropeptide neurotransmitter, is expressed in dorsal paired medial (DPM) neurons. Cyclic AMP-dependent protein kinase (PKA) is also expressed in the MBs. Flies mutated in genes for the catalytic or regulatory subunits of PKA are deficient in learning. Transposon-induced mutant screening identified more new genes involved in olfactory learning: *latheo, nalyot, linotte, leonardo, volado,* and *fasciclin II* (see Waddell & Quinn, 2001 for review).

The MBs have an embryonic origin and continue to grow during development to the adult stages (Armstrong, de Belle, Wang, & Kaiser, 1998; Technau & Heisenberg, 1982). During metamorphosis, the MBs display a remarkable structural plasticity involving neural degeneration, birth, and regrowth. The volume of MBs is influenced by genetic and environmental factors during development (Balling, Technau, & Heisenberg, 1987; Technau, 1984). Heisenberg, Heusipp, & Wanke (1995) observed the effect of social experience on the volume of the *D. melanogaster* brain during adulthood and found that the volume of the MBs was significantly increased in the flies held in groups compared to those held in isolation (see also Barth & Heisenberg, 1997; Barth, Hirsch, Meinertzhagen, & Heisenberg, 1997 for visual experiences). In a series of investigations into the development of *D. paulistorum* discrimination abilities, Kim (2008) also showed that flies reared in a total isolation during development had a significantly smaller MB size than did flies reared in groups, and the isolates were more aggressive. Such results suggest that changes in MB structure result from social interactions during development and may play roles in drosophilid aggressiveness (Baier et al., 2002).

Neurotransmitters dopamine and octopamine are known to be involved in courtship conditioning and olfactory learning in *Drosophila* (Dubai, Buxbaum, Corfas, & Ofarim, 1987; O'Dell, 1994), and their receptors have been identified in MBs (Han, Millar, & Davis, 1998; Han, Millar, Grotewiel, & Davis, 1996). Baier et al. (2002) observed that the aggressiveness of *D. melanogaster* males was significantly reduced when dopamine and octopamine were genetically and pharmacologically depleted. Further, when synaptic outputs from MBs were blocked, males were not aggressive. Similarly, MB-ablated males were significantly less aggressive than controls with intact MBs (Kim et al., 2007). Such data indicate that the MBs may also be involved in *Drosophila* aggressive behavior. Meanwhile, serotonin (5-hydroxytryptophan, 5-HT) was reported to have no effects on aggression in *D. melanogaster* (Baier et al., 2002; but see Dierick & Greenspan, 2007), although it has been known that 5-HT is associated with aggression and social dominance status in a variety of organisms (Kravitz, 2000; Kravitz & Huber, 2003; Murakami & Itoh, 2001, 2003; Popova, 2006). Instead, 5-HT influences male courtship behavior; when 5-HT levels in the brain were reduced, male homosexual courtship was induced (Hing & Carlson, 1996; Zhang & Odenwald, 1995; see Liu et al., 2008 for the effect of dopamine on male-male courtship). Lee and Hall (2000) reported that serotonergic neurons in the brain were not different between more aggressive *fruitless* mutant males and wild-type ones.

Genetics of Aggressive Behavior

Recent dissections of interactions between wildtype *D. melanogaster* males have identified several genes involved in *Drosophila* aggressiveness (see Robin, Daborn, & Hoffmann, 2006 for review; Table 22.2). Jacob (1978) first reported that *ebony* mutants were more aggressive than wild types while *black* mutants were less aggressive and that aggressiveness was influenced by β-alanine. Two genes involved in the sex-determination hierarchy, *fruitless (fru)* and *dissatisfaction (dsf)*, were also reported to be associated with aggression (Lee & Hall, 2000). The *fru* gene, which plays a prominent role in male courtship behavior, was associated with high levels of head-to-head interactions between males. Males homozygous for any of the *fru* mutations displayed vigorous head-buttings. The frequency of the head-buttings was significantly higher between *fru* mutant males than between *fru* male and a wild-type male or female. Aging *dsf* mutant males led to a high level of head-to-head interactions compared to *fru* mutant males. Generating alleles for *fru* that are constitutively spliced in the male (*fru*^{M}) or female (*fru*^{F}) mode (Demir & Dickson, 2005), Vrontou et al. (2006) further analyzed the crucial role of the *fru* gene in aggression and found that *fru*^{F} males were more inclined to fight females than to court them, whereas *fru*^{M} females courted other females and fought males. In a pairing between *fru*^{F} males and *fru*^{M} females, the behavior of *fru*^{M} females was indistinguishable from that of control *fru*^{F} males; and *fru*^{F} males behaved like *fru*^{C} females. They also found that *fru*^{F} males, like normal females, did not establish strong dominance relationships; fights between *fru*^{F} males were significantly less frequent than those in *fru*^{C} males.

Boake et al. (1998) crossed *D. silvestris* (S) and *D. heteroneura* (H) and observed aggression (initial posture, wing extension, head position, and leg posture) in heterospecific pairs, a hybrid male and either an S or an H male. Parental males performed differently as regards wing extensions and leg postures, and the hybrids resembled S males for early postures and extended leg postures, but were like H males for wing extensions. They hypothesized that aggressive behavior is controlled by a single gene with major effects.

Utilizing classical selection experiment procedures widely employed in quantitative genetics (Falconer & Mackay, 1996; Harshman & Hoffmann, 2000), Dierick and Greenspan (2006) produced four lines of *D. melanogaster*, two selected for high aggression and two for "neutral" lines in which highly aggressive males were removed from the breeding pool of each generation, and they showed that aggressiveness in the selected lines significantly increased over generations. After 21 generations of selection, a fighting index increased more than 30-fold in the selected lines. This indicates that multiple genes contribute to aggressiveness. The neutral lines displayed decreased aggressiveness since selection was exercised against aggressive flies every generation. Selected lines dominated the neutral lines when paired with them. However, in mating competition assays, the selected lines mated significantly less than the neutral lines (but see Dow & von Schilcher, 1975; Hoffmann, 1987). The selected lines were also lighter in weight than the neutral ones. Using microarray analysis, these authors evaluated gene expression in the brains of representative flies from their two lines and identified approximately 80 genes that were significantly differentially expressed. The selected lines expressed higher levels of these genes than did the neutral ones. Note that 42 genes showed differences in transcription abundances between these lines. Mutation analysis, inserting these candidate

genes into wild types, showed that *Cyp6a20*, which encodes a cytochromosome P450 (Feyereisen, 2005), is directly involved in changes in aggressive behavior during selection experiments. No differences in levels of those genes related to serotonin (5-HT) metabolism or in total 5-HT levels were recorded between the two lines. Subsequently, Dierick and Greenspan (2007) found that both pharmacological and genetic manipulations that increase 5HT synthesis also increase aggression. This effect was the same in selected and neutral lines, indicating that selection had not affected 5HT levels or receptors. They also found that destroying or feminizing brain cells that produce neuropeptide F (NPF) in a male-specific pattern increased aggression, which parallels what has been found with neuropeptide Y (NPY) in *Mus* (knocking out an NPY receptor gene increases territorial aggression in mice, which appears to be stimulated by 5HT). Flies whose 5HT synthesis in Ddc-producing cells was shut off nonetheless showed low levels of aggression, leading Dierick and Greenspan to postulate an "aggression circuit" in the brain, which is modulated by 5HT and by NPF but which does not require 5HT. The modulatory roles of 5HT (which stimulates aggression) and NPF (which inhibits aggression) appear to be additive, and this suggests that the two molecules act independently.

Edwards, Rollmann, Morgan, and Mackay (2006) used artificial selection as did Dierick and Greenspan (2006) when selecting for high and low levels of male aggression. A realized heritability (h^2) of 0.10 over 28 generations was reported for aggressive behavior in *D. melanogaster*, indicating considerable environmental variance. Selection for aggressive behavior did not affect other *Drosophila* behaviors or physiology, such as mating, locomotion, stress resistance, temperature tolerance, and longevity. Using whole-genome microarrays, these authors assessed aggressiveness levels for 19 mutations in candidate genes and identified 15 genes that form the genetic architecture of *Drosophila* aggressive behavior: 7 genes (*trantrack, longitudinals lacking, scribbler, Male-specific RNA 87F, kismet, muscleblind, and Darkener of apricot*) were involved in other biological processes and 8 genes could not be characterized.

Future Directions

The function of aggressive behavior is to defend territories against intruders in nature. Considerable variation in territorial behavior has been observed among *Drosophila* species and territoriality is positively correlated with male mating success. Aggressive behavior is reported in many picture-winged Hawaiian *Drosophila* species and *D. melanogaster* in the wild and in the laboratory, which demonstrates that more aggressive individuals are more territorial and increase their reproductive fitness. However, these relationships are not always true and need to be studied more with different species. Aggressive behavior is also given during courtship but it has not received attentions from evolutionary biologists because its frequency is relatively low in the laboratory where flies are routinely placed with small amounts of food in *plastic* mating chambers. Roles of aggressive behavior in sexual isolation between strains and species need to be investigated.

Using single-gene studies of induced mutations, several genes have been reported to affect aggressive behavior in *D. melanogaster* (see Table 22.2). However, there are no genes controlling behavior directly. Rather, many genes have pleiotropic effects and influence other behaviors as well (e.g., Anholt & Mackay, 2004; Kyriacou, 2002). Aggressive behavior is polygenic and regulated by interactions between multiple genes with small effects as well as by brain neurotransmitters. Studies of QTLs responsible for natural variation in aggression could be combined with single-gene studies of biochemical mutations providing additional candidate genes or gene regions (Greenspan, 2004). Aggressive behavior is also significantly influenced by environment. Especially, various experiences that individuals obtain during development will affect their subsequent behaviors including aggression. Studies of antisocial personality disorders in humans show that the magnitude of genetic and environmental influences on the traits substantially changes during development stages: these traits are more influenced by shared environments during adolescence but more genetically influenced during adulthood. Studies of social experience/learning in *Drosophila* aggressive behavior are limited to the observations of experience-dependent modification of behavior, memory, structural plasticity, and quantitative levels of neurotransmitters in the brain. Using mutant flies, studies of gene–environment interactions and correlations on aggressive behavior will provide insights into complex gene–behavior pathways in aggression.

Acknowledgment I am very grateful to Drs. W. Anderson, L. Ehrman, J.-F. Ferveur, P. A. Gowaty, J.-M. Jallon, T. Markow, and J. Ringo for helpful comments on the manuscript. This work was partially supported by UGA Faculty Grant (10-21-RR-064-069) and NIH Grant 5R01 GM048528-06 (via W. Anderson).

Table 22.2 Factors involved in *Drosophila* aggressive behavior

Factor	References
1) Genes	
ebony	Jacob (1960)
black	Jacob (1960)
fruitless (fru)	Lee and Hall (2000)
fru^M	Vrontou et al. (2006)
fru^F	Vrontou et al. (2006)
dissatisfaction (dsf)	Lee and Hall (2000)
neuropeptide F (npf)	Dierick and Greenspan (2007)
Cyp6a20	Dierick and Greenspan (2006)
15 *P*-element mutations	Edwards et al. (2006)
2) Neurotransmitters	
Dopamine	Baier et al. (2002)
Octopamine	Baier et al. (2002)
Serotonin	Baier et al. (2002); Dierick and Greenspan (2007)
3) Environment	
Food resource	Hoffmann (1990); Papaj & Messing (1998)
Social experience	Hoffmann (1990); Kim and Ehrman (1998); Kamyshev et al. (2002); Ueda & Kidokoro (2002); Svetec, Cobb, & Ferveur (2005); Yurkovic et al. (2006); Kim et al. (2007)

References

Alves, H., Rouault, J.-D., Kondoh, Y., Nakano, Y., Yamamoto, D., Kim, Y.-K., & Jallon, J.-M. (2008). Evolution of cuticular hydrocarbons of Hawaiian *Drosophilidae* (submitted).

Andersson, M. (1994). *Sexual selection*. New Jersey: Princeton University Press.

Anderson, W. W., Kim, Y.-K., & Gowaty, P. A. (2007). Experimental constraints on mate preferences in *Drosophila pseudoobscura* decrease offspring viability and fitness of mated pairs. *Proceedings of the National Academy of Sciences USA, 104*, 4484–4488.

Anholt, R. R. H., & Mackay, T. F. C. (2004). Quantitative genetic analyses of complex behaviors in *Drosophila. Nature Review/Genetics, 5*, 838–849.

Armstrong, J. D., de Belle J. S., Wang, Z., & Kaiser, K. (1998). Metamorphosis of the mushroom bodies: Large-scale rearrangements of the neural substrates for associative learning and memory in *Drosophila. Learning & Memory, 5*, 102–114.

Aspi, J., & Hoikkala, A. (1995). Male mating success and survival in the field with respect to size and courtship song character in *Drosophila littoralis* and *D. montana* (Diptera: Drosophilidae). *Journal of Insect Behavior, 8*, 67–87.

Baier, A., Wittek, B., & Brembs, B. (2002). *Drosophila* as a new model organisms for the neurobiology of aggression? *Journal of Experimental Biology, 205*, 1233–1240.

Balling, A., Technau, G. M., & Heisenberg, M. (1987). Are the structural changes in the adult *Drosophila* mushroom bodies memory traces? Studies on biochemical learning mutants. *Journal of Neurogenetics, 4*, 65–73.

Barth, M., & Heisenberg, M. (1997). Vision affects mushroom bodies and central complex in *Drosophila melanogaster. Learning & Memory, 4*, 219–229.

Barth, M., Hirsch, H. V. B., Meinertzhagen, I. A., & Heisenberg, M. (1997). Experience-dependent developmental plasticity in the optic lobe of *Drosophila melanogaster. Journal of Neuroscience, 17*, 1493–1504.

Blows, M. W. (2002). Interaction between natural and sexual selection during the evolution of mate recognition. *Proceedings of the Royal Society of London, 269*, 1113–1118.

Blows, M. W., & Allan, R. A. (1998). Levels of mate recognition within and between two *Drosophila* species and their hybrids. *American Naturalist, 152*, 826–837.

Boake, C. R. B. (1989). Correlations between courtship success, aggressive success and body size in a picture-winged fly, *Drosophila silvestris. Ethology, 80*, 318–329.

Boake, C. R. B., DeAngelis, M. P., & Andreadis, D. K. (1997). Is sexual selection and species recognition a continuum? Mating behavior of the stalk-eyed *Drosophila heteroneura*. *Proceedings of the National Academy of Sciences USA, 94*, 12442–12445.

Boake, C. R. B., & Konigsberg, L. (1998). Inheritance of male courtship behavior, aggressive success, and body size in *Drosophila silvestris*. *Evolution, 52*, 1487–1492.

Boake, C. R. B., Price, D. K., & Andreadis, D. K. (1998). Inheritance of behavioural differences between two interfertile, sympatric species, *Drosophila silvestris* and *D. heteroneura*. *Heredity, 80*, 642–650.

Carson, H. L. (1997). Sexual selection: A driver of genetic change in Hawaiian *Drosophila*. *Journal of Heredity, 88*, 343–352.

Carson, H. L. (2002). Female choice in *Drosophila*: Evidence from Hawaii and implications for evolutionary biology. *Genetica, 116*, 383–393.

Carson, H. L. Hardy, D. E., Spieth, H. T., & Stone, W. S. (1970). The evolutionary biology of the Hawaiian Drosophilidae. In M. K. Hecht & W. C. Steere (Eds.), *Essays in evolution and genetics in honor of Theodosius Dobzhansky* (pp. 437–543). New York: Appleton-Centry-Crofts.

Chen, S., Lee, A. Y., Bowens, N., Huber, R., & Kravitz, E. A. (2002). Fighting fruit flies: A model system for the study of aggression. *Proceedings of the National Academy of Sciences USA, 99*, 5664–5668.

Chenoweth, S. F., & Blows, M. W. (2005). Contrasting mutual sexual selection on homologous signal traits in *Drosophila serrata*. *American Naturalist, 165*, 281–289.

Cobb, M., & Jallon, J.-M. (1990). Pheromones, mate recognition and courtship stimulation in the *Drosophila melanogaster* species subgroup. *Animal Behaviour, 39*, 1058–1067.

Connolly, J. B., Roberts, I., J. H., Amstrong, J. D., Kaiser, K., Forte, M., Tully, T., & O'Kane, C. J. (1996). Associative learning disrupted by impaired G$_s$ signaling in *Drosophila* mushroom bodies. *Science, 274*, 2104–2107.

Coyne, J. A., & Charlesworth, B. (1997). Genetics of a pheromonal difference affecting sexual isolation between *Drosophila mauritiana* and *D. sechellia*. *Genetics, 145*, 1015–1030.

Coyne, J. A., Crittenden, A. P., & Mah, K. (1994). Genetics of a pheromonal difference contributing to reproductive isolation in *Drosophila*. *Science, 265*, 1461–1464.

Crossley, S., & Wallace, B. (1987). The effects of crowding on courtship and mating success in *Drosophila melanogaster*. *Behavior Genetics, 17*, 513–522.

Darwin, C. (1871). *The descent of man, and selection in relation to sex.* London: J. Murray.

David, J. R., Allemand, R., Van Herrewege, J., & Cohert, Y. (1983). Ecophysiology: Abiotic factors. In M. Ashburner, H. L. Carson, & J. N. Thompson (Eds.), *The genetics and biology of Drosophila* (pp. 106–109). London: Academic Press.

Davis, R. L. (1993). Mushroom bodies and *Drosophila* learning. *Neuron, 11*, 1–14.

de Belle, J. S., & Heisenberg, M. (1994). Associative odor learning in *Drosophila* abolished by chemical ablation of mushroom bodies. *Science, 263*, 692–695.

Demir, E., & Dickson, B. J. (2005). *fruitless* splicing specifies male courtship behavior in *Drosophila*. *Cell, 121*, 785–794.

Dierick, H. A., & Greenspan, R. J. (2006). Molecular analysis of flies selected for aggressive behavior. *Nature Genetics, 38*, 1023–1031.

Dierick, H. A., & Greenspan, R. J. (2007). Serotonin and neuropeptide F have opposite modulatory effects on fly aggression. *Nature Genetics, 39*, 678–682.

Dow, M. A., & von Schilcher, F. (1975). Aggression and mating success in *Drosophila melanogaster*. *Nature, 254*, 511–512.

Dubai, Y., Buxbaum, J., Corfas, G., & Ofarim, M. (1987). Formamidines interact with *Drosophila* octopamine receptors alter the flies' behavior and reduce their learning ability. *Journal of Comparative Physiology, 161*, 739–746.

Dubai, Y., Jan, Y.-N., Byers, D., Quinn, W., & Benzer, S. (1976). *dunce*, a mutant of *Drosophila* deficient in learning. *Proceedings of the National Academy of Sciences USA, 73*, 1684–1688.

Dubnau, J., Chiang, A.-S., & Tully, T. (2003). Neural substrates of memory: From synapses to system. *Journal of Neurobiology, 54*, 238–253.

Dubnau, J., & Tully, T. (2001). Functional anatomy: From molecule to memory. *Current Biology, 11*, R24–R243.

Edwards, A. C., Rollmann, S. M., Morgan, T. J., & Mackay, T. F. C. (2006). Quantitative genomics of aggressive behavior in *Drosophila melanogaster*. *PLoS Genetics, 2*, 1386–1395.

Ellis, L. B., & Kessler, S. (1975). Differential posteclosion housing experiences and reproduction in *Drosophila*. *Animal Behaviour, 23*, 949–952.

Ewing, L. S., & Ewing, A. W. (1984). Courtship in *Drosophila melanogaster*: Behaviour of mixed sex groups in large observation chambers. *Behaviour, 90*, 184–202.

Falconer, D. S., & Mackay, T. F. C. (1996). *Introduction to quantitative genetics* (4th ed.). London: Longmans Green.

Ferveur, J.-F. (2005). Cuticular hydrocarbons: The evolution and roles in *Drosophila* pheromonal communication. *Behavior Genetics, 35*, 279–295.

Feyereisen, R. (2005). Insect cytochrome P450. In L. I. Gilbert, K. Iatrou, & S. S. Gill (Eds.), *Comprehensive molecular insect science* (Vol. 4, pp. 1–77). Amsterdam: Elsevier.

Fisher, R. A. (1930). *The genetical theory of natural selection*. Oxford: Clarendon Press.

Folkers, E., Drain, P. F., & Quinn, W. G. (1993). *radish*, a *Drosophila* mutant deficient in consolidated memory. *Proceedings of the National Academy of Sciences USA, 90*, 8123–8127.

Fulker, D. W. (1966). Mating speed in male *Drosophila melanogaster*: A psychogenetic analysis. *Science, 153*, 203–205.

Gleason, J. M. (2005). Mutations and natural genetic variation in the courtship song of *Drosophila*. *Behavior Genetics, 35*, 265–277.

Gowaty, P. A., Anderson, W. W., Bluhm, C. K., Drickamer, L. C., Kim, Y.-K., & Moore, A. (2007). The hypothesis of reproductive compensation for lowered offspring viability: Tests of assumptions and predictions. *Proceedings of the National Academy of Sciences USA, 104*, 15023–15027.

Greenspan, R. J. (2004). E pluribus unum, ex uno plura: Quantitative- and single-gene perspectives on the study of behavior. *Annual Reviews of Neuroscience, 27*, 79–105.

Greenspan, R. J., & Ferveur, J.-F. (2000). Courtship in *Drosophila*. *Annual Reviews of Genetics, 34*, 205–232.

Grillet M., Dartevelle, L., & Ferveur, J.-F. (2006). A *Drosophila* male pheromone affects female sexual receptivity. *Proceedings of the Royal Society of London, 273, 315–323*.

Hall, J. C. (1994). The mating of a fly. *Science, 264*, 1702–1714.

Han, K.-A., Millar, N. S., & Davis, R. L. (1998). A novel octopamine receptor with preferential expression in *Drosophila* mushroom bodies. *Journal of Neuroscience, 18*, 3650–3658.

Han, K.-A., Millar, N. S., Grotewiel, M. S., & Davis, R. L. (1996). DAMB, a novel dopamine receptor expressed specifically in *Drosophila* mushroom bodies. *Neuron, 16*, 1127–1135.

Harshman, L. G., & Hoffmann, A. A. (2000). Laboratory selection experiments using *Drosophila*: What do they really tell us? *Trends in Ecology and Evolution, 15*, 32–36.

Heisenberg, M. (2003). Mushroom body memoir: From maps to models. *Nature Review/Neuroscience, 4*, 266–275.

Heisenberg, M., Borst, A., Wagner, S., & Byers, D. (1985). *Drosophila* mushroom body mutants are deficient in olfactory learning. *Journal of Neurogenetics, 2*, 1–30

Heisenberg, M., Heusipp, M., & Wanke, C. (1995). Structural plasticity in the *Drosophila* brain. *Journal of Neuroscience, 15*, 1951–1960.

Hing, A. L., & Carlson, J. R. (1996). Male-male courtship behavior induced by ectopic expression of the *Drosophila* white gene: Role of sensory function and age. *Journal of Neurobiology, 30,* 454–464.

Hoffmann, A. A. (1987a). A laboratory study of male territoriality in the sibling species *Drosophila melanogaster and Drosophila simulans. Animal Behaviour, 35,* 807–818.

Hoffmann, A. A. (1987b). Territorial encounters between *Drosophila* males of different sizes. *Animal Behaviour, 35,* 1899–1901.

Hoffmann, A. A. (1988). Heritable variation for territorial success in two *Drosophila melanogaster* populations. *Animal Behaviour, 36,* 1180–1189.

Hoffmann, A. A. (1989). Georgraphic variation in the territorial success of *Drosophila melanogaster* males. *Behavior Genetics, 19,* 241–255.

Hoffmann, A. A. (1990). The influence of age and experience with conspecifics on territorial behavior in *Drosophila melanogaster. Journal of Insect Behavior, 3,* 1–12.

Hoffmann, A. A. (1991). Heritable variation for territorial success in field-collected *Drosophila melanogaster. American Naturalist, 138,* 668–679.

Hoffmann, A. A., & Cacoyianni, Z. (1989). Selection for territoriality in *Drosophila melanogaster*: Correlated responses in mating success and other fitness components. *Animal Behaviour, 38,* 23–34.

Hoikkala, A. (2005). Inheritance of male sound characteristics in *Drosophila* species. In S. Drosopoulos & M. F. Claridge (Eds.), *Insect sounds and communication: Physiology, behaviour, ecology and evolution* (pp. 167–177). Boca Raton, FL: CRC Taylor & Francis.

Hoikkala, A., Aspi, J., & Suvanto, L. (1998). Male courtship song frequency as an indicator of male genetic quality in an insect species, *Drosophila montana. Proceedings of the Royal Society of London, 265,* 503–508.

Howard, R. W., Jackson, L. L., Banse H., & Blows, M. W. (2003). Cuticular hydrocarbons of *Drosophila birchii* and *D. serrata*: Identification and role in mate choice in *D. serrata. Journal of Chemical Ecology, 29,* 961–976.

Hoy, R. R., Hoikkala, A., & Kaneshiro, K. Y. (1988). Hawaiian courtship songs: Evolutionary innovation in communication signals in *Drosophila. Science, 240,* 217–219.

Irwin, D. E., & Price, T. (1999). Sexual imprinting, learning and speciation. *Heredity, 82,* 347–354.

Jacob, M. E. (1960). Influence of light on mating of *Drosophila melanogaster. Ecology, 41,* 182–188.

Jacob, M. E. (1978). Influence of β-alanine on mating and territorialism in *Drosophila melanogaster. Behavior Genetics, 8,* 487–502.

Jallon, J. M. (1984). A few chemical words exchanged by *Drosophila* during courtship and mating. *Behavior Genetics, 14,* 441–478.

Jallon, J.-M., Antony, C., & Benemar, O. (1981). Un antiaphrodisiac produit par les males de *Drosophila melanogaster* et transfere aux femelles lors de la copulation. *Comptes rendus de l'Académie des sciences. Paris, 292,* 1147–1149.

Kamyshev, N. G., Smirnova, G. P., Kamysheva, E. A., Nikiforov, O. N., Parafenyuk, I. V., & Ponomarenko, V. V. (2002). Plasticity of social behavior in *Drosophila. Neuroscience and Behavioral Physiology, 32,* 401–408.

Kaneshiro, K. Y., & Boake, C. R. B. (1987). Sexual selection and speciation: Issue raised by *Hawaiian Drosophila. Trends in Ecology and Evolution, 2,* 207–212.

Kaul, D., & Parsons, P. A. (1965). The genotypic control of mating speed and duration of copulation in *Drosophila pseudoobscura. Heredity, 20,* 381–392.

Kessler, S. (1969). The genetics of *Drosophila* mating behavior. II. The genetic architecture of mating speed in *Drosophila pseudoobscura. Genetics, 62,* 421–433.

Kido, A., & Ito, K. (2002). Mushroom bodies are not required for courtship behavior by normal and sexually mosaic *Drosophila. Journal of Neurobiology, 52,* 302–311.

Kim, Y.-K. (2008). Developmental isolation and subsequent adult behavior of *Drosophila paulistorum*. VIII. Quantitative variation of mushroom bodies (submitted).

Kim, Y.-K., Abramowicz, K., & Anderson, W. W. (2008). A role of aggressive behavior in sexual isolation between *Drosophila pseudoobscura* and *D. persimilis. Behavior Genetics, 38,* 632.

Kim, Y.-K., Alvarez, D., Barber, J., Brock, A., & Jeon, J. (2007). Genetic and environmental influence on the *Drosophila* aggressive behavior. *Behavior Genetics, 37,* 766.

Kim, Y.-K., Basset, C., Laverentz, J., & Anderson, W. W. (2005). Female choice in sexual selection of *Drosophila pseudoobscura. Behavior Genetics, 35,* 808.

Kim, Y.-K., & Ehrman, L. (1998). Developmental isolation and subsequent adult behavior of *D. paulistorum*: IV. Courtship. *Behavior Genetics, 28,* 57–65.

Kim, Y.-K., Ehrman, L., & Koepfer, H. R. (1992). Developmental isolation and subsequent adult behavior of *Drosophila paulistorum*. I. Survey of the six semispecies. *Behavior Genetics, 22,* 545–556.

Kim, Y.-K., Ehrman, L., & Koepfer, H. R. (1996). Developmental isolation and subsequent adult behavior of *Drosophila paulistorum*. II. Prior experience. *Behavior Genetics, 26,* 15–25.

Kim, Y.-K., Gowaty, P. A., & Anderson, W. W. (2005). Testing for mate preference and mate choice in *Drosophila pseudoobscura*. In L. Noldus, J. J. F. Grieco, L. W. S. Loijens, & P. H. Zimmerman (Eds.), *Proceedings of the 5th International Conference on Methods and Techniques in Behavioral Research* (pp. 533–535). The Netherlands: Wageningen.

Kim, Y.-K., Koepfer, H. R., & Ehrman, L. (1996). Developmental isolation and subsequent adult behavior of *Drosophila paulistorum*. III. Alternative rearing. *Behavior Genetics, 26,* 27–37.

Kim, Y.-K., Phillips, D., Chao, T., & Ehrman, L. (2004). Developmental isolation and subsequent adult behavior of *Drosophila paulistorum*. V. Quantitative variation of cuticular hydrocarbons. *Behavior Genetics, 34,* 385–394.

Kirkpatrick, M. (1987). Sexual selection by female choice in polygynous animals. *Annual Reviews of Ecology and Systematics, 18,* 43–70.

Kravitz, E. A. (2000). Serotonin and aggression: Insights gained from a lobster model system and speculations on the role of amine neurons in a complex behavior like aggressions. *Journal of Comparative Physiology, 186,* 221–238.

Kravitz, E. A., & Huber, R. (2003). Aggression in invertebrates. *Current Opinion in Neurobiology, 13,* 736–743.

Kyriacou, C. P. (2002). Single gene mutations in *Drosophila*: What can they tell us about the evolution of sexual behaviour? *Genetica, 116,* 197–203.

Kyriacou, C. P., & Hall, J. C. (1982). The function of courtship song rhythms in *Drosophila. Animal Behaviour, 30,* 794–801.

Lacy, R. C., & Sherman, P. W. (1983). Kin recognition by phenotype matching. *American Naturalist, 121,* 489–512.

Lande, R. (1981). Models of speciation by sexual selection on polygenic traits. *Proceedings of the National Academy of Sciences USA, 78,* 3721–3725.

Lee, G., & Hall, J. C. (2000). A newly uncovered phenotype associated with the fruitless gene of *Drosophila melanogaster*: Aggression-like head interactions between mutant males. *Behavior Genetics, 30,* 263–275.

Liu, T., Dartevelle, L., Yuan, C., Wei, H., Wang, Y., & Ferveur, J.-F., et al., (2008). Increased dopamine level enhances male-male courtship in *Drosophila. Journal of Neuroscience, 28,* 5539–5546.

Mackay, T. F. C., Heinsohn, S. L., Lyman, R. F., Moehring, A. J., Morgan, T. J., & Rollmann, S. M. (2005). *Proceedings of the National Academy of Sciences USA, 102,* 6622–6629.

Mane, S. D., Tompkins, L., & Richmond, R. C. (1983). Male esterase 6 catalyzes the synthesis of a sex pheromone in *Drosophila melanogaster* females. *Science, 222*, 419–421.

Manning, A. (1961). The effects of artificial selection for mating speed in *Drosophila melanogaster. Animal Behaviour, 9*, 82–92.

Marcillac, F., Bousquet, F., Alabouvette, J., Savarit, F., & Ferveur, J-F. (2005). A mutation with major effects on *Drosophila melanogaster* sex pheromones. *Genetics, 171*, 1617–1628.

Markow, T. A. (1988). Reproductive behavior of *Drosophila melanogaster* and *D. nigrospiracula* in the field and in the laboratory. *Journal of Comparative Psychology, 102*, 169–173.

Markow, T. A. (1996). Evolution of *Drosophila* mating systems. *Evolutionary Biology, 29*, 73–106

Markow, T. A. (2000). Forced matings in natural populations of *Drosophila. American Naturalist, 156*, 100–103.

Markow, T. A., & O'Grady, P. M. (2005). Evolutionary genetics of reproductive behavior in *Drosophila*: Connecting the dots. *Annual Reviews of Genetics, 39*, 263–291.

McBride, S. M. J., Giuliani, G., Choi, C., Krause, P., Correale, D., Watson, K., Baker, G., & Siwicki, K. K. (1999). Mushroom body ablation impairs short-term memory and long-term memory of courtship conditioning in *Drosophila melanogaster. Neuron, 24*, 967–977.

McGuire, S. E., Le, P. T., & Davis, R. L. (2001). The role of *Drosophila* mushroom body signaling in olfactory memory. *Science, 293*, 1330–1333.

McRobert, S. P., Tompkins, L., Barr, N. B., Bradner, J., Lucas, D., Rattigan, D. M., & Tannous, A. F. (2003). Mutations in *raised Drosophila melanogaster* affect experience-dependent aspects of sexual behavior in both sexes. *Behavior Genetics, 33*, 347–356.

Milani, R. (1956). Relations between courting and fighting behaviour in some *Drosophila* species (*obscura* group). *1st. Sup. di Sanita, 1*, 213–224.

Moehring, A. J., & Mackay, T. F. C. (2004). The quantitative genetic basis of male mating behavior in *Drosophila melanogaster. Genetics, 167*, 1249–1263.

Murakami, S., & Itoh, M. T. (2001). Effects of aggression and wing removal on brain serotonin levels in male crickets, *Gryllus bimaculatus. Journal of Insect Physiology, 47*, 1309–1312.

Murakami, S., & Itoh, M. T. (2003). Removal of both antennae influences the courtship and aggressive behaviors in male crickets. *Journal of Neurobiology, 57*, 110–118.

Nilsen, S. P., Chan, Y.-B., Huber, R., & Kravitz, E. A. (2004). Gender-selective patterns of aggressive behavior in *Drosophila melanogaster. Proceedings of the National Academy of Sciences USA, 101*, 12342–12347.

Noor, M. A. (1995). Speciation driven by natural selection in *Drosophila. Nature, 375*, 674–675.

Noor, M. A. F., & Ortíz-Barrientos, D. (2006). Simulating natural conditions in the laboratory: A re-examination of sexual isolation between sympatric and allopatric populations of *Drosophila pseudoobscura* and *D. persimilis. Behavior Genetics, 36*, 322–327.

O'Dell, K. M. C. (1994). The inactive mutation leads to abnormal experience-dependent courtship modification in male *Drosophila melanogaster. Behavior Genetics, 24*, 381–388.

O'Donald, P. (1983). Sexual selection by female choice. In P. Bateson (Ed.), *Mate Choice* (pp. 53–66). Cambridge: Cambridge University Press.

Paillette, M., Ikeda, H., & Jallon, J.-M. (1991). A new acoustic signal of the fruit-flies *Drosophila simulans* and *D. melanogaster. Bioacoustics, 3*, 247–254.

Papaj, D. R., & Messing, R. H. (1998). Asymmetries in physiological state as a possible cause of resident advantage in contests. *Behaviour, 135*, 1013–1030.

Partridge, L. Ewing, A., & Chandler, A. (1987). Male size and mating success in *Drosophila melanogaster*: The roles of male and female behavior. *Animal Behaviour, 35*, 555–562.

Partridge, L., & Farquhar. M. (1983). Lifetime mating success of male fruitflies (*Drosophila melanogaster*) is related to their size. *Animal Behaviour, 31*, 871–877.

Partridge, L., Hoffmann, A., & Jones, S. (1987). Male size and mating success in *Drosophila melanogaster* and *Drosophila pseudoobscura* under field conditions. *Animal Behaviour, 35*, 468–476.

Partridge, L., Mackay, T. F. C., & Aitken, S. (1985). Male mating success and fertility in *Drosophila melanogaster, Genetical Research Cambridge, 46*, 279–285.

Pascual, A., & Préat, T. (2001). Localization of long-term memory within the *Drosophila* mushroom body. *Science, 294*, 1115–1117.

Paterson, H. E. H. (1978). More evidence against speciation by reinforcement. *South African Journal of Science, 74*, 369–371.

Paterson, H. E. H. (1985). The recognition concept of species. In E. S. Vrba (Ed.), *Species and speciation. Transvaal Museum Monograph, 4* (pp. 21–29). Pretoria: Transvaal Museum.

Popova, N. K. (2006). From genes to aggressive behavior: The role of serotonin system. *BioEssays, 28*, 495–503.

Price, D. K., & Boake, C. R. B. (1995). Behavioral reproductive isolation in *Drosophila silvestris, D. heteroneura* and their F_1 hybrids (Diptera: Drosophilaidae). *Journal of Insect Behavior, 8*, 595–616.

Quinn, W. G., Sziber, P. P., & Booker, R. (1979). The *Drosophila* memory mutant amnesiac. *Nature, 277*, 212–214.

Ringo, J. M. (1977). Why 300 species of Hawaiian *Drosophila*? The sexual selection hypothesis. *Evolution, 31*, 694–696.

Ringo, J., Kananen, M. K., & Wood, D. (1983). Aggression and mating success in three species of *Drosophila. Journal of Comparative Ethology, 61*, 341–350.

Ringo, J. M., & Hodosh, R. J. (1978). A multivariate analysis of behavioral divergence among closely related species of endemic Hawaiian *Drosophila. Evolution, 32*, 389–397.

Ritchie, M. G., & Gleason, J. M. (1995). Rapid evolution of courtship song pattern in *Drosophila willistoni* sibling species. *Journal of Evolutionary Biology, 8*, 463–479.

Ritchie, M. G., & Kyriacou, C. P. (1996). Artificial selection for a courtship signal in *Drosophila melanogaster. Animal Behaviour, 52*, 603–611.

Ritchie, M. G., Saarikettu, M., Livingstone, S., & Hoikkala, A. (2001). Characterisation of female preference functions for a sexually selected acoustic signal in *D. montana*, and a test of the "temperature coupling" hypothesis. *Evolution, 55*, 721–727.

Ritchie, M. G. Townhill, R. M., & Hoikkala, A. (1998). Female preference for fly song: Playback experiments confirm the targets of sexual selection. *Animal Behaviour, 56*, 713–717.

Robertson, F. W., & Reeve, E. (1952). Studies in quantitative inheritance. I. The effects of selection on wing and thorax length in *Drosophila melanogaster. Journal of Genetics, 50*, 414–448.

Robin, C., Daborn, P. J., & Hoffmann, A. A. (2006). Fighting fly genes. *Trends in Genetics, 23*, 51–54.

Roman, G., & Davis, R. L. (2001). Molecular biology and anatomy of *Drosophila* olfactory associative learning. *Bioessays, 23*, 571–581.

Saarikettu, M., Liimatainen, J., & Hoikkala, A. (2005). The role of male courtship song in species recognition in *Drosophila montana. Behavior Genetics, 35*, 257–263.

Santos, M., Ruiz, A., Barbadilla, A., Hasson, E., & Fontdevila, A. (1988). The evolutionary history of *Drosophila*. XIV. Larger flies mate more often in nature. *Heredity, 61*, 255–262.

Savarit, F., Sureau, G., Cobb, M., & Ferveur, J.-F. (1999). Genetic elimination of known pheromones reveals the fundamental chemical bases of mating and isolation in *Drosophila. Proceedings of the National Academy of Sciences USA, 96*, 9015–9020.

Singh, R. N., & Singh, K. (1984). Fine-structure of the sensory organs of *Drosophila melanogaster* meigen larva. *International Journal of Insect Morphology and Embryology, 13*, 255–273.

Skrzipek, K. H., Kroner, B., & Hager, H. (1979). Aggression bei *Drosophila melanogaster* – Laboruntersuchungen. *Journal of Comparative Psychology, 49*, 87–103.

Snook, R. R., Robertson, A., Crudgington, H. S., & Ritchie, M. G. (2005). Experimental manipulation of sexual selection and the evolution of courtship song in *Drosophila pseudoobscura. Behavior Genetics, 35*, 245–255.

Spiess, E. B. (1987). Discrimination among prospective mates in *Drosophila*. In D. J. C. Fletcher & C. D. Michener (Eds.), *Kin recognition in animals* (pp. 75–119). New York: John Wiley & Sons.

Spiess, E. B., & Langer, B. (1964). Mating speed control by gene arrangements in *Drosophila pseudoobscura* homokaryotypes. *Proceedings of the National Academy of Sciences USA, 51*, 1015–1019.

Spieth, H. T. (1966). Courtship behavior of endemic Hawaiian *Drosophila. Studies in Genetics III. University of Texas Publication 6615*, 245–313.

Spieth, H. T. (1968). Evolutionary implications of sexual behavior in *Drosophila. Evolutionary Biology, 2*, 157–193.

Spieth, H. T. (1974). Courtship behavior in *Drosophila. Annual Reviews of Entomology, 19*, 385–405.

Spieth, H. T. (1981). *Drosophila heteroneura* and *Drosophila silvestris*: Head shapes, behavior and evolution. *Evolution, 35*, 921–930.

Spieth, H. T. (1982). Behavioral biology and evolution of the Hawaiian picture-winged species group of *Drosophila. Evolutionary Biology, 14*, 351–437.

Spieth, H. T., & Heed, W. B. (1975). The *Drosophila pinicola* species group. *Pan-Pacific Entomologist, 51*, 287–295.

Strausfeld, N. J., Hansen, L., Li, Y., Gomez, R. S., & Ito. K. (1998). Evolution, discovery and interpretations of arthropod mushroom bodies. *Leaning & Memory, 5*, 11–37.

Svetec, N., Cobb, M., & Ferveur, J.-F., (2005). Chemical stimuli induce courtship dominance in *Drosophila. Current Biology, 15(19)*, R790–R792.

Svetec, N., & Ferveur, J.-F. (2005). Social experience and pheromonal perception can exchange male-male interactions in *Drosophila melanogaster. Journal of Experimental Biology, 208*, 891–898.

Tauber, E., & Eberl, D. F. (2002). The effect of male competition on the courtship song of *Drosophila melanogaster. Journal of Insect Behavior, 15*, 109–120.

Tauber, E., & Eberl, D. F. (2003). Acoustic communication in *Drosophila. Behavioural Processes, 64*, 197–210.

Technau, G. M. (1984). Fiber number in the mushroom bodies of adult *Drosophila melanogaster* depends on age, sex and experience. *Journal of Neurogenetics, 1*, 113–126.

Technau, G. M., & Heisenberg, M. (1982). Neural reorganization during metamorphosis of the corpora pedunculata in *Drosophila melanogaster. Nautre, 295*, 405–407.

Templeton, A. R. (1977). Analysis of head shape differences between two interfertile species of Hawaiian *Drosophila. Evolution, 31*, 630–641.

Ueda, A., & Kidokoro, Y. (2002). Aggressive behaviours of female *Drosophila melanogaster* are influenced by their social experience and food resources. *Phsiological Entomology, 27*, 21–28.

Vrontou, E., Nilsen, S. P., Kravitz, E. A., & Dickson, B. J. (2006). *fruitless* regulates aggression and dominance in *Drosophila. Nature Neuroscience, 9*, 1469–1471.

Waddell, S., & Quinn, W. G. (2001). What can we teach *Drosophila*? What can they teach us? *Trends in Genetics, 17*, 719–726.

Wallace, B. (1974). Studies on intra- and inter-specific competition in *Drosophila. Ecology, 55*, 227–244.

Wheeler, C. J., Fields, W. L., & Hall, J. C. (1988). Spectral analysis of *Drosophila* courtship songs: *D. melanogaster, D. simulans*, and their interspecific hybrids. *Behavior Genetics, 18*, 675–703.

Wilkinson, G. S., & Johns, P. M. (2005). Sexual selection and the evolution of mating systems in flies. In D. K. Yates & B. M. Wiegmann (Eds.), *The evolutionary biology of flies* (pp. 312–339). New York: Columbia University Press.

Williams, M. A., Blouin, A. G., & Noor, M. A. F. (2001). Courtship songs of *Drosophila pseudoobscura* and *D. persimilis*. II. Genetics of species differences. *Heredity, 86*, 68–77.

Yurkovic, A., Wang, O., Basu, A. C., & Kravitz, E. A. (2006). Learning and memory associated with aggression in *Drosophila melanogaster. Proceedings of the National Academy of Sciences USA, 103*, 17519–17524.

Zamudio, K. R., Huey, R. B., & Crill, W. D. (1995). Bigger isn't always better: Body size, developmental and parental temperature and male territorial success in *Drosophila melanogaster. Animal Behaviour, 49*, 671–677.

Zars, T., Fischer, M., Schulz, R., & Heisenberg, M. (2000). Localization of a short-term memory in *Drosophila. Science, 288*, 672–675.

Zhang, S.-D., & Odenwald, W. F. (1995). Misexpression of the white (w) gene triggers male-male courtship in *Drosophila. Proceedings of the National Academy of Sciences USA, 92*, 5525–5529.

Chapter 23

Handedness: A Behavioral Laterality Manifestation

Ira B. Perelle and Lee Ehrman

Apologia

Because we are, respectively, a psychologist and a geneticist, though only one of us is a lefthander, we view handedness as an adapted and adaptive behavioral phenotype. *Handbook* utilizers are referred to relevant chapters herein by Neal, Cherny, Posthuma, Ulbricht, and Neiderhiser covering:

Biometrical Models
QTL Methodology
Multivariate Analysis
Genotype–Environment Interaction

Introduction

Why study handedness? In Western cultures hardly anyone pays attention to whether a person is right- or lefthanded. This was not always true: Not very long ago, in terms of human evolution, to be lefthanded was to be evil, mentally deficient, sickly, possessed, clumsy, oafish, or any combination of these. Now, most of the developed world knows better: Lefthanders, approximately 10% of the population, write with their left hand. Does it matter? Not really, but it is different and we, as scientists, are interested in lefthanders because we are particularly interested in things that differ from the norm.

I.B. Perelle (✉)
Mercy College, Dobbs Ferry, NY 10522, USA
e-mail: iperelle@mercy.edu

L. Ehrman (✉)
Purchase College, Purchase, NY 10588, USA
e-mail: lehrman@purchase.edu

Historical Perspective

The point at which man became primarily righthanded is difficult to determine. No artifacts located to date provide any reason to believe *Homo erectus*, as a group, were lateralized for hand use, although it is probable that individuals were either righthanded or lefthanded in approximately equal proportions, as are chimpanzees, gorillas, and other nonhuman primates today (Finch, 1941; Kummer & Goodall, 1985; McGrew & Marchant, 2001). Stone Age implements, discovered in various parts of what is now Europe, lead to the conclusion that early hominids did not express a species-wide hand preference. A study of the crafting of 426 Stone Age implements found in southern France indicated that some were made to be used by the left hand, some by the right hand, and some by either hand, in almost equal numbers (Sarasin, 1884).

It is probable that utterances were a part of the behavioral repertoire of *Homo erectus;* however these almost certainly were extremely crude, not unlike the utterances of nonhuman higher primates existing today. It is also probable that the tendency toward righthandedness appeared first in *Homo sapiens* (with Neanderthals considered the first in the proto-human lineage, existing between 80,000 and 35,000 BCE) shortly after the time crude utterances developed into simple speech. Cro-Magnon man, closely resembling modern man, came on the scene approximately 35,000 BCE and, though there is no indication in the fossil record of direct evolutionary continuity between Neanderthals and Cro-Magnon, there undoubtedly was a common intermediate form. It is believed that interbreeding occurred between Neanderthals and Cro-Magnon (Coyne & Orr, 2004; Futuyma, 2005).

Language, however, is qualitatively different than simple speech. Although other animals have been taught to communicate with humans in various formats and in some cases with dramatic results, this communication cannot be considered language. Alex, the African Grey parrot trained by Irene Pepperberg, could answer, in clear spoken English, many spoken questions relating to quantity, material, difference, and sameness, and Alex could also initiate requests

Y.-K. Kim (ed.), *Handbook of Behavior Genetics*,
DOI 10.1007/978-0-387-76727-7_23, © Springer Science+Business Media, LLC 2009

(Pepperberg, 2002). We maintain that as impressive as Alex's speech ability was it was not language in the strict sense of the word. Language is systematic communication by vocal or written symbols used and understood by a group of people. It must include words for objects and actions, rules for the placement of various word types in sentences, and rules for sentence structure. In short, the use of language is a highly complex and sophisticated behavior. As early hominids evolved into *Homo sapiens*, language became an important acquisition, perhaps *the* most important. It was surely the feature that made advanced cultural transmission possible, and most certainly there was strong evolutionary pressure on *H. sapiens* for the capability to learn and use language. Prior to language development, the bilateral hominid brain would have had no dedicated area for this function. As language became such an important part of human existence, a section of the brain evolved for processing this function, and this verbal processing area, located by Paul Broca, is in the left hemisphere for most humans (Broca, 1861; Broca, 1865). This is now known as Broca's area. The reason verbal processing developed in the left hemisphere is not known, nor may it ever be known. It has been suggested by Corballis (1997) and others that the cause was a chance mutation in the left hemisphere and could just have easily been in the right hemisphere, but Nottebohm and Nottebohm (1976) found that songbirds, which depend on audible communication for successful courtship and territory maintenance, also process their "language" in the left hemisphere. (For more on bird laterality, see Ehrman & Perelle, 1983.)

Gesturing

Primates gesture. Gorillas and chimpanzees in the wild and in naturalistic habitats gesture to each other in what appears to be intentional communication. Gorillas in the San Francisco Zoo have been observed to use at least 30 gestures, most of them subtle and difficult for human observers to detect (Tanner & Byrne, 1996). Chimpanzees have also been seen to use gestures in what appeared to be a form of conversation with another chimpanzee (Tomasello et al., 1997). De Wall (1982) believes that such primate gestures originate as behavior toward some object and then become standardized for communication. Since nonhuman primates use gestures for communication and there is no reason to believe that this is a recently developed behavior, it is reasonable to presume that primate gesturing occurred in prehistoric times and remained in the behavioral repertoire of any intermediate forms that evolved into early *Homo*, then descended into *Homo sapiens*. Gesturing is a form of communication, and because humans developed a specific part of the brain for communication purposes, it is almost certain that the communicative aspect of gesturing is initiated and interpreted in

Broca's area, part of the left hemisphere (see also Corballis, 1999). The importance of the left hemisphere cannot be underestimated. Luria (1973) was quite aware of this when he wrote:

> The left hemisphere (in right handers) begins to play an essential role not only in the cerebral organization of speech, but also in the cerebral organization of all higher forms of cognitive activity connected with speech – perception organized into logical schemes, active verbal memory, logical thought – whereas the right hemisphere begins to play a subordinate role
> ...(Luria, 1973, p.78).

Recently it has been observed that most baboons (78%) use their right hand for communicative gesturing, but not for object manipulation (Meguerditchian & Vauclair, 2006). From an evolutionary viewpoint, this finding implies that an early form of left hemisphere specialization for communication may have existed in a common primate ancestor that predates early human forms.

Motor Control and Righthanders

Although the evolutionary origin is unknown, it is recognized that control of both fine and coarse muscles is located in the hemisphere contralateral to the muscle being controlled (Corballis & Morgan, 1978; Geschwind, 1975a). If, therefore, humans process language in their left cerebral hemisphere and the left hemisphere controls the right side of the body, it would be most efficient for gesturing and, eventually writing, to be performed with the right arm and/or hand, as is true of most humans. Because the right hand of most modern humans is used for higher forms of cognitive activities (e.g., writing and gesturing), it is the hand initially selected for many other tasks and eventually becomes adept at tool use and other common behaviors such as ball throwing, tooth brushing, etc. Since our modern world was designed by righthanders, many of the everyday activities we perform are very difficult to do with the left hand (e.g., scissor use), which further reinforces the use of the right hand. Why, then, are there any lefthanders at all?

The existence of lefthanders presents many problems: superstition (are lefthanders inherently sinister?), proportion (how many lefthanders are there?), difference (other than preferred hand, is a lefthander different from a righthander?), determination (how do you determine a lefthander?), and etiology (how does one become a lefthander?).

Superstitions and Myths

In the past, and even in some places even today, lefthandedness has been associated with evil. The Catholic Church once declared that lefthanded people were servants of the devil, and up to the latter part of the 20th century, Catholic schools forced lefthanded children to switch to their right hand for writing and other activities (Winters, 2004). Some African tribes equate lefthandedness with evil spirits; in Japan, until approximately 40 years ago, discovering one's wife to be lefthanded was cause for divorce; and in the past, Jewish priests had to be free of any bodily defects, including blindness, lameness, dwarfism, and, of course, lefthandedness.

In many Middle Eastern countries, social customs prohibit public use of the left hand. In these countries food is eaten from a common vessel, and bodily functions, usually without the luxury of toilet tissue, are performed with the left hand. It is obvious why reaching into the common food bowl with the left hand is prohibited. In these countries, it is a major insult to hand an object to someone with the left hand. In response to a query from a 12-year-old lefthander, who had been told many times that he will literally "go to hell" for eating and writing with his left hand, the following response appeared in *Arab News*, "If a left-handed person trains himself patiently to eat and drink with his right hand until this becomes quite easy for him, and he does so in order to follow the Prophet's example, he is sure to earn God's reward" (Arab News, 2005). In other words, God abhors a lefthander.

In India, lefthanders have been deemed dirty, evil, and/or sinister. In India's major cities and among the educated, but not in rural areas, this has changed and lefthanders are now accepted as normal and are welcomed on sports teams and in general society. For religious purposes, however, the use of the left hand is prohibited and there are restrictions against its use for worship ritual (Pooja), holy food (Prasad), as well as other ceremonial functions. In rural areas, the old ways still prevail. Lefthanded potential daughters-in-law or sons-in-law are not accepted, and lefthanded children are first discouraged, then forcibly prohibited from using their left hand (Chaugule, 2005).

Many Eskimos believe all lefthanded people are potential sorcerers, and in Morocco lefthanders are thought to be devils and cursed. In colonial America, lefthandedness was ample reason to be accused of being a witch. Trials consisted of, among other things, an examination of the naked body of the accused, and a blemish or a mole, or a third nipple, found on the left side of the body was sufficient cause for conviction and execution. Joan d'Arc is believed to have been convicted of witchcraft in this manner and in various pictures she is shown holding her sword in her left hand. There are historians, however, who believe she was really righthanded but was shown lefthanded in pictures to justify her execution.

Are Lefthanders Different?

British educational psychologist Cyril Burt, describing lefthanders, opined:

They squint, they stammer, they shuffle and shamble, they flounder about like seals out of water. Awkward in the house and clumsy in their games, they are fumblers and bunglers in whatever they do. (Burt, 1937, p 287)

Are lefthanders really that different? Sir Thomas Browne, as far back as the mid 1600s, seemed to think so. He became interested in handedness and asked, among other things, whether it is good to stop children from using their left hands (Wilkin, 1852). Some lefthanders indeed are different, but Burt's diatribe is extreme. The frequency of lefthanders is significantly greater than the global average in several populations and later in this chapter we will explain why. Lefthanders are overrepresented in populations of mentally retarded individuals, children with learning deficits, and those with reading difficulties, and it has been found that the proportion of lefthanders increases as IQ decreases (Gregory & Paul, 1980; McBurney & Dunn, 1976; Pirozzolo & Rayner, 1979; Springer & Eisenson, 1977). Lefthanders, however, are also overrepresented among the gifted. In a study of the handedness of members of the Mensa Society, all members of which are in the upper 2% of the IQ range, we (Ehrman & Perelle, 1983; Granville, Ehrman, & Perelle, 1979) found the proportion of lefthanders to be 20%, double that in the general population (p < 0.001). Studies of unselected school children, however, show no difference in cognitive ability between left- and righthanders (Hardyk & Petrinovich, 1977; Swanson, Kinsbourne, & Horn, 1980), nor were any differences between left- and righthanders found in articulation, stammering, speech, writing productivity, or syntactic maturity among a representative population of 11-year-olds (Calnan & Richardson, 1976).

In one of the most flawed studies of lefthanders, but one which received much publicity, lefthanders were found to be considerably different: On the average, they were found to die 9 years earlier than righthanders (Coren & Halpern, 1999). Lefthanders do not die younger than righthanders. One thing the authors neglected to consider was hand switching among the subjects in their study, most of whom were senior citizens. It is well known that in the first half of the 20th century many, perhaps most, lefthanders were switched to right hand use by teachers, parents, and others who believed all the myths and negative events, discussed above, that would plague lefthanders. There are far fewer older lefthanders because there were far fewer people who were permitted to be lefthanded (Wolf, D'Agostino, & Coss, 1999), and there are actually far fewer living lefthanders simply because there are far fewer lefthanders in older cohorts. Several other studies have also shown the early death hypothesis to be incorrect (Kuhlemeier, 1991), and in one, Basso et al. (2000), it was reported that in 60% of dizygotic twin pairs, the righthanded twin died first (50% in the monozygotic twin pairs). In another recent study, 50,000 Swedish military conscripts, aged 18–21, inducted into the military from 1969 to 1970, were tracked through 1989. Of this cohort, 954 died, and of the 954, 82 or 8.6% were lefthanded, slightly less than the population proportion (Person & Allebeck, 1994). The Iowa Women's Health Study determined, among other things, that lefthanded women showed no increased mortality compared to righthanded women (Cerhab et al., 1994). The second major problem with Coren's research (Coren & Halpern, 1999) was that the deceased's handedness was determined by telephone inquiry to a relative. Reports of handedness by relatives are notoriously unreliable (Bryden, 1982).

In the mid 1900s, it was fairly common for researchers to conduct studies that showed many physical problems associated with lefthandedness. Geschwind and Behan (1982, 1984) managed to find lefthanders significantly deficient relative to righthanders for all manner of ills, including migraine, allergies, dyslexia, stuttering, skeletal malformations, and thyroid disorders. Others found evidence of increased schizophrenia (Krynicki & Nahas, 1979) as well as other malfunctions (Benson & Geschwind, 1972; Carter-Saltzman, 1979). None of this is true. In 1984, McManus and his colleagues performed a meta-analysis incorporating 89 studies that included 21,000 patients and 34,000 controls. The result showed no indication that lefthanders had a greater tendency toward disorders than did righthanders (described in McManus, 2002). Erlenmeyer-Kimling and her colleagues analyzed data from three studies and found that "there are no significant findings for *anything* related to hand preference or hand skills" (Erlenmeyer-Kimling et al., 2005, p. 355), and this result was consistent within each study as well as across all studies.

Are lefthanders different? Yes, they write with the left hand. Aside from that, lefthanders as a group are not different. They are overrepresented in some populations but that overrepresentation is not skewed in a single direction (for example, there is a higher proportion of lefthanders in both the gifted and the retarded populations). What has been found is that lefthanders are more varied as a group than righthanders (Perelle & Ehrman, 1982). Many of the early negative findings were the result of faulty determination of handedness, which will be discussed below, as well as the self-fulfilling prophecy effect, well known in psychology: If it is bad to be a lefthander and an individual is a lefthander, that individual must be bad.

Handedness Proportions

Currently, there is some agreement, but not a universal consensus, that the proportion of lefthanders in the population is approximately 10%. Several factors affect this proportion including how you determine a "lefthander," how you measure "lefthandedness," where you measure handedness, and what you are looking for. We believe the oldest written record of handedness proportion occurs in the Old Testament *Book of Judges, 20:12*. One passage states that the Benjamite Army of 26,000 men included an elite sling shot unit composed of 700 lefthanders, " …every one who could sling a stone at a hair and not miss" (The Holy Bible, 1611/1962, p. 234). This special unit comprised 2.7% of the total army. There is no indication in this passage, however, that these were the only lefthanders in the army.

There are almost as many estimated proportions of lefthanders reported as there are papers written on handedness. Hecaen and Ajuriaguerra (1964) reviewed a number of studies and found the proportions of lefthanders specified ranged from 1 to 30%. Bryden (1979), investigating young adults in Canada, determined the proportion of lefthanders to be 10.39%. In Japan, however, reported proportions of lefthanders are considerably lower than in the United States and much of Europe: Hatta and Nakatsuka (1976) indicated that 3.1% of their subjects were lefthanders. Dawson (1977) provides lefthander proportions of approximately 10% for Alaskan Eskimo, but between 1 and 3% for the Hong Kong Hakka, the Katanganese, and the Temme of Sierra Leone, these last three being highly conservative and conforming societies.

We (Perelle & Ehrman, 1994) implemented a large-scale investigation of handedness in 32 countries. We developed a self-administered questionnaire, partially based on the same questions that other researchers were asking at the time (more about this later). The questionnaire was translated from English into French, Spanish, Italian, Dutch, German, Swedish, Russian, and Portuguese. (The method of translation, to assure validity across languages, was standard translate/back translate with iterations until the back translation exactly matched the original.) We contacted colleagues in 32 countries in which the language spoken was one of those into which we had the questionnaire translated and solicited their cooperation. If an affirmative reply was given, and it almost always was, we forwarded questionnaires, return envelopes, and payment for return postage. We received 12,000 completed questionnaires of which 11,074 were useable. Countries from which less than 100 questionnaires were received were grouped into an "other" category and include Haiti and Cyprus, from which one or two questionnaires were returned, and Israel and Venezuela, each of which returned approximately 90 questionnaires. The final results are shown in Table 23.1, where it can be seen that, based on writing hand, lefthanders' proportions ranged from 2.5 (Mexico) to 12.8% (Canada), with the overall proportion being 9.5%.

Because of the social bias against lefthandedness in most countries prior to mid 1900s, many handedness researchers expect the proportion of lefthandedness in a cohort to decrease as older cohorts are selected, and our results

Table 23.1 Frequency, sex distribution, writing hand, self-perceived handedness, and switching attempt of subjects by country of residence (after and augmented from Perelle & Ehrman, 1994)

Country	Australia	Belgium	Canada	England	France	Italy	Mexico	Netherlands	New Zealand	Nigeria	Spain	Turkey	United States	Other	Total
Subjects															
N	642	489	656	179	941	214	446	346	231	276	1220	115	4885	434	11,074
% Female	51.1	51.3	60.7	46.4	52.7	40.2	46.9	44.8	48.5	47.1	56	67	56.8	54.6	54.4
Writing hand															
% Left	10.7	7.5	12.8	11.2	7.8	6.5	2.5	10.8	6.5	5.5	3.6	2.6	12.2	6.3	9.5
% Right	88.3	92.3	86.3	88.3	91.1	92.5	96.2	88.7	93.1	92.6	96	95.7	86.6	93	89.6
% Either	0.9	0.2	0.9	0.6	1.1	0.9	1.3	0.6	0.4	0.7	0.4	1.7	1	0.7	0.9
Self-perceived handedness															
% Strong left	5.6	7.1	5.4	8	5 8	4.7	2.3	9.4	3.1	4.1	2.2	4.5	7.2	5.6	5.9
% Mod. left	4.6	8.6	6.9	4	4.9	3.3	1.6	6.4	3.1	0.8	3.8	4.5	4.6	3.1	4.5
% Ambidextrous	3.2	5.6	3.8	3.4	5.4	3.3	4.3	6.7	2.2	1.1	1.5	1.8	4	1.6	3.7
% Mod. right	22.3	20.8	33.8	16.1	24.1	29.4	26.8	16.1	23.1	13.5	24.7	13.4	17.7	12.7	20.6
% Strong right	64.3	57.9	50.2	68.4	59.8	59.3	65.1	61.4	68.4	80.5	67.9	75.9	66.5	76.9	65.2
Attempt to switch															
% Yes	8.8	12.5	6.5	4.5	7	4.3	5.8	10.3	5.2	15.5	6.5	10.8	6	7.3	7

Because of rounding, totals may not total 100%

confirmed this. Dawson (1977) found a negative relationship between the cultural formality of a population and the proportion of lefthanders in that population. Medland (Medland et al., 2004), using our data (Perelle & Ehrman, 1994[1]), divided the countries in which the subjects learned to write into two categories: Formal (strict/conformist) and Non Formal (permissive/tolerant) cultures, using Hofstede's (1983) characterization. Medland found that, with one exception (Italy), in countries classified as Formal, the proportion of lefthanded writers was significantly less than in countries classified as Non Formal, as indicated by a standard contingency test ($\mathcal{X}^2 = 63.35$; p < 0.0001; Medland et al., 2004, p. 292) and concluded that "...the context in which one learns to write may have a lasting effect on adult handedness" (p. 295).

Research and Issues

One of the most contentious topics in the study of handedness is the method used to determine the handedness of individuals. This is not a frivolous topic since, depending on the method used, subjects may be placed in a hand category different from the category in which they believe they belong. As Geschwind observed in his review of a book on the topic, "A striking omission is the failure to define the subject. We never really find out how one defines the left-hander or how one constructs reliable measures for testing this trait. The problem is not trivial." (Geschwind, 1975b, p. 23). In the early years of the 20th century, there was little concern about the importance of this topic. The determination of handedness was more or less informal: "... a righthander is a human being who, when only an infant, uses his right hand as his main instrument, whereby the left serves only as assistance." (Schaefer, 1911, in Wile, 1934, p. 85).

In order to determine differences between two groups, there must be a reliable and valid test to determine membership in each group. Many researchers have used, and some are still using, the Edinburgh Handedness Inventory (Oldfield, 1971) or some variation of it, a questionnaire to be completed by the subject or the researcher. Most, if not all, researchers using this test make up their own scoring protocol, based on the following: If the subject claimed to use the right hand for all tasks listed, the subject is righthanded. If, however, the subject used the left hand for one or more tasks, classification depended on the researcher's preference. Some researchers classified subjects more or less non-righthanded (rather than lefthanded) depending on the number of tasks performed with the left hand (e.g., Rife, 1940). Some devel-

oped a numerical scale such as 0 to 10 or -5 to $+5$ and based subjects' handedness on an arbitrary cutoff. Several included the categories of "slightly lefthanded" and "slightly righthanded." Others considered the middle scores as indicative of ambidextrality. This lack of consistency meant (and still means) studies could not be compared since lefthandedness (non-righthandedness) could mean whatever the researcher wanted it to mean. Bishop's 1990 paper, "How to increase your chances of obtaining a significant association between handedness and disorder" was an attempt to bring this problem to the attention of the handedness research community. Holder's unpublished thesis, "Hand preference questionnaires: One gets what one asks for," was another (Holder, 1992). In an Interim Report, Holder stated that the term "handedness" does not have a precise standard: There are no empirical means by which to determine handedness; handedness questionnaire designs and analysis are based on assumptions; and the results from a set of data may be predetermined by the questionnaire, the scoring system, and/or the subject selection methods (Holder, 2005).

Quite obviously we agree, and we propose a very simple solution: In literate populations, handedness will be determined by the hand with which a person writes. Except for an exceedingly small proportion of the population, probably less than 0.5%, who actually can write equally easily and clearly with both hands, this will quickly and unequivocally provide an immediate dichotomous handedness classification. In our international study (Perelle & Ehrman, 1994) we found that 85.5% of the subjects who wrote with the left hand considered themselves to be lefthanded and 85.8% of the subjects who wrote with the right hand considered themselves righthanded (see Table 23.2).

In Table 23.2, it can also be observed that 4.8% of the subjects who wrote with the left hand considered themselves righthanded (column 4: 3.1% + 1.7%) and 2.7% who wrote with the right hand considered themselves lefthanded (column 7: 0.8% + 1.9%). This latter group may be switched lefthanders who still considered themselves lefthanders, but it is highly unlikely that the first group represents righthanders who were switched to lefthanded writing. It is possible that because of the social stigma of lefthandedness in their society, some lefthanders actually call themselves righthanded. The remaining subjects considered themselves ambidextrous, but if our solution is adopted, this problem is eliminated since less than 1% of the subjects claimed that they wrote with either hand (Table 23.1).

Among the problems arising when asking the hand used for various tasks is that many individuals have easily learned to perform a task with either hand. Automobile mechanics, for example, who sometimes have to work in very tight places under a car, learn to use a wrench or screwdriver with whichever hand most conveniently reaches the nut or bolt to be adjusted. Consider pianists and other musicians

---

[1] The Perelle & Ehrman (1994) data are available to qualified researchers, in ASCII format, for further analyses.

Table 23.2 Self-classification of handedness by writing hand (after and adapted from Perelle & Ehrman, 1994)

Self-classification of handedness	Writing hand								
	Left			Right			Total		
	N	% of row	% of column	N	% of row	% of column	N	% of row	% of column
Strongly left	568	88.5	55.9	74	11.5	0.8	642	100.0	5.9
Moderately left	301	61.3	29.6	190	38.7	1.9	491	100.0	4.5
Ambidextrous	99	24.6	9.7	304	75.4	3.1	403	100.0	3.7
Moderately right	31	1.4	3.1	2208	98.6	22.5	2239	100.0	20.6
Strongly right	17	0.2	1.7	7054	99.8	71.8	7071	100.0	65.2
Total	1016	9.4	100.0	9830	90.6	100.0	10846	100.0	100.0

who must use both hands simultaneously and equally well to perform their craft. There have been at least two lefthanded (reversed) pianos built, one in 1826 and the other in 1998 (The Piano, 1999). This was probably a matter of idiosyncrasy, rather than musical necessity, since whether it be a lefthanded piano or a righthanded piano, the pianist must still use both hands. Puentes (2004) found no atypical handedness distribution among undergraduate music and visual arts majors. Typists and other keyboard operators also must use both hands with equal proficiency.

We (Perelle et al., 1981) investigated the speed with which an individual's nonwriting hand could be trained to perform a task relative to their writing hand (sometimes called "the preferred hand"). We used the Crawford Small Parts Dexterity Test (Crawford & Crawford, 1956), which requires subjects, using tweezers, to place tiny pegs in holes and then to place tiny collars on these pegs. As would be expected for both lefthanders and righthanders, in the pretest trial there was a highly significant difference between the speed of the writing hand and the nonwriting hand. After performing six learning trials a day for 5 days, a post-test trial was administered. For both lefthanders and righthanders, performance speeds increased significantly, but the speed of performance for the nonwriting hand increased to such an extent that for several subjects it was actually faster than the performance of the writing hand. In less than 1 week all subjects had learned to use their nonwriting hand with equal or greater proficiency than their writing hand (Perelle, Ehrman, & Manowitz, 1981).

People tend to use their writing hand for all manner of tasks because it is the hand that gets the most training. There are, however, situations in which the critical hand is not necessarily the writing hand. Military personnel and hunters, when asked, will say they fire a rifle righthanded. This means they place the butt of the rifle against the right shoulder with the index finger of the right hand on the trigger. The critical hand, though, is the left hand. That is the hand that supports the barrel of the rifle and must coordinate with the eye to aim the rifle. If the left hand is not smooth in its movement, steady and accurate, the target will be missed. As another example, show dog handlers must keep the dog on their left during show trials and control the dog with the left hand. Handlers have no trouble learning these behaviors.

It is obvious that, unless we have only one functional arm, we all use both hands and, except for writing, can quickly learn to use the nonwriting hand as efficiently as the writing hand. Beyond a doubt, the writing hand, as handedness determinant, is, therefore, the ideal solution to the classification problem.

Handedness Etiology

From the time serious research in handedness began, the Holy Grail of handedness research was to find the sole cause, or etiology, of this "aberrant" behavior, lefthandedness. In 1902, Cunningham, in a short commentary titled "Lefthandedness and Right-brainedness" stated that lefthandedness is hereditary (Cunningham, 1902, p.273). What makes his comments particularly interesting is that Cunningham discussed the relationship between brain hemispheres and handedness, but not many paid attention to his foresight until recently. Cunningham attributed the dominance of the left hemisphere for verbal processing to natural selection, which was becoming popular as a theory at the turn of the century. A sampling of the researchers proposing that handedness is heritable or genetically based includes Annett (1973), Chamberlain (1928), Levy and Nagylaki (1972), Merrell (1957), and Rife (1940). These researchers suggested single dominant genes for righthandedness, two genes (Annett's "right shift theory"), and others. A fairly detailed analysis of these theories can be found in Bradshaw and Nettleton (1983). All used family handedness data to support their theory but, as Bryden (1982) said,

> There is really no wholly satisfactory study of handedness in families. The problems associated with the measurement of handedness have already been discussed. Family studies of handedness invariably violate one or more of the[se] caveats... self report as a means of assessing handedness, questionnaires not standardized, accept statements about relatives' handedness, criteria for considering a person lefthanded vary from study to study, and so on. (Bryden, 1982, p. 183).

There are those who disagree with genetic theories. Blau believed that lefthandedness was indicative of a learning failure, not genetics, and arose from "an inherent deficiency, faulty education, or emotional negativism." (Blau, 1946, p.182). We find this rather excessive. Corballis & Morgan (1978) determined that handedness is not genetic in origin, and Collins (1977), after years of attempting to selectively breed mice for pawedness direction, is convinced that although strength of laterality may be genetic, genes are not lateralized, that is, direction of laterality has no genetic basis. We (Ehrman & Perelle, 1983) investigated *Drosophila melanogaster* and *D. paulistorum* to determine laterality, if any, within and across subjects, and within and across families. Using *Drosophila* courtship behaviors, circling, tapping, and wing extension, we found no evidence that left- or right-oriented behaviors had any genetic component. Approximately 20% of our subjects were highly lateralized either left or right; lateralization had no effect on courtship success nor did it in any way predict the lateralization of F1 or F2 generations. Females displayed no preference for the side from which they were courted, or the wing with which they were touched.

The Three Etiologies of Handedness

We strongly believe that any theory of handedness etiology must include a consideration of the known facts from various disciplines, and we also strongly believe in the principle of parsimony: If there are several ways to explain a phenomenon, the simplest is usually the correct way. Any theory must consider why, among other things, relative to the general population there is a greater proportion of lefthanders among both the lower and higher ends of the IQ distribution, why a higher proportion of lefthanders have a lower birth weight, why many twin pairs, particularly monozygotic twins, are discordant for handedness, why a higher proportion of lefthanders stutter when they are young but cease as they approach puberty, and why lefthanders perform a greater proportion of nonwriting tasks with their right hand than righthanders do with their left. We introduced our theory of the etiology of lefthandedness and briefly answered some of these questions in Perelle and Ehrman (2001) and will provide a more detailed explanation here.

We believe it is clear that no single etiology can account for the various behavioral manifestations associated with lefthandedness, and we are pleased that Linke & Kersebaum (2005) agree with us. We posit there are three etiologies that result in lefthandedness. The first is the "pathological lefthander" [an untoward designation coined by Satz (1972)]. These individuals are lefthanded as the result of a traumatic event prenatally, perinatally, or shortly postnatally, which, if prenatally, interfered with the normal development of their left hemisphere or, if peri- or postnatally, caused damage to an otherwise normal left hemisphere through oxygen deprivation or through physical trauma. It is possible that excessive maternal alcohol consumption is also a factor. Sperry (1971) determined that some lefthanders were processing verbal material in their right hemisphere, and this would be typical of pathological lefthanders. Their normal verbal processing area has been destroyed or badly damaged and they must develop a new area, probably after birth, in their right hemisphere. It is, therefore, easy to understand why their verbal development is slower than in the general population. These lefthanders are those who are at the low end of the IQ distribution, whose reaction time is somewhat slower than the population mean, and who may be somewhat more clumsy and involved in a higher proportion of accidents than the population mean. Some of them would exhibit low birth weight as a result of prenatal trauma. They may very well be the lefthanders that Blau (1946) and Burt (1937) had in mind when they penned their unfavorable comments, mentioned above. Brain scans show pathological lefthanders are using the right hemisphere for verbal processing and most other processing, with very little activity in the left hemisphere. The pathological lefthander can hardly be the result of genetic control for handedness since the etiologies of the pathological lefthander are not under genetic control. These individuals forcibly resist attempts to switch their handedness.

The second type of lefthander can be called the "normal lefthander." These individuals are normal in every way except that their brain function (and other characteristics such as hair whorl and fingerprint patterns) is the reverse of normal righthanders. These lefthanders are usually the result of twinning, primarily monozygotic twinning. It is known that in monozygotic twin pairs, quite often one twin is the reverse, or mirror image, of the other (Torgersen, 1950; Boklage, 1980). In these individuals, brain function is also reversed with the verbal processing area developing in the right hemisphere. If the verbal processing area is in the right hemisphere, for the same reason left hemisphere verbal processors use the right hand, efficiency dictates that these individuals use their left hand for writing and other manual tasks. In many, perhaps most, twin pregnancies one twin dies in utero and is absorbed (Hall & Lopez-Rangel, 1996). The remaining twin, if it is the mirror image (right hemisphere verbal processing) twin, is then born as a lefthanded singleton. Investigation using magnetic resonance imaging would show normal lefthanders have both hemispheres intact, with functions reversed from that of normal righthanders. If there is any genetic influence to lefthandedness it would appear in the normal lefthander, since they are a result of monozygotic (and possibly dizygotic) twinning, and some twinning is believed to be under genetic control. These individuals can be switched to righthand writing but at a high cost to their educational achievement. Many who have been switched have been classified as learning disabled and have trouble reading and writing. One of us (IBP) has had several switched normal lefthanders in class and has been told by them that they were diagnosed as learning disabled. When their background was probed, they recall having their writing hand switched. It was suggested that they try writing with the left hand. In most cases, in less than 1 week their "learning disability" disappeared, although their lefthanded writing was a bit shaky at first.

The third type of lefthander is the "learned lefthander." These individuals were born as normal righthanders, with the normal verbal processing area in the left hemisphere. Sometime very early in life they were handed or reached for a bottle, some toy, or other object and, by chance, grasped it with the left hand (infants do not show strong lateralization until about age 3). Since they were successful using the left hand ("reinforced" in behavioral terms), the probability was high that they would again use the left hand when reaching for or grasping another toy. By the time they are old enough to be given crayons or other writing implements they have already become comfortable using the left hand and will continue to use it. When they start to write, they start to have some of the problems listed above. Since they are processing verbal material in the left hemisphere, this verbal information must be transmitted across the *corpus callosum* so it can be used in the right hemisphere which controls the left hand. This takes a small but finite amount of time which interferes with writing skill and manifests itself as poor handwriting. Having switched their primary verbal processing areas also accounts for the stuttering in these lefthanders. Eventually these individuals develop a second verbal processing area in the right hemisphere, which accounts for Sperry's (1971) finding that verbal processing areas exist in both hemispheres in some lefthanders. As they mature they gain control over verbal function and their stuttering disappears. These lefthanders are rarely, if ever, found in authoritarian or conformist cultures. At the first sign of left-hand preference, parents would have immediately removed the object from the left hand, placed it in the right hand, and (perhaps) praised the child for right-hand use or punished it for left-hand use. If left hand use is not noticed until children reach school age, they can learn righthanded writing and use with very little pressure since they already possess their verbal processing area in their left hemisphere.

Dysfluency

Dysfluency (stuttering), defined as an intermittent disturbance in the rhythm of speech, is an interesting behavioral phenomena. Anderson (1932) found grade school stutterers to be inferior to grade school normals in hand coordination, but college stutterers did not differ from college normals in this behavior. Spadino (1972) found no differences between stutterers and non-stutterers in quality of composition, quality of handwriting, or any other written verbal task. Most researchers agree that while non-stutters show a pattern of primarily left hemisphere activation during speech, stutters show activation of the right or both hemispheres. Modern investigative techniques, such as MRI and PET scans, confirm these findings (De Nil & Kroll, 2001; Foundas, Corey, Hurley, & Heilman, 2004). Both right- and lefthanders stutter, usually for the same reason. Learned lefthanders, as described above, initially have their verbal processing area in the left hemisphere, but they have learned, through their own persistence, to write with their left hand. For some of these individuals the involvement of both hemispheres creates an interference that results in stuttering. Both pathological and normal lefthanders process verbal material in the right hemisphere. If they are forced to write with the right hand they experience the same interference as learned lefthanders, but in the opposite direction. If they are allowed to revert to left-hand use their stuttering will likely disappear (Clairborne, 1916).

Writing Position

There have been several theories proposed for the differences in handwriting postures among lefthanders (and a few righthanders). Some lefthanders write in what looks like a very contorted position, with their left hand "hooked" over the paper and the writing implement pointed downward, toward the writer. Other lefthanders write in a more or less normal posture. One theory involves the relationship between the gene or genes controlling the laterality of the "dominant" hand and the gene or genes controlling the laterality of the brain or speech function (Levy & Reid, 1976). Another theory of handwriting posture suggests that it is related to visual control of movement (Moscovitch & Smith, 1979). Both theories are wrong. Lefthanders who write in a hooked position do so because they had a poorly trained second grade teacher. In second grade, when most schools teach handwriting, some teachers do not take the time to determine whether there are any lefthanders in their class. They show the entire class the "correct" way to place the paper on the desk, the "correct" way to hold writing implement, and the "correct" way to position the arm. If lefthanders position the paper as is correct for righthanders, they must hook their arm to see what they are writing. They also must be careful not to smear ink if they are writing with pens, since lefthanders write over their writing, not ahead of it as do righthanders (Clark, 1961).

If lefthanders are taught to position their paper in the opposite direction from that which righthanders use, they will be able to use a normal writing posture (although they still must remove their hand to see what they have just written).

Future Directions

Lefthanders in the developed world have to live in an overwhelmingly righthanded environment. Tools, entertainment equipment, office equipment, notebooks, and many other articles that righthanders take for granted were designed by righthanders for righthanders. Most lefthanders, early in their life, learn to cope, and many normal and learned lefthanders become better able to adapt to difficult situations because of their early learning experiences. They are particularly successful in occupations in which they are on their own much of the time and have to make decisions based on their own evaluation of a problem. The question, then, is why Behavior Genetics, as a discipline, should be involved with handedness at all?

There are complex areas of handedness which would benefit from rigorous research, the first being the etiology of individuals' lefthandedness. Since we know there are at least three etiologies of lefthandedness, early identification, particularly in the instance of the pathological lefthander, could mean early academic intervention for youngsters now often categorized as learning disabled. A non-invasive test of young individuals showing a proclivity for left-hand use could alert parents to the possibility of left hemisphere problems, and, if further measures showed left hemisphere dysfunction, suitable educational protocols could be instituted. If the individual displayed a functional left hemisphere, indicating a normal or learned lefthander, other intervention to preclude stuttering and other undesirable verbal behaviors, would be appropriate. In fact, behavior geneticists, working in concert with educators, could develop a series of tests and interventions that would reduce or eliminate the verbal communication problems of many young lefthanders.

There is also the possibility of devising a measure to determine the etiology of older lefthanders to ascertain the proportion of each type in our society. This might not provide any benefit to the left-handed population, but, as we stated above, scientists are interested in differences and, as good scientists, we want to quantify whatever can be quantified.

There is also the possibility, we believe extremely small, that yet another etiology of lefthandedness exists. We are confident that the three etiologies we have described account for all the behavioral and hemispheric manifestations thus far found in lefthanders, but we want to leave the door open for future scientific endeavors. As to the future, we defer to our editor's final chapter: Future Directions for Behavior Genetics.

Acknowledgment We wish to thank, once again, Diane A. Granville whose advice, corrections, and suggestions significantly improved our writing, and Lauren Lazar, Librarian at Purchase College, State University of New York, who helped gather our widely dispersed sources, some of which are old enough to be considered historic.

References

Anderson, L. O. (1932). Stuttering and allied disorders. *Comparative Psychology Monographs, 1.*

Annett, M. (1973). Handedness in families. *Annals Of Human Genetics, 37*, 93–105.

Arab News. (2005, July 24). Left-handedness [http://www. islamicity/dialog]. *Arab News*, p. 2.

Basso, O., Olsen, J., Skytthe, A., Vaupel, J., & Christensen, K. (2000). Handedness and mortality, A follow-up study of Danish twins born between 1900 and 1910. *Epidemiology, 11*, 576–580.

Benson, D. F., & Geschwind, N. (1972). Cerebral dominance and its disturbances. In R. Paine, H. Myklebust, & D. Weiss (Eds.), *Dyslexia and Reading Disabilities* (pp. 108–118). New York: MSS Inf. Corp.

Bishop, D. V. (1990). How to increase your chances of obtaining a significant association between handedness and disorder. *Journal Of Clinical Experimental Psychology, 12*, 812–816.

Blau, A. (1946). The master hand. *American Orthopsychiatric Association, 5.*

Boklage, C. (1980). The sinistral blastocyst: An embryologic perspective in the development of brain-function asymmetries. In J. Herron (Ed.), *Neuropsychology of Left-Handedness* (pp. 115–137). New York: Academic Press.

Bradshaw, J. L., & Nettleton, N. C. (1983). *Human Cerebral Asymmetry*. Englewood Cliffs, NJ: Prentice-Hall.

Broca, P. (1861). Remarques sur le siège da la faculté du langage articulé, suivie d'une observation d'aphémie [Remarks on the source of the faculty of the articulated language, followed by an observation of its importance]. *Bulletin de la Societe d'Anthropologie, 2*, 330–357.

Broca, P. (1865). Sur de la faculté du langage articulé [On the faculty of articulated language]. *Bulletin De La Societe D'anthopologie, 6*, 337–393.

Bryden, M. P. (1979). *Possible genetic mechanisms of handedness and laterality*. Paper presented at the Canadian Psychological Association, Quebec City, Canada.

Bryden, M. P. (1982). *Laterality, Functional asymmetry in the intact brain*. New York: Academic Press.

Burt, C. (1937). *The backward child*. New York: Appleton Century.

Calnan, M., & Richardson, K. (1976). Development correlates of handedness in a national sample of 11- year-olds. *Annual Of Human Biology, 3*, 329–342.

Carter-Saltzman, L. (1979). Patterns of cognitive functioning in relation to handedness and sex-related differences. In M. A. Wittig & A. C. Petersen (Eds.), *Sex-related differences in cognitive functioning* (pp. 97–114). New York: Academic Press.

Cerhab, J. R., Folsum, A. R., Potter, J. D., & Prineas, R. J. (1994). Handedness and mortality risk in older women. *American Journal Of Epidemiology, 140*, 368–374.

Chamberlain, H. D. (1928). The inheritance of left-handedness. *Journal Of Heredity, 19*, 557–559.

Chaugule, B. (2005). India Perspective. *India Association of Left-Handers*. Retrieved from http://www.geocities.com/ lefthandersind/indpersp.htm

Clairborne, J. H. (1916). The psychology and physiology of mirror writing. *University Of California Publications In Psychology, 2*, 199–265.

Clark, M. M. (1961). *Teaching left-handed children*. New York: Philosophical Library.

Collins, R. L. (1977). Origins of the sense of asymmetry: mendelian and non-mendelian models of inheritance. *Annals Of The New York Academy Of Science, 299*, 283–305.

Corballis, M. C. (1997). The genetics and evolution of handedness. *Psychological Review, 104*, 714–727.

Corballis, M. C. (1999). The gestural origins of language. *American Scientist, 87*, 138–145.

Corballis, M. C., & Morgan, M. J. (1978). On the biological basis of human laterality. *Behavioral Brain Science, 2*, 261–336.

Coren, S., & Halpern, D. F. (1999). Left-handedness: A marker for decreased survival fitness. *Psychological Bulletin, 109*, 90–106.

Coyne, J., & Orr, H. A. (2004). *Speciation*. Sunderland, MA: Sinauer.

Crawford, J. E., & Crawford, D. M. (1956). *Crawford Small Parts Dexterity Test, Part 1*. New York: Psychological Corp.

Cunningham, D. J. (1902). Right-handedness and left brainedness. *Journal Of The Anthropological Institute Of Great Britain And Ireland, 32*, 273.

Dawson, J. L. M. B. (1977). Alaskan Eskimo hand, eye, auditory dominance, and cognitive style. *Psychologia, 20*, 121–135.

De Nil, L. F., & Kroll, R. M. (2001). Searching for the neural basis of stuttering treatment outcome: Recent neuroimaging studies. *Clinical Linguistics & Phonetics, 15*, 163–168

De Wall, F. (1982). *Chimpanzee politics*. New York: Harper and Row.

Ehrman, L., & Perelle, I. B. (1983). Laterality. *Mensa Research Journal, 16*, 3–31.

Erlenmeyer-Kimling, L., Hans, S., Ingraham, L., Marcus, J., Wynne, L., Rehman, A., et al. (2005). Handedness in children of schizophrenic parents: Data from three high risk studies. *Behavior Genetics, 35*(3), 351–358.

Finch, G. (1941). Chimpanzee handedness. *Science, 94*, 117–118.

Foundas, A. L., Corey, D. M., Hurley, M. M., & Heilman, K. M. (2004). Verbal dichotic listening in developmental subgroups with atypical auditory processing. *Cognitive and Behavioral Neurology, 17*(4), 224–232

Futuyama, D. (2005). *Evolution*. Sunderland, MA: Sinauer.

Geschwind, N. (1975a). On the other hand. *The Sciences, 63*, 22–24.

Geschwind, N. (1975b). The apraxias: neural mechanisms of disorders of learned movements. *American Scientist, 63*, 188–195.

Geschwind, N., & Behan, O. P. (1982). Left-handedness: Association with immune disease, migraine, and developmental learning disorders. *Proceedings Of The National Academy Of Science, 79*, 5097–5100.

Geschwind, N., & Behan, O. P. (1984). Laterality, hormones, and immunity. In N. Geshwind & A. M. Galaburda (Eds.), *Cerebral Dominance* (pp. 221–224). Cambridge, MA: Harvard University Press.

Granville, D., Ehrman, L., & Perelle, I. B. (1979). Laterality survey. *Mensa Bulletin, 224*, 211–214.

Gregory, P., & Paul, J. (1980). The effects of handedness and writing posture on neuropsychological test results. *Neuropsychologia, 18*, 231–285.

Hall, J. G., & Lopez-Rangel, E. (1996). Embryologic development and monozygotic twinning. *Acta Genetica Med Gemellol, 45*, 53–57.

Hardyk, C., & Petrinovich, L. F. (1977). Left-handedness. *Psychological Bulletin, 84*, 385–404.

Hatta, T., & Nakatsuka, Z. (1976). Note on hand preference of Japanese people. *Perceptual and Motor Skills, 42*, 530.

Hecaen, H., & Ajuriaguerra, J. (1964). *Left-handedness: Manual superiority and cerebral dominance*. New York: Grune & Stratton.

Hofstede, G. (1983). National cultures revisited. *Behavior Science Research, 18*, 285–305.

Holder, M. K. (1992). *Hand preference questionnaires: One gets what one asks for*. Unpublished master's thesis, Rutgers University, New Brunswick, NJ.

Holder, M. K. (2005). *Hand preference questionnaires: Interim report*. [http://www.indiana.edu/~primate/forms/debrief.html] Bloomington: University of Indiana.

The Holy Bible: Revised Standard Version (1962). Cleveland, OH: The World Publishing Co. (Original work published 1611)

Krynicki, V. E., & Nahas, A. D. (1979). Differing lateralized perceptual-motor patterns in schizophrenic and non-schizophrenic children. *Perceptual and Motor Skills, 49*, 603–610.

Kuhlemeier, K. V. (1991). Longevity and lefthandedness. *American Journal Of Public Health, 81*, 513.

Kummer, H., & Goodall, J. (1985). Conditions of innovative behaviour in primates. *Transactions of The Royal Society B, 308*, 203–214.

Levy, J., & Nagylaki, T. (1972). A model for the genetics of handedness. *Genetics, 72*, 117–128.

Levy, J., & Reid, M. (1976). Variations in writing posture and cerebral organization. *Science, 194*, 337–339.

Linke, D. B., & Kersebaum, S. (2005). Left Out. *Scientific American Mind, 16*(4), 79–83.

Luria, A. R. (1973). *The working brain* (B. Haigh, Trans.). New York: Basic Books.

McBurney, A., & Dunn, H. G. (1976). Handedness, footedness, eyedness: A prospective study with special reference to the development of speech and language skills. In R. M. Knights & D. J. Bakker (Eds.), *The Neuropsychology of Learning Disorders*. London: University Park Press.

McGrew, W. C., & Marchant, L. F. (2001). Ethological study of manual laterality in chimpanzees of the Mahale Mountains, Tanzania. *Behaviour, 138*, 329–358.

McManus, C. (2002). *Right hand, left hand*. Cambridge, MA: Harvard University Press.

Medland, S. E., Perelle, I. B., De Monte, V., & Ehrman, L. (2004). Effects of culture, sex, and age on the distribution of handedness: An evaluation of the sensitivity of three measures of handedness. *Laterality, 9*, 287–297.

Meguerditchian, A., & Vauclair, J. (2006) Baboons communicate with their right hand. *Behavioral Brian Research, 171*, 170–174.

Merrell, D. J. (1957). Dominance of hand and eye. *Human Biology, 29*, 314–328.

Moscovitch, M., & Smith, L. (1979). Differences in neural organization between individuals with inverted and noninverted hand postures during writing. *Science, 205*, 710–712.

Nottebohm, F., & Nottebohm, M. E. (1976). Left hypoglossal dominance in the control of canary and white-crowned sparrow song. *Journal Of Comparative Psychology, 108*, 171–192.

Oldfield, R. C. (1971). The assessment and analysis of handedness: The Edinburgh Inventory. *Neuropsychology, 9*, 97–113.

Pepperberg, I. M. (2002). *The Alex Studies*. Cambridge, MA: Harvard University Press. (Original work published 1999)

Perelle, I. B., & Ehrman, L. (1982). What is a lefthander? *Experientia, 38*, 1257–1258.

Perelle, I. B., & Ehrman, L. (1983). The development of laterality. *Behavioral Science, 28*, 284–297.

Perelle, I. B., & Ehrman, L. (1994). An international study of human handedness. *Behavior Genetics, 24*(3), 217–227.

Perelle, I. B., & Ehrman, L. (2001). Handedness: A look at laterality. In E. C. R. Reeve (Ed.), *Encyclopedia of Genetics* (pp. 573–576). London: Fitzroy Dearborn.

Perelle, I. B., Ehrman, L., & Manowitz, J. W. (1981). Human handedness: The influence of learning. *Perceptual And Motor Skills, 53*, 967–977.

Persson, P. G., & Allebeck, P. (1994). Do left-handers have increased mortality? *Epidemiology, 5*, 337–340.

The Piano. The first left-handed piano. (1999). Retrieved July 24, 2005, from http://?/?www.lefthandedpiano.co.uk/?about.html

Pirozzolo, F. J., & Rayner, K. (1979). Cerebral organization and reading disability. *Neuropsychologia, 17*, 485–491.

Puentes, F. (2004). *Laterality in the fine arts.* Senior thesis, Purchase College, State University of New York, Purchase.

Rife, D. C. (1940). Handedness with special reference to twins. *Genetics, 25,* 178–186.

Sarasin, P. (1884). Ueber Rechts- und Links-Haendigkeit in der Praehistorie, und Rechtschandigkeit in der historischen Zeit [On right and left-handedness in Prehistory, and Righthandedness in historical times]. *Naturforschende Gesellschaft, 29,* 122–196.

Satz, P. (1972). Pathological left-handedness: An explanatory model. *Cortex, 11,* 121–135.

Schaefer, M. (1911). Die linkshander in den Berliner Gemeindeschulen [The lefthander in the Berlin Public Schools]. *Berlin Kleiner Wochenscher, 48*(7). Quoted by Wile (1934) p. 85.

Spadino, E. J. (1972). Writing and laterality characteristics of stuttering children (Doctoral dissertation, Columbia University, 1941). *Dissertation Abstracts,* 1–82.

Sperry, R. W. (1971). How the developing brain gets itself wired for adaptive function. In E. Tobach, L. R. Aronson, & E. Shaw (Eds.), *The biopsychology of development* New York: Academic Press.

Springer, S. P., & Eisenson, J. (1977). Hemispheric specialization for speech in language-disordered children. *Neuropsychologia, 15,* 287–293.

Swanson, J. M., Kinsbourne, M., & Horn, J. M. (1980). Cognitive deficit and left-handedness: A cautionary note. In Jeannine Herron (Ed.), *Neuropsychology of left-handedness* (pp. 281–291). New York: Academic Press.

Tanner, J. E., & Byrne, R. W. (1996). Representation of action through iconic gestures in a captive lowland gorilla. *Current Anthropology, 37,* 162–173.

Tomasello, M., Call, J., Warren, J., Frost, G. T., Carpenter, M., & Nagell, K. (1997). The ontogeny of chimpanzee gestural signals: A comparison across groups and generations. *Evolution Of Communication, 1,* 223–259.

Torgersen, J. (1950). Situs inversus, asymmetry, and twinnings. *American Journal Of Human Genetics, 2,* 361–370.

Wile, A. (1934). *Handedness right and left.* Boston: Lothrop, Lee and Shepard.

Wilkin, S. (1852). *The works of Sir Tomas Browne.* London: Henry Bohn. (Original work published 1836)

Winters, D. (2004). *Lefthandedness.* Retrieved from http://www. uncle-tax.com/at/janfeb04/lefthandedness.html

Wolf, P. A., D'Agostino, R. B., & Coss, L. (1999). Letter to the Editor. *New England Journal Of Medicine, 325,* 1042.

Chapter 24

Genetics of Exercise Behavior

Janine H. Stubbe and Eco J.C. de Geus

Introduction

A sedentary lifestyle has been cited as one of the main causes of the explosive rise in obesity that starts at an increasingly younger age (Martinez-Gonzalez, Martinez, Hu, Gibney, & Kearney, 1999). Furthermore, regular exercisers have lower risks for cardiovascular disease (CVD) and type 2 diabetes than non-exercisers (Albright et al., 2000; Kaplan, Strawbridge, Cohen, & Hungerford, 1996; Kesaniemi et al., 2001) and the percentage of people at risk because of inactivity is higher than for hypertension, smoking, and cholesterol (Caspersen, 1987; Stephens & Craig, 1990). Despite these well-documented benefits of exercise, a large proportion of adults in the Western world do not exercise on a regular basis (Crespo, Keteyian, Heath, & Sempos, 1996; Haase, Steptoe, Sallis, & Wardle, 2004; Stephens & Craig, 1990). As a consequence, a sedentary lifestyle – and the accompanying risk for obesity – remains a major threat to health in today's society. This is reflected in public health recommendations which unanimously include an encouragement to a more active lifestyle (WHO/FIMS Committee on Physical Activity for Health, 1995; U.S. Department of Health and Human Services, 2005).

To increase the success of intervention on this important health behavior, much research has been devoted to the determinants of exercise behavior. The bulk of these studies have attempted to explain low exercise prevalence in terms of social and environmental barriers. These include, amongst others, poor access to facilities (Matson-Koffman, Brownstein, Neiner, & Greaney, 2005; Varo et al., 2003), low socioeconomic status (Haase et al., 2004; Varo et al., 2003), non-Caucasian race (Kaplan, Lazarus, Cohen, & Leu, 1991), high job strain (Payne, Jones, & Harris, 2005; Van Loon, Tijhuis, Surtees, & Ormel, 2000), subjective "lack of time"

(Shephard, 1985; Sherwood & Jeffery, 2000), inadequate health beliefs (Haase et al., 2004), and low social support by family, peers, or colleagues (King et al., 1992; Orleans, Kraft, Marx, & McGinnis, 2003; Sherwood & Jeffery, 2000). Despite their face validity, none of these factors has emerged as a strong causal determinant of exercise behavior (Dishman, Sallis, & Orenstein, 1985; Seefeldt, Malina, & Clark, 2002). Increasingly, therefore, biological factors have been invoked to explain why exercisers exercise and why non-exercisers do not (Rowland, 1998; Thorburn & Proietto, 2000; Tou & Wade, 2002). As will become evident in this chapter, these factors should prominently include a genetic disposition to exercise.

Before examining in detail the existing behavior genetics work on this topic, we will briefly go into the definition of exercise behavior and review a number of large-scale studies that give insight into the current prevalence of exercise behavior.

Definition of Exercise Behavior

Operational definitions of exercise behavior have differed strongly across studies. First, a distinction can be made between studies querying "pure" exercise activities (jogging, gymnasia, and all individual or team sports) versus studies including all physical activities which may improve cardiorespiratory health but are not primarily intended that way (gardening, walking the dog, or bicycling to school/work) (Caspersen, Powell, & Christenson, 1985). Even when we restrict ourselves to pure exercise activities in leisure time, exercise definitions differ across studies (Table 24.1). Only two very specific phenotypes have been defined in a highly comparable way. *Sedentary* subjects simply do not engage in any type of leisure time physical activity, whereas *vigorous exercisers* perform activities above the intensity and frequency thresholds required to maintain a continued increase in aerobic fitness above their sedentary level. To achieve such an increase, subjects need to engage in large muscle dynamic

E.J.C. de Geus (✉)
Department of Biological Psychology, Vrije Universiteit, Amsterdam 1081 BT, The Netherlands
e-mail: JCN.de.Geus@psy.vu.nl

Y.-K. Kim (ed.), *Handbook of Behavior Genetics*,
DOI 10.1007/978-0-387-76727-7_24, © Springer Science+Business Media, LLC 2009

Table 24.1 Three exercise levels used across studies to categorize exercise behavior

Exercise level	Definition
Sedentary	Does not engage in any type of leisure time exercise behavior
Vigorous exercise	Performs leisure time exercise activities above the intensity and frequency thresholds required to maintain a continued increase in aerobic fitness above their sedentary level, i.e., engage in large muscle dynamic exercise activities requiring more than 50% of their maximal oxygen consumption for at least three times a week for 20 min or more per occasion
Light-to-moderate exercise	All exercise behavior in between sedentary and vigorous exercise. A further distinction can be made between light exercise (less than 60 min a week or intensity below 4 METs) and moderate exercise (at least 60 min weekly with a minimum intensity of 4 METs), but note that not all studies collect data on intensity and frequency

exercise activities requiring more than 50% of their maximal oxygen consumption for at least three times a week for 20 min or more per occasion (Blair et al., 1996; Pate et al., 1995).

Measures of *light-to-moderate exercise*, i.e., all activity levels in between sedentary and vigorous exercise, are much harder to compare across studies. Studies use different criteria for the minimum frequency and the minimum intensity that is required to classify participants as "regular exercisers". Criteria for frequency have varied from once per 2 weeks (Haase et al., 2004; Steptoe et al., 1997, 2002) to five or more times a week (Caspersen, Pereira, & Curran, 2000). In some studies the reported specific exercise activities were coded for intensity and had to meet a certain minimal intensity (De Geus, Boomsma, & Snieder, 2003; Perusse, Tremblay, Leblanc, & Bouchard, 1989; Stubbe, Boomsma, & De Geus, 2005), whereas in others no specific exercise activities were reported or no minimum intensity was specified (Haase et al., 2004; Steptoe et al., 1997).

The differences in the operational definition of regular exercise are compounded by the varying methods of assessment of regular exercise. Some studies use surveys with only a single YES/NO question (Boomsma, Vandenbree, Orlebeke, & Molenaar, 1989; Koopmans, Van Doornen, & Boomsma, 1994) whereas others query the type, duration, frequency, and intensity in great detail (Martinez-Gonzalez et al., 2001). Some studies use an interview strategy (Caspersen et al., 2000) rather than a survey-based approach, or even direct measurements of energy expenditure with accelerometry or physiological recording (Pate et al., 2002; Sirard & Pate, 2001). This makes it difficult to either pool or compare the prevalence of exercise behavior across studies. Fortunately, there are five very

large studies that, together, provide a reasonable insight into the prevalence of exercise in industrialized societies (Caspersen et al., 2000; Haase et al., 2004; Martinez-Gonzalez et al., 2001; Steptoe et al., 1997, 2002). We will describe their assessment strategies and outcomes in more detail in the next paragraph.

Prevalence of Exercise Behavior

The European Health and Behavior Study (EHBS) (Steptoe et al., 1997) and the International Health and Behavior Study (IHBS1/IHBS2) (Haase et al., 2004; Steptoe et al., 2002) are two large survey studies assessing the prevalence of leisure time physical activity in 18- to 30-year-old university students. The EHBS survey was carried out in 16,483 students from 21 European countries in 1990 (Steptoe et al., 1997). The 2000 IHBS1 and IHBS2 studies used the same measures as the EHBS study and partly the same sample. The IHBS1 study (Steptoe et al., 2002) included 10,336 participants from 13 of the 21 European countries included by the EHBS. The IHBS2 study (Haase et al., 2004) extended the sample by using more countries worldwide resulting in a final sample of 19,298 university students from 23 countries.

In all three studies, leisure time exercise participation was assessed by responses to three items. The first item asked whether an individual had participated in any exercise (e.g., sports activities, physically active pastime) over the past 2 weeks. Those who responded positively were asked what kind of activity they carried out. The most reported forms of activity were jogging/running, swimming, football (soccer), and aerobics. Furthermore, participants were asked how many times they had exercised in the past 2 weeks. Data were analyzed by dividing the sample into three groups. Inactive subjects (i.e., sedentary subjects) did not engage in any exercise at all; subjects who engaged one to four times per 2 weeks in exercise were considered regular exercisers at "low-frequent activity" (corresponding to light to moderate as defined in the previous paragraph); subjects who exercised more than five times per 2 weeks were considered "frequent" exercisers (this corresponds roughly to vigorous exercise in Table 24.1, but note that no intensity was coded). Regular exercise was defined as exercising at least once over the past 2 weeks.

In the 1990 study, 73% of the men and 68% of the women exercised regularly, suggesting that 27% of the male and 32% of the female students are sedentary. A total of 36% of men and 30% of women were vigorous exercisers, i.e., had exercised on five or more occasions during the previous 2 weeks (Steptoe et al., 1997). In the EHBS study (Steptoe et al., 2002), the survey was repeated 10 years later for 13 of the 21 countries (IHBS1). Figure 24.1 shows that the

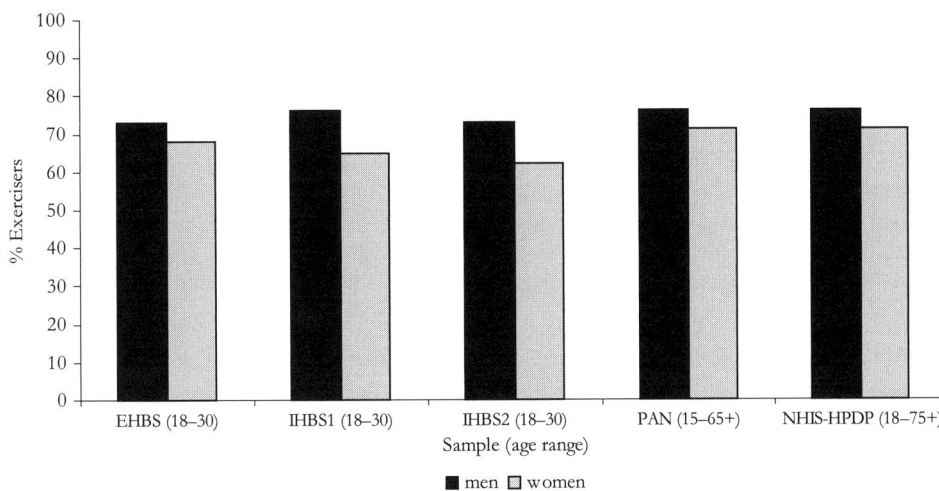

Fig. 24.1 Prevalence of regular exercise in five different studies, the Health and Behavior Study (EHBS, IHBS1, and IHBS2), the pan-European study of adults from 15 member states of the Euro- pean Union (PAN), and the National Health Interview Survey-Health Promotion/Disease Prevention (NHIS-HPDP)

prevalence of regular exercise remained fairly stable over a 10-year time period. Extending the sample with students from countries worldwide (IHBS2) again resulted in compa- rable prevalences (Haase et al., 2004).

All three studies showed that men were more likely than women to have exercised in the previous 2 weeks. In the third and largest study, for instance, more women than men reported to be sedentary (38% versus 27%), whereas the proportion engaged in vigorous exercise was larger in men (28%) than that in women (19%). There was no overall dif- ference in the proportion of men (45%) and women (43%) active at light to moderate levels (active one to four times per 2 weeks).

The samples used in the EHBS and IHBS studies may not be representative for the whole population, because it was conducted in students 18–30 year old. A pan-European study (PAN) of adult exercise participation by Martinez-Gonzalez and colleagues (2001) used a population-based sample of more than 15,000 adults from 15 member states of the Euro- pean Union. Subjects were divided in age bins of 10 years, starting at age 15 and leading up to a final category of sub- jects aged 65 and over. To assess activity levels, subjects were asked to select the activities in which they participated from a list of 17 activities (i.e., athletics, cycling, dancing, equestrian sports, fishing, football, gardening, golf, hill walk- ing, climbing, keep fit, aerobic, jogging, martial arts, rac- quet sports, rowing, canoeing, skiing, skating, swimming, team sports, walking, and water sports). Metabolic equiva- lents (METs) assigned to each activity were used to quan- tify the amount of leisure time physical activity, with one MET representing the rate of energy expenditure of an indi- vidual at rest which is approximately 1 kcal/kg/h (Ainsworth et al., 1993, 2000). Participants also indicated the number of

hours a week they participated in each activity. Regular exer- cise was defined as engaging in any of the queried exercise activities, with constraints on intensity and frequency.

Across the entire age range studied, an average of 76% of the male and 71% of the female EU population partici- pated in some kind of exercise activity. With increasing age, exercise participation decreased, ranging from 83% in 15–24 year old to 65% in people aged 65 years and over. A wide variability was found in the prevalence of exercise activi- ties among European countries. Northern European coun- tries showed higher exercise prevalences than southern ones. Figure 24.1 shows that, across all countries, the overall per- centage of regular exercisers is in close agreement with the estimates of prevalence by the EHBS and IHBS studies (Haase et al., 2004; Steptoe et al., 1997; Steptoe et al., 2002). As in the EHBS and IHBS studies, a higher percentage of men engaged in any leisure time exercise activities, and the average intensity of their activities (in METs) was higher than in women.

In the 1991 National Health Interview Survey-Health Pro- motion/Disease Prevention (NHIS-HPDP) study, physical activity levels were assessed in 43,732 men and women from the USA, aged 18 years and over (Caspersen et al., 2000). Frequency and duration were assessed of gardening and exer- cise activities (i.e., walking for exercise, stretching exercises, weightlifting, jogging, aerobics, bicycling, stair climbing for exercise, swimming for exercise, play tennis, golf, baseball, basketball, volleyball, handball, soccer, football, racquet- ball or squash, bowling, and skiing (downhill, cross-country, and water). To get information about the intensity level, questions were asked about increases in breathing or heart rate. According to the Healthy people 2000 objectives (U.S. Department of Health and Human Services, 2005) exercise behavior of the participants was categorized into three activity patterns: physically inactive (i.e., no participation in any leisure time physical activity), engaging in regular, sustained light to moderate activities (five or more times a week and 30 min or more per occasion of any activity), and engaging in regular, vigorous activities (three or more times per week and 20 min or more per occasion of any activity performed at ≥50% of maximal oxygen consumption). These activity patterns correspond closely to the three exercise levels listed in Table 24.1.

The average prevalence across adulthood for any form of exercise behavior (i.e., light to moderate or vigorous) varied between 73% for women and 79% for men. Women not only had a significantly higher prevalence of being sedentary than men; they also reported less engagement in light to moderate exercise (27% versus 21%). For vigorous activity the difference between men and women was small at 18–29 years (4% more males), moderate at 65–74 years (9% more males), but very large at ≥75 years (16% more males).

A clear picture arises from these five studies. Despite the well-documented benefits of exercise, a large group of young adolescents and adults do not engage in exercise on a regular basis. Worldwide, the prevalence for sedentary behavior varies between 21 and 27% for males and between 27 and 38% for females. Prevalence for light to moderate exercise ranges between 27 and 45% for males and between 21 and 43% for females. Finally, between 28 and 36% of males are engaged in vigorous activities and this percentage varies between 19 and 30% for females.

What factors cause exercisers to exercise and, more importantly, what keeps non-exercisers from doing the same? The remainder of this chapter will review evidence from behavioral genetics for a significant genetic contribution to voluntary exercise.

Family Studies on Exercise Behavior

As with many other traits, exercise behavior appears to run in the family. Familial resemblance in exercise behavior has been tested by the correlation of exercise behavior in parent–child, sister–sister, brother–brother, and sister–brother pairings. Significant familial resemblance in exercise behavior between parents and their offspring has been reported in various studies. Parent–offspring correlations have ranged from low ($r = 0.09$–0.13) for participation defined as activities requiring at least five times the resting metabolic rate (Perusse et al., 1989) or weekly time spent on the main exercise activity during the previous year (Simonen et al., 2002) to moderate ($r = 0.29$–0.37) for exercise participation coded as a dichotomous variable using the single question "Do you participate in sports?" (Koopmans et al., 1994).

In the Canadian Fitness Survey (Perusse, Leblanc, & Bouchard, 1988), the degree of familial resemblance for leisure time energy expenditure, total time spent on leisure time activities and the activity level (derived from total time spent on leisure time activities and total number of months for the reported activities) was assessed in 16,477 subjects, aged 10 years and older. Siblings and parent–offspring pairs were formed to compute familial correlations in energy expenditure, time spent on activities and activity level. These familial correlations ranged between 0.12 and 0.62 for the three variables, suggesting evidence for familial resemblance. However, familial correlations were higher within generations (siblings) than across generations (parent–offspring). Also a significant correlation between spouses was found and, within the same generation, correlations for spouses and siblings were of the same magnitude. This suggested to the authors that familial resemblance resulted primarily from environmental factors common to members of the same generation (i.e., family, neighborhood, facilities, and general cultural attitudes on exercise). However, parent–offspring studies underestimate heritability if different genes are expressed at different ages, and spousal correlations may also partly represent assortative mating.

Twin Studies on Exercise Behavior

Twin studies can directly decompose familial resemblance into shared genetic and shared environmental influences by comparing the resemblance in exercise behavior between monozygotic (MZ) and dizygotic (DZ) twins. As opposed to parent–offspring family designs they do so within members of the same generation. A variety of twin studies have shown that genetic factors contribute to individual differences in exercise participation and measures of exercise frequency, duration, and/or intensity (Aarnio, Winter, Kujala, & Kaprio, 1997; Beunen & Thomis, 1999; Boomsma et al., 1989; De Geus et al., 2003; Frederiksen & Christensen, 2003; Heller et al., 1988; Koopmans et al., 1994; Kujala et al., 2002; Lauderdale et al., 1997; Maia, Thomis, & Beunen, 2002; Perusse et al., 1989; Stubbe et al., 2005, 2006). The main results of these studies are summarized in Table 24.2. Studies were included only if estimates of genetic (a^2 or d^2) or shared environmental (c^2) contribution to total variance were given in the paper or if the correlations of MZ and DZ twins were supplied. The latter makes it possible to calculate the contribution of additive ($a^2 = 2(r_{MZ}-r_{DZ})$) or non-additive ($d^2 = 4r_{DZ}-r_{MZ}$) genetic factors or of shared environmental ($c^2 = 2r_{DZ}-r_{MZ}$) factors (Plomin, DeFries, McClearn, & McGuffin, 2000). Table 24.2 shows these various estimates to range widely across studies. The large range in these estimates may be caused in part by the use of various definitions of exercise, but as we will argue in detail below, also by the vastly different age ranges studied.

Five twin samples have been used to address the heritability of exercise participation in adolescents. In a large family cohort based on the Quebec family study, a three-day activity record was used to determine the activity level of young adolescent twins (mean age 14.6) (Perusse et al., 1989). Each day was divided into 96 periods of 15 min, and for each 15-min period subjects were asked to note, on a scale from one to nine, the energy expenditure of the dominant physical activity of that period. Regular vigorous exercise behavior was assessed from the number of periods in which exercise activities or moderate to intense manual work (i.e., tree cutting, snow shoveling, etc.) were reported that were rated 6 or higher on the nine-point scale (i.e., activities requiring 4.8 times the resting oxygen consumption). The average value of the ratings across these periods was used as the measure of regular exercise. Monozygotic and dizygotic twin correlations did not differ significantly from each other, indicating that genetic factors explained 0% of the variation in regular exercise behavior. Individual differences in regular exercise were attributed to common environmental (74%) and unique environmental factors (26%).

In a Dutch twin study, exercise participation was assessed in 2,628 young complete twin pairs aged between 13 and 20 (Stubbe et al., 2005). Ainsworth's Compendium of physical activity was used to recode the reported exercise activities into METs. Subjects were classified as regular exercisers if they engaged in competitive or non-competitive leisure time exercise activities with a minimal intensity of four METs for at least 60 min per week. In the classification scheme of Table 24.1 this would include both light to moderate and vigorous exercisers. Genetic and common environmental contributions to exercise participation were computed separately within age groups 13–14 years, 15–16 years, 17–18 years, and 19–20 years. Very large familial resemblance was found at all ages. In agreement with the study by Perusse and colleagues (1989), genes were of no importance to exercise participation in 13- to 16-year-old children, whereas environmental factors shared by children from the same family largely accounted for 78% (15–16 years) to 84% (13–14 years) of the individual differences in participation. Genetic influences started to appear (36%) at the age of 17–18 years with the role of common environment rapidly decreasing (47%). After the age of 18 years, genes almost entirely explain individual differences in exercise participation (85%) and common environmental factors do not contribute at all.

The large shift from common environmental to genetic influences on exercise habits in adolescence implies that studies collapsing twin data across this age range will arrive at "mixture" estimates. That this indeed happens is illustrated by two other studies on the Dutch twins that had pre-

viously estimated the genetic and environmental influences on individual differences in exercise participation in Dutch adolescents using smaller samples with larger age ranges (Boomsma et al., 1989; Koopmans et al., 1994). Both studies defined exercise participation by the response to the single question "Have you been involved in exercise activities during the last 3 months?". In 90 adolescent Dutch twin pairs aged 14–20 years (average age = 17 years old) heritability was estimated at 64% for both males and females but evidence for common environment was also suggested (Boomsma et al., 1989). In 1,587 13- to 22-year-old Dutch twins (mean age of 18 years), Koopmans et al. (1994) estimated heritability and common environmental influences to be 48 and 38%, respectively.

A combination of common environmental and genetic influences in adolescence has also been reported by other studies, which additionally suggest a sex difference such that the common environment loses its importance earlier in boys than in girls. In the Leuven Longitudinal Twin Study (Beunen & Thomis, 1999), 92 Flemish male twins and 91 female twins aged 15 years reported the number of hours they exercised each week. For girls, 44% of the variation in exercise participation was explained by genetic factors and 54% by common environmental factors. For boys, genetic factors already explained about 83% of the total variance at age 15. In a study based on 411 Portuguese twins aged 12–25 years (mean age was approximately 17 years) an exercise participation index was computed as a composite score of items that takes into account the expected energy expenditure for a given exercise activity, number of hours practiced per week, and number of months per year (Maia et al., 2002). In agreement with Beunen and Thomis (1999), larger heritability estimates were found for males (68%) compared to females (40%). Finally, Aarnio and colleagues (1997) found substantially lower opposite-sex twin pair correlations than dizygotic same-sex twin pair correlations in 16-year-old Finnish twins, which is again in keeping with a different genetic architecture for males and females in this age range.

To our knowledge, five studies have investigated the influences of genes and environment on exercise behavior in adults (Frederiksen & Christensen, 2003; Heller et al., 1988; Kujala et al., 2002; Lauderdale et al., 1997; Stubbe et al., 2006). An Australian study of 200 twin pairs assessed genetic influences on several lifestyle risk factors, including a single exercise question, "vigorous exercise in the past 2 weeks" (Heller et al., 1988). Ages ranged from 17 to 66 years with the mean ages of MZ and DZ twins being 36.9 (SD = 13.2) and 35.6 (SD = 11.5) years, respectively. Heritability was estimated at 39% for this question. In 3,344 male twin pairs aged 33–51 years from the Vietnam Era Twin Registry (Lauderdale et al., 1997), regular exercise was assessed with five questions about vigorous forms of exercise (>4.5 METs) performed in the last 3 months: (1) jog

Table 24.2 Twin studies on exercise participation

Study	Sample	Phenotype	Categorization	Results
Stubbe et al. (2006)[1]	13,676 MZ and 23,375 DZ pairs from seven different countries participating in the GenomEUtwin project (aged 19–40)[2]	Engage in leisure time exercise activities with a minimal intensity of 4 METs for at least 60 min per week (yes/no)	Light to moderate plus vigorous exercise	$a^2 = 27$–67%; $c^2 = 0$–37% for males $a^2 = 48$–71%; $c^2 = 0\%$ for females
Stubbe et al. (2005)[1]	2,628 Complete Dutch twin pairs (aged 13–14, 15–16, 17–18, 19–20)[1]	Engage in leisure time exercise activities with a minimal intensity of 4 METs for at least 60 min per week (yes/no)	Light to moderate plus vigorous exercise	$a^2 = 0\%$; $c^2 = 84\%$ for 13- to 14-year-old twins $a^2 = 0\%$; $c^2 = 78\%$ for 15- to 16-year-old twins $a^2 = 36\%$; $c^2 = 47\%$ for 17- to 18-year-old twins $a^2 = 85\%$; $c^2 = 0\%$ for 19- to 20-year-old twins
Beunen et al. (2003)[1]	92 Male and 91 female Belgium twin pairs (aged 15)	Number of hours spent on sports each week	Moderate plus vigorous exercise	$a^2 = 83\%$; $c^2 = 0\%$ for males $a^2 = 44\%$; $c^2 = 54\%$ for females
De Geus et al. (2003)[1]	157 Adolescent (aged 13–22) and 208 middle-aged Dutch twin pairs (aged 35–62)[2]	Average weekly METs spent on sports or other vigorous activities in leisure time in the last 3 months (≥4 METS)	Moderate plus vigorous exercise	$a^2 = 79\%$; $c^2 = 0\%$ for adolescent twins $a^2 = 41\%$; $c^2 = 0\%$ for middle-aged twins
Frederiksen et al. (2003)[1]	616 MZ and 642 same-sex DZ twin pairs (aged 45–68)	Engage in leisure time in any of the 11 different exercise activities (yes/no)	Moderate plus vigorous exercise (jogging, gym, swim, tennis, badminton, football, handball, aerobics, rowing, table tennis, volleyball)	$a^2 = 49\%$; $c^2 = 0\%$ for males and females
Kujala et al. (2002)[3]	Data on both members of 1,772 MZ and 3,551 dizygotic same-sex twin pairs (aged 24–60)	Participation in vigorous physical activity based on the question: "is your physical activity during leisure time about as strenuous, on average, as (1) walking, (2) alternatively walking and jogging, (3) jogging (light running), or (4) running?". Those who chose alternative 2, 3, or 4 were classified as participating in vigorous activity	Vigorous exercise	$a^2 = 56\%$; $c^2 = 4\%$ for vigorous activity
Maia et al. (2002)[1]	411 Portuguese twin pairs (aged 12–25)	A composite sports participation index (SPI), that takes into account the energy expenditure for a given sport, number of hours practiced per week, and number of months per year	Moderate plus vigorous exercise	$a^2 = 68\%$; $c^2 = 20\%$ for males $a^2 = 40\%$; $c^2 = 26\%$ for females

Table 24.2 (continued)

Study	Sample	Phenotype	Exercise type	Heritability
Aarnio et al. (1997)[3]	3,254 Twins at age 16, their parents, and grandparents	The categorical phenotype consisted of five physical activity categories ranging from very active to hardly active based on two questions about the – frequency of leisure time PA – intensity of leisure time PA	Light to moderate plus vigorous exercise	$a^2 = 54\%; c^2 = 18\%$ for males $a^2 = 46\%; c^2 = 18\%$ for females
Lauderdale et al. (1996)[3]	3,344 Male twin pairs of the Vietnam Era Twin Registry (aged 33–51)	Five questions assessed regular participation in specific, intense athletic activities (running, bicycling, swimming, racquet, and other sports) (yes/no)	Vigorous exercise	$a^2 = 0\%; d^2 = 53\%$ for jogging $a^2 = 48\%; c^2 = 4\%$ for racquet sports $a^2 = 30\%; c^2 = 17\%$ for strenuous sports $a^2 = 0\%; d^2 = 58\%$ for bicycling $a^2 = 8\%; c^2 = 31\%$ for swimming
Koopmans et al. (1994)[1]	1,587 Adolescent Dutch twin pairs (aged 13–22)	Do you participate in leisure time exercise? (yes/no)	Light to moderate plus vigorous exercise	$a^2 = 48\%; c^2 = 38\%$ for males and females
Boomsma et al. (1989)[1]	44 MZ and 46 DZ Dutch adolescent twin pairs (aged 14–20)	Do you participate in leisure time exercise? (yes/no)	Light to moderate plus vigorous exercise	$a^2 = 64\%$
Heller et al. (1988)[3]	200 Twin pairs (aged 17–66)	Engaged in vigorous exercise in the past 2 weeks (yes/no)	Vigorous exercise	$a^2 = 39\%$
Perusse et al. (1989)[3]	55 Monozygotic and 56 dizygotic Canadian twin pairs (aged 15)[2]	A 3-day activity record was used to determine the activity level of the subjects. The number of periods corresponding to activities with an intensity of ≥ 4.8 METs was counted each day and the average value was used as an indicator of exercise participation	Moderate plus vigorous exercise	$a^2 = 0\%; c^2 = 78\%$ for males and females

[1] Heritability was estimated using variance component methods.
[2] The variable age is used as a regressor.
[3] Heritability was estimated using formulas to calculate the percentage by hand.

or run at least 10 miles per week, (2) play strenuous racquet sports at least 5 h per week, (3) play other strenuous sports (basketball, soccer, etc.), (4) ride a bicycle at least 50 miles per week, (5) swim at least 2 miles per week. For all of the measures, MZ correlations were higher than DZ correlations, suggesting that genes play a role in explaining individual differences in regular exercise. For running or jogging, racquet sports, and bicycling, broad-sense heritability was estimated between 48 and 58%. For bicycling and jogging, MZ correlations exceeded the DZ correlations by more than a factor of 2, making this the only study to report significant non-additive effects. In a Finnish twin study, heritability was estimated in 3,551 dizygotic same-sex twin pairs and 1,772 monozygotic same-sex twin pairs aged 24–60 years (Kujala et al., 2002). Participation in vigorous physical activity was based on the question: "is your physical activity during leisure time about as strenuous, on average, as (1) walking, (2) alternatively walking and jogging, (3) jogging (light running), or (4) running?". Those who chose alternative 2, 3, or 4 were classified as participating in vigorous activity. Heritability was estimated at 56%.

Recently we conducted the largest twin study on exercise behavior ever (Stubbe et al., 2006). The GenomEUtwin project ("Genome-wide analyses of European twin and population cohorts to identify genes predisposing to common diseases") entails one of the largest research consortia in genetic epidemiology in the world with a collection of over 0.8 million twins. Self-reported data on frequency, duration, and intensity of exercise behavior from Australia, Denmark, Finland, Norway, the Netherlands, Sweden, and the UK were used to create an index of exercise participation in each country. Participants had to be engaged in exercise activities for at least 60 min per week with a minimum intensity of about four METs to be classified as regular exercisers. Results obtained in 85,198 twins aged 19–40 years showed an average percentage of male and female exercisers of 44 and 35%, respectively.

Per country, the estimates of the heritability of regular exercise participation are depicted in Table 24.3. The median heritability of exercise participation was 62% across the seven countries and ranged, in males, from 27% in Norway

to 67% in the Netherlands and, in females, from 48% in Australia to 71% in the UK. Shared environmental effects played a role only in exercise participation of the Norwegian males (37%), but were of no importance in the other countries.

Frederiksen and Christensen (2003) were the only ones to report the influence of genetic factors on exercise participation in a group of middle-aged to elderly twins. Information on leisure time exercise participation of people aged 45–68 years was assessed through the questions: "Do you in your leisure time participate in any of the following sports: jogging, gymnastics, swimming, tennis, badminton, football, handball, aerobics, rowing, table tennis, or volleyball?" The exercisers were defined as those indicating participation in any of these activities, whereas the sedentary participants did not report any participation. Genes explained 49% of the variance in exercise participation.

Twin Studies on Physical Activity

Since the innate drive to exercise will be most obvious in leisure time we have focused above on voluntary leisure time exercise behavior. A number of twin studies have quantified regular total physical activity rather than exercise activities limited to leisure time only. Since a large part of regular physical activity in these studies could effectively be attributed to voluntary exercise activities in leisure time, we briefly review these studies here. Slightly more caution is needed in the interpretation of these studies, because the heterogeneity in the definition of regular physical activity will be larger than that in the definition of regular leisure time exercise.

Table 24.4 summarizes the relevant twin studies, again including only those where heritability and "environmentability" estimates or correlations of MZ and DZ twins were given in this chapter. Common environmental influences were again almost completely restricted to children and young adolescents. In adults, reported heritability estimates vary between 46 and 56%. In spite of the larger heterogeneity in the phenotype, Table 24.4 confirms the overall finding that genetic factors contribute significantly to individual differences in physical activity of adults.

Table 24.3 Heritability of exercise participation in subjects aged 21–40 in seven countries participating in the collaborative GenomEUtwin project

Country	No. of complete twin pairs	Percentage exercisers		Heritability estimates	
		Male (%)	Female (%)	Male (%)	Female (%)
Australia	2, 728	64	56	48	48
Denmark	9, 456	43	33	52	52
Finland	8, 842	37	29	62	62
The Netherlands	2, 681	58	55	67	67
Norway	3, 995	55	51	27	56
Sweden	8, 927	37	23	62	62
United Kingdom	422	–	53	–	70

Table 24.4 Twin studies on physical activity level (PA)

Study	Sample	Phenotype	Results
Franks et al. (2005)[1]	100 Same-sex dizygotic ($n = 38$) and monozygotic ($n = 62$) twin pairs (aged 4–10)	(1) Physical activity energy expenditure (PAEE) (2) Total energy expenditure (TEE) labeled water	$a^2 = 0\%; c^2 = 69\%$ $a^2 = 19\%; c^2 = 59\%$
Maia et al. (2002)[1]	411 Portuguese twin pairs (aged 12–25)	The continuous variable leisure time PA is a composite score based on the following four items: – hours watching TV – frequency of walking in leisure time – minutes spent walking per day – frequency of cycling	$a^2 = 63\%; c^2 = 0\%$ for males $a^2 = 32\%; c^2 = 38\%$ for females
Kujala et al. (2002)[2]	Data on both members of 1,772 MZ and 3,551 dizygotic same-sex twin pairs (aged 24–60)	Assessment of leisure time physical activity was based on a series of structured questions on leisure PA (frequency, duration, and intensity of PA sessions) and PA during journey to and from work. The activity MET index was expressed as the summary score of leisure MET-hours per day. Subjects whose volume of activity was ≥ 2 MET-hours per day were classified as physically active at leisure	$a^2 = 46\%; c^2 = 0\%$
Perusse et al. (1989)[1]	55 Monozygotic and 56 dizygotic Canadian twin pairs (aged 15)	A 3-day activity record was used to determine physical activity of the subjects. Each day was divided into 96 periods of 15 min, and for each 15-min period the subjects were asked to note on a scale from 1 to 9 the energy expenditure of the dominant activity of that period. The categorical scores were summed over the 96 15-min periods of each day and the mean sum of the 3 days was used as an indicator of the level of habitual PA	$a^2 = 20\%; c^2 = 52\%$
Perusse et al. (1989)[1]	55 Monozygotic and 56 dizygotic Canadian twin pairs (aged 15)	A 3-day activity record was used to determine physical activity of the subjects. Each day was divided into 96 periods of 15 min, and for each 15-min period the subjects were asked to note on a scale from 1 to 9 the energy expenditure of the dominant activity of that period. The categorical scores were summed over the 96 15-min periods of each day and the mean sum of the 3 days was used as an indicator of the level of habitual PA	$a^2 = 20\%; c^2 = 52\%$
Kaprio et al. (1981)[2]	1,537 MZ and 3,507 DZ male twin pairs (aged ≥ 18 years)	Leisure time PA was based on the amount of physical activity currently engaged in, its intensity and duration, and number of years of physical activity engaged in the adult life. The intensity and duration scores were multiplied together to obtain an activity score	$a^2 = 46\%; d^2 = 11\%$ (age adjusted)

[1] Heritability was estimated using variance component methods.
[2] Heritability was estimated using formulas to calculate the percentage by hand.

Differences in Genetic Architecture of Exercise Behavior Across the Life Span

When we summarize the studies reviewed in Tables 24.2, 24.3 and 24.4, two striking findings stand out: (1) the genetic architecture of exercise behavior is vastly different across the life span with the largest differences seen between the ages 15 and 20 and (2) all studies in *adult* twins consistently suggest a significant genetic contribution to adult exercise participation. Figure 24.2 plots the heritability estimates from the twin studies in Tables 24.2 and 24.3 as a function of the mean age of the sample. Up till age 13–14, genes are of no importance in explaining individual differences in exercise participation, whereas a huge familial resemblance is found through common environmental effects. In late adolescence (from approximately age 17–18 onward), genetic factors start to appear and the role of common environment decreases. Genetic factors peak in their contribution to exercise behavior around age 19–20 to decrease again from young adulthood onward to reach a stable value of about 50% in middle-aged subjects.

The tentative curve drawn through this plot clearly shows that the genetic architecture is different at various points in the life span. These differences have direct bearing on studies assessing heritability using parent–offspring correlations or younger–older sibling correlations. Such studies have systematically yielded lower heritability estimates than twin studies. This may be due to a violation of the assumption that the genetic architecture is the same in younger and older members of the family. If this is not the case, e.g.,

when different genes are expressed in parents and offspring, the parent–offspring correlation does not estimate the heritability in either parental or offspring generations correctly. We suggest, therefore, that the lower heritability estimates from these family studies may partly reflect the comparison of "apples and oranges".

The Lack of Common Environmental Influences on Adult Exercise Behavior

A number of studies show low to moderate tracking from childhood exercise behavior to adult exercise behavior (Beunen et al., 2004; Fortier, Katzmarzyk, Malina, & Bouchard, 2001; Malina, 1996; Simonen, Levalahti, Kaprio, Videman, & Battie, 2004; Twisk, Kemper, & van Mechelen, 2000). Tracking, or stability, refers to the maintenance of relative rank or position over time. Inter-age correlations between repeated measures of the trait are generally used to estimate stability. It has been suggested that correlations <0.30 are considered to be indicative of low stability, whereas those ranging from 0.30 to 0.60 are moderate, and those >0.60 are high (Malina, 1996). A review by Malina (1996) shows that, although different indicators of physical activity and different methods of analysis are used, it appears that physical activity tracks low to moderately from adolescence to adulthood. This is consistent with results from the longitudinal Amsterdam Growth and Health Study (Twisk et al., 2000). In subjects with a mean age of 13.1 (±0.8) years, total time spent on all habitual physical activities in relation to school, work, sports, and other leisure time activities was measured with an interviewer-administered activity questionnaire. During the first 4 years of the study, yearly measurements were carried out. Later on, two follow-up measurements took place after 8 and 14 years, respectively. The stability coefficient, summarizing tracking across all intervals, was 0.34 (95% CI = 0.19–0.49) for daily activity, indicating that there was low to moderate tracking.

Using data from the Netherlands Twin Registry (NTR), we essentially replicated this finding. Table 24.5 shows 7-year tracking of exercise participation from ages 13–16 to ages 20–23. Again low to moderate tracking coefficients were found ranging from 0.22 to 0.44. Model-fitting results showed that these correlations did not significantly differ from each other ($p = 0.56$), resulting in an overall tracking coefficient of 0.37 from ages 13–20 to 16–23, which is in keeping with the stability coefficient of 0.34 found in the Amsterdam Growth and Health Study, even though our cohort was born more than 10 years later.

In view of the striking shift in genetic architecture during adolescence, this tracking is puzzling. If common environmental factors influence exercise behavior among children and their exercise behavior tracks into adulthood,

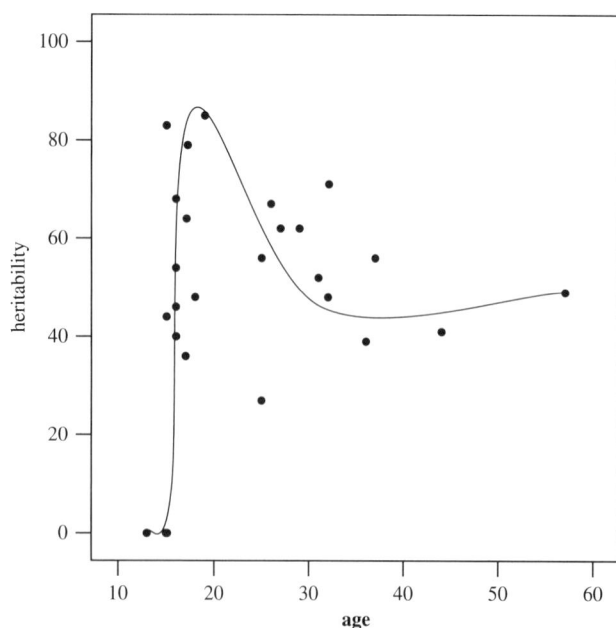

Fig. 24.2 Heritability estimates for exercise participation as a function of the mean age of the twin sample

Table 24.5 Seven-year tracking of exercise participation from adolescence into young adulthood in the Dutch Twin Survey

Initial age and age at follow-up	Number of subjects participating in two surveys	Tetrachoric correlation
From age 13 to 20	169	0.41
From age 14 to 21	184	0.22
From age 15 to 22	181	0.44
From age 16 to 23	214	0.36

one would expect to find enduring effects of the environment they shared as youngsters even after they reach adulthood. In spite of this expectation, most of the studies in adults do not find evidence for common environment at all, including six of the seven samples in the GenomEUtwin study. A first potential explanation for the failure to detect C in adult samples is a lack of power to detect common environment in smaller-sized twin studies. Most studies measured exercise behavior as a dichotomy, and at heritabilities between 30 and 70%, large samples are needed to detect additional common environmental influences of modest size as is shown in Table 24.6 (Neale, Eaves, & Kendler, 1994). However, at least three samples of the GenomEUtwin study (with heritability at 50%) easily exceed this sample size number and yet did not detect common environment.

A second potential explanation is that in adulthood common environmental factors interact with genetic make-up. Since twin studies cannot discriminate between main effects of genes and their interaction with common environmental influences ($C \times G$), in the classical twin model any $C \times G$ interaction would end up as a main effect of genetic factors (Purcell, 2001). There is, in fact, a straightforward theoretical account for a $C \times G$ interaction on exercise behavior that would be compatible with such a scenario. It has been suggested that genetic influences on exercise ability, which are very strong both for strength and endurance phenotypes (Arden & Spector, 1997; Bouchard et al., 1998; Thomis et al., 1997), may explain part of the heritability of exercise behavior (Stubbe et al., 2006).

The basic idea is that people will seek out the activities that they excel in. This is particularly true in adolescence. Being "good in sports" is an important source of self-esteem for teenagers and the athletic role model is continuously reinforced by the media (Field et al., 1999; Pope, Olivardia, Borowiecki, & Cohane, 2001). Hence, genes coding for exercise ability may well become genes for adolescent exercise behavior. The parents and older siblings may be helpful to

make sure the youngsters regularly get to the playing field in the first place, and to provide positive feedback on their performance. The extent of positive feedback, however, may depend on their (exercise ability) genotypes. This is even truer for feedback by peers and colleagues, who will base their judgment entirely on performance rather than family ties. The family environment, in short, determines exposure and encouragement in early adolescence, but actual ability will determine whether they like exercising enough (by excelling in it) to maintain the behavior when the perception of peers and colleagues increases in importance in late adolescence.

A final possibility is that the estimates of common environmental influences in early adolescence include genetic effects that are correlated with the family environment. Such a correlation would come about if the parents that most encourage their children to become engaged in exercise were themselves of above-average athletic ability. If they pass on these genes and create a family environment that encourages sports, a positive correlation between common environmental and genetic influences would arise and in the twin samples genes for exercise ability would then become correlated to an encouraging environment shared by the twins. It has been shown that a correlation of genes and shared environment inflates the estimates of common environmental effects in twin studies (Purcell, 2001).

Different Genes at Different Ages?

The above $C \times G$ scenario would still leave unexplained why there is a peak in heritability around age 18–25. This peak was most clearly demonstrated in a study that assessed exercise behavior as weekly energy expenditure in an identical way in a cohort of 17-year-old and a cohort of 45-year-old twins (De Geus et al., 2003). Heritability was found to be much higher in adolescents (79%) than in adults (41%). Does the impact of the unique environment on exercise habits increase after young adulthood, for instance, due to factors like work stress and child care load? That is entirely possible, and would fit with data indicating that the most often reported barrier to exercise is "lack of time" (King et al., 1992; Sallis & Hovell, 1990). However, total variance in leisure time energy expenditure was also seen to go down in the same study (De Geus et al., 2003). This does not rule out an

Table 24.6 Sample size in subjects (N) needed to detect common environmental influences (V_C) in full ACE models under varying levels of variation due to additive genetic sources (V_A)

	$V_A = 30\%$		$V_A = 40\%$		$V_A = 50\%$		$V_A = 60\%$		$V_A = 70\%$	
V_C (%)	10	20	10	20	10	20	10	20	10	20
N	13,681	3,152	12,908	2,918	12,007	2,661	11,000	2,387	9,919	2,108

Note: MZ/DZ ratio = 1/1; significance level $\alpha = 0.05$; power $(1-\beta) = 0.80$.

increase in environmental variance, but it does mean that a decrease in genetic variance must have occurred. Another possibility, therefore, is that different genes play a role in exercise behavior in adolescence than in adulthood.

As stated above, core components of exercise ability like aerobic endurance and muscular strength show large heritability, and the genes influencing exercise ability may play an important role in the choices of adolescents. Exercise ability, however, may start to loose significance in adulthood when recreational exercise start to become more prominent than competitive exercise. In this phase, genes that determine personality may increasingly start to influence exercise behavior. In adult samples a modest but highly significant association between neuroticism and extraversion and exercise participation is found (De Moor, Beem, Stubbe, Boomsma, & De Geus, 2005), for instance, whereas this link is absent in adolescence (Allison et al., 2005). Physical and mental health benefits of exercise may also become increasingly important motives in adulthood. If there are differences in the genetic sensitivity to these health benefits of exercise, genes coding for this differential sensitivity may well become genes for adult exercise behavior.

In favor of this hypothesis, standardized training programs have already shown some persons to be more responsive to the same exercise regime than others in terms of increased aerobic fitness (Bouchard et al., 1999), increased muscular strength (Thomis et al., 1998), reduced body fat (Perusse et al., 2000), increased HDL/LDL cholesterol ratio (Rice et al., 2002), decreased C-reactive protein (Lakka et al., 2005), increased insulin sensitivity and glucose effectiveness (Boule et al., 2005; Teran-Garcia, Rankinen, Koza, Rao, & Bouchard, 2005), and decreased heart rate and blood pressure (An et al., 2003; Rice et al., 2002). By studying the response of family members to an identical exercise program, these differential health effects were shown to largely reflect differences in genetic make-up (An et al., 2003; Bouchard et al., 1999; Boule et al., 2005; Lakka et al., 2005; Perusse et al., 2000; Rice et al., 2002; Teran-Garcia et al., 2005; Thomis et al., 1997). Although currently unknown, psychological benefits may well show a similar dependency on genotype.

As it stands, the idea that different genes influence exercise behavior across the lifespan remains hypothetical. This hypothesis can be fully tested, however, in longitudinal twin data. Ideally these should span the crucial period between age 18 and 30.

Assortative Mating

So far, we have suggested that twin studies of exercise behavior in adolescence may be complicated by the presence of C × G interaction and by adolescence-specific genetic effects. Additional complexity may derive from assortative mating. In a three-generation Finnish study (Aarnio et al., 1997), intra- and inter-generational associations of leisure time physical activity among family members were examined. The sample consisted of 3,254 twins at the age of 16, their parents and grandparents. The correlation was 0.19 between parents, 0.33 between paternal grandparents and 0.43 between maternal grandparents, suggesting that assortative mating is present. In the Quebec family study, familial aggregation of physical activity phenotypes was investigated in 696 subjects from 200 families (Simonen et al., 2002). For moderate to strenuous physical activity, the parental correlation was 0.22. Similar assortment was found in our own sample (Willemsen, Vink, & Boomsma, 2003). Tetrachoric correlations between exercise participation of spouses as a function of the duration of the relationship were 0.45, 0.42, and 0.49 for relations lasting <5 years, ≥5 years, and >15 years, respectively.

How will assortment for exercise participation affect the estimates in twin studies? If the environment causes assortment no effects on genetic variance will be seen. If the assortment is phenotypic, as we expect, it will act to both increase total genetic variance and heritability (Falconer & Mackay, 1996). In the classical twin design, however, phenotypic assortment will look like common environmental influences when fitting an ACE model because it also increases the average amount of shared genes of DZ pairs above the theoretical 50%. Thus, the heritability in the population increases as a consequence of phenotypic assortment but use of the classical twin design will increase the estimate of common environmental influences. Does the common environmental influence on exercise found in studies on adolescent twins in part reflect assortative mating?

At first sight, the finding that common environmental effects disappear in later adulthood seems to argue against assortative mating since the higher than 50% genetic resemblance should stay in effect throughout the lifespan. However, in the above we argued that genes that are expressed in early adulthood may partly differ from the genes that influence exercise later in life. If the assortment is phenotypic, it will exclusively operate on the genes that are in effect during the main mating period, e.g., in late adolescence and young adulthood. In this case, the genes that affect exercise in later stages of life may still be under random mating. Future modeling of exercise data obtained in twins as well as their parents and their spouses may shed more light on these issues. As it stands, early conclusions based on the resemblance of young sibling–sibling and spousal correlations that "familial resemblance is the result of environmental factors shared by members of the same generation rather than inherited factors" (Perusse et al., 1988) seem premature in retrospect.

Future Directions

The prevailing theoretical perspective in preventive medicine now holds that social and environmental factors largely account for voluntary lifestyle choices. Here, in contrast, it is shown that in adulthood some of the choices for a healthy lifestyle reflect differences in genetic make-up, although potentially in interaction with shared environment. This requires a change in our perspective, such that we change from "population-based" intervention strategies to "personalized" intervention strategies. Currently, this concept of "personalized medicine" is increasingly being applied to curative medicine and pharmacotherapeutic intervention. We suggest extending this concept to preventive medicine.

Crucial to such personalized preventive medicine is a mechanistic understanding of the genetic pathways that underlie the genetic contribution to individual variation in this behavior. Such understanding may not only help to improve intervention strategies but may impact on research on health in general. Randomized controlled training trials have clearly shown that regular exercise has a causal effect on mental (Babyak et al., 2000; Moore & Blumenthal, 1998; Steptoe, Edwards, Moses, & Mathews, 1989) and physical health (Berlin & Colditz, 1990). It is possible, therefore, that the well-known heritability of many health parameters like depression (Kendler & Aggen, 2001), obesity (Schousboe et al., 2003), thrombosis (Dunn et al., 2004), hypertension (Kupper et al., 2005), diabetes (De Lange et al., 2003), and even cardiovascular mortality (Zdravkovic et al., 2004) may partly reflect the genetic factors causing the adoption and maintenance of regular exercise behavior. In that case, finding the "genes for exercise behavior" immediately translates to finding genes that contribute to the heritability of mental and physical health.

So which genes could explain the heritability of exercise behavior? Unfortunately, this is a vastly under-explored question. For exercise, *ability* coordinated efforts exist worldwide and successful association has been reported for a number of genes influencing endurance or strength phenotypes, some of which have been replicated in independent samples. Specifically, a systematic and yearly update of the Human Gene Map for Performance and Health-Related Fitness Phenotypes is published in *Medicine and Science in Sports and Exercise* (Perusse et al., 2003; Rankinen et al., 2001, 2002, 2004; Wolfarth et al., 2005). For exercise *behavior* no such coordinated effort exists, although the most recent version of this *MSSE* Gene Map included for the first time a new section on this topic (Wolfarth et al., 2005). Fortunately, as we have argued above, genetic variation in exercise ability may partly overlap with genetic variation in exercise behavior, which means that many of the genes on the Human Gene Map for Performance and Health-Related Fitness Phenotypes can be considered promising candidate genes.

One example that illustrates this is the insertion/deletion (I/D) polymorphism in the angiotensin-converting enzyme (ACE) gene. Montgomery and colleagues (Williams et al., 2000; Woods, Humphries, & Montgomery, 2000) determined the ACE I/D genotype in British army recruits who were tested for a number of fitness traits before and after a 10-week training program. Efficiency of the muscles, or delta efficiency, computed as the increase in power output for a given increase in oxygen consumption, was found to increase almost ninefold more in subjects homozygous for the I allele. Almost no training effect was found in those homozygous for the D allele. As previously noted people generally like doing what they are good at, and will pursue those activities in leisure time as much as possible. Taking this one step further, we may reasonably assume that people feel specifically competent when they notice themselves to gain more in performance compared to others who nonetheless follow the same exercise regime. In support of this "competence hypothesis" a multicenter study in Italian borderline hypertensives (Winnicki et al., 2004) showed that the ACE polymorphism accounted for 21% of the variance in exercise participation. The most sedentary group had a clear excess of the genotype (DD) that caused the lowest increase in muscle efficiency after training in the British recruits. Here, at least, a genetic effect on exercise ability indeed coincided with reduced amounts of exercise behavior.

Apart from the link between the ACE polymorphism and exercise behavior, association for other candidate genes has been reported. In women but not in men, physical activity levels were associated with polymorphisms in the dopamine D2 receptor gene, which is proposed to play a role in rewarding mechanisms (Simonen, Rankinen, Perusse, Leon, et al., 2003). In 331 early postmenopausal women, physical activity was associated with a polymorphism in the CYP19 (aromatase) gene (Salmen et al., 2003). In the Quebec Family study, the Melanocortin-4 receptor gene (MC4R-C-2745T variant) showed significant associations with moderate-to-strenuous activity scores and with inactivity scores (Loos et al., 2005). Finally, in 97 healthy girls, physical activity was associated with polymorphisms in a calcium-sensing receptor gene (Lorentzon, Lorentzon, Lerner, & Nordstrom, 2001). To our knowledge, only one whole genome scan based on linkage analysis exists for physical exercise (Simonen, Rankinen, Perusse, Rice, et al., 2003). A few putative genomic regions were identified that might harbor genes influencing participation in regular exercise, but the evidence was only suggestive, as the power for linkage in this relatively small and unselected sample was small.

Combining the importance of exercise for health to the strong evidence for its heritability makes it paramount that large-scale gene finding studies start targeting this crucial behavior.

Acknowledgment This study was supported by the Netherlands Organization for Scientific Research (NWO-MW 904-61-193, NWO 575-25-006, and NWO 985-10-002) and the European Commission under the program "Quality of Life and Management of the Living Resources" of 5th Framework Program (GenomEUtwin QLG2-CT-2002-01254).

References

Aarnio, M., Winter, T., Kujala, U. M., & Kaprio, J. (1997). Familial aggregation of leisure-time physical activity: A three generation study. *International Journal of Sports Medicine, 18*, 549–556.

Ainsworth, B. E., Haskell, W. L., Leon, A. S., Jacobs, D. R., Jr., Montoye, H. J., Sallis, J. F., et al. (1993). Compendium of physical activities: Classification of energy costs of human physical activities. *Medicine and Science in Sports and Exercise, 25*, 71–80.

Ainsworth, B. E., Haskell, W. L., Whitt, M. C., Irwin, M. L., Swartz, A. M., Strath, S. J., et al. (2000). Compendium of physical activities: An update of activity codes and MET intensities. *Medicine and Science in Sports and Exercise, 32*, S498–S504.

Albright, A., Franz, M., Hornsby, G., Kriska, A., Marrero, D., Ullrich, I., et al. (2000). Exercise and type 2 diabetes. *Medicine and Science in Sports and Exercise, 32*, 1345–1360.

Allison, K. R., Adlaf, E. M., Irving, H. M., Hatch, J. L., Smith, T. F., Dwyer, J. J. M., et al. (2005). Relationship of vigorous physical activity to psychologic distress among adolescents. *Journal of Adolescent Health, 37*, 164–166.

An, P., Perusse, L., Rankinen, T., Borecki, I. B., Gagnon, J., Leon, A. S., et al. (2003). Familial aggregation of exercise heart rate and blood pressure in response to 20 weeks of endurance training: The HERITAGE Family Study. *International Journal of Sports Medicine, 24*, 57–62.

Arden, N. K., & Spector, T. D. (1997). Genetic influences on muscle strength, lean body mass, and bone mineral density: A twin study. *Journal of Bone and Mineral Research, 12*, 2076–2081.

Babyak, M., Blumenthal, J. A., Herman, S., Khatri, P., Doraiswamy, M., Moore, K., et al. (2000). Exercise treatment for major depression: Maintenance of therapeutic benefit at 10 months. *Psychosomatic Medicine, 62*, 633–638.

Berlin, J. A., & Colditz, G. A. (1990). A meta-analysis of physical activity in the prevention of coronary heart disease. *American Journal of Epidemiology, 132*, 612–628.

Beunen, G., & Thomis, M. (1999). Genetic determinants of sports participation and daily physical activity. *International Journal of Obesity, 23*, S55–S63.

Beunen, G., Lefevre, J., Philippaerts, R. M., Delvaux, K., Thomis, M., Claessens, A. L., et al. (2004). Adolescent correlates of adult physical activity: A 26-year follow-up. *Medicine and Science in Sports and Exercise, 36*, 1930–1936.

Blair, S. N., Booth, M., Gyarfas, I., Iwane, H., Mati, B., Matsudo, V., et al. (1996). Development of public policy and physical activity initiatives internationally. *Sports Medicine, 21*, 157–163.

Boomsma, D. I., Vandenbree, M. B. M., Orlebeke, J. F., & Molenaar, P. C. M. (1989). Resemblances of parents and twins in sports participation and heart-rate. *Behavior Genetics, 19*, 123–141.

Bouchard, C., Daw, E. W., Rice, T., Perusse, L., Gagnon, J., Province, M. A., et al. (1998). Familial resemblance for VO2max in the sedentary state: The HERITAGE Family Study. *Medicine and Science in Sports and Exercise, 30*, 252–258.

Bouchard, C., An, P., Rice, T., Skinner, J. S., Wilmore, J. H., Gagnon, J., et al. (1999). Familial aggregation of VO2max response to exercise training: Results from the HERITAGE Family Study. *Journal of Applied Physiology, 87*, 1003–1008.

Boule, N. G., Weisnagel, S. J., Lakka, T. A., Tremblay, A., Bergman, R. N., Rankinen, T., et al. (2005). Effects of exercise training on glucose homeostasis. *Diabetes Care, 28*, 108–114.

Caspersen, C. J., Powell, K. E., & Christenson, G. M. (1985). Physical activity, exercise, and physical fitness: Definitions and distinctions for health-related research. *Public Health Reports, 100*, 126–131.

Caspersen, C. J. (1987). Physical inactivity and coronary heart-disease. *Physician in Sports Medicine, 15*, 43–44.

Caspersen, C. J., Pereira, M. A., & Curran, K. M. (2000). Changes in physical activity patterns in the United States, by sex and cross-sectional age. *Medicine and Science in Sports and Exercise, 32*, 1601–1609.

Crespo, C. J., Keteyian, S. J., Heath, G. W., & Sempos, C. T. (1996). Leisure-time physical activity among US adults. Results from the Third National Health and Nutrition Examination Survey. *Archives of Internal Medicine, 156*, 93–98.

De Geus, E. J. C., Boomsma, D. I., & Snieder, H. (2003). Genetic correlation of exercise with heart rate and respiratory sinus arrhythmia. *Medicine and Science in Sports and Exercise, 35*, 1287–1295.

De Lange, M., Snieder, H., Ariens, R. A. S., Andrew, T., Grant, P. J., & Spector, T. D. (2003). The relation between insulin resistance and hemostasis: Pleiotropic genes and common environment. *Twin Research, 6*, 152–161.

De Moor, M., Beem, A. L., Stubbe, J. H., Boomsma, D., & De Geus, E. J. C. (2005). Regular exercise, anxiety, depression and personality: A population-based study. *Preventive Medicine, 42*, 273–279.

Dishman, R. K., Sallis, J. F., & Orenstein, D. R. (1985). The determinants of physical activity and exercise. *Public Health Reports, 100*, 158–171.

Dunn, E. J., Ariens, R. A., De Lange, M., Snieder, H., Turney, J. H., Spector, T. D., et al. (2004). Genetics of fibrin clot structure: A twin study. *Blood, 103*, 1735–1740.

Falconer, D. S., & Mackay, T. F. C. (1996). *Introduction to quantitative genetics.* (4th ed.). Essex: Pearson Education Limited.

Field, A. E., Cheung, L., Wolf, A. M., Herzog, D. B., Gortmaker, S. L., & Colditz, G. A. (1999). Exposure to the mass media and weight concerns among girls. *Pediatrics, 103*, 1911.

Fortier, M. D., Katzmarzyk, P. T., Malina, R. M., & Bouchard, C. (2001). Seven-year stability of physical activity and musculoskeletal fitness in the Canadian population. *Medicine and Science in Sports and Exercise, 33*, 1905–1911.

Franks P. W., Ravussin E., Hanson R. L., Harper I. T., Allison D. B., Knowler W. C., Tataranni P. A., & Salbe A. D. (2005). Habitual physical activity in children: the role of genes and the environment. *The American Journal of Clinical Nutrition, 82*, 901–908.

Frederiksen, H., & Christensen, K. (2003). The influence of genetic factors on physical functioning and exercise in second half of life. *Scandinavian Journal of Medicine and Science in Sports, 13*, 9–18.

Haase, A., Steptoe, A., Sallis, J. F., & Wardle, J. (2004). Leisure-time physical activity in university students from 23 countries: Associations with health beliefs, risk awareness, and national economic development. *Preventive Medicine, 39*, 182–190.

Heller, R. F., O'Connell, D. L., Roberts, D. C. K., Allen, J. R., Knapp, J. C., Steele, P. L., et al. (1988). Lifestyle factors in monozygotic and dizygotic twins. *Genetic Epidemiology, 5*, 311–321.

Kaplan, G. A., Lazarus, N. B., Cohen, R. D., & Leu, D. J. (1991). Psychosocial factors in the natural history of physical activity. *American Journal of Preventive Medicine, 7*, 12–17.

Kaplan, G. A., Strawbridge, W. J., Cohen, R. D., & Hungerford, L. R. (1996). Natural history of leisure-time physical activity and its correlates: Associations with mortality from all causes and cardiovascular disease over 28 years. *American Journal of Epidemiology, 144*, 793–797.

Kaprio J., Koskenvuo M., & Sarna S. (1981). Cigarette smoking, use of alcohol, and leisure time physical activity among same-sexed adult male twins. *Progress in Clinical and Biological Research, 69*, 37–46.

Kendler, K. S., & Aggen, S. H. (2001). Time, memory and the heritability of major depression. *Psychological Medicine, 31*, 923–928.

Kesaniemi, Y. A., Danforth, E., Jensen, M. D., Kopelman, P. G., Lefebvre, P., & Reeder, B. A. (2001). Dose-response issues concerning physical activity and health: An evidence-based symposium. *Medicine and Science in Sports and Exercise, 33*, S351–S358.

King, A. C., Blair, S. N., Bild, D. E., Dishman, R. K., Dubbert, P. M., Marcus, B. H., et al. (1992). Determinants of physical activity and interventions in adults. *Medicine and Science in Sports and Exercise, 24*, S221–S236.

Koopmans, J. R., Van Doornen, L. J. P., & Boomsma, D. I. (1994). Smoking and sports participation. In U. Goldbourt & U. De Faire (Eds.), *Genetic factors in coronary heart disease*. Dordrecht: Kluwer Academic Publisher.

Kujala U. M., Kaprio J., & Koskenvuo M. (2002). Modifiable risk factors as predictors of all-cause mortality: The roles of genetics and childhood environment. *American Journal of Epidemiology, 156*, 985–993.

Kupper, N., Willemsen, G., Riese, H., Posthuma, D., Boomsma, D. I., & De Geus, E. J. C. (2005). Heritability of daytime ambulatory blood pressure in an extended twin design. *Hypertension, 45*, 80–85.

Lakka, T. A., Lakka, H. M., Rankinen, T., Leon, A. S., Rao, D. C., Skinner, J. S., et al. (2005). Effect of exercise training on plasma levels of C-reactive protein in healthy adults: The HERITAGE Family Study. *European Heart Journal, 26*, 2018–2025.

Lauderdale, D. S., Fabsitz, R., Meyer, J. M., Sholinsky, P., Ramakrishnan, V., & Goldberg, J. (1997). Familial determinants of moderate and intense physical activity: A twin study. *Medicine and Science in Sports and Exercise, 29*, 1062–1068.

Loos, R. J. F., Rankinen, T., Tremblay, A., Perusse, L., Chagnon, Y., & Bouchard, C. (2005). Melanocortin-4 receptor gene and physical activity in the Quebec Family Study. *International Journal of Obesity, 29*, 420–428.

Lorentzon, M., Lorentzon, R., Lerner, U. H., & Nordstrom, P. (2001). Calcium sensing receptor gene polymorphism, circulating calcium concentrations and bone mineral density in healthy adolescent girls. *European Journal of Endocrinology, 144*, 257–261.

Maia, J. A. R., Thomis, M., & Beunen, G. (2002). Genetic factors in physical activity levels: A twin study. *American Journal of Preventive Medicine, 23*, 87–91.

Malina, R. M. (1996). Tracking of physical activity and physical fitness across the lifespan. *Research Quarterly for Exercise and Sports, 67*, S48–S57.

Martinez-Gonzalez, M. A., Martinez, J. A, Hu, F. B., Gibney, M. J., & Kearney, J. (1999). Physical inactivity, sedentary lifestyle and obesity in the European Union. *International Journal of Obesity and////////// elated Metabolic Disorders, 23*, 1192–1201.

Martinez-Gonzalez, M. A., Varo, J. J., Santos, J. L., De Irala, J., Gibney, M., Kearney, J., et al. (2001). Prevalence of physical activity during leisure-time in the European Union. *Medicine and Science in Sports and Exercise, 33*, 1142–1146.

Matson-Koffman, D. M., Brownstein, J. N., Neiner, J. A., & Greaney, M. L. (2005). A site-specific literature review of policy and environmental interventions that promote physical activity and nutrition for cardiovascular health: What works? *American Journal of Health Promotion, 19*, 167–193.

Moore, K. A., & Blumenthal, J. A. (1998). Exercise training as an alternative treatment for depression among older adults. *Alternative Therapies in Health and Medicine, 4*, 48–56.

Neale, M. C., Eaves, L. J., & Kendler, K. S. (1994). The power of the classical twin study to resolve variation in threshold traits. *Behavior Genetics, 24*, 239–258.

Orleans, C. T., Kraft, M. K., Marx, J. F., & McGinnis, J. M. (2003). Why are some neighborhoods active and others not? Charting a new course for research on the policy and environmental determinants of physical activity. *Annals of Behavioral Medicine, 25*, 77–79.

Pate, R. R., Pratt, M., Blair, S. N., Haskell, W. L., Macera, C. A., Bouchard, C., et al. (1995). Physical activity and public health: A recommendation from the centers for disease control and prevention and the American College of Sports Medicine. *Journal of the American Medical Association, 273*, 402–407.

Pate, R. R., Freedson, P. S., Sallis, J. F., Taylor, W. C., Sirard, J., Trost, S. G., et al. (2002). Compliance with physical activity guidelines: Prevalence in a population of children and youth. *Annals of Epidemiology, 12*, 303–308.

Payne, N., Jones, F., & Harris, P. R. (2005). The impact of job strain on the predictive validity of the theory of planned behaviour: An investigation of exercise and healthy eating. *British Journal of Health Psychology, 10*, 115–131.

Perusse, L., Leblanc, C., & Bouchard, C. (1988). Familial resemblance in lifestyle components: Results from the Canada Fitness Survey. *Canadian Journal of Public Health, 79*, 201–205.

Perusse, L., Tremblay, A., Leblanc, C., & Bouchard, C. (1989). Genetic and environmental influences on level of habitual physical activity and exercise participation. *American Journal of Epidemiology, 129*, 1012–1022.

Perusse, L., Rice, T., Province, M. A., Gagnon, J., Leon, A. S., Skinner, J. S., Wilmore, J. H., Rao, D. C., & Bouchard, C. (2000). Familial aggregation of amount and distribution of subcutaneous fat and their responses to exercise training in the HERITAGE Family Study. *Obesity Research, 8*, 140–150.

Perusse, L., Rankinen, T., Rauramaa, R., Rivera, M. A., Wolfarth, B., & Bouchard, C. (2003). The human gene map for performance and health-related fitness phenotypes: The 2002 update. *Medicine and Science in Sports and Exercise, 35*, 1248–1264.

Plomin, R., DeFries, J. C., McClearn, G. E., & McGuffin, P. (2000). *Behavioral genetics* (4th ed.). New York: Worth Publishers.

Pope, H. G., Olivardia, R., Borowiecki, J. J., & Cohane, G. H. (2001). The growing commercial value of the male body: A longitudinal survey of advertising in women's magazines. *Psychotherapy and Psychosomatics, 70*, 189–192.

Purcell, S. (2001). Gene-by-environment interaction in twin and sib-pair analysis. *Behavior Genetics, 31*, 466.

Rankinen, T., Perusse, L., Rauramaa, R., Rivera, M. A., Wolfarth, B., & Bouchard, C. (2001). The human gene map for performance and health-related fitness phenotypes. *Medicine and Science in Sports and Exercise, 33*, 855–867.

Rankinen, T., Perusse, L., Rauramaa, R., Rivera, M. A., Wolfarth, B., & Bouchard, C. (2002). The human gene map for performance and health-related fitness phenotypes: The 2001 update. *Medicine and Science in Sports and Exercise, 34*, 1219–1233.

Rankinen, T., Perusse, L., Rauramaa, R., Rivera, M. A., Wolfarth, B., & Bouchard, C. (2004). The human gene map for performance and health-related fitness phenotypes: The 2003 update. *Medicine and Science in Sports and Exercise, 36*, 1451–1469.

Rice, T., Despres, J. P., Perusse, L., Hong, Y. L., Province, M. A., Bergeron, J., et al. (2002). Familial aggregation of blood lipid response to exercise training in the health, risk factors, exercise training, & genetics (HERITAGE) family study. *Circulation, 105*, 1904–1908.

Rowland, T. W. (1998). The biological basis of physical activity. *Medicine and Science in Sports and Exercise, 30*, 392–399.

Sallis, J. F., & Hovell, M. F. (1990). Determinants of exercise behavior. *Exercise and Sport Sciences Reviews, 18*, 307–330.

Salmen, T., Heikkinen, A. M., Mahonen, A., Kroger, H., Komulainen, M., Pallonen, H., et al. (2003). Relation of aromatase gene polymorphism and hormone replacement therapy to serum estradiol levels, bone mineral density, and fracture risk in early postmenopausal women. *Annals of Medicine, 35,* 282–288.

Schousboe K., Willemsen, G., Kyvik, K. O., Mortensen, J., Boomsma, D. I., Cornes, B. K, et al. (2003). Sex differences in heritability of BMI: A comparative study of results from twin studies in eight countries. *Twin Research, 6,* 409–421.

Seefeldt, V., Malina, R. M., & Clark, M. A. (2002). Factors affecting levels of physical activity in adults. *Sports Medicine, 32,* 143–168.

Shephard, R. J. (1985). Factors influencing the exercise behavior of patients. *Sports Medicine, 2,* 348–366.

Sherwood, N. E., & Jeffery, R. W. (2000). The behavioral determinants of exercise: Implications for physical activity interventions. *Annual Review of Nutrition, 20,* 21–44.

Simonen, R. L., Perusse, L., Rankinen, T., Rice, T., Rao, D. C., & Bouchard, C. (2002). Familial aggregation of physical activity levels in the Quebec family study. *Medicine and Science in Sports and Exercise, 34,* 1137–1142.

Simonen, R. L., Rankinen, T., Perusse, L., Leon, A. S., Skinner, J. S., Wilmore, J. H., et al. (2003). A dopamine D2 receptor gene polymorphism and physical activity in two family studies. *Physiology and Behavior, 78,* 751–757.

Simonen, R. L., Rankinen, T., Perusse, L., Rice, T., Rao, D. C., Chagnon, Y., et al. (2003). Genome-wide linkage scan for physical activity levels in the Quebec family study. *Medicine and Science in Sports and Exercise, 35,* 1355–1359.

Simonen, R. L., Levalahti, E., Kaprio, J., Videman, T., & Battie, M. C. (2004). Multivariate genetic analysis of lifetime exercise and environmental factors. *Medicine and Science in Sports and Exercise, 36,* 1559–1566.

Sirard, J. R., & Pate, R. R. (2001). Physical activity assessment in children and adolescents. *Sports Medicine, 31,* 439–454.

Stephens, T., & Craig, C. L. (1990). *The well-being of Canadians: Highlights of the 1988 Campbell's Survey.* Ottawa: Canadian Fitness and Lifestyle Research Institute.

Steptoe, A., Edwards, S., Moses, J., & Mathews, A. (1989). The effects of exercise training on mood and perceived coping ability in anxious adults from the general-population. *Journal of Psychosomatic Research, 33,* 537–547.

Steptoe, A., Wardle, J., Fuller, R., Holte, A., Justo, J., Sanderman, R., et al. (1997). Leisure-time physical exercise: Prevalence, attitudinal correlates, and behavioral correlates among young Europeans from 21 countries. *Preventive Medicine, 26,* 845–854.

Steptoe, A., Wardle, J., Cui, W. W., Bellisle, F., Zotti, A. M., Baranyai, R., et al. (2002). Trends in smoking, diet, physical exercise, and attitudes toward health in European university students from 13 countries, 1990–2000. *Preventive Medicine, 35,* 97–104.

Stubbe, J. H., Boomsma, D. I., & De Geus, E. J. C. (2005). Sports participation during adolescence: A shift from environmental to genetic factors. *Medicine and Science in Sports and Exercise, 37,* 563–570.

Stubbe, J. H., Boomsma, D. I., Vink, J. M., Cornes, B. K., Martin, N. G., Skythe, A., et al. (2006) Genetic influence on exercise participation in 37,051 twin pairs from seven countries. *PLoS One, 1,* e22.

Teran-Garcia, M., Rankinen, T., Koza, R. A., Rao, D. C., & Bouchard, C. (2005). Endurance training-induced changes in insulin sensitivity and gene expression. *American Journal of Physiology, Endocrinology and Metabolism, 288,* E1168–E1178.

Thomis, M. A., Van Leemputte, M., Maes, H. H., Blimkie, C. J. R., Claessens, A. L., Marchal, G., et al. (1997). Multivariate genetic analysis of maximal isometric muscle force at different elbow angles. *Journal of Applied Physiology, 82,* 959–967.

Thomis, M. A., Beunen, G. P., Maes, H. H., Blimkie, C. J. R., Van Leemputte, M., Claessens, A. L., et al. (1998). Strength training: Importance of genetic factors. *Medicine and Science in Sports and Exercise, 30,* 724–731.

Thorburn A. W., & Proietto. J. (2000). Biological determinants of spontaneous physical activity. *Obesity Reviews, 1,* 87–94.

Tou, J. C. L., & Wade, C. E. (2002). Determinants affecting physical activity levels in animal models. *Experimental Biology and Medicine, 227,* 587–600.

Twisk, J. W. R., Kemper, H. C. G., & van Mechelen, W. (2000). Tracking of activity and fitness and the relationship with cardiovascular disease risk factors. *Medicine and Science in Sports and Exercise, 32,* 1455–1461.

U.S. Department of Health and Human Services (2005). *Healthy people 2000: National health promotion and disease prevention objectives.* Washington, DC: U.S. Department of Health and Human Services.

Van Loon, A. J. M., Tijhuis, M., Surtees, P. G., & Ormel, J. (2000). Lifestyle risk factors for cancer: The relationship with psychosocial work environment. *International Journal of Epidemiology, 29,* 785–792.

Varo, J. J., Martinez-Gonzalez, M. A., Irala-Estevez, J., Kearney, J., Gibney, M., & Martinez, J. A. (2003). Distribution and determinants of sedentary lifestyles in the European Union. *International Journal of Epidemiology, 32,* 138–146.

WHO/FIMS Committee on Physical Activity for Health. (1995). *Exercise for health. Bulletin of the World Health Organization, 73,* 135–136.

Willemsen, G., Vink, J. M., & Boomsma, D. I. (2003). Assortative mating may explain spouses' risk of same disease. *British Medical Journal, 326,* 396.

Williams A. G., Rayson, M. P., Jubb, M., World, M., Woods, D. R., Hayward, M., et al. (2000). The ACE gene and muscle performance. *Nature, 403,* 614.

Winnicki, M., Accurso, V., Hoffmann, M., Pawlowski, R., Dorigatti, F., Santonastaso, M., et al. (2004). Physical activity and angiotensin-converting enzyme gene polymorphism in mild hypertensives. *American Journal of Medical Genetics, 125A,* 38–44.

Wolfarth, B., Bray, M. S., Hagberg, J. M., Perusse, L., Rauramaa, R., Rivera, M. A., et al. (2005). The human gene map for performance and health-related fitness phenotypes: The 2004 update. *Medicine and Science in Sports and Exercise, 37,* 881–903.

Woods, D. R., Humphries, S. E., & Montgomery, H. E. (2000). The ACE I/D polymorphism and human physical performance. *Trends in Endocrinology and Metabolism, 11,* 416–420.

Zdravkovic, S., Wienke, A., Pedersen, N. L., Marenberg, M. E., Yashin, A. I., & De Faire, U. (2004). Genetic influences on CHD-death and the impact of known risk factors: Comparison of two frailty models. *Behavior Genetics, 34,* 585–592.

Chapter 25

Genetics of ADHD, Hyperactivity, and Attention Problems

Eske M. Derks, James J. Hudziak, and Dorret I. Boomsma

Introduction

Attention deficit hyperactivity disorder (ADHD) is characterized by symptoms of inattention, and/or hyperactivity-impulsivity. Inattention symptoms are present when an individual fails to pay attention and has difficulty in concentrating. Children or adults who are hyperactive fidget, squirm and move about constantly and can not sit still for any length of time. Impulsivity can be described as acting or speaking too quickly without first thinking of the consequences. Children with ADHD face developmental and social difficulties. As adults, they may face problems related to employment, driving a car, or relationships (Barkley, 2002). As is the case for many other psychiatric disorders, the diagnosis of ADHD is not based on a specific pathological agent, such as a microbe, a toxin, or a genetic mutation, but instead on the collection of signs and symptoms that occur together more frequently than expected by chance (Todd, Constantino, & Neuman, 2005). Genetic studies of psychiatric disorders are complicated by this lack of clear diagnostic tests (Hudziak, 2001). Heritability estimates in epidemiological genetic studies and the results of gene-finding studies may vary as a consequence of the instrument that is used to assess ADHD, and of other factors such as the specific population that is investigated. In the current chapter we will focus on behavioral measures of ADHD, and not on endophenotypes (i.e., phenotypes that form a link between the biological pathway and the behavioral outcome, for example, executive functioning). An excellent overview of endophenotypes for ADHD can be found in Castellanos and Tannock (2002). In this overview, we will first present epidemiological studies on the prevalence of ADHD (Section *Prevalence of ADHD*). Next, the results of studies reporting the heritability of ADHD and related phenotypes will be dis-

cussed (Section *Genetic Epidemiological Studies on ADHD in Children*). We concentrate on variation in these statistics as a result of the specific characteristics of the samples (e.g., age and sex of the children) and as a result of variation in the assessment methods and informants. Finally, we give an overview of studies reporting on the agreement between questionnaire data and diagnostic interviews (Section *The Relation Between Questionnaire Data and Diagnostic Interviews*).

Prevalence of ADHD

The current guidelines for the diagnosis of ADHD in the fourth edition of the Diagnostic and Statistical Manual of Mental Disorders (DSM-IV) describe three different subtypes of ADHD: (i) ADHD of the inattentive type, which requires the presence of six out of nine symptoms related to inattention; (ii) ADHD of the hyperactive/impulsive type, which requires the presence of six out of nine hyperactive/impulsive symptoms; and (iii) ADHD of the combined type, which requires the presence of six out of nine inattention symptoms and six out of nine hyperactive/impulsive symptoms (American Psychiatric Association, 1994). Additional criteria are the presence of some hyperactive/impulsive or inattentive symptoms before age 7 years, and impairment from the symptoms in two or more settings.

In research settings, the diagnosis of ADHD is not always based on these formal criteria. In some studies, the diagnosis is based on behavior checklists, whose items are summed into a total score. ADHD is then assumed to be present when a child scores above a certain diagnostic cutoff criterion. Diagnoses based on checklists usually do not incorporate additional requirements such as age of onset before age 7 years, or impairment.

Prevalence estimates of ADHD may vary as a result of instrument variance (e.g., DSM diagnoses versus checklists) and as a function of sex and age of the children. We summarize epidemiological studies that report prevalence estimates

D.I. Boomsma
Department of Biological Psychology, Vrije Universiteit, Van der Boechorststraat 1, 1081 BT, Amsterdam, The Netherlands
e-mail: DI.Boomsma@psy.vu.nl

Y.-K. Kim (ed.), *Handbook of Behavior Genetics*,
DOI 10.1007/978-0-387-76727-7_25, © Springer Science+Business Media, LLC 2009

Table 25.1 Prevalence estimates based on clinical diagnosis in community-based samples

Study	N	Method	Any ADHD Boys/girls (sex ratio)	Inattentive Boys/girls (sex ratio)	Hyperactive Boys/girls (sex ratio)	Combined Boys/girls (sex ratio)	Age
Lavigne et al. (1996)*	1150	DSM-III-R diagnosis by clinician	2.4/1.3 (1.8)	–	–	–	2–5
Breton et al. (1999)**	2400	DSM-III-R clinical interview with child 6-month prevalence Impairment not included	3.3	–	–	–	6–14
	2400	Clinical interview with teacher 6-month prevalence Impairment not included	8.9	–	–	–	6–14
	2400	Clinical interview with parent 6-month prevalence Impairment not included	5.0	–	–	–	6–14
Rohde et al. (1999)	1013	DSM-IV clinical interview with parent and child	5.5/6.1 (.90)	2.0	0.8	3.0	12–14
Cuffe et al. (2001)	490	DSM-III-R clinical interview with adolescent and parent	2.6/0.5 (4.9)	–	–	–	16–22
Graetz, Sawyer, Hazell, Arney, and Baghurst (2001)	3597	DSM-IV clinical interview with parent	11.0/4.0 (2.8)	5.1/2.3 (2.2)	2.4/1.4 (1.7)	3.1/0.7 (4.4)	6–17
Costello et al. (2003)	1420	DSM-IV clinical interview with parent 3-month prevalence	1.5/0.3 (5)	–	–	–	9–13
Ford, Goodman, and Meltzer (2003)	10438	DSM-IV clinical interview with parent, teacher, and self (diagnosis based on judgment by clinician)	3.6/0.9 (4.3)	1.0/0.3 (3.0)	0.3/0.04 (7)	2.3/0.5 (4.6)	5–15
Graetz et al. (2005)	2375	DSM-IV clinical interview with parent Impairment not included	19.0/8.8 (2.2)	8.9/4.4 (2.0)	3.4/1.8 (1.9)	6.7/2.6 (2.6)	6–13
Neuman et al. (2005)	1472	DSM-IV clinical interview with parent	7.4/3.9 (1.9)	4.5/0.6 (7.5)	0.5/1.2 (.4)	2.3/2.1 (1.1)	7–19

*This is the weighted N which is calculated based on the information provided in the paper. The weighted prevalence of ADHD is 2%, the number of subjects is 23, so the weighted total number of subjects is 23/0.02=1150.

**Breton et al. do not give the prevalences by sex, but do report the odds ratio's for male:female. These are 4.0 in self-reports, 5.1 in teacher reports, and 2.9 in parental reports.

for ADHD based on DSM criteria in Table 25.1. These prevalences can be compared with the prevalences based on checklist data which are shown in Table 25.2. In both tables, information on the assessment method and on the age and sex of the children has been included.

The prevalences based on *diagnostic interview studies* varied between 1.5 and 19.0% in boys, and between 0.3 and 8.8% in girls. In both boys and girls, the lowest prevalence was reported in a study that used a 3-month prevalence instead of the usual 1 year prevalence which may explain the discrepancy with other findings (Costello, Mustillo, Erkanli, Keeler, & Angold, 2003). The highest prevalence was reported in a study that did not include impairment criteria (Graetz, Sawyer, & Baghurst, 2005). Breton et al. (1999) also excluded impairment criteria. Excluding the results of these three studies, the prevalences are in the range of 2.4–11% in boys and 1.3–4% in girls. The prevalences based on *checklist data* range between 2.9 and 23.1% in boys and between 1.4 and 13.6% in girls. Baumgaertel, Wolraich, and Dietrich (1995), who did not show the prevalences by sex, reported a prevalence of 17.8, which is in the upper range for both sexes.

Clearly, higher prevalences are reported when diagnosis is based on questionnaire data compared to clinical diagnoses. How can this discrepancy be explained? Wolraich, Hannah, Baumgaertel, and Feurer (1998) showed that the rate of overall ADHD (i.e., irrespective of subtype) based on checklist data in a sample of 698 boys and girls drops from 16.1 to 6.8% when impairment is required for diagnosis. Similarly, in the study of Breton et al. (1999), the prevalence based on parental reports dropped from 5.0 to 4.0% when including impairment criteria. Because impairment criteria are usually included in diagnostic interview studies and not in studies using questionnaire data, it is likely that the higher prevalence in questionnaire data is the result of the exclusion of impairment criteria.

In Tables 25.1 and 25.2, higher prevalences for ADHD are reported in boys than in girls. The mean sex ratios were calculated by taking the average of the sex ratios across studies. For overall ADHD, the ratio of boys:girls ranges from 0.9:1 to 5:1 with a mean sex ratio of about 2.5:1. The sex ratio is lowest in young children (3–5 years; mean sex ratio is 1.7:1) and highest in older children (5–13 years; mean sex ratio is about 3:1). In adolescents (13–17 years), the sex ratio is about 2.5:1. The sex ratio's do not vary much by subtype. The sex ratio's are 2.5:1, 2.5:1, and 3.5:1 for the inattentive type, the hyperactive-impulsive type, and the combined type, respectively. The male:female ratio is not very high in epidemiological studies (about 3:1), but is clearly higher (about 9:1) in clinical settings (Gaub & Carlson, 1997).

In two studies, the prevalence of ADHD was estimated separately in three age groups (Cuffe, Moore, & McKeown, 2005; Nolan, Gadow, & Sprafkin, 2001). Both studies show a relatively low prevalence of ADHD in young children, an increased prevalence in older children, and again a relatively low prevalence in adolescents. A recent epidemiological study in adults showed that ADHD may be common in adulthood. Broad screening DSM-IV criteria (symptom occurred sometimes or often) identified 16.4% of a population of 966 adults as having ADHD, while 2.9% of the adults met narrow screening criteria (symptom occurred often) (Faraone & Biederman, 2005).

Genetic Epidemiological Studies on ADHD in Children

Many studies report the heritability of ADHD from a comparison of the covariance structure in monozygotic (MZ) and dizygotic (DZ) twins. In these studies, variation in the vulnerability for ADHD is decomposed into genetic and environmental components. The decomposition of variance takes place by comparing the similarity (covariance or correlation) between MZ twins, who are nearly always genetically identical, and DZ twins, who on average share half of their segregating alleles. MZ twins share all additive genetic and non-additive genetic variance. DZ twins on average share half of the additive genetic and one quarter of the non-additive genetic variance (Plomin, DeFries, McClearn, & McGuffin, 2001). The environmental decomposition of the phenotypic variance is into shared environmental variance and non-shared, or specific, environmental variance. The environmental effects shared in common by two members of a twin-pair (C) are by definition perfectly correlated in both monozygotic and dizygotic twins. The non-shared environmental effects (E) are by definition uncorrelated in twinpairs. A first estimate of additive genetic heritability based on twin data is obtained from comparing MZ and DZ correlations: $a^2 = 2(r_{MZ} - r_{DZ})$. The importance of non-additive genetic influence is obtained from $d^2 = 4(r_{DZ} - r_{MZ})$ and of shared environmental factors $c^2 = 2r_{DZ} - r_{MZ}$. Finally, the estimate of the non-shared environmental component is obtained from $e^2 = 1 - r_{MZ}$. In the classic twin design, one cannot estimate D and C simultaneously and usually the choice for an ADE or ACE model is based on the pattern of MZ and DZ twin correlations. Parameters a^2, c^2, d^2, and e^2 are then obtained with, e.g., maximum likelihood estimation using software packages as Mx (Neale, Boker, Xie, & Maes, 2003) or Mplus (Muthén & Muthén, 2000).

Papers reporting on the heritability of ADHD find large genetic influences, irrespective of the choice of instrument, informant, or sex and age of the child. Another general finding is the non-significant influence of the shared environment. We summarize these results by measurements

Table 25.2 Prevalence estimates based on behavioral checklist data in community based samples

Study	N	Method	Any ADHD Boys/girls (sex ratio)	Inattentive Boys/girls (sex ratio)	Hyperactive Boys/girls (sex ratio)	Combined Boys/girls (sex ratio)	Age
Szatmari, Offord, and Boyle (1989)	1486	DSM-III-R rating scale by parent, teacher, and self. Prevalences based on hierarchical log-linear models	10.1/3.3 (3.1)				4–11
	1236	DSM-III-R rating scale by parent, teacher, and self. Prevalences based on hierarchical log-linear models	7.3/3.4 (2.1)	–	–	–	12–16
Baumgaertel et al. (1995)	1077	DSM-IV rating scale by teacher	17.8	9.0	3.9	4.8	5–12
Wolraich, Hannah, Pinnock, Baumgaertel, and Brown (1996)	8258	DSM-IV rating scale by teacher	16.2/6.1 (2.7)	7.2/3.5 (2.1)	3.8/0.9 (4.2)	5.3/1.6 (3.3)	Kinder garten through 5th grade
Nolan et al. (2001)	413	DSM-IV rating scale by teacher	21.5/13.6 (1.6)	3.8/4.0 (.95)	7.6/5.1 (1.5)	10.1/4.6 (2.2)	3–5
	1520	DSM-IV rating scale by teacher	23.1/8.2 (2.8)	14.4/6.0 (2.4)	3.4/1.1 (3.1)	5.3/1.1 (4.8)	5–12
	1073	DSM-IV rating scale by teacher	20.1/8.8 (2.3)	14.5/8.0 (1.8)	1.6/0.0 (incalculable)	4.0/0.8 (5.0)	12–18
Larsson et al. (2004)	2063	DSM-III-R rating scale by parent	4.7			–	8–9
	2055	DSM-III-R rating scale by parent	3.1		–	–	13–14
Levy, Hay, Bennett, and McStephen (2005)	1550	DSM-IV rating scale by mother	–	1/4.3 (2.3)	3.1/1.7 (1.8)	5.8/2.0 (2.9)	4–12
Cuffe et al. (2005)	6933	Strengths and Difficulties Questionnaire	3.1/1.4 (2.2)	–	–	–	4–8
	7431	Strengths and Difficulties Questionnaire	6.3/2.1 (3.0)	–	–	–	9–13
	5636	Strengths and Difficulties Questionnaire	2.9/1.8 (1.6)	–	–	–	14–17

of: (i) ADHD symptoms (i.e., instrument includes both hyperactivity–impulsivity and attention problem symptoms (Table 25.3); (ii) hyperactivity (Table 25.4); and (iii) attention problems (Table 25.5). In the tables, we included information on the instrument that was used to assess ADHD. It should be noted that the majority of the studies used symptom counts rather than categorical diagnosis. If a research group published more than one paper based on the same sample, we included only the study with the largest sample size. The broad-band heritability of ADHD ranges between 35 and 89%. For hyperactivity, the broad-band heritability ranges between 42 and 100%. Finally, for attention problems, the broad-band heritability ranges between 39 and 81%.

Longitudinal studies show that symptom ratings of attention problems are stable between ages 7 and 12 (Rietveld, Hudziak, Bartels, Beijsterveldt van, & Boomsma, 2004). The same is true for symptom ratings of ADHD between 8 and 13 years of age (Larsson, Larsson, & Lichtenstein, 2004). These two studies report remarkably similar correlations of about 0.5 for 5-year test–retest correlations. Likewise, both studies report that the stability of symptom ratings of attention problems is mainly explained by additive genetic effects, but that the genetic effects are far from perfectly stable. Only a subset of the genes that operate at one age does so at a later age.

Although shared environmental influences on ADHD seem to be absent, a number of recent studies have shown that the genetic effects may be mediated by environmental factors (Brookes et al., 2006; Kahn et al., 2003; Seeger et al., 2004). Interaction between genetic and shared environmental influences inflate the estimate of the genetic effects. The finding of significant gene by environment interaction in these studies highlights the importance of considering the effects of both environmental and genetic factors, and their interactions in future studies on ADHD.

Sex Differences in Genetic Influences on ADHD

When examining the genetic architecture of a trait, two different kinds of sex differences can be distinguished. *Quantitative sex differences* reflect sex differences in the magnitude of the genetic influences: do genes explain the same or different amounts of variation in boys and girls? *Qualitative sex differences* reflect differences in the specific genes that are expressed in boys and girls. Below, we discuss quantitative and qualitative sex differences in ADHD.

Thirteen of the studies reported in Tables 25.3, 25.4, and 25.5 tested for quantitative sex differences in ADHD (see Tables 25.3, 25.4, and 25.5). Seven of these studies reported the absence of significant sex differences. In the remaining six studies, the presence of sex differences varied by informant and age. The effect sizes of the statistically significant sex differences were small, and the pattern of sex differences was inconsistent over studies. In some studies heritability was higher in boys, while in other studies heritability was higher in girls. The small effect sizes and the inconsistent pattern of results support the conclusion that the magnitudes of the etiological factors influencing variation in ADHD do not vary much as a function of the child's sex.

Nine studies investigated if different genes are expressed in boys and girls. Eight studies did not find qualitative sex differences. One study reported on different genes in boys and girls, but only for twins who were rated by the same teacher and not for twins rated by parents or different teachers (Saudino, Ronald, & Plomin, 2005). Future studies should reveal if this finding of qualitative sex differences in teacher ratings can be replicated.

Informant Differences

The heritabilities for ADHD rated by father and mother appear to be similar in most studies (Beijsterveldt van, Verhulst, Molenaar, & Boomsma, 2004; Derks, Hudziak, Beijsterveldt van, Dolan, & Boomsma, 2004; Eaves et al., 1997), but not in others (Goodman & Stevenson, 1989). Heritabilities for teacher ratings range between 39 and 81% and are usually lower than the heritabilities based on parental ratings in the same sample (Eaves et al., 1997; Kuntsi & Stevenson, 2001; Simonoff et al., 1998; Vierikko, Pulkkinen, Kaprio, & Rose, 2004, but see Martin, Scourfield, & McGuffin, 2002).

A complexity encountered when teacher ratings are analyzed is that both members of a twin-pair may be rated by the same teacher or by different teachers. Twin correlations are usually higher in children rated by the same teacher than in children rated by different teachers (Saudino et al., 2005; Simonoff et al., 1998; Towers et al., 2000; Vierikko et al., 2004) but not in Sherman, Iacono, and McGue (1997). Simonoff et al. (1998) developed two different models to explore this finding. One model was based on the assumption that teachers have difficulty distinguishing the two children ("twin confusion model"). The other model was based on the assumption that ratings by the same teacher are correlated because (i) raters have their own subjective views on which behaviors are appropriate and which are not or (ii) raters influence the behavior of the child because of the rater's own personality characteristics ("correlated errors model"). Although Simonoff et al. (1998) were not able to differentiate between these two models, Derks, Hudziak, Beijsterveldt van et al. (2006) reported a better fit of the correlated errors model in a large sample of Dutch twins rated by their teacher.

Table 25.3 Heritability estimates based on epidemiological studies of ADHD

Study	Age	N pairs	Sample and assessment instrument	Rater	Heritability (A+D)	Quantitative sex difference in heritability	Contrast or dominance effect
Eaves et al. (1997)	8–16	1412	Virginia Twin Study of Adolescent Behavioral Development Diagnostic interview	Mother and father	71(boys)/74 (girls) (Mother) 78 (boys)/55 (girls) (Father)	No (mother) Yes (father)	AE$_S$ model fits better than ADE model
	8–16	1412	Virginia Twin Study of Adolescent Behavioral Development Questionnaire	Mother, father, and teacher	62 (boys) /54 (girls) (Teacher) 75 (boys) /63 (girls) (Mother) 82 (boys) /76 (girls) (Father)	No	Mother and father: AE$_S$ model fits better than ADE model. Teacher: AE model
Sherman, Iacono, and McGue (1997)	11–12	288	Minnesota Twin Family Study Mother: Diagnostic interview Teacher: Rating Form	Mother and teacher	89 (Mother) 73 (Teacher)	–	–
Coolidge, Thede, and Young (2000)	Mean=9	112	Coolidge Personality and Neuropsychological Inventory	Parent	82	–	AE model
Thapar et al. (2000)	5–17	1321*	Greater Manchester Twin Register DuPaul rating scale (parent and teacher), and Rutter A scale (parent)	Parent and teacher	78 (DuPaul, parent) 84 (Rutter, parent) 77 (DuPaul, teacher)	No	(Parent, DuPaul): ADE (var MZ=varDZ) Parent, Rutter: AE$_S$ (var DZ > var MZ) Teacher: AE model
Young, Stallings, Corley, Krauter, and Hewitt (2000)	12–18	334	Colorado Drug Research Center Diagnostic interview	Self	49	–	AE model
Burt, Krueger, McGue, and Iacono (2001)	10–12	753	Minnesota Twin Family Study Diagnostic Interview	Symptom present when mother or self report it	57	No	–
Price et al. (2001)	2, 3, and 4	3118 (2) 2796(3) 2452(4)	Twins Early Development Study Conners Rating Scale	Parent	83 (boys)/79(girls) (age 2) 81 (age 3) 79 (age 4)	Yes (age 2) No (ages 3,4)	AE$_S$ model fits better than ADE model (ages 2, 3, and 4)
Larsson et al. (2004)	8–9 and 13–14	1106 (8–9) 1063 (13–14)	Young Twins Study DSM-III-R based questionnaire	Parent	35(boys)/68 (girls) (age 8–9) 74 (boys)/61(girls) (age 13–14)	Yes	–
Dick, Viken, Kaprio, Pulkkinen, and Rose (2005)	14	631	FinnTwin12 Diagnostic interview	Self	70	–	–
Kuntsi, Rijsdijk, Ronald, Asherson, and Plomin (2005)	Mean=8	3541	Twins Early Development Study Conners Rating Scale	Parent	72	No	–
Hudziak, Derks, Althoff, Rettew, and Boomsma (2005)	7	1585	Netherlands Twin Register Conners Rating Scale	Mother	78	No	ADE model (var MZ = var DZ)

*The authors report in the abstract that data from 2082 pairs are included, but 162 pairs with unknown zygosity and 599 opposite sex pairs are excluded from statistical analyses. AE$_S$ model includes additive genetic effects, non-shared environmental effects and contrast effects; ADE model includes additive genetic effects, non-additive genetic effects, and non-shared environmental effects; var, variance.

Table 25.4 Heritability estimates based on epidemiological studies of hyperactivity

Study	Age	N pairs	Sample and assessment instrument	Rater	Heritability (A+D)	Quantitative sex difference in heritability	Contrast or dominance effect
Goodman and Stevenson (1989)*	13	213	Medical Research Counsel and Inner London Education Authority (both London) Rutter Behavior Questionnaire	Mother, father and teacher	>100% (mother) 42% (father) 42% (teacher)	–	–
Thapar, Hervas, and McGuffin (1995)	8–16	376	Cardiff Births Survey Rutter A scale	Mother	88	–	AE_S model (var DZ >var MZ)
Sherman et al. (1997)	11–12	288	Minnesota Twin Family Study Teacher: rating form Mother: diagnostic interview	Teacher and mother	69 (teacher) 91 (mother)	–	–
Simonoff et al. (1998)	8–16	1044	Virginia Twin Study of Adolescent Behavioral Development Questionnaire	Mother and teacher	75 (mother same teacher) 69 (mother different teacher) 52 (teacher)	No	Mother: AE_S model fits better than ADE model Teacher: AE model
Kuntsi and Stevenson (2001)	7–11	268	Twins from primary schools in Southern England Conners Rating Scale	Parent and teacher	71 (parent) 57 (teacher)	–	Parent: AE_S model fits better than ADE model Teacher: AE model
Martin et al. (2002)	5–16	682 (Parent) 443 (Teacher)	Cardiff Birth Survey Conners Rating Scale (CRS), and Strengths and Difficulties Questionnaire (SDQ)	Parent and teacher	74 (parent, CRS) 72 (parent, SDQ) 80 (teacher, CRS) 81 (teacher, SDQ)	–	Parent (CRS and SDQ): ADE model fits better than AE_S model Teacher (CRS and SDQ): AE model
Derks et al. (2004)	3	6250 (Father) 9445 (Mother)	Netherlands Twin Register Child Behavior Checklist 2/3	Father and mother	66 (father) 70 (mother)	No	ADE model (var MZ=var DZ)
Vierikko et al. (2004)	11–12	1636	FinnTwin12 Multidimensional Peer Nomination Inventory	Parent and teacher	78 (boys, parent) 81 (girls, parent) 49 (boys, teacher) 55 (girls, teacher)	No (parent) Yes (teacher)	Parent (girls): AE_S model (but var MZ=var DZ) Parent (boys): AE model Teacher: AE model
Saudino et al. (2005)	7	3714	Twins Early Development Study Strengths and Difficulties Questionnaire	Parent and teacher	77 (parent, boys) 75 (parent, girls) 74 (same teacher, boys) 76 (same teacher, girls) 66 (different teacher, boys) 55 (different teacher, girls)	No	–

*The heritability is more than 100% while the authors calculated A, C, and E based on the MZ and DZ correlations, whereas the DZ correlations were lower than half the MZ correlations. AE_s model includes additive genetic effects, non-shared environmental effects, and contrast effects; ADE model includes additive genetic effects, non-additive genetic effects, and non-shared environmental effects; var=variance.

Table 25.5 Heritability estimates based on epidemiological studies of attention problems

Study	Age	N pairs	Sample and assessment instrument	Rater	Heritability (A+D)	Quantitative sex difference in heritability	Contrast or dominance effect
Edelbrock, Rende, Plomin, and Thompson (1995)	7–15	181	Western Reserve Twin Project Child Behavior Checklist	Parent	66	–	–
Gjone, Stevenson, and Sundet (1996)	5–9; 12–15	390 (5–9); 526 (12–15)	Norwegian Medical Birth Registry Child Behavior Checklist	Parent	73–76 (age 5–9); 75–79 (age 12–15)	–	–
Sherman et al. (1997)	11–12	288	Minnesota Twin Family Study Teacher: Rating form Mother: Diagnostic interview	Mother and teacher	69 (mother); 39 (teacher)	–	–
Hudziak, Rudiger, Neale, Heath, and Todd (2000)	8–12	492	Missouri Twin Study Child Behavior Checklist	Parent	60–76	–	No contrast effect, D was not tested
Schmitz and Mrazek (2001)	4–11	207	Colorado Department of Health Statistics Child Behavior Checklist	Mother	54	–	–
Rietveld et al. (2003)*	7; 10; 12	3373; 2485; 1305	Netherlands Twin Register Child Behavior Checklist	Mother	71 (age 7); 70 (boys, age 10); 71 (girls, age 10); 69 (boys, age 12); 73 (girls, age 12)	No (7) Yes (10); Yes (12)	An ADE and an AEs model both provide a good fit to the data
Beijsterveldt van et al. (2004)	5	7679 (mother); 6999 (father)	Netherlands Twin Register Devereux Child Behavior Rating Scale	Mother and father	79 (mother boys); 81 (mother girls); 76 (father boys and girls)	Yes (mother) No (father)	AE$_S$ model (var DZ>var MZ)

*Both the ADE and the AEs model provided a good fit at some ages. The heritability estimates are based on the ADE model because this model provided a good fit at all ages. AE$_s$ model includes additive genetic effects, non-shared environmental effects and contrast effects; ADE model includes additive genetic effects, non-additive genetic effects, and non-shared environmental effects.

Selected Samples (DeFries–Fulker Regression)

Several twin studies have based heritability estimates for ADHD on data from subjects who were selected on a high vulnerability for ADHD. In some of these studies, the subjects with a high vulnerability were selected based on a clinical diagnosis of ADHD, in others they obtained a high behavior checklist score. DeFries and Fulker (1985) developed a multiple regression model that is especially appropriate for the analysis of data in twin-pairs in which one member of a pair has been selected because of a deviant score. The rationale of this method is based on the fact that when probands are selected based on high scores on a heritable trait, MZ cotwins are expected to obtain higher scores on the trait than DZ cotwins because of a lower degree of regression to the mean. In the regression model, the cotwin's score is predicted from a proband's score (P) and the coefficient of relationship (R). The coefficient of relationship equals 0.5 and 1 in DZ and MZ twins, respectively. The basic regression model is as follows: $C = B_1 P + B_2 R + A$, where C is a cotwin's predicted score; B_1 is the partial regression of the cotwin's score on the proband's score; B_2 is the partial regression of the cotwin's score on the coefficient of relationship; and A is the regression constant. B_1 is a measure of twin resemblance that is independent of zygosity. A significant regression coefficient B_2 indicates that being a member of the affected group is heritable. The extreme group heritability (h_g^2) equals: $h_g^2 = B_2/(\text{mean score proband−mean score cotwin})$. After establishing the heritability of the condition by testing the significance of B_2, direct estimates of h^2 (the extent to which individual differences in the unselected population are heritable) and c^2 (the extent to which individual differences in the unselected population are explained by shared environmental factors) can be obtained by fitting the following extended regression model: $C = B_3 P + B_4 R + B_5 PR + A$, where PR is the product of the proband's score and the coefficient of relationship R. B_5 is a direct estimate of h^2, while B_3 is a direct estimate of c^2. DeFries and Fulker (1985) note that if affected individuals represent the lower end of a normal distribution of individual differences, the estimate of h^2 (heritability of the trait in the unselected sample) should be similar to the estimate of h_g^2 (heritability of extreme group membership).

The DeFries–Fulker regression model has been used to estimate h_g^2 and h^2 in a number of studies (Gillis, Gilger, Pennington, & DeFries, 1992; Rhee, Waldman, Hay, & Levy, 1999; Stevenson, 1992). Gillis et al. studied the heritability of ADHD in a sample of 74 twin-pairs in which at least one of the twin members was diagnosed with ADHD. They report an estimate of 0.98 (±0.26) for h_g^2. This is in agreement with an estimate of 0.81 (±0.51) for h_g^2 based on hyperactivity scores in a sample of 196 13-year-old twin-pairs (Stevenson, 1992), although this latter estimate did not reach significance.

A number of studies showed that h_g^2 does not vary as a function of the diagnostic cutoff score that is used for assessing ADHD (Levy, Hay, McStephen, Wood, & Waldman, 1997; Price, Simonoff, & Waldman, 2001; Willcutt, Pennington, & DeFries, 2000). Gjone, Stevenson, Sundet, and Eilertsen (1996) also report an absence of change in group heritability with increasing severity, but a slight tendency toward decreased heritability in the more severely affected groups. This suggests that the extreme group heritability does not vary as a function of the diagnostic cutoff score, although there may be a somewhat lower heritability of ADHD at the extreme of the distribution.

An interesting application of DeFries–Fulker regression was shown in Willcutt et al. (2000) who studied ADHD in 373 8- to 18-year-old twin-pairs. They investigated if h_g^2 of inattention varies as a function of the level of hyperactivity/impulsivity, and vice versa, if h_g^2 of hyperactivity/impulsivity varies as a function of the level of inattention. The etiology of extreme inattention was similar whether the proband exhibited low or high levels of hyperactivity/impulsivity. In contrast, the heritability of extreme hyperactivity/impulsivity was high in individuals who show high levels of inattention, while it was low and non-significant in individuals with low levels of inattention.

The Relation Between Questionnaire Data and Diagnostic Interviews

Derks et al. (2006) reviewed studies that investigated the relation between behavior checklist scores on attention problems and the clinical diagnosis for ADHD and reported on the positive and negative predictive power, sensitivity, and specificity. Many of these studies used the attention problem scale of the Child Behavior Checklist to predict ADHD. Despite its name, the scale also contains items related to hyperactivity–impulsivity. Positive predictive power (PPP) refers to the proportion of children with a high checklist score who obtain a positive DSM diagnosis (i.e., affected), and negative predictive power (NPP) refers to the proportion of children with a low checklist score who obtain a negative DSM diagnosis (i.e., unaffected). Sensitivity and specificity refer to the proportion of children with a positive DSM diagnosis, who score high on the checklist, and the proportion of children with a negative DSM diagnosis, who score low on the checklist, respectively. Table 25.6 summarizes the results of the studies that used these Diagnostic Efficiency Measures (DES). A negative feature of the DES is their dependence on the baseline prevalence of the disorder. Therefore, the baseline prevalence was also included in Table 25.6. On the basis of the results, we can conclude that the association between behavior checklist scores and clinical diagnoses for ADHD

Table 25.6 Diagnostic efficiency statistics of studies that examined the association between behavior checklist scores and ADHD

Study	Sample boys/girls	N	Cutpoint	Prevalence (%)	PPP	NPP	SE	SP
Gould, Bird, and Staghezza Jaramillo (1993)	NR	157	T > 65	23	0.36	0.96	0.46	0.95
Chen, Faraone, Biederman, and Tsuang (1994)	SR	111/108	T ≥ 65	16/8	1.00 (boys) 0.67 (girls)	0.86 (boys) 0.93 (girls)	0.17 (boys) 0.22 (girls)	1.00 (boys) 0.99 (girls)
Eiraldi, Power, Karustis, and Goldstein (2000)	R	192/50	T ≥ 65	83	0.93	0.37	0.78	0.69
Lengua, Sadowski, Friedrich, and Fisher (2001)	R	203	Based on regression	29	0.50	0.71	0.02	0.99
Sprafkin, Gadow, Salisbury, Schneider, and Loney (2002)	R	247/0	T ≥ 60	71	0.78	0.83	0.97	0.33
Hudziak, Copeland, Stanger, and Wadsworth (2004)	SR	101/82	T ≥ 65	36	0.97	0.76	0.47	0.99
Derks et al. (2006)	NR	192/216	Longitudinal	14/12	0.59/0.36	0.96/0.97	0.74/0.80	0.92/0.81

R=clinically referred sample, NR=non-referred sample, SR=siblings of referred children, PPP=Positive Predictive Power, NPP=Negative Predictive Power, SE=Sensitivity, SP=Specificity.

is strong. However, in population-based studies, a low score on the behavior checklist is highly predictive of the absence of ADHD, while a high score is less predictive of ADHD. Derks et al. (2006) further showed that a boy with a high CBCL score has a higher chance of obtaining a positive diagnosis for ADHD than a girl with a high CBCL score. In other words, questionnaire scores better predict clinical diagnosis in boys than girls.

In the field of behavioral genetics, the focus of interest is not only on the genetic and environmental influences on the variance of a trait but also on the genetic and environmental influences on the covariance of two traits. Future studies should investigate the aetiology of the covariance between behavior checklist scores and DSM-IV diagnoses of ADHD. An important issue that needs to be addressed is the overlap of the genetic factors that explain variation in different measures of ADHD.

Current Topics

In the previous sections we gave an overview of the results of epidemiological studies on ADHD. A few general findings emerged, among which a higher prevalence of ADHD in boys than girls, and a high heritability of ADHD in children irrespective of sex, age, or informant. In Section *Measurement Invariance with Respect to Sex*, *Genetic Dominance or Rater Bias/Sibling Interaction*, *Multiple Informants*, *Are the Subtypes of ADHD Genetically Heterogeneous?*, *Is Liability to ADHD Continuous or Categorical?*, and *Molecular Genetic Studies of ADHD*, we discuss current topics in the

research field of ADHD. Section *Measurement Invariance with Respect to Sex* addresses the question if measurement instruments assess ADHD equally well in boys and girls. Section *Genetic Dominance or Rater Bias/Sibling Interaction* discusses the controversy between studies claiming the presence of contrast effects versus non-additive genetic effects on individual differences in ADHD. In Section *Multiple Informants* we report on the results of genetic analyses in which the ratings from multiple informants are analyzed simultaneously. Sections *Are the Subtypes of ADHD Genetically Heterogeneous?* and *Is Liability to ADHD Continuous or Categorical?* show two applications of latent class analyses: examination of genetic heterogeneity of the ADHD subtypes and investigation of the categorical versus continuous distribution of the liability for ADHD. Finally, in Section *Molecular Genetic Studies of ADHD*, we provide a brief overview of the results obtained in gene-finding studies on ADHD.

Measurement Invariance with Respect to Sex

The prevalence of ADHD is about 2.5 times higher in boys than girls, and there are sex differences in the association between checklist scores and clinical diagnoses. Heritability seems not to vary much as a function of the child's sex, and only one out of nine studies suggests that different genes are expressed in boys and girls.

Before any sex differences in ADHD can be interpreted, we should first establish if the measurement instrument is not biased with respect to sex. Stated differently, the instrument should measure the same construct, i.e., latent variable of interest, in boys and girls (Mellenbergh, 1989; Meredith, 1993). If this is the case then we expect the observed score (i.e., the score obtained on the measurement instrument) of a person to depend on that person's score on the latent construct, but not on that person's sex. If this is not the case, a boy and a girl with identical levels of problem behavior may obtain systematically (i.e., regardless of measurement error) different scores on the instrument. This is undesirable because obviously we wish our measurements to reflect accurate and interpretable differences between cases in different groups. If the measurement instrument is not biased with respect to sex, we say that it is measurement invariant (MI) with respect to sex.

The criteria of MI are empirically testable in the common factor model (Meredith, 1993). Factor analysis may be viewed as a regression model in which observed variables (e.g., item scores) are regressed on a latent variable or common factor. In terms of this regression, the MI criteria are (1) equality of regression coefficients (i.e., factor loadings) over groups; (2) equality of item intercepts over groups (i.e., differences in item means can only be the result of differences in factor means), and (3) equality of residual variances (i.e., variance in the observed variables, not explained by the common factor) over groups. When satisfied, these restrictions ensure that any group differences in the mean and variance of the observed variables are due to group differences in the mean and variance of the latent factor.

In a sample of 800 boys and 851 girls rated by their teacher, Derks, Dolan, Hudziak, Neale, and Boomsma (2007) established measurement invariance with respect to sex for the Cognitive problems-inattention scale, the Hyperactive scale, and the ADHD-index of the Conners Teacher Rating Scale-Revised. This implies that teacher ratings on ADHD are not biased as a result of the child's sex. Although future studies should show if measurement invariance is also tenable for parental ratings on ADHD, the results in teacher ratings suggest that sex differences in the prevalence of ADHD, and on the predictive value of questionnaire scores are not the result of measurement bias.

Genetic Dominance or Rater Bias/Sibling Interaction

When reviewing the literature on ADHD, it is remarkable that many studies report very low DZ correlations for parental ratings but not for teacher ratings on ADHD. Low DZ correlations can be explained either by the presence of non-additive genetic effects (Lynch & Walsh, 1998) or by social interaction. The effects of social interaction among siblings were discussed by Eaves (1976) and others

(Boomsma, 2005; Carey, 1986). Social interactions between siblings may create an additional source of variance and can either be cooperative (imitation) or competitive (contrast). Cooperation implies that behavior in one sibling leads to similar behavior in the other siblings. In the case of competition, the behavior in one child leads to the opposite behavior in the other child.

In the classical twin design, cooperation, or positive interaction, leads to increased twin correlations for both monozygotic (MZ) and dizygotic (DZ) twins. The relative increase is larger for DZ than for MZ correlations, and the pattern of correlations thus resembles the pattern which is seen if a trait is influenced by the shared environment. Negative sibling interaction, or competition, will result in MZ correlations which are more than twice as high as DZ correlations, a pattern also seen in the presence of non-additive genetic effects.

In data obtained from parental ratings on the behavior of their children, the effects of cooperation and competition may be mimicked (Simonoff et al., 1998). When parents are asked to evaluate and report upon their children's phenotype, they may compare the behavior of siblings. Parents may either stress similarities or differences between children, resulting in an apparent cooperation or competition effect.

The presence of a contrast effect, caused by either social interaction or rater bias, is indicated by differences in MZ and DZ variances. If there is a contrast effect the variances of MZ and DZ twins are both decreased, and this effect is greatest on the MZ variance. Contrast and non-additive genetic effects can theoretically be distinguished by making use of the fact that contrast effects lead to differences in variances in MZ and DZ twins, while non-additive genetic effects do not. However, Rietveld, Posthuma, Dolan, and Boomsma (2003) showed that the statistical power to separate these effects is low in the classical twin design.

In Tables 25.3, 25.4, and 25.5, we included information on the influence of non-additive genetic effects and contrast effects on individual differences in ADHD. In the 14 studies testing for the presence of these effects, a consistent finding was the absence of non-additive genetics and contrast effects in teacher ratings. In parental ratings, nine studies reported significant contrast effects. However, one of these studies did not report larger variances in DZ than MZ twins, and the presence of non-additive genetic effects was not considered (Vierikko et al., 2004). Another study reported significant contrast effects on the Rutter scale, but significant non-additive genetic effects on the DuPaul rating scale (Thapar, Harrington, Ross, & McGuffin, 2000). The authors argue that rater contrast effects may be more pronounced for some scales, as a result of differences in the number of items or in the format of the questionnaires. The influence of non-additive genetic effects was also reported in two other studies on hyperactivity. Furthermore, Rietveld, Hudziak, Bartels, Beijsterveldt van, and Boomsma (2003) reported that a model with non-additive genetic effects and a model with contrast effects both provided a good fit to the data. Finally, two studies found no significant influences of either contrast or non-additive genetic effects. Teacher ratings do not indicate the presence of either one of these influences, suggesting that rater bias rather than genetic dominance plays a role in parental ratings. However, this is contradicted by the non-significant variance differences in MZ and DZ twins in some studies. So far, the results on the presence of non-additive genetic effects or contrast effects in parental ratings on ADHD are inconclusive. The issue may be resolved by including ratings from other family members which increases the statistical power to detect genetic dominance.

Multiple Informants

When investigating genetic and environmental influences on individual differences in problem behavior, we should acknowledge the fact that ratings of problem behavior may be influenced by the rater's personal values and by the unique settings in which the rater and child co-exist. Agreement between raters shows that some aspects of the behavior can be reliably assessed across settings and by different informants. Disagreement may reflect the fact that different raters assess unique aspects of the behavior, which are apparent in a particular set of circumstances, but not in others. For example, a child's inability to concentrate or sit still may be obvious in the classroom setting, but less evident in other settings, where sustained attention is less important (e.g., at play or at home with family members). For CBCL-AP scores, paternal and maternal ratings correlate 0.73, while parent and teacher correlations show a lower correlation of 0.44 (Achenbach & Rescorla, 2001).

Different models for twins rated by multiple informants have been developed. In this chapter, we will restrict the discussion to the psychometric model (Hewitt, Silberg, Neale, Eaves, & Erickson, 1992; Neale & Cardon, 1992).

In the psychometric model (see Fig. 25.1), the ratings of the child's behavior are allowed to be influenced by aspects of the child's behavior that are perceived both by raters (common factor) and uniquely by each rater (rater-specific factors). Unique perceptions could arise if the child behaves differentially toward his or her parents, or if the parents observe the child in different situations. The common and unique aspects are both allowed to be influenced by genetic and environmental factors.

Maternal and *paternal* ratings on overactive behavior in 3-year-olds correlate between 0.66 and 0.68 in boys, girls, and opposite-sex twins. Bivariate analyses showed that 68% of the variance is explained by a factor that is stable across informant (Derks et al., 2004). The remaining variance is explained by rater-specific factors. The heritability of the common factor is high (72%). In addition, genes explain more than half of the variation of the rater-specific factors (55% for fathers and 67% for mothers). The fact that variation in the rater-specific factors is not completely explained by environmental factors, implies that disagreement between parents is not only the result from rater-specific views (i.e., measurement error). In contrary, paternal and maternal ratings are influenced by aspects of the child's behavior that are uniquely perceived by each parent.

To determine how much of the variation in *parent* and *teacher* ratings is due to rating similar versus situation-specific components of behavior, some investigators employed bivariate model fitting analyses, which revealed that maternal and teacher ratings partly reflect a common latent phenotype (Derks et al., 2006; Martin et al., 2002; Simonoff et al., 1998). In Martin et al., 42% of the variation in the Strengths and Difficulties Questionnaire (SDQ) is explained by a factor that is common to parent and teacher ratings, the heritability of this factor is 90%. The heritability of the rater-specific factors is 22% in parent ratings and 65% in teacher ratings. The authors also obtained parental and teacher Conners Rating Scale (CRS) scores. Variation in parent and teacher's CRS scores was for 38% explained by a common factor. This factor showed a heritability of 82%. The rater-specific factors showed heritabilities of 65 and 79% for parent and teacher ratings, respectively. Simonoff et al. reported a heritability of 89% for the common factor. The genetic component of this common factor was greater than in the univariate models (52 and 69–75% in teacher and maternal ratings, respectively). Derks et al. (2006) also showed a higher heritability of the common factor (78%) than of the rater-specific factors (76 and 39% for maternal and teacher ratings, respectively). In summary, all three studies report a higher heritability of the common factor than of the rater-specific factors. This can be explained by the fact that when multiple indicators for a latent phenotype are used (e.g., over time or across raters), only a proportion of the measurement error of the individual ratings is passed on to the latent phenotype (Simonoff et al., 1998). Therefore, future gene finding studies could increase statistical power by focusing on the highly heritable common factor because it is less subject to measurement error.

Are the Subtypes of ADHD Genetically Heterogeneous?

ADHD is a disorder that may include symptoms of inattention, hyperactivity/impulsivity, or both. Because of this heterogeneity in symptom profiles, concerns have been raised over the validity of the DSM-IV subtypes (Todd, 2000). In

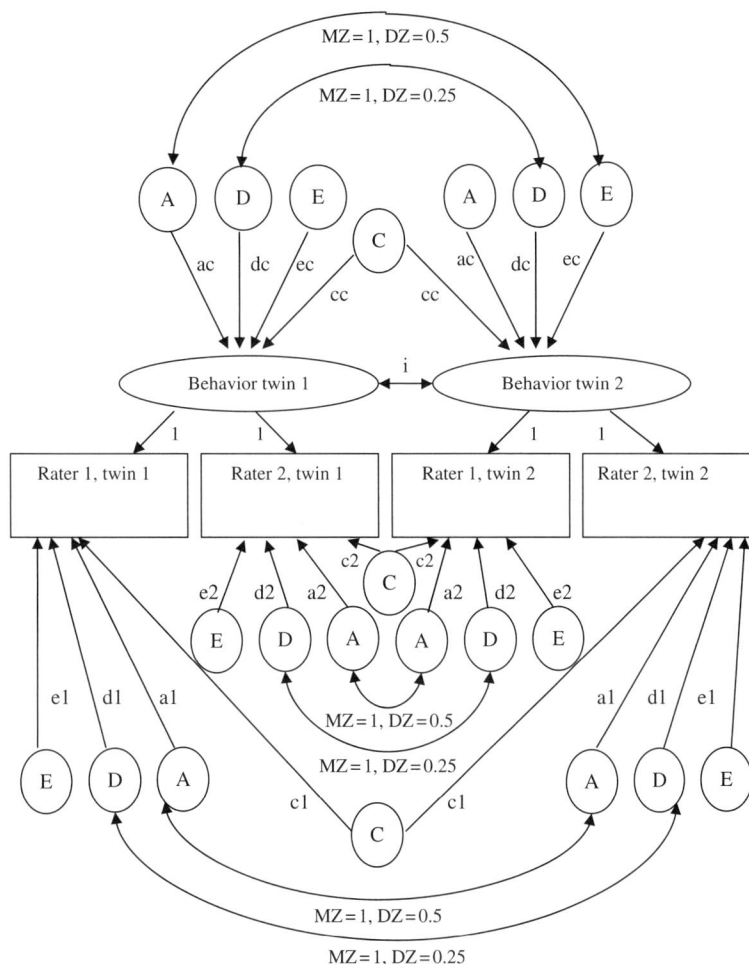

Fig. 25.1 Rater model

Note: The illustrated model is a psychometric model. Both twins are rated by two informants (rater 1 and rater 2). Variation in behavior is explained by common A, C or D, and E (shown in the upper part of the figure), and rater-specific A, C or D, and E (shown in the lower part of the figure). A=additive genetic factor; D=dominant genetic factor; C=shared environmental factor; E=non-shared environmental factor; ac=additive genetic common; dc=dominant genetic common; ec=non-shared environment common; cc=shared environment common; a1=additive genetic rater 1; d1=dominant genetic rater 1; e1=non-shared environment rater 1; c1=shared environment rater 1; a2=additive genetic rater 2; d2=dominant genetic rater 2; e2=non-shared environment rater 2; c2=shared environment rater 2, i=social interaction path

this section, we address the question if the different subtypes of ADHD are genetically heterogeneous. In other words, is the variability in symptoms profiles explained by different genetic influences on the inattentive type, the hyperactive/impulsive type, and the combined type? A number of papers have looked at the familiality and heritability of the DSM-IV subtypes of ADHD. These studies failed to identify significant familial (i.e., genetic or shared environmental) clustering of the subtypes and concluded that symptom variability is largely a function of non-familial causes (Faraone, Biederman, & Friedman, 2000; Faraone, Biederman, Mick, et al., 2000; Smalley, McCracken, & McGough, 2001).

Todd et al. (2001) used latent class analysis (LCA; McCutcheon, 1987) to examine if the clustering of symptoms can be described with more meaningful subtypes. LCA assumes the presence of a number of latent classes with a categorical rather than a continuous distribution. Estimates are provided for (i) the number of latent classes; (ii) the prevalence of each class; and (iii) the item endorsement probabilities conditional on latent class membership. Todd et al. (2001) applied LCA to parent reports on 2018 female adolescent twin pairs from the state of Missouri and investigated if the original DSM-IV subtypes and the derived latent classes represent independent genetic entities. The DSM-IV combined type and inattentive type showed a lack of familial specificity (e.g., a proband with the inattentive type has a higher chance of having a cotwin with either the inattentive or the combined type, but does not have a higher chance of having a cotwin with the hyperactive/impulsive type). The hyperactive/impulsive type did show familial specificity (e.g., a proband with the hyperactive/impulsive type has a higher chance of having a cotwin with the hyperactive/impulsive type, but does not have a higher chance of having a cotwin with the inattentive or combined type). This suggests that the hyperactive/impulsive type is independent of the other two subtypes. The LCA resulted in an eight-class solution. This eight-class solution was replicated in a sample of Australian twins (Rasmussen et al., 2002) and a similar (7-class) solution was found in an independent sample from Missouri (Volk, Neuman, Joyner, & Todd, 2005). In contrast to the DSM-IV subtypes, the eight latent classes appeared to represent pure genetic categories. The authors conclude that "these results are most compatible with the presence of independent, familial forms of ADHD that are approximated by latent-class analysis and are imperfectly operationalized by DSM-IV criteria".

Is Liability to ADHD Continuous or Categorical?

Another interesting feature of LCA is that it can help clarify whether ADHD shows a categorical or a continuous distribution. If the underlying nature of the phenotype is a continuum of problems with inattention, hyperactivity/impulsivity, or both, then symptoms endorsement profiles of the observed classes will reflect differences in severity or frequency of the reporting of symptoms only (Hudziak et al., 1998). Analyzing data on 1549 female twin-pairs, Hudziak et al. (1998) showed symptom profiles that indicated the presence of three separate continua of severity of problems: inattention, hyperactivity/impulsivity, and combined type. Thus, within the domains, the symptoms are better described as existing on a continuum rather than as discrete disease entities. Future studies should reveal if there are indeed significant cross-class heritabilities among the mild and severe latent classes, as would be expected if the distribution of ADHD is continuous.

Molecular Genetic Studies of ADHD

Molecular genetic studies address the question which genes explain the high heritability of AHDH. It is beyond the scope of this paper to provide an extensive overview of the results of molecular genetic studies. Recently, a number of review studies on the molecular genetics of ADHD have been published (Asherson, 2004; Bobb, Castellanos, Addington, & Rapoport, 2005; Faraone et al., 2005; Thapar, O'Donovan, & Owen, 2005).

Faraone et al. (2005) reviewed candidate gene studies of ADHD and computed pooled odds ratio's (ORs)

across studies for gene variants examined in three or more case–control or family-based studies. Seven gene variants showed a pooled OR that is significantly larger than 1: dopamine receptor D4 (DRD4), dopamine receptor D5 (DRD5), dopamine rransporter (DAT), dopamine β-hydroxylase (DBH), synaptosomal-associated protein 25 (SNAP-25), serotonin transporter (5-HTT), and serotonin receptor (HTR1B). These small ORs are consistent with the idea that the genetic vulnerability to ADHD is mediated by many genes of small effect.

Five groups have conducted genome-wide linkage scans in an attempt to find regions of chromosomes that are involved with ADHD. We will discuss the regions for which LOD scores higher than 2 (p<~0.002) were found. The first genome-wide scan on ADHD was published in 2002 by Fisher et al. (2002) who analyzed data from 126 affected sibling pairs in 104 families. In 2003, the sample was extended and contained 204 families with 207 affected sibling pairs (Ogdie et al., 2003). In the extended sample, LOD > 2 was found at chromosome 16p13 and 17p11. Bakker et al. (2003) performed a genome scan on 238 children from 164 Dutch affected sib pairs with ADHD. They report a LOD score of 3.04 at chromosome 7p and of 3.54 at chromosome 15q. Arcos-Burgos et al. (2004) analyzed data from 16 genetically isolated families in Columbia. They reported linkage peaks (LOD score > 2) at chromosomes 4q, 8q, and 11q in specific families. The fourth genome-wide scan was performed in a sample of 102 families encompassing a total of 229 affected children (Hebebrand et al., 2006). For clinical diagnosis of ADHD, the highest LOD score of 2.74 was reported on chromosome 5p. A LOD score > 2 was also found at chromosome 12q. For quantitative DSM-IV measures, the highest LOD scores were observed on chromosome 5p (total an inattentive scores) and chromosome 12q (inattentive scores). For hyperactivity, no LOD scores > 2 were reported. Finally, Gayan et al. (2005) reported linkage for ADHD at chromosomes 14q32 and 20q11.

The linkage peaks of these four studies do not show much overlap. An interesting resemblance between the studies is that four genome-wide scans report modest evidence (LOD > 1) for linkage at chromosome 5p. An obvious candidate gene at chromosome 5p, is the DAT gene, but in the study of Hebebrand et al., allelic variation at the DAT1 was not responsible for the linkage signal. Furthermore, the gene with the largest pooled OR as reported by Faraone et al., DRD4, is located at chromosome 11p. None of the genome-wide scans reported a linkage peak at this location.

The results of these four studies are inconsistent. This may be due to the different sampling procedures which are applied to select subjects or to differences in the definition of the phenotype. Furthermore, because each gene is expected to show a small effect and because a correction to the type-I error

(α) has to be made because of multiple testing, the statistical power in each study is low.

Some Directions for Future Research

Phenotype Definitions: Application of Item Response Theory (IRT)

In many instances, heritability of a trait is estimated for sum scores (e.g., of items or symptoms) and the distribution of sum scores often displays a large degree of skewness and kurtosis. Especially when analyzing symptom data on psychopathology, the distribution of sum scores is usually L-shaped, due to the fact that the vast majority of subjects displays a few or no symptoms (Oord van den et al., 2003). Derks et al. (2004) showed in simulated data with such an L-shaped distribution that if the true model is an ADE model, and parameters are estimated with normal theory maximum likelihood, the additive genetic component is underestimated, and the non-additive genetic component and the non-shared environmental component are overestimated. They recommend the use of a liability threshold model when analyzing sum scores with an L-shaped distribution (Lynch & Walsh, 1998).

Another concern when analyzing sum scores, that is not resolved by using a threshold model for the sum scores, is that some of the information that is contained in the original item scores is ignored when analyzing sum scores. The fact that the relationship between the latent trait and the observed item score may well be probabilistic (i.e., a person who is below the threshold on the latent trait, has a relatively low probability to score positive on the item), instead of deterministic (i.e., a person who is below the threshold, has a zero probability to score positive on the item) may also cause bias in the heritability estimate. Within the item response theory (IRT) framework, item scores are modeled as a function of one or more latent factors. Two recent papers show the advantages of IRT in the behavior genetic research field (Eaves et al., 2005; Berg van den, Glas, & Boomsma, 2007). According to Berg van den et al., advantages of using an IRT framework include (i) IRT provides a model for the relation between item scores and the latent phenotype; (ii) it supports the use of incomplete item administration and handling of missing data; (iii) it accounts for measurement error both in dependent and in independent variables, and (iv) it handles the problem of L-shaped distributed data. The application of this approach in future studies on ADHD is particularly interesting for gene finding studies while it may increase statistical power to detect the influence of genes with small effects.

Heritability of ADHD in Adults

The heritability of ADHD has been studied extensively in children. In contrast, not much is known on the magnitude of the genetic and environmental influences on individual differences in ADHD in adults. This may partly be explained by the fact that some of the earlier work suggested that ADHD is rare in adulthood. However, Faraone et al. (2005) performed a meta-analysis of follow-up studies on ADHD. They show that syndromatic persistence (i.e., the maintenance of full diagnostic status) is low ($\sim$15%), but that symptomatic persistence (i.e., the maintenance of partial diagnostic status with impairment) is much higher with a persistence rate of 40–60% (the higher estimate excludes two outlying observations). Therefore, future research should focus on the identification of the genetic and environmental influences on individual differences in ADHD in adults.

The only study that investigates the etiological influences on attention problems in adults estimates genetic and environmental influences based on self-report data from The Netherlands Twin Register at three different time waves (Berg van den, Willemsen, Geus de, & Boomsma, 2006). The mean age of the young adults is 19.6, 21.3, and 22.8 years at wave 1, 2, and 3, respectively. Irrespective of measurement wave, the heritability of attention problems is about 40%. The authors further showed that the stability in attention problems is largely explained by genetic factors. In addition, variation in ADHD at different ages in young adulthood is mainly explained by the same genes. It is unclear if the lower heritabilities in adults compared to children can be explained by age effects or by the fact that ratings of ADHD are usually based on parental or teacher reports in children and on self-reports in adults. Future studies of The Netherlands Twin Register will look into genetic and environmental influences on stability of the attention problems from early childhood (parent and teacher reports), through adolescence (parent, teacher, and self-reports) into adulthood (self-reports).

Acknowledgment This work was supported by NWO Grant numbers 575-25-006, 575-25-012, and 904-57-94 (Boomsma, P.I.), NIMH Grant number MH58799 (Hudziak, P.I.), and by the Centre for Neurogenomics and Cognition Research (CNCR) of the Vrije Universiteit, Amsterdam.

References

Achenbach, T. M., & Rescorla, L. A. (2001). *Manual for the ASEBA preschool forms & profiles*. Burlington, VT: University of Vermont, Department of Psychiatry.

American Psychiatric Association. (1994). *Task force on DSM-IV. Diagnostic and statistical manual of mental disorder: DSM-IV* (4th ed.). Washington, DC: American Psychiatric Association.

Arcos-Burgos, M., Castellanos, F. X., Pineda, D., Lopera, F., Palacio, J. D., Palacio, L. G., et al. (2004). Attention-deficit/hyperactivity disorder in a population isolate: Linkage to loci at 4q13.2, 5q33.3, 11q22, and 17p11. *American Journal of Human Genetics, 75,* 998–1014.

Asherson, P. (2004). Attention-deficit hyperactivity disorder in the post-genomic era. *European Child and Adolescent Psychiatry, 13 Suppl 1,* I50–I70.

Bakker, S. C., Meulen van der, E. M., Buitelaar, J. K., Sandkuijl, L. A., Pauls, D. L., Monsuur, A. J., et al. (2003). A whole-genome scan in 164 Dutch sib-pairs with attention-deficit hyperactivity disorder: Suggestive evidence for linkage on chromosomes 7p and 15q. *American Journal of Human Genetics, 72,* 1251–1260.

Barkley, R. A. (2002). Major life activity and health outcomes associated with attention-deficit/hyperactivity disorder. *Journal of Clinical Psychiatry, 63,* 10–15.

Baumgaertel, A., Wolraich, M. L., & Dietrich, M. (1995). Comparison of diagnostic-criteria for attention-deficit disorders in a German elementary-school sample. *Journal of the American Academy of Child and Adolescent Psychiatry, 34,* 629–638.

Beijsterveldt van, C. E. M., Verhulst, F. C., Molenaar, P. C. M., & Boomsma, D. I. (2004). The genetic basis of problem behavior in 5-year-old Dutch twin pairs. *Behavior Genetics, 34,* 229–242.

Berg van den, S. M., Glas, C. A., & Boomsma, D. I. (2007). Variance decomposition using an IRT measurement model. *Behavior Genetics, 37,* 604–616.

Berg van den, S. M., Willemsen, G., Geus de, E. J., & Boomsma, D. I. (2006) Genetic etiology of stability of attention problems in young adulthood. *American Journal of Medical Genetics Part B-Neuropsychiatric Genetics, 141,* 55–60.

Bobb, A. J., Castellanos, F. X., Addington, A. M., & Rapoport, J. L. (2005). Molecular genetic studies of ADHD: 1991 to 2004. *American Journal of Medical Genetics Part B-Neuropsychiatric Genetics, 132,* 109–125.

Boomsma, D. I. (2005). Sibling interaction effects. In B. Everitt & D. Howell (Eds.), *Encyclopedia of statistics in behavioral science* (pp. 1831–1832). Chisester: John Wiley & Sons.

Breton, J. J., Bergeron, L., Valla, J. P., Berthiaume, C., Gaudet, N., Lambert, J., et al. (1999). Quebec child mental health survey: Prevalence of DSM-III-R mental health disorders. *Journal of Child Psychology and Psychiatry and Allied Disciplines, 40,* 375–384.

Brookes, K. J., Mill, J., Guindalini, C., Curran, S., Xu, X., & Knight, J., et al. (2006). A common haplotype of the dopamine transporter gene associated with attention-deficit/hyperactivity disorder and interacting with maternal use of alcohol during pregnancy. *Archives of General Psychiatry 2006; 63,* 74–81.

Burt, S. A., Krueger, R. F., McGue, M., & Iacono, W. G. (2001). Sources of covariation among attention-deficit/hyperactivity disorder, oppositional defiant disorder, and conduct disorder: The importance of shared environment. *Journal of Abnormal Psychology, 110,* 516–525.

Carey, G. (1986). Sibling imitation and contrast effects. *Behavior Genetics, 16,* 319–340.

Castellanos, F. X., & Tannock, R. (2002). Neuroscience of attention-deficit/hyperactivity disorder: The search for endophenotypes. *Nature Reviews Neuroscience, 3,* 617–628.

Chen, W. J., Faraone, S. V., Biederman, J., & Tsuang, M. T. (1994). Diagnostic accuracy of the child behavior checklist scales for attention-deficit hyperactivity disorder: A receiver-operating characteristic analysis. *Journal of Consulting and Clinical Psychology, 62,* 1017–1025.

Coolidge, F. L., Thede, L. L., & Young, S. E. (2000). Heritability and the comorbidity of attention deficit hyperactivity disorder with behavioral disorders and executive function deficits: A preliminary investigation. *Developmental Neuropsychology, 17,* 273–287.

Costello, E. J., Mustillo, S., Erkanli, A., Keeler, G., & Angold, A. (2003). Prevalence and development of psychiatric disorders in childhood and adolescence. *Archives of General Psychiatry, 60,* 837–844.

Cuffe, S. P., McKeown, R. E., Jackson, K. L., Addy, C. L., Abramson, R., & Garrison, C. Z. (2001). Prevalence of attention-deficit/hyperactivity disorder in a community sample of older adolescents. *Journal of the American Academy of Child and Adolescent Psychiatry, 40,* 1037–1044.

Cuffe, S. P., Moore, C. G., & McKeown, R. E. (2005). Prevalence and correlates of ADHD symptoms in the National Health Interview Survey. *Journal of Attention Disorders, 9,* 392–401.

DeFries, J. C., & Fulker, D. W. (1985). Multiple regression analysis of twin data. *Behavior Genetics, 15,* 467–473.

Derks E. M., Dolan C. V., Hudziak J. J., Neale M. C., and Boomsma D. I. (2007). Teacher reports and the role of sex and genetics in attention deficit hyperactivity disorder (ADHD) and oppositional defiant disorder (ODD). *Behavior Genetics, 37,* 559–566.

Derks, E. M., Hudziak, J. J., Beijsterveldt van, C. E. M., Dolan, C. V., & Boomsma, D. I. (2004). A study of genetic and environmental influences on maternal and paternal CBCL syndrome scores in a large sample of 3-year-old Dutch twins. *Behavior Genetics, 34,* 571–583.

Derks, E. M., Hudziak, J. J., Beijsterveldt van, C. E. M., Dolan, C. V., & Boomsma, D. I. (2006). Genetic Analyses of Maternal and Teacher Ratings on Attention Problems in 7-year-old Dutch Twins. *Behavior Genetics, 36,* 833–844.

Derks, E. M., Hudziak, J. J., Dolan, C. V., Ferdinand, R. F., & Boomsma, D. I. (2006). The relations between DISC-IV DSM diagnoses of ADHD and multi-informant CBCL-AP syndrome scores. *Comprehensive Psychiatry, 47,* 116–122.

Dick, D. M., Viken, R. J., Kaprio, J., Pulkkinen, L., & Rose, R. J. (2005). Understanding the covariation among childhood externalizing symptoms: Genetic and environmental influences on conduct disorder, attention deficit hyperactivity disorder, and oppositional defiant disorder symptoms. *Journal of Abnormal Child Psychology, 33,* 219–229.

Eaves, L. J. (1976). A model for sibling effects in man. *Heredity, 36,* 205–214.

Eaves, L. J., Silberg, J. L., Meyer, J. M., Maes, H. H., Simonoff, E., Pickles, A., et al. (1997). Genetics and developmental psychopathology .2. The main effects of genes and environment on behavioral problems in the Virginia twin study of adolescent behavioral development. *Journal of Child Psychology and Psychiatry and Allied Disciplines, 38,* 965–980.

Eaves, L., Erkanli, A., Silberg, J., Angold, A., Maes, H. H., & Foley, D. (2005). Application of Bayesian inference using Gibbs sampling to item-response theory modeling of multi-symptom genetic data. *Behavior Genetics 35,* 765–780.

Edelbrock, C., Rende, R., Plomin, R., & Thompson, L. A. (1995). A twin study of competence and problem behavior in childhood and early adolescence. *Journal of Child Psychology and Psychiatry and Allied Disciplines, 36,* 775–785.

Eiraldi, R. B., Power, T. J., Karustis, J. L., & Goldstein, S. G. (2000). Assessing ADHD and comorbid disorders in children: The Child Behavior Checklist and the Devereux scales of mental disorders. *Journal of Clinical Child Psychology, 29,* 3–16.

Faraone, S. V., & Biederman, J. (2005). What is the prevalence of adult ADHD? Results of a population screen of 966 adults. *Journal of Attention Disorders, 9,* 384–391.

Faraone, S. V., Biederman, J., & Friedman, D. (2000). Validity of DSM-IV subtypes of attention-deficit/hyperactivity disorder: A family study perspective. *Journal of the American Acadamy of Child and Adolescent Psychiatry, 39,* 300–307.

Faraone, S. V., Biederman, J., Mick, E., Williamson, S., Wilens, T., Spencer, T., et al. (2000). Family study of girls with attention deficit hyperactivity disorder. *American Journal of Psychiatry, 157,* 1077–1083.

Faraone, S. V., Perlis, R. H., Doyle, A. E., Smoller, J. W., Goralnick, J. J., Holmgren, M. A., et al. (2005). Molecular genetics of attention-deficit/hyperactivity disorder. *Biological Psychiatry, 57,* 1313–1323.

Fisher, S. E., Francks, C., McCracken, J., McGough, J. J., Marlow, A. J., MacPhie I. L., et al. (2002). A genomewide scan for loci involved in attenion-deficit/hyperactivity disorder. *American Journal of Human Genetics, 70,* 1183–1196.

Ford, T., Goodman, R., & Meltzer, H. (2003). The British child and adolescent mental health survey 1999: The prevalence of DSM-IV disorders. *Journal of the American Academy of Child and Adolescent Psychiatry, 42,* 1203–1211.

Gaub, M., & Carlson, C. L. (1997). Gender differences in ADHD: A meta-analysis and critical review. *Journal of the American Academy of Child and Adolescent Psychiatry, 36,* 1036–1045.

Gayan, J., Willcutt, E. G., Fisher, S. E., Francks, C., Cardon, L. R., Olson, R. K., et al. (2005). Bivariate linkage scan for reading disability and attention-deficit/hyperactivity disorder localizes pleiotropic loci. *Journal of Child Psychology and Psychiatry and Allied Discipline, 46,* 1045–1056.

Gillis, J. J., Gilger, J. W., Pennington, B. F., & DeFries, J. C. (1992). Attention deficit disorder in reading-disabled twins: Evidence for a genetic etiology. *Journal of Abnormal Child Psychology, 20,* 303–315.

Gjone, H., Stevenson, J., & Sundet, J. M. (1996). Genetic influence on parent-reported attention-related problems in a Norwegian general population twin sample. *Journal of the American Academy of Child and Adolescent Psychiatry, 35,* 588–596.

Gjone, H., Stevenson, J., Sundet, J. M., & Eilertsen, D. E. (1996). Changes in heritability across increasing levels of behavior problems in young twins. *Behavior Genetics, 26,* 419–426.

Goodman, R., & Stevenson, J. (1989). A twin study of hyperactivity .2. The etiological role of genes, family relationships and perinatal adversity. *Journal of Child Psychology and Psychiatry and Allied Disciplines, 30,* 691–709.

Gould, M. S., Bird, H., & Staghezza Jaramillo, B. (1993). Correspondence between statistically derived behavior problem syndromes and child psychiatric diagnoses in a community sample. *Journal of Abnormal Child Psychology, 21,* 287–313.

Graetz, B. W., Sawyer, M. G., & Baghurst, P. (2005). Gender differences among children with DSM-IV ADHD in Australia. *Journal of the American Academy of Child and Adolescent Psychiatry, 44,* 159–168.

Graetz, B. W., Sawyer, M. G., Hazell, P. L., Arney, F., & Baghurst, P. (2001). Validity of DSM-IV ADHD subtypes in a nationally representative sample of Australian children and adolescents. *Journal of the American Academy of Child and Adolescent Psychiatry, 40,* 1410–1417.

Hebebrand, J., Dempfle, A., Saar, K., Thiele, H., Herpertz-Dahlmann, B., Linder, M., et al. (2006). A genome-wide scan for attention-deficit/hyperactivity disorder in 155 german sib-pairs. *Molecular Psychiatry, 11,* 196–205.

Hewitt, J. K., Silberg, J. L., Neale, M. C., Eaves L. J., & Erickson, M. (1992). The analysis of parental ratings of children's behavior using LISREL. *Behavior Genetics, 22,* 293–317.

Hudziak, J. J. (2001). The role of phenotypes (diagnoses) in genetic studies of attention-deficit/hyperactivity disorder and related child psychopathology. *Child and Adolescent Psychiatric Clinics of North America, 10,* 279–97, viii.

Hudziak, J. J., Copeland, W., Stanger, C., & Wadsworth, M. (2004). Screening for DSM-IV externalizing disorders with the Child Behavior Checklist: A receiver-operating characteristic analysis. *Journal of Child Psychology and Psychiatry and Allied Disciplines, 45,* 1299–1307.

Hudziak, J. J., Derks, E. M., Althoff, R. R., Rettew, D. C., & Boomsma, D. I. (2005). The genetic and environmental contributions to attention deficit hyperactivity disorder as measured by the Conners' Rating Scales – Revised. *American Journal of Psychiatry, 162,* 1614–1620.

Hudziak, J. J., Heath, A. C., Madden, P. F., Reich, W., Bucholz, K. K., Slutske, W., et al. (1998). Latent class and factor analysis of DSM-IV ADHD: A twin study of female adolescents. *Journal of the American Academy of Child and Adolescent Psychiatry, 37,* 848–857.

Hudziak, J. J., Rudiger, L. P., Neale, M. C., Heath, A. C., & Todd, R. D. (2000). A twin study of inattentive, aggressive and anxious/depressed behaviors. *Journal of the American Academy of Child and Adolescent Psychiatry, 39,* 469–476.

Kahn, R. S., Khoury, J., Nichols, W. C., & Lanphear, B. P. (2003). Role of dopamine transporter genotype and maternal prenatal smoking in childhood hyperactive-impulsive, inattentive, and oppositional behaviors. *Journal of Pediatrics, 143,* 104–110.

Kuntsi, J., Rijsdijk, F., Ronald, A., Asherson, P., & Plomin, R. (2005). Genetic influences on the stability of attention-deficit/hyperactivity disorder symptoms from early to middle childhood. *Biological Psychiatry, 57,* 647–654.

Kuntsi, J., & Stevenson, J. (2001). Psychological mechanisms in hyperactivity: II. The role of genetic factors. *Journal of Child Psychology and Psychiatry and Allied Disciplines, 42,* 211–219.

Larsson, J., Larsson, H., & Lichtenstein, P. (2004). Genetic and environmental contributions to stability and change of ADHD symptoms between 8 and 13 years of age: A longitudinal twin study. *Journal of the American Academy of Child and Adolescent Psychiatry, 43,* 1267–1275.

Lavigne, J. V., Gibbons, R. D., Kaufer Christoffel, K., Arend, R., Rosenbaum, D., Binns, H., et al. (1996). Prevalence rates and correlates of psychiatric disorders among preschool children. *Journal of the American Academy of Child and Adolescent Psychiatry, 35,* 204–214.

Lengua, L. J., Sadowski, C. A., Friedrich, W. N., & Fisher, J. (2001). Rationally and empirically derived dimensions of children's symptomatology: Expert ratings and confirmatory factor analyses of the CBCL. *Journal of Consulting and Clinical Psychology, 69,* 683–698.

Levy, F., Hay, D. A., Bennett, K. S., & McStephen, M. (2005). Gender differences in ADHD subtype comorbidity. *Journal of the American Academy of Child and Adolescent Psychiatry, 44,* 368–376.

Levy, F., Hay, D. A., McStephen, M., Wood, C., & Waldman, I. (1997). Attention-deficit hyperactivity disorder: A category or a continuum? Genetic analysis of a large-scale twin study. *Journal of the American Academy of Child and Adolescent Psychiatry, 36,* 737–744.

Lynch, M., & Walsh, B. (1998). *Genetics and analysis of quantitative traits.* Sunderland, MA: Sinauer Associates.

Martin, N., Scourfield, J., & McGuffin, P. (2002). Observer effects and heritability of childhood attention-deficit hyperactivity disorder symptoms. *British Journal of Psychiatry, 180,* 260–265.

McCutcheon, A. L. (1987). *Latent class analysis.* Newbury Park: Sage Publications.

Mellenbergh, G. J. (1989). Item bias and item response theory. *International Journal of Educational Research, 13,* 127–143.

Meredith, W. (1993). Measurement invariance, factor-analysis and factorial invariance. *Psychometrika, 58,* 525–543.

Muthén, L. K., & Muthén, B. O. (2000). *Mplus. Statistical analyses with latent variables. User's guide (Version 2).* Los Angeles: Muthén & Muthén.

Neale, M. C., Boker, S. M., Xie, G., & Maes, H. H. (2003). *Mx: Statistical modeling* (6th ed.). Richmond, VA: Department of Psychiatry.

Neale, M. C., & Cardon, L. R. (1992). *Methodology for genetic studies of twins and families.* Dordrecht: Kluwer Academic Publisher.

Neuman, R. J., Sitdhiraksa, N., Reich, W., Ji, T. H. C., Joyner, C. A., Sun, L. W., et al. (2005). Estimation of prevalence of DSM-IV and latent class-defined ADHD subtypes in a population-based sample of child and adolescent twins. *Twin Research and Human Genetics, 8,* 392–401.

Nolan, E. E., Gadow, K. D., & Sprafkin, J. (2001). Teacher reports of DSM-IV ADHD, ODD, and CD symptoms in schoolchildren.

Journal of the American Academy of Child and Adolescent Psychiatry, 40, 241–249.

Ogdie, M. N., MacPhie, I. L., Minassian, S. L., Yang, M., Fisher, S. E., Francks, C., et al. (2003). A genomewide scan for attention-deficit/hyperactivity disorder in an extended sample: Suggestive linkage on 17p11. *American Journal of Human Genetics, 72*, 1268–1279.

Oord van den, E. J., Pickles, A., & Waldman, I. D. (2003). Normal variation and abnormality: an empirical study of the liability distributions underlying depression and delinquency. *Journal of Child Psychology & Psychiatry, 44*, 180–192.

Plomin, R., DeFries, J. C., McClearn, G. E., & McGuffin, P. (2001). *Behavioral genetics.* New York: Worths Publishers.

Price, T. S., Simonoff, E., & Waldman, I. (2001). Hyperactivity in preschool children is highly heritable. *Journal of the American Academy of Child and Adolescent Psychiatry, 40*, 1362–1364.

Rasmussen, E. R., Neuman, R. J., Heath, A. C., Levy, F., Hay, D. A., & Todd, R. D. (2002). Replication of the latent class structure of attention-deficit hyperactivity disorder (ADHD) subtypes in a sample of Australian twins. *Journal of Child Psychology and Psychiatry and Allied Disciplines, 43*, 1018–1028.

Rhee, S. H., Waldman, I. D., Hay, D. A., & Levy, F. (1999). Sex differences in genetic and environmental influences on DSM-III-R attention-deficit/hyperactivity disorder. *Journal of Abnormal Psychology, 108*, 24–41.

Rietveld, M. J. H., Hudziak, J. J., Bartels, M., Beijsterveldt van, C. E. M., & Boomsma, D. I. (2003). Heritability of attention problems in children: I. Cross-sectional results from a study of twins, age 3–12 years. *American Journal of Medical Genetics Part B-Neuropsychiatric Genetics, 117B*, 102–113.

Rietveld, M. J. H., Hudziak, J. J., Bartels, M., Beijsterveldt van, C. E. M., & Boomsma, D. I. (2004). Heritability of attention problems in children: Longitudinal results from a study of twins, age 3 to 12. *Journal of Child Psychology and Psychiatry and Allied Disciplines, 45*, 577–588.

Rietveld, M. J., Posthuma, D., Dolan, C. V., & Boomsma, D. L. (2003). ADHD: Sibling interaction or dominance: An evaluation of statistical power. *Behavior Genetics, 33*, 247–255.

Rohde, L. A., Biederman, J., Busnello, E. A., Zimmermann, H., Schmitz, M., Martins, S., et al. (1999). ADHD in a school sample of Brazilian adolescents: A study of prevalence, comorbid conditions, and impairments. *Journal of the American Academy of Child and Adolescent Psychiatry, 38*, 716–722.

Saudino, K. J., Ronald, A., & Plomin, R. (2005). The etiology of Behavior problems in 7-year-old twins: Substantial genetic influence and negligible shared environmental influence for parent ratings and ratings by same and different teachers. *Journal of Abnormal Child Psychology, 33*, 113–130.

Seeger, G., Schloss, P., Schmidt, M. H., Ruter-Jungfleisch, A., & Henn, F. A. (2004). Gene-environment interaction in hyperkinetic conduct disorder (HD + CD) as indicated by season of birth variations in dopamine receptor (DRD4) gene polymorphism. *Neuroscience Letters, 366*, 282–286.

Schmitz, S., & Mrazek, D. A. (2001). Genetic and environmental influences on the associations between attention problems and other problem behaviors. *Twin Research, 4*, 453–458.

Sherman, D. K., Iacono, W. G., & McGue, M. K. (1997). Attention-deficit hyperactivity disorder dimensions: A twin study of inattention and impulsivity-hyperactivity. *Journal of the American Academy of Child and Adolescent Psychiatry, 36*, 745–753.

Sherman, D. K., McGue, M. K., & Iacono, W. G. (1997). Twin concordance for attention deficit hyperactivity disorder: A comparison of teachers' and mothers' reports. *American Journal of Psychiatry, 154*, 532–535.

Simonoff, E., Pickles, A., Hervas, A., Silberg, J. L., Rutter, M., & Eaves, L. (1998). Genetic influences on childhood hyperactivity:

Contrast effects imply parental rating bias, not sibling interaction. *Psychological Medicine, 28*, 825–837.

Smalley, S. L., McCracken, J., & McGough, J. (2001). Refining the ADHD phenotype using affected sibling pair families. *American Journal of Medical Genetics, 105*, 31–33.

Sprafkin, J., Gadow, K. D., Salisbury, H., Schneider, J., & Loney, J. (2002). Further evidence of reliability and validity of the Child Symptom Inventory-4: Parent checklist in clinically referred boys. *Journal of Clinical Child and Adolescent Psychology, 31*, 513–524.

Stevenson, J. (1992). Evidence for a genetic etiology in hyperactivity in children. *Behavior Genetics, 22*, 337–344.

Szatmari, P., Offord, D. R., & Boyle, M. H. (1989). Ontario child health study – Prevalence of attention deficit disorder with hyperactivity. *Journal of Child Psychology and Psychiatry and Allied Disciplines, 30*, 219–230.

Thapar, A., Harrington, R., Ross, K., & McGuffin, P. (2000). Does the definition of ADHD affect heritability? *Journal of the American Academy of Child and Adolescent Psychiatry, 39*, 1528–1536.

Thapar, A., Hervas, A., & McGuffin, P. (1995). Childhood hyperactivity scores are highly heritable and show sibling competition effects – Twin study evidence. *Behavior Genetics, 25*, 537–544.

Thapar, A., O'Donovan, M., & Owen, M. J. (2005). The genetics of attention deficit hyperactivity disorder. *Human Molecular Genetics, 14 Spec No. 2*, R275–R282.

Todd, R. D. (2000). Genetics of attention deficit/hyperactivity disorder: Are we ready for molecular genetic studies? *American Journal of Medical Genetics, 96*, 241–243.

Todd, R. D., Constantino, J. N., & Neuman, R. J. (2005). Redefining early-onset disorders for genetic studies: Attention-deficit/hyperactiviy disorder and autism. In C. F. Zorumski & E. H. Rubin (Eds.), *Psychopathology in the genome and neuroscience era* (1st ed., pp. 33–49). Arlington, VA: American Psychiatric Publishing, Inc.

Todd, R. D., Rasmussen, E. R., Neuman, R. J., Reich, W., Hudziak, J. J., Bucholz, K. K., et al. (2001). Familiality and heritability of subtypes of attention deficit hyperactivity disorder in a population sample of adolescent female twins. *American Journal of Psychiatry, 158*, 1891–1898.

Towers, H., Spotts, E., Neiderhiser, J. M., Hetherington, E. M., Plomin, R., & Reiss, D. (2000). Genetic and environmental influences on teacher ratings of the Child Behavior Checklist. *International Journal of Behavioral Development, 24*, 373–381.

Vierikko, E., Pulkkinen, L., Kaprio, J., & Rose, R. J. (2004). Genetic and environmental influences on the relationship between aggression and hyperactivity-impulsivity as reated by teachers and parents. *Twin Research, 7*, 261–274.

Volk, H. E., Neuman, R. J., Joyner, C., & Todd, R. D. (2005). Replication and validation of population derived ADHD phenotypes. *American Journal of Medical Genetics Part B-Neuropsychiatric Genetics, 138B*, 62–63.

Willcutt, E. G., Pennington, B. F., & DeFries, J. C. (2000). Etiology of inattention and hyperactivity/impulsivity in a community sample of twins with learning difficulties. *Journal of Abnormal Child Psychology, 28*, 149–159.

Wolraich, M. L., Hannah, J. N., Baumgaertel, A., & Feurer, I. D. (1998). Examination of DSM-IV criteria for attention deficit/hyperactivity disorder in a county-wide sample. *Journal of Developmental and Behavioral Pediatrics, 19*, 162–168.

Wolraich, M. L., Hannah, J. N., Pinnock, T. Y., Baumgaertel, A., & Brown, J. (1996). Comparison of diagnostic criteria for attention-deficit hyperactivity disorder in a county-wide sample. *Journal of the American Academy of Child and Adolescent Psychiatry, 35*, 319–324.

Young, S. E., Stallings, M. C., Corley, R. P., Krauter, K. S., & Hewitt, J. K. (2000). Genetic and environmental influences on behavioral disinhibition. *American Journal of Medical Genetics, 96*, 684–695.

Chapter 26

Depression and Anxiety in Childhood and Adolescence: Developmental Pathways, Genes and Environment

Frances Rice and Anita Thapar

Introduction

Depression and anxiety are amongst the most common mental health problems experienced in the general population (Kessler et al., 1994). It is now well established that these traits and disorders often have their origins in childhood. Nevertheless, relatively little is known about which etiological factors are important, how risk factors work together and what are the mechanisms that lead to depression and anxiety. In the last 15 years, there has been a large increase in epidemiological and behavior genetic research on childhood depression and anxiety. In this chapter we review the evidence from this research and highlight new directions for future research.

Family studies have used a categorical approach to define depression and anxiety and have mainly included clinically referred samples. In contrast, twin and adoption studies of childhood anxiety and depression have generally adopted a dimensional approach to define psychopathology in community samples. There is good evidence to support a dimensional approach. For instance, symptoms of depression and anxiety that fall below the diagnostic threshold are associated with functional impairment and future episodes (Angold, Costello, Farmer, Burns, & Erkanli, 1999; Pickles, Rowe, Simonoff, Foley, Rutter, & Silberg, 2001; Van den Oord, Pickles, & Waldman, 2003). Nevertheless, it also needs to be borne in mind that high symptom scores cannot be equated with clinical disorder. A problem for both research and clinical practice in defining childhood anxiety and depression is that it is not clear whether it is preferable to rely on parent or child ratings of symptoms. In particular, it is unclear how to interpret findings for different raters. Behavior genetic studies have also had to deal with this issue.

F. Rice (✉)
Child & Adolescent Psychiatry Section, Department of Psychological Medicine, School of Medicine, Cardiff University, Heath Park, Cardiff CF14 4XN, UK
e-mail: ricef@cardiff.ac.uk

Epidemiology of Anxiety

Estimates of the prevalence of any childhood anxiety disorder are in the order of 3–12% (Costello et al., 1996; Simonoff et al., 1997) and rise to as high as 40% or over if impairment is not required for a diagnosis (Simonoff et al., 1997). In general, epidemiological studies show that rates of any anxiety disorder are higher in children than adolescents (Costello, Mustillo, Erkanli, Keeler, & Angold, 2003) and are higher in females than males (Lewinsohn, Hops, Roberts, Seeley, & Andrews, 1993; Lewinsohn, Zinbarg, Seeley, Lewinsohn, & Sack, 1997).

Outcomes of Anxiety During Childhood and Adolescence

High levels of anxiety symptoms during childhood (Goodwin, Fergusson, & Horwood, 2004) and adolescence (Pine, Cohen, & Brook, 2001) have been found to be linked with increased rates of adult anxiety disorders (Goodwin et al., 2004) and with increased risk of depressive disorder in adulthood (Pine et al., 2001). However, it has been suggested that children with early onset anxiety disorders show relatively well-adjusted outcomes in young adult life unless they also had a comorbid depressive disorder (Last, Hansen, & Franco, 1997; Last, Perrin, Hersen, & Kazdin, 1996).

Epidemiology of Depression

The prevalence of depressive disorder in children ranges from 0.4 to 2.7% (Birmaher et al., 1996; Harrington, 1994). Adolescence is associated with a sharp increase in prevalence with 12-month prevalence estimates ranging from 2 to 8.3% (Birmaher et al., 1996; Harrington, 1994; Meltzer, Gatward, & Ford, 2000). In childhood, studies report an equal proportion of boys and girls affected or a slight excess of boys. Conversely, in adolescence, the rate of depression is higher in females, with a sex ratio in the order of 2:1, which mirrors the pattern seen in adult life (Harrington, 1994). As with clinical disorder, adolescence is associated with a marked increase in symptoms of depression (Angold, Erkanli, Silberg, Eaves, & Costello, 2002).

Outcomes of Depression During Childhood and Adolescence

Depression in children and adolescents has immediate and long-term detrimental effects. Both depressive disorder and sub-clinical depressive symptoms are associated with psychosocial impairment, use of medical services (Angold, Costello, Farmer, et al., 1999; Lewinsohn, Solomon, Seeley, & Zeiss, 2002) and deliberate self-harm (Harrington, 1994; Johnson, Weissman, & Klerman, 1992). Depression in childhood and adolescence also shows strong continuities with clinical depression in adulthood (Fombonne, Wostear, Cooper, Harrington, & Rutter, 2001; Harrington, Fudge, Rutter, Pickles, & Hill, 1990; Lewinsohn, Hoberman, & Rosenbaum, 1988; Lewinsohn et al., 2002). Similarly, high levels of depressive symptoms are strongly predictive of future clinical depressive episodes in adulthood (Pine, Cohen, Cohen, & Brook, 1999).

Comorbidity of Anxiety and Depression

Depression and anxiety co-occur more commonly than would be expected by chance in children and adolescents with rates of anxiety disorders in adolescents with depression ranging from 20 to 75% (Angold, Costello, & Erkanli, 1999; Kovacs & Devlin, 1998). This co-occurrence has been identified both in clinical studies of children and adolescents and general population samples that have examined sub-clinical levels of depression and anxiety symptoms (Brady & Kendall, 1992; Kovacs & Devlin, 1998). More specifically, anxiety symptoms or disorders most often precede depressive symptoms or disorders (Avenevoli, Stolar, Li, Dierker, & Merikangas, 2001; Kovacs, Gatsonis, Paulauskas, & Richards, 1989). For example, in a referred sample of depressed children, Kovacs and colleagues (1989) found that in those children with a comorbid anxiety disorder, the anxiety disorder preceded the depressive disorder in two thirds of cases. Similar evidence that anxiety disorders tend to precede depression has been reported in longitudinal epidemiological studies (Goodwin et al., 2004; Merikangas et al., 2003; Orvaschel, Lewinsohn,

& Seeley, 1995; Woodward & Fergusson, 2001). Panic disorder is, however, a notable exception and is more likely to follow than precede depression, and there is some evidence that obsessive compulsive disorder does not commonly co-occur with depression (Lewinsohn et al., 1997). Despite these exceptions, the observations that anxiety commonly precedes depression have led to suggestions that anxiety may be a developmental precursor of depression (Merikangas, 1993). Results from some family studies suggest that this may particularly be the case in young people who are at increased familial risk of depression indexed by a history of parental depression (Rende, Warner, Wickramarante, & Weissman, 1999; Warner, Weissman, Mufson, & Wickramaratne, 1999). The familial and genetic contribution to comorbidity is an important area of research that will be dealt with in this chapter.

In summary, anxiety and depression during childhood and adolescence are associated with negative outcomes including disorder during adult life. However, there is a larger evidence base for the long-term deleterious effects of childhood and adolescent depression. There is substantial comorbidity between the two syndromes with most of the evidence suggesting that anxiety precedes depression.

Family Studies

Anxiety

An important first step of any quantitative genetic study is to consider evidence for the familiality of the trait under study. Family studies of childhood and adolescent disorder can be differentiated according to their strategy: (1) 'Bottom-up' studies of the relatives of child probands with a psychiatric diagnosis or (2) 'top-down' studies of the children of parents with a psychiatric diagnosis. There have been a number of family studies of anxiety disorders in childhood and adolescence although a pooled or meta-analysis of these studies has not been published and there are too few data to establish to what extent specific types of anxiety disorder (e.g., over-anxious disorder versus separation anxiety) are familially distinct. Overall, the relatives of children with anxiety disorders show significantly higher rates of these disorders than do relatives of controls (Last, Hersen, Kazdin, Orvaschel, & Perrin, 1991) and children of parents with anxiety disorders show elevated rates of anxiety disorder when compared to children whose parents have no disorder (Beidel & Turner, 1997). However, these studies have not included control groups affected by other types of psychopathology. A number of other top-down studies have examined the influence of parental anxiety disorders comorbid with depression (Biederman, Rosenbaum, Bolduc, Faraone, & Hirshfeld, 1991; Warner, Mufson, & Weissman, 1995; Weissman, Leckman, Merikangas, Gammon, & Prusoff, 1984) and these are discussed below.

Depression

A large body of research has examined the familiality of depressive disorder in children and adolescents (e.g., Rice, Harold, & Thapar, 2002a). On the whole, both 'bottom-up' and 'top-down' family studies have reported an elevated risk of major depressive disorder (MDD) in these groups in comparison to controls. These studies have calculated odds ratios, which in this case is the ratio of the odds of being exposed (i.e., having a family history of MDD) if affected with MDD to those of being exposed if unaffected with MDD (Silman, 1995). An odds ratio of 1 would denote no difference in the odds of MDD between the exposed and unexposed groups, while an odds ratio of 2, for example, would indicate a doubling of the odds of MDD in the exposed group. Studies of both types have also been consistent in showing that the risk is not depression-specific and that there is familial clustering of other types of psychopathology, for example antisocial behavior (Harrington, Fudge, Rutter, Pickles, & Hill, 1991). Family studies of depressive symptoms have also used the top-down approach to examine depression symptom scores and internalizing symptoms according to the Child Behavior Checklist (CBCL) (Achenbach, 1991) in the children of depressed parents. 'Top-down' studies have most often examined the offspring of depressed mothers rather than fathers. It should also be noted that the age range of the offspring included in 'top-down' studies is wide.

Bottom-Up Studies

Studies of the relatives of child probands with major depressive disorder (MDD) (e.g., Goodyer, Cooper, Vize, & Ashby, 1993; Harrington et al., 1997; Klein, Shankman, Lewinsohn, Rohde, & Seeley, 2004; Kutcher & Marton 1991; Puig-Antich et al., 1989; Weissman et al., 1999; Wickramaratne, Greenwald, & Weissman, 2000) report an increased prevalence of MDD of around twofold in first-degree relatives (FDR) compared to non-affective psychiatric controls and to the relatives of never psychiatrically ill controls. Two studies, however, reported no significant difference in the familial rate of MDD in young MDD probands compared to young non-affective psychiatric controls (Mitchell, McCauley, Burke, Calderon, & Schloredt, 1989; Puig-Antich et al., 1989). Rice and colleagues (2002a) pooled estimates of familial risk from published family studies meeting certain inclusion criteria and obtained an

odds ratio of 2.12 for MDD in the first-degree relatives of child probands with MDD in comparison to the family members of children with no psychiatric diagnosis. An odds ratio of 1.89 was obtained when the comparison group was a non-affective psychiatric control group.

Top-Down Studies

Studies of the offspring of parents with MDD (e.g., Biederman et al., 1991; Keller et al., 1986; Mufson, Weissman, & Warner, 1992; Orvaschel, Walsh-Allis, & Ye, 1989; Radke-Yarrow, Nottelmann, Martinez, Fox, & Belmont, 1992; Weissman, Leckman et al., 1984; Wickramaratne, Warner, & Weissman, 2000) have reported higher levels of MDD in the offspring of depressed parents compared to offspring of parents with no psychopathology (relative risk/odds ratio range = 0.9–8.8, median = 2.75). Compared to psychiatric (Biederman et al., 1991; Mufson, 1992 or medical control groups (Hammen, Burge, Burney, & Adrian, 1990) an increased risk in disorder of around twofold to offspring of depressed parents has been reported although few 'top-down' studies have included such control groups. An odds ratio of 3.98 was obtained from a pooled analysis of published studies when the comparison group was children whose parents had never had a psychiatric diagnosis and 1.70 when the comparison group was a psychiatric control group (Rice et al., 2002a). Interestingly, the odds ratio of 1.70 was not significant which suggests that the children of depressed parents are no more likely to experience depression than children of parents with a different psychiatric disorder. However, only three 'top-down' studies that met inclusion criteria for the pooled analysis had included a psychiatric or medically ill comparison group, and therefore the power to detect significant familial effects was low.

Top-Down Studies of Depressive Symptoms in Offspring

There have been two meta-analyses examining the strength of the link between depression in parents and depressive symptoms in offspring (Connell & Goodman, 2002; Kane & Garber, 2004). The first examined the link with internalizing and externalizing problems in children whose mothers or fathers were depressed. The second examined the relationship between depression in fathers and children's psychopathology. Both of these studies included parents with clinical depression and dimensional measures of depressive symptoms from parents in community samples and examined the magnitude of the relationship with children's internalizing symptoms. Both studies found evidence for a significant positive correlation between parental depression and internalizing problems in children (mean effect size range r = 0.14–0.24). Connell & Goodman (2002) found evidence that the link was stronger for mothers and children's internalizing symptoms (regardless of child gender). They also reported an interesting effect of the child's age, which differed for maternal and paternal depressions. Maternal depression had a greater effect on younger children's symptoms while paternal depression had a greater effect on adolescent children's symptoms. This effect was observed for both children's internalizing and externalizing symptoms.

There has been a large amount of research examining the links between parental depression specifically in the post-natal period and childhood emotional symptoms (Marks, Hipwell, & Kumar, 2002). These studies have consistently shown that parental depression during this time, whether maternal or paternal (Ramchandani, Stein, Evans, & O'Connor, 2005), is associated with later childhood problems. Finally, one community-based study that used a dimensional approach to assessing parental depression found that the association between depression in parents and their adolescent children was greatest when the parents were vulnerable to depression but also had no formal qualifications (Eley, Liang, et al., 2004). This result suggests that the risk of depression to adolescents is increased when parental depression is accompanied by psychosocial adversity and also underlines the fact that many risk factors for psychopathology act in a cumulative manner (Rutter, Pickles, Murray, & Eaves, 2001).

Issues Arising from Family Studies

Age of Onset

'Top-down' studies of the Yale cohort (Weissman, Wickramaratne, et al., 1984; Wickramaratne & Weissman, 1998) have shown that familial aggregation of depression in offspring is greatest when the onset of the parents' depression occurred at a relatively young age (Weissman, Wickramaratne, et al., 1984 – under 20 years; Wickramaratne & Weissman – under 30 years). These findings, coupled with observations from other conditions that an early age of onset is more often the result of a more strongly genetically influenced condition (e.g., diabetes, Alzheimer's disease), have been interpreted as implying that childhood onset depressive disorder may have a stronger genetic component than adult onset depression. However, the evidence is not consistent and greater familial loading could arise from both common environmental and genetic effects. For example, exposure to negative cognitions or parental rejection could be conditions that might increase familial aggregation of depressive disorder (e.g., Whitbeck et al., 1992). Other studies have compared depression in children with adult and adolescent groups. Again findings are conflicting. Neuman,

Geller, Rice, & Todd (1997) found evidence for greater familial loading in childhood onset cases compared to adult onset cases. On the other hand, two studies showed no difference in the familiality of childhood and adolescent depression: Harrington and colleagues (1997) found no evidence that familiality differed between childhood (14.9%) and adolescent onset (16.8%) cases, although higher levels of criminality and family discord were found in the relatives of the childhood onset cases. Similarly, Wickramaratne, Greenwald, et al. (2000) found no evidence that the rates of MDD in first-degree relatives of childhood onset cases (40.6%) and adolescent onset cases (46.9%) differed (odds ratio 0.7, 95% CI 0.4–1.2). Looking more closely, Wickramaratne and colleagues (1998, 2000) found some evidence that recurrent childhood onset MDD may be more familial than either adolescent or adult onset MDD. Finally, odds ratios obtained from a pooled analysis of family studies of depression in children and adolescents (odds ratios = 1.70, 1.83, 2.12, 3.98) depending on the method and control group (Rice et al., 2002a) are similar (and not substantially greater) to that obtained from a meta-analysis of family studies of depression in adults (odds ratio = 2.84) (Sullivan, Neale, & Kendler, 2000). Thus, there is little consistent evidence to substantiate the claim that childhood onset MDD is more familial than adolescent or adult onset MDD. However, onset in early adulthood rather than later in adulthood might be more familial (Wickramaratne, Warner, et al., 2002) although the evidence is not conclusive. Nonetheless, increased familiality does not necessarily imply greater genetic loading – twin and adoption studies are needed to disentangle similarity due to genetic and environmental factors. However, it may be important to identify individuals with an onset in early adulthood for clinical reasons: There is some evidence that individuals with an onset of MDD in late adolescence/young adulthood have a worse prognosis and shorter time to relapse than those with an onset of depression later in adulthood (Gollan, Raffety, Gortner, & Dobson, 2005; Parker, Roy, Hadzi-Pavlovic, Mitchell, & Wilhelm, 2003).

Specificity of Transmission

Including psychiatric as well as healthy comparison groups allows assessment of whether there is specificity in transmission of disorder. That is, does the disorder 'breed-true' or does having a depressed parent increase offspring risk of a wide range of psychopathology. The former explanation would be more consistent with a genetic transmission hypothesis and the latter explanation would be more consistent with an environmental transmission hypothesis although the two are not mutually exclusive. Processes involved in the generational transmission of depression might include parenting style, family conflict, a negative cognitive style and endocrine factors such as dysfunction of the hypothalamic pituitary adrenal (HPA) axis (e.g., Goodman & Gotlib, 1999). Rice and colleagues (2002a) found some evidence for specificity of transmission as the odds ratio obtained from bottom-up studies was significant with a psychiatric comparison group. However, the equivalent odds ratio from 'top-down' studies was not significant although few 'top-down' studies included such comparison groups. More recent evidence (Kim-Cohen, Moffitt, Taylor, Pawlby, & Caspi, 2005) suggests that the increased rate of antisocial behavior in offspring of depressed parents is explained by both genetic and social mechanisms.

As there are few family studies of anxiety that have included psychiatric control groups, the evidence of specific familial aggregation for anxiety is less clear. The likelihood of psychopathology among relatives of anxious children extends to types of psychopathology other than anxiety including depression and alcohol abuse (Bell-Dolan, Last, & Strauss, 1990).

The Familial Relationship Between Anxiety and Depression

Some 'top-down' family studies of depression have found evidence for familial aggregation of anxiety rather than depression in children who have a depressed parent (Rende et al., 1999; Warner et al., 1999). It is not clear whether this is due to comorbidity of the parent's illness, in particular panic disorder (Biederman et al., 1991; Warner et al., 1995; Weissman, Wickramaratne, et al., 1984), or whether anxiety may be a genetic precursor to depression in children. Thus, it may be that in children, familial risk for depression manifests itself as anxiety. Alternatively, anxiety may tend to precede depression in children regardless of familial risk for depression (Wickramaratne, Warner, et al., 2002). That is, given the developmental differences in the onset of anxiety and depressive disorders – the former are more common in children and the latter in adolescents – it may simply be that children tend to exhibit symptoms of anxiety rather than depression regardless of genetic risk (Kovacs & Devlin, 1998).

Familial Transmission from Mothers and Fathers to Offspring

The vast majority of top-down studies have examined the association between depressive illness in the mother and rates of depression in offspring without examining the influence of the father. Nevertheless, there is preliminary evidence that associations may be stronger between mothers and children than fathers and children (Connell & Goodman, 2002; Klein et al., 2004). This is an intriguing difference, which might come about for several reasons. It has been suggested that the difference might arise because of the tendency for mothers to be the parent most responsible for child-rearing activities (Pleck, 1997) and evidence that depression in mothers may influence their ability and sensitivity in these activities, which in turn impacts upon children's socio-emotional development (Connell & Goodman, 2002). However, an alternative explanation for the greater similarity between mothers and children than fathers and children is that intrauterine environmental factors may be important in the familial transmission of depression. Consistent with this possibility, a large number of animal studies have found evidence for the role of maternal exposure to stress during pregnancy and anxiety/depressive-like traits in the resulting offspring (Seckl & Meaney, 2004). Similarly, children whose mothers have high levels of depressive symptoms during pregnancy exhibit high levels of emotional and behavioral problems even when potential confounders such as post-natal depressive symptoms are included (O'Connor, Heron, Golding, & Glover, 2003).

Twin Studies

Identical (monozygotic; MZ) twins share all of their genes in common, while fraternal (dizygotic; DZ) twins share half of their genes in common on average. These differences in the level of genetic relatedness between MZ and DZ twins mean that studying twins allows inferences to be made regarding the genetic and environmental origins of a trait or behavior. As with all methods, twin studies have a number of strengths and limitations and although a discussion of these is beyond the scope of this chapter, they are reviewed elsewhere (e.g., Plomin, DeFries, McClearn, & McGuffin, 2001).

In contrast to the family studies which have been based on clinically referred populations, nearly all the twin studies of anxiety and depression have been based on non-clinical samples and have used questionnaire measures of anxiety and depression. A few notable exceptions have used symptom scores derived from psychiatric interviews (Eaves et al., 1997) and one twin study has examined diagnoses of depressive disorder in adolescent girls (Glowinski, Madden, Bucholz, Lynskey, & Heath, 2003).

In general, there has been less behavioral genetic research on anxiety in young people than on depression. In fact, the literature on the role of genetic influences on anxiety in children is relatively sparse. However, quite a number of twin and adoption studies have examined symptoms of internalizing problems according to the Child Behavior Checklist (Achenbach, 1991), which includes components of anxiety, depression and withdrawal.

Estimates of the magnitude of genetic influences on variation in depression and anxiety are shown in Tables 26.1, 26.2, and 26.3. For the most part, the estimates illustrate that symptoms of both depression and anxiety are influenced by genetic factors. However, the figures also highlight some of the inconsistencies between different studies, in particular, differences that depend on the rater (parent or child) of the symptoms. Table 26.1 shows estimates of genetic effects on depressive symptoms in children and adolescents. Table 26.2 illustrates genetic parameter estimates for twin studies of anxiety, and Table 26.3 gives estimates from twin studies of internalizing symptom scores. These estimates have been taken from univariate twin analyses. Estimates from bivariate twin models (which include two phenotypes) have not been included since they may be influenced by the ordering of variables and covariation between the two phenotypes (e.g., Loehlin, 1996). Thus, including only estimates from univariate studies allows more direct comparison across different studies. Where possible, estimates obtained from full (including genetic, shared and non-shared environmental estimates of variation) rather than nested models are presented (although these were not always presented in published articles). Where multiple reports on the same sample have been published, results from one

paper have been reported except where different assessment tools have been used.

Twin Studies of Depression

Table 26.1 illustrates that the vast majority of these studies find evidence for a significant genetic component to depressive symptomatology. However, it is apparent that there is wide variation in the genetic parameter estimates across different studies (range 11–72%). Some variation is to be expected since estimates of heritability are specific to the population under study. Moreover studies have been based on depression symptoms that have been reported on by different informants. However, these substantial differences in etiology are in contrast to twin studies of other types of childhood psychopathology such as symptoms of ADHD and conduct disorder where estimates are relatively consistent across different raters (Arseneault et al., 2003; Scourfield, Van den Bree, Martin, & McGuffin, 2004; Thapar, Harrington, Ross, & McGuffin, 2000). Thus, it seems that measurement and rater effects must account for some of the differences between studies. It is widely known

Table 26.1 Genetic parameter estimates from twin studies of childhood and adolescent depression

Authors	Details	Genetic estimate %
Glowinski et al. (2003)	Adolescent interview girls	
	Major depressive disorder	42
	Broader phenotype	24
Happonen, Pulkkinen, Kaprio, Van der Meere, Viken, and Rose (2002)	Child questionnaire	45
	Parent questionnaire	43
Rice et al. (2002a)	Child questionnaire girls	31
	Child questionnaire boys	43
	Child questionnaire	55
	Parent questionnaire	25
Eaves et al. (1997)	Child questionnaire girls	15
	Child questionnaire boys	16
	Father questionnaire girls	60
	Father questionnaire boys	60
	Mother questionnaire girls	64
	Mother questionnaire boys	65
	Child interview girls	19
	Child interview boys	11
	Father interview girls	54
	Father interview boys	72
	Mother interview girls	66
	Mother interview boys	64
Eley 1997	Child questionnaire	48
Murray and Sines (1996)	Parent questionnaire girls	27
	Parent questionnaire boys	35
Thapar and McGuffin (1994)	Child questionnaire	70
	Parent questionnaire	48
Rende et al. (1993)	Child questionnaire	34

Broader phenotype = 2 weeks of depressed mood, irritability, or anhedonia.

Table 26.2 Genetic parameter estimates from twin studies of childhood and adolescent anxiety

Authors	Details	Genetic estimate %
Rice et al. (2004)	Parent questionnaire	46
Eley et al. (2003)	Pre-schoolers	
	Parent questionnaire girls – shy/inhibited	66
	Parent questionnaire boys – shy/inhibited	76
	Parent questionnaire – obsessions	65
	Parent questionnaire – fears	44
	Parent questionnaire – separation anxiety	39
Legrand, McGue, and Iacono (1999)	Child questionnaire – state anxiety	11
	Child questionnaire – trait anxiety	40
Eaves et al. (1997)	Child questionnaire girls	37
	Child questionnaire boys	0
	Father questionnaire girls	69
	Father questionnaire boys	72
	Mother questionnaire girls	52
	Mother questionnaire boys	57
	Child interview girls – over anxious	46
	Child interview boys – over anxious	30
	Father interview girls – over anxious	59
	Father interview boys – over anxious	54
	Mother interview girls – over anxious	66
	Mother interview boys – over anxious	31
	Child interview girls – separation anxiety	31
	Child interview boys – separation anxiety	19
	Father interview girls – separation anxiety	74
	Father interview boys – separation anxiety	0
	Mother interview girls – separation anxiety	74
	Mother interview boys – separation anxiety	4
Thapar and McGuffin (1995)	Child questionnaire	0
	Parent questionnaire	48

Table 26.3 Genetic parameter estimates from twin studies of childhood and adolescent internalizing

Authors	Details	Genetic estimate %
Boomsma et al. (2005)	Father questionnaire girls – age 12 anx/dep	39
	Father questionnaire boys – age 12 anx/dep	45
	Mother questionnaire girls – age 12 anx/dep	37
	Mother questionnaire boys– age 12 anx/dep	40
	Father questionnaire girls – age 10 anx/dep	41
	Father questionnaire boys – age 10 anx/dep	52
	Mother questionnaire girls – age 10 anx/dep	45
	Mother questionnaire boys– age 10 anx/dep	47
	Father questionnaire girls – age 7 anx/dep	47
	Father questionnaire boys – age 7 anx/dep	50
	Mother questionnaire girls – age 7 anx/dep	48
	Mother questionnaire boys– age 7 anx/dep	51
Deater-Deckard et al. (1997)	Father questionnaire	52
	Mother questionnaire	62
Gjone and Stevenson (1997)	Mother questionnaire	34
Schmitz et al. (1995)	Mother questionnaire – preschoolers	17
	Mother questionnaire	37
Edelbrock et al. (1995)	Mother questionnaire – anx/dep	50
	Mother questionnaire	34
Silberg et al. (1994)	Mother questionnaire	23

Anx/dep = anxious depressed subscale of the Child Behavior Checklist. The study of Boomsma et al. (2005) also includes genetic estimates for children aged 3 and 5. The estimates for older children are presented here as the version of the CBCL used at ages 3 and 5 differed from that used at ages 7, 10, and 12.

that correlations between different informants (for depressive symptoms) are modest (Cantwell, Lewinsohn, Rohde, & Seeley, 1997; Verhulst, Dekker, & vanderEnde, 1997) and it seems that each informant provides meaningful information from their own perspective (e.g., Boomsma, van Beijsterveldt, & Hudziak, 2005; Verhulst et al., 1997). Thus, it may be that different informants rate slightly different phenotypes or are influenced by different factors. For instance Hay and colleagues (1999) reported that mothers are influenced by their own mental state when reporting on their child's psychopathology, while fathers are influenced by characteristics of the child namely cognitive ability. Cole and colleagues (2002) found that agreement between mothers and children about the child's depressive symptoms improved when rates of change over time were examined rather than absolute values. The rater differences for twin studies of depression are particularly puzzling as their direction differs across different studies. So, for instance, in The Virginia Twin Study of Adolescent and Behavior Development (VTSABD) child ratings were consistently less heritable than parent ratings (Eaves et al., 1997), while in The Cardiff Study of all Wales and North West of England Twins (CASTANET) adolescent ratings were more heritable than parent ratings (Rice, Harold, & Thapar, 2002b; Thapar and McGuffin, 1994). One other possibility for the wide variation in genetic estimates across twin studies is developmental differences in the etiology of depressive symptoms. A number of studies have found substantial age/developmental differences in etiology when examining data more closely. Thapar and McGuffin (1994) in The Cardiff Twin Study observed that parental ratings of children under the age of 11 were not significantly genetically influenced while those of children and adolescents aged 11–16 were significantly heritable (also reported by Rice et al., 2002b; Scourfield et al., 2003 in independent cohorts). A number of other studies have reported similar effects. Eley & Stevenson (1999) reported a similar pattern of results for boys' self-ratings in an English sample of twins. Silberg and colleagues (1999) in the VTSABD found that self-rated symptoms of depression assessed by the Child and Adolescent Psychiatric Assessment (CAPA; Angold & Costello, 2000) were heritable only in pubertal girls. These observed age-related differences in the contribution of genes and environment present the possibility that such effects might contribute, in part, to differences in parameter estimates between studies.

Twin Studies of Anxiety

Table 26.2 illustrates genetic parameter estimates from twin studies of anxiety in childhood and adolescence. It can be seen that questionnaire measures of anxiety show sig-

nificant genetic influence. Some interesting findings arise from the few studies that have separately examined different dimensions of anxiety disorder symptoms. There is fairly consistent evidence from these studies that for separation anxiety, shared environmental influences are present (Eaves et al., 1997; Eley et al., 2003; Silberg, Rutter, Neale, & Eaves, 2001). This is in contrast to results from other dimensions of anxiety such as generalized anxiety disorder, which is substantially genetically influenced (Silberg et al., 2001). One twin study found that paternal absence was associated with separation anxiety in girls suggesting that this might be an important shared environmental factor (Cronk, Slutske, Madden, Bucholz, & Heath, 2004). Again, rater effects appear to be important – three studies of self-rated anxiety found no evidence for a genetic component to anxiety symptoms (Eaves et al., 1997; Thapar and McGuffin, 1995) – for boys' ratings on a modified version of the Revised Children's Manifest Anxiety Scale (RCMAS; Reynolds & Richards, 1978) and boys' symptoms of separation anxiety as assessed by the CAPA (Angold & Costello, 2000).

Twin Studies of Internalizing Symptoms

Table 26.3 shows that there have been fewer studies of internalizing symptoms than of depressive and anxiety symptoms. However, results are quite consistent across studies, showing modest genetic influences between 30 and 40% (e.g., Deater-Deckard, Reiss, Hetherington, & Plomin, 1997; Gjone, 1997; Schmitz, Fulker, & Mrazek, 1995). Boomsma and colleagues (2005) did not find any marked changes in the contribution of genetic factors to variation in the anxiety/depression scale of the CBCL between the ages of 7 and 12 years. This is in contrast to the age-related findings in the genetic etiology of depression between childhood and adolescence.

In summary, symptoms of depression and anxiety are influenced by genetic factors to a moderate degree although there is wide variability in the magnitude of genetic estimates. Measurement and rater differences appear to account for some of this variability. Differences in the genetic etiology of depression have been reported according to age with genetic factors more important for adolescents than for children.

High Levels of Depressive and Anxiety Symptoms

There has only been one study of depressive disorder in young people (Glowinski et al., 2003) and no twin study of anxiety disorders in children and adolescents. However, a number of studies have used the De Fries and Fulker regression method (DeFries & Fulker, 1985) to examine the etiology of high levels of depressive symptoms (Deater-Deckard et al., 1997; Eley, 1997; Gjone, Stevenson, Sundet, & Eilertsen, 1996; Rende, Plomin, Reiss, & Hetherington, 1993; Rice et al., 2002b). The vast majority of these studies (Deater-Deckard et al., 1997; Eley, 1997; Rende et al., 1993; Rice et al., 2002b) have found that high levels of depressive symptoms when rated by the adolescents (the study of Deater-Deckard et al., 1997, is an exception in that parent-rated symptoms were used) are less heritable (group heritability; h^2_g) than symptoms within the normal range (heritability for individual differences; h^2). In fact, shared environmental factors (group-shared environment; c^2_g) were of the greatest importance in the etiology of high levels of symptoms. These findings are consistent across different samples and different methods of estimating the c^2_g statistic (see Eley, 1997, for a description of the different methods used). However, these findings contrast with those of Glowinski and colleagues (2003) who performed the only twin study of depressive disorder and found no evidence of a shared environmental contribution (best fitting model was an AE model A = 40%, CI = 24, 66; E = 60%, CI = 50, 76). This might reflect measurement differences (high scores on questionnaire are not synonymous with depressive disorder) or differences in the age distribution between the samples. The individuals in the study of Glowinski and colleagues (2003) ranged from 12 to 23 years, while those included in the studies that used the DF method ranged from 5 to 18 years. Thus, the individuals included in the study of Glowinski and colleagues (2003) were older. Finally, the classical twin design is known to have low power to detect shared environmental effects especially when data are categorical (Neale & Cardon, 1992); therefore, this may account for the observed differences. Nonetheless, Glowinski and colleagues (2003) also performed a similar analysis of a broader depressive phenotype (rather than diagnosis of MDD) and found evidence for shared environmental effects (best fitting model was an ACE model; A = 24%; C = 30%, E = 45%). As these authors pointed out, this suggests that shared environmental factors may be important in the etiology of a broad depressive phenotype but not in the diagnosis or syndrome of MDD.

In summary, the one twin study of depressive disorder in adolescents showed that genetic and non-shared environmental factors were important influences. This result is in line with findings from the adult literature (Sullivan et al., 2000). In contrast, high levels of depressive symptoms are mainly influenced by shared environmental factors. It may be that this difference in etiology reflects subtle differences between the phenotype of depressive disorder and high levels of depressive symptoms assessed by questionnaire.

Covariation Between Anxiety and Depression

In comparison to many of the findings from univariate analyses, results from studies of the covariation between anxiety and depression are relatively consistent. Two cross-sectional studies (Eley & Stevenson, 1999; Thapar & McGuffin, 1997) have shown that anxiety and depression symptoms are associated mainly because they share a common genetic liability. Silberg and colleagues (2001) used a longitudinal design to examine the genetic and environmental architecture of the association between three dimensions of anxiety (over-anxious disorder (OAD), simple phobias and separation anxiety) and depression in girls using self-reported symptoms from the CAPA (Angold & Costello, 2000). They found evidence for a common set of genes influencing early (between ages 8 and 13) over-anxious disorder and simple phobias and later depression (between ages 14 and 17) although later OAD and phobias also had unique genetic influences. Of course, because this study used a classical twin approach – inferring the contribution of genes through the use of a genetically sensitive design – it was not possible to identify the specific genetic variants that were included in this common set of genes. Another longitudinal study again found that early anxiety and later depression shared a common genetic etiology and that the link could not be explained by earlier depression (Rice, van den Bree, & Thapar, 2004). In addition, the link between early anxiety and later depression was not explained by a phenotypic causal path (Rice et al., 2004).

In summary, both cross-sectional and longitudinal studies suggest that anxiety and depression in childhood and adolescence are associated because they share a common genetic liability. Studies that have included parent and child-reported symptoms have reported this same pattern of results.

Longitudinal Studies of Depressive and Internalizing Symptoms

Given the importance of developmental change and continuity in depression, it is surprising that there have not been more longitudinal twin studies. O'Connor, McGuire, Reiss, Hetherington, and Plomin (1998) used the NEAD cohort (a mixed-twin family design) to examine factors influencing the stability and change of depressive symptoms over a 3-year period. A composite measure of depressive symptoms was used comprising adolescent questionnaire reports and an observational assessment of depressed mood. Genetic factors were most important in accounting for the stability of symptoms over time (64% of the stability of depression was attributed to genetic influences and the remainder to nonshared influences). Schmitz and colleagues (1995) examined the continuity of internalizing symptoms in a sample of 95 twin pairs from toddler-hood (ages 2–3) to middle childhood. Shared environmental effects accounted for the greatest proportion of stability in symptoms over time. Scourfield et al. (2003) examined continuity of parent-rated depressive symptoms as assessed by the short version of the Mood and Feelings Questionnaire (MFQ) (Angold et al., 1995). Again, shared environmental factors accounted for most of the covariation between symptoms across a 3-year time lag. This may have been due to shared rater effects (mothers reported on their twins' symptoms at both time points). In an older sample of twins and siblings (G1219), Lau and Eley (2006) examined continuity and change in depressive symptoms again assessed by the short MFQ (adolescent self-reported symptoms). Symptoms were assessed at three time points (baseline, on average 8 months later and on average 25 months after the second assessment). A common genetic influence at time 1 affected depressive symptoms at each time point and a 'new' genetic influence emerged at time 2 which additionally affected symptoms at time 3. There was a shared environmental influence that was common to depressive symptoms at all three time points but this was not significant. Non-shared environmental factors were significant and specific to each time point but were not associated with stability of symptoms over time. Finally, Bartels and colleagues (2004) examined the continuity of internalizing symptoms from ages 3, 7, 10 and 12 in the Netherlands Twin Registry. There were strong genetic (43%) and shared environmental (47%) contributions to the stability of internalizing symptoms. There was genetic variance at age 3 that continued to contribute to symptoms at all ages studied, although there were additional specific genetic influences at ages 7, 10 and 12. Thus, in general, genetic influences appear to be important in the continuity of depressive symptoms. Shared environmental effects also appear to play a role in continuity (at least in children and younger adolescents). There are, however, relatively few longitudinal studies of depressive symptoms and results are not entirely consistent over existing studies.

In summary, there have been few longitudinal twin studies of depression and anxiety. To date, the evidence suggests that both genetic and shared environmental factors contribute to the stability of depression and anxiety over time.

Adoption Studies of Depression/Anxiety

Adoption studies provide another approach to disentangling genetic and environmental influences on behavior. As with twin studies, adoption studies depend on comparisons of the similarity between pairs of relatives who differ in their degree of genetic relatedness. For example, comparisons of the similarity between biological parents and children and adoptive parents and children can be made, with greater similarity between biological parents and children consistent with a role for genetic factors, while greater similarity between adoptive parents and children is consistent with a role for shared environmental factors. The strengths and weaknesses of adoption studies are described elsewhere (e.g., Plomin et al., 2001; Rutter et al., 2001)

There have been two adoption studies of depression/anxiety in childhood. Van den Oord, Boomsma, and Verhulst (1994) examined internalizing symptoms in an international adoptee sample using a sibling design. Correlations between non-biological (adoptive) siblings were as high as those between biological siblings suggesting strong shared environmental influences and no evidence for substantial genetic effects.

Eley, Deater-Deckard, Fombonne, Fulker, and Plomin (1998) studied the Colorado Adoption Project cohort using both a sibling and a parent–offspring design and again found no evidence for genetic effects. In fact correlations between parents and children were very low, suggesting non-shared environmental influences. However, in line with the results of van den Oord and colleagues, when mothers rated their children's internalizing problems, correlations were higher, and suggested some shared environmental (or shared rater) influences.

Thus, the results from adoption studies are at odds with those from twin studies in that adoption studies find that genetic influences are unimportant in the etiology of depressive symptoms. First, twin studies rely on comparisons being made between siblings of exactly the same age, thus any developmental differences in the etiology and phenomenology of depression will be controlled for. The study of Eley and colleagues (1998) included a direct test of genetic effects by including a parent–offspring component to the study design. Therefore, as they note, any developmental differences between the phenotype in the two generations as well as potential genetic heterogeneity may have influenced results. Passive gene–environment correlation may account for some of the differences in results between twin and adoption studies of depression in childhood and adolescence. Passive gene–environment correlation occurs when children are exposed to family environments provided by their parents that are correlated with their genetic characteristics. It does not seem unlikely that this may occur with depression – for example, it is well known that children of depressed parents often experience a disharmonious family environment (e.g., inconsistent parenting, conflict between parents) (see Downey & Coyne, 1990) in addition to presumably inheriting genes that increase vulnerability to depression. In a classic twin design, passive gene–environment correlation would be subsumed within the genetic parameter estimate. The adoption design is thought to remove passive gene–environment correlation. Thus, this might account for differences in the findings of adoption and twin studies and also suggests that passive gene–environment correlation may play a role in the etiology of depression in young people.

In summary, there have only been two adoption studies of depression/anxiety during childhood. Neither has found evidence for the role of genetic factors. This contrasts with results from twin studies. A possible reason for the discrepancy in findings from twin and adoption studies is the influence of passive gene–environment correlation. In addition, study designs differ in their pattern of strength and weakness and it is possible that this may account for differences in the pattern of results from twin and adoption studies.

Gene–Environment Interplay

Gene–Environment Correlation

Two main distinctions between types of gene–environment interplay have been made: gene–environment correlation and gene–environment interaction. We first discuss the evidence for gene–environment correlation. Although we consider gene–environment correlation and interaction separately, we recognize that gene–environment correlation and interaction may not be independent and may simultaneously influence risk for depression and anxiety (e.g., Eaves, Silberg, & Erkanli, 2003). Gene–environment correlation is defined as genetic influences on exposure to the environment and refers to observations that genetic and environmental risks are not independent and often go hand in hand. This type of effect (at a phenotypic level) has also been termed person–environment correlation.

Three broad dimensions of gene–environment correlation have been proposed: passive, evocative and active (Plomin et al., 2001). Passive gene–environment correlation is thought to come about as (in most cases) parents both pass on genes and provide a rearing environment for their children, and their genes and the environment they provide may be correlated. An example of this is that children of parents with mental illness often experience environmental adversity such as hostile or inconsistent parenting (Rutter & Quinton, 1984). Thus, parental characteristics are strongly associated with the rearing environment they provide for their children – and parental characteristics will be influenced by the parents' genes, which in turn are correlated with their child's genes. Evocative gene–environment correlation refers to an individual evoking a response from another person because of his/her behavior (which is partly influenced by his/her genes). Finally, active gene–environment correlation refers

to an individual actively seeking out an environment that is correlated with his/her genes – sometimes known as 'niche-fitting' (Scarr & McCartney, 1983).

Twin studies have used structural equation modeling to examine the impact of gene–environment correlation on depression and anxiety by including environmental factors as a phenotype in a bivariate genetic model (Neiderhiser, Reiss, Hetherington, & Plomin, 1999; Pike, McGuire, Hetherington, Reiss, & Plomin, 1996; Rice, Harold, & Thapar, 2003; Silberg et al., 1999). Using this approach, the presence of gene–environment correlation is statistically inferred and will include the effects of many different genes.

Gene–environment correlation with depression and stressful life events has been examined in several studies of children and adolescents. Stressful life events have been recognized for many years as being robustly associated with depression in adults (Brown & Harris, 1978) as well as in young people (Goodyer, Kolvin, & Gatzanis, 1985). A distinction has long been made between independent and dependent life events (Brown & Harris, 1978). Independent life events are thought to be largely random events outside an individual's control (such as death of a loved one). On the other hand, dependent life events are stressful events, which an individual's own behavior may play some role in precipitating (such as getting into a fight and being injured or arguing and falling out with a friend). Dependent life events may therefore involve either active or evocative gene–environment correlation as they may depend upon an individual's behavior, which in turn is partly influenced by their genes. Thapar, Harold, and McGuffin (1998) found substantial gene–environment correlation with parent-rated negative life events and depression in a sample of twins aged 8–17 years. Silberg and colleagues (1999) found a similar result looking at self-reported depressive symptoms from the CAPA and parent-reported dependent life events in adolescent girls. Rice and colleagues (2003) examined whether gene–environment correlation with life events could account for the age-related differences in the etiology of depressive symptoms observed in The Cardiff Twin Study. Gene–environment correlation with dependent life events was substantially greater in adolescents than children and seemed to account for much of the age-related differences in the genetic etiology of depression. Taken together these results suggest the importance of either active or evocative processes with stressful life events in adolescent depression. In summary, there is good evidence for gene–environment correlation with negative life events and depression in adolescents, with some evidence suggesting stronger effects in adolescents than in children (Rice et al., 2003).

One study found evidence of gene–environment correlation with a different environmental stressor. Pike and colleagues (1996) found that a composite measure of family negativity (defined as anger/hostility, coercion and transactional conflict) showed gene–environment correlation with depressive symptoms in the NEAD study. Gene–environment correlation accounted for a substantial proportion of the phenotypic correlation between depressive symptoms and mother, father and sibling negativity. However, there were still environmentally mediated effects of family negativity on depression symptoms. Neiderhiser and colleagues (1999) examined this same sample longitudinally over a 3-year time period and again found evidence for gene–environment correlation with maternal and paternal conflict negativity and depressive symptoms.

Gene–Environment Interaction

Gene–environment interaction refers to situations where the effects of an environmental risk factor on health or behavior are contingent upon an individual's genotype (Moffitt, Caspi, & Rutter, 2005). Or, put another way, genes can have different effects on a trait when the environment differs. Where gene–environment interaction exists, this means that genes, rather than directly influencing a trait or illness, are having influences as a result of individual susceptibility to an environmental hazard (Rutter, 2006).

A number of twin studies have used structural equation modeling to examine the impact of gene–environment interaction on depression and anxiety (Eaves et al., 2003; Rice, Harold, Shelton, & Thapar, 2006; Silberg et al., 2001). Again, using this approach the genetic variants (at a molecular level) that influence susceptibility to environmental stress are not known and gene–environment interaction will include the effect of many different genes.

Two studies of the Virginia Twin Cohort have shown evidence for gene–environment interaction with stressful life events and depression. First, Silberg and colleagues (2001) examined independent life events that were not influenced by genetic factors in order to identify potential gene–environment interaction (i.e., by minimizing gene–environment correlation). They found evidence for gene–environment interaction with depression and generalized anxiety in adolescent girls. A later study by Eaves and colleagues (2003) used a Monte Carlo Markov Chain (MCMC) statistical approach to simultaneously estimate gene–environment correlation and gene–environment interaction with anxiety and depressive symptoms in adolescent girls. They found a main effect of genes that influence both early anxiety and later depression. They also found evidence for substantial gene–environment interaction and correlation. First, genes that influenced early anxiety increased sensitivity to later stressful life events (gene–environment interaction) and also increased exposure to depression-inducing life events (gene–environment correlation).

Examining a different environmental stressor, Rice and colleagues (2006) found evidence for gene–environment interaction with a measure of overt family conflict and depressive symptoms. Specifically, those at genetic risk of depression showed stronger depressogenic effects to family conflict and the genetic variance of depressive symptoms was increased at higher levels of family conflict.

However, recent studies have included direct measures of genetic polymorphisms (variants in a specific gene) in assessing gene–environment interaction with depression, in particular a variable nucleotide tandem repeat (VNTR) in the serotonin transporter gene (5HTT; SERT). Thus, these studies have assessed genetic risk according to naturally occurring variation at a single functional polymorphism. Caspi and colleagues (2002, 2003) provided the first clear-cut evidence of gene–environment interaction with a single measured genetic variant (the VNTR of the 5HTT gene). In this study of the Dunedin cohort, individuals (young adults) who possessed a copy of the risk genotype (the short allele) were significantly more likely to experience clinical depression following stressful life events than individuals with the long allele. This finding has been partially replicated in adolescents (Eley, Sugden et al., 2004) and adults (Kendler, Kuhn, Vittum, Prescott, & Riley, 2005).

In summary, there has been consistent evidence to suggest the importance of indirect genetic influences on adolescent depression and anxiety. These indirect genetic influences affect exposure to environmental risk (gene–environment correlation) and susceptibility to environmental hazards (gene–environment interaction). To date the most consistent evidence points to gene–environment correlation and interaction with stressful life events and depression.

Conclusion and Future Directions

There is a large volume of research on the genetic and environmental etiology of internalizing problems in children and adolescents, in particular on the etiology of depression. Depression (assessed categorically as a disorder and dimensionally by symptoms) is familial and most evidence suggests the importance of genetic influences as well as environmental factors. Recent behavior genetic evidence on gene–environment correlation and interaction has underlined the fact that genetic and environmental factors are not distinct but act together to influence depression and anxiety. Research questions remain with regard to rater, developmental change and overlap with other types of psychopathology. We next highlight some potential areas for future research.

Generational Transmission

We know very little about the factors that account for the transmission of depression between different generations and the specific mechanisms involved despite the huge numbers of twin and family studies of child and adolescent depression. This is a clinically important area as it may help targeted interventions of children at high risk of depression. The discrepancy between results from twin and adoption studies of depression suggests that passive gene–environment correlation may play a role in this process. The children of twins design may be a useful paradigm for assessing the importance of passive gene–environment correlation in the transmission of depression between parents and children (e.g., Silberg & Eaves, 2004).

Environmental Factors in Very Early Life

Across internal medicine, there has been much interest in the 'fetal origins of adult disease' for many complex multifactorial diseases including cardiovascular disease and type II diabetes (Barker, 1998). A large body of research in animals has suggested the importance of maternal stress during pregnancy and resulting anxiety-like traits in the offspring as adults (Seckl & Meaney, 2004). Similarly, population-based studies have reported that gestational stress is associated with behavioral problems in children including anxiety and depression (O'Connor et al., 2003). Other pre-natal environmental factors such as birth weight have also been implicated (e.g., Patton, Coffey, Carlin, Olsson, & Morley, 2004). However, to date, very few studies have examined the impact of pre-natal environmental factors in genetically sensitive designs. The potential role of pre-natal environmental factors in conjunction with genetic risk is thus an avenue for future research. Pre-natal cross fostering studies in animals are able to identify whether pre-natal environmental factors exert effects on offspring independently of genes shared between mother and child. The use of in vitro fertilization with donated gametes and surrogacy as a method of conception now means that similar pre-natal cross fostering studies are possible in humans (Thapar et al., 2007).

Indirect Genetic Mechanisms and Intermediate Phenotypes

Many genetic influences on psychopathology could come about indirectly rather than through direct changes in the coding sequence or changes in gene expression. Genetic influences could occur via behavior where gene–environment correlation exists, via susceptibility to environmental factors where gene–environment interaction exists or through an intermediate phenotype (endophenotype) which itself carries risk for disorder. To take the short allele of the 5HTT gene variant as an example: First, gene–environment interaction has been reported which suggests a modifying effect of the short allele on risk for depression following stressful life events. However, in general, there does not appear to be a main effect of the short allele on depressive disorder although some studies have found the 5HTT short allele to be linked with neurotic and anxious personality traits/stress-responsivity (Hariri et al., 2005; Lesch et al., 1996). Neuroimaging work has also illustrated that the 5HTT allele may be strongly related to the engagement of neuronal systems that underlie emotional processing (Hariri et al., 2005). Taken together, this evidence suggests that the influence of the 5HTT short allele on risk for depressive disorder might be mediated through neuronal systems involved in emotional processing. Examining potential indirect genetic mechanisms, in particular intermediate phenotypes, may be a fruitful avenue for future research to elucidate developmental pathways to disorder.

Person–Environment Correlation

There has been much interest in gene–environment interplay and there is good evidence to suggest it exists for child and adolescent depression and stressful life events. Thus, individuals play a role in the level of their exposure to certain types of environmental risk (Jaffee & Price, 2007). It seems likely that more examples of this phenomenon will be found. One important point that Rutter and his colleagues note is that it cannot always be assumed that an individual's effect on his/her own environment (including the others in it) is entirely due to genes (Rutter & Silberg, 2002; Rutter et al., 2001). This is because the selection of environments stems from the characteristics of the individual rather than of genes. That is, individual exposure to certain environments stems from the person; for example, factors such as personality traits and past experience of adversity may influence a person's exposure to risk environments and only a proportion of these 'person effects' will be influenced by genes. This point is well illustrated by the results of a study by O'Connor, Deater-Deckard, et al., (1998). O'Connor and colleagues identified evocative gene–environment correlation between disruptive child behavior and harsh parenting using an adoption design. Children whose biological parents were antisocial were themselves more likely to be disruptive and, in turn, to evoke negative parenting from their adoptive parents. That is, the child's disruptive behavior evoked a harsh parenting response from adoptive parents. However, not all of this effect was due to genes – a similar effect was observed even in children whose biological parents were not antisocial. This result suggests that the important influence was the child's own behavior and only part of this was genetically influenced. One important step for future research is thus to identify how much of observed 'person effects' are in fact due to genes and, furthermore, to identify what features of the individual and the processes account for 'person effects'.

Conclusion

Much has been learnt from behavioral genetics research about the etiology and comorbidity of childhood and adolescent depression and anxiety. The field is now well positioned to extend this research and to move toward examining the complex interplay between genes and environment, as well as the mechanisms involved in the pathogenesis and development of psychopathology.

Acknowledgment This work was supported by the Wellcome Trust.

References

Achenbach, T. (1991). Manual for the child behavior checklist/ 4–18 and 1991 profile. Burlington, VT: University of Vermont Department of Psychiatry.

Angold, A., & Costello, E. J. (2000). The child and adolescent psychiatric assessment (CAPA). *Journal of the American Academy of Child and Adolescent Psychiatry, 39*, 39–48.

Angold, A., Costello, E. J., & Erkanli, A. (1999). Comorbidity. *Journal of Child Psychology and Psychiatry, 40*, 57–87.

Angold, A., Costello, E. J., Farmer, E. M., Burns, B. J., & Erkanli, A. (1999). Impaired but undiagnosed. *Journal of the American Academy of Child Adolescent Psychiatry, 38*, 129–137.

Angold, A., Costello, E. J., Messer, S. C., Pickles, A., Winder, F., & Silver, D. (1995). Development of a short questionnaire for use in epidemiological studies of depression in children and adolescents. *International Journal of Methods in Psychiatric Research, 5*, 237–249.

Angold, A., Erkanli, A., Silberg, J., Eaves, L., & Costello, E. J. (2002). Depression scale scores in 8–17 year olds: Effects of age and gender. *Journal of Child Psychology and Psychiatry, 43*, 1052–1063.

Arseneault, L., Moffitt, T. E., Caspi, A., Taylor, A., Rijsdijk, F. V., Jaffee, S. R., et al. (2003). Strong genetic effects on cross-situational antisocial behaviour among 5-year-old children according to mothers, teachers, examiner-observers, and twins' self-reports. *Journal of Child Psychology and Psychiatry, 44*, 832–848.

Avenevoli, S., Stolar, M., Li, J., Dierker, L., & Merikangas, K. R. (2001). Comorbidity of depression in children and adolescents: Models and evidence form a prospective high-risk study. *Biological Psychiatry, 49*, 1071–1081.

Barker, D. J. P. (1998.) *Mothers, babies and health in later life.* Edinburgh: Chruchill Livingston.

Bartels, M., van den Oord, E. J., Hudziak, J. J., Rietveld, M. J., van Beijsterveldt, C. E., & Boomsma, D. I. (2004). Genetic and environmental mechanisms underlying stability and change in problem behaviors at ages 3, 7, 10, and 12. *Developmental Psychology, 40*, 852–867.

Beidel, D. C., & Turner, S. M. (1997). At risk for anxiety: I. Psychopathology in the offspring of anxious parents. *Journal of the American Academy of Child Adolescent Psychiatry, 36*, 918–924.

Bell-Dolan, D. J., Last, C. G., & Strauss, C. C. (1990). Symptoms of anxiety disorders in normal children. *Journal of the American Academy of Child Adolescent Psychiatry, 29*, 759–765.

Biederman, J., Rosenbaum, J. F., Bolduc, E. A., Faraone, S. V., & Hirshfeld, D. R. (1991). A high risk study of young children of parents with panic disorder and agoraphobia with and without comorbid major depression. *Psychiatry Research, 37*, 333–348.

Birmaher, B., Ryan, N. D., Williamson, D. E., Brent, D. A., Kaufman, J., Dahl, R. E., et al. (1996). Childhood and adolescent depression: A review of the past 10 years. Part I. *Journal of the American Academy of Child Adolescent Psychiatry, 35*, 1427–1439.

Boomsma, D. I., van Beijsterveldt, C. E., & Hudziak, J. J. (2005). Genetic and environmental influences on anxious/depression during childhood: A study from the Netherlands Twin Register. *Genes Brain and Behavior, 4*, 466–481.

Brady, E. U., & Kendall, P. C. (1992). Comorbidity of anxiety and depression in children and adolescents. *Psychological Bulletin, 111*, 244–255.

Brown, G. W., & Harris, T. O. (1978). *Social origins of depression: A study of psychiatric disorder in women.* London: Tavistock.

Cantwell, D. P., Lewinsohn, P. M., Rohde, P., & Seeley, J. R. (1997). Correspondence between adolescent report and parent report of psychiatric diagnostic data. *Journal of the American Academy of Child Adolescent Psychiatry, 36*, 610–619.

Caspi, A., Sugden, K., Moffitt, T. E., Taylor, A., Craig, I. W., Harrington, H., et al. (2003). Influence of life stress on depression: moderation by a polymorphism in the 5-HTT gene. *Science, 301*(5631), 386–389.

Cole, D. A., Tram, J. M., Martin, J. M., Hoffman, K. B., Ruiz, M. D., Jacquez, F. M., et al. (2002). Individual differences in the emergence of depressive symptoms in children and adolescents: A longitudinal investigation of parent and child reports. *Journal of Abnormal Psychology, 111*, 156–165.

Connell, A. M., & Goodman, S. H. (2002). The association between psychopathology in fathers versus mothers and children's internalizing and externalizing behavior problems: A meta-analysis. *Psychological Bulletin, 128*, 746–773.

Costello, E. J., Angold, A., Burns, B. J., Erkanli, A., Stangl, D. K., & Tweed, D. L. (1996). The Great Smoky Mountains Study of youth. Functional impairment and serious emotional disturbance. *Archives of General Psychiatry, 53*(12), 1137–1143.

Costello, E. J., Mustillo, S., Erkanli, A., Keeler, G., & Angold, A. (2003). Prevalence and development of psychiatric disorders in childhood and adolescence. *Archives of General Psychiatry, 60*, 837–844.

Cronk, N. J., Slutske, W. S., Madden, P. A., Bucholz, K. K., & Heath, A. C. (2004). Risk for separation anxiety disorder among girls: Paternal absence, socioeconomic disadvantage, and genetic vulnerability. *Journal of Abnormal Psychology, 113*, 237–247.

Deater-Deckard, K., Reiss, D., Hetherington, E. M., & Plomin, R. (1997). Dimensions and disorders of adolescent adjustment: A quantitative genetic analysis of unselected samples and selected extremes. *Journal of Child Psychology and Psychiatry, 38*, 515–525.

DeFries, J. C., & Fulker, D. W. (1985). Multiple regression analysis of twin data. *Behavior Genetics, 15*, 467–473.

Downey, G., & Coyne, J. C. (1990). Children of depressed parents: An integrative review. *Psychological Bulletin, 108*, 50–76.

Eaves, L., Silberg, J., & Erkanli, A. (2003). Resolving multiple epigenetic pathways to adolescent depression. *Journal of Child Psychology and Psychiatry, 44*, 1006–1014.

Eaves, L. J., Silberg, J. L., Meyer, J. M., Maes, H. H., Simonoff, E., Pickles, A., et al. (1997). Genetics and developmental psychopathology: 2. The main effects of genes and environment on behavioral problems in the Virginia Twin Study of Adolescent Behavioral Development. *Journal of Child Psychology and Psychiatry, 38*, 965–980.

Edelbrock, C., Rende, R., Plomin, R., & Thompson, L. (1995). A twin study of competence and problem behavior in childhood and early adolescence. *Journal of Child Psychology and Psychiatry, 36*, 775–785.

Eley, T. C. (1997). Depressive symptoms in children and adolescents: Etiological links between normality and abnormality: A research note. *Journal of Child Psychology and Psychiatry, 38*, 861–865.

Eley, T. C., Bolton, D., O'Connor, T. G., Perrin, S., Smith, P., & Plomin, R. (2003). A twin study of anxiety-related behaviours in preschool children. *Journal of Child Psychology and Psychiatry, 44*, 945–960.

Eley, T. C., Deater-Deckard, K., Fombonne, E., Fulker, D. W., & Plomin, R. (1998). An adoption study of depressive symptoms in middle childhood. *Journal of Child Psychology and Psychiatry, 39*, 337–345.

Eley, T.C., Liang, H., Plomin, R., Sham, P., Sterne, A., Williamson, R., et al. (2004). Parental familial vulnerability, family environment, and their interactions as predictors of depressive symptoms in adolescents. *Journal of the American Academy of Child and Adolescent Psychiatry, 43*, 298–306.

Eley, T. C., & Stevenson, J. (1999). Exploring the covariation between anxiety and depression symptoms: A genetic analysis of the effects of age and sex. *Journal of Child Psychology and Psychiatry, 40*, 1273–1282.

Eley, T. C., Sugden, K., Corsico, A., Gregory, A. M., Sham, P., McGuffin, P., et al. (2004). Gene–environment interaction analysis of serotonin system markers with adolescent depression. *Molecular Psychiatry, 9*, 908–915.

Fombonne, E., Wostear, G., Cooper, V., Harrington, R., & Rutter, M. (2001). The Maudsley long-term follow-up of child and adolescent depression. 1. Psychiatric outcomes in adulthood. *British Journal of Psychiatry, 179*, 210–217.

Gjone, H., Stevenson, J., Sundet, J. M., & Eilertsen, D. E. (1996). Changes in heritability across increasing levels of behavior problems in young twins. *Behavior Genetics, 26*, 419–426.

Gjone, H., & Stevenson, J. (1997). The association between internalizing and externalizing behavior in childhood and early adolescence: Genetic or environmental common influences? *Journal of Abnormal Child Psychology 25*, 277–286.

Glowinski, A. L., Madden, P. A., Bucholz, K. K., Lynskey, M. T., & Heath, A. C. (2003). Genetic epidemiology of self-reported lifetime DSM-IV major depressive disorder in a population-based twin sample of female adolescents. *Journal of Child Psychology and Psychiatry, 44*, 988–996.

Goodman, S. H., & Gotlib, I. H. (1999). Risk for psychopathology in the children of depressed mothers: A developmental model for understanding mechanisms of transmission. *Psychological Review, 106*, 458–490.

Goodwin, R. D., Fergusson, D. M., & Horwood, L. J. (2004). Early anxious/withdrawn behaviors predict later internalizing disorders. *Journal of Child Psychology and Psychiatry, 45*, 874–883.

Goodyer, I. M., Cooper, P. J., Vize, C. M., & Ashby, L. (1993). Depression in 11–16-year-old girls: The role of past parental psychopathology and exposure to recent life events. *Journal of Child Psychology and Psychiatry, 34*, 1103–1115.

Goodyer, I., Kolvin, I., & Gatzanis, S. (1985). Recent undesirable life events and psychiatric-disorder in childhood and adolescence. *British Journal of Psychiatry 147*, 517–523.

Gollan, J., Raffety, B., Gortner, E., & Dobson, K. (2005). Course profiles of early- and adult-onset depression. *Journal of Affective Disorders, 86*, 81–86.

Hammen, C., Burge, D., Burney, E., & Adrian, C. (1990). Longitudinal study of diagnoses in children of women with unipolar and bipolar affective disorder. *Archives of General Psychiatry, 47*, 1112–1117.

Happonen, M., Pulkkinen, L., Kaprio, J., Van der Meere, J., Viken, R. J., & Rose, R. J. (2002). The heritability of depressive symptoms: Multiple informants and multiple measures. *Journal of Child Psychology and Psychiatry, 43*, 471–480.

Hariri, A. R., Drabant, E. M., Munoz, K. E., Kolachana, L. S., Mattay, V. S., Egan, M. F., et al. (2005). A susceptibility gene for affective disorders and the response of the human amygdala. *Archives of General Psychiatry, 62*, 146–152.

Harrington, R. (1994). Affective disorders. In M. Rutter, E. Taylor, & L. Hersov, (Eds.), *Child and adolescent psychiatry: Modern approaches* (pp. 330–345). Oxford: Blackwell Scientific Publications.

Harrington, R., Fudge, H., Rutter, M., Pickles, A., & Hill, J. (1990). Adult outcomes of childhood and adolescent depression. I. Psychiatric status. *Archives of General Psychiatry, 47*, 465–473.

Harrington, R., Fudge, H., Rutter, M., Pickles, A., & Hill, J. (1991). Adult outcomes of childhood and adolescent depression. 2. Links with antisocial disorders. *Journal of the American Academy of Child and Adolescent Psychiatry, 30*, 434–439.

Harrington, R., Rutter, M., Weissman, M., Fudge, H., Groothues, C., Bredenkamp, D., et al. (1997). Psychiatric disorders in the relatives of depressed probands. I. Comparison of prepubertal, adolescent and early adult onset cases. *Journal of Affective Disorders, 42*, 9–22.

Hay, D. F., Pawlby, S., Sharp, D., Schmucker, G., Mills, A., Allen, H., et al. (1999). Parents' judgements about young children's problems: Why mothers and fathers might disagree yet still predict later outcomes. *Journal of Child Psychology and Psychiatry, 40*, 1249–1258.

Jaffee, S. R., & Price, T. S. (2007). Gene environment correlations: A review of the evidence and implications for prevention of mental illness. *Molecular Psychiatry, 12*, 432–442.

Johnson, J., Weissman, M. M., & Klerman, G. L. (1992). Service utilization and social morbidity associated with depressive symptoms in the community. *JAMA, 267*, 1478–1483.

Kane, P., & Garber, J. (2004). The relations among depression in fathers, children's psychopathology, and father–child conflict: A meta-analysis. *Clinical Psychology Review, 24*, 339–360.

Keller, M. B., Beardslee, W. R., Dorer, D. J., Lavori, P. W., Samuelson, H., & Klerman, G. R. (1986). Impact of severity and chronicity of parental affective illness on adaptive functioning and psychopathology in children. *Archives of General Psychiatry, 43*, 930–937.

Kendler, K. S., Kuhn, J. W., Vittum, J., Prescott, C. A., & Riley, B. (2005). The interaction of stressful life events and a serotonin transporter polymorphism in the prediction of episodes of major depression: A replication. *Archives of General Psychiatry, 62*, 529–535.

Kessler, R. C., McGonagle, K. A., Zhao, S., Nelson, C. B., Hughes, M., Eshleman, S., et al. (1994). Lifetime and 12-month prevalence of DSM-III-R psychiatric disorders in the United States. Results from the National Comorbidity Survey. *Archives of General Psychiatry, 51*, 8–19.

Kim-Cohen, J., Moffitt, T. E., Taylor, A., Pawlby, S. J., & Caspi, A. (2005). Maternal depression and children's antisocial behavior – Nature and nurture effects. *Archives of General Psychiatry, 62*, 173–181.

Klein, D. N., Shankman, S. A., Lewinsohn, P. M., Rohde, P., & Seeley, J. R. (2004). Family study of chronic depression in a community sample of young adults. *American Journal of Psychiatry, 161*, 646–653.

Kovacs, M., & Devlin, B. (1998). Internalizing disorders in childhood. *Journal of Child Psychology and Psychiatry, 39*, 47–63.

Kovacs, M., Gatsonis, C., Paulauskas, S. L., & Richards, C. (1989). Depressive disorders in childhood. IV. A longitudinal study of comorbidity with and risk for anxiety disorders. *Archives of General Psychiatry, 46*, 776–782.

Kutcher, S., & Marton, P. (1991). Affective disorders in first-degree relatives of adolescent onset bipolars, unipolars, and normal controls. *Journal of the American Academy of Child Adolescent Psychiatry, 30*, 75–78.

Last, C. G., Hansen, C., & Franco, N. (1997). Anxious children in adulthood: A prospective study of adjustment. *Journal of the American Academy of Child Adolescent Psychiatry, 36*, 645–652.

Last, C. G., Hersen, M., Kazdin, A., Orvaschel, H., & Perrin, S. (1991). Anxiety disorders in children and their families. *Archives of General Psychiatry, 48*, 928–934.

Last, C. G., Perrin, S., Hersen, M., & Kazdin, A. E. (1996). A prospective study of childhood anxiety disorders. *Journal of the American Academy of Child Adolescent Psychiatry, 35*, 1502–1510.

Lau, J. Y. F., & Eley, T. C. (2006). Changes in genetic and environmental influences on depressive symptoms across adolescence and young adulthood. *British Journal of Psychiatry, 189*, 422–427.

Legrand, L. N., McGue, M., & Iacono, W. G. (1999). A twin study of state and trait anxiety in childhood and adolescence. *Journal of Child Psychology and Psychiatry, 40*, 953–958.

Lesch, K. P., Bengel, D., Heils, A., Sabol, S. Z., Greenberg, B. D., Petri, S., et al. (1996). Association of anxiety-related traits with a polymorphism in the serotonin transporter gene regulatory region. *Science, 274*, 1527–1531.

Lewinsohn, P. M., Hoberman, H. M., & Rosenbaum, M. (1988). A prospective study of risk factors for unipolar depression. *Journal of Abnormal Psychology, 97*, 251–264.

Lewinsohn, P. M., Hops, H., Roberts, R. E., Seeley, J. R., & Andrews, J. A. (1993). Adolescent psychopathology: I. Prevalence and incidence of depression and other DSM-III-R disorders in high school students. *Journal of Abnormal Psychology, 102*, 133–144.

Lewinsohn, P. M., Solomon, A., Seeley, J. R., & Zeiss, A. (2002). Clinical implications of 'subthreshold' depressive symptoms. *Journal of Abnormal Psychology, 109*, 345–351.

Lewinsohn, P. M., Zinbarg, R., Seeley, J. R., Lewinsohn, M., & Sack, W. H (1997). Lifetime comorbidity among anxiety disorders and between anxiety disorders and other mental disorders in adolescents. *Journal of Anxiety Disorders, 11*, 377–394.

Loehlin, J. C. (1996). The Cholesky approach: A cautionary note. *Behavior Genetics, 26*, 65–69.

Marks, M., Hipwell, A. E., & Kumar, R. C. (2002). Implications for the infant of maternal puerperal psychiatric disorders. In M. Rutter, & E. Taylor (Eds.), *Child and adolescent psychiatry* (pp. 858–880). Oxford: Blackwell Scientific Publications.

Meltzer, H., Gatward, R., & Ford, T. (2000). Mental health of children and adolescents in Great Britain. London: The Stationery Office.

Merikangas, K. R. (1993). Genetic epidemiologic studies of affective disorders in childhood and adolescence. *European Archives of Psychiatry and Clinical Neuroscience, 243*, 121–130.

Merikangas, K. R., Zhang, H. P., Avenevoli, S., Acharyya, S., Neuenschwander, M., & Angst, J. (2003). Longitudinal trajectories of depression and anxiety in a prospective community study – The Zurich cohort study. *Archives of General Psychiatry, 60*, 993–1000.

Mitchell, J., McCauley, E., Burke, P., Calderon, R., & Schloredt, K. (1989). Psychopathology in parents of depressed children and adolescents. *Journal of the American Academy of Child Adolescent Psychiatry, 28*, 352–357.

Moffitt, T. E., Caspi, A., & Rutter, M. (2005). Strategy for investigating interactions between measured genes and measured environments. *Archives of General Psychiatry, 62*, 473–481.

Mufson, L., Weissman, M. M., & Warner, V. (1992). Depression and anxiety in parents and children: A direct interview study. *Journal of Anxiety Disorders, 6*, 1–13.

Murray, K. T., & Sines, J. O. (1996). Parsing the genetic and nongenetic variance of children's depressive behavior. *Journal of Affective Disorders, 38*, 23–34.

Neale, M. C., & Cardon. L. R. (1992). *Methodology for genetic studies of twins and families*. Series D: Behavioural and social sciences. Dordrecht: Kluwer Academic Publishers.

Neiderhiser, J. M., Reiss, D., Hetherington, E. M., & Plomin, R. (1999). Relationships between parenting and adolescent adjustment over time: Genetic and environmental contributions. *Developmental Psychology, 35*, 680–692.

Neuman, R. J., Geller, B., Rice, J. P., & Todd, R. D. (1997). Increased prevalence and earlier onset of mood disorders among relatives of prepubertal versus adult probands. *Journal of the American Academy of Child Adolescent Psychiatry, 36*, 466–473.

O'Connor, T. G., Deater-Deckard, K., Fulker, D., Rutter, M., & Plomin, R. (1998). Genotype-environment correlations in late childhood and early adolescence: Antisocial behavioral problems and coercive parenting. *Developmental Psychology, 34*, 970–981.

O'Connor, T. G., Heron, J., Golding, J., & Glover, V. (2003). Maternal antenatal anxiety and behavioural/emotional problems in children: A test of a programming hypothesis. *Journal of Child Psychology and Psychiatry, 44*, 1025–1036.

O'Connor, T. G., McGuire, S., Reiss, D., Hetherington, M., & Plomin, R. (1998). Co-occurrence of depressive symptoms and antisocial behavior in adolescence: A common genetic liability. *Journal of Abnormal Psychology, 107*, 27–37.

Orvaschel, H., Lewinsohn, P. M., & Seeley, J. R. (1995). Continuity of psychopathology in a community sample of adolescents. *Journal of the American Academy of Child Adolescent Psychiatry, 34*, 1525–1535.

Orvaschel, H., Walsh-Allis, G., & Ye, W. (1989). Psychopathology in children of parents with recurrent depression. *Journal of Abnormal Child Psychology, 16*, 17–28.

Parker, G., Roy, K., Hadzi-Pavlovic, B., Mitchell, D., & Wilhelm, K. (2003). Distinguishing early and late onset non-melancholic unipolar depression. *Journal of Affective Disorders, 74*, 131–138.

Patton, G. C., Coffey, C., Carlin, J. B., Olsson, C. A., & Morley, R. (2004). Prematurity at birth and adolescent depressive disorder. *British Journal of Psychiatry, 184*, 446–447.

Pickles, A., Rowe, R., Simonoff, E., Foley, D., Rutter, M., & Silberg, J. (2001). Child psychiatric symptoms and psychosocial impairment: Relationship and prognostic significance. *British Journal of Psychiatry, 179*, 230–235.

Pike, A., McGuire, S., Hetherington, E. M., Reiss, D., & Plomin, R. (1996). Family environment and adolescent depressive symptoms and antisocial behavior: A multivariate genetic analysis. *Developmental Psychology, 32*, 590–603.

Pine, D. S., Cohen, P., & Brook, J. (2001). Adolescent fears as predictors of depression. *Biological Psychiatry, 50*, 721–724.

Pine, D. S., Cohen, E., Cohen, P., & Brook, J. (1999). Adolescent depressive symptoms as predictors of adult depression: Moodiness or mood disorder? *American Journal of Psychiatry, 156*, 133–135.

Pleck, J. (1997). Paternal involvement: Levels, sources and consequences. In M. E. Lamb (Ed.), *The role of the father in child development* (3rd ed., pp. 66–103). New York: Wiley.

Plomin, R., DeFries, J. C., McClearn, G. E., & McGuffin, P. (2001). Behavioral genetics (4th ed.). New York: Worth.

Puig-Antich, J., Goetz, D., Davies, M., Kaplan, T., Davies, S., Ostrow, L., et al. (1989). A controlled family history study of prepubertal major depressive disorder. *Archives of General Psychiatry, 46*, 406–418.

Radke-Yarrow, M., Nottelmann, E., Martinez, P., Fox, M. B., & Belmont, B. (1992). Young children of affectively ill parents: A longitudinal study of psychosocial development. *Journal of the American Academy of Child Adolescent Psychiatry, 31*, 68–77.

Ramchandani, P., Stein, A., Evans, J., & O'Connor, T. G. (2005). ALSPAC study team. Paternal depression in the postnatal period and child development: A prospective population study (2005). *Lancet, 365*, 2201–2205.

Rende, R. D., Plomin, R., Reiss, D., & Hetherington, E. M. (1993). Genetic and environmental influences on depressive symptomatology in adolescence: Individual differences and extreme scores. *Journal of Child Psychology and Psychiatry, 34*, 1387–1398.

Rende, R., Warner, V., Wickramarante, P., & Weissman, M. M. (1999). Sibling aggregation for psychiatric disorders in offspring at high and low risk for depression: 10-year follow-up. *Psychological Medicine, 29*, 1291–1298.

Reynolds, C. R., & Richards B. O. (1978). What I think and feel: A revised measure of children's manifest anxiety. *Journal of Abnormal Child Psychology, 6*, 271–280.

Rice, F., Harold, G. T., Shelton, K. H., & Thapar, A. (2006). Family conflict interacts with genetic liability in predicting childhood and adolescent depression. *Journal of the American Academy of Child and Adolescent Psychiatry, 45*, 841–848.

Rice, F., Harold, G., & Thapar, A. (2002a). The genetic aetiology of childhood depression: A review. *Journal of Child Psychology and Psychiatry, 43*, 65–79.

Rice, F., Harold, G. T., & Thapar, A. (2002b). Assessing the effects of age, sex and shared environment on the genetic aetiology of depression in childhood and adolescence. *Journal of Child Psychology and Psychiatry, 43*, 1039–1051.

Rice, F., Harold, G. T., & Thapar, A. (2003). Negative life events as an account of age-related differences in the genetic aetiology of depression in childhood and adolescence. *Journal of Child Psychology and Psychiatry, 44*, 977–987.

Rice, F., van den Bree, M. B., & Thapar, A. (2004). A population-based study of anxiety as a precursor for depression in childhood and adolescence. *BMC Psychiatry, 4*, 43.

Rutter, M. (2006). *Genes and behavior: Nature–nurture interplay explained.* Oxford: Blackwell Publishing.

Rutter, M., Pickles, A., Murray, R., & Eaves, L. (2001). Testing hypotheses on specific environmental causal effects on behavior. *Psychological Bulletin, 127*, 291–324.

Rutter, M., & Quinton, D. (1984). Parental psychiatric disorder: Effects on children. *Psychological Medicine, 14*, 853–880.

Rutter, M., & Silberg, J. (2002). Gene-environment interplay in relation to emotional and behavioral disturbance. *Annual Review of Psychology, 53*, 463–490.

Scarr, S., & McCartney, K. (1983). How people make their own environments – A theory of genotype-environment effects. *Child Development, 54*, 424–435.

Schmitz, S., Fulker, D. W., & Mrazek, D. A. (1995). Problem behavior in early and middle childhood: An initial behavior genetic analysis. *Journal of Child Psychology and Psychiatry, 36*, 1443–1458.

Scourfield, J., Rice, F., Thapar, A., Harold, G. T., Martin, N., & McGuffin, P. (2003). Depressive symptoms in children and adolescents: Changing aetiological influences with development. *Journal of Child Psychology and Psychiatry, 44*, 968–976.

Scourfield, J., Van den Bree, M., Martin, N., & McGuffin, P. (2004). Conduct problems in children and adolescents – A twin study. *Archives of General Psychiatry, 61*, 489–496.

Seckl, J. R., & Meaney, M. J. (2004). Glucocorticoid programming. *Annals of the New York Academy of Science, 1032*, 63–84.

Silberg, J., & Eaves, L. J. (2004). Analysing the contributions of genes and parent–child interaction to childhood behavioural and emotional problems: A model for the children of twins. *Psychological Medicine, 34*, 347–356.

Silberg, J., Erikson, M. T., Meyer, J. M., Eaves, L. J., Rutter, M. L., & Hewitt, J. K. (1994). The application of structural equation modeling to maternal ratings of twins' behavioural and emotional problems. *Journal of Consulting and Clinical Psychology, 62*, 510–521.

Silberg, J., Pickles, A., Rutter, M., Hewitt, J., Simonoff, E., Maes, H., et al. (1999). The influence of genetic factors and life stress on depression among adolescent girls. *Archives of General Psychiatry, 56*, 225–232.

Silberg, J., Rutter, M., Neale, M., & Eaves, L. (2001). Genetic moderation of environmental risk for depression and anxiety in adolescent girls. *British Journal of Psychiatry, 179*, 116–121.

Silman, A. J. (1995). *Epidemiological studies: A practical guide.* Cambridge: Cambridge University Press.

Simonoff, E., Pickles, A., Meyer, J. M., Silberg, J. L., Maes, H. H., Loeber, R., et al. (1997). The Virginia twin study of adolescent behavioral development. Influences of age, sex, and impairment on rates of disorder. *Archives of General Psychiatry, 54*, 801–808.

Sullivan, P. F., Neale, M. C., & Kendler, K. S. (2000). Genetic epidemiology of major depression: Review and meta-analysis. *American Journal of Psychiatry, 157*, 1552–1562.

Thapar, A., Harold, G., & McGuffin, P. (1998). Life events and depressive symptoms in childhood – Shared genes or shared adversity? A research note. *Journal of Child Psychology and Psychiatry, 39*, 1153–1158.

Thapar, A., Harrington, R., Ross, K., & McGuffin, P. (2000). Does the definition of ADHD affect heritability? *Journal of the American Academy of Child Adolescent Psychiatry, 39*, 1528–1536.

Thapar, A., & McGuffin, P. (1994). A twin study of depressive symptoms in childhood. *British Journal of Psychiatry, 165*, 259–265.

Thapar, A., & McGuffin, P. (1995). Are anxiety symptoms in childhood heritable? *Journal of Child Psychology and Psychiatry, 36*, 439–4475.

Thapar, A., & McGuffin, P. (1997). Anxiety and depressive symptoms in childhood – A genetic study of comorbidity. *Journal of Child Psychology and Psychiatry, 38*, 651–656.

Van den Oord, E., Boomsma, D. I., & Verhulst, F. C. (1994). A study of problem behaviors in 10 to 15 year old biologically related and unrelated international adoptees. *Behavior Genetics, 24*, 193–205.

Van den Oord, E. J., Pickles, A., & Waldman, I. D. (2003). Normal variation and abnormality: An empirical study of the liability distributions underlying depression and delinquency. *Journal of Child Psychology and Psychiatry, 44*, 180–192.

Verhulst, F. C., Dekker, M. C., & vanderEnde, J. (1997). Parent, teacher and self-reports as predictors of signs of disturbance in adolescents: Whose information carries the most weight? *Acta Psychiatrica Scandinavica, 96*, 75–81.

Warner, V., Mufson, L., & Weissman, M. M. (1995). Offspring at high and low risk for depression and anxiety: Mechanisms of psychiatric disorder. *Journal of the American Academy of Child Adolescent Psychiatry, 34*, 786–797.

Warner, V., Weissman, M. M., Mufson, L., & Wickramaratne, P. J. (1999). Grandparents, parents, and grandchildren at high risk for depression: A three-generation study. *Journal of the American Academy of Child Adolescent Psychiatry, 38*, 289–296.

Weissman, M. M., Leckman, J. F., Merikangas, K. R., Gammon, G. D., & Prusoff, B. A. (1984). Depression and anxiety disorders in parents and children. Results from the Yale family study. *Archives of General Psychiatry, 41*(9), 845–852.

Weissman, M. M., Wickramaratne, P., Merikangas, K. R., Leckman, J. F., Prusoff, B. A., Caruso, K. A., et al. (1984). Onset of major depression in early adulthood. Increased familial loading and specificity. *Archives of General Psychiatry, 41*, 1136–1143.

Weissman, M. M., Wolk, S., Goldstein, R. B., Moreau, D., Adams, P., Greenwald, S., et al. (1999). Depressed adolescents grown up. *JAMA, 281*, 1707–1713.

Whitbeck, L. B., Hoyt, D. R., Simons, R. L., Conger, R. D., Elder, G. H., Lorenz, F. O., et al. (1992). Intergenerational continuity of parental rejection and depressed affect. *Journal of Personality and Social Psychology, 63*, 1036–1045.

Wickramaratne, P. J., Greenwald, S., & Weissman, M. M. (2000). Psychiatric disorders in the relatives of probands with prepubertal-onset or adolescent-onset major depression. *Journal of the American Academy of Child and Adolescent Psychiatry, 39*, 1396–1405.

Wickramaratne, P. J., Warner, V., & Weissman, M. M. (2000). Selecting early onset MDD probands for genetic studies: Results from a longitudinal high-risk study. *American Journal of Medical Genetics, 96*, 93–101.

Wickramaratne, P. J., & Weissman, M. M. (1998). Onset of psychopathology in offspring by developmental phase and parental depression. *Journal of the American Academy of Child Adolescent Psychiatry, 37*(9), 933–942.

Woodward, L. J., & Fergusson, D. M. (2001). Life course outcomes of young people with anxiety disorders in adolescence. *Journal of the American Academy of Child Adolescent Psychiatry, 40*, 1086–1093.

Chapter 27

Genetics of Autism

Sarah Curran and Patrick Bolton

Introduction

Autism is the prototypical form of a group of disorders that are referred to as the pervasive developmental disorders (PDD). It is a behaviorally defined syndrome characterized by the presence of qualitative abnormalities in the development of reciprocal social interaction and communication, coupled with restricted, repetitive and stereotyped patterns of behavior and interests. The syndrome has, by definition, an onset before the age of 3 years. The definition and diagnostic criteria for autism in the main international classification systems are closely comparable (i.e., the International Classification of Diseases version-10 (ICD-10) of the World Health Organization and the Diagnostic and Statistical Manual version IV (DSM-IV) of the USA). However, the two classification schemes take a rather different approach to the categorization of other forms of pervasive developmental disorder. Some of the other PDD subtypes seem very likely to be closely related conditions that represent variants of autism, although this is not yet firmly established. The main subtypes recognized in ICD-10 include Asperger's syndrome, atypical autism and other pervasive developmental disorders. In DSM-IV only Asperger's syndrome and pervasive developmental disorder – not otherwise specified – PDD-NOS (DSM-IV) are separately classified. Currently these syndromes are thought to be genetically related. Accordingly, many research groups are collectively referring to this group of conditions as autism spectrum disorders (ASD) and combining them for inclusion in genetic studies.

Another condition, called Rett's syndrome, is also categorized as a PDD. However, this has a phenotypically different presentation to autism. It is largely confined to females, who appear to develop normally during the first 6 months, but subsequently there is a regression in development with a partial or complete loss of acquired skills, together with deceleration of head growth and the emergence of hand-wringing motor stereotypies, hyperventilation and truncal ataxia. Social interest, however, seems to be maintained despite these other losses. This condition appears to be genetically distinct from ASD and many cases have been found to have mutations in the gene encoding methyl-CpG-binding protein-2 (MECP2) [MIM 300005] on the X chromosome (X q28); however, there have been approximately 200 different mutations identified. Up until recently, MECP2 mutations were thought to be lethal in males, but over the last few years a few males who have survived birth, despite having a hemizygous mutation, have been identified and they have been reported to present with a neonatal encephalopathy.

Childhood disintegrative disorder (sometimes referred to as Heller's syndrome in the early literatures) is also classified under the rubric of a pervasive developmental disorder (PDD). Disintegrative disorders are characterized by an apparently normal period of development for at least the first 2 years of life, followed by a marked or dramatic and general loss of skills, coupled with the emergence of an autistic syndrome. There is often a very poor prognosis and many children deteriorate to such a degree that they end up severely or profoundly mentally impaired. It is an extremely uncommon disorder (0.2/10,000) and sometimes rare neurological conditions are identified as the cause of the regression. In some cases the child may die. Classic cases are fairly easy to distinguish from cases of typical autism, but in cases with less typical or dramatic presentations it can be difficult to make a differential diagnosis from cases with late-onset autistic regression. In genetic studies of the autism spectrum, the rare cases of Rett's syndrome and childhood disintegrative disorder are usually excluded in order to reduce heterogeneity.

Epidemiology

Early epidemiological studies conducted between the mid-1960s and mid-1970s yielded prevalence rates for autism of

S. Curran (✉)
Department of Psychologist Medicine, Institute of Psychiatry, King's College London, De Crespigny Park, London SE5 8AF, UK
e-mail: s.curran@iop.kcl.ac.uk

Y.-K. Kim (ed.), *Handbook of Behavior Genetics*,
DOI 10.1007/978-0-387-76727-7_27, © Springer Science+Business Media, LLC 2009

around 4–5 per 10,000 (Lotter, 1967). These early studies were based on classical criteria for autism, which were largely derived from the original description of Leo Kanner, who in 1943 first described the condition. Variants of classically defined autism, such as Asperger's syndrome, atypical autism and other PDD, were introduced into the ICD-10 in 1992 and Asperger's disorder and PDD-NOS were introduced into the DSM-IV in 1994. Thus, there has been a general broadening of the definition of PDD over the last decade. A recent comprehensive review by Fombonne (2005) examined all studies published in English, including early studies. The review suggested that the current rate of all PDDs is about 60 per 10,000, with autism estimated at 13 per 10,000, Asperger's disorder approximately 3 per 10,000 and other PDDs approximately 44 per 10,000. The latest epidemiological research conducted in the UK has suggested that prevalence for autism spectrum disorders may be over 1% (Baird et al., 2006). The increase in the prevalence of these disorders over the last 30–40 years has led to a heightened public awareness and concern that some environmental factor underlies what some have called the 'epidemic of autism'. Although such major and quick shifts in prevalence are incompatible with genetic explanations, this does not necessarily mean that the change is due to some specific environmental factor. Instead, much of the change seems to be attributable to the change and broadening of diagnostic concepts that have taken place, along with the improvements in methods of case identification and diagnosis. It still remains unclear, however, whether these factors account for all the change in prevalence. The inevitable uncertainty over the explanation for the increase in prevalence figures has fuelled debate and argument about the possible causal role and hence safety of childhood vaccines. Currently, however, there is no good evidence to implicate a causal role for any specific environmental risk factor, including vaccines (see below).

One of the most consistent findings from all the epidemiological reports concerns the sex difference in prevalence. Males have been found to outnumber females by an order of around 3–4:1. This difference in sex ratio seems to be even greater amongst people with Asperger's syndrome. As yet, however, the basis for the sex differences in prevalence has defied explanation.

Associated Features

Overall, relatively little is found on physical examination of children with autism, apart from the occasional pathognomic signs of a comorbid medical condition (discussed below) such as fragile X syndrome or tuberous sclerosis, which at most account for just a few percent of cases.

However, macrocephaly (head circumference greater than the 97th percentile) has been found to be present in a proportion of children with autism and Asperger's syndrome (~25%) and their family members (Fidler, Bailey, & Smalley, 2000; Gillberg & de Souza 2002; Miles, Hadden, Takahashi, & Hillman, 2000; Woodhouse et al., 1996). The presence of macrocephaly does not clearly correlate with other features of autism. It seems to develop in the early postnatal period (Courchesne, 2004) and derive from rapid early postnatal growth of the brain (Hazlett et al., 2005).

Abnormal movement kinematics are commonly found, with abnormal posturing in those with autism and gross motor problems in those with Asperger's syndrome (Rinehart et al., 2006). The precise cause for these differences in motor development is not yet understood.

Epilepsy develops in approximately 25–33%, but may not begin until adolescence or early adult life. This late age of onset is different to the typical distribution of age of onset seen in epilepsy cases without autism.

A slight increased risk of nonspecific minor congenital abnormalities has also been reported and they may index a reduced likelihood of familial recurrence (Pickles et al., 2000). There have been some studies (Rodier, Bryson, & Welch, 1997) that have suggested that abnormalities of the ears are the most common form of minor anomalies associated with autism, and these findings have been used to argue that events during specific periods of fetal brain development give rise to autism.

There is a well-established increased risk of minor obstetric and perinatal complications in those diagnosed with autism. Although at first sight these seem to be promising contenders for environmental risk factors, this is not necessarily the case. Current evidence suggests that they may represent epiphenomena or reflect a gene–environment correlation (Bolton, Murphy et al., 1997; Glasson et al., 2004; Zwaigenbaum et al., 2002).

Course and Prognosis

Long-term outcome studies of individuals with autism or autism spectrum disorders have shown that language ability and IQ are the key predictors of later functioning. However, even able people with high functioning autism and Asperger's syndrome can have problems functioning in adult life with difficulties in educational and employment progress, independent living, social relationships, behavioral and psychiatric problems (Howlin, Mawhood, & Rutter 2000).

Genetic and Environmental Determinants

A few single gene disorders have been reported in association with autism or the autism spectrum disorders. The two best established genetic conditions as causes of autism are tuberous sclerosis (TS) and fragile X syndrome. People with TS frequently develop autism or a variant of the classic syndrome. Approximately 30–50% of people with TS develop an autism spectrum disorder, a rate that is considerably higher than the rate of autism spectrum disorders in the general population (Bolton, Park, Higgins, Griffiths, & Pickles, 2002; Harrison & Bolton 1997). The rate of tuberous sclerosis in children diagnosed as suffering from an autism spectrum disorder is around 1–2%. Although this is a relatively low rate, it is still much higher than the rate of TS in the general population (~ 1 in 10,000). Therefore, the overlap between autism spectrum disorders and TS is very clear. TS is a Mendelian autosomal dominant genetic disorder that is due to mutation in one of two genes (TSC1 on chromosome 9q or TSC2 on chromosome 16p). Only one of the genes needs to be affected for TSC to be present. The TSC1 gene produces a protein called *hamartin*. The TSC2 gene produces the protein *tuberin*. These proteins form a protein complex that operates in the insulin signaling pathway to regulate cell proliferation and differentiation – the processes in which nerve cells divide to form new generations of cells and acquire individual characteristics (McCall, Chin, Salzman, & Fults, 2006). Studies are beginning to explore the risk mechanisms that lead to autism spectrum disorders in tuberous sclerosis (Bolton, Park et al., 2002).

Approximately 15–20% of those with fragile X syndrome exhibit autistic-type behaviors, such as poor eye contact, hand-flapping or odd gesture movements, hand-biting, poor sensory skills and speech/language delay (Hagerman, 2006; Hatton et al., 2006). Fragile X is a sex-linked genetic abnormality caused by mutations in the FMR1 gene on the X chromosome. The most common mutation is a triplet repeat expansion. The syndrome only usually becomes manifest once the number of triplet repeats exceeds a threshold, but progression to the full mutation may occur through a permutation stage, characterized by a subthreshold but nevertheless increased number of triplet repeats. Fragile X affects approximately 1 in every 1,000–2,000 male individuals, and the female carrier frequency may be substantially higher. Males afflicted with this syndrome typically have a moderate-to-severe form of intellectual handicap. Females may also be affected but generally have a mild form of impairment. Premutation carriers are more common and there is some evidence to suggest that they may sometimes manifest phenotypic abnormalities, such as learning difficulties and possible autism (Aziz et al., 2003; Hagerman, 2006).

The association between both these genetic disorders, yet the marked variability in the expression of the autism phenotype within them, provides an important window into potential pathophysiological mechanisms that can lead to autism.

Apart from these known genetic disorders, the only other major medical causes of autism currently seen in developed countries comprise various forms of chromosomal abnormality. In most instances a definitive causal role for these abnormalities is presumptive, but there is clear evidence that various abnormalities of chromosome 15 that involve the Prader–Willi/Angelman syndrome critical region between 15q11-13 can give rise to an increased risk for autism spectrum disorder (Bolton, Dennis et al., 2001; Cook et al., 1997; Milner et al., 2005).

Evidently though, in the vast majority of cases of autism (about 95%), there is no known medical condition to account for the syndrome (Rutter, Bailey, Bolton, & Le Conteur 1994). In the remaining apparently idiopathic cases, there has been much conjecture about etiology. Evidence for the potential importance of genetic influences in causation initially came from family studies and the recognition that a 2–4% rate of autism in siblings, although low in absolute terms, nevertheless, indicated strong familial aggregation (Rutter, 1968; Rutter & Lockyer, 1967). The sibling recurrence risk has been reported to be between ~2 and 6% (August, Stewart, & Tsai, 1981; Baird & August, 1985; Bolton, Macdonald et al., 1994; Boutin et al., 1997; Fombonne & du Mazaubrun 1992; Minton, Campbell, Green, Jennings, & Samit, 1982; Piven, Gayle et al., 1990; Ritvo, Jorde et al., 1989). However, it may be underestimated as a result of 'stoppage rules' (the tendency for families to curtail having further children, after the birth of a handicapped child (Jones & Szatmari 1988)).

The first landmark twin study to investigate the basis for this familial aggregation was reported in 1977 and it clearly demonstrated that autism was genetically determined (Folstein and Rutter 1977). However, the findings also suggested that genetic factors could not account for all the risk and that nongenetic factors were also probably operating. Subsequent twin studies on highly selected clinical populations have largely confirmed these conclusions and the concordance rates amongst monozygotic (MZ) and dizygotic (DZ) twins for autism have been estimated at (respectively) 36–91 and 0–24% (Bailey et al., 1995; Folstein & Rutter, 1977; Ritvo, Freeman, Mason-Brothers, Mo, & Ritvo, 1985; Steffenburg et al., 1989). The results indicate a complex mode of inheritance.

Environmental Risk Factors

Early theories proposed that autism was a consequence of abnormal rearing of the child by cold, aloof, 'refrigerator' parents. These and similar psychosocial theories are now discredited. But what evidence is there to implicate specific environmental risk factors? At present, it has to be concluded that there is no compelling evidence that any postulated specific environmental risk factor is involved in pathogenesis, despite many claims to the contrary. Numerous hypotheses have been proposed, concerning the role of a wide range of possible environmental exposures, including infections (prenatal and postnatal), dietary factors, vaccines and medicines as well as allergens and toxins (thimerosal, lead, mercury, etc.). Presently, however, none of the research evidence is very persuasive regarding a role for any of these. The use of vaccines in early childhood has been particularly controversial over recent years, yet the vast majority of evidence reported has failed to support the notion that vaccines play a significant role in etiology (Rutter 2005; Taylor 2006). Nevertheless, the twin data indicate that nongenetic factors play a role in shaping the expression of the phenotype and may play a role in etiology, so nongenetic factors remain to be identified. This includes, but not exclusively, environmental risks. We shall return to this issue later.

Current Issues

Broadening the Definition of the Autism Phenotype

As well as the evidence that autism is familial, with a sibling recurrence risk of around 2–4%, a number of family studies have also addressed the question as to whether other forms of pervasive developmental disorder aggregate in the families of individuals with autism. Thus, in the family study conducted by Bolton and his colleagues, the rate of other pervasive developmental disorders such as Asperger's syndrome and atypical autism and other PDDs was found to be 2.9% which is higher than the population rate for these conditions (Bolton, Macdonald et al., 1994). Rates of Asperger's syndrome have also been reported to be elevated in the first-degree relatives of children with autism compared with children with other neuropsychiatric disorders and typically developing children (Gillberg, Gillberg, & Steffenburg, 1992).

A few studies have looked at the familiality of autism spectrum disorders in the relatives of individuals with Asperger's syndrome. Thus, Ghaziuddin and colleagues (Ghaziuddin, 2005) investigated a sample of 58 children with Asperger's syndrome and compared them to a group of children with high functioning autism. They found that 5% of the Asperger's syndrome probands had a first-degree relative with Asperger's syndrome, a rate that is also likely to be higher than the rate in the general population (Ghaziuddin, 2005). A similar elevation in the rates of autism spectrum disorders in the families of individuals with Asperger's syndrome was reported by Gillberg and Cederlund (2005).

Family studies of children with autism spectrum disorders (i.e., children with autism or Asperger's syndrome or atypical autism or other forms of pervasive developmental disor-

der) have also been reported to show elevated rates of autism spectrum disorders amongst siblings (Micali, Chakrabarti, & Fombonne, 2004; Szatmari, Jones et al., 1993).

These studies all support the view that the distinction between some of the subtypes of the pervasive developmental disorders does not appear to be warranted on behavior genetic grounds. However, the low absolute rate of pervasive developmental disorders in the relatives of individuals with autism spectrum disorders means that formal statistical proof of the increased familial aggregation of these conditions is often lacking. It is possible, for example, that heterogeneity does exist, but that at the moment we have not yet hit on the right approach to subtyping. Nevertheless, subgroups of pervasive developmental disorders may be meaningful in terms of distinguishing different degrees of severity.

Apart from the issue of the familial aggregation of autism spectrum disorders, the original twin study of autism conducted by Folstein and Rutter (1977) raised another important issue concerning the phenotypic definition of autism. The principle observation was that the nonautistic co-twins of identical twins with autism seem to exhibit more subtle problems in development and educational attainment. In a series of studies, this possibility has now been more systematically investigated. To begin with, detailed evaluation of the first-, second- and third-degree relatives of carefully diagnosed probands with autism was undertaken using family history study methods. The rates of hypothetically linked impairments in social interaction, communication and play behaviors were evaluated and compared with the rates in a controlled group of relatives of individuals with Down's syndrome due to trisomy 21. The findings, illustrated in Fig. 27.1, clearly demonstrated that the siblings of individuals with autism were at increased risk for social communication impairments and restricted and repetitive patterns of interests and activities that extended well beyond the traditional diagnostic boundaries of the pervasive developmental disorders. A narrow definition of this broadened phenotypic concept was defined as a combination of impairments in any two of the three key, operationally defined

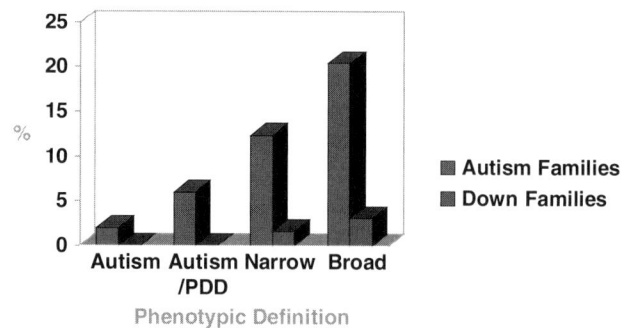

Fig. 27.1 Rates of impairments in social interaction, communication and play behaviors in siblings of subjects with autism compared to siblings of subjects with Down's syndrome

domains (social, communication, interest patterns/repetitive behaviors). Around 12% or so of siblings exhibited impairments in a combination of these areas. Relaxing the definition of the phenotype further, to include impairments in just one of these domains, led to an estimated rate of this broader phenotype in over 20% of siblings of children with autism. The comparable rate in the siblings of Down's syndrome children was around 5% (Bolton, Macdonald et al., 1994). Since then, a number of other family studies have confirmed that the relatives of individuals with autism spectrum disorders are at increased risk from subtle impairments in social communication skills and unusual interest patterns (Baron-Cohen, Wheelwright, Skinner, Martin, & Clubley, 2001; Bishop, Maybery, Maley et al., 2004; Bishop, Maybery, Wong, Maley, & Hallmayer, 2006; Constantino, Lajonchere et al., 2006; Piven, Palmer, Jacobi, Childress, & Arndt, 1997a, 1997b; Szatmari, MacLean et al., 2000).

Although the use of Down's syndrome controls and other contrast groups in the family studies that have been reported to date have demonstrated that the familial liability to this broader phenotype is largely specific to autism, the basis of the familiality of these more subtle impairments can only be addressed formally using the twin study methodology. Accordingly, the original set of twins studied by Folstein and Rutter and a new sample of twins born in the UK since the original twin study were investigated, using comparable methods of assessment undertaken in the family study by Bolton and colleagues (Bailey et al., 1995). The results of this study are illustrated in Fig. 27.2. The findings clearly demonstrated that as the definition of the phenotype was successively broadened and relaxed to include more and more subtle impairments, the concordance rates in the MZ twins steadily increased whereas the concordance rates in the DZ twin pairs changed to a very limited extent (Bailey, et al., 1995). The implications were clear: the genetic liability to autism confers a risk for a broader range of impairments in social communication skills and unusual patterns of interests and activities. As with previous research on the genetic epidemiology of autism it was also evident that male relatives of

probands with autism were much more likely to exhibit these impairments than female relatives.

The evidence for broadening the phenotypic definition beyond the traditional diagnostic boundaries of autism has been a key advance and has led to a number of reconceptualization of diagnostic criteria and approaches to classification. In addition, it has raised the important question as to what elements comprise this broader phenotype and where the boundaries of the phenotype should be drawn. Some more in-depth studies of the features of the broader phenotype have indicated difficulties in the formation of friendships and intimate, confiding relationships (Le Couteur et al., 1996; Piven et al., 1997a, 1997b; Szatmari, Volkmar, & Walter, 1995). In the communication domain, impairments in the pragmatics of communication (understanding and responding to the needs of the listener in communication) have also been found to characterize the broader phenotype (Landa, Folstein, & Isaacs, 1991; Landa, Piven et al., 1992). The evidence regarding other aspects of language development and function has not been studied in much depth, but it appears that phonological short-term memory does not show the same familiality observed in normal and language-disordered families (Bishop, Maybery, Wong et al., 2004). From a conceptual stance, it makes intuitive sense that the phenotype should include autistic-like impairments, but various other lines of evidence have raised the question as to whether other psychiatric disorders or personality differences may also stem from the genetic liability to autism. The issue with regard to other psychiatric disorders was also investigated in the families studied by Bolton, Macdonald et al. (1994). Systematic and standardized evaluations of psychiatric disorder were undertaken using the lifetime version of the Schedule for Affective Disorders and Schizophrenia in addition to family history methods (Bolton, Pickles, Murphy, & Rutter, 1998). Contrary to earlier speculations that autism might reflect a form of early-onset psychotic disorder of childhood, there was no suggestion of an increased rate of schizophrenia or other psychoses in the families of the children with autism compared with the Down's syndrome control families. There was however evidence for increased familial aggregation of affective disorders including anxiety and depression. However, the pattern of findings did not clearly suggest that these affective disturbances constituted part of the broader phenotype. Rather, the results raised the possibility that the increased propensity of relatives to suffer from these problems reflected the stresses and demands of raising a child with a serious developmental disorder (Bolton, Pickles et al., 1998). The question could not be resolved entirely in that investigation and warrants further study in the future. Interestingly, in this and other studies, the relatives of people with autism appeared to be at a greater risk of developing obsessive compulsive disorders (Bolton, Macdonald

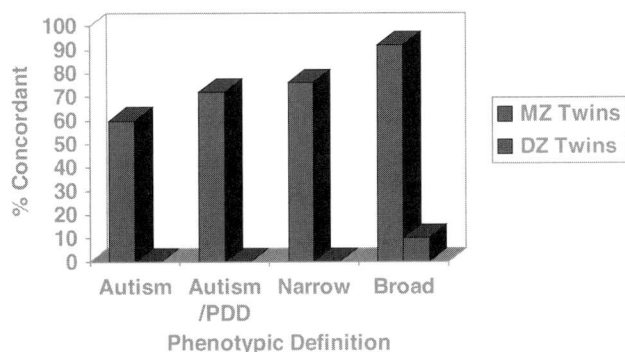

Fig. 27.2 Concordance rates in MZ and DZ twins according to definitions of the autism phenotype

et al., 1994; Micali et al., 2004; Wilcox, Tsuang, Schnurr, & Braida-Fragoso 2003).

Studies of the personality attributes of the relatives of autistic children have also been undertaken. The results of one of the more comprehensive study indicated that first-degree relatives were significantly more anxious, impulsive, aloof, shy, oversensitive, irritable and eccentric than controls (Murphy et al., 2000). Factor analysis of this data suggested that the personality attributes fell into three main areas: withdrawn, tense and difficult (Murphy, et al.). Other studies of the personality traits of relatives have also supported the notion that certain personality styles seem to be more common in parents and first-degree relatives of children with autism, although the precise pattern of findings has not been entirely consistent (Narayan, Moyes, & Wolff 1990; Piven, Wzorek et al.,1994; Piven et al.). In some respects it seems that the associated personality traits probably represent the adult expression of the social and communication difficulties and unusual interest patterns that have been reported in siblings early in their development (Murphy et al., 2000).

Variable Expression

Another key question raised by the evidence for a broader phenotype of autism concerns the determinants of the variation in phenotypic manifestation. The data to answer this question are limited and the findings are somewhat difficult to interpret. First, it is clear that amongst identical twins, one twin may have severe autism yet the co-twin may just exhibit subtle signs of the broader phenotype (Bailey et al., 1995). To this extent, the same genetic liability can be associated with marked variation in phenotypic expression. The question still remains, however, as to whether more severely affected individuals are more likely to have more affected relatives who are more severely affected, as predicted by the multi-factorial threshold model. Le Couteur and colleagues studied MZ twin pairs concordant for a diagnosis of an autism spectrum disorder. They found no evidence for greater within- than between-twin-pair similarity for the autism phenotype, suggesting that symptom severity is not correlated with liability and that differences in clinical manifestations are not an indication of genetic heterogeneity (Le Couteur et al., 1996). The number of twin pairs was, however, small and the power of the study limited. By contrast however a more recent study found significantly reduced within-family variants for all the main domains of the autism phenotype in their study of MZ twin pairs (Kolevzon, Smith, Schmeidler, Buxbaum, & Silverman, 2004). These latter findings are more in keeping with the evidence from the family study data, where familiality of the broader autism phenotype was correlated with the severity of the autistic symptoms

in the proband, most clearly amongst those probands that had useful speech (Bolton, Macdonald et al., 1994; Pickles et al., 2000) and the reports that severity is correlated in affected sibling pairs (Spiker, Lotspeich, Dimiceli, Myers, & Risch, 2002). These finding are broadly consistent with the predictions of the multi-factorial threshold model. However, intriguingly, in the family studies that have examined the issue, there was no evidence that the less frequently affected sex (females) were at higher liability for autism spectrum disorders (Bolton, Macdonald et al., 1994; Pickles, et al.).

Dimensions Versus Categories

Another issue raised by the findings of a broader autism phenotype concerns the conceptualization of the autism spectrum disorders and whether they are best considered as categorically distinct or whether they represent the extreme end of normally distributed traits. In support of the dimensional conceptualization, a number of questionnaire-based studies of families of individuals with autism spectrum disorders and investigations of general population-based samples have indicated that autistic-like traits are 'quasi'normally distributed in the general population and that the distribution amongst siblings is shifted upwards, partway between the general population mean and the mean for individuals with autism spectrum disorders (Bishop, Maybery, Wong, Maley, & Hallmayer, 2006; Constantino, Lajonchere et al., 2006; Constantino & Todd, 2003, 2005). Moreover, the distribution in the sibs of autism spectrum disorder probands does not appear to be bimodal in form, although the numbers of individuals are too small for a definitive conclusion about this. These findings raised the important question as to whether or not the factors that determine variation in autistic-like traits within the normal range are involved in the etiology of autism spectrum disorder. As yet, a definitive answer is not available and will only come once susceptibility genes for autism are identified (Ronald, Happe, Bolton et al., 2006).

Compound Phenotypes

Although the classic syndrome of autism is characterized by a combination of impairments in reciprocal social interaction and communication, coupled with restricted and stereotyped patterns of behavior and interest, there has been a long-standing interest in the nature of the relationship between these different aspects of the syndrome. Quite a few investigations have been undertaken in an attempt to examine the relationship. For the most part, the reports examining this question have represented secondary analyses of data collected for other purposes and as a consequence there are inevitable difficulties with the interpretation of the findings. This is perhaps particularly the case with respect to the investigation of multiplex families collected and recruited for the purpose of molecular genetic linkage studies. The problem lies in the fact that selection bias could have a very significant impact on the results. Moreover, the investigation of clinically defined populations gives rise to an inherent problem in the circularity of the logic of the argument: examining the inter-relationship between the triad of impairments in a population that is defined by the presence of impairment in all three areas will inevitably lead to considerable restriction in the variance in each domain and a high likelihood to conclude that the domains are inter-related.

The typical approach has usually entailed some form of factor analysis of data from clinical samples. In a number of studies the results have indicated that a large proportion of the variance in autistic behaviors loads on an unrotated first principle component (Constantino, Gruber et al., 2004; Szatmari, Merette et al., 2002; Volkmar et al., 1988; Wadden, Bryson, & Rodger 1991). The results suggest that the triad of impairments should be construed as a unitary phenomenon. However, not all studies have found this factor structure, and several other reports have found that between 3- and 6-factor solutions provide the best explanation for the data (Berument, Rutter, Lord, Pickles, & Bailey, 1999; DiLalla & Rogers, 1994; Miranda-Linne & Melin 2002; Stella, Mundy, & Tuchman, 1999; Tadevosyan-Leyfer et al., 2003; Wadden, et al.). The differences in findings may reflect the differences in the approach to assessment and measurement of the autistic behaviors as well as the differences between studies in the methods of case ascertainment and the age and characteristics of the populations investigated.

A number of investigators have used cluster analysis as an alternative approach to this question. In general, these cluster analytic studies have not been particularly informative about the relationship between components of the triad of impairments but a few studies have suggested that symptom severity may define clusters and that within each group ordered by severity the children exhibit a combination of social impairments, communication impairments and repetitive stereotype behaviors (Constantino, Gruber et al., 2004; Sevin et al., 1995; Spiker et al., 2002; Szatmari, Bartolucci, & Bremner, 1989; Waterhouse et al., 1996).

Two studies have used factor analytic approaches to look at the structure of the triad in the general population. Although this strategy has the benefits of less bias and larger sample size it is inevitably constrained by the fact that very few individuals in the general population will be scoring in the clinical range. The first reports represented results from the social responsiveness scale which is a measure of autistic-like traits (Constantino, Davis et al., 2003). The authors undertook a confirmatory factor analysis which produced a first factor that explained a large proportion of the variance (Constantino, Gruber et al., 2004). In general the findings were interpreted as supporting the notion of there being a unitary factor. It must be noted however that the social responsiveness scale predominantly includes items that are about social behaviors, and there are relatively few items that tap into other aspects of the triad of impairments. It is of note therefore that somewhat different results were obtained by an analysis of another questionnaire called the Autism Spectrum Quotient (AQ) which is a self-report measure of autistic symptomatology for adults. Analysis of data from this questionnaire produced a 3-factor solution with the factors described as pertaining to social skills, details and patterns and communication and mind reading abilities (Austin, 2005).

Another approach to the issue, that circumvents the pitfalls of studying clinical groups and the problems associated with low rates of the more severe difficulties in population-based samples, has been to investigate the factor structure in the relatives of carefully diagnosed and characterized children with autism as well as the relatives of individuals with other conditions such as Down's syndrome. The advantage of this strategy is that the samples are not selected clinically but they are enriched with individuals at risk for autistic-like impairments (Bolton, Macdonald et al., 1994). The findings from a confirmatory factor analysis supported the notion of a hierarchical model with three factors relating to social impairments, communication impairments and restrictive repetitive patterns of interest and activity, all of which loaded on one single latent factor pertaining to autism. The disadvantage of this approach was the potential of a reporting bias, because the information regarding the relatives was obtained from a parent, using the family history method.

Yet another approach to the issue has been to employ more genetically informed approaches, by studying the co-segregation of traits according to the degree of genetic resemblance of relatives. A number of studies have examined the pattern of familial aggregation of aspects of the phenotype. A detailed and systematic examination of the question has been undertaken by Ronald et al. (Ronald, Happe, Bolton et al., 2006; Ronald, Happe, & Plomin, 2005). They studied a large general population cohort of UK twin pairs that were originally recruited at 18 months and that have then been studied over the intervening years using a number of different assessment tools. When they reach the age of 8 years, the parents were sent the CAST (Williams et al., 2005) questionnaire (childhood Asperger's syndrome test). This test was developed for primary school-age children and covers a number of autistic-like traits. The strength of using a twin sample for this purpose was that it allowed for multivariate model fitting to test the extent to which different compo-

Males/females

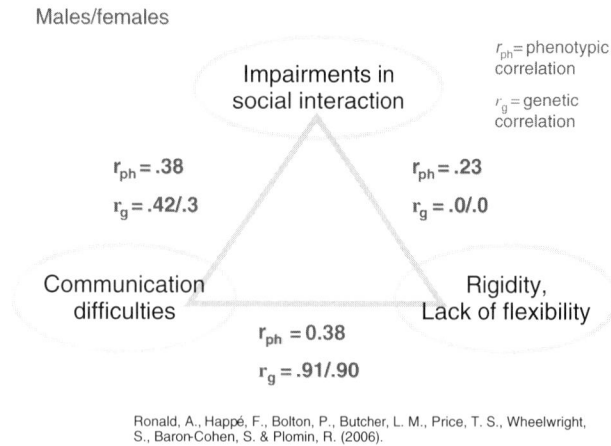

Ronald, A., Happé, F., Bolton, P., Butcher, L. M., Price, T. S., Wheelwright,
S., Baron-Cohen, S. & Plomin, R. (2006).

Fig. 27.3 Population survey of autistic traits; phenotype- genotype correlations in dimensional autistic traits

nents of the phenotype had distinct as well as shared genetic and environmental risk factors. The results of this study are summarized in Fig. 27.3. The findings are noteworthy for showing that each component of the autism phenotype was highly heritable, but that the phenotypic and genetic correlations between the different triad of impairments were relatively modest (Ronald, Happe, Bolton et al., 2006). The findings supported the notion that the autism phenotype is more complex than a unitary model would suggest and that there may be a need to search for genetic susceptibility loci that relate to different components of the phenotype as well as for loci that contribute to all aspects of the syndrome.

Heterogeneity and Genetic Architecture

There is clear evidence of etiological and genetic heterogeneity in autism spectrum disorder (cf. tuberous sclerosis, fragile X, chromosome abnormalities); however, the extent of heterogeneity is unknown and this potentially possesses a major challenge in the search for susceptibility genes. The limited available behavior genetic data have suggested that the presence of useful speech may index heterogeneity (Bolton, Macdonald et al., 1994; Pickles et al., 2000) and this possibility is partly supported by evidence suggesting that linkage signals differ when sib pairs without speech are analyzed separately (Alarcon, Cantor, Liu, Gilliam, & Geschwind, 2002; Buxbaum, Silverman, Smith, Kilifarski, et al., 2001). Other putative markers of heterogeneity include the presence of minor congenital anomalies (Pickles, et al., 2000), savant skills (Nurmi, Dowd et al., 2003) and repetitive compulsive behaviors/insistence on sameness (Shao et al., 2003; Sutcliffe et al., 2005).

The reported recurrence risk for autism amongst siblings of 2–4% taken with the early prevalence estimates for autism of 2–4/10,000 led to the view that the relative risk to siblings (λs) was 50–100 fold higher than the rate in the general population. This estimate suggested that traditional molecular linkage investigations of affected sib pairs should have a fair chance of identifying susceptibility genes. However, now that recent epidemiological studies have shown that the prevalence is much higher, it is clear that λs was significantly overestimated. It is hard to accurately estimate λs from the currently available data, because reported prevalence rates in siblings may be biased by stoppage effects and the general population rates vary considerably by time, place and method. As a result, the range of estimates is wide and the confidence interval broad, but it now seems that λs is more likely to lie somewhere from 5 to 25. This might partly explain why studies in the largest samples so far reported have failed to identify strong linkage signals (Yonan, Alarcon et al., 2003).

The published twin studies of autism have not been amenable or subjected to contemporary model fitting approaches, partly because the samples were too small in size and highly selected. Nevertheless, all the studies have indicated a very high MZ concordance rate (although less than 100%) and a DZ concordance rate that, in the more rigorous studies, was less than half the MZ concordance rate. Indeed, as far as it was possible to tell with such small samples, the rate of autism in DZ co-twins appears to be no greater than the sibling recurrence rate (Bailey et al., 1995; Folstein and Rutter, 1977; Ritvo, Freeman et al., 1985; Steffenburg et al., 1989;). The findings therefore indicate that heritability is very high and that the involvement of shared environmental risk factors is unlikely. The results have also suggested the presence of epistasis (gene–gene interaction). This represents another challenge for molecular genetic studies.

Three twin studies of autistic-like traits in general population samples of twins have been published, and they have been analyzed using multivariate model fitting approaches. All have found that the large majority of the variance is attributable to genetic factors and two have found no evidence for a shared environmental contribution (Constantino, Hudziak, & Todd, 2003; Constantino & Todd, 2003; Ronald, Happe, Bolton et al., 2006; Ronald, Happe, & Plomin, 2005; Scourfield, McGuffin, & Thapar, 1997). In the largest study to date (Ronald, Happe, Bolton, et al.), there was some indication of nonadditive genetic effects. Defries–Fulker extreme analysis of these data suggested that the pattern of findings was similar at the extreme of the distribution, although the DZ twin correlations were lower at the extremes. However, as outlined above, the findings also suggested that the genetic and environmental influences for the three main components of the triad differed (Ronald, Happe, Bolton et al., 2006).

Molecular Genetics

The evidence from the behavior genetic analyses for strong genetic determination of autism has prompted a large number of molecular genetic studies aimed at identifying susceptibility genes. Genome-wide linkage scans of multiplex autism families have identified several genomic regions that may harbor susceptibility genes, and across studies there is some emerging convergence in findings. Evidence for linkage on 7q and 2q has been amongst the more consistent findings, but there is also suggestive evidence of linkage on 3q, 16p, 17q and 13q (Veenstra-VanderWeele & Cook 2004). Guided by these and other findings, positional candidate genetic association studies have been conducted and a number of positive association have been reported. For the most part, the findings remain largely unreplicated at present, but association with markers in gamma-aminobutyric acid GABA$_A$ receptor subunit genes (Ashley-Koch et al., 2006; Curran et al., 2006; McCauley et al., 2004) and the serotonin (5-HT) pathway genes are beginning to be replicated (e.g., Ashley-Koch, et al.; Betancur et al., 2002; Curran, Purcell, Craig, Asherson, & Sham, 2005; Kim et al., 2002; Klauck, Poustka, Benner, Lesch, & Poustka, 1997; McCauley, et al.; Mulder, Anderson, Anderson, Kema, Brugman et al., 2005; Sutcliffe et al., 2005). Whilst these results have not yet identified a single causative mutation of major effect, they do implicate these pathways in the genetic underpinnings of autism. Another approach has been to study genes disrupted in the cases of autism with rare co-occurring cytogenetic abnormalities (Castermans, Wilquet, Parthoens et al., 2003; Vorstman et al., 2006). This approach has helped identify neuroligin genes (Jamain et al., 2003) and the neurobeachin gene (Castermans, Wilquet, Steyaert et al., 2004). Is there any evidence that different aspects of the autism phenotype are determined by different susceptibility genes? At this stage there are relatively limited data to answer this question, because most of the evidence comes from linkage studies which can only provide an indirect support for the idea. Nevertheless, a number of linkage scans which have been reported have undertaken subsidiary analyses to see whether or not the strength and pattern of the linkage signal are correlated with different features/traits. There is some suggestion that this may be the case with reports that linkage signals are increased in certain regions when the analyses focus on more obsessive compulsive features of autism and in different regions when one focuses more on language and communication impairments (Alarcon et al., 2002; Buxbaum, Silverman, Smith, Kilifarski, et al., 2001; Nurmi, Dowd et al., 2003; Shao et al., 2003; Sutcliffe et al., 2005). However, many of these analyses were not predicated on strong a priori grounds, and linkage studies of relatively small samples are notoriously sensitive to minor changes in phenotypic definition, so for the moment, the results can only be considered as partially supportive.

Imprinted Genes

Analysis of linkage data has indicated the parent of origin effects and hence the possibility that imprinted susceptibility genes may be involved in etiology (i.e., genes being expressed preferentially from either father or mother). Further support for this notion has come from evidence that maternally rather than paternally derived duplications of chromosome 15q11-13 seem to be associated with an increased risk for autism spectrum disorder (Bolton, Dennis et al., 2001; Cook et al., 1997; Milner et al., 2005; Vorstman et al., 2006). Genes that are maternally or possibly maternally expressed in this region, such as UBE3A and GABRB3, have been implicated (Buxbaum, Silverman, Smith, Greenberg, et al., 2002; Curran et al., 2006; McCauley et al., 2004; Nurmi, Amin et al., 2003; Nurmi, Bradford et al., 2001). Based on these data and studies of postmortem brain tissue, a mixed epigenetic and genetic model for autism has been proposed with both de novo and inherited contributions (Jiang et al., 2004).

Environmental Risks

Although in general no clear environmental risk factors have been implicated in etiology, there is evidence that extreme early deprivation, as experienced by, for example, Romanian orphans, may give rise to 'quasi'autistic-like traits (Rutter, Andersen-Wood et al., 1999). In addition, there is some evidence to suggest that severe early-onset epilepsy may be an intrinsic environmental risk factor (Bolton, Park et al., 2002; Humphrey, Neville, Clarke, & Bolton, 2006). In neither case, however, do these factors play a role in the vast majority of cases of autism spectrum disorder.

Future Directions

First, it is clear that more genetic epidemiological research is required in order to clarify questions concerning the genetic architecture of autism. These will need to investigate population-based samples and to integrate molecular with behavior genetic approaches. An obvious gap at present is the absence of any adoption studies of autism spectrum disorders or autistic-like traits. In addition, however, twin and family studies of autism spectrum disorders and the broader pheno-type are required, as our current knowledge is almost entirely based on studies of more traditionally defined autism. Some caution is needed here, because it is evident that the current findings may be sensitive to the measurement properties of the instruments used for assessing the phenotype and, in addition, there is still a good deal of uncertainty concerning the elements that make up the trait(s). As such, further work on the development of measurement tools will be required. In particular, it will be necessary to develop measures that gather information from different informants and sources (parents, teachers, etc.). Moreover, these studies will have to adopt more developmental perspectives on the phenotype, which changes over time in the nature of its expression. One obvious area for behavior genetics research in this regard will be to examine the inter-relationship between correlated features of the phenotype longitudinally, such as language development (Dworzhinsky et al., 2007) and imitation skills (McEwen et al., 2007), in order to explore the basis for the ontogeny of associations.

Second, there is a need to investigate the potential of various putative endophenotypes in autism and related traits in order to clarify unresolved issues about the nature of the genetic architecture and pathophysiological processes. Endophenotypes are conceptualized as phenotypes that are more proximal to the underlying biological processes of more distal behavioral measures. They are traits that share a substantial genetic component with a clinical disorder, but which are also found in apparently unaffected relatives (Gottesman & Gould 2003; Gould & Gottesman 2006). Studying the inheritance of endophenotypes can increase the power to detect the genes involved and clarify the pathways leading from genetic predisposition to clinical disorder. Candidate endophenotypes for ASD are neurocognitive differences in executive function, theory of mind abilities and central coherence (Baron-Cohen, 1992; Baron-Cohen & Hammer, 1997; Briskman, Happe, Frith, 2001; Happe, Briskman, & Frith, 2001; Hughes, Plumet, & Leboyer, 1999; Hughes et al., 1997; Ozonoff, Rogers, Farnham, & Pennington, 1993; Piven & Palmer 1997; Szatmari, Jones et al., 1993); neuroanatomical brain changes (as identified by in vivo neuroimaging, e.g., McAlonan et al., 2005) and neurochemical measures of platelet serotonin (Mulder, Anderson, Kema, de Bildt et al., 2004; Perry, Cook, Leventhal, Wainwright, & Freedman,1991). Neurophysiological measures may also hold promise (Grice, Halit et al., 2005; Grice, Spratling et al., 2001).

Third, large-scale molecular genetic studies are now underway, addressing the problems of the earlier underpowered studies. Recently, these have identified copy number variation as a potential risk factor for ASD and coupled with genome-wide association studies, genes are being identified (Cook & Scherer, 2008). These approaches, coupled with advances in bioinformatics (Yonan, Palmer et al., 2003) will help identify the network of genes that create the risk for autism. In addition, further investigations of the mechanisms that give rise to autism spectrum disorder in individuals with known Mendelian disorders will compliment these studies and throw light on developmental processes.

Fourth, it is evident that there is a need to identify the nongenetic factors that are involved in risk and variable expression of the phenotype. A number of large-scale epidemiological case–control (Hertz-Picciotto et al., 2006) and cohort studies are underway that should help test various environmental risk theories, but this line of enquiry is limited by the lack of any strong guiding principles to help narrow the field of search.

Fifth, one of the major unexplained enigmas of autism spectrum disorders concerns the explanation for the sex differences in prevalence. An explanation for this would represent a major step forward, but at present it remains rather elusive with no clear evidence to implicate any factor, although X-linked genes (Jamain et al., 2003) and fetal testosterone levels have been proposed to account for the differences (Knickmeyer, Baron-Cohen, Fane et al., 2006; Knickmeyer, Baron-Cohen, Raggatt, Taylor, & Hackett, 2006).

A sixth area for future research will be animal studies. At present, a behavioral model of autism in animals has not been convincingly demonstrated. This is not surprising given that the condition is primarily characterized by abnormalities in social behavior and communication. A number of studies of potentially relevant traits are being investigated in animal models, but it is too early to tell how informative they will be in furthering our understanding of the human condition. Much more fruitful and promising at the moment, however, are the investigations of animals with gene mutations that are implicated in the etiology of autism (e.g., fragile X; Mineur, Huynh, & Crusio, 2006). As the genetic architecture of autism becomes clearer and susceptibility genes are identified, investigations of their role in brain and behavioral development in animals will provide fundamental insights into pathophysiological mechanisms. Patrick Bolton was supported by the UK NIHR Biomedical Research Centre for Mental Health at the Institute of Psychiatry, Kings College London and The South London and Maudsley NHS Foundation Trust.

References

Alarcon, M., Cantor, R. M., Liu, J., Gilliam, T. C., & Geschwind, D. H. (2002). Evidence for a language quantitative trait locus on chromosome 7q in multiplex autism families. *American Journal of Human Genetics, 70*(1), 60–71.

Ashley-Koch, A. E., Mei, H., Jaworski, J., Ma, D. Q., Ritchie, M. D., Menold, M. M., et al. (2006). An analysis paradigm for investigating multi-locus effects in complex disease: Examination of three GABA receptor subunit genes on 15q11-q13 as risk factors for autistic disorder. *Annals of Human Genetics, 70*(Pt 3), 281–292.

August, G. J., Stewart, M. A., & Tsai, L. (1981). The incidence of cognitive disabilities in the siblings of autistic children. *The British Journal of Psychiatry, 138*, 416–422.

Aziz, M., Stathopulu, E., Callias, M., Taylor, C., Turk, J., Oostra, B., et al. (2003). Clinical features of boys with fragile X premutations and intermediate alleles. *American Journal of Medical Genetics Part B: Neuropsychiatric Genetics, 121*(1), 119–127.

Bailey, A., Le Couteur, A., Gottesman, I., Bolton, P., Simonoff, E., Yuzda E., et al. (1995). Autism as a strongly genetic disorder: Evidence from a British twin study. *Psychological Medicine, 25*(1), 63–77 issn: 0033-2917.

Baird, G., Simonoff, E., Pickles, A., Chandler, S., Loucas, T., Meldrum, D., et al. (2006). Prevalence of disorders of the autism spectrum in a population cohort of children in South Thames: The Special Needs and Autism Project (SNAP). *Lancet, 368*(9531), 210–215.

Baird, T. D., & August, G. J. (1985). Familial heterogeneity in infantile autism. *Journal of Autism and Developmental Disorders, 15*(3), 315–321 issn: 0162-3257.

Baron-Cohen, S. (1992). Out of sight or out of mind? Another look at deception in autism. *Journal of Child Psychology and Psychiatry, and Allied Disciplines, 33*(7), 1141–1155 issn: 0021-9630.

Baron-Cohen, S., & Hammer, J. (1997). Parents of children with Asperger syndrome: What is the cognitive phenotype? *Journal of Cognitive Neuroscience, 9*(4), 548–554.

Baron-Cohen, S., Wheelwright, S., Skinner, R., Martin, J., & Clubley, E. (2001). The autism-spectrum quotient (AQ): Evidence from Asperger syndrome/high-functioning autism, males and females, scientists and mathematicians. *Journal of Autism and Developmental Disorders, 31*(1), 5–17.

Berument, S. K., Rutter, M., Lord, C., Pickles, A., & Bailey, A. (1999). Autism screening questionnaire: Diagnostic validity. *The British Journal of Psychiatry, 175*, 444–451.

Betancur, C., Corbex, M., Spielewoy, C., Philippe, A., Laplanche, J.-L., Launay, J.-M., et al. (2002). Serotonin transporter gene polymorphisms and hyperserotonemia in autistic disorder. *Molecular Psychiatry, 7*(1), 67–71.

Bishop, D. V., Maybery, M., Maley, A., Wong, D., Hill, W., & Hallmayer, J. (2004). Using self-report to identify the broad phenotype in parents of children with autistic spectrum disorders: A study using the Autism-Spectrum Quotient. *Journal of Child Psychology and Psychiatry, 45*(8), 1431–1436.

Bishop, D. V., Maybery, M., Wong, D., Maley, A., & Hallmayer, J. (2006). Characteristics of the broader phenotype in autism: A study of siblings using the children's communication checklist-2. *American Journal of Medical Genetics Part B: Neuropsychiatric Genetics, 141*(2), 117–122.

Bishop, D. V., M. Maybery, Wong, D. Maley, A., Hill, W. & Hallmayer, J. (2004). Are phonological processing deficits part of the broad autism phenotype? *American Journal of Medical Genetics Part B: Neuropsychiatric Genetics, 128*(1), 54–60.

Bolton, P. F., Dennis, N. R., Browne, C. E., Thomas, N. S., Veltman, M. W. M., Thompson, R. J., et al. (2001). The phenotypic manifestations of interstitial duplications of proximal 15q with special reference to the autistic spectrum disorders. *American Journal of Medical Genetics, 105*(8), 675–685.

Bolton, P., Macdonald, H., Pickles, A., Rios, P., Goode, S., Crowson, M., et al. (1994). A case-control family history study of autism. *Journal of Child Psychology and Psychiatry, 35*, 877–900.

Bolton, P. F., Murphy, M., Macdonald, H., Whitlock, B., Pickles, A., & Rutter, M. (1997). Obstetric complications in autism: Consequences or causes of the condition? *Journal of the American Academy of Child and Adolescent Psychiatry, 36*(2), 272–281 issn: 0890-8567.

Bolton, P. F., Park, R. J., Higgins, J. N. P., Griffiths, P. D., & Pickles, A. (2002). Neuro-epileptic determinants of autism spectrum disorders in tuberous sclerosis complex. *Brain, 125*(Pt 6), 1247–1255.

Bolton, P., Pickles, A., Murphy, M., & Rutter, M. (1998). Autism, affective and other psychiatric disorders: Patterns of familial aggregation. *Psychological Medicine* 28: 385–395.

Boutin, P., Maziade, M., Merette, C., Mondor, M., Bedard, C., & Thivierge, J. (1997). Family history of cognitive disabilities in first-degree relatives of autistic and mentally retarded children. *Journal of Autism and Developmental Disorders, 27*(2), 165–176 issn: 0162-3257.

Briskman, J., Happe, F., & Frith, U. (2001). Exploring the cognitive phenotype of autism: Weak "central coherence" in parents and siblings of children with autism: II. Real-life skills and preferences. *Journal of Child Psychology and Psychiatry, 42*(3), 309–316.

Buxbaum, J. D., Silverman, J. M., Smith, C. J., Greenberg, D. A., Kilifarski, M., Reichert, J., et al. (2002). Association between a GABRB3 polymorphism and autism. *Molecular Psychiatry, 7*(3), 311–316.

Buxbaum, J. D., Silverman, J. M., Smith, C. J., Kilifarski, M., Reichert, J., Hollander, E., et al. (2001). Evidence for a susceptibility gene for autism on chromosome 2 and for genetic heterogeneity. *American Journal of Human Genetics, 68*(6), 1514–1520.

Castermans, D., Wilquet, V., Parthoens, E., Huysmans, C., Steyaert, J., Swinnen, L., et al. (2003). The neurobeachin gene is disrupted by a translocation in a patient with idiopathic autism. *Journal of Medical Genetics, 40*(5), 352–356.

Castermans, D., Wilquet, V., Steyaert, J., Van de Ven, W., Fryns, J. P., & Devriendt, K. (2004). Chromosomal anomalies in individuals with autism: A strategy towards the identification of genes involved in autism. *Autism, 8*(2), 141–161.

Constantino, J. N., Davis, S. A., Todd, R. D., Schindler, M. K., Gross, M. M., Brophy, S. L., et al. (2003). Validation of a brief quantitative measure of autistic traits: Comparison of the social responsiveness scale with the autism diagnostic interview-revised. *Journal of Autism and Developmental Disorders, 33*(4), 427–433.

Constantino, J. N., Gruber, C. P., Davis, S., Hayes, S., Passanante, N., & Przybeck. T. (2004). The factor structure of autistic traits. *Journal of Child Psychology and Psychiatry, 45*(4), 719–726.

Constantino, J. N., Hudziak, J. J., & Todd, R. D. (2003). Deficits in reciprocal social behavior in male twins: Evidence for a genetically independent domain of psychopathology. *Journal of the American Academy of Child and Adolescent Psychiatry, 42*(4), 458–467.

Constantino, J. N., Lajonchere, C., Lutz, M., Gray, T., Abbacchi, A., McKenna, K., et al. (2006). Autistic social impairment in the siblings of children with pervasive developmental disorders. *The American Journal of Psychiatry, 163*(2), 294–296.

Constantino, J. N., & Todd, R. D. (2003). Autistic traits in the general population: A twin study. *Archives of General Psychiatry, 60*(5), 524–530.

Constantino, J. N. & Todd, R. D. (2005). Intergenerational transmission of subthreshold autistic traits in the general population. *Biological Psychiatry, 57*(6), 655–660.

Cook, E. H., Jr., Lindgren, V., Leventhal, B. L., Courchesne, R., Lincoln, A., Shulman, C., et al. (1997). Autism or atypical autism in maternally but not paternally derived proximal 15q duplication. *American Journal of Human Genetics, 60*(4), 928–934 issn: 0002-9297.

Cook, E. H., Jr., Scherer, S. W. (2008). Copy-number variations associated with neuropsychiatric conditions. *Nature, 455*, 919–923.

Courchesne, E. (2004). Brain development in autism: Early overgrowth followed by premature arrest of growth. *Mental Retardation and Developmental Disabilities Research Reviews, 10*(2), 106–111.

Curran, S., Powell, J., Neale, B. M., Dworzynski, K., Li, T., Murphy, D., et al. (2006). An association analysis of candidate genes on chromosome 15 q11-13 and autism spectrum disorder. *Molecular Psychiatry, 11*(8), 709–713.

Curran, S., Purcell, S., Craig, I., Asherson, P., & Sham, P. (2005). The serotonin transporter gene as a QTL for ADHD. *American Journal of Medical Genetics B, Neuropsychiatric Genetics, 134B,* 42–47.

DiLalla, D. L., & S. J. Rogers (1994). Domains of the childhood autism rating scale: Relevance for diagnosis and treatment. *Journal of Autism and Developmental Disorders, 24*(2), 115–128.

Dworzynski, K., Ronald, A., Hayiou-Thomas, M., Rijsdijk, F., Happé, F., Bolton, P. F., & Plomin, R. (2007). Aetiological relationship between language performance and autistic-like traits in childhood: a twin study. *International Journal of Language & Communication Disorders, 42,* 273–292.

Fidler, D. J., Bailey, J. N., & Smalley, S. L. (2000). Macrocephaly in autism and other pervasive developmental disorders. *Developmental Medicine and Child Neurology, 42*(11), 737–740.

Folstein, S., & Rutter, M. (1977). Infantile autism: A genetic study of 21 twin pairs. *Journal of Child Psychology and Psychiatry, 18,* 297–321.

Fombonne, E. (2005). Epidemiology of autistic disorder and other pervasive developmental disorders. *The Journal of Clinical Psychiatry, 66*(Suppl. 10), 3–8.

Fombonne, E. & du Mazaubrun, C. (1992). Prevalence of infantile autism in four French regions. *Social Psychiatry and Psychiatric Epidemiology, 27,* 203–210.

Ghaziuddin, M. (2005). A family history study of Asperger syndrome. *Journal of Autism and Developmental Disorders, 35*(2), 177–182.

Gillberg, C., & Cederlund, M. (2005). Asperger syndrome: Familial and pre- and perinatal factors. *Journal of Autism and Developmental Disorders, 35*(2), 159–166.

Gillberg, C., & de Souza, L. (2002). Head circumference in autism, Asperger syndrome, and ADHD: A comparative study. *Developmental Medicine and Child Neurology, 44*(5), 296–300.

Gillberg, C., Gillberg, I. C., & Steffenburg, S. (1992). Siblings and parents of children with autism: A controlled population-based study. *Developmental Medicine and Child Neurology, 34*(5), 389–98 issn: 0012-1622.

Glasson, E. J., Bower, C., Petterson, B., de Klerk, N., Chaney, G. & Hallmayer, J. F. (2004). Perinatal factors and the development of autism: A population study. *Archives of General Psychiatry, 61*(6), 618–627.

Gottesman, I. I., & Gould, T. D. (2003). The endophenotype concept in psychiatry: Etymology and strategic intentions. *The American Journal of Psychiatry, 160*(4), 636–645.

Gould, T. D., & Gottesman, I. I. (2006). Psychiatric endophenotypes and the development of valid animal models. *Genes, Brain, and Behaviour, 5*(2): 113–119.

Grice, S. J., Halit, H., Farroni, T., Baron-Cohen, S.,. Bolton, P., & Johnson, M. H. (2005). Neural correlates of eye-gaze detection in young children with autism. *Cortex, 41*(3), 342–353.

Grice, S. J., Spratling, M. W., Karmiloff-Smith, A., Halit, H., Csibra, G., de Haan, M. et al. (2001). Disordered visual processing and oscillatory brain activity in autism and Williams syndrome. *Neuroreport, 12*(12), 2697–2700.

Hagerman, R. J. (2006). Lessons from fragile X regarding neurobiology, autism, and neurodegeneration. *Journal of Developmental and Behavioral Pediatrics, 27*(1), 63–74.

Happe, F., Briskman, J., & Frith, U. (2001). Exploring the cognitive phenotype of autism: Weak "central coherence" in parents and siblings of children with autism: I. Experimental tests. *Journal of Child Psychology and Psychiatry, 42*(3), 299–307.

Harrison, J. E. & Bolton, P. F. (1997). Annotation: Tuberous sclerosis. *Journal of Child Psychology and Psychiatry, and Allied Disciplines, 38*(6), 603–614 issn: 0021-9630.

Hatton, D. D., Sideris, J., Skinner, M., Mankowski, J., Bailey D. B., Jr, Roberts, J., et al. (2006). Autistic behavior in children with fragile X syndrome: Prevalence, stability, and the impact of FMRP. *American Journal of Medical Genetics A, 140*(17), 1804–1813.

Hazlett, H. C., Poe, M., Gerig, G., Smith, R. G., Provenzale, J., Ross, A., et al. (2005). Magnetic resonance imaging and head circumference study of brain size in autism: Birth through age 2 years. *Archives of General Psychiatry, 62*(12), 1366–1376.

Hertz-Picciotto, I., Croen, L. A., Hansen, R., Jones, C. R., van de Water, J. & Pessah, I. N. (2006). The CHARGE study: An epidemiologic investigation of genetic and environmental factors contributing to autism. *Environmental Health Perspectives, 114*(7), 1119–1125.

Howlin, P., Mawhood, L., & Rutter, M. (2000). Autism and developmental receptive language disorder–a follow-up comparison in early adult life. II: Social, behavioural, and psychiatric outcomes. *Journal of Child Psychology and Psychiatry, 41*(5), 561–578.

Hughes, C., Leboyer, M., & Bouvard, M. (1997). Executive function in parents of children with autism. *Psychological Medicine, 27*(1), 209–220 issn: 0033-2917.

Hughes, C., Plumet, M. H., & Leboyer, M. (1999). Towards a cognitive phenotype for autism: Increased prevalence of executive dysfunction and superior spatial span amongst siblings of children with autism. *Journal of Child Psychology and Psychiatry, 40*(5), 705–718.

Humphrey, A., Neville, B. G., Clarke, A., & Bolton, P. F. (2006). Autistic regression associated with seizure onset in an infant with tuberous sclerosis. *Developmental Medicine and Child Neurology, 48*(7), 609–611.

IMGSAC (2001). A genomewide screen for autism: Strong evidence for linkage to chromosomes 2q, 7q, and 16p. *American Journal of Human Genetics, 69*(3), 570–581.

Jamain, S., Quach, H., Betancur, C., Rastam, M., Colineaux, C., Gillberg, I. C., et al. (2003). Mutations of the X-linked genes encoding neuroligins NLGN3 and NLGN4 are associated with autism. *Nature Genetics, 34*(1), 27–29.

Jiang, Y. H., Sahoo, T., Michaelis, R. C., Bercovich, D., Bressler, J., Kashork, C. D., et al. (2004). A mixed epigenetic/genetic model for oligogenic inheritance of autism with a limited role for UBE3A. *American Journal of Medical Genetics A, 131*(1), 1–10.

Jones, M. B., & Szatmari, P. (1988). Stoppage rules and genetic studies of autism [published erratum appears in *Journal of Autism and Developmental Disorders* 1988 Sep;18(3):477]. *Journal of Autism and Developmental Disorders, 18*(1), 31–40 issn: 0162-3257.

Kim, S. J., Cox, N., Courchesne, R., Lord, C., Corsello, C., Akshoomoff, N., et al. (2002). Transmission disequilibrium mapping at the serotonin transporter gene (SLC6A4) region in autistic disorder. *Molecular Psychiatry, 7*(3), 278–288.

Klauck, S. M., Poustka, F., Benner, A., Lesch, K. P., & Poustka, A. (1997). Serotonin transporter (5-HTT) gene variants associated with autism? *Human Molecular Genetics, 6*(13), 2233–2238.

Knickmeyer, R., Baron-Cohen, S., Fane, B. A., Wheelwright, S., Mathews, G. A., Conway, G. S., et al. (2006). Androgens and autistic traits: A study of individuals with congenital adrenal hyperplasia. *Hormones and Behavior, 50*(1), 148–153.

Knickmeyer, R., Baron-Cohen, S., Raggatt, P., Taylor, K., & Hackett, G. (2006). Fetal testosterone and empathy. *Hormones and Behavior, 49*(3), 282–292.

Kolevzon, A., Smith, C. J., Schmeidler, J., Buxbaum, J. D., & Silverman, J. M. (2004). Familial symptom domains in monozygotic siblings with autism. *American Journal of Medical Genetics Part B: Neuropsychiatric Genetics, 129*(1), 76–81.

Landa, R., Folstein, S. E., & Isaacs, C. (1991). Spontaneous narrative-discourse performance of parents of autistic individuals. *Journal of Speech and Hearing Research, 34*(6), 1339–1345.

Landa, R., Piven, J., Wzorek, M. M., Gayle, J. O., Chase, G. A., & Folstein, S. E. (1992). Social language use in parents of autistic individuals. *Psychological Medicine, 22*(1): 245–254 issn: 0033-2917.

Le Couteur, A., Bailey, A., Goode, S., Pickles, A., Gottesman, I., Robertson, S., et al. (1996). A broader phenotype of autism: The clinical spectrum in twins. *Journal of Child Psychology and Psychiatry, and Allied Disciplines, 37*(7), 785–801 issn: 0021-9630.

Lotter, V. (1967). Epidemiology of Autistic conditions in young children. II. Some characteristics of parents and children. *Social Psychiatry, 1,* 163–173.

McAlonan, G. M., Cheung, V., Cheung, C., Suckling, J., Lam, G. Y., Tai, K. S., et al. (2005). Mapping the brain in autism. A voxel-based MRI study of volumetric differences and intercorrelations in autism. *Brain, 128*(Pt 2), 268–276.

McCall, T., Chin, S. S., Salzman, K. L. & Fults, D. W.. (2006). Tuberous sclerosis: A syndrome of incomplete tumor suppression. *Neurosurg Focus, 20*(1), E3.

McCauley, J. L., Olson, L. M., Delahanty, R., Amin, T., Nurmi E. L., Organ E. L.,et al. (2004). A linkage disequilibrium map of the 1-Mb 15q12 GABA(A) receptor subunit cluster and association to autism. *American Journal of Medical Genetics Part B: Neuropsychiatric Genetics, 131*(1), 51–9.

McEwen, F., Happé, F., Bolton, P., Rijsdijk, F., Ronald, A., Dworzynski, K., & Plomin, R. (2007). Origins of individual differences in imitation: links with language, pretend play, and socially insightful behaviour in two-year-old twins. *Child Development, 78,* 474–492.

Micali, N., Chakrabarti, S., & Fombonne, E. (2004). The broad autism phenotype: Findings from an epidemiological survey. *Autism, 8*(1), 21–37.

Miles, J. H., Hadden, L. L., Takahashi T. N., & Hillman R. E.. (2000). Head circumference is an independent clinical finding associated with autism. *American Journal of Medical Genetics, 95*(4), 339–350.

Milner, K. M., Craig, E. E., Thompson, R. J., Veltman, M. W. M., Thomas, S., Roberts, S., et al. (2005). Prader–Willi syndrome: Intellectual abilities and behavioural features by genetic subtype. *Journal of Child Psychology and Psychiatry, 46*(10), 1089–1096.

Mineur, Y. S., Huynh, L. X., & Crusio, W. E. (2006). Social behavior deficits in the Fmr1 mutant mouse. *Behavioural Brain Research, 168*(1), 172–175.

Minton, J., Campbell, M., Green, W., Jennings, S., & Samit, C. (1982). Cognitive assessment of siblings of autistic children. *Journal of the American Academy of Child Psychiatry, 21*(3), 256–261 issn: 0002-7138.

Miranda-Linne, F. M. & Melin, L. (2002). A factor analytic study of the autism behavior checklist. *Journal of Autism and Developmental Disorders, 32*(3), 181–188.

Mulder, E. J., Anderson, G. M., Kema, I. P., Brugman, A. M., Ketelaars C. E., de Bildt, A. et al. (2005). Serotonin transporter intron 2 polymorphism associated with rigid-compulsive behaviors in Dutch individuals with pervasive developmental disorder. *American Journal of Medical Genetics Part B: Neuropsychiatric Genetics, 133*(1), 93–96.

Mulder, E. J., Anderson, G. M., Kema, I. P., de Bildt, A., van Lang,. N. D. J., den Boer, J. A., et al. (2004). Platelet serotonin levels in pervasive developmental disorders and mental retardation: Diagnostic group differences, within-group distribution, and behavioral correlates. *Journal of the American Academy of Child and Adolescent Psychiatry, 43*(4), 491–499.

Murphy, M., Bolton, P. F., Pickles, A., Fombonne, E., Piven, J., & Rutter, M. (2000). Personality traits of the relatives of autistic probands. *Psychological Medicine, 30*(6): 1411–1424.

Narayan, S., Moyes, B., & Wolff, S (1990). Family characteristics of autistic children: A further report [see comments]. *Journal of Autism and Developmental Disorders, 20*(4), 523–535 issn: 0162-3257.

Nurmi, E. L., Amin, T., Olson, L. M., Jacobs, M. M., McCauley, J. L., Lam, A. Y., et al. (2003). Dense linkage disequilibrium mapping in the 15q11-q13 maternal expression domain yields evidence for association in autism. *Molecular Psychiatry, 8*(6), 624–634, 570.

Nurmi, E. L., Bradford, Y., Chen, Y., Hall, J., Arnone, B., Gardiner, M. B., et al. (2001). Linkage disequilibrium at the Angelman syndrome gene UBE3A in autism families. *Genomics, 77*(1–2), 105–113.

Nurmi, E. L., Dowd, M., Tadevosyan-Leyfer, O., Haines, J. L., Folstein, S. E., Sutcliffe, J. S. (2003). Exploratory subsetting of autism families based on savant skills improves evidence of genetic linkage to 15q11-q13. *Journal of the American Academy of Child and Adolescent Psychiatry, 42*(7), 856–863.

Ozonoff, S., Rogers, S. J., Farnham, J. M., & Pennington, B. F. (1993). Can standard measures identify subclinical markers of autism? *Journal of Autism and Developmental Disorders, 23*(3), 429–441 issn: 0162-3257.

Perry, B. D., Cook, E. H., Jr., Leventhal, B. L., Wainwright, M. S., & Freedman, D. X., (1991). Platelet 5-HT2 serotonin receptor binding sites in autistic children and their first-degree relatives. *Biological Psychiatry, 30*(2), 121–130 issn: 0006-3223.

Pickles, A., Starr, E. Kazak, S., Bolton, P., Papanikolaou, K., Bailey, A., et al. (2000). Variable expression of the autism broader phenotype: Findings from extended pedigrees. *Journal of Child Psychology and Psychiatry, 41*(4), 491–502.

Piven, J., Gayle, J., Chase, J., Fink, B., Landa, R., Wzorek, M. M., et al. (1990). A family history study of neuropsychiatric disorders in the adult siblings of autistic individuals [see comments]. *Journal of the American Academy of Child and Adolescent Psychiatry, 29*(2), 177–183 issn: 0890-8567.

Piven, J., & Palmer, P. (1997). Cognitive deficits in parents from multiple-incidence autism families. *Journal of Child Psychology and Psychiatry, 38*(8), 1011–1021.

Piven, J., Palmer, P., Jacobi, D., Childress, D., & Arndt, S. (1997a). Broader autism phenotype: Evidence from a family history study of multiple-incidence autism families. *The American Journal of Psychiatry, 154*(2), 185–190.

Piven, J., Palmer, P., Jacobi, D., Childress, D. & Arndt, S. (1997b). Personality and language characteristics in parents from multiple-incidence autism families. *American Journal of Medical Genetics, 74*(4), 398–411.

Piven, J., Wzorek, M., Landa, R., Lainhart, J., Bolton, P., Chase, G. A., et al. (1994). Personality characteristics of the parents of autistic individuals. *Psychological Medicine, 24*(3), 783–795.

Rinehart, N. J., Bellgrove, M. A., Tonge, B. J., Brereton, A. V., Howells-Rankin, D., & Bradshaw, J. L. (2006). An examination of movement kinematics in young people with high-functioning autism and asperger's disorder: Further evidence for a motor planning deficit. *Journal of Autism and Developmental Disorders, 36*(6), 757–767.

Ritvo, E. R., Freeman, B. J., Mason-Brothers, A., Mo, A. & Ritvo, A. M. (1985). Concordance for the syndrome of autism in 40 pairs of afflicted twins. *The American Journal of Psychiatry, 142*(1): 74–77 issn: 0002-953x.

Ritvo, E. R., Jorde, L. B., Mason-Brothers, A., Freeman, B. J., Pingree, C., Jones, M. B., et al. (1989). The UCLA-University of Utah epidemiologic survey of autism: Recurrence risk estimates and genetic counseling. *The American Journal of Psychiatry, 146*(8), 1032–1036 issn: 0002-953x.

Rodier, P. M., Bryson, S. E., & Welch, J. P. (1997). Minor malformations and physical measurements in autism: Data from Nova Scotia. *Teratology 55*(5): 319–325 issn: 0040-3709.

Ronald, A., Happe, F., Bolton, P., Butcher, L. M., Price, T., Wheelwright, S., et al. (2006). Genetic heterogeneity between the three components of the autism spectrum: A twin study. *Journal of the American Academy of Child and Adolescent Psychiatry, 45*(6), 691–699.

Ronald, A., Happe, F., & Plomin, R. (2005). The genetic relationship between individual differences in social and nonsocial behaviours characteristic of autism. *Dev Sci* 8(5): 444–458.

Rutter, M. (1968). Concepts of autism: A review of research. *Journal of Child Psychology and Psychiatry, and Allied Disciplines, 9*(1), 1–25 issn: 0021-9630.

Rutter, M. (2005). Incidence of autism spectrum disorders: Changes over time and their meaning. *Acta Paediatr, 94*(1), 2–15.

Rutter, M., Andersen-Wood, L., Beckett, C., Bredenkamp, D., Castle, J., Groothues, C., et al. (1999). Quasi-autistic patterns following severe early global privation. English and Romanian Adoptees (ERA) study team. *Journal of Child Psychology and Psychiatry, 40*(4), 537–549.

Rutter, M., Bailey, A. Bolton, P., & Le Conteur, A (1994). Autism and known medical conditions: Myth and substance. *Journal of Child Psychology and Psychiatry, 35*, 311–322.

Rutter, M., & Lockyer, L. (1967). A five to fifteen year follow-up study of infantile psychosis. I. Description of sample. *The British Journal of Psychiatry, 113*(504), 1169–1182 issn: 0960-5371.

Scourfield, J., McGuffin, P., & Thapar, A. (1997). Genes and social skills. *Bioessays, 19*(12), 1125–1127.

Sevin, J. A., Matson, J. L., Coe, D., Love, S. R., Matese, M. T., & Benavidez, D. A. (1995). Empirically derived subtypes of pervasive developmental disorders: A cluster analytic study. *Journal of Autism and Developmental Disorders, 25*(6), 561–578 issn: 0162-3257.

Shao, Y., Cuccaro, M. L., Hauser, E. R., Raiford, K. L., Menold, M. M., Wolpert, C. M., et al. (2003). Fine mapping of autistic disorder to chromosome 15q11-q13 by use of phenotypic subtypes. *American Journal of Human Genetics, 72*(3), 539–548.

Spiker, D., Lotspeich, L. J., Dimiceli, S., Myers, R. M., & Risch, N. (2002). Behavioral phenotypic variation in autism multiplex families: Evidence for a continuous severity gradient. *American Journal of Medical Genetics, 114*(2), 129–136.

Steffenburg, S., Gillberg, C. Hellgren, L., Andersson, L., Gillberg, I. C., Jakobsson, G., et al. (1989). A twin study of autism in Denmark, Finland, Iceland, Norway and Sweden. *Journal of Child Psychology and Psychiatry, and Allied Disciplines, 30*(3), 405–416 issn: 0021-9630.

Stella, J., Mundy, P., & Tuchman, R. (1999). Social and nonsocial factors in the Childhood Autism Rating Scale. *Journal of Autism and Developmental Disorders, 29*(4), 307–317.

Sutcliffe, J. S., Delahanty, R. J., Prasad, H. C., McCauley, J. L., Han, Q., Jiang, L., et al. (2005). Allelic heterogeneity at the serotonin transporter locus (SLC6A4) confers susceptibility to autism and rigid-compulsive behaviors. *American Journal of Human Genetics, 77*(2), 265–279.

Szatmari, P., Bartolucci, G., & Bremner, R. (1989). Asperger's syndrome and autism: Comparison of early history and outcome. *Developmental Medicine and Child Neurology, 31*(6), 709–720 issn: 0012-1622.

Szatmari, P., Jones, M. B., Tuff, L., Bartolucci, G., Fisman, S., &. Mahoney, W. J. (1993). Lack of cognitive impairment in first-degree relatives of children with pervasive developmental disorders. *Journal of the American Academy of Child and Adolescent Psychiatry, 32*(6): 1264–1273 issn: 0890-8567.

Szatmari, P., MacLean, J. E., Jones, M. B., Bryson, S. E.,. Zwaigenbaum, L., Bartolucci, G.,et al. (2000). The familial aggregation of

the lesser variant in biological and nonbiological relatives of PDD probands: A family history study. *Journal of Child Psychology and Psychiatry, 41*(5), 579–586.

Szatmari, P., Merette, C., Bryson, S. E., Thivierge, J., Roy, M. A., Cayer, M., et al. (2002). Quantifying dimensions in autism: A factor-analytic study. *Journal of the American Academy of Child and Adolescent Psychiatry, 41*(4), 467–474.

Szatmari, P., Volkmar, F., & Walter, S. (1995). Evaluation of diagnostic criteria for autism using latent class models. *Journal of the American Academy of Child and Adolescent Psychiatry, 34*(2): 216–222 issn: 0890-8567.

Tadevosyan-Leyfer, O., Dowd, M., Mankoski, R., Winklosky, B.,. Putnam, S., McGrath, L., et al. (2003). A principal components analysis of the autism diagnostic interview-revised. *Journal of the American Academy of Child and Adolescent Psychiatry, 42*(7), 864–872.

Taylor, B. (2006). Vaccines and the changing epidemiology of autism. *Child: Care, Health and Development, 32*(5), 511–519.

Veenstra-VanderWeele, J., & Cook, E. H., Jr. (2004). Molecular genetics of autism spectrum disorder. *Molecular Psychiatry, 9*(9), 819–832.

Volkmar, F. R., Cicchetti, D. V., Dykens, E., Sparrow, S. S., Leckman, J. F., & Cohen, D. J. (1988). An evaluation of the autism behavior checklist. *Journal of Autism and Developmental Disorders, 18*(1), 81–97 issn: 0162-3257.

Vorstman, J. A., Staal, W. G., van Daalen, E., van Engeland, H., Hochstenbach, P. F. R., & Franke, L.. (2006). Identification of novel autism candidate regions through analysis of reported cytogenetic abnormalities associated with autism. *Molecular Psychiatry, 11*(1), 1, 18–28.

Wadden, N. P., Bryson, S. E., & Rodger, R. S. (1991). A closer look at the autism behavior checklist: Discriminant validity and factor structure. *Journal of Autism and Developmental Disorders, 21*(4), 529–541 issn: 0162-3257.

Waterhouse, L., Morris, R., Allen, D., Dunn, M., Fein, D., Rapin, L., et al. (1996). Diagnosis and classification in autism. *Journal of Autism and Developmental Disorders, 26*(1), 59–86 issn: 0162-3257.

Wilcox, J. A., Tsuang, M. T., Schnurr, T., & Braida-Fragoso, N. (2003). Case-control family study of lesser variant traits in autism. *Neuropsychobiology* 47(4): 171–177.

Williams, J., Scott, F., Stott, C., Allison, C., Bolton, P., Baron-Cohen,. S., et al. (2005). The CAST (Childhood Asperger Syndrome Test): Test accuracy. *Autism* 9(1): 45–68.

Woodhouse, W., Bailey, A., Rutter, M., Bolton, P., Baird, G., & Le Couteur, A. (1996). Head circumference in autism and other pervasive developmental disorders. *Journal of Child Psychology and Psychiatry, 37*(6), 665–671.

Yonan, A. L., Alarcon, M., Cheng, R., Magnusson, P.K., Spence, S. J., Palmer, A. A., et al. (2003). A genomewide screen of 345 families for autism-susceptibility loci. *American Journal of Human Genetics, 73*(4), 886–897.

Yonan, A. L., Palmer, A. A., Smith, K. C., Feldman, L., Lee, H. K., Yonan, J. M., et al. (2003). Bioinformatic analysis of autism positional candidate genes using biological databases and computational gene network prediction. *Genes, Brain, and Behaviour, 2*(5): 303–320.

Zwaigenbaum, L., Szatmari, P., Jones, M. B., Bryson, S. E., Maclean, J. E., Mahoney, W. J., et al. (2002). Pregnancy and birth complications in autism and liability to the broader autism phenotype. *Journal of the American Academy of Child and Adolescent Psychiatry, 41*(5), 572–579.

Chapter 28

Genetics of Smoking Behavior

Richard J. Rose, Ulla Broms, Tellervo Korhonen, Danielle M. Dick, and Jaakko Kaprio

Introduction

The public health significance of sustained smoking is difficult to overstate. Worldwide, every other current smoker will prematurely die from tobacco-related diseases (Doll, Peto, Wheatley, Gray, & Sutherland, 1994; Neubauer et al., 2006). Should current trends continue, annual deaths attributable to smoking will exceed 10 million by 2025 (Mackay, Eriksen, & Shafey, 2006). Persistent smoking is the most preventable cause of disability and death; it is associated with wide-ranging adverse health effects, including heart disease, pulmonary disease, and lung and other cancers (Doll, Peto, Boreham, & Sutherland, 2005; Risch et al., 1993) in both industrialized and developing countries (Mackay, et al.).

Smoking behaviors aggregate in families and in peer networks, due to genetic dispositions and familial and extra-familial environmental influences. In a recent study of adult twins, achieving a high school education halved the likelihood of persistent smoking, parental smoking almost doubled it and having an MZ co-twin who ever smoked increased the likelihood nearly 10-fold (Hamilton et al., 2006). Smoking, like drinking, may be understood within a developmental framework – behavior for which precursors are found in early childhood with causal modifiers evident throughout lifetime. Risk of nicotine dependence is conditional on initiating smoking; some at high risk never smoke, and the factors underlying individual differences in smoking initiation, like those underlying drinking/abstaining, are multiple in nature and include factors outside, as well as within, family households. A developmental perspective is necessary to appreciate the associations of smoking behaviors with drinking, with other patterns of substance use/abuse, and with behav-

ioral and mental disorders. And as is true for the initiation and escalation of drinking, smoking patterns illustrate gene–environment interactions and correlations. For such reasons, smoking behaviors invite behavior genetic study. This chapter offers an overview of that research with a focus on recent twin studies and with illustrations drawn from research conducted within Finland.

Finland offers an ideal living laboratory for such study. Finland maintains a Population Register Center (PRC) into which each newborn is added with a unique identifier code that incorporates date of birth and a link to the biological mother. The PRC updates information on residential addresses of all citizens, so Finnish twins from a given birth cohort can be exhaustively identified and followed throughout their lives. Loss to follow-up is minimized by individualized linkage of each Finn to health and institutional outcome registries. Neither incomplete ascertainment nor self-selection biases create inescapable problems. Finns have a long history of participation in nationwide epidemiological research, encouraging extended twin designs including those that use classrooms containing twins and families of twins as sampling units. These and other aspects of Finnish society make it a superb setting for twin studies (Rose, 2006), including those directed to smoking.

One in four Finnish adults (26% of men, 18% of women) is a daily smoker (Helakorpi, Patja, Prättälä, & Uutela, 2005). Smoking among Finns at ages 15–16 is more prevalent than in most other European societies. And in Finland, as elsewhere, smoking patterns show within-culture variation across levels of education, income, and social status (Giskes et al., 2005; Helakorpi, Patja, Prättälä, & Uutela, 2007). Intervention efforts have had little effect in reducing smoking within some SES groups and larger effects within others. Smoking prevalence remains unchanged in some Western countries (Giskes, et al.), and within-culture variation across age, birth cohort, and gender is common (Cavelaars et al., 2000; Platt, Amos, Gnich, & Parry, 2002; Regidor, Gutierrez-Fisac, Calle, Navarro, & Dominguez, 2001). In some Western societies, the prevalence of smoking is decreasing among both men and women

R.J. Rose (✉)
Department of Psychological and Brain Sciences, Indiana University, Bloomington; Department of Public Health, University of Helsinki, Finland
e-mail: rose@indiana.edu

Y.-K. Kim (ed.), *Handbook of Behavior Genetics*,
DOI 10.1007/978-0-387-76727-7_28, © Springer Science+Business Media, LLC 2009

of advantaged socioeconomic status; such change is evident in the United States and in Finland and its Nordic neighbors.

Smoking Behaviors

We begin with a brief overview of the multiple ways phenotypes of smoking behavior and nicotine dependence are defined and assessed for research purposes. In the research literature, the development of nicotine dependence is studied within the context of reward mechanisms, the development of tolerance, difficulty in smoking cessation even among those motivated to attempt it, and withdrawal effects.

Smoking Initiation

In Finland, most experimental smoking is initiated during ages 11–15 (Rimpelä, Rainio, Pere, Lintonen, & Rimpelä, 2005). The typical self-report of initial experience is in taking a couple of puffs and then, often, an entire cigarette (Lerman & Berrettini, 2003), usually with peers. Those who experience positive subjective (and often, social) effects and find it rewarding in biological, social, and/or psychosocial ways are those who will continue smoking and escalate to become regular users. And as smoking continues, acute effects of nicotine accompany the influence of social factors in reinforcing smoking behavior and dispositional differences in susceptibility to nicotine dependency become evident (Lerman & Berrettini).

Environmental influences of smoking among family members (de Vries, Engels, Kremers, Wetzels, & Mudde, 2003; Kestilä et al., 2006; Sasco, Merrill, Benhaim-Luzon, Gerard, & Freyer, 2003), peers and friends of same age (Johnson et al., 2002; Vink, Willemsen, & Boomsma, 2003), and social and cultural context are clearly important in starting to smoke (Flay, Petraitis, & Hu, 1999; Lerman & Berrettini, 2003). Smoking initiation is influenced by imitative modeling effects from peers, but analyses of peer affiliation among young Finnish twins illustrate active selection processes (assortative pairing) in their friendship choices made at ages 11–12 (Rose, 2002; Rose, 2007), so these effects are not exclusively passive. Very large between-school differences in smoking rates and correlations for smoking initiation among unrelated school classmates (Rose, Viken et al., 2003; Siddiqui, Hedeker, Flay, & Hu, 1996) underscore the importance of extra-familial environmental effects.

Genetically influenced dispositional differences in temperament and personality are relevant also to smoking initiation; research indicates that greater inattentiveness (Barman, Pulkkinen, Kaprio, & Rose, 2004) and neuroticism (Terracciano & Costa, 2004) associate with earlier smoking initiation. Pharmacological factors become of increasing importance relative to social factors in maintaining smoking behavior once regular smoking becomes more established.

Behavior genetic research has defined smoking initiation (SI) in several ways: age of onset of regular smoking (however defined), age of initial experimentation (first puff on a cigarette), or lifetime ever/never smoking, but with varying requirements of what constitutes "regular" smoking; under some definitions "chippers" and very infrequent smokers are classified along with those who have never initiated.

Persistent Smoking

Similarly, there are multiple definitions for smoking persistence (SP): quantity measures (number of cigarettes smoked per day), current tobacco use, or even nicotine dependence according to standard DSM criteria. About one in five Finnish adolescents is a daily smoker, a prevalence only slightly less than that among Finnish adults. Among young men (15–24 years old), 21% are daily smokers and an additional 9% smoke on occasion; patterns are similar (18 and 8%) among age-matched Finnish women (Helakorpi et al., 2007). In research with such regular smokers, a frequently used marker of dependence is daily cigarette consumption, on the assumption that the more cigarettes smoked per day, the more difficult the smoker will find it to stop. But matters may be more complicated, for a monotonic relationship between smoking quantity and adverse health outcomes is not always observed. Smoking behaviors include frequency and duration of puffs, duration of inhalation/exhalation, lung retention time, and it may be nicotine dose, rather than the number of cigarettes smoked per day that is more important. A high level of nicotine intake should foster more pronounced neuronal adaptation, may reflect a greater constitutional need for nicotine and a more readily learned, extinction-resistant behavior. Whatever the mechanism(s), there is clear evidence that those who smoke more cigarettes per day are less likely able to stop (Etter, Duc, & Perneger, 1999; Hymowitz et al., 1997; Kaprio & Koskenvuo, 1988; Senore et al., 1998). Evidence (Broms, Silventoinen, Lahelma, Koskenvuo, & Kaprio, 2004; DiFranza, Savageau, Fletcher, Ockene et al., 2002; DiFranza, Savageau, Fletcher, O'Loughlin et al., 2007) suggests that nicotine dependence symptoms can develop soon after initiation and that these symptoms may lead to smoking intensification.

DiFranza and colleagues (2002) showed that approximately 20% of adolescents reported nicotine dependence symptoms within 1 month of initiating regular smoking. Adolescent smokers quickly learn to appreciate the rewarding effects of nicotine and become skilled at manipulating dosage. A recent study (Rubinstein, Thompson, Benowitz,

Shiffman, & Moscicki, 2007) concluded that adolescents may be more sensitive to nicotine effects than adults (a concern mirroring that for age differences in acute sensitivity to ethanol), and some adolescents may develop dependence with low levels of nicotine: perhaps 70% of adolescent smokers have tried to quit but only 10% succeed (Lerman & Berrettini, 2003; Prokhorov, Pallonen, Fava, Ding, & Niaura, 1996).

Nicotine Tolerance

Smoking may produce a pleasant feeling almost immediately. Within seconds after starting to smoke a cigarette, nicotine travels through the bloodstream to the brain, causing changes in brain chemicals that often produce a pleasant effect. But the effect is short lasting, and smokers usually need to smoke repeatedly, throughout the day to maintain it (O'Brien, Childress, McLellan, & Ehrman, 1992; West & Hardy, 2006).

Tolerance refers to adaptation to repeated drug exposure (Benowitz, 1996). Over time, tolerance is signaled by increased intake of the drug to attain the same effect. Tolerance to some of the effects of nicotine develops fairly rapidly and several mechanisms are involved. Acute tolerance is illustrated when a few equal doses of nicotine are applied in close succession, and the later doses have diminished effect (Perkins, Fonte, Sanders, Meeker, & Wilson, 2001). Nicotine levels are lowest in the blood of a smoker in the morning and acute tolerance develops during the day (Le Houezec, 2003). Presumably, tissue resistance to later doses of nicotine in this case is due to the persistent occupation of nicotine receptors by nicotine molecules (Perkins, Fonte et al., 2001).

Tolerance after quitting is important for preventing relapse, but the process is not well understood. The neuronal pathways that have been activated earlier are more prone to activate again in the future, because of some facilitation process (Leonard & Bertrand, 2001; Perkins, Fonte et al., 2001; Perkins, Gerlach, Broge, Fonte, & Wilson, 2001; Perkins, Gerlach, Broge, Grobe et al., 2001; Perkins, Gerlach, Broge, Sanders et al., 2001).

Smoking Cessation

Smoking cessation can be distinguished from smoking persistence and studied as a separate phenotype. Quitting is difficult for many smokers, and smoking is a chronically relapsing behavior. Mark Twain observed that giving up smoking is easy; he had done it a hundred times. Many who have sustained long periods of abstinence eventually

return to smoking. Relapse rates among those initially succeeding decline over time but remain substantial for some years: about 50% of those abstinent for 6 weeks relapse by 6 months, and 20% of those abstinent for 6 months relapse by 1 year (Stapleton, 1998). There is evidence that 30–50% of those abstinent for 1 year will relapse by 5 years (Stapleton; Blondal, Gudmundsson, Olafsdottir, Gustavsson, & Westin, 1999). Six of ten Finnish smokers report that they would like to quit smoking (Helakorpi et al., 2007).

Nicotine dependence (ND) as employed in research studies has had different definitions, as well, changing with new diagnostic frameworks and changing as new instruments for epidemiological assessment become available. For ND, however, explicit evaluation of informative phenotypic definition for genetic research has begun (Lessov et al., 2004).

Demographics of Smoking Behavior and Nicotine Dependence

Developmental Age and Birth Cohort

Smoking is a developmental phenomenon set within familial, school, neighborhood, and cultural contexts. Age variation is critical in understanding development of smoking behavior and nicotine dependence. Adolescents likely experience similar kinds of withdrawal as do adults if they try to quit. Because attempts may finally be successful, and because the number of quit attempts increases with duration of the smoking career, cessation rates relate to age (Jarvis, 1997). And age at smoking initiation is a risk factor for later dependence. Compelling evidence of a strong association between age at drinking initiation and risk for alcohol dependency has been reported, and early onset age of smoking appears to be an analogous risk factor for ND (Choi, Gilpin, Farkas, & Pierce, 2001).

The prevalence of adolescent and adult smoking reflects societal and cultural attitudes toward the behavior. Cohort differences are evident in changes in the prevalence of smoking in years subsequent to publicized reports of its adverse health effects and increasingly common legislative efforts to prohibit public smoking.

Gender

Gender differences in smoking behaviors and in their risk factors have been reported. Some of the differences between men and women may make it more difficult for women to quit smoking (Perkins, Donny, & Caggiula, 1999). Women are not as successful with nicotine replacement therapy, perhaps because of their hypothesized lower sensitivity to nicotine (Cepeda-Benito, Reynoso, & Erath, 2004; Shiffman & Paton, 1999). Women tend to worry more about weight gain if they quit smoking (Saarni, Silventoinen, Rissanen, Sarlio-Lahteenkorva, & Kaprio, 2004), and women more often report starting and maintaining smoking as a way to control weight (Perkins, 2001). Women are encouraged not to smoke during pregnancy (Windsor, Boyd, & Orleans, 1998), and such injunctions likely influence smoking cessation among fertile women.

Gender differences (Perkins, Sanders, D'Amico, & Wilson, 1997) in nicotine discrimination have been reported; male smokers may be better at discriminating and titrating nicotine doses, so that women's smoking may be less driven by nicotine reinforcement and more influenced by other sensory and behavioral effects of smoking (Perkins, Donny et al., 1999).

Gender differences in smoking vary widely across culture and over time, but gender differences are narrowest in the developed countries, while developing countries demonstrate very wide gender gaps (Crofton, 1990; McLellan, 1995). Scandinavian countries show nearly equal smoking prevalence among men and women (Crofton). While differences in prevalence have shifted over time, differences in smoking rate among smokers have not (Shiffman & Paton, 1999).

Some data suggest that women are less nicotine dependent than men, as measured by standard assessment instruments such as the Fagerström Scale or FTND (Fagerström et al., 1996). Paradoxically, however, women report more severe signs of nicotine withdrawal (Hatsukami, Skoog, Allen, & Bliss, 1995). Surprisingly, given that women seemingly have lower nicotine dependence, they are, as noted, less successful with nicotine replacement therapy, a finding attributed to their hypothesized lower sensitivity to nicotine (Shiffman & Paton, 1999).

Family and Context

Family and marital/partner status are contexts in which smoking behaviors develop. Smoking by family members (de Vries et al., 2003; Kestilä et al., 2006; Sasco et al., 2003) and friends of similar age (de Vries, et al.; Johnson et al., 2002; Sasco, et al.; Vink, Willemsen, Engels, & Boomsma, 2003) influence adolescent smoking initiation. Smoking prevalence is very similar among married or cohabiting (23% men, 20% women) and single (22% men and 20% women) adults, but those separated smoke more (46% men and 35% women) in Finnish data (Helakorpi et al., 2007). Living with a spouse is a predictor of smoking cessation, especially among men (Broms, Silventoinen, Lahelma, et al., 2004) and a review (Haustein, 2006) shows that tobacco use is greater among

people living alone. Children with smoking parents are more likely to become smokers themselves (Huurre, Aro, & Rahkonen, 2003). Parents caring for children are more likely to quit than adults without children (Jarvis, 1996). Marriage to a nonsmoker or to an ex-smoker may be associated with an increase in successful cessation.

Socioeconomic Status (SES)

Among lower SES groups, smoking has shown little decline in recent years in Finland. Smoking has become increasingly associated with lower levels of education and income (Giskes et al., 2005). A 20-year survey of Finnish adults illustrates this trend: Finns with limited education comprised the largest group of smokers (Lahelma et al., 1997). Daily smoking is found in 27% of Finnish men and 24% of women in the lowest education group compared to 19 and 17% in the most educated groups (Helakorpi et al., 2007). Children of less educated parents are more likely to smoke than children of better educated parents (Huurre et al., 2003; Kestilä et al., 2006), and non-smoking adolescents were more likely to reach higher levels of education in early adulthood (Glendinning, Hendry, & Shucksmith, 1995; Koivusilta, Rimpela, & Vikat, 2003). A review by Haustein (2006) found that tobacco is used more among unemployed people. A Finnish prospective study (Kaprio & Koskenvuo, 1988) found that young adult Finns who had quit smoking during a 6-year follow-up period were better educated than those who continued to smoke, a finding confirmed in a later analyses of the same cohort (Broms, Silventoinen, Madden et al., 2004).

Early Twin Studies of Smoking Behavior

The first twin studies of smoking behavior were reported during the late 1950s and 1960s as an integral part of ongoing debate on the causal role of smoking in lung cancer. The debate engaged notable statisticians, including R.A. Fisher, who raised critical argument, in the journal *Nature,* that the first twin cohort studies of smoking and lung cancer demonstrated an association, but had not proven causality (Fisher, 1958). Fisher argued that studies of smoking-discordant twin pairs would provide a test of causality. Small (Conterio & Chiarelli, 1962; Friberg, Kaij, Dencker, & Jonsson, 1959) and larger sample studies (Cederlöf, 1966; Cederlöf, Friberg, & Lundman, 1977; Eysenck, 1980; Friedman, King, Klatsky, & Hulley, 1981; Hannah, Hopper, & Mathews, 1985 Hrubec, Cederlöf, & Friberg, 1976; Kaprio, Sarna, Koskenvuo, & Rantasalo, 1978; Partanen, Bruun, & Markkanen, 1966; Raaschou-Nielsen, 1960), reviewed by Kaprio (1984), generally demonstrated that concordance for smoking was higher in MZ pairs than DZ pairs, but these early studies did not provide quantitative estimates of heritability. Many of these Nordic and US data sets have been later re-analyzed using more sophisticated analyses (vida infra). Early twin studies on co-twins reared apart supported the general inference of genetic effects (Farber, 1981; Kaprio, Koskenvuo, & Langinvainio, 1984; Shields, 1962); other studies of twins reared apart (see also Eaves & Eysenck, 1980; Osler, Holst, Prescott, & Sorensen, 2001; Sullivan & Kendler, 1999) failed to find substantial familial environmental influences, a finding confirmed in Danish adoption data on smoking behavior (Shenassa et al., 2003).

Recent Behavior Genetic Studies of Smoking

Subsequent to these historical studies, larger, population-based samples and new analytic techniques have permitted parameter estimation based on model-fitting genetic and environmental effects on smoking behaviors and nicotine dependence. Many of the newer research studies test multistage models of smoking behaviors to evaluate whether the same genetic and environmental factors contribute to various stages of smoking. For example, two-stage models can test the extent to which the factors that influence smoking initiation also impact smoking persistence. These models can be extended to examine overlap of factors influencing smoking initiation, smoking persistence and regular tobacco use, and nicotine dependence. Further, multi-stage models can accommodate the fact that liability for subsequent smoking dimensions (e.g., regular use, dependence) is unknown for individuals who have never initiated smoking. In an early discussion of the utility of multi-stage models, Heath and colleagues (Heath, Martin, Lynskey, Todorov, & Madden, 2002) suggested that many risk factors (e.g., antisocial personality) could plausibly affect both initiation and outcome. And, they argued, it is equally plausible that either a genetic effect (e.g., one that regulates nicotine sensitivity or metabolism) or an environmental exposure (e. g., spousal smoking) could affect smoking outcomes, but have no impact on smoking initiation. Most of the studies on the genetic architecture of smoking behavior have examined smoking initiation (SI) and smoking persistence (SP), while quantitative genetic studies on other smoking behaviors and outcomes including ND are fewer in number.

Genetic and Environmental Influences on Smoking Initiation

We begin with twin studies of smoking initiation (SI), and these studies are given more detailed attention, because there are more of them and because causal factors in SI are of critical importance, given the conditional nature of all other smoking behaviors and the association of early onset smoking with elevated risk for ND.

The first entry in Table 28.1 provides a summary of a meta-analysis of twin studies of SI reported during the 1990's (Li, Cheng, Ma, & Swan, 2003). Included are 17 twin cohorts from six studies of SI conducted in three countries (Australia, Finland, and the USA) in research reported from 1993 to 1999. In meta-analysis of the eight cohorts of twin brothers, the weighted estimate of h^2 (mean $\pm$ SEM) was 0.37 ± 0.04, ranging from 0.11 to 0.64 across cohorts; for the nine cohorts of twin sisters, estimated h^2 ranged from 0.32 to 0.78 with a mean of 0.55 ± 0.04. Estimated effects of common environment varied across these cohorts of male and female twins, as well; in half of the cohorts of twin brothers, c^2 was estimated at or above 0.5, while a similarly high estimate of c^2 was found for only one of the cohorts of twin sisters, and in five of the nine cohorts of female twins, c^2 estimates were 0.0–0.18. Their meta-analysis led Li et al. to conclude that genetic factors play a more significant role in smoking initiation for adult women than for men, and conversely, that common environmental factors have greater impact on SI among adult men. No significant gender differences were found for effects of e^2 on SI.

Li et al. included multiple analyses of the same (or overlapping) data sets and, accordingly, their meta-analysis may have given undue weight to some (e.g., Australian) twin data. Inferences drawn about gender differences in parameter estimates may reflect weighted effects of specific samples. Yet, the major inference drawn from the Li et al. meta-analysis, that genetic factors are substantial across culture and gender and account for no less than one-third and perhaps one-half of the observed variance, is consistent with other reviews. Sullivan & Kendler (1999) reviewed a larger (and earlier) set of studies and estimated h^2 of liability for SI at 60%. And inferences of "a substantial genetic component to smoking" were drawn from a third review of the 1990s twin research literature (Batra, Patkar, Berrettini, Weinstein, & Leone, 2003).

The remaining entries in Table 28.1 summarize parameter estimates from eight recent twin studies reported during 2002–2007. The estimates derive from data from twin cohorts in four countries (twin samples from the Netherlands, as well as Australia, Finland and USA); most of the studied twins are adults, and three of the samples, those studied by Morley et al. in Australia, by Broms, et al. in Finland, and by Hamilton et al. in the USA, include elderly twins > age 60. In contrast, twins in the two Netherlands reports included Table 28.1 are younger (mean ages 25 in Vink, Beem et al., 2004 and <31 in Vink, Willemsen, & Boomsma, 2005), and age seemingly makes for differences in parameter estimates. All ACE parameter estimates could be constrained equal across gender in the two Dutch data

Table 28.1 Parameter estimates for smoking initiation (SI)

Study	Year	Sample		Men			Women		
				a^2	c^2	e^2	a^2	c^2	e^2
Li	2003	Meta-analysis	Age range	0.37	0.49	0.17	0.55	0.24	0.16
Heath	2002	AUS; 4,342 pairs	30–50[1]	0.22	0.42	0.37	0.63	0.11	0.27
Madden	2004	21,883 pairs[2]	18–46	0.46–0.68	0.00–0.35	0.11–0.33	0.43–0.59	0.16–0.48	0.10–0.25
Vink	2004	NL; 3,657 pairs	25±11	0.36	0.56	0.07	0.36	0.56	0.07
Maes	2004	USA; >6,800 twins	35 ± 8	0.72	0	0.28	0.63	0.17	0.21
Hamilton	2006	USA; 32,000 pairs	19–60+	0.71	0.12	0.17	0.32	0.48	0.21
Vink	2005	NL; 1,572 pairs	31±12	0.44	0.51	0.05	0.44	0.51	0.05
Broms	2006	FIN; 9,880 pairs	23–88	0.59	0.19	0.22	0.35	0.51	0.14
Hardie	2006	USA; 546 pairs	25–75	0.48[3]	0	0.52	0.48	0	0.52
Morley	2007	AUS: >5,000 pairs	16–88	0.63	0.18[3]	0.19	0.54	0.24[4]	0.17

[1]Age range or mean age ± st. dev. as given in published report. [2]Madden et al. studied three twin samples (from AUS, FIN and SWE) and divided each sample into three age groups; entries for this study are the range of estimates across samples and age groups. Estimates of h^2 for lifetime smoking could be set invariate across the three samples and the three age groups. For women, estimates of c^2 could be set equal across the three samples by each of the three age groups (45% for the youngest age group, 35% for the next youngest and 26% for the oldest). For male twins, however, estimates of c^2 were higher for both Nordic samples (33, 29 and 19% by increasing age) than for the Australian males (26, 9 and 11% by increasing age). [3]Hardie et al. do not report sample sizes or parameter estimates separately by gender. [4]Morley et al. added siblings of twins into their sample enabling their analysis to distinguish a twin-specific environmental effect from the usual c^2 estimate; in so doing, they estimate that the twin-specific environmental factors accounted for 19% of the variance in SI among females and 12% of variance in SI among males.

sets, a result not true of any other data in Table 28.1. Note also that the genetic and environmental estimates for SI obtained in data from Heath et al. (>4000 Australian twins) and Madden et al. (nearly 22,000 twin pairs from three countries divided into three age groups) are consistent with the inferences drawn by Li et al. from their meta-analysis of earlier research: greater h^2 among adult women, greater c^2 among adult men. What is, perhaps, more interesting in the cross-cultural comparisons made by Madden and her colleagues is that the estimates of c^2 in women could be equated across twins from the three countries (AUS, FIN, and SWE) and across the three age groups (18–25, 26–35, and 36–46) used for analysis. But that was not so for men, where estimates of c^2 were consistently greater for twins from the two Nordic countries than for twin brothers from Australia. Madden and her colleagues interpreted their results to imply that the risk of becoming a smoker can be modified by experi-

ences shared by co-twins and that these modifying effects likely differ across age and culture.

The large population-based twin data set studied by Maes, et al. in Virginia was of twins aged 20–59 with mean age of 36; nearly 80% of the twins had initiated tobacco use during their lifetime, and univariate estimates of heritability for SI were slightly greater for men than women, with modest c^2 for women but none for men. In the report by Morley et al. (>5, 000 adult Australian twin pairs), SI was assessed as age-of-onset of smoking, and the obtained estimates of h^2 were very similar in magnitude for men and women. But, interestingly, these researchers report significant effects from common environment, and because their sample included many of the twins' non-twin siblings, shared environmental effects specific to twins could be estimated. Estimates of the influence from twin-specific c^2 were substantial for variation in SI (0.19 in female twins and 0.12 in male twins).

This effect could not be dropped from models without significant loss of fit. Prevalence of ever smoking by twins and siblings was similar for both men and women, so the difference in the twin-specific c^2 is not a function of different prevalence.

The large-sample study of California twins (Hamilton et al., 2006) offers additional insights into twin-specific environments and gender differences in the magnitude of genetic and environmental effects on SI. And, again, the gender differences were in the direction contrary to that inferred by Li et al.'s meta-analysis of twin data from the 1990s. The California data (impressively large, but reflecting a response rate of under 40%) were obtained from one or both co-twins in >32,000 twin pairs, ages 19–60+, including >10,000 MZ, >11,000 same-sex DZ, and nearly 10,000 opposite-sex DZ pairs in parameter estimation of SI. Estimated genetic effects on SI accounted for >70% of the total variance in men, but less than a third of the variance among women; conversely, effects attributed to common environment accounted for nearly half the variance in smoking initiation among women but only 12% for men.

Setting aside gender differences in the parameter estimates and considering the twins as individuals, odds ratios (ORs) obtained from this large data set enhance our understanding of genetic and environmental factors causally relevant to SI including possible twin-specific environments. The OR for SI linearly increased with twins' age at assessment, from 1.0 for twins < age 25 to 3.4 for those aged 60+, while twins born in later cohorts, after health consequences of smoking were publicized by the Surgeon General, had reduced odds of SI relative to twins born in early cohorts in an era when health benefits of smoking were widely touted by the tobacco industry. But the most dramatic association with SI in these data from California twins was smoking by a twin's co-twin: with a never-smoked MZ co-twin as the reference, the OR was 5.7 (95% CI, 5.2–6.2) for SS DZ twins with an ever-smoking co-twin and 9.7 (95% CI 8.8–10.6) for MZ twins with ever-smoking co-twins. Parental smoking, in contrast, had a more modest association, elevating OR less than 2-fold. These California data add information on twin-specific environmental effects. Closeness to co-twin (assessed by frequency of communication between co-twins; twin pairs communicating with one another weekly or more often versus those in less frequent interaction) varied across both gender and zygosity, and closeness to co-twin was protective, yielding an OR of 0.86; that finding may reflect cooperative interaction of twin siblings as suggested in analysis of drinking/abstaining among Finnish adolescents, where prevalence of abstinence was associated with zygosity (Rose, Dick, Viken, Pulkkinen, & Kaprio, 2001). Perhaps the reduced risk of SI among twin pairs who maintain frequent contact is another reflection of the special twin-specific environment to which significant variance was

attributed in the Australian data of Morley et al. Among adolescent Finnish twins, an influence of sibling interactions associated with differences in co-twin dependency on abstinence/alcohol use has been reported (Penninkilampi-Kerola, Kaprio, Moilanen, & Rose, 2005).

Table 28.1 also includes an analysis of Finnish adult twin data (Broms et al., 2006); the data were collected by questionnaire in 1981, but only now (as part of doctoral dissertation research) have the data been analyzed with modern multivariate techniques. The Finnish sample includes 9,880 same-sex twin pairs born prior to 1958, and aged 23–88 (mean age = 40) at assessment. SI was assessed as age-at-initiation, and the analysis classified those never smoking, as well as those who reported smoking only occasionally as non-initiators. So defined, nearly two-thirds of the twin women and slightly more than one-third of the twin men had never smoked regularly. Of those who did smoke, about as many started before age 18 as later. In univariate model-fitting these data, h^2 for SI (defined as age at initiating regular smoking) was estimated at 0.59 (95% CI, 0.49–0.69) for men and 0.36 (0.28–0.43) for women – a result at odds with Li et al.'s inference from other earlier data sets.

Smoking initiation was studied recently in an adult twin sample recruited nationwide from the United States (Hardie, Moss, & Lynch, 2006). Here, SI was studied as the reported age at first cigarette, for which data from 546 twin pairs, ages 25–75, were available. An AE model fit these smoking initiation data from adult twins with estimates of $h^2 = 0.48$ and $e^2 = 0.52$.

Summary of Smoking Initiation Studies

As this review makes clear, twin studies of SI constitute a substantial body of behavior genetic research based on smoking data reported by >60,000 twin pairs from countries in Europe, North America, Australia, and Asia. So, what conclusions can confidently be drawn? Among adult twins, a confident conclusion is that additive genetic effects are significant, typically explaining half or more of the variance. Effects from shared experience are found in most studies, but c^2 effects are not always significant, and AE models fit much of the observed data from adults. Significant effects of common environment may be age specific, for such effects are found in several studies of adolescent twins. And genetic variance may be age modulated, as well, absent in some parameter models fit to SI among adolescents (Stallings, Hewitt, Beresford, Heath, & Eaves, 1999). Some data suggest twin-specific shared environmental effects, but documenting such effects require adding non-twin siblings in extended twin designs, or measured interactions of contact frequency or perceived closeness of co-twins, and those designs are uncommon. Gender differences in relative magnitude of h^2 and c^2 are evident in some studies but not in a consistent direction.

Given that the most of the studies we reviewed are large-sample studies, why are their estimates so inconsistent – with estimated h^2 ranging from 0.22 to 0.75 for men and 0.32 to 0.63 for women? Perhaps such variation is to be expected for smoking behaviors, given influences of diverse between- and within-family factors including age, gender, birth cohort, religiosity, family structure and status, parental smoking history, and the likelihood that the magnitude of genetic effect varies with time and place (Kendler, Neale, Sullivan et al., 1999) as well as across development. Add, as well, that the manner in which the phenotype is defined for research purposes must affect parameter estimation. Most of the studies we have reviewed examined the phenotype of lifetime ever/never smoking (albeit with varying requirements of quantity/frequency to define "regular" smoking) and the phenotype was treated as a dichotomy; other studies used age at initial experimentation (first cigarette), because it is a risk factor for regular smoking; still other studies have used age at initiation of regular smoking, treating that phenotype as quasi-continuous.

Smoking Persistence

The meta-analysis of Li, Cheng et al. (2003) included studies of smoking persistence (SP) from 11 adult male twin cohorts and 6 adult female twin cohorts. As in the meta-analysis of SI, these studies of SP from four countries (AUS, FIN, SWE, and USA) had been reported during the 1990s. The measure of SP varied across study, and three phenotypes were included: quantity (number of cigarettes smoked per day), current tobacco use or not, and dependence; while acknowledging that these measured behaviors "may not" be identical, Li, et al. collapsed across them on the rationale that each of these smoking phenotypes correlated strongly with nicotine dependency.

So defined, the weighted estimates of h^2 from meta-analysis of SP in these adult twin studies were 0.59 ± 0.02 for males and 0.46 ± 0.12 for females; for c^2 the estimates were 0.08 ± 0.04 and 0.28 ± 0.08, respectively. Estimates of e^2 were 0.37 ± 0.03 for males and 0.24 ± 0.07 for females. For smoking persistence, in contrast to their inferences for smoking initiation, Li and colleagues concluded that genetic factors play a less significant role in female adults than among adult males, and, again in contrast to SI, shared environmental factors play a greater role among adult women than men.

We add a summary of results from recent twin studies of SP, focusing on twin studies we reviewed for SI and included in Table 28.1, adding a recent report from China.

And, for possible clarity, we distinguish studies of smoking persistence or SP from studies of smoking amount or SA. Of the eight recent studies of SI tabulated in Table 28.1, six also report analyses of SP, and some test for genetic correlations between initiation and persistence of smoking using multivariate, multi-stage models. While the heritability of dependence can be assessed only among those who have initiated smoking, genetic variance in smoking initiation can be assessed among all persons in the population. Two-stage modeling permits inclusion of information from never smokers in the analysis (Heath, Martin et al., 2002; Morley et al., 2007) and allows estimation of the degree to which genetic influences on smoking initiation overlap those on dependence. The first detailed consideration of a two-stage model of SP (Heath, Martin, et al.) confirmed substantial genetic effects on smoking persistence, but those genetic effects had minimal overlap with the genetic influences on smoking initiation. That finding has been replicated in subsequent twin research.

Data on SP reported by Madden and her colleagues are from 2,284 pairs from Australia, 8,651 from Sweden pairs, and 10,948 from Finland, aged 18–46. Models fit to SP did not require c^2, and estimates shown in Table 28.2 were constrained to equality across samples from the three countries, both genders, and the three age groups within each gender, a quite different model fit than was found for SI from this three-country sample.

More than half the adult USA twin sample studied by Maes and her colleagues reported regular tobacco use, defined as averaging at least seven cigarettes a week for a month or more. Using that definition of SP, a bivariate analysis of SI and SP was fit to a causal, contingent, common-pathways model. Effects of common environment could be set to zero for both male and female samples to yield AE estimates with equal causal paths for all zygosity/gender groups: $h^2 = 0.80$ and $e^2 = 0.20$. The common-pathways model fit better than a single liability model, implying that the g and e paths that contribute to liability for SI and SP are not fully overlapping.

The California twin sample studied by Hamilton et al. contained 8,625 pairs concordant for having initiated smoking for their analysis of SP. A univariate ACE model, fit to the combined sample, apportioned 54% to additive genes, 9% to common environments, and 37% to unshared experience. But the lower bound of 95% CIs for the c^2 parameter went to zero, and neither twins reported closeness to co-twin, their birth cohort, or nor their age at time of assessment had significant influence on SP, a very different result than was found for SI.

In their analysis of smoking persistence in Australian twins, Morley et al. dichotomously defined whether or not twins who had reported smoking initiation were, at the time of questionnaire assessment, a current or ex-smoker. About

Table 28.2 Smoking persistence (SP)

Study	Year	Sample	Men			Women		
			a^2	c^2	e^2	a^2	c^2	e^2
Li	2003	Meta-analysis	0.59	0.08	0.37	0.46	0.28	0.24
Heath	2002	AUS: 4,382 pairs	0.42	0.09	0.49	0.42	0.10	0.48
Madden	2004	10,000 pairs[1]	0.39–0.49	0.01–0.19	0.39–0.58	0.42–0.45	0.01–0.22	0.30–0.53
Vink	2004	NL; 3,657 pairs	0.36	0.56	0.07	0.36	0.56	0.07
Maes	2004	USA; 6,805 pairs	0.80	0	0.20	0.80	0	0.20
Hamilton	2006	USA; 8,625 pairs[2]	0.54	0.09	0.37	0.55	0.09	0.37
Broms	2006	FIN: 9,880 pairs	0.58	0	0.42	0.50	0	0.50
Lessov-Schlaggar	2006	CHN; 192 ♂ pairs	0.75	0	0.25			
Morley	2007	AUS:>5,000 pairs	0.50	0.03	0.47	0.41	0.10	0.49

[1]Madden et al. studied three twin samples (from AUS, FIN and SWE) and divided each sample into three age groups; entries for this study are the range of estimates across samples and age groups. [2]Hamilton et al. studied SP in all twin pairs concordant for ever having initiated smoking.

44% of the female twins and 49% of male twins were current or ex-smokers. Multivariate model-fitting included a three-stage common-effects, sex-limitation model that added to ACE parameters the additional T parameter here fit for shared experience unique to twin siblings. Results were consistent with that found by Hamilton et al.: twin-specific factors although substantially important for SI were *not* found for SP and are ignored in Table 28.2.

The USA adult twins studied by Hardie and his colleagues yielded a very small sample for study of SP; modeling those small-sample data, two-thirds of the variance in SP was attributed to genetic factors. The genetic correlation between reported age at first cigarette and SP (duration of regular use since onset) was essentially zero.

Broms and her colleagues dichotomized their Finnish twins into current smokers and lifetime smokers who were not current smokers, ignoring for their analysis of SP never smokers and those who reported never smoking more than occasionally. To that categorical measure, a two-stage bivariate model was fit to SI and SP. For men, the c^2 estimate from the full model was bounded by zero, so it was dropped in favor of an AE model yielding estimates of $h^2 = 0.58$ and $e^2 = 0.43$. An AE model offered the best fit to SP for women, estimating both h^2 and e^2 at 0.50. Correlations for additive genetic variance between SI and SP were 0.22 for men and 0.17 for women, suggesting that genetic factors influencing SI account, at most, for 2–4% of the variance in SP.

An AE model fit dichotomously scored current smoking status of twin brothers in China, as well (Lessov-Schlaggar et al., 2006); fully three-fourths of the variance were attributed to additive genes. Although of limited size, the Chinese sample is of interest, as most (58%) of the stud-

ied twins were current smokers, and the results extend cross-cultural replication of AE effects on SI.

To offer a summary of results from these recent twin studies of smoking persistence: Substantial additive genetic variance, little or no effect from common environment, no twin-specific environmental effect, and, in most studies and most interestingly, little or no overlap in genetic effects for SI with those for SP.

Smoking Amount

Quantity or smoking amount (SA) has been studied both as a trait in itself and as a proxy measure for nicotine dependence. Analyses of SA have been made for the highest number of cigarettes ever smoked per day, average number of cigarettes smoked daily during the time of heaviest smoking, a dichotomy of light or heavy smokers, and as the maximum number of cigarettes ever smoked in a 24-h period; in earlier studies, (Carmelli, Swan, Robinette, & Fabsitz, 1990; Swan, Carmelli, & Cardon, 1996; Swan, Carmelli, & Cardon, 1997; Swan, Carmelli, Rosenman, Fabsitz, & Christian, 1990) these phenotypes were found moderately heritable (40–56%).

Table 28.3 offers a summary of results from recent studies of SA. In analysis of adult Finnish twins, Broms et al. dichotomized their sample into those smoking less than 20 cigarettes daily versus those reporting 20 or more and reported heritability estimates of 0.54 for men and 0.61 for women. A two-stage bivariate analysis of SI (age at initiation) and this measure of SA found very little genetic correlation.

Gene–Environment Interactions in Smoking Initiation and Persistence

Gene–environment interactions may be ubiquitous in smoking behaviors, but have not, as yet, been studied widely. Because smoking initiation is influenced by environmental factors shared within families and extra-familial environmental factors shared with peers and birth cohorts, one expects substantial gene–environment correlations and interactions, likely sources of variability in parameter estimates across studies.

One example of moderation of genetic and environmental effects on smoking comes from data from a younger Finnish twin cohort, *FT12*, that show how parenting may moderate the expression of genetic predispositions on adolescent smoking. Significant moderating effects were associated with two dimensions of parenting (parental monitoring and time spent in activities with parents). Genetic influences on adolescent smoking decreased, and common environmental influences increased, at higher levels of parental monitoring (Dick, Viken et al., 2007). Limited parental monitoring may offer an environment that allows greater opportunity for the expression of adolescents' genetic predispositions. In contrast, significant moderating effects for time spent engaged in activities with parents were observed, but this parenting characteristic operated in a different manner: as adolescents reported spending more time with their parents, genetic variance on smoking increased. These findings suggest that spending more time with parents may restrict expression of individual genetic predispositions. Interestingly, the effects associated with parenting on adolescent smoking remained significant from age 14 to age 17, consistent with other findings from Finnish longitudinal analyses that common environmental influences on adolescent smoking vary little across adolescence.

Religiosity is another household factor of interest, as it has been shown to significantly modulate heritable influences on initiation of drinking (Koopmans, Slutske, van Baal, & Boomsma, 1999; Winter, Karvonen, & Rose, 2002). And religiosity modulates genetic effects on smoking initiation in a similar manner. In analysis of data from the AddHealth Study, Timberlake and colleagues have reported that high levels of self-rated religiousness were associated with lower rates of SI, and that religiousness significantly attenuated genetic variance (Timberlake et al., 2005).

Recall that significant effects of a special "twin environment" were found for SI among adult Australian twin pairs. Reciprocal sibling interactions would be expected to impact on prevalence of SI, at least at early ages when smoking is experimental and social in character and prevalence is low. And the differences in prevalence for individual twins would be ordered MZ<SSDZ<OSDZ. Evidence of such ordering, suggestive of cooperative sibling interaction effects, has been found for drinking (Rose, Dick et al., 2001), but effects are inconsistent in the published twin studies of smoking initiation.

Gene × environment interactions on SI have been examined for educational attainment. McCaffery, Papandonatos et al. (2008) found educational level correlated with SI in part due to G–E correlation and interaction, but after taking SI into account, no such effects on ND remained.

Perhaps effects are evident, also, from measured differences in frequency of co-twin contact or perceived closeness to co-twin or self-reported dependency on co-twin. Using data from *FinnTwin16*, Penninkilampi-Kerola et al. (2005) found that co-twin dependence (assessed by questionnaire self-report) modulated h^2 effects on abstinence and alcohol use. Within dependent twin pairs, genetic effects on alcohol-related variables were reduced. In subsequent study, similar effects have been found for smoking initiation (Penninkilampi et al., 2007; see Kaprio, 2007). Most of the studies of interaction effects have been conducted in adolescents, and, importantly, significant evidence of a changing degree of genetic and environmental influences has been demonstrated across several of the studied dimensions. This may suggest that the determinants of smoking are particularly sensitive to external influences, and it may explain variability in estimates of genetic and environmental influences across different studies. Additional work on the extent to which gene–environment interactions may exist on adult smoking phenotypes seems warranted.

Table 28.3 Parameter estimates for smoking amount (SA)

Study	Year	Sample	Men			Women		
			a^2	c^2	e^2	a^2	c^2	e^2
Vink	2004	NL; 3,657 prs	0.51	0.30	0.18	0.51	0.30	0.18
Hardie	2006	USA; 94 prs	0.40	0	0.51	0.49	0	0.51
Broms	2006	FIN: 9,880 prs	0.54	0	0.46	0.61	0	0.39
Morley	2007	AUS:>5,000 prs	0.40	0.12	0.48	0.41	0.20	0.39
Haberstick	2007	USA; 1,078 prs	0.50	0	0.50	0.50	0	0.50

SA was defined differently across studies; e.g., Vink et al uses maximum number of cigarettes smoked daily as a categorical variable; Morley et al used average daily cigarette consumption; Broms et al dichotomized smoking twins into those smoking 20 more cigarettes daily versus those smoking less.

Nicotine Dependence

Adult twin and family studies have suggested high heritability for nicotine dependence (Kendler, Neale, Sullivan et al., 1999; Lessov et al., 2004; Li, Cheng et al., 2003; Maes, Sullivan et al., 2004; Vink, Beem et al., 2004; True et al., 1999). Estimates of heritability have varied from 0.60 among men (True, et al.) to 0.72 among women (Kendler, Neale, Sullivan, et al.) and 0.56 among both men and women (Lessov, et al.). Genetic influences on nicotine dependence (Table 28.4) have been studied using DSM-IV and III-R criteria and as defined by Fagerström's questionnaires (FTQ and FTND). Measured by DSM-III-R criteria, heritability was estimated among US male veterans as 0.60 (True, et al.), and similarly, as 0.56 using a measure of DSM-IV among Australian men and women (Lessov, et al.). Using FTQ, heritability was estimated at 0.62 among US men and women (Maes, Sullivan et al., 2004) and at 0.72 in a separate sample of US women (Kendler, Neale, Sullivan et al., 1999). In a study of Dutch men and women, ND was assessed with the FTND, and heritability was estimated at 0.75 (Vink, Willemsen et al., 2005).

Thus, twin studies on nicotine dependence have shown fairly high and reasonably consistent heritability estimates by different measurements in different cultures. However, a recent family study from Finland showed lower heritability estimates (Broms et al., 2007), whether assessed by the Nicotine Dependence Syndrome Scale ($h^2 = 0.30$) or by FTND ($h^2 = 0.40$) with no evidence for sex-specific genetic effects.

Smoking in Adolescence

Although the research literature is less extensive than that for adults, a number of recent studies have investigated how genetic and environmental factors influence smoking during adolescence. An early study of 1,600 Dutch adolescent twin pairs attributed 31% of the variance in smoking behavior to genetic factors, with a substantial influence (59%)

from shared environments; results from opposite-sex DZ pairs suggested some gender-specific environmental effects (Boomsma, Koopmans, van Doornen, & Orlebeke, 1994). Subsequent analyses of the Dutch twin sample divided it by age groups, with analyses examining 12–14 year olds, 15–16 year olds, and 17–25 year olds separately. Genetic influences were nonsignificant for "ever smoking" in the younger two age groups, but substantial in the oldest group, with genes accounting for 66% of variance in boys and 33% in girls (Koopmans, van Doornen, & Boomsma, 1997).

In an Australian study, 414 pairs of MZ and DZ twins had complete data across three waves, with initial assessments from ages 13–18, a follow-up 3 years later, and a second follow-up 4 years later, when the twins were between 20 and 25. As in other longitudinal studies, mean levels of tobacco involvement increased across assessment waves from adolescence into early adulthood, and consistent with the Dutch study, genetic influences increased over time, accounting for 15% of the variance at wave 1, 20% at wave 2, and 35% at wave 3. Common environmental influences showed corresponding decreases in importance across time, accounting for 55% of the variance at wave 1, 50% at wave 2, and 35% at wave 3. When models were fit to the sexes separately, genetic factors were stronger in females than in males, although it was not clear if this difference was statistically significant. Interestingly, adjusting genetic and environmental estimates for peer and parental smoking caused a significant decrease in genetic influence, particularly at the first wave of data collection. Some of the genetic influence on smoking in early adolescence may be an indirect influence of genetic influences on choice of peers (White, Hopper, Wearing, & Hill, 2003).

A Colorado data set was used to examine genetic and environmental influences across several classes of substances. The sample consisted of 345 MZ twin pairs, 337 DZ twin pairs, 306 biological sibling pairs, and 74 adoptive sibling pairs, assessed at ages 12–19 (mean ∼16 years). Moderate genetic and common environmental influences on tobacco initiation (38% A; 34% C) were found, with no evidence of significant sex differences. There were, however, significant sex differences in influences on tobacco use and

Table 28.4 Parameter estimates for nicotine dependency (ND)

Study	Year	Sample	Men			Women		
			a^2	c^2	e^2	a^2	c^2	e^2
True	1999	USA; 2,356 pairs	0.60	0	0.40			
Kendler	1999	USA: 851 pairs				0.72	0	0.28
Lessov	2004	AUS: 2,293 pairs	0.56	0	0.44	0.56	0	0.44
Maes	2004	USA; 6,805 twins	0.67	0	0.33	0.67	0	0.33
Vink	2005	NL; 1,572 pairs	0.75	0	0.25	0.75	0	0.25
Broms	2007	FIN; 291 pairs	0.40	0	0.60	0.40	0	0.60
Haberstick	2007	USA: 1,078 pairs	0.61	0	0.39	0.61	0	0.39

DSM criteria of ND employed in studies by True, Kendler and Lessov; FTND used in studies by Maes, Vick, and Broms; HSI by Haberstick in a twin sample to which 1,305 full sibs and 384 half-sibs were added.

problem use, with considerably higher genetic influence in girls (with no evidence of C), and substantially lower genetic influences and higher common environmental influences in boys (Rhee et al., 2003).

Other studies have found no evidence of gender differences in the importance of genetic and environmental effects on SI. There was a suggestion of gender differences in early data from the Minnesota Twin-Family Study, where model fitting to tetrachoric correlations of ever/never lifetime tobacco use at age 17 yielded estimates of 0.59 and 0.18 for h^2 and c^2 among adolescent males, but 0.11 and 0.71 for adolescent females (Han, McGue, & Iacono, 1999); but when split by gender, this Minnesota twin sample was of modest power, and a gender-invariant ACE model could be fit to the data. In updated analyses of the sample, based on 626 male and female twin pairs, there was no evidence of gender differences, and the heritability of tobacco use and nicotine dependence was substantial, on the order of ~50%.

A cohort-sequential design studied twins in Virginia (Maes, Woodfard, et al., 1999) and reported baseline data from >1, 400 twin pairs, about evenly distributed across ages 8–16+. The twins were asked whether or not they had ever consumed more than one cigarette per day. So defined, ever smoking rose from about 1.0% prevalence at age 12 to 36% for boys and 22% for girls at age 16. Prevalence varied with family structure and status, parental religious affiliation, and other variables. Individual differences in tobacco use from reports of twins ages 13–16 were largely explained by genetic factors, estimated at 84% for lifetime use. No gender differences in genetic and environmental effects were found.

Data from adolescent Finnish twins have been used to examine genetic and environmental influences on smoking behavior. In the *FinnTwin16* sample, baseline questionnaires were administered across a 60-month period to age-standardize enrollment at age 16. By that age, about half of Finnish twins reported that they had initiated smoking, a prevalence about equal in males and females. And it remained virtually unchanged from age 16 to the second follow-up assessment at age 18 1/2, suggesting that most individuals who will initiate smoking in adolescence have begun by age 16. Analyses of smoking frequency suggested that the factors influencing smoking across this age range are very consistent, as well. Genetic factors accounted for ~50% of the variation at all time points; common environmental factors accounted for ~30% of the variation; and unique environmental influences accounted for the remaining 20% of the variation. There were no gender differences in the factors influencing smoking frequency across this age (Dick, Barman, & Pitkänen, 2006). These findings replicate the Minnesota and Virginia studies in suggesting gender invariance of genetic and environmental influences. Unlike

the Australian and Dutch longitudinal studies, studies that expanded the age range into young adulthood, there was no evidence of increasing genetic influences across adolescence in Finnish twins. Notably, however, these findings were very different from studies of adolescent drinking conducted on the same twin birth cohorts across the same ages: drinking showed a steady increase in the importance of genetic effects across this age range (Rose, Dick et al., 2001).

Data from the younger twin study, *FinnTwin12*, permit study of factors that impact smoking initiation earlier in development. To 1,262 same-sex twins ages 11–12, we yoked a gender- and age-matched classmate control; twins and controls in each pair attended the public school serving their residential area and each twin–control dyad was in the same classroom. The classmate controls are genetic strangers reared in separate households but sharing schools and neighborhoods with their yoked twin; adding them into the classic twin model teases apart effects of familial and extra-familial influences. The twins were studied with supervised in-classroom procedures; their teachers were instructed to select a same-sex classmate closest in age to each twin (for some cohorts) or, in other cohorts, to select same-sex classmates adjacent to the twins in the alphabetized class roster of children's surnames (Rose, Viken et al., 2003). The classmate controls then completed questionnaires at age 12 identical to those completed by the twins. Included among items asked in the questionnaire were two concerning smoking: "Have you ever smoked cigarettes"? And "Do you have friends who smoke cigarettes"? Most of the variance (73%) in the item on smoking initiation at age 12 was again accounted for by common environmental effects. And of that substantial shared environment component, familial factors were more important (accounting for 49% of the variation in early onset smoking) than extra-familial influences (24% of the variance). In contrast, 54% of the variance in having friends who are smoking by age 12 was attributed to common experience; and of that, school-based shared environment was a bit more influential than familial environment. These results suggest that variation across communities, neighborhoods, and schools exert causal influences on early initiation of substance use and on the likelihood of being exposed to smoking peers at an early age.

In summary, studies of adolescent twins demonstrate the importance of genetic factors on smoking behaviors at this earlier developmental stage. There is evidence that the influence of genetic effects increases from early adolescence into adulthood. Whether gender differences modulate the magnitude of genetic and common environmental influences is less certain. Data from the Australia and Colorado samples suggest that genetic influences may be greater in girls; Dutch data found it higher in boys; data from Minnesotan, Virginia, and Finnish twin samples suggest gender invariance. A potentially interesting role of the environment emerges across multiple studies, with Australian results suggesting that genetic effects on early smoking operate through peers, the Dutch study suggestion of sex-specific environmental effects, and the Virginia sample finding evidence that smoking prevalence varied with a number of environmental factors.

Comorbidities of Smoking Behaviors

Abuse of other substances (alcohol, drugs) and mental disorders (depression, schizophrenia) are major comorbid conditions associated with smoking (Jane-Llopis & Matytsina, 2006). Competing hypotheses have been offered to explain these comorbidities. Cigarette smoking may serve as a gateway to illicit drug use (Kandel & Yamaguchi, 1993). Depressive symptoms may lead to self-medication and foster initiation of smoking and drinking: results of a four-wave longitudinal study of American adolescents found persistent depressive symptoms were prospective predictors of increased smoking across time, after controlling for baseline smoking levels (Windle & Windle, 2001). Or, conversely, long-term exposure to persistent cigarette smoking or heavy drinking may have a role in the etiology of depression (Morrell & Cohen, 2006). And there is evidence that health compromising life styles, including sedentary behavior patterns are linked to tobacco use: Finnish data show that persistent physical inactivity in adolescence predicts subsequent smoking initiation, even after controlling for third-variable familial confounds in contrasts of activity-discordant co-twins (Kujala, Kaprio, & Rose, 2007). There is evidence, as well, that early adverse life events are associated with subsequent tobacco addiction. Childhood sexual abuse predicts nicotine dependence in early adulthood (Al Mamun et al., 2007). Further, shared genetic or environmental factors may enhance vulnerability to smoking co-occurring with drinking, drug abuse, depression, or schizophrenia (Williams & Ziedonis, 2004). But a caveat: comorbidities may be confounded by methodological factors leading to overestimation of that dual diagnosis (Kessler, 2004). We briefly review literature on the major smoking comorbidities of cannabis use, alcohol use, depression, and schizophrenia.

Smoking and Cannabis

Cannabis use exhibits strong comorbidity with cigarette use (Agrawal, Grant et al., 2006; Guxens, Nebot, Ariza, & Ochoa, 2007). Finnish data find early smoking initiation is a potent predictor of cannabis use initiation. Twins who

had their first cigarette by the age of 12 had more than a 20-fold risk of subsequent cannabis use compared to twins who had never initiated smoking. That risk attenuated among smoking-discordant twin pairs, but it remained significant (Korhonen, Huizink et al., 2008). Cannabis use shows modest to moderate (20–30%) genetic influence among adolescents and more substantial influence in adults (Agrawal & Lynskey, 2006), so potential mechanisms explaining its association with smoking behaviors are of interest.

Two alternative models have been suggested. The gateway hypothesis (Kandel & Yamaguchi, 1993) suggests a causal pathway from smoking (licit drug) to cannabis (illicit drug). An alternative (Kandel, Yamaguchi, & Klein, 2006; Maccoun, 2006) is a common liability model suggesting shared genetic or environmental influences affecting both phenotypes. According to Agrawal & Lynskey (2006), comorbid tobacco and cannabis use is due partly to shared genetic risk factors and to a modest but significant overlap of environmental influences. Modeling initiation and progression of nicotine and cannabis use with two-stage models, where progression is conditional on initiation, allows studying genetic and environmental influences on both stages (Neale, Harvey, Maes, Sullivan, & Kendler, 2006). For tobacco and marijuana use, the relation between initiation and progression suggests substantial overlap in genetic/environmental etiologies. Common environmental influences tend to be greater for initiation, while genetic influences are stronger for heavier use. To account for relationships between stages of comorbid phenotypes, such as the pathway from smoking initiation to cannabis abuse, Neale and his co-authors (2006) introduced a multivariate two-stage model. The first application was conducted among adult female twins, where the multivariate models, i.e., the Cholesky model and the reciprocal causal model, were fitted. The two models generated similar fit, but the causal model was more parsimonious. Liability to initiation and progression for nicotine and cannabis use were closely related, especially within each substance. The Cholesky model demonstrated a high genetic correlation ($r_A = 0.82$) between nicotine and cannabis initiation. A novel finding was the negative unique environment correlation for nicotine and cannabis dependence, indicating that the two traits have more risk factors in common than their phenotypic correlation suggests, but some specific environment risk factors may act in opposite ways on each. The causal model indicated that liability to initiate accounts for a substantial proportion of variance in progression. A novel finding was that liability to initiate smoking increased liability to initiate cannabis use (path = 0.85), but liability to initiate cannabis use decreased liability to initiate smoking (path = −0.40) (Neale et al., 2006).

Huizink et al. (2007) tested both the causal gateway model and the common liability model with data from *FinnTwin12–17* adolescents. The two models generated similar fit, and, as in adult data, the causal model was more parsimonious. But the Cholesky model suggested a lower genetic correlation between nicotine and cannabis initiation ($r_A = 0.57$) than had been found in adult females, suggesting less shared genetic vulnerability in their covariation. The causal model suggested that the liability to initiate smoking increased the liability to initiate cannabis use (path $= 0.63$).

Smoking and Alcohol

Although smoking rates have decreased in the developed world, they remain high in individuals with alcohol use disorders (Durazzo & Meyerhoff, 2007). In the United States, prevalence of nicotine dependence is very high among alcohol abusers (Grant, Hasin, Chou, Stinson, & Dawson, 2004), and there is a dose–response relationship between amount of cigarettes smoked and amount of alcohol consumed (Dani & Harris, 2005). And early onset smoking persistence is associated with an increased risk of alcohol dependence (John et al., 2003) and hazardous alcohol use later in life, such as drunk driving (Riala, Hakko, Isohanni, Jarvelin, & Rasanen, 2004).

It might seem plausible that the high rate of comorbid tobacco and alcohol use merely reflects the high prevalence of each addictive behavior in the general population, or that similar social situations trigger both behaviors. However, several neurobiological effects have been suggested to modulate the behaviors of nicotine and alcohol abusers. For example, ethanol administration can increase cigarette smoking (Li, Volkow, Baler, & Egli, 2007). Further, twin/family studies demonstrate shared genetic influences on the comorbidity of nicotine and alcohol dependence (Dani & Harris, 2005), the genetic correlation high ($r_A = 0.68$) for severe cases of lifetime dependence (True et al., 1999). Initiation of both behaviors during adolescence may be substantially influenced by shared environmental factors, but regular use of both substances in adulthood may be more influenced by shared genetic factors (Tyndale, 2003). Significant familial association between risk of regular smoking and alcohol dependence, due at least in part by common genetic mechanisms, remained after controlling for personality and psychiatric factors, common risk factors for both phenotypes and both genetically influenced (Madden & Heath, 2002). Assortative mating among comorbid smokers and drinkers (Agrawal, Heath et al., 2006) occurs, as well.

Smoking and Depression

Smokers have a greater likelihood of lifetime depression or current depressive symptoms than non-smokers and those with depressive disorders tend more often to be smokers than are controls (Dani & Harris, 2005; Jane-Llopis & Matytsina, 2006; Morrell & Cohen, 2006); prevalence of current smokers among US adults with history of major depression doubled that of those without history of depression (Lasser et al., 2000). The association is well established, but there are competing hypotheses for its explanation. First, pre-existing depressiveness may predict onset of smoking and/or progression to nicotine dependence (Breslau, Novak, & Kessler, 2004; Fergusson, Goodwin, & Horwood, 2003; Murphy et al., 2003). Among adolescents, depressive symptoms prospectively predict smoking escalation, and the likelihood of rapid escalation interacts with the *DRD2 A1* allele (Audrain-McGovern, Lerman, Wileyto, Rodriguez, & Shields, 2004). And adolescent smoking may increase development of nicotine dependence (Kassel, Stroud, & Paronis, 2003). Conversely, long-term persistent smoking may increase the risk of depression, as reported by longitudinal studies of adolescents (Goodman & Capitman, 2000; Wu & Anthony, 1999) and adults (Klungsoyr, Nygard, Sorensen, & Sandanger, 2006; Korhonen, Broms et al., 2007). Bidirectional predictive associations have been demonstrated in adolescent samples (Breslau, Peterson, Schultz, Chilcoat, & Andreski, 1998; Windle & Windle, 2001). Smoking and depression may share common etiologic risks (Bergen & Caporaso, 1999; Williams & Ziedonis, 2004). Among adolescents, conduct disorders (Breslau, Peterson, et al.) and delinquent peers (Fergusson, Lynskey, & Horwood, 1996) may be such shared risk factors.

Correlations between genetic components of smoking and depression (r_A) have been investigated among adolescents and adults with conflicting results. In adolescents, Silberg, Rutter, D'Onofrio, and Eaves (2003) suggested that early experimental smoking and depression were genetically correlated in girls, but environmentally correlated among boys. McCaffery, Papandonatos, Stanton, Lloyd-Richardson, & Niaura, (2008) replicated shared genetic vulnerability in girls, but found significant unique environmental correlations in both boys and girls. In adults, Kendler, Neale, MacLean et al. (1993) reported that comorbidity of smoking and MDD largely arose from familial factors ($r_A = 0.56$) in adult twin sisters. But other twin studies have reported lower genetic correlations. In male twins McCaffery and co-authors (2003) found that unique environmental factors accounted for most covariation between liability to smoking and depression, with a very modest genetic correlation ($r_A = 0.17$). In a recent longitudinal twin study, Korhonen and colleagues (2007) found that, after controlling for familial factors, smoking remained a gender-sensitive predictor of depressive symptoms. The stronger association in men was modestly accounted for by underlying shared genes ($r_A = 0.25$). Finally, Fu et al. (2007) reported significant shared genetic vulnerability to nicotine dependence and MDD in cross-sectional study of twin brothers. However, after controlling for genetic influences on conduct and antisocial disorders, that genetic correlation approached zero ($r_A = 0.06$).

Effects of specific genes on the smoking-depression comorbidity have been investigated in a limited number of studies with small sample sizes, and mostly for dopamine genes (Lerman & Niaura, 2002). A *DRD4* genotype × depression interaction was found for negative-affect reduction smoking, suggesting that the acute rewarding effects of nicotine partly depend on genetic factors involved in dopamine transmission (Lerman et al., 1998). Audrain-McGovern, Lerman et al. (2004) reported association of the *DRD2 A1* allele with smoking progression among adolescents, an effect potentiated by depressive symptoms. Serotonin transporter genes have received attention, as well but results are inconsistent. Both genetic and environmental correlations may be sensitive to variation in phenotypic definitions of smoking behavior (smoking initiation, persistent smoking, nicotine dependence) and depression (diagnosed MDD, self-reported symptoms) and sensitive as well as to study design (cross-sectional or longitudinal) and sample characteristics.

Smoking and Schizophrenia

Schizophrenics have high (70–90%) smoking rates (Dani & Harris, 2005; de Leon & Diaz, 2005) and high nicotine dependence prevalence. Why? Nicotine affects both expression of schizophrenic symptoms and antipsychotic medications used to treat it (Lyon, 1999). Schizophrenics smoke to self-medicate, and nicotine is hypothesized to normalize deficits in sensory processing associated with the disease (Dani & Harris). Smoking temporarily improves processing of auditory stimuli and may lessen negative symptoms by increasing dopamine in the nucleus accumbens and prefrontal and frontal cortex (Lyon). Cigarette smoking reduces adverse reactions to drug therapy (Haustein, Haffner, & Woodcock, 2002), perhaps enhancing metabolism of antipsychotics (Lyon). And vulnerability to schizophrenia may be associated with vulnerability to smoking (de Leon & Diaz), perhaps via a shared neurobiology, related to altered dopaminergic and cholinergic transmission in the mesolimbic systems (Williams & Ziedonis, 2004). Finally, shared genetic and/or environmental familial factors may underlie smoking and schizophrenia (de Leon & Diaz). There is some evidence linking nicotine receptors (nAChRs) and schizophrenia: the chromosomal region containing the nAChR alpha7 subunit may be linked to genetic risk for this disease, and alpha7 expression may be reduced in these patients (Dani & Harris, 2005).

That the evidence of shared genetic vulnerability for smoking or nicotine dependence and its studied comorbidities is inconsistent is not surprising. Findings are sensitive to definitions of phenotypes, study designs, and sampling used in each analysis. The assessment of causal relations requires a design allowing the inclusion of temporal and concurrent factors contributing to the association. Explorations of common genetic liability for comorbid conditions such as nicotine dependence and alcohol abuse are at an early stage.

Linkage and Association Studies

Molecular Genetic Analyses of Smoking Behavior

To date, genome-wide scans and candidate gene studies of smoking behavior and nicotine dependence have yielded inconsistent, non-replicated results – in common with molecular studies of diseases and behavioral traits with complex genetics. Ho & Tyndale (2007) reviewed published literature as of mid-2006. Eleven different projects had reported genome-wide linkage scans of smoking-related phenotypes, but few had been explicitly designed to study genetics of smoking. LOD scores greater than 3.0 were reported for chromosome 5 (number of cigarettes), chromosome 6 (smoking status), chromosome 7 (DSM-IV dependence), chromosome 11 (comorbid habitual smoking and alcohol dependence), and chromosome 16 (phenotype: short-term cessation). But these had not been consistently replicated. Since then, the Nicotine Addiction Genetics (NAG) Project has reported significant linkage in two independent family data sets from Finland and Australia, with a combined LOD score for the combined data set exceeding 5 on chromosome 22 for the maximum number of cigarettes ever smoked in a 24-h period (Saccone et al., 2007). In contrast to many earlier studies, the NAG study samples were ascertained for heavy cigarette use and nicotine dependence with a very large sample size and consistent methods. In the Finnish arm of the NAG, novel results and replication of several earlier linkage findings were recently reported (Loukola et al., 2007), notably for findings on 10q (maxLOD of 3.12) for a smoker phenotype, and on 7q and 11p (max LOD 2.50 and 2.25, respectively) for the DSM-IV nicotine dependence phenotype. Other work is in progress with fine-mapping of regions identified in earlier scans, which will indicate possible candidate genes to evaluate (Gelernter, Panhuysen et al., 2007; Gelernter, Yu et al., 2006; Lou, Ma, Sun, Payne, & Li, 2007; Zhang et al., 2006).

Ho & Tyndale (2007) also reviewed research literature on candidate genes, both for the neurotransmitter genes in the CNS and the role of metabolic pathways, in particular for CYP2A6, which metabolizes most nicotine to cotinine. Of interest among these neurotransmitter genes are the nicotinic receptor genes, such as those coding for the alpha 4 (CHRNA4), alpha 7 (CHRNA7), and beta 2 (CHRNB2) subunits of the five-unit receptor, but also several other nicotinic receptors. The alpha4-beta2 receptor is a well-known binding site for nicotine and the site of action of varenicline, a new medication for nicotine dependence. Yet, further replication of the somewhat inconsistent association finding to date is needed. As for alcohol and other substances, genetic variation in the dopamine genes, in particular DRD2, and in the gamma-aminobutyric acid neurotransmitter genes, such as GABAB2, has been inconsistently related to the studied phenotypes. However, recent evidence suggests that genetic variants of the CYP2A6 gene are functionally important and probably play a role in the development of nicotine dependence (Audrain-McGovern, Al Koudsi et al., 2007; Benowitz, Swan, Jacob, Lessov-Schlaggar, & Tyndale, 2006; Kubota et al., 2006).

Very recently, large-scale whole genome-wide association studies (GWAS) have identified novel genes in many complex diseases. In these studies, cases (for example persons with nicotine dependence) are compared to controls (healthy subjects) with respect to genotype frequencies of a large number of SNPs covering much of the major sequence variation in the genome. Low-frequency variants are not included. Current technologies, which are rapidly evolving, cover some three hundred thousand to a million SNPs. Because of the very large number of statistical associations that arise, even very small p-values are to be treated cautiously, for the possibility of bias exists, especially for underpowered studies (Garner, 2007). True associations need to be replicated in several data sets, as has been the case for obesity and type 2 diabetes (Fraying & McCarthy, 2007) Three GWAS have been published for nicotine dependence (Bierut et al., 2007; Swan et al., 2006; Uhl et al., 2007), each were based on relatively small samples. Nonetheless, some novel and some expected candidate genes were exposed such as the Neurexin (NRXN) 1 gene (Berrettini, 2008). These efforts represent but the first applications of this novel technique to smoking behavior and nicotine dependence; clearly, larger data sets will be needed to find novel genes for nicotine dependence and various aspects of smoking behavior.

Summary and Future Directions

Twin, adoption, and family studies have clearly established that genetic effects are important in smoking-related behaviors. However, the complexity and evolution of these behaviors over the time-course of an individual's smoking history from experimentation to established use to ultimate cessation pose challenges for genetic study. Most research to date has used fairly crude assessments of smoking behav-

ior, and a critical goal is to refine phenotypic assessments to permit better identification of genetic variants underlying specific aspects of smoking behavior.

Thus, in terms of specific phases or types of smoking behavior, it is important to address the causes of variation in initial susceptibility to smoking and the rapid development of nicotine dependence in adolescents. Likewise, genetic contributions to characteristics of social peer relationships, which strongly predict smoking patterns, should be examined in genetically informative designs combined with studies of friends and peers in relevant subcultures, which segment current societies and provide very specific environments.

It would be important to look at those who are able to control their smoking, so-called light and intermittent smokers. Most research so far has primarily targeted daily and regular smokers, seeking genes underlying this preponderant behavior in smokers. However, in countries where the overall smoking prevalence is decreasing, the proportion of non-daily smokers has been increasing. As of now, there is little consensus as to whether intermittent smoking represents a transitional stage toward smoking cessation on one hand or toward daily smoking on the other hand, or whether intermittent smokers consistently maintain their low frequency of tobacco use. Nor is there research on whether genetic influences are different for intermittent smokers than for regular smokers. Finally, the question could arise whether there are genetic influences "protecting" some individuals from becoming regular smokers, such as genes involved in nicotine metabolism.

Smoking cessation is a process that often takes place after multiple attempts and relapses. Few studies have examined the possible genetics of relapse after a quit attempt, and why some persons can quit "cold turkey" while others make countless quit attempts without success. Pharmacogenomic studies are beginning to examine the role of specific genetic variants, such as those related to dopamine, serotonin, and nicotinic acetylcholine systems, and their interactions with specific medications for smoking cessations such as brupropion or varenicline.

Linkage studies in families have located some chromosomal areas potentially linked to smoking behaviors, but these have been hampered by the fact that many analyses have been done on families collected for other reasons than smoking, and smoking phenotypes were relatively crude. A handful of large-scale genome-wide association studies have been conducted on relatively modest sample sizes. More linkage and association studies focused on smoking are needed to provide the power to detect and replicate reliably genes accounting for some fraction of the variation in smoking behavior. In addition to detecting sequence variants (such as SNPs), epigenetic changes and other genetic mechanisms underlying the interindividual variation in smoking behavior need to be evaluated.

Powerful neuroimaging tools such as functional MRI and PET scans have been used in recent years to look at many aspects of smoking and its effect on the structure and function of the central nervous system, but only occasionally has this approach been combined with genetic informed designs. Twin studies of smoking-discordant pairs using neuroimaging would be a powerful design for disentangling genes from environment.

More research is needed on the role of genetic factors in non-nicotine aspects of tobacco dependence and on other tobacco products than cigarettes, smokeless tobacco in particular. Finally, assessing heritable variance in hypothesized endophenotypes relevant to ND is another promising approach illustrated in a laboratory paradigm (Ray, Rhee, & Stallings, 2007) reporting heritability estimates of 0.47–0.68 for self-reports of tension reduction after smoking. Genetic studies of smoking behaviors offer great challenge but much opportunity.

Acknowledgment Preparation of this chapter was supported by NIAAA grant R37 AA12502. Portions of the literature review and tables were adapted from a dissertation submitted by Ulla Broms to the Faculty of the University of Helsinki (Broms, 2008).

References

Agrawal, A., Grant, J. D., Waldron, M., Duncan, A. E., Scherrer, J. F., & Lynskey, M. T. (2006). Risk for initiation of substance use as a function of age of onset of cigarette, alcohol and cannabis use: Findings in a midwestern female twin cohort. *Preventive Medicine, 43*, 125–128.

Agrawal, A., Heath, A. C., Grant, J. D., Pergadia, M. L., Statham, D. J., Bucholz, K. K., et al. (2006). Assortative mating for cigarette smoking and for alcohol consumption in female Australian twins and their spouses. *Behavior Genetics, 36*, 553–566.

Agrawal, A., & Lynskey, M. T. (2006). The genetic epidemiology of cannabis use, abuse and dependence. *Addiction, 101*, 801–812.

Al Mamun, A., Alati, R., O'Callaghan, M., Hayatbakhsh, M. R., O'Callaghan, F. V., Najman, J. M., et al. (2007). Does childhood sexual abuse have an effect on young adults' nicotine disorder (dependence or withdrawal)? Evidence from a birth cohort study. *Addiction, 102, 647–654.*

Audrain-McGovern, J., Al Koudsi, N., Rodriguez, D., Wileyto, E. P., Shields, P. G., & Tyndale, R. F. (2007). The role of CYP2A6 in the emergence of nicotine dependence in adolescents. *Pediatrics, 119*, e264–e274.

Audrain-McGovern, J., Lerman, C., Wileyto, E. P., Rodriguez, D., & Shields, P. G. (2004). Interacting effects of genetic predisposition and depression on adolescent smoking progression. *The American Journal of Psychiatry, 161*, 1224–1230.

Barman, S. K., Pulkkinen, L., Kaprio, J., & Rose, R. J. (2004). Inattentiveness, parental smoking and adolescent smoking initiation. *Addiction,* 99, 1049–1061.

Batra, V. Patkar, A. A., Berrettini, W. H. Weinstein, S. P., & Leone, F. T. (2003). The genetic determinants of smoking. *Chest, 123*, 1730–1739.

Benowitz, N. L. (1996). Pharmacology of nicotine: addiction and therapeutics. *Annual Review of Pharmacology & Toxicology, 36*, 597–613.

Benowitz, N. L., Swan, G. E., Jacob, P., Lessov-Schlaggar, C. N., & Tyndale, R. F. (2006). CYP2A6 genotype and the metabolism and disposition kinetics of nicotine. *Clinical Pharmacol.Therapy, 80*, 457–467.

Bergen, A. W., & Caporaso, N. (1999). Cigarette smoking. *Journal of the National Cancer Institute, 91*, 1365–1375.

Berrettini, W. (2008). Nicotine addiction. *American Journal of Psychiatry, 65*,1089–92.

Bierut, L. J., Madden, P. A., Breslau, N., Johnson, E. O., Hatsukami, D., & Pomerleau, O. F., et al. (2007). Novel genes identified in a high-density genome wide association study for nicotine dependence, *Human Molecular Genetics, 16*, 24–35.

Blondal, T., Gudmundsson, L. J., Olafsdottir, I., Gustavsson, G., & Westin, A. (1999). Nicotine nasal spray with nicotine patch for smoking cessation: randomized trial with six year follow up. *British Medical Journal, 318*, 285–288.

Boomsma, D. I., Koopmans, J. R., van Doornen, L. J. & Orlebeke, J. F. (1994). Genetic and social influences on starting to smoke: a study of Dutch adolescent twins and their parents. *Addiction.* 89:219–226.

Breslau, N., Novak, S. P., & Kessler, R. C. (2004). Psychiatric disorders and stages of smoking. *Biological Psychiatry, 55*, 69–76.

Breslau, N., Peterson, E. L., Schultz, L. R., Chilcoat, H. D., & Andreski, P. (1998). Major depression and stages of smoking. A longitudinal investigation. *Archives of General Psychiatry, 55*, 161–166.

Broms, U. (2008). *Nicotine Dependence and Smoking Behaviour: A genetic and epidemiological study*. Publications of Public Health M 193:2008. University of Helsinki, Multiprint Oy. Helsinki.

Broms, U., Madden, P. A. F., Heath, A. C., Pergadia, M. L., Shiffman, S., & Kaprio J. (2007). The nicotine dependence syndrome scale in finnish smokers. *Drug and Alcohol Dependence, 89*, 42–51.

Broms, U., Silventoinen, K., Lahelma, E., Koskenvuo, M, & Kaprio J. (2004). Smoking cessation by socioeconomic status and marital status: the contribution of smoking behavior and family background. *Nicotine & Tobacco Research, 6*, 447–455.

Broms, U., Silventoinen, K., Madden, P. A., Heath, A. C., & Kaprio, J. (2006). Genetic architecture of smoking behavior: A study of Finnish adult twins. *Twin Research and Human Genetics, 9*, 64–72.

Carmelli, D., Swan, G. E., Robinette, D. & Fabsitz, R. R. (1990). Heritability of substance use in the NAS-NRC twin registry. *Acta Geneticae Medicae et Gemellologiae, 39*, 91–98.

Cavelaars, A. E., Kunst, A. E., Geurts, J. J., Crialesi, R., Grotvedt, L., & Helmert, U.et al. (2000). Educational differences in smoking: international comparison. *British Medical Journal, 320*, 1102–1107.

Cederlöf, R. (1966). *The twin method in epidemiological studies on chronic disease*. Stockholm: University of Stockholm.

Cederlöf, R., Friberg, L., & Lundman, T. (1977). The interactions of smoking, environment and heredity and their implications for disease etiology. *Acta Medica Scandinavica – Supplementum, 612*, 1–128.

Cepeda-Benito, A., Reynoso, J. T., & Erath, S. (2004). Meta-analysis of the efficacy of nicotine replacement therapy for smoking cessation: Differences between men and women. *Journal of Consulting & Clinical Psychology, 72*, 712–722.

Choi, W. S., Gilpin, E. A., Farkas, A. J., & Pierce, J. P. (2001). Determining the probability of future smoking among adolescents. *Addiction, 96*, 313–323.

Conterio, F., & Chiarelli, B. (1962). Study of the inheritance of some daily life habits. *Heredity, 17, 347–359.*

Crofton, J. (1990). Tobacco and the Third World. *Thorax, 45*, 164–169.

Dani, J. A., & Harris, R. A. (2005). Nicotine addiction and comorbidity with alcohol abuse and mental illness. *Nature Neuroscience, 8*, 1465–1470.

de Leon, J., & Diaz, F. J. (2005). A meta-analysis of worldwide studies demonstrates an association between schizophrenia and tobacco smoking behaviors. *Schizophrenia Research, 76*, 135–157.

de Vries, H., Engels, R., Kremers, S., Wetzels, J., & Mudde, A. (2003). Parents' and friends' smoking status as predictors of smoking onset: findings from six European countries, *Health Education Research, 18*, 627–636.

Dick, D. M., Barman, S., & Pitkänen. T. (2006). Genetic and environmental influences on the initiation and continuation of smoking and drinking. In L. Pulkkinen, J. Kaprio, & R. J. Rose (Eds.), *Socioemotional Development and Health from Adolescence to Adulthood*, Cambridge: Cambridge University Press, pp. 126–145.

Dick, D. M., Viken, R., Purcell, S., Kaprio, J., Pulkkinen, L., & Rose, R. J. (2007). Parental monitoring moderates the importance of genetic and environmental influences on adolescent smoking. *Journal of Abnormal Psychology, 116*, 213–218.

DiFranza, J. R., Savageau, J. A., Fletcher, K., Ockene, J. K., Rigotti, N. A., McNeill, A. D., et al. (2002). Measuring the loss of autonomy over nicotine use in adolescents: the DANDY (Development and Assessment of Nicotine Dependence in Youths) study. *Archives of Pediatrics & Adolescent Medicine, 156*, 397–403.

DiFranza, J. R., Savageau, J. A., Fletcher, K., O'Loughlin, J., Pbert, L., Ockene, J. K., et al. (2007). Symptoms of tobacco dependence after brief intermittent use: the Development and Assessment of Nicotine Dependence in Youth-2 study. *Archives of Pediatrics & Adolescent Medicine, 161*, 704–710.

Doll, R., Peto, R., Boreham, J., & Sutherland, I. (2005). Mortality from cancer in relation to smoking: 50 years observations on British doctors. *British Journal of Cancer, 92*, 426–429.

Doll, R., Peto, R., Wheatley, K., Gray, R., & Sutherland, I. (1994). Mortality in relation to smoking: 40 years' observations on male British doctors, *British Medical Journal, 309*, 901–911.

Durazzo, T. C., & Meyerhoff, D. J. (2007). Neurobiological and neurocognitive effects of chronic cigarette smoking and alcoholism. *Frontiers in Bioscience: a Journal and Virtual Library, 12*, 4079–4100.

Eaves, L. J., & Eysenck, H. J. (1980). Are twins enough? The analysis of family and adoption data. In H. J. Eysenck (Ed.), *The causes and effects of smoking* (pp. 236–282). London: Maurice Temple Smith.

Etter, J. F., Duc, T. V. & Perneger, T. V. (1999). Validity of the Fagerström test for nicotine dependence and of the heaviness of smoking index among relatively light smokers, *Addiction, 94*, 269–281.

Eysenck, H. J. (1980). *The Causes and Effects of Smoking.* London: Maurice Temple Smith.

Fagerström, K. O., Kunze, M., Schoberberger, R., Breslau, N., Hughes, J. R., Hurt, R. D., et al. (1996). Nicotine dependence versus smoking prevalence: comparisons among countries and categories of smokers. *Tobacco Control, 5*, 52–56.

Farber, S. L. (1981). *Identical Twins Reared Apart.* New York: Basic Books.

Fergusson, D. M., Goodwin, R. D., & Horwood, L. J. (2003). Major depression and cigarette smoking: Results of a 21-year longitudinal study. *Psychological Medicine, 33*, 1357–1367.

Fergusson, D. M., Lynskey, M. T., & Horwood, L. J. (1996). Comorbidity between depressive disorders and nicotine dependence in a cohort of 16-year-olds. *Archives of General Psychiatry, 53*, 1043–1047.

Fisher, R. A. (1958). Cancer and smoking. *Nature, 182*, 596.

Flay, B. R., Petraitis, J., & Hu, F. B. (1999). Psychosocial risk and protective factors for adolescent tobacco use. *Nicotine & Tobacco Research, 1*, Suppl 1, S59–S65.

Fraying T. M., & McCarthy M. I. (2007). Genetic studies of diabetes following the advent of the genome-wide association study: Where do we go from here? *Diabetologia. 11*, 2229–2233.

Friberg, L., Kaij, L., Dencker, S. J., & Jonsson, E. (1959). Smoking habits of monozygotic and dizygotic twins. *British Medical Journal, 1*, 1090–1092.

Friedman, G. D., King, M. C., Klatsky, A. L., & Hulley, S. B. (1981). Characteristics of smoking- discordant monozygotic twins. *Progress in Clinical and Biological Research, 69 Part C*, 17–22.

Fu, Q., Heath, A. C., Bucholz, K. K., Lyons, M. J., Tsuang, M. T., True, W. R., et al. (2007). Common genetic risk of major depression and nicotine dependence: The contribution of antisocial traits in a United States veteran male twin cohort. *Twin Research Human Genetics, 10*, 470–478.

Garner, C. (2007). Upward bias in odds ratio estimates from genome-wide association studies. *Genetic Epidemiology, 31*, 288–295.

Gelernter, J., Panhuysen, C., Weiss, R., Brady, K., Poling, J., Krauthammer, M. et al. (2007). Genomewide linkage scan for nicotine dependence: Identification of a chromosome 5 risk locus. *Biological Psychiatry, 61*, 119–126.

Gelernter, J., Yu, Y., Weiss, R., Brady, K., Panhuysen, C., Yang, B. Z. et al. (2006). Haplotype spanning TTC12 and ANKK1, flanked by the DRD2 and NCAM1 loci, is strongly associated to nicotine dependence in two distinct American populations. *Human Molecular Genetics, 15*, 3498–3507.

Giskes, K., Kunst, A. E., Benach, J., Borrell, C., Costa, G., Dahl, E. et al (2005). Trends in smoking behaviour between 1985 and 2000 in nine European countries by education. *Journal of Epidemiology & Community Health, 59*, 395–401

Glendinning, A., Hendry, L., & Shucksmith, J. (1995). Lifestyle, health and social class in adolescence. *Social Science and Medicine, 41*, 235–248.

Goodman, E., & Capitman, J. (2000). Depressive symptoms and cigarette smoking among teens. *Pediatrics, 106*, 748–755.

Grant, B. F., Hasin, D. S., Chou, S. P., Stinson, F. S., & Dawson, D. A. (2004). Nicotine dependence and psychiatric disorders in the united states: Results from the national epidemiologic survey on alcohol and related conditions. *Archives of General Psychiatry, 61*, 1107–1115.

Guxens, M., Nebot, M., Ariza, C., & Ochoa, D. (2007). Factors associated with the onset of cannabis use: a systematic review of cohort studies. *Gaceta anitaria, 21*, 252–260.

Haberstick, B. C., Timberlake, D., Ehringer, M. A., Lessem, J. M., Hopfer, C. J., Smolen, A. et al. (2007). Genes, time to first cigarette and nicotine dependence in a general population sample of young adults. *Addiction, 102*, 655–665.

Hamilton, A. S., Lessov-Schlaggar, C. N., Cockburn, M. G., Unger, J. B., Cozen, W., & Mack, T. M. (2006). Gender differences in determinants of smoking initiation and persistence in California twins. *Cancer Epidemiology, Biomarkers & Prevention, 15*, 1189–1197.

Han C., McGue, M. K., & Iacono W. G. (1999). Lifetime tobacco, alcohol and other substance use in adolescent Minnesota twins: univariate and multivariate behavioral genetic analyses. *Addiction, 94*, 981–993.

Hannah, M. C., Hopper, J. L., & Mathews, J. D. (1985). Twin concordance for a binary trait. II. Nested analysis of ever-smoking and ex-smoking traits and unnested analysis of a "committed-smoking" trait. *American Journal of Human Genetics, 37*, 153–165.

Hardie, T.L., Moss, H. B., & Lynch, K. G. (2006). Genetic correlations between smoking initiation and smoking behaviors in a twin sample. *Addictive behaviors, 31*, 2030–2037.

Hatsukami, D., Skoog, K., Allen, S., & Bliss, R. (1995). Gender and the effects of different doses of nicotine gum on tobacco withdrawal symptoms, *Experimental and Clinical Psychopharmacology, 3*, 163–173.

Haustein, K. O. (2006). Smoking and poverty. *European Journal of Cardiovascular Prevention & Rehabilitation, 13*, 312–318.

Haustein, K. O., Haffner, S., & Woodcock, B. G. (2002). A review of the pharmacological and psychopharmacological aspects of smoking and smoking cessation in psychiatric patients. *International Journal of Clinical Pharmacology and Therapeutics, 40*, 404–418.

Heath, A.C., Martin, N.G., Lynskey, M.T., Todorov, A.A., & Madden, P.A. (2002). Estimating two-stage models for genetic influences on alcohol, tobacco or drug use initiation and dependence vulnerability in twin and family data. *Twin Research, 5,* 113–124.

Helakorpi, S., Patja, K., Prättälä, R., & Uutela, A. (2005). *Health Behaviour and Health Among the Finnish Adult Population, Spring 2005.* Publications of the National Health Institute. Helsinki.

Helakorpi, S., Patja, K., Prättälä, R., & Uutela, A. (2007). *Health behaviour and health among the Finnish adult population, Spring 2006.* Publications of the National Health Institute. Helsinki.

Ho, M. K., & Tyndale, R. F. (2007). Overview of the pharmacogenomics of cigarette smoking. *Pharmacogenomics Journal, 7,* 81–98.

Hrubec, Z., Cederlöf, R., & Friberg, L. (1976). Background of angina pectoris: social and environmental factors in relation to smoking. *American Journal of Epidemiology, 103,* 16–29.

Huizink, A. C., Korhonen, T., Levälahti, E., Dick, D., Rose, R. J., Pulkkinen, L.et al. (2007). Testing the gateway model in Finnish adolescent twins: Progression from cigarette smoking to cannabis use. *Behavior Genetics. 37,* 734–809 (abstr.)

Huurre, T., Aro, H., & Rahkonen, O. (2003). Well-being and health behaviour by parental socioeconomic status: a follow-up study of adolescents aged 16 until age 32 years. *Social Psychiatry & Psychiatric Epidemiology, 38,* 249–255.

Hymowitz, N., Risch, H. A., Howe, G. R., Jain, M., Burch, J. D., Holowaty, E. J.et al. (1997). Predictors of smoking cessation in a cohort of adult smokers followed for five years. *Tobacco Control, 6,* Suppl 2, S57–S62.

Jane-Llopis, E., & Matytsina, I. (2006). Mental health and alcohol, drugs and tobacco: A review of the comorbidity between mental disorders and the use of alcohol, tobacco and illicit drugs. *Drug and Alcohol Review, 25,* 515–536.

Jarvis, M. J. (1996). The association between having children, family size and smoking cessation in adults. *Addiction, 91,* 427–434.

Jarvis, M. (1997). Patterns and predictors of smoking cessation in the general population. In C. T. Bolliger & K. O. Fagerström (Eds.). *The tobacco epidemic* (pp. 151–164). Basel: Karger.

John, U., Meyer, C., Rumpf, H. J., Schumann, A., Thyrian, J. R., & Hapke, U. (2003). Strength of the relationship between tobacco smoking, nicotine dependence and the severity of alcohol dependence syndrome criteria in a population-based sample. *Alcohol and Alcoholism, 38,* 606–612.

Johnson, C. C., Li, D., Perry, C. L., Elder, J. P., Feldman, H .A., Kelder, S. H. et al. (2002). Fifth through eighth grade longitudinal predictors of tobacco use among a racially diverse cohort: CATCH. *Journal of School Health, 72,* 58–64.

Kandel, D. B., Yamaguchi, K., & Klein, L. C. (2006). Testing the gateway hypothesis. *Addiction, 101,* 470–472.

Kandel, D., & Yamaguchi, K. (1993). From beer to crack: Developmental patterns of drug involvement. *American Journal of Public Health, 83,* 851–855.

Kaprio, J. (1984). *The incidence of coronary heart disease in twin pairs discordant for cigarette smoking. A six-year follow-up of adult like-sexed male twin pairs.* Helsinki: Kansanterveystieteen julkaisuja M84:1984.

Kaprio, J. (2007). Differences in smoking habits of MZ and DZ twins; A commentary on Tishler and Carey. *Twin Research and Human Genetics, 10,* 718–720.

Kaprio, J. & Koskenvuo, M. (1988). A prospective study of psychological and socioeconomic characteristics, health behavior and morbidity in cigarette smokers prior to quitting compared to persistent smokers and non-smokers. *Journal of Clinical Epidemiology, 41,* 139–150.

Kaprio, J., Koskenvuo, M., & Langinvainio, H. (1984). Finnish twins reared apart. IV: Smoking and drinking habits. A preliminary analysis of the effect of heredity and environment. *Acta Geneticae Medicae et Gemellologiae, 33,* 425–433.

Kaprio, J., Sarna, S., Koskenvuo, M., & Rantasalo, I. (1978). *The Finnish Twin registry: baseline characteristics : Section II: History of symptoms and illnesses, use of drugs, physical characteristics, smoking, alcohol and physical activity.* (vols. M37) Helsinki: Kansanterveystieteen julkaisuja (Publications in Public Health Science).

Kassel, J. D., Stroud, L. R., & Paronis, C. A. (2003). Smoking, stress, and negative affect: Correlation, causation, and context across stages of smoking. *Psychological Bulletin, 129,* 270–304.

Kendler, K. S., Neale, M. C., MacLean, C. J., Heath, A. C., Eaves, L. J., & Kessler, R. C. (1993). Smoking and major depression. A causal analysis. *Archives of General Psychiatry, 50,* 36–43.

Kendler, K. S., Neale, M. C., Sullivan, P., Corey, L. A., Gardner, C. O., & Prescott, C. A. (1999). A population-based twin study in women of smoking initiation and nicotine dependence. *Psychological Medicine, 29,* 299–308.

Kessler, R. C. (2004). The epidemiology of dual diagnosis. *Biological psychiatry, 56* 730–737.

Kestilä, L., Koskinen, S., Martelin, T., Rahkonen, O., Pensola, T., Pirkola, S. et al. (2006). Influence of parental education, childhood adversities, and current living conditions on daily smoking in early adulthood. *European Journal of Public Health, 16,* 617–626

Klungsoyr, O., Nygard, J. F., Sorensen, T., & Sandanger, I. (2006). Cigarette smoking and incidence of first depressive episode: An 11-year, population-based follow-up study. *American Journal of Epidemiology, 163,* 421–432.

Koivusilta L., Rimpela A., & Vikat, A. (2003). Health behaviors and health in adolescence as predictors of educational level in adulthood: A follow-up study from Finland. *Social Science and Medicine,* 57, 577–593.

Koopmans, J. R., Slutske,W.S., van Baal, G. C., & Boomsma, D. I. (1999). The influence of religion on alcohol use initiation: evidence for genotype X environment interaction. *Behavior Genetics,* 29, 445–453.

Koopmans J. R., van Doornen L .J., & Boomsma, D. I. (1997). Association between alcohol use and smoking in adolescent and young adult twins: A bivariate genetic analysis. *Alcoholism: Clinical & Experimental Research, 21,* 537–546.

Korhonen, T., Broms, U., Varjonen, J., Romanov, K., Koskenvuo, M., Kinnunen, T. et al. (2007). Smoking behavior as a predictor for depression among Finnish men and women: A prospective study of adult twins. *Psychological Medicine, 37,* 705–715.

Korhonen, T., Huizink, A. C., Dick, D., Pulkkinen, L., Rose, R. J., & Kaprio, J. (2008). Role of individual, peer and family factors in the use of cannabis and other illicit drugs: A longitudinal analysis among Finnish adolescent twins. *Drug and Alcohol Dependence, 97,* 33–43.

Kubota, T., Nakajima-Taniguchi, C., Fukuda, T., Funamoto, M., Maeda, M., Tange, E. et al. (2006). CYP2A6 polymorphisms are associated with nicotine dependence and influence withdrawal symptoms in smoking cessation. *Pharmacogenomics Journal., 6,* 115–119.

Kujala, U. M., Kaprio, J., & Rose, R. J. (2007). Physical activity in adolescence and smoking in young adulthood: A prospective twin cohort study. *Addiction, 102,* 1151–1157.

Lahelma, E., Rahkonen, O., Berg, M. A., Helakorpi, S., Prattala, R., Puska, P., et al. (1997). Changes in health status and health behavior among Finnish adults 1978–1993. *Scandinavian Journal of Work, Environment & Health, 23,* Suppl 3, 85–90.

Lasser, K., Boyd, J. W., Woolhandler, S., Himmelstein, D. U., McCormick, D., & Bor, D. H. (2000). Smoking and mental illness: A population-based prevalence study. *JAMA, 284,* 2606–2610.

Le Houezec, J. (2003). Role of nicotine pharmacokinetics in nicotine addiction and nicotine replacement therapy: A review. *International Journal of Tuberculosis & Lung Disease, 7,* 811–819.

Leonard, S., & Bertrand, D. (2001). Neuronal nicotinic receptors: from structure to function. *Nicotine & Tobacco Research, 3*, 203–223.

Lerman, C., & Berrettini, W. (2003). Elucidating the role of genetic factors in smoking behavior and nicotine dependence. *American Journal of Medical Genetics. Part B, 118*, 48–54.

Lerman, C., Caporaso, N., Main, D., Audrain, J., Boyd, N. R., Bowman, E. D., et al. (1998). Depression and self-medication with nicotine: The modifying influence of the dopamine D4 receptor gene. *Health Psychology, 17*, 56–62.

Lerman, C., & Niaura, R. (2002). Applying genetic approaches to the treatment of nicotine dependence. *Oncogene, 21*, 7412–7420.

Lessov, C. N., Martin, N. G., Statham, D. J., Todorov, A. A., Slutske, W. S., Bucholz, K. K., et al. (2004). Defining nicotine dependence for genetic research: Evidence from Australian twins. *Psychological Medicine, 34*, 865–879.

Lessov-Schlaggar, C. N., Pang, Z., Swan, G. E., Guo, Q., Wang, S., Cao, W.et al. (2006). Heritability of cigarette smoking and alcohol use in Chinese male twins: The Qingdao twin registry. *International Journal of Epidemiology, 35*, 1278–1285.

Li, M.D., Cheng, R., Ma, J.Z., & Swan, G.E. (2003). A meta-analysis of estimated genetic and environmental effects on smoking behavior in male and female adult twins. *Addiction, 98*, 23–31.

Li, T. K., Volkow, N. D., Baler, R. D., & Egli, M. (2007). The biological bases of nicotine and alcohol co-addiction. *Biological Psychiatry, 61*, 1–3.

Lou, X. Y., Ma, J. Z., Sun, D., Payne, T. J., & Li, M. D. (2007). Fine mapping of a linkage region on chromosome 17p13 reveals that GABARAP and DLG4 are associated with vulnerability to nicotine dependence in European-Americans. *Human Molecular Genetics, 16*, 142–153.

Loukola, A., Broms, U., Maunu, H., Widen, E., Heikkila, K., Siivola, M., et al. (2007). Linkage of nicotine dependence and smoking behavior on 10q, 7q and 11p in twins with homogeneous genetic background. *The Pharmacogenomics Journal, 8*, 209–219.

Lyon, E. R. (1999). A review of the effects of nicotine on schizophrenia and antipsychotic medications. *Psychiatric services (Washington, D.C.), 50*, 1346–1350.

Mackay, J., Eriksen, M., & Shafey, O. (2006). *The Tobacco Atlas*. The American Cancer Society: Atlanta

Maccoun, R. J. (2006). Competing accounts of the gateway effect: The field thins, but still no clear winner. *Addiction, 101*, 473–474.

Madden, P. A. & Heath, A. C. (2002). Shared genetic vulnerability in alcohol and cigarette use and dependence. *Alcoholism: Clinical & Experimental Research, 26*, 1919–1921.

Madden, P. A., Pedersen, N. L., Kaprio, J., Koskenvuo, M. J., & Martin, N. G.(2004). The epidemiology and genetics of smoking initiation and persistence: crosscultural comparisons of twin study results. *Twin Research, 7*, 82–97.

Maes, H. H., Sullivan, P. F., Bulik, C. M., Neale, M. C., Prescott, C. A., Eaves, L. J., et al. (2004). A twin study of genetic and environmental influences on tobacco initiation, regular tobacco use and nicotine dependence. *Psychological medicine, 34*, 1251–1261.

Maes, H. H., Woodard C. E., Murrelle L., Meyer J. M., Silberg J. L., Hewitt J. K., *et al.* (1999). Tobacco, alcohol and drug use in eight- to sixteen-year-old twins: the Virginia twin study of adolescent behavioral development. *Journal of Studies on Alcohol, 60*, 293–305

McCaffery, J. M., Niaura, R., Swan, G. E., & Carmelli, D. (2003). A study of depressive symptoms and smoking behavior in adult male twins from the NHLBI twin study. *Nicotine & Tobacco Research, 5*, 77–83.

McCaffery, J. M., Papandonatos, G. D., Lyons, M. J., Koenen, K. C., Tsuang, M. T., & Niaura, R. (2008). Educational attainment, smoking initiation and lifetime nicotine dependence among male Vietnam-era twins. *Psychological Medicine, 38*, 1287–1297.

McCaffery, J. M., Papandonatos, G. D., Stanton, C., Lloyd-Richardson, E. E., & Niaura, R. (2008). Depressive symptoms and cigarette smoking in twins from the National Longitudinal Study of Adolescent Health. *Health Psychology, 27*, S207–S215.

McLellan, D. (1995). Women and diversity. In K. Slama (Ed.) *Tobacco and Health* (pp. 19–25). Plenum, New York.

Morley, K. I., Lynskey, M. T., Madden, P. A., Treloar, S. A., Heath, A. C., & Martin, N. G. (2007). Exploring the inter-relationship of smoking age-at-onset, cigarette consumption and smoking persistence: genes or environment? *Psychological Medicine, 37*, 1357–1367.

Morrell, H. E. R. & Cohen, L. M. (2006). Cigarette smoking, anxiety and depression. *Journal of Psychopathology and Behavioral Assessment, 28*, 283–297.

Murphy, J. M., Horton, N. J., Monson, R. R., Laird, N. M., Sobol, A. M., & Leighton, A. H. (2003). Cigarette smoking in relation to depression: Historical trends from the Stirling county study. *The American Journal of Psychiatry, 160*, 1663–1669.

Neale, M. C., Harvey, E., Maes, H. H., Sullivan, P. F., & Kendler, K. S. (2006). Extensions to the modeling of initiation and progression: Applications to substance use and abuse. *Behavior Genetics, 36*, 507–524.

Neubauer, S., Welte, R., Beiche, A., Koenig, H. H., Buesch, K., & Leidl, R. (2006). Mortality, morbidity and costs attributable to smoking in Germany: update and a 10-year comparison. *Tobacco Control, 15*, 464–471.

O'Brien, C. P., Childress, A. R., McLellan, A. T., & Ehrman, R. (1992). A learning model of addiction. *Research Publications – Association for Research in Nervous & Mental Disease, 70*, 157–177.

Osler, M., Holst, C., Prescott, E., & Sorensen, T .I. (2001). Influence of genes and family environment on adult smoking behavior assessed in an adoption study. *Genetic Epidemiology, 21*, 193–200.

Partanen, J., Bruun, K., & Markkanen, T. (1966). *Inheritance of drinking behavior. A study of intelligence, personality, and use of alcohol of adult twins*. Helsinki: The Finnish Foundation for Alcohol Studies.

Penninkilampi-Kerola, V., Kaprio, J., Moilanen, I., & Rose, R. J. (2005). Co-twin dependence modifies heritability of abstinence and alcohol use: A population-based study of Finnish twins. *Twin Research and Human Genetics, 8*: 232–244.

Penninkilampi-Kerola, V., Rose, R. J., & Kaprio, J. (2007). *Co-twin relationship and heritability of smoking behavior*. Presentation at International Society of Twin Studies Meetings, Ghent; cited in Kaprio (2007).

Perkins, K.A. (2001). Smoking cessation in women. Special considerations, *CNS Drugs, 15*, 391–411.

Perkins, K.A., Donny, E., & Caggiula, A.R. (1999). Sex differences in nicotine effects and self-administration: review of human and animal evidence, *Nicotine & Tobacco Research, 1*, 301–315.

Perkins, K.A., Fonte, C., Sanders, M., Meeker, J., & Wilson, A. (2001a). Threshold doses for nicotine discrimination in smokers and non-smokers. *Psychopharmacology, 155*, 163–170.

Perkins, K.A., Gerlach, D., Broge, M., Fonte, C., & Wilson, A. (2001b). Reinforcing effects of nicotine as a function of smoking status, *Experimental & Clinical Psychopharmacology, 9*, 243–250.

Perkins, K.A., Gerlach, D., Broge, M., Grobe, J. E., Sanders, M., Fonte, C., et al. (2001c) Dissociation of nicotine tolerance from tobacco dependence in humans, *Journal of Pharmacology & Experimental Therapeutics, 296*, 849–856.

Perkins, K. A., Gerlach, D., Broge, M., Sanders, M., Grobe, J., Fonte, C., et al. (2001d). Quitting cigarette smoking produces minimal loss of chronic tolerance to nicotine, *Psychopharmacology, 158*, 7–17.

Perkins, K.A., Sanders, M., D'Amico, D., & Wilson, A. (1997). Nicotine discrimination and self-administration in humans as a function of smoking status. *Psychopharmacology, 131*, 361–370.

Platt, S., Amos, A., Gnich, W., & Parry, O. (2002). Reducing inequalities in health: a European perspective. In J. P. Mackenbach & M. J. Bakker (Eds.), *Smoking Policies* (pp. 125–143). London: Routledge.

Prokhorov, A. V., Pallonen, U. E., Fava, J. L., Ding, L., & Niaura, R. (1996). Measuring nicotine dependence among high-risk adolescent smokers. *Addictive Behaviors, 21*, 117–127.

Raaschou-Nielsen, E. (1960). Smoking habits in twins. *Danish Medical Bulletin, 7*, 82–88.

Ray, L. A., Rhee, S H., & Stallings, M. C. (2007). Examining the heritability of a laboratory-based smoking endophenotype. *Twin Research Human Genetics, 10*, 546–553.

Regidor, E., Gutierrez-Fisac, J. L., Calle, M. E., Navarro, P., & Dominguez, V. (2001). Trends in cigarette smoking in Spain by social class. *Preventive Medicine, 33*, 241–248.

Rhee, S. H., Hewitt, J. K., Young, S. E., Corley, R. P., Crowley, T. J., & Stallings, M. C. (2003). Genetic and environmental influences on substance initiation, use, and problem use in adolescents. *Archives of General Psychiatry, 60*, 1256–1264.

Riala, K., Hakko, H., Isohanni, M., Jarvelin, M. R., & Rasanen, P. (2004). Teenage smoking and substance use as predictors of severe alcohol problems in late adolescence and in young adulthood. *The Journal of Adolescent Health, 35*, 245–254.

Rimpelä, A., Rainio, S., Pere, L., Lintonen, T., & Rimpelä, M. (2005). Use of tobacco products and substance use in 1977–2005. Adolescent Health and Lifestyle Survey. Ministry of Social and Health Affairs, Helsinki. (In Finnish). 118 pps. (Reports of the Ministry of Social Affairs and Health, ISSN 1236-2115).

Risch, H. A., Howe, G. R., Jain, M., Burch, J. D., Holowaty, E. J., & Miller, A. B. (1993). Are female smokers at higher risk for lung cancer than male smokers? A case-control analysis by histologic type. *American Journal of Epidemiology, 138*, 281–293.

Rose, R. J. (2002). How do adolescents select their friends? A behavior-genetic perspective. In L. Pulkkinen & A. Caspi (Eds.), *Paths to Successful Development: Personality in the Life Course.* (pp. 106–125) Cambridge: Cambridge University Press.

Rose, R. J. (2006). Introduction. In L. Pulkkinen, J. Kaprio, & R. J. Rose (Eds.) *Socioemotional development and health from adolescence to adulthood.* (pp. 1–25) Cambridge: Cambridge University Press.

Rose, R. J. (2007). Peers, parents, and processes of adolescent socialization: A twin-study perspective. In H. Stattin, M. Kerr, & R. Engels (Eds.) *Friends, lovers & groups: Who is important and Why?* (pp. 105–124) London/New York: John Wiley.

Rose, R. J., Dick, D. M., Viken, R. J., Pulkkinen, L. & Kaprio, J. (2001). Drinking or abstaining at age 14? A genetic epidemiological study. *Alcoholism: Clinical & Experimental Research, 25*, 1594–1604.

Rose, R. J., Viken, R. J., Dick, D. M., Bates, J. E., Pulkkinen, L., & Kaprio, J. (2003). It *does* take a village: Non-familial environments and childrens behavior. *Psychological Science, 14*: 273–277.

Rubinstein, M. L., Thompson, P. J., Benowitz, N. L., Shiffman, S., & Moscicki, A. B. (2007). Cotinine levels in relation to smoking behavior and addiction in young adolescent smokers. *Nicotine & Tobacco Research, 9*, 129–135.

Saarni, S. E., Silventoinen, K., Rissanen, A., Sarlio-Lahteenkorva, S., & Kaprio, J. (2004). Intentional weight loss and smoking in young adults. *International Journal of Obesity & Related Metabolic Disorders, 28*, 796–802.

Saccone, S. F., Pergadia, M. L., Loukola, A., Broms, U., Montgomery, G. W., Wang, J. C. et al. (2007). Genetic linkage to chromosome 22q12 for a heavy-smoking quantitative trait in two independent samples. *American Journal of Human Genetics, 80*, 856–866.

Sasco, A. J., Merrill, R. M., Benhaim-Luzon, V., Gerard, J. P., & Freyer, G. (2003). Trends in tobacco smoking among adolescents in Lyon, France. *European Journal of Cancer, 39*, 496–504.

Senore, C., Battista, R. N., Shapiro, S. H., Segnan, N., Ponti, A., Rosso, S., et al. (1998). Predictors of smoking cessation following physicians' counseling. *Preventive Medicine, 27*, 412–421.

Shenassa, E. D., McCaffery, J. M., Swan, G. E., Khroyan, T. V., Shakib, S., Lerman, C., et al. (2003). Intergenerational transmission of tobacco use and dependence: a transdisciplinary perspective. *Nicotine & Tobacco Research, 5*, Suppl 1, S55–S69.

Shields, J. (1962). *Monozygotic twins brought up apart and brought up together.* London: Oxford University Press.

Shiffman, S., & Paton, S. M. (1999). Individual differences in smoking: gender and nicotine addiction. *Nicotine & Tobacco Research, 1*, Suppl 2, S153–S157.

Siddiqui, O., Hedeker, D., Flay, B. R., & Hu, F. B. (1996). Intraclass correlation estimates in a school-based prevention study. *American Journal of Epidemiology, 144*, 425–433.

Silberg, J., Rutter, M., D'Onofrio, B., & Eaves, L. (2003). Genetic and environmental risk factors in adolescent substance use. *Journal of Child Psychology and Psychiatry, 44*, 664–676.

Stallings, M. C., Hewitt, J. K., Beresford, T., Heath, A. C., & Eaves, L. J. (1999). A twin study of drinking and smoking onset milestones and latencies from first use to regular use. *Behavior Genetics, 29*, 409–421.

Stapleton, J. (1998). Cigarette smoking prevalence, cessation and relapse. *Statistical Methods in Medical Research, 7*, 187–203.

Sullivan, P.F., & Kendler, K.S. (1999). The genetic epidemiology of smoking. *Nicotine & Tobacco Research, 1*, Suppl 2, S51–S57.

Swan, G. E., Carmelli, D., & Cardon, L.R. (1996). The consumption of tobacco, alcohol, and coffee in Caucasian male twins: a multivariate genetic analysis. *Journal of Substance Abuse, 8*, 19–31.

Swan, G. E., Carmelli, D., & Cardon, L.R. (1997). Heavy consumption of cigarettes, alcohol and coffee in male twins. *Journal of Studies on Alcohol, 58*, 182–190.

Swan, G.E., Carmelli, D., Rosenman, R. H., Fabsitz, R. R., & Christian, J. C. (1990). Smoking and alcohol consumption in adult male twins: genetic heritability and shared environmental influences. *Journal of Substance Abuse, 2*, 39–50.

Swan, G. E., Hops, H., Wilhelmsen, K. C., Lessov-Schlaggar, C. N., Cheng, L. S., Hudmon, K. S., et al. (2006). A genome-wide screen for nicotine dependence susceptibility loci. *American Journal of Medical Genetics. Part B, Neuropsychiatric Genetics, 141*, 354–360.

Terracciano, A., & Costa, P. T., Jr. (2004). Smoking and the Five-Factor Model of personality. *Addiction, 99*, 472–481.

Timberlake, D. S., Rhee, S. H., Haberstick, B. C., Hopfer, C., Ehringer, M., Lessem, J. M., et al. (2005). The moderating effect of religiosity on the genetic and environmental determinants of smoking initiation. *Nicotine and Tobacco Research, 8*, 123–133.

True, W. R., Xian, H., Scherrer, J. F., Madden, P. A., Bucholz, K. K., Heath, A. C., et al. (1999). Common genetic vulnerability for nicotine and alcohol dependence in men. *Archives of General Psychiatry, 56*, 655–661.

Tyndale, R. F. (2003). Genetics of alcohol and tobacco use in humans. *Annals of Medicine, 35*, 94–121.

Uhl, G. R., Liu, Q. R., Drgon, T., Johnson, C., Walther, D., & Rose, J. E. (2007). Molecular genetics of nicotine dependence and abstinence: whole genome association using 520,000 SNPs. *BMC Genetics 8*, 10.

Vink, J. M., Beem,A. L., Posthuma, D., Neale, M. C., Willemsen, G., Kendler, K .S.et al. (2004). Linkage analysis of smoking initiation and quantity in Dutch sibling pairs. *Pharmacogenomics Journal, 4*, 274–282.

Vink, J. M., Willemsen, G., & Boomsma, D. I. (2003). The association of current smoking behavior with the smoking behavior of parents, siblings, friends and spouses. *Addiction, 98*, 923–931.

Vink, J. M., Willemsen, G., & Boomsma, D. I. (2005). Heritability of smoking initiation and nicotine dependence. *Behavior Genetics, 35*, 397–406.

Vink, J. M., Willemsen, G., Engels, R. C., & Boomsma, D. I. (2003). Smoking status of parents, siblings and friends: predictors of regular smoking? Findings from a longitudinal twin-family study. *Twin Research, 6*, 209–217

West, R., & Hardy, A. (2006). *Theory of addiction,* Blackwell/Addiction Press, Oxford: Malden, MA.

White, V. M., Hopper, J .L., Wearing, A. J., & Hill, D. J. (2003). The role of genes in tobacco smoking during adolescence and young adulthood: a multivariate behaviour genetic investigation. *Addiction, 98,* 1087–1100.

Williams, J. M., & Ziedonis, D. (2004). Addressing tobacco among individuals with a mental illness or an addiction. *Addictive Behaviors, 29,* 1067–1083.

Windle, M, & Windle, R. C. (2001). Depressive symptoms and cigarette smoking among middle adolescents: prospective associations and intrapersonal and interpersonal influences. *Journal of Consulting and Clinical Psychology, 69,* 215–226.

Windsor, R. A., Boyd, N. R., & Orleans, C. T. (1998). A meta-evaluation of smoking cessation intervention research among pregnant women: improving the science and art. *Health Education Research, 13,* 419–438.

Winter, T., Karvonen, S., & Rose, R.J. (2002). Does religiousness explain regional differences in alcohol use in Finland? *Alcohol & Alcoholism, 37,* 330–339.

Wu, L. T., & Anthony, J. C. (1999). Tobacco smoking and depressed mood in late childhood and early adolescence. *American Journal of Public Health, 89,* 1837–1840.

Zhang, H., Ye, Y., Wang, X., Gelernter, J., Ma, J. Z., & Li, M. D. (2006). DOPA decarboxylase gene is associated with nicotine dependence. *Pharmacogenomics, 7,* 1159–1166

Chapter 29

The Genetics of Substance Use and Substance Use Disorders

Danielle M. Dick, Carol Prescott, and Matt McGue

Introduction

The abuse of licit and illicit drugs constitutes one of the leading public health problems in the world. There are tens of millions of alcohol or drug abusers and more than a billion smokers in the world (WHO, 2002). Each year, substance abuse results in the loss of tens of millions of dollars due to health-care costs and lost productivity (Cartwright, 2008; Rehm, Taylor, & Room, 2006). Substance abuse also shortens lives, increases risk for chronic disabling illness, and results in untold social costs in terms of broken families, ruined careers, and violent victimization (Goldman, Oroszi, & Ducci, 2005). While effective treatments for addiction exist, the amelioration of the societal burden of substance abuse will require more effective prevention. Behavioral genetic research in this area is part of a larger effort aimed at improving prevention and intervention efforts by bringing about a better understanding of the origins of substance use disorders (O'Brien, 2008).

Behavioral genetic research has helped to transform the addictions research field. No longer are substance use disorders considered a form of moral weakness or an outgrowth of an unresolved dynamic conflict. Rather, there is now widespread recognition that substance use disorders have a neurological basis and are genetically influenced. The present chapter provides an overview of the behavioral genetic research that has helped to achieve this transition. This is an extremely active area for behavioral genetic research, especially over the past 10 years, and it is well beyond the scope of the present review to comprehensively cover all relevant research. This review is consequently selective but intended to be representative. The scope of our review includes behavioral genetic research on the use and abuse of both common licit (i.e., alcohol) and illicit

drugs. Because behavioral genetic research on smoking is comprehensively reviewed in another chapter in this volume, we limit our discussion of smoking research to instances where we can draw parallels to findings on alcohol or illicit substance use and abuse. In this chapter, we review (1) twin and adoption research that has established the existence of genetic and environmental influences on substance use phenotypes and helped to identify intermediate phenotypes that appear to underlie these influences; (2) molecular genetic investigations that are beginning to identify the specific genes underlying risk for substance use disorders; and (3) developmental behavioral genetic research aimed at investigating the joint contribution of genetic and environmental factors to the development of substance use disorders. Before reviewing behavioral genetic research, we briefly discuss what is known about the clinical epidemiology of substance use and abuse.

Substance Use and Substance Use Disorders: The Nature of the Phenotype

Five core features of the epidemiology of substance use and abuse are central to understanding behavioral genetic approaches in this area. First, substance use disorders are relatively common. Consequently, large numbers of individuals with clinically relevant levels of substance abuse will be included in community-based samples and population twin surveys. Second, both clinical and sub-clinical patterns of substance use behavior have public health significance. Consequently, behavioral genetic research is not limited to the investigation of substance use disorders, but rather has considered a broad range of substance use phenotypes. Third, substance use disorders can be usefully conceptualized within a developmental framework. This framework traces the roots of substance use disorders from childhood, prior to substance use initiation, to late adolescence and early adulthood, when patterns of maladaptive substance use typically emerge. Fourth, substance use

D. M. Dick (✉)
Virginia Institute for Psychiatric and Behavior Genetics, Virginia Commonwealth University
e-mail: ddick@vcu.edu

Y.-K. Kim (ed.), *Handbook of Behavior Genetics*,
DOI 10.1007/978-0-387-76727-7_29, © Springer Science+Business Media, LLC 2009

disorders rarely occur in isolation but rather typically co-occur with each other and with other mental health problems. Finally, access to substances is a necessary precondition for the development of substance use disorders. Although this observation may seem trivial, we will show it has significant consequences for understanding the interplay of genetic and environmental factors in the development of substance use disorders.

The DSM and ICD distinguish two forms of substance use disorders (Hasin, 2003). Substance dependence refers to a maladaptive pattern of substance use characterized by impaired control, neglect of major social and occupational responsibilities, and physical signs of dependence. Substance abuse refers to a recurrent pattern of substance use despite physical, social, or psychological harm. Although substance abuse is likely to be associated with lower levels of impairment than substance dependence, substance abuse has a clinical course that suggests it is not merely a milder form of dependence (Hasin, Grant, & Endicott, 1990). Substance use disorders vary in prevalence across cultures, but are among the most common of mental health problems in nearly all Western countries (Rehm, Room, van den Brink, & Jacobi, 2005; Rehm, Room, van den Brink, & Jacobi, 2005). Epidemiological studies in the United States, for example, indicate that approximately 25% of the adult population is nicotine dependent (Breslau, Johnson, Hiripi, & Kessler, 2001), and that 15–20% of the population has a lifetime diagnosis of a substance use disorder other than nicotine dependence (Kessler, Berglund, Demler, Jin, & Walters, 2005). The rate of nicotine dependence is similar in men and women (Breslau, et al.), although males are approximately twice as likely as females to have other substance use disorders (Kessler, et al.).

The public health burden of substance use behavior is not due entirely to the existence of diagnosable cases. Subclinical levels of substance use are associated with increased risk of driving accidents and violence, and decreased worker productivity (Gmel & Rehm, 2003; Hingson & Winter, 2003). Substance use in adolescence, and especially in early adolescence, is also a major risk factor for substance use disorders in adulthood (McGue & Iacono, 2005; McGue, Iacono, Legrand, & Elkins, 2001a). Moreover, substance use problems can be conceptualized as existing along a dimension of problematic use rather than as simple discrete entities (Gillespie, Neale, Prescott, Aggen, & Kendler, 2007; Krueger et al., 2004). For these reasons, rather than focus exclusively on substance use disorders, behavioral geneticists have investigated a range of substance use phenotypes including the initiation of substance use in adolescence, quantity–frequency indices of substance use in adulthood, and symptom count scales as quantitative markers of problem use severity.

Although there is individual variation (Maggs & Schulenberg, 2004), the typical course of the development of substance use and abuse in adolescence and early adulthood has been characterized by epidemiological research. Initiation of substance use typically begins in early to middle adolescence with experimentation with "licit" drugs like tobacco and alcohol. The majority of individuals will try tobacco and alcohol sometime during adolescence (Johnston, O'Malley, Bachman, & Schulenberg, 2005). Consequently, more important than whether an adolescent has ever used tobacco or alcohol is when they start and how rapidly they escalate in their use of these substances. Relative to the percentage of adolescents who use tobacco or alcohol, a smaller percentage of adolescents will ever use marijuana and other illicit drugs. Initiation of substance use after the mid-twenties is uncommon (Chen & Kandel, 1995). The prevalence of both licit and illicit substance use continues to increase throughout adolescence and into early adulthood, and substance use disorders typically onset in late adolescence or early adulthood (Kessler et al., 2005). For most individuals, however, substance use other than tobacco begins to moderate later in adulthood as they marry, initiate careers, and assume other adult roles (Bachman et al., 2002). For a minority, however, the attainment of adulthood does not result in a moderation of youthful insobriety.

While substance availability and social context are important contributors, individual-level factors exert a powerful influence on the development of substance use behavior (Sher & Trull, 1994; Tarter & Vanyukov, 1994). Early initiation of substance use, for example, is not entirely circumstantial, as children who are characterized as impulsive and oppositional or who suffer from mental health problems like attention-deficit/hyperactivity disorder (ADHD) and conduct disorder (CD) are more likely to initiate substance use early and develop a substance use disorder than children not having these disorders (Elkins, McGue, & Iacono, 2007). These developmentally early dispositional and mental health markers of risk may contribute to another salient feature of substance use disorders – their substantial comorbidity. Substance abusers typically abuse multiple rather than a single substance. Smoking is markedly elevated among alcoholics (Istvan & Matarazzo, 1984), who also show significantly elevated rates of illicit drug use disorders (Stinson et al., 2006). Substance use problems are also highly comorbid with mental health problems, especially those characterized by disinhibition or poor impulse control (Krueger, Hicks et al., 2002), and, to a lesser degree, internalizing disorders like anxiety and depression (Grant et al., 2006). This has led some to hypothesize the existence of one or more generalized dimensions of risk that convey vulnerability not only to substance use disorders but also to associated mental health problems.

Twin and Adoption Studies: Characterizing the Nature of Genetic and Environmental Influences on Substance Use Phenotypes

Twin and Adoption Research on Substance Use and Abuse

Overview: Substance use and abuse is one of the most active areas of behavioral genetic research. It is not within the scope of this chapter to review or summarize every published twin and adoption study in this area. Consequently, our focus will be on large-scale community-based twin and adoption studies covering two substances: alcohol and cannabis. We focus on community-based samples because clinically ascertained samples, which overrepresent individuals with severe and comorbid forms of substance use disorders, may yield biased estimates (Prescott, Aggen, & Kendler, 2000). We focus on studies with large samples since they are likely to yield statistically stable estimates of effects. Finally, we restrict our analysis to phenotypes related to alcohol and cannabis use because, along with tobacco, these are the most widely studied substances from a behavioral genetic perspective. Findings from the limited number of twin studies on other licit (e.g., caffeine) and illicit (e.g., cocaine, hallucinogens) substances are generally consistent with those from research on the two substances we focus on here.

Twin and Adoption Research on Drinking and Alcoholism: Table 29.1 gives standardized variance component estimates (i.e., h^2, c^2, and e^2) reported in large-scale, community-based twin studies of alcohol-related phenotypes. The table also gives the sample size weighted-average variance component estimates. Twin studies of adult samples are overwhelmingly consistent: Genetic and non-shared environmental factors each account for approximately 50% of the variance in alcohol outcomes, while shared environmental factors appear to have little or no impact. Heritability estimates for quantity/frequency measures of alcohol consumption in adults appear comparable to those for problem use or dependence. Although early twin and adoption research suggested stronger heritable influences on risk of alcoholism in men than women (McGue, Pickens, & Svikis, 1992), this gender difference has not emerged in larger and more recent twin studies (Heath et al., 1997; Prescott, 2002). The finding of moderate genetic influences is consistent with earlier reviews of twin studies on alcoholism (McGue, 1991, 1994) and is further supported by the repeated finding from adoption studies of significant elevations in drinking problems and alcoholism in the adult reared-away offspring of alcoholic biological parents (Cloninger, Bohman, et al., 1981; Goodwin et al., 1973).

As noted in the recent review by Hopfer et al. (2005), behavioral genetic studies of drinking behavior in adoles-cent populations have focused primarily on initiation and quantity/frequency measures rather than with problem or abusive patterns of alcohol use. Variance component estimates from adolescent twin samples are more variable across studies than estimates based on adult twin samples. More-over, the weighted-average estimates of h^2 are lower and c^2 are higher in adolescent as compared to adult twin samples. Significantly, this conclusion is supported by longitudinal as well as cross-sectional studies. In a large sample of adolescent Finnish twins followed longitudinally, for exam-ple, Rose, Dick, Viken, and Kaprio (2001) reported signifi-cant increases in the heritability of drinking and significant decreases in the influence of shared environmental factors from age 16 to 18.5 years. The findings from this study are reproduced in Fig. 29.1.

Multi-stage biometric models that seek to characterize the shared and unique contributions to substance use initiation versus substance use progression have been fit to adoles-cent twin data on alcohol use. In a sample of 1,214 adoles-cent twin pairs, for example, Fowler et al. (2007) reported a heritability estimate of 0.26 and a shared environmental influence of 0.65 for drinking initiation. Various indices of drinking progression were assessed in this study (e.g., quan-tity, binge drinking), and estimates of heritability for these measures were consistently higher (0.27–0.64) and shared environmental effects consistently lower (0.00–0.36) than for drinking initiation. Importantly, the association of drinking initiation with drinking progression was only moderate in magnitude, suggesting that there are unique causal factors underlying the two types of substance use phenotypes. Stud-ies of adolescent adopted siblings provide additional evi-dence for the existence of strong shared environmental influ-ence on adolescent drinking behavior. McGue, Sharma, & Benson (1996) reported a correlation of 0.24 for an index of drinking behavior in a sample of 255 adopted adolescent sibling pairs. The sibling correlation was, however, substan-tially greater for like-sex adopted siblings who were no more than 2 years apart in age (r = 0.45) versus unlike sibling pairs who were more than 2 years apart in age (r = 0.05). The finding of substantial similarity in drinking behavior among only demographically similar siblings, along with the failure in this study to observe significant adopted parent–offspring resemblance for drinking outcomes, led these researchers to posit that sibling factors may be a major contributor to the shared environmental influence on adolescent drinking.

Although measures of alcohol use clearly show genetic influence in adolescence, the more limited number of stud-ies investigating alcohol dependence symptoms in early ado-lescence suggests a very different picture than studies of alcohol dependence in adults. Analyses of alcohol depen-dence symptoms at age 14 in a large sample of Finnish twins found no evidence of genetic effects in either girls or boys at this age (Rose, Dick, Viken, & Pulkkinen, 2004). Data

Table 29.1 Large-scale, community-based twin studies of alcohol-related phenotypes

Studies of adult populations

Study	Country	Sample	N	Smoking phenotype	Biometric estimates h^2	c^2	e^2
Jardine & Martin (1984)	Australia	Male	2,745 indivs	Weekly consumption	0.36	0.23	0.41
		Female	4,875 indivs	Weekly consumption	0.56	0.00	0.44
Kaprio et al. (1992)	Finland	Younger Male	898 pairs	Weekly consumption	0.64	0.00	0.36
		Younger Female	987 pairs	Weekly consumption	0.58	0.00	0.42
		Older Male	1,388 pairs	Weekly consumption	0.48	0.00	0.52
		Older Female	1,199 pairs	Weekly consumption	0.49	0.00	0.51
Kendler et al. (1994)	US	Female	1,030 pairs [a]	Alcohol dependence	0.59	0.00	0.41
Swan et al. (1996)	US	Male Vets	356 pairs	Weekly consumption	0.49	0.00	0.51
Heath et al. (1997)	Australia	Combined Male and Female	5,889 indivs	Alcohol dependence	0.64	0.01	0.35
Kendler et al. (1997)	Sweden	Male	8,935 Pairs	Temperance board registration	0.54	0.14	0.32
Hettema et al. (1999)	US	Male	1,295 Indivs	Weekly consumption	0.60	0.00	0.40
		Female	1,871 Indivs	Weekly consumption	0.47	0.00	0.53
Prescott & Kendler (1999)	US	Male	3,516 Indivs	Alcohol dependence	0.48	0.00	0.52
Weighted average			10,269 pairs [b]	Weekly consumption	0.52	0.03	0.45
			18,073 pairs [b]	Dependence/temperance	0.55	0.08	0.37
Han et al. (1999)	US	Male	274 pairs	Initiation	0.60	0.23	0.17
		Female	225 pairs	Initiation	0.10	0.68	0.22
Koopmans et al. (1999)	Netherlands	Male	1553 indivs	Initiation	0.00	0.92	0.08
		Female	1849 indivs	Initiation	0.41	0.54	0.05
Maes, Woodfard, et al. (1999)	US	Combined Male and Female	1412 pairs	Initiation	0.54	0.17	0.29
Rose et al. (2001)	Finland	Male	1330 indivs	Initiation	0.00	0.76	0.24
		Female	1380 indivs	Initiation	0.18	0.76	0.06
Rhee et al. (2003)	US	Male	1148 indivs [c]	Initiation	0.41	0.36	0.23
				Problem use	0.41[e]	0.33	0.26
		Female	976 indivs [c]	Initiation	0.41	0.22	0.37
				Problem use	0.60[e]	0.17	0.23
Rose et al. (2004)	Finland	Male	916 indivs	Alcohol dependence	0.00	0.76	0.24
		Female	694 indivs	Alcohol dependence	0.00	0.83	0.17
Hopfer et al. (2005)	US	Combined Male and Female	2427 pairs [d]	Weekly consumption	0.52	0.00	0.48
Weighted average			8455 pairs [b]	Initiation/consumption	0.37	0.36	0.27
			1867 pairs [b]	Dependence/problem use	0.28	0.49	0.23

h^2 = proportion of variance due to genetic effects; c^2 = shared environmental effects; e^2 = non-shared environmental effects.
[a] Parents of the twins are included in the sample; [b] single individuals counted as half a pair in weighted average; [c] study includes biological and adoptive sibling pairs as well as twin pairs; [d] includes full and half-sibling pairs in addition to twins; [e] heritability estimate includes combined additive and non-additive (dominance) effects.

from the Missouri Adolescent Female Twin Study showed a similar pattern of results, with alcohol dependence symptoms in adolescence largely influenced by environmental factors (Knopik, 2005). These studies suggest that the factors influencing alcohol dependence symptoms that manifest very early in adolescence may differ from the etiological causes of adult alcohol dependence. Longitudinal studies of dependence symptoms from adolescence into adulthood will be necessary to further explore this question.

The behavioral genetic findings on alcohol use and abuse are quite similar to those with smoking. As reviewed elsewhere in this volume as well as in the meta-analyses of twin studies of smoking published by Sullivan and

Kendler (1999) and Li, Cheng, Ma, and Swan (2003), smoking progression is more strongly heritable than smoking initiation. Conversely there is greater evidence of shared environmental effects on smoking initiation than on smoking progression. Also, as is found with alcohol, heritable effects on smoking appear to increase while shared environmental effects appear to decrease during the transition from early adolescence to early adulthood (Koopmans, vanDoornen, & Boomsma, 1997). The similar pattern of developmental behavioral genetic findings for the two substances are consistent with, although clearly not proof of, the existence of common mechanisms of risk, a possibility we address below.

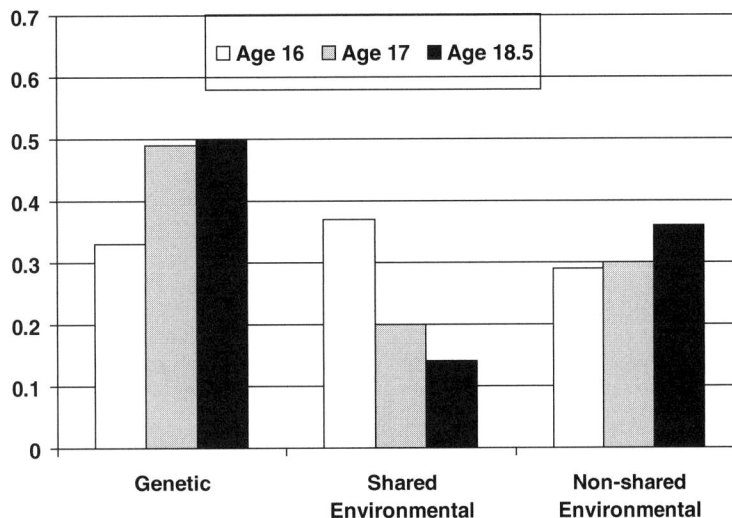

Fig. 29.1 Proportion of variance in drinking frequency associated with genetic and environmental factors at three ages in a longitudinal study of Finnish twins (Rose et al., 2001). The figure shows that the relative influence of genetic factors increases while that of shared environmental factors decreases in importance

Twin Research on Cannabis-Related Phenotypes: A summary of large-scale population-based twin studies of cannabis-related phenotypes is given in Table 29.2; this literature was also recently reviewed by Agrawal and Lynskey (2006). Studies of adult twins implicate moderate to strong levels of heritable influence. Interestingly, and largely unlike adult twin studies of alcohol and tobacco use, a moderate level of shared environmental influence is also implicated in adult twin studies of cannabis-related phenotypes. As with the other two substance phenotypes, heritability estimates arc generally lower while shared environmental influences are generally higher in adolescent as compared to adult twin samples. Also, as has been found in multi-stage modeling of smoking and alcohol use, heritable influences are weaker and shared environmental influences are stronger on marijuana use initiation than progression. To our knowledge there have been no systematic adoption studies of cannabis-related phenotypes.

Multivariate Biometric Models of Multiple Substance Use Phenotypes

Research reviewed thus far has focused on the use and abuse of individual substances. One of the most salient features of substance use disorders, however, is that they rarely occur in isolation. Substance use disorders are highly comorbid with each other as well as with other forms of disinhibitory psychopathology (Goldman & Bergen, 1998). The existence of substantial genetic contributions to each of the individual substance use and disinhibitory psychopathology disorders raises the question of whether common genetic factors are the primary source of disorder comorbidity. Behavioral genetic methodology is well suited to address this question (Neale & Kendler, 1995).

In a sample of over 3000 pairs of adult male twins from the Vietnam Era Twin Registry, Tsuang, Lyons, Meyer et al. (1998) reported evidence for the existence of a common vulnerability factor that accounted for most of the genetic effects on abuse of marijuana, sedatives, stimulants, and hallucinogens. Only abuse of heroin/opiates showed evidence of substantial unique genetic effects not accounted for by the common liability. Kendler, Myers, & Prescott (2007) extended this line of investigation by investigating dependence on both licit (alcohol, caffeine, and nicotine) and illicit (cocaine and cannabis) substances in a sample of 4,865 like-sex male and female twin pairs from the Virginia Adult Twin Study of Psychiatric and Substance Use Disorders. As in the study by Tsuang and colleagues, Kendler et al. found evidence for a general genetic liability to cocaine and cannabis. They also found evidence for a second general genetic liability factor that loaded primarily on symptoms of alcohol and nicotine dependence. Genetic effects on symptoms of caffeine dependence were largely independent of the two liability factors. Interestingly, the genetic liability to licit and illicit substance dependence were highly correlated (r=0.82), implicating the existence of a general process underlying dependence on a broad range of both licit and illicit substances.

Table 29.2 Large-scale, community-based twin studies of cannabis-related phenotypes

Studies of adult populations

Study	Country	Sample	N	Cannabis phenotype	h^2	c^2	e^2
Tsuang et al. (1996)	US	Male veterans	3,372 pairs	Abuse/dependence [a]	0.33	0.29	0.38
Kendler and Prescott (1998)	US	Female	1,934 indivs	Use	0.40	0.35	0.25
				Dependence	0.43	0.18	0.39
Kendler et al. (2000)	US	Male	1,198 pairs	Use	0.33	0.34	0.33
				Abuse Dependence	0.58	0.00	0.42
Lynskey et al. (2002)	Australia	Male	2,779 indivs	Use	0.72	0.00	0.28
				Dependence	0.50	0.12	0.38
		Female	3,444 indivs	Use	0.63	0.17	0.21
				Dependence	0.40	0.18	0.46
Weighted average			5,326 pairs [b]	Use	0.54	0.20	0.26
			8,698 pairs [b]	Abuse/dependence	0.42	0.19	0.40
Maes, Woodfard, et al. (1999)	US	Combined male and female	1,412 pairs	Initiation	0.22	0.68	0.09
McGue et al. (2000)	US	Male	289 pairs	Initiation	0.26	0.56	0.18
				Abuse/dependence	0.54	0.27	0.19
		Female	337 pairs	Initiation	0.13	0.61	0.26
				Abuse/dependence	0.06	0.68	0.26
Miles et al. (2001)	US	Combined male and female	738 pairs	Initiation	0.31	0.47	0.22
Rhee et al. (2003)	US	Male	1,148 indivs[b]	Initiation	0.39	0.44	0.17
				Problem use	0.44	0.35	0.21
		Female	976 indivs[b]	Initiation	0.72	0.24	0.04
				Problem use	0.22	0.33	0.45
Weighted average			3,838 pairs [b]	Initiation	0.32	0.53	0.14
			1,688 pairs [b]	Abuse/dependence/ problem use	0.32	0.40	0.29

h^2 = proportion of variance due to genetic effects; c^2 = shared environmental effects; e^2 = non-shared environmental effects.
[a]DSM-III-R diagnosis of cannabis dependence or abuse; [b]single individuals counted as half a pair in weighted average.

The comorbidity of substance use disorders with other behavioral disorders may provide insights into the nature of this general liability. Specifically, substance use disorders appear to be strongly comorbid with other forms of disinhibitory psychopathology, including antisocial personality and conduct disorder (Krueger & Markon, 2006), while substance use and misuse in adolescence often co-occurs with other indicators of adolescent disinhibitory behavior (Young, Rhee, Stallings, Corley, & Hewitt, 2006). In a sample of 1,048 male and female 17-year old twins, Krueger, Hicks et al. (2002) found that the phenotypic associations among multiple substance use and disinhibitory disorders, as well as a personality measure of behavioral constraint (the converse of impulsivity) could be accounted for by a general liability to externalizing psychopathology (Fig. 29.2). Importantly, this externalizing factor was highly heritable (81%), with non-significant estimates of residual genetic effects for all of the specific indicators except the personality measure. Collectively, these findings suggest that while there are substance-specific genetic influences, a large portion of the genetic influences underlying substance use disorders owes to a general and highly heritable vulnerability to disinhibitory psychopathology.

Evidence from studies of adult twins indicates there is also genetic overlap between substance use disorders and internalizing psychopathology, including depression (Prescott, Aggen, & Kendler, 2000) and anxiety (Kendler, Prescott, Myers, & Neale, 2003). The co-occurrence of these disorders may represent self-medication of internalizing symptoms. Internalizing disorders are often found in individuals whose substance use disorders have relatively later onset and are possibly secondary to the internalizing disorders (Kuo, Gardner, Kendler, & Prescott, 2006). There is also evidence for depressive disorders being secondary to chronic substance use (e.g., Schuckit, 2006). These findings are consistent with results from some twin and adoption studies supporting the existence of partially distinct etiology of early and later onset substance use disorders (Cloninger, Bohman, et al., 1981; McGue, Pickens, et al., 1992).

Summary of Twin and Adoption Research on Substance Use Phenotypes

Our review of relevant twin and family research on substance use and abuse leads us to several general conclusions. First, substance use phenotypes are uniformly moderately to highly heritable. Indeed, our survey of twin and adoption research on alcohol and cannabis use and abuse as well as other

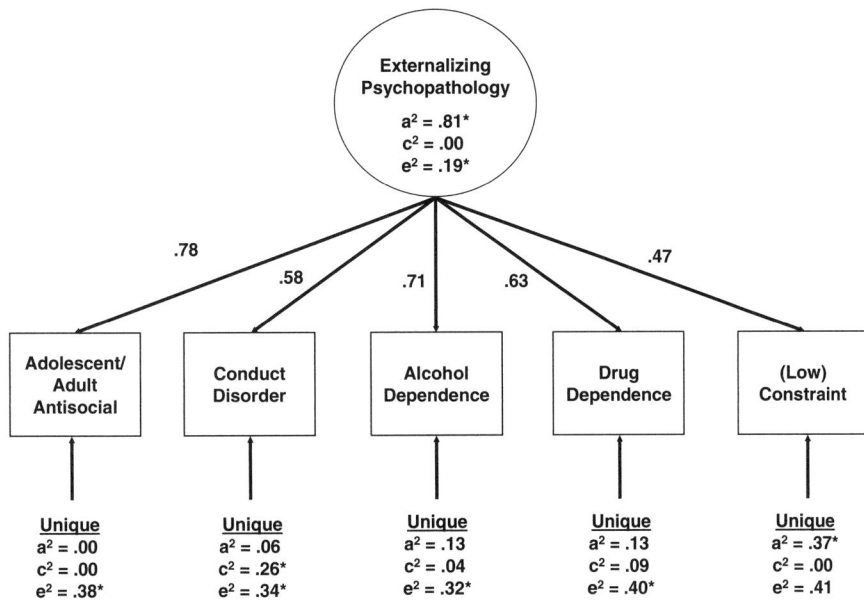

Fig. 29.2 General externalizing model of substance use disorders, disinhibitory behavior, and personality markers of risk. Variance component parameters designated with an * are significantly different from 0 at $p < 0.05$; a^2 = proportion of variance due to additive genetic effects, c^2 = proportion due to shared environmental effects, and e^2 = proportion due to non-shared environmental effects. Findings are reproduced from Krueger et al. (2002)

reviews of behavioral genetic research on tobacco suggest comparable levels of heritability across these three distinct classes of substances. Second, the magnitude of heritable effects on substance use phenotypes appears to be developmentally conditioned, with heritability estimates being consistently higher in adulthood than in adolescence. This conclusion is consistent with one of the most robust findings to emerge from the behavioral genetic literature: the heritability of a wide range of behavioral phenotypes increases during the transition from adolescence to adulthood (Bergen, Gardner, & Kendler, 2007).

Our third general conclusion concerns the substantial comorbidity that exists among substance use disorders and between substance use disorders and other mental health disorders. To a large degree these comorbidities appear to reflect a general and highly heritable vulnerability to disinhibitory psychopathology. This conclusion has substantial implications for gene identification research on substance use phenotypes in that it implies the existence of genetic polymorphisms that convey risk across a wide range of substance use and mental health disorders. Nonetheless, it is important to recognize that this common genetic vulnerability does not account entirely for genetic influences on specific substances. Consequently, we also expect there to be substance-specific genetic influences, possibly related to the metabolism or receptor targets of individual substances (Goldman et al., 2005).

Gene Identification for Substance Use Phenotypes

Methodologies for Gene Identification

Historically, two primary strategies have been used for gene identification: linkage and association studies. Linkage studies have the advantage that they enable researchers to scan the entire genome for possible regions containing genes influencing a trait of interest, without any a priori knowledge of where these genes may reside. This can be accomplished by testing genetic markers approximately evenly spaced throughout the genome. Traditionally, this was accomplished with microsatellite markers spaced every ~10cM, necessitating ~400 markers to cover the whole genome. Early studies using linkage methods in psychiatric disorders were largely unsuccessful (Egeland et al., 1987) and led to much early disappointment and frustration. This was based, in part, on the use of linkage methods (called parametric) that required specification of a disease model (e.g., mode of inheritance, penetrance, disease allele frequency in the population). Although parametric linkage had been used successfully to map genes for hundreds of Mendelian disorders (http://www3.ncbi.nlm.nih.gov/Omim/), studying complex phenotypes introduced a number of new complications, including the involvement of many genes of small effect, genetic heterogeneity, environmental influences, and phenotypic imprecision. Subsequently, linkage methods were created that were better suited for the many complexities inherent in complex phenotypes such as substance use disorders. These methods, called nonparametric linkage, are based on a concept called "identity by descent" (IBD) marker allele sharing. If siblings inherit the same marker allele from the same parent, the allele is called IBD. If the marker being tested is in close physical proximity to a gene influencing the disease or trait under study, then siblings who are similar for the trait would be expected to share more IBD marker alleles. Conversely, siblings who are dissimilar would be expected to exhibit fewer IBD marker alleles near the gene influencing the trait. Nonparametric linkage methods also allow the inclusion of more extended families beyond sibling pairs in the genetic analysis, as basic statistical genetic probability can be used to calculate deviations from expected allele sharing across different relative types.

Although linkage analyses are useful for identifying chromosomal regions likely to harbor genes influencing the phenotype of interest, linkage peaks are often broad, containing hundreds of genes, and linkage is imprecise in its ability to localize the underlying susceptibility variant. In addition, we expect that many genes involved in complex behavioral phenotypes may have effect sizes too small to be detected in linkage analyses. Accordingly, the second method of analysis, association analysis, has become increasingly popular.

Association analyses provide a useful tool for several purposes, including testing the role of potential candidate genes, fine-mapping in regions of linkage, and more recently, genome-wide analyses. Allelic association refers to a significantly increased or decreased frequency of a particular marker allele with a disease trait. There are two basic types of association studies: case–control studies and family-based studies. Case–control studies compare the frequency of alleles between a group of unrelated, affected individuals and a group of matched controls. The controls should be matched to the cases with respect to numerous factors, such as age, gender, and ethnicity, so that they differ only in disease status. In this way, differences in allele frequencies between the two groups are interpreted as evidence that the gene is involved in disease status.

The advantage of this approach lies in its relative simplicity. The necessary statistics are straightforward and the method is generally powerful for detecting genes of smaller effect than those that can be identified with linkage analysis. Case–control studies are also fairly easy to implement in terms of sample collection. The primary disadvantage of this approach is its sensitivity to the existence of population stratification. Population stratification refers to the mismatching of cases and controls for population substructure (i.e., ethnicity). If cases and controls are mismatched on ethnicity and ethnic groups differ in disease risk, then spurious genetic

associations can result. However, because large numbers of single nucleotide polymorphisms (SNPs; i.e., genetic variation at the level of a single base of DNA) are now available to allow for testing and correcting for population substructure, case–control studies have become increasingly popular and concerns about population stratification have lessened.

The other method used in association studies is the family-based method. The most basic form of the family-based test is called the transmission disequilibrium test (TDT). This method compares the frequency of alleles transmitted to affected children from heterozygous parents with the frequency of alleles not transmitted. The sample needed for the TDT analysis consists of an affected individual and his/her parents. The rationale behind the TDT is that for a diallelic marker, there is a 50/50 chance that a child will inherit each of the two possible alleles from a heterozygous parent. Accordingly, in the event of no association, across many trios of affected individuals (with heterozygous parents), we expect that ~50% of the individuals will have inherited allele 1 and ~50% of individuals should have inherited allele 2. However, if a particular allele is found more often than expected by chance in the sample of affected individuals, this indicates that the allele is associated with disease status. In this way, we are testing for over-transmission of a particular allele to affected individuals.

The primary advantage of family-based methods is that they eliminate the need for matched samples, thereby avoiding potential problems associated with population stratification. However, they are generally less powerful than case–control methods, and the necessary sample structure may be more difficult to collect. Affected individuals with both parents are necessary, and only heterozygous parents are informative. Accordingly, the number of trios that will be informative will vary as a function of the allele frequencies at the marker of interest (e.g., if one allele is very rare, there will be fewer heterozygous parents, requiring more trios to be collected). In addition, for late onset disorders, such as Alzheimer's disease, it may be impractical or impossible to ascertain parents of affected individuals. There are many extensions of the basic TDT, including methods that use data from additional affected or unaffected siblings or from even more extended pedigrees. In addition, the test has been extended to the analysis of quantitative traits in addition to qualitative traits. There are many excellent references describing methods for linkage and association in great detail (e.g., Neale, Ferreira, Medland, & Posthuma, 2008)

Gene Identification for Alcohol Phenotypes

Linkage Studies of Alcohol Phenotypes: The first large-scale study aimed at identifying genes contributing to alcohol dependence was the Collaborative Study on the Genetics of

Alcoholism (COGA), a multi-site collaboration initiated in 1989 that ascertained families densely affected with alcohol dependence from treatment centers across the United States. An initial genome-wide linkage scan for alcohol dependence was reported on 291 markers, genotyped in 987 individuals from 105 densely affected families, with an average inter-marker distance of 13.8 cM. Nonparametric linkage analyses yielded evidence of linkage on chromosomes 1 and 7, with more modest evidence of linkage on chromosome 2. In addition, there was evidence for a protective locus on chromosome 4 (Reich et al., 1998). A subsequent report was published on an additional sample of 157 alcohol-dependent families, ascertained using identical criteria (Foroud, Edenberg, et al., 2000). Additional genotyping was also conducted on the initial sample, yielding a total of 351 markers genotyped in both samples. Evidence for linkage on chromosomes 1 and 7 was present in the second sample, and increased in the combined sample. The chromosome 2 finding was specific to the initial data set only and a new finding of linkage on chromosome 3 was identified in the second sample.

Since these initial reports of linkage with alcohol dependence phenotypes, a number of additional papers have been published using the COGA data, reporting analyses using novel analytic techniques (Williams et al., 1999), and using other alcohol-related phenotypes of interest, such as maximum number of drinks in a 24-hour period (Saccone et al., 2000), subjective response to alcohol (Schuckit et al., 2001), alcohol dependence comorbid with depression (Nurnberger et al., 2001), and conduct disorder (Dick, Li et al., 2004). These papers have identified other chromosomal regions of interest that may contain genes conferring risk to specific components of the alcohol dependence phenotype, subgroups of alcoholics, and related traits.

COGA has been joined by a number of other genome-wide scans for alcohol dependence. A microsatellite scan was conducted on a sample of 330 individuals from multiplex alcohol-dependent families recruited through treatment centers in the Pittsburgh area. Linkage analyses were conducted incorporating information about clinical, personality, and event-related potential characteristics, and yielded evidence for loci on chromosomes 1, 2, 6, 7, 10, 12, 14, 16, and 17 (Hill et al., 2004). A sample of 152 alcohol-dependent individuals from 32 extended families from a Southwestern American Indian tribe produced evidence for linkage on chromosomes 11p and 4p (Long et al., 1998). In a sample of Mission Indian families, a genome-wide scan was performed on a panel of 791 microsatellite markers. Although there were no findings with dichotomous alcohol dependence phenotypes, chromosomes 4 and 12 showed evidence of linkage with a severity phenotype, and chromosomes 6, 15, and 16 showed linkage with a withdrawal phenotype (Ehlers et al., 2004). Finally, in the Irish Affected Sib Pair Study of Alcohol Dependence (IASPSAD), a 4cM genome

scan was conducted on a sample of 474 families ascertained via probands in treatment facilities in Ireland. The strongest evidence of linkage was observed with number of alcohol dependence symptoms on chromosome 4. There was weaker evidence of association observed with alcohol dependence on several other regions, including 1q, 13q, and 22q for alcohol dependence, and 2q, 9q, and 18p for alcohol dependence symptoms (Prescott, Sullivan et al., 2006).

In summary, a number of different chromosomal regions have been identified as potentially containing genes conferring risk for alcohol dependence and related phenotypes. Only the primary reports from the major existent studies have been reviewed here, but these findings underscore two important points from the linkage literature. First, the magnitude of the lod scores (i.e., the standard test statistic used in a linkage analysis) has generally been modest, rarely meeting recommended levels for genome-wide significance (Lander & Kruglyak, 1995). Second, many different chromosomal regions have been implicated, and different studies often identify different regions.

There are plausible explanations for these results. Modest lod scores may reflect the small effect size of individual genes involved in susceptibility to alcohol dependence and the general lack of power of linkage analyses to detect genes of small effect. Findings across multiple regions may reflect the fact that numerous genes are thought to be involved in pathways related to risk for dependence. Different findings emerging across different studies may reflect genetic heterogeneity, whereby a different subset of susceptibility genes happens to be operating in different samples. Another consideration is that different studies have used different definitions of alcohol dependence, including DSMIII-R, DSMIV, and ICD10 criteria. Finally, linkage peaks are often broad and have imprecise localization (Roberts, MacLean, Neale, Eaves, & Kendler, 1999), making it difficult to determine what constitutes a replication. These considerations make it very difficult to distinguish valid reasons for inconsistent linkage findings across samples from false-positive reports. It is thus challenging to determine when to invest the necessary time and resources to follow up linkage peaks to attempt to identify specific susceptibility genes in the region.

Despite this, at least two chromosomal regions have been followed up with association analyses that have successfully led to gene identification. These regions are on chromosomes 4 and 7 (Dick, Jones et al., 2006). As reviewed above, several independent studies found evidence of linkage to chromosome 4, including with multiple phenotypes in the COGA sample (Saccone et al., 2000), as well as in the Southwest American Indian study (Long et al., 1998) and IASPSAD (Prescott, Sullivan et al., 2006). Chromosome 4 contains many candidate genes of interest, including a cluster of alcohol dehydrogenase (ADH; involved in ethanol metabolism) genes and a cluster of GABA-A receptor genes. Specific genes that have been identified on these chromosomes are reviewed below.

Association Studies of Alcohol Phenotypes: The second broad approach to gene identification is the association study. Two major strategies for association have been used successfully in the substance abuse field thus far. The first involves investigating genetic variation in genes at or near significant linkage peaks, sometimes called positional candidates. The second involves investigating genetic variation in genes that affect the biological systems thought to underlie substance use and abuse outcomes.

Positional Candidates: Replicated linkage of alcoholism to regions on chromosome 4 has resulted in the targeted investigation of several genes that map to that chromosome, including genetic polymorphisms that affect alcohol metabolism. The primary pathway of ethanol metabolism involves oxidation to acetaldehyde, catalyzed by alcohol dehydrogenases (ADHs), followed by further oxidation to acetate, catalyzed by aldehyde dehydrogenases (ALDHs) (Hurley, Edenberg, & Li, 2002). Humans have seven *ADH* genes tightly clustered on chromosome 4q22 in a head-to-tail array extending over approximately 365 kb: (from 5′ to 3′) *ADH7-ADH1C-ADH1B-ADH1A-ADH6-ADH4-ADH5*. The ADH class I isozymes play the major role in ethanol metabolism, and genetic polymorphisms have been detected in two of these genes that code for subunits of proteins which, in vitro, have greater enzymatic activity, suggesting faster conversion of ethanol to acetaldehyde in individuals carrying these alleles. A number of studies have reported lower frequencies of both the ADH1B*2 and ADH1C*1 alleles among alcoholics, as compared to non-alcoholics, in a variety of East Asian populations (e.g., Shen et al., 1997). More recently, a study of Jewish men living in Israel found that the ADH1B*2 allele was related to a reduced level of peak weekly alcohol intake (Neumark, Friedlander, Thomasson, & Li, 1998). Additionally, ADH1B*2 was associated with lower levels of alcohol consumption in men in a European population (Whitfield et al., 1998).

Repeated linkage evidence on chromosome 4 has also led to more extensive examination of the other ADH genes on chromosome 4. In the COGA study, markers were genotyped across all genes on the chromosome 4 cluster, and association was reported between DSMIV alcohol dependence and *ADH4* (Edenberg, Xuei, Chen, et al., 2006). This has also been reported in an independent sample of European Americans (Luo et al., 2005). In the Irish sample, association has been reported between alcohol dependence and markers in several of the ADH genes in the cluster (Kuo et al., 2008).

Chromosome 4 also contains a cluster of GABA-A receptor genes in the vicinity of reported linkage peaks. Several lines of evidence suggest that GABA, the major inhibitory neurotransmitter in the human central nervous system, is involved in many of the behavioral effects of alcohol, including motor incoordination, anxiolysis, sedation, withdrawal signs, and ethanol preference (Grobin, Matthews, Devaud, & Morrow, 1998). This suggests that the GABA-A receptor genes may be good candidates for contributing to the genetic risk for alcohol dependence. Chromosome 4 contains the genes *GABRA2, GABRA4, GABRB1*, and *GABRG1*. COGA systematically genotyped SNPs across each of these genes and identified strong evidence of association with one of the genes, *GABRA2*, and alcohol dependence (Edenberg, Dick, et al., 2004). This finding has now been replicated in a number of independent samples (Covault, Gelernter, Hesselbrock, Nellissery, & Kranzler, 2004; Fehr et al., 2006; Lappalainen et al., 2005; Soyka et al., 2008).

Several other genes on chromosome 4 that were considered good potential candidates have also been followed up in the COGA sample with more extensive genotyping to test for potential association with alcohol dependence. Association has recently been reported with *NFKB1*, which encodes a subunit of a transcription factor, and alcohol dependence. NF-κB, the associated protein, regulates many genes relevant to brain function, and its actions can be potentiated by ethanol; thus *NFKB1* was considered an excellent candidate gene for alcoholism (Edenberg, Xuei, Wetherill, et al., 2008). Additional potential candidates on chromosome 4 remain under investigation in COGA based on the rationale that since chromosome 4 showed evidence of linkage across multiple phenotypes and multiple samples, there are likely to be multiple genes in the region related to susceptibility to alcohol dependence and related traits.

The IASPSAD study has used a systematic approach to identify genes contributing to the linkage peak on chromosome 4 observed in its sample. TagSNPs (i.e., SNPS selected to systematically cover common genetic variation in the targeted regions) were genotyped across the 65 genes located in the 1-lod interval surrounding the linkage peak associated with alcohol dependence symptoms in the Irish sample. Association with a number of novel genes was identified, including association between number of alcohol dependence symptoms among alcoholics and *DKK2*, a gene thought to be involved in metabolic signaling pathways (Kalsi et al., 2008).

Chromosome 7 has also been followed-up in the COGA project with more extensive genotyping and association analyses, as this was the strongest linkage signal in the original genome scan (lod = 3.49), and subsequent genotyping conducted as part of the Genetic Analysis Workshop 14 (GAW14) further increased the evidence for linkage (Lod = 4.1; Dunn et al., 2005). Association has been detected with multiple candidate genes that have been tested around the linkage region on chromosome 7, including a muscarinic cholinergic receptor gene, *CHRM2* (Wang, Hinrichs, Stock et al., 2004), and two bitter taste receptors, *TAS2R16* and

TAS2R38 (Wang, Hinrichs, Bertelsen et al., 2007). In addition to testing markers across candidate genes in the region, a systematic approach to follow-up was applied, in which tagSNPs were genotyped across a 2-lod support interval surrounding the peak lod score from the GAW analyses. This systematic screen of genetic variation across the linkage peak led to the detection of a novel candidate gene, *ACN9*, which was associated with alcohol dependence (Dick, Aliev, Wang, Grucza et al., 2008). Little is known about the function of *ACN9* in humans; however, data from yeast indicate involvement in the assimilation of ethanol or acetate into carbohydrate, suggesting a plausible role for its involvement in alcohol dependence.

Biologically Based Candidates: Alcohol Metabolism. The genes that have been most consistently associated with alcohol dependence are polymorphisms in the alcohol-metabolizing enzymes. As described above in the section on positional candidates, multiple ADH genes have been associated with alcohol dependence; however, the effects associated with variation in ALDH are more powerful because it is the rate-limiting step in the metabolic pathway. Specifically, genetic variation in the gene that codes for the mitochondrial version of the ALDH enzyme, designated *ALDH2*, is associated with a markedly reduced capacity to eliminate acetaldehyde. The mutant form of the *ALDH2* gene (designated the *ALDH2*2* allele) results in a deficient form of the enzyme and slow rates of acetaldehyde clearance. Since acetaldehyde levels are associated with many of the dysphoric effects of alcohol (e.g., dizziness, nausea, flushing), inheritance of the mutant allele is expected to be protective against heavy drinking and the development of alcoholism. The *ALDH2*2* allele is nearly absent in whites and blacks, but it is considerably more common in Asians, with up to 43% of the Japanese population carrying this allele (Higuchi, Matsushita, Murayama, Tagaki, & Hayashida, 1995).

A potential role for the involvement of *ALDH2* in alcohol dependence was detected as early as 1982, when Harada and colleagues reported that *ALDH2* deficiency was substantially lower among Japanese alcoholics, suggesting that the deficient *ALDH2*2* allele may play a protective role by reducing the risk of alcohol dependence (Harada, Agarwal, Goedde, Tagaki, & Ishikawa, 1982). Subsequent studies have also reported reduced rates of the *ALDH2*2* allele among alcoholics in Asian populations (Shen et al., 1997). *ALDH2*2* confers up to a 10-fold reduction in the risk of alcohol dependence, giving it a stronger protective effect than either the *ADH1B* or *ADH1C* genes (Shen, et al.). The effect of the *ADH1B* genotype appears to be independent of, and additive with, that of the *ALDH2* locus (Chen et al., 1996).

Dopamine System: Dopamine is believed to play an important role in reward behavior. It is thought that the effects of alcohol are mediated through the reward pathway in the mesolimbic system, making genes involved in dopaminergic transmission potential candidates for involvement in alcohol dependence. There are five dopamine receptors, although the dopamine D2 receptor gene (*DRD2*) has received the greatest study in relation to alcohol-related phenotypes. In 1990, Blum et al. reported an association between *DRD2* and alcoholism (Blum et al., 1990). This association subsequently has been replicated by several groups (e.g., Neiswanger, Hill, & Kaplan, 1995; Noble, 1998). However, the association remains controversial, as many studies have failed to replicate an association between *DRD2* and alcohol dependence (e.g., Gelernter et al., 1991; Kidd, 1993). A recent meta-analysis suggests that *DRD2* is significantly, albeit modestly (pooled odds ratio [OR] of ∼ 1.3), associated with alcoholism (Smith, Watson, Gates, Ball, & Foxcroft, 2008). It has been suggested that *DRD2* may contribute to a "reward deficiency syndrome", a collection of addictive, impulsive, or compulsive behaviors, including alcoholism, polysubstance abuse, smoking, obesity, attention-deficit disorder, and gambling. More recently, it has been recognized that the Taq1A polymorphism that is commonly genotyped in studies investigating *DRD2* is actually located in a neighboring kinase gene, named ankyrin repeat and kinase domain containing 1 (*ANKK1*), a member of a family of proteins involved in signal transduction pathways. Although initial genotyping of the Taq1A polymorphism in the COGA sample was negative, more systematic genotyping across the *DRD2/ANKK1* region suggests evidence of association between SNPs in the region and alcohol dependence and antisociality in the Collaborative Study of the Genetics of Alcoholism sample (Dick, Wang, et al., 2007)

The Serotonin System. Genes involved in the regulation of the serotonin (5-hydroxytryptamine; 5-HT) system provide plausible candidates for involvement in many neuropsychiatric phenotypes, including alcohol consumption, abuse, and dependence, because of the role of serotonin in mood regulation, sleep, aggression, and appetite. Pharmacological agents that increase 5-HT cause a reduction in alcohol self-administration in humans. The gene encoding the serotonin transporter (5-HTT) on chromosome 17q11.2 exhibits functional polymorphism in the promoter region of the gene (the polymorphism is designated 5-HTTLPR), with the shorter allele demonstrating lower transcriptional efficiency. There is currently mixed evidence for association of genetic variants in 5-HTT with alcohol use. A recent meta-analysis concluded that there is significant evidence of association between alcohol dependence and the short allele of 5-HTT (OR=1.18), with somewhat greater association among alcohol dependence comorbid with another psychiatric condition or characterized by greater severity (Feinn, Nellissery, & Kranzler, 2005). Additionally, the short allele has been associated with alcohol consumption in a sample of social drinkers (Munafo & Johnstone, 2005).

Endophenotypes

An important strategy that has been used as a means to deal with the complications introduced in gene identification efforts by the heterogeneity that characterizes alcohol dependence is to study endophenotypes. Endophenotypes are characteristics that represent more basic underlying biologic features in the gene-to-behavior pathways. This was a strategy first applied to psychiatric disorders by Gottesman and Shields in 1972, and more recently reviewed in Gottesman and Gould (2003). It seems probable that genes act more directly on an endophenotype than on a diagnostic classification (particularly for disorders like substance abuse that involve behavioral components), and therefore the study of endophenotypes may more efficiently lead to the identification of genes. In addition, endophenotypes may provide information about the pathways leading from genes to the manifestation of clinical disorder.

There is a substantial body of literature suggesting that electrophysiological measures represent relevant endophenotypes for alcohol dependence. It has been proposed that the genetic predisposition to alcohol dependence involves central nervous system (CNS) disinhibition/hyperexcitability (Begleiter & Porjesz, 1999) and electrophysiological measures can reflect this CNS disinhibition. Alcoholic individuals evidence several abnormalities in electrophysiological functioning. In the COGA sample, increased beta power and theta power in all three bands of resting EEG has been observed in alcohol-dependent individuals, as compared to controls. An increase in beta power has also been observed in the offspring of male alcoholics, further suggesting this may be a marker of an inherited predisposition to alcohol dependence, rather than an outcome of chronic alcohol consumption. In addition, the frequency bands of EEG are highly heritable; the averaged heritabilities for the delta, theta, alpha, and beta frequencies are 76, 89, 89, and 86%, respectively (vanBeijsterveldt, Molenaar, deGeus, & Boomsma, 1996).

Perhaps the most extensively studied psychophysiological endophenotype in the substance abuse field is the amplitude of the P3 event-related brain potential (ERP). P3 is a late positive deflection in the ERP wave-form that is thought to reflect allocation of attentional resources during memory update (Polich, Pollock, & Bloom, 1994). Although the precise biological mechanisms underlying the generation of P3 are not known, it is believed to reflect neural processes related to disinhibition (Begleiter & Porjesz, 1999). In terms of its utility as an endophenotype, P3 amplitude evoked in a visual ERP paradigm is on average markedly lower among abstinent alcoholics as compared to non-alcoholics (Porjesz & Begleiter, 1998). This association does not appear to be a consequence of alcohol toxicity on brain function as P3 amplitude reduction has been consistently observed among the preadolescent sons of alcoholics (Begleiter, Porjesz, Bihari, & Kissin, 1984; Polich, et al.). Moreover, individual differences in P3 amplitude are substantially heritable (Katsanis, Iacono, McGue, & Carlson, 1997; O'Connor, Morzorati, Christian, & Li, 1994), and longitudinal studies report that reduced P3 amplitude predicts early onset alcohol abuse (Berman, Whipple, Fitch, & Noble, 1993; Hill, Steinhauer, Lowers, & Locke, 1995).

Consistent with its characterization as an indicator of vulnerability to generalized disinhibitory psychopathology rather than being a specific marker of alcoholism risk, P3 amplitude reduction has been associated with early adolescent problem behavior (Iacono and McGue, 2006), early onset substance use disorders (Iacono, Carlson, Malone, & McGue, 2002), adolescent substance use and misuse (Yoon, Iacono, Malone, & McGue, 2006), and general externalizing psychopathology (Iacono, et al.). Moreover, Patrick and colleagues (Patrick et al., 2006) showed that P3 amplitude reduction was associated with a latent disinhibitory psychopathology factor, while Hicks et al. (2007) showed that this association was predominantly genetically mediated.

Taken together, these findings suggest that electrophysiological endophenotypes are appropriate biological markers of the genetic predisposition to alcohol dependence and would have utility in genetic analyses. This has proven to be the case in the COGA project, where electrophysiological endophenotypes have been used as a complement to clinical diagnoses to hunt for genes involved in the predisposition to alcohol dependence. In fact, chromosomes 4 and 7 were selected preferentially for follow-up based on (loosely converging) evidence of linkage with both dependence diagnoses and electrophysiological endophenotypes. On both chromosomes, the evidence for linkage was considerably stronger with the electrophysiological endophenotypes (Jones et al., 2004; Porjesz et al., 2002), and the linkage peaks with the electrophysiological endophenotypes were narrower and more directly located over the genes that were subsequently identified (GABRA2 and CHRM2). These findings underscore the utility of using endophenotypes to identify genes involved in the predisposition to alcohol dependence (as reviewed in (Dick, Jones et al., 2006)).

Extending Gene Identification Efforts Beyond Alcohol Dependence Diagnoses

Another exciting development in gene identification projects directed at understanding how genes predispose to alcohol problems has been the expansion of analyses beyond adult alcohol dependence. Although early gene identification efforts focused largely on alcohol dependence diagnoses, there has been a new effort to incorporate a developmental perspective into gene identification efforts and to expand analyses to other clinically relevant phenotypes suggested by twin studies. For example, the role of *GABRA2*, first associated with adult alcohol dependence, has been studied across a sample of individuals ranging in age from 7 to 65+, with evidence that *GABRA2* is associated with conduct disorder in childhood/adolescence, and the association with alcohol dependence does not emerge until the early twenties (Dick, Bierut et al., 2006). Additional analyses suggest that *GABRA2* is also associated with antisocial personality disorder (Dick, Agrawal et al., 2006) and with other forms of illicit drug dependence (Agrawal et al., 2006; Dick, Bierut, et al.), suggesting that this may be a gene involved in a variety of forms of disinhibitory psychopathology. A similar pattern has been found for *CHRM2*, where the association is strongest with a latent externalizing factor score consisting of symptoms of alcohol dependence, other drug dependence, conduct disorder, adult antisocial behavior, and disinhibitory personality traits (Dick, Aliev, Wang, Saccone et al., 2008). Thus, *GABRA2* and *CHRM2* may be examples of genes that are involved in alcohol dependence through a general externalizing pathway, a pattern originally suggested by multivariate twin analyses.

In addition to broadening the phenotype, other association analyses have been directed at analyzing aspects of alcohol use and dependence phenotypes. For example, although initial analyses of the GABA$_A$ receptor genes on chromosome 5 showed no evidence of association with alcohol dependence diagnoses in the COGA sample (Dick, Edenberg et al., 2005), subsequent analyses of drinking-related phenotypes, including a broader definition of problematic use, age of first intoxication, having a history of blackouts and level of response to alcohol, detected association with *GABRA1* (Dick, Plunkett et al., 2006). These findings are of particular interest since the evidence for the GABA$_A$ receptor genes on chromosome 5 came largely from the animal literature where drinking phenotypes, rather than alcohol dependence diagnoses per se, are necessarily modeled. Similarly, association has been reported between the alpha synuclein gene (*SNCA*), another candidate gene from the animal literature, and alcohol craving, despite no association originally detected with dependence diagnoses (Foroud, Wetherill, et al., 2007). These findings demonstrate the importance of examining alcohol use variables as phenotypes, in addition to a diagnosis of alcohol dependence, in order to better understand the genetic susceptibility toward alcohol problems.

Gene–Environment Interplay

Although classic twin models partition variance into genetic and environmental sources of variance, we know that this separation into distinct sources of variance is an oversimpli-

fication. For substance dependence and related phenotypes, both genes and the environment play an important role (as reviewed above), and these genetic and environmental influences are likely to combine in complex ways.

Gene–Environment Interaction in the Etiology of Substance Use Disorders

One form of such interplay is gene–environment interaction (G×E). G×E can be thought of as the effects of the environment being dependent on a person's genotype, or, equivalently, as the effects of genotype being dependent on (or moderated by) the environment.

Studies of twins have demonstrated that the magnitude of importance of genetic effects on substance use can vary as a function of numerous factors, including marital status (Heath, Jardine, & Martin, 1989), religiosity (Koopmans, Slutske, van Baal, & Boomsma, 1999), regional residency (Dick, Rose, Viken, Kaprio, & Koskenvuo, 2001; Rose et al., 2001), peer deviance (Dick, Viken et al., 2007), and parental monitoring (Dick, Viken, et al.). This suggests that some environments exacerbate the expression of genetic predispositions, whereas others are protective. Furthermore, the importance of different environmental factors may vary across developmental periods (Dick, Viken, et al.).

Results from adoption studies are more limited, but also suggest some environmental interactions. In studies of Swedish adoptees, having a biological father with criminality interacted with unstable rearing environment to increase risk of antisocial alcoholism among males (Cloninger, Sigvardsson, Bohman, & von Knorring, 1982). In the Iowa adoption studies, among females, a biological predisposition to alcoholism interacted with conflict in the rearing environment or adoptive parent psychopathology to increase risk for subsequent development of alcoholism (Cutrona et al., 1994).

Gene identification efforts are now attempting to test for moderation of effects associated with specific genes as a function of environmental risk factors. For example, a recent study of the influences of the serotonin transporter 5-HTTLPR polymorphism on college students' drinking and drug use found an interaction between 5-HTTLPR genotypes and stressful life events (Covault et al., 2007). The short allele of 5-HTTLPR was associated with increased drinking and drug use only among individuals who had experienced multiple negative life events, suggesting these individuals may be particularly susceptible to stress. There is also preliminary evidence of gene–environment interaction associated with *GABRA2*, whereby the effect of the genotype differed by marital status (Dick, Agrawal et al., 2006). Even the effect of the inherited *ALDH2* deficiency, which on the surface appears to be direct and straightforward (i.e., those with ALDH2 deficiency get sick when they drink and so they drink less), is complex and culturally conditioned. Higuchi et al. (1994) reported that the proportion of *ALDH2*2* heterozygotes among Japanese alcoholics has increased over time. In 1979, 2.5% of Japanese alcoholics were carriers of the *ALDH2*2* allele, suggesting a powerful protective effect since roughly 40–50% of Japanese are carriers. However, the proportion of *ALDH2*2* carriers among Japanese alcoholics rose, however, to 8% by 1986, and 13% by 1992. Although these latter rates still reflect a protective effect of *ALDH2* deficiency, this effect has clearly diminished over time. Higuchi et al. (1994) attributed the increasing frequency of *ALDH2* deficiency among Japanese alcoholics to cultural factors – during the period covered by the study, Japanese culture had become increasingly encouraging of heavy drinking. So, the consequences of inheriting even a relatively straightforward genetic variant – an allele that can make alcohol toxic – will still depend on context.

Gene–Environment Correlation in the Etiology of Substance Use Disorders

As indicated by the examples described above, behavior genetic studies of substance use and abuse are moving beyond latent variable models to include measured genes and measured contextual risk factors. However, some caution is warranted when interpreting these studies as simply reflecting gene–environment interactions. Factors such as marital status, religious affiliation, peer deviance, and stressful life events are often assumed to reflect "environmental" risk, but there is evidence that each of these is also influenced by genetic factors (e.g., Eaves, Eysenck, & Martin, 1989; Kendler, Jacobson et al., 2007; Kendler & Prescott, 2006). Thus, it is plausible that another form of gene–environment interplay is relevant: gene–environment correlation.

Gene–environment correlation is when individuals whose genotypes predispose them to develop a condition are at increased risk to experience predisposing environments. Scarr and McCartney (1983) described three forms of gene–environment correlations: passive, active, and evocative, all of which are likely to come into play in the development of substance use behaviors. Passive gene–environment correlations arise when parental genotypes influence their children's genetic risk and their environments. A striking example of this is provided by children of alcoholic parents, who are at increased risk to experience many socio-cultural and psychological factors associated with the development of substance use disorders, including family conflict, financial difficulties, inadequate parenting and supervision, and heavy drinking role models (McGue, 1997).

Active gene–environment correlations occur when individuals' genotypes lead them to create or seek out environments. For example, individuals with low vulnerability to develop substance abuse may also be predisposed to select protective social situations, such as non-deviant peers and stable marital relationships. Or, individuals with genetically influenced characteristics, such as risk-taking or impulsive personality traits, may create stressful life events which then provide a trigger for substance abuse.

Finally, evocative gene–environment correlations occur when the genotypes of individuals lead to behavior which evokes reactions in others or the environment that predispose to substance use. In the previously described example of the Swedish adoptees (Cloninger, Sigvardsson et al., 1982), behavioral deviance of the adolescents could have contributed to their unstable rearing environment, increasing their risk to develop antisocial alcoholism.

Identifying gene–environment interactions in the context of such gene–environment correlations is one of the challenges of behavior genetic research. Approaches such as extended kinship designs have been applied to substance use (e.g., Maes, Neale et al., 2006) but it is not possible to test simultaneously all possible mechanisms for parent–offspring transmission of risk. Eventually, the use of measured gene studies will help to disentangle these processes, but currently available studies examine variation at single alleles, and these are likely to represent only a small fraction of the genetically transmitted risk for substance use disorders. A complete understanding of these processes awaits elucidation of all the genetic and environmental processes contributing to the development of these disorders.

Using Genetic Information to Address Causal Hypotheses

The contributions of behavioral genetic studies to understanding the etiology of substance use disorders extend beyond providing estimates of heritability. Twin studies are being used to address a number of causal hypotheses about substance use. Although these studies have their limitations, they provide information that helps provide a more nuanced context for interpreting the results from epidemiological studies of risk factors for substance use disorders. A complete description of all the work done in this area is beyond the scope of our review. Here we provide several examples illustrating how behavior genetic studies are being used to evaluate the causality of risk factors for substance abuse.

Risk Factors: Causal or Correlated?

From a logical perspective, the existence of causality is easier to disprove than prove. The approach used in these studies is to estimate the basis for the association between a risk factor and substance abuse. If the basis is familial (either genetic or shared environmental) this suggests that the risk factor is correlated with, rather than causal for, the outcome. If the outcome and risk factors are associated due to individual-specific factors, this is consistent with, but not proof of, a causal relation. The behaviors could be causally related or associated through some unmeasured process.

Age at First Drink. Early age of onset of alcohol use is associated with increased risk for developing alcoholism (Grant & Dawson, 1997). In some studies, the prevalence of alcoholism among individuals who drink before age 15 is as high as 50%. Some researchers have interpreted this association as causal – that early drinking *directly* increases risk for later alcohol problems, leading to a call for prevention efforts aimed at delaying age at first drink. However, an alternative explanation is that early drinking is just a manifestation of liability to deviance, and delaying alcohol use would not alter the underlying liability to adolescent problem behavior or to adult alcoholism.

The information to test causality comes from twin pairs discordant for early drinking. Under the causal hypothesis, twins with earlier drinking onset are expected to have higher risk for alcoholism than their later-drinking cotwins. But if early drinking is an index of behavioral deviance (which has been found to be highly heritable), we would expect that the prevalence of alcoholism would be similar for members of discordant-onset pairs. The "unexposed" twins (with a later onset of drinking) would be expected to share their cotwins' risk for behavioral deviance and thus have a higher risk for alcoholism than observed in pairs in which neither twin drank early.

Results from the VATSPSUD study were inconsistent with the causal hypothesis (Prescott & Kendler, 1999a, 1999b). The analyses, which examined discordant pair information and conducted latent variable modeling in the whole sample, estimated the overlap between early drinking and alcohol dependence as due almost entirely to shared genetic liability. For example, among 282 MZ pairs discordant for early drinking, there was only a slight increase in the prevalence of alcoholism among twins who drank early relative to their cotwins who did not. This familial pattern of results was replicated in a sample of adolescent twins from Minnesota (McGue, Iacono, Legrand, & Elkins, 2001b). Adolescents whose parents used alcohol early were more likely to begin drinking at a young age and showed strong twin-pair resemblance for early drinking. It is also noteworthy that the pair resemblance for early drinking was largely attributable to pair similarity for externalizing behaviors, consistent with

early drinking being a manifestation for behavioral deviance. Although there are good reasons to try to delay and reduce adolescent alcohol use, these results suggest this targeting drinking onset age may not be an effective way to reduce the prevalence of adult alcoholism.

Drinking and Marital Status. A significant predictor of alcohol consumption among adults is marital status, with married individuals tending to drink less than those who are single or divorced (Temple et al., 1991). A longitudinal study of female twin pairs found limited support for a causal association between divorce and drinking (Prescott & Kendler, 2001). Women who would later get divorced drank more heavily in the several years prior to their divorce, and their cotwins drank more than other women, regardless of cotwin marital status. These results suggest a model of a causal association between divorce and drinking is too simplistic and there are familial factors that contribute to the association between drinking and divorce.

The Gateway Hypothesis

As discussed earlier, there is a strong association between the use of one substance and the subsequent use of other substances. The common liability model posits that the comorbidity that exists among multiple substance abuse phenotypes and other indicators of disinhibitory psychopathology is due to the existence of a common underlying liability. An alternative is the gateway model, which posits that these associations arise because the use/abuse of any specific substance increases the likelihood of the use/abuse of other, stronger, substances as well as disinhibited behavior (Kandel, Yamaguchi, & Chen, 1992). Consequently, interventions targeted at reducing use of substances early in the progression would under the gateway model reduce other substance abuse and disinhibited behavior by impeding the progression of externalizing behavior, but would not be expected to produce general benefits under the common liability model if the common liability was not the target of the intervention. Both twin and measured gene studies have been used to address these alternatives.

Discordant Twin-Pair Studies: Lynskey, Heath, et al., (2003) used the discordant twin-pair design to address the gateway model for the progression from use of cannabis to use and abuse of other drugs. Among 311 Australian young adult twin pairs discordant for trying cannabis before age 17, progression to other drugs and development of substance abuse were more than twice as common among the twins who were early cannabis users than their cotwins who did not use cannabis early. The results suggest that for these pairs, the key variables distinguishing their trajectories were individual-specific environmental factors, consistent with early use being in the causal chain for later substance use. However, it is also possible that other factors, such as behavioral deviance, are responsible for both the early cannabis use and the subsequent use of other substances. It is noteworthy that the pairs discordant for early cannabis use were atypical, representing only 11% of pairs in the sample. This suggests that genetic and shared environmental factors contribute to these behaviors as well. A subsequent study in the Virginia adult twin sample was less supportive of a causal explanation; the association between early cannabis use and subsequent drug involvement was due primarily to genetic and family environmental factors (Agrawal, Neale, Prescott, & Kendler, 2004).

ALDH2 Deficiency: The existence of *ALDH2* deficiency provides an opportunity to test one of the key implications of the common liability model of substance use disorders. *ALDH2* deficiency provides a natural analog to an experimental drinking intervention, something the epidemiologist George Davey-Smith has called Mendelian randomization (Davey-Smith & Ebrahim, 2004). That is, by chance some individuals inherit the deficient *ALDH2*2* allele while others will not. We know that those with *ALDH2* deficiency will drink less. The key question is whether they will, as the gateway model predicts, have reduced rates of smoking, other drug use, and disinhibited behavior relative to individuals not inheriting ALDH2 deficiency. Using a unique sample of Asian-American adopted adolescents, Irons et al. (2007) addressed this question. As expected, the *ALDH2*2* allele was associated with lower rates of drinking; it was not, however, associated with lower rates of smoking, illicit drug use, or delinquent behavior. These findings thus suggest that preventions that target use/abuse of specific substances (e.g., raising the price of cigarettes), even when successful, may have only limited effects on the full range of adolescent problem behavior.

Conclusion and Future Directions

Use and misuse of substances occurs within a complex matrix of social, familial, and biological factors. Behavior genetic research has established that a substantial portion of the risk for developing substance use disorders can be attributed to genetic variation. However, it is also clear that environmental factors are an important influence on initiation of substance use and may interact with genetic risk.

Contemporary behavior genetic studies of substance use have moved beyond a focus on heritability estimates and are contributing to understanding the etiology of these disorders by identifying specific genes that alter risk, by investigating how genetic influences are moderated by contextual

influences, and by evaluating putative causal risk factors. A variety of challenges remain, including clarifying the etiological heterogeneity in the development of these disorders and unraveling the specific mechanisms of shared genetic risk for substance and comorbid conditions. We expect that over the next decade, molecular genetic studies of susceptibility loci will combine with findings from genetic epidemiologic research and studies of endophenotypes in high-risk populations to add substantially to our knowledge about the matrix of risk.

References

Agrawal, A., Edenberg, H. J., Foroud, T., Bierut, L. J., Dunne, G., Hinrichs, A. L., et al. (2006). Association of GABRA2 with drug dependence in the collaborative study of the genetics of alcoholism sample. *Behavior Genetics, 36*(5), 640–650.

Agrawal, A., & Lynskey, M. T. (2006). The genetic epidemiology of cannabis use, abuse and dependence. *Addiction, 101*(6), 801–812.

Agrawal, A., Neale, M. C., Prescott, C. A., & Kendler, K. S. (2004). A twin study of early cannabis use and subsequent use and abuse/dependence of other illicit drugs. *Psychological Medicine, 34*(7), 1227–1237.

Bachman, J. G., O'Malley, P. M., Schulenberg, J. E., Johnston, L. D., Bryant, A. L., & Merline, A. C. (2002). *The decline of substance use in young adulthood.* Mahwah, New Jersey: Erlbaum.

Begleiter, H., & Porjesz, B. (1999). What is inherited in the predisposition toward alcoholism? A proposed model. *Alcoholism: Clinical and Experimental Research, 23*(7), 1125–1135.

Begleiter, H., Porjesz, B., Bihari, B., & Kissin, B. (1984). Event-related brain potentials in boys at risk for alcoholism. *Science, 225*(4669), 1493–1496.

Bergen, S. E., Gardner, C. O., & Kendler, K. S. (2007). Age-related changes in heritability of behavioral phenotypes over adolescence and young adulthood: A meta-analysis. *Twin Research and Human Genetics, 10*(3), 423–433.

Berman, S. M., Whipple, S. C., Fitch, R. J., & Noble, E. P. (1993). P3 in young boys as a predictor of adolescent substance use. *Alcohol, 10*(1), 69–76.

Blum, K., Noble, E. P., Sheridan, P. J., Montgomery, A., Ritchie, T., Jagadeeswaran, P., et al. (1990). Allelic association of human dopamine D2 receptor gene in alcoholism. *JAMA, 263*(15), 2055–2060.

Breslau, N., Johnson, E. O., Hiripi, E., & Kessler, R. (2001). Nicotine dependence in the United States: Prevalence, trends and smoking persistence. *Archives of General Psychiatry, 58*, 810–816.

Cartwright, W. S. (2008). Economic costs of drug abuse: Financial, cost of illness, and services. *Journal of Substance Abuse Treatment, 34*(2), 224–233.

Chen, K., & Kandel, D. B. (1995). The natural history of drug use from adolescence to the mid-thirties in a general population sample. *American Journal of Public Health, 85*(1), 41–47.

Chen, W. J., Loh, E. W., Hsu, Y. P. P., Chen, C. C., Yu, J. M., & Cheng, A. T. A. (1996). Alcohol-metabolising genes and alcoholism among Taiwanese Han men: Independent effect of ADH2, ADH3 and ALDH2. *British Journal of Psychiatry, 168*(6), 762–767.

Cloninger, C. R., Bohman, M., & Sigvardsson, S. (1981). Inheritance of alcohol abuse. Cross-fostering analysis of adopted men. *Archives of General Psychiatry, 38*(8), 861–868.

Cloninger, C. R., Sigvardsson, S., Bohman, M., & von Knorring, A. L. (1982). Predisposition to petty criminality in Swedish adoptees: II.

Cross-fostering analysis of gene-environment interactions. *Archives of General Psychiatry, 39*, 1242–1247.

Covault, J., Gelernter, J., Hesselbrock, V., Nellissery, M., & Kranzler, H. R. (2004). Allelic and haplotypic association of GABRA2 with alcohol dependence. *American Journal of Medical Genetics Part B-Neuropsychiatric Genetics, 129B*(1), 104–109.

Covault, J., Tennen, H., Armeli, S., Conner, T. S., Herman, A. I., Cillessen, A. H. N., et al. (2007). Interactive effects of the serotonin transporter 5-HTTLPR polymorphism and stressful life events on college student drinking and drug use. *Biological Psychiatry, 61*(5), 609–616.

Cutrona, C. E., Cadoret, R. J., Suhr, J. A., Richards, C. C., Troughton, E., Schutte, K., et al. (1994). Interpersonal variables in the prediction of alcoholism among adoptees: Evidence for gene-environment interactions. *Comprehensive Psychiatry, 35*(3), 171–179.

Davey-Smith, G., & Ebrahim, S. (2004). Mendelian randomization: prospects, potentials, and limitations. *International Journal of Epidemiology, 33*(1), 30–42.

Dick, D. M., Agrawal, A., Schuckit, M. A., Bierut, L., Hinrichs, A., Fox, L., et al. (2006). Marital status, alcohol dependence, and GABRA2: Evidence for gene-environment correlation and interaction. *Journal of Studies on Alcohol, 67*(2), 185–194.

Dick, D. M., Aliev, F., Wang, J. C., Grucza, R. A., Schuckit, M., Kuperman, S., et al. (2008). Using dimensional models of externalizing psychopathology to aid in gene identification. *Archives of General Psychiatry, 65*(3), 310–318.

Dick, D. M., Aliev, F., Wang, J. C., Saccone, S., Hinrichs, A., & Bertelsen, S. (2008). A systematic single nucleotide polymorphism screen to fine-map alcohol dependence genes on chromosome 7 identifies a novel susceptibility gene ACN9. *Biological Psychiatry, 63*, 1047–1053.

Dick, D. M., Bierut, L., Hinrichs, A., Fox, L., Bucholz, K. K., Kramer, J., et al. (2006). The role of GABRA2 in risk for conduct disorder and alcohol and drug dependence across developmental stages. *Behavior Genetics, 36*(4), 577–590.

Dick, D. M., Edenberg, H. J., Xuei, X. L., Goate, A., Hesselbrock, V., Schuckit, M., et al. (2005). No association of the GABA(A) receptor genes on chromosome 5 with alcoholism in the collaborative study on the genetics of alcoholism sample. *American Journal of Medical Genetics Part B-Neuropsychiatric Genetics, 132B*(1), 24–28.

Dick, D. M., Jones, K., Saccone, N., Hinrichs, A., Wang, J. C., Goate, A., et al. (2006). Endophenotypes successfully lead to gene identification: Results from the collaborative study on the genetics of alcoholism. *Behavior Genetics, 36*(1), 112–126.

Dick, D. M., Li, T. K., Edenberg, H. J., Hesselbrock, V., Kramer, J., Kuperman, S., et al. (2004). A genome-wide screen for genes influencing conduct disorder. *Molecular Psychiatry, 9*(1), 81–86.

Dick, D. M., Plunkett, J., Wetherill, L. F., Xuei, X. L., Goate, A., Hesselbrock, V., et al. (2006). Association between GABRA1 and drinking behaviors in the collaborative study on the genetics of alcoholism sample. *Alcoholism-Clinical and Experimental Research, 30*(7), 1101–1110.

Dick, D. M., Rose, R. J., Viken, R. J., Kaprio, J., & Koskenvuo, M. (2001). Exploring gene-environment interactions: Socioregional moderation of alcohol use. *Journal of Abnormal Psychology, 110*(4), 625–632.

Dick, D. M., Viken, R., Purcell, S., Kaprio, J., Pulkkinen, L., & Rose, R. J. (2007). Parental monitoring moderates the importance of genetic and environmental influences on adolescent smoking. *Journal of Abnormal Psychology, 116*(1), 213–218.

Dick, D.M., Wang, J. C., Plunkett, J., Hinrichs, A., Bertelsen, S., Budde, J. P., et al. (2007). Family-based analyses of alcohol dependence yield association with neighboring gene ANKK1 rather than DRD2. *Alcoholism: Clinical and Experimental Research*, 31, 1645–1653.

Dunn, G., Hinrichs, A. L., Bertelsen, S., Jin, C. H., Kauwe, J. S. K., Suarez, B. K., et al. (2005). Microsatellites versus single-nucleotide polymorphisms in linkage analysis for quantitative and qualitative measures. *Bmc Genetics, 6*.

Eaves, L. J., Eysenck, H. J., & Martin, N. G. (1989). *Genes, culture, and personality: an empirical approach.* London: Academic Press.

Edenberg, H. J., Dick, D. M., Xuei, X. L., Tian, H. J., Almasy, L., Bauer, L. O., et al. (2004). Variations in GABRA2, encoding the alpha 2 subunit of the GABA(A) receptor, are associated with alcohol dependence and with brain oscillations. *American Journal of Human Genetics, 74*(4), 705–714.

Edenberg, H. J., Xuei, X. L., Chen, H. J., Tian, H. J., Wetherill, L. F., Dick, D. M., et al. (2006). Association of alcohol dehydrogenase genes with alcohol dependence: a comprehensive analysis. *Human Molecular Genetics, 15*(9), 1539–1549.

Edenberg, H. J., Xuei, X. L., Wetherill, L. F., Bierut, L., Bucholz, K., Dick, D. M., et al. (2008). Association of NFKB1, which encodes a subunit of the transcription factor NF-kappa B, with alcohol dependence. *Human Molecular Genetics, 17*(7), 963–970.

Egeland, J. A., Gerhard, D. S., Pauls, D. L., Sussex, J. N., Kidd, K. K., Allen, C. R., et al. (1987). Bipolar affective disorders linked to DNA markers on chromosome 11. *Nature, 325*(6107), 783–787.

Ehlers, C. L., Gilder, D. A., Wall, T. L., Phillips, E., Feiler, H., & Wilhelmsen, K. C. (2004). Genomic screen for loci associated with alcohol dependence in mission Indians. *American Journal of Medical Genetics Part B-Neuropsychiatric Genetics, 129B*(1), 110–115.

Elkins, I. J., McGue, M., & Iacono, W. G. (2007). Prospective effects of attention-deficit/hyperactivity disorder, conduct disorder, and sex on adolescent substance use and abuse. *Archives of General Psychiatry, 64*(10), 1145–1152.

Fehr, C., Sander, T., Tadic, A., Lenzen, K. P., Anghelescu, I., Klawe, C., et al. (2006). Confirmation of association of the GABRA2 gene with alcohol dependence by subtype-specific analysis. *Psychiatric Genetics, 16*(1), 9–17.

Feinn, R., Nellissery, M., & Kranzler, H. R. (2005). Meta-analysis of the association of a functional serotonin transporter promoter polymorphism with alcohol dependence. *American Journal of Medical Genetics Part B-Neuropsychiatric Genetics, 133B*(1), 79–84.

Foroud, T., Edenberg, H. J., Goate, A., Rice, J., Flury, L., Koller, D. L., et al. (2000). Alcoholism susceptibility loci: Confirmation studies in a replicate sample and further mapping. *Alcoholism-Clinical and Experimental Research, 24*(7), 933–945.

Foroud, T., Wetherill, L. F., Liang, T., Dick, D. M., Hesselbrock, V., Kramer, J., et al. (2007). Association of alcohol craving with alpha-synuclein (SNCA). *Alcoholism-Clinical and Experimental Research, 31*(4), 537–545.

Fowler, T., Lifford, K., Shelton, K., Rice, F., Thapar, A., Neale, M. C., et al. (2007). Exploring the relationship between genetic and environmental influences on initiation and progression of substance use. *Addiction, 102*(3), 413–422.

Gelernter, J., O'Malley, S., Risch, N., Kranzler, H. R., Krystal, J., Merikangas, K., et al. (1991). No association between an allele at the D2 dopamine receptor gene (DRD2) and alcoholism [see comments]. *JAMA, 266*(13), 1801–1807.

Gillespie, N. A., Neale, M. C., Prescott, C. A., Aggen, S. H., & Kendler, K. S. (2007). Factor and item-response analysis DSM-IV criteria for abuse of and dependence on cannabis, cocaine, hallucinogens, sedatives, stimulants and opioids. *Addiction, 102*(6), 920–930.

Gmel, G., & Rehm, J. (2003). Harmful alcohol use. *Alcohol Research and Health, 27*(1), 52–62.

Goldman, D., & Bergen, A. (1998). General and specific inheritance of substance abuse and alcoholism. *Archives of General Psychiatry, 55*, 964–965.

Goldman, D., Oroszi, G., & Ducci, F. (2005). The genetics of addictions: Uncovering the genes. *Nature Reviews Genetics, 6*, 521–532.

Goodwin, D. W., Schulsinger, F., Hermansen, L., Guze, S. B., & Winokur, G. (1973). Alcohol problems in adoptees raised apart from alcoholic biological parents. *Archives of General Psychiatry, 28*(2), 238–243.

Gottesman, I. I., & Gould, T. D. (2003). The endophenotype concept in psychiatry: etymology and strategic intentions. *American Journal of Psychiatry, 160*(4), 636–645.

Grant, B. F., & Dawson, D. A. (1997). Age at onset of alcohol use and its association with DSM-IV alcohol abuse and dependence: Results from the National Longitudinal Alcohol Epidemiologic Survey. *Journal of Substance Abuse, 9*, 103–110.

Grant, B. F., Stinson, F. S., Dawson, D. A., Chou, S. P., Dufour, M., Compton, W., et al. (2006). Prevalence and co-occurrence of substance use disorders and independent mood and anxiety disorders. *Archives of General Psychiatry, 61*, 807–816.

Grobin, A. C., Matthews, D. B., Devaud, L. L., & Morrow, A. L. (1998). The role of GABA(A) receptors in the acute and chronic effects of ethanol. *Psychopharmacology, 139*(1–2), 2–19.

Han, C., McGue, M. K., & Iacono, W. G. (1999). Lifetime tobacco, alcohol, and other substance use in adolescent Minnesota twins: Univariate and multivariate behavioral genetic analyses. *Addiction, 7*, 981–993.

Harada, S., Agarwal, D. P., Goedde, H. W., Tagaki, S., & Ishikawa, B. (1982). Possible protective role against alcoholism for aldehyde dehydrogenase isozyme deficiency in Japan [letter]. *Lancet, 2*(8302), 827.

Hasin, D. (2003). Classification of alcohol use disorders. *Alcohol Research and Health, 27*(1), 5–17.

Hasin, D. S., Grant, B., & Endicott, J. (1990). The natural-history of alcohol-abuse – implications for definitions of alcohol-use disorders. *American Journal of Psychiatry, 147*(11), 1537–1541.

Heath, A. C., Bucholz, K. K., Madden, P. A. F., Dinwiddie, S. H., Slutske, W. S., Bierut, L. J., et al. (1997). Genetic and environmental contributions to alcohol dependence risk in a national twin sample: consistency of findings in women and men. *Psychological Medicine, 27*(6), 1381–1396.

Heath, A. C., Jardine, R., & Martin, N. G. (1989). Interactive effects of genotype and social environment on alcohol consumption in female twins. *Journal of Studies on Alcohol, 50*, 38–48.

Hettema, J. M., Corey, L. A., & Kendler, K. S. (1999). A multivariate genetic analysis of the use of tobacco, alcohol, and caffeine in a population based sample of male and female twins. *Drug and Alcohol Dependence, 57*(1), 69–78.

Hicks, B. M., Bernat, E., Malone, S. M., Iacono, W. G., Patrick, C. J., Krueger, R. F., et al. (2007). Genes mediate the association between P3 amplitude and externalizing disorders. *Psychophysiology, 44*(1), 98–105.

Higuchi, S., Matsushita, S., Imazeki, H., Kinoshita, T., Takagi, S., & Kono, H. (1994). Aldehyde dehydrogenase genotypes in Japanese alcoholics [letter]. *Lancet, 343*(8899), 741–742.

Higuchi, S., Matsushita, S., Murayama, M., Tagaki, S., & Hayashida, M. (1995). Alcohol and alcohol dehydrogenase polymorphisms and the risk of alcoholism. *American Journal of Psychiatry, 152* (1219–1221).

Hill, S. Y., Shen, S., Zezza, N., Hoffman, E. K., Perlin, M., & Allan, W. (2004). A genome wide search for alcoholism susceptibility genes. *American Journal of Medical Genetics Part B-Neuropsychiatric Genetics, 128B*(1), 102–113.

Hill, S. Y., Steinhauer, S., Lowers, L., & Locke, J. (1995). Eight-year longitudinal follow-up of P300 and clinical outcome in children from high-risk for alcoholism families. *Biol Psychiatry, 37*(11), 823–827.

Hingson, R., & Winter, M. (2003). Epidemiology and consequences of drinking and driving. *Alcohol Research and Health, 27*(1), 63–78.

Hopfer, C. J., Timberlake, D., Haberstick, B., Lessem, J. M., Ehringer, M. A., Smolen, A., et al. (2005). Genetic influences on quantity of

alcohol consumed by adolescents and young adults. *Drug and Alcohol Dependence, 78*(2), 187–193.

Hurley, T. D., Edenberg, H., & Li, T. K. (2002). The pharmacogenomics of alcoholism. In J. Licinio & M. L. Wong (Eds.), *Pharmacogenomics: The search for individualized therapies* (pp. 417–441). Weinheim, Germany: Wiley.

Iacono, W. G., Carlson, S. R., Malone, S. M., & McGue, M. (2002). P3 event-related potential amplitude and the risk for disinhibitory disorders in adolescent boys. *Archives of General Psychiatry, 59*(8), 750–757.

Iacono, W. G., & McGue, M. (2006). Association between P3 event-related brain potential amplitude and adolescent problem behavior. *Psychophysiology, 43*, 465–469.

Irons, D. E., McGue, M., Iacono, W. G., & Oetting, W. S. (2007). Mendelian randomization: A novel test of the gateway hypothesis. *Development and Psychopathology, 19*, 1181–1195.

Istvan, J., & Matarazzo, J. D. (1984). Tobacco, alcohol, and caffeine use: a review of their interrelationships. *Psychological Bulletin, 95*(2), 301–326.

Jardine, R., & Martin, N. G. (1984). Causes of variation in drinking habits in a large twin sample. *Acta Gemellogicae et Medicae, 33*, 435–450.

Johnston, L. D., O'Malley, P. M., Bachman, J. G., & Schulenberg, J. E. (2005). *Teen drug use down but progress halts among youngest teens.* Retrieved 07/18/2006, from www.monitoringthefuture.org

Jones, K. A., Pojesz, B., Almasy, L., Bierut, L., Goate, A., Wang, J. C., et al. (2004). Linkage and linkage disequilibrium of evoked EEG oscillations with CHRM2 receptor gene polymorphisms: implications for human brain dynamics and cognition. *International Journal of Psychophysiology, 53*(2), 75–90.

Kalsi, G., Kuo, P-H., Alexander, J., Aliev, F., Sullivan, P. F., Patterson, D. G., et al. (2008). Identification of novel candidate genes on 4q22-q32 in the Irish affected sib-pair study of alcohol dependence. Conference Presentation at the Annual Meeting of the Research Society of Alcoholism, June 2008, Washington, DC.

Kandel, D. B., Yamaguchi, K., & Chen, K. (1992). Stages of progression in drug involvement from adolescence to adulthood: Further evidence for the gateway theory. *Journal of Studies on Alcohol, 53*, 447–457.

Kaprio, J., Viken, R., Koskenvuo, M., Romanov, K., & Rose, R. J. (1992). Consistency and change in patterns of social drinking: a 6-year follow-up of the Finnish Twin Cohort. *Alcoholism, Clinical and Experimental Research, 16*(2), 234–240.

Katsanis, J., Iacono, W. G., McGue, M. K., & Carlson, S. R. (1997). P300 event-related potential heritability in monozygotic and dizygotic twins [published erratum appears in *Psychophysiology* 1998 Jan;35(1):133]. *Psychophysiology, 34*(1), 47–58.

Kendler, K. S., Jacobson, K. C., Gardner, C. O., Gillespie, N., Aggen, S. A., & Prescott, C. A. (2007). Creating a social world – A developmental twin study of peer-group deviance. *Archives of General Psychiatry, 64*(8), 958–965.

Kendler, K. S., Karkowski, L. M., Neale, M. C., & Prescott, C. A. (2000). Illicit psychoactive substance use, heavy use, abuse, and dependence in a US population-based sample of male twins. *Archives of General Psychiatry, 57*(3), 261–269.

Kendler, K. S., Myers, J., & Prescott, C. A. (2007). Specificity of genetic and environmental risk factors for symptoms of cannabis, cocaine, alcohol, caffeine, and nicotine dependence. *Archives of General Psychiatry, 64*(11), 1313–1320.

Kendler, K. S., Neale, M. C., Heath, A. C., Kessler, R. C., & Eaves, L. J. (1994). A twin-family study of alcoholism in women. *American Journal of Psychiatry, 151*(5), 707–715.

Kendler, K. S., & Prescott, C. A. (1998). Cannabis use, abuse, and dependence in a population-based sample of female twins. *American Journal of Psychiatry, 155*(8), 1016–1022.

Kendler, K. S., & Prescott, C. A. (2006). *Genes, environment and psychopathology: understanding the causes of psychiatric and substance use disorders*. New York: Guilford.

Kendler, K. S., Prescott, C. A., Myers, J., & Neale, M. C. (2003). The structure of genetic and environmental factors for common psychiatric and substance use disorders in men and women. *Archives of General Psychiatry, 60*, 929–937.

Kendler, K. S., Prescott, C. A., Neale, M. C., & Pedersen, N. L. (1997). Temperance board registration for alcohol abuse in a national sample of Swedish male twins, born 1902 to 1949. *Archives of General Psychiatry, 54*(2), 178–184.

Kessler, R. C., Berglund, P., Demler, O., Jin, R., & Walters, E. E. (2005). Lifetime prevalence and age-of-onset distributions' of DSM-IV disorders in the national comorbidity survey replication. *Archives of General Psychiatry, 62*(6), 593–602.

Kidd, K. K. (1993). Associations of disease with genetic markers; deja vu all-over again. *American Journal of Medical Genetics (Neuropsychiatric Genetics), 48*, 71–73.

Knopik, V. (2005) Alcoholism and externalizing disorders: Comorbidity risk mediated via intrauterine exposure. Paper presented at the Fifth Annual Guze Symposium on Alcoholism, St. Louis, MO, 2005.

Koopmans, J. R., Slutske, W. S., van Baal, G. C., & Boomsma, D. I. (1999). The influence of religion on alcohol use initiation: Evidence for genotype X environment interaction. *Behavior Genetics, 29*, 445–453.

Koopmans, J. R., vanDoornen, L. J. P., & Boomsma, D. I. (1997). Association between alcohol use and smoking in adolescent and young adult twins: A bivariate genetic analysis. *Alcoholism: Clinical and Experimental Research, 21*(3), 537–546.

Krueger, R. F., Hicks, B. M., Patrick, C. J., Carlson, S. R., Iacono, W. G., & McGue, M. (2002). Etiologic connections among substance dependence, antisocial behavior, and personality: modeling the externalizing spectrum. *Journal of Abnormal Psychology, 111*(3), 411–424.

Krueger, R. F., & Markon, K. E. (2006). Understanding psychopathology: Melding behavior genetics, personality, and quantitative psychology to develop an empirically based model. *Current Directions in Psychological Science, 15*(3), 113–117.

Krueger, R. F., Nichol, P. E., Hicks, B. M., Markon, K. E., Patrick, C. J., Iacono, W. G., et al. (2004). Using latent trait modeling to conceptualize an alcohol problems continuum. *Psychological Assessment, 16*(2), 107–119.

Kuo, P.-H., Gardner, C. O., Kendler, K. S., & Prescott, C. A. (2006). The temporal relationship of the onsets of alcohol dependence and major depression: Using a genetically informative study design. *Psychological Medicine, 36*, 1153–1162.

Kuo, P. H., Kalsi, G., Prescott, C. A., Hodgkinson, C. A., Goldman, D., & Van den Oord, E. J. J. (2008). Association of ADH and ALDH genes with alcohol dependence in the Irish Affected Sib Pair Study of Alcohol Dependence (IASPSAD) Sample. *Alcoholism: Clinical and Experimental Research, 32*, 785–795.

Lander, E., & Kruglyak, L. (1995). Genetic dissection of complex traits – Guidelines for interpreting and reporting linkage results. *Nature Genetics, 11*(3), 241–247.

Lappalainen, J., Krupitsky, E., Remizov, M., Pchelina, S., Taraskina, A., Zvartau, E., et al. (2005). Association between alcoholism and gamma-amino butyric acid alpha 2 receptor subtype in a Russian population. *Alcoholism-Clinical and Experimental Research, 29*(4), 493–498.

Li, M. D., Cheng, R., Ma, J. Z., & Swan, G. E. (2003). A meta-analysis of estimated genetic and environmental effects on smoking behavior in male and female adult twins. *Addiction, 98*(1), 23–31.

Long, J. C., Knowler, W. C., Hanson, R. L., Robin, R. W., Urbanek, M., Moore, E., et al. (1998). Evidence for genetic linkage to alcohol dependence on chromosomes 4 and 11 from an autosome-wide scan in an American Indian population. *American Journal of Medical Genetics, 81*(3), 216–221.

Luo, X. G., Kranzler, H. R., Zuo, L. J., Yang, B. Z., Lappalainen, J., & Gelernter, J. (2005). ADH4 gene variation is associated with alcohol and drug dependence: results from family controlled and population-structured association studies. *Pharmacogenetics and Genomics, 15*(11), 755–768.

Lynskey, M. T., Heath, A. C., Bucholz, K. K., Slutske, W. S., Madden, P. A. F., Nelson, E. C., et al. (2003). Escalation of drug use in early-onset cannabis users vs co-twin controls. *Jama-Journal of the American Medical Association, 289*(4), 427–433.

Lynskey, M. T., Heath, A. C., Nelson, E. C., Bucholz, K. K., Madden, P. A. F., Slutske, W. S., et al. (2002). Genetic and environmental contributions to cannabis dependence in a national young adult twin sample. *Psychological Medicine, 32*, 195–207.

Maes, H. H., Neale, M. C., Kendler, K. S., Martin, N. G., Heath, A. C., & Eaves, L. J. (2006). Genetic and cultural transmission of smoking initiation: An extended twin kinship model. *Behavior Genetics, 36*(6), 795–808.

Maes, H. H., Woodfard, C. E., Murrelle, L., Meyer, J. M., Silberg, J. L., Hewitt, J. K., et al. (1999). Tobacco, alcohol and drug use in eight- to sixteen-year-old twins: the Virginia Twin Study of Adolescent Behavioral Development. *Journal of Studies on Alcohol, 60*(3), 293–305.

Maggs, J. L., & Schulenberg, J. E. (2004). Trajectories of alcohol use during the transition to adulthood. *Alcohol Research and Health, 28*(4), 195–201.

McGue, M. (1991). Behavioral genetic models of alcoholism and drinking. In K. E. Leonard & H. T. Blane (Eds.), *Psychological theories of drinking and alcoholism* (pp. 372–421). New York: Guilford.

McGue, M. (1994). Genes, environment, and the etiology of alcoholism. In R. Zucker, G. Boyd & J. Howard (Eds.), *The development of alcohol problems: Exploring the biopsychosocial matrix of risk* (Vol. NIAAA Research Monograph No. 26, NIH Publication No. 94–3495, pp. 1–40). Bethesda, MD: National Institutes of Health.

McGue, M. (1997). A behavioral-genetic perspective on children of alcoholics. *Alcohol Health and Research World, 21*, 210–217.

McGue, M., Elkins, I., & Iacono, W. G. (2000). Genetic and environmental influences on adolescent substance use and abuse. *American Journal of Medical Genetics (Neuropsychiatric Genetics), 96*, 671–677.

McGue, M., & Iacono, W. G. (2005). The association of early adolescent problem behavior with adult psychopathology. *American Journal of Psychiatry, 162*(6), 1118–1124.

McGue, M., Iacono, W. G., Legrand, L. N., & Elkins, I. (2001a). The origins and consequences of age at first drink. I. Associations with substance-use disorders, disinhibitory behavior and psychopathology, and P3 amplitude. *Alcoholism: Clinical and Experimental Research, 25*, 1156–1165.

McGue, M., Iacono, W. G., Legrand, L. N., & Elkins, I. (2001b). Origins and consequences of age at first drink. II. Familial risk and heritability. *Alcoholism: Clinical and Experimental Research, 25*, 1166–1173.

McGue, M., Pickens, R. W., & Svikis, D. S. (1992). Sex and age effects on the inheritance of alcohol problems: a twin study. *Journal of Abnormal Psychology, 101*(1), 3–17.

McGue, M., Sharma, A., & Benson, P. (1996). Parent and sibling influences on adolescent alcohol use and misuse: Evidence from a U.S. adoption cohort. *Journal of Studies on Alcohol, 57*, 8–18.

Miles, D. R., van den Bree, M. B. M., Gupman, A. E., Newlin, D. B., Glantz, M. D., & Pickens, R. W. (2001). A twin study on sensation seeking, risk taking behavior and marijuana use. *Drug and Alcohol Dependence, 62*(1), 57–68.

Munafo, M., & Johnstone, E. C. (2005). Association between the serotonin transporter gene and alcohol consumption in social drinkers: evidence for mediation by anxiety-related personality. *European Neuropsychopharmacology, 15*, S591-S591.

Neale, B. M., Ferreira, M. A. R., Medland, S. E., & Posthuma, D. (2008). *Statistical genetics: gene mapping through linkage and association*. New York: Taylor and Francis.

Neale, M. C., & Kendler, K. S. (1995). Models of comorbidity for multifactorial disorders. *American Journal of Human Genetics, 57*, 935–953.

Neiswanger, K., Hill, S. Y., & Kaplan, B. B. (1995). Association and linkage studies of the TAQI A1 allele at the dopamine D2 receptor gene in samples of female and male alcoholics [see comments]. *American Journal of Medical Genetics, 60*(4), 267–271.

Neumark, Y. D., Friedlander, Y., Thomasson, H. R., & Li, T. K. (1998). Association of the ADH2*2 allele with reduced ethanol consumption in Jewish men in Israel: A pilot study. *Journal of Studies on Alcohol, 59*(2), 133–139.

Noble, E. P. (1998). The D2 dopamine receptor gene: a review of association studies in alcoholism and phenotypes. *Alcohol, 16*(1), 33–45.

Nurnberger, J. I., Foroud, T., Flury, L., Su, J., Meyer, E. T., Hu, K. L., et al. (2001). Evidence for a locus on chromosome 1 that influences vulnerability to alcoholism and affective disorder. *American Journal of Psychiatry, 158*(5), 718–724.

O'Brien, C. P. (2008). Prospects for a genomic approach to the treatment of alcoholism. *Archives of General Psychiatry, 65*(2), 132–133.

O'Connor, S., Morzorati, S., Christian, J. C., & Li, T. K. (1994). Heritable features of the auditory oddball event-related potential: peaks, latencies, morphology and topography. *Electroencephalography and Clinical Neurophysiology, 92*(2), 115–125.

Patrick, C. J., Bernat, E. M., Malone, S. M., Iacono, W. G., Krueger, R. F., & McGue, M. (2006). P300 amplitude as an indicator of externalizing in adolescent males. *Psychophysiology, 43*(1), 84–92.

Polich, J., Pollock, V. E., & Bloom, F. E. (1994). Meta-analysis of P300 amplitude from males at risk for alcoholism. *Psychological Bulletin, 115*(1), 55–73.

Porjesz, B., & Begleiter, H. (1998). Genetic basis of event-related potentials and their relationship to alcoholism and alcohol use. *Journal of Clinical Neurophysiology, 15*(1), 44–57.

Porjesz, B., Begleiter, H., Wang, K. M., Almasy, L., Chorlian, D. B., Stimus, A. T., et al. (2002). Linkage and linkage disequilibrium mapping of ERP and EEG phenotypes. *Biological Psychology, 61*(1–2), 229–248.

Prescott, C. A. (2002). Sex differences in the genetic risk for alcoholism. *Alcohol Research & Health, 26*(4), 264–273.

Prescott, C. A., Aggen, S. H., & Kendler, K. S. (2000). Sex-specific genetic influences on the comorbidity of alcoholism and major depression in a population-based sample of U.S. twins. *Archives of General Psychiatry, 57*, 803–811.

Prescott, C. A., & Kendler, K. S. (1999a). Age at first drink and risk for alcoholism: A noncausal association. *Alcoholism: Clinical and Experimental Research, 23*, 101–107.

Prescott, C. A., & Kendler, K. S. (1999b). Genetic and environmental contributions to alcohol abuse and dependence in a population-based sample of male twins. *American Journal of Psychiatry, 156*(1), 34–40.

Prescott, C. A., & Kendler, K. S. (2000). Influence of ascertainment strategy on finding sex differences in genetic estimates from twin studies of alcoholism. *American Journal of Medical Genetics, 96*(6), 754–761.

Prescott, C. A., & Kendler, K. S. (2001). Associations between marital status and alcohol consumption in a longitudinal study of female twins. *Journal of Studies on Alcohol, 62*(5), 589–604.

Prescott, C. A., Sullivan, P. F., Kuo, P. H., Webb, B. T., Vittum, J., Patterson, D. G., et al. (2006). Genomewide linkage study in the Irish affected sib pair study of alcohol dependence: evidence for a susceptibility region for symptoms of alcohol dependence on chromosome 4. *Molecular Psychiatry, 11*(6), 603–611.

Rehm, J., Room, R., van den Brink, W., & Jacobi, F. (2005). Alcohol use disorders in EU countries and Norway: An overview of the epidemiology. *European Neuropsychopharmacology, 15*, 377–388.

Rehm, J., Room, R., van den Brink, W., & Kraus, L. (2005). Problematic drug use and drug use disorders in EU countries and Norway: An overview of the epidemiology. *European Neuropsychopharmacology, 15*, 389–397.

Rehm, J., Taylor, B., & Room, R. (2006). Global burden of disease from alcohol, illicit drugs and tobacco. *Drug and Alcohol Review, 25*(6), 503–513.

Reich, T., Edenberg, H. J., Goate, A., Williams, J. T., Rice, J. P., Van Eerdewegh, P., et al. (1998). Genome-wide search for genes affecting the risk for alcohol dependence. *American Journal of Medical Genetics, 81*(3), 207–215.

Rhee, S. H., Hewitt, J. K., Young, S. E., Corley, R. P., Crowley, T. J., & Stallings, M. C. (2003). Genetic and environmental influences on substance initiation, use, and problem use in adolescents. *Archives of General Psychiatry, 60*(12), 1256–1264.

Roberts, S. B., MacLean, C. J., Neale, M. C., Eaves, L. J., & Kendler, K. S. (1999). Replication of linkage studies of complex traits: An examination of variation in location estimates. *American Journal of Human Genetics, 65*(3), 876–884.

Rose, R. J., Dick, D. M., Viken, R. J., & Kaprio, J. (2001). Gene-environment interaction in patterns of adolescent drinking: Regional residency moderates longitudinal influences on alcohol use. *Alcoholism: Clinical and Experimental Research, 25*(5), 637–643.

Rose, R. J., Dick, D. M., Viken, R. J., Pulkkinen, L., Nurnberger Jr., J. I. and Kaprio, J. (2004). Genetic and environmental effects on conduct disorder, alcohol dependence symptoms, and their covariation at age 14. *Alcoholism: Clinical and Experimental Research, 28*, 1541–1548.

Saccone, N. L., Kwon, J. M., Corbett, J., Goate, A., Rochberg, N., Edenberg, H. J., et al. (2000). A genome screen of maximum number of drinks as an alcoholism phenotype. *American Journal of Medical Genetics, 96*, 632–637.

Scarr, S., & McCartney, K. (1983). How people make their own environments: A theory of genotype => environment effects. *Child Development, 54*, 424–435.

Schuckit, M. A. (2006). Comorbidity between substance use disorders and psychiatric conditions. *Addiction, 101*(Supplement 1), 76–88.

Schuckit, M. A., Edenberg, H. J., Kalmijn, J., Flury, L., Smith, T. L., Reich, T., et al. (2001). A genome-wide search for genes that relate to a low level of response to alcohol. *Alcoholism-Clinical and Experimental Research, 25*(3), 323–329.

Shen, Y. C., Fan, J. H., Edenberg, H. J., Li, T. K., Cui, Y. H., Wang, Y. F., et al. (1997). Polymorphism of ADH and ALDH genes among four ethnic groups in China and effects upon the risk for alcoholism. *Alcoholism-Clinical and Experimental Research, 21*(7), 1272–1277.

Sher, K. J., & Trull, T. J. (1994). Personality and disinhibitory psychopathology: Alcoholism and antisocial personality disorder. *Journal of Abnormal Psychology, 103*, 92–102.

Smith, L., Watson, M., Gates, S., Ball, D., & Foxcroft, D. (2008). Meta-analysis of the association of the Taq1A polymorphism with the Risk of alcohol dependency: A HuGE gene-disease association review. *American Journal of Epidemiology, 167*(2), 125–138.

Soyka, M., Preuss, U. W., Hesselbrock, V., Zill, P., Koller, G., & Bondy, B. (2008). GABA-A2 receptor subunit gene (GABRA2) polymorphisms and risk for alcohol dependence. *Journal of Psychiatric Research, 42*(3), 184–191.

Stinson, F. S., Grant, B. F., Dawson, D. A., Ruan, W. J., Huang, B., & Saha, T. (2006). Comorbidity between DSM-IV alcohol and specific drug use disorders in the United States. *Alcohol Research and Health, 29*(2), 94–106.

Sullivan, P. F., & Kendler, K. S. (1999). The genetic epidemiology of smoking. *Nicotine and Tobacco Research, 1*, S51-S57.

Swan, G. E., Carmelli, D., & Cardon, L. R. (1996). The consumption of tobacco, alcohol, and coffee in Caucasian Male twins: A multivariate genetic analysis. *Journal of Substance Abuse, 8*(1), 19–31.

Tarter, R. E., & Vanyukov, M. (1994). Alcoholism: A developmental disorder. *Journal of Consulting and Clinical Psychology, 62*, 1096–1107.

Temple, M. T., Fillmore, K. M., Hartka, E., Johnstone, B., Leino, E. V., & Motoyoshi, M. (1991). A meta-analysis of change in marital and employment status as predictors of alcohol consumption on a typical occasion. *British Journal of Addiction, 86*(10), 1269–1281.

Tsuang, M. T., Lyons, M. J., Eisen, S. A., Goldberg, J., True, W., Lin, N., et al. (1996). Genetic influences on DSM-III-R drug abuse and dependence: A study of 3,372 twin pairs. *American Journal of Medical Genetics, 67*(5), 473–477.

Tsuang, M. T., Lyons, M. J., Meyer, J. M., Doyle, T., Eisen, S. A., Goldberg, J., et al. (1998). Co-occurrence of abuse of different drugs in men: the role of drug- specific and shared vulnerabilities [see comments]. *Archives of General Psychiatry, 55*(11), 967–972.

vanBeijsterveldt, C. E. M., Molenaar, P. C. M., deGeus, E. J. C., & Boomsma, D. I. (1996). Heritability of human brain functioning as assessed by electroencephalography. *American Journal of Human Genetics, 58*(3), 562–573.

Wang, J. C., Hinrichs, A. L., Bertelsen, S., Stock, H., Budde, J. P., Dick, D. M., et al. (2007). Functional variants in TAS2R38 and TAS2R16 influence alcohol consumption in high-risk families of African-American origin. *Alcoholism-Clinical and Experimental Research, 31*(2), 209–215.

Wang, J. C., Hinrichs, A. L., Stock, H., Budde, J., Allen, R., Bertelsen, S., et al. (2004). Evidence of common and specific genetic effects: Association of the muscarinic acetylcholine receptor M2 (CHRM2) gene with alcohol dependence and major depressive syndrome. *Human Molecular Genetics, 13*(17), 1903–1911.

Whitfield, J. B., Nightingale, B. N., Bucholz, K. K., Madden, P. A. F., Heath, A. C., & Martin, N. G. (1998). ADH genotypes and alcohol use and dependence in Europeans. *Alcoholism-Clinical and Experimental Research, 22*(7), 1463–1469.

Williams, J. T., Begleiter, H., Porjesz, B., Edenberg, H. J., Foroud, T., Reich, T., et al. (1999). Joint multipoint linkage analysis of multivariate qualitative and quantitative traits. II. Alcoholism and event-related potentials. *American Journal of Human Genetics., 65*(4), 1148–1160.

Yoon, H. H., Iacono, W. G., Malone, S. M., & McGue, M. (2006). Using the brain P300 response to identify novel phenotypes reflecting genetic vulnerability for adolescent substance misuse. *Addictive Behaviors, 31*(6), 1067–1087.

Young, S. E., Rhee, S. H., Stallings, M. C., Corley, R. P., & Hewitt, J. K. (2006). Genetic and environmental vulnerabilities underlying adolescent substance use and problem use: General or specific? *Behavior Genetics, 36*(4), 603–615.

Chapter 30

Genetic Analysis of Conduct Disorder and Antisocial Behavior

Soo Hyun Rhee and Irwin D. Waldman

Introduction

In this chapter, we examine the evidence for genetic influences on conduct disorder and antisocial behavior. First, we present results from a meta-analysis of twin and adoption studies estimating the relative magnitude of genetic and environmental influences on antisocial behavior (Rhee & Waldman, 2002). Second, we discuss recent studies that have examined several interesting issues in the etiology of antisocial behavior, including genotype × environment interactions, co-occurrence with other psychiatric disorders, the etiology of psychopathy, and the etiology of adolescent-limited versus life-course-persistent antisocial behavior. Third, we review association studies examining the influence of specific candidate genes on antisocial behavior and linkage studies conducting genome-wide screens for quantitative trait loci (QTLs) influencing antisocial behavior.

In this review, we focused on phenotypes directly related to conduct disorder or antisocial behavior. Studies were included if they clearly examined antisocial personality disorder, conduct disorder, criminality, or aggression; if there was empirical evidence suggesting that the measure of antisocial behavior used in the study successfully discriminated an antisocial group from a control group; or if there was empirical evidence suggesting that the measure used in the study is significantly related to a more established operationalization of antisocial behavior.

Studies examining four operationalizations of antisocial behavior were included. First, studies examining psychiatric diagnoses of antisocial personality disorder (ASPD) and conduct disorder (CD) were included. The Diagnostic and Statistical Manual of Mental Disorders, fourth edition (DSM-IV; American Psychiatric Association [APA], 1994), describes the essential features of ASPD as "a pervasive pattern of dis-

regard for, and violation of, the rights of others that begins in childhood or early adolescence and continues into adulthood" (p. 645). A diagnosis of ASPD requires a history of CD before the age of 15 and three or more of the following criteria: failure to conform to social norms with respect to lawful behaviors (i.e., as indicated by repeatedly performing acts that are grounds for arrest), deceitfulness, impulsivity, irritability and aggressiveness, reckless disregard for safety, consistent irresponsibility, and lack of remorse. Conduct disorder, a criterion for the diagnosis of ASPD, is described by the DSM-IV as "a repetitive and persistent pattern of behavior in which the basic rights of others or major age-appropriate societal norms or rules are violated" (p. 90, APA, 1994). It usually occurs in childhood or early adolescence and is manifested as aggression toward people and animals, destruction of property, deceitfulness or theft, and serious violations of rules. Second, studies examining criminality (an unlawful act that leads to arrest, conviction, or incarceration) and delinquency (unlawful acts committed as a juvenile) were included. Third, we included studies examining aggression, which is usually studied as a personality characteristic and assessed with measures such as the Adjective Checklist (Gough & Heilbrun, 1972). Fourth, studies examining an omnibus operationalization that includes aggression and delinquency items, such as the externalizing scale from the Child Behavior Checklist (Achenbach & Edelbrock, 1983), were included.

Current Issues and Research

A Meta-analysis of Twin and Adoption Studies Examining Antisocial Behavior

More than a hundred twin and adoption studies of antisocial behavior have been published. Nonetheless, it is difficult to draw clear conclusions regarding the magnitude of genetic and environmental influences on antisocial behavior given the current literature. The main reason for this

S.H. Rhee (✉)
Institute for Behavioral Genetics, University of Colorado, Boulder, CO 80309, USA
e-mail: Soo.Rhee@colorado.edu

Y.-K. Kim (ed.), *Handbook of Behavior Genetics*,
DOI 10.1007/978-0-387-76727-7_30, © Springer Science+Business Media, LLC 2009

difficulty is the considerable heterogeneity of the results in this area of research, with published heritability estimates (i.e., the magnitude of genetic influences) ranging from very low (e.g., 0.00; Plomin, Foch, & Rowe, 1981) to very high (e.g., 0.71; Slutske et al., 1997). It is important to remember that there is not a fixed, absolute heritability for antisocial behavior. The heritability estimate describes the magnitude of genetic influences in a particular population at a particular time (Plomin, DeFries, McClearn, & McGuffin, 2001), and it is possible that the heterogeneity in the results of behavior genetic studies of antisocial behavior may reflect real, substantive differences as well as those due to methodological variations across studies. In the literature, various hypotheses have been proposed to explain these heterogeneous results across studies, including differences in the age of the sample (e.g., Cloninger & Gottesman, 1987), the age of onset of antisocial behavior (e.g., Moffitt, 1993), and the measurement of antisocial behavior (e.g., Plomin, Nitz, & Rowe, 1990).

We conducted a meta-analysis of twin and adoption studies in order to provide a clearer and more comprehensive picture of the magnitude of genetic and environmental influences on antisocial behavior. Given previous hypotheses proposed to explain the heterogeneity in the results, we examined the possible moderating effects of three study characteristics (i.e., the operationalization of antisocial behavior, assessment method, and zygosity determination method) and two participant characteristics (i.e., the age and sex of the participants) on the magnitude of genetic and environmental influences on antisocial behavior.

We tested the operationalization of antisocial behavior as a possible moderator given the evidence that antisocial personality disorder, conduct disorder, criminality, and aggression are distinct but related constructs (e.g., Robins & Regier, 1991). We examined four levels of operationalization, which included diagnosis (ASPD or CD), criminality, aggression, and antisocial behavior (an omnibus operationalization that included aggression and delinquency items). We tested assessment method and zygosity determination as moderators because of evidence suggesting that these are potential methodological confounders (e.g., Plomin, 1981; McCartney, Harris, & Bernieri, 1990). We compared five assessment methods, including self-report, report by others (i.e., parent and teacher report), official records, objective measures, and reactions to aggressive material as well as three zygosity determination methods, including blood grouping, questionnaires, and a combination of blood grouping and questionnaires. Sex was examined given the consistent evidence that antisocial behavior is more prevalent in males than females (e.g., Hyde, 1984; Wilson & Herrnstein, 1985). Age was examined because of the potential to test an interesting hypothesis regarding the development of antisocial behavior by comparing studies that included children, adolescents, and adults. DiLalla and Gottesman (1989)

and Moffitt (1993) have suggested that antisocial individuals can be divided into a smaller group whose antisocial behavior is persistent throughout the life course and caused predominantly by genetic influences and a larger group whose antisocial behavior is limited to adolescence and caused predominantly by environmental influences. If their hypothesis is correct, the magnitude of genetic influences on antisocial behavior should be lower in adolescence than in childhood or adulthood. We also compared the results of twin and adoption studies. Twin and adoption studies have unique assumptions or biases that can make interpretations of their results difficult. Comparing the results of twin and adoption studies can help determine whether the results of behavior genetic studies have been influenced by these unique assumptions or biases. To the degree that the results of twin and adoption studies are similar, it is more likely that the results reflect the true magnitude of genetic and environmental influences. One cannot rule out the possibility, however, that the results of twin and adoption studies are similar because they share similar biases to some extent that influence their results in the same direction.

Two types of adoption studies were included in the present meta-analysis: (1) parent–offspring adoption studies (i.e., comparing the correlation between adoptees and their adoptive parents with the correlation between adoptees and their biological parents) and (2) sibling adoption studies (i.e., comparing the correlation between adoptive siblings with the correlation between biological siblings). When interpreting the results of parent–offspring data, it is important to consider the possibility that the correlations between the parents and the offspring may be reduced by the age difference between the two generations and that the magnitude of familial (i.e., genetic and shared environmental) influences may be underestimated. Genetic influences on a trait may differ from one generation to another because the genes affecting the same trait may differ in their expression across age due to genotype–environment interaction. For example, genetic influences in the younger generation may be increased because of environmental facilitation of antisocial behavior, e.g., via secular increases in substance use and less-stringent parenting practices (e.g., Lykken, 1997). Also, there may be cohort-specific shared environmental influences other than the cultural transmission from parents to offspring. Therefore, each type of adoption study was compared to the twin studies separately.

One hundred forty-one twin and adoption studies of antisocial behavior were identified by examining the PsycInfo and Medline databases and contacting authors of unpublished manuscripts identified by abstracts of the Behavior Genetics Association meetings and searching the Dissertation Abstracts and ERIC databases. After excluding studies that were unsuitable given inadequate construct validity, inability to calculate effect sizes given lack of information, and simultaneous assessment of related disorders, such as alcoholism, 96 studies remained. Non-independence among these studies was addressed by either choosing the effect size from the largest sample when non-independent samples had differing sample sizes or by averaging the effect sizes across the samples when non-independent samples had the same sample size. After addressing the problem of non-independence, 10 independent adoption samples and 42 independent twin samples remained. The meta-analysis was conducted on these 52 independent samples. The studies included in the meta-analysis are marked with an asterisk (*) in the References section.

Effect sizes were determined in one of three ways. First, some adoption and twin studies used a continuous variable to measure antisocial behavior and reported either Pearson product moment or intraclass correlations, which were the effect sizes used from these studies in the meta-analysis. Second, a dichotomous variable was used, and concordances, percentages, or a contingency table (including the number of twin pairs with both twins affected, one twin affected, and neither twin affected) was reported. The information from the concordances or percentages was transformed into a contingency table, which was then used to estimate the tetrachoric correlation (i.e., the correlation between the latent continuous variables that are assumed to underlie the observed dichotomous variables). For these studies, the tetrachoric correlation was the effect size used in the meta-analysis. Third, we were able to directly estimate the tetrachoric correlation from the raw data for some studies because we had access to the data (e.g., Slutske et al., 1997; Waldman, Levy, & Hay, 1995).

Alternative models positing that antisocial behavior is affected by additive genetic influences (A), shared environmental influences (C), non-additive genetic influences (D), and nonshared environmental influences (E) were tested. The ACE model, the AE model, the CE model, and the ADE model were compared. Given that the ACDE model can be tested only when both twin and adoption studies are included in the analysis, it was only possible to estimate c^2 and d^2 simultaneously when analyzing all of the data included in the meta-analysis. For other analyses (i.e., the comparison between twin and adoption studies and the tests of moderators), both twin and adoption studies were not always available across different types of studies. Therefore, we were limited to comparing the ACE, AE, CE, and ADE models for analyses other than those that included all data included in the meta-analysis.

Overview of the Results

When all available data from both twin and adoption studies were analyzed together and the magnitude of non-additive genetic influences was estimated in addition to the magnitude of shared environmental influences, the best fitting model was the ACDE model. Based on this analysis, there were moderate additive genetic ($a^2 = 0.32$), non-additive genetic ($d^2 = 0.09$), shared environmental ($c^2 = 0.16$), and nonshared environmental ($e^2 = 0.43$) influences on antisocial behavior.

Operationalization was a significant moderator of the magnitude of genetic and environmental influences. This means that there was a statistically significant difference between a model that constrained the parameter estimates to be the same across the different levels of the moderator and a model that allowed the parameter estimates to differ across the different levels of the moderator. The ACE model was the best fitting model for diagnosis ($a^2 = 0.44$, $c^2 = 0.11$, $e^2 = 0.45$), aggression ($a^2 = 0.44$, $c^2 = 0.06$, $e^2 = 0.50$), and antisocial behavior ($a^2 = 0.47$, $c^2 = 0.22$, $e^2 = 0.31$), whereas the ADE model was the best fitting model for criminality ($a^2 = 0.33$, $d^2 = 0.42$, $e^2 = 0.25$). Within the operationalization of diagnosis, the a^2 estimate was higher in studies examining CD ($a^2 = 0.50$, $c^2 = 0.11$, $e^2 = 0.39$), whereas the e^2 estimate was higher in studies examining ASPD ($a^2 = 0.36$, $c^2 = 0.10$, $e^2 = 0.54$).

Assessment method also was a significant moderator, with the ACE model fitting best for self-report ($a^2 = 0.39$, $c^2 = 0.06$, $e^2 = 0.55$) and report by others ($a^2 = 0.53$, $c^2 = 0.22$, $e^2 = 0.25$), whereas the AE model fit best for reaction to aggressive stimuli ($a^2 = 0.52$, $e^2 = 0.48$). All of the studies using the assessment method of records were also studies examining criminality, for which the ADE model fits best ($a^2 = 0.33$, $d^2 = 0.42$, $e^2 = 0.25$).

Age was a significant moderator, with the ACE model fitting best for children ($a^2 = 0.46$, $c^2 = 0.20$, $e^2 = 0.34$), adolescents ($a^2 = 0.43$, $c^2 = 0.16$, $e^2 = 0.41$), and adults ($a^2 = 0.41$, $c^2 = 0.09$, $e^2 = 0.50$). The magnitude of familial influences (a^2 and c^2) decreased with age, whereas the magnitude of non-familial influences (e^2) increased with age.

Operationalization, assessment method, and age were frequently confounded in the studies included in the meta-analysis. For example, parent report was more frequently used in studies examining antisocial behavior than in studies examining diagnosis or aggression, as well as in studies examining children than in studies examining adolescents or adults. The results of analyses examining age as a significant moderator were not consistent with DiLalla and Gottesman's hypothesis that the magnitude of genetic influences should be higher in childhood and adulthood than in adolescence. In contrast, the magnitude of genetic influence was lower in both adolescence and adulthood than in childhood. Nonetheless, the presence of confounding among the moderators must be considered, implying that the higher heritability for childhood may actually reflect the higher heritability for parent report.

Zygosity determination method was a significant moderator, such that the ADE model was the best fitting model for studies using blood grouping ($a^2 = 0.14$, $d^2 = 0.33$, $e^2 = 0.53$), whereas the ACE model was the best fitting model for studies using the questionnaire method ($a^2 = 0.43$, $c^2 = 0.27$, $e^2 = 0.30$) and a combination of the two methods ($a^2 = 0.39$, $c^2 = 0.11$, $e^2 = 0.50$). These parameter estimates are difficult to interpret, given that studies using the most stringent method of zygosity determination (i.e., blood grouping) and the least stringent method of zygosity determination (i.e., questionnaire) yielded higher estimates of genetic influences (broad $h^2 = 0.43$–0.47) than studies using a combination of the two methods (broad $h^2 = 0.39$). Although sex was a significant moderator when data from all studies were analyzed, there were no statistically significant sex differences in studies that included both sexes (males: $a^2 = 0.43$, $c^2 = 0.19$, $e^2 = 0.38$; females: $a^2 = 0.41$, $c^2 = 0.20$, $e^2 = 0.39$).

Parent–offspring adoption studies found a lower magnitude of familial influences on antisocial behavior (i.e., lower a^2 and c^2 and higher e^2) than the twin and sibling adoption studies. There are several possible reasons for this result. First, the age difference between the children and their parents may lead to lower correlations, given that there may be age- or cohort-specific genetic and/or environmental influences. This age difference is absent in the twin studies and smaller in the sibling adoption studies, thus supporting this explanation. Second, because of the practical obstacles involved in conducting an adoption study, in several studies different operationalizations and methods of assessment were used in the adoptees and their parents (e.g., criminality via official records in the parents and aggression via self-report in the adoptees).

There was no statistically significant difference between the results of twin studies and the sibling adoption studies. This result should be interpreted cautiously considering the fact that 42 independent twin samples were compared with only 3 independent sibling adoption samples. Although the power to detect a statistically significant difference between the two types of studies may have been limited by the small number of sibling adoption studies, the parameter estimates for the twin studies ($a^2 = 0.45$, $c^2 = 0.12$, $e^2 = 0.43$) and the sibling adoption studies ($a^2 = 0.48$, $c^2 = 0.13$, $e^2 = 0.39$) were very similar.

Additional Issues in the Etiology of Antisocial Behavior

Several interesting issues in the etiology of antisocial behavior were beyond the scope of the meta-analysis, or we were unable to conduct a quantitative review of these issues given the small number of studies addressing them in the literature. Below, studies examining genotype × environment interaction, co-occurrence with other psychiatric disorders, the etiology of psychopathy, and the etiology of adolescent-limited versus life-course-persistent antisocial behavior are reviewed.

Genotype × Environment Interaction

In addition to estimating the main effects of genes and environments on various forms of antisocial behavior, researchers also have examined whether genes and the environment interact to influence antisocial behavior using both the adoption design and the twin design, which have differing advantages and disadvantages. The adoption study is the ideal method for testing genotype–environment interactions because the genetic and environmental influences on a trait are disentangled and can be measured distinctly. Unfortunately, the power to test the genotype × environment interaction term may be reduced in adoption studies of antisocial behavior because of range restriction in the variables used to indicate the environmental and/or genetic influences on antisocial behavior. McClelland & Judd (1993) demonstrated that restricting the range of the predictor variables will reduce the residual variance of the product of the two predictors and, in turn, the statistical power to detect an interaction. The problem with range restriction is especially a concern in adoption studies of antisocial behavior because the chance of adoptees being placed in adoptive homes with criminal or antisocial adoptive parents is very low. Therefore, the statistical difficulties of detecting interactions should be considered in interpreting adoption studies examining gene–environment interactions. In contrast, genotype–environment interactions are more difficult to test in twin studies because the genetic and environmental influences on a trait are likely to be correlated. On the other hand, range restriction is less of a problem in twin studies, where the samples are more representative of the general population.

Data from several adoption studies (Cadoret, Cain, & Crowe, 1983; Cloninger, Sigvardsson, Bohman, & von Knorring, 1982; Mednick, Gabrielli, & Hutchings, 1983) show evidence of genotype–environment interaction for antisocial behavior. Mednick et al. (1983) conducted a cross-fostering analysis of Danish adoptees. Among adoptees who had a criminal background in both their biological and adoptive parents, 24.5% of them became criminal themselves. This is in comparison to 20% of adoptees who have a criminal background only in their biological parents, 14.7% of adoptees who have a criminal background only in their adoptive parents, and 13.5% of adoptees with no criminal background. Cloninger et al. (1982) found similar results for petty criminality in Swedish adoptees when they considered both biological variables (i.e., criminal biological parents) and environmental variables (i.e., negative rearing experiences and adoptive placement). Among adoptees with both biological and environmental risks, 40% were criminal. This is in comparison to 12.1% of those with only biological risk factors, 6.7% of those with only environmental risk factors, and 2.9% of those with neither biological nor environmental risk factors. Also, in a sample of adoptees from Iowa, Cadoret et al. (1983) found that when both genetic and environmental risk factors were present, they accounted for a greater number of antisocial behaviors than an additive combination of the two kinds of risk factors acting independently. The genotype–environment interactions were not statistically significant in Cloninger et al. (1982) or Mednick et al. (1983), most likely given reduced power to test an interaction in the presence of range restriction.

More recently, several twin studies have examined the interaction between genes and specific environmental influences. Rowe, Almeida, and Jacobson (1999) examined the interaction between genetic influences and family warmth on aggression and reported that heritability increased and the magnitude of shared environmental influences decreased with greater family warmth. Rowe et al. suggested the possibility that greater genetic influences are required for the expression of aggression in more benign environments, whereas social norms and peer models may lead to the expression of aggression in individuals without genetic predisposition in more adverse environments. A recent study by Button, Scourfield, Martin, Purcell, and McGuffin (2005) supports Rowe et al.'s findings. They examined the interaction between genes and family dysfunction and concluded that the heritability of antisocial behavior decreased and the magnitude of both shared environmental and nonshared environmental influences increased as family dysfunction increased.

Jaffee et al. (2005) examined the effect of the interaction between genetic influences and physical maltreatment on conduct problems in a different way. They divided their twin sample into four groups based on levels of genetic risk as a function of their co-twin's conduct disorder status and the pair's zygosity (i.e., lowest risk – MZ co-twin with no CD diagnosis; low risk – DZ co-twin with no CD diagnosis; high risk – DZ co-twin with CD diagnosis; highest risk – MZ co-twin with CD diagnosis) and absence/presence of maltreatment. The maltreated group with the highest genetic risk had the highest probability of a CD diagnosis. There was a significant interaction between genetic risk and maltreatment, but the conclusions are different from those reached by Rowe et al. (1999) and Button et al. (2005). In the non-maltreated group, there was a 44.2% increase in the probability of a CD diagnosis from the lowest (1.9%) to the highest (46.1%) genetic risk group. In the maltreated group, there was a 66.1% increase in the probability of a CD diagnosis

from the lowest (3.5%) to the highest (69.6%) genetic risk group. That is, genetic risk had less effect on conduct problems in the non-maltreated group (i.e., the more benign environment) than in the maltreated group (i.e., the less benign environment).

The Jaffee et al. (2005) study differs from the Rowe et al. (1999) and Button et al. (2005) studies in several ways, including the specific environmental variable examined and the analytical method used. It is possible that the effects of the interaction between genes and family warmth/dysfunction on antisocial behavior are different from those of the interaction between genes and physical maltreatment. Additional studies examining genotype × specific environmental influence interactions on antisocial behavior are needed.

Co-occurrence with Other Psychiatric Disorders

Antisocial behavior co-occurs with several other psychiatric disorders such as major depressive disorders, attention-deficit/hyperactivity disorder (ADHD), and drug dependence. In recent years, many researchers have examined the degree to which the covariance between antisocial behavior and other psychiatric disorders is due to common genetic or environmental influences.

There is evidence of significant co-occurrence between conduct disorder (CD) and ADHD, which occur together in 30–50% of the cases in both epidemiological and clinical samples (Biederman, Newcorn, & Sprich, 1991). The results of several multivariate behavior genetic studies examining the co-occurrence between ADHD and CD are largely consistent, with most studies (Dick, Viken, Kaprio, Pulkkinen, & Rose, 2005; Nadder, Rutter, Silberg, Maes, & Eaves, 2002; Nadder, Silberg, Eaves, Maes, & Meyer, 1998; Silberg et al., 1996; Thapar, Harrington, & McGuffin, 2001; Waldman, Rhee, & Levy, 2001) reporting a substantial overlap between the genetic influences on ADHD and the genetic influences on CD. An exception was Burt, Krueger, McGue, and Iacono's (2001) examination of ADHD, oppositional defiant disorder (ODD), and CD, in which a single shared environmental factor was the largest contributor to the covariation among ADHD, ODD, and CD.

Researchers also have noted the high rate of co-occurrence between CD and MDD, with reported rates of co-occurrence between depression and CD/ODD ranging from 21 to 83% (Angold & Costello, 1993). There are three published studies examining the etiology of the occurrence between CD and MDD or the broader constructs of externalizing and internalizing behavior/disorders, and the results of these studies are conflicting. Gjone and Stevenson (1997) examined the covariance between parent reports of internalizing and externalizing behavior problems in adolescent twins and concluded that although there are both common genetic and shared environmental influences on the covariance between internalizing and externalizing behavior, there is more consistent evidence for common shared environmental influences. O'Connor, McGuire, Reiss, Hetherington, and Plomin (1998) examined the covariation between antisocial behavior and depressive symptoms assessed using a composite of parent report, self-report, and observers' report in adolescents and found that there were moderate genetic, shared environmental, and nonshared environmental influences on the covariation between the two symptom domains. In Kendler, Prescott, Myers, and Neale's (2003) examination of self-reported internalizing and externalizing disorders in adults, the authors found that genetic factors influencing externalizing disorders had little influence on internalizing disorders, and genetic factors influencing internalizing disorders had little influence on externalizing disorders. (They did not find a clear distinction for shared environmental influences and nonshared environmental influences on externalizing and internalizing disorders.) Several methodological differences among these studies may explain their differing conclusions, including the specific constructs examined, the assessment methods used, and the age of the participants.

Conduct disorder also commonly co-occurs with substance use and substance use disorders. For example, Reebye, Moretti, and Lessard (1995) reported that 52% of adolescent inpatients with CD also had a substance use disorder. Researchers have concluded that there are genetic influences on the co-occurrence between antisocial behavior and substance use/substance use disorders across a range of substances. Silberg, Rutter, D'Onofrio, and Eaves (2003) found significant common genetic influences on smoking and conduct disturbance. Slutske et al. (1998) reported that genetic influences accounted for most of the covariance (with nonshared environmental influences accounting for the remainder) between CD and alcohol dependence. Miles, van den Bree, and Pickens (2002) found evidence of moderate common genetic influences and low common nonshared environmental influences on CD and marijuana use. Grove et al. (1990) found significant genetic correlations between illegal drug problems and both childhood and adult antisocial behavior. Recently, Button et al. (2006) reported that there were moderate genetic, shared environmental, and nonshared environmental influences on the covariance between CD and polysubstance dependence vulnerability. These studies are consistent in suggesting that common genetic influences play a significant role in the substantial co-occurrence of conduct problems and substance use and/or dependence.

The Etiology of Psychopathy

Psychopathy is a personality-based construct characterized by superficial charm, lack of empathy, lack of guilt, egocen-

tricity, and dishonesty. Personality characteristics are emphasized in this condition, rather than overt antisocial acts. Researchers hypothesize that psychopathic individuals, who represent a subset of individuals diagnosed with CD or ASPD, may be those who are most likely to engage in life-course-persistent antisocial behavior (Viding, Blair, Moffitt, & Plomin, 2005) and instrumental aggression (Blair, Peschardt, Budhani, Mitchell, & Pine, 2006), both of which are particularly heritable.

Recently, Waldman and Rhee (2006) presented the results of a meta-analysis of the small number of behavior genetic studies examining psychopathy. These included the studies reviewed in the earlier Rhee & Waldman (2002) meta-analysis (Brandon & Rose, 1995; DiLalla, Carey, Gottesman, & Bouchard, 1996; Gottesman, 1963, 1965; Loehlin & Nichols, 1976; Loehlin, Willerman, & Horn, 1987; Taylor, Iacono, and McGue, 2000; Torgersen, Skre, Onstad, Edvardsen, & Kringlen, 1993) as well more recent results from Taylor, Loney, Bobadilla, Iacono, and McGue (2003) and Blonigen, Carlson, Krueger, and Patrick (2003). In contrast to the aforementioned overall results for antisocial behavior ($a^2 = 0.32$, $d^2 = 0.09$, $c^2 = 0.16$, $e^2 = 0.43$), there is little evidence for shared environmental influences on psychopathy, and the AE model fits best ($h^2 = 0.49$, $e^2 = 0.51$).

The results from an interesting study by Viding et al. (2005) were not included in the meta-analysis because it examined the magnitude of genetic and environmental influences on extreme status on psychopathy (whereas the other studies examined the etiology of psychopathy in the general population). Viding et al.'s study is also the first examination of the etiology of psychopathy in children. They examined a sample of 7-year-old twin pairs in the general population and then selected probands based on extremely high scores on antisocial behavior and callous-unemotional traits. Their results are similar to those from the Waldman and Rhee meta-analysis. Extreme antisocial behavior accompanied by extreme callous-unemotional traits was highly heritable ($h^2_g = 0.81$) with no evidence of shared environmental influences, but extreme antisocial behavior unaccompanied by extreme callous-unemotional traits was less heritable ($h^2_g = 0.30$) and affected by shared environmental influences ($c^2_g = 0.34$).

Recently, Larsson, Andershed, and Lichtenstein (2006) published a twin study examining the etiology of psychopathy in adolescents. Twins aged 16–17 years were assessed via the Youth Psychopathic Traits Inventory (YPI). Their results are also consistent with those of previous studies examining the etiology of psychopathy, as they found moderate genetic and nonshared environmental influences and little evidence of shared environmental influences on all three YPI dimensions (grandiose/manipulative – $a^2 = 0.51$, $c^2 = 0.03$, $e^2 = 0.46$; callous/unemotional – $a^2 = 0.43$, $c^2 = 0.00$, $e^2 = 0.57$; impulsive/irresponsible – $a^2 = 0.56$, $c^2 = 0.00$, $e^2 = 0.44$). A common pathway model (where the covariance among the three YPI dimensions is represented by an intermediate latent variable) fits the data well, and the heritability of the latent psychopathic personality factor was slightly higher ($h^2 = 0.63$, $c^2 = 0.00$, $e^2 = 0.37$) than those of the three YPI dimensions.

Another recent paper by Blonigen, Hicks, Krueger, Patrick, and Iacono (2005) examined the co-occurrence between psychopathic traits and internalizing and externalizing psychopathology. They found evidence of genetic influences on two distinct psychopathic traits, fearless dominance (which is characterized by social potency, stress immunity, and fearlessness) and impulsive antisociality (which is characterized by negative emotionality and low behavioral constraint). Genetic influences on fearless dominance were negatively correlated with those on internalizing psychopathology, whereas genetic influences on impulsive antisociality were positively correlated with those on externalizing psychopathology.

Genetic Influences on Adolescent-Limited Versus Life-Course-Persistent Antisocial Behavior

In a seminal review paper in 1993, Moffitt hypothesized that there are two developmental types of antisocial behavior that have different etiologies. Life-course-persistent antisocial behavior begins in early childhood, continues throughout the child's lifetime, and is posited to show stronger genetic influences and greater deficits in neuropsychological functioning. In contrast, adolescent-limited antisocial behavior is posited to be more affected by environmental influences such as antisocial peer models. Therefore, Moffitt hypothesizes that the magnitude of genetic influences should be higher for antisocial behavior occurring in childhood or adulthood than in adolescence.

The results of the meta-analysis do not support this hypothesis, as the magnitude of genetic influences on antisocial behavior was lower in both adolescence and adulthood than in childhood. However, it is possible that this result is due to the presence of confounding between age and assessment method (i.e., most studies that examined the heritability of antisocial behavior in childhood used parent reports, and most studies that examined the heritability of antisocial behavior in adolescence and adulthood used self-reports). Given that studies that use parent reports have yielded higher heritabilities than studies that use self-reports, the higher heritability for antisocial behavior in childhood may thus reflect the higher heritability for parent report.

Several studies that have examined adolescent-limited and life-course-persistent antisocial behavior in the same sample using the same assessment method have reported results supporting Moffitt's hypothesis. Lyons et al. (1995) contrasted the etiology of adult and juvenile antisocial traits and reported that adult antisocial traits had significant genetic influences, whereas juvenile antisocial traits were largely influenced by shared environmental influences. Taylor, Iacono, and McGue (2000) examined antisocial behavior in the co-twins of probands with early-onset and late-onset delinquency. Monozygotic co-twins of probands with early-onset delinquency were more antisocial than dizygotic co-twins of probands with early-onset delinquency, but the monozygotic and dizygotic co-twins of probands with late-onset delinquency had similar levels of antisocial behavior. Finally, Jacobson, Neale, Prescott, and Kendler (2001) reported that adolescent antisocial behavior had a lower heritability and a greater magnitude of shared environmental influences in their subsample of youth whose antisocial behavior did not persist into adulthood than in their subsample of youth whose antisocial behavior did persist into adulthood.

Examination of the Influence of Specific Genes and QTLs on Antisocial Behavior

Given the evidence of moderate genetic influences on conduct disorder and antisocial behavior, many researchers have investigated the relations of specific genes with antisocial behavior in linkage and association studies. The number of association and linkage studies examining antisocial behavior is far fewer than those examining disorders or phenotypes that are often comorbid with antisocial behavior, such as ADHD, alcoholism, and suicidality. Thus, many researchers examining these overlapping phenotypes also have examined the evidence for association of specific genes with antisocial behavior in clinical samples with ADHD (e.g., Thapar et al., 2005), alcoholism and drug abuse (e.g., Soyka, Preuss, Koller, Zill, & Bondy, 2004), suicidality (e.g., Zalsman et al., 2001), and schizophrenia (e.g., Strous et al., 2003). In the present review, we focused on studies of antisocial behavior in the general population, clinical samples with antisocial behavior, or general psychiatric samples, given the possibility that some of the conflicting results and failures to replicate across studies may be due to sample differences.

Broadly speaking, there are two general strategies for identifying quantitative trait loci (QTLs) that contribute to the etiology of a disorder or trait, namely linkage and association. The purpose of a linkage study is to test whether the putative risk-inducing gene is at a particular chromosomal locus, whereas the purpose of an association study is to test whether a particular allele of a candidate gene is associated with the disorder or trait.

In genome scans, evidence for linkage between a disorder or trait and evenly spaced DNA markers distributed across the entire genome is evaluated. Evidence for linkage between any of these DNA markers and the trait or disorder of interest implicates a broad segment of the genome that may contain hundreds of genes, and lack of evidence for linkage can, in some cases, be used to exclude genomic segments. Subsequent fine-grained linkage analyses can then use a new set of more tightly grouped markers within the implicated genomic region to locate the functional mutation. Thus, genome scans may be thought of as exploratory searches for putative genes that contribute to the etiology of a disorder. The fact that major genes have been found for many medical diseases via genome scans without a priori knowledge of those genes' function or etiological significance is a testament to the usefulness of this method. Unfortunately, the power of linkage analyses in genome scans is typically quite low, making it very difficult, if not impossible, to detect QTLs that account for less than $\sim$15% of the variance in a disorder. Therefore, the promise for linkage-based genome scans of complex traits remains largely unknown.

In contrast to linkage, association has the statistical power to detect QTLs that account for a relatively small percentage of the variance in a disorder. A disadvantage of association is that a QTL cannot be detected unless the marker being examined is very close to the QTL. Therefore, many more DNA markers would be needed to conduct a genome scan using association, which has been made possible by recent advances in array-based genotyping technologies. Until recently, researchers have used association solely to conduct candidate gene studies.

In many ways, candidate gene studies are polar opposites of genome scans. In contrast to the exploratory nature of genome scans, well-conducted candidate gene studies represent a targeted test of the role of specific genes in the etiology of a disorder as the location, function, and etiological relevance of candidate genes are most often known or strongly hypothesized a priori. Thus, an advantage of well-conducted candidate gene studies in comparison with genome scans is that positive findings are easily interpretable because one already knows the gene's location, function, and etiological relevance, even if the specific polymorphism(s) chosen for study in the candidate gene is not functional and the functional mutation(s) in the candidate gene is as yet unidentified. There are also disadvantages to the candidate gene approach given that only previously identified genes can be studied. Thus, one cannot find genes that one has not looked for or have yet to be discovered, and because there are relatively few strong candidate genes for psychiatric disorders, the same genes are examined as candidates for almost all psychiatric disorders, regardless of how disparate the disorders may be in terms of their symptomatology or conjectured pathophysiology.

In well-designed studies, however, knowledge regarding the biology of the disorder is used to select genes based on the known or hypothesized involvement of their gene product in the etiology of the trait or disorder (i.e., its pathophysiological function and etiological relevance). With respect to antisocial behavior, genes underlying various aspects of the dopaminergic and serotonergic neurotransmitter pathways may be conjectured based on several lines of converging evidence suggesting a role for these neurotransmitter systems in the etiology and pathophysiology of these traits and their relevant disorders.

Association Studies

Dopamine transmission. There is significant co-occurrence between CD and ADHD, with 30–50% of cases having both disorders in both epidemiological and clinical samples (Biederman et al., 1991). Also, several twin studies examining the co-occurrence between the two disorders suggest that there are genetic influences common to both (Dick et al., 2005; Nadder et al., 1998, 2002; Silberg et al., 1996; Waldman et al., 2001). Therefore, specific genes associated with ADHD also may influence CD.

There are several lines of evidence suggesting that the dopamine transporter gene is associated with ADHD. The dopamine transporter is the site of action of stimulant medications effective in the treatment of ADHD (Winsberg & Comings, 1999), and dopamine transporter levels are approximately 70% higher in the brains of individuals with ADHD than those of controls (Dougherty et al., 1999). "Knockout" mice with a deletion of both copies of the dopamine transporter are hyperactive (Giros, Jaber, Jones, Wightman, & Caron, 1996) and in humans, a VNTR polymorphism in the 3' untranslated region of the dopamine transporter gene has shown association with ADHD in previous studies, with the more common 10-repeat allele being the risk allele (e.g., Cook et al., 1995; Faraone et al., 2005; Mill et al., 2005; Waldman et al., 1998).

Given such evidence, two studies have examined the evidence of association between this polymorphism of the dopamine transporter gene and externalizing behavior in children. Jorm, Prior, Sanson, Smart, Zhang, and Easteal (2001) examined the association between the DAT1 gene and externalizing behavior problems and associated temperament traits in a longitudinal study of children from age 3 to 16 in the general population. They examined whether children with the 10-repeat/10-repeat, 10-repeat/non 10-repeat, and non 10-repeat/non 10-repeat genotypes differed in the mean level of behavior problems but found no evidence of association between the DAT1 gene and any of the phenotypes examined, including parent reports of hostile-aggressive behavior, conduct disorder, socialized aggression, or oppositional behavior, at any age.

Young et al. (2002) examined the association between the same VNTR polymorphism of the DAT1 gene and parent reports of externalizing behavior at ages 4, 7, and 9. A community sample of children from twin and adoption studies was examined, using a sibling-based association method (Fulker, Cherny, Sham, & Hewitt, 1999) that separates the allelic effect into between-family effects (that may include possible population stratification effects) and within-family effects (where population stratification is controlled by the inclusion of siblings). In contrast to evidence from ADHD studies suggesting that the 10-repeat allele is the risk-inducing allele, Young et al. found that the 9-repeat was the risk-inducing allele for externalizing behavior at ages 4 and 7, but not at age 9. Young et al. cite that the less frequent 9-repeat allele has been shown to be the risk-inducing allele for other phenotypes correlated with externalizing behavior, such as alcohol dependence (Sander et al., 1997; Schmidt, Harms, Kuhn, Rommelspacher, & Sander, 1998) and cocaine intoxication (Gelernter, Kranzler, Satel, & Rao, 1994).

Serotonin transmission. Several lines of evidence suggest that genes influencing serotonergic neurotransmission may be associated with conduct disorder and antisocial behavior. Low cerebrospinal fluid concentrations of 5-hydroxyindoleacetic acid, the major metabolite of serotonin, are found in aggressive subjects (e.g., Kruesi et al., 1990; Lidberg, Tuck, Asberg, Scalia-Tomba, & Bertilsson, 1985). Also, pharmacological challenge trials suggest an association between low serotonin functioning and violence or aggression. For example, aggression is associated with blunted prolactin (PRL) responses to the treatment of 5-HT releaser, d-fenfluramine (Coccaro, Kavoussi, Cooper, & Hauger, 1997; Coccaro et al., 1987). Given this evidence, researchers have examined the association between antisocial behavior and several genes involved in serotonergic transmission, including the serotonin transporter gene, the HTR1B gene, and the tryptophan hydroxylase gene.

The serotonin transporter is known to influence serotonergic transmission and several researchers have examined a polymorphism in the promoter area with two allelic variants that differ in transcriptional activity (5-HTTLPR). The S or short allele of the polymorphism has lower transcriptional activity (Heils et al., 1996) and is associated with a blunted response in fenfluramine-induced prolactin release pharmacological challenges (e.g., Reist, Mazzanti, Vu, Tran, & Goldman, 2001), whereas serotonin transporter availability is higher in individuals with the L or long allele than in those with the S allele (Heinz et al., 2000).

Baca-Garcia et al. (2004) examined the association between the serotonin transporter gene promoter linked region and impulsivity and aggressive behavior among suicide attempters and a control group of blood donors and found no difference among the SS, SL, and LL genotypes in the level of impulsivity or history of aggressive behavior in either group. In contrast, Retz, Retz-Junginger, Supprian, Thome, and Rösler (2004) found a significant association between the serotonin transporter gene and violent behavior in a sample of adult males referred for a forensic

examination. Retz et al. divided the subjects into a violent and nonviolent group and found that the S allele was significantly over-represented in the violent group. More recently, Sakai et al. (2006) found that the S allele is a risk allele for conduct disorder with aggression in a case–control study comparing conduct-disordered adolescents and non-conduct-disordered controls as well as in a transmission disequilibrium test examining conduct-disordered adolescents and their parents. Beitchman et al. (2006) reported similar results in a study examining three alleles of the 5-HTTLPR polymorphism (i.e., S, Lg, and La; La is "high transcribing"; Lg is "low transcribing", similar to S); they found that the "low-expressing" genotypes of the 5-HTTLPR polymorphism (S/S, Lg/S, Lg/Lg) are associated with childhood aggression in a case–control study.

In an adoption study sample, Cadoret et al. (2003) found a complex pattern of association between the serotonin transporter gene and externalizing behavior (i.e., aggressivity, conduct disorder, and attention-deficit/hyperactivity disorder). In the presence of antisocial biological parents, the L allele was associated with increased externalizing behavior, whereas in the presence of alcoholism in the biological parents, the S allele was associated with increased externalizing behavior. The authors also found a significant genotype by gender interaction, with the S allele being associated with higher externalizing behavior in males and the S allele being associated with lower externalizing behavior in females. Haberstick, Smolen, and Hewitt (2006) also reported mixed findings. Family-based association tests were conducted in a twin sample, examining parent and teacher reports of aggressive behavior at ages 7, 9, 10, 11, and 12, with teacher reports also available at age 8. A statistically significant result was obtained only for teacher reports of aggressive behavior at age 9, with the S allele being the risk allele.

Given research suggesting that genes involved in serotonin functioning may be candidate genes for aggression and the evidence that 5-HT1B knockout mice show increased aggression (Saudou et al., 1994), New et al. (2001) examined the association between a common polymorphism caused by a silent G to C substitution in the HTR1B gene and self-reported impulsive aggression in a sample of subjects with one or more personality disorders but found no evidence of association.

Tryptophan hydroxylase (TPH) is the rate-limiting enzyme in the serotonin pathway in the conversion of tryptophan into 5-hydroxytryptophan, the direct precursor of 5-HT. Researchers have examined two common SNPs in intron 7 of the TPH gene, A218C and A779C, that are in strong linkage disequilibrium. The genotypes of these two SNPs were found to be identical in nearly 100% of the sample in one Japanese study (Kunugi et al., 1999), with the 218A and the 779A being linked and the 218C and 779C alleles being linked.

The results of studies examining the association between the TPH gene and antisocial behavior are conflicting. Two studies suggest that the A allele of the A218C/A779C polymorphism is associated with increased antisocial behavior. Manuck et al. (1999) examined association of the A218C polymorphism of the TPH gene with aggression and anger-related traits and found that the A allele was associated with higher aggression, a tendency to experience unprovoked anger, and a tendency to express anger outwardly. The A allele was also associated with attenuated peak prolactin response to fenfluramine, although this relationship was only found in men. Hennig, Reuter, Netter, Burk, and Landt (2005) examined the association between the TPH A779C polymorphism and aggression in a community sample. They examined two different types of aggression assessed by the Buss–Durkee hostility inventory, neurotic hostility and aggressive hostility, and found that the A allele of the A779C polymorphism of the TPH gene was associated with aggressive hostility but not neurotic hostility. In contrast to Hennig et al. and Manuck et al., Staner et al. (2002) found that the C allele of the A218C polymorphism was associated with a higher level of impulsive aggression in a sample of nonpsychotic, impulsive inpatients. Similarly, New et al. (1998) examined the association between the A218C polymorphism and impulsive aggression in a sample of patients with personality disorders and found that the C allele was associated with higher levels of impulsive aggression.

Staner et al. (2002) suggest that the differences in sample composition may account for the discrepancies in these results, as the association with the C allele was found in psychiatric samples (New et al., 1998; Staner et al., 2002) and not in community samples (Manuck et al., 1999; Staner et al., 2002). Hennig et al. (2005) suggest that differences in the conclusions may be due to differences in the type of aggression being examined, as they found evidence for association of the A allele with aggressive hostility, but not with neurotic hostility, which was more similar to the measure of impulsive aggression examined in New et al. (1998).

MAOA. Monoamine oxidase A is a degradative enzyme that catalyzes deamination of serotonin, norepinephrine, and dopamine. Two lines of evidence suggest that the gene coding for MAOA may be related to antisocial behavior. First, deletion of the gene encoding MAOA led to heightened aggression in transgenic mice (Cases et al. 1995). Second, a rare MAOA point mutation resulting in no MAOA enzyme was associated with mild mental retardation and impulsive aggression in males in a large Dutch kindred (Brunner, Nelson, et al., 1993; Brunner, Nelson, Breakefield, Ropers, & van Oost, 1993). Although this mutation is rare and unlikely to predict aggressive behavior in the general population, other common allelic variations of the MAOA gene may be associated with aggression. Several researchers have examined a functional 30-bp variable number of tandem repeat polymorphism in the promoter region of the MAOA gene, for example, in which alleles 2 and 3 (i.e., 3.5 or 4 repeats) have shown greater transcriptional activity than alleles 1 and 4 (i.e., 3 or 5 repeats; Sabol, Hu, & Hamer, 1998).

Manuck, Flory, Ferrell, Mann, and Muldoon (2000) examined the association between the VNTR polymorphism of the MAOA gene and aggression and impulsivity in a male community sample. In contrast to the results expected based on the findings from the Dutch kindred study (Brunner, Nelson, et al., 1993; Brunner, Nelen, et al., 1993) and the transgenic mouse study (Cases et al., 1995), individuals with the 1 and 4 alleles (i.e., with lower transcriptional activity) scored lower on the composite measure of aggression and impulsivity than individuals with the 2 and 3 alleles (i.e., with higher transcriptional activity). Also, individuals in the 1/4 allele group also showed more pronounced central nervous system serotonergic responsivity than individuals in the 2/3 allele group in a neuropsychopharmacologic challenge (i.e., prolactin response to fenfluramine hydrochloride).

Caspi et al. (2002) examined the interaction between the MAOA gene and maltreatment early in life, given evidence that maltreatment in early life alters norepinephrine, serotonin, and dopamine systems (e.g., Bremner & Vermetten, 2001). They hypothesized that maltreated children with a genotype that confers high levels of MAOA expression would be less likely to develop antisocial behavior, and tested this hypothesis in a longitudinal community sample of male adolescents. The main effect of MAOA on antisocial behavior was not significant, although the main effect of maltreatment was significant. The effect of maltreatment on antisocial behavior was much stronger in males with the genotype conferring low MAOA activity than in males with the genotype conferring high MAOA activity. This was a significant finding, as it is one of the first demonstrations of an interaction between a measured gene and a measured environmental influence on a behavioral trait in humans. Caspi et al.'s results are not necessarily in contrast to those of Manuck et al., given that among individuals with no maltreatment experience, those with the genotype conferring low MAOA activity had lower antisocial behavior than those with the genotype conferring high MAOA activity.

Since the publication of Caspi et al.'s results (2002), several researchers have attempted to replicate the MAOA-by-maltreatment interaction on antisocial behavior, with varying results. Huang et al. (2004) found a significant interaction in a psychiatric sample, reporting that the high-activity MAOA allele was associated with lower impulsivity in those reporting early childhood abuse. However, this finding did not extend to measures of aggression or hostility, and similar results were not found in females. In males, Huang et al. found that childhood abuse was more common in individuals with the low-activity MAOA allele, suggesting evidence of evocative gene–environment correlation.

Foley et al. (2004) examined the interaction between the MAOA gene and adverse childhood environment (i.e., parental neglect, exposure to interparental violence, and inconsistent parental discipline) on conduct disorder in a community twin sample. They replicated Caspi et al.'s (2002) findings, reporting that the low MAOA activity allele was a risk factor for conduct disorder only in the presence of adverse childhood environment. Most of the power to detect this interaction came from the individuals with extremely adverse childhood environment. Foley et al. (2004) also examined the evidence for evocative gene–environment correlation, but in contrast to Huang et al. (2004), found that the low MAOA activity allele did not predict exposure to childhood adversity. Nilsson et al. (2005) found support for the initial Caspi et al. (2002) finding in a study examining the interaction between MAOA and a broader construct of psychosocial risk (i.e., dwelling in multi-family housing and experiences with violent victimization). They reported a stronger effect of psychosocial risk on antisocial behavior in boys with the low-activity MAOA genotype than in boys with the high-activity MAOA genotype. Widom and Brzustowicz (2006) attempted to replicate the Caspi et al. (2002) finding in participants in a prospective cohort design study. The interaction between MAOA genotype and abuse/neglect was replicated in Caucasians but not in non-white participants.

Haberstick et al. (2005) also examined the interaction between MAOA and maltreatment on antisocial behavior in a community sample. Despite having adequate power, they were not able to replicate Caspi et al.'s (2002) finding. However, they found a nonsignificant trend in the predicted direction, with individuals with the high-activity MAOA allele having lower antisocial behavior in the maltreated group and individuals with the low-activity MAOA allele having a lower level of antisocial behavior in the non-maltreated group. Huizinga et al. (2006) examined the interaction between MAOA and self-reported physical abuse and violent victimization in participants in the National Youth Survey Family Study. Maltreatment by a parent was significantly related to adolescent and adult antisocial and violence-related behavioral problems, but the main effect of MAOA and the MAOA-by-maltreatment interaction was not significant. Similar results were found for violent victimization. Young et al. (2006) examined the interaction between MAOA and maltreatment in a group of adolescents being treated for significant conduct and substance use problems. They were unable to replicate Caspi et al.'s (2002) findings, reporting that there were no significant differences in the relationship between maltreatment and conduct disorder between the low- and high-activity MAOA genotypes.

Recently, Kim-Cohen et al. (2006) reported replication of the Caspi et al. (2002) findings in children assessed at age 5 and 7. As in Caspi et al., maltreatment had a stronger effect on antisocial behavior in boys with the low MAOA activity genotype than in boys with the high-activity genotype. In addition, Kim-Cohen conducted a meta-analysis of five studies examining the maltreatment by MAOA interaction on antisocial behavior (Caspi et al., 2002; Foley et al., 2004; Haberstick et al., 2005; Kim-Cohen et al., 2006; Nilsson et al., 2005) and found support for the original Caspi et al. finding.

Linkage Studies

There are no linkage studies of conduct disorder examining a general population sample or a clinical sample being treated only for conduct problems. However, two recent linkage-based genome-wide screens have been conducted in samples ascertained for substance use problems to search for loci influencing conduct disorder.

Dick et al. (2004) published the first genome-wide linkage scan of conduct disorder in an adult sample collected as part of the Collaborative Study on Genetics of Alcoholism (COGA). They examined retrospectively reported childhood conduct disorder and conduct disorder symptoms in a sample of alcoholic probands and their first-degree relatives. They conducted nonparametric, multipoint linkage analyses using affected sibling pairs, reporting lod scores equal to or greater than 1.5 for CD diagnosis at chromosome 19p13 at 35 cM (lod score = 2.14), chromosome 2p11 at 136 cM (lod score = 1.65), chromosome 12q13 at 78 cM (lod score = 1.79), and chromosome 3q12 at 134 cM (lod score = 1.60). For CD symptoms as a quantitative trait, suggestive evidence for linkage was found at chromosome 1q32 at 34 cM (lod score = 2.17) and at chromosome 19q12 at 46 cM (lod score = 2.1).

Stallings et al. (2005) examined evidence for linkage for conduct disorder symptoms in a clinical sample being treated for comorbid substance use disorders and antisocial behavioral problems. Using DeFries–Fulker linkage analysis, a regression-based sibling pair linkage method, they found two peaks yielding lod scores greater than 1 for conduct disorder symptoms, one on chromosome 9q34 at 162 cM (lod score = 1.76) and the other on chromosome 17q12 at 54 cM (lod score = 1.26). Unfortunately, there are no overlaps between the suggestive linkage peaks reported in Dick et al. (2004) and Stallings et al. (2005). Although Stallings et al. (2005) found a suggestive peak on chromosome 3q24-3q25 at 173 cM with a lod score of 1.63 for substance dependence vulnerability, a similar peak was not found for conduct disorder symptoms.

Future Directions

Twin and Adoption Studies

The results of Rhee & Waldman's (2002) meta-analysis suggest several future directions for twin and adoption studies of antisocial behavior. First, there may be meaningful distinctions in the operationalizations of antisocial behavior that need further examination, such as violent versus nonviolent crime, criminality versus delinquency, relational versus overt aggression, and instrumental versus reactive aggression. Second, multivariate analyses examining the extent to which the different operationalizations of antisocial behavior have common or specific genetic and environmental influences are needed. An example of such an analysis is Cloninger & Gottesman's (1987) finding that there is little genetic overlap between violent and nonviolent crime. Such results will be very important to consider in the search for specific genes influencing antisocial behavior. Related to this goal is the clarification of the relation between antisocial behavior and possible endophenotypes. Third, the results of the meta-analysis show that there is significant confounding between assessment method and operationalization and age (e.g., criminality always being assessed using official records and childhood externalizing behavior always being assessed using parent report), suggesting that the assessment methods used in future behavior genetic studies of antisocial behavior should be diversified given the possibility that results may reflect the assessment method rather than the operationalization of antisocial behavior being assessed. One such example is Arseneault et al.'s (2003) finding that there were strong genetic influences on antisocial behavior when ratings agreed across mother, teacher, observer, and self-report. Fourth, the power to test genotype × environment interactions may be limited by range restriction in adoption studies. Therefore, future behavior genetic studies should consider alternative research design strategies, such as over-sampling extreme observations (McClelland & Judd, 1993). For example, such studies may over-sample children with a low genetic predisposition to antisocial behavior who are reared in environments that predispose them to antisocial behavior. Fifth, studies examining specific shared and nonshared environmental influences while controlling for genetic influences are needed.

Association and Linkage Studies

In comparison to other comorbid disorders such as ADHD and alcoholism, there are few association and linkage studies of conduct disorder and antisocial behavior. So far, there is little consensus in the findings of these studies. There are several possible reasons for conflicting results and failures to replicate, including small sample sizes, differences in the types of samples employed, differences in the range of phenotypes examined, differences in analytical approaches, and the possibility of population stratification. Future directions for association studies include the need to increase our understanding regarding how the various operationalizations of antisocial behavior are related to each other and apply this knowledge to the search for specific genes influencing antisocial behavior. For example, a highly heritable common latent phenotype or endophenotype for antisocial behavior may be a better phenotype than conduct disorder symptoms, which generally show moderate shared environmental influences. The differences across studies suggest a need for increased collaboration among researchers investigating the genetics of antisocial behavior. Such collaboration among researchers examining attention-deficit/hyperactivity disorder has been productive (e.g., Kent, 2004). There are many possible candidate genes for antisocial behavior that have yet to be examined, and technological advances will allow more intensive, systematic studies involving dense mapping or sequencing of specific candidate genes and whole genome association studies.

Knowledge regarding the neuro-cognitive dysfunctions related to antisocial behavior from neuropsychological and functional imaging studies has been increasing. For example, Blair et al. (2006) recently reviewed the evidence for two neural systems implicated in psychopathy, one involving the amygdala and the other involving the orbital/ventrolateral frontal cortex. As the evidence regarding specific genetic effects on antisocial behavior becomes clearer, the application of imaging genomics (i.e., combining neuroimaging and genetic analysis to examine genes' effects on brain information processing) in the etiology of antisocial behavior will become an important future direction.

The first evidence of a specific gene × measured environment interaction was found for antisocial behavior 4 years ago (Caspi et al., 2002), resulting in considerable interest in measured gene × environment interactions. Given the large number of possible candidate genes and environmental risk factors for antisocial behavior, the possibility of type I errors and spurious findings should be considered carefully. Recently, Moffitt, Caspi, and Rutter (2005) have encouraged careful measured gene × environment interaction hypothesis testing and have emphasized the importance of specifying a priori theoretically plausible triads of a gene, an environmental pathogen, and a behavioral phenotype. They described seven strategic steps needed to organize future hypothesis-driven studies of measured gene × environment interaction, which include (1) consulting the quantitative behavioral-genetic literature, (2) identifying a candidate environmental pathogen for the disorder in question, (3) optimizing environmental risk measurement, (4) identifying candidate susceptibility genes, (5) testing for an interaction, (6) evaluating whether a measured gene × environment interaction extends beyond the initially hypothesized triad of gene, environmental pathogen, and disorder, and (7) replication and meta-analysis.

Acknowledgment This work was supported in part by NIDA DA-13956 and NIMH MH-01818.

Note: The asterisk denotes a study that was considered for inclusion in the meta-analysis.

References

Achenbach, T. M., & Edelbrock, C. S. (1983). *Manual for the Child Behavior Checklist and Revised Behavior Profile*. Burlington, VT: University of Vermont.

American Psychiatric Association. (1994). *Diagnostic and statistical manual of mental disorders* (4th ed.). Washington, DC: American Psychiatric Association.

Angold, A, & Costello, E. J. (1993). Depressive comorbidity in children and adolescents: Empirical, theoretical and methodological issues. *American Journal of Psychiatry, 150,* 1779–1791.

Arseneault, L., Moffitt, T. E., Caspi, A., Taylor, A., Rijsdijk, F. V., Jaffee, S. R., et al. (2003). Strong genetic effects on cross-situational antisocial behaviour among 5-year-old children according to mothers, teachers, examiner-observers, and twins' self-reports. *Journal of Child Psychology and Psychiatry, 44,* 832–848.

Baca-Garcia, E., Vaquero, C., Diaz-Sastre, C., García-Resa, E., Saiz-Ruiz, J., Fernández-Piqueras, J., et al. (2004). Lack of association between the serotonin transporter promoter gene polymorphism and impulsivity or aggressive behavior among suicide attempters and healthy volunteers. *Psychiatry Research, 126,* 99–106.

*Baker, L., Mack, W., Moffitt, T., & Mednick, S. (1989). Sex differences in property crime in a Danish adoption cohort. *Behavior Genetics, 19,* 355–370.

Beitchman, J. H., Baldassarra, L., Mik, H., De Luca, V., King, N., Bender, D., et al. (2006). Serotonin transporter polymorphisms and persistent, pervasive childhood aggression. *American Journal of Psychiatry, 163,* 1103–1105.

Biederman, J., Newcorn, J., & Sprich, S. (1991). Comorbidity of attention deficit hyperactivity disorder with conduct, depressive, anxiety, and other disorders. *American Journal of Psychiatry, 148,* 564–577.

Blair, R. J. R., Peschardt, K. S., Budhani, S., Mitchell, D. G. V., & Pine, D. S. (2006). The development of psychopathy. *Journal of Child Psychology and Psychiatry, 47,* 262–275.

*Blanchard, J. M., Vernon, P. A., & Harris, J. A. (1995). A behavior genetic investigation of multiple dimensions of aggression. *Behavior Genetics, 25,* 256.

Blonigen, D. M., Carlson, S. R., Krueger, R. F., & Patrick, C. J. (2003). A twin study of self-reported psychopathic personality traits. *Personality and Individual Differences, 35,* 179–197.

Blonigen, D. M., Hicks, B. M., Krueger, R. F., Patrick, C. J., & Iacono, W. G. (2005). Psychopathic personality traits: Heritability and genetic overlap with internalizing and externalizing psychopathology. *Psychological Medicine, 35,* 637–648.

*Bohman, M. (1978). Some genetic aspects of alcoholism and criminality: A population of adoptees. *Archives of General Psychiatry, 35,* 269–276.

*Brandon, K, & Rose, R. J. (1995). A multivariate twin family study of the genetic and environmental structure of personality, beliefs, and alcohol use. *Behavior Genetics, 25,* 257.

Bremner, J. D., & Vermetten, E. (2001). Stress and development: behavioral and biological consequences. *Development and Psychopathology, 13,* 473–489.

Brunner, H. G., Nelson, M., Breakefield, X. O., Ropers, H. H., & van Oost, B. A. (1993). Abnormal behavior associated with a point mutation in the structural gene for monoamine oxidase. *Science, 262,* 578–580.

Brunner, H. G., Nelen, M. R., van Zandvoort, P., Abeling, N. G., van Gennip, A. H., Wolters, E. C., et al. (1993). X-linked borderline mental retardation with prominent behavioral disturbance: phenotype, genetic localization, and evidence for disturbed monoamine metabolism. *American Journal of Human Genetics, 52,* 1032–1039.

Burt, S. A., Krueger, R. F., McGue, M., & Iacono, W. G. (2001). Sources of covariation among attention-deficit/hyperactivity disorder, oppositional defiant disorder, and conduct disorder: The importance of shared environment. *Journal of Abnormal Psychology, 110,* 516–525.

Button, T. M. M., Hewitt, J. K., Rhee, S. H., Young, S. E., Corley, R. P., & Stallings, M. C. (2006). Examination of the causes of covariation between conduct disorder symptoms and vulnerability to drug dependence. *Twin Research and Human Genetics, 9,* 38–45.

Button, T. M. M., Scourfield, J., Martin, N., Purcell, S., & McGuffin, P. (2005). Family dysfunction interacts with genes in the causation of antisocial symptoms. *Behavior Genetics, 35,* 115–120.

Cadoret, R. J., Cain, C. A., & Crowe, R. R. (1983). Evidence for gene-environment interaction in the development of adolescent antisocial behavior. *Behavior Genetics, 13,* 301–310.

Cadoret, R. J., Langbehn, D., Caspers, K., Troughton, E. P., Yucuis, R., Sandhu, H. K., et al. (2003). Associations of the serotonin transporter promoter polymorphism with aggressivity, attention deficit, and conduct disorder in an adoptee population. *Comprehensive Psychiatry, 44,* 88–101.

*Cadoret, R. J., Troughton, E., O'Gorman, T. W., & Heywood, E. (1986). An adoption study of genetic and environmental factors in drug abuse. *Archives of General Psychiatry, 43,* 1131–1136.

*Cadoret, R. J., Yates, W. R., Troughton, E., Woodworth, G., & Stewart, M. A. (1995). Adoption study demonstrating two genetic pathways to drug abuse. *Archives of General Psychiatry, 52,* 42–52.

*Cadoret, R. J., Yates, W. R., Troughton, E., Woodworth, G., & Stewrat, M. A. (1996). An adoption study of drug abuse/dependency in females. *Comprehensive Psychiatry, 37,* 88–94.

*Carey, G. (1992). Twin imitation for antisocial behavior: Implications for genetic and family environment research. *Journal of Abnormal Psychology, 101,* 18–25.

Cases, O., Seif, I., Grimsby, J., Gaspar, P., Chen, K., Pournin, S., et al. (1995). Aggressive behavior and altered amounts of brain serotonin and norepinephrine in mice lacking MAOA. *Science, 268,* 1763–1766.

Caspi, A., McClay, J., Moffitt, T. E., Mill, J., Martin, J., Craig, I. W., et al. (2002). Role of genotype in the cycle of violence in maltreated children. *Science, 297,* 851–854.

*Cates, D. S., Houston, B. K., Vavak, C. R., Crawford, M. H., & Uttley, M. (1993). Heritability of hostility-related emotions, attitudes, and behaviors. *Journal of Behavioral Medicine, 16,* 237–256.

*Centerwall, B. S., & Robinette, C. D. (1989). Twin concordance for dishonorable discharge from the military: With a review of genetics of antisocial behavior. *Comprehensive Psychiatry, 30,* 442–446.

Cloninger, C. R, & Gottesman, I. I. (1987). Genetic and environmental factors in antisocial behavior disorders. In S. A. Mednick, T. E. Moffitt, & S. A. Stack (Eds.), *The causes of crime: New biological approaches* (pp. 92–109). New York: Cambridge University Press.

Cloninger, C. R., Sigvardsson, S., Bohman, M., & von Knorring, A. -L. (1982). Predisposition to petty criminality in Swedish adoptees. II.

Cross-fostering analysis of gene-environment interaction. *Archives of General Psychiatry, 39*, 1242–1247.

Coccaro, E. F., Kavoussi, R. J., Cooper, T. B., & Hauger, R. L. (1997). Central serotonin activity and aggression: inverse relationship with prolactin response to D-fenfluramine, but not CSF 5-HIAA concentration, in human subjects. *American Journal of Psychiatry, 154*, 1430–1435.

Coccaro, E. F., Siever, L. J., Klar, H., Rubenstein, K., Benjamin, E., & Davis, K. L. (1987). Diminished prolactin responses to repeated fenfluramine challenge in man. *Psychiatry Research, 22*, 257–259.

*Coid, B., Lewis, S. W., & Reveley, A. M. (1993). A twin study of psychosis and criminality. *British Journal of Psychiatry, 162*, 87–92.

Cook, E. H., Stein, M. A., Krasowski, M. D., Cox, N. J., Olkon, D. M., Kieffer, J. E., et al. (1995). Association of attention-deficit disorder and the dopamine transporter gene. *American Journal of Human Genetics, 56*, 993–998.

*Cunningham, L., Cadoret, R. J., Loftus, R., & Edwards, J. E. (1975). Studies of adoptees from psychiatrically disturbed biological parents: Psychiatric conditions in childhood and adolescence. *British Journal of Psychiatry, 126*, 534–549.

*Deater-Deckard, K., Reiss, D., Hetherington, E. M., & Plomin, R. (1997). Dimensions and disorders of adolescent adjustment: A quantitative genetic analysis of unselected samples and selected extremes. *Journal of Child Psychology and Psychiatry, 38*, 515–525.

Dick, D. M., Li, T.-K., Edenberg, H. J., Hesselbrock, V., Kramer, J., Kuperman, S., et al. (2004). A genome-wide screen for genes influencing conduct disorder. *Molecular Psychiatry, 9*, 81–86.

Dick, D. M., Viken, R. J., Kaprio, J., Pulkkinen, L., & Rose, R. J. (2005). Understanding the covariation among childhood externalizing symptoms: Genetic and environmental influences on conduct disorder, attention deficit hyperactivity disorder, and oppositional defiant disorder symptoms. *Journal of Abnormal Child Psychology, 33*, 219–229.

*DiLalla, D. L., Carey, G., Gottesman, I. I., & Bouchard, T. J. (1996). Heritability of MMPI personality indicators of psychopathology in twins reared apart. *Journal of Abnormal Psychology, 105*, 491–499.

DiLalla, L. F., & Gottesman, I. I. (1989). Heterogeneity of causes for delinquency and criminality: Lifespan perspectives. *Development and Psychopathology, 1*, 339–349.

Dougherty, D., Bonab, A. A., Spencer, T. J., Rauch, S. L., Madras, B. K., & Fischman, A. J. (1999). Dopamine transporter density in patients with attention deficit hyperactivity disorder. *Lancet, 354*, 2132–2133.

*Eaves, L. J., Silberg, J. L., Meyer, J. M., Maes, H. H., Simonoff, E., Pickles, A., et al. (1997). Genetics and developmental psychopathology: 2. The main effects of genes and environment on behavioral problems in the Virginia Twin Study of Adolescent Behavioral Development. *Journal of Child Psychology and Psychiatry, 38*, 965–980.

*Edelbrock, C., Rende, R., Plomin, R., & Thompson, L. A. (1995). A twin study of competence and problem behavior in childhood and early adolescence. *Journal of Child Psychology and Psychiatry, 36*, 775–785.

*Eley, T. C., Lichtenstein, P., & Stevenson, J. (1999). Sex differences in the etiology of aggressive and nonaggressive antisocial behavior: Results from two twin studies. *Child Development, 70*, 155–168.

Faraone, S. V., Perlis, R. H., Doyle, A. E., Smoller, J. W., Goralnick, J. J., Holmgren, M. A., et al. (2005). Molecular genetics of attention-deficit/hyperactivity disorder. *Biological Psychiatry, 57*, 1313–1323.

*Finkel, D., & McGue, M. (1997). Sex differences and nonadditivity in heritability of the multidimensional personality questionnaire scales. *Journal of Personality and Social Psychology, 72*, 929–938.

Foley, D. L., Eaves, L. J., Wormley, B., Silberg, J. L., Maes, H. H., Kuhn, J., et al. (2004). Childhood adversity, monoamine oxidase A genotype, and risk for conduct disorder. *Archives of General Psychiatry, 61*, 738–744.

Fulker, D. W., Cherny, S. S., Sham, P. C., & Hewitt, J. K. (1999). Combined linkage and association sib-pair analysis for quantitative traits. *American Journal of Human Genetics, 64*, 259–267.

Gelernter, J., Kranzler, H. R., Satel, S., & Rao, P. A. (1994). Genetic association between dopamine transporter protein alleles and cocaine-induced paranoid. *Neuropsychopharmacology, 11*, 195–200.

*Ghodsian-Carpey, J., & Baker, L. A. (1987). Genetic and environmental influences on aggression in 4- to 7-year-old twins. *Aggressive Behavior, 13*, 173–186.

Giros, B., Jaber, M., Jones, S. R., Wightman, R. M., & Caron, M. G. (1996). Hyperlocomotion and indifference to cocaine and amphetamine in mice lacking the dopamine transporter. *Nature, 379*, 606–612.

Gjone, H., & Stevenson, J. (1997). The association between internalizing and externalizing behavior in childhood and early adolescence: Genetic or environmental common influences? *Journal of Abnormal Child Psychology, 25*, 277–286.

*Gottesman, I. I. (1963). Heritability of personality: A demonstration. *Psychological Monographs, 77*(9, Whole No. 572).

*Gottesman, I. I. (1965). Personality and natural selection. In S. G. Vandenberg (Ed.), *Methods and goals in human behavior genetics* (pp. 63–80). New York: Academic Press.

Gough, H. G., & Heilbrun, A. B. (1972). *The adjective checklist manual.* Palo Alto, CA: Consulting Psychologists Press.

Grove, W. M., Eckert, E. D., Heston, L., Bouchard, T. J., Segal, N., & Lykken, D. T. (1990). Heritability of substance abuse and antisocial behavior: A study of monozygotic twins reared apart. *Biological Psychiatry, 27*, 1293–1304.

*Gustavsson, J. P., Pedersen, N. L., Åsberg, M., & Schalling, D. (1996). Exploration into the sources of individual differences in aggresion-, hostility-, and anger-related (AHA) personality traits. *Personality and Individual Differences, 21*, 1067–1071.

Haberstick, B. C., Lessem, J. M., Hopfer, C. J., Smolen, A., Ehringer, M. A., Timberlake, D., et al. (2005). Monoamine oxidase A (MAOA) and antisocial behaviors in the presence of childhood and adolescent maltreatment. *American Journal of Medical Genetics, 135*, 59–64.

Haberstick, B. C., Smolen, A., & Hewitt, J. K. (2006). Family-based association test of the 5HTTLPR and aggressive behavior in a general population sample of children. *Biological Psychiatry, 59*, 836–843.

Heils, A., Teufel, A., Petri, S., Stöber, G., Riederer, P., Bengel, D., et al. (1996). Allelic variation of human serotonin transporter gene expression. *Journal of Neurochemistry, 66*, 2621–2624.

Heinz, A., Jones, D. W., Mazzanti, C., Goldman, D., Ragan, P., Hommer, D., et al. (2000). A relationship between serotonin transporter genotype and in vivo protein expression and alcohol neurotoxicity. *Biological Psychiatry, 47*, 643–649.

Hennig, J., Reuter, M., Netter, P., Burk, C., & Landt, O. (2005). Two types of aggression are differentially related to serotonergic activity and the A779C TPH polymorphism. *Behavioral Neuroscience, 119*, 16–25.

Huang, Y., Cate, S. P., Battistuzi, C., Oquendo, M. A., Brent, D., & Mann, J. J. (2004). An association between a functional polymorphism in the monoamine oxidase A gene promoter, impulsive traits and early abuse experiences. *Neuropsychopharmacology, 29*, 1498–1505.

Huizinga, D., Haberstick, B. C., Smolen, A., Menard, S., Young, S. E., Corley, R. P., et al. (2006). Childhood maltreatment, subsequent antisocial behavior, and the role of monoamine oxidase A genotype. *Biological Psychiatry, 60*, 677–683.

Hyde, J. S. (1984). How large are gender differences in aggression? A developmental meta-analysis. *Developmental Psychology, 20,* 722–736.

Jacobson, K. C., Neale, M. C., Prescott, C. A., & Kendler, K. S. (2001). Behavioral genetic confirmation of a life-course perspective on antisocial behavior: Can we believe the results? *Behavior Genetics, 31,* 456.

Jaffee, S. R., Caspi, A., Moffitt, T. E., Dodge, K. A., Rutter, M., Taylor, A., et al. (2005). Nature×nurture: Genetic vulnerabilities interact with physical maltreatment to promote conduct problems. *Development and Psychopathology, 17,* 67–84.

Jorm, A. F., Prior, M., Sanson, A., Smart, D., Zhang, Y., & Easteal, S. (2001). Association of a polymorphism of the dopamine transporter gene with externalizing behavior problems and associated temperament traits: A longitudinal study from infancy to the mid-teens. *American Journal of Medical Genetics, 105,* 346–350.

Kendler, K. S., Prescott, C. A., Myers, J., & Neale, M. C. (2003). The structure of genetic and environmental risk factors for common psychiatric and substance use disorders in men and women. *Archives of General Psychiatry, 60,* 929–937.

Kent, L. (2004). Recent advances in the genetics of attention deficit hyperactivity disorder. *Current Psychiatry Reports, 6,* 143–148.

Kim-Cohen, J., Caspi, A., Taylor, A., Williams, B., Newcombe, R., Craig, I. W., et al. (2006). MAOA, maltreatment, and gene-environment interaction predicting children's mental health: New evidence and a meta-analysis. *Molecular Psychiatry, 11,* 903–913.

Kruesi, M. J., Rapoport, J. L., Hamburger, S., Hibbs, E., Potter, W., Z., Lenane, M., et al. (1990). Cerebrospinal fluid monoamine metabolites, aggression, and impulsivity in disruptive behavior disorders of children and adolescents. *Archives of General Psychiatry, 55,* 989–994.

Kunugi, H., Ishida, S., Kato, T., Sakai, S., Tatsumi, M., Hirose, T., et al. (1999). No evidence for an association of polymorphisms of the tryptophan hydroxylase gene with affective disorders or attempted suicide among Japanese patients. *American Journal of Psychiatry, 156,* 774–776.

Larsson, H., Andershed, H., & Lichtenstein, P. (2006). A genetic factor explains most of the variation in the psychopathic personality. *Journal of Abnormal Psychology, 115,* 221–230.

Lidberg, L., Tuck, J. R., Asberg, M., Scalia-Tomba, G. P., & Bertilsson, L. (1985). Homicide, suicide and CSF 5-HIAA. *Acta Psychiatrica Scandinavica, 71,* 230–236.

*Livesley, W. J., Jang, K. L., Jackson, D. N., & Vernon, P. A. (1993). Genetic and environmental contributions to dimensions of personality disorder. *American Journal of Psychiatry, 150,* 1826–1831.

*Loehlin, J. C., & Nichols, R. C. (1976). *Heredity, environment, and personality.* Austin, TX: University of Texas Press.

*Loehlin, J. C., Willerman, L., & Horn, J. M. (1987). Personality resemblance in adoptive families: A 10-year follow-up. *Journal of Personality and Social Psychology, 53,* 961–969.

Lykken, D. T. (1997). Incompetent parenting: its causes and cures. *Child Psychiatry and Human Development, 27,* 129–137.

*Lyons, M. J., True, W. R., Eisen, S. A., Goldberg, J., Meyer, J. M., Faraone, S. V., et al. (1995). Differential heritability of adult and juvenile antisocial traits. *Archives of General Psychiatry, 52,* 906–915.

*Lytton, H., Watts, D., & Dunn, B. E. (1988). Stability of genetic determination from age 2 to age 9: A longitudinal twin study. *Social Biology, 35,* 62–73.

Manuck, S. B., Flory, J. D., Ferrell, R. E., Dent, K. M., Mann, J. J., & Muldoon, M. F. (1999). Aggression and anger-related traits associated with a polymorphism of the tryptophan hydroxylase gene. *Biological Psychiatry, 45,* 603–614.

Manuck, S. B., Flory, J. D., Ferrell, R. E., Mann, J. J., & Muldoon, M. F. (2000). A regulatory polymorphism of the monoamine oxidase-

A gene may be associated with variability in aggression, impulsivity, and central nervous system serotonergic responsivity. *Psychiatry Research, 95,* 9–23.

McCartney, K., Harris, M. J., & Bernieri, F. (1990). Growing up and growing apart: A developmental meta-analysis of twin studies. *Psychological Bulletin, 107,* 226–237.

McClelland, G. H, & Judd, C. M. (1993). Statistical difficulties of detecting interactions and moderator effects. *Psychological Bulletin, 114,* 376–390.

*McGue, M., Sharma, A., & Benson, P. (1996). The effect of common rearing on adolescent adjustment: Evidence from a U.S. adoption cohort. *Developmental Psychology, 32,* 604–613.

Mednick, S. A., Gabrielli, W. F, & Hutchings, B. (1983). Genetic influences in criminal behavior: Evidence from an adoption cohort. In K. T. Van Dusen & S. A. Mednick (Eds.), *Prospective studies of crime and delinquency* (pp. 39–56). Boston, MA: Kluwer-Nijhof.

*Meininger, J. C., Hayman, L. L., Coates, P. M, & Gallagher, P. (1988). Genetics or environment? Type A behavior and cardiovascular risk factors in twin children. *Nursing Research, 37,* 341–346.

Miles, D. R., van den Bree, M. B., & Pickens, R. W. (2002). Sex differences in shared genetic and environmental influences between conduct disorder symptoms and marijuana use in adolescents. *American Journal of Medical Genetics, 114,* 159–168.

Mill, J., Xu, X., Ronald, A., Curran, S., Price, T., Knight, J., et al. (2005). Quantitative trait locus analysis of candidate gene alleles associated with attention deficit hyperactivity disorder (ADHD) in five genes: DRD4, DAT1, DRD5, SNAP-25, and 5HT1B. *American Journal of Medical Genetics, 133,* 68–73.

Moffitt, T. E. (1993). Adolescence-limited and life-course-persistent antisocial behavior: A developmental taxonomy. *Psychological Review, 100,* 674–701.

Moffitt, T. E., Caspi, A., & Rutter, M. (2005). Strategy for investigating interactions between measured genes and measured environments. *Archives of General Psychiatry, 62,* 473–481.

Nadder, T. S., Rutter, M., Silberg, J. L., Maes, H. H., & Eaves, L. J. (2002). Genetic effects on the variation and covariation of attention deficit-hyperactivity disorder (ADHD) and oppositional-defiant disorder/conduct disorder (Odd/CD) symptomatologies across informant and occasion of measurement. *Psychological Medicine, 32,* 39–53.

Nadder, T. S., Silberg, J. L., Eaves, L. J., Maes, H. H., & Meyer, J. M. (1998). Genetic effects on ADHD symptomatology in 7- to 13-year-old twins: Results from a telephone survey. *Behavior Genetics, 28,* 83–99.

*Nathawat, S. S., & Puri, P. (1995). A comparative study of MZ and DZ twins on level I and level II mental abilities and personality. *Journal of the Indian Academy of Applied Psychology, 21,* 87–92.

New, A. S., Gelernter, J., Goodman, M., Mitropoulou, V., Koenigsberg, H., Silverman, J., et al. (2001). Suicide, impulsive aggression, and HTR1B genotype. *Biological Psychiatry, 50,* 62–65.

New, A. S., Gelernter, J., Yovell, Y., Trestman, R. L., Nielsen, D. A., Silvermann, J., et al. (1998). Tryptophan hydroxylase genotype is associated with impulsive-aggression measures. A preliminary report. *American Journal of Medical Genetics, 81,* 13–17.

Nilsson, K. W., Sjöberg, R. L., Damberg, M., Leppert, J., Öhrvik, J., Alm, P. O., et al. (2005). Role of monoamine oxidase A genotype and psychosocial factors in male adolescent criminal activity. *Biological Psychiatry, 59,* 121–127.

*O'Connor, T. G., McGuire, S., Reiss, D., Hetherington, E. M., & Plomin, R. (1998). Co-occurrence of depressive symptoms and antisocial behavior in adolescence: A common genetic liability. *Journal of Abnormal Psychology, 107,* 27–37.

*Owen, D., & Sines, J. O. (1970). Heritability of personality in children. *Behavior Genetics, 1,* 235–248.

Plomin, R. (1981). Heredity and temperament: A comparison of twin data for self-report questionnaires, parental ratings, and objectively assessed behavior. In L. Gedda, P. Parisi, & W. E. Nance (Eds.), *Twin research 3: Intelligence, personality, and development* (pp. 269–278). New York: Alan R. Liss, Inc.

Plomin, R., DeFries, J. C., McClearn, G. E., & McGuffin, P. (2001). *Behavioral genetics* (4th ed.). New York: Worth Publishers.

*Plomin, R., Foch, T. T., & Rowe, D. C. (1981). Bobo clown aggression in childhood: Environment, not genes. *Journal of Research in Personality, 15*, 331–342.

Plomin, R., Nitz, K., & Rowe, D. C. (1990). Behavioral genetics and aggressive behavior in childhood. In M. Lewis & S. M. Miller (Eds.), *Handbook of developmental psychopathology* (pp. 119–133). New York: Plenum Press.

*Rahe, R. H., Hervig, L., & Rosenman, R. H. (1978). Heritability of type A behavior. *Psychosomatic Medicine, 40*, 478–486.

Reebye, P., Moretti, M. M., & Lessard, J. C. (1995). Conduct disorder and substance use disorders: Comorbidity in a clinical sample of preadolescents and adolescents. *Canadian Journal of Psychiatry, 40*, 313–319.

Reist, C., Mazzanti, C., Vu, R., Tran, D., & Goldman, D. (2001). Serotonin transporter promoter polymorphism is associated with attenuated prolactin response to fenfluramine. *American Journal of Medical Genetics, 105*, 363–368.

Retz, W., Retz-Junginger, P., Supprian, T., Thome, J., & Rösler, M. (2004). Association of serotonin transporter promoter gene polymorphism with violence: Relation with personality disorders, impulsivity, and childhood ADHD psychopathology. *Behavioral Sciences and the Law, 22*, 415–425.

Rhee, S. H., & Waldman, I. D. (2002). Genetic and environmental influences on antisocial behavior: A meta-analysis of twin and adoption studies. *Psychological Bulletin, 128*, 490–529.

Robins, L. N., & Regier, D. A. (1991). *Psychiatric disorders in America*. New York: The Free Press.

*Rowe, D. C. (1983). Biometrical genetic models of self-reported delinquent behavior: A twin study. *Behavior Genetics, 13*, 473–489.

Rowe, D. C., Almeida, D. M., & Jacobson, K. C. (1999). School context and genetic influences on aggression in adolescence. *Psychological Science, 10*, 277–280.

*Rushton, J. P., Fulker, D. W., Neale, M. C., Nias, D. K. B, & Eysenck, H. J. (1986). Altruism and aggression: The heritability of individual differences. *Journal of Personality and Social Psychology, 50*, 1192–1198.

Sabol, S. Z., Hu, S., & Hamer, D. (1998). A functional polymorphism in the monoamine oxidase A gene promoter. *Human Genetics, 103*, 273–279.

Sakai, J. T., Young, S. E., Stallings, M. C., Timberlake, D., Smolen, A., Stetler, G. L., et al. (2006). Case-control and within-family tests for an association between conduct disorder and 5HTTLPR. *American Journal of Medical Genetics, 141*, 825–832.

Sander, T., Harms, H., Podschus, J., Finckh, U., Nickel, B., Rolfs, A., et al. (1997). Alellic association of a dopamine transporter gene polymorphism in alcohol dependence with withdrawal seizures or delirium. *Biological Psychiatry, 41*, 299–304.

Saudou, F., Amara, D. A., Dierich, A., LeMeur, M., Ramboz, S., Segu, L., et al. (1994). Enhanced aggressive behavior in mice lacking 5-HT1B receptor. *Science, 265*, 1875–1878.

*Scarr, S. (1966). Genetic factors in activity motivation. *Child Development, 37*, 663–673.

Schmidt, L. G., Harms, H., Kuhn, S., Rommelspacher, H., & Sander, T. (1998). Modification of alcohol withdrawal by the A9 allele of the dopamine transporter gene. *American Journal of Psychiatry, 155*, 474–478.

*Schmitz, S., Fulker, D. W., & Mrazek, D. A. (1995). Problem behavior in early and middle childhood: An initial behavior genetic analysis. *Journal of Child Psychology and Psychiatry, 36*, 1443–1458.

*Seelig, K. J., & Brandon, K. O. (1997). Rater differences in gene-environment contributions to adolescent problem behavior. *Behavior Genetics, 27*, 605.

Silberg, J., Rutter, M., D'Onofrio, B., & Eaves, L. (2003). Genetic and environmental risk factors in adolescent substance use. *Journal of Child Psychology and Psychiatry, 44*, 664–676.

Silberg, J., Rutter, M., Meyer, J., Maes, H., Hewitt, J., Simonoff, E., et al. (1996). Genetic and environmental influences on the covariation between hyperactivity and conduct disturbance in juvenile twins. *Journal of Child Psychology and Psychiatry, 37*, 803–816.

*Slutske, W. S., Heath, A. C., Dinwiddie, S. H., Madden, P. A. F., Bucholz, K. K., Dunne, M. P., et al. (1997). Modeling genetic and environmental influences in the etiology of conduct disorder: A study of 2,682 adult twin pairs. *Journal of Abnormal Psychology, 106*, 266–279.

Slutske, W. S., Heath, A. C., Dinwiddie, S. H., Madden, P. A., Bucholz, K. K., Dunne, M. P., et al. (1998). Common genetic risk factors for conduct disorder and alcohol dependence. *Journal of Abnormal Psychology, 107*, 363–374.

Soyka, M., Preuss, U. W., Koller, G., Zill, P., & Bondy, B. (2004). Association of 5-HT1B receptor gene and antisocial behavior in alcoholism. *Journal of Neural Transmission, 111*, 101–109.

Stallings, M. C., Corley, R. P., Dennehey, B., Hewitt, J. K., Krauter, K. S., Lessem, J. M., et al. (2005). A genome-wide search for quantitative trait loci influencing antisocial drug dependence in adolescence. *Archives of General Psychiatry, 62*, 1042–1051.

Staner, L., Uyanik, G., Correa, H., Tremeau, F., Monreal, J., Crocq, M.-A., et al. (2002). A dimensional impulsive-aggressive phenotype is associated with the A218C polymorphism of the tryptophan hydroxylase gene: A pilot study in well-characterized impulsive inpatients. *American Journal of Medical Genetics, 114*, 553–557.

*Stevenson, J, & Graham, P. (1988). Behavioral deviance in 13-year-old twins: An item analysis. *Journal of the American Academy of Child and Adolescent Psychiatry, 27*, 791–797.

Strous, R. D., Nolan, K. A., Lapidus, R., Diaz, L., Saito, T., & Lachman, H. M. (2003). Aggressive behavior in schizophrenia is associated with the low enzyme activity COMT polymorphism: A replication study. *American Journal of Medical Genetics, 120*, 29–34.

Taylor, J., Iacono, W. G., & McGue, M. (2000). Evidence for a genetic etiology of early-onset delinquency. *Journal of Abnormal Psychology, 109*, 634–643.

Taylor, J., Loney, B. R., Bobadilla, L., Iacono, W. G., & McGue, M. (2003). Genetic and environmental influences on psychopathy trait dimensions in a community sample of male twins. *Journal of Abnormal Child Psychology, 31*, 633–645.

*Taylor, J., McGue, M., Iacono, W. G., & Lykken, D. T. (2000). A behavioral genetic analysis of the relationship between the socialization scale and self-reported delinquency. *Journal of Personality, 68*, 29–50.

*Tellegen, A., Lykken, D. T., Bouchard, T. J., Wilcox, K., Segal, N., & Rich, S. (1988). Personality similarity in twins reared apart and together. *Journal of Personality and Social Psychology, 54*, 1031–1039.

Thapar, A., Harrington, R., & McGuffin, P. (2001). Examining the comorbidity of ADHD-related behaviours and conduct problems using a twin study design. *British Journal of Psychiatry, 179*, 224–229.

Thapar, A., Langley, K., Fowler, T., Rice, F., Turic, D., Whittinger, N., et al. (2005). Catechol *o*-methyltransferase gene variant and birth weight predict early-onset antisocial behavior in children with attention-deficit/hyperactivity disorder. *Archives of General Psychiatry, 62*, 1275–1278.

*Thapar, A., & McGuffin, P. (1996). A twin study of antisocial and neurotic symptoms in childhood. *Psychological Medicine, 26*, 1111–1118.

*Torgersen, S. Skre, I., Onstad, S., Edvardsen, J., & Kringlen, E. (1993). The psychometric-genetic structure of DSM-III-R personality disorder criteria. *Journal of Personality Disorders, 7*, 196–213.

*van den Oord, E. J. C. G., Boomsma, D. I., & Verhulst, F. C. (1994). A study of problem behaviors in 10- to 15-year-old biologically related and unrelated international adoptees. *Behavior Genetics, 24,* 193–205.

*van den Oord, E. J. C. G., Verhulst, F. C., & Boomsma, D. I. (1996). A genetic study of maternal and paternal ratings of problem behaviors in 3-year-old twins. *Journal of Abnormal Psychology, 105,* 349–357.

Viding, E., Blair, R. J. R., Moffitt, T. E., & Plomin, R. (2005). Evidence for substantial genetic risk for psychopathy in 7-year-olds. *Journal of Child Psychology and Psychiatry, 46,* 592–597.

*Waldman, I. D., Levy, F., & Hay, D. A. (1995). Multivariate genetic analyses of the overlap among DSM-III-R disruptive behavior disorder symptoms. *Behavior Genetics, 25,* 293–294.

*Waldman, I. D., Pickens, R. W., & Svikis, D. S. (1989). Sex differences in genetic and environmental components of childhood conduct problems. *Behavior Genetics, 19,* 779–780.

Waldman, I. D, & Rhee, S. H. (2006). Genetic and environmental influences on psychopathy and antisocial behavior. In C. Patrick (Ed.), *Handbook of psychopathy* (pp. 205–228). New York: The Guilford Press.

Waldman, I. D., Rhee, S. H., & Levy, F. (2001). Causes of overlap among symptoms of ADHD, oppositional defiant disorder, and conduct disorder. In F. Levy & D. A. Hay (Eds.), *Attention, genes, and ADHD* (pp. 115–138). New York: Brunner-Routledge.

Waldman, I. D., Rowe, D. C., Abramowitz, A., Kozel, S. T., Mohr, J. H., Sherman, S. L., et al. (1998). Association and linkage of the dopamine transporter gene and attention-deficit hyperactivity disorder in children: heterogeneity owing to diagnostic subtype and severity. *American Journal of Human Genetics, 63,* 1767–1776.

Widom, C. S, & Brzustowicz, L. M. (2006). MAOA and the "cycle of violence": childhood abuse and neglect, MAOA genotype, and risk for violent and antisocial behavior. *Biological Psychiatry, 60,* 684–689.

*Willcutt, E. G., Shyu, V., Green, P., & Pennington, B. F. (1995, April). *A twin study of the comorbidity of the disruptive behavior disorders of childhood.* Paper presented at the annual meeting of the Society for Research in Child Development. Indianapolis, IN.

Wilson, J. Q., & Herrnstein, R. J. (1985). *Crime and human nature.* New York: Simon and Schuster.

*Wilson, G. D., Rust, J., & Kasriel, J. (1977). Genetic and family origins of humor preferences: A twin study. *Psychological Reports, 41,* 659–660.

Winsberg, B. G, & Comings, D. E. (1999). Association of the dopamine transporter gene (DAT1) with poor methylphenidate response. *Journal of the American Academy of Child and Adolescent Psychiatry, 38,* 1474–1477.

Young, S. E., Smolen, A., Corley, R. P., Krauter, K. S., DeFries, J. C., Crowley, T. J., et al. (2002). Dopamine transporter polymorphism associated with externalizing behavior problems in children. *American Journal of Medical Genetics, 114,* 144–149.

Young, S. E., Smolen, A., Hewitt, J. K., Haberstick, B. C., Stallings, M. C., Corley, R. P., et al. (2006). Interaction between MAO-A genotype and maltreatment in the risk for conduct disorder: Failure to confirm in adolescent patients. *American Journal of Psychiatry, 163,* 1019–1025.

*Young, S. E., Stallings, M. C., Corley, R. P., Hewitt, J. K., & Fulker, D. W. (1997). Sibling resemblance for conduct disorder and attention deficit-hyperactivity disorder in selected, adoptive, and control families. *Behavior Genetics, 27,* 612.

Zalsman, G., Frisch, A., Bromberg, M., Gelernter, J., Michaelovsky, E., Campino, A., et al. (2001). Family-based association study of serotonin transporter promoter in suicidal adolescents: no association with suicidality but possible role in violence traits. *American Journal of Medical Genetics, 105,* 239–245.

Chapter 31

Schizophrenia and Affective Psychotic Disorders – Inputs from a Genetic Perspective

Daniel R. Hanson

Introduction

A person once walked into my inner city, county hospital, psychiatry office wearing clothing that he had sewn together out of old rubber inner tubes. He explained that he made this outfit to protect himself from infections – a suit of armor for the age of communicable disease. It was a hot day; the person appeared flushed and dehydrated with a dry mouth, absence of perspiration, and pulse of 130. His efforts to protect himself had actually placed him in jeopardy of heat stroke, even more probable because his psychiatric medications had the side effect of altering temperature regulation. How did this person's behavior become so maladaptive? What happens to the human brain/mind during the development of psychotic behaviors remains one of greatest puzzles confronting both the sciences and the humanities. The human genome and nervous system were "designed" by evolution and by experience to perceive the environment accurately and respond to the environment to enhance the adaptation of both the individual and the species. With so much at stake, how can the brain and mind go so far awry? This chapter will explore some of the behavioral genetic data and theory about the origins of psychoses, termed insanity by the legal system and the public.

The key feature of psychoses involves a loss of contact with reality for long periods while remaining fully alert and oriented to one's surroundings. Delirium, by contrast, involves loss of contact with reality but also involves disturbances where consciousness waxes and wanes, and people become confused. The distinction is crucial since deliria are often the result of severe physiological problems that can be rapidly medically dangerous or fatal. Deliria and psychoses both may involve hallucinations, delusions, and maladaptive behaviors, so the distinction can be difficult without adequate historical information about the ill person. This chapter focuses on the two most common forms of psychoses – schizophrenia and the affective psychoses of bipolar and unipolar disorders (formerly known as "manic-depressive insanity"). These categories were first demarcated in the late 19th and early 20th centuries to help distinguish these disorders from the degenerative brain disorders of old age such as senility and Alzheimer's disease, as well as the disease induced by the syphilis spirochete infecting the brain. The diagnostic foundations for schizophrenia and bipolar illness as laid down by Emil Kraepelin (1919), Bleuler (1911/1950) and Kurt Schneider (1959) still form the basis for modern-day diagnostic practices described in the American Psychiatric Association's *Diagnostic and Statistical Manual* edition IV (DSM-IV-TR, 2000). The classic and simplified conceptualizations of schizophrenia describe a disorder of thought processes (illogical, incoherent) inferred from language, disrupted expression of emotion (typically flattened or inappropriate), and a deteriorating course after onset in late teenage years or early adulthood. Bipolar illness, by contrast, involves wide swings in emotion from profound depression to euphoric excitement with a return to normal moods between episodes. We now realize these capsule descriptions are greatly oversimplified and even misleading. The course of either of these illnesses may range from malignant deterioration to lengthy social recovery. Both syndromes may involve hallucinations and delusions. Formal thought disorder occurs in the mood disorders as well as schizophrenia (Lake, 2007).

For over a 100 years, schizophrenia and bipolar illness were conceptualized as separate entities. However, there have been minority dissenters along the way (Ødegaard, 1972; Strömgren, 1994) and recently there has been an upsurge in discussions about the overlap between schizophrenia and bipolar illness using molecular genetic information (Craddock, O'Donovan, & Owen, 2005) – more on this topic later. Consequently, this chapter does not automatically assume that schizophrenia and bipolar illness are completely separate entities but, instead, attempts to prepare the reader

D.R. Hanson (✉)
Departments of Psychiatry & Psychology, University of Minnesota, F282/2A West, 2450 Riverside Avenue, Minneapolis, MN 55454, 612-273-9765, USA
e-mail: drhanson@umn.edu

Y.-K. Kim (ed.), *Handbook of Behavior Genetics*,
DOI 10.1007/978-0-387-76727-7_31, © Springer Science+Business Media, LLC 2009

to maintain an open mind about this question. Therefore, rather than including the details of the diagnostic criteria for schizophrenia and bipolar disorders, the reader can consult DSM-IV-TR for the current diagnostic concepts. These nosologies are certain to change (Craddock, O'Donovan, & Owen, 2006; Nordgaard, Arnfred, Handest, & Parnas, 2008) as the field moves away from diagnostic categories based on behavioral signs and symptoms and moves closer to diagnosing mental illness based on endophenotypes (Gottesman & Gould, 2003) and, ultimately, genomic etiologies.

A third category of psychoses, often described as cyclical psychoses (Leonhard, 1961; Perris & Brockington, 1981), are mentioned only in passing. Typically, such psychoses involve an acute onset and an admixture of schizophrenic and affective symptoms followed by recovery though episodes may reoccur. This syndrome has received scant attention in the American psychiatric literature (the DSM-IV category of Brief Psychotic Disorders captures some of the features of the cyclical psychoses) but is more fully integrated into European and Scandinavian diagnostic practices. There is a consensus that these psychoses are separate from schizophrenia (Peralta, Cuesta, & Zandio, 2007). It is less certain that they are separate from bipolar disorder though recent family study data suggest that cyclic psychoses are also distinct from bipolar illness (Jabs et al., 2006; Pfuhlmann et al., 2004).

What We Know About the Common Psychoses

Before entering a detailed discussion of genetic factors in schizophrenia and bipolar illness, we need to set the stage with some basic facts about these disorders. A clear understanding of the frequency (base rates) of these illnesses in the general population provides the basis for comparison of the rates in relatives of affected individuals in a search for familiality. It is upon these comparisons that behavioral geneticists build the case for the important roles and causal weights of genetic, environmental, epigenetic, and stochastic (chance) factors in the development of these illnesses. The population rates also give us clues as to how common are the causal factors (genetic and environmental) that contribute to the development of these disorders. Secondly, exploring the phenomena that precede overt illness gives important insights as to what kind of causal factors we need to search for and when in development these factors make their impact.

Frequency of Common Psychoses

Schizophrenia and bipolar mood disorders are relatively common illness. Rates of an illness can be expressed as an incidence (typically, new cases per year per 100,000 population), prevalence (total cases observed – both new and old – per year per unit population), or morbid risk (age-adjusted) likelihood of any one individual developing the disorder over the course of a lifetime. Behavioral geneticists prefer the latter statistic for its conceptual simplicity and because it reminds us that schizophrenia is a lifelong process affecting *individuals*. About 1% of the general population will experience schizophrenia and another 1% will experience bipolar illness sometime over the course of their lifetimes. These 1% rates are remarkably constant around the globe regardless of culture, geography, or climate. The onset of the illness is earlier in men but by the end of the risk period the rates are equal for men and women. There have been some speculations that the occurrence of schizophrenia is declining (Brewin et al., 1997; Geddes, Black, Whalley, & Eagles, 1993; Suvisaari, Haukka, Tanskanen, & Lönnqvist, 1999). Whether there is a true decline or the changing rates is an artifact of sampling or changing diagnostic practices remains debated (Osby et al., 2001). The issue is important for behavioral geneticists because the time frame for this proposed decrease is too short for any evolutionary change in the frequency of the genes involved in these illnesses. If such a decline is real, and if we understood why the change occurred, we would have a valuable insight into modifying environmental factors to change or prevent the development of illness.

The general population 1% morbid risks are generally derived using operational definitions of these syndromes that are based on specified diagnostic criteria. However, an additional factor – impairment of function – also enters into the ascertainment of diagnosable cases. Most of the statistics describing general population rates of mental illness have relied on counting individuals who have entered into treatment and thus imply that there has been some degree of disruption of function. It is possible that there are people with the syndromes but whose functional status is sufficient in isolated or supportive environments such that they do not present for treatment. Such cases would not come to professional attention and would escape enumeration. There have been efforts to screen large and representative samples of the population (Kessler et al., 1994) and in such studies the rates of schizophrenia and bipolar illness are higher than the typical 1% rates by as much as twofold. Sometimes these undiagnosed cases in the population are referred to as "sub-threshold" versions of the syndromes. There are many controversies surrounding these population studies (Murray, Jones, Susser, Van Os, & Cannon, 2003). To make large population-based epidemiological studies feasible, trained research assistants apply questionnaires and standardized interviews to gather the data. However, the data are not based on observations by experienced psychiatrists or psychologists. Concerns have been raised that the survey method leads to an over-diagnosis of mental health syndromes (Helzer et al., 1985) among other distortions. To illustrate, the diagnosis of osteoarthritis is typically made when wear and tear on the body's joints become so bad that pain and disability bring the affected person to a healthcare provider for treatment. However, if you went from door to door asking people if they had a sore joint in recent weeks, the researcher would find a vastly larger number of people with sore joints compared to the number who actually seek treatment. Which strategy provides the best answer to the question of how common is osteoarthritis? There is no clear answer – much depends on why one is asking the question. If the question is asked to guide health care policy for providing adequate services to help people with disabling osteoarthritis, the first strategy may be most helpful. If one is asking the question to find the largest possible market for selling a treatment for joint pain, the latter strategy may be emphasized. It is beyond the scope of this chapter to say any more about economic and political forces that influence the process of enumerating diseases – just be aware that such forces exist that can either raise or reduce criteria-based results.

The general population rates of illness also provide important information about how common the causal factors are that give rise to these illnesses. When searching for the factors that contribute to the development of schizophrenia or bipolar illness, we need to know if we are looking for rare and abnormal genetic or environmental factors or common and ordinary events that are cumulatively toxic. Among studies of identical twins, if one twin has schizophrenia, the genetically identical co-twin has the disorder only about 50% of the time at the end of the risk period. The less-than-100% concordance tells us that there must be something genetic (G) *and* something environmental (E) about the causes of this schizophrenia. Many environmental factors have been proposed including such things as in utero exposure to infection (Meyer, Yee, & Feldon, 2007; Smith, Li, Garbett, Mirnics, & Patterson, 2007), prenatal malnutrition (Hoek, Brown, & Susser, 1998; Hulshoff Pol et al., 2000; St Clair et al., 2005), birth and pregnancy complications (Byrne, Agerbo, Bennedsen, Eaton, & Mortensen, 2007; M. Cannon & Clarke, 2005; Clarke, Harley, & Cannon, 2006), season of birth (Bersani et al., 2006; McGrath, Saha, Lieberman, & Buka, 2006), and urban and disadvantaged social–economic situations (Byrne, Agerbo, Eaton, & Mortensen, 2004; Cantor-Graae & Selten, 2005; Cooper, 2005). Figure 31.1 provides a summary of the impact (relative risk) of some of these environmental influences and contrasts them with the relative risk conferred by a genetic relationship to a sibling who has a schizophrenia-related psychosis or an identical twin with schizophrenia. So far, all of the posited environmental factors have only weak impact on risk compared to genetic factors. That is not to say that there are no highly impactful environmental factors; there probably are, but we have not identified them yet.

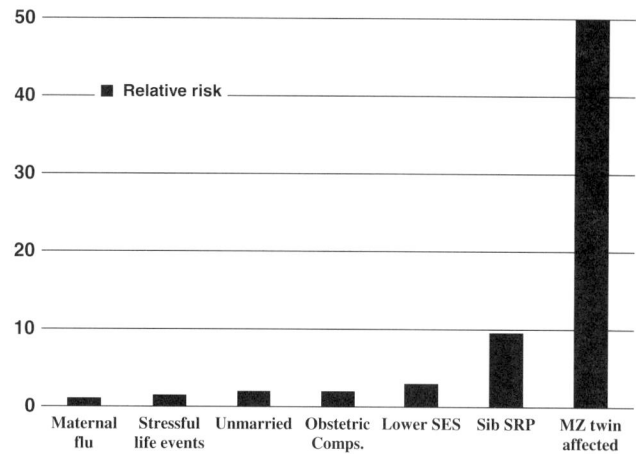

Fig. 31.1 Relative "potency" of factors associated with an increased rate of schizophrenia. The vertical axis is odds ratios – the degree to which the risk factor increases the risk above the general population risk. On the low end of the scale, children exposed in utero to maternal influenza have a slight increased risk compared to the general population (odds ratio = 1.1). At the other extreme, individuals with an affected identical twin have a vastly increased risk (odds ratio = 50). SRP = schizophrenia-related psychosis

Knowing that schizophrenia and bipolar illness each affect about 1% of the population, and with a few assumptions about the factors being independent of each other, we can calculate approximate base rates of the individual causal contributors. Details and elaborations of the calculations can be found in Hanson (2004) and Hanson and Gottesman (2007). The surprising result is that the environmental and the genetic risk factors for acquiring schizophrenia or bipolar illness are individually likely to be relatively common events. Figure 31.2 illustrates the phenomenon for any illness with a population rate of 1%. If there are, say, four (horizontal axis) independent causal factors, then we would expect that each one of these factors, individually, would impact about a third (vertical axis) of the general population. The purpose of Fig. 31.2 is to offer rough "ball-park" estimates of the population frequencies of risk factors and it is not intended as a rigorous epidemiological model. However, Fig. 31.2 serves as a point of departure in thinking about the frequency in the population of causal factors. Somewhere along the curves in Fig. 31.2, we would stop thinking of the risk factors (genetic or environmental) as rare pathological events and start thinking of them as common individual differences within the normal range (Thomson & Esposito, 1999). If this turns out to be true, illness results more from the bad luck of experiencing a *combination* of risk factors that jointly cause illness but individually are normal and without deleterious consequences (Becker, 2004; Hanson and Gottesman, 2007). Additionally, if the causal factors are rare it would be easier to imagine schizophrenia and bipolar illness as separate entities, each with their own infrequent

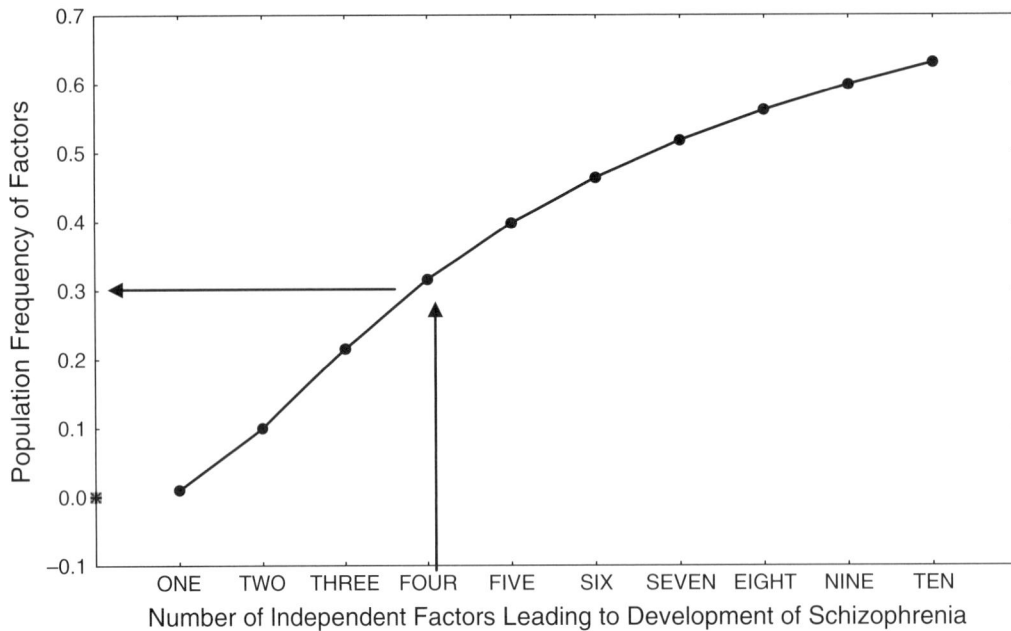

Fig. 31.2 Possible relationships between the number of factors that contribute to the development of schizophrenia (with a population base rate of 1%) and the frequency of causal factors in the general population.

This illustration provides a guess as to how common the causal factors could be in the general population assuming each factor is independent of the others and that all factors are about equally common

and separate etiologies. However, if the causal factors are common, then it is much more likely that these common factors would overlap across both illnesses.

Psychotic Disorders Develop Over Time

The range in age of onset of schizophrenia is unusually broad. Symptoms usually start in the late teenage years or early twenties, but the illness can start in middle childhood (Hanson & Gottesman, 1976) and may rarely start in old age (Bridge & Wyatt, 1980; Howard, Rabins, Seeman, & Jeste, 2000; Slater & Cowie, 1971). The DSM III statement that schizophrenia could not start after age 45 is clearly false and this claim was expunged from subsequent editions. Likewise, bipolar illness is increasingly recognized in children (Correll, Penzner, Frederickson et al., 2007; Correll, Penzner, Frederickson et al., 2007; Somanath, Jain, & Reddy, 2002; Tillman & Geller, 2006) and may onset late in life (Almeida & Fenner, 2002) though, like schizophrenia, the peak age of onset is in the late teens or early twenties (Loranger & Levine, 1978). This variable age of onset presents complications to psychiatric geneticists because many of the individuals in a sample of people at risk (e.g., relatives of a bipolar proband) will still be young. These younger folks may not have the illness at the time of the study but some will have the potential to go on to develop the illness at a later age. The net result would be to underestimate the true rates of

illness in the study sample. To get around this problem, the sizes of study samples are sometimes statistically adjusted for the age structure of the sample (sometimes referred to by the German term Bezugsziffer – or BZ). To provide a simple example, if a person had lived through only half of the empirically known risk period, he/she would be counted as only 0.5 in the denominator of a risk calculation that lists the number affected in the numerator, divided by the total number of (age-corrected) individuals in the denominator.

The common psychotic syndromes often show evidence of precursors early in the life of those who eventually break down. From the time of Bleuler (1911/1950) and Kraepelin (1919) "It is certain that many a schizophrenia can be traced back into the early years of the patient's lives..." (Bleuler, 1911/1950 p. 252). Likewise, a prodrome of unstable affect, sometimes appearing like an attention-deficit disorder, is not rare in the lives of individuals who eventually develop bipolar illness and there is growing interest in defining the bipolar prodrome (Correll, Penzner, Frederickson et al., 2007). It is these early forerunners of illness that motivated the studies of individuals at high risk (Chapter 32 in this volume) in an effort to better understand the natural history of these illnesses. Not everyone who develops schizophrenia or bipolar illness has a clear prodrome and not all cases have a gradual slide into illness. The natural history can vary from gradual, undulating, or rapid progression into illness and the subsequent course can, likewise, be progressive deterioration, undulating, or return to health (Bleuler, 1978; Tsuang & Winokur, 1975; Winokur et al., 1994). The realization of such a variable array of onsets and course has led to a shift in thinking about degenerative models vs. developmental models of these illnesses.

Degenerative models for the major psychoses imply that after a period of normal development, the brain, or one of its parts, begins to deteriorate with a commensurate perturbation of behavior. However, stating that an illness is degenerative is not particularly helpful. What needs to be done is to determine when in the life course the degeneration begins and how the degeneration is initiated and maintained. Answers to the "when?" and "how?" questions would then describe the *degenerative* process in *developmental* terms.

Developmental models of psychoses implicate abnormalities of early brain assembly that predispose to future malfunction. The proponents of the model further argue that the perturbations of development are limited to the early times of development and are discontinuous. Without this qualifier, developmental models are indistinguishable from degenerative models where the degeneration commences early in the life span. The early abnormalities are not necessarily the cause of psychoses, but, instead, create a state of risk for a future episode. That is, a diathesis or predisposition is not a disease. Consequently, there must be factors later in life that convert the vulnerability to an illness. These additional factors are presumed to damage development in such a way that a predisposition becomes actualized. To gain a complete understanding of the syndrome, we must return to the question of "what happens?."

Following this line of reasoning, the distinction between degenerative and developmental models blurs. In fact, a medical–behavioral condition can be both developmental and degenerative as exemplified by Down syndrome (Head & Lott, 2004; Kornberg et al., 1990; Opitz & Gilbert-Barness, 1990). Some individuals born with Down syndrome (trisomy 21) exhibit a number of developmental anomalies including cardiac malformations, abnormal dermatoglyphics, skeletal changes, and muscular hypotonia, to name a few. As trisomy 21 infants mature, most exhibit degrees of mental retardation. By about age 50, these individuals invariably develop Alzheimer-like CNS degenerative pathology and it is observed at autopsy (Head & Lott, 2004).

The Genetic Evidence

Family and Twin Studies

The evidence supporting a role for genetic factors in schizophrenia and bipolar disorder comes from multiple strategies that include studies of families, twins, adoptees, genetic linkages, and searches for candidate SNPs and genes. Table 31.1 summarizes the classic epidemiological studies of schizophrenia and bipolar illness. These pooled data come from scores of studies that cross decades, countries, methodological strategies, and diagnostic practices. Yet, in spite of the differences there is, with rare exception, a consistent demonstration of a higher risk for illness in relatives of affected individuals compared to general population risks and the risk varies consistently with degree of genetic relationship. Furthermore, the risks are higher for relatives of severely affected (early onset) probands and for relatives from families with many affected individuals. The gender of the affected proband does not alter risks.

The 50% (approx.) concordance rate in identical twins is also the best evidence that genetics is not the whole story. Environmental, epigenetic, and stochastic factors, yet to be identified, must also play a role. The study of offspring of discordant identical twins is especially illuminating, though such opportunities are rare. In one such study the risk to children of the parents affected with schizophrenia was 17%; that was the same as the risk to children born to the unaffected but genetically identical co-twins (Gottesman & Bertelsen, 1989). The conclusion is that it is possible for a person to have the gene(s) for psychosis, show no illness, but still pass those genes on to offspring. If we understood how this works (Wong, Gottesman, & Petronis, 2005), we would be better able to prevent or treat these illnesses.

Twin data also help us answer other questions. There has been a longstanding debate about the separateness of unipolar depression and bipolar illness. Bertelsen's twin data from Denmark (Bertelsen, 1978; Bertelsen, Harvald, & Hauge, 1977) clearly show that the identical twins of bipolars can have lifelong unipolar syndromes and unipolars

Table 31.1 The risks for schizophrenia or bipolar illness in various categories of genetic relationship

	Risk for schizophrenia in relatives of probands with schizophrenia (Gottesman et al., 1982) (%)	Risk for bipolar illness in relatives of probands with bipolar illness (Craddock & Jones, 1999; Jones, Kent, & Craddock, 2004)
First degree	10–15	~7%
Second degree	2–3	No data
Identical (MZ) twins	40–50	40–60%
Fraternal (DZ) twins	10–15	0–19%
Adoptees	10–18	Two studies: increased risk

can have bipolar identical co-twins. Subsequent data (Bertelsen, 2005) have confirmed these observations. The clinical implications are that there are individuals who appear to have a unipolar depression but who, in fact, are intrinsically bipolar (with unexpressed genotypes) and would benefit from the same mood-stabilizing medications used for bipolar illness. There is, for example, a growing literature on using the mood-stabilizing medication lithium to *augment* antidepressants when severely depressed patient does not respond to antidepressants alone. In such cases, the lithium is probably not simply boosting the antidepressant; it is treating the underlying bipolar illness. The genetic-epidemiological data also help us understand the boundaries of these illnesses. There is substantial evidence that people with bipolar illness have increased rates of panic disorder, obsessionality, and general anxiety and there is an increase of these syndromes in the relatives of bipolars (Dilsaver, Akiskal, Akiskal, & Benazzi, 2006; Edmonds et al., 1998; Joyce et al., 2004; Kessler et al., 1994; Merikangas et al., 2007; Merikangas & Low, 2004; Rzhetsky, Wajngurt, Park, & Zheng, 2007; Schulze, Hedeker, Zandi, Rietschel, & McMahon, 2006; Simon et al., 2003). One way to conceptualize this phenomenon is that the core of these syndromes does not lie in any one of these behavioral manifestations but, instead, there is a basic disruption of brain centers that modulate affect/emotion. In other words, the problem with depression (substitute mania, panic, etc.) is not that the brain is depressed (excited, aroused) but that the mechanisms that regulate this spectrum of emotion are broken or dysregulated. If so, it then makes some sense why medications such as the so-called "antidepressants" not only help with low mood but also with the anxiety syndromes such as panic or obsessive compulsive disorders. This formulation also helps us understand the syndrome of mixed mania where there is an admixture of negative emotions (depression, irritability, and anger) combined with a high-energy state. If bipolar illness were truly a swing from one pole of depression through a mid-point of normal to an opposite pole of mania it would be hard to explain the mixed state. The bipolar syndrome is a state of emotional dysregulation where any and all emotions may simultaneously exist in excess. Einat and Manji (2006) have postulated that the faulty regulation occurs at the level of cellular plasticity – see also Gottesman and Hanson (2005) for more on plasticity.

The genetic-epidemiological data also speak to the question of separateness of schizophrenia and bipolar illness. As alluded to previously, these syndromes were originally conceptualized as distinct entities on clinical grounds. The majority of the genetic-epidemiology studies are consistent with the distinction by failing to show increased rates of bipolar illness in the families of schizophrenia probands or schizophrenia in the families of bipolar probands. However, most of the genetic studies were never designed to specifically test the question. Investigators typically set out to test the genetics of just one of these disorders and did so already assuming the illnesses were separate. Thus, there may have been a built-in bias where, for example, if schizophrenia were the focus of the study, and psychotic relatives were found, there may have been unwitting and unknown factors that led to affected relatives being perceived as having the same diagnosis as the proband. By contrast, Ødegaard (1972) set out to study psychoses with few preconceived notions about the separateness of the illnesses. His data from Norwegian national registers (Table 31.2) suggest that schizophrenia and bipolar illness can occur within the same families. A few other historical observations have reached similar conclusions (Reed, Hartley, Anderson, Phillips, & Johnson, 1973; Tsuang, Winokur, & Crowe, 1980). In more recent times, there has been a reappraisal of the overlap between schizophrenia and bipolar illness (Blackwood et al., 2007; Craddock et al., 2006; Maier, Hofgen, Zobel, & Rietschel, 2005; Palo et al., 2007), especially as the field has moved into linkage studies and the search for candidate genes.

Table 31.2 Diagnostic distribution of Norwegian index patients and their psychotic relatives (Ødegaard, 1972)

Proband diagnosis	No. of psychotic relatives	Percent of psychotic relatives diagnosed as having (not = to risk figures)		
		Schizophrenia	Reactive psychoses	Affective psychoses
Schizophrenia, severe defect	109	78	7	15
Schizophrenia, slight defect	368	71	16	14
Schizophrenia, no defect (schizoaffective)	179	46	23	31
Reactive psychoses	82	28	48	24
Atypical affective psychoses	39	36	28	36
Bipolar psychoses	47	19	11	70

Linkage Analyses and Candidate Genes

Linkage analyses search for genes or regions of chromosomes that are consistently passed along in families hand in hand with illness – i.e., the chromosomal regions are "linked" to the disease. By knowing what genes lie on the linked segment of a chromosome, research could hone in on specific genes in that chromosomal region that contribute to risk. The history of linkage studies of schizophrenia and bipolar illness has been long and characterized by exciting announcements of linkage followed by a disappointing series of failures to replicate (Crow, 2007). Linkage analysis for psychiatric syndromes is particularly precarious because, as we have seen with discordant identical twin data, the relevant genes (whatever they are) are often *not* associated with illness. Furthermore, as we have also seen, there are many uncertainties about what constitutes "illness." If we are studying linkage in bipolar illness, should we count unipolar depression, panic, anxiety, or phobias as cases? This is one of the stickiest issues confronting the field of human behavior genetics. Hanson, Gottesman, and Meehl (1977) offer further discussion of the issues of diagnoses and symptom expression as it relates to prediction in psychiatric genetic endeavors, including the study of children at risk. In spite of all of the difficulties, there have been several loci linked to either schizophrenia, bipolar illness, or both, as listed in Table 31.3. This table is derived from the up-to-date and rigorous review by Craddock et al. (2005). However, given the shifting sands of this line of research, the reader is well advised to search for subsequent reviews for updates (http://www.schizophreniaforum.org). It is difficult to draw any firm conclusions about specific genes impacting our illnesses of interest. However, the evidence from linkage studies is shaping up to confirm that the contributing genes to schizophrenia and bipolar illness are likely to be several to many in number, each with small effects and

that, to a considerable extent, there is some genetic overlap across these syndromes.

Candidate genes are genes with known chromosomal locations that are suspected to be involved in a disease and whose protein product suggests that it could be the disease gene in question. One way to search for candidate genes starts with some theory about the illness and then looks to see if genes related to that theory are linked to the illness. For example, the neurotransmitter dopamine has long been implicated in schizophrenia, so investigators have been looking at candidate genes coding for aberrant dopamine receptors (of which there are several). Unfortunately, this approach has had little in the way of replicable success. Alternatively, candidate genes are identified when highly detailed linkage studies lead investigators to a specific gene that is linked to illness. Table 31.3 summarizes the most promising candidate genes to date but the reader is cautioned that none of these genes are solidly proven to be associated with psychoses (Sanders et al., 2008). Table 31.4 provides a brief description of some of these candidate genes.

The striking commonality among these candidate genes is that, when known, their functions are fundamental and widespread throughout the brain, if not the entire body. They involve the kinds of biological mechanisms that fit what we know clinically, entailing early developmental perturbations followed by later compounding of problems leading to overt illness. Yet, given the fundamental nature of the candidate genes actions, they must be only a tiny bit "off" from normal. While it is true that schizophrenia and bipolar illness have a profound effect on those with the illness, the vast majority of brain function comes close to working normally. People with these illnesses walk, talk, think, and feel – all systems are operative to some degree and can normalize – unlike brain diseases such as stroke, or degenerative disorders such as Alzheimer and Huntington. Whatever is wrong is pervasive

Table 31.3 Linkage sites and candidate genes for schizophrenia and/or bipolar illness (based on Craddock, O'Donovan, & Owen, 2005)

	Schizophrenia	Bipolar illness	Both
Linkage sites	6p24-22	6q16-q22	1q21-q22
	1q21-22	12q23-q24	1q42
	13q32-34	9p22-p21	2p13-p16
	8p21-22	10q21-22	4p-16
	6q16-25	14q24-q32	4q-32
	22q11-12	13q-32-q3422	6q21-q25
	5q21-q33	q11-q22	10q25-26
	10p15-p11	Chromosome 18	12q23-q24
	1q42		13q32-q34
			15q14
			17p11-q25
Candidate genes	NRG1	DAOA(G72	DAOA(G72)
	DTNBP1	BDNF	DISC1
	DISC1		NRG1
	DAOA(G72)		
	BDNF		
Chromosome abnormalities	22q11		1q42

Table 31.4 Brief description of candidate genes for schizophrenia and bipolar illness

Candidate gene	Function
Neuregulin 1 (NRG1)	Synaptogenesis, axon guidance and maintenance, glial cell development and others
D-Aminoacid oxidase and its activator (DAOA)	Uncertain
Dysbindin (DTNBP1)	Glutamate neurotransmitter systems, biogenesis of lysosome-related organelles
Disrupted in schizophrenia (DISC1)	Intracellular transport, neurite structure, neuronal migration, very active in embryogenesis
Regulator of G-protein signaling 4 (RGS4)	Modulates cellular production of "second messengers" within the cells in response to external signals (neurotransmitters, hormones)
Brain-derived neurotropic factor (BDNF)	Modify growth, development, survival of neurons, likely involved in CNS plasticity

and profound, yet subtle and mild. No wonder we still have not solved this mystery in more than a 100 years of effort. Because the common psychoses are undoubtedly multifactorial in cause, tasks for the future will be to study arrays of candidate genes in relation to each other and in relation to environmental factors that might influence these genes to lead to illness.

A Framework for Putting It all Together: Epigenesis, Reaction Surface, and Endophenotypes

Epigenesis

Human development is more than a simple summation of genetic and environmental factors. The concept of epigenesis moves us closer to understanding the developmental processes that lead to such disorders as schizophrenia and bipolar illness. The term epigenesis originated with embryological theories suggesting that complex organisms originate from relatively simple undifferentiated cells. Broadly speaking, the term currently is used to include all the forces that lead to the phenotypic expression of an individual's genotype (Petronis, 2004; Waddington, 1957). In the early 1970s, Gottesman and Shields (1972) introduced the concept of epigenesis into psychiatric genetics followed by later elaborations (Gottesman, Shields, & Hanson, 1982). The definition of "epigenetic" continues to evolve and, to many molecular biologists, the term refers to the mechanisms by which cells change form or function and then transmit that form or function to future cells in that cell line (Jablonka & Lamb, 2002; Jaenisch & Bird, 2003; Morange, 2002). Transformation of an undifferentiated embryo cell into a liver cell and transformation of a normal liver cell into a cancerous cell are examples of epigenesis.

The best studied mechanisms for the epigenetic regulation of mammalian gene expression involves the addition of a methyl group to cytosine that, along with adenine, thiamine and guanine, form the four-letter alphabet of DNA (Petronis, 2003; Petronis et al., 2003). This methylation of cytosine changes the configuration of the DNA such that the genetic information encoded in that area cannot be read and is nullified (Jaenisch & Bird, 2003; Jones & Takai, 2001) – the gene is essentially turned off. Conversely, removing DNA methylation allows expression of the gene. Failure of methylation systems leads to clinical syndromes such as Rett syndrome that involves mental retardation, autistic-like behaviors, and other neuro-developmental anomalies in girls (Shahbazian & Zoghbi, 2002). Such epigenetic mechanisms may account for why, in a rodent model, maternal behavior toward young offspring affects the size of the offspring's hippocampus in adulthood, depending on the offspring's genotype (Weaver, Grant, & Meaney, 2002). The nurturing-induced effects on brain development and stress response persists into maturity (Sapolsky, 2004; Weaver et al., 2004). It is tempting to speculate about the power of such findings to help explain how very early life experiences/exposures may help set the stage for schizophrenia (Petronis et al., 2003) and mood disorders (Caspi et al., 2003; Charney & Manji, 2004) many years later in life.

Epigenetic perspectives grapple with complexities of how multiple genetic factors integrate over time with multiple environmental factors through dynamic, often non-linear, sometimes non-reversible, processes to produce behaviorally relevant endophenotypes and phenotypes. A key epigenetic question is how identical twins can be discordant for psychoses (Cannon et al., 2002; Hulshoff Pol et al., 2004; Kuratomi et al., 2007; van Erp et al., 2004; van Haren et al., 2004; Wong, Gottesman, & Petronis, 2005).

Reaction Surface

At any time, any one genotype may have a wide array of potential phenotypes, referred to as a "reaction range" or "reaction surface" (Turkheimer, Goldsmith, & Gottesman, 1995). The actual phenotype will depend on the influence of the individual's other genes and on the specific contexts of environments experienced among a wide array of possible environments. Which environment is experienced may be stochastic (chance) or may be a function of the individual's past phenotypes. Indeed, an individual's phenotype (which is partially a result of his/her genotype) may lead him/her to select certain environments, thereby establishing a correlation between genotype and environment (Carey, 2003).

The array of possible outcomes for any developmental process could, in theory, be plotted in multidimensional space, as functions of genotypes, environments, and time. The plot would produce an undulating surface that would represent the phenotype for that unique combination of genotype, environment, and time. Such a surface has been referred to as a "reaction surface" (Gottesman & Gould, 2003; Sing, Stengard, & Kardia, 2003) or "phenotypic surface" (Nijhout, 2003) and these articles provide informative graphics. Figure 31.3 provides such an example (Gottesman & Gould, 2003) applied to the ontogenesis of schizophrenia. The illustration demonstrates a changing reaction surface, a threshold for illness, suggested endophenotypes (some already connected to candidate genes), and a dimension of environmental inputs.

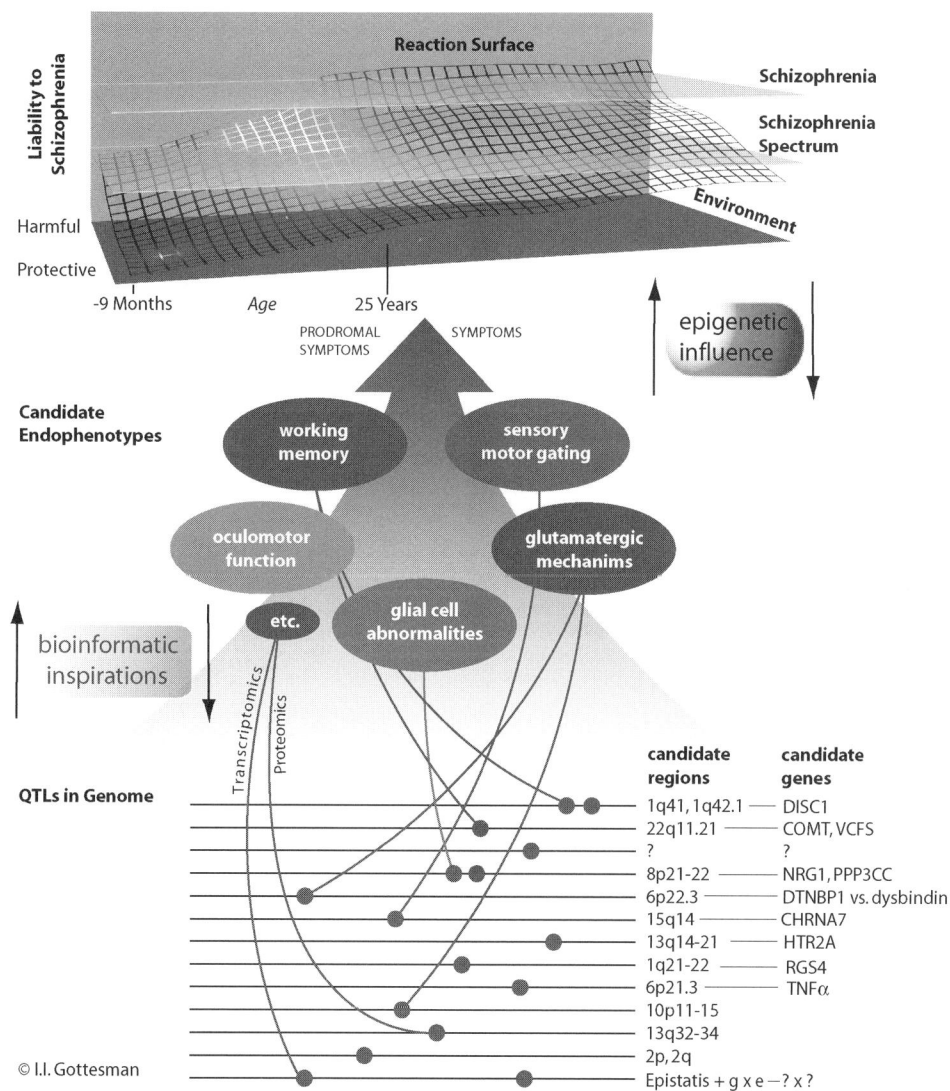

Fig. 31.3 Illustration of a systems biology approach toward explaining complex behavior incorporating dynamic interplay among candidate genes and gene regions, endophenotypes, and pre- and postnatal environmental and epigenetic influences (protective or harmful) over the course of development. Question marks indicate gaps in our knowledge. Two planes intersect the reaction surface for the liability to developing schizophrenia over time indicating levels above which clinical diagnoses are manifest (cf. Gottesman & Gould, 2003; Manji, Gottesman, & Gould, 2003). Copyright 2005 by I.I. Gottesman (used with permission)

Questions for the future include whether or not schizophrenia and bipolar belong to the same reaction surface or not.

Endophenotypes

It is a long road from genotype through epigenetic pathways to endpoint phenotypes such as schizophrenia and bipolar illness. As we have seen, these behavioral phenotypes are of uncertain value in specifying the limits and bounds of the biological entities that contribute to these phenomena. The clinical outcome may be too distant in time and development to define the origins – explorers should not look backward through their telescopes. What are needed are intermediate traits that sit closer to the genotype end of the developmental trajectory. In spite of the best efforts to improve the reliability of psychiatric classification, the diagnoses in the official nomenclatures are still syndromal and lack validating biological markers. The missing links have been referred to as "endophenotypes" (Gottesman & Gould, 2003; Gottesman and Shields, 1972, Shields & Gottesman 1973). Alternative concepts with similar but different meanings include "biological markers," "intermediate phenotypes," "risk factors," "vulnerability markers," and "sub-clinical traits." These terms can be confusing because they are used in different ways by different authors. Table 31.5 attempts to compare and contrast some of the different nuances associated with

these terms. Attempts to identify those characteristics that are genetically mediated – endophenotypes – would require the following:

1. Endophenotypes that would be associated with the trait in the population.
2. The endophenotype would be demonstrably heritable.
3. The endophenotype is present whether the trait/disease is present or not (e.g., vulnerability marker) but may require a non-harmful triggering/eliciting test analogous to a glucose tolerance test or a cardiac stress test.
4. Within families, endophenotype and trait co-segregate (but not perfectly – see 3 above).
5. The endophenotype found in families with the trait (especially an illness) is found in non-affected family members at a higher rate than in the general population.

An instructive example comes from cardiology and the long QT syndrome. Every normal heartbeat is associated with a specific pattern of electrical polarization and depolarization of cardiac muscles. These shifts in polarization can be measured with an electrocardiogram (EKG) and the letters P, Q, R, S, and T have been arbitrarily assigned to identify specific shifts in polarization. The Q wave indicates the start of depolarization of the ventricles (contraction) and the T wave represents ventricular repolarization (relaxation) in preparation for the next beat. If there is a failure of repolarization (inherited, medication side effects) the QT time interval gets

Table 31.5 Terms frequently used in behavioral genetics

Term	Definition	Example/Elaboration
Endophenotype	A heritable and quantitative measure of neurobiological function associated with illness (see additional elaboration in text)	Long QT interval as a marker for Jervell and Lange-Nielsen cardio-auditory syndrome
Genetic linkage	A gene or chromosomal site statistically associated with illness but not necessarily a causal factor	Illness may be linked to a specific gene site because the identified gene is located close on the chromosome to some other gene that is the actual relevant gene
Biological marker	A biological trait associated with being at increased risk but not necessarily inherited nor causal	Elevation in the levels of prostatic-specific antigen is associated with prostate cancer (or prostate inflammation)
Genetic marker	A genetic modification associated with increased risk for illness	CAG nucleotide repeats identify Huntington disease
Intermediate phenotype	A quantitative measure both associated with the illness and believed critical to the underlying disease pathophysiology (see distinction with endophenotypes in text)	Increased rate of colonic polyps associated with colon cancer
Subclinical trait	Classification of disease based on specific clinical features	Fatigue associated with diabetes
Liability indicator	A factor or test that indicates a person is at increased risk for illness	These may be endophenotypes or genetic/biologic markers
Risk factor	Any thing, environmental or genetic, that is associated with increased risk if a person has, or is exposed to, the factor	Smoking is a risk factor for lung cancer
Candidate gene	A gene that , based on current knowledge and theory, may be a causal factor for illness	See Table 31.4

longer, and, in the extreme, the heartbeat can become dangerously erratic. It was known that phenotypes including syncope, ventricular arrhythmias, and sudden death aggregated in families. The common denominator turned out to be QT elongation on EKG. Using QT elongation as the endophenotype, and by excluding or including family members with this finding, genetic linkage studies were successful in identifying the associated genes (Keating et al., 1991; Keating & Sanguinetti, 2001).

Reprise

Schizophrenia and bipolar illness have profound impact on those affected, their families, and society. With good treatment, the impact is modified and affected individuals can experience productive lives. Recalling examples such as polio, we know that low tech but patient and caring physical therapy restored paralyzed limbs to function long before we understood the biological basis of polio. Likewise, we do not have to wait to uncover the secrets of the cause of these illnesses to move forward in minimizing the toll these illnesses can take. However, prevention would be better and, as the twins discordant for these illnesses prove to us, there must be a way to prevent genetic predispositions from becoming manifest. Based on the kinds of data summarized above plus the insights offered by epigenetic thinking, new theories are emerging (Hanson & Gottesman, 2005) that may move us closer to the goal. There is room for more ideas – any takers?

References

Almeida, O. P., & Fenner, S. (2002). Bipolar disorder: similarities and differences between patients with illness onset before and after 65 years of age. *International Psychogeriatrics, 14*(3), 311–322.

Becker, K. G. (2004). The common variants/multiple disease hypothesis of common complex genetic disorders. *Medical Hypotheses, 62*(2), 309–317.

Bersani, G., Pucci, D., Gherardelli, S., Conforti, F., Bersani, I., Osborn, J. F., et al. (2006). Excess in the spring and deficit in the autumn in birth rates of male schizophrenic patients in Italy: potential role of perinatal risk factors. *Journal of Maternal-Fetal & Neonatal Medicine, 19*(7), 425–431.

Bertelsen, A. (1978). A Danish twin study of manic-depressive disorders. *Progress in Clinical and Biological Research, 24A*, 119–124.

Bertelsen, A. (2005). Contributions of Danish Registers to Understanding Psychopathology (A Lifetime of 30 Years' Collaboration with Irving I. Gottesman). In L. DiLalla (Ed.), *Behavior genetics principles: Perspectives in development, personality, and psychopathology*. Washington, DC: American Psychological Press.

Bertelsen, A., Harvald, B., & Hauge, M. (1977). A Danish twin study of manic-depressive disorders. *British Journal of Psychiatry, 130*, 330–351.

Blackwood, D. H., Pickard, B. J., Thomson, P. A., Evans, K. L., Porteous, D. J., & Muir, W. J. (2007). Are some genetic risk factors com-

mon to schizophrenia, bipolar disorder and depression? Evidence from DISC1, GRIK4 and NRG1. *Neurotoxicity Research, 11*(1), 73–83.

Bleuler, E. (1911/1950). *Dementia praecox or the group of Schizophrenias* (J. Zinkin, Trans.). New York: International University Press.

Bleuler, M. (1978). *The schizophrenic disorders: Long-term patient and family studies*. New Haven: Yale University Press.

Brewin, J., Cantwell, R., Dalkin, T., Fox, R., Medley, I., Glazebrook, C., et al. (1997). Incidence of schizophrenia in Nottingham. A comparison of two cohorts, 1978–1980 and 1992–1994. *British Journal of Psychiatry, 171*, 140–144.

Bridge, T., & Wyatt, R. (1980). Paraphrenia: Paranoid states of late life I. European Research. *American Geriatrics Society, 28*(5), 193–200.

Byrne, M., Agerbo, E., Bennedsen, B., Eaton, W. W., & Mortensen, P. B. (2007). Obstetric conditions and risk of first admission with schizophrenia: A Danish national register based study. *Schizophrenia Research, 97*(3), 51–59.

Byrne, M., Agerbo, E., Eaton, W. W., & Mortensen, P. B. (2004). Parental socio-economic status and risk of first admission with schizophrenia- a Danish national register based study. *Social Psychiatry and Psychiatric Epidemiology, 39*(2), 87–96.

Cannon, M., & Clarke, M. C. (2005). Risk for schizophrenia – broadening the concepts, pushing back the boundaries. *Schizophrenia Research, 79*(1), 5–13.

Cannon, T. D., Thompson, P. M., van Erp, T. G., Toga, A. W., Poutanen, V. P., Huttunen, M., et al. (2002). Cortex mapping reveals regionally specific patterns of genetic and disease-specific gray-matter deficits in twins discordant for schizophrenia. *Proceedings of the National Academy of Sciences of the United States of America, 99*(5), 3228–3233.

Cantor-Graae, E., & Selten, J. P. (2005). Schizophrenia and migration: a meta-analysis and review. *American Journal of Psychiatry, 162*(1), 12–24.

Carey, G. (2003). *Human genetics for the social sciences*. Thousand Oaks CA: Sage.

Caspi, A., Sugden, K., Moffitt, T. E., Taylor, A., Craig, I. W., Harrington, H., et al. (2003). Influence of life stress on depression: moderation by a polymorphism in the 5-HTT gene. *Science, 301*(5631), 386–389.

Charney, D. S., & Manji, H. K. (2004). Life stress, genes, and depression: multiple pathways lead to increased risk and new opportunities for intervention. *Science's STKE, 2004*(225), re5.

Clarke, M. C., Harley, M., & Cannon, M. (2006). The role of obstetric events in schizophrenia. *Schizophrenia Bulletin, 32*(1), 3–8.

Cooper, B. (2005). Schizophrenia, social class and immigrant status: the epidemiological evidence. *Epidemiologia e Psichiatria Sociale, 14*(3), 137–144.

Correll, C. U., Penzner, J. B., Frederickson, A. M., Richter, J. J., Auther, A. M., Smith, C. W., et al. (2007). Differentiation in the preonset phases of schizophrenia and mood disorders: evidence in support of a bipolar mania prodrome. *Schizophrenia Bulletin, 33*(3), 703–714.

Correll, C. U., Penzner, J. B., Lencz, T., Auther, A., Smith, C. W., Malhotra, A. K., et al. (2007). Early identification and high-risk strategies for bipolar disorder. *Bipolar Disorders, 9*(4), 324–338.

Craddock, N., & Jones, I. (1999). Genetics of bipolar disorder. *Journal of Medical Genetics, 36*, 585–594.

Craddock, N., O'Donovan, M., & Owen M. J. (2005). The genetics of schizophrenia and bipolar disorder: dissecting psychosis. *Journal of Medical Genetics, 42*(3), 193–204.

Craddock, N., O'Donovan, M., & Owen, M. (2006). Genes for schizophrenia and bipolar disorder? Implications for psychiatric nosology. *Schizophrenia Bulletin, 32*(1), 9–16.

Crow, T. (2007). How and why genetic linkage has not solved the problem of psychosis. *American Journal of Psychiatry, 164*(1), 13–21.

Dilsaver, S., Akiskal, H., Akiskal, K., & Benazzi, F. (2006). Dose-response relationship between number of comorbid anxiety disorders in adolescent bipolar/unipolar disorders, and psychosis, suicidality, substance abuse and familiality. *Journal of Affective Disorders, 96*(3), 249–258.

DSM-IV-TR. (2000). *Diagnostic and statistical manual of mental disorders, Fourth Edition, text revision*. Washington, D C: American Psychiatric Association.

Edmonds, L. K., Mosley, B. J., Admiraal, A. J., Olds, R. J., Romans, S. E., Silverstone, T., et al. (1998). Familial bipolar disorder: preliminary results from the Otago Familial Bipolar Genetic Study. *Australian and New Zealand Journal of Psychiatry, 32*(6), 823–829.

Einat, H., & Manji, H. (2006). Cellular plasticity cascades: genes-to-behavior pathways in animal models of bipolar disorder. *Biological Psychiatry, 59*(16), 1160–1171.

Geddes, J., Black, R., Whalley, L., & Eagles, J. (1993). Persistence of the decline in the diagnosis of schizophrenia among first admissions to Scottish hospitals from 1969 to 1988. *British Journal of Psychiatry, 164*(4), 620–626.

Gottesman, I. I., & Bertelsen, A. (1989). Confirming unexpressed genotypes for schizophrenia. Risks in the offspring of Fischer's Danish identical and fraternal discordant twins. *Archives of General Psychiatry, 46*(10), 867–872.

Gottesman, I. I., & Gould, T. D. (2003). The endophenotype concept in psychiatry: etymology and strategic intentions. *American Journal of Psychiatry, 160*(4), 1–10.

Gottesman, I. I., & Hanson, D. R. (2005). Human development: biological and genetic processes. *Annual Review of Psychology, 56*, 263–286.

Gottesman, I. I., & Shields, J. (1972). *Schizophrenia and genetics: a twin study vantage point*. New York: Academic Press.

Gottesman, I. I., Shields, J., & Hanson, D. R. (1982). *Schizophrenia: the epigenetic puzzle*. Cambridge: Cambridge University Press.

Hanson, D. R. (2004). Getting the bugs into our genetic theories of schizophrenia. In L. DiLalla (Ed.), *Behavior genetics principles: Perspective in development, personality and psychopathology* (pp. 205–216). Washington, DC: American Psychological Press.

Hanson, D. R., & Gottesman, I. I. (1976). The genetics, if any, of infantile autism and childhood schizophrenia. *Journal of Autism and Childhood Schizophrenia, 6*(3), 209–234.

Hanson, D. R., & Gottesman, I. I. (2005). Theories of schizophrenia: a genetic-inflammatory-vascular synthesis. *BMC Medical Genetics, 6*, 7.

Hanson, D. R., & Gottesman, I. I. (2007). Choreographing genetic, epigenetic, and stochastic steps in the dances of psychopathology. In A. Maston (Ed.), *Multilevel dynamics in developmental psychopathology* (pp. 27–44). Mahwal, NJ: Erlbaum Associates.

Hanson, D. R., Gottesman, I. I., & Meehl, P. E. (1977). Genetic theories and the validation of psychiatric diagnoses: implications for the study of children of schizophrenics. *Journal of Abnormal Psychology, 86*(6), 575–588.

Head, E., & Lott, I. (2004). Down syndrome and beta-amyloid deposition. *Current Opinion in Neurology, 17*, 95–100.

Helzer, J., Robins, L., McEvoy, L, T, Spitznagel, E. L., Stoltzman, R., Farmer, A., et al. (1985). A comparison of clinical and diagnostic interview schedule diagnoses. Physician reexamination of lay-interviewed cases in the general population. *Archives of General Psychiatry, 41*(7), 657–666.

Hoek, H. W., Brown, A. S., & Susser, E. (1998). The Dutch famine and schizophrenia spectrum disorders. *Social Psychiatry and Psychiatric Epidemiology, 33*(8), 373–379.

Howard, R., Rabins, P. V., Seeman, M. V., & Jeste, D. V. (2000). Late-onset schizophrenia and very-late-onset schizophrenia-like psychosis: an international consensus. The International Late-Onset Schizophrenia Group. *American Journal of Psychiatry, 157*(2), 172–178.

Hulshoff Pol, H. E., Brans, R. G. H., van Haren, N. E. M., Schnack, H. G., Langen, M., Baare, W. F. C., et al. (2004). Gray and white

matter volume abnormalities in monozygotic and same-gender dizygotic twins discordant for schizophrenia. *Biological Psychiatry, 55*, 126–130.

Hulshoff Pol, H. E., Hoek, H. W., Susser, E., Brown, A. S., Dingemans, A., Schnack, H. G., et al. (2000). Prenatal exposure to famine and brain morphology in schizophrenia. *American Journal of Psychiatry, 157*(7), 1170–1172.

Jablonka, E., & Lamb, M. (2002). The changing concept of epigenetics. *Annals of the New York Academy of Sciences, 981*(Dec), 82–96.

Jabs, B., Althaus, G., Bartsch, A., Schmidtke, A., Stöber, G., Beckmann, H., et al. (2006). Cycloid psychoses as atypical manic-depressive disorders. Results of a family study. *Nervenarzt, 77*(9), 1096–1100, 1102–1104.

Jaenisch, R., & Bird, A. (2003). Epigenetic regulation of gene expression: how the genome integrates intrinsic and environmental signals. *Nature Genetics, 33 Suppl* (Mar), 245–254.

Jones, I., Kent, L., & Craddock, N. (2004). Genetics of affective disorders. In P. McGuffin, O. Palo, M. Owen & I. Gottesman (Eds.), *Psychiatric genetics & genomics* (pp. 211–245). Oxford: Oxford University Press.

Jones, P., & Takai, D. (2001). The role of DNA methylation in mammalian epigenetics. *Science, 263*, 1068–1070.

Joyce, P. R., Doughty, C. J., Wells, J. E., Walsh, A. E., Admiraal, A., Lill, M., et al. (2004). Affective disorders in the first-degree relatives of bipolar probands: results from the South Island Bipolar Study. *Comprehensive Psychiatry, 45*(3), 168–174.

Keating, M., Dunn, C., Atkinson, D., Timothy, K., Vincent, G. M., & Leppert, M. (1991). Consistent linkage of the long-QT syndrome to the Harvey ras-1 locus on chromosome 11. *American Journal of Medical Genetics, 49*(6), 1335–1339.

Keating, M., & Sanguinetti, M. (2001). Molecular and cellular mechanisms of cardiac arrhythmias. *Cell, 104*(659–580).

Kessler , R., McGonagle, K., Zhao, S., Nelson, C., Hughes, M., Eshleman, S., et al. (1994). Lifetime and 12-month prevalence of DSM-III-R psychiatric disorders in the United States. Results from the National Comorbidity Survey. *Archives of General Psychiatry, 51*, 8–19.

Kornberg, J., Kawashima, H., Pulst, S., Allen, L., Magenis, E., & Epstein, C. (1990). Down syndrome: toward a molecular definition of the phenotype. *American Journal of Medical Genetics Suppl, 7*, 91–97.

Kraepelin, E. (1919). *Dementia praecox and paraphrenia* (R. Barclay, Trans.). Edinburgh: E & S Livingston.

Kuratomi, G., Iwamoto, K., Bundo, M., Kusumi, I., Kato, N., Iwata, N., et al. (2007). Aberrant DNA methylation associated with bipolar disorder identified from discordant monozygotic twins. *Molecular Psychiatry, 13*(4), 429–441.

Lake, C. (2007). Disorders of thought are severe mood disorders: the selective attention defect in mania challenges the Kraepelinian dichotomy-a review. *Schizophrenia Bulletin* (e-pub).

Leonhard, K. (1961). Cycloid psychoses-endogenous psychoses which are neither schizophrenic nor manic depressive. *Journal of Mental Science, 197*, 632–648.

Loranger, A. W., & Levine, P. M. (1978). Age at onset of bipolar affective illness. *Archives of General Psychiatry, 35*(11), 1345–1348.

Maier, W., Hofgen, B., Zobel, A., & Rietschel, M. (2005). Genetic models of schizophrenia and bipolar disorder: overlapping inheritance or discrete genotypes? *European Archives of Psychiatry and Clinical Neuroscience, 255*(3), 159–166.

Manji, H. K., Gottesman, I. I., & Gould, T. D. (2003). Signal transduction and genes-to-behaviors pathways in psychiatric diseases. *Science's STKE, 2003*(207), pe49.

McGrath, J. J., Saha, S., Lieberman, D. E., & Buka, S. (2006). Season of birth is associated with anthropometric and neurocognitive outcomes during infancy and childhood in a general population birth cohort. *Schizophrenia Research, 81*(1), 91–100.

Merikangas, K. R., Akiskal, H. S., Angst, J., Greenberg, P. E., Hirschfeld, R. M., Petukhova, M., et al. (2007). Lifetime and 12-month prevalence of bipolar spectrum disorder in the National Comorbidity Survey replication. *Archives of General Psychiatry, 64*(5), 543–552.

Merikangas, K. R., & Low, N. C. (2004). The epidemiology of mood disorders. *Current Psychiatry Reports, 6*(6), 411–421.

Meyer, U., Yee, B., & Feldon, J. (2007). The neurodevelopmental impact of prenatal infections at different times of pregnancy: the earlier the worse?*Neuroscientist, 13*(3), 241–256.

Morange, M. (2002). The relations between genetics and epigenetics: a historical point of view. *Annals of the New York Academy of Sciences, 981*(Dec), 50–60.

Murray, R., Jones, P., Susser, E., Van Os, J., & Cannon, M. (Eds.). (2003). *The epidemiology of Schizophrenia.* Cambridge: Cambridge University Press.

Nijhout, H. (2003). The importance of context in genetics. *American Scientist, 91*, 416–423.

Nordgaard, J., Arnfred, S., Handest, P., & Parnas, J. (2008). The diagnostic status of first-rank symptoms. *Schizophrenia Bulletin, 34*(1), 137–154.

Ødegaard, Ø. (1972). The multifactorial theory of inheritance in predisposition to schizophrenia. In A. Kaplan (Ed.), *Genetic factors in "schizophrenia"* (pp. 256–275). Springfield: Charles C Thomas.

Opitz, J., & Gilbert-Barness, E. (1990). Reflections on the pathogenesis of Down syndrome. *American Journal of Medical Genetics Suppl, 7*, 38–51.

Osby, U., Hammar, N., Brandt, L., Wicks, S., Thinsz, Z., Ekbom, A., et al. (2001). Time trends in first admissions for schizophrenia and paranoid psychosis in Stockholm County, Sweden. *Schizophrenia Research, 47*(2–3), 247–254.

Palo, O. M., Antila, M., Silander, K., Hennah, W., Kilpinen, H., Soronen, P., et al. (2007). Association of distinct allelic haplotypes of DISC1 with psychotic and bipolar spectrum disorders and with underlying cognitive impairments. *Human Molecular Genetics, 16*(20), 3517–3528.

Peralta, V., Cuesta, M., & Zandio, M. (2007). Cycloid psychoses: an examination of the validity of the concept. *Current Psychiatry Reports, 9*(3), 184–192.

Perris, C., & Brockington, I. (1981). Cycloid psychoses and their relation to the major psychoses. In C. Perris, G. Struwe & B. Jansson (Eds.), *Biological Psychiatry* (pp. 447–450). Amsterdam: Elsevier.

Petronis, A. (2003). *Epigenetics: influence on behavioral disorders.* London: Nature Publishing Group.

Petronis, A. (2004). The origin of schizophrenia: genetic thesis, epigenetic antithesis, and resolving synthesis. *Biological Psychiatry, 55*(10), 965–970.

Petronis, A., Gottesman, I. I., Kan, P., Kennedy, J. L., Basile, V. S., Paterson, A. D., et al. (2003). Monozygotic twins exhibit numerous epigenetic differences: clues to twin discordance? *Schizophrenia Bulletin, 29*(1), 169–178.

Pfuhlmann, B., Jabs, B., Althaus, G., Schmidtke, A., Bartsch, A., Stöber, G., et al. (2004). Cycloid psychoses are not part of a bipolar affective spectrum: results of a controlled family study. *Journal of Affective Disorders, 83*(1), 11–19.

Reed, S., Hartley, C., Anderson, V., Phillips, V., & Johnson, N. (1973). *The psychoses: family studies.* Philadelphia: W.B. Saunders.

Rzhetsky, A., Wajngurt, D., Park, N., & Zheng, T. (2007). Probing genetic overlap among complex human phenotypes. *Proceedings of the National Academy of Sciences of the United States of America, 104*(28), 11694–11699.

Sanders, A. R., Duan, J., Levinson, D. F., Shi, J., He, D., Hou, C., et al. (2008). No significant association of 14 candidate genes with schizophrenia in a large European ancestry sample: Implications for

psychiatric genetics. *American Journal of Psychiatry* (e-pub).

Sapolsky, R. M. (2004). Mothering style and methylation. *Nature Neuroscience, 7*(8), 791–792.

Schneider, K. (1959). *Clinical Psychopathology.* New York: Grune & Stratton.

Schulze, T. G., Hedeker, D., Zandi, P., Rietschel, M., & McMahon, F. J. (2006). What is familial about familial bipolar disorder? Resemblance among relatives across a broad spectrum of phenotypic characteristics. *Archives of General Psychiatry, 63*(12), 1368–1376.

Shahbazian, M., & Zoghbi, H. (2002). Rett syndrome and the MeCP2: Linking epigenetics and neuronal function. *American Journal of Medical Genetics, 71*(6), 1259–1272.

Shields, J., & Gottesman, I. I. (1973). Genetic Studies of schizophrenia as signposts to biochemistry. *Biochemical Society 1*, 165–174 (Special Publication).

Simon, N., Smoller, J., Fava, M., Sachs, G., Racette, S., Perlis, R., et al. (2003). Comparing anxiety disorders and anxiety-related traits in bipolar disorder and unipolar depression. *Journal of Psychiatric Research, 37*(3), 187–192.

Sing, C., Stengard, J., & Kardia, S. (2003). Genes, environment, and cardiovascular disease. *Arteriosclerosis, Thrombosis, and Vascular Biology, 23*(7), 1190–1196.

Slater, E., & Cowie, V. (1971). *The genetics of mental disorders.* London: Oxford University Press.

Smith, S., Li, J., Garbett, K., Mirnics, K., & Patterson, P. (2007). Maternal immune activation alters fetal brain development through interleukin-6. *Journal of Neuroscience, 27*(40), 10695–10702.

Somanath, C., Jain, S., & Reddy, Y. (2002). A family study of early-onset bipolar I disorder. *Journal of Affective Disorders, 70*(1), 91–94.

St Clair, D., Xu, M., Wang, P., Yu, Y., Fang, Y., Zhang, F., et al. (2005). Rates of adult schizophrenia following prenatal exposure to the Chinese famine of 1959–1961. *Journal of the American Medical Association, 294*(5), 557–562.

Strömgren, E. (1994). The unitary psychosis (Einheitpsychose) concept: past and present. *Neurology, Psychiatry, and Brain Research, 2*, 201–205.

Suvisaari, J., Haukka, J., Tanskanen, A. J., & Lönnqvist, J. K, (1999). Decline in the incidence of schizophrenia in Finnish cohorts born from 1954 to 1965. *Archives of General Psychiatry, 56*(8), 733–740.

Thomson, G., & Esposito, M. S. (1999). The genetics of complex diseases. *Trends in Cell Biology, 9*(12), M17–20.

Tillman, R., & Geller, B. (2006). Controlled study of switching from attention-deficit/hyperactivity disorder to a prepubertal and early adolescent bipolar I disorder phenotype during 6-year prospective follow-up: rate, risk, and predictors. *Development and Psychopathology, 18*(4), 1037–1053.

Tsuang, M. T., & Winokur, G. (1975). The Iowa 500: field work in a 35-year follow-up of depression, mania, and schizophrenia. *Canadian Psychiatric Association Journal, 20*(5), 359–365.

Tsuang, M. T., Winokur, G., & Crowe, R. R. (1980). Morbidity risks of schizophrenia and affective disorders among first degree relatives of patients with schizophrenia, mania, depression and surgical conditions. *British Journal of Psychiatry, 137*, 497–504.

Turkheimer, E., Goldsmith, H., & Gottesman, I. (1995). Commentary – some conceptual deficiencies in 'developmental' behavioral genetics. *Human Development, 38*, 142–153.

van Erp, T. G., Saleh, P. A., Huttunen, M., Lonnqvist, J., Kaprio, J., Salonen, O., et al. (2004). Hippocampal volumes in schizophrenic twins. *Archives of General Psychiatry, 61*(4), 346–353.

van Haren, N. E., Picchioni, M. M., McDonald, C., Marshall, N., Davis, N., Ribchester, T., et al. (2004). A controlled study of brain structure in monozygotic twins concordant and discordant for schizophrenia. *Biological Psychiatry, 56*(6), 454–461.

Waddington, C. (1957). *The strategy of the genes.* London: George Allen & Unwin LTD.

Weaver, I. C., Cervoni, N., Champagne, F. A., D'Alessio, A. C., Sharma, S., Seckl, J. R., et al. (2004). Epigenetic programming by maternal behavior. *Nature Neuroscience, 7*(8), 847–854.

Weaver, I. C., Grant, R. J., & Meaney, M. J. (2002). Maternal behavior regulates long-term hippocampal expression of BAX and apoptosis in the offspring. *Journal of Neurochemistry, 82*(4), 998–1002.

Winokur, G., Coryell, W., Akiskal, H. S., Endicott, J., Keller, M., & Mueller, T. (1994). Manic-depressive (bipolar) disorder: the course in light of a prospective ten-year follow-up of 131 patients. *Acta Psychiatrica Scandinavica, 89*(2), 102–110.

Wong, A. H., Gottesman, I. I., & Petronis, A. (2005). Phenotypic differences in genetically identical organisms: the epigenetic perspective. *Human Molecular Genetics, 14 Spec No 1*, R11–18.

Chapter 32

Genetic Risks in Schizophrenia: Cross-National Prospective Longitudinal High-Risk Studies

Judith G. Auerbach, L. Erlenmeyer-Kimling, Barbara Fish, Sydney L. Hans, Loring J. Ingraham, Joseph Marcus, Thomas F. McNeil, and Erland Schubert

Introduction

Prospective longitudinal studies are a powerful means to identify the causal chains of biological and environmental factors that underlie the development of serious mental disorders. Schizophrenia is one of these disorders. Schizophrenia is a multifactorial neurodevelopmental disorder whose specific molecular genetic, epigenetic, stochastic, and environmental bases remain elusive. The chronic debilitating nature of the disorder places a heavy emotional and financial burden on the individual, family, and society. Although the lifetime prevalence of schizophrenia is 1% in the general population that has passed through the risk period, it accounts for up to 3% of total national health-care expenditures in Western countries (Knapp, Mangalore, & Simon, 2004). The hypothetical ability to intervene in the developmental progression of the disorder at a point before breakdown is contingent upon the elucidation of early behavioral and other markers of genetic liability to the disorder and their triggers. In this chapter, we focus on research aimed at early detection of markers that may predict to the later onset of schizophrenia. Once these markers are identified, intervention can, at least in principle, be targeted to those individuals most at risk for the disorder.

Background

Strong evidence for genetic factors in schizophrenia comes from family, adoption, and twin studies and more recent molecular approaches (Hanson, this volume). Lifetime risk for schizophrenia averages 13% for the offspring of schizophrenia probands, 10% for their siblings, and 6% for their parents (Owen, O'Donovan, & Gottesman, 2002). The risk to adopted-away children of biological parents with schizophrenia is in the range of that for children brought up with their own schizophrenic parents (Kety, Wender, Jacobsen, & Ingraham, 1994; Tienari et al., 2004). Twin studies are usually presented as evidence for the role of heredity in schizophrenia because of the fivefold or greater concordance rate for the disorder in monozygotic compared with dizygotic twins. Nevertheless, the consistent finding that concordance in genetically identical monozygotic twins is considerably below 100% also constitutes evidence for the influence of environmental and epigenetic factors – many acting during gestation, as well as throughout later developmental periods – that may impinge differentially on the expression of the initially identical genomes and thus lead to discordance for the illness in some monozygotic twin pairs (Wong, Gottesman, & Petronis, 2005). What is inherited, therefore, is not the disorder itself but a liability to develop the disorder. Whether schizophrenia is the ultimate outcome depends not only on this liability but also on other causal, mediating, moderating, and/or chance factors.

Models of schizophrenia incorporating biological and environmental factors were first formally proposed by Gottesman and Shields (1967, 1972), Meehl (1962), and Rosenthal (1970). These diathesis–stressor models conceptualize schizophrenia as the outcome of gene–environment interactions. According to such models, individuals differ not only in their degree of genetic liability to schizophrenia but also in the type and strength of environmental/epigenetic stressors to which they may be exposed. The result is a broad continuum ranging from low to high risks to schizophrenia, depending on the joint impact of genetic and environmental/epigenetic factors. The diathesis–stressor models have gradually contributed to the current conceptualization of a spectrum ranging from less to more disabling, with schizophrenia as the most extreme expression (Kendler, McGuire, Gruenberg, & Walsh, 1995). Recognition of this spectrum of less malignant disorders helps to clarify some of the conditions that earlier investigators (e.g., Kallmann, 1938; Slater, 1968) struggled to define in twin and family studies and casts light on the old idea of "incomplete

J.G. Auerbach (✉)
Department of Psychology, Ben-Gurion University, Beer-Sheva, Israel
e-mail: judy@bgu.ac.il

Y.-K. Kim (ed.), *Handbook of Behavior Genetics*,
DOI 10.1007/978-0-387-76727-7_32, © Springer Science+Business Media, LLC 2009

penetrance" when thought was focused only on single major locus models (either recessive or dominant). Underlying genetic liability is not always expressed phenotypically as schizophrenia (Faraone, Green, Seidman, & Tsuang, 2001; Meehl, 1990), or even as major spectrum disorders, as illustrated by Gottesman and Bertelsen's (1989) study of offspring of twins discordant for schizophrenia spectrum disorders. In monozygotic pairs, the rate of these disorders was the same in offspring of the schizophrenic twins as in offspring of the fully discordant cotwins. In dizygotic pairs, however, offspring of the phenotypically unaffected twin showed much lower rates similar to those found for second-degree relatives (e.g., nieces/nephews) of schizophrenic probands. Thus, the phenotypically unaffected twins in the monozygotic pairs, but not the dizygotic pairs, shared the same genetic liability as the schizophrenic cotwins but it remained unexpressed.

Nevertheless, first-degree relatives of schizophrenic patients are usually at increased risk not only for schizophrenia itself, but also for several personality disorders, especially schizotypal and perhaps paranoid personality disorders (e.g., Baron et al., 1985; Erlenmeyer-Kimling et al., 1997; Kendler & Gardner, 1997; Kety et al., 1994; Maier, Lichtermann, Minges, & Heun, 1994; Webb & Levinson, 1993) and sometimes schizoid (Kendler et al., 1993) or avoidant personality disorder (Hans, Auerbach, Styr, & Marcus, 2004). In fact, first-degree relatives often also display schizotypal symptoms that may be insufficient in number or severity to reach diagnostic criteria for schizotypal personality disorder and, yet, stand out from control or general population profiles (Clementz, Grove, Katsanis, & Iacono, 1991; Kendler et al., 1995; Squires-Wheeler et al., 1997). The common underpinning for these various clinical manifestations of genetic liability may relate to deficient neurobehavioral functioning.

Since Kraepelin (1919) first defined *dementia praecox*, neurological signs and neurobehavioral deficits (e.g., disturbances in cognitive, motor, psychophysiological performance) have been a recurring finding in empirical studies of schizophrenia (Cox & Ludwig, 1979; Manschreck & Ames, 1984; Quitkin, Rifkin, & Klein, 1976). These deficits have been called "endophenotypes" by Gottesman and colleagues (cf., Gottesman & Shields, 1972; Gottesman & Gould, 2003) because they are more directly expressive of gene action than are, for example, hallucinations and delusions. Not only are such features seen in individuals with schizophrenia, but they also are more prevalent in first-degree relatives (Clementz, Sweeney, Hirt, & Haas, 1992; Faraone et al., 1995; Holzman et al., 1974; Keefe et al., 1997; Pogue-Geile, Garrett, Brunke, & Hall, 1991) than in controls. Meehl (1962) suggested that the underlying basis of these deficits is a neurointegrative defect, a term introduced by Fish (1957), which Meehl called "schizotaxia". In Meehl's

view, schizotaxia is expressed in most circumstances as "schizotypy", a nonpsychotic personality organization that develops into schizophrenia in certain environmental conditions (cf., Lenzenweger, Maher, & Manschreck, 2005).

Recently Tsuang, Faraone and colleagues (Faraone et al., 2001; Tsuang, Stone, & Faraone, 2000; Tsuang, Stone, Tarbox, & Faraone, 2002) have "reformulated" the concept of schizotaxia, proposing that schizotaxia is expressed through neuropsychological deficits as well as negative schizotypal symptoms that are qualitatively similar to, but milder than, those seen in schizophrenia. The investigators' studies suggest that 20–50% of first-degree relatives of schizophrenic probands manifest a "schizotaxic" behavior profile, not unlike Heston's (1970) earlier defense of a dominant gene model.

Thus, schizophrenia is currently conceptualized as a neurodevelopmental disorder grounded in developmental processes early in life (e.g., Cicchetti & Cannon, 1999; Gooding & Iacono, 1995; Murray & Fearon, 1999; Walker & Neumann, 1995; Weinberger, 1986). Fish (Fish, 1957, 1977; Fish, Marcus, Hans, Auerbach, & Perdue, 1992; Fish & Kendler, 2005) was the first to provide evidence that genetic liability to the disorder could be detected as a neurointegrative defect in the first weeks of life in infants of mothers with schizophrenia. Her research set the stage for later prospective, longitudinal "high-risk" studies that were intended to identify endophenotypic deficiencies in individuals considered to be genetically susceptible to schizophrenia. Pearson & Kley (1957) formalized the underlying theory of high-risk research, but it was Norman Garmezy (cf., Garmezy & Streitman, 1974, Part I; Garmezy, 1974, Part II) who undertook the role of cheerleader and mentor to the emerging field of studies, at least those that started in the United States.

Approaches Used in Studies of Early Development in Schizophrenia

For many years, retrospective accounts – usually unstructured – obtained from schizophrenic individuals and/or their relatives provided the only source of information about early development of the disorder. Usually, the retrospective accounts were aimed at possible person-based and/or environment-based precursors of the illness, in line with then-contemporary hypotheses about the origins of schizophrenia. Clearly, however, such sources were likely to be unreliable and tainted by knowledge of the eventual psychiatric outcome and, thus needed to be replaced by newer research approaches (Mednick & McNeil, 1968). The main longitudinal approaches that followed represent an improvement in objectivity and have enhanced our knowledge about possible early behavioral precursors to schizophrenia spectrum disorders.

Regardless of recruitment approach and type of high-risk study, expectations for high-risk research are that some relatively specific deficits in neurobehavioral traits will differentiate the high risk for schizophrenia (HRS) group from the no-parental-mental-illness (NMI) comparison group. Moreover, when a psychiatric comparison group is included, the neurobehavioral deficits must differentiate between the HRS group and a group at high risk for other mental disorders (HRO), to classify the marker as specific to the liability to schizophrenia spectrum disorders. Important also is the fact that the deficit must be present only in a subgroup of the HRS group because not all of these children will inherit the genetic liability to the disorder, and those who do not are not expected to show neurobehavioral impairment. Further cross-sectional differences between offspring of schizophrenic and offspring of other parent groups used as control do not necessarily represent antecedents of schizophrenia or "schizophrenicity" (McNeil & Kaij, 1979). Such differences may simply reflect combined genetic and environmental consequences of having, as opposed to not having, a schizophrenic parent. Identification of predictive antecedents depends on prospective, longitudinal data, rather than cross-sectional comparisons.

Longitudinal studies differ in their methods of subject recruitment, and in the extent to which some or all of the data come from investigations specifically focused on the development of mental illness or from research initially intended to examine other issues. Three longitudinal approaches can be distinguished, although in practice many studies use some combination of the second and third approaches. A less common approach sometimes taken follows psychiatric outcomes in individuals selected for some behaviors thought to reflect schizophrenia proneness – e.g., anhedonia, perceptual aberrations (Chapman, Edell, & Chapman, 1980), or a given type of biological response pattern (e.g., autonomic nervous system responses), as in a study of preschool children on the Island of Mauritius (Schulsinger, Mednick, Venables, Raman, & Bell, 1975).

(1) **Follow-back studies** examine some aspects of early development in known schizophrenic patients and case controls, for whom archival records such as teachers' reports, school grades, home movies, etc., are available (Niemi, Suvisaari, Tuulio-Henriksson, & Lönnqvist, 2003; Walker, Grimes, & Davis, 1993; Walker & Lewine, 1990; Watt, 1978). This type of study, however, is limited by the fact that the available records were not originally intended to provide information about pre-schizophrenic states or about precursors of the disorder. Also, follow-back studies cannot typically contribute information about the genetics of schizophrenia. (2) **Large epidemiological data sets,**

obtained on birth cohorts, for example, have been used, with linkage to population registers or other record systems, to identify individual members of the cohort who either have become schizophrenic patients in adulthood or are classified as being at risk because record searches have tagged them as offspring of affected parents. Data from the (US) National Collaborative Perinatal Project and from British and New Zealand birth cohort studies have been used in both of these ways (Cannon et al., 2000; Cannon et al., 2002; Jones, 1995; Jones, Rodgers, Murray, & Marmot, 1994; Rosso et al., 2000). Data from these studies are valuable because the samples are relatively representative of the population of individuals who develop schizophrenia. Again, however, the available data on early development may be less complete than desired because the original epidemiological investigations were not designed with schizophrenia in mind and typically do not have measures selected specifically for their relevance to theories about the development of this disorder.

(3) Another main approach employs a **prospective, longitudinal high-risk strategy.** In this paradigm, individuals at high genetic risk to a disorder – usually first-degree relatives of patients – are targeted as the subject population in infancy or childhood and then followed over time to identify possible endophenotypic markers as predictors of the disorder. The high-risk research paradigm is of particular value in the investigation of disorders with low prevalence in the general population (Garmezy & Streitman, 1974; Pearson & Kley, 1957). An advantage of this approach, compared with the other two methods, is that the investigators can select measures based on their likely relevance to contemporary conceptualizations of schizophrenia. In fact, over the course of the prospective high-risk studies, a major paradigm shift occurred in the causal models proposed to account for schizophrenia. Psychosocial and psychodynamic hypotheses, which had dictated the types of material collected retrospectively about the early backgrounds of adult patients, began to yield their dominance to a growing interest among researchers in the possible biological roots of schizophrenia. As it happened, most of the investigators who succeeded in carrying out longitudinal follow-up of high-risk subjects held a biological viewpoint from the beginning and thus tackled questions about early development with assessments of biological and neurobehavioral characteristics. To a large extent, then, the prospective studies were in the vanguard of the trend toward biological psychiatry that was emerging and that has since shaped the current "mainstream" of hypotheses and assessment in schizophrenia research. While many measures available when the first high-risk studies began (Fish, 1957, 1977; Mirsky, Silberman, Latz, & Nagler, 1985) are considered less sophisticated than those introduced later on, they nevertheless tapped important trait constructs that

have shown logical consistency over developmental stages and across studies – even those started at much later dates.

An important part of being able to select one's measures is that they can be adjusted to allow similar trait constructs (e.g., selective attention, working memory) to be tested at different ages. This flexibility in modifying measures is crucial because it permits assessments of the same construct across different developmental stages. Moreover, an index of deviance on measures of multiple constructs can be derived and is likely to prove more reliable than a single measure as a predictor of later schizophrenia (Erlenmeyer-Kimling, 2000; Erlenmeyer-Kimling et al., 2000).

Like the other approaches, a prospective longitudinal high-risk strategy is not without its disadvantages. The samples are often small and it is difficult to maintain the sample over time. Selective dropout may affect the validity of both cross-sectional and predictive findings. There is also the question of the generalizability to the general population when findings are based on groups at known genetic risk for schizophrenia. A biobehavioral marker with high predictive value for individuals at genetic risk for schizophrenia may not be predictive for individuals without genetic risk.

Ultimately, the goal of prospective, longitudinal studies of high-risk populations concerns the two intertwined aspects of prediction, as related to intervention and prevention: the search for early indicators/markers that point to later emergence of the disorder and offer clues as to **what** needs to be targeted by interventions, and the search for at-risk individuals **who**, by displaying these markers, appear to be the most likely candidates to warrant interventions. When "risk" is based a genetic relationship to a schizophrenic proband, the "what" is likely to be an endophenotype (Gottesman & Gould, 2003), which is more proximal to the biological origins of the disorder than are clinical symptoms and represents an intermediary state between the genetic liability and a possible outcome in the schizophrenia spectrum. As noted previously, the scope of "who" is at highest risk may not be entirely clear, however, because even individuals with the genetic liability may not always develop schizophrenia or major signs of the spectrum personality disorders. For example, the study noted earlier by Gottesman and Bertelsen (1989) clearly demonstrated that the schizophrenic and phenotypically unaffected members of discordant monozygotic twin pairs share the same genetic liability, as evidenced by their transmission of schizophrenia risk to their offspring at the same rate.

The Prospective High-Risk Studies

This chapter highlights contributions to the search for possible genetic liability markers and putative behavioral predictors of the disorder from prospective, longitudinal studies meeting four specific criteria: (1) the subjects under study are offspring of parents with schizophrenia; (2) recruitment and first assessment occurred before adolescence, thus avoiding possible confusion of pre-schizophrenic predictors with signs occurring in the prodrome (the disturbed period immediately preceding psychosis onset in many individuals); (3) neurobehavioral (endophenotypic) assessments were a critical part of the assessment battery; and (4) the samples were followed at least until late adolescence or early adulthood. Of the six studies considered here, four started with infants, many of whom are only now coming of age, as they mature into late adolescence or young adulthood, during which schizophrenia typically emerges. Several well-known, frequently referenced studies are mentioned where appropriate but are not highlighted because they did not meet the above criteria.

The highlighted studies are two from the United States – the New York Infant Study (**NYIS**) and the New York High-Risk Project (**NYHRP**), two from Israel – the Israeli High-Risk Study (**IHRS**) and the Jerusalem Infant Development Study (**JIDS**), one from Sweden – the Swedish High-Risk Study (**SHRP**), and one from Denmark – the Danish Cohort Study (**DCS**). Along with the high risk for schizophrenia (HRS) group, each study includes a no-parental-mental-illness (NMI) comparison group, and four studies (**NYHRP, JIDS, SHRP and DCS**) also include a comparison group of children at high risk for other mental disorders (HRO). In general, the studies espoused Paul Meehl's notion of schizotaxia as a neurointegrative defect that would be manifested both early in life and in a number of neurodevelopmental domains. Table 32.1 lists the principal investigators and a brief description of each study. Table 32.2 presents the ages (mean and range) and the number of subjects in each parental diagnostic group at each main assessment period for each study.

Abbreviations

Individual study names

DCS	Danish Cohort Study
IHRS	Israeli High-Risk Study
JIDS	Jerusalem Infant Development Study
NYIS	New York Infant Study
NYHRP	New York High-Risk Project
SHRP	Swedish High-Risk Project

Subject subgroup names

HRS	High risk to schizophrenia
HRO	High risk to other mental disorders
NMI	No parental mental illness

Table 32.1 Prospective studies of offspring of schizophrenic parents

Study name (abbreviation)	Year started	Current principal investigator(s)	Description
New York Infant Study (NYIS)	1952	B. Fish	Sample of schizophrenic and normal women giving birth at a New York City hospital. Infants followed from birth, with examinations at days 1 and 3, months 1–4, 6, 9, 13, 18, and 24, one examination each in childhood and adolescence, two examinations in adulthood up to mean age 28
Israeli High-Risk Study (IHRS)	1967	L. J. Ingraham (initiated by D. Rosenthal and continued by A. Mirsky)	Sample of offspring of schizophrenic and normal parents, half of each parent group kibbutz reared, half town reared. Offspring followed from mean age 11–31 years, with examinations in childhood, adolescence, and adulthood
New York High-Risk Project (NYHRP)	Sample A: 1971 Sample B: 1978 Siblings: 1985	L. Erlenmeyer-Kimling	Samples A and B: schizophrenic and affective parents screened from consecutive admissions at New York state psychiatric hospitals, and offspring with normal parents from schools or population sampling. Children followed from mean age 9–40 years, with one examination in childhood and three examinations in adolescence, and three examinations in adulthood. Siblings from both samples followed from age 18 to –60 years
Jerusalem Infant Development Study (JIDS)	Targets: 1973 Sibs: 1983 New Recruits: 1991	S. L. Hans J. Marcus J. G. Auerbach	Original offspring sample followed from birth with examinations at days 3 and 14, months 4, 8, and 12, one examination each in childhood and adolescence. In childhood, siblings recruited and followed from mean age 10–17 years. New recruits and additional siblings were examined once only at adolescent ages 12–22
Swedish High-Risk Project (SHRP)	1973	T. F. McNeil	Sample of mentally ill and normal women identified as pregnant and followed to birth of infants. Prenatal testing of mothers, researchers present at birth, examinations of infants at day 3, weeks 3 and 6, months 3, 6, and 12, one examination each in childhood, adolescence, and early adulthood
Danish Cohort Study (DCS)	1968	S. A. Mednick	Selected from the National Danish Birth Cohort Study (9,148 births) and followed from birth with multiple examinations to 12 months by the epidemiological birth cohort study; 183 offspring of mentally ill parents identified through the national psychiatric register and 82 offspring of normal parents followed for the high-risk study with one examination in childhood ages 11–13 years and adulthood ages 31–32 years

Findings from the Prospective, Longitudinal High-Risk Studies

Neurobehavioral Deficits as Possible Endophenotypes

Neurobehavioral deficits in infancy, childhood, and adolescence are presented in Table 32.3.

Infancy: Neurobehavioral functioning in HRS infants was first assessed in Fish's, (1957, 1977) ground-breaking **NYIS**. Fish extensively examined a number of aspects of infant development during the first 2 years of life. A subset of these infants showed a simultaneous occurrence of retarded cranial growth, as well as retarded and erratic postural-motor development on one or more examinations between 2 and 8 months of age. Fish termed this picture of disorder "pandysmaturation" to reflect the several domains of abnormal development. She regarded pandysmaturation as a disorder of the timing and integration of neurological maturation, which reflected genetic schizotypal traits and predicted a schizotypal outcome (Fish, 1977; Fish et al., 1992; Fish & Kendler, 2005; Fish, Hans, Marcus, & Auerbach, 2005).

Three other studies, the **SHRP**, **JIDS**, and **DCS**, with data from birth, also found that HRS infants were characterized by developmental retardation during this period, and showed a neurobehavioral profile differentiating them from infants in the NMI and HRO comparison groups. Again, the behaviors characterized a subset of HRS infants, rather than all of them. In the **SHRP**, a subset of the HRS infants showed decreased arousal, combined with neurological abnormalities, and deviant sensitivity to stimulation (Schubert, Blennow, & McNeil, 1996). During the first 4 years of life, HRS offspring also showed a higher frequency of developmental deviations – including delayed walking, visual dysfunction, language delay, enuresis, poor social competence, and social withdrawal – than offspring in the comparison groups (Henriksson & McNeil, 2004).

In the **JIDS**, in addition to group differences reflecting developmental delay, a subset of HRS infants were characterized by poor functioning in motor and sensorimotor areas during the first year of life (Marcus, Auerbach, Wilkinson, & Burack, 1981). As neonates, these infants showed weaknesses in upper-torso control and motor immaturity. At 4,

Table 32.2 Prospective high-risk studies: Number of subjects by parental groups and ages at different developmental stages

Study	Infancy		Childhood		Adolescence		Adulthood	
	Age[a] (months)	Sample size	Mean age (range)	Sample size	Mean age (range)	Sample size	Mean age[b] (range)	Sample size
NYIS	<24	12 HRS 12 NMI	10 (9–11)	12 HRS 12 NMI	16 (15–17)	12 HRS 11 NMI	25 (21–35)	12 HRS 11NMI
IHRS			11 (8–15)	50 HRS 50 NMI			31 (26–34)	44 HRS 43 NMI
NYHRP								
Sample A + B			9 (7–12)	109 HRS 81 HR 162 ONMI	15 (10–20)	71 HRS 67 HRO 138 NMI	40 (32–46)	96 HRS 73 HRO 154 NMI
Siblings (A + B)							42(23–60)	46 HRS 47 HRO 76 NMI
JIDS								
Original sample	<12	19 HRS 20 HRO 19 NMI	(8–12)	15 HRS 18 HR 12 ONMI	(12–22)[b]	15 HRS 15 HRO 10 NMI		
Sibling additions			(8–12)	10 HRS 10 HR 8 ONMI	(12–22)[b]	9 HRS 10 HRO 6 NMI		
New recruitment					(12–22)[b]	17 HRS 14 HRO 20 NMI		
SHRP	<24[c,d]	44 HRS 44 HRO 88 NMI	6 (5–7)	31 HRS 33 HR 97 ONMI			22 (19–25)	38 HRS 37 HRO 91 NMI
DCS	<12	90 HRS 93 HRO 82 NMI	(11–13)	90 HRS 93 HR 82 ONMI			(31–32)	81 HRS 87 HRO 74 NMI

[a]Testing at multiple times within age span. [b]Age at final testing. [c]Evaluations also conducted during pregnancy. [d]Well-baby clinic data (0–4 years) for HRS 43, HRO 41, NMI 100. NYIS, New York Infant Study; IHRS, Israeli High-Risk Study; NYHRP, New York High-Risk Project; JIDS, Jerusalem Infant Development Study; SHRP, Swedish High-Risk Project; DCS, Danish Cohort Study; HRS, High risk for schizophrenia group; HRO, High risk for other psychiatric disorders group; NMI, no parental mental illness group.

8, and 12 months, their levels of fine motor control, eye–hand coordination, and gross motor skills were below the age level. In a partial replication of Fish's study, pandysmaturation in the **JIDS** sample was related to parental diagnosis of schizophrenia and to poor functioning on the **JIDS** cognitive battery in childhood (Fish et al., 1992). Pediatric records of HRS infants in the **DCS** showed weaker or absent Moro reflexes at birth and delays in head control and walking with support than infants born to parents with a character disorder or NMI parents (Mednick, Mura, Schulsinger, & Mednick, 1971).

In another study, The Rochester Longitudinal Study (Sameroff, Seifer, Zax, & Barocas, 1987), indices of 4-month mental and psychomotor development were significantly lower for HRS infants than for HRO and NMI infants but by 12 months, only the difference between the HRS and HRO infants was significant.

School age: The HRS group, or subgroup, in each of the highlighted studies differed from the NMI group as well as from the HRO subjects in the **NYIS**, **NYHRP**, **SHRP**, and **JIDS**, with respect to a number of neurobehavioral measures,

including poorer neurological functioning and perceptual-motor deficits (Fish & Hagin, 1973), gross neuromotor skills (Erlenmeyer-Kimling et al., 2000; Marcus, 1974; Marcus, Hans, Auerbach, & Auerbach, 1993; McNeil, Cantor-Graae, & Blennow, 2003), perceptual-cognitive functioning (Marcus et al., 1993), and perceptual and attentional functioning, particularly in tasks with high-processing demands (Erlenmeyer-Kimling & Cornblatt, 1992; Sohlberg & Yaniv, 1985). On a summary measure of performance on several attentional tasks calculated for the **NYHRP** (Erlenmeyer-Kimling & Cornblatt, 1992), about 25% of the HRS offspring were consistently impaired across several assessment rounds. In contrast, 10 and 6% of the HRO and NMI groups, respectively, showed such impairments at the first assessment round, in childhood, with lower rates of impairment at later rounds. Short-term memory deficits, especially in distraction conditions, were found in the **NYHRP** and the **IHRS** (Erlenmeyer-Kimling & Cornblatt, 1987; Lifshitz, Kugelmass, & Karov, 1985), although in the latter study impairments were not found on other memory tasks. In the **DCS** and the **JIDS**, multiple signs of

Table 32.3 Deficits in neurobehavioral functioning characterizing subgroups of offspring of schizophrenics

Study age range	Infancy <24 months	Childhood 5–13 years	Adolescence 10–22 years
NYIS	• Cranial growth retardation plus postural-motor and/or visual-motor delay by 10 months (=pandysmaturation)	• Perceptual motor deficits	
IHRS		• Poor neurological functioning • Perceptual, attentional, motor deficits	• Neuromotor deficits
NYHRP		• Deficits in visual sustained attention, auditory distractibility, auditory–visual short-term verbal memory • Poor neuromotor functioning • Attention deficits • Lower IQ	• Deviant in verbal sustained attention, auditory selective attention under overload, auditory short-term verbal memory • Attention deficits • Lower IQ
JIDS	• Motor and sensorimotor deficits in neonatal period and in first year	• Motor and perceptual motor deficits	• Motor cognitive-attentional deficits
SHRP	• Reduced arousal, neurological abnormalities, deviant response to stimulation in neonatal period • Developmental deviations	• Deviant neurological and motoric functioning	
DCS	• Weak or absent reflexes • Gross motor delays	• Poor neurological functioning	

Note: NYIS, New York Infant Study; IHRS, Israeli High-Risk Study; NYHRP, New York High-Risk Project; JIDS, Jerusalem Infant Development Study; SHRP, Swedish High-Risk Project; DCS, Danish Cohort Study.

neurological dysfunctions characterized a subgroup of HRS subjects in both studies (Marcus, Hans, Lewow, Wilkinson, & Burack, 1985).

Two high-risk studies in the US reporting data from childhood only, the Minnesota High-Risk Studies (Nuechterlein, 1983) and the Stony Brook High-Risk Study (Winters, Stone, Weintraub, & Neale, 1981), also found attentional impairments in a higher percentage of the HRS children than in the NMI children. In the Minnesota study, these differences also held between HRS and HRO offspring whose mothers were nonpsychotic, but in the Stony Brook study, children of depressed mothers did not differ significantly from HRS children.

Supportive evidence for neurobehavioral deficits in high-risk children also comes from the Helsinki High-Risk follow-back study (Niemi et al., 2003) and subsamples of children from the epidemiological US National Collaborative Perinatal Project (Rieder & Nichols, 1979; Rosso et al., 2000). These studies found that a greater number of HRS children had neurological soft signs than did the NMI children, although in the Collaborative Perinatal Project (Rieder & Nichols, 1979) this was true only for boys. In the Minnesota subsample of the US Collaborative Perinatal Project (Hanson, Gottesman, & Heston, 1976), a subset of HRS children were characterized by poor motor skills, large intra-

individual variability on cognitive measures, and social emotional difficulties.

Adolescence: During adolescence, assessments were carried out in the **JIDS, NYHRP,** and **IHRS** on the same trait constructs that had been measured at younger ages. In both the **IHRS** (Marcus, Hans, Lewow, Wilkinson, & Burack, 1985; Marcus, Hans, Mednick, Schulsinger, & Michelsen, 1985) and **NYHRP** (Erlenmeyer-Kimling & Cornblatt, 1992), the HRS subjects continued to show poorer performance than the NMI or HRO adolescents with respect to neuromotor functioning (**IHRS**) and to attentional impairment (**NYHRP**). In the **JIDS**, poor motoric and cognitive/attentional functioning was found in 42% of the HRS adolescents contrasted with 22% of the HRO and 4% of the NMI. These performance deficits continued to be seen in the HRS subjects from one assessment time to the next (Hans et al., 1999), as was true also in the **NYHRP** (Winters, Cornblatt, & Erlenmeyer-Kimling, 1991).

Although not universally observed in the several high-risk studies, gender acted as an effect moderator in the **JIDS**, with male children being nearly 4 times more likely to function poorly than females. Gender plays an important role in the developmental trajectory of schizophrenia. In males, schizophrenia is characterized by more neurobehavioral and neuroanatomical signs and earlier onset (Castle, Sham, & Murray, 1998).

Summary of neurobehavioral deficits: Across ages and studies, the strongest candidates for neurobehavioral markers of the genetic liability to schizophrenia seem to be hypoarousal in the neonatal period combined with neuromotor impairments, and pandysmaturation from infancy onwards. Attentional impairments, short-term memory deficits, and neuromotor dysfunctions have usually characterized a larger subgroup of HRS children and adolescents than subjects in the comparison groups.

Neurobehavioral Predictors

Diagnostic assessments were carried out at adolescence and/or adulthood for the participants in all six highlighted studies. The predictors suggested by each of the six high-risk studies are summarized in Table 32.4. The neurobehavioral markers indicative of genetic liability were for the most part also predictive of a schizophrenia spectrum diagnosis.

Infancy: In the **NYIS** and **JIDS**, pandysmaturation in infancy predicted all the spectrum disorders with onsets in childhood (**NYIS**), adolescence (**NYIS, JIDS**), or adulthood (**NYIS**) (Fish & Kendler, 2005; Fish et al., 2005). In both studies, poor motor and sensorimotor functioning during infancy were more prominent in those at-risk infants who went on to develop a schizophrenia spectrum disorder than in those who did not (Fish, 1987; Hans et al., 2004). In the **SHRP**, visual dysfunction at age 4 was not only predictive of neurological abnormalities at age 6 (McNeil, Cantor-Graae, & Blennow, 2003) but also of schizophrenia spectrum disorder in young adulthood (Henriksson & McNeil, 2004; Schubert, Henriksson, & McNeil, 2005).

Further support for the predictive value of developmental impairments in infancy comes from two studies using different methodologies to investigate the development of schizophrenia. The New Zealand (Dunedin) birth cohort study (Cannon et al., 2002) found that schizophreniform disorder at age 26 was predicted by persistent, pan-developmental impairment from early childhood. In a follow-back study making use of home movies taken in the first 2 years of life, Walker and her colleagues (Walker & Lewine, 1990) found more signs of motor delay and movement abnormalities in those children who went on to develop adult-onset schizophrenia than in their unaffected siblings.

School age and adolescence: At school age and adolescence, neuromotor impairment predicted an adolescent or adult diagnosis within the schizophrenia spectrum. In these age periods, cognitive/attentional impairment was predictive of schizophrenia. In the **NYHRP**, global attention deficits, gross motor impairment, deficient verbal short-term memory in childhood, and the effect of having a schizophrenic versus normal or affectively ill parent predicted a later diagnosis of schizophrenia-related psychosis (Erlenmeyer-Kimling et al., 2000). In regression models on each of the neurobehavioral measures, the schizophrenic parent effect was strong (indicating a strong genetic effect on the neurobehavioral measures, themselves), thus highly supporting the suggestion that these measures may be phenotypic indicators of the genetic liability to schizophrenia. Childhood verbal short-term memory and gross neuromotor deficits, each, had reasonably high prediction (*sensitivity*, 83% and 75%, respectively) to future schizophrenia in HRS subjects, but also predicted the disorder in some subjects who did not develop it. A deviance index based on deviance on the three neurobehavioral measures together, however, predicted 50%

Table 32.4 Neurobehavioral functioning predictive of schizophrenia spectrum disorders

Study age range	Infancy <24 months	Childhood 5–13 years	Adolescence 10–22 years
NYIS	• Pandysmaturation • Visual motor deficits	• Perceptual motor deficits	
IHRS		• Focused attention deficit • Hyporesponsivity (EDA – best predicts affective and next schizophrenia spectrum)	
NYHRP		• Gross neuromotor deficits • Verbal memory deficits • Attention deficit • Lower IQ	• Poor neuromotor functioning • Poor attention
JIDS	• Poor neurobehavioral functioning (motor and sensorimotor functioning) • Pandysmaturation	• Poor neurobehavioral functioning (motor and cognitive-perceptual functioning)	• Poor neurobehavioral functioning (motor and cognitive-attentional functioning)
SHRP **DCS**		• Poor neurobehavioral functioning • Neuromotor functioning • Global deficits in laterality • Minor physical anomalies	

Note: NYIS, New York Infant Study; IHRS, Israeli High-Risk Study; NYHRP, New York High-Risk Project; JIDS, Jerusalem Infant Development Study; SHRP, Swedish High-Risk Project; DCS, Danish Cohort Study.

of the future schizophrenics and showed a relatively low false-positive rate of 10% in the HRS group, but in the HRO or NMI groups, no children were deviant on all three measures at once. The measures showed no relationship to other forms of later illness.

The **IHRS** also noted that attentional measures from age 11 were predictive of a schizophrenia spectrum diagnosis in adulthood (Mirsky, Ingraham, & Kugelmass, 1995). In the **JIDS**, the four HRS individuals diagnosed with a schizophrenia spectrum disorder in adolescence (Hans et al., 1999) showed a stable pattern of poor neurobehavioral functioning (motor and cognitive/attentional) at school age and adolescence.

Other Neurobehavioral Deficits

Numerous attempts have been made to identify deficits in other neurobehavioral domains; we consider two of them here. **Smooth-pursuit-eye movement** dysfunctions have been assessed in the NYHRP in adulthood (Rosenberg et al., 1997) and in childhood in a Colorado study of children of schizophrenic parents (Ross, Hommer, Radant, Roath, & Freedman, 1996). In both studies, HRS individuals showed more dysfunctions and more frequent anticipatory saccades as has been found with schizophrenic patients and a number of their unaffected first-degree relatives (Levy, Holzman, Matthysse, & Mendell, 1994). **Brain imaging assessments**, based on *event-related potentials* (ERPs) in the NYHRP (Friedman & Squires-Wheeler, 1994; Squires-Wheeler, Erlenmeyer-Kimling, & Friedman, 1999), *computed tomography* (CT) scans in the Copenhagen High-Risk Project (Cannon et al., 1994), and *magnetic resonance imaging* (MRI) in the Edinburgh High-Risk Study (Job, Whalley, Johnstone, & Lawrie, 2005; Job et al., 2006) and the Pittsburgh Risk Evaluation Program (Keshavan, Diwadkar, Montrose, Rajarethinam, & Sweeney, 2005; Keshavan, Diwadkar, Montrose, Stanley, & Pettegrew, 2005) have produced increasingly interesting results with advancing technology. For example, although event-related potentials in the NYHRP showed an association between reduced amplitudes of the P3 wave with poor global adjustment, negative symptoms, and working memory deficit in all three risk groups (Squires-Wheeler, Erlenmeyer-Kimling, & Friedman, 1999), there were no differences among the risk (distinguished by parental mental illness) or among the children's diagnostic outcomes in adulthood (Friedman & Squires-Wheeler, 1994). CT scan in abnormalities in the Copenhagen High-Risk Project were more frequent in HRS subjects who had already developed schizophrenia than in the HRS subjects who had not, or in NMI subjects. MRI scans – arguably the most advanced of the brain imaging technologies considered here – have yielded a number of important results

in the Edinburgh High-Risk Study and the Pittsburgh Risk Evaluation Program. A recent report from the former (Job et al., 2006), showing decreased gray matter in the temporal gyrus in scans taken about 1.5 years after the initial ones, has yielded the largest *plausible* positive predictive value for development of schizophrenia in individuals at high genetic risk that has emerged from high-risk research thus far. Similar reductions in temporal gyrus gray matter were also seen in young HRS subjects of the Pittsburg study (Rajarethinam, Sahni, Rosenberg, & Keshavan, 2004). Other findings based on MRI data from these two studies support the neurodevelopmental model of schizophrenia and may be expected to provide a convergent picture of brain development.

Social Maladjustment as a Possible Endophenotype

Problems of social adjustment characterize adolescents and adults with schizophrenia (Asarnow & Ben-Meir, 1988; Bellack, Morrison, Wixted, & Mueser, 1990) are predictive of later hospitalization for schizophrenia (Davidson et al., 1999), and are also associated with poor prognosis in schizophrenic patients (Eggers & Bunk, 1997; Werry, McClellan, & Chard, 1991). Whether these problems are expressions of genetic liability, whether they are associated with certain neurobehavioral indicators, and whether they predict later disorder in individuals at genetic risk are questions of interest. Further questions about the way in which growing up with a mentally ill parent may affect social adjustment, either independently or jointly in interaction with genetic effects, are of interest but cannot be addressed by the highlighted studies, which contain no systematic subgroups of children reared apart from their biological parents.

Social adjustment has been observed to be poor in all of the highlighted studies. In the **SHRP**, 4-year-old HRS children showed poorer social competence and greater social withdrawal than HRO or NMI children (Henriksson & McNeil, 2004). The same was true of school-age children in both the **JIDS** and **IHRS** where social withdrawal was especially characteristic of *male* HRS subjects (Hans, Marcus, Henson, Auerbach, & Mirsky, 1992; Sohlberg & Yaniv, 1985). Social isolation was also noted in four of five pre-schizotypal children and one of the two ADHD children in the **NYIS** (Carlson & Fish, 2005). In the **DCS**, Danish 11- to 13-year-old children who later developed schizophrenia demonstrated lower sociability in videotaped interactions, as well as more neuromotor deficits, compared to children who did not develop a psychiatric disorder or who developed other psychiatric disorders (Schiffman et al., 2004). In adolescent HRS subjects, poor social adjustment was also more common than in either HRO or NMI children in the **NYHRP** (Cornblatt, Lenzenweger, Dworkin, & Erlenmeyer-Kimling, 1992; Dworkin et al., 1990), the **IHRS** (Nagler &

Glueck, 1985), and the **JIDS** (Hans, Auerbach, Asarnow, Styr, & Marcus, 2000). In the **JIDS**, social problems were most apparent in impaired relationships with the opposite sex (Hans et al., 2000), but lack of peer engagement and social problems in adolescence were related only to parent diagnostic group and not to the adolescents' own diagnostic status.

Examination of a possible association between neurobehavioral markers and social adjustment problems in a combined analysis of data from **JIDS** and **IHRS** (Hans et al., 1992) showed correlations of perceptual-cognitive and motor indices with social adjustment, social withdrawal, and aggression. However, positive correlation between neuromotor signs and social withdrawal was significant only for boys, i.e., HRS boys with signs of neuromotor dysfunction were also more socially withdrawn. In the **NYHRP**, attention deviance in HRS children predicted social deficits in adolescence and adulthood (Cornblatt et al., 1992), whereas childhood neuromotor dysfunction predicted affective flattening in adolescence (Dworkin et al., 1990).

Gene–Environment Interactions

The role of the childrearing environment as a contributing factor in the development of serious mental illness is often given no more than lip service in most high-risk studies. Notable exceptions are the **IHRS** (Mirsky et al., 1985) and the SHRS (McNeil & Kaij, 1987). In the IHRS, the childrearing setting was part of the research design. Fifty percent of the children in each group (HRS and NMI) came from a kibbutz setting and 50% from a town setting. During the period in which the study was conducted, kibbutz children were raised in group homes and spent a limited amount of time with their own parents. One of the major findings of the study was that the frequency of *overall* psychopathology at age 25 was higher for the HRS children who had been raised in the kibbutz. Schizophrenia and schizophrenia spectrum disorders, however, occurred only in the HRS group, regardless of childrearing setting. Thus, the setting in which the child was raised did not in and of itself increase the probability of a schizophrenia spectrum disorder outcome.

The picture is different when quality of parenting behavior is considered. In the **IHRS**, poor parenting behavior in the form of over involvement, inconsistency, and hostility was predictive of an adult spectrum diagnosis in HRS offspring, particularly when they had early neurobehavioral impairment (Marcus et al., 1987). In the **SHRS**, mental disturbance as early as age 6 was significantly related to the psychotic condition of the mother 6–24 months post-partum (McNeil & Kaij, 1987). During the first 12 months of life, negative interaction was more characteristic of the HRS mothers than NMI mothers, with HRO mothers, for the most part, more similar to the HRS

group. Infant interactive behaviors were remarkably similar across the groups although decreased social behavior toward the mother at 3 days and 3.5 months was noted for the HRS infants (Näslund, Persson-Blennow, McNeil, & Kaij, 1985; Näslund, Persson-Blennow, McNeil, Kaij, & Malmquist-Larsson, 1985; Persson-Blennow, Näslund, McNeil, & Kaij, 1986).

More direct evidence of gene–environmental interplay comes from a Finnish study and a Danish study. Tienari et al. (2004) studied the adopted-away offspring of Finnish mothers with schizophrenia and found that the adoptive-family environment had a significant impact on the diagnostic outcome of HRS adoptees compared with NMI adoptees. HRS adoptees exposed to an environment that could be described as critical/conflictual, constricted, or presenting boundary problems were significantly more likely to be diagnosed with a schizophrenia spectrum disorder at later ages than were HRS adoptees in more optimal family environments. In the Danish study, HRS adolescents' perception of their relationships with their parents was associated with psychiatric outcome in adulthood (Schiffman et al., 2002). Those HRS offspring who reported good relations with both parents were less likely to receive an adult diagnosis of schizophrenia than those HRS offspring who reported poor relations with both parents (7.0% versus 23.4%, respectively) suggesting that good relations with both parents may act as a protective factor in the development of schizophrenia, or indicate a low genetic risk.

Future Directions

The era of prospective longitudinal high-risk studies in schizophrenia starting in childhood or earlier may well be over. This is unfortunate, as these studies have defined many neurobehavioral endophenotypes expressed by the underlying genetic liability, some of which independently or in combination with other factors also show strong promise as *predictors* of future psychiatric outcome. The overall consistency of the findings across the studies appears to give them a face validity that is reassuring. Nevertheless, except for Fish's NYIS, subjects in the other studies (**SHRP, JIDS,** and **DCS**) starting in infancy have not yet been followed into adulthood through the schizophrenia risk period. Further follow-up of the subjects in these studies is essential to strengthen and add power to a consolidated database that confirms predictive relationships between later development of schizophrenia and earlier neurobehavioral deficits, problem behaviors and exposure to environmental stressors, as others (e.g., Niemi, Suvisaari, Tuulio-Henriksson, & Lönnqvist, 2003) have noted.

A new generation of high-risk studies could then capitalize on such a database containing the accumulated information imparted by the first-generation studies, while using newer methodologies to identify gene functions directly, investigate pathological connections in the brain, and explore gene–environment interaction. To a certain extent, this is what the Edinburgh and Pittsburgh studies, and some others, have already done. It would be particularly helpful if new studies were organized by a central control to ensure use of a common protocol across multiple sites to deal with power issues from the start. In an ongoing epidemiologic birth cohort study from New Zealand which has attempted to identify children at risk for various mental disorders, Caspi and colleagues (2002) have shown the fruitfulness of searching for gene–environment interaction in predicting behavioral outcome in individuals subjected to severe early maltreatment. In schizophrenia research, the role of the environment as a causal factor in the development of the disorder has not been greatly emphasized. In part, this is a result of the difficulty of defining precisely which environmental factors are likely to increase the probability of an eventual diagnosis. Many authors, over many years, have pointed out that "environment" is unlikely to consist of a single factor, especially a single factor that is the same for a large majority of people who develop the illness. Instead, environment probably consists of the cumulative effect of multiple idiosyncratic epigenetic interactions, handily covered under the rubric of "stress" that increases the odds of a pathological outcome. In any event, a new generation of longitudinal studies should take up the challenge of viewing "environment" more complexly and considering the possibilities of its interactions in biological pathways at numerous points along the way. This is an undertaking similar to what needs to be addressed in pharmacogenetics – if it is to play a vital part in therapeutic medicine – i.e., evaluating the role of numerous (but limited for practicality) exogenous and endogenous influences upon each unique genotype's responsivity to different therapeutic components. Clearly these multisite studies would include the collection of DNA, and all the other technological advances that many of the earlier studies missed.

In hindsight, high-risk researchers would probably agree that more progress might have been made if the multisite design had been applied to the studies, with all of them tapping the same domains of measures. Even so it is impressive that the studies concur in their delineation of neurobehavioral risk.

References

Asarnow, J. R., & Ben-Meir, S. (1988). Children with schizophrenia spectrum and depressive disorders: A comparative study of premorbid adjustment, onset pattern and severity of impairment. *Journal of Child Psychology and Psychiatry and Allied Disciplines, 29*, 477–488.

Baron, M., Gruen, R., Rainer, J. D., Kane, J., Asnis, L., & Lord, S. (1985). A family study of schizophrenic and normal control probands: Implications for the spectrum concept of schizophrenia. *American Journal of Psychiatry, 142*, 447–455.

Bellack, A. S., Morrison, R. L., Wixted, J. T., & Mueser, K. T. (1990). An analysis of social competence in schizophrenia. *British Journal of Psychiatry, 156*, 809–818.

Cannon, T. D., Bearden, C. E., Hollister, J. M., Rosso, I. M., Sanchez, L. E., & Hadley, T. (2000). Childhood cognitive functioning in schizophrenia patients and their unaffected siblings: A prospective cohort study. *Schizophrenia Bulletin, 26*, 379–393.

Cannon, M., Caspi, A., Moffitt, T. E., Harrington, H., Taylor, A., Murray, R. M., et al. (2002). Evidence for early-childhood pandevelopmental impairment specific to schizophreniform disorder: Results from a longitudinal birth cohort. *Archives of General Psychiatry, 59*, 449–457.

Cannon T. D., Mednick S. A., Parnas J., Schulsinger F., Praestholm J., & Vestergaard A. (1994). Developmental brain abnormalities in the offspring of schizophrenic mothers. II. Structural brain characteristics of schizophrenia and schizotypal personality disorder. *Archives of General Psychiatry, 51*, 955–962.

Carlson, G. A., & Fish, B. (2005). Longitudinal course of schizophrenia spectrum symptoms in offspring of psychiatrically hospitalized mothers. *Journal of Child and Adolescent Psychopharmacology, 15*, 362–383.

Caspi, A., McClay, J., Moffitt, T. E., Mill, J., Martin, J., Craig, I. W., et al. (2002). Role of genotype in the cycle of violence in maltreated children. *Science, 297*, 851–854.

Castle, D. J., Sham, P., & Murray, R. M. (1998). Differences in distribution of ages of ones in males and females with schizophrenia. *Schizophrenia Research, 33*, 179–183.

Chapman, L. J., Edell, W. S., & Chapman, J. P. (1980). Physical anhedonia, perceptual aberration, and psychosis proneness. *Schizophrenia Bulletin, 6*, 639–653.

Cicchetti, D., & Cannon, T. D. (1999). Neurodevelopmental processes in the ontogenesis and epigenesis of psychopathology. *Development and Psychopathology, 11*, 375–393.

Clementz, B. A., Grove, W. M., Katsanis, J., & Iacono, W. G. (1991). Psychometric detection of schizotypy: Perceptual aberration and physical anhedonia in relatives of schizophrenics. *Journal of Abnormal Psychology, 100*, 607–612.

Clementz, B. A., Sweeney, J. A., Hirt, M., & Haas, G. (1992). Pursuit gain and saccadic intrusions in first-degree relatives of probands with schizophrenia. *Journal of Abnormal Psychology, 99*, 327–335.

Cornblatt, B. A., Lenzenweger, M. F., Dworkin, R. H., & Erlenmeyer-Kimling, L. (1992). Childhood attentional dysfunctions predict social deficits in unaffected adults at risk for schizophrenia. *British Journal of Psychiatry, 161*(Suppl. 18), 59–64.

Cox, S. M., & Ludwig, A. M. (1979). Neurological soft signs and psychopathology: Incidence in diagnostic groups. *Canadian Journal of Psychiatry, 24*, 668–673.

Davidson, M., Reichenberg, A., Rabinowitz, J., Weiser, M., Kaplan, Z., & Mark, M. (1999). Behavioral and intellectual markers for schizophrenia in apparently healthy male adolescents. *American Journal of Psychiatry, 156*, 1328–1335.

Dworkin, R. II., Green, S. R., Small, N. E., Warner, M. L., Cornblatt, B. A., & Erlenmeyer-Kimling, L. (1990). Positive and negative symptoms and social competence in adolescents at risk for schizophrenia and affective disorder. *American Journal of Psychiatry, 147*, 1234–1236.

Eggers, C., & Bunk, D. (1997). The long-term course of childhood-onset schizophrenia: A 42-year follow up. *Schizophrenia Bulletin, 23*, 105–117.

Erlenmeyer-Kimling, L. (2000). Neurobehavioral deficits in offspring of schizophrenic parents: Liability indicators and predictors of illness. *American Journal of Medical Genetics, 97,* 65–71.

Erlenmeyer-Kimling, L., & Cornblatt, B. (1987). High risk research in schizophrenia: A summary of what has been learned. *Journal of Psychiatric Research, 21,* 401–411.

Erlenmeyer-Kimling, L., & Cornblatt, B. A. (1992). A summary of attentional findings in the New York High-Risk Project. *Journal of Psychiatric Research, 26,* 405–426.

Erlenmeyer-Kimling, L., Hilldoff Adamo, U., Rock, D., Roberts, S. A., Bassett, A. S., Squires-Wheeler, E., et al. (1997). The New York High-Risk Project: Prevalence and comorbidity of Axis I disorders in offspring of schizophrenic parents at 25 years of follow-up. *Archives of General Psychiatry, 54,* 1096–1102.

Erlenmeyer-Kimling, L., Rock, D., Roberts, S. A., Janal, M., Kestenbaum, C., Cornblatt, B., et al. (2000). Attention, memory, and motor skills as childhood predictors of schizophrenia-related psychoses: The New York High-Risk Project. *American Journal of Psychiatry, 157,* 1416–1422.

Faraone, S. V., Green, A. I., Seidman, L. J., & Tsuang, M. T. (2001). "Schizotaxia": Clinical implications and new directions for research. *Schizophrenia Bulletin, 27,* 1–18.

Faraone, S. V., Seidman, L. J., Kremen, W. S., Pepple, J. R., Lyons, M. J., & Tsuang, M. T. (1995). Neuropsychological functioning among the nonpsychotic relatives of schizophrenic patients: A diagnostic efficiency analysis. *Journal of Abnormal Psychology, 104,* 286–304.

Fish, B. (1957). The detection of schizophrenia in infancy. *Journal of Nervous and Mental Disorders. 125,* 1–24.

Fish, B. (1977). Neurobiologic antecedents of schizophrenia in children. *Archives of General Psychiatry, 34,* 1297–313.

Fish, B. (1987). Infant predictors of the longitudinal course of schizophrenic development. *Schizophrenia Bulletin, 13,* 395–409.

Fish, B., & Hagin, R. (1973). Visual-motor disorders in infants at risk for schizophrenia. *Archives of General Psychiatry, 28,* 900–904.

Fish, B., Hans, S. L., Marcus, J., & Auerbach, J. G. (2005). Life courses of schizophrenia spectrum disorders observed from birth. *Schizophrenia Bulletin, 31,* 199–200.

Fish, B., & Kendler, K. S. (2005). Abnormal infant neurodevelopment predicts schizophrenia spectrum disorders. *Journal of Child and Adolescent Psychopharmacology, 15,* 348–361.

Fish, B., Marcus, J., Hans, S. L., Auerbach, J. G., & Perdue, S. (1992). Infants at risk for schizophrenia: Sequelae of a genetic neurointegrative defect. A review and replication analysis of pandysmaturation in the Jerusalem Infant Development Study. *Archives of General Psychiatry, 49,* 221–235.

Friedman, D., & Squires-Wheeler, E. (1994). Event-related potentials (ERPs) as indicators of risk for schizophrenia. *Schizophrenia Bulletin, 20,* 63–74.

Garmezy, N. (1974). Children at risk: The search for the antecedents of schizophrenia. Part II: Ongoing research programs, issues, and intervention. *Schizophrenia Bulletin, 9,* 55–125.

Garmezy, N., & Streitman, S. (1974). Children at risk: The search for the antecedents of schizophrenia. Part I. Conceptual models and research models. *Schizophrenia Bulletin, 8,* 14–90.

Gooding, D. C., & Iacono, W. G. (1995). Schizophrenia through the lens of a developmental psychopathology perspective. In D. Cicchetti & D. J. Cohen (Eds.), *Manual of developmental psychopathology, Vol. II: Risk, disorder and adaptation* (pp. 535–580). New York: John Wiley & Sons.

Gottesman, I. I., & Bertelsen, A. (1989). Confirming unexpressed genotypes for schizophrenia. *Archives of General Psychiatry, 46,* 867–872.

Gottesman, I. I., & Gould, T. D. (2003). The endophenotype concept in psychiatry: Etymology and strategic intentions. *American Journal of Psychiatry, 160,* 636–645.

Gottesman, I. I., & Shields, J. (1967). A polygenic theory of schizophrenia. *Proceedings of the National Academy of Sciences, 58,* 199–205.

Gottesman, I. I., & Shields, J. (1972). A polygenic theory of schizophrenia. *International Journal of Mental Health, 1,* 107–115.

Hans, S. L., Auerbach, J. G., Asarnow, J. R., Styr, B., & Marcus, J. (2000). Social adjustment of adolescents at risk for schizophrenia: The Jerusalem Infant Development Study. *Journal of the American Academy of Child and Adolescent Psychiatry, 39,* 1406–1414.

Hans, S. L., Auerbach, J. G., Styr, B., & Marcus, J. (2004). Offspring of parents with schizophrenia: Mental disorders during childhood and adolescence. *Schizophrenia Bulletin, 30,* 303–315.

Hans, S. L., Marcus, J., Henson, L., Auerbach, J. G., & Mirsky, A. F. (1992). Interpersonal behavior of children at risk for schizophrenia. *Psychiatry, 55,* 314–335.

Hans, S. L., Marcus, J., Nuechterlein, K. H., Asarnow, R. F., Styr, B., & Auerbach, J. G. (1999). Neurobehavioral deficits at adolescence in children at risk for schizophrenia: The Jerusalem Infant Development Study. *Archives of General Psychiatry, 56,* 741–748.

Hanson, D. R. (this book). Psychoses: Schizophrenia, bipolar disorder and their nonpsychotic spectra. In Kim, Y. (Ed.), *Handbook of behavior genetics.* New York: Springer.

Hanson, D. R., Gottesman, I. I., & Heston, L. L. (1976). Some possible childhood indicators of adult schizophrenia inferred from children of schizophrenics. *British Journal of Psychiatry, 129,* 142–154.

Henriksson, K. M., & McNeil, T. F. (2004). Health and development in the first 4 years of life of offspring of women with schizophrenia and affective psychoses: Well-Baby Clinic information. *Schizophrenia Research, 70,* 39–48.

Heston, L. L. (1970). The genetics of schizophrenic and schizoid diseases. *Science, 167,* 249–256.

Holzman, P. S., Proctor, L. R., Levy, D. L., Yasillo, N. J., Meltzer, H. Y., & Hurst, S. W. (1974). Eye-tracking dysfunctions in schizophrenic patients and their relatives. *Archives of General Psychiatry, 31,* 143–151.

Job, D. E., Whalley, H. C., Johnstone, E. C., & Lawrie, S. M. (2005). Grey matter changes over time in high risk subjects developing schizophrenia. *Neuroimage, 25,* 1023–1030.

Job, D. E., Whalley, H. C., McIntosh, A. M., Owens, D. G., Johnstone, E. C., & Lawrie, S. M. (2006). Grey matter changes can improve the prediction of schizophrenia in subjects at high risk. *BMC Medicine, 7,* 4–29.

Jones, P. (1995). Childhood motor milestones and IQ prior to adult schizophrenia: Results from a 43 year old British birth cohort. *Psychiatria Fennica, 26,* 63–80.

Jones, P., Rodgers, B., Murray, R., & Marmot, M. (1994). Child developmental risk factors for adult schizophrenia in the British 1946 birth cohort. *Lancet, 344,* 1398–1402.

Kallmann, F. J. (1938). *The genetics of schizophrenia.* New York: Augustin.

Keefe, R. S., Silverman, J. M., Mohs, R. C., Siever, L. J., Harvey, P. D., Friedman, L., et al. (1997). Eye tracking, attention, and schizotypal symptoms in nonpsychotic relatives of patients with schizophrenia. *Archives of General Psychiatry, 54,* 169–176.

Kendler, K. S., & Gardner, C. O. (1997). The risk for psychiatric disorders in relatives of schizophrenic and control probands: A comparison of three independent studies. *Psychological Medicine, 27,* 411–419.

Kendler, K. S., McGuire, M., Gruenberg, A. M., Spellman, M., O'Hare, A., & Walsh, D. (1993). The Roscommon Family Study: II. The risk of nonschizophrenic nonaffective psychoses in relatives. *Archives of General Psychiatry, 50,* 645–652.

Kendler, K. S., McGuire, M., Gruenberg, A. M., & Walsh, D. (1995). Schizotypal symptoms and signs in the Roscommon Family Study: Their factor structure and familial relationship with psychotic and affective disorders. *Archives of General Psychiatry, 52,* 296–303.

Keshavan, M. S., Diwadkar, V. A., Montrose, D. M., Rajarethinam, R., & Sweeney, J. A. (2005). Premorbid indicators and risk for schizophrenia: A selective review and update. *Schizophrenia Research, 79,* 45–57.

Keshavan, M. S., Diwadkar, V. A., Montrose, D. M., Stanley, J. A., & Pettegrew, J. W. (2005). Premorbid characterization in schizophrenia: The Pittsburgh High Risk Study. *World Psychiatry, 3,* 163–168.

Kety, S. S., Wender, P. H., Jacobsen, B., & Ingraham, L. J. (1994). Mental illness in the biological and adoptive relatives of schizophrenic adoptees: Replication of the Copenhagen Study in the rest of Denmark. *Archives of General Psychiatry, 51,* 442–455.

Knapp, M., Mangalore, R., & Simon, J. (2004). The global cost of schizophrenia. *Schizophrenia Bulletin, 30,* 279–293.

Kraepelin, E. (1919). *Dementia praecox and paraphenia.* Edinburgh, Scotland: E. S. Livingstone.

Lenzenweger, M. F., Maher, B. A., & Manschreck, T. C. (2005). Paul E. Meehl's influence on experimental psychopathology: Fruits of the nexus of schizotypy and schizophrenia, neurology and methodology. *Journal of Clinical Psychology, 61,* 1295–1315.

Levy, D. L., Holzman, P. S., Matthysse, S., & Mendell, N. R. (1994). Eye tracking and schizophrenia: A selective review. *Schizophrenia Bulletin, 20,* 47–62.

Lifshitz, M., Kugelmass, S., & Karov, M. (1985). Perceptual-motor and memory performance of high-risk children. *Schizophrenia Bulletin, 11,* 74–84.

Maier, W., Lichtermann, D., Minges, J. R., & Heun, R. (1994). Personality disorders among the relatives of schizophrenia patients. *Schizophrenia Bulletin, 20,* 481–493.

Manschreck, T. C., & Ames, D. (1984). Neurologic features and psychopathology. *Biological Psychiatry, 19,* 703–719.

Marcus, J. (1974). Cerebral functioning in offspring of schizophrenics: A possible genetic factor. *International Journal of Mental Health, 3,* 57–73.

Marcus, J., Auerbach, J., Wilkinson, L., & Burack, C. M. (1981). Infants at risk for schizophrenia: The Jerusalem Infant Development Study. *Archives of General Psychiatry, 38,* 703–713.

Marcus, J., Hans, S. L., Auerbach, J. G., & Auerbach, A. G. (1993). Children at risk for schizophrenia: The Jerusalem Infant Development Study. II. Neurobehavioral deficits at school age. *Archives of General Psychiatry, 50,* 797–809.

Marcus, J., Hans, S. L., Lewow, E., Wilkinson, L., & Burack, C. M. (1985). Neurological findings in high-risk children: Childhood assessment and 5-year follow up. *Schizophrenia Bulletin, 11,* 85–100.

Marcus, J., Hans, S. L., Mednick, S., Schulsinger, F., & Michelsen, N., (1985). Neurological dysfunctioning in offspring of schizophrenics in Israel and Denmark: A replication analysis. *Archives of General Psychiatry, 42,* 753–761.

Marcus, J., Hans, S. L., Nagler, S., Auerbach, J. G., Mirsky, A. F., & Aubrey, A. (1987). Review of the NIMH Israeli Kibbutz-City Study and the Jerusalem Infant Development Study. *Schizophrenia Bulletin, 13,* 425–437.

McNeil, T. F., Cantor-Graae, E., & Blennow, G. (2003). Mental correlates of neuromotor deviation in 6-year-olds at heightened risk for schizophrenia. *Schizophrenia Research, 60,* 219–228.

McNeil, T. F., & Kaij, L. (1979). Etiological relevance of comparisons of high-risk and low-risk groups. *Acta Psychiatrica Scandinavia, 59,* 545–560.

McNeil, T. F., & Kaij, L. (1987). Swedish high-risk study: Sample characteristics at age 6. *Schizophrenia Bulletin, 13,* 373–381.

Mednick, S. A., & McNeil, T. (1968). Current methodology in research on the etiology of schizophrenia: Serious difficulties which suggest the use of the high-risk group method. *Schizophrenia Bulletin, 70,* 681–693.

Mednick, S. A., Mura, M., Schulsinger, F., & Mednick, B. (1971). Perinatal conditions and infant development in children with schizophrenic parents. *Social Biology, 18,* S103–113.

Meehl, P. E. (1962). Schizotaxia, schizotypy, schizophrenia. *American Psychologist, 17,* 827–838.

Meehl, P. E. (1990). Toward an integrated theory of schizotaxia, schizotypy, and schizophrenia. *Journal of Personality Disorders, 4,* 1–99.

Mirsky, A. F., Ingraham, L. J., & Kugelmass, S. (1995). Neuropsychological assessment of attention and its pathology in the Israeli cohort *Schizophrenia Bulletin, 21,* 193–204.

Mirsky, A. F., Silberman, E. K., Latz, A., & Nagler, S. (1985). Adult outcomes of high risk children: Differential effects of town and kibbutz rearing. *Schizophrenia Bulletin, 11,* 150–154.

Murray, R., & Fearon, P. (1999). The developmental "risk factor" model of schizophrenia. *Journal of Psychiatric Research, 33,* 497–499.

Nagler, S., & Glueck, Z. (1985). The clinical interview. *Schizophrenia Bulletin, 11,* 38–47.

Näslund, B., Persson-Blennow, I., McNeil, T. F., & Kaij, L. (1985). Offspring of women with nonorganic psychosis: Mother–infant interaction at three and six weeks of age. *Acta Psychiatrica Scandinavia, 71,* 441–450.

Näslund, B., Persson-Blennow, I., McNeil, T. F., Kaij, L., & Malmquist-Larsson, A. (1985). Offspring of women with nonorganic psychosis: Infant attachment to mother at one year of age. *Acta Psychiatrica Scandinavia, 69,* 231–241.

Niemi, L. T., Suvisaari, J. M., Haukka, J. K., & Lonnqvist, J. K. (2005). Childhood predictors of future psychiatric morbidity in offspring of mothers with psychotic disorder: Results from the Helsinki High-Risk Study. *British Journal of Psychiatry, 186,* 108–114.

Niemi, L. T., Suvisaari, J. M., Tuulio-Henriksson, A., & Lönnqvist, J. K. (2003). Childhood developmental abnormalities in schizophrenia: evidence from high-risk studies. *Schzophrenia Research, 60,* 239–258.

Nuechterlein, K. H. (1983). Signal detection in vigilance tasks and behavioral attributes among offspring of schizophrenic mothers and among hyperactive children. *Journal of Abnormal Psychology, 92,* 4–28.

Owen, M. J., O'Donovan, M. C., & Gottesman, I. I. (2002). Schizophrenia. In P. McGuffin, M. J. Owen, & II. Gottesman (Eds.), *Psychiatric genetics and genomics* (pp. 247–266). New York: Oxford University Press.

Pearson, J. S., & Kley, I. B. (1957). On the application of genetic expectancies on age-specific rates in the study of human behavior disorders. *Psychological Bulletin, 54,* 406–420.

Persson-Blennow, I., Näslund, B., McNeil, T. F., & Kaij, L. (1986). Offspring of women with nonorganic psychosis: Mother–infant interaction at one year of age. *Acta Psychiatrica Scandinavia, 73,* 207–213.

Pogue-Geile, M. F., Garrett, A. H., Brunke, J. J., & Hall, J. K. (1991). Neuropsychological impairments are increased in siblings of schizophrenic patients. *Schizophrenia Research, 4,* 390.

Quitkin, F., Rifkin, A., & Klein, D. F. (1976). Neurological soft signs in schizophrenia and character disorder. *Archives of General Psychiatry, 33,* 845–853.

Rajarethinam, R., Sahni, S., Rosenberg, D. R., & Keshavan, M. S. (2004). Reduced superior temporal gyrus volume in young offspring of patients with schizophrenia. *American Journal of Psychiatry, 161,* 1121–1124.

Rieder, R. O., & Nichols, P. L. (1979). Offspring of schizophrenics. III. Hyperactivity and neurological soft signs. *Archives of General Psychiatry, 36,* 665–674.

Rosenberg, D. R., Sweeney, J. A., Squires-Wheeler, E., Keshavan, M. S., Cornblatt, B. A., & Erlenmeyer-Kimling, L. (1997). Eye-tracking dysfunction in offspring from the New York High-Risk Project:

Diagnostic specificity and the role of attention. *Psychiatry Research,* 66, 121–130.

Rosenthal, D. (1970). *Genetic theory and abnormal behavior.* New York, NY: McGraw-Hill International Book Co.

Ross, R. G., Hommer, D., Radant, A., Roath, M., & Freedman, R. (1996). Early expression of smooth-pursuit eye movement abnormalities in children of schizophrenic parents. *Journal of the American Academy of Child and Adolescent Psychiatry, 35,* 941–949.

Rosso, I. M., Bearden, C. E., Hollister, J. M., Gasperoni, T. L., Sanchez, L. E., Hadley, T., et al. (2000). Childhood neuromotor dysfunction in schizophrenia patients and their unaffected siblings: A prospective cohort study. *Schizophrenia Bulletin, 26,* 367–378.

Sameroff, A. J., Seifer, R., Zax, M., & Barocas, R. (1987). Early indicators of developmental risk: Rochester Longitudinal Study. *Schizophrenia Bulletin, 13,* 383–394.

Schiffman, J., LaBrie, J., Carter, J., Cannon, T., Schulsinger, F., Parnas, J., et al. (2002). Perception of parent–child relationships in high-risk families, and adult schizophrenia outcome of offspring. *Journal of Psychiatric Research, 36,* 41–47.

Schiffman, J., Walker, E., Ekstrom, M., Schulsinger, F., Sorensen, H., & Mednick, S. (2004). Childhood videotaped social and neuromotor precursors of schizophrenia: A prospective investigation. *American Journal of Psychiatry, 161,* 2021–2027.

Schubert, E. W., Blennow, G., & McNeil, T. F. (1996). Wakefulness and arousal in neonates born to women with schizophrenia: Diminished arousal and its association with neurological deviations. *Schizophrenia Research, 22,* 49–59.

Schubert, E. W., Henriksson, K. M., & McNeil, T. F. (2005). A prospective study of offspring of women with psychosis: Visual dysfunction in early childhood predicts schizophrenia-spectrum disorders in adulthood. *Acta Psychiatrica Scandinavica, 112,* 385–393.

Schulsinger, F. Mednick, S. A., Venables, P. H., Raman, A. C., & Bell, B. (1975). Early detection and prevention of mental illness: The Mauritius project. *Neuropsychobiology, 1,* 166–179.

Slater, E. (1968). A review of earlier evidence on genetic factors in schizophrenia. In D. Rosenthal & S. S. Kety (Eds.), *The transmission of schizophrenia.* Oxford: Pergamon.

Sohlberg, S. C., & Yaniv, S. (1985). Social adjustment and cognitive performance of high-risk children. *Schizophrenia Bulletin, 11,* 61–65.

Squires-Wheeler, E., Erlenmeyer-Kimling, L., & Friedman, D. (1999). Event-related potential evaluation of working memory function. *Schizophrenia Research, 36,* 259.

Squires-Wheeler, E., Friedman, D., Amminger, G. P., Skodol, A., Looser-Ott, S., Roberts, S., et al. (1997). Negative and positive dimensions of schizotypal personality disorder. *Journal of Personality Disorders, 11,* 285–300.

Tienari, P., Wynne, L. C., Sorri, A., Lahti, I., Laksy, K., Moring, J., et al. (2004). Genotype-environment interaction in schizophrenia-spectrum disorder. Long-term follow-up study of Finnish adoptees. *British Journal of Psychiatry, 184,* 216–222.

Tsuang, M. T., Stone, W. S., & Faraone, S. V. (2000). Toward reformulating the diagnosis of schizophrenia. *American Journal of Psychiatry, 157,* 1041–1050.

Tsuang, M. T., Stone, W. S., Tarbox, S. I., & Faraone, S. V. (2002). An integration of schizophrenia with schizotypy: Identification of schizotaxia and implications for research on treatment and prevention. *Schizophrenia Research, 54,* 169–175.

Walker, E. F., Grimes, K. E., & Davis, D. M. (1993). Childhood precursors of schizophrenia: Facial expressions of emotion. *American Journal of Psychiatry, 150,* 1654–1660.

Walker, E. F., & Lewine, R. J. (1990). Prediction of adult-onset schizophrenia from childhood home movies of the patients. *American Journal of Psychiatry, 147,* 1052–1056.

Walker, E. F., & Neumann, C. S. (1995). Neurodevelopmental models of schizophrenia: The role of central nervous system maturation in the expression of neuropathology. In J. L. Waddington & P. F. Buckley (Eds.), *The neurodevelopmental hypothesis of schizophrenia.* Texas: R. G. Landes.

Watt, N. F. (1978). Patterns of childhood social development in adult schizophrenics. *Archives of General Psychiatry, 35,* 160–165.

Webb, C. T., & Levinson, D. F. (1993). Schizotypal and paranoid personality disorder in the relatives of patients with schizophrenia and affective disorders: A review. *Schizophrenia Research, 11,* 117–137.

Weinberger, D. R. (1986). The pathogenesis of schizophrenia: A neurodevelopmental theory. In H. A. W. Nasrallah & D. R. Weinberger (Eds.), *The neurology of schizophrenia.* Amsterdam: Elsevier.

Werry, J. S., McClellan, J. M., & Chard, L. (1991). Childhood and adolescent schizophrenic, bipolar, and schizoaffective disorders: A clinical and outcome study. *Journal of the American Academy of Child and Adolescent Psychiatry, 30,* 457–465.

Winters, L., B. Cornblatt, B. A., & Erlenmeyer-Kimling, L. (1991). The prediction of psychiatric disorders in late adolescence. In E. Walker (Ed.), *Schizophrenia: A life-course developmental perspective* (pp. 123–137). New York: Academic Press.

Winters, K. C., Stone, A. S., Weintraub, S., & Neale, J. M. (1981). Cognitive and attentional deficits in children vulnerable to psychopathology. *Journal of Abnormal Child Psychology, 9,* 435–453.

Wong, A. H., Gottesman, I. I., & Petronis, A. (2005). Phenotypic differences in genetically identical organisms: The epigenetic perspective. *Human Molecular Genetics, 14,* R11–R18.

Chapter 33

Attention and Working Memory: Animal Models for Cognitive Symptoms of Schizophrenia – Studies on D2-Like Receptor Knockout Mice

Claudia Schmauss

Introduction

The ability of humans and other higher vertebrates to implement and orchestrate behavioral strategies that are geared towards defined goals is governed by fundamental cognitive functions of attentional control and working memory that control lower-level sensory, memory, and motor operations for the purpose of achieving these goals. Studies on patients with distinct brain lesions have provided the foundation to delineate a structure–function relationship of such cognitive functions. Moreover, recent advances in brain imaging using positron emission tomography (PET) and functional magnetic resonance imaging (fMRI) have added a new dimension to the study of this structure–function relationship, namely the potential of investigating functional interactions among various brain regions in subjects performing defined cognitive tasks. Hence, the last decade of research on higher cognitive functions yielded a wealth of new knowledge about the multiplicity of brain regions that are activated at different stages of information processing.

To date, almost all studies on higher cognitive function are conducted on non-human and human primates, and they mostly employ psychological and neuroimaging/electrophysiological techniques. Psychological studies yielded a number of interesting results regarding the deficits of cognitive performance in various psychiatric illnesses. A limitation of these studies is, however, that they cannot examine the neural mechanisms of these cognitive processes. Neuroimaging and electrophysiological studies have powerfully illuminated the anatomic areas involved in attention and working memory, and they have provided important insight into functional aspects of the neural processes underlying these cognitive functions. It is, however, also recognized that these studies are correlational in nature, and that it is difficult to draw causal conclusions about the function of the various brain regions based on their activation (Raz, 2004).

The majority of animal studies on higher cognitive function relies on research on primates because of the anatomic relatedness of their brains to the human brain, the high degree of genetic homology between both species, and the high visual acuity of both species that enables the application of highly similar cognitive tasks that are based on visual features. Moreover, primates can implement complex plans and exhibit sophisticated goal-directed behaviors. Nevertheless, rodent models of higher cognitive functions are being increasingly explored because such models have the potential to complement primate studies in at least three important ways. (1) Investigations of cognitive phenotypes of mice carrying mutations in distinct neuronal genes allow probing isolated sets of genes for their role in higher cognitive function. Such studies are of significance because very little is presently known about the genes involved in cognitive network activation. Although some genetic variations have been shown to correlate with impaired performances in cognitive tasks requiring attention and/or working memory (Blasi et al., 2005; Bertolino et al., 2006; Egan et al., 2001; Fan, Fossella, Sommer, Wu, & Posner, 2003; Glickstein, Hof, & Schmauss, 2002; Glickstein, DeSteno, Hof, & Schmauss, 2005), the molecular biological underpinnings of these cognitive functions are largely unknown. (2) A comparative analysis of gene expression profiles can be performed on lines of mice (or rats) selectively bred for high or low cognitive task performance and thus facilitate the identification of genes involved in modulating cognitive functioning. (3) Brains of rodents are readily available for high-resolution anatomic studies on cognitive network activation that can span from global network identification to the cellular functions within any given network.

The objective of this chapter is to provide a brief overview of the anatomic regions and neurotransmitter systems that support working memory and attention and to describe

C. Schmauss (✉)

Department of Psychiatry and Molecular Therapeutics, Columbia University and New York State Psychiatric Institute, New York, NY 10032, USA
e-mail: cs581@columbia.edu

Y.-K. Kim (ed.), *Handbook of Behavior Genetics*,
DOI 10.1007/978-0-387-76727-7_33, © Springer Science+Business Media, LLC 2009

several behavioral paradigms that probe different cognitive domains of attentional control in rodents. Moreover, results of studies that applied some of these paradigms to mice lacking dopamine D2 and D3 receptors are discussed in view of the extent to which these studies could provide new knowledge of the role of dopamine and its receptors in working memory and other domains of attentional controls.

Anatomic Structures and Neurotransmitter Systems Supporting Attention and Working Memory

Attentional control is governed by a system of anatomical areas carrying out the function of three different control systems: selecting, orienting, and alerting (Posner & Petersen, 1990). These three functions are differently modulated by different neurotransmitter systems: Cholinergic systems originating in the basal forebrain play an important role in orienting, the norepinephrine system originating in the locus coeruleus plays a role in alerting, and the mesocortical dopamine system targets those prefrontal cortical and anterior cingulate areas involved in executive ("selecting") attention (for reviews see Posner & Petersen, 1990; Sarter, Givens, & Bruno, 2001; Raz, 2004; Raz & Buhle, 2006). Attention can be "selective" or "sustained". Selective attention requires directing attention to a specific (external or internal) object and suppressing attention to irrelevant objects. Related to selective attention is attentional shifting, a process involving engagement of attention on a stimulus, disengagement from this stimulus, and engagement on another relevant one. Sustained attention (duration, once an object has been selected) requires that task performance continues to be accurate over an extended period of time, i.e., that relevant stimuli continue to be selected and irrelevant stimuli continue to be ignored (Kindlon, 1998; Sarter et al., 2001).

Imaging studies in humans revealed that the orienting network of attention relies heavily on activation of the parietal system (superior and temporal parietal, frontal eye fields, superior colliculus), the selecting network involves activation of the anterior cingulate cortex, the lateral ventral prefrontal cortex (PFC), and the basal ganglia, and the alerting network involves activation of the locus coeruleus and the right frontal and parietal cortex (for reviews see Posner & Petersen, 1990; Raz, 2004; Raz & Buhle, 2006).

At difference to the complex and multidimensional constructs of attention, working memory is defined as the ability to hold relevant information "on-line" over a short period of time and subject it to further processing to enable accurate executive functioning. As such, working memory can be viewed as the interface between perception, long-term memory, and action. Theoretical concepts of working memory assume a limited capacity system, termed "the central executive", whose multidimensional coding allows different systems to be integrated. This "central executive" is assisted by two subsidiary storage systems, termed phonological loop (which is based on sound and language and temporarily represents new phoneme sequences) and visuospatial sketchpad (which maintains and manipulates visuospatial representations) (for review see Baddeley, 2003).

Working memory has at least four components: (1) encoding, (2) control of attention, (3) maintenance of information, and (4) resistance to interference. Hence, working memory is clearly distinct from the permanent inscription on neuronal circuitry due to learning and, in contrast to long-term memory (associated with proper hippocampal function), working memory is primarily a prefrontal cortical function (Goldman-Rakic, 1995) which, as further discussed below, is highly susceptible to disruptions under conditions of altered dopamine-mediated signaling. Furthermore, in contrast to tasks that test "attention", correct responses in working memory tests are guided by a representation of the *prior* stimulus, information must be updated on a trial-to-trial basis, and the execution of correct responses relies on the memory of the most recent response (Goldman-Rakic, 1995).

In the primate, working memory tasks lead to activation of not only the prefrontal cortex, but also the hippocampus and posterior parietal cortex, and it is thought that working memory relies on a re-entrant network organization that enables the prefrontal cortex and the hippocampus to operate with other cortical and subcortical structures as an integrated unit (Friedman & Goldman-Rakic, 1994). Hence, some anatomic structures support both attention and working memory, and an overlapping anatomic network is particularly evident for selecting the network of attention, i.e., a higher-level metacognitive attentional system (executive attention) that is also concerned with working memory.

Working Memory and Attention Deficits in Major Mental Disorders

Deficits in working memory and other domains of attentional controls are symptoms of a variety of psychiatric disorders, most notably schizophrenia, mood disorders, autism, attention-deficit hyperactivity disorder, and obsessive–compulsive disorders. Although neither attention nor working memory is implemented in a unitary fashion, attention and working memory processes interact in several ways: (1) attention acts as a "gatekeeper" that determines which item occupies the limited space of working memory, (2) attention facilitates the early identification of new sensory information and recruits it for active maintenance in working memory, and (3) attention functions as an "executive process" that actively manipulates and updates the content of working memory. Hence, working memory and (at least "selective") attention are closely intertwined to enable goal-driven processing by increasing accessibility of relevant over irrelevant information (Awh, Vogel, & Oh, 2006). It is thus not surprising that disorders affecting fundamental mechanisms of working memory can give rise to numerous difficulties, including difficulties in focusing attention (Etchepareborda & Abas-Mas, 2005). For example, a recent meta-analysis of studies on children with ADHD revealed that these children had, in fact, prominent working memory impairments (Martinussen, Hayden, Hogg-Johnson, & Tannock, 2005). Moreover, as discussed further below, working memory deficits are prominent in schizophrenia (Park & Holzman, 1992; Weinberger, Berman, & Zec, 1986), and schizophrenic patients often exhibit impairments of attention/vigilance (Nuechterlein et al., 2004). Although a causal relationship between the working memory and attention deficits in many of the psychiatric disorders appears plausible, direct proof of such a functional relationship remains elusive (Kindlon, 1998).

To date, cognitive dysfunctions in psychiatric disorders are perhaps the most extensively studied in schizophrenic patients (Barch, 2006). Dysfunctions of working memory and attentional processes occur prior to the onset of clinically manifest schizophrenia, and they persist throughout the course of the disease (Erlenmeyer-Kimling, 2000; Peuskens, Demily, & Thibaut, 2005). Whereas previous research has primarily focused on the positive symptoms of the disease (delusions, hallucinations, and thought disorders) and on the treatment of these symptoms with antipsychotic drugs, it is now recognized that cognitive dysfunctions are core psychopathological features of the disease that are largely refractory to conventional antipsychotic treatment.

Patients with schizophrenia exhibit impairments in a variety of cognitive domains, most notably perception and attention, short- and long-term memory, executive and motor function (Peuskens et al., 2005). Working memory deficits, however, are the most commonly observed cognitive dysfunction and, as a core deficit, the extent of working memory impairment is thought to be the best predictor of social integration and the propensity for relapse (for review see Goldman-Rakic, Castner, Svensson, Siever, & Williams, 2004).

The Role of Dopamine in Working Memory

Optimal working memory performance is critically dependent upon the integrity of dopaminergic innervation of the prefrontal cortex. In a seminal study, Brozoski, Brown, Rosvold, and Goldman-Rakic (1979) showed that lesions of the dorsolateral prefrontal cortex of monkeys with 6-hydroxydopamine (6-OHDA) elicited a delay-dependent impairment in the performance on a spatial delayed response task. Moreover, Collins, Wilkinson, Everitt, Robbins, and Roberts (2000) showed that such lesions also impaired the acquisition of this task. Additional evidence for a role of dopamine in cognition came from studies of patients with Parkinson's disease. Early in the course of the disease, i.e., prior to pharmacological intervention, such patients exhibit several cognitive deficits, including spatial working memory deficits, and subsequent L-DOPA treatment reverses some of these deficits (Owens et al., 1992). Conversely, excessive dopamine signaling caused by stress also produces working memory deficits, and these deficits are reversed by D1-antagonist treatment (Arnsten & Goldman-Rakic, 1998). Indeed, evidence suggests that the activity of prefrontal cortical dopamine D1-receptors plays a critical role in the control of working memory. Microinjection of dopamine D1- (but not D2-) receptor antagonists into the dorsolateral prefrontal cortex disrupts memory-guide saccades in the monkey (Sawaguchi & Goldman-Rakic, 1991) and iontophoretic application of D1-selective ligands revealed a nonlinear relationship between D1 receptor activation and the strengths of memory fields in prefrontal pyramidal neurons of monkeys (Williams & Goldman-Rakic, 1995). The latter data have let the authors to postulate that there is not just an optimal window for D1-receptor stimulation to enable optimal working memory performance, but an inverted U-shaped relationship between the strengths of memory fields and the level of D1-receptor activation. This postulate is supported by findings that subjects with low baseline performance in working memory tasks improve upon indirect dopamine-receptor stimulation via amphetamine administration whereas those with optimal performance deteriorate, and these phenomena correlate with signal intensities in the right dorsolateral prefrontal cortex measured with functional magnetic resonance imaging (Mattay et al., 2000). Similar results were obtained in experiments on rodents that exhibit different levels of baseline performance in the five-choice serial reaction time task (a test of sustained attention further described below) (Granon et al., 2000), suggesting that suitable doses of D1 agonists improve working memory and higher doses disrupt it, and that the efficacy of D1-agonist stimulation depends on the baseline working memory performance. Moreover, since one characteristic of working memory is its resistance to interference, it has been hypothesized that that D1-receptor activation prevents distraction during delay periods of working memory tasks and thus stabilizes stimulus representations (Robbins, 2005).

Finally, several human studies have shown that genetic variances of dopamine-related genes (catecholamine-O-

methyltransferase, dopamine transporter) differentially affect higher cognitive functions and prefrontal cortical activation (Blasi et al., 2005; Bertolino et al., 2006; Egan et al., 2001), and much has been learned about the limits of cognitive functions such as working memory and attention and the role of dopamine from the study of schizophrenic patients (Barch, 2006). Such patients are almost exclusively treated with neuroleptic drugs that block either D2-like dopamine receptors (typical neuroleptics like haloperidol) or both D2- and 5-HT2-like receptors (atypical neuroleptics like clozapine). It is now recognized that typical neuroleptics, although efficacious in ameliorating the positive symptoms of schizophrenia, are ineffective in the treatment of deficits in working memory and attentional control. Moreover, as further discussed below, three animal studies have shown clearly that chronic blockade of D2-like dopamine receptors (either with haloperidol or in knockout mice) induces working memory deficits (Castner, Williams, & Goldman-Rakic, 2000; Glickstein et al., 2002) and that inactivation of D2 receptors in knockout mice impairs their performance in an attention-set-shifting task (Glickstein et al., 2005). It is evident from earlier studies (Glickstein et al., 2002, 2005) that compromised D1-receptor function in the mPFC alone is insufficient to account for the deficits in attentional control and working memory, and that more research is needed to understand how inactivation of D2-like receptors leads to disrupted cognitive functioning.

Cognitive Functions Studied in Wild-Type and Mutant Mouse Models: Dopamine D1 Receptor Function in Mice Lacking D2 and D3 Receptors

Genetic, molecular, and high-resolution anatomic studies on networks that support major cognitive processes in primates are hampered by the limited availability of brain tissue. Hence, there is a need for complementary studies on lower vertebrates, including studies on genetically modified mice with altered expression of genes implicated in the control of higher cognitive function.

As outlined above, studies on non-human and human primates have provided substantial insight into the role of prefrontal cortical dopamine and dopamine D1 receptors in the control of working memory. These studies have extended the conceptual thinking about the role of dopamine in schizophrenia beyond dopamine D2-like receptors, the main targets of neuroleptic drugs, and the subcortical hyperdopaminergia that justifies neuroleptic treatment. They have opened avenues for considering that prefrontal cortical hypodopaminergia also contributes to the core psychopathology of schizophrenia and that decreased dopamine

D1-receptor function plays a pivotal role therein. However, while these studies focused exclusively on dopamine D1 receptors, they have seldom addressed the role of D2-like receptors in modulating prefrontal cortical D1-receptor activation. For example, the study of Castner et al. (2000) showed that monkeys chronically treated with the typical neuroleptic drug haloperidol also exhibited working memory deficits and that intermittent application of a D1 agonist leads to long-lasting improvement of this dysfunction. This finding suggested that normal cognitive function requires a balanced activity of D1- and D2-like dopamine receptors. Indeed, subsequent studies on knockout mice lacking D2 and D3 receptors illustrated that chronic D2-like receptor blockade alters agonist-promoted activity of D1 receptors (Schmauss, 2000; Glickstein et al., 2002). As further discussed below, these studies have shown that inactivation of D2- and D3-receptors (as achieved by chronic neuroleptic treatment) decreases agonist-stimulated D1 receptor activation in the medial prefrontal cortex and that mice lacking D2 and D3 receptors have spatial working memory deficits.

D1 Receptor Function in the Forebrain of Mice Lacking Dopamine D2 and D3 Receptors

Studies on knockout mice lacking individual dopamine receptors revealed clearly that normal motor and cognitive function requires an intricately balanced activity of all dopamine receptors (Schmauss, 2000; Glickstein et al., 2002; Glickstein & Schmauss, 2004). The conclusion that genetic manipulations that selectively targeted either dopamine D2 or D3 receptors also lead to decreased agonist-stimulated D1 receptor function originated from studies on D1-receptor-dependent induction of expression of mRNA encoding the immediate early gene c-*fos*. In these studies, D2 and D3 receptor knockout mice exhibited blunted neocortical c-*fos* responses to D1-agonist stimulation although their D1-receptor expression levels were unaltered. A single pharmacological stimulation of D1 receptors in these mutants with either a selective D1 agonist or the indirect dopamine receptor agonist methamphetamine, however, led to a long-term (as much as 2 weeks) rescue of D1-receptor function in the forebrain neocortex of these mutants (Schmauss, 2000). Hence, although chronic inactivation of D2 and D3 receptors blunts neocortical D1-receptor function in vivo, this situation is not irreversible. It is noteworthy, however, that the long-term increase in agonist-promoted neocortical D1-receptor function induced by either a single dose of methamphetamine or a selective D1 agonist is only detected in brains with either naturally low (preadolescent mice) or abnormally blunted (D2 and D3 knockout mice) c-*fos* expression in response to D1 agonist stimulation (Schmauss, 2000).

D1 Receptor Activation in the Medial Prefrontal Cortex of Mice Lacking Dopamine D2 and D3 Receptors

A stereological assessment of the number of neurons expressing Fos immunoreactivity in response to D1-agonist stimulation showed that the blunted neocortical c-fos response of D2 and D3 mutants was most prominent in the medial prefrontal cortex (mPFC; Glickstein et al., 2002). This was of particular interest because the rodent mPFC is the functional homologue of the dorsolateral PFC of the primate. Like the primate PFC, the rodent mPFC is anatomically defined by its reciprocal connections to the mediodorsal thalamus as well as reciprocal cortico-cortical connections, and several functional properties mediated by the rodent mPFC are homologous to the dorsolateral PFC of the primate, most notably the control of working memory, attention, and attention shifts (Uylings, Groenewegen, & Kolb, 2003). In fact, the recognition of this functional homology between the primate dorsolateral PFC and the mPFC stimulated studies on higher cognitive functions in rats and mice.

Figure 33.1 schematically illustrates the anatomic topography of the three subregions of the mouse mPFC, the anterior cingulate (AC), prelimbic (PL), and infralimbic (IL) cortex, and their neuronal cytoarchitecture revealed by labeling with an antibody directed against the neuronal protein NeuN. In the mPFC of D2- and D3-receptor mutant mice, the number of neurons expressing Fos immunoreactivity in response to a D1 agonist challenge was reduced in superficial (II/III) and deeper (V/VI) laminar territories of all three subregions, and this reduction was largest in the AC, inter-

mediate in the PL, and lowest in the IL subregions (Glickstein et al., 2002). A single dose of methamphetamine, however, administered 1 week prior to the D1-agonist challenge, led to robustly increased expression of Fos immunoreactivity in both D2 and D3 mutants (but not in wild type), and this increase was highest in the AC and lowest in the IL subregions, and it was larger in D2 mutants compared with D3 mutants. In fact, in the AC and PL, D1-agonist-stimulated c-*fos* responses of methamphetamine-pretreated D2 mutants were indistinguishable from wild type. Hence, methamphetamine pre-treatment let to a long-term rescue of agonist-promoted D1-receptor function in the mPFC.

Given the well-documented role of prefrontal cortical D1 receptors in the control of working memory, it was of interest to test whether D2 and D3 mutants also exhibit working memory deficits and whether methamphetamine pre-treatment, known to rescue D1-receptor function in the mPFC, would also restore normal working memory performance. As shown below, D2 and D3 mutants have indeed working memory deficits. Interestingly, however, the ability of methamphetamine pre-treatment to restore normal working memory function differs between both mutants.

Cognitive Functions Studied in Wild-Type and Mutant Mouse Models: Working Memory

Delayed Alternation: Tests of Spatial Working Memory of Rodents

In rodents, working memory tests are primarily testing spatial working memory and such tests are conducted in two- (T- and Y-maze) or eight-arm (8-arm radial maze) test chambers. The most commonly used test paradigm is the delayed alternation task performed in a T-maze which has been shown to be a valuable tool for evaluating spatial working memory deficits associated with prefrontal cortical dysfunction in all mammalian species (Markowitsch & Pritzel, 1977; Moran, 1993; Van Haaren, De Bruin, Heinsbroek, & Van de Poll, 1985). This test was originally developed for rats, but it has been used successfully in studies on knockout mice (see, for example, Glickstein et al., 2002). Briefly, mice are trained (over a period of 10–14 days) for alternate arm entries in the T-maze until they reach ~80% correct arm entries on two consecutive days with no inter-trial delay (10 trials per test). On subsequent days, they are tested using a variable delay (5 s to 1 min) until a delay period is reached that impairs the performance to chance levels of correct arm entry (i.e., 50%).

As expected from results of the studies described above, both D2 and D3 mutants exhibited working memory deficits with increasing impairments at increasing delay periods.

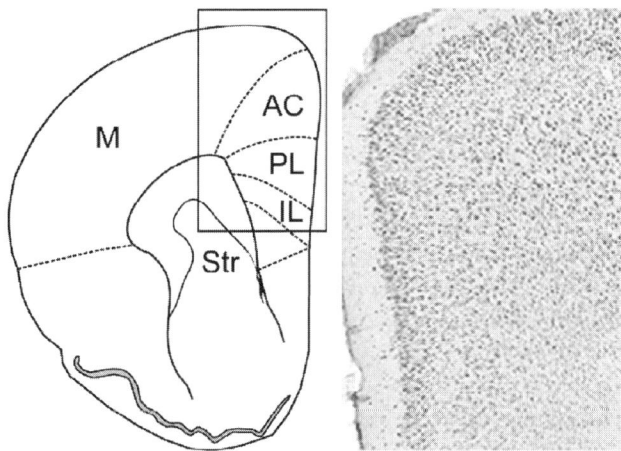

Fig. 33.1 *The mouse medial prefrontal cortex.* Schematic illustration of the regional extent of the mouse medial prefrontal cortex in a coronal section taken 5.5 mm rostral from the interaural line and its neuronal cytoarchitecture revealed by immunolabeling with an antibody directed against NeuN. M, motor cortex; Str, striatum; AC, anterior cingulate cortex; PL, prelimbic cortex; IL, infralimbic cortex. Adopted from Glickstein et al. (2002)

These deficits were more severe in D2 mutants which exhibited significantly lower performance compared with wild type already at the 5-s delay intervals. These behavioral differences paralleled the above-described differences between both mutants in the magnitude of D1-agonist-stimulated c-*fos* responses in the AC and PL subregions of the mPFC. Pre-treatment of these mutants with a single dose of methamphetamine rescued their working memory performance within 1 week following methamphetamine administration. However, the same treatment had no effect on the working memory test performance of D3 mutants. These findings illustrate that the working memory deficits of D2 mutants can be solely attributed to their decreased D1-receptor function. However, since diminished D1-receptor function alone cannot explain the resistance of D3 mutants to methamphetamine pre-treatment it appears that signaling through D3 receptors is also essential for proper working memory performance. The extent to which D3-receptor inactivation alters the activation of those cognitive networks that support working memory remains to be further investigated.

Because attention and working memory are fundamental cognitive processes that interact with one another it was of interest to test whether the working memory deficits of D2 and D3 mutant mice are also accompanied by deficits in other domains of attentional control. The following section addresses this question along with a brief description of experimental paradigms that are suitable for testing different domains of attentional control in rodents.

Attention

At present, very few studies of attentional control in rodents have used the mouse. The behavioral paradigms successfully developed for the rat, however, are generally applicable to the mouse, and some of the most commonly employed experimental paradigms are briefly described below.

A Two-Choice Discrimination Attentional Set-Shifting Tasks Designed for Rodent

Attention-set-shifting tasks typically involve a series of compound discriminations that require the subject to learn task rules, maintain and shift attentional sets, and to reverse previously acquired rules. When applied to humans or primates, such tasks are usually based on different visual features of a stimulus, such as filled shapes versus configurations of lines superimposed on these shapes.

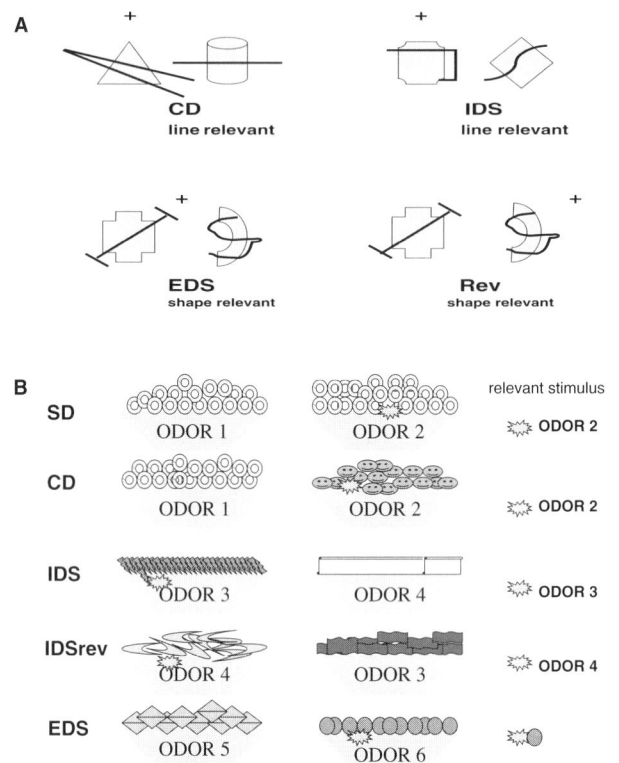

Fig. 33.2 *Attention set-shifting tests designed for humans* (**A**) *and rodents* (**B**). Whereas the human test relies on visual stimuli (*lines* and *shapes*), the rodent version of the test uses olfactory (odors) and somatosensory (texture of digging medium) stimulus dimensions. In (**A**), the + sign indicates the reward option. In (**B**), the relevant stimuli on the right indicates reward location. SD, simple discrimination; CD, compound discrimination; IDS, intradimensional set shift; EDS, extradimensional set shift; Rev, reversal

As illustrated in Fig. 33.2A, a stimulus pair can consist of a shape and a line but, at any given time, only one aspect of the pair (shape or line) is relevant. In such a continuous attentional performance test, subjects first complete a compound discrimination (CD) in which, for example, the line represents the relevant (guiding) stimulus dimension. In the following intradimensional set-shifting phase (IDS), new shapes and lines are presented but lines remain the relevant dimension. In an extradimensional set-shifting phase (EDS), however, the previously relevant dimension (line) becomes irrelevant and the previously irrelevant dimension (shape) guides the correct response selection. Finally, set rules are reversed (Rev), i.e., the previously relevant stimulus property (a distinct line type) is no longer relevant, but the previously irrelevant one guides correct response selection.

In contrast to primates, the ability of rodents to discriminate visual stimuli by pattern and shape is greatly diminished. Thus, the application of attention-set-shifting tasks to rodents necessitates the use of stimulus dimensions that are appropriate for this species. Hence, Birrell and Brown (2000) have adopted the general principles of the attention-set-shifting task to rats by using two new stimulus dimensions, odor and texture. In this paradigm, terra cotta pots are used as digging bowls and their rims are scented with perfumed oil to produce a lasting odor. The bowls contain a food pellet buried underneath different digging media. As shown in Fig. 33.2B, two pots are presented in each trial. Animals are first trained to dig in unscented small bowls filled with familiar bedding to retrieve a food reward that is deeply buried underneath the bedding. Then, animals are trained in two simple discriminations (SD) of either odor (patchouli/jasmine, mango/vanilla, tea rose/dewberry, fuzzy peach/woody sandalwood, etc.) or digging media of different textures (wood chips/alpha dry bedding, glass beads/Eppendorf tube lids, ribbon/yarn, shredded paper/shredded rubber, etc.) to a criterion of six consecutive correct trials. The order of the two SDs and relevant stimulus dimensions (odor or digging medium) are randomized across animals. After an SD between two odors or two digging media, a compound discrimination (CD) follows in which a new dimension is added to the stimuli presented in the initial SD. This new dimension, however, is not a reliable indicator of the food location. The next test phase requires an intradimensional shift of attention. The IDS is another CD in which both relevant and irrelevant stimuli are changed, but the previously relevant stimulus dimension (odor or digging medium) remains the same. This IDS is then subjected to reversal rules, i.e., the previously negative stimulus becomes a positive one but the irrelevant stimulus dimension is still not predicting the reward location. Finally, in a task requiring an extradimensional shift of attention (EDS), the formerly irrelevant dimension becomes relevant and the originally guiding dimension loses its predictive value. Also this test phase is subjected to reversal rules. In all test phases, animals have to reach a criterion of six consecutive correct trials, and each sensory stimulus is used in only one test phase.

The attention-set-shifting paradigm described above engages processes from several cognitive domains, ranging from associative (procedural) learning (CD), selective attention (maintaining sets), shifting between perceptual stimuli or dimension (IDS, EDS), and reversal learning. Studies of Brown and colleagues (Birrell and Brown, 2000; McAlonan & Brown, 2003) have shown that different phases of the test are differently affected by lesions of the orbital and medial prefrontal cortex of the rat: Whereas lesions of the medial prefrontal cortex (mainly confined to IL and PL subregions) impair EDS performance but not reversal learning (Birrell and Brown, 2000), reversal learning, but not set shifting, is impaired in rats with orbital prefrontal lesions (McAlonan & Brown, 2003), suggesting that different neocortical

structures control cognitive processes and/or performance in the different phases of the task.

Performance of D2 and D3 Mutant Mice in the Attention-Set-Shifting Task

Because working memory is required across the different cognitive domains of attentional control, Glickstein et al. (2005) tested whether the working memory deficits of D2 and D3 mutants would lead to impairments in the performance of these mutants in the attention-set-shifting task illustrated in Fig. 33.2B. Interestingly, the cognitive phenotype revealed by this task differed quite substantially. D2 mutants exhibited significant impairment in the first compound discrimination (CD) of the task, indicating difficulties in the acquisition of the rules that governed the task. Once they have learned the rules after a significantly increased number of trials to criterion in the CD, they proceeded through the remaining phases of the test in a manner indistinguishable from wild type. Interestingly, in contrast to the effect of methamphetamine on the working memory performance of these mutants, pre-treatment with methamphetamine did not improve the performance of these mutants in the CD phase of the test.

At difference to D2 mutants, D3 mutants outperformed wild-type controls in set-shifting phases of the task, and their response accuracy was significantly better under reversal conditions (IDSRev). These differences in task performance were also reflected in differences in the number of mPFC neurons that expressed Fos immunoreactivity in tested animals. After completion of either CD or EDS phases of the test (see Fig. 33.2B), neuronal activation (as indicated by Fos expression) was highest in D3 mutants, intermittent in wild type, and lowest in D2 mutants. Hence, Fos expression levels appear to correlate directly with the response accuracies of the three genotypes.

Interestingly, in wild-type animals, test-induced activation of the AC subregion of the mPFC exceeded that of PL/IL subregions. This difference, however, was never detected in D3 mutants. It is of interest to note that the rodent mPFC also has features of other cortical regions of primates, most notably the anterior cingulate cortex (Uylings et al., 2003). In primates, one interpretation of neuronal activation in the AC during tasks requiring attention and response selection is that the AC monitors conflict between different action plans ("performance monitoring"; Botvinick, Braverm, Barch, Carter, & Cohen, 2001; MacDonald, Cohen, Stenger, & Carter, 2000) to signal greater cognitive control to the dorsolateral PFC. This decreases conflict, and in subsequent trial with correct response selection, activation of the AC decreases and activation of the dorsolateral PFC increases (Kerns et al., 2004). It is thus tempting to speculate that the AC subregion of the rodent mPFC has similar influences on the IL and PL subregions, and that the higher response accuracy of D3 mutants is partly due to their higher PL/IL activation that signals less conflict and hence greater cognitive control (see Glickstein et al., 2005).

Finally, it was noted that D3, but not D2, mutants exhibited prolonged response latencies in all test phases of the attention set-shifting task illustrated in Fig. 33.2B. The reasons for this prolonged response latency is presently unclear. It is possible that D3 mutants have subtle abnormalities in information processing that cannot be uncovered with the attention set-shifting task. For example, the higher neuronal activation in the mPFC of these mutants could also be due to the inability of these mutants to ignore irrelevant stimuli so that both relevant and irrelevant information are continuously co-processed. Such potential deficit in optimizing cognitive control of selective attention could perhaps be unraveled with some of the additional cognitive tests that are described below.

Rodent Tests of Selective Attention

In contrast to the complexity of cognitive domains engaged by the attention-set-shifting task described above, several other behavioral paradigms have been developed that focus on a more limited set of cognitive domains. Among them is the *cued target detection task* (CTD), a task of selective (also called transient, exogenous, or reflexive) visuospatial attention developed by Posner and colleagues (Posner, 1980). This test examines the orienting and alerting components of attention and eliminates sustained attentional components. In the CTD, attention is evoked by an abrupt, peripheral visual stimulus that serves as a cue for the location of a subsequent visual target. The test requires that the animals maintain fixation during the presentation of the cue and target. The orienting reaction to the cue is mediated by shifts in attention (called covert orienting; Posner (1980)). As schematically illustrated in Fig. 33.3A, four types of trials are presented in this test.

One is a "valid" trial in which both cue and target are located on the same side of the fixation. The other is an "invalid" trial that positions cue and target in opposite sides of the fixation. These two trials test the orienting component of attention. In the valid trial, the benefits of spatial information for facilitating target detection or discrimination is measured and the invalid trial estimates the cost of misorienting. The time to detect the target when it is congruent with the cue is shorter than the time to detect incongruent cues and targets, and this difference in the reaction times is referred

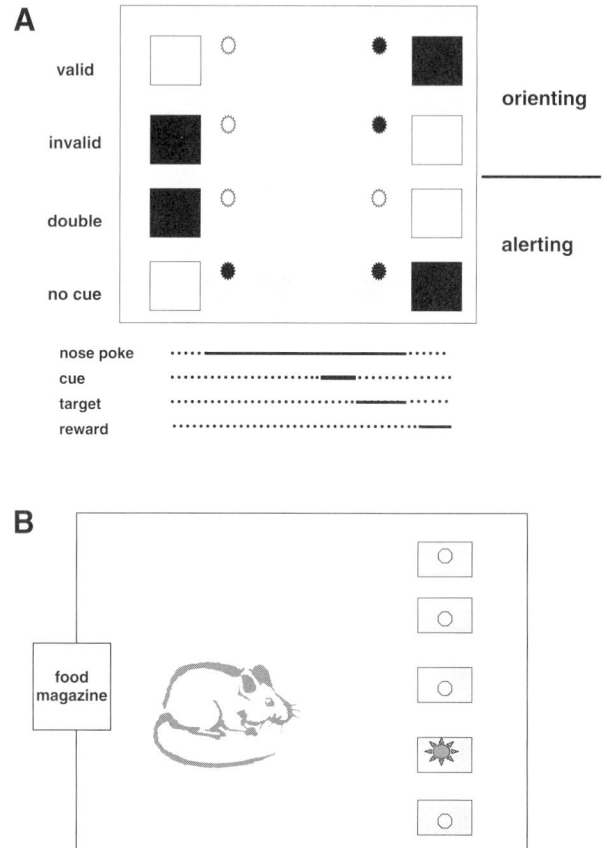

Fig. 33.3 *Rodent tests of selective and sustained attention.* (**A**) The cued target detection test (selective attention). (**B**) The 5-choice serial reaction time task (sustained attention). In (**A**), rectangles represent dispenser enclosures and cycles represent cue lights. Unfilled rectangles and cue lights indicate that they are illuminated. The timing of CTD events is indicated on the bottom. Nose-pokes trigger the cue and dispenser light after a variable interval. Once the cue is presented, the animal withdraws its nose and moves to the reward dispenser. Adopted from Stewart et al. (2001). For the 5-CSRTT schematically illustrated in (**B**), a test chamber with a rear magazine equipped with a pellet dispenser and nose-poke detector is used. The front of the chamber has an array of five equally spaced holes, each containing a light bulb and an infrared nose-poke detector. During the test duration, the animal scans the array of holes and, when the animal responds correctly (nose poke) to the light stimulus presented in one of them, it obtains a food reward.

to a "validity effect" (VE). The other two components of the task measure stimulus "alerting" effects defined as the temporal benefits of response time provided by the abrupt stimulus onset. In one test component, double cues are presented, but the target appears subsequently at only one of them. Hence, the cue lacks explicit spatial information and requires no orienting response. However, the cue provides temporal information as it informs the subject of the target appearance. In the fourth component, no cues are presented prior to target presentations and the difference between reaction times in the double cued and non-cued target tests is termed "alerting affect" (AE).

This test proved valuable in studies on rats that examined the role of nicotinic cholinergic neurotransmission in modulating the orienting component of attention (Stewart, Burke, & Marrocco, 2001) and it is likely that similar studies on wild-type and knockout mice will provide novel insight into the role of individual molecules in the control of orienting and alerting components of attention.

Another process related to selective attention is termed *latent inhibition* (LI), a negative transfer phenomenon characterized by an impaired generation of a conditioned response to a stimulus that had been pre-exposed without consequences. LI reflects a stimulus-specific decline of attention as a function of repeated irrelevant (non-reinforced) stimulus pre-exposure (Lubow, 2005). This acquired irrelevance of the stimulus reduces this stimulus' ability to enter into associative relationships compared with novel stimuli or interferes with the subsequent expression of such associative relationships. Tests of latent inhibition are thought to be a valid model for dysfunctions of selective attention in schizophrenia that are due to high distractibility, i.e., in schizophrenia, irrelevant stimuli attract attention and, when repeated, these stimuli continue to be subjected to cognitive processing and thus, the normal shift of the processing from a controlled to an automatic mode is inhibited (for review see Lubow, 2005).

Weiner (2003) proposes a "two-headed" LI model that describes two extremes of deficient cognitive switching: disrupted LI and persistent LI. In animal models, disrupted LI occurs under conditions that lead to LI in normal animals, and persistent LI is observed under conditions that disrupt LI in normal animals. These two extremes reflect enhanced and delayed switching between associations acquired during non-reinforced preexposure (stimulus-no event) and during subsequent conditioning (stimulus reinforcement), they are differentially modulated by dopamine, and they result from dysfunction in different brain circuitries. Enhanced dopaminergic neurotransmission as well as stress disrupt LI and thus promote rapid switching of responses according to the stimulus-reinforcement contingency. On the contrary, blockade of dopaminergic neurotransmission enhances the control of the stimulus-no event contingency and retards switching to a new stimulus-reinforcement association (Shalev & Weiner, 2001; Weiner, 2003; Lubow, 2005; Young, Moran, & Joseph, 2005). Studies in rats have shown that the dopaminergic innervation of the nucleus accumbens plays a critical role in LI (for review see Young et al., 2005). Damage to the shell of the nucleus accumbens disrupts LI, while damage to the core leads to abnormally persistent LI. The effects of shell lesions are also paralleled by lesions in the entorhinal cortex, and lesions of the hippocampus and the basolateral nucleus of the amygdala produce abnormal LI phenotypes resembling that of core lesions of the nucleus accumbens (Weiner, 2003).

To date, studies on LI employed procedures typically used with rats, including conditioned suppression, conditioned avoidance, conditioned taste aversion, and classical defensive and appetitive conditioning (Lubow, 2005), and these LI paradigms have been successfully used in wild-type and knockout mice (Caldarone, Duman, & Piccioto, 2000; Miyakawa et al., 2003). It is plausible that appropriate modifications of the SD and CD components of the attention-set-shifting task described above will also provide a suitable test of LI in the mouse. For example, during the SD, a novel stimulus could be a scented bowl that does not contain a food pellet and is paired with the presentation of the non-scented bowl containing the food pellet. The test phase is completed when the animal disregards the scented bowl over several consecutive trials. Then, in the CD, the same scented bowl would signal reward, but it is presented along with new stimulus properties of the second dimension (digging media) and another scented bowl that does not contain the food reward. The number of trials necessary to induce inattentiveness to the irrelevant stimulus in the SD and the number of trials necessary to attend the previously irrelevant stimulus as the relevant in the CD should provide an estimate of the extent of LI in wild type and different dopamine receptor knockout mice when these data are compared with corresponding results obtained from animals not exposed to a negative stimulus in the SD that becomes the positive stimulus in the CD.

Rodent Tests of Sustained Attention

The CTD task described above tests reflexive attention, i.e., the most primitive form of attention that is engaged with shortest response latency whenever an abrupt sensory stimulus is presented. At difference, voluntary, sustained attention requires conscious mental activity, is more slowly activated than reflexive attention, and is controlled by cognitive demand (Beane & Marrocco, 2004). Two tests of sustained attention have been used in rodents. One is the *sustained attention task* (SAT; McGaughy & Sarter, 1995) in which rats press one of two levers if a target occurs in a defined time window and targets are presented in about half of the trials and omitted from the remaining trails. This test may not find a broad applicability to mice due to the difficulty of mice to engage the motor act of lever pressing.

Another test of sustained attention is the *five-choice serial reaction time task* (5-CSRTT) developed by Robbins and colleagues (Carli, Robbins, Evenden, & Everitt, 1983; for review see Robbins, 2002). In this test, rats must attend to an array of potential target locations, but respond only to the one that is illuminated. Typically, rodents scan a horizontal array of five spatial apertures for the location of a brief visual target stimulus over a large number of trials (see Fig. 33.3B).

Studies employing this test have shown that ablations of different neurotransmitters affect different task-related measures. For example, cholinergic lesions affect response accuracy, noradrenergic lesions increase distractibility, serotonergic lesions increase impulsivity, and dopaminergic lesions slow responding (for reviews see Dalley, Cardinal, & Robbins, 2004; Robbins, 2002).

The 5-CSRTT has been widely used in studies on rats (for review see Robbins, 2002). However, Humby, Laird, Davies, and Wilkinson (1999) were the first to demonstrate the successful application of this test to mice, and they showed that systematic manipulations of task parameters could uncover behavioral differences between F1 generations of C57Bl/6xDBA/2 and C57Bl/6x129Sv inbred strains of mice in response to pharmacological manipulations of central cholinergic neurotransmission. Moreover, Young et al. (2004) successfully used the 5-CSRTT in wild-type mice and mutant littermates lacking the α7 nicotinic acetylcholine receptor (nAChR) and showed that nicotine improves sustained attention in normal mice and that the α7 nAChR plays a significant role in attentional control processes. Thus, the 5-CSRTT has already made its entry into cognitive studies on knockout mice and comparative cognitive studies of different inbred strains.

Conclusion and Future Directions

Despite the predominant focus of research on primates, there is a role for mice in studies of higher cognitive function. The study of the mouse promises a unique advantage, namely the availability of genetically engineered knockout lines with targeted disruptions of genes thought to be involved in shaping higher cognitive function. As illustrated here, some progress has been made, for example, with studies on the cognitive functions of knockout mice lacking dopamine D2 and D3 receptors, and this progress was possible by the ability to combine genetic engineering with quantitative anatomic, molecular, and behavioral analyses. As a result, these studies could provide a plausible link between D2-like receptor inactivation and decreased D1-receptor function in the mPFC that ultimately leads to working memory deficits. Moreover, these studies revealed differential effects of individual subtypes of D2-like receptors in other domains of attentional control, and they begun to illuminate the reasons why chronic administration of neuroleptic drugs that block D2-like receptors cannot improve cognitive deficits.

It is evident that studies on cognitive functions of mice have benefited tremendously from the behavioral paradigms that were systematically developed for the rat prior to the era of knockout mice. It is possible that a standard battery of cognitive behavioral tests (optimized for the mouse) will be recommended for studies that probe the role of novel genes in cognitive function in the mouse. Such future studies will not only be delimited to knockout mice but also take further advantage of the impressive repertoire of genetically tractable inbred strains of mice to conduct comparative studies of cognitive function between them to ultimately determine the genetic underpinnings of efficient or disrupted cognitive control.

Finally, high-resolution anatomic studies on brains of mice will complement studies on primates to facilitate the process of identifying anatomic networks, their interactions, and the cellular contributions to specific network activation necessary for efficient attentional control. Thus, although current studies of cognitive functions in knockout and genetically distinct strains of mice are still in their earliest stage, studies on these mice promise a powerful contribution to the field of cognitive neuroscience.

References

Arnsten, A. F., & Goldman-Rakic, P. S. (1998). Noise stress impairs prefrontal cortical cognitive function in monkeys: Evidence for a hyperdopaminergic mechanism. *Archives of General Psychiatry, 55*, 362–368.

Awh, E., Vogel, E. K., & Oh, S.-H. (2006). Interactions between attention and working memory. *Neuroscience, 139*, 201–208.

Baddeley, A. (2003). Working memory: Looking back and looking forward. *Nature Reviews. Neuroscience, 4*, 829–839.

Barch, D. M. (2006). What can research on schizophrenia tell us about the cognitive neuroscience of working memory? *Neuroscience, 139*, 73–84.

Beane, M., & Marrocco, R. T. (2004). Norepinephrine and acetylcholine mediation of the components of reflexive attention: Implications for attention deficit disorders. *Progress in Neurobiology, 74*, 167–181.

Bertolino, A., Blasi, G., Latorre, V., Rubino, V., Rampino, A., Sinibaldi, L., et al. (2006). Additive effects of genetic variation in dopamine regulating genes on working memory cortical activity in human brain. *Journal of Neuroscience, 26*, 3918–3922.

Birrell, J. M., & Brown, V. J. (2000). Medial frontal cortex mediates perceptual attention set shifting in the rat. *Journal of Neuroscience, 20*, 4320–4324.

Blasi, G., Mattay, V. S., Bertolino, A., Elvervåg, B., Callicott, J. H., Das, S., et al. (2005). Effect of catechol-O-methyltransferase val[158]met genotype on attentional control. *Journal of Neuroscience, 25*, 5038–5045.

Botvinick, M. M., Braverm T. S., Barch, D. M., Carter, C. S., & Cohen, J. D. (2001). Conflict monitoring and cognitive control. *Psychological Review, 108*, 624–652.

Brozoski, T. J., Brown, R. M., Rosvold, H. E., & Goldman-Rakic, P. S. (1979). Cognitive deficits caused by regional depletion of dopamine in prefrontal cortex of rhesus monkey. *Science, 205*, 929–932.

Caldarone, B. J., Duman, C. H., & Picciota, M. R. (2000). Fear conditioning and latent inhibition in mice lacking the high affinity subclass of nicotinic acetylcholine receptors in the brain. *Neuropharmacology, 39*, 2779–2784.

Carli, M., Robbins, T. W., Evenden, J. L., & Everitt, B. J. (1983). Effects of lesions to ascending noradrenergic neurons on performance of a 5-choice serial reaction task in rats; implications for theories of dorsal noradrenergic bundle function based on selective attention and arousal. *Behavioural Brain Research, 9*, 361–380.

Castner, S. A., Williams, G. V., & Goldman-Rakic, P. S. (2000). Reversal of antipsychotic-induced working memory deficits by short-term dopamine D1-receptor stimulation. *Science, 287*, 2020–2022.

Collins, P., Wilkinson, L. S., Everitt, B. J., Robbins, T. W., & Roberts, A. C. (2000). The effect of dopamine depletion from the caudate nucleus of the common marmoset (Callithrix jacchus) on tests of prefrontal cognitive function. *Behavioural Neuroscience, 114*, 3–17.

Dalley, J. W., Cardinal, R. N., & Robbins, T. W. (2004). Prefrontal executive and cognitive functions in rodents: Neural and neurochemical substrates. *Neuroscience and Biobehavioural Reviews, 28*, 771–784.

Egan, M. F., Goldberg, T. E., Kolachana, B. S., Callicott, J. H., Mazzanti, C. M., Straub, R. E., et al. (2001). Effect of COMT Val[108/158] Met genotype on frontal lobe function and risk for schizophrenia. *Proceedings of the National Academy of Sciences of the United States of America, 98*, 6917–6922.

Erlenmeyer-Kimling, L. (2000). Neurobiological deficits in offspring of schizophrenic parents: Liability indicators and predictors of illness. *American Journal of Medical Genetics, 97*, 65–71.

Etchepareborda, M. C., & Abas-Mas, L. (2005). Working memory in basic learning processes. *Reviews of Neurology, 15*(Suppl. 1), S79–S83.

Fan, J., Fossella, J., Sommer, T., Wu, Y., & Posner, M. I. (2003). Mapping the genetic variation of executive attention onto brain activation. *Proceedings of the National Academy of Sciences of the United States of America, 100*, 7406–7411.

Friedman, H. R., & Goldman-Rakic, P. S. (1994). Coactivation of prefrontal cortex and inferior parietal cortex in working memory tasks revealed by 2DG functional mapping in the rhesus monkey. *Journal of Neuroscience, 14*, 2775–2788.

Glickstein, S. B., DeSteno, D. A., Hof, P. R., & Schmauss, C. (2005). Mice lacking dopamine D$_2$ and D$_3$ receptors exhibit different differential activation of prefrontal cortical neurons during tasks requiring attention. *Cerebral Cortex, 15*, 1016–1024.

Glickstein, S. B., Hof, P. R., & Schmauss, C. (2002). Mice lacking dopamine D2 and D3 receptors have spatial working memory deficits. *Journal of Neuroscience, 22*, 5619–5629.

Glickstein, S. B., & Schmauss, C. (2004). Focused motor stereotypies do not require enhanced activation of neurons in striosomes. *Journal of Comparative Neurology, 469*, 227–238.

Goldman-Rakic, P. S. (1995). Cellular basis of working memory. *Neuron, 14*, 477–485.

Goldman-Rakic, P. S., Castner, S. A., Svensson, T. H., Siever, L. J., & Williams, G. V. (2004). Targeting the dopamine D1 receptor in schizophrenia: Insights for cognitive dysfunctions. *Psychopharmacology, 174*, 3–16.

Granon, S., Passetti, F., Thomas, K. L., Dalley, J. W., Everitt, B. J., & Robbins, T. W. (2000). Enhanced and impaired attentional performance after infusion of D1 dopaminergic agents into rat prefrontal cortex. *Journal of Neuroscience, 20*, 1208–1215.

Humby, T., Laird, F. M., Davies, W., & Wilkinson, L. S. (1999). Visuospatial attentional functioning in mice: Interactions between cholinergic manipulations and genotype. *European Journal of Neuroscience, 11*, 2813–2823.

Kerns, J. G., Cohen, J. D., MacDonald, A. W., III, Cho, R. Y., Stenger, V. A., & Carter, C. S. (2004). Anterior cingulate conflict monitoring and adjustments in control. *Science, 303*, 1023–1026.

Kindlon, D. J. (1998). The measurements of attention. *Child Psychology and Psychiatry Review, 3*, 72–78.

Lubow, R. E. (2005). Construct validity of the animal latent inhibition model of selective attention deficits in schizophrenia. *Schizophrenia Bulletin, 31*, 139–153.

MacDonald, A. W., III, Cohen, J. D., Stenger, V. A., & Carter, C. S. (2000). Dissociating the role of the dorsolateral prefrontal and anterior cingulate cortex in cognitive control. *Science, 288*, 1835–1838.

Markowitsch, H. J., & Pritzel, M. (1977). Comparative analysis of prefrontal learning functions in rats, cats, and monkeys. *Psychological Bulletin, 84*, 817–837.

Martinussen, R., Hayden, J., Hogg-Johnson, S., & Tannock, R. (2005). A meta-analysis of working memory impairments in children with attention-deficit/hyperactivity disorder. *Journal of the American Academy of Child and Adolescent Psychiatry, 44*, 377–384.

Mattay, V. S., Callicott, J. H., Bertolino, A., Heaton, I., Frank, J. A., Coppola, R., et al. (2000). Effects of dextroamphetamine on cognitive performance and cortical activation. *Neuroimage, 12*, 268–275.

McAlonan, K., & Brown, V. J. (2003). Orbital prefrontal cortex mediates reversal learning and not attentional set shifting in the rat. *Behavioural Brain Research, 146*, 97–103.

Mcgaughy, J., & Sarter, M. (1995). Behavioral vigilance in rats: Task validation and effects of age, amphetamine, and benzodiazepine receptor ligands. *Psychopharmacology, 117*, 340–357.

Miyakawa, T., Leiter, L. M., Gerber, D. J., Gainetdinov, R. R., Sotnikova, T. D., Zeng, H., et al. (2003). Conditional calcineurin knockout mice exhibit multiple abnormal behaviors related to schizophrenia. *Proceedings of the National Academy of Sciences of the United States of America, 100*, 8987–8992.

Moran, P. M. (1993). Differential effects of scopolamine and mecamylamine on working and reference memory in the rat. *Neuroscience Letters, 138*, 157–160.

Nuechterlein, K. H., Barch, D. M., Gold, J. M., Goldberg, J. M., Green, M. F., & Heaton, R. K. (2004). Identification of separable cognitive factors in schizophrenia. *Schizophrenia Research, 15*, 29–39.

Owens, A. M., James, M., Leigh, P. N., Summers, B. A., Marsden, C. D., Quinn, N. P., et al. (1992.) Fronto-striatal cognitive deficits at different stages of Parkinson's disease. *Brain, 115*, 1727–1751.

Park, S., & Holzman, P. S. (1992). Schizophrenics show spatial working memory deficits. *Archives of General Psychiatry, 49*, 975–982.

Peuskens, J., Demily, C., & Thibaut, F. (2005). Treatment of cognitive dysfunctions in schizophrenia. *Clinical Therapeutics, 27*, S25–S37.

Posner, M. I. (1980). Orienting of attention. *Quarterly Journal of Experimental Psychology, 32*, 3–25.

Posner, M. I., & Petersen, S. E. (1990). The attention system of the human brain. *Annual Review of Neuroscience, 13*, 25–42.

Raz, A. (2004). Anatomy of attentional networks. *The American Record (Part B: New Anat), 281B*, 21–26.

Raz, A., & Buhle, J. (2006). Typologies of attentional networks. *Nature Reviews. Neuroscience, 7*, 367–379.

Robbins, T. W. (2002). The 5-choice serial reaction time task: Behavioral pharmacology and functional neurochemistry. *Psychopharmacology, 163*, 362–380.

Robbins, T. W. (2005). Chemistry of the mind: Neurochemical modulation of prefrontal cortical function. *Journal of Comparative Neurology, 493*, 140–146.

Sarter, M., Givens, B., & Bruno, J. P. (2001). The cognitive neuroscience of sustained attention: Where top-down meets bottom-up. *Brain Research Reviews, 35*, 146–160.

Sawaguchi, T., & Goldman-Rakic, P. S. (1991). D1 dopamine receptors in prefrontal cortex: Involvement in working memory. *Science, 251*, 947–950.

Schmauss, C. (2000). A single dose of methamphetamine leads to a long-term reversal of the blunted dopamine D$_1$-receptor-mediated neocortical c-*fos* responses in mice deficient for D$_2$ and D$_3$ receptors. *Journal of Biological Chemistry, 275*, 38944–38948.

Shalev, U., & Weiner, I. (2001). Gender-dependent differences in latent inhibition following prenatal stress and corticosterone administration. *Behavioural Brain Research, 126*, 57–63.

Stewart, C., Burke, S., & Marrocco, R. (2001). Cholinergic modulation of covert attention in the rat. *Psychopharmacology, 155*, 210–218.

Uylings, H. B. M., Groenewegen, H. J., & Kolb, B. (2003). Do rats have a prefrontal cortex? *Behavioural Brain Research, 146*, 3–17.

Van Haaren, F., De Bruin, J. P., Heinsbroek, R. P., & Van de Poll, N. E. (1985). Delayed spatial response alternation: Effects of delayed-interval duration and lesions of the medial prefrontal cortex on response accuracy of male and female Wistar rats. *Behavioural Brain Research, 18*, 41–49.

Weinberger, D. R., Berman, K. F., & Zec, R. F. (1986). Physiological dysfunction of dorsolateral prefrontal cortex in schizophrenia: I. Regional cerebral blood flow (rCBF) evidence. *Archives of General Psychiatry, 43*, 114–125.

Weiner, I. (2003). The "two-headed" latent inhibition model of schizophrenia: Modeling positive and negative symptoms and their treatment. *Psychopharmacology, 169*, 257–297.

Williams, G. V., & Goldman-Rakic, P. S. (1995). Modulation of memory fields by dopamine D1 receptors in prefrontal cortex. *Nature, 376*, 549–550.

Young, J. W., Finlayson, K., Spratt, C., Marston, H. M., Crawford, N., Kelly, J. S., et al. (2004). Nicotine improves sustained attention in mice: Evidence for involvement of the a7 nicotinic acetylcholine receptor. *Neuropsychopharmacology, 29*, 891–900.

Young, A. M. J., Moran, P. M., & Joseph, M. H. (2005). The role of dopamine in conditioning and latent inhibition: What, when, where and how? *Neuroscience and Biobehavioral Reviews, 29*, 963–976.

Conclusion

Chapter 34

Future Directions for Behavior Genetics

Yong-Kyu Kim

The Handbook has been organized into (1) an introduction; (2) quantitative methods and models in behavior genetics; (3) genetics of cognition; (4) genetics of personality; (5) genetics of psychopathology; and (6) conclusion. The Handbook has selected 33 current topics and issues in behavior genetics. Twenty-four studies of human behaviors and nine studies of animal models have been presented.

Directions for future research on the selected topics have been highlighted in each chapter. Complex disorders or traits that have been presented here are polygenic and multifactorial, and thus they are influenced by both genes and environment, as well as by interactions between these two factors, and probably epigenetic factors that regulate the expression of implicated genes (a field still in its infancy). The quantitative genetic approaches using twin, adoption, or family studies determine the influences of genetic and environmental factors (a^2, d^2, c^2, and e^2) on the behaviors of interest. For most disorders or traits discussed in the Handbook, MZ twins almost always show higher correlations than DZ twins, suggesting that there is some genetic influence on the traits. However, the presence of the discordant MZ co-twins of probands with the disorders indicates that (1) non-shared environment is important; (2) unique environment can differentially influence expression of the same genes that the co-twins share; (3) multiple genes, interaction of the genes, and/or interaction of genes and environment play roles in the development of the traits; and (4) epigenetic factors not yet identified play a role in masking gene expression. This list suggests that (1) future research in behavior genetics will be enhanced by studying non-shared environmental sources of variation; (2) there is a need to identify more genes directly involved in the behavior with the advance in molecular genetics; (3) interactions between genes and environment, as well as gene-environment correlation, are to be further explored;

(4) longitudinal studies are warranted to see how specific environments affect gene expression with age; (5) multivariate analysis showing how specific environmental influence interacts with genetic effects needs to be further developed; (6) knowledge of endophenotypes will increase the power to detect the genes involved and clarify the pathways between genetic predisposition and the disorders; (7) powerful technologies, i.e., neuroimaging, are warranted to look at the effects of genes in the structures and function of the brains; and (8) further development of animal models will contribute to our understanding of human behavioral research.

When I outlined the Handbook, I focused upon the complex disorders in humans and animal models of the disorders. It was extremely difficult to put all of the broad domains in behavior genetics into a single book. Given the necessity of page limits, several important domains in behavior genetics have been missed.

Epigenetics. The theory of epigenesis suggests that complex organisms form from relatively simple undifferentiated cells through a series of radical transformations. Various tissues or organs contain the same set of genes, but perform very different functions. Epigenetics involves modifications of the activation of certain genes during the development of complex organisms, and thus describes any aspect other than DNA sequence that influences the development of an organism. The epigenetic factors are not static, but change over time (Gottesman, 2004). Such instability of epigenetic modification leads to differences in gene activity even within pairs of MZ twins. For example, Wong, Gottesman, and Petronis (2005) suggest that epigenetic factors play a role in substantial phenotypic variation observed in genetically identical organisms, i.e., MZ twins and inbred animals, *in the absence of* either genetic background differences or identifiable environmental variation. Specific epigenetic processes include paramutation, genomic imprinting, gene silencing, X chromosome inactivation, etc.

Endophenotypes. Endophenotypes are measurable internal phenotypes and fill the gap (epigenetic pathways) between genes and complex disorders. The identification of endophenotypes can help to resolve the etiologies of

Y.-K. Kim (✉)
Department of Genetics, University of Georgia, Athens, GA 30602, USA
e-mail: yongkyu@uga.edu

the disorders, for example, multiple genes. Recently, a number of attempts to determine candidate endophenotypes have been successfully made in several complex disorders, i.e., schizophrenia, mood disorders, AD, ADHD, alcoholism, and personality disorders (see Gottesman, 2004 Gottesman & Gould, 2003; for further information, see also Chapters 31 in this volume).

Circadian rhythm. A circadian rhythm is a roughly 24-hour cycle in the biochemical, physiological, or behavioral processes of living organisms including bacteria, fungi, plants, and animals (see Dunlap, 1999 for review). It is driven by a feedback loop of gene transcription in the suprachiasmatic nucleus in the hypothalamus. *Clock* and related genes have been isolated from many species. Several human complex disorders are known to bear a striking resemblance to the phenotypes of mutant animals (Wagner-Smith & Kay, 2000). Currently neural and genetic networks to explain how genes and their products, as well as environments, interact to determine such a complex trait are being studied.

Learning and memory. Learning is an adaptive behavior, and memory resulting from prior experience is stored in the hippocampus. Using animal models, especially *Drosophila*, behavioral screens for learning mutants isolated from chemical mutagenesis or targeted mutations have identified many genes that influence the development of neuroanatomical structure and the biochemical pathways underlying behavioral plasticity. Further, neuroscientists have attempted to identify neuronal mechanisms by which genes have their effects in the brain. For example, biochemical analysis of *dunce* and *rutabaga* in *D. melanogaster* suggests that cyclic AMP (cAMP) is crucial for learning and memory, as well as for synaptic plasticity and that several mutations affect the cAMP signaling and block the memory processing (Dubnau & Tully, 1998). Currently, the QTL analyses for learning disorders have been reported for reading disabilities and speech/language disorders (see this volume). Molecular findings for dementia come from research on Alzheimer's disease (AD), i.e., *APOE4* gene. Knock-out mouse models for AD-related genes are generated for studies of cognitive neuroscience. Powerful neuroimaging tools facilitate the process of identifying anatomic networks of cognitive decline and the changes of brain structure and function with age.

Chromosomal abnormality. A single-base deletion or insertion in the DNA codon produces different amino acids and proteins, and subsequently alters the function of the proteins. Chromosomal abnormality is caused by mutations during cell division and involves gross genetic imbalance which contributes to multiple defects including behavioral problems. For instance, Huntington's disease and Fragile X syndrome are caused by expanded triplet repeats in the *Htt* gene and *FMR1* gene, respectively, and these disorders cause mental retardation and other neurodegenerative disorders. Angelman syndrome and Prader–Willi syndrome are also caused by a small deletion in chromosome 15 depending on which parental chromosome is affected. Williams syndrome has a small deletion in chromosome 7. These syndromes are associated with mental disorders. Turner syndrome (TS) is the most common chromosomal abnormality (XO) in females, and TS patients are susceptible to a range of disorders, including hypertension and poor social activity during adolescence and adulthood (Elsheikh, Dunger, Conway, & Wass, 2002). Research on the behavioral disorders associated with chromosomal abnormality is required.

Obesity. Although the Handbook has one chapter on exercise behavior in health psychology, health-related issues such as obesity or wellbeing have not been discussed. Obesity is a major risk factor for several medical disorders. Current studies show that not only is it heritable and runs in families, but it is also influenced by environment such as stress. A few candidate genes for obesity have been identified. Obesity is also highly comorbid with eating disorders and psychopathological diseases. Behavioral genetic research on other health medicine issues, i.e., blood pressure, hypertension, diabetes, and cardiovascular disease are warranted.

Aging. Senescence means a decline in health and function associated with aging. Thus, when we age, we are changed physically, psychologically, and socially. Aging is one of the universal concerns in human beings. Chapter 7 has discussed changes in cognitive abilities in the aging process. Longitudinal studies of cognitive aging show a decrease in heritability in late adulthood, which results from increasing non-shared environmental variance, i.e., lifestyle variables or disease. Alzheimer's disease is the most common form of dementia, representing approximately two-thirds of dementia cases, and prevalence rates for AD increase with age. Considerable variation in longevity exists between individuals. Longevity is known to be associated with low rates of chronic illness and healthy lifestyle. Sibling studies show moderate genetic influence on longevity. Currently, genome-wide linkage and association studies to find candidate genes for longevity from the populations with high longevity rate are in progress. Research on the loss of expression of specific genes at both the mRNA level and the protein level is warranted.

Social behavior. Behavior genetics focuses upon the nature and origins of individual differences in human or animal behaviors. In mammals, altruistic behavior, i.e., alarming calls or postures, has been well studied from the ecological and evolutionary points of view. Although individuals displaying such altruistic behavior are directly subject to predation, their genes indirectly pass onto future generations via surviving relatives, and thus increase inclusive fitness (Hamilton, 1968). Another well-known social behavior is Parental care. Females of most animal species, including humans, provide offspring with *unselfish* parental care. Parental investment plays a role in enhancing the fitness of individuals. Although ecological studies demonstrate that it depends upon the mating systems in nature, investigations into specific genes contributing to the social behavior are almost entirely lacking.

Evolutionary psychology. Darwin (1871) stated that conspicuous traits such as large body size, bright colorful plumage, or elaborate courtship are favored to attract mates and enhance the reproductive success of individuals. The exaggerated traits supposedly indicate the physical and genetic quality of individuals that possess the traits. Individuals of the same sex vigorously compete to acquire mating with opposite-sex individuals by displaying the conspicuous traits or by directly fighting with each other. The genetic mechanisms of sexually selected traits or behaviors have been intensively studied in animal models. There is evidence in *Drosophila melanogaster* that males who mate the fastest also copulate more often, more successfully, and leave more progeny (Fulker, 1966). More aggressive males are also more successful in mating with females (see this volume). Similarly, in humans, men prefer attractive women, and women prefer males with high social status (Buss, 1989). Little, however, is known about the genetic architecture of this adaptive behavior.

Behavior and evolution. Genetics concerns the genetic constitution of organisms and the laws governing the transmission of hereditary information from one generation to the next generation at the individual levels. However, population genetics concern the genetic constitution of populations (of a given species) and deal with the genetic variation of particular genes from generation to generation. The genetic variants include morphological characters, chromosomal inversions, gene polymorphisms, blood group polymorphisms, DNA sequences, mating behavior, and so on. In the Biological Species Concept, species is defined as members of a group that are interbreeding with each other, but are reproductively isolated from those of other groups (Dobzhansky, 1940; Mayr, 1963). There are two types of reproductive isolating mechanisms: (1) premating isolation and (2) postmating isolation. When two different species mate with each other in nature or in the laboratory, they usually produce reproductively sterile hybrids and, thus, genes cannot pass onto the next generation. Therefore, premating isolation, that is sexual isolation or behavioral isolation, is evolved. Sexual isolation is known to be a by-product of genetic differentiation resulting from adaptation to new environments. Individuals in the same species share species-specific courtship and pheromone profiles (Paterson, 1985) and new species are formed when gene flow between populations is blocked. Considerable genetic variation in courtship and pheromone profiles exists in natural populations. The

QTL approaches to sexual isolation between populations or between species are to be encouraged.

Readers can get additional information on behavior genetics from thoughtfully organized books that must be in any selected reading list (Carey, 2003; DiLalla, 2004; Ehrman, Maxson, & Kim, 2009; Jang, 2005; McGuffin, Owen, & Gottesman, 2004; Plomin, DeFries, Craig, & McGuffin, 2003; Plomin, DeFries, McClearn, & McGuffin, 2008). The following sources can also reveal the most recent discoveries and efforts in behavior genetics: http://www.faseb.org for the American Society of Human Genetics; http://www.bga.org for the Behavior Genetics Association; http://ww.ists.qimr.edu.au for the International Society for Twin Studies; http://www.ispg.net for the International Society of Psychiatric Genetics; http://www.brain.com for research on the brain; http://www.ncbi.nlm.nih.gov for the database of single-base nucleotide substitutions and short deletion and insertion polymorphism; http://www.nhgri.nih.gov for the database of human genes and genetic disorders; http://www.biomedcentral.com; and http://www.jbiol.com for access to recently published articles.

In conclusion, behavior genetics is now becoming an even more important domain in human and animal behavior research with new findings, methods, and designs. The advances in molecular genetics and quantitative genetics will improve our understanding of specific genes associated with complex traits or disorders and of gene–environment interaction and correlation for traits. The goal of behavior genetics is to find specific genes associated with a particular disorder and to provide environmental prevention of the disorder before it appears. This is the motivation of the *Handbook of Behavior Genetics*. I hope that the Handbook will be a useful reference for young scientists of behavior genetics. The Handbook will contribute to our understanding of behavior genetics and future research endeavors in the twenty-first century.

References

Buss, D. M. (1989). Sex differences in human mate preferences: Evolutionary hypotheses tested in 37 cultures. *Behavioral and Brain Sciences, 12,* 1–49.

Carey, G. (2003). *Human genetics for the social sciences.* Thousand Oaks, CA: Sage.

Darwin, C. (1871). *The descent of man, and selection in relation to sex.* London: J. Murray.

DiLalla, L. F. (2004). *Behavior genetics principles: Perspectives in development, personality, and psychopathology.* Washington, DC: American Psychological Association.

Dubnau, J., & Tully, T. (1998). Gene discovery in *Drosophila*: New insights for learning and memory. *Annual Review of Neuroscience, 21,* 407–444.

Dunlap, J. C. (1999). Molecular bases for circadian clocks. *Cell, 96,* 271–290.

Dobzhansky, Th. (1940). Speciation as a stage in evolutionary divergence. *American Naturalist, 74*, 312–321.

Ehrman, L., Maxson, S., & Kim, Y.-K. (2009). Behavior Genetics and Evolution (3rd ed.) Oxford: Oxford University Press.

Elsheikh, M., Dunger, D. B., Conway, G. S., & Wass, J. A. H. (2002). Turner's syndrome in adulthood. *Endocrine Reviews, 23*, 120–140.

Fulker, D. W. (1966). Mating speed in male *Drosophila melanogaster*: A psychogenetic analysis. *Science, 153*, 203–205.

Gottesman, I. I. (2004). Postscript: Eyewitness to the maturation of behavioral genetics. In L. F. DiLalla (Ed.), *Behavior genetics principles: Perspectives in development, personality, and psychopathology* (pp. 217–223). Washington, DC: American Psychological Association.

Gottesman, I. I., & Gould, T. D. (2003). The endophenotype concept in psychiatry: Etymology and strategic intentions. *American Journal of Psychiatry, 160*, 636–645.

Hamilton, D. W. (1968). The genetical theory of social behaviour. *Journal of Theoretical Biology, 7*, 1–52.

Jang, K. L. (2005). *The behavioral genetics of psychopathology: A clinical guide*. Mahwah, NJ: Lawrence Erlbaum Associates.

Mayr, E. (1963). *Animal species and evolution*. Cambridge, MA: Harvard University Press.

McGuffin, P., Owen, M. J., & Gottesman, I. I. (2004). *Psychiatric genetics and genomics*. Oxford: Oxford University Press.

Paterson, H. E. H. (1985). The recognition concept of species. In E. S. Vrba (Ed.), *Species and speciation. Transvaal Museum Monograph, 4* (pp. 21–29). Pretoria: Transvaal Museum.

Plomin, R., DeFries, J. C., Craig, I. W., & McGuffin, P. (2003). *Behavioral genetics in the postgenomic era*. Washington, DC: American Psychological Association.

Plomin, R., DeFries, J. C., McClearn, G. E., & McGuffin, P. (2008). *Behavioral genetics* (5th ed.). New York: Worth.

Wagner-Smith, K., & Kay, S. A. (2000). Circadian rhythm genetics: From flies to mice to humans. *Nature genetics, 26*, 23–27.

Wong, A. H. C., Gottesman, I. I., & Petronis, A. (2005). Phenotypic differences in genetically identical organisms: The epigenetic perspective. *Human Molecular Genetics, 14*, R11–R18.

Index

Printed in the United States of America